Eleventh Edition

Introductory Algebra

Margaret L. Lial
American River College

John Hornsby
University of New Orleans

Terry McGinnis

VP, Courseware Portfolio Management:	Chris Hoag
Director, Courseware Portfolio Management:	Michael Hirsch
Courseware Portfolio Manager:	Matthew Summers
Courseware Portfolio Assistant:	Shannon Bushee
Content Producer:	Sherry Berg
Managing Producer:	Karen Wernholm
Producer:	Shana Siegmund
Manager, Courseware Quality Assurance:	Mary Durnwald
Manager, Content Development:	Rebecca Williams
Product Marketing Manager:	Alicia Frankel
Product Marketing Assistant:	Hanna Lafferty
Field Marketing Managers:	Jennifer Crum, Lauren Schur
Senior Author Support/Technology Specialist:	Joe Vetere
Manager, Rights and Permissions:	Gina Cheselka
Manufacturing Buyer:	Carol Melville, LSC Communications
Associate Director of Design:	Blair Brown
Program Design Lead:	Barbara Atkinson
Text Design, Production Coordination, Composition, and Illustrations:	Cenveo® Publisher Services
Cover Design:	Studio Montage
Cover Image:	ML Harris/The Image Bank/ Getty Images

Library of Congress Cataloging-in-Publication Data

Names: Lial, Margaret L. | Hornsby, John, 1949- | McGinnis, Terry.
Title: Introductory algebra / Margaret Lial, John Hornsby, Terry McGinnis.
Description: 11th edition. | Boston : Pearson, [2018]
Identifiers: LCCN 2016003620 | ISBN 9780134474083 (pbk. : alk. paper)
Subjects: LCSH: Algebra-Textbooks.
Classification: LCC QA152.3 .L557 2018 | DDC 512.9—dc23
LC record available at http://lccn.loc.gov/2016003620

ISBN 13: 978-0-13-447408-3
ISBN 10: 0-13-447408-2

To Dotty, Puddles, and Gus. You loved us unconditionally.

E.J.H. and T.R.M.

Contents

Preface

It is with great pleasure that we offer the eleventh edition of *Introductory Algebra*. We have remained true to the original goal that has guided us over the years—to provide the best possible text and supplements package to help students succeed and instructors teach. This edition faithfully continues that process through enhanced explanations of concepts, new and updated examples and exercises, student-oriented features like Pointers, Cautions, Problem-Solving Hints, Margin Problems, and Study Skills, as well as an extensive package of helpful supplements and study aids.

This text is part of a series that also includes the following books:

- *Basic College Mathematics,* Tenth Edition, by Lial, Salzman, and Hestwood
- *Prealgebra,* Sixth Edition, by Lial and Hestwood
- *Intermediate Algebra,* Eleventh Edition, by Lial, Hornsby, and McGinnis
- *Introductory and Intermediate Algebra,* Sixth Edition, by Lial, Hornsby, and McGinnis
- *Developmental Mathematics: Basic Mathematics and Algebra,* Fourth Edition, by Lial, Hornsby, McGinnis, Salzman, and Hestwood

WHAT'S NEW IN THIS EDITION

We are pleased to offer the following new textbook features and supplements.

- ***Revised Exposition*** With each edition of the text, we continue to polish and improve discussions and presentations of topics to increase readability and student understanding. We believe this edition is the best yet in this regard.
- ***More Figures and Diagrams*** For visual learners, we have made a concerted effort to add mathematical figures, diagrams, tables, and graphs whenever possible.
- ***Improved Study Skills*** Most of these special activities now include a ***Now Try This*** section to increase student involvement. Each is designed independently to allow flexible use with individuals or small groups of students, or as a source of material for in-class discussions.
- ***More What Went Wrong? Exercises*** We have increased the number of these popular **CONCEPT CHECK** exercises, which highlight common student errors.
- ***More Relating Concepts Exercises*** We have increased the number of these flexible groups of exercises, located at the end of many exercise sets. Specially written to help students tie concepts together, as well as compare and contrast ideas, identify and describe patterns, and extend concepts to new situations, these sets of problems may be used with individual students or collaboratively with pairs or small groups.
- ***Dedicated Mixed Review Exercises*** Each chapter review has been expanded to include a one-page set of Mixed Review Exercises to help students further synthesize concepts.
- ***Enhanced MyMathLab Resources*** Exercise coverage has been refined with new videos and homework problems, including new Relating Concepts questions added throughout the course. See pages x and xi for more details.
- ***Learning Catalytics*** This interactive student response tool uses students' own devices to engage them in the learning process. Learning Catalytics is accessible through MyMathLab and can be customized to an instructor's specific needs. Instructors can employ this tool to generate class discussion, promote peer-to-peer learning, and use real-time data to adjust instructional strategy. As an introduction to this exciting new tool, we have provided prerequisite skills questions at the beginning of each section to check students' preparedness for the new section. Learn more about Learning Catalytics in the Instructor Resources tab in MyMathLab.

- ***Data Analytics*** We analyzed aggregated student usage and performance data from MyMathLab for the previous edition of this text. The results of this analysis helped us improve the quality and quantity of exercises that matter the most to instructors and students.

CONTENT CHANGES

Specific content changes include the following:

- **Exercise sets** have been updated with a renewed focus on conceptual understanding, skill development, and review. New or revised figures are included wherever possible.
- **Real-world data** in the examples and exercises has been updated.
- **More "word equations"** are included in application examples to help students translate words into equations.
- **Expanded Chapter R** includes new figures and exposition on fractions, as well as new discussion, examples, and exercises on converting between fractions, decimals, and percents.
- **Expanded Mid-Chapter Summary Exercises** in Chapter 2 continue our emphasis on the difference between simplifying expressions and solving equations. The mid-chapter Summary Exercises in Chapters 4, 6, and 9 include new examples that illustrate and distinguish between solution methods.
- **Separate sections on slope-intercept form and point-slope form** now appear in Chapter 3 and include enhanced discussion and new examples and exercises.
- **Reorganized Chapter 5** introduces the rules for exponents and application to scientific notation at the beginning of the chapter, followed by the sections on polynomials and their operations.
- **The following topics are among those that have been enhanced and/or expanded:**

  Operations with signed numbers (Sections 1.4–1.6)
  Order of operations involving absolute value expressions (Sections 1.5 and 1.6)
  Solving linear equations in one variable (Sections 2.1 and 2.2)
  Solving linear inequalities with fractions (Section 2.7)
  Graphing linear equations in two variables using intercepts (Section 3.2)
  Solving linear systems of equations using elimination (Section 4.3)
  Dividing a polynomial by a polynomial (Section 5.7)
  Discussion of sums of squares and factoring perfect square trinomials (Section 6.5)
  Solving variation problems (Section 7.8)
  Adding and subtracting radicals (Section 8.3)
  Solving quadratic equations by completing the square (Section 9.2)

HALLMARK FEATURES

We have enhanced the following popular features, each of which is designed to increase ease-of-use by students and/or instructors.

- ***Emphasis on Problem-Solving*** We introduce our six-step problem-solving method in Chapter 2 and integrate it throughout the text. The six steps, *Read, Assign a Variable, Write an Equation, Solve, State the Answer,* and *Check,* are emphasized in boldface type and repeated in examples and exercises to reinforce the problem-solving process for students. We also provide students with **Problem-Solving Hint** boxes that feature helpful problem-solving tips and strategies.
- ***Helpful Learning Objectives*** We begin each section with clearly stated, numbered objectives, and the included material is directly keyed to these objectives so that students and instructors know exactly what is covered in each section.

- ***Popular Cautions and Notes*** One of the most popular features of previous editions, we include information marked CAUTION and Note to warn students about common errors and emphasize important ideas throughout the exposition. The updated text design makes them easy to spot.

- ***Comprehensive Examples*** The new edition features a multitude of step-by-step, worked-out examples that include pedagogical color, helpful side comments, and special pointers. We give special attention to checking example solutions—more checks, designated using a special CHECK tag and ✓, are included than in past editions.

- ***More Pointers*** Because they were so well received by both students and instructors in the previous edition, we incorporate more pointers in examples and discussions throughout this edition of the text. They provide students with important on-the-spot reminders and warnings about common pitfalls.

- ***Ample Margin Problems*** Margin problems, with answers immediately available at the bottom of the page, are found in every section of the text. This key feature allows students to immediately practice the material covered in the examples in preparation for the exercise sets. Many include guided solutions.

- ***Updated Figures, Photos, and Hand-Drawn Graphs*** Today's students are more visually oriented than ever. As a result, we include appealing mathematical figures, diagrams, tables, and graphs, including a "hand-drawn" style of graphs, whenever possible. We have incorporated depictions of well-known mathematicians as well as photos to accompany applications in examples and exercises.

- ***Relevant Real-Life Applications*** We include many new or updated applications from fields such as business, pop culture, sports, technology, and the health sciences that show the relevance of algebra to daily life.

- ***Extensive and Varied Exercise Sets*** The text contains a wealth of exercises to provide students with opportunities to practice, apply, connect, review, and extend the skills they are learning. Numerous illustrations, tables, graphs, and photos help students visualize the problems they are solving. Problem types include skill building and writing exercises, as well as applications, matching, true/false, multiple-choice, and fill-in-the-blank problems.

  In the Annotated Instructor's Edition of the text, the writing exercises are marked with an icon so that instructors may assign these problems at their discretion. Students can watch an instructor work through the complete solution for all exercises marked with a Play Button icon in MyMathLab.

- ***Special Summary Exercises*** We include a set of these popular in-chapter exercises in every chapter beginning in Chapter 1. They provide students with the all-important ***mixed review problems*** they need to master topics and often include summaries of solution methods and/or additional examples.

- ***Step-by-Step Solutions to Selected Exercises*** Exercise numbers enclosed in a blue square, such as **11.**, indicate that a worked-out solution for the problem is available in MyMathLab. These solutions are given for selected exercises that most commonly cause students difficulty.

Resources for Success

MyMathLab Online Course for Lial/Hornsby/McGinnis *Introductory Algebra*, 11th edition

The corresponding MyMathLab course tightly integrates the authors' approach, giving students a learning environment that encourages conceptual understanding and engagement.

NEW! Learning Catalytics

Integrated into MyMathLab, Learning Catalytics use students' mobile devices for an engagement, assessment, and classroom intelligence system that gives instructors real-time feedback on student learning. LC annotations for instructors in the text provide corresponding questions that they can use to engage their classrooms.

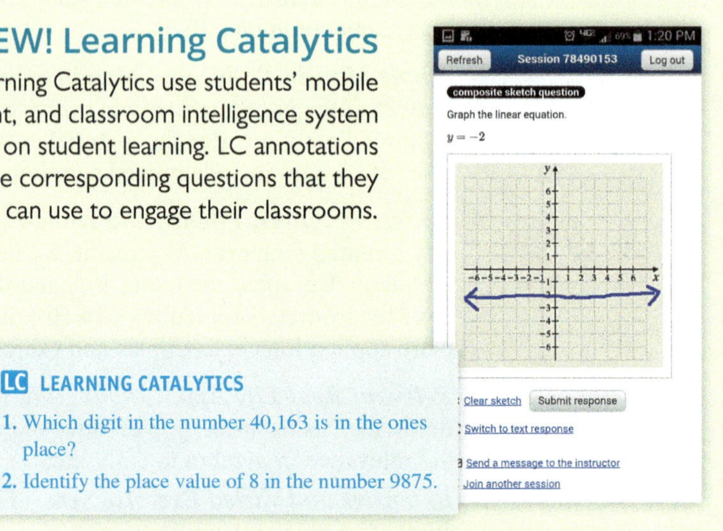

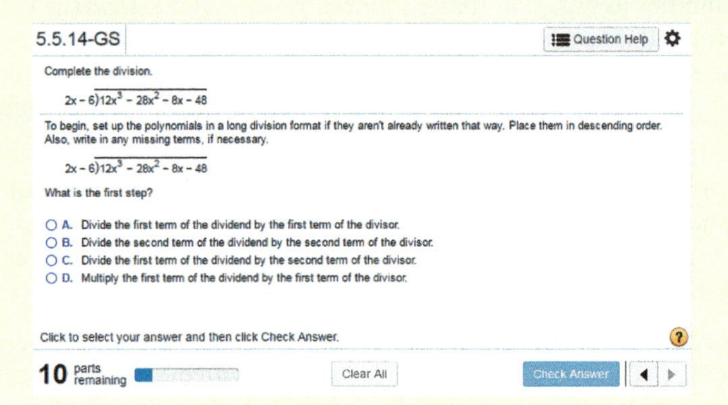

Expanded! Conceptual Exercises

In addition to MyMathLab's hallmark interactive exercises, the Lial team provides students with exercises that tie concepts together and help students problem-solve. Guided Solutions exercises, marked with a "GS" in the Assignment Manager, test student understanding of the problem-solving steps while guiding them through the solution process. Relating Concepts exercises in the text help students make connections and problem-solve at a higher level. These sets are assignable in MyMathLab, with expanded coverage.

NEW! Workspace Assignments

These new assignments allow students to naturally write out their work by hand, step-by-step, showing their mathematical reasoning as they receive instant feedback at each step. Each student's work is captured in the MyMathLab gradebook so instructors can easily pinpoint exactly where in the solution process students struggled.

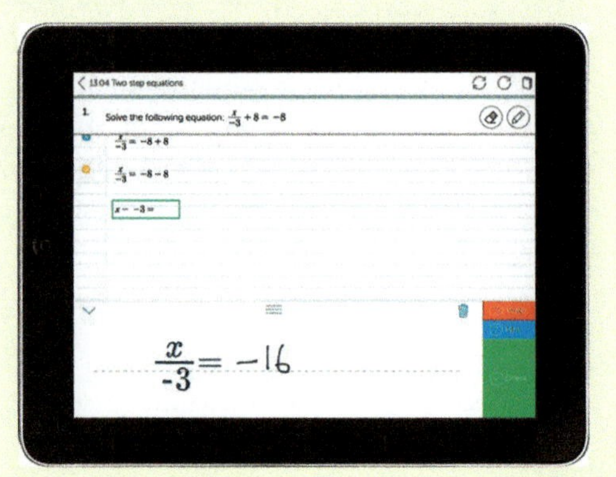

www.mymathlab.com

Resources for Success

NEW! Adaptive Skill Builder

When students struggle on an exercise, Skill Builder assignments provide just-in-time, targeted support to help them build on the requisite skills needed to complete their assignment. As students progress, the Skill Builder assignments adapt to provide support exercises that are personalized to each student's activity and performance throughout the course.

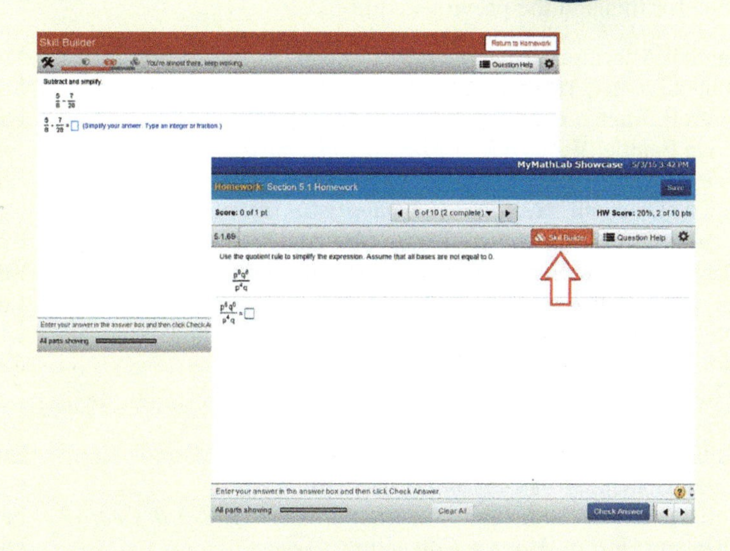

Instructor Resources

Annotated Instructor's Edition

ISBN 10: 0-13-449580-2 **ISBN 13:** 978-0-13-449580-4

The AIE provides annotations for instructors, including answers, Learning Catalytics suggestions, and vocabulary and teaching tips.

The following resources can be downloaded from www.pearsonhighered.com or in MyMathLab:

Instructor's Solutions Manual

This manual provides solutions to all exercises in the text.

Instructor's Resource Manual

This manual includes Mini-Lectures to provide new instructors with objectives, key examples, and teaching tips for every section of the text.

PowerPoints

These slides, which can be edited, present key concepts and definitions from the text.

TestGen

TestGen® (www.pearsoned.com/testgen) enables instructors to build, edit, print, and administer tests using a computerized bank of questions developed to cover all the objectives of the text.

Student Resources

Student Solutions Manual

ISBN 10: 0-13- 450922-6 **ISBN 13:** 978-0-13- 450922-8

This manual contains completely worked-out solutions for all the odd-numbered exercises in the text.

Lial Video Workbook

ISBN 10: 0-13-450923-4 **ISBN 13:** 978-0-13-450923-5

This workbook/note-taking guide helps students develop organized notes as they work along with the videos. The notebook includes

- Guided Examples to be used in conjunction with the Lial Section Lecture Videos and/or Objective-Level Video clips, plus corresponding Now Try This Exercises for each text objective.
- Extra practice exercises for every section of the text, with ample space for students to show their work.
- Learning objectives and key vocabulary terms for every text section, along with vocabulary practice problems.

ACKNOWLEDGMENTS

The comments, criticisms, and suggestions of users, nonusers, instructors, and students have positively shaped this text over the years, and we are most grateful for the many responses we have received. The feedback gathered for this revision of the text was particularly helpful, and we especially wish to thank the following individuals who provided invaluable suggestions for this and the previous editions:

Randall Allbritton, *Daytona State College*
Jannette Avery, *Monroe Community College*
Sarah E. Baxter, *Gloucester County College*
Linda Beattie, *Western New Mexico University*
Jean Bolyard, *Fairmont State College*
Tim C. Caldwell, *Meridian Community College*
Russell Campbell, *Fairmont State University*
Shawn Clift, *Eastern Kentucky University*
Bill Dunn, *Las Positas College*
Lucy Edwards, *Las Positas College*
Morris Elsen, *Cape Fear Community College*
J. Lloyd Harris, *Gulf Coast State College*
Terry Haynes, *Eastern Oklahoma State College*
Edith Hays, *Texas Woman's University*
Karen Heavin, *Morehead State University*
Christine Heinecke Lehmann, *Purdue University—North Central*
Elizabeth Heston, *Monroe Community College*
Sharon Jackson, *Brookhaven College*
Harriet Kiser, *Georgia Highlands College*
Valerie Lazzara, *Palm Beach State College*
Valerie H. Maley, *Cape Fear Community College*
Susan McClory, *San Jose State University*
Pam Miller, *Phoenix College*
Jeffrey Mills, *Ohio State University*
Linda J. Murphy, *Northern Essex Community College*
Celia Nippert, *Western Oklahoma State College*
Elizabeth Olgilvie, *Horry-Georgetown Technical College*
Enyinda Onunwor, *Stark State College*
Larry Pontaski, *Pueblo Community College*
Diann Robinson, *Ivy Tech State College—Lafayette*
Rachael Schettenhelm, *Southern Connecticut State University*
Jonathan Shands, *Cape Fear Community College*
Lee Ann Spahr, *Durham Technical Community College*
Carol Stewart, *Fairmont State University*
Fariheh Towfiq, *Palomar College*
Diane P. Veneziale, *Burlington County College*
Cora S. West, *Florida State College at Jacksonville*
Johanna Windmueller, *Seminole State College*
Gabriel Yimesghen, *Community College of Philadelphia*

Over the years, we have come to rely on an extensive team of experienced professionals. Our sincere thanks go to these dedicated individuals at Pearson Arts & Sciences, who worked hard to make this revision a success: Chris Hoag, Michael Hirsch, Sherry Berg, Shana Siegmund, Matt Summers, Alicia Frankel, and Ruth Berry.

We are especially pleased to welcome Callie Daniels to our team. She thoroughly reviewed all chapters and helped extensively with manuscript preparation. Special thanks to Shannon d'Hemecourt, who assisted once again with updating real data applications.

We are also grateful to Carol Merrigan and Marilyn Dwyer of Cenveo, Inc., for their excellent production work; Connie Day for her copyediting expertise; Cenveo for their photo research; and Lucie Haskins for producing another accurate, useful index. Jack Hornsby, Paul Lorczak, and Sarah Sponholz did a thorough, timely job accuracy checking page proofs and Jack Hornsby checked the index.

We particularly thank the many students and instructors who have used this text over the years. You are the reason we do what we do. It is our hope that we have positively impacted your mathematics journey. We would welcome any comments or suggestions you might have via email to math@pearson.com.

John Hornsby
Terry McGinnis

Prealgebra Review

R.1 Fractions

OBJECTIVES

1. Identify prime numbers.
2. Write numbers in prime factored form.
3. Write fractions in lowest terms.
4. Convert between improper fractions and mixed numbers.
5. Multiply and divide fractions.
6. Add and subtract fractions.
7. Solve applied problems that involve fractions.
8. Interpret data in a circle graph.

The numbers used most often in everyday life are the **natural (counting) numbers,**

$$1, 2, 3, 4, \ldots,$$

the **whole numbers,**

$$0, 1, 2, 3, 4, \ldots,$$

The three dots, or *ellipsis points,* indicate that each list of numbers continues in the same way indefinitely.

and **fractions,** such as

$$\frac{1}{2}, \quad \frac{2}{3}, \quad \text{and} \quad \frac{11}{12}.$$

The parts of a fraction are named as follows.

$$\text{Fraction bar} \rightarrow \frac{3}{8} \begin{array}{l} \leftarrow \text{Numerator} \\ \leftarrow \text{Denominator} \end{array}$$

The fraction bar represents division $\left(\frac{a}{b} = a \div b\right)$.

Note

Fractions are a way to represent parts of a whole. In a fraction, the **numerator** gives the number of parts being represented. The **denominator** gives the total number of equal parts in the whole. See **Figure 1.**

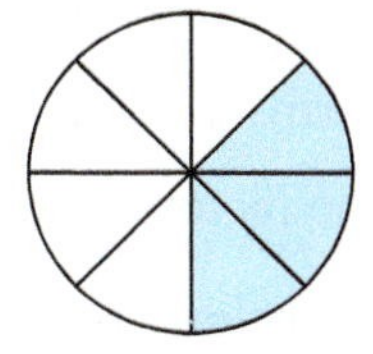

The shaded region represents $\frac{3}{8}$ of the circle.

Figure 1

A fraction is classified as being either a **proper fraction** or an **improper fraction.**

Proper fractions	$\frac{1}{5}, \frac{2}{7}, \frac{9}{10}, \frac{23}{25}$	Numerator is **less than** denominator. Value is less than 1.
Improper fractions	$\frac{3}{2}, \frac{5}{5}, \frac{11}{7}, \frac{28}{4}$	Numerator is **greater than or equal to** denominator. Value is greater than or equal to 1.

OBJECTIVE 1 Identify prime numbers. In work with fractions, we will need to write numerators and denominators as *products*. A **product** is the answer to a multiplication problem. When 12 is written as the product 2×6, for example, 2 and 6 are **factors** of 12. Other factors of 12 are 1, 3, 4, and 12.

A natural number greater than 1 is **prime** if it has only itself and 1 as factors. "Factors" are understood here to mean natural number factors.

2, 3, 5, 7, 11, 13, 17, 19, 23, 29, 31, 37 First dozen prime numbers

A natural number greater than 1 that is not prime is a **composite number.**

4, 6, 8, 9, 10, 12, 14, 15, 16, 18, 20, 21 First dozen composite numbers

By agreement, the number 1 is neither prime nor composite.

EXAMPLE 1 Distinguishing between Prime and Composite Numbers

Identify each number as *prime* or *composite*.

(a) 33 Since 33 has factors of 3 and 11, as well as 1 and 33, it is composite.

(b) 43 There are no numbers other than 1 and 43 itself that divide *evenly* into 43, so the number 43 is prime.

(c) 9832 Since 9832 can be divided by 2, giving 2×4916, it is composite.

◀ **Work Problem 1 at the Side.**

1 Identify each number as *prime* or *composite*.

(a) 13

(b) 27

(c) 59

(d) 1806

OBJECTIVE 2 Write numbers in prime factored form. We *factor* a number by writing it as the product of two or more numbers.

Multiplication	Factoring
$6 \cdot 3 = 18$	$18 = 6 \cdot 3$
Factors Product	Product Factors

Factoring is the reverse of multiplying two numbers to obtain the product.

Note

In algebra, a raised dot $\cdot$ is often used instead of the $\times$ symbol to indicate multiplication because $\times$ may be confused with the letter x.

A composite number written using factors that are all prime numbers is in **prime factored form.**

EXAMPLE 2 Writing Numbers in Prime Factored Form

Write each number in prime factored form.

(a) 35 We factor 35 using the prime factors 5 and 7 as $35 = 5 \cdot 7$.

(b) 24 We use a factor tree, as shown below. The prime factors are circled.

Divide by the least prime factor of 24, which is 2. $24 = 2 \cdot 12$

Divide 12 by 2 to find two factors of 12. $24 = 2 \cdot 2 \cdot 6$

Now factor 6 as $2 \cdot 3$. $24 = 2 \cdot 2 \cdot 2 \cdot 3$

All factors are prime.

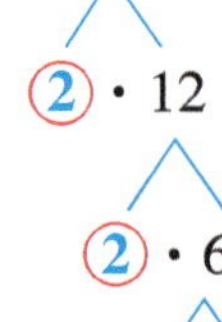

◀ **Work Problem 2 at the Side.**

2 Write each number in prime factored form.

(a) 39

(b) 70

(c) 72

(d) 135

Answers

1. (a) prime (b) composite (c) prime (d) composite
2. (a) $3 \cdot 13$ (b) $2 \cdot 5 \cdot 7$ (c) $2 \cdot 2 \cdot 2 \cdot 3 \cdot 3$ (d) $3 \cdot 3 \cdot 3 \cdot 5$

Note

No matter which prime factor we start with when factoring, we will *always* obtain the same prime factorization. Verify that if we start with 3 instead of 2 in **Example 2(b)**, we obtain

$$24 = 3 \cdot 2 \cdot 2 \cdot 2.$$

The order of the factors is different, but the same prime factors result.

OBJECTIVE 3 **Write fractions in lowest terms.** The following properties are useful when writing a fraction in *lowest terms*.

Properties of 1

Any nonzero number divided by itself is equal to 1. *Example:* $\frac{3}{3} = 1$

Any number multiplied by 1 remains the same. *Example:* $\frac{2}{5} \cdot 1 = \frac{2}{5}$

A fraction is in **lowest terms** when the numerator and denominator have no factors in common (other than 1).

Writing a Fraction in Lowest Terms

Step 1 Write the numerator and denominator in factored form.

Step 2 Replace each pair of factors common to the numerator and denominator with 1.

Step 3 Multiply the remaining factors in the numerator and in the denominator.

(This procedure is sometimes called **"simplifying the fraction."**)

EXAMPLE 3 **Writing Fractions in Lowest Terms**

Write each fraction in lowest terms.

(a) $\frac{10}{15} = \frac{2 \cdot 5}{3 \cdot 5} = \frac{2}{3} \cdot \frac{5}{5} = \frac{2}{3} \cdot 1 = \frac{2}{3}$ Use the first property of 1 to replace $\frac{5}{5}$ with 1.

(b) $\frac{15}{45}$ By inspection, the greatest common factor of 15 and 45 is 15.

$$\frac{15}{45} = \frac{15}{3 \cdot 15} = \frac{1}{3 \cdot 1} = \frac{1}{3}$$ Remember to write 1 in the numerator.

If the greatest common factor is not obvious, factor the numerator and denominator into prime factors.

$$\frac{15}{45} = \frac{3 \cdot 5}{3 \cdot 3 \cdot 5} = \frac{1 \cdot 1}{3 \cdot 1 \cdot 1} = \frac{1}{3}$$ The same answer results.

(c) $\frac{150}{200} = \frac{3 \cdot 50}{4 \cdot 50} = \frac{3}{4} \cdot 1 = \frac{3}{4}$ 50 is the greatest common factor of 150 and 200.

Another strategy is to choose *any* common factor and work in stages.

$$\frac{150}{200} = \frac{15 \cdot 10}{20 \cdot 10} = \frac{3 \cdot 5 \cdot 10}{4 \cdot 5 \cdot 10} = \frac{3 \cdot 1 \cdot 1}{4 \cdot 1 \cdot 1} = \frac{3}{4}$$ The same answer results.

Work Problem 3 at the Side. ▶

3 Write each fraction in lowest terms.

(a) $\frac{8}{14}$

(b) $\frac{10}{70}$

(c) $\frac{72}{120}$

Answers

3. (a) $\frac{4}{7}$ (b) $\frac{1}{7}$ (c) $\frac{3}{5}$

4 Write $\frac{92}{5}$ as a mixed number.

5 Write $11\frac{2}{3}$ as an improper fraction.

OBJECTIVE 4 Convert between improper fractions and mixed numbers.

A **mixed number** is a single number that represents the sum of a natural number and a proper fraction.

$$\text{Mixed number} \rightarrow 5\frac{3}{4} = 5 + \frac{3}{4}$$

EXAMPLE 4 Converting an Improper Fraction to a Mixed Number

Write $\frac{59}{8}$ as a mixed number.

The fraction bar represents division. We divide the numerator of the improper fraction by the denominator.

$$\begin{array}{r} 7 \\ 8\overline{)59} \\ 56 \\ \hline 3 \end{array}$$

7 ← Quotient

Denominator of fraction (divisor) → 8; 59 ← Numerator of fraction (dividend)

3 ← Remainder

$$\frac{59}{8} = 7\frac{3}{8}$$

◀ Work Problem 4 at the Side.

EXAMPLE 5 Converting a Mixed Number to an Improper Fraction

Write $6\frac{4}{7}$ as an improper fraction.

We multiply the denominator of the fraction by the natural number and then add the numerator to obtain the numerator of the improper fraction.

$$7 \cdot 6 = 42 \quad \text{and} \quad 42 + 4 = 46$$

The denominator of the improper fraction is the same as the denominator in the mixed number, which is 7 here. Thus, $6\frac{4}{7} = \frac{46}{7}$.

◀ Work Problem 5 at the Side.

Multiplying Fractions

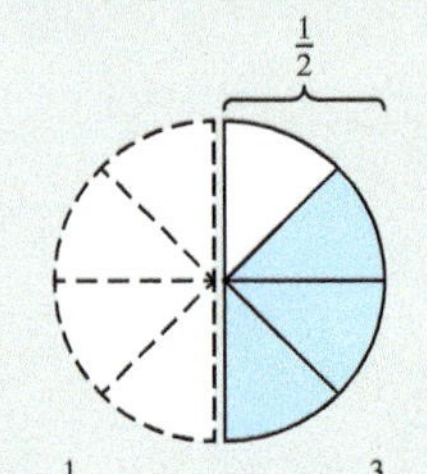

$\frac{3}{4}$ of $\frac{1}{2}$ is equivalent to $\frac{3}{4} \cdot \frac{1}{2}$, which equals $\frac{3}{8}$ of the circle.

Figure 2

OBJECTIVE 5 Multiply and divide fractions. See **Figure 2.**

Multiplying Fractions

To multiply two fractions, multiply the numerators to obtain the numerator of the product. Multiply the denominators to obtain the denominator of the product. ***The product should be written in lowest terms.***

EXAMPLE 6 Multiplying Fractions

Find each product, and write it in lowest terms as needed.

(a) $\frac{3}{8} \cdot \frac{4}{9}$

$= \frac{3 \cdot 4}{8 \cdot 9}$ Multiply numerators. Multiply denominators.

$= \frac{3 \cdot 4}{2 \cdot 4 \cdot 3 \cdot 3}$ Factor the denominator.

$= \frac{1}{2 \cdot 3}$ $\frac{3}{3} = 1$ and $\frac{4}{4} = 1$; Remember to write 1 in the numerator.

$= \frac{1}{6}$ Write in lowest terms.

Continued on Next Page

Answers

4. $18\frac{2}{5}$

5. $\frac{35}{3}$

(b)

$$2\frac{1}{3} \cdot 5\frac{1}{4}$$

$$= \frac{7}{3} \cdot \frac{21}{4}$$ Write each mixed number as an improper fraction.

$$= \frac{7 \cdot 21}{3 \cdot 4}$$ Multiply numerators. Multiply denominators.

$$= \frac{7 \cdot 3 \cdot 7}{3 \cdot 4}$$ Factor the numerator.

Think: $\frac{49}{4}$ means $49 \div 4$.

$$= \frac{49}{4}, \text{ or } 12\frac{1}{4}$$ Write in lowest terms and as a mixed number.

Work Problem 6 at the Side. ▶

Two fractions are **reciprocals** of each other if their product is 1. See the table.

RECIPROCALS

Number	Reciprocal
$\frac{3}{4}$	$\frac{4}{3}$
$\frac{11}{7}$	$\frac{7}{11}$
$\frac{1}{5}$	5, or $\frac{5}{1}$
9, or $\frac{9}{1}$	$\frac{1}{9}$

Example: $\frac{3}{4} \cdot \frac{4}{3} = \frac{12}{12} = 1$

Because division is the inverse, or opposite, of multiplication, we use reciprocals to divide fractions.

Figure 3 illustrates dividing fractions.

Dividing Fractions

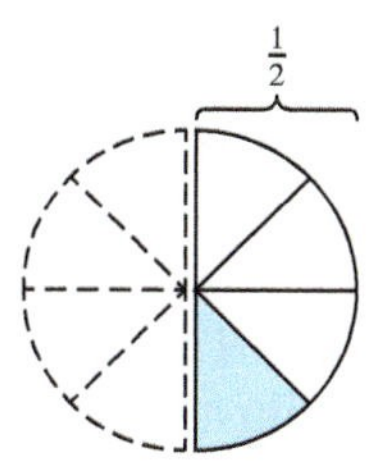

$\frac{1}{2} \div 4$ is equivalent to $\frac{1}{2} \cdot \frac{1}{4}$, which equals $\frac{1}{8}$ of the circle.

Figure 3

Dividing Fractions

To divide two fractions, multiply the first fraction by the reciprocal of the second. The result or **quotient** should be written in lowest terms.

As an example of why this procedure works, we know that

$$20 \div 10 = 2 \quad \text{and also that} \quad 20 \cdot \frac{1}{10} = 2.$$

---

6 Find each product, and write it in lowest terms as needed.

(a) $\frac{5}{6} \cdot \frac{3}{10}$

(b) $\frac{4}{7} \cdot \frac{5}{8}$

(c) $3\frac{1}{3} \cdot 1\frac{3}{4}$

Answers

6. (a) $\frac{1}{4}$ (b) $\frac{5}{14}$ (c) $\frac{35}{6}$, or $5\frac{5}{6}$

7 Find each quotient, and write it in lowest terms as needed.

(a) $\frac{2}{7} \div \frac{3}{10}$

(b) $\frac{3}{4} \div \frac{7}{16}$

(c) $\frac{4}{3} \div 6$

(d) $3\frac{1}{4} \div 1\frac{2}{5}$

Answers

7. (a) $\frac{20}{21}$ **(b)** $\frac{12}{7}$, or $1\frac{5}{7}$ **(c)** $\frac{2}{9}$

(d) $\frac{65}{28}$, or $2\frac{9}{28}$

EXAMPLE 7 Dividing Fractions

Find each quotient, and write it in lowest terms as needed.

(a) $\frac{3}{4} \div \frac{8}{5}$

$= \frac{3}{4} \cdot \frac{5}{8}$ Multiply by the reciprocal of the second fraction.

$= \frac{3 \cdot 5}{4 \cdot 8}$ Multiply numerators. Multiply denominators.

$= \frac{15}{32}$ Make sure the quotient is in lowest terms.

(b) $\frac{3}{4} \div \frac{5}{8}$

$= \frac{3}{4} \cdot \frac{8}{5}$ Multiply by the reciprocal.

$= \frac{3 \cdot 4 \cdot 2}{4 \cdot 5}$ Multiply and factor.

$= \frac{6}{5}$, or $1\frac{1}{5}$

(c) $\frac{5}{8} \div 10$ Think of 10 as $\frac{10}{1}$ here.

$= \frac{5}{8} \cdot \frac{1}{10}$ Multiply by the reciprocal.

$= \frac{5 \cdot 1}{8 \cdot 2 \cdot 5}$ Multiply and factor.

$= \frac{1}{16}$ Remember to write 1 in the numerator.

(d) $1\frac{2}{3} \div 4\frac{1}{2}$

$= \frac{5}{3} \div \frac{9}{2}$ Write each mixed number as an improper fraction.

$= \frac{5}{3} \cdot \frac{2}{9}$ Multiply by the reciprocal of the second fraction.

$= \frac{10}{27}$ Multiply numerators and denominators. The quotient is in lowest terms.

◀ **Work Problem 7 at the Side.**

OBJECTIVE 6 **Add and subtract fractions.** The result of adding two numbers is the **sum** of the numbers. **Figure 4** illustrates adding fractions.

Adding Fractions

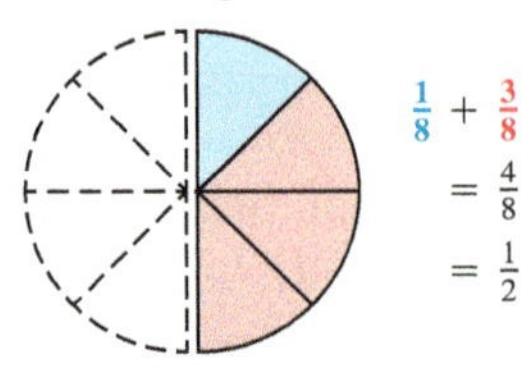

Figure 4

Adding Fractions

To find the sum of two fractions with the *same* denominator, add their numerators and ***keep the same denominator.***

EXAMPLE 8 Adding Fractions (Same Denominator)

Find each sum, and write it in lowest terms as needed.

(a) $\frac{3}{7} + \frac{2}{7}$

$= \frac{3+2}{7}$ Add numerators. Keep the same denominator.

$= \frac{5}{7}$ The answer is in lowest terms.

(b) $\frac{2}{10} + \frac{3}{10}$

$= \frac{2+3}{10}$ Add numerators. Keep the same denominator.

$= \frac{5}{10}$

$= \frac{1}{2}$ Write in lowest terms.

Work Problem 8 at the Side. ▶

If the fractions to be added do not have the same denominator, we must first rewrite them with a common denominator. For example, to rewrite $\frac{3}{4}$ as a fraction with a denominator of 12, think as follows.

$$\frac{3}{4} = \frac{?}{12}$$

We must find the number that can be multiplied by 4 to give 12. Because $4 \cdot 3 = 12$, by the second property of 1, we multiply the numerator and the denominator by 3.

$$\frac{3}{4} = \frac{3}{4} \cdot 1 = \frac{3}{4} \cdot \frac{3}{3} = \frac{3 \cdot 3}{4 \cdot 3} = \frac{9}{12}$$

$\frac{3}{4}$ is equivalent to $\frac{9}{12}$. See **Figure 5.**

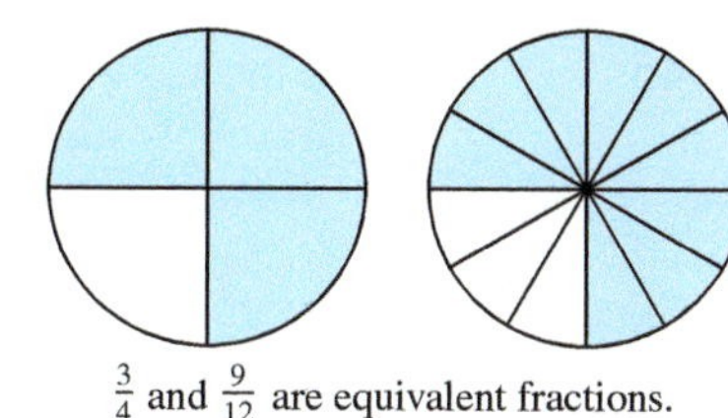

$\frac{3}{4}$ and $\frac{9}{12}$ are equivalent fractions.

Figure 5

Note

The process of writing an equivalent fraction is the reverse of writing a fraction in lowest terms.

Finding the Least Common Denominator (LCD)

To add or subtract fractions with different denominators, find the **least common denominator (LCD)** as follows.

Step 1 Write each denominator in prime factored form.

Step 2 The LCD is the product of every (different) factor that appears in any of the factored denominators. If a factor is repeated, use the greatest number of repeats as factors of the LCD.

Step 3 Write each fraction with the LCD as the denominator, using the second property of 1.

8 Find each sum, and write it in lowest terms as needed.

(a) $\frac{5}{11} + \frac{3}{11}$

(b) $\frac{5}{14} + \frac{3}{14}$

(c) $\frac{3}{5} + \frac{4}{5}$

Answers

8. **(a)** $\frac{8}{11}$ **(b)** $\frac{4}{7}$ **(c)** $\frac{7}{5}$, or $1\frac{2}{5}$

9 Find each sum, and write it in lowest terms as needed.

(a) $\frac{5}{12} + \frac{3}{8}$

(b) $\frac{7}{30} + \frac{2}{45}$

(c) $2\frac{1}{8} + 1\frac{2}{3}$

(d) $4\frac{5}{6} + 2\frac{1}{3}$

Answers

9. (a) $\frac{19}{24}$ (b) $\frac{5}{18}$ (c) $\frac{91}{24}$, or $3\frac{19}{24}$

(d) $\frac{43}{6}$, or $7\frac{1}{6}$

EXAMPLE 9 Adding Fractions (Different Denominators)

Find each sum, and write it in lowest terms as needed.

(a) $\frac{4}{15} + \frac{5}{9}$

Step 1 To find the LCD, write each denominator in prime factored form.

$$15 = 5 \cdot 3 \quad \text{and} \quad 9 = 3 \cdot 3$$

3 is a factor of both denominators.

Step 2

$$\text{LCD} = 5 \cdot 3 \cdot 3 = 45$$

(15 = 5 · 3; 9 = 3 · 3)

In this example, the LCD needs one factor of 5 and two factors of 3 because the second denominator has two factors of 3.

Step 3 Now we can use the second property of 1 to write each fraction with 45 as the denominator.

$$\frac{4}{15} = \frac{4}{15} \cdot \frac{3}{3} = \frac{12}{45} \quad \text{and} \quad \frac{5}{9} = \frac{5}{9} \cdot \frac{5}{5} = \frac{25}{45}$$

At this stage, the fractions are *not* in lowest terms.

$$\frac{4}{15} + \frac{5}{9}$$

$$= \frac{12}{45} + \frac{25}{45}$$ Use the equivalent fractions with the common denominator.

$$= \frac{37}{45}$$ Add numerators. Keep the same denominator.

Make sure the sum is in lowest terms.

(b) $3\frac{1}{2} + 2\frac{3}{4}$

Method 1

$$3\frac{1}{2} + 2\frac{3}{4}$$

$$= \frac{7}{2} + \frac{11}{4}$$ Write each mixed number as an improper fraction.

$$= \frac{14}{4} + \frac{11}{4}$$ Find a common denominator. The LCD is 4.

Think: $\frac{7 \cdot 2}{2 \cdot 2} = \frac{14}{4}$

$$= \frac{25}{4}, \text{ or } 6\frac{1}{4}$$ Add. Write as a mixed number.

Method 2

$$3\frac{1}{2} = 3\frac{2}{4}$$

$$+\,2\frac{3}{4} = 2\frac{3}{4}$$

Write $3\frac{1}{2}$ as $3\frac{2}{4}$. Then add vertically. Add the natural numbers and the fractions separately.

$$5\frac{5}{4} = 5 + 1\frac{1}{4} = 6\frac{1}{4}, \text{ or } \frac{25}{4}$$ The same answer results.

◀ Work Problem 9 at the Side.

The result of subtracting one number from another number is the **difference** of the numbers. **Figure 6** illustrates subtracting fractions.

Subtracting Fractions

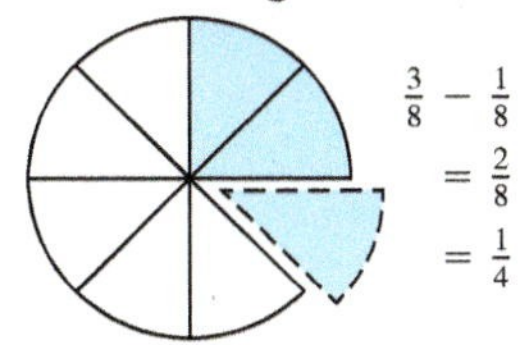

Figure 6

Subtracting Fractions

To find the difference of two fractions with the *same* denominator, subtract their numerators and ***keep the same denominator.***

If the fractions have *different* denominators, write them with a common denominator first.

EXAMPLE 10 Subtracting Fractions

Find each difference, and write it in lowest terms as needed.

(a) $\frac{15}{8} - \frac{3}{8}$

$= \frac{15-3}{8}$ Subtract numerators. Keep the same denominator.

$= \frac{12}{8}$

$= \frac{3}{2}$, or $1\frac{1}{2}$ Write in lowest terms and as a mixed number.

(b) $\frac{15}{16} - \frac{4}{9}$

$= \frac{15}{16} \cdot \frac{9}{9} - \frac{4}{9} \cdot \frac{16}{16}$ Because 16 and 9 have no common factors except 1, the LCD is $16 \cdot 9 = 144$.

$= \frac{135}{144} - \frac{64}{144}$ Write equivalent fractions.

$= \frac{71}{144}$ Subtract numerators. Keep the common denominator.

(c) $\frac{7}{18} - \frac{4}{15}$

$= \frac{7}{2 \cdot 3 \cdot 3} \cdot \frac{5}{5} - \frac{4}{3 \cdot 5} \cdot \frac{2 \cdot 3}{2 \cdot 3}$ $18 = 2 \cdot 3 \cdot 3$ and $15 = 3 \cdot 5$, so the LCD is $2 \cdot 3 \cdot 3 \cdot 5 = 90$.

$= \frac{35}{90} - \frac{24}{90}$ Write equivalent fractions.

$= \frac{11}{90}$ Subtract. The answer is in lowest terms.

Continued on Next Page

10 Find each difference, and write it in lowest terms as needed.

(a) $\frac{9}{11} - \frac{3}{11}$

(b) $\frac{5}{11} - \frac{2}{9}$

(c) $\frac{13}{15} - \frac{5}{6}$

(d) $2\frac{3}{8} - 1\frac{1}{2}$

(e) $10\frac{1}{4} - 2\frac{2}{3}$

11 Solve the problem.

To make a three-piece outfit from the same fabric, Jen needs $1\frac{1}{4}$ yd for the blouse, $1\frac{2}{3}$ yd for the skirt, and $2\frac{1}{2}$ yd for the jacket. How much fabric does she need?

Answers

10. (a) $\frac{6}{11}$ (b) $\frac{23}{99}$ (c) $\frac{1}{30}$ (d) $\frac{7}{8}$ (e) $7\frac{7}{12}$

11. $5\frac{5}{12}$ yd

(d) $4\frac{1}{2} - 1\frac{3}{4}$

Method 1

$4\frac{1}{2} - 1\frac{3}{4}$

$= \frac{9}{2} - \frac{7}{4}$ Write each mixed number as an improper fraction.

Think: $\frac{9 \cdot 2}{2 \cdot 2} = \frac{18}{4}$

$= \frac{18}{4} - \frac{7}{4}$ Find a common denominator. The LCD is 4.

$= \frac{11}{4}$, or $2\frac{3}{4}$ Subtract. Write as a mixed number.

Method 2

$4\frac{1}{2} = 4\frac{2}{4} = 3\frac{6}{4}$ The LCD is 4. $4\frac{2}{4} = 3 + 1 + \frac{2}{4} = 3 + \frac{4}{4} + \frac{2}{4} = 3\frac{6}{4}$

$-1\frac{3}{4} = 1\frac{3}{4} = 1\frac{3}{4}$

$2\frac{3}{4}$, or $\frac{11}{4}$ The same answer results.

◀ Work Problem **10** at the Side.

OBJECTIVE 7 Solve applied problems that involve fractions.

EXAMPLE 11 Solving an Applied Problem with Fractions

The diagram in **Figure 7** appears with directions for a woodworking project. Find the height of the desk to the top of the writing surface.

We must add these measures (" means inches.)

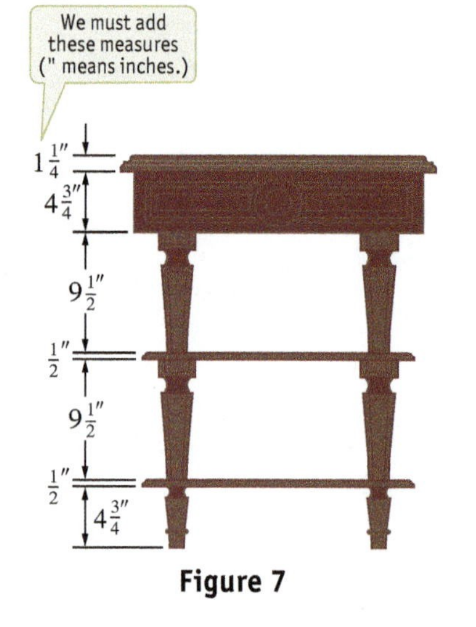

Figure 7

$$\begin{aligned} 1\tfrac{1}{4} &\rightarrow 1\tfrac{1}{4} \\ 4\tfrac{3}{4} &\rightarrow 4\tfrac{3}{4} \\ 9\tfrac{1}{2} &= 9\tfrac{2}{4} \\ \tfrac{1}{2} &= \tfrac{2}{4} \\ 9\tfrac{1}{2} &= 9\tfrac{2}{4} \\ \tfrac{1}{2} &= \tfrac{2}{4} \\ +\,4\tfrac{3}{4} &\rightarrow 4\tfrac{3}{4} \\ \hline & \;\; 27\tfrac{15}{4} \end{aligned}$$

Add the section and writing surface heights to obtain the total height. The common denominator is 4.

Think: $\frac{15}{4}$ means $15 \div 4$.

Because $\frac{15}{4}$ is an improper fraction, this is not the final answer.

Because $\frac{15}{4} = 3\frac{3}{4}$, we have $27\frac{15}{4} = 27 + 3\frac{3}{4} = 30\frac{3}{4}$. The height is $30\frac{3}{4}$ in.

◀ Work Problem **11** at the Side.

EXAMPLE 12 Solving an Applied Problem with Fractions

An upholsterer needs $2\frac{1}{4}$ yd from a bolt of fabric to cover a chair. How many chairs can be covered with $23\frac{2}{3}$ yd of fabric?

To better understand the problem, we replace the fractions with whole numbers. Suppose each chair requires 2 yd, and we have 24 yd of fabric. Dividing 24 by 2 gives 12, the number of chairs that can be covered. To solve the original problem, we must divide $23\frac{2}{3}$ by $2\frac{1}{4}$.

$$23\frac{2}{3} \div 2\frac{1}{4}$$

$$= \frac{71}{3} \div \frac{9}{4} \quad \text{Convert each mixed number to an improper fraction.}$$

$$= \frac{71}{3} \cdot \frac{4}{9} \quad \text{Multiply by the reciprocal.}$$

$$= \frac{284}{27}, \text{ or } 10\frac{14}{27} \quad \text{Multiply numerators. Multiply denominators. Write as a mixed number.}$$

Thus, 10 chairs can be covered, with some fabric left over.

Work Problem 12 at the Side. ▶

OBJECTIVE 8 Interpret data in a circle graph. In a **circle graph,** or **pie chart,** a circle is used to indicate the total of all the data categories represented. The circle is divided into *sectors,* or wedges, whose sizes show the relative magnitudes of the categories. The sum of all the fractional parts must be 1 (for 1 whole circle).

EXAMPLE 13 Using a Circle Graph to Interpret Information

In September 2015, there were about 3300 million (3.3 billion) Internet users worldwide. The circle graph in **Figure 8** shows the fractions of these users living in various regions of the world.

Worldwide Internet Users by Region

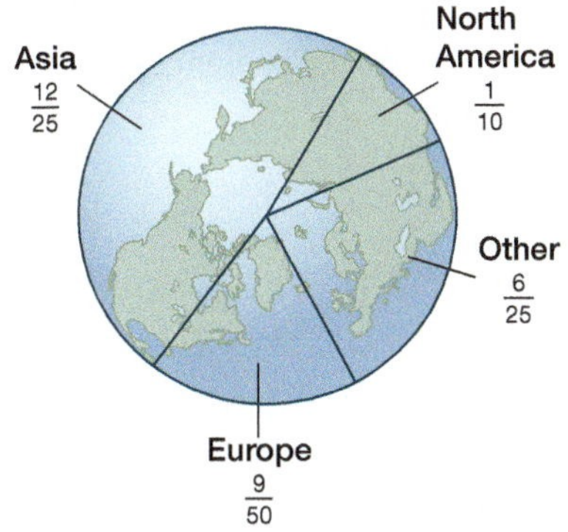

Asia $\frac{12}{25}$

North America $\frac{1}{10}$

Other $\frac{6}{25}$

Europe $\frac{9}{50}$

Data from www.internetworldstats.com

Figure 8

(a) Which region had the largest share of Internet users? What was that share?

In the circle graph, the sector for Asia is the largest, so Asia had the largest share of Internet users, $\frac{12}{25}$.

Continued on Next Page

12 Solve the problem.

A gallon of paint covers 500 ft^2. (ft^2 means square feet.) To paint his house, Tram needs enough paint to cover 4200 ft^2. How many gallons of paint should he buy?

Answer

12. $8\frac{2}{5}$ gal are needed, so he should buy 9 gal.

13 Refer to the circle graph in **Figure 8.**

(a) Which region had the second-largest number of Internet users in September 2015?

(b) *Estimate* the number of Internet users in Asia.

(c) How many *actual* Internet users were there in Asia?

(b) *Estimate* the number of Internet users in Europe.

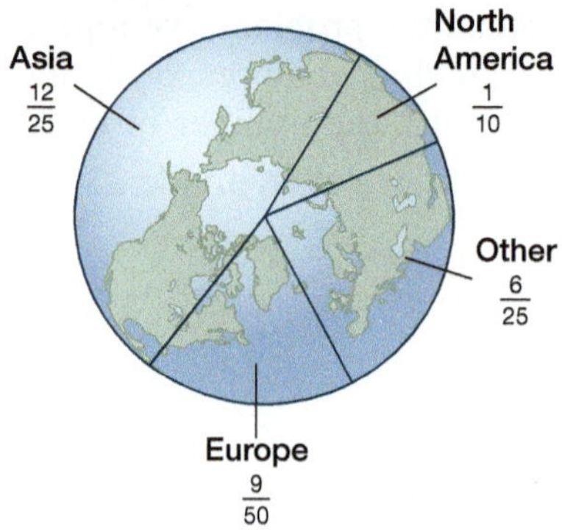

Data from www.internetworldstats.com

Figure 8 (repeated)

From the graph, a share of $\frac{9}{50}$ for Europe can be rounded to $\frac{10}{50}$, or $\frac{1}{5}$. The total number of Internet users, 3300 million, can be rounded to 3000 million (3 billion).

$$\frac{1}{5} \cdot 3000 = 600 \text{ million}$$ Multiply to estimate the number of Internet users in Europe.

(c) How many *actual* Internet users were there in Europe?

$$\frac{9}{50} \cdot \mathbf{3300}$$ Multiply the actual fraction from the graph for Europe by the number of Internet users.

$$= \frac{9}{50} \cdot \frac{\mathbf{3300}}{\mathbf{1}}$$ $a = \frac{a}{1}$ for all a.

$$= \frac{29{,}700}{50}$$ Multiply numerators. Multiply denominators.

$$= 594$$ Divide.

Thus, 594 million, or 594,000,000 people in Europe used the Internet.

◀ **Work Problem 13 at the Side.**

Answers

13. **(a)** Other **(b)** 1500 million **(c)** 1584 million

R.1 Exercises

Go to MyMathLab for worked-out, step-by-step solutions to exercises enclosed in a square and video solutions to exercises.

CONCEPT CHECK *Decide whether each statement is* true *or* false. *If it is* false, *explain why.*

1. In the fraction $\frac{3}{7}$, 3 is the numerator and 7 is the denominator.

2. The mixed number equivalent of the improper fraction $\frac{41}{5}$ is $8\frac{1}{5}$.

3. The fraction $\frac{7}{7}$ is a proper fraction.

4. The number 1 is prime.

5. The fraction $\frac{17}{51}$ is in lowest terms.

6. The reciprocal of $\frac{8}{2}$ is $\frac{4}{1}$.

7. The product of 8 and 2 is 10.

8. The difference of 12 and 2 is 6.

Identify each number as prime *or* composite. ***See Example 1.***

9. 19 **10.** 61 **11.** 52 **12.** 99

13. 2468 **14.** 3125 **15.** 97 **16.** 83

Write each number in prime factored form. ***See Example 2.***

17. 30 **18.** 50 **19.** 57 **20.** 51 **21.** 124

22. 165 **23.** 252 **24.** 168 **25.** 500 **26.** 700

Write each fraction in lowest terms. ***See Example 3.***

27. $\frac{15}{18}$ **28.** $\frac{16}{20}$ **29.** $\frac{8}{16}$ **30.** $\frac{4}{12}$ **31.** $\frac{15}{50}$

32. $\frac{24}{64}$ **33.** $\frac{18}{90}$ **34.** $\frac{16}{64}$ **35.** $\frac{90}{150}$ **36.** $\frac{100}{140}$

Write each improper fraction as a mixed number. ***See Example 4.***

37. $\frac{77}{12}$ **38.** $\frac{101}{15}$ **39.** $\frac{83}{11}$ **40.** $\frac{67}{13}$ **41.** $\frac{12}{7}$ **42.** $\frac{16}{9}$

Write each mixed number as an improper fraction. ***See Example 5.***

43. $2\frac{3}{5}$ **44.** $5\frac{6}{7}$ **45.** $10\frac{3}{8}$ **46.** $12\frac{2}{3}$ **47.** $10\frac{1}{5}$ **48.** $18\frac{1}{6}$

CONCEPT CHECK *Choose the letter of the correct response.*

49. For the fractions $\frac{p}{q}$ and $\frac{r}{s}$, which can serve as a common denominator?

A. $q \cdot s$ **B.** $q + s$ **C.** $p \cdot r$ **D.** $p + r$

50. Which fraction is *not* equal to $\frac{5}{9}$?

A. $\frac{15}{27}$ **B.** $\frac{30}{54}$ **C.** $\frac{40}{74}$ **D.** $\frac{55}{99}$

Find each product or quotient, and write it in lowest terms as needed.
See Examples 6 and 7.

51. $\frac{4}{5} \cdot \frac{6}{7}$ **52.** $\frac{5}{9} \cdot \frac{2}{7}$ **53.** $\frac{5}{12} \cdot \frac{3}{10}$ **54.** $\frac{3}{4} \cdot \frac{2}{15}$

55. $\frac{1}{10} \cdot \frac{12}{5}$ **56.** $\frac{1}{8} \cdot \frac{10}{7}$ **57.** $\frac{15}{4} \cdot \frac{8}{25}$ **58.** $\frac{21}{8} \cdot \frac{4}{7}$

59. $3\frac{1}{4} \cdot 1\frac{2}{3}$ **60.** $2\frac{2}{3} \cdot 1\frac{3}{5}$ **61.** $2\frac{3}{8} \cdot 3\frac{1}{5}$ **62.** $3\frac{3}{5} \cdot 7\frac{1}{6}$

63. $\frac{2}{5} \div \frac{7}{9}$ **64.** $\frac{2}{3} \div \frac{5}{7}$ **65.** $\frac{5}{4} \div \frac{3}{8}$ **66.** $\frac{7}{5} \div \frac{3}{10}$

67. $\frac{32}{5} \div \frac{8}{15}$ **68.** $\frac{24}{7} \div \frac{6}{21}$ **69.** $\frac{3}{4} \div 12$ **70.** $\frac{2}{5} \div 30$

71. $1\frac{3}{5} \div 2\frac{1}{3}$ **72.** $1\frac{1}{3} \div 2\frac{1}{2}$ **73.** $2\frac{5}{8} \div 1\frac{15}{32}$ **74.** $2\frac{3}{10} \div 1\frac{4}{5}$

Find each sum or difference, and write it in lowest terms as needed.
See Examples 8–10.

75. $\frac{7}{15} + \frac{4}{15}$ **76.** $\frac{2}{9} + \frac{5}{9}$ **77.** $\frac{7}{12} + \frac{1}{12}$ **78.** $\frac{3}{16} + \frac{5}{16}$

79. $\frac{5}{9} + \frac{1}{3}$ **80.** $\frac{4}{15} + \frac{1}{5}$ **81.** $\frac{3}{8} + \frac{5}{6}$ **82.** $\frac{5}{6} + \frac{2}{9}$

83. $3\frac{1}{8}+2\frac{1}{4}$

84. $4\frac{2}{3}+2\frac{1}{6}$

85. $\frac{7}{9}-\frac{2}{9}$

86. $\frac{8}{11}-\frac{3}{11}$

87. $\frac{13}{15}-\frac{3}{15}$

88. $\frac{11}{12}-\frac{3}{12}$

89. $\frac{7}{12}-\frac{1}{9}$

90. $\frac{11}{16}-\frac{1}{12}$

91. $6\frac{1}{4}-5\frac{1}{3}$

92. $5\frac{1}{3}-4\frac{1}{2}$

93. $\frac{5}{3}+\frac{1}{6}-\frac{1}{2}$

94. $\frac{7}{15}+\frac{1}{6}-\frac{1}{10}$

Solve each problem. ***See Examples 11 and 12.***

Use the chart to work Exercises 95 and 96.

95. How many cups of dry grits would be needed for eight microwave servings of Quaker Quick Grits?

96. How many teaspoons of salt would be needed for five stove-top servings? (*Hint:* 5 is halfway between 4 and 6.)

	Microwave	Stove Top		
Servings	**1**	**1**	**4**	**6**
Water or Milk	1 cup	1 cup	4 cups	6 cups
Grits	$\frac{1}{4}$ cup	$\frac{1}{2}$ cup	1 cup	$1\frac{1}{2}$ cups
Salt (optional)	Dash	Dash	$\frac{1}{4}$ tsp	$\frac{1}{2}$ tsp

Data from www.quakeroats.com

97. A piece of property has an irregular shape, with five sides as shown in the figure. Find the total distance around the piece of property. This distance is the **perimeter** of the figure.

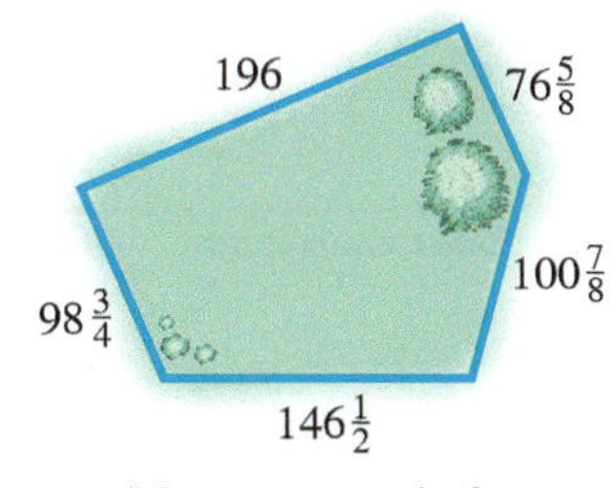

Measurements in feet

98. A triangle has sides of lengths $5\frac{1}{4}$ ft, $7\frac{1}{2}$ ft, and $10\frac{1}{8}$ ft. Find the perimeter of the triangle. **See Exercise 97.**

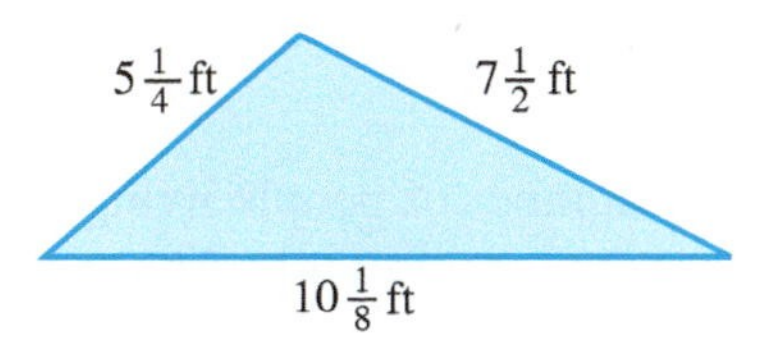

99. A hardware store sells a 40-piece socket wrench set. The measure of the largest socket is $\frac{3}{4}$ in., while the measure of the smallest socket is $\frac{3}{16}$ in. What is the difference between these measures?

100. Two sockets in a socket wrench set have measures of $\frac{9}{16}$ in. and $\frac{3}{8}$ in. What is the difference between these two measures?

101. Under existing standards, most of the holes in Swiss cheese must have diameters between $\frac{11}{16}$ and $\frac{13}{16}$ in. To accommodate new high-speed slicing machines, the U.S. Department of Agriculture wants to reduce the minimum size to $\frac{3}{8}$ in. How much smaller is $\frac{3}{8}$ in. than $\frac{11}{16}$ in.? (Data from U.S. Department of Agriculture.)

102. The Pride Golf Tee Company, the only U.S. manufacturer of wooden golf tees, has created the Professional Tee System. Two lengths of tees are the ProLength Max and the Shortee, as shown in the figure. How much longer is the ProLength Max than the Shortee? (Data from *The Gazette*.)

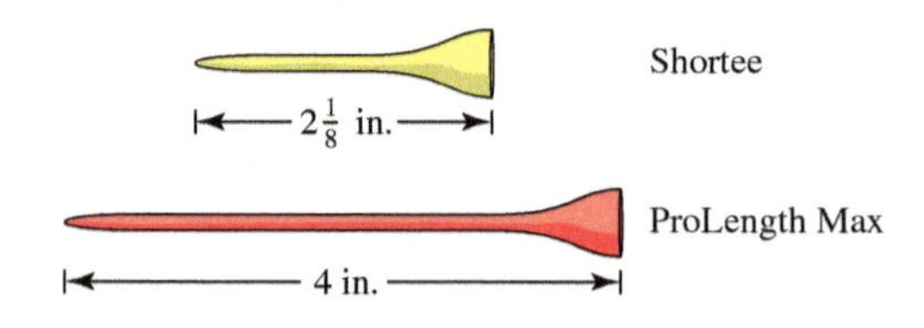

103. A cake recipe calls for $1\frac{3}{4}$ cups of sugar. A caterer has $15\frac{1}{2}$ cups of sugar on hand. How many cakes can he make?

104. Kyla needs $1\frac{1}{8}$ yd of fabric to make a pillow. How many pillows can she make with $8\frac{3}{4}$ yd of fabric?

105. It takes $2\frac{3}{8}$ yd from a bolt of fabric to make a costume for a school play. How much fabric would be needed for seven costumes?

106. A cookie recipe calls for $2\frac{2}{3}$ cups of sugar. How much sugar would be needed to make four batches of cookies?

Approximately 40 *million people living in the United States were born in other countries. The circle graph gives the fractional number from each region of birth for these people. Use the graph to work each problem.* ***See Example 13.***

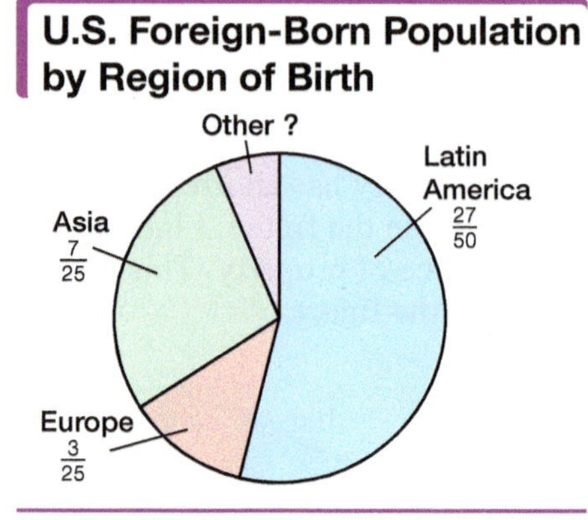

Data from U.S. Census Bureau.

107. Estimate the number of people living in the United States who were born in Asia.

108. Estimate the number of people living in the United States who were born in Latin America.

109. How many people (in millions) were born in Europe?

110. How many people (in millions) were born in Latin America?

111. What fractional part of the foreign-born population was from other regions?

112. What fractional part of the foreign-born population was from Latin America or Asia?

R.2 Decimals and Percents

OBJECTIVES

1. Write decimals as fractions.
2. Add and subtract decimals.
3. Multiply and divide decimals.
4. Write fractions as decimals.
5. Write percents as decimals and decimals as percents.
6. Write percents as fractions and fractions as percents.
7. Solve applied problems that involve percents.

Fractions are one way to represent parts of a whole. Another way is with a decimal fraction or **decimal,** a number written with a decimal point.

9.4, 14.001, 0.25 Decimal numbers

Each digit in a decimal number has a place value, as shown below.

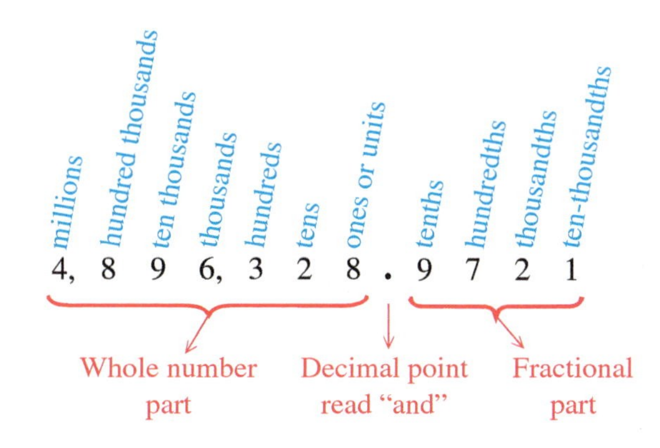

Each successive place value is ten times greater than the place value to its right and one-tenth as great as the place value to its left.

OBJECTIVE 1 Write decimals as fractions. Place value is used to write a decimal number as a fraction.

Converting a Decimal to a Fraction

Read the decimal using the correct place value. Write it in fractional form just as it is read.

- The numerator will be the digits to the right of the decimal point.
- The denominator will be a power of 10—that is, 10 for tenths, 100 for hundredths, and so on.

EXAMPLE 1 Writing Decimals as Fractions

Write each decimal as a fraction. (Do not write in lowest terms.)

(a) 0.95 We read 0.95 as "ninety-five hundredths."

$$0.95 = \frac{95}{100}$$ ← For hundredths

(b) 0.056 We read 0.056 as "fifty-six thousandths."

$$0.056 = \frac{56}{1000}$$

↑ For thousandths

Do not confuse **0.056** with **0.56**, read "fifty-six *hundredths,*" which is the fraction $\frac{56}{100}$.

Continued on Next Page

(c) 4.2095 We read this decimal number, which is greater than 1, as "Four ***and*** two thousand ninety-five ten-thousandths."

$$4.2095 = 4\frac{2095}{10{,}000}$$ Write the decimal number as a mixed number.

$$= \frac{42{,}095}{10{,}000}$$ Write the mixed number as an improper fraction.

◄ **Work Problem 1 at the Side.**

OBJECTIVE 2 Add and subtract decimals.

EXAMPLE 2 Adding and Subtracting Decimals

Add or subtract as indicated.

(a) 6.92 + 14.8 + 3.217

Place the digits of the decimal numbers in columns by place value, so that tenths are in one column, hundredths in another column, and so on.

Be sure to line up decimal points.

$$\begin{array}{r} 6.92 \\ 14.8 \\ +\ 3.217 \\ \hline 24.937 \end{array}$$ Decimal points are aligned.

To avoid errors, attach zeros as placeholders so there are the same number of places to the right of each decimal point.

$$\begin{array}{r} 6.92 \\ 14.8 \\ +\ 3.217 \\ \hline \end{array} \quad \text{becomes} \quad \begin{array}{r} 6.920 \\ 14.800 \\ +\ 3.217 \\ \hline 24.937 \end{array}$$ Attach 0s.

6.92 is equivalent to 6.920.
14.8 is equivalent to 14.800.

(b) 47.6 − 32.509

$$\begin{array}{r} 47.6 \\ -\ 32.509 \\ \hline \end{array} \quad \text{becomes} \quad \begin{array}{r} 47.600 \\ -\ 32.509 \\ \hline 15.091 \end{array}$$ Write the decimal numbers in columns, attaching 0s as needed.

(c) 3 − 0.253

$$\begin{array}{r} 3.000 \\ -\ 0.253 \\ \hline 2.747 \end{array}$$ A whole number is assumed to have the decimal point at the right of the number. Write 3 as 3.000.

◄ **Work Problem 2 at the Side.**

OBJECTIVE 3 Multiply and divide decimals. To multiply decimals, follow these steps.

Multiplying Decimals

Step 1 Ignore the decimal points and multiply as if the numbers were whole numbers.

Step 2 Add the number of **decimal places** (digits to the *right* of the decimal point) in each factor. Place the decimal point that many digits from the right in the product.

---

1 Write each decimal as a fraction. (Do not write in lowest terms.)

(a) 0.8

(b) 0.431

(c) 2.58

2 Add or subtract as indicated.

(a) 68.9 + 42.72 + 8.973

(b) 351.8 − 2.706

(c) 32 − 21.72

Answers

1. (a) $\frac{8}{10}$ **(b)** $\frac{431}{1000}$ **(c)** $\frac{258}{100}$

2. (a) 120.593 **(b)** 349.094 **(c)** 10.28

EXAMPLE 3 **Multiplying Decimals**

Multiply.

(a) 29.3 × 4.52

```
   29.3     1 decimal place
 × 4.52     2 decimal places
    586       1 + 2 = 3
  1465
 1172
 132.436    3 decimal places
```

(b) 31.42 × 65

```
   31.42    2 decimal places
 ×    65    0 decimal places
   15710      2 + 0 = 2
  18852
 2042.30    2 decimal places
```

The final 0 can be dropped and the product written 2042.3.

(c) 0.05 × 0.3

Here 5 × 3 = 15. Be careful placing the decimal point.

2 decimal places 1 decimal place

0.05 × 0.3

= 0.015 — Do *not* write 0.150.

2 + 1 = 3 decimal places

Attach 0 as a placeholder in the tenths place.

Work Problem 3 at the Side. ▶

3 Multiply.

(a) 9.32 × 1.4

(b) 2.13 × 51

(c) 0.06 × 0.004

To divide decimals, follow these steps.

Dividing Decimals

Step 1 Change the **divisor** (the number we are dividing *by*) into a whole number by moving the decimal point as many places as necessary to the right.

Step 2 Move the decimal point in the **dividend** (the number we are dividing *into*) to the right by the same number of places.

Step 3 Move the decimal point straight up, and then divide as with whole numbers to find the **quotient.**

$$25\overline{)125} = 5$$

Divisor → 25; Dividend → 125; Quotient → 5

Remember this terminology for the parts of a division problem.

EXAMPLE 4 **Dividing Decimals**

Divide.

(a) 233.45 ÷ 11.5

Write the problem as follows. $11.5\overline{)233.45}$

$11.5\overline{)233.45}$ To change the divisor 11.5 into a whole number, move *each* decimal point one place to the right.

To see why this works, write the division in fraction form and multiply by $\frac{10}{10}$. The result is the same as when we moved the decimal point one place to the right in the divisor and the dividend.

$$\frac{233.45}{11.5} \cdot \frac{10}{10} = \frac{2334.5}{115}$$ Multiplying by $\frac{10}{10}$ is equivalent to multiplying by 1.

Continued on Next Page

Answers

3. (a) 13.048 (b) 108.63 (c) 0.00024

4 Divide.

(a) $451.47 \div 14.9$

(b) $5.476 \div 0.37$

(c) $7.334 \div 1.3$
(Round the quotient to two decimal places.)

5 Multiply or divide as indicated.

(a) 294.72×10

(b) 19.5×1000

(c) $4.793 \div 100$

(d) $960.1 \div 10$

Answers

4. (a) 30.3 (b) 14.8 (c) 5.64
5. (a) 2947.2 (b) 19,500 (c) 0.04793 (d) 96.01

Move the decimal point straight up, and divide as with whole numbers.

```
      20.3
115)2334.5     Move the decimal point straight up.
    230
     345       115 does not divide into 34, so we used
     345       0 as a placeholder in the quotient.
       0
```

(b) $8.949 \div 1.25$ (Round the quotient to two decimal places.)

```
1.25)8.949    Move each decimal point two places to the right.

       7.159
125)894.900    Move the decimal point straight up, and divide
    875        as with whole numbers. Attach 0s as placeholders.
     199
     125
      740
      625
      1150
      1125
        25
```

We carried out the division to three decimal places so that we could round to two decimal places, obtaining the quotient 7.16.

◄ Work Problem 4 at the Side.

Note

To round 7.159 in **Example 4(b)** to two decimal places (that is, to the nearest hundredth), we look at the digit to the *right* of the hundredths place. **If this digit is 5 or greater, we round up. If it is less than 5, we drop the digit(s) beyond the desired place.**

Hundredths place ↓

7.159 — 9, the digit to the right of the hundredths place, is 5 or greater.

≈ 7.16 — Round 5 up to 6. $\approx$ means "is approximately equal to."

Multiplying or Dividing by Powers of 10 (Shortcuts)

- To ***multiply*** by a power of 10, ***move the decimal point to the right*** as many places as the number of zeros.
- To ***divide*** by a power of 10, ***move the decimal point to the left*** as many places as the number of zeros.

In both cases, insert 0s as placeholders if necessary.

EXAMPLE 5 Multiplying and Dividing by Powers of 10

Multiply or divide as indicated.

(a) 48.731×100

$= 48.73\,1$ or 4873.1 — Move the decimal point two places to the right because 100 has two 0s.

(b) $48.7 \div 1000$

$= 048.7$ or 0.0487 — Move the decimal point three places to the left because 1000 has three 0s. Insert a 0 in front of the 4 to do this.

◄ Work Problem 5 at the Side.

OBJECTIVE 4 Write fractions as decimals.

Writing a Fraction as a Decimal

Because a fraction bar indicates division, write a fraction as a decimal by dividing the numerator by the denominator.

EXAMPLE 6 Writing Fractions as Decimals

Write each fraction as a decimal.

(a) $\frac{19}{8}$

$$\begin{array}{r} 2.375 \\ 8\overline{)19.000} \\ \underline{16} \\ 3\,0 \\ \underline{2\,4} \\ 60 \\ \underline{56} \\ 40 \\ \underline{40} \\ 0 \end{array}$$

Divide 19 by 8. Add a decimal point and as many 0s as necessary.

$\frac{19}{8} = 2.375$ ← Terminating decimal

(b) $\frac{2}{3}$

$$\begin{array}{r} 0.6666\ldots \\ 3\overline{)2.0000\ldots} \\ \underline{1\,8} \\ 20 \\ \underline{18} \\ 20 \\ \underline{18} \\ 20 \\ \underline{18} \\ 2 \end{array}$$

$\frac{2}{3} = 0.6666\ldots$ ← Repeating decimal

- The remainder in the division in part (a) is 0, so this quotient is a **terminating decimal.**
- The remainder in the division in part (b) is never 0. Because a number, in this case 2, is always left after the subtraction, this quotient is a **repeating decimal.** A convenient notation for a repeating decimal is a bar over the digit (or digits) that repeats.

$$\frac{2}{3} = 0.6666\ldots, \quad \text{or} \quad 0.\overline{6}$$

We often round repeating decimals to as many places as needed.

$$\frac{2}{3} \approx 0.667 \quad \text{An approximation to the nearest thousandth}$$

Work Problem 6 at the Side. ▶

6 Write each fraction as a decimal. For repeating decimals, write the answer by first using bar notation and then rounding to the nearest thousandth.

(a) $\frac{2}{9}$

(b) $\frac{17}{20}$

(c) $\frac{1}{11}$

(d) $\frac{13}{5}$

OBJECTIVE 5 Write percents as decimals and decimals as percents.

The word **percent** means **"per 100."** Percent is written with the symbol **%**. ***One percent means "one per one hundred," or "one one-hundredth."***

Percent, Fraction, and Decimal Equivalents

$$1\% = \frac{1}{100}, \quad \text{or} \quad 1\% = 0.01$$

EXAMPLE 7 Writing Percents as Decimals

Write each percent as a decimal.

(a) 73%   We use the fact that 1% = 0.01 and convert as follows.

$$73\% = 73 \cdot 1\% = 73 \cdot 0.01 = 0.73$$

Continued on Next Page

Answers

6. **(a)** $0.\overline{2}$, 0.222 **(b)** 0.85 **(c)** $0.\overline{09}$, 0.091 **(d)** 2.6

7 Write each percent as a decimal.

(a) 23% (b) 310%

(c) $5\frac{1}{4}\%$ (d) 40%

8 Write each decimal as a percent.

(a) 0.71 (b) 1.32

(c) 0.06 (d) 0.685

9 Convert each percent to a decimal and each decimal to a percent.

(a) 52% (b) 2%

(c) 0.45 (d) 3.5

Answers

7. (a) 0.23 (b) 3.10, or 3.1 (c) 0.0525 (d) 0.4

8. (a) 71% (b) 132% (c) 6% (d) 68.5%

9. (a) 0.52 (b) 0.02 (c) 45% (d) 350%

(b) $125\% = 125 \cdot 1\% = 125 \cdot 0.01 = 1.25$ $\quad 1\% = 0.01$

A percent greater than 100 represents a number greater than 1.

(c) $3\frac{1}{2}\%$

First write the fractional part as a decimal.

$$3\frac{1}{2}\% = (3 + 0.5)\% = 3.5\%$$

Now write the percent in decimal form.

$$3.5\% = 3.5 \cdot 1\% = 3.5 \cdot 0.01 = 0.035$$

Be careful placing the decimal point.

◀ Work Problem **7** at the Side.

EXAMPLE 8 Writing Decimals as Percents

Write each decimal as a percent.

(a) 0.32

This conversion is the opposite of what we did in **Example 7** when we wrote percents as decimals. We use $1\% = 0.01$ in reverse.

$$0.32 = 32 \cdot 0.01 = 32 \cdot 1\% = 32\% \quad 0.01 = 1\%$$

(b) $0.05 = 5 \cdot 0.01 = 5 \cdot 1\% = 5\%$ $\quad 0.01 = 1\%$

(c) $2.63 = 263 \cdot 0.01 = 263 \cdot 1\% = 263\%$

A number greater than 1 is more than 100%.

◀ Work Problem **8** at the Side.

Converting Percents and Decimals (Shortcuts)

- To convert a percent to a decimal, move the decimal point two places to the *left* and drop the % symbol.
- To convert a decimal to a percent, move the decimal point two places to the *right* and attach a % symbol.

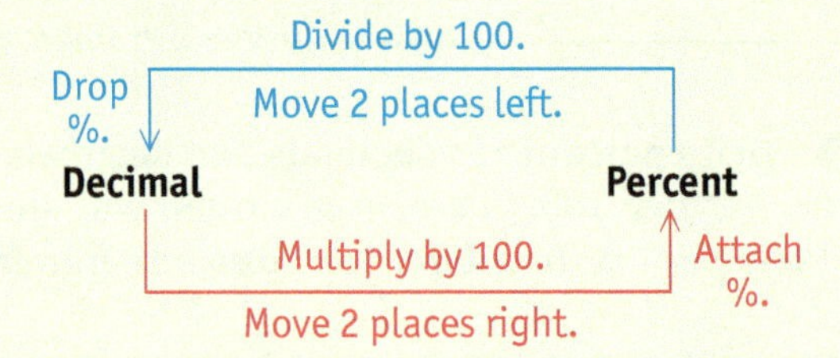

EXAMPLE 9 Converting Percents and Decimals by Moving the Decimal Point

Convert each percent to a decimal and each decimal to a percent.

(a) $45\% = 0.45$ **(b)** $250\% = 2.50$ **(c)** $9\% = 09\% = 0.09$

(d) $0.57 = 57\%$ **(e)** $1.5 = 1.50 = 150\%$ **(f)** $0.007 = 0.007 = 0.7\%$

◀ Work Problem **9** at the Side.

OBJECTIVE 6 Write percents as fractions and fractions as percents.

EXAMPLE 10 Writing Percents as Fractions

Write each percent as a fraction. Give answers in lowest terms as needed.

(a) 8%

We use the fact that $1\% = \frac{1}{100}$, and convert as follows.

$$8\% = 8 \cdot 1\% = 8 \cdot \frac{1}{100} = \frac{8}{100}$$

In lowest terms, $\frac{8}{100} = \frac{2 \cdot 4}{25 \cdot 4} = \frac{2}{25}.$

Thus, $8\% = \frac{2}{25}$.

(b) $175\% = 175 \cdot 1\% = 175 \cdot \frac{1}{100} = \frac{175}{100}$

In lowest terms, $\frac{175}{100} = \frac{7 \cdot 25}{4 \cdot 25} = \frac{7}{4}$, or $1\frac{3}{4}$.

A number greater than 1 is more than 100%.

(c) 13.5%

$= 13\frac{1}{2} \cdot 1\%$ Write 13.5 as a mixed number.

$= \frac{27}{2} \cdot \frac{1}{100}$ Write $13\frac{1}{2}$ as an improper fraction. Use the fact that $1\% = \frac{1}{100}$.

$= \frac{27}{200}$ Multiply the fractions.

Work Problem 10 at the Side. ▶

We know that 100% of something is the whole thing. One way to convert a fraction to a percent is to multiply by 100%, which is equivalent to 1.

EXAMPLE 11 Writing Fractions as Percents

Write each fraction as a percent.

(a) $\frac{2}{5}$

$= \frac{2}{5} \cdot 100\%$ Multiply by 1 in the form 100%.

$= \frac{2}{5} \cdot \frac{100}{1}\%$ $a = \frac{a}{1}$

$= \frac{2 \cdot 5 \cdot 20}{5 \cdot 1}\%$ Multiply and factor.

$= \frac{2 \cdot 20}{1}\%$ Divide out the common factor.

$= 40\%$ Simplify.

(b) $\frac{1}{6}$

$= \frac{1}{6} \cdot 100\%$

$= \frac{1}{6} \cdot \frac{100}{1}\%$

$= \frac{1 \cdot 2 \cdot 50}{2 \cdot 3 \cdot 1}\%$

$= \frac{50}{3}\%$

$= 16\frac{2}{3}\%$, or $16.\overline{6}\%$

Work Problem 11 at the Side. ▶

---

10 Write each percent as a fraction. Give answers in lowest terms as needed.

(a) 20%

(b) 160%

(c) 1.5%

11 Write each fraction as a percent.

(a) $\frac{6}{25}$

(b) $\frac{2}{9}$

Answers

10. (a) $\frac{1}{5}$ (b) $\frac{8}{5}$, or $1\frac{3}{5}$ (c) $\frac{3}{200}$

11. (a) 24% (b) $22\frac{2}{9}\%$, or $22.\overline{2}\%$

12 Solve the problem.

A pair of jeans that regularly sells for $69 is on sale at 30% off. Find the amount of the discount and the sale price of the jeans.

OBJECTIVE 7 Solve applied problems that involve percents. The decimal form of a percent is generally used in calculations.

EXAMPLE 12 Using Percent to Solve an Applied Problem

A DVD with a regular price of $18 is on sale this week at 22% off. Find the amount of the discount and the sale price of the DVD.

The discount is 22% *of* 18. The word *of* here means multiply.

$$\begin{array}{ccc} 22\% & \text{of} & 18 \\ \downarrow & \downarrow & \downarrow \\ 0.22 & \cdot & 18 \end{array} \quad \text{Write 22\% as a decimal.}$$

$$= 3.96 \quad \text{Multiply.}$$

The discount is $3.96. The sale price is found by subtracting.

$$\$18.00 - \$3.96 = \$14.04 \quad \text{Original price} - \text{discount} = \text{sale price}$$

The sale price is $14.04.

◀ Work Problem **12** at the Side.

Answer

12. $20.70; $48.30

R.2 Exercises

FOR EXTRA HELP — *Go to MyMathLab for worked-out, step-by-step solutions to exercises enclosed in a square and video solutions to ▶ exercises.*

CONCEPT CHECK *Provide the correct response.*

1. For the decimal number 367.9412, name the digit that has each place value.

(a) tens **(b)** tenths **(c)** thousandths

(d) ones or units **(e)** hundredths

2. For the decimal number 46.249, round to the place value indicated.

(a) hundredths **(b)** tenths

(c) ones or units **(d)** tens

3. Round each decimal to the nearest thousandth.

(a) $0.\overline{8}$ **(b)** $0.\overline{5}$

(c) 0.9762 **(d)** 0.8642

4. Find each product or quotient.

(a) 25.4×10 **(b)** 25.4×100

(c) $25.4 \div 100$ **(d)** $25.4 \div 1000$

Write each decimal as a fraction. (Do not write in lowest terms.) **See Example 1.**

5. 0.4 **6.** 0.6 **7.** 0.64 **8.** 0.82 **9.** 0.138

10. 0.104 **11.** 0.043 **12.** 0.087 **13.** 3.805 **14.** 5.166

Add or subtract as indicated. **See Example 2.**

15. $25.32 + 109.2 + 8.574$ **16.** $90.527 + 32.43 + 589.8$ **17.** $28.73 - 3.12$ **18.** $46.88 - 13.45$

19. $43.5 - 28.17$ **20.** $345.1 - 56.31$ **21.** $3.87 + 15 + 2.9$ **22.** $8.2 + 1.09 + 12$

23. $32.56 + 47.356 + 1.8$ **24.** $75.2 + 123.96 + 3.897$ **25.** $18 - 2.789$ ▶ **26.** $29 - 8.582$

Multiply or divide as indicated. **See Examples 3–5.**

27. 12.8×9.1 **28.** 34.04×0.56 **29.** 0.2×0.03 **30.** 0.07×0.004

31. $78.65 \div 11$ ▶ **32.** $73.36 \div 14$ **33.** $19.967 \div 9.74$ **34.** $44.4788 \div 5.27$

35. 57.116×100 **36.** 82.053×100 **37.** 0.094×1000 **38.** 0.025×1000

39. $1.62 \div 10$ **40.** $8.04 \div 10$ **41.** $24.03 \div 100$ **42.** $490.35 \div 100$

CONCEPT CHECK *Complete the table of fraction, decimal, and percent equivalents.*

	Fraction in Lowest Terms (or Whole Number)	Decimal	Percent
43.	$\frac{1}{100}$	0.01	
44.	$\frac{1}{50}$		2%
45.		0.05	5%
46.	$\frac{1}{10}$		
47.	$\frac{1}{8}$	0.125	
48.			20%
49.	$\frac{1}{4}$		
50.	$\frac{1}{3}$		
51.			50%
52.	$\frac{2}{3}$		$66\frac{2}{3}\%$, or $66.\overline{6}\%$
53.		0.75	
54.	1	1.0	

Write each fraction as a decimal. For repeating decimals, write the answer by first using bar notation and then rounding to the nearest thousandth. ***See Example 6.***

55. $\frac{1}{8}$ **56.** $\frac{7}{8}$ **57.** $\frac{5}{4}$ **58.** $\frac{9}{5}$

59. $\frac{5}{9}$ **60.** $\frac{8}{9}$ **61.** $\frac{1}{6}$ **62.** $\frac{5}{6}$

Write each percent as a decimal. ***See Examples 7 and 9(a)–(c).***

63. 54% **64.** 39% **65.** 7% **66.** 4% **67.** 90%

68. 10% **69.** 117% **70.** 189% **71.** 2.4% **72.** 3.1%

73. $6\frac{1}{4}\%$ **74.** $5\frac{1}{2}\%$ **75.** 0.8% **76.** 0.5%

Write each decimal or whole number as a percent. ***See Examples 8 and 9(d)–(f).***

77. 0.73 **78.** 0.83 **79.** 0.02 **80.** 0.08 **81.** 0.004 **82.** 0.005

83. 1.28 **84.** 2.35 **85.** 0.3 **86.** 0.6 **87.** 6 **88.** 10

Write each percent as a fraction. Give answers in lowest terms as needed. **See Example 10.**

89. 51% **90.** 47% **91.** 15% **92.** 35% **93.** 2%

94. 8% **95.** 140% **96.** 180% **97.** 7.5% **98.** 2.5%

Write each fraction as a percent. **See Example 11.**

99. $\frac{4}{5}$ **100.** $\frac{3}{25}$ **101.** $\frac{7}{50}$ **102.** $\frac{9}{20}$ **103.** $\frac{2}{11}$

104. $\frac{4}{9}$ **105.** $\frac{9}{4}$ **106.** $\frac{8}{5}$ **107.** $\frac{13}{6}$ **108.** $\frac{31}{9}$

Solve each problem. **See Example 12.**

109. What is 50% of 320? **110.** What is 25% of 120? **111.** What is 6% of 80?

112. What is 5% of 70? **113.** What is 14% of 780? **114.** What is 26% of 480?

Solve each problem. **See Example 12.**

115. Elwyn's bill for dinner at a restaurant was $89. He wants to leave a 20% tip. How much should he leave for the tip? What is his total bill for dinner and tip?

116. Gary earns $15 per hour at his job. He recently received a 7% raise. How much per hour was his raise? What is his new hourly rate?

117. Find the discount on a leather recliner with a regular price of $795 if the recliner is on sale at 15% off. What is the sale price of the recliner?

118. A laptop computer with a regular price of $597 is on sale at 20% off. Find the amount of the discount and the sale price of the computer.

In a recent year, approximately 60 *million people from other countries visited the United States. The circle graph shows the distribution of these international visitors by country or region. Use the graph to work each problem.*

119. How many travelers visited the United States from Canada?

120. How many travelers visited the United States from Mexico?

121. What percent of travelers visited the United States from places other than Canada, Mexico, Europe, or Asia? (*Hint:* The sum of the parts of the graph must equal 1 whole, that is, 100%.)

122. Use the answer from **Exercise 121** to find how many travelers visited the United States from places other than Canada, Mexico, Europe, or Asia.

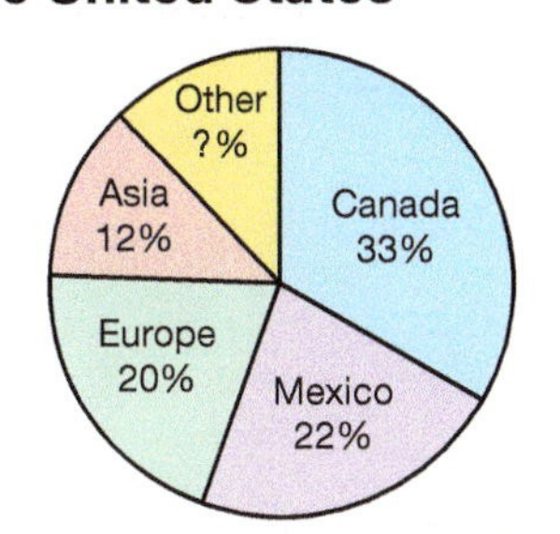

Data from U.S. Department of Commerce.

Study Skills

USING YOUR MATH TEXT

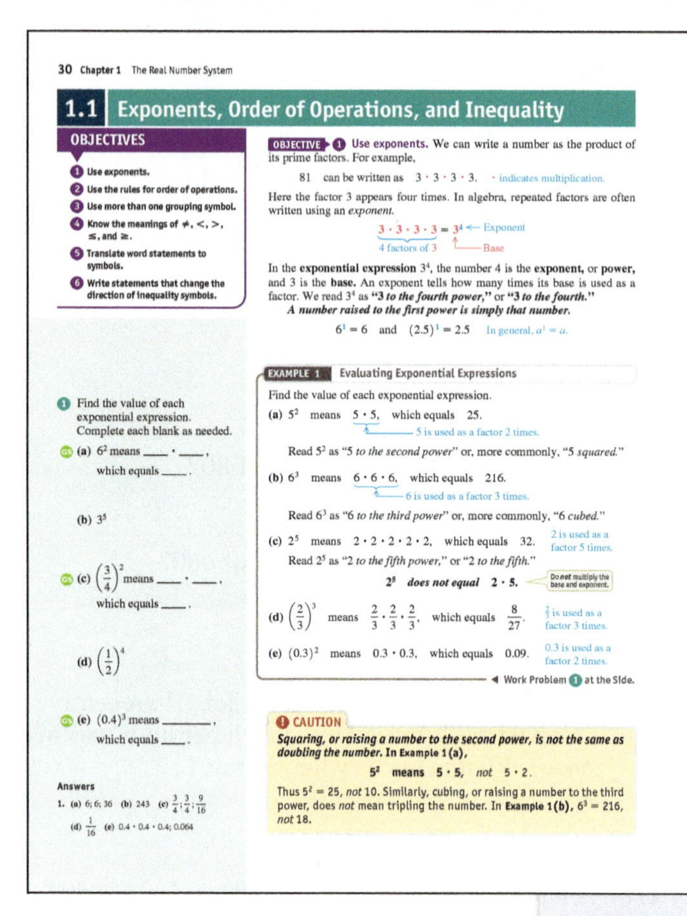

30 Chapter 1 The Real Number System

1.1 Exponents, Order of Operations, and Inequality

OBJECTIVES

1. Use exponents.
2. Use the rules for order of operations.
3. Use more than one grouping symbol.
4. Know the meanings of ≠, <, >, ≤, and ≥.
5. Translate word statements to symbols.
6. Write statements that change the direction of inequality symbols.

OBJECTIVE 1 **Use exponents.** We can write a number as the product of its prime factors. For example,

81 can be written as $3 \cdot 3 \cdot 3 \cdot 3$. · indicates multiplication.

Here the factor 3 appears four times. In algebra, repeated factors are often written using an *exponent.*

$3 \cdot 3 \cdot 3 \cdot 3 = 3^4$ ← Exponent; 4 factors of 3; Base

In the **exponential expression** 3^4, the number 4 is the **exponent,** or **power,** and 3 is the **base.** An exponent tells how many times its base is used as a factor. We read 3^4 as **"3 *to the fourth power,"*** or **"3 *to the fourth.*"**

A number raised to the first power is simply that number.

$6^1 = 6$ and $(2.5)^1 = 2.5$ In general, $a^1 = a$.

EXAMPLE 1 Evaluating Exponential Expressions

Find the value of each exponential expression.

(a) 5^2 means $5 \cdot 5$, which equals 25. — 5 is used as a factor 2 times.

Read 5^2 as "5 *to the second power*" or, more commonly, "5 *squared.*"

(b) 6^3 means $6 \cdot 6 \cdot 6$, which equals 216. — 6 is used as a factor 3 times.

Read 6^3 as "6 *to the third power*" or, more commonly, "6 *cubed.*"

(c) 2^5 means $2 \cdot 2 \cdot 2 \cdot 2 \cdot 2$, which equals 32. — 2 is used as a factor 5 times.

Read 2^5 as "2 *to the fifth power,*" or "2 *to the fifth.*"

2^5 *does not equal* $2 \cdot 5$. — Do *not* multiply the base and exponent.

(d) $\left(\frac{2}{3}\right)^3$ means $\frac{2}{3} \cdot \frac{2}{3} \cdot \frac{2}{3}$, which equals $\frac{8}{27}$. — $\frac{2}{3}$ is used as a factor 3 times.

(e) $(0.3)^2$ means $0.3 \cdot 0.3$, which equals 0.09. — 0.3 is used as a factor 2 times.

◄ Work Problem 1 at the Side.

CAUTION

Squaring, or raising a number to the second power, is not the same as doubling the number. In **Example 1 (a),**

5^2 **means** $5 \cdot 5$, *not* $5 \cdot 2$.

Thus $5^2 = 25$, *not* 10. Similarly, cubing, or raising a number to the third power, does *not* mean tripling the number. In **Example 1 (b),** $6^3 = 216$, *not* 18.

1 Find the value of each exponential expression. Complete each blank as needed.

(a) 6^2 means ___ · ___, which equals ___.

(b) 3^5

(c) $\left(\frac{3}{4}\right)^2$ means ___ · ___, which equals ___.

(d) $\left(\frac{1}{2}\right)^4$

(e) $(0.4)^3$ means ______, which equals ___.

Answers

1. (a) 6; 6; 36 (b) 243 (c) $\frac{3}{4}; \frac{3}{4}; \frac{9}{16}$ (d) $\frac{1}{16}$ (e) $0.4 \cdot 0.4 \cdot 0.4$; 0.064

Your text is a valuable resource. You will learn more if you make full use of the features it offers.

Now Try This

General Features

Locate each feature, and complete any blanks.

- **Table of Contents** This is located at the front of the text.
  Find it and mark the chapters and sections you will cover, as noted on your course syllabus.
- **Answer Section** ***Tab this section at the back of the text.***
  Refer to it frequently when doing homework. Answers to odd-numbered section exercises are provided. Answers to ALL Concept Check, writing, Relating Concepts, summary, chapter review, test, and cumulative review exercises are given.
- **List of Formulas** Use this helpful list of geometric formulas, along with review information on triangles and angles, throughout the course.
  Find this information at the back of the text.
  The formula for the volume of a cube is ______.

Specific Features

Look through Chapter 2 and give the number of a page that includes an example of each of the following specific features.

- **Objectives** The objectives are listed at the beginning of each section and again within the section as the corresponding material is presented. Once you finish a section, ask yourself if you have accomplished them. ***See page*** ______.
- **Margin Problems** These exercises allow you to immediately practice the material covered in the examples and prepare you for the exercise set. Check your results using the answers at the bottom of the page. ***See page*** ______.
- **Pointers** These small shaded balloons provide on-the-spot warnings and reminders, point out key steps, and give other helpful tips. ***See page*** ______.
- **Cautions** These provide warnings about common errors that students often make or trouble spots to avoid. ***See page*** ______.
- **Notes** These provide additional explanations or emphasize important ideas. ***See page*** ______.
- **Problem-Solving Hints** These boxes give helpful tips or strategies to use when you work applications. ***See page*** ______.

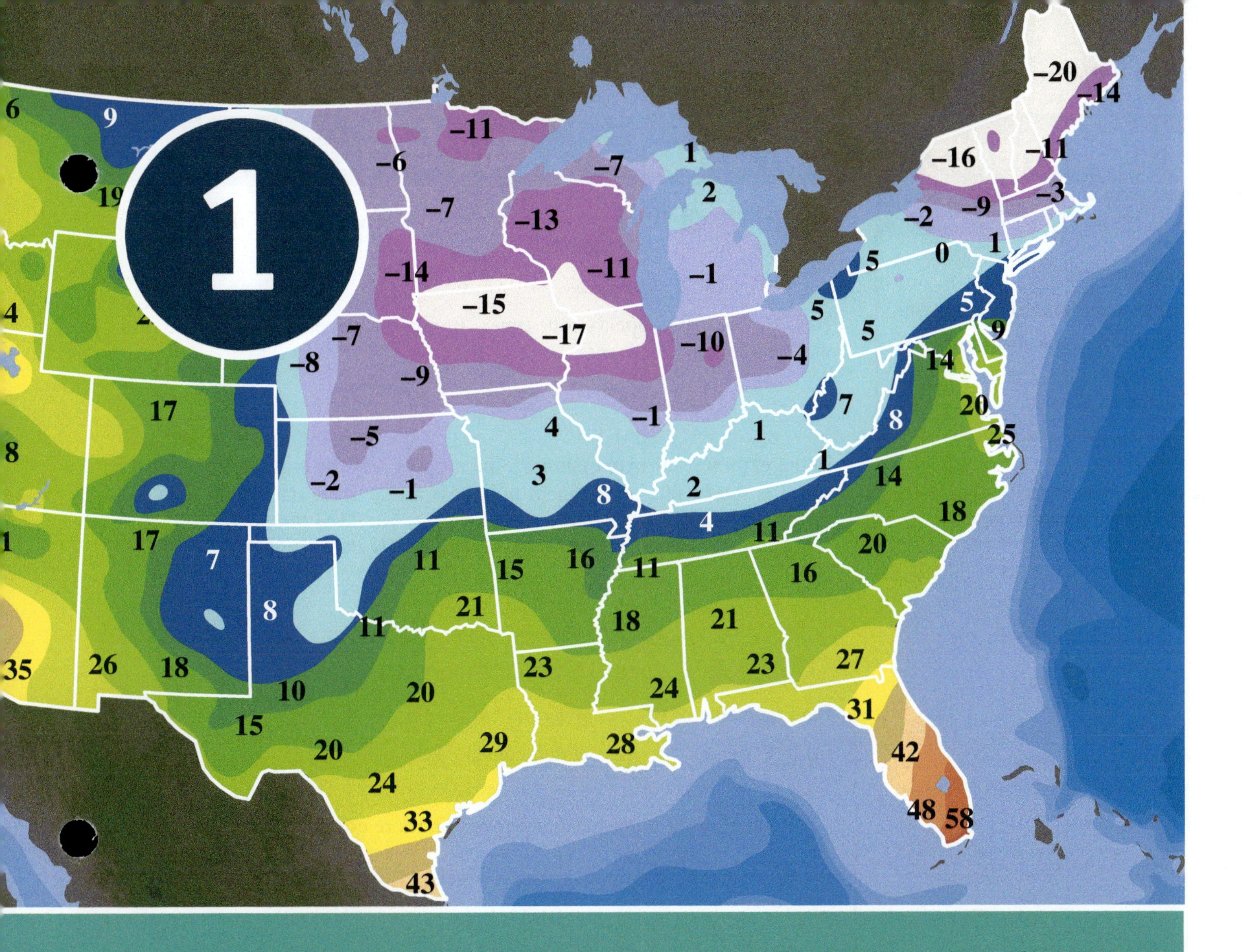

The Real Number System

Positive and *negative numbers,* used to indicate temperatures above and below zero, elevations above and below sea level, and gains and losses in the stock market or on a football field, are examples of *real numbers,* the subject of this chapter.

1.1 Exponents, Order of Operations, and Inequality

OBJECTIVES

1. Use exponents.
2. Use the rules for order of operations.
3. Use more than one grouping symbol.
4. Know the meanings of $\neq$, $<$, $>$, $\leq$, and $\geq$.
5. Translate word statements to symbols.
6. Write statements that change the direction of inequality symbols.

OBJECTIVE 1 Use exponents. We can write a number as the product of its prime factors. For example,

$$81 \quad \text{can be written as} \quad 3 \cdot 3 \cdot 3 \cdot 3. \qquad \cdot \text{ indicates multiplication.}$$

Here the factor 3 appears four times. In algebra, repeated factors are often written using an *exponent.*

$$\underbrace{3 \cdot 3 \cdot 3 \cdot 3}_{\text{4 factors of 3}} = 3^4 \leftarrow \text{Exponent}$$

(3 is the Base.)

In the **exponential expression** 3^4, the number 4 is the **exponent,** or **power,** and 3 is the **base.** An exponent tells how many times its base is used as a factor. We read 3^4 as ***"3 to the fourth power,"*** or ***"3 to the fourth."***

A number raised to the first power is simply that number.

$$6^1 = 6 \quad \text{and} \quad (2.5)^1 = 2.5 \qquad \text{In general, } a^1 = a.$$

EXAMPLE 1 Evaluating Exponential Expressions

Find the value of each exponential expression.

(a) 5^2 means $5 \cdot 5$, which equals 25.

5 is used as a factor 2 times.

Read 5^2 as "5 *to the second power*" or, more commonly, "5 *squared.*"

(b) 6^3 means $6 \cdot 6 \cdot 6$, which equals 216.

6 is used as a factor 3 times.

Read 6^3 as "6 *to the third power*" or, more commonly, "6 *cubed.*"

(c) 2^5 means $2 \cdot 2 \cdot 2 \cdot 2 \cdot 2$, which equals 32. 2 is used as a factor 5 times.

Read 2^5 as "2 *to the fifth power,*" or "2 *to the fifth.*"

$$2^5 \textit{ does not equal } 2 \cdot 5.$$

Do **not** multiply the base and exponent.

(d) $\left(\frac{2}{3}\right)^3$ means $\frac{2}{3} \cdot \frac{2}{3} \cdot \frac{2}{3}$, which equals $\frac{8}{27}$. $\frac{2}{3}$ is used as a factor 3 times.

(e) $(0.3)^2$ means $0.3 \cdot 0.3$, which equals 0.09. 0.3 is used as a factor 2 times.

◀ Work Problem 1 at the Side.

1 Find the value of each exponential expression. Complete each blank as needed.

GS **(a)** 6^2 means ____ • ____, which equals ____.

(b) 3^5

GS **(c)** $\left(\frac{3}{4}\right)^2$ means ____ • ____, which equals ____.

(d) $\left(\frac{1}{2}\right)^4$

GS **(e)** $(0.4)^3$ means ________, which equals ____.

! CAUTION

Squaring, or raising a number to the second power, is not the same as doubling the number. In **Example 1(a),**

$$5^2 \text{ means } 5 \cdot 5, \textit{ not } 5 \cdot 2.$$

Thus $5^2 = 25$, *not* 10. Similarly, cubing, or raising a number to the third power, does *not* mean tripling the number. In **Example 1(b),** $6^3 = 216$, *not* 18.

Answers

1. (a) 6; 6; 36 (b) 243 (c) $\frac{3}{4}$; $\frac{3}{4}$; $\frac{9}{16}$ (d) $\frac{1}{16}$ (e) $0.4 \cdot 0.4 \cdot 0.4$; 0.064

OBJECTIVE 2 Use the rules for order of operations. When an expression involves more than one operation, we often use **grouping symbols,** such as parentheses ( ), to indicate the order in which the operations should be performed.

Consider the following expression.

$$5 + 2 \cdot 3$$

To show that the multiplication should be performed before the addition, we use parentheses to group $2 \cdot 3$.

$$5 + (2 \cdot 3) \quad \text{equals} \quad 5 + 6, \quad \text{which equals} \quad 11.$$

If addition is to be performed first, the parentheses should group $5 + 2$.

$$(5 + 2) \cdot 3 \quad \text{equals} \quad 7 \cdot 3, \quad \text{which equals} \quad 21.$$

Other grouping symbols are brackets [ ], braces { }, and fraction bars. (For example, in $\frac{8-2}{3}$, the expression $8 - 2$ is "grouped" in the numerator.)

To simplify an expression involving more than one operation, we use the following rules for **order of operations.** This order is used by most calculators and computers.

Order of Operations

If grouping symbols are present, work within them, innermost first (and above and below fraction bars separately), in the following order.

Step 1 Apply all **exponents.**

Step 2 Do any **multiplications** or **divisions** in order from left to right.

Step 3 Do any **additions** or **subtractions** in order from left to right.

If no grouping symbols are present, start with Step 1.

Note

Multiplication is understood in expressions with parentheses.

Examples: $3(7)$, $(6)2$, $(5)(4)$, $3(4 + 1)$

EXAMPLE 2 Using the Rules for Order of Operations

Find the value of each expression.

(a) $24 - 12 \div 3$ (Be careful. Divide first.)

$= 24 - 4$ Divide.

$= 20$ Subtract.

(b) $9(6 + 11)$

$= 9(17)$ Work inside the parentheses.

$= 153$ Multiply.

(c) $6 \cdot 8 + 5 \cdot 2$

$= 48 + 10$ Multiply, working from left to right.

$= 58$ Add.

(d) $48 \div 2 \cdot 3$

$= 24 \cdot 3$ Divide.

$= 72$ Multiply.

Multiplications and divisions are done from left to right as they appear. Additions and subtractions are then done, again working from left to right.

Continued on Next Page

2 **GS** Label the order in which each operation should be performed. Then find the value of each expression.

(a) $7 + 3 \cdot 8$

② ①

$= 7 + \_\_\_\_$

$= \_\_\_\_$

(b) $7 \cdot 3 - 2 \cdot 9$

○ ○ ○

$= \_\_\_\_ - \_\_\_\_$

$= \_\_\_\_$

(c) $8 + 2(5 - 1)$

○ ○ ○

(d) $7 \cdot 6 - 3(8 + 1)$

○ ○ ○ ○

(e) $10 - 3^2 + 5$

○ ○ ○

3 Find the value of each expression.

(a) $9[(4 + 8) - 3]$

(b) $\dfrac{2(7 + 8) + 2}{3 \cdot 5 + 1}$

Answers

2. **(a)** 24; 31
(b) ①, ③, ②; 21; 18; 3
(c) ③, ②, ①; 16
(d) ②, ④, ③, ①; 15
(e) ②, ①, ③; 6

3. **(a)** 81 **(b)** 2

(e) $16 - 3(2 + 3)$ — Do *not* subtract 16 − 3 first.

$= 16 - 3(5)$ Add inside the parentheses.

$= 16 - 15$ Multiply.

$= 1$ Subtract.

(f) $2(5 + 6) + 7 \cdot 3$

Start here. $= 2(11) + 7 \cdot 3$ Work inside the parentheses.

$= 22 + 21$ Multiply.

$= 43$ Add.

(g) $9 + 2^3 - 5$ — $2^3 = 2 \cdot 2 \cdot 2$, not $2 \cdot 3$

$= 9 + 8 - 5$ Apply the exponent.

$= 12$ Add, and then subtract.

◀ Work Problem **2** at the Side.

OBJECTIVE 3 **Use more than one grouping symbol.** In an expression like

$$2(8 + 3(6 + 5)),$$

we often use brackets, [ ], in place of the outer pair of parentheses.

EXAMPLE 3 Using Brackets and Fraction Bars as Grouping Symbols

Find the value of each expression.

(a) $2[8 + 3(6 + 5)]$ — Start here.

$= 2[8 + 3(11)]$ Add inside the parentheses.

$= 2[8 + 33]$ Multiply inside the brackets.

$= 2[41]$ Add inside the brackets.

$= 82$ Multiply.

(b) $\dfrac{4(5 + 3) + 3}{3 \cdot 2 - 1}$ Simplify the numerator and denominator separately.

$= \dfrac{4(8) + 3}{3 \cdot 2 - 1}$ Work inside the parentheses.

$= \dfrac{32 + 3}{6 - 1}$ Multiply.

$= \dfrac{35}{5}$ Add and subtract.

$= 7$ Divide.

◀ Work Problem **3** at the Side.

Note

The expression $\frac{4(5+3)+3}{3 \cdot 2 - 1}$ in **Example 3(b)** can be written as a quotient.

$$[4(5+3)+3] \div [3 \cdot 2 - 1]$$

The fraction bar "groups" the numerator and denominator separately.

OBJECTIVE 4 Know the meanings of ≠, <, >, ≤, and ≥. So far, we have used the equality symbol $=$. The symbols

$$\neq, \quad <, \quad >, \quad \leq, \quad \text{and} \quad \geq \qquad \text{Inequality symbols}$$

are used to express an **inequality,** a statement that two expressions may not be equal. The equality symbol with a slash through it, ≠, means "is *not* equal to."

$$7 \neq 8 \qquad \text{7 is not equal to 8.}$$

If two numbers are not equal, then one of the numbers must be less than the other. Reading from left to right, the symbol < means "is less than."

$$7 < 8 \qquad \text{7 is less than 8.}$$

Reading from left to right, the symbol > means "is greater than."

$$8 > 2 \qquad \text{8 is greater than 2.}$$

To keep the meanings of the symbols < and > clear, remember that the symbol always points to the lesser number.

$$\text{Lesser number} \rightarrow 8 < 15$$

$$15 > 8 \leftarrow \text{Lesser number}$$

Reading from left to right, the symbol ≤ means "is less than or equal to."

$$5 \leq 9 \qquad \text{5 is less than or equal to 9.}$$

If either the < part or the = part is true, then the inequality ≤ is true. Thus, $5 \leq 9$ is true because $5 < 9$ is true. Also, $8 \leq 8$ is true because $8 = 8$ is true.

The symbol ≥ means "is greater than or equal to."

$$9 \geq 5 \qquad \text{9 is greater than or equal to 5.}$$

EXAMPLE 4 Using Inequality Symbols

Determine whether each statement is *true* or *false*.

(a) $6 \neq 5 + 1$ This statement is false because $6 = 5 + 1$.

(b) $5 + 3 < 19$ The statement $5 + 3 < 19$ is true because $8 < 19$.

(c) $15 \leq 20 \cdot 2$ The statement $15 \leq 20 \cdot 2$ is true because $15 < 40$.

(d) $25 \geq 30$ Both $25 > 30$ and $25 = 30$ are false, so $25 \geq 30$ is false.

(e) $12 \geq 12$ Because $12 = 12$, this statement is true.

(f) $\frac{6}{15} \geq \frac{2}{3}$ Find a common denominator.

$\frac{6}{15} \geq \frac{10}{15}$ The statements $\frac{6}{15} > \frac{10}{15}$ and $\frac{6}{15} = \frac{10}{15}$ are false. Because at least one of them is false, $\frac{6}{15} \geq \frac{2}{3}$ is also false.

Work Problem 4 at the Side. ▶

4 Determine whether each statement is *true* or *false*.

(a) $28 \neq 4 \cdot 7$

(b) $5 > 4 + 2$

(c) $25 \geq 10$

(d) $21 \leq 21$

(e) $9 \cdot 3 \geq 28$

(f) $\frac{4}{7} \leq \frac{5}{8}$

Answers

4. (a) false **(b)** false **(c)** true **(d)** true **(e)** false **(f)** true

5 Write each word statement in symbols.

(a) Nine is equal to eleven minus two.

(b) Seventeen is less than thirty.

(c) Eight is not equal to ten.

(d) Fourteen is greater than twelve.

(e) Thirty is less than or equal to fifty.

(f) Two is greater than or equal to two.

6 Write each statement as another true statement with the inequality symbol reversed.

(a) $8 < 10$

(b) $3 > 1$

(c) $\frac{2}{3} \geq \frac{3}{5}$

(d) $0.5 \leq 1.2$

Answers

5. (a) $9 = 11 - 2$ (b) $17 < 30$
(c) $8 \neq 10$ (d) $14 > 12$
(e) $30 \leq 50$ (f) $2 \geq 2$

6. (a) $10 > 8$ (b) $1 < 3$
(c) $\frac{3}{5} \leq \frac{2}{3}$ (d) $1.2 \geq 0.5$

OBJECTIVE 5 Translate word statements to symbols.

EXAMPLE 5 Translating from Words to Symbols

Write each word statement in symbols.

(a) Twelve **is equal to** ten **plus** two. $12 = 10 + 2$

(b) Nine **is less than** ten. $9 < 10$

Compare this with "9 less than 10," which is written $10 - 9$.

(c) Fifteen **is not equal to** eighteen. $15 \neq 18$

(d) Seven **is greater than** four. $7 > 4$

(e) Thirteen **is less than or equal to** forty. $13 \leq 40$

(f) Six **is greater than or equal to** six. $6 \geq 6$

◀ Work Problem 5 at the Side.

OBJECTIVE 6 Write statements that change the direction of inequality symbols.

Any statement with $<$ can be converted to one with $>$, and any statement with $>$ can be converted to one with $<$. ***We do this by reversing both the order of the numbers and the direction of the symbol.***

$$6 < 10 \quad \text{becomes} \quad 10 > 6.$$

Interchange numbers.

Reverse symbol.

EXAMPLE 6 Converting between Inequality Symbols

Write each statement as another true statement with the inequality symbol reversed.

(a) $5 > 2$ is equivalent to $2 < 5$. **(b)** $\frac{1}{2} \leq \frac{3}{4}$ is equivalent to $\frac{3}{4} \geq \frac{1}{2}$.

◀ Work Problem 6 at the Side.

SUMMARY OF EQUALITY AND INEQUALITY SYMBOLS

Symbol	Meaning	Example
$=$	Is equal to	$0.5 = \frac{1}{2}$ means 0.5 is equal to $\frac{1}{2}$.
$\neq$	Is not equal to	$3 \neq 7$ means 3 is not equal to 7.
$<$	Is less than	$6 < 10$ means 6 is less than 10.
$>$	Is greater than	$15 > 14$ means 15 is greater than 14.
$\leq$	Is less than or equal to	$4 \leq 8$ means 4 is less than or equal to 8.
$\geq$	Is greater than or equal to	$1 \geq 0$ means 1 is greater than or equal to 0.

! CAUTION

Equality and inequality symbols are used to write mathematical ***sentences.*** Operation symbols ($+$, $-$, $\cdot$, and $\div$) are used to write mathematical ***expressions*** that represent a number. Compare the following.

Sentence: $4 < 10$ ← Gives the relationship between 4 and 10

Expression: $4 + 10$ ← Tells how to operate on 4 and 10 to get 14

1.1 Exercises

Go to MyMathLab for worked-out, step-by-step solutions to exercises enclosed in a square and video solutions to ▶ exercises.

CONCEPT CHECK *Decide whether each statement is* true *or* false. *If it is false, explain why.*

1. $3^2 = 6$

2. $1^3 = 3$

3. $3^1 = 1$

4. The expression 6^2 means that 2 is used as a factor 6 times.

Find the value of each exponential expression. ***See Example 1.***

5. 7^2

6. 4^2

7. 3^2

8. 8^2

9. 12^2

10. 14^2

11. 4^3

12. 5^3

13. 10^3

14. 11^3

15. 3^4

16. 6^4

17. 4^5

18. 3^5

19. $\left(\frac{2}{3}\right)^4$

20. $\left(\frac{3}{4}\right)^3$

21. $\left(\frac{1}{6}\right)^2$

22. $\left(\frac{1}{3}\right)^2$

23. $(0.04)^3$

24. $(0.05)^4$

Find the value of each expression. ***See Examples 2 and 3.***

25. $13 + 9 \cdot 5$
○ ○
$= 13 +$ ____
$=$ ____

26. $11 + 7 \cdot 6$
○ ○
$= 11 +$ ____
$=$ ____

27. $20 - 4 \cdot 3 + 5$
○ ○ ○
$= 20 -$ ____ $+ 5$
$=$ ____ $+ 5$
$=$ ____

28. $18 - 7 \cdot 2 + 6$
○ ○ ○
$= 18 -$ ____ $+ 6$
$=$ ____ $+ 6$
$=$ ____

29. $9 \cdot 5 - 13$
○ ○

30. $7 \cdot 6 - 11$
○ ○

31. $18 - 2 + 3$
○ ○

32. $22 - 8 + 9$
○ ○

33. $64 \div 4 \cdot 2$

34. $250 \div 5 \cdot 2$

35. $9 \cdot 4 - 8 \cdot 3$

36. $11 \cdot 4 + 10 \cdot 3$

37. $\frac{1}{4} \cdot \frac{2}{3} + \frac{2}{5} \cdot \frac{11}{3}$

38. $\frac{9}{4} \cdot \frac{2}{3} + \frac{4}{5} \cdot \frac{5}{3}$

39. $25.2 - 12.6 \div 4.2$

40. $12.4 - 9.3 \div 3.1$

41. $10 + 40 \div 5 \cdot 2$

42. $12 + 35 \div 7 \cdot 3$

43. $18 - 2(3 + 4)$

44. $30 - 3(4 + 2)$

45. $3(4 + 2) + 8 \cdot 3$

46. $9(1 + 7) + 2 \cdot 5$

47. $18 - 4^2 + 3$

48. $22 - 2^3 + 9$

49. $5[3 + 4(2^2)]$

50. $6[2 + 8(3^3)]$

51. $3^2[(11 + 3) - 4]$

52. $4^2[(13 + 4) - 8]$

53. $2 + 3[5 + 4(2)]$

54. $5 + 4[1 + 7(3)]$

55. $\dfrac{6(3^2 - 1) + 8}{3 \cdot 2 - 2}$

56. $\dfrac{2(8^2 - 4) + 8}{4 \cdot 3 - 10}$

57. $\dfrac{4(7 + 2) + 8(8 - 3)}{6(4 - 2) - 2^2}$

58. $\dfrac{6(5 + 1) - 9(1 + 1)}{5(8 - 4) - 2^3}$

CONCEPT CHECK *Insert one pair of parentheses in each expression so that the given value results when the operations are performed.*

59. $3 \cdot 6 + 4 \cdot 2 = 60$

60. $2 \cdot 8 - 1 \cdot 3 = 42$

61. $10 - 7 - 3 = 6$

62. $15 - 10 - 2 = 7$

63. $8 + 2^2 = 100$

64. $4 + 2^2 = 36$

Simplify each expression involving an operation as needed. Then determine whether the given statement is true *or* false. ***See Examples 2–4.***

65. $8 \geq 17$

66. $10 \geq 41$

67. $\dfrac{1}{2} \leq \dfrac{2}{4}$

68. $\dfrac{3}{9} \leq \dfrac{1}{3}$

69. $17 \leq 18 - 1$

70. $12 \geq 10 + 2$

71. $9 \cdot 3 - 11 \leq 16$

72. $6 \cdot 5 - 12 \leq 18$

73. $6 \cdot 8 + 6 \cdot 6 \geq 0$

74. $4 \cdot 20 - 16 \cdot 5 \geq 0$

75. $6[5 + 3(4 + 2)] \leq 70$

76. $6[2 + 3(2 + 5)] \leq 135$

77. $\dfrac{9(7 - 1) - 8 \cdot 2}{4(6 - 1)} > 3$

78. $\dfrac{2(5 + 3) + 2 \cdot 2}{2(4 - 1)} > 1$

79. $8 \leq 4^2 - 2^2$

80. $10^2 - 8^2 > 6^2$

Write each word statement in symbols. ***See Example 5.***

81. Fifteen is equal to five plus ten.

82. Twelve is equal to twenty minus eight.

83. Nine is greater than five minus four.

84. Ten is greater than six plus one.

85. Sixteen is not equal to nineteen.

86. Three is not equal to four.

87. Two is less than or equal to three.

88. Five is less than or equal to six.

Write each statement in words, and decide whether it is true *or* false.
See Examples 4 and 5.

89. $7 < 19$

90. $9 < 10$

91. $8 \geq 11$

92. $4 \leq 2$

93. $\dfrac{1}{3} \neq \dfrac{3}{10}$

94. $\dfrac{10}{7} \neq \dfrac{3}{2}$

Write each statement as another true statement with the inequality symbol reversed. ***See Example 6.***

95. $5 < 30$ **96.** $8 > 4$ **97.** $12 \geq 3$ **98.** $25 \leq 41$

99. $2.5 \geq 1.3$ **100.** $4.1 \leq 5.3$ **101.** $\frac{4}{5} > \frac{3}{4}$ **102.** $\frac{8}{3} < \frac{11}{4}$

One way to measure a person's cardiofitness is to calculate how many METs, or metabolic units, he or she can reach at peak exertion. One MET is the amount of energy used when sitting quietly. To calculate ideal METs, we can use the following expressions.

$14.7 - \text{age} \cdot 0.13$ For women

$14.7 - \text{age} \cdot 0.11$ For men

(Data from *New England Journal of Medicine.*)

103. A 40-yr-old woman wishes to calculate her ideal MET.

(a) Write the expression, using her age.

(b) Calculate her ideal MET. (*Hint:* Use the rules for order of operations.)

(c) Researchers recommend that a person reach approximately 85% of his or her MET when exercising. Calculate 85% of the ideal MET from part (b). Then refer to the following table. What activity can the woman do that is approximately this value?

Activity	METs	Activity	METs
Golf (with cart)	2.5	Skiing (water or downhill)	6.8
Walking (3 mph)	3.3	Swimming	7.0
Mowing lawn (power)	4.5	Walking (5 mph)	8.0
Ballroom or square dancing	5.5	Jogging	10.2
Cycling	5.7	Skipping rope	12.0

Data from Harvard School of Public Health.

(d) Repeat parts (a)–(c) for a 55-yr-old man.

104. Repeat parts (a)–(c) of **Exercise 103** for your age and gender. For yourself 5 yr from now.

The table shows the number of pupils per teacher in U.S. public schools in selected states. Use the table to answer each question.

State	Pupils per Teacher
Alaska	15.3
Texas	14.7
California	22.2
Wyoming	12.7
Maine	11.8
Idaho	18.4
Missouri	14.1

Data from National Center for Education Statistics.

105. Which states had a number greater than 14.1?

106. Which states had a number that was at most 14.7?

107. Which states had a number not less than 14.1?

108. Which states had a number greater than 22.2?

Study Skills

TAKING LECTURE NOTES

Study the set of sample math notes given here.

- **Include the date and title** of the day's lecture topic.
- **Include definitions,** written here in parentheses—don't trust your memory.
- **Skip lines and write neatly** to make reading easier.
- **Emphasize direction words** (like *simplify*) with their explanations.
- **Mark important concepts with stars, underlining, etc.**
- **Use two columns,** which allows an example and its explanation to be close together.
- **Use brackets and arrows** to clearly show steps, related material, etc.

January 12

Exponents

Exponents used to show repeated multiplication.

$3 \cdot 3 \cdot 3 \cdot 3$ can be written 3^4 ← exponent (how many times it's multiplied); base (the number being multiplied)

Read 3^2 as 3 to the 2nd power or 3 squared

3^3 as 3 to the 3rd power or 3 cubed

3^4 as 3 to the 4th power

etc.

Simplifying an expression with exponents → actually do the repeated multiplication

2^3 means $2 \cdot 2 \cdot 2$ and $2 \cdot 2 \cdot 2 = 8$

☆ Careful! 5^2 means $5 \cdot 5$ NOT $5 \cdot 2$

so $5^2 = 5 \cdot 5 = 25$ BUT $5^2 \neq 10$

Example	Explanation
Simplify $2^4 \cdot 3^2$	Exponents mean multiplication.
$2 \cdot 2 \cdot 2 \cdot 2 \cdot 3 \cdot 3$	Use 2 as a factor 4 times. Use 3 as a factor 2 times.
$16 \cdot 9$	$2 \cdot 2 \cdot 2 \cdot 2$ is 16; $3 \cdot 3$ is 9; $16 \cdot 9$ is 144
144	Simplified result is 144 (no exponents left)

Now Try This

With a partner or in a small group, compare lecture notes. Then answer each question.

1. What are you doing to show main points in your notes (such as boxing, using stars, etc.)? ______
2. In what ways do you set off explanations from worked problems and subpoints (such as indenting, using arrows, circling, etc.)? ______
3. What new ideas did you learn by examining your classmates' notes? ______
4. What new techniques will you try when taking notes in future lectures? ______

1.2 Variables, Expressions, and Equations

OBJECTIVES

1. Evaluate algebraic expressions, given values for the variables.
2. Translate word phrases to algebraic expressions.
3. Identify solutions of equations.
4. Translate sentences to equations.
5. Distinguish between equations and expressions.

A **constant** is a fixed, unchanging number. A **variable** is a symbol, usually a letter, used to represent an unknown number.

$5, \frac{3}{4}, 8\frac{1}{2}, 10.8$ Constants | a, x, y, z Variables

An **algebraic expression** is a sequence of constants, variables, operation symbols, and/or grouping symbols, such as parentheses, square brackets, or fraction bars.

$x + 5, \quad 2m - 9, \quad 8p^2 + 6(p - 2)$ Algebraic expressions

$2m$ means $2 \cdot m$, the product of 2 and m.

$6(p - 2)$ means the product of 6 and $p - 2$.

OBJECTIVE 1 Evaluate algebraic expressions, given values for the variables. To *evaluate* an expression is to find its *value*. An algebraic expression can have different numerical values for different values of the variables.

EXAMPLE 1 Evaluating Algebraic Expressions

Evaluate each expression for $x = 5$.

(a) $8 + x$

$= 8 + 5$ Let $x = 5$.

$= 13$ Add.

(b) $8x$ ← Multiplication is understood.

$= 8 \cdot x$

$= 8 \cdot 5$ Let $x = 5$.

$= 40$ Multiply.

(c) $2x - 9$

$= 2 \cdot x - 9$ $2x = 2 \cdot x$

$= 2 \cdot 5 - 9$ Let $x = 5$.

$= 10 - 9$ Multiply.

$= 1$ Subtract.

(d) $3x^2$

$= 3 \cdot x^2$

$= 3 \cdot 5^2$ Let $x = 5$. ($5^2 = 5 \cdot 5$)

$= 3 \cdot 25$ Square 5.

$= 75$ Multiply.

Work Problem 1 at the Side. ▶

! CAUTION

$3x^2$ means $3 \cdot x^2$, ***not*** $3x \cdot 3x$. See **Example 1(d).**

Unless parentheses are used, the exponent refers only to the variable or number just before it. We would need to use parentheses to write $3x \cdot 3x$ with exponents.

$(3x)^2$ means $3x \cdot 3x$.

1 Evaluate each expression for $x = 3$.

GS **(a)** $x + 12$

$= \_\_\_\_ + 12$

$= \_\_\_\_$

GS **(b)** $6x$

$= \_\_\_\_ \cdot x$

$= 6 \cdot \_\_\_\_$

$= \_\_\_\_$

GS **(c)** $5x^2$

$= \_\_\_\_ \cdot x^2$

$= 5 \cdot \_\_\_\_^2$

$= 5 \cdot \_\_\_\_$

$= \_\_\_\_$

(d) $16x - 10$

Answers

1. (a) 3; 15 **(b)** 6; 3; 18 **(c)** 5; 3; 9; 45 **(d)** 38

2 Evaluate each expression for $x = 6$ and $y = 9$.

GS **(a)** $4x + 7y$

$= 4 \cdot \_\_\_\_ + 7 \cdot \_\_\_\_$

$= \_\_\_\_ + \_\_\_\_$

$= \_\_\_\_$

(b) $\dfrac{4x - 2y}{x + 1}$

(c) $2x^2 + y^2$

Answers

2. (a) 6; 9; 24; 63; 87 (b) $\frac{6}{7}$ (c) 153

EXAMPLE 2 Evaluating Algebraic Expressions

Evaluate each expression for $x = 5$ and $y = 3$.

(a) $2x + 5y$ — We could use parentheses and write $2(5) + 5(3)$.

$= 2 \cdot 5 + 5 \cdot 3$ Let $x = 5$ and $y = 3$.

$= 10 + 15$ Multiply. (Follow the rules for order of operations.)

$= 25$ Add.

(b) $\dfrac{9x - 8y}{2x - y}$

$= \dfrac{9 \cdot 5 - 8 \cdot 3}{2 \cdot 5 - 3}$ Let $x = 5$ and $y = 3$.

$= \dfrac{45 - 24}{10 - 3}$ Multiply.

$= \dfrac{21}{7}$ Subtract.

$= 3$ Divide.

(c) $x^2 - 2y^2$

$= 5^2 - 2 \cdot 3^2$ Let $x = 5$ and $y = 3$. ($3^2 = 3 \cdot 3$)

$= 25 - 2 \cdot 9$ Apply the exponents. ($5^2 = 5 \cdot 5$)

$= 25 - 18$ Multiply.

$= 7$ Subtract.

◀ Work Problem **2** at the Side.

OBJECTIVE 2 Translate word phrases to algebraic expressions.

EXAMPLE 3 Using Variables to Write Word Phrases as Algebraic Expressions

Write each word phrase as an algebraic expression, using x as the variable.

(a) The **sum** of a number and 9

$x + 9$, or $9 + x$ "Sum" is the answer to an addition problem.

(b) 7 **minus** a number

$7 - x$ "Minus" indicates subtraction.

The expression $x - 7$ is incorrect. We cannot subtract in either order and obtain the same result.

(c) A number **subtracted from 12**

$12 - x$ — Be careful with order.

Compare this result with "12 subtracted from a number," which is $x - 12$.

(d) The **product** of 11 and a number

$11 \cdot x$, or $11x$

Continued on Next Page

(e) 5 divided by a number

$$5 \div x, \quad \text{or} \quad \frac{5}{x}$$

$\frac{x}{5}$ is *not* correct here.

(f) The product of 2 and the difference of a number and 8

We are multiplying 2 times "something." This "something" is the difference of a number and 8, written $x - 8$. We use parentheses around this difference.

$$2 \cdot (x - 8), \quad \text{or} \quad 2(x - 8)$$

$8 - x$, which means the difference of 8 and a number, is *not* correct.

Work Problem 3 at the Side. ▶

OBJECTIVE 3 Identify solutions of equations. An **equation** is a statement that two expressions are equal. ***An equation always includes the equality symbol, =.***

$$x + 4 = 11, \quad 2y = 16, \quad 4p + 1 = 25 - p,$$
$$\frac{3}{4}x + \frac{1}{2} = 0, \quad z^2 = 4, \quad 4(m - 0.5) = 2m$$

Equations

To **solve an equation,** we must find all values of the variable that make the equation true. Such a value of the variable is a **solution** of the equation.

EXAMPLE 4 Deciding Whether a Number Is a Solution of an Equation

Decide whether each equation has the given number as a solution.

(a) $5p + 1 = 36$; 7

$5p + 1 = 36$

$5 \cdot 7 + 1 \stackrel{?}{=} 36$ Let $p = 7$.

$35 + 1 \stackrel{?}{=} 36$ Multiply.

$36 = 36$ ✓ True—the left side of the equation equals the right side.

We could use parentheses and write 5 (7) here.

Be careful. Multiply first.

The number 7 is a solution of the equation.

(b) $9m - 6 = 32$; $\frac{14}{3}$

$9m - 6 = 32$

$9 \cdot \frac{14}{3} - 6 \stackrel{?}{=} 32$ Let $m = \frac{14}{3}$.

$42 - 6 \stackrel{?}{=} 32$ Multiply.

$36 = 32$ False—the left side does *not* equal the right side.

The number $\frac{14}{3}$ is not a solution of the equation.

Work Problem 4 at the Side. ▶

3 Write each word phrase as an algebraic expression, using x as the variable.

(a) The sum of 5 and a number

(b) A number minus 4

(c) A number subtracted from 48

(d) The product of 6 and a number

(e) A number divided by 7

(f) 9 multiplied by the sum of a number and 5

4 Decide whether each equation has the given number as a solution.

(a) $x - 1 = 3$; 2

(b) $2k + 3 = 15$; 7

(c) $7p - 11 = 5$; $\frac{16}{7}$

Answers

3. **(a)** $5 + x$ **(b)** $x - 4$ **(c)** $48 - x$ **(d)** $6x$ **(e)** $\frac{x}{7}$ **(f)** $9(x + 5)$

4. **(a)** no **(b)** no **(c)** yes

5 Write each word sentence as an equation. Use x as the variable.

GS **(a)** Three times the sum of a number and 13 is 19.

___(___ + ___) = ___

(b) Five times a number subtracted from 21 is 15.

(c) Five less than six times a number is equal to nineteen.

(d) Fifteen divided by a number equals the number minus two.

6 Decide whether each of the following is an *equation* or an *expression*.

(a) $2x + 5y - 7$

(b) $\frac{3x - 1}{5}$

(c) $2x + 5 = 7$

(d) $\frac{x - 3}{2} = 4x$

Answers

5. **(a)** 3; x; 13; 19 **(b)** $21 - 5x = 15$
(c) $6x - 5 = 19$ **(d)** $\frac{15}{x} = x - 2$

6. **(a)** expression **(b)** expression
(c) equation **(d)** equation

OBJECTIVE 4 **Translate sentences to equations.**

EXAMPLE 5 **Translating Sentences to Equations**

Write each word sentence as an equation. Use x as the variable.

(a) Twice the sum of a number and four is six.

"Twice" means two times. The word *is* suggests equals.

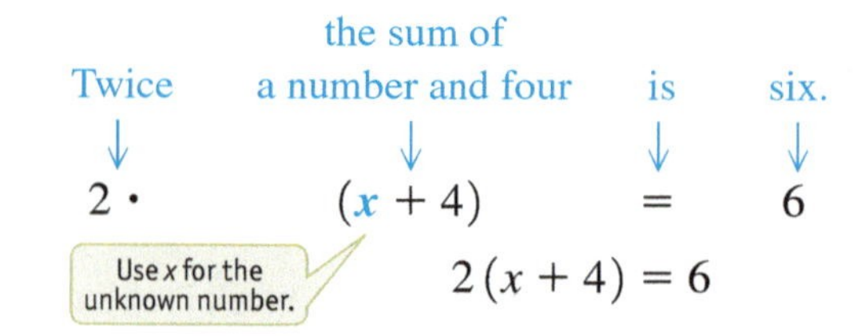

(b) Nine more than five times a number is 49.

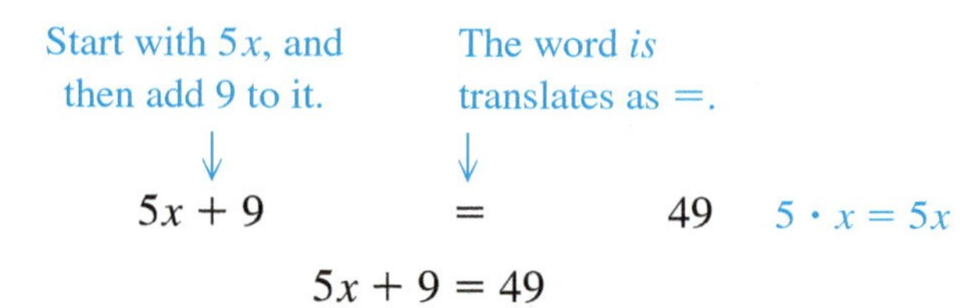

(c) Seven less than three times a number is equal to eleven.

Three times a number	less	seven	is equal to	eleven.
↓	↓	↓	↓	↓
$3x$	$-$	7	$=$	11

$$3x - 7 = 11$$

◀ Work Problem **5** at the Side.

OBJECTIVE 5 **Distinguish between equations and expressions.**

Distinguishing between an Equation and an Expression

An **equation** is a sentence—it has something on the left side, an = symbol, and something on the right side.

An **expression** is a phrase that represents a number.

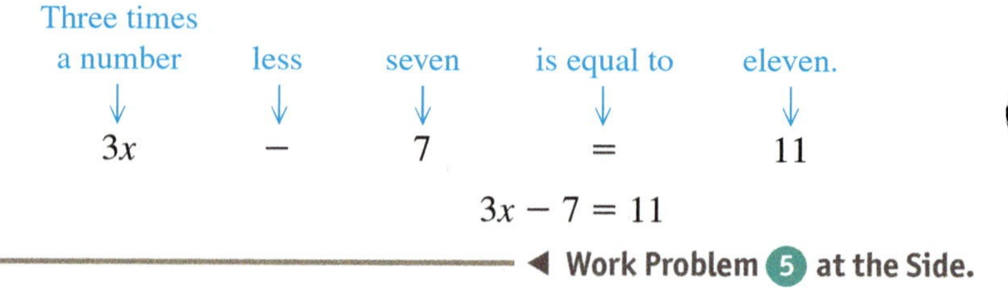

EXAMPLE 6 **Distinguishing between Equations and Expressions**

Decide whether each of the following is an *equation* or an *expression*.

(a) $2x - 3 - 8$ Ask, "*Is there an equality symbol?*" The answer is no, so this is an *expression*.

(b) $2x - 3 = 8$ Because there is an equality symbol with something on either side of it, this is an *equation*.

(c) $5x^2 + 2y^2$ There is no equality symbol. This is an *expression*.

◀ Work Problem **6** at the Side.

1.2 Exercises

FOR EXTRA HELP — *Go to* MyMathLab *for worked-out, step-by-step solutions to exercises enclosed in a square and video solutions to ▶ exercises.*

CONCEPT CHECK *Choose the letter(s) of the correct response.*

1. The expression $8x^2$ means ______.

A. $8 \cdot x \cdot 2$ **B.** $8 \cdot x \cdot x$ **C.** $8 + x^2$ **D.** $8x \cdot 8x$

2. For $x = 2$ and $y = 1$, the value of xy is ______.

A. $\frac{1}{2}$ **B.** 1 **C.** 2 **D.** 3

3. The sum of 15 and a number x is represented by ______.

A. $15 + x$ **B.** $15 - x$ **C.** $x - 15$ **D.** $15x$

4. Which of the following are expressions?

A. $6x = 7$ **B.** $6x + 7$ **C.** $6x - 7$ **D.** $6x - 7 = 0$

CONCEPT CHECK *Complete each statement.*

5. For $x = 3$, the value of $x + 8$ is ______.

6. For $x = 1$ and $y = 2$, the value of $5xy$ is ______.

7. "The sum of 13 and x" is represented by the expression ______. For $x = 3$, the value of this expression is ______.

8. $2x + 6$ is an (*equation* / *expression*), while $2x + 6 = 8$ is an (*equation* / *expression*).

Evaluate each expression for **(a)** $x = 4$ *and* **(b)** $x = 6$. ***See Example 1.***

9. $x + 7$

10. $x - 3$

11. $4x$

12. $6x$

13. $5x + 2$

14. $7x - 8$

15. $4x^2$ ▶

16. $5x^2$

17. $\frac{x + 1}{3}$

18. $\frac{x - 2}{5}$

19. $\frac{3x - 5}{2x}$

20. $\frac{4x - 1}{3x}$

21. $6.459x$

22. $3.275x$

23. $3x^2 + x$

24. $2x + x^2$

Evaluate each expression for **(a)** $x = 2$ *and* $y = 1$ *and* **(b)** $x = 1$ *and* $y = 5$.
See Example 2.

25. $13x - 2y$

26. $8x + 3y$

27. $8x + 3y + 5$

28. $4x + 2y + 7$

29. $3(x + 2y)$

30. $2(2x + y)$

31. $\frac{x}{2} + \frac{y}{3}$

32. $\frac{x}{5} + \frac{y}{4}$

33. $\frac{2x + 4y - 6}{5y + 2}$

34. $\frac{4x + 3y - 1}{2x + y}$

35. $2y^2 + 5x$

36. $6x^2 + 4y$

37. $\frac{3x + y^2}{2x + 3y}$

38. $\frac{x^2 + 1}{4x + 5y}$

39. $0.841x^2 + 0.32y^2$

40. $0.941x^2 + 0.2y^2$

Write each word phrase as an algebraic expression, using x as the variable.
See Example 3.

41. Twelve times a number

42. Fifteen times a number

43. Thirteen added to a number

44. Six added to a number

45. Two subtracted from a number

46. Eight subtracted from a number

47. The difference of twice a number and 6

48. The difference of 6 and three times a number

49. One-third of a number, subtracted from seven

50. One-fifth of a number, subtracted from fourteen

51. 12 divided by the sum of a number and 3

52. The difference of a number and 5, divided by 12

53. The product of 6 and four less than a number

54. The product of 9 and five more than a number

Decide whether each equation has the given number as a solution. ***See Example 4.***

55. $x - 5 = 12$; 7

56. $x + 6 = 15$; 10

57. $5x + 2 = 7$; 1

58. $3x + 5 = 8$; 1

59. $6x + 4x + 9 = 11$; $\frac{1}{5}$

60. $2x + 3x + 8 = 20$; $\frac{12}{5}$

61. $2y + 3(y - 2) = 14; \quad 3$

62. $6x + 2(x + 3) = 14; \quad 2$

63. $2x^2 + 1 = 19; \quad 3$

64. $3r^2 - 2 = 46; \quad 4$

65. $\frac{z + 4}{2 - z} = \frac{13}{5}; \quad \frac{1}{3}$

66. $\frac{x + 6}{x - 2} = \frac{37}{5}; \quad \frac{13}{4}$

Write each word sentence as an equation. Use x as the variable. ***See Example 5.***

67. The sum of a number and 8 is 18.

68. A number minus three equals 1.

69. Five more than twice a number is 5.

70. The product of a number and 3 is 6.

71. Sixteen minus three-fourths of a number is 13.

72. The sum of six-fifths of a number and 2 is 14.

73. Three times a number is equal to 8 more than twice the number.

74. Triple a number plus six equals five times the number.

75. A number divided by 3 equals four subtracted from the number.

76. Twelve divided by a number equals $\frac{1}{3}$ times the number.

Decide whether each of the following is an equation *or an* expression. ***See Example 6.***

77. $3x + 2(x - 4)$

78. $5y - (3y + 6)$

79. $7t + 2(t + 1) = 4$

80. $9r + 3(r - 4) = 2$

81. $x + y = 9$

82. $x + y - 9$

83. $\frac{3x - 8}{2}$

84. $\frac{3x - 8}{2} = 11$

Relating Concepts (Exercises 85–88) For Individual or Group Work

A ***mathematical model*** *is an equation that describes the relationship between two quantities. For example, the life expectancy of Americans at birth can be approximated by the equation*

$$y = 0.180x - 283,$$

where x is a year between 1960 and 2010 and y is age in years. (Data from Centers for Disease Control and Prevention.)

Use this model to approximate life expectancy (to the nearest year) in each of the following years.

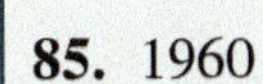

85. 1960

86. 1975

87. 1995

88. 2010

Take time to read each section and its examples before doing your homework. You will learn more and be better prepared to work the exercises.

Approaches to Reading Your Math Text

Student A learns best by listening to his teacher explain things. He "gets it" when he sees the instructor work problems. He previews the section before the lecture, so he knows generally what to expect. **Student A carefully reads the section in his text *AFTER* he hears the classroom lecture on the topic.**

Student B learns best by reading on her own. She reads the section and works through the examples before coming to class. That way, she knows what the teacher is going to talk about and what questions she wants to ask. **Student B carefully reads the section in her text *BEFORE* she hears the classroom lecture on the topic.**

Which reading approach works better for you—that of Student A or that of Student B? ________

Tips for Reading Your Math Text

- **Turn off your cell phone and the TV.** You will be able to concentrate more fully on what you are reading.
- **Survey the material.** Glance over the assigned material to get an idea of the "big picture." Look at the list of objectives to see what you will be learning.
- **Read slowly.** Read only one section—or even part of a section—at a sitting, with paper and pencil in hand.
- **Pay special attention to important information given in colored boxes or set in boldface type.** Highlight any additional information you find helpful.
- **Study the examples carefully.** Pay particular attention to the blue side comments and any pointer balloons.
- **Do the margin problems in the workspace provided or on separate paper as you go.** These problems mirror the examples and prepare you for the exercise set. Check your answers with those given at the bottom of the page.
- **Make study cards as you read.** Make cards for new vocabulary, rules, procedures, formulas, and sample problems.
- **Mark anything you don't understand. *ASK QUESTIONS*** in class—everyone will benefit. Follow up with your instructor, as needed.

Now Try This

Think through and answer each question.

1. Which two or three reading tips will you try this week? ________________

   ____

   ____

   ____

2. Did the tips you selected improve your ability to read and understand the material? Explain. ________________

   ____

   ____

   ____

1.3 Real Numbers and the Number Line

OBJECTIVES

1. Classify numbers and graph them on number lines.
2. Tell which of two real numbers is less than the other.
3. Find the additive inverse of a real number.
4. Find the absolute value of a real number.

A **set** is a collection of objects. In mathematics, these objects are usually numbers. The objects that belong to a set are its **elements.** They are written between **braces** { }.

{1, 2, 3, 4, 5} ← The set containing the elements 1, 2, 3, 4, and 5

OBJECTIVE 1 Classify numbers and graph them on number lines. The set of numbers used for counting is the *natural numbers*. The set of *whole numbers* includes 0 with the natural numbers.

Natural Numbers and Whole Numbers

{1, 2, 3, 4, 5, . . . } is the set of **natural numbers** (or **counting numbers**).

{0, 1, 2, 3, 4, 5, . . . } is the set of **whole numbers.**

We can represent numbers on a **number line** like the one in **Figure 1.**

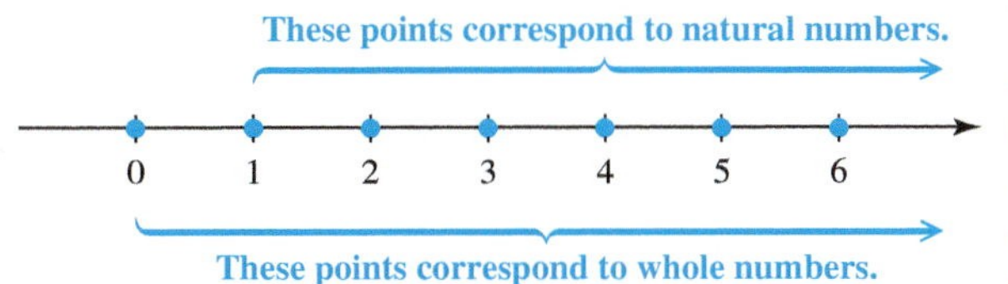

Figure 1

To draw a number line, choose any point on the line and label it 0. Then choose any point to the right of 0 and label it 1. Use the distance between 0 and 1 as the scale to locate, and then label, other points.

The natural numbers are located to the right of 0 on the number line. For each natural number, we can place a corresponding number to the left of 0, labeling the points −1, −2, −3, and so on, as shown in **Figure 2.** Each is the **opposite,** or **negative,** of a natural number. The natural numbers, their opposites, and 0 form the set of *integers*.

Integers

{. . . , −3, −2, −1, 0, 1, 2, 3, . . . } is the set of **integers.**

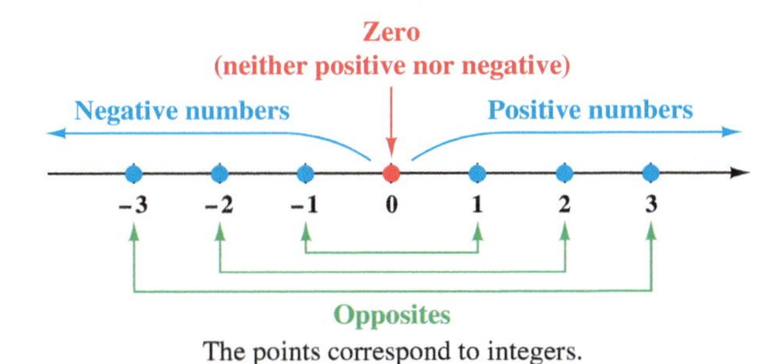

The points correspond to integers.

Figure 2

Positive numbers and *negative numbers* are **signed numbers.**

EXAMPLE 1 Using Negative Numbers

Use an integer to express the boldface italic number in each statement.

(a) The lowest Fahrenheit temperature ever recorded in meteorological records was ***129°*** below zero at Vostok, Antarctica, on July 21, 1983. (Data from *The World Almanac and Book of Facts.*)

Use -129 because "below zero" indicates a negative number.

(b) The shore surrounding the Dead Sea is ***1348*** ft below sea level. (Data from *The World Almanac and Book of Facts.*)

Here "below sea level" indicates a negative number, -1348.

◀ Work Problem 1 at the Side.

1 Use an integer to express the boldface italic number(s) in each statement.

(a) Erin discovers that she has spent $***53*** more than she has in her checking account.

(b) The record high Fahrenheit temperature in the United States was ***134°*** in Death Valley, California, on July 10, 1913. (Data from *The World Almanac and Book of Facts.*)

(c) A football team gained ***5*** yd, then lost ***10*** yd on the next play.

Fractions are *rational numbers.*

Rational Numbers

$\{x \mid x$ is a quotient of two integers, with denominator not $0\}$ is the set of **rational numbers.**

(Read the part in the braces as *"the set of all numbers x such that x is a quotient of two integers, with denominator not* 0.")

Note

The set symbolism used in the definition of rational numbers,

$\{x \mid x$ has a certain property$\}$,

is **set-builder notation.** This notation is convenient to use when it is not possible to list all the elements of a set.

Because any number that can be written as the quotient of two integers (that is, as a fraction) is a rational number, ***all integers, mixed numbers, terminating (or ending) decimals, and repeating decimals are rational.*** The table gives examples.

RATIONAL NUMBERS

Rational Number	Equivalent Quotient of Two Integers
-5	$\frac{-5}{1}$ (means $-5 \div 1$)
$1\frac{3}{4}$	$\frac{7}{4}$ (means $7 \div 4$)
0.23 (terminating decimal)	$\frac{23}{100}$ (means $23 \div 100$)
0.3333…, or $0.\overline{3}$ (repeating decimal)	$\frac{1}{3}$ (means $1 \div 3$)
4.7	$\frac{47}{10}$ (means $47 \div 10$)

To **graph** a number, we place a dot on the number line at the point that corresponds to the number. The number is the **coordinate** of the point.

Answers

1. (a) -53 (b) 134 (c) $5, -10$

EXAMPLE 2 Graphing Rational Numbers

Graph each rational number on a number line.

$$-\frac{3}{2}, \quad -\frac{2}{3}, \quad 0.5, \quad 1\frac{1}{3}, \quad \frac{23}{8}, \quad 3.25, \quad 4$$

To locate the improper fractions on a number line, write them as mixed numbers or decimals. The graph is shown in **Figure 3.**

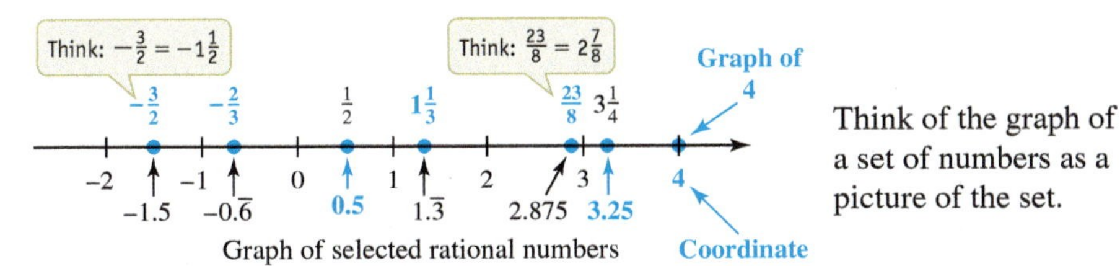

Graph of selected rational numbers

Think of the graph of a set of numbers as a picture of the set.

Figure 3

Work Problem 2 at the Side. ▶

2 Graph each rational number on a number line.

$$-3, \quad \frac{17}{8}, \quad -2.75, \quad 1\frac{1}{2}, \quad -\frac{3}{4}$$

-3 -2 -1 0 1 2 3

Not all numbers are rational. For example, the square root of 2, written $\sqrt{2}$, cannot be written as a quotient of two integers. Because of this, $\sqrt{2}$ is an *irrational number*. See **Figure 4.**

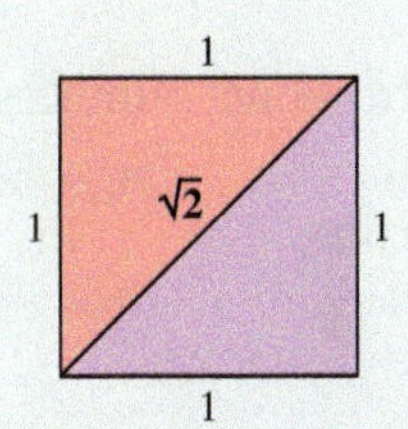

This square has diagonal of length $\sqrt{2}$. The number $\sqrt{2}$ is an irrational number.

Figure 4

Irrational Numbers

$\{x \mid x$ is a nonrational number represented by a point on a number line$\}$ is the set of **irrational numbers.**

The decimal form of an irrational number neither terminates nor repeats.

Both rational and irrational numbers can be represented by points on a number line and together form the set of **real numbers.** See **Figure 5.**

Real Numbers

$\{x \mid x$ is a rational or an irrational number$\}$ is the set of **real numbers.**

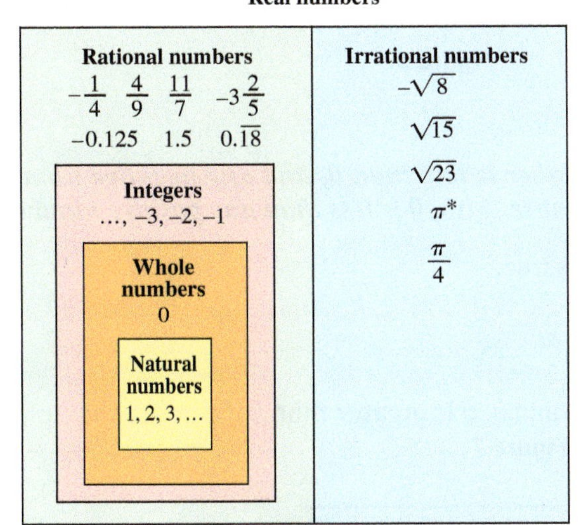

Figure 5

An example of a number that is not a real number is the square root of a negative number, such as $\sqrt{-5}$.

*The value of the irrational number π (pi) is approximately 3.141592654. The decimal digits continue forever with no repeated pattern.

Answer

2.
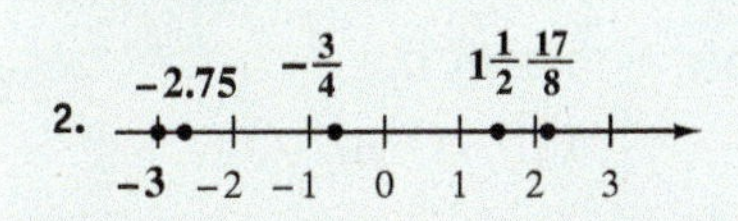

3 List the numbers in the following set that belong to each set of numbers.

$$\left\{-7, -\frac{4}{5}, 0, \sqrt{3}, 2.7, 13\right\}$$

(a) Whole numbers

(b) Integers

(c) Rational numbers

(d) Irrational numbers

Answers

3. (a) 0, 13 (b) −7, 0, 13
(c) $-7, -\frac{4}{5}, 0, 2.7, 13$ (d) $\sqrt{3}$

EXAMPLE 3 Determining Whether a Number Belongs to a Set

List the numbers in the following set that belong to each set of numbers.

$$\left\{-5, -\frac{2}{3}, 0, 0.\overline{6}, \sqrt{2}, \pi, 3\frac{1}{4}, 5, 5.8\right\}$$

(a) Natural numbers: 5

(b) Whole numbers: 0 and 5
The whole numbers consist of the natural (counting) numbers and 0.

(c) Integers: −5, 0, and 5

(d) Rational numbers: $-5, -\frac{2}{3}, 0, 0.\overline{6}\left(\text{or } \frac{2}{3}\right), 3\frac{1}{4}\left(\text{or } \frac{13}{4}\right), 5,$ and $5.8\left(\text{or } \frac{58}{10}\right)$
Each of these numbers can be written as the quotient of two integers.

(e) Irrational numbers: $\sqrt{2}$ and π

(f) Real numbers: All the numbers in the set are real numbers.

◄ Work Problem **3** at the Side.

OBJECTIVE 2 Tell which of two real numbers is less than the other.
Given any two different positive numbers, we can determine which number is less than the other. Positive numbers decrease as the corresponding points on the number line go to the left. For example,

$8 < 12$ because 8 is to the left of 12 on the number line.

This ordering is extended to all real numbers by definition.

Ordering of Real Numbers

For any two real numbers *a* and *b*, ***a* is less than *b*** if *a* lies to the left of *b* on a number line. See **Figure 6.**

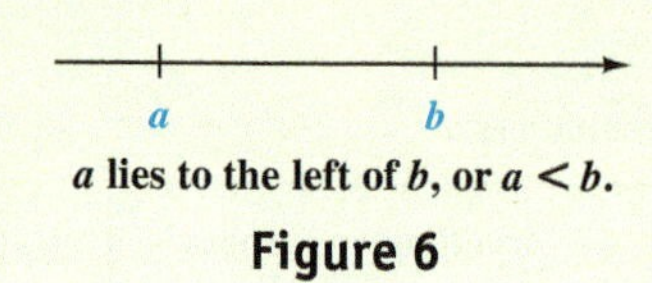

***a* lies to the left of *b*, or $a < b$.**

Figure 6

Thus, any negative number is less than 0, and any negative number is less than any positive number. Also, 0 is less than any positive number.

The following also holds true.

Ordering of Real Numbers

For any two real numbers *a* and *b*, ***a* is greater than *b*** if *a* lies to the right of *b* on a number line. See **Figure 7.**

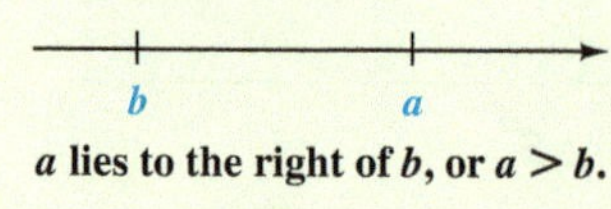

***a* lies to the right of *b*, or $a > b$.**

Figure 7

EXAMPLE 4 **Determining the Order of Real Numbers**

Is the statement $-3 < -1$ *true* or *false*?

Locate -3 and -1 on a number line. See **Figure 8.** Because -3 lies to the *left* of -1, -3 is less than -1. The statement $-3 < -1$ is true.

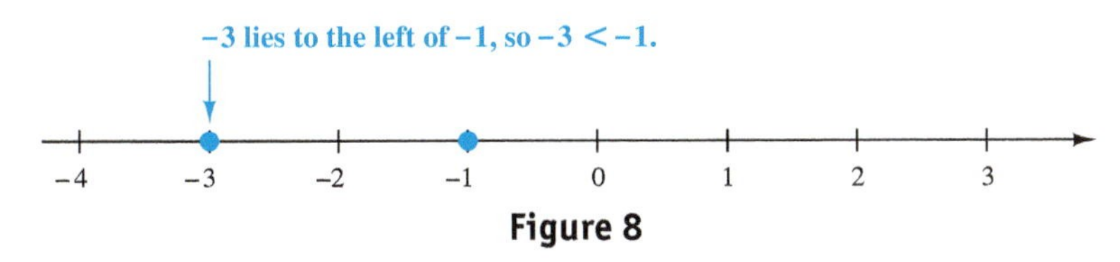

Figure 8

Also, the statement $-1 > -3$ is true because -1 lies to the *right* of -3 on the number line. See **Figure 8.**

Work Problem 4 at the Side. ▶

OBJECTIVE 3 Find the additive inverse of a real number. For any real number x (except 0), there is exactly one number on a number line the same distance from 0 as x, but on the *opposite* side of 0. See **Figure 9.** Such pairs of numbers are *additive inverses,* or *opposites,* of each other.

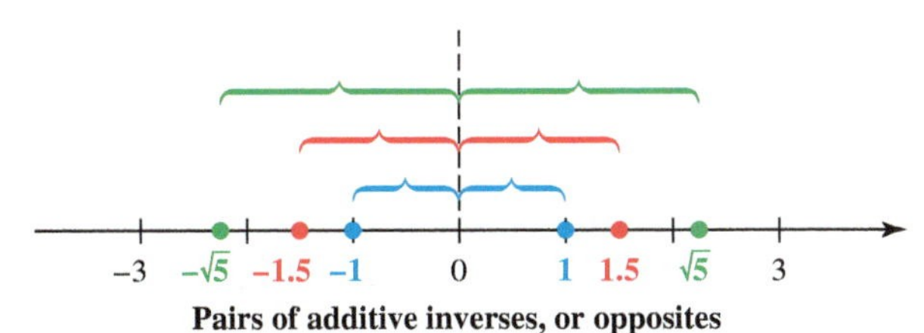

Pairs of additive inverses, or opposites

Figure 9

Additive Inverse

The **additive inverse** of a nonzero number x is the number that is the same distance from 0 on a number line as x, but on the *opposite* side of 0.

We indicate the additive inverse of a number by writing the symbol $-$ in front of the number. For example, the additive inverse of 7 is -7 (read "*negative* 7"). We could write the additive inverse of -4 as $-(-4)$, but we know that 4 is the opposite of -4. Because a number can have only one additive inverse, $-(-4)$ and 4 must represent the same number.

$$-(-4) = 4$$

Double Negative Rule

For any real number x, $\quad -(-x) = x.$

The table in the margin gives examples of additive inverses.

Finding an Additive Inverse

The additive inverse of a nonzero number is found by changing the sign of the number.

A nonzero number and its additive inverse have opposite signs.

Work Problem 5 at the Side. ▶

4 Tell whether each statement is *true* or *false.*

GS **(a)** $-2 < 4$

-2 is to the (*left* / *right*) of 4 on a number line, so the statement is (*true* / *false*).

GS **(b)** $6 > -3$

6 is to the (*left* / *right*) of -3 on a number line, so the statement is (*true* / *false*).

(c) $-9 < -12$

(d) $-4 \geq -1$

(e) $-6 \leq 0$

ADDITIVE INVERSES

Number	Additive Inverse
-4	$-(-4)$, or 4
0	0
5	-5
$-\frac{2}{3}$	$\frac{2}{3}$
0.52	-0.52

Note that the number 0 is its own additive inverse.

5 Find the additive inverse of each number.

(a) 15 **(b)** -9

(c) -12 **(d)** $\frac{1}{2}$

(e) 0 **(f)** -0.1

Answers

4. **(a)** left; true **(b)** right; true **(c)** false **(d)** false **(e)** true

5. **(a)** -15 **(b)** 9 **(c)** 12 **(d)** $-\frac{1}{2}$ **(e)** 0 **(f)** 0.1

OBJECTIVE 4 Find the absolute value of a real number. Because additive inverses are the same distance from 0 on a number line, a number and its additive inverse have the same *absolute value*. The **absolute value** of a real number x, written $|x|$ and read ***"the absolute value of x,"*** can be defined as the distance between 0 and the number on a number line.

$|2| = 2$ The distance between 2 and 0 on a number line is 2 units.

$|-2| = 2$ The distance between -2 and 0 on a number line is also 2 units.

Distance is a physical measurement, which is never negative. ***Therefore, the absolute value of a number is never negative.***

Absolute Value

For any real number x,

$$|x| = \begin{cases} x & \text{if } x \geq 0 \\ -x & \text{if } x < 0. \end{cases}$$

By this definition, if x is a positive number or 0, then its absolute value is x itself.

$$|8| = 8 \quad \text{and} \quad |0| = 0$$

If x is a negative number, then its absolute value is the additive inverse of x.

$|-8| = -(-8) = 8$ The additive inverse of -8 is 8.

EXAMPLE 5 Finding Absolute Value

Find each absolute value and simplify if needed.

(a) $|0| = 0$ **(b)** $|5| = 5$ **(c)** $|-5| = -(-5) = 5$

(d) $-|5| = -(5) = -5$ **(e)** $-|-5| = -(5) = -5$

(f) $|8-2| = |6| = 6$ **(g)** $-|8-2| = -|6| = -6$

Absolute value bars are grouping symbols. In parts (f) and (g), we perform any operations inside the absolute value bars *before* finding the absolute value.

◀ Work Problem 6 at the Side.

6 Find each absolute value and simplify if needed.

(a) $|-9|$

(b) $|9|$

(c) $-|15|$

GS (d) $-|-9|$
$= -$____
$=$____

(e) $|9-4|$

GS (f) $-|32-2|$
$= -|$____$|$
$=$____

Answers

6. (a) 9 (b) 9 (c) -15
(d) 9; -9 (e) 5 (f) 30; -30

1.3 Exercises

FOR EXTRA HELP — Go to MyMathLab for worked-out, step-by-step solutions to exercises enclosed in a square and video solutions to ▶ exercises.

CONCEPT CHECK *Complete each statement.*

1. The number __________ is a whole number, but not a natural number.

2. The natural numbers, their additive inverses, and 0 form the set of __________.

3. The additive inverse of every negative number is a (*negative* / *positive*) number.

4. If x and y are real numbers with $x > y$, then x lies to the (*left* / *right*) of y on a number line.

5. A rational number is the __________ of two integers, with the __________ not equal to 0.

6. Decimals that neither terminate nor repeat are __________ numbers.

Use an integer to express each boldface italic number representing a change. **See Example 1.**

7. Between 2012 and 2013, the male population in the United States increased by ***1,212,795.*** (Data from U.S. Census Bureau.)

8. From 2010 to 2015, the mean SAT critical reading score for Ohio students increased by ***19.*** (Data from The College Board.)

9. From 2009 to 2014, newspaper circulation in the United States on Sundays went from 46,164 thousand to 42,751 thousand, a decrease of ***3413*** thousand papers. (Data from *Editor and Publisher International Yearbook*.)

10. In 1935, there were 15,295 banks in the United States. By 2015, the number was 6242, a decrease of ***9053*** banks. (Data from Federal Deposit Insurance Corporation.)

Graph each rational number on a number line. **See Example 2.**

11. $0, 3, -5, -6$

12. $2, 6, -2, -1$

13. $-0.5, -6, -4, \frac{7}{4}, 4$

14. $-5, -\frac{7}{2}, -2, 0, 4.5$

15. $\frac{1}{4}, 2\frac{1}{2}, -3\frac{4}{5}, -4, -\frac{13}{8}$

16. $5\frac{1}{4}, \frac{41}{9}, -2\frac{1}{3}, 0, -3\frac{2}{5}$

List all numbers from each set that are the following. **See Example 3.**

(a) natural numbers *(b) whole numbers* *(c) integers*

(d) rational numbers *(e) irrational numbers* *(f) real numbers*

17. $\left\{-9, -\sqrt{7}, -1\frac{1}{4}, -\frac{3}{5}, 0, \sqrt{5}, 3, 5.9, 7\right\}$

18. $\left\{-5.3, -5, -\sqrt{3}, -1, -\frac{1}{9}, 0, 1.2, 4, \sqrt{12}\right\}$

19. $\left\{\frac{7}{9}, -2.\overline{3}, \sqrt{3}, 0, -8\frac{3}{4}, 11, -6, \pi\right\}$

20. $\left\{1\frac{5}{8}, -0.\overline{4}, \sqrt{6}, 9, -12, 0, \sqrt{10}, 0.026\right\}$

CONCEPT CHECK *Exercises 21–38 check understanding of the various sets of numbers.*

Give a number that satisfies the given condition.

21. An integer between 3.6 and 4.6

22. A rational number between 2.8 and 2.9

23. A whole number that is not positive and is less than 1

24. A whole number greater than 3.5

25. An irrational number that is between $\sqrt{12}$ and $\sqrt{14}$

26. A real number that is neither negative nor positive

Decide whether each statement is true *or* false.

27. Every natural number is positive.

28. Every whole number is positive.

29. Every integer is a rational number.

30. Every rational number is a real number.

31. Some numbers are both rational and irrational.

32. Every terminating decimal is a rational number.

Give three numbers between -6 *and* 6 *that satisfy each given condition.*

33. Positive real numbers but not integers

34. Real numbers but not positive numbers

35. Real numbers but not whole numbers

36. Rational numbers but not integers

37. Real numbers but not rational numbers

38. Rational numbers but not negative numbers

Select the lesser of the two given numbers. ***See Example 4.***

39. $-11, -4$

40. $-9, -16$

41. $-21, 1$

42. $-57, 3$

43. $0, -100$

44. $-215, 0$

45. $-\frac{2}{3}, -\frac{1}{4}$

46. $-\frac{3}{8}, -\frac{9}{16}$

Decide whether each statement is true *or* false. ***See Example 4 and* Figures 6–8.**

47. $8 < -16$

48. $12 < -24$

49. $-5 < -2$

50. $-10 < -9$

51. $-4 > 0$

52. $-9 > 0$

53. $-11 > -10$

54. $-8 > -2$

For each number, find **(a)** *its additive inverse and* **(b)** *its absolute value.* ***See Objectives 3 and 4.***

55. -2 **56.** -8 **57.** 6 **58.** 11

59. $-\frac{3}{4}$ **60.** $-\frac{1}{3}$ **61.** 4.95 **62.** 8.1

63. **CONCEPT CHECK** Match each expression in Column I with its value in Column II. Choices in Column II may be used once, more than once, or not at all.

I	II
(a) $\lvert -9 \rvert$	**A.** 9
(b) $-(-9)$	**B.** -9
(c) $-\lvert -9 \rvert$	**C.** Neither A nor B
(d) $-\lvert -(-9) \rvert$	**D.** Both A and B

64. **CONCEPT CHECK** Fill in each blank with the correct value.

The opposite of -5 is ____, while the absolute value of -5 is ____.

The additive inverse of -5 is ____, while the additive inverse of the absolute value of -5 is ____.

Find each absolute value and simplify if needed. ***See Example 5.***

65. $\lvert -6 \rvert$ **66.** $\lvert -3 \rvert$ **67.** $-\lvert 12 \rvert$ **68.** $-\lvert 23 \rvert$ **69.** $-\left| -\frac{2}{3} \right|$

70. $-\left| -\frac{4}{5} \right|$ **71.** $\lvert 13 - 4 \rvert$ **72.** $\lvert 8 - 7 \rvert$ **73.** $-\lvert 6 - 3 \rvert$ **74.** $-\lvert 9 - 4 \rvert$

Decide whether each statement is true *or* false.

75. $\lvert -8 \rvert < 7$ **76.** $\lvert -6 \rvert \geq -\lvert 6 \rvert$ **77.** $4 \leq \lvert 4 \rvert$ **78.** $-\lvert -3 \rvert > 2$

79. $-4 \leq -(-5)$ **80.** $\lvert -6 \rvert < \lvert -9 \rvert$ **81.** $-\lvert 8 \rvert > \lvert -9 \rvert$ **82.** $-\lvert -5 \rvert \geq -\lvert -9 \rvert$

The table shows the percent change in the Consumer Price Index (CPI) for selected categories of goods and services from 2013 to 2014 and from 2014 to 2015. Use the table to answer each question.

Category	Change from 2013 to 2014	Change from 2014 to 2015
Apparel	0.7	−1.9
Food	3.1	1.6
Energy	7.1	−17.1
Medical care	1.4	3.0
Transportation	−1.1	−7.9

Data from U.S. Bureau of Labor Statistics.

83. Which category in which year represents the greatest percent decrease?

84. Which category in which year represents the greatest percent increase?

85. Which category in which year represents the least change?

86. Which category represents a decrease for both years?

Study Skills

USING STUDY CARDS

You may have used "flash cards" in other classes. In math, "study cards" can help you remember terms and definitions, procedures, and concepts. Use study cards to

- Help you understand and learn the material;
- Quickly review when you have a few minutes;
- Review before a quiz or test.

One of the advantages of study cards is that you learn the material while you are making them.

Vocabulary Cards

Put the word and a page reference on the front of the card. On the back, write the definition, an example, any related words, and a sample problem (if appropriate).

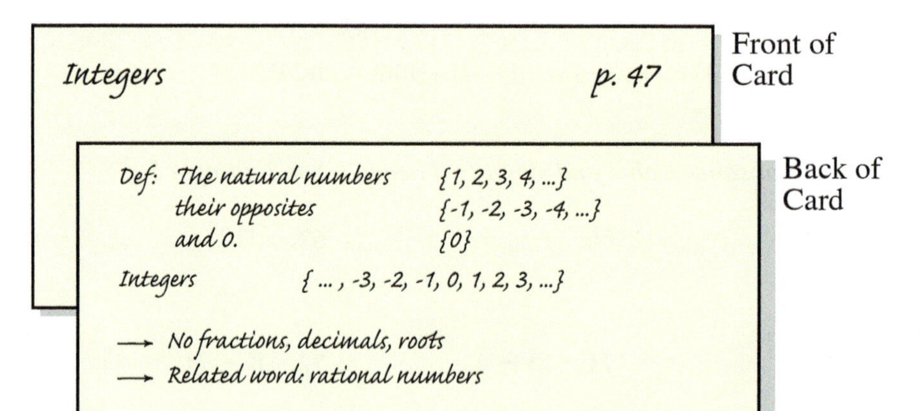

Integers
p. 47
Front of Card
Def: The natural numbers {1, 2, 3, 4, ...}
their opposites {-1, -2, -3, -4, ...}
and 0. {0}
Integers {..., -3, -2, -1, 0, 1, 2, 3, ...}
→ No fractions, decimals, roots
→ Related word: rational numbers
Back of Card

Procedure ("Steps") Cards

Write the name of the procedure on the front of the card. Then write each step in words. On the back of the card, put an example showing each step.

Evaluating Absolute Value (Simplifying) *p. 52*

Front of Card

1. *Work inside absolute value bars first (like working inside parentheses).*
2. *Find the absolute value (never negative).*
3. *A negative sign in front of the absolute value bar is NOT affected, so keep it!*

Back of Card

Examples:

Simplify $|10-6|$ → $|4| = 4$ — *Work inside: 10 − 6 = 4. Absolute value of 4 is 4.*

Simplify $-|-12|$ → -12 — *Absolute value of −12 is 12. Keep negative sign that was in front.*

Now Try This

Make a vocabulary card and a procedure card for material you are learning now.

1.4 Adding Real Numbers

OBJECTIVES

1. Add two numbers with the same sign.
2. Add two numbers with different signs.
3. Use the rules for order of operations when adding real numbers.
4. Translate words and phrases that indicate addition.

OBJECTIVE 1 Add two numbers with the same sign. Recall that the answer to an addition problem is a **sum.** The numbers being added are the **addends.**

$$x + y = z \leftarrow \text{Sum}$$

Addends (x and y)

EXAMPLE 1 Adding Numbers (Same Sign) on a Number Line

Use a number line to find each sum.

(a) $2 + 3$

Step 1 Start at 0 and draw an arrow 2 units to the *right*. See **Figure 10.**

Step 2 From the right end of that arrow, draw another arrow 3 units to the *right* to represent the addition of a *positive* number.

The number below the end of the second arrow is 5, so $2 + 3 = 5$.

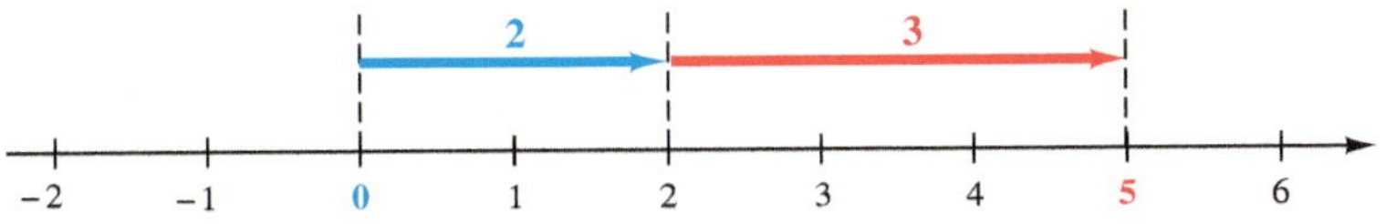

Figure 10

(b) $-2 + (-4)$

(We write parentheses around -4 due to the $+$ and $-$ symbols next to each other.)

Step 1 Start at 0 and draw an arrow 2 units to the *left*. See **Figure 11.**

Step 2 From the left end of the first arrow, draw a second arrow 4 units to the *left* to represent the addition of a *negative* number.

The number below the end of the second arrow is -6, so $-2 + (-4) = -6$.

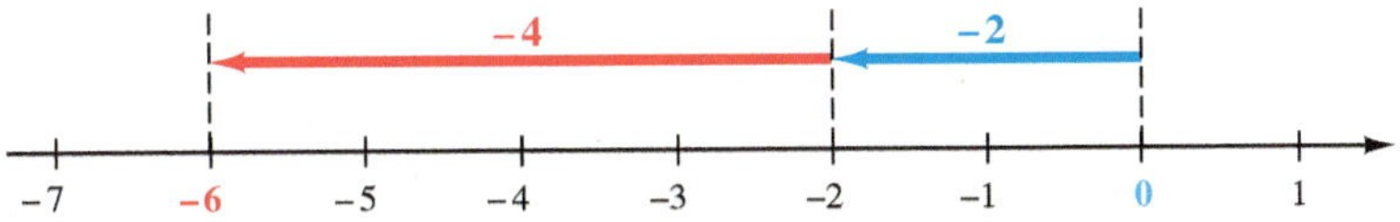

Figure 11

Work Problem 1 at the Side. ▶

In **Example 1(b),** the sum of the two negative numbers -2 and -4 is a negative number whose distance from 0 is the sum of the distance of -2 from 0 and the distance of -4 from 0. ***That is, the sum of two negative numbers is the opposite of the sum of their absolute values.***

Adding Signed Numbers (Same Sign)

To add two numbers with the *same* sign, add their absolute values. The sum has the same sign as the addends.

Examples: $2 + 4 = 6$ and $-2 + (-4) = -6$

1 Use a number line to find each sum.

(a) $1 + 4$

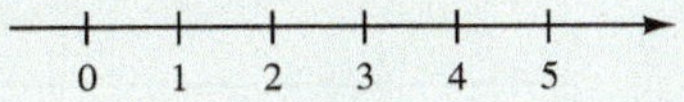

(b) $-2 + (-5)$

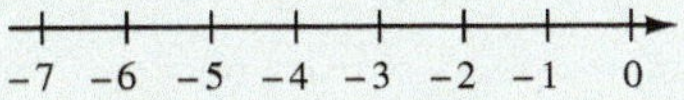

Answers

1. (a) $1 + 4 = 5$

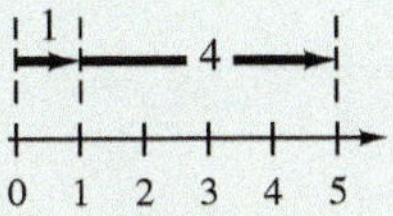

(b) $-2 + (-5) = -7$

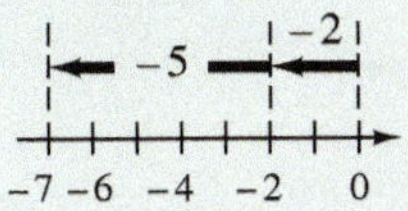

2 Find each sum.

(a) $-7 + (-3)$

(b) $-\frac{2}{5} + \left(-\frac{1}{2}\right)$

(c) $-1.27 + (-5.46)$

3 Use a number line to find each sum.

(a) $6 + (-3)$

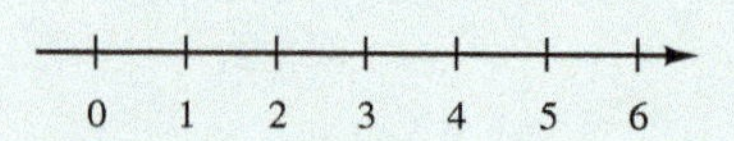

(b) $-5 + 1$

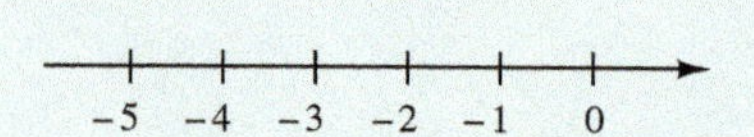

Answers

2. (a) -10 (b) $-\frac{9}{10}$ (c) -6.73

3. (a) $6 + (-3) = 3$

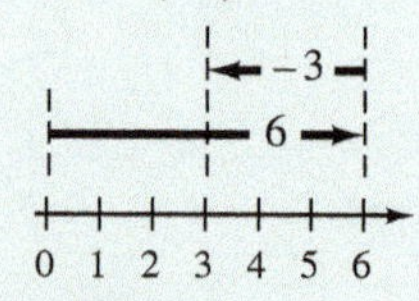

(b) $-5 + 1 = -4$

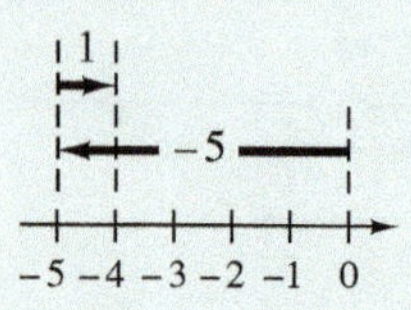

EXAMPLE 2 Adding Two Negative Numbers

Find each sum.

(a) $-9 + (-2)$ — Both addends are negative.

$= -(|-9| + |-2|)$ — Add the absolute values of the addends.

(Sign of each addend)

$= -(9 + 2)$ — Take the absolute values.

$= -11$ — The sum of two negative numbers is negative.

(b) $-\frac{1}{4} + \left(-\frac{2}{3}\right)$ — Both addends are negative.

$= -\frac{3}{12} + \left(-\frac{8}{12}\right)$ — Write equivalent fractions using the LCD, 12.

Think: $\left|-\frac{3}{12}\right| = \frac{3}{12}$ and $\left|-\frac{8}{12}\right| = \frac{8}{12}$

$= -\frac{11}{12}$ — Add the absolute values of the addends. Use the common negative sign.

(c) $-2.6 + (-4.7)$ — Both addends are negative.

$= -7.3$ — Add the absolute values of the addends. Use the common negative sign.

◀ Work Problem **2** at the Side.

OBJECTIVE 2 Add two numbers with different signs.

EXAMPLE 3 Adding Numbers (Different Signs) on a Number Line

Use a number line to find the sum $-2 + 5$.

Step 1 Start at 0 and draw an arrow 2 units to the left. See **Figure 12.**

Step 2 From the left end of this arrow, draw a second arrow 5 units to the *right* to represent the addition of a *positive* number.

The number below the end of the second arrow is 3, so $-2 + 5 = 3$.

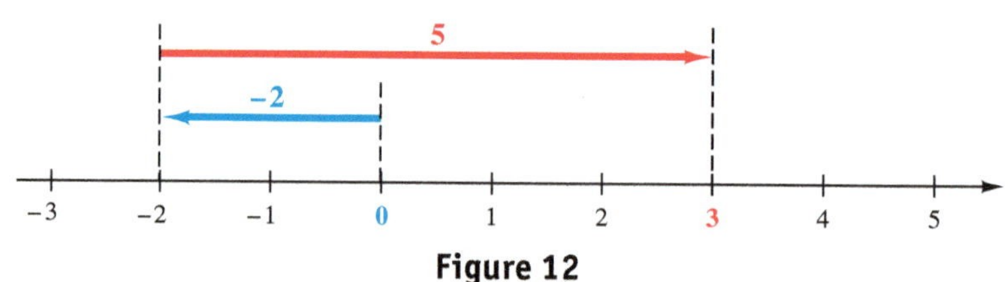

Figure 12

◀ Work Problem **3** at the Side.

Adding Signed Numbers (Different Signs)

To add two numbers with *different* signs, find their absolute values and subtract the lesser absolute value from the greater. The sum has the same sign as the addend with greater absolute value.

Examples: $-2 + 6 = 4$ and $2 + (-6) = -4$

EXAMPLE 4 Adding Signed Numbers (Different Signs)

Find each sum.

(a) $-12 + 5$

$|-12|$ $|5|$ — Find the absolute value of each addend, and subtract the lesser from the greater.

$= -(12 - 5)$

Use the sign of the addend with greater absolute value.

$= -7$

(b) $-8 + 12$ — Find the absolute value of each addend, and subtract the lesser from the greater.

$= +(12 - 8)$

Use the sign of the addend with greater absolute value.

$= 4$ — The + symbol is understood.

(c) $\frac{5}{6} + \left(-1\frac{1}{3}\right)$

$= \frac{5}{6} + \left(-\frac{4}{3}\right)$ — Write the mixed number as an improper fraction.

$= \frac{5}{6} + \left(-\frac{8}{6}\right)$ — Find a common denominator.

$= -\left(\frac{8}{6} - \frac{5}{6}\right)$ — $\left|\frac{5}{6}\right| = \frac{5}{6}$ and $\left|-\frac{8}{6}\right| = \frac{8}{6}$; Subtract the lesser absolute value from the greater.

Use a − symbol because $\left|-\frac{8}{6}\right| > \left|\frac{5}{6}\right|$.

$= -\frac{3}{6}$ — Subtract the fractions.

$= -\frac{1}{2}$ — Write in lowest terms.

(d) $8.1 + (-4.6)$

$= +(8.1 - 4.6)$

$|8.1| > |-4.6|$

$= 3.5$

(e) $-16 + 16$

$= 0$

(f) $42 + (-42)$

$= 0$

In parts (e) and (f), the difference of the absolute values is 0. ***In general, when additive inverses are added, the sum is 0.***

Work Problem 4 at the Side. ▶

OBJECTIVE 3 Use the rules for order of operations when adding real numbers.

EXAMPLE 5 Adding with Grouping Symbols

Find each sum.

(a) $-3 + [4 + (-8)]$ (Start here.)

$= -3 + [-4]$ (We could write (-4).)

$= -7$

(b) $8 + [(-2 + 6) + (-3)]$

$= 8 + [4 + (-3)]$

$= 8 + 1$

$= 9$

Continued on Next Page

---

4 Find each sum.

GS (a) $-8 + 2$

$= (+/-)(\_\_\_\_ - \_\_\_\_)$

$\uparrow |-8|$ $\uparrow |2|$

$= \_\_\_\_$

(b) $-15 + 4$

(c) $17 + (-10)$

(d) $\frac{3}{4} + \left(-1\frac{3}{8}\right)$

(e) $-9.5 + 3.8$

(f) $37 + (-37)$

Answers

4. (a) $-$; 8; 2; -6 **(b)** -11 **(c)** 7 **(d)** $-\frac{5}{8}$ **(e)** -5.7 **(f)** 0

5 Label the order in which each **GS** operation should be performed. Then find each sum.

(a) $2 + [7 + (-3)]$
② ①
$= 2 + \_\_\_\_$
$= \_\_\_\_$

(b) $6 + [(-2 + 5) + 7]$
○ ○ ○

(c) $-9 + [-4 + (-8 + 6)]$
○ ○ ○

(d) $|-8 + (-10)| + |7 + (-7)|$
○ ○ ○

6 Write a numerical expression for each phrase, and simplify the expression.

(a) 4 more than -12

(b) The sum of 6 and -7

(c) 7 increased by the sum of 8 and -3

7 Solve the problem.

A football team lost 8 yd on first down, lost 5 yd on second down, and then gained 7 yd on third down. Find the total net yardage for the plays.

Answers

5. **(a)** 4; 6
(b) ③, ①, ②; 16
(c) ③, ②, ①; -15
(d) ①, ③, ②; 18

6. **(a)** $-12 + 4$; -8 **(b)** $6 + (-7)$; -1
(c) $7 + [8 + (-3)]$; 12

7. -6 yd

(c) $|-9 + 3| + |7 + (-11)|$ — The absolute value bars serve as grouping symbols.

$= |-6| + |-4|$ — Add inside the absolute value bars.

$= 6 + 4$ — Find each absolute value.

$= 10$ — Add.

◀ Work Problem **5** at the Side.

OBJECTIVE 4 Translate words and phrases that indicate addition. Problem solving requires translating words and phrases into symbols.

WORDS AND PHRASES THAT INDICATE ADDITION

Word or Phrase	Example	Numerical Expression and Simplification
Sum of	The *sum of* -3 and 4	$-3 + 4$, which equals 1
Added to	5 *added to* -8	$-8 + 5$, which equals -3
More than	12 *more than* -5	$(-5) + 12$, which equals 7
Increased by	-6 *increased by* 13	$-6 + 13$, which equals 7
Plus	3 *plus* 14	$3 + 14$, which equals 17

EXAMPLE 6 Translating Words and Phrases (Addition)

Write a numerical expression for each phrase, and simplify the expression.

(a) The **sum of** -8 and 4 and 6

$-8 + 4 + 6$ simplifies to $-4 + 6$, which equals 2.

Add in order from left to right.

(b) 3 **more than** -5, **increased by** 12

$(-5 + 3) + 12$ simplifies to $-2 + 12$, which equals 10.

Here we *simplified* each expression by performing the operations.

◀ Work Problem **6** at the Side.

EXAMPLE 7 Solving an Application Involving Addition

A football team **gained 3 yd** on first down, **lost 12 yd** on second down, and then **gained 13 yd** on third down. Find the total net yardage for the plays.

$3 + (-12) + 13$

$= [3 + (-12)] + 13$

$= (-9) + 13$

$= 4$

Represent gains with positive numbers, losses with negative numbers. Add from left to right.

The total net yardage on these plays was 4 yd.

◀ Work Problem **7** at the Side.

1.4 Exercises

FOR EXTRA HELP

Go to MyMathLab for worked-out, step-by-step solutions to exercises enclosed in a square and video solutions to ▶ exercises.

CONCEPT CHECK *Complete each of the following.*

1. The sum of two negative numbers will always be a (*positive / negative*) number. Give a number-line illustration using the sum $-2 + (-3) =$ ____.

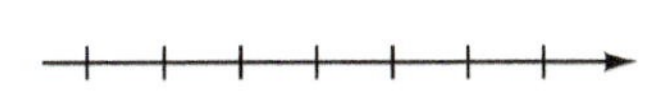

2. When adding a positive number and a negative number, where the negative number has the greater absolute value, the sum will be a (*positive / negative*) number. Give a number-line illustration using the sum $-4 + 2 =$ ____.

3. The sum of a number and its opposite will always be ______.

4. To simplify the expression $8 + [-2 + (-3 + 5)]$, begin by adding ____ and ____, according to the rules for order of operations.

Find each sum. ***See Examples 1–5.***

5. $-7 + (-3)$
6. $-11 + (-4)$
7. $-10 + (-3)$
8. $-16 + (-7)$
9. $6 + (-4)$
10. $8 + (-5)$
11. $12 + (-15)$
12. $4 + (-8)$
13. $-16 + 7$
14. $-13 + 6$
15. $6 + (-6)$
16. $-11 + 11$
17. $-\frac{1}{3} + \left(-\frac{4}{15}\right)$
18. $-\frac{1}{4} + \left(-\frac{5}{12}\right)$
19. $-\frac{1}{6} + \frac{2}{3}$
20. $-\frac{6}{25} + \frac{19}{20}$
21. $\frac{9}{10} + \left(-1\frac{3}{8}\right)$
22. $\frac{5}{8} + \left(-1\frac{5}{12}\right)$
23. $2\frac{1}{2} + \left(-3\frac{1}{4}\right)$
24. $1\frac{3}{8} + \left(-2\frac{1}{4}\right)$
25. $-12.4 + (-3.5)$
26. $-21.3 + (-2.5)$
27. $7.8 + (-9.4)$
28. $14.7 + (-10.1)$
29. $10 + [-3 + (-2)]$
30. $13 + [-4 + (-5)]$
31. $5 + [14 + (-6)]$
32. $7 + [3 + (-14)]$
33. $-3 + [5 + (-2)]$
34. $-7 + [10 + (-3)]$
35. $-8 + [(-1 + 3) + (-2)]$
36. $-7 + [(-8 + 5) + 3]$
37. $-7.1 + [3.3 + (-4.9)]$
38. $-9.5 + [-6.8 + (-1.3)]$
39. $[-8 + (-3)] + [-7 + (-7)]$
    $=$ ____ $+$ (____)
    $=$ ____
40. $[-5 + (-4)] + [9 + (-2)]$
    $=$ ____ $+$ ____
    $=$ ____

41. $|-7+5|+|8+(-12)|$

42. $|-8+2|+|4+(-9)|$

43. $|-12+(-4)|+|-4+10|$

44. $|-10+(-2)|+|-8+17|$

45. $\left(-\frac{1}{2}+0.25\right)+\left(-\frac{3}{4}+0.75\right)$

46. $\left(-\frac{3}{2}+0.75\right)+\left(-\frac{1}{2}+2.25\right)$

Perform each operation, and then determine whether the statement is true *or* false. *Try to do all work mentally.*

47. $-10+6+7=-3$

48. $-12+8+5=-1$

49. $\frac{7}{3}+\left(-\frac{1}{3}\right)+\left(-\frac{6}{3}\right)=0$

50. $-\frac{3}{2}+1+\frac{1}{2}=0$

51. $|-8+10|=-8+(-10)$

52. $|-4+6|=-4+(-6)$

53. $2\frac{1}{5}+\left(-\frac{6}{11}\right)=-\frac{6}{11}+2\frac{1}{5}$

54. $-1\frac{1}{2}+\frac{5}{8}=\frac{5}{8}+\left(-1\frac{1}{2}\right)$

55. $-7+[-5+(-3)]=[(-7)+(-5)]+3$

56. $6+[-2+(-5)]=[(-4)+(-2)]+5$

Write a numerical expression for each phrase, and simplify the expression. ***See Example 6.***

57. The sum of -5 and 12 and 6

58. The sum of -3 and 5 and -12

59. 14 added to the sum of -19 and -4

60. -2 added to the sum of -18 and 11

61. The sum of -4 and -10, increased by 12

62. The sum of -7 and -13, increased by 14

63. $\frac{2}{7}$ more than the sum of $\frac{5}{7}$ and $-\frac{9}{7}$

64. 0.85 more than the sum of -1.25 and -4.75

The table gives scores (above or below par—that is, above or below the score "standard") for selected golfers during the 2015 PGA Tour Championship. Write a signed number that represents the total score (above or below par) for the four rounds for each golfer.

	Golfer	Round 1	Round 2	Round 3	Round 4	Total Score
65.	Dustin Johnson	−6	+1*	−4	−3	
66.	Billy Horschel	0	−4	−4	+3	
67.	Jason Dufner	−1	+3	−3	+5	
68.	Charles Howell III	−2	−2	+5	+2	

*Golf scoring commonly includes a + symbol with a score over par.
Data from www.pga.com

Solve each problem. ***See Example 7.***

69. The surface, or rim, of a canyon is at altitude 0. On a hike down into the canyon, a party of hikers stops for a rest at 130 m below the surface. They then descend another 54 m. What is their new altitude? (Write the altitude as a signed number.)

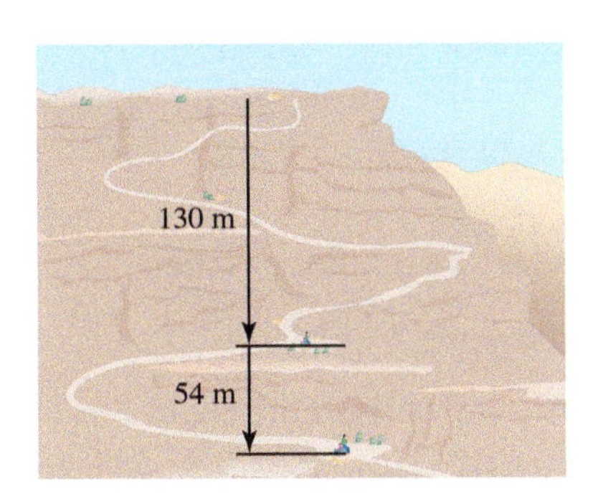

70. A pilot announces to the passengers that the current altitude of their plane is 34,000 ft. Because of some unexpected turbulence, the pilot is forced to descend 2100 ft. What is the new altitude of the plane? (Write the altitude as a signed number.)

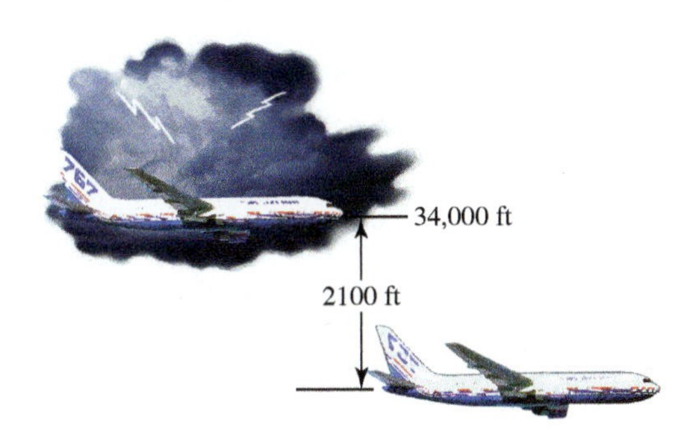

71. Based on 2020 population projections, New York will lose 5 seats in the U.S. House of Representatives, Pennsylvania will lose 4 seats, and Ohio will lose 3. Write a signed number that represents the total number of seats these three states are projected to lose. (Data from Population Reference Bureau.)

72. In 2020, Michigan is projected to lose 3 seats in the U.S. House of Representatives and Illinois 2. Projected to gain the most seats are California with 9, Texas with 5, Florida with 3, Georgia with 2, and Arizona with 2. Write a signed number that represents the algebraic sum of these changes. (Data from Population Reference Bureau.)

73. J. D. enjoys playing Triominoes every Wednesday night. Last Wednesday, on four successive turns, his scores were -19, 28, -5, and 13. What was his final score for the four turns?

74. Gail also enjoys playing Triominoes. On five successive turns, her scores were -13, 15, -12, 24, and 14. What was her total score for the five turns?

75. On three consecutive passes, a quarterback passed for a gain of 6 yd, was sacked for a loss of 12 yd, and passed for a gain of 43 yd. Find the total net yardage for the plays.

76. On a series of three consecutive running plays, a running back gained 4 yd, lost 3 yd, and lost 2 yd. Find his total net yardage for the plays.

77. The lowest temperature ever recorded in Arkansas was $-29°$F. The highest temperature ever recorded there was 149°F more than the lowest. What was this highest temperature? (Data from National Climatic Data Center.)

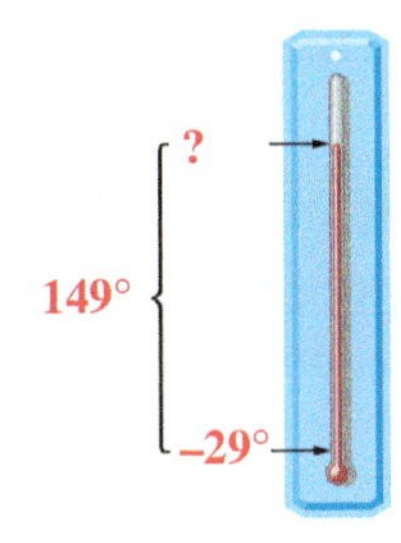

78. On January 23, 1943, the temperature rose 49°F in two minutes in Spearfish, South Dakota. If the starting temperature was $-4°$F, what was the temperature two minutes later? (Data from *Guinness World Records*.)

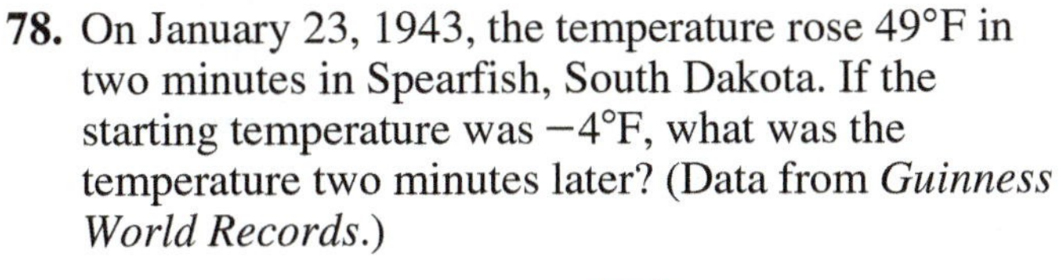

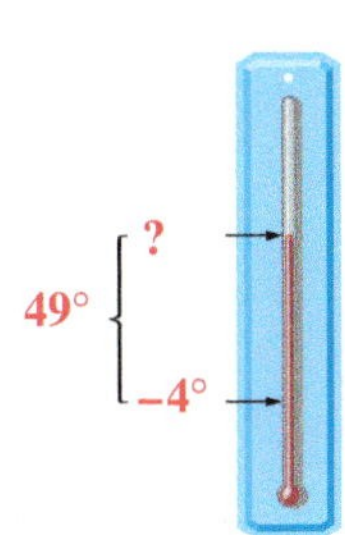

1.5 Subtracting Real Numbers

OBJECTIVES

1. Subtract two numbers on a number line.
2. Use the definition of subtraction.
3. Use the rules for order of operations when subtracting real numbers.
4. Translate words and phrases that indicate subtraction.

OBJECTIVE 1 Subtract two numbers on a number line. Recall that the answer to a subtraction problem is a **difference.** In the subtraction $x - y$, x is the **minuend** and y is the **subtrahend.**

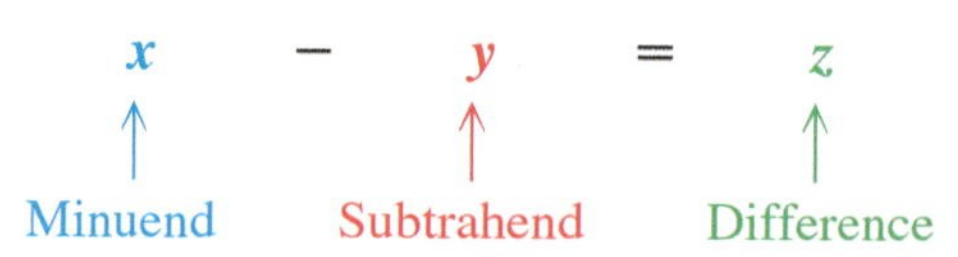

EXAMPLE 1 Subtracting Numbers on a Number Line

Use a number line to find the difference $7 - 4$.

Step 1 Start at 0 and draw an arrow 7 units to the *right*. See **Figure 13.**

Step 2 From the right end of the first arrow, draw a second arrow 4 units to the *left* to represent the subtraction.

The number below the end of the second arrow is 3, so $7 - 4 = 3$.

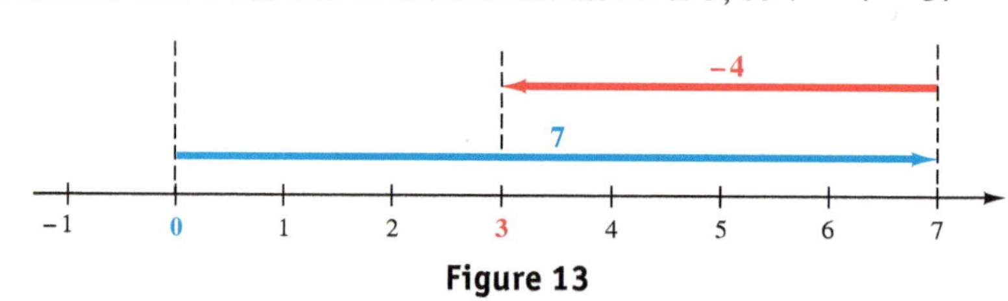

Figure 13

◀ Work Problem 1 at the Side.

1 Use a number line to find the difference $6 - 2$.

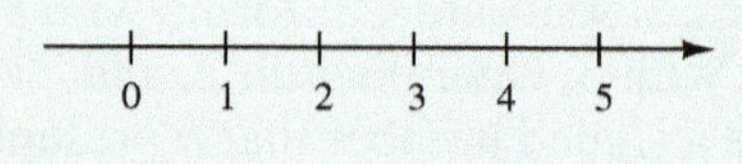

OBJECTIVE 2 Use the definition of subtraction. The procedure used in **Example 1** to find $7 - 4$ is the *same* procedure for finding $7 + (-4)$.

$$7 - 4 \quad \text{is equal to} \quad 7 + (-4).$$

This shows that *subtracting* a positive number from a greater positive number is the same as *adding* the opposite of the lesser number to the greater.

Definition of Subtraction

For any real numbers x and y,

$$x - y \quad \text{is defined as} \quad x + (-y).$$

To subtract y from x, add the additive inverse (or opposite) of y to x. In words, change the subtrahend to its opposite and add.

Example: $4 - 9$

$= 4 + (-9)$

$= -5$

Subtracting Signed Numbers

Step 1 Change the subtraction symbol to an addition symbol, and change the sign of the subtrahend.

Step 2 Add the signed numbers.

Answers

1. $6 - 2 = 4$

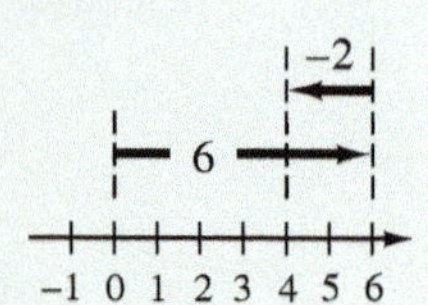

EXAMPLE 2 Subtracting Signed Numbers

Find each difference.

(a) $12 - 3$

Change − to +.

$= 12 + (-3)$

No change; Additive inverse of 3

$= 9$ — 12 has the greater absolute value, so the sum is positive.

(b) $5 - 7$

Change − to +.

$= 5 + (-7)$

No change; Additive inverse of 7

$= -2$ — −7 has the greater absolute value, so the sum is negative.

(c) $-8 - 15$

Change − to +.

$= -8 + (-15)$

No change; Additive inverse of 15

$= -23$ — The sum of two negative numbers is negative.

(d) $-3 - (-5)$

Change − to +.

$= -3 + 5$

No change; Additive inverse of −5

$= 2$ — 5 has the greater absolute value, so the sum is positive.

(e) $\frac{3}{8} - \left(-\frac{4}{5}\right)$

$= \frac{15}{40} - \left(-\frac{32}{40}\right)$ Write equivalent fractions using the LCD, 40.

$= \frac{15}{40} + \frac{32}{40}$ Definition of subtraction

$= \frac{47}{40}$ Add the fractions.

(f) $-8.75 - (-2.41)$

$= -8.75 + 2.41$ Definition of subtraction

$= -6.34$ Add the decimals.

Work Problem 2 at the Side. ▶

2 Find each difference.

GS **(a)** $6 - 10$

$= 6 + (\_\_\_\_)$

$= \_\_\_\_$

(b) $-2 - 4$

GS **(c)** $3 - (-5)$

$= 3 + \_\_\_\_$

$= \_\_\_\_$

(d) $-8 - (-12)$

(e) $\frac{5}{4} - \left(-\frac{3}{7}\right)$

(f) $7.5 - 9.2$

Answers

2. **(a)** -10; -4 **(b)** -6 **(c)** 5; 8 **(d)** 4 **(e)** $\frac{47}{28}$ **(f)** -1.7

3 GS Label the order in which each operation should be performed. Then perform the operations.

(a) $2-[(-3)-(4+6)]$

(b) $[(5-7)+3]-8$

(c) $\left[-\frac{1}{6}-\left(-\frac{1}{3}\right)\right]-\frac{1}{4}$

(d) $3|6-9|-|4-12|$

Uses of the Symbol –

We use the symbol – for three purposes.

1. ***It can represent subtraction,*** as in $9-5=4$.
2. ***It can represent negative numbers,*** such as -10, -2, and -3.
3. ***It can represent the additive inverse (or opposite) of a number,*** as in "the additive inverse (or opposite) of 8 is -8."

We may see more than one use of – in the same expression, such as $-6-(-9)$, where -9 is subtracted from -6. The meaning of the symbol depends on its position in the algebraic expression.

OBJECTIVE 3 Use the rules for order of operations when subtracting real numbers. As before, first perform any operations inside the grouping symbols.

EXAMPLE 3 Subtracting with Grouping Symbols

Perform each operation.

(a) $-6-[2-(8+3)]$ Work from the inside out.

$=-6-[2-11]$ Add inside the parentheses.

$=-6-[2+(-11)]$ Definition of subtraction

$=-6-(-9)$ Add inside the brackets.

$=-6+9$ Definition of subtraction

$=3$ Add.

(b) $\frac{2}{3}-\left[\frac{1}{12}-\left(-\frac{1}{4}\right)\right]$

$=\frac{8}{12}-\left[\frac{1}{12}-\left(-\frac{3}{12}\right)\right]$ Write equivalent fractions using the LCD, 12.

$=\frac{8}{12}-\left[\frac{1}{12}+\frac{3}{12}\right]$ Work inside the brackets.

$=\frac{8}{12}-\frac{4}{12}$ Add inside the brackets.

$=\frac{4}{12}$ Subtract.

$=\frac{1}{3}$ Write in lowest terms.

(c) $|4-7|-2|6-3|$ $2|6-3|$ means $2\cdot|6-3|$.

$=|-3|-2|3|$ Work within the absolute value bars.

$=3-2\cdot 3$ Find each absolute value.

Be careful. Multiply first.

$=3-6$ Multiply.

$=-3$ Subtract.

◄ Work Problem 3 at the Side.

Answers

3. (a) ③, ②, ①; 15
(b) ①, ②, ③; -7
(c) ①, ②; $-\frac{1}{12}$
(d) ③, ①, ④, ②; 1

OBJECTIVE 4 Translate words and phrases that indicate subtraction.

WORDS AND PHRASES THAT INDICATE SUBTRACTION

Word, Phrase, or Sentence	Example	Numerical Expression and Simplification
Difference of	The *difference of* -3 and -8	$-3 - (-8)$ simplifies to $-3 + 8$, which equals 5
Subtracted from*	12 *subtracted from* 18	$18 - 12$, which equals 6
From . . . , subtract	*From* 12, *subtract* 8.	$12 - 8$ simplifies to $12 + (-8)$, which equals 4
Less	6 *less* 5	$6 - 5$, which equals 1
Less than*	6 *less than* 5	$5 - 6$ simplifies to $5 + (-6)$, which equals -1
Decreased by	9 *decreased by* -4	$9 - (-4)$ simplifies to $9 + 4$, which equals 13
Minus	8 *minus* 5	$8 - 5$, which equals 3

*Be careful with order when translating.

CAUTION

When subtracting two numbers, be careful to write them in the correct order, because, in general,

$$x - y \neq y - x. \quad \text{For example, } 5 - 3 \neq 3 - 5.$$

Think carefully before interpreting an expression involving subtraction.

EXAMPLE 4 Translating Words and Phrases (Subtraction)

Write a numerical expression for each phrase, and simplify the expression.

(a) The **difference of** -8 and 5

When "difference of" is used, write the numbers in the order given.

$$-8 - 5 \quad \text{simplifies to} \quad -8 + (-5), \quad \text{which equals} \quad -13.$$

(b) 4 **subtracted from** the **sum of** 8 and -3

Here the operation of addition is also used, as indicated by the words *sum of*. First, add 8 and -3. Next, subtract 4 from this sum.

$$[8 + (-3)] - 4 \quad \text{simplifies to} \quad 5 - 4, \quad \text{which equals} \quad 1.$$

(c) 4 **less than** -6

Here 4 must be taken *from* -6, so write -6 first.

Be careful with order.

$$-6 - 4 \quad \text{simplifies to} \quad -6 + (-4), \quad \text{which equals} \quad -10.$$

Notice that "4 less than -6" differs from "4 *is less than* -6." The statement "4 is less than -6" is symbolized $4 < -6$ (which is a false statement).

(d) 8, **decreased by** 5 **less than** 12

First, write "5 less than 12" as $12 - 5$. Next, subtract $12 - 5$ from 8.

$$8 - (12 - 5) \quad \text{simplifies to} \quad 8 - 7, \quad \text{which equals} \quad 1.$$

Work Problem 4 at the Side. ▶

4 Write a numerical expression for each phrase, and simplify the expression.

(a) The difference of -5 and -12

(b) -2 subtracted from the sum of 4 and -4

(c) 7 less than -2

(d) 9, decreased by 10 less than 7

Answers

4. **(a)** $-5 - (-12)$; 7
(b) $[4 + (-4)] - (-2)$; 2
(c) $-2 - 7$; -9
(d) $9 - (7 - 10)$; 12

5 Solve the problem.

The highest elevation in Argentina is Mt. Aconcagua, which is 6960 m above sea level. The lowest point in Argentina is the Valdes Peninsula, 40 m below sea level. Find the difference between the highest and lowest elevations.

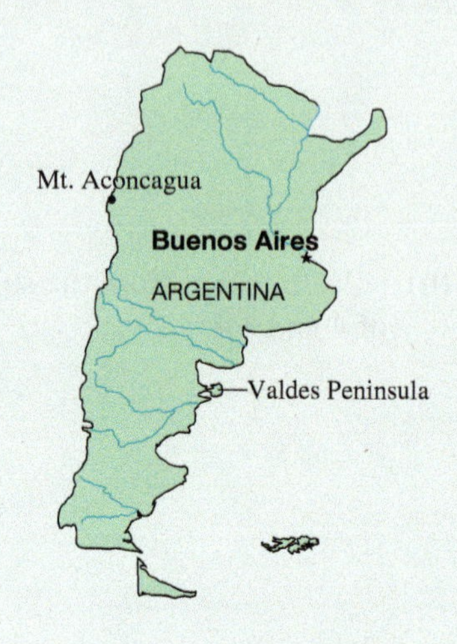

Answer

5. 7000 m

EXAMPLE 5 Solving an Application Involving Subtraction

The record high temperature of 134°F in the United States was recorded at Death Valley, California, in 1913. The record low was −80°F, at Prospect Creek, Alaska, in 1971. See **Figure 14.** What is the difference between these highest and lowest temperatures? (Data from National Climatic Data Center.)

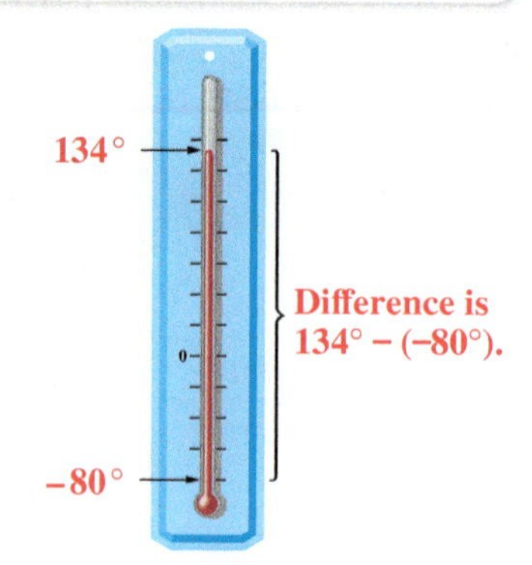

Figure 14

We must subtract the lowest temperature from the highest temperature.

$$134 - (-80)$$
$$= 134 + 80 \quad \text{Definition of subtraction}$$
$$= 214 \quad \text{Add.}$$

Order of numbers matters in subtraction.

The difference between the two temperatures is 214°F.

◀ Work Problem **5** at the Side.

1.5 Exercises

FOR EXTRA HELP — *Go to MyMathLab for worked-out, step-by-step solutions to exercises enclosed in a square and video solutions to ▶ exercises.*

CONCEPT CHECK *Fill in each blank with the correct response.*

1. By the definition of subtraction, in order to perform the subtraction $-6-(-8)$, we add the opposite of _______ to _______ to obtain _______.

2. To subtract y from x, add the _______ _______, or _______, of y to x.

3. By the rules for order of operations, to simplify $8-[3-(-4-5)]$, the first step is to subtract _______ from _______.

4. "The difference of 7 and 12" translates as _______, while "the difference of 12 and 7" translates as _______.

Find each difference. **See Examples 1 and 2.**

5. $5-9$ ▶

6. $6-11$

7. $4-7$

8. $8-13$

9. $-7-3$ GS ▶
$=-7+(\_\_\_\_)$
$=\_\_\_\_$

10. $-12-5$ GS
$=-12+(\_\_\_\_)$
$=\_\_\_\_$

11. $-10-6$

12. $-13-16$

13. $7-(-4)$ GS
$=7+\_\_\_\_$
$=\_\_\_\_$

14. $9-(-6)$ GS
$=9+\_\_\_\_$
$=\_\_\_\_$

15. $6-(-13)$ ▶

16. $13-(-3)$

17. $-7-(-3)$ ▶

18. $-8-(-6)$

19. $-3-(-12)$

20. $-7-(-16)$

21. $\frac{1}{2}-\left(-\frac{1}{4}\right)$

22. $\frac{1}{3}-\left(-\frac{4}{3}\right)$

23. $-\frac{3}{4}-\frac{5}{8}$ ▶

24. $-\frac{5}{6}-\frac{1}{2}$

25. $4.4-(-9.2)$

26. $6.7-(-12.6)$

27. $-7.4-4.5$

28. $-5.4-9.6$

Perform the indicated operations. **See Example 3.**

29. $(4-6)+12$ ▶

30. $(3-7)+4$

31. $(8-1)-12$

32. $(9-3)-15$

33. $3-(4-6)$

34. $6-(7-14)$

35. $-3-(6-9)$

36. $-4-(5-12)$

37. $8-(-3)-9+6$

38. $12-(-8)-25+9$

39. $-9-[1-(4+2)]$

40. $-8-[-5-(9-2)]$

41. $[(-5-8)-4]-3$

42. $[(-9+1)-11]-4$

43. $|-5-6|+|9+2|$

44. $|-4+8|+|6-1|$

45. $|-8-2|-|-9-3|$

46. $|-4-2|-|-8-1|$

47. $|6-9|-4|8-2|$

48. $|-2-5|-3|9-3|$

49. $\frac{5}{8}-\left[\frac{1}{2}-\left(-\frac{3}{4}\right)\right]$

50. $\frac{9}{10}-\left[\frac{3}{10}-\left(-\frac{1}{8}\right)\right]$

51. $\left(-\frac{3}{8}-\frac{2}{3}\right)-\left(-\frac{9}{8}-3\right)$

52. $\left(-\frac{3}{4}-\frac{5}{2}\right)-\left(-\frac{1}{8}-1\right)$

53. $-5.2-(8.4-10.8)$

54. $-9.6-(3.5-12.6)$

55. $[(-3.1)-4.5]-(0.8-2.1)$

56. $[(-7.8)-9.3]-(0.6-3.5)$

57. $-12-[(9-2)-(-6-3)]$

58. $-4-[(6-9)-(-7-4)]$

59. $[-12.25-(8.34+3.57)]-17.88$

60. $[-34.99+(6.59-12.25)]-8.33$

CONCEPT CHECK *Suppose that x represents a positive number and y represents a negative number. Determine whether the given expression must represent a* positive *or* negative *number.*

61. $y-x$

62. $x-y$

63. $x+|y|$

64. $y-|x|$

65. $|x|+|y|$

66. $-(|x|+|y|)$

Write a numerical expression for each phrase, and simplify the expression. ***See Example 4.***

67. The difference of 4 and -8

68. The difference of 7 and -14

69. 8 less than -2

70. 9 less than -13

71. The sum of 9 and -4, decreased by 7

72. The sum of 12 and -7, decreased by 14

73. 12 less than the difference of 8 and -5

74. 19 less than the difference of 9 and -2

Solve each problem. ***See Example 5.***

75. The lowest temperature ever recorded in Illinois was -36°F on January 5, 1999. The lowest temperature ever recorded in Utah was on February 1, 1985, and was 33°F lower than Illinois's record low. What is the record low temperature in Utah? (Data from National Climatic Data Center.)

76. The lowest temperature ever recorded in South Carolina was -19°F. The lowest temperature ever recorded in Wisconsin was 36° lower than South Carolina's record low. What is the record low temperature in Wisconsin? (Data from National Climatic Data Center.)

77. The top of Mount Whitney, visible from Death Valley, has an altitude of 14,494 ft above sea level. The bottom of Death Valley is 282 ft below sea level. Using 0 as sea level, find the difference between these two elevations. (Data from *The World Almanac and Book of Facts*.)

78. The height of Mount Pumasillo in Peru is 20,492 ft. The depth of the Peru-Chile Trench in the Pacific Ocean is 26,457 ft below sea level. Find the difference between these two elevations. (Data from *The World Almanac and Book of Facts*.)

79. A chemist is running an experiment under precise conditions. At first, she runs it at $-174.6°F$. She then lowers the temperature by $2.3°F$. What is the new temperature for the experiment?

80. At 2:00 A.M., a plant worker found that a dial reading was 7.904. At 3:00 A.M., she found the reading to be -3.291. Find the difference between these two readings.

81. In August, Kari began with a checking account balance of \$904.89. She made the following withdrawals and deposits.

Withdrawals	Deposits
\$35.84	\$85.00
\$26.14	\$120.76
\$3.12	

Assuming no other transactions, what was her new account balance?

82. In September, Derek began with a checking account balance of \$537.12. He made the following withdrawals and deposits.

Withdrawals	Deposits
\$41.29	\$80.59
\$13.66	\$276.13
\$84.40	

Assuming no other transactions, what was his new account balance?

83. Kim owes \$870.00 on her MasterCard account. She returns two items costing \$35.90 and \$150.00 and receives credit for these on the account. Next, she makes a purchase of \$82.50, and then two more purchases of \$10.00 each. She makes a payment of \$500.00. She then incurs a finance charge of \$37.23. How much does she still owe?

84. Charles owes \$679.00 on his Visa account. He returns three items costing \$36.89, \$29.40, and \$113.55 and receives credit for these on the account. Next, he makes purchases of \$135.78 and \$412.88, and two purchases of \$20.00 each. He makes a payment of \$400. He then incurs a finance charge of \$24.57. How much does he still owe?

85. José enjoys diving in Lake Okoboji. He dives to 34 ft below the surface of the lake. His partner, Sean, dives to 40 ft below the surface, but then ascends 20 ft. What is the vertical distance between José and Sean?

86. Rhonda also enjoys diving. She dives to 12 ft below the surface of False River. Her sister, Sandy, dives to 20 ft below the surface, but then ascends 10 ft. What is the vertical distance between Rhonda and Sandy?

87. The federal budget had a surplus of \$236 billion in 2000 and projected a deficit of \$474 billion in 2016. Express the difference between these amounts as a positive number. (Data from U.S. Treasury Department.)

88. In 2008, undergraduate college students had an average (mean) credit card balance of \$3173. The average balance decreased \$2076 in 2011, then increased \$366 in 2012, and then decreased \$538 in 2013. What was the average credit card balance of undergraduate college students in 2013? (Data from Sallie Mae.)

Median sales prices for existing single-family homes in the United States for selected years are shown in the table. Complete the table, determining the change from one year to the next by subtraction.

	Year	Median Sales Price	Change from Previous Year
	2008	\$198,100	—
89.	2009	\$175,500	
90.	2010	\$173,100	
91.	2011	\$166,200	
92.	2012	\$177,200	

Data from National Association of Realtors.

The two tables show the heights of some selected mountains and the depths of some selected trenches. Use the information given to answer each question.

Mountain	Height (in feet, as a positive number)
Foraker	17,400
Wilson	14,246
Pikes Peak	14,110

Trench	Depth (in feet, as a negative number)
Philippine	−32,995
Cayman	−24,721
Java	−23,376

Data from *The World Almanac and Book of Facts.*

93. What is the difference between the height of Mt. Foraker and the depth of the Philippine Trench?

94. What is the difference between the height of Pikes Peak and the depth of the Java Trench?

95. How much deeper is the Cayman Trench than the Java Trench?

96. How much deeper is the Philippine Trench than the Cayman Trench?

97. How much higher is Mt. Wilson than Pikes Peak?

98. If Mt. Wilson and Pikes Peak were stacked one on top of the other, how much higher would they be than Mt. Foraker?

Study Skills

COMPLETING YOUR HOMEWORK

You are ready to do your homework ***AFTER*** you have read the corresponding text section and worked through the examples and margin problems.

Homework Tips

- **Survey the exercise set.** Take a few minutes to glance over the problems that your instructor has assigned to get a general idea of the types of exercises you will be working. Skim directions, and note any references to section examples.
- **Work problems neatly.** Use pencil and write legibly, so others can read your work. Skip lines between steps. Clearly separate problems from each other.
- **Show all your work.** It is tempting to take shortcuts. Include ALL steps.
- **Check your work frequently to make sure you are on the right track.** It is hard to unlearn a mistake. For all odd-numbered problems and other selected exercises, answers are given in the back of the text.

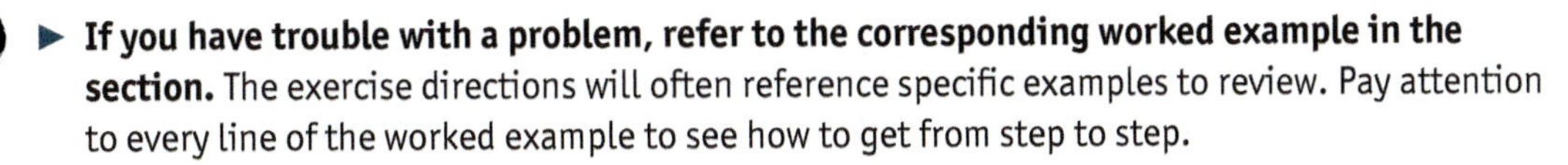

- **If you have trouble with a problem, refer to the corresponding worked example in the section.** The exercise directions will often reference specific examples to review. Pay attention to every line of the worked example to see how to get from step to step.

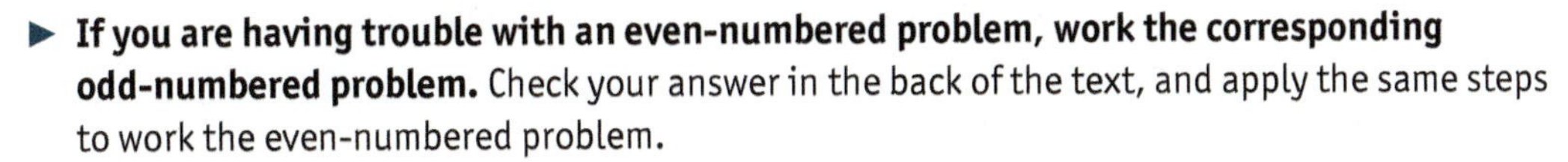

- **If you are having trouble with an even-numbered problem, work the corresponding odd-numbered problem.** Check your answer in the back of the text, and apply the same steps to work the even-numbered problem.
- **Do some homework problems every day.** This is a good habit, even if your math class does not meet each day.
- **Mark any problems you don't understand.** Ask your instructor about them.

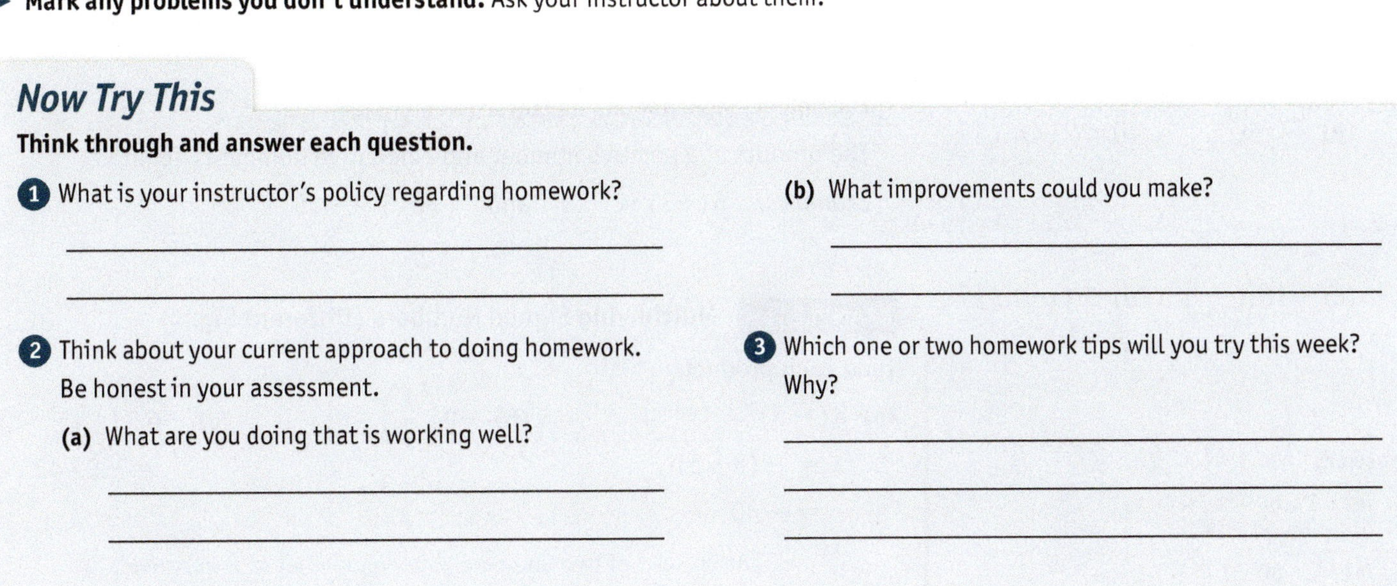

Now Try This

Think through and answer each question.

1. What is your instructor's policy regarding homework?

____

____

2. Think about your current approach to doing homework. Be honest in your assessment.

   (a) What are you doing that is working well?

   ____

   ____

   (b) What improvements could you make?

   ____

   ____

3. Which one or two homework tips will you try this week? Why?

____

____

____

1.6 Multiplying and Dividing Real Numbers

OBJECTIVES

1. Find the product of a positive number and a negative number.
2. Find the product of two negative numbers.
3. Use the reciprocal of a number to apply the definition of division.
4. Use the rules for order of operations when multiplying and dividing real numbers.
5. Evaluate expressions involving variables.
6. Translate words and phrases that indicate multiplication and division.
7. Translate simple sentences into equations.

The result of multiplication is a **product.** We know that the product of two positive numbers is positive. And we also know that the product of 0 and any positive number is 0, so we extend that property to all real numbers.

Multiplication Property of 0

For any real number x, the following hold.

$$x \cdot 0 = 0 \quad \text{and} \quad 0 \cdot x = 0$$

OBJECTIVE 1 Find the product of a positive number and a negative number. Observe the following pattern.

$$3 \cdot 5 = 15$$
$$3 \cdot 4 = 12$$
$$3 \cdot 3 = 9$$
$$3 \cdot 2 = 6$$
$$3 \cdot 1 = 3$$
$$3 \cdot 0 = 0$$
$$3 \cdot (-1) = ?$$

The products decrease by 3.

What should $3 \cdot (-1)$ equal? Multiplication can also be considered repeated addition. As a result, the product $3 \cdot (-1)$ represents the sum

$$-1 + (-1) + (-1), \quad \text{which equals} \quad -3,$$

so the product $3 \cdot (-1)$ should be -3. This fits the pattern above. Also, $3 \cdot (-2)$ represents the sum

$$-2 + (-2) + (-2), \quad \text{which equals} \quad -6.$$

◀ Work Problem 1 at the Side.

These results maintain the pattern above and suggest the following rule.

Multiplying Signed Numbers (Different Signs)

The product of a positive number and a negative number is negative.

Examples: $6(-3) = -18$ and $(-6)3 = -18$

EXAMPLE 1 Multiplying Signed Numbers (Different Signs)

Find each product.

(a) $8(-5)$
$= -(8 \cdot 5)$
$= -40$

(b) $-9\left(\frac{1}{3}\right)$
$= -3$

(c) $-6.2(4.1)$
$= -25.42$

The product of two numbers with *different* signs is *negative.*

◀ Work Problem 2 at the Side.

---

1 Find each product by finding the sum of three numbers.

(a) $3 \cdot (-3)$ (b) $3 \cdot (-4)$

(c) $3 \cdot (-5)$

2 Find each product.

(a) $7(-8)$ (b) $-9(2)$

(c) $-16\left(\frac{5}{32}\right)$ (d) $3.1(-2.5)$

Answers

1. (a) -9 (b) -12 (c) -15
2. (a) -56 (b) -18 (c) $-\frac{5}{2}$ (d) -7.75

OBJECTIVE 2 Find the product of two negative numbers. Look at another pattern.

$$-5(4) = -20$$
$$-5(3) = -15$$
$$-5(2) = -10$$
$$-5(1) = -5$$
$$-5(0) = 0$$
$$-5(-1) = ?$$

The products increase by 5.

The numbers in color on the left side of the equality symbols decrease by 1 for each step down the list. The products on the right increase by 5 for each step down the list. To maintain this pattern, $-5(-1)$ should be 5 more than $-5(0)$, or 5 more than 0, so

$$-5(-1) = 5.$$

The pattern continues with

$$-5(-2) = 10$$
$$-5(-3) = 15$$
$$-5(-4) = 20, \quad \text{and so on.}$$

These results suggest the next rule.

Multiplying Two Negative Numbers

The product of two negative numbers is positive.

Example: $-6(-3) = 18$

EXAMPLE 2 Multiplying Two Negative Numbers

Find each product.

(a) $-9(-2)$
$= 18$

(b) $-\frac{2}{3}\left(-\frac{3}{2}\right)$
$= 1$

(c) $-0.5(-1.25)$
$= 0.625$

The product of two numbers with the *same* sign is *positive*.

Work Problem 3 at the Side. ▶

Multiplying Signed Numbers

The product of two numbers with the *same* sign is *positive*.

The product of two numbers with *different* signs is *negative*.

OBJECTIVE 3 Use the reciprocal of a number to apply the definition of division. The definition of division depends on the idea of a *reciprocal,* or *multiplicative inverse,* of a number.

Reciprocals

Pairs of numbers whose product is 1 are **reciprocals,** or **multiplicative inverses,** of each other.

3 Find each product.

(a) $-5(-6)$

(b) $-7(-3)$

(c) $-\frac{1}{7}\left(-\frac{5}{2}\right)$

(d) $-1.2(-1.1)$

Answers

3. (a) 30 (b) 21 (c) $\frac{5}{14}$ (d) 1.32

4 Complete the table.

	Number	Reciprocal
(a)	6	____
(b)	-2	____
(c)	$\frac{2}{3}$	____
(d)	$-\frac{1}{4}$	____
(e)	0.75	____
(f)	0	____

RECIPROCALS

Number	Reciprocal
4	$\frac{1}{4}$
-5	$\frac{1}{-5}$, or $-\frac{1}{5}$
0.3, or $\frac{3}{10}$	$\frac{10}{3}$
$-\frac{5}{8}$	$-\frac{8}{5}$

A number and its reciprocal have a product of 1.

Example: $4 \cdot \frac{1}{4} = \frac{4}{4}$, or 1

0 has no reciprocal because the product of 0 and any number is 0, not 1.

◀ Work Problem 4 at the Side.

Recall that the answer to a division problem is a **quotient.** For example, we can write the quotient of 15 and 3 as $15 \div 3$, which equals 5. We obtain the same answer if we multiply $15 \cdot \frac{1}{3}$, the reciprocal of 3. This suggests the next definition.

Definition of Division

For any real numbers x and y, where $y \neq 0$,

$$x \div y = x \cdot \frac{1}{y}.$$

To divide two numbers, multiply the first number (the **dividend**) by the reciprocal, or multiplicative inverse, of the second number (the **divisor**).

Example: $15 \div 3 = 15 \cdot \frac{1}{3} = 5$

Recall that an equivalent form of $x \div y$ is $\frac{x}{y}$, where the fraction bar represents division. ***In algebra, quotients are usually represented with a fraction bar.*** For example,

$$15 \div 3 \quad \text{is equivalent to} \quad \frac{15}{3}.$$

Note

The following forms all represent division, where $y \neq 0$.

$$x \div y, \quad \frac{x}{y}, \quad x/y, \quad \text{and} \quad y\overline{)x}$$

Example: $15 \div 3$, $\frac{15}{3}$, $15/3$, and $3\overline{)15}$ are equivalent forms.

Because division is defined in terms of multiplication, the rules for multiplying signed numbers also apply to dividing them.

Dividing Signed Numbers

The quotient of two numbers with the *same* sign is *positive.*

The quotient of two numbers with *different* signs is *negative.*

Examples: $\frac{15}{3} = 5$, $\frac{-15}{-3} = 5$, $\frac{15}{-3} = -5$, and $\frac{-15}{3} = -5$

Answers

4. (a) $\frac{1}{6}$ (b) $\frac{1}{-2}$, or $-\frac{1}{2}$ (c) $\frac{3}{2}$ (d) -4 (e) $\frac{4}{3}$ (f) none

EXAMPLE 3 Dividing Signed Numbers

Find each quotient.

(a) $\frac{8}{-2} = -4$ **(b)** $\frac{-100}{5} = -20$ **(c)** $\frac{-4.5}{-0.09} = 50$

(d) $-\frac{1}{8} \div \left(-\frac{3}{4}\right)$

$= -\frac{1}{8} \cdot \left(-\frac{4}{3}\right)$ Multiply by the reciprocal of the divisor.

$= \frac{1}{6}$ Multiply the fractions. Write in lowest terms.

Work Problem 5 at the Side. ▶

Consider the quotient $\frac{12}{3}$.

$$\frac{12}{3} = 4 \quad \text{because} \quad 4 \cdot 3 = 12.$$

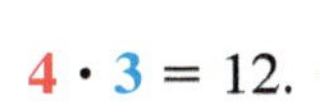

This relationship between multiplication and division allows us to investigate division involving 0. Consider the quotient $\frac{0}{3}$.

$$\frac{0}{3} = 0 \quad \text{because} \quad 0 \cdot 3 = 0.$$

Now consider $\frac{3}{0}$.

$$\frac{3}{0} = ?$$

We need to find a number that when multiplied by 0 will equal 3, that is,

$$? \cdot 0 = 3.$$

No real number satisfies this equation because the product of any real number and 0 must be 0. ***Thus, division by 0 is undefined.***

Division Involving 0

For any real number x, with $x \neq 0$,

$$\frac{0}{x} = 0 \quad \text{and} \quad \frac{x}{0} \textbf{ is undefined.}$$

Examples: $\frac{0}{-10} = 0$ and $\frac{-10}{0}$ is undefined.

Work Problem 6 at the Side. ▶

From the definitions of multiplication and division of real numbers,

$$\frac{-40}{8} = -5 \quad \text{and} \quad \frac{40}{-8} = -5, \quad \text{so} \quad \frac{-40}{8} = \frac{40}{-8}.$$

Based on this example, the quotient of a positive number and a negative number can be expressed in different, yet equivalent, forms.

5 Find each quotient.

(a) $\frac{-8}{-2}$

(b) $\frac{12}{-4}$

(c) $\frac{-1.44}{-0.12}$

(d) $\frac{1}{4} \div \left(-\frac{2}{3}\right)$

6 Find each quotient if possible.

(a) $\frac{-3}{0}$

(b) $\frac{0}{-53}$

Answers

5. **(a)** 4 **(b)** -3 **(c)** 12 **(d)** $-\frac{3}{8}$

6. **(a)** undefined **(b)** 0

Also, $\frac{-40}{-8} = 5$ and $\frac{40}{8} = 5$, so $\frac{-40}{-8} = \frac{40}{8}$.

Equivalent Forms

For any positive real numbers x and y, the following are equivalent.

$\frac{-x}{y}$, $\frac{x}{-y}$, and $-\frac{x}{y}$ ← We generally use this form for negative final answers.

$\frac{-x}{-y}$ and $\frac{x}{y}$ ← We generally use this form for positive final answers.

OBJECTIVE 4 Use the rules for order of operations when multiplying and dividing real numbers.

EXAMPLE 4 Using the Rules for Order of Operations

Perform the indicated operations.

(a) $-9(2) - (-3)(2)$

$= -18 - (-6)$ Multiply.

$= -18 + 6$ Definition of subtraction

$= -12$ Add.

(b) $-6 + 2(3 - 5)$ ← Begin inside the parentheses.

Do *not* add first.

$= -6 + 2(-2)$ Subtract inside the parentheses.

$= -6 + (-4)$ Multiply.

$= -10$ Add.

(c) $|3 - 2(4)| - 2(-6)$

$= |3 - 8| - 2(-6)$ Multiply inside the absolute value bars.

$= |-5| - 2(-6)$ Subtract inside the absolute value bars.

$= 5 - 2(-6)$ Find the absolute value.

$= 5 - (-12)$ Multiply.

$= 17$ Subtract.

(d) $\frac{5(-2) - 3(4)}{2(1 - 6)}$ Simplify the numerator and denominator separately.

$= \frac{-10 - 12}{2(-5)}$ Multiply in the numerator. Subtract in the denominator.

$= \frac{-22}{-10}$ Subtract in the numerator. Multiply in the denominator.

$= \frac{11}{5}$ Write in lowest terms.

◀ Work Problem 7 at the Side.

7 Label the order in which each operation should be performed. Then perform the operations.

GS **(a)** $-3(4) - 2(6)$

○ ○○

$=$ ____ $-$ ____

$=$ ____

GS **(b)** $-8[-1 - (-4)(-5)]$

○ ○ ○

$= -8[-1 -$ ____ $]$

$= -8[$ ____ $]$

$=$ ____

(c) $-4 - 2(7 - 10)$

(d) $6(-2) + |3(3) - 11|$

(e) $\frac{6(-4) - 2(5)}{3(2 - 7)}$

GS **(f)** $\frac{2^2 - 5^2}{3(7 - 8)}$

$= \frac{\_\_\_ - \_\_\_}{3(\_\_\_)}$

$= \frac{\_\_\_}{-3}$

$=$ ____

Answers

7. (a) ①, ③, ②; −12; 12; −24
(b) ③, ②, ①; 20; −21; 168
(c) 2 (d) −10 (e) $\frac{34}{15}$
(f) 4; 25; −1; −21; 7

OBJECTIVE 5 Evaluate expressions involving variables.

EXAMPLE 5 Evaluating Algebraic Expressions

Evaluate each expression for $x = -1$, $y = -2$, and $m = -3$.

(a) $(3x + 4y)(-2m)$

Use parentheses around substituted negative values to avoid errors.

$= [3(-1) + 4(-2)][-2(-3)]$ Substitute the given values for the variables.

$= [-3 + (-8)][6]$ Multiply.

$= [-11]6$ Add inside the brackets.

$= -66$ Multiply.

(b) $2x^2 - 3y^2$

Think: $(-2)^2 = -2(-2)$

$= 2(-1)^2 - 3(-2)^2$ Substitute -1 for x and -2 for y.

Think: $(-1)^2 = -1(-1)$

$= 2(1) - 3(4)$ Apply the exponents.

$= 2 - 12$ Multiply.

$= -10$ Subtract.

(c) $\dfrac{4y^2 + x}{m}$

$= \dfrac{4(-2)^2 + (-1)}{-3}$ Substitute -2 for y, -1 for x, and -3 for m.

$= \dfrac{4(4) + (-1)}{-3}$ Apply the exponent.

$= \dfrac{16 + (-1)}{-3}$ Multiply.

$= \dfrac{15}{-3}$ Add.

$= -5$ Divide.

(d) $\left(\frac{3}{4}x + \frac{5}{8}y\right)\left(-\frac{1}{2}m\right)$

$= \left[\frac{3}{4}(-1) + \frac{5}{8}(-2)\right]\left[-\frac{1}{2}(-3)\right]$ Substitute for x, y, and m.

$= \left[-\frac{3}{4} + \left(-\frac{5}{4}\right)\right]\left[\frac{3}{2}\right]$ Multiply inside the brackets.

$= \left[-\frac{8}{4}\right]\left[\frac{3}{2}\right]$ Add inside the brackets.

$= -2\left(\frac{3}{2}\right)$ Divide.

$= -3$ Multiply.

Work Problem 8 at the Side. ▶

8 Evaluate each expression.

GS **(a)** $2x - 7(y + 1)$,
for $x = -4$ and $y = 3$

$2x - 7(y + 1)$

$= 2(\_\_\_) - 7(\_\_\_ + 1)$

$= \_\_\_ - 7(\_\_\_)$

$= -8 - \_\_\_$

$= \_\_\_$

(b) $2x^2 - 4y^2$,
for $x = -2$ and $y = -3$

(c) $\dfrac{4x - 2y}{-3x}$,
for $x = 2$ and $y = -1$

(d) $\left(\frac{2}{5}x - \frac{5}{6}y\right)\left(-\frac{1}{3}z\right)$,
for $x = 10$, $y = 6$, and $z = -9$

Answers

8. **(a)** -4; 3; -8; 4; 28; -36
(b) -28 **(c)** $-\frac{5}{3}$ **(d)** -3

OBJECTIVE 6 Translate words and phrases that indicate multiplication and division.

9 Write a numerical expression for each phrase, and simplify the expression.

(a) The product of 6 and the sum of -5 and -4

(b) Three times the difference of 4 and -6

(c) Three-fifths of the sum of 2 and -7

(d) 20% of the sum of 9 and -4

(e) Triple the product of 5 and 6

WORDS AND PHRASES THAT INDICATE MULTIPLICATION

Word or Phrase	Example	Numerical Expression and Simplification
Product of	The *product of* -5 and -2	$-5(-2)$, which equals 10
Times	13 *times* -4	$13(-4)$, which equals -52
Twice (meaning "2 times")	*Twice* 6	$2(6)$, which equals 12
Triple (meaning "3 times")	*Triple* 4	$3(4)$, which equals 12
Of (used with fractions)	$\frac{1}{2}$ *of* 10	$\frac{1}{2}(10)$, which equals 5
Percent of	12*% of* -16	$0.12(-16)$, which equals -1.92
As much as	$\frac{2}{3}$ *as much as* 30	$\frac{2}{3}(30)$, which equals 20

EXAMPLE 6 Translating Words and Phrases (Multiplication)

Write a numerical expression for each phrase, and simplify the expression.

(a) The **product of** 12 and the sum of 3 and -6

$12[3+(-6)]$ simplifies to $12[-3]$, which equals -36.

(b) **Twice** the difference of 8 and -4

$2[8-(-4)]$ simplifies to $2[12]$, which equals 24.

(c) Two-thirds **of** the sum of -5 and -3

$\frac{2}{3}[-5+(-3)]$ simplifies to $\frac{2}{3}[-8]$, which equals $-\frac{16}{3}$.

(d) 15**% of** the difference of 14 and -2

$0.15[14-(-2)]$ simplifies to $0.15[16]$, which equals 2.4.

Remember that 15% = 0.15.

(e) **Double** the product of 3 and 4

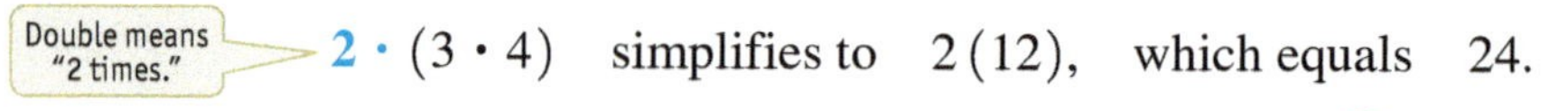

Double means "2 times."

$2\cdot(3\cdot 4)$ simplifies to $2(12)$, which equals 24.

◀ Work Problem 9 at the Side.

PHRASES THAT INDICATE DIVISION

Phrase	Example	Numerical Expression and Simplification
Quotient of	The *quotient of* -24 and 3	$\frac{-24}{3}$, which equals -8
Divided by	-16 *divided by* -4	$\frac{-16}{-4}$, which equals 4
Ratio of	The *ratio of* 2 to 3	$\frac{2}{3}$

When translating a phrase involving division into a fraction, we write the first number named as the numerator and the second as the denominator.

Answers

9. (a) $6[(-5)+(-4)]$; -54
(b) $3[4-(-6)]$; 30
(c) $\frac{3}{5}[2+(-7)]$; -3
(d) $0.20[9+(-4)]$; 1
(e) $3(5\cdot 6)$; 90

EXAMPLE 7 Translating Words and Phrases (Division)

Write a numerical expression for each phrase, and simplify the expression.

(a) The **quotient of** 14 and the sum of -9 and 2

"Quotient" indicates division.

$$\frac{14}{-9+2} \text{ simplifies to } \frac{14}{-7}, \text{ which equals } -2.$$

(b) The product of 5 and -6, **divided by** the difference of -7 and 8

$$\frac{5(-6)}{-7-8} \text{ simplifies to } \frac{-30}{-15}, \text{ which equals } 2.$$

Work Problem 10 at the Side. ▶

OBJECTIVE 7 Translate simple sentences into equations. We can use words and phrases to translate sentences into equations.

EXAMPLE 8 Translating Sentences into Equations

Write each sentence as an equation, using x as the variable.

(a) Three **times** a number **is** -18.

The word *times* indicates multiplication. The word *is* translates as =.

$$3 \cdot x = -18, \quad \text{or} \quad 3x = -18 \qquad 3 \cdot x = 3x$$

(b) The **sum** of a number and 9 **is** 12.

$$x + 9 = 12$$

(c) The **difference of** a number and 5 **is** 0.

$$x - 5 = 0$$

(d) The **quotient of** 24 and a number **is** -2.

$$\frac{24}{x} = -2$$

Work Problem 11 at the Side. ▶

CAUTION

In **Examples 6 and 7,** the *phrases* translate as *expressions*, while in **Example 8,** the *sentences* translate as *equations*.

- ***An expression is a phrase.***
- ***An equation is a sentence with something on the left side, an = symbol, and something on the right side.***

$$\frac{5(-6)}{-7-8} \qquad 3x = -18$$

↑ Expression ↑ Equation

10 Write a numerical expression for each phrase, and simplify the expression.

(a) The quotient of 20 and the sum of 8 and -3

(b) The product of -9 and 2, divided by the difference of 5 and -1

11 Write each sentence as an equation, using x as the variable.

(a) Twice a number is -6.

(b) The difference of -8 and a number is -11.

(c) The sum of 5 and a number is 8.

(d) The quotient of a number and -2 is 6.

Answers

10. (a) $\frac{20}{8+(-3)}$; 4 **(b)** $\frac{-9(2)}{5-(-1)}$; -3

11. (a) $2x = -6$ **(b)** $-8 - x = -11$ **(c)** $5 + x = 8$ **(d)** $\frac{x}{-2} = 6$

1.6 Exercises

FOR EXTRA HELP Go to MyMathLab for worked-out, step-by-step solutions to exercises enclosed in a square and video solutions to ▶ exercises.

CONCEPT CHECK *Fill in each blank with one of the following.*

greater than 0, less than 0, *or* equal to 0

1. The product or the quotient of two numbers with the same sign is ___________.

2. The product or the quotient of two numbers with different signs is ___________.

3. If three negative numbers are multiplied together, the product is ___________.

4. If two negative numbers are multiplied and then their product is divided by a negative number, the result is ___________.

5. If a negative number is squared and the result is added to a positive number, the final answer is ___________.

6. The reciprocal of a negative number is ___________.

CONCEPT CHECK *Work each problem.*

7. Complete this statement: The quotient formed by any nonzero number divided by 0 is ________, and the quotient formed by 0 divided by any nonzero number is ________. Give an example of each quotient.

8. Which expression is undefined?

A. $13 \div 0$ **B.** $13 \div 13$ **C.** $0 \div 13$ **D.** $13 \cdot 0$

Find each product. ***See Examples 1 and 2.***

9. $-7(4)$
10. $-8(5)$
11. $-5(-6)$
12. $-4(-20)$
13. $-8(0)$
14. $0(-12)$
15. $-\frac{3}{8}\left(-\frac{20}{9}\right)$
16. $-\frac{5}{4}\left(-\frac{6}{25}\right)$
17. $-6.8(0.35)$
18. $-4.6(0.24)$
19. $-6\left(-\frac{1}{4}\right)$
20. $-8\left(-\frac{1}{2}\right)$

Find each quotient. ***See Example 3*** *and the discussion of division involving 0.*

21. $\frac{-15}{5}$
22. $\frac{-18}{6}$
23. $\frac{96}{-16}$
24. $\frac{38}{-19}$
25. $\frac{-8.8}{2.2}$
26. $\frac{-4.6}{0.23}$
27. $\frac{-160}{-10}$
28. $\frac{-260}{-20}$

29. $\frac{0}{-3}$ **30.** $\frac{0}{-5}$ **31.** $\frac{11.5}{0}$ **32.** $\frac{15.2}{0}$

33. $-\frac{5}{6} \div \frac{8}{9}$ **34.** $-\frac{7}{10} \div \frac{3}{4}$ **35.** $-\frac{3}{4} \div \left(-\frac{1}{2}\right)$ **36.** $-\frac{3}{16} \div \left(-\frac{5}{8}\right)$

Perform the indicated operation. ***See Example 4.***

37. $7 - 3 \cdot 6$ **38.** $8 - 2 \cdot 5$ **39.** $-2(5) - (-4)(2)$ **40.** $-4(3) - (-3)(6)$

41. $-7(3 - 8)$ **42.** $-5(4 - 7)$ **43.** $7 + 2(4 - 1)$ **44.** $5 + 3(6 - 4)$

45. $-4 + 3(2 - 8)$ **46.** $-8 + 4(5 - 7)$ **47.** $3(-5) + |3 - 10|$ **48.** $4(-8) + |4 - 15|$

49. $|8 - 7(2)| - 6(-2)$ **50.** $|5 - 3(9)| - 7(-4)$ **51.** $\frac{-5(-6)}{9 - (-1)}$ **52.** $\frac{-12(-5)}{7 - (-5)}$

53. $\frac{-21(3)}{-3 - 6}$ **54.** $\frac{-40(3)}{-2 - 3}$ **55.** $\frac{-10(2) + 6(2)}{-3 - (-1)}$

56. $\frac{8(-1) + 6(-2)}{-6 - (-1)}$ **57.** $\frac{-6 - |-9 + 5|}{2 - (-3)}$ **58.** $\frac{-8 - |-3 + 2|}{-3 - (-6)}$

59. $\frac{-27(-2) - (-12)(-2)}{-2(3) - 2(2)}$ **60.** $\frac{-13(-4) - (-8)(-2)}{(-10)(2) - 4(-2)}$ **61.** $\frac{3^2 - 4^2}{7(-8 + 9)}$

62. $\frac{5^2 - 7^2}{2(3 + 3)}$ **63.** $\frac{4(2^3 - 5) - 5(-3^3 + 21)}{3[6 - (-2)]}$ **64.** $\frac{-3(-2^4 + 10) + 4(2^5 - 12)}{-2[8 - (-7)]}$

Evaluate each expression for $x = 6$, $y = -4$, *and* $a = 3$. ***See Example 5.***

65. $6x - 5y + 4a$ **66.** $5x - 2y + 3a$ **67.** $(2x + y)(3a)$

68. $(5x - 2y)(-2a)$ **69.** $\left(\frac{5}{6}x + \frac{3}{2}y\right)\left(-\frac{1}{3}a\right)$ **70.** $\left(\frac{1}{3}x - \frac{4}{5}y\right)\left(-\frac{1}{5}a\right)$

71. $(6 - x)(5 + y)(3 + a)$

72. $(-5 + x)(-3 + y)(3 - a)$

73. $5x - 4a^2$

74. $-2y^2 + 3a$

75. $\dfrac{2y - x}{a - 3}$

76. $\dfrac{xy + 8a}{x - 6}$

Write a numerical expression for each phrase, and simplify the expression.
See Examples 6 and 7.

77. The product of 4 and -7, added to -12

78. The product of -9 and 2, added to 9

79. Twice the product of -1 and 6, subtracted from -4

80. Twice the product of -8 and 2, subtracted from -1

81. The product of -3 and the difference of 3 and -7

82. The product of 12 and the difference of 9 and -8

83. Three-tenths of the sum of -2 and -28

84. Four-fifths of the sum of -8 and -2

85. 20% of the product of -5 and 6

86. 30% of the product of -8 and 5

87. The quotient of -12 and the sum of -5 and -1

88. The quotient of -20 and the sum of -8 and -2

89. The sum of -18 and -6, divided by the product of 2 and -4

90. The sum of 15 and -3, divided by the product of 4 and -3

91. The product of $-\frac{2}{3}$ and $-\frac{1}{5}$, divided by $\frac{1}{7}$

92. The product of $-\frac{1}{2}$ and $\frac{3}{4}$, divided by $-\frac{2}{3}$

Write each sentence as an equation, using x as the variable. ***See Example 8.***

93. Nine times a number is -36.

94. Seven times a number is -42.

95. The quotient of a number and 4 is -1.

96. The quotient of a number and 3 is -3.

97. $\frac{9}{11}$ less than a number is 5.

98. $\frac{1}{2}$ less than a number is 2.

99. When 6 is divided by a number, the result is -3.

100. When 15 is divided by a number, the result is -5.

A few years ago, the following question and expression appeared on boxes of Swiss Miss Hot Cocoa Mix: On average, how many mini-marshmallows are in one serving?

$$3 + 2 \times 4 \div 2 - 3 \times 7 - 4 + 47$$

101. The box gave 92 as the answer. What is the *correct* answer?

102. ***What Went Wrong?*** Explain the error that somebody at the company made in calculating the answer.

Relating Concepts (Exercises 103–108) For Individual or Group Work

To find the ***average (mean)*** *of a group of numbers, we add the numbers and then divide the sum by the number of terms added.* ***Work Exercises 103–106 in order,*** *to find the average of* 23, 18, 13, -4, *and* -8.

103. Find the sum of the given group of numbers.

104. How many numbers are in the group?

105. Divide the answer for **Exercise 103** by the answer for **Exercise 104.** Give the quotient as a mixed number.

106. What is the average of the given group of numbers?

Find the average of each group of numbers.

107. All integers between -10 and 14, including both -10 and 14

108. All integers between -15 and -10, including both -15 and -10

Summary Exercises *Performing Operations with Real Numbers*

Operations with Signed Numbers

Addition

Same sign Add the absolute values of the numbers. The sum has the same sign as the addends.

Different signs Find the absolute values of the numbers, and subtract the lesser absolute value from the greater. The sum has the same sign as the addend with greater absolute value.

Subtraction

Add the additive inverse (or opposite) of the subtrahend to the minuend.

Multiplication and Division

Same sign The product or quotient of two numbers with the same sign is positive.

Different signs The product or quotient of two numbers with different signs is negative.

Division by 0 is undefined.

Perform operations with signed numbers using the rules for order of operations.

Order of Operations

If grouping symbols are present, work within them, innermost first (and above and below fraction bars separately), in the following order.

Step 1 Apply all **exponents.**

Step 2 Do any **multiplications** or **divisions** in order from left to right.

Step 3 Do any **additions** or **subtractions** in order from left to right.

If no grouping symbols are present, start with Step 1.

Perform the indicated operations.

1. $14 - 3 \cdot 10$

2. $-3(8) - 4(-7)$

3. $(3 - 8)(-2) - 10$

4. $-6(7 - 3)$

5. $7 + 3(2 - 10)$

6. $-4[(-2)(6) - 7]$

7. $(-4)(7) - (-5)(2)$

8. $-5[-4 - (-2)(-7)]$

9. $40 - (-2)[8 - 9]$

10. $\dfrac{5(-4)}{-7 - (-2)}$

11. $\dfrac{-3 - (-9 + 1)}{-7 - (-6)}$

12. $\dfrac{5(-8 + 3)}{13(-2) + (-7)(-3)}$

13. $\dfrac{6^2 - 8}{-2(2) + 4(-1)}$

14. $\dfrac{16(-8 + 5)}{15(-3) + (-7 - 4)(-3)}$

15. $\frac{9(-6)-3(8)}{4(-7)+(-2)(-11)}$

16. $\frac{2^2+4^2}{5^2-3^2}$

17. $\frac{(2+4)^2}{(5-3)^2}$

18. $\frac{4^3-3^3}{-5(-4+2)}$

19. $\frac{-9(-6)+(-2)(27)}{3(8-9)}$

20. $|-4(9)|-|-11|$

21. $\frac{6(-10+3)}{15(-2)-3(-9)}$

22. $\frac{(-9)^2-9^2}{3^2-5^2}$

23. $\frac{(-10)^2+10^2}{-10(5)}$

24. $-\frac{3}{4}\div\left(-\frac{5}{8}\right)$

25. $\frac{1}{2}\div\left(-\frac{1}{2}\right)$

26. $\frac{8^2-12}{(-5)^2+2(6)}$

27. $\left[\frac{5}{8}-\left(-\frac{1}{16}\right)\right]+\frac{3}{8}$

28. $\left(\frac{1}{2}-\frac{1}{3}\right)-\frac{5}{6}$

29. $-0.9(-3.7)$

30. $-5.1(-0.2)$

31. $|-2(3)+4|-|-2|$

32. $40+2(-5-3)$

Evaluate each expression for $x=-2$, $y=3$, and $a=4$.

33. $-x+y-3a$

34. $(x-y)-(a-2y)$

35. $\left(\frac{1}{2}x+\frac{2}{3}y\right)\left(-\frac{1}{4}a\right)$

36. $\frac{2x+3y}{a-xy}$

37. $\frac{x^2-y^2}{x^2+y^2}$

38. $-x^2+3y$

39. $\frac{-x+2y}{2x+a}$

40. $\frac{2x+a}{-x+2y}$

1.7 Properties of Real Numbers

OBJECTIVES

1. Use the commutative properties.
2. Use the associative properties.
3. Use the identity properties.
4. Use the inverse properties.
5. Use the distributive property.

In the basic properties in this section, a, b, and c represent real numbers.

OBJECTIVE 1 Use the commutative properties. The word *commute* means to go back and forth. We might commute to work or to school. If we travel from home to work and follow the same route from work to home, we travel the same distance each time.

The **commutative properties** say that if two numbers are added or multiplied in either order, the result is the same.

Commutative Properties

$a + b = b + a$ Addition

$ab = ba$ Multiplication

EXAMPLE 1 Using the Commutative Properties

Use a commutative property to complete each statement.

(a) $-8 + 5 = 5 + \underline{\ ?\ }$

Notice that the "order" changed.

$-8 + 5 = 5 + (-8)$ Commutative property of addition

(b) $(-2)7 = \underline{\ ?\ }(-2)$

$-2(7) = 7(-2)$ Commutative property of multiplication

◀ Work Problem 1 at the Side.

1 Use a commutative property to complete each statement.

(a) $x + 9 = 9 + \underline{\qquad}$

(b) $5x = x \cdot \underline{\qquad}$

OBJECTIVE 2 Use the associative properties. When we *associate* one object with another, we think of those objects as being grouped together.

The **associative properties** say that when we add or multiply three numbers, we can group the first two together or the last two together and obtain the same answer.

Associative Properties

$(a + b) + c = a + (b + c)$ Addition

$(ab)c = a(bc)$ Multiplication

EXAMPLE 2 Using the Associative Properties

Use an associative property to complete each statement.

(a) $-8 + (1 + 4) = (-8 + \underline{\ ?\ }) + 4$

The "order" is the same. The "grouping" changed.

$-8 + (1 + 4) = (-8 + 1) + 4$ Associative property of addition

(b) $[2 \cdot (-7)] \cdot 6 = 2 \cdot \underline{\ ?\ }$

$[2 \cdot (-7)] \cdot 6 = 2 \cdot [(-7) \cdot 6]$ Associative property of multiplication

◀ Work Problem 2 at the Side.

2 Use an associative property to complete each statement.

(a) $-5 + (2 + 8)$
$= (\underline{\qquad\quad}) + 8$

(b) $10 \cdot [-8 \cdot (-3)]$
$= \underline{\qquad\quad}$

Answers

1. (a) x (b) 5
2. (a) $-5 + 2$ (b) $[10 \cdot (-8)] \cdot (-3)$

By the associative property, the sum (or product) of three numbers will be the same no matter how the numbers are "associated" in groups. Parentheses can be left out if a problem contains only addition (or multiplication). For example,

$(-1 + 2) + 3$ and $-1 + (2 + 3)$ can be written as $-1 + 2 + 3$.

EXAMPLE 3 **Distinguishing between Properties**

Decide whether each statement is an example of a *commutative property,* an *associative property,* or *both.*

(a) $(2 + 4) + 5 = 2 + (4 + 5)$

The order of the three numbers is the same on both sides of the equality symbol. The only change is in the *grouping*, or association, of the numbers. This is an example of the *associative property*.

(b) $6 \cdot (3 \cdot 10) = 6 \cdot (10 \cdot 3)$

The same numbers, 3 and 10, are grouped on each side. On the left, the 3 appears first, but on the right, the 10 appears first. The only change involves the *order* of the numbers, so this is an example of the *commutative property*.

(c) $(8 + 1) + 7 = 8 + (7 + 1)$

Both the order and the grouping are changed. On the left, the order of the three numbers is 8, 1, and 7. On the right, it is 8, 7, and 1. On the left, the 8 and 1 are grouped. On the right, the 7 and 1 are grouped. Therefore, *both* properties are used.

Work Problem 3 at the Side. ▶

We can sometimes use the commutative and associative properties to rearrange and regroup numbers to simplify calculations.

EXAMPLE 4 **Using the Commutative and Associative Properties**

Find each sum or product.

(a) $23 + 41 + 2 + 9 + 25$

$= (41 + 9) + (23 + 2) + 25$

$= 50 + 25 + 25$ Use the commutative and associative properties.

$= 100$

(b) $25(69)(4)$

$= 25(4)(69)$

$= 100(69)$

$= 6900$

Work Problem 4 at the Side. ▶

OBJECTIVE 3 **Use the identity properties.** If a child wears a costume on Halloween, the child's appearance is changed, but his or her *identity* is unchanged. The identity of a real number is left unchanged when identity properties are applied.

The **identity properties** say that the sum of 0 and any number equals that number, and the product of 1 and any number equals that number.

Identity Properties

$a + 0 = a$ and $0 + a = a$ Addition

$a \cdot 1 = a$ and $1 \cdot a = a$ Multiplication

3 Decide whether each statement is an example of a *commutative property*, an *associative property*, or *both*.

(a) $2 \cdot (4 \cdot 6) = (2 \cdot 4) \cdot 6$

(b) $(2 \cdot 4) \cdot 6 = (4 \cdot 2) \cdot 6$

(c) $(2 + 4) + 6 = 4 + (2 + 6)$

4 Find each sum or product.

(a) $5 + 18 + 29 + 31 + 12$

(b) $5(37)(20)$

Answers

3. **(a)** associative property **(b)** commutative property **(c)** both

4. **(a)** 95 **(b)** 3700

5 Use an identity property to complete each statement.

(a) $9 + 0 =$ ____

(b) ____ $+ (-7) = -7$

(c) ____ $\cdot 1 = 5$

6 In part (a), write in lowest terms. In part (b), perform the operation.

(a) $\frac{85}{105}$

(b) $\frac{9}{10} - \frac{53}{50}$

Answers

5. (a) 9 (b) 0 (c) 5

6. (a) $\frac{17}{21}$ (b) $-\frac{4}{25}$

The number 0 leaves the identity, or value, of any real number unchanged by addition, so 0 is the **identity element for addition,** or the **additive identity.** Because multiplication by 1 leaves any real number unchanged, 1 is the **identity element for multiplication,** or the **multiplicative identity.**

EXAMPLE 5 Using the Identity Properties

Use an identity property to complete each statement.

(a) $-3 + \underline{\ ?\ } = -3$

$-3 + 0 = -3$

Identity property of addition

(b) $\underline{\ ?\ } \cdot \frac{1}{2} = \frac{1}{2}$

$1 \cdot \frac{1}{2} = \frac{1}{2}$

Identity property of multiplication

◀ Work Problem **5** at the Side.

We use the identity property of multiplication to write fractions in lowest terms and to find common denominators.

EXAMPLE 6 Using the Identity Property to Simplify Expressions

In part (a), write in lowest terms. In part (b), perform the operation.

(a) $\frac{49}{35}$

$= \frac{7 \cdot 7}{5 \cdot 7}$ Factor.

$= \frac{7}{5} \cdot \frac{7}{7}$ Write as a product.

$= \frac{7}{5} \cdot 1$ Property of 1

$= \frac{7}{5}$ Identity property

(b) $\frac{3}{4} + \frac{5}{24}$

$= \frac{3}{4} \cdot 1 + \frac{5}{24}$ Identity property

$= \frac{3}{4} \cdot \frac{6}{6} + \frac{5}{24}$ Use $1 = \frac{6}{6}$ to obtain a common denominator.

$= \frac{18}{24} + \frac{5}{24}$ Multiply.

$= \frac{23}{24}$ Add.

◀ Work Problem **6** at the Side.

OBJECTIVE 4 Use the inverse properties. Each day before we go to work or school, we likely put on our shoes. Before we go to sleep at night, we likely take them off. These operations from everyday life are examples of *inverse* operations.

The **inverse properties** of addition and multiplication lead to the additive and multiplicative identities, respectively. Recall that $-a$ is the **additive inverse,** or **opposite,** of a, and $\frac{1}{a}$ is the **multiplicative inverse,** or **reciprocal,** of the nonzero number a.

Inverse Properties

$a + (-a) = 0$ and $-a + a = 0$ Addition

$a \cdot \frac{1}{a} = 1$ and $\frac{1}{a} \cdot a = 1 \quad (a \neq 0)$ Multiplication

EXAMPLE 7 Using the Inverse Properties

Use an inverse property to complete each statement.

(a) $\underline{\ ?\ } + \frac{1}{2} = 0$

$-\frac{1}{2} + \frac{1}{2} = 0$

(b) $4 + \underline{\ ?\ } = 0$

$4 + (-4) = 0$

(c) $-0.75 + \frac{3}{4} = \underline{\ ?\ }$

$-0.75 + \frac{3}{4} = 0$

The inverse property of addition is used in parts (a)–(c).

(d) $\underline{\ ?\ } \cdot \frac{5}{2} = 1$

$\frac{2}{5} \cdot \frac{5}{2} = 1$

(e) $-5(\underline{\ ?\ }) = 1$

$-5\left(-\frac{1}{5}\right) = 1$

(f) $4(0.25) = \underline{\ ?\ }$

$4(0.25) = 1$

The inverse property of multiplication is used in parts (d)–(f).

Work Problems 7 and 8 at the Side. ▶

OBJECTIVE 5 **Use the distributive property.** The word *distribute* means "to give out from one to several." Consider the following expressions.

$2(5 + 8)$ equals $2(13)$, or 26.

$2(5) + 2(8)$ equals $10 + 16$, or 26.

Both expressions equal 26.

Thus, $2(5 + 8) = 2(5) + 2(8)$.

This result is an example of the *distributive property of multiplication with respect to addition,* the only property involving *both* addition and multiplication. With this property, a product can be changed to a sum or difference. This idea is illustrated by the divided rectangle in **Figure 15.**

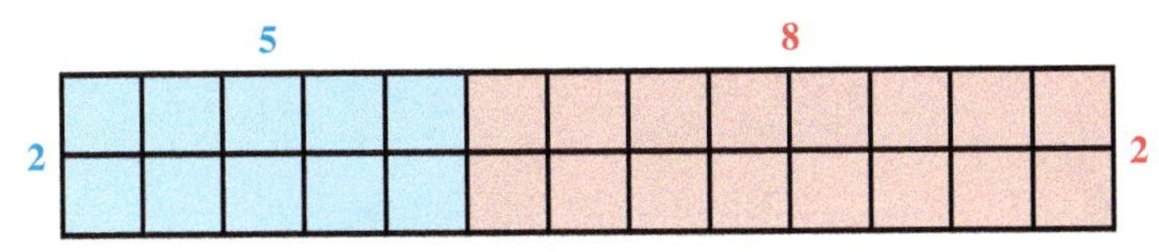

The area of the left part is 2(5) = 10.
The area of the right part is 2(8) = 16.
The total area is 2(5 + 8) = 26, or the total area is 2(5) + 2(8) = 10 + 16 = 26.
Thus, 2(5 + 8) = 2(5) + 2(8).

Figure 15

The **distributive property** says that multiplying a number a by a sum of two numbers $b + c$ gives the same result as multiplying a by b and a by c and then adding the two products.

Distributive Property

$$a(b + c) = ab + ac \quad \text{and} \quad (b + c)a = ba + ca$$

The a outside the parentheses is "distributed" over the b and c inside. The distributive property is valid for multiplication over subtraction.

$$a(b - c) = ab - ac \quad \text{and} \quad (b - c)a = ba - ca$$

7 Use an inverse property to complete each statement.

(a) $-6 + \underline{\quad\quad} = 0$

(b) $\frac{4}{3} \cdot \underline{\quad\quad} = 1$

(c) $-\frac{1}{9} \cdot (\underline{\quad\quad}) = 1$

8 Complete each statement so that it is an example of either an identity property or an inverse property. Tell which property is used.

(a) $275 + \underline{\quad\quad} = 275$

(b) $-0.6 + \frac{3}{5} = \underline{\quad\quad}$

(c) $0.2(5) = \underline{\quad\quad}$

(d) $\frac{2}{5} \cdot \underline{\quad\quad} = \frac{2}{5}$

Answers

7. (a) 6 (b) $\frac{3}{4}$ (c) -9

8. (a) 0; identity property
(b) 0; inverse property
(c) 1; inverse property
(d) 1; identity property

9 Use the distributive property to rewrite each expression. Simplify if possible.

GS **(a)** $2(p + 5)$

$= 2 \cdot \_\_\_\_ + 2 \cdot \_\_\_\_$

$= \_\_\_\_ + \_\_\_\_$

(b) $-4(y + 7)$

(c) $5(m - 4)$

(d) $-5(4x - 3)$

(e) $-\frac{2}{3}(6x - 3)$

(f) $7(2y + 7k - 9m)$

Answers

9. **(a)** p; 5; $2p$; 10 **(b)** $-4y - 28$
(c) $5m - 20$ **(d)** $-20x + 15$
(e) $-4x + 2$ **(f)** $14y + 49k - 63m$

The distributive property can also be extended to more than two numbers.

$$a(b + c + d) = ab + ac + ad$$

EXAMPLE 8 Using the Distributive Property

Use the distributive property to rewrite each expression. Simplify if possible.

(a) $5(9 + 6)$

We could write $5(9) + 5(6)$ here.

$= 5 \cdot 9 + 5 \cdot 6$ The factor 5 is "distributed" over the 9 and 6.

$= 45 + 30$ Multiply.

$= 75$ Add.

Multiply first.

(b) $4(x + 5 + y)$

$= 4x + 4 \cdot 5 + 4y$

$= 4x + 20 + 4y$

(c) $-2(x + 3)$

$= -2x + (-2)(3)$

$= -2x + (-6)$

$= -2x - 6$

(d) $-\frac{1}{2}(4x + 3)$

Think: $-\frac{1}{2}(4x) = \left(-\frac{1}{2} \cdot 4\right)x = \left(-\frac{1}{2} \cdot \frac{4}{1}\right)x$

$= -\frac{1}{2}(4x) + \left(-\frac{1}{2}\right)(3)$ Distributive property

$= -2x + \left(-\frac{3}{2}\right)$ Multiply.

This step is often omitted.

$= -2x - \frac{3}{2}$ Definition of subtraction

(e) $3(k - 9)$

Be careful here.

$= 3[k + (-9)]$ Definition of subtraction

$= 3k + 3(-9)$ Distributive property

$= 3k - 27$ Multiply.

(f) $-2(3x - 4)$

$= -2[3x + (-4)]$ Definition of subtraction

$= -2(3x) + (-2)(-4)$ Distributive property

$= (-2 \cdot 3)x + (-2)(-4)$ Associative property

$= -6x + 8$ Multiply.

(g) $8(3r + 11t + 5z)$

$= 8(3r) + 8(11t) + 8(5z)$ Distributive property

$= (8 \cdot 3)r + (8 \cdot 11)t + (8 \cdot 5)z$ Associative property

This step is often omitted.

$= 24r + 88t + 40z$ Multiply.

◀ **Work Problem 9 at the Side.**

The expression $-a$ may be interpreted as $-1 \cdot a$. Using this result and the distributive property, we can *clear* (or *remove*) *parentheses.*

EXAMPLE 9 **Using the Distributive Property to Clear Parentheses**

Write each expression without parentheses.

(a) $-(2y + 3)$

The − symbol indicates a factor of −1.

$= -1 \cdot (2y + 3)$ $-a = -1 \cdot a$

$= -1 \cdot 2y + (-1) \cdot 3$ Distributive property

$= -2y - 3$ Multiply; definition of subtraction

(b) $-(-9w - 2)$

$= -1(-9w - 2)$

$= -1(-9w) - 1(-2)$

$= 9w + 2$

We can also interpret the negative sign in front of the parentheses to mean the *opposite* of each of the terms within the parentheses.

$-1(-9w - 2)$
$= +9w + 2$

(c) $-(-x - 3y + 6z)$

$= -1(-1x - 3y + 6z)$ Be careful with signs.

$= -1(-1x) - 1(-3y) - 1(6z)$ Distributive property

$= x + 3y - 6z$ Multiply.

$-1(-1x) = 1x = x$

Work Problem 10 at the Side. ▶

Summary of the Properties of Addition and Multiplication

For any real numbers a, b, and c, the following properties hold true.

Commutative properties $a + b = b + a$ and $ab = ba$

Associative properties $(a + b) + c = a + (b + c)$

$(ab)c = a(bc)$

Identity properties   There is a real number 0 such that

$a + 0 = a$ and $0 + a = a$.

There is a real number 1 such that

$a \cdot 1 = a$ and $1 \cdot a = a$.

Inverse properties   For each real number a, there is a single real number $-a$ such that

$a + (-a) = 0$ and $(-a) + a = 0$.

For each nonzero real number a, there is a single real number $\frac{1}{a}$ such that

$$a \cdot \frac{1}{a} = 1 \quad \text{and} \quad \frac{1}{a} \cdot a = 1.$$

Distributive property $a(b + c) = ab + ac$

$(b + c)a = ba + ca$

10 Write each expression without parentheses.

GS **(a)** $-(3k - 5)$

$= \_\_\_\_ \cdot (3k - 5)$

$= \_\_\_\_ \cdot 3k + (-1)(\_\_\_\_)$

$= \_\_\_\_ + \_\_\_\_$

(b) $-(2 - r)$

(c) $-(-5y + 8)$

(d) $-(-z + 4)$

(e) $-(-t - 4u + 5v)$

Answers

10. (a) -1; -1; -5; $-3k$; 5
(b) $-2 + r$ **(c)** $5y - 8$
(d) $z - 4$ **(e)** $t + 4u - 5v$

1.7 Exercises

FOR EXTRA HELP Go to MyMathLab for worked-out, step-by-step solutions to exercises enclosed in a square and video solutions to ▶ exercises.

1. **CONCEPT CHECK** Match each item in Column I with the correct choice(s) from Column II. Choices may be used once, more than once, or not at all.

I	II
(a) Identity element for addition	**A.** $(5 \cdot 4) \cdot 3 = 5 \cdot (4 \cdot 3)$
(b) Identity element for multiplication	**B.** 0
(c) Additive inverse of a	**C.** $-a$
(d) Multiplicative inverse, or reciprocal, of the nonzero number a	**D.** -1
(e) The number that is its own additive inverse	**E.** $5 \cdot 4 \cdot 3 = 60$
(f) The two numbers that are their own multiplicative inverses	**F.** 1
(g) The only number that has no multiplicative inverse	**G.** $(5 \cdot 4) \cdot 3 = 3 \cdot (5 \cdot 4)$
(h) An example of the associative property	**H.** $5(4 + 3) = 5 \cdot 4 + 5 \cdot 3$
(i) An example of the commutative property	**I.** $\frac{1}{a}$
(j) An example of the distributive property	

2. **CONCEPT CHECK** Fill in the blanks: The commutative property allows us to change the _______ of the terms in a sum or the factors in a product. The associative property allows us to change the _______ of the terms in a sum or the factors in a product.

CONCEPT CHECK *Tell whether or not the following everyday activities are commutative.*

3. Washing your face and brushing your teeth
4. Putting on your left sock and putting on your right sock
5. Preparing a meal and eating a meal
6. Starting a car and driving away in a car
7. Putting on your socks and putting on your shoes
8. Getting undressed and taking a shower
9. **CONCEPT CHECK** Use parentheses to show how the associative property can be used to give two different meanings to the phrase "foreign sales clerk."
10. **CONCEPT CHECK** Use parentheses to show how the associative property can be used to give two different meanings to the phrase "defective merchandise counter."

Use a commutative or an associative property to complete each statement. State which property is used. ***See Examples 1 and 2.***

11. $-15 + 9 = 9 +$ _____
12. $6 + (-2) = -2 +$ _____
13. $-8 \cdot 3 =$ _____ $\cdot (-8)$
14. $-12 \cdot 4 = 4 \cdot$ _____
15. $(3 + 6) + 7 = 3 + ($_____ $+ 7)$
16. $(-2 + 3) + 6 = -2 + ($_____ $+ 6)$
17. $7 \cdot (2 \cdot 5) = ($_____ $\cdot 2) \cdot 5$
18. $8 \cdot (6 \cdot 4) = (8 \cdot$ _____$) \cdot 4$

19. **CONCEPT CHECK** Evaluate $25 - (6 - 2)$ and evaluate $(25 - 6) - 2$. Does it appear that subtraction is associative?

20. **CONCEPT CHECK** Evaluate $180 \div (15 \div 3)$ and evaluate $(180 \div 15) \div 3$. Does it appear that division is associative?

21. **CONCEPT CHECK** Complete the table and each statement beside it.

Number	Additive Inverse	Multiplicative Inverse
5		
-10		
$-\frac{1}{2}$		
$\frac{3}{8}$		
x		
$-y$		

In general, a number and its additive inverse have (*the same/opposite*) signs.

A number and its multiplicative inverse have (*the same/opposite*) signs.

22. **CONCEPT CHECK** The following conversation took place between one of the authors of this text and his son, Jack, when Jack was 4 years old.

DADDY: "Jack, what is $3 + 0$?"
JACK: "3."
DADDY: "Jack, what is $4 + 0$?"
JACK: "4. And Daddy, *string* plus zero equals *string*!"

What property of addition did Jack recognize?

Decide whether each statement is an example of a commutative, *an* associative, *an* identity, *an* inverse *or the* distributive property. ***See Examples 1, 2, 3, and 5–8.***

23. $\frac{2}{3}(-4) = -4\left(\frac{2}{3}\right)$

24. $6\left(-\frac{5}{6}\right) = \left(-\frac{5}{6}\right)6$

25. $-6 + 6 = 0$

26. $12 + (-12) = 0$

27. $\frac{2}{3}\left(\frac{3}{2}\right) = 1$

28. $\frac{5}{8}\left(\frac{8}{5}\right) = 1$

29. $2.34 \cdot 1 = 2.34$

30. $-8.456 \cdot 1 = -8.456$

31. $6(x + y) = 6x + 6y$

32. $14(t + s) = 14t + 14s$

33. $-\frac{5}{9} = -\frac{5}{9} \cdot \frac{3}{3} = -\frac{15}{27}$

34. $\frac{13}{12} = \frac{13}{12} \cdot \frac{7}{7} = \frac{91}{84}$

35. $-6 + (12 + 7) = (-6 + 12) + 7$

36. $(-8 + 13) + 2 = -8 + (13 + 2)$

37. $(4 + 17) + 3 = 3 + (4 + 17)$

38. $(-8 + 4) + (-12) = -12 + (-8 + 4)$

39. $5(2x) + 5(3y) = 5(2x + 3y)$

40. $3(5t) - 3(7r) = 3(5t - 7r)$

Write a new expression that is equal to the given expression, using the given property. Then simplify the new expression if possible. **See Examples 1, 2, 5, 7, and 8.**

41. $r + 7$; commutative

42. $t + 9$; commutative

43. $s + 0$; identity

44. $w + 0$; identity

45. $-6(x + 7)$; distributive

46. $-5(y + 2)$; distributive

47. $(w + 5) + (-3)$; associative

48. $(b + 8) + (-10)$; associative

Use the properties of this section to perform the operations. **See Example 4.**

49. $1999 + 2 + 1 + 8$

50. $2998 + 3 + 2 + 17$

51. $50(67)2$

52. $5(47)(2)$

53. $43 - 31 + 7 + 31$

54. $26 + 8 - 26 + 12$

55. $-\frac{3}{8} + \frac{2}{5} + \frac{8}{5} + \frac{3}{8}$

56. $-\frac{5}{12} - \frac{3}{7} + \frac{5}{12} + \frac{10}{7}$

57. $-\frac{8}{5}(0.77)\left(-\frac{5}{8}\right)$

58. $\frac{9}{7}(-0.38)\left(\frac{7}{9}\right)$

59. $6t + 8 - 6t + 3$

60. $9r + 12 - 9r + 1$

61. CONCEPT CHECK A student used the distributive property to rewrite the expression $-3(4 - 6)$ as shown.

$$\begin{aligned} &-3(4 - 6) \\ &= -3(4) - 3(6) \\ &= -12 - 18 \\ &= -30 \end{aligned}$$

What Went Wrong? Rewrite the expression correctly.

62. CONCEPT CHECK A student wrote the expression $-(3x + 4)$ without parentheses as shown.

$$\begin{aligned} &-(3x + 4) \\ &= -1(3x + 4) \\ &= -1(3x) + 4 \\ &= -3x + 4 \end{aligned}$$

What Went Wrong? Rewrite the expression correctly.

63. Explain how the procedure of changing $\frac{3}{4}$ to $\frac{9}{12}$ requires the use of the multiplicative identity element, 1.

64. Explain how the procedure for changing $\frac{9}{12}$ to $\frac{3}{4}$ requires the use of the multiplicative identity element, 1.

Use the distributive property to rewrite each expression. Simplify if possible. **See Example 8.**

65. $5(9 + 8)$

66. $6(11 + 8)$

67. $4(t + 3)$

68. $5(w + 4)$

69. $7(z - 8)$

70. $8(x - 6)$

71. $-8(r + 3)$

72. $-11(x + 4)$

73. $-\frac{1}{4}(8x + 3)$

74. $-\frac{1}{3}(9x + 5)$

75. $-5(y - 4)$
GS $= -5(\_\_\_\_) - 5(\_\_\_\_)$
$= \_\_\_\_\_\_\_\_$

76. $-9(g - 4)$
GS $= -9(\_\_\_\_) - 9(\_\_\_\_)$
$= \_\_\_\_\_\_\_\_$

77. $2(6x + 5)$

78. $3(3x + 4)$

79. $-3(2x - 5)$

80. $-4(3x - 2)$

81. $-6(8x + 1)$

82. $-5(4x + 1)$

83. $-\frac{4}{3}(12y + 15z)$

84. $-\frac{2}{5}(10b + 20a)$

85. $8(3r + 4s - 5y)$

86. $2(5u - 3v + 7w)$

87. $-3(8x + 3y + 4z)$

88. $-5(2x - 5y + 6z)$

Write each expression without parentheses. **See Example 9.**

89. $-(6x + 5)$

90. $-(8y + 7)$

91. $-(4t + 3m)$

92. $-(9x + 12y)$

93. $-(-5c - 4d)$

94. $-(-13x - 15y)$

95. $-(-3q + 5r - 8s)$

96. $-(-4z + 5w - 9y)$

1.8 Simplifying Expressions

OBJECTIVES

1. Simplify expressions.
2. Identify terms and numerical coefficients.
3. Identify like terms.
4. Combine like terms.
5. Simplify expressions from word phrases.

OBJECTIVE 1 Simplify expressions. We now simplify expressions using the properties introduced in the previous section.

EXAMPLE 1 Simplifying Expressions

Simplify each expression.

(a) $4x + 8 + 9$ simplifies to $4x + 17$.

(b) $4(3m - 2n)$ (To simplify, we clear the parentheses.)

$= 4(3m) - 4(2n)$ Distributive property

$= (4 \cdot 3)m - (4 \cdot 2)n$ Associative property

$= 12m - 8n$ Multiply.

(c) $6 + 3(4k + 5)$ (Do *not* start by adding.)

$= 6 + 3(4k) + 3(5)$ Distributive property

$= 6 + (3 \cdot 4)k + 3(5)$ Associative property

$= 6 + 12k + 15$ Multiply.

$= 6 + 15 + 12k$ Commutative property

$= 21 + 12k$ Add.

(d) $5 - (2y - 8)$

$= 5 - 1(2y - 8)$ $-a = -1 \cdot a$

$= 5 - 1(2y) - 1(-8)$ Distributive property

$= 5 - 2y + 8$ Multiply. (Be careful with signs.)

$= 5 + 8 - 2y$ Commutative property

$= 13 - 2y$ Add.

◀ Work Problem 1 at the Side.

1 Simplify each expression.

(a) $9k + 12 - 5$

GS **(b)** $7(3p + 2q)$

$= 7(\_\_\_\_) + 7(\_\_\_\_)$

$= \_\_\_\_ + \_\_\_\_$

(c) $2 + 5(3z - 1)$

(d) $-3 - (2 + 5y)$

Note

The steps using the commutative and associative properties will not be shown in the rest of the examples. However, be aware that they are usually involved.

OBJECTIVE 2 Identify terms and numerical coefficients. A **term** is a number (constant), a variable, or a product or quotient of a number and one or more variables raised to powers.

$$9x, \quad 15y^2, \quad -3, \quad -8m^2n, \quad \frac{2}{p}, \quad k \qquad \text{Terms}$$

In the term $9x$, the **numerical coefficient,** or simply the **coefficient,** of the variable x is 9. Additional examples are shown in the table on the next page.

Answers

1. (a) $9k + 7$ (b) $3p$; $2q$; $21p$; $14q$ (c) $15z - 3$ (d) $-5 - 5y$

TERMS AND THEIR COEFFICIENTS

Term	Numerical Coefficient
8	8
$-7y$	-7
$34r^3$	34
$-26x^5yz^4$	-26
$-k$, or $-1k$	-1
r, or $1r$	1
$\frac{3x}{8}$, or $\frac{3}{8}x$	$\frac{3}{8}$
$\frac{x}{3} = \frac{1x}{3}$, or $\frac{1}{3}x$	$\frac{1}{3}$

Work Problem 2 at the Side. ▶

CAUTION

It is important to be able to distinguish between ***terms*** and ***factors.***

$8x^3 + 12x^2$ — This expression has **two terms,** $8x^3$ and $12x^2$. **Terms** are separated by a + or − symbol.

$(8x^3)(12x^2)$ — This is a **one-term** expression. The **factors** $8x^3$ and $12x^2$ are multiplied.

OBJECTIVE 3 Identify like terms. Terms with exactly the same variables that have the same exponents on the variables are **like terms.**

Like Terms			Unlike Terms			
$9t$	and	$4t$	$4y$	and	$7t$	Different variables
$6x^2$	and	$-5x^2$	$17x$	and	$-8x^2$	Different exponents
$-2pq$	and	$11pq$	$4xy^2$	and	$4xy$	Different exponents
$3x^2y$	and	$5x^2y$	$-7wz^3$	and	$2xz^3$	Different variables

Work Problem 3 at the Side. ▶

OBJECTIVE 4 Combine like terms. Recall the distributive property.

$\boldsymbol{a(b+c) = ab + ac}$ can be written "in reverse" as $\boldsymbol{ba + ca = (b+c)a}$.

This last form provides justification for **combining like terms.**

EXAMPLE 2 Combining Like Terms

Combine like terms in each expression.

(a) $-9m + 5m$

$= (-9 + 5)m$ — Distributive property in reverse

$= -4m$

(b) $6r + 3r + 2r$

$= (6 + 3 + 2)r$

$= 11r$

(c) $4x + x$

$= 4x + 1x$ — $x = 1x$

$= (4 + 1)x$

$= 5x$

(d) $16y^2 - 9y^2$

$= (16 - 9)y^2$

$= 7y^2$

(e) $32y + 10y^2$

These unlike terms cannot be combined.

Work Problem 4 at the Side. ▶

2 Complete the table.

	Term	Numerical Coefficient
(a)	$15q$	____
(b)	$-2m^3$	____
(c)	$-18m^7q^4$	____
(d)	$-r$	____
(e)	$\frac{5x}{4}$	____

3 Identify each pair of terms as *like* or *unlike.*

(a) $9x,\ 4x$

(b) $-8y^3,\ 12y^2$

(c) $5x^2y^4,\ 5x^4y^2$

(d) $7x^2y^4,\ -7x^2y^4$

(e) $13kt,\ 4tk$

4 Combine like terms.

(a) $4k + 7k$

(b) $4r - r$

(c) $5z + 9z - 4z$

(d) $8p + 8p^2$

Answers

2. (a) 15 (b) −2 (c) −18 (d) −1 (e) $\frac{5}{4}$

3. (a) like (b) unlike (c) unlike (d) like (e) like

4. (a) $11k$ (b) $3r$ (c) $10z$ (d) cannot be combined

CAUTION
Remember that only like terms may be combined.

Simplifying an Expression

An expression has been simplified when the following conditions have been met.

- All grouping symbols have been removed.
- All like terms have been combined.
- Operations have been performed, when possible.

EXAMPLE 3 Simplifying Expressions Involving Like Terms

Simplify each expression.

(a) $14y + 2(6 + 3y)$ — Start by distributing the 2.

$= 14y + 2(6) + 2(3y)$ Distributive property

$= 14y + 12 + 6y$ Multiply.

$= 20y + 12$ Combine like terms.

$14y + 6y = (14 + 6)y = 20y$

(b) $9k - 6 - 3(2 - 5k)$ — Be careful with signs.

$= 9k - 6 - 3(2) - 3(-5k)$ Distributive property

$= 9k - 6 - 6 + 15k$ Multiply.

$= 24k - 12$ Combine like terms.

(c) $-(2 - r) + 10r$

$= -1(2 - r) + 10r$ $\quad -(2 - r) = -1(2 - r)$

$= -1(2) - 1(-r) + 10r$ Distributive property

$= -2 + r + 10r$ Multiply. (Be careful with signs.)

$= -2 + 11r$ Combine like terms; $r = 1r$.

Alternatively, $-(2 - r)$ can be thought of as the *opposite* of $(2 - r)$—that is, $-2 + r$—which can then be added to $10r$ to obtain $-2 + 11r$.

(d) $-\frac{2}{3}(x - 6) - \frac{1}{6}x$

$= -\frac{2}{3}x - \frac{2}{3}(-6) - \frac{1}{6}x$ Distributive property

$= -\frac{2}{3}x + 4 - \frac{1}{6}x$ Multiply.

$= -\frac{4}{6}x + 4 - \frac{1}{6}x$ Get a common denominator.

$= -\frac{5}{6}x + 4$ Combine like terms.

Continued on Next Page

(e) $5(2a - 6) - 3(4a - 9)$

$= 5(2a) + 5(-6) - 3(4a) - 3(-9)$ Distributive property twice

$= 10a - 30 - 12a + 27$ Multiply.

$= -2a - 3$ Combine like terms.

Work Problem 5 at the Side. ▶

Note

Examples 2 and 3 suggest that like terms may be combined by adding or subtracting the coefficients of the terms and keeping the same variable factors.

OBJECTIVE ▶ 5 Simplify expressions from word phrases.

EXAMPLE 4 Translating Words into a Mathematical Expression

Write the phrase as a mathematical expression using x as the variable, and simplify.

The sum of 9, five times a number,
four times the number, and
six times the number

The word "sum" indicates that the terms should be added. Use x for the number.

$9 + 5x + 4x + 6x$ simplifies to $9 + 15x$. Combine like terms.

This is an expression to be simplified, *not* an equation to be solved.

Work Problem 6 at the Side. ▶

5 Simplify each expression.

GS **(a)** $10p + 3(5 + 2p)$

$= 10p + \_\_\_\_(5) + 3(\_\_\_\_)$

$= 10p + \_\_\_\_ + \_\_\_\_$

$= \_\_\_\_\_$

(b) $7z - 2 - (1 + z)$

(c) $-(3k^2 + 5k) + 7(k^2 - 4k)$

(d) $-\frac{3}{4}(x - 8) - \frac{1}{3}x$

(e) $2(5r + 3) - 3(2r - 3)$

6 Write each phrase as a mathematical expression using x as the variable, and simplify.

(a) Three times a number, subtracted from the sum of the number and 8

(b) Twice a number added to the sum of 6 and the number

Answers

5. (a) 3; $2p$; 15; $6p$; $16p + 15$
(b) $6z - 3$ (c) $4k^2 - 33k$
(d) $-\frac{13}{12}x + 6$ (e) $4r + 15$

6. (a) $(x + 8) - 3x$; $-2x + 8$
(b) $(6 + x) + 2x$; $3x + 6$

1.8 Exercises

FOR EXTRA HELP — Go to MyMathLab for worked-out, step-by-step solutions to exercises enclosed in a square and video solutions to ▶ exercises.

CONCEPT CHECK *Choose the letter of the correct response.*

1. Which expression is a simplified form of $-(6x-3)$?

A. $-6x-3$ **B.** $-6x+3$ **C.** $6x-3$ **D.** $6x+3$

2. Which is an example of a pair of like terms?

A. $6t, 6w$ **B.** $-8x^2y, 9xy^2$ **C.** $5ry, 6yr$ **D.** $-5x^2, 2x^3$

3. Which is an example of a term with numerical coefficient 5?

A. $5x^3y^7$ **B.** x^5 **C.** $\frac{x}{5}$ **D.** 5^2xy^3

4. Which is a correct translation for "six times a number, subtracted from the product of eleven and the number" (if x represents the number)?

A. $6x-11x$ **B.** $11x-6x$ **C.** $(11+x)-6x$ **D.** $6x-(11+x)$

Simplify each expression. **See Example 1.**

5. $4r+19-8$ ▶

6. $7t+18-4$

7. $7(3x-4y)$

8. $8(2p-9q)$

9. $5+2(x-3y)$ ▶

10. $8+3(s-6t)$

11. $-2-(5-3p)$ ▶

12. $-10-(7-14r)$

13. $6+(4-3x)-8$

14. $-12+(7-8x)+6$

In each term, give the numerical coefficient of the variable(s). **See Objective 2.**

15. $-12k$ ▶

16. $-23y$

17. $5m^2$

18. $-3n^6$

19. xw

20. pq

21. $-x$

22. $-t$

23. 10

24. 15

25. $28xy^2$

26. $17a^2b$

27. $-\frac{3}{8}x$

28. $-\frac{5}{4}z$

29. $\frac{x}{2}$

30. $\frac{x}{6}$

31. $\frac{2x}{5}$

32. $\frac{8x}{9}$

33. $-1.28r^2$

34. $-2.985t^3$

Identify each group of terms as like *or* unlike. **See Objective 3.**

35. $8r, -13r$ ▶

36. $-7a, 12a$

37. $3x, 3y$

38. $9m, 9n$

39. $5z^4, 9z^3$

40. $8x^5, -10x^3$

41. $4, 9, -24$

42. $7, 17, -83$

43. x, y

44. t, s

Simplify each expression. **See Examples 1–3.**

45. $3x+12x$

46. $4y+9y$

47. $-6x-3x$

48. $-4z-8z$

49. $12b+b$ ▶

50. $19x+x$

51. $3k+8+4k+7$

52. $15z+1+4z+2$

53. $-2x+3+4x-17+20$

54. $r - 6 - 12r - 4 + 16$

55. $-\frac{4}{3} + 2t + \frac{1}{3}t - 8 - \frac{8}{3}t$

56. $-\frac{5}{6} + 8x + \frac{1}{6}x - 7 - \frac{7}{6}$

57. $6y^2 + 11y^2 - 8y^2$

58. $-9m^3 + 3m^3 - 7m^3$

59. $2p^2 + 3p^2 - 8p^3 - 6p^3$

60. $5y^3 + 6y^3 - 3y^2 - 4y^2$

61. $2y^2 - 7y^3 - 4y^2 + 10y^3$

62. $9x^4 - 7x^6 + 12x^4 + 14x^6$

63. $2(4x + 6) + 3$

64. $4(6y + 9) + 7$

65. $-\frac{5}{6}(y + 12) - \frac{1}{2}y$

66. $-\frac{2}{3}(w + 15) - \frac{1}{4}w$

67. $-\frac{4}{3}(y - 12) - \frac{1}{6}y$

68. $-\frac{7}{5}(t - 15) - \frac{1}{2}t$

69. $-5 - 2(x - 3)$
$= -5 - 2(\_\_\_\_) - 2(\_\_\_\_)$
$= -5 - \_\_\_\_ + \_\_\_\_$
$= \_\_\_\_\_\_$

70. $-8 - 3(2x + 4)$
$= -8 - 3(\_\_\_\_) - 3(\_\_\_\_)$
$= -8 - \_\_\_\_ - \_\_\_\_$
$= \_\_\_\_\_\_$

71. $13p + 4(4 - 8p)$

72. $5x + 3(7 - 2x)$

73. $-5(5y - 9) + 3(3y + 6)$

74. $-3(2t + 4) + 8(2t - 4)$

75. $-3(2r - 3) + 2(5r + 3)$

76. $-4(5y - 7) + 3(2y - 5)$

77. $8(2k - 1) - (4k - 3)$

78. $6(3p - 2) - (5p + 1)$

79. $\frac{1}{2}(2x + 4) - \frac{1}{3}(9x - 6)$

80. $\frac{1}{4}(8x + 16) - \frac{1}{5}(20x - 15)$

81. $-\frac{2}{3}(5x + 7) - \frac{1}{3}(4x + 8)$

82. $-\frac{3}{4}(7x + 9) - \frac{1}{4}(5x + 7)$

83. $-2(-3k + 2) - (5k - 6) - 3k - 5$

84. $-2(3r - 4) - (6 - r) + 2r - 5$

85. $-4(-3x+3)-(6x-4)-2x+1$

86. $-5(8x+2)-(5x-3)-3x+17$

87. $-5.3r+4.9-(2r+0.7)+3.2r$

88. $2.7b+5.8-(3b+0.5)-4.4b$

89. $-7.5(2y+4)-2.9(3y-6)$

90. $8.4(6t-6)+2.4(9-3t)$

91. **CONCEPT CHECK** A student simplified the expression $7x-2(3-2x)$ as shown.

$$
\begin{aligned}
&7x-2(3-2x)\\
&=7x-2(3)-2(2x)\\
&=7x-6-4x\\
&=3x-6
\end{aligned}
$$

What Went Wrong? Find the correct simplified answer.

92. **CONCEPT CHECK** A student simplified the expression $3+2(4x-5)$ as shown.

$$
\begin{aligned}
&3+2(4x-5)\\
&=5(4x-5)\\
&=5(4x)+5(-5)\\
&=20x-25
\end{aligned}
$$

What Went Wrong? Find the correct simplified answer.

Write each phrase as a mathematical expression using x as the variable, and simplify. ***See Example 4.***

93. Five times a number, added to the sum of the number and three

94. Six times a number, added to the sum of the number and six

95. A number multiplied by -7, subtracted from the sum of 13 and six times the number

96. A number multiplied by 5, subtracted from the sum of 14 and eight times the number

97. Six times a number added to -4, subtracted from twice the sum of three times the number and 4

98. Nine times a number added to 6, subtracted from triple the sum of 12 and 8 times the number

Relating Concepts (Exercises 99–102) For Individual or Group Work

A manufacturer has fixed costs of \$1000 *to produce gizmos. Each gizmo costs* \$5 *to make. The fixed cost to produce gadgets is* \$750, *and each gadget costs* \$3 *to make.*
Work Exercises 99–102 in order.

99. Write an expression for the cost to make x gizmos. (*Hint:* The cost will be the sum of the fixed cost and the cost per item times the number of items.)

100. Write an expression for the cost to make y gadgets.

101. Write an expression for the total cost to make x gizmos and y gadgets.

102. Simplify the expression from **Exercise 101.**

Study Skills

REVIEWING A CHAPTER

Your text provides material to help you prepare for quizzes or tests in this course. Refer to the **Chapter Summary** as you read through the following techniques.

414 Chapter 5 Exponents and Polynomials

Chapter 5 Summary

Key Terms

5.1

exponential expression A number written with an exponent (or power) is an exponential expression.

5.3

scientific notation A number written in the form $a \times 10^n$, where $1 \le |a| < 10$ and n is an integer, is in scientific notation.

standard notation We refer to a number such as 125 as the standard notation of 1.25×10^2.

5.4

term A term is a number (constant), a variable, or a product or quotient of a number and one or more variables raised to powers.

like terms Terms with exactly the same variables (including the same exponents) are like terms.

polynomial in x A term or the sum of a finite number of terms of the form ax^n, for any real number a and any whole number n, is a polynomial in x.

descending powers A polynomial in x is written in descending powers if the exponents on x in its terms decrease from left to right.

degree of a term The degree of a term is the sum of the exponents on the variables.

degree of a polynomial The degree of a polynomial is the greatest degree of any nonzero term of the polynomial.

monomial A monomial is a polynomial with exactly one term.

binomial A binomial is a polynomial with exactly two terms.

trinomial A trinomial is a polynomial with exactly three terms.

5.5

FOIL method The FOIL method is used to find the product of two binomials. The letters of the word **FOIL** originate as follows: Multiply the **F**irst terms, multiply the **O**uter terms (to obtain the outer product), multiply the **I**nner terms (to obtain the inner product), and multiply the **L**ast terms.

outer product The outer product of $(a + b)(c + d)$ is ad.

inner product The inner product of $(a + b)(c + d)$ is bc.

5.6

conjugate The conjugate of $x + y$ is $x - y$.

New Symbols

x^{-n} x to the negative n power

Test Your Word Power

See how well you have learned the vocabulary in this chapter.

1. A **polynomial** is an algebraic expression made up of
   A. a term or a finite product of terms with positive coefficients and exponents
   B. a term or a finite sum of terms with real coefficients and whole number exponents
   C. the product of two or more terms with positive exponents
   D. the sum of two or more terms with whole number coefficients and exponents.
2. The **degree of a term** is the
   A. number of variables in the term
   B. product of the exponents on the variables
   C. least exponent on the variables
   D. sum of the exponents on the variables.
3. A **trinomial** is a polynomial with
   A. exactly one term
   B. exactly two terms
   C. exactly three terms
   D. more than three terms.
4. A **binomial** is a polynomial with
   A. exactly one term
   B. exactly two terms
   C. exactly three terms
   D. more than three terms.
5. A **monomial** is a polynomial with
   A. exactly one term
   B. exactly two terms
   C. exactly three terms
   D. more than three terms.

Chapter Reviewing Techniques

- **Review the Key Terms.** Make a study card for each. Include a definition, an example, a sketch (if appropriate), and a section or page reference.
- **Review any New Symbols.** Cover the column with the symbol meanings and confirm that you know each symbol.
- **Take the Test Your Word Power quiz** to check your understanding of new vocabulary. The answers immediately follow.
- **Read the Quick Review.** Pay special attention to the headings. Study the explanations and examples given for each concept. Try to think about the whole chapter.
- **Reread your lecture notes.** Focus on what your instructor has emphasized in class, and review that material in your text.
- **Look over your homework and any quizzes.** Pay special attention to any trouble spots.
- **Work the Review Exercises.** They are grouped by section.
  - ✓ Pay attention to direction words, such as *simplify*, *solve*, and *evaluate*.
  - ✓ After you've done each section of exercises, check your answers in the answer section.
  - ✓ Are your answers exact and complete? Did you include the correct labels, such as \$, cm^2, ft, etc.?
  - ✓ Make study cards for difficult problems.
- **Work the Mixed Review Exercises.** They are in mixed-up order. Check your answers in the answer section.
- **Take the Chapter Test under test conditions.**
  - ✓ Time yourself.
  - ✓ Use a calculator or notes (if your instructor permits them on tests).
  - ✓ Take the test in one sitting.
  - ✓ Show all your work.
  - ✓ Check your answers in the back of the book.

Reviewing a chapter will take some time. Avoid rushing through your review in one night. Use the suggestions over a few days or evenings to better understand the material and remember it longer.

Now Try This

Follow these reviewing techniques to prepare for your next test. Then answer the questions below.

1. How much time did you spend reviewing for your test? Was it enough?

   ____

2. How did the reviewing techniques work for you?

   ____

3. What will you do differently when reviewing for your next test?

   ____

Chapter 1 Summary

Key Terms

1.1

exponent An exponent, or **power**, is a number that indicates how many times a factor is repeated.

3^4 ← Exponent, Base — Exponential expression

base A base is the number that is a repeated factor when written with an exponent.

exponential expression A number written with an exponent is an exponential expression.

inequality An inequality is a statement that two expressions may not be equal.

1.2

constant A constant is a fixed, unchanging number.

variable A variable is a symbol, usually a letter, used to represent an unknown number.

algebraic expression An algebraic expression is a sequence of numbers (constants), variables, operation symbols, and/or grouping symbols.

equation An equation is a statement that two expressions are equal.

solution A solution of an equation is any value of the variable that makes the equation true.

1.3

set A set is a collection of objects.

element The objects that belong to a set are its elements.

natural numbers The set of natural numbers is $\{1, 2, 3, 4, \ldots\}$.

whole numbers The set of whole numbers is $\{0, 1, 2, 3, 4, \ldots\}$.

number line A number line shows the ordering of real numbers on a line.

additive inverse The additive inverse, or **opposite,** of a number x is the number that is the same distance from 0 on a number line as x, but on the opposite side of 0.

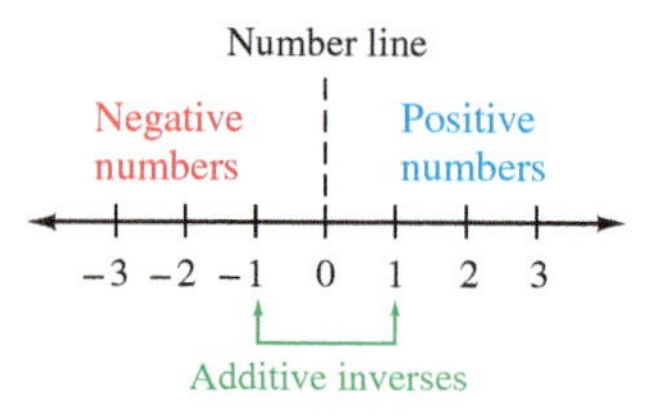

integers The set of integers is $\{\ldots, -3, -2, -1, 0, 1, 2, 3, \ldots\}$.

negative number A negative number is located to the *left* of 0 on a number line.

positive number A positive number is located to the *right* of 0 on a number line.

signed numbers Positive numbers and negative numbers are signed numbers.

rational numbers A rational number is a number that can be written as the quotient of two integers, with denominator not 0.

set-builder notation Set-builder notation uses a variable and a description to describe a set. It is often used to describe sets whose elements cannot easily be listed.

coordinate The number that corresponds to a point on a number line is the coordinate of that point.

irrational numbers An irrational number is a real number that is not a rational number.

real numbers Real numbers are numbers that can be represented by points on a number line (that is, all rational and irrational numbers).

absolute value The absolute value of a number is the distance between 0 and the number on a number line.

1.4

sum The answer to an addition problem is the sum.

addends The numbers being added are the addends.

$8 + 5 = 13$ ← Sum; 8 and 5: Addends

1.5

minuend In the operation $x - y$, x is the minuend.

subtrahend In the operation $x - y$, y is the subtrahend.

difference The answer to a subtraction problem is the difference.

$15 - 7 = 8$ ← Difference; 15: Minuend; 7: Subtrahend

1.6

product The answer to a multiplication problem is the product.

quotient The answer to a division problem is the quotient.

dividend In the division $x \div y$, x is the dividend.

divisor In the division $x \div y$, y is the divisor.

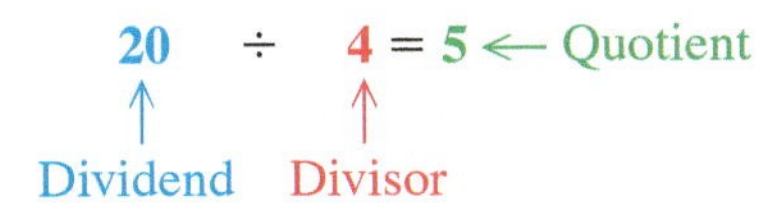

reciprocal Pairs of numbers whose product is 1 are reciprocals, or **multiplicative inverses,** of each other.

1.7

identity element for addition (additive identity) When the identity element for addition, which is 0, is added to a number, the number is unchanged.

identity element for multiplication (multiplicative identity) When a number is multiplied by the identity element for multiplication, which is 1, the number is unchanged.

1.8

term A term is a number (constant), a variable, or a product or quotient of a number and one or more variables raised to powers.

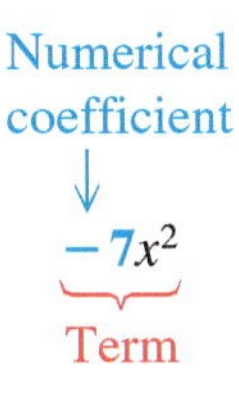

numerical coefficient The numerical factor of a term is its numerical coefficient, or **coefficient.**

like terms Terms with exactly the same variables (that have the same exponents on the variables) are like terms.

New Symbols

Symbol	Meaning
a^n	n factors of a
( ), [ ], { }	grouping symbols (parentheses, square brackets, braces)
$=$	is equal to
$\neq$	is not equal to
$<$	is less than
$\leq$	is less than or equal to
$>$	is greater than
$\geq$	is greater than or equal to
$x(y)$, $(x)y$, $(x)(y)$, $x \cdot y$, or xy	x times y
$x \div y$, $\frac{x}{y}$, x/y, or $y\overline{)x}$	x divided by y
$\{x \mid x$ **has a certain property**$\}$	set-builder notation
$\lvert x \rvert$	absolute value of x
$-x$	additive inverse, or opposite, of x
$\frac{1}{x}$	multiplicative inverse, or reciprocal, of x (where $x \neq 0$)

Test Your Word Power

See how well you have learned the vocabulary in this chapter.

1 A **product** is
- **A.** the answer in an addition problem
- **B.** the answer in a multiplication problem
- **C.** one of two or more numbers that are added to obtain another number
- **D.** one of two or more numbers that are multiplied to obtain another number.

2 An **exponent** is
- **A.** a symbol that tells how many numbers are being multiplied
- **B.** a number raised to a power
- **C.** a number that tells how many times a factor is repeated
- **D.** one of two or more numbers that are multiplied.

3 A **variable** is
- **A.** a symbol used to represent an unknown number
- **B.** a value that makes an equation true
- **C.** a solution of an equation
- **D.** the answer in a division problem.

4 An **integer** is
- **A.** a positive or negative number
- **B.** a natural number, its opposite, or zero
- **C.** any number that can be graphed on a number line
- **D.** the quotient of two numbers.

5 A **coordinate** is
- **A.** the number that corresponds to a point on a number line
- **B.** the graph of a number
- **C.** any point on a number line
- **D.** a distance on a number line.

6 The **absolute value** of a number is
- **A.** the graph of the number
- **B.** the reciprocal of the number
- **C.** the opposite of the number
- **D.** the distance between 0 and the number on a number line.

7 A **term** is
- **A.** a numerical factor
- **B.** a number, a variable, or a product or quotient of numbers and variables raised to powers
- **C.** one of several variables with the same exponents
- **D.** a sum of numbers and variables raised to powers.

8 The **subtrahend** in $a - b = c$ is
- **A.** a
- **B.** b
- **C.** c
- **D.** $a - b$.

Answers to Test Your Word Power

1. B; *Example:* The product of 2 and 5, or 2 times 5, is 10.
2. C; *Example:* In 2^3, the number 3 is the exponent (or power), so 2 is a factor three times; $2^3 = 2 \cdot 2 \cdot 2 = 8$.
3. A; *Examples:* x, y, z
4. B; *Examples:* $-9, 0, 6$
5. A; *Example:* The point graphed three units to the right of 0 on a number line has coordinate 3.
6. D; *Examples:* $|2| = 2$ and $|-2| = 2$
7. B; *Examples:* $6, \frac{x}{2}, -4ab^2$
8. B; *Example:* In $5 - 3 = 2$, 3 is the subtrahend.

Quick Review

Concepts	Examples
1.1 Exponents, Order of Operations, and Inequality	
Order of Operations Work within any parentheses or brackets and above and below fraction bars in the following order. ***Step 1*** Apply all exponents. ***Step 2*** Multiply or divide from left to right. ***Step 3*** Add or subtract from left to right.	Simplify. $36 - 4(2^2 + 3)$ $= 36 - 4(4 + 3)$ Apply the exponent. $= 36 - 4(7)$ Add inside the parentheses. $= 36 - 28$ Multiply. $= 8$ Subtract.
1.2 Variables, Expressions, and Equations	
To evaluate an expression means to find its value. Evaluate an expression with a variable by substituting a given number for the variable.	Evaluate $2x + y^2$ for $x = 3$ and $y = -4$. $2x + y^2$ $= 2(3) + (-4)^2$ Substitute. $= 6 + 16$ Multiply. Apply the exponent. $= 22$ Add.
A value of a variable that makes an equation true is a solution of the equation.	Is 2 a solution of $5x + 3 = 18$? $5(2) + 3 \stackrel{?}{=} 18$ Let $x = 2$. $13 = 18$ False 2 is not a solution.
1.3 Real Numbers and the Number Line	
Ordering Real Numbers a is less than b if a lies to the left of b on a number line. a is greater than b if a lies to the right of b on a number line. The additive inverse, or opposite, of x is $-x$. The absolute value of x, written $\lvert x \rvert$, is the distance between x and 0 on a number line.	Graph -2, 0, and 3. −3 −2 −1 0 1 2 3 4 $-2 < 3$ $3 > 0$ $0 < 3$ $-(5) = -5$ $-(-7) = 7$ $-0 = 0$ $\lvert 13 \rvert = 13$ $\lvert 0 \rvert = 0$ $\lvert -5 \rvert = 5$
1.4 Adding Real Numbers	
Adding Two Signed Numbers ***Same sign*** Add their absolute values. The sum has the same sign as the addends. ***Different signs*** Subtract their absolute values. The sum has the sign of the addend with greater absolute value.	Add. $9 + 4 = 13$ $-8 + (-5) = -13$ $7 + (-12) = -5$ $-5 + 13 = 8$

Concepts	Examples
1.5 Subtracting Real Numbers	
Definition of Subtraction For any real numbers x and y, $x - y$ is defined as $x + (-y)$.	Subtract. $5 - (-2) = 5 + 2 = 7$ $-3 - 4 = -3 + (-4) = -7$ $-2 - (-6) = -2 + 6 = 4$
1.6 Multiplying and Dividing Real Numbers	
Multiplying and Dividing Two Signed Numbers ***Same sign*** The product (or quotient) is *positive.* ***Different signs*** The product (or quotient) is *negative.*	Multiply or divide. $6 \cdot 5 = 30$ $-7(-8) = 56$ $\frac{-24}{-6} = 4$ $-6(5) = -30$ $\frac{-18}{9} = -2$ $\frac{49}{-7} = -7$
Definition of Division For any real numbers x and y, $x \div y = x \cdot \frac{1}{y}$ (where $y \neq 0$).	$10 \div 2 = \frac{10}{2} = 10 \cdot \frac{1}{2} = 5$
0 divided by a nonzero number equals 0. ***Division by 0 is undefined.***	$\frac{0}{5} = 0$ $\frac{5}{0}$ is undefined.
1.7 Properties of Real Numbers	
Commutative Properties $a + b = b + a$ $ab = ba$	$7 + (-1) = -1 + 7$ $5(-3) = (-3)5$
Associative Properties $(a + b) + c = a + (b + c)$ $(ab)c = a(bc)$	$(3 + 4) + 8 = 3 + (4 + 8)$ $[-2(6)]4 = -2[6(4)]$
Identity Properties $a + 0 = a$ $0 + a = a$ $a \cdot 1 = a$ $1 \cdot a = a$	$-7 + 0 = -7$ $0 + (-7) = -7$ $9 \cdot 1 = 9$ $1 \cdot 9 = 9$
Inverse Properties $a + (-a) = 0$ $-a + a = 0$ $a \cdot \frac{1}{a} = 1$ $\frac{1}{a} \cdot a = 1$ $(a \neq 0)$	$7 + (-7) = 0$ $-7 + 7 = 0$ $-2\left(-\frac{1}{2}\right) = 1$ $-\frac{1}{2}(-2) = 1$
Distributive Properties $a(b + c) = ab + ac$ and $(b + c)a = ba + ca$ $a(b - c) = ab - ac$ and $(b - c)a = ba - ca$	$5(x + 2) = 5x + 5(2)$ $(x + 2)5 = x \cdot 5 + 2(5)$ $9(5y - 4) = 9(5y) - 9(4)$ $(5y - 4)9 = 5y(9) - 4(9)$
1.8 Simplifying Expressions	
Only like terms may be combined. We use a distributive property. $ba + ca = (b + c)a$ and $a(b + c) = ab + ac$	Simplify each expression. $-3y^2 + 6y^2 = (-3 + 6)y^2 = 3y^2$ $4(3 + 2x) - 6(5 - x) = 4(3) + 4(2x) - 6(5) - 6(-x) = 12 + 8x - 30 + 6x = 14x - 18$

Chapter 1 Review Exercises

If you need help with any of these Review Exercises, look in the section indicated.

1.1 *Find the value of each expression.*

1. 5^4

2. $(0.03)^4$

3. 0.21^3

4. $\left(\frac{5}{2}\right)^3$

Evaluate each expression.

5. $8 \cdot 5 - 13$

6. $5[4^2 + 3(2^3)]$

7. $16 + 12 \div 4 - 2$

8. $20 - 2(5 + 3)$

9. $\frac{7(3^2 - 5)}{16 - 2 \cdot 6}$

10. $\frac{5(6^2 - 2^4)}{3 \cdot 5 + 10}$

Write each word statement in symbols.

11. Thirteen is less than seventeen.

12. Five plus two is not equal to ten.

Write each statement in words, and decide whether it is true *or* false.

13. $6 < 15$

14. $\frac{2}{4} \neq \frac{3}{6}$

1.2 *Evaluate each expression for $x = 6$ and $y = 3$.*

15. $2x + 6y$

16. $4(3x - y)$

17. $\frac{x}{3} + 4y$

18. $\frac{x^2 + 3}{3y - x}$

Write each word phrase as an algebraic expression, using x as the variable.

19. Six added to a number

20. A number subtracted from eight

21. Nine subtracted from six times a number

22. Three-fifths of a number added to 12

Decide whether each equation has the given number as a solution.

23. $5x + 3(x + 2) = 22;\quad 2$

24. $\frac{x + 5}{3x} = 1;\quad 6$

Write each word sentence as an equation. Use x as the variable.

25. Six less than twice a number is 10.

26. The product of a number and 4 is 8.

Decide whether each of the following is an equation *or an* expression.

27. $5r - 8(r + 7) = 2$

28. $2y + (5y - 9) + 2$

1.3 *Graph each number on a number line.*

29. $-4, -\frac{1}{2}, 0, 2.5, 5$

−6 −4 −2 0 2 4 6

30. $-3\frac{1}{4}, \frac{14}{5}, -1\frac{1}{8}, \frac{5}{6}$

−6 −4 −2 0 2 4 6

Classify each number, using the sets natural numbers, whole numbers, integers, rational numbers, irrational numbers, *and* real numbers.

31. $\frac{4}{3}$

32. 19

Select the lesser of the two given numbers.

33. $-10, 5$

34. $-8, -9$

35. $-\frac{2}{3}, -\frac{3}{4}$

36. $0, -|23|$

Decide whether each statement is true *or* false.

37. $12 > -13$

38. $0 > -5$

39. $-9 < -7$

40. $-13 > -13$

Find each absolute value and simplify if needed.

41. $-|3|$

42. $-|-19|$

43. $-|9 - 2|$

44. $|15 - 6|$

1.4 *Find each sum.*

45. $-10 + 4$

46. $14 + (-18)$

47. $-8 + (-9)$

48. $\frac{4}{9} + \left(-\frac{5}{4}\right)$

49. $-5 + [-6 + (-8) + 19]$

50. $|-12 + 3| + |-2 + (-4)|$

Write a numerical expression for each phrase, and simplify the expression.

51. 19 added to the sum of -31 and 12

52. 13 more than the sum of -4 and -8

Solve each problem.

53. Otis found that his checking account balance was $-\$23.75$, so he deposited \$50.00. What is his new balance?

54. The low temperature in Yellowknife, in the Canadian Northwest Territories, one January day was $-26°F$. It rose 16° that day. What was the high temperature?

1.5 *Find each difference.*

55. $-7 - 4$

56. $-12 - (-11)$

57. $5 - (-2)$

58. $-\frac{3}{7} - \frac{4}{5}$

59. $2.56 - (-7.75)$

60. $(-10 - 4) - (-2)$

61. $(-3 + 4) - (-1)$

62. $|5 - 9| - |-3 + 6|$

Write a numerical expression for each phrase, and simplify the expression.

63. The difference of -4 and -6

64. Five less than the sum of 4 and -8

65. The difference of 18 and -23, decreased by 15

66. Nineteen, decreased by 12 less than -7

Solve each problem.

67. A quarterback passed for a gain of 8 yd, was sacked for a loss of 12 yd, and then threw a 42 yd touchdown pass. Find the total net yardage for the plays.

68. On December 4, 2015, the Dow Jones Industrial Average closed at 17,847.63, up 370 points from the previous day. What was the closing price on December 3, 2015? (Data from *The Wall Street Journal*.)

The table shows the number of people naturalized in the United States (that is, made citizens of the United States) for the years 2008 through 2014. Use a signed number to represent the change in the number of people naturalized for each time period.

Year	Number of People (in thousands)
2008	1050
2009	742
2010	619
2011	691
2012	763
2013	777
2014	655

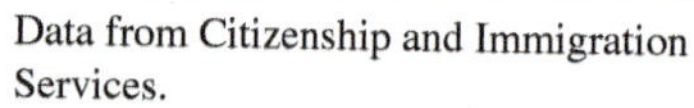
Data from Citizenship and Immigration Services.

69. 2008 to 2009

70. 2010 to 2011

71. 2012 to 2013

72. 2013 to 2014

1.6 *Perform the indicated operations.*

73. $(-12)(-3)$

74. $15(-7)$

75. $-\frac{4}{3}\left(-\frac{3}{8}\right)$

76. $-4.8(-2.1)$

77. $5(8-12)$

78. $(5-7)(8-3)$

79. $2(-6)-(-4)(-3)$

80. $3(-10)-5$

81. $\frac{-36}{-9}$

82. $\frac{220}{-11}$

83. $-\frac{1}{2} \div \frac{2}{3}$

84. $-33.9 \div (-3)$

85. $\frac{-5(3)-1}{8-4(-2)}$

86. $\frac{5(-2)-3(4)}{-2[3-(-2)]+10}$

87. $\frac{10^2-5^2}{8^2+3^2-(-2)}$

88. $\frac{4^2-8 \cdot 2}{(-1.2)^2-(-0.56)}$

Evaluate each expression for $x=-5$, $y=4$, and $z=-3$.

89. $6x-4z$

90. $5x+y-z$

91. $5x^2$

92. $z^2(3x-8y)$

Write a numerical expression for each phrase, and simplify the expression.

93. Nine less than the product of -4 and 5

94. Five-sixths of the sum of 12 and -6

95. The quotient of 12 and the sum of 8 and -4

96. The product of -20 and 12, divided by the difference of 15 and -15

Write each sentence as an equation, using x as the variable.

97. The quotient of a number and the sum of the number and 5 is -2.

98. 3 less than 8 times a number is -7.

1.7 *Decide whether each statement is an example of a* commutative, *an* associative, *an* identity, *an* inverse, *or the* distributive property.

99. $6 + 0 = 6$

100. $5 \cdot 1 = 5$

101. $-\frac{2}{3}\left(-\frac{3}{2}\right) = 1$

102. $17 + (-17) = 0$

103. $3x + 3y = 3(x + y)$

104. $w(xy) = (wx)y$

105. $5 + (-9 + 2) = [5 + (-9)] + 2$

106. $(1 + 2) + 3 = 3 + (1 + 2)$

Use the distributive property to rewrite each expression. Simplify if possible.

107. $7(y + 2)$

108. $-12(4 - t)$

109. $3(2s + 5y)$

110. $-(-4r + 5s)$

1.8 *Simplify each expression.*

111. $7y + y$

112. $16p^2 - 8p^2 + 9p^2$

113. $4r^2 - 3r + 10r + 12r^2$

114. $-8(5k - 6) + 3(7k + 2)$

115. $2s - (-3s + 6)$

116. $-7(2t - 4) - 4(3t + 8) - 19(t + 1)$

Write each phrase as a mathematical expression using x as the variable, and simplify.

117. Seven times a number, subtracted from the product of -2 and three times the number

118. A number multiplied by 8, added to the sum of 5 and four times the number

Chapter 1 Mixed Review Exercises

Complete the table.

	Number	Absolute Value	Additive Inverse	Multiplicative Inverse
1.	-3	____	____	____
2.	12	____	____	____
3.	____	____	____	$-\frac{3}{2}$
4.	____	____	-0.2	____

5. To which of the following sets does $0.\overline{6}$ belong: natural numbers, whole numbers, integers, rational numbers, irrational numbers, real numbers?

6. Evaluate $(x + 6)^3 - y^3$ for $x = -2$ and $y = 3$.

Perform the indicated operations.

7. $\frac{15}{2} \cdot \left(-\frac{4}{5}\right)$

8. $\left(-\frac{5}{6}\right)^2$

9. $-|(-7)(-4)| - (-2)$

10. $\frac{6(-4) + 2(-12)}{5(-3) + (-3)}$

11. $\frac{3}{8} - \frac{5}{12}$

12. $\frac{12^2 + 2^2 - 8}{10^2 - (-4)(-15)}$

13. $\frac{8^2 + 6^2}{7^2 + 1^2}$

14. $-16(-3.5) - 7.2(-3)$

15. $2\frac{5}{6} - 4\frac{1}{3}$

16. $-8 + [(-4 + 17) - (-3 - 3)]$

17. $-\frac{12}{5} \div \frac{9}{7}$

18. $(-8 - 3) - 5(2 - 9)$

Solve each problem.

19. The highest temperature ever recorded in Iowa was 118°F. The lowest temperature ever recorded in the state was 165°F lower than the highest temperature. What is the record low temperature for Iowa? (Data from National Climatic Data Center.)

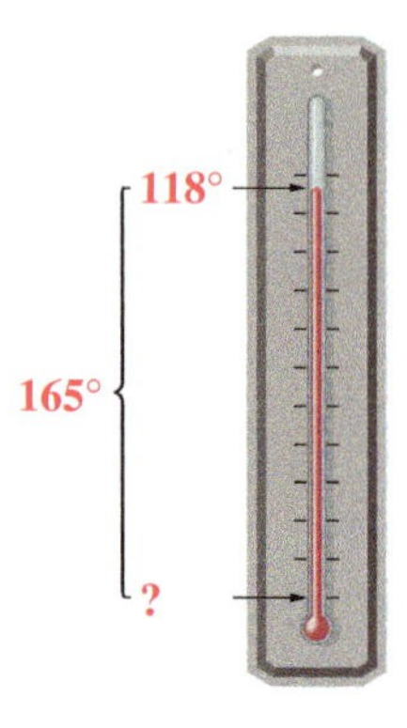

20. Humpback whales love to heave their 45-ton bodies out of the water. This is called *breaching*. A whale breached 15 ft above the surface of the ocean, while her mate cruised 12 ft below the surface. What is the difference between these two heights?

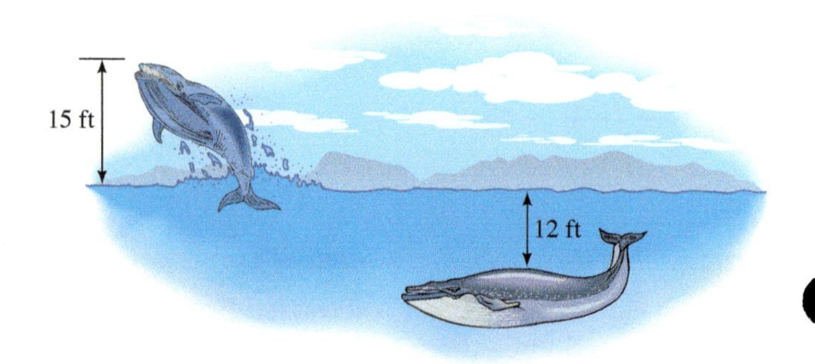

Chapter 1 Test

The Chapter Test Prep Videos with step-by-step solutions are available in MyMathLab or on YouTube at https://goo.gl/8aAsWP

Decide whether each statement is true *or* false.

1. $4[-20+7(-2)] \leq -135$

2. $\left(\frac{1}{2}\right)^2 + \left(\frac{2}{3}\right)^2 = \left(\frac{1}{2}+\frac{2}{3}\right)^2$

3. Graph the numbers $-1, -3, |-4|$, and $|-1|$ on the number line.

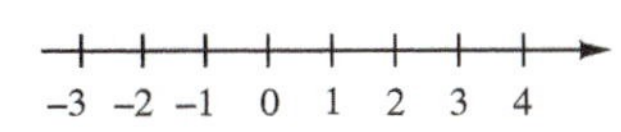

4. To which of the following sets does $-\frac{2}{3}$ belong: natural numbers, whole numbers, integers, rational numbers, irrational numbers, real numbers?

Select the lesser of the two given numbers.

5. $6, -|-8|$

6. $-0.742, -1.277$

7. Write a numerical expression for the phrase, and simplify the expression.

The quotient of -6 and the sum of 2 and -8

8. If a and b are both negative, is $\frac{a+b}{a \cdot b}$ positive or negative?

Perform the indicated operations.

9. $-2-(5-17)+(-6)$

10. $-5\frac{1}{2}+2\frac{2}{3}$

11. $4^2+(-8)-(2^3-6)$

12. $-6.2-[-7.1+(2.0-3.1)]$

13. $(-5)(-12)+4(-4)+(-8)^2$

14. $\dfrac{-7-|-6+2|}{-5-(-4)}$

15. $\dfrac{30(-1-2)}{-9[3-(-2)]-12(-2)}$

Evaluate each expression for $x=-2$ and $y=4$.

16. $3x-4y^2$

17. $\dfrac{5x+7y}{3(x+y)}$

Solve each problem.

18. The highest Fahrenheit temperature ever recorded in Idaho was 118°F, while the lowest was −60°F. What is the difference between these highest and lowest temperatures? (Data from *The World Almanac and Book of Facts*.)

19. For 2014, the U.S. federal government collected \$3.02 trillion in revenues but spent \$3.51 trillion. Write the federal budget deficit as a signed number. (Data from Office of Management and Budget.)

20. For a certain system of rating relief pitchers, 3 points are awarded for a save, 3 points are awarded for a win, 2 points are subtracted for a loss, and 2 points are subtracted for a blown save. If a pitcher has 4 saves, 3 wins, 2 losses, and 1 blown save, how many points does he have?

Match each statement in Column I with the property it illustrates in Column II.

I	II
21. $3x + 0 = 3x$	**A.** Commutative property
22. $(5 + 2) + 8 = 8 + (5 + 2)$	**B.** Associative property
23. $-3(x + y) = -3x + (-3y)$	**C.** Inverse property
24. $-5 + (3 + 2) = (-5 + 3) + 2$	**D.** Identity property
25. $-\frac{5}{3}\left(-\frac{3}{5}\right) = 1$	**E.** Distributive property

Simplify each expression.

26. $8x + 4x - 6x + x + 14x$

27. $-(3x - 1)$

28. $-2(3x^2 + 4) - 3(x^2 + 2x)$

29. $5(2x - 1) - (x - 12) + 2(3x - 5)$

30. Consider the expression $-6[5 + (-2)]$.

(a) Evaluate it by first working within the brackets.

(b) Evaluate it using the distributive property.

(c) Why must the answers in parts (a) and (b) be the same?

Equations, Inequalities, and Applications

Solving *linear equations*, the subject of this chapter, can be thought of in terms of the concept of balance.

2.1 The Addition Property of Equality

OBJECTIVES

1. Identify linear equations.
2. Use the addition property of equality.
3. Simplify, and then use the addition property of equality.

An *equation* is a statement asserting that two algebraic expressions are equal.

CAUTION

Remember that an equation always includes an equality symbol.

Equation	Expression
Left side → $x - 5 = 2$ ← Right side	$x - 5$
An equation can be solved.	An expression **cannot** be solved. (It can be *evaluated* for a given value or *simplified*.)

OBJECTIVE 1 **Identify linear equations.**

Linear Equation in One Variable

A **linear equation in one variable** (here x) can be written in the form

$$Ax + B = C,$$

where A, B, and C are real numbers and $A \neq 0$.

Examples: $4x + 9 = 0$, $2x - 3 = 5$, $x = 7$ Linear equations

$x^2 + 2x = 5$, $x^3 = -1$, $\frac{1}{x} = 6$, $|2x + 6| = 0$ *Non*linear equations

Recall that a **solution** of an equation is a number that makes the equation true when it replaces the variable. An equation is solved by finding its **solution set,** the set of all solutions. Equations that have exactly the same solution sets are **equivalent equations.**

A linear equation in x is solved by using a series of steps to produce a simpler equivalent equation of the form

$$x = \textbf{a number} \quad \text{or} \quad \textbf{a number} = x.$$

OBJECTIVE 2 **Use the addition property of equality.** In the linear equation $x - 5 = 2$, both $x - 5$ and 2 represent the same number because that is the meaning of the equality symbol. To solve the equation, we change the left side from $x - 5$ to just x, as follows.

$x - 5 = 2$ Given equation

$x - 5 + 5 = 2 + 5$ Add 5 to *each* side to keep them equal.

$x + 0 = 7$ Additive inverse property

$x = 7$ Additive identity property

Add 5. It is the opposite (additive inverse) of -5, and $-5 + 5 = 0$.

To check that 7 is the solution, we replace x with 7 in the original equation.

CHECK $x - 5 = 2$ Original equation

$7 - 5 \stackrel{?}{=} 2$ Let $x = 7$.

$2 = 2$ ✓ True

The left side equals the right side.

We write a solution set using set braces.

The final equation is true, so 7 is the solution and $\{7\}$ is the solution set.

To solve $x - 5 = 2$, we used the **addition property of equality.**

Addition Property of Equality

If A, B, and C represent real numbers, then the equations

$$A = B \quad \text{and} \quad A + C = B + C \quad \text{are equivalent.}$$

In words, the same number may be added to each side of an equation without changing the solution set.

In this property, any quantity that represents a real number C can be added to each side of an equation to obtain an equivalent equation.

Note

Equations can be thought of in terms of a balance. Thus, adding the *same* quantity to each side does not affect the balance. See **Figure 1.**

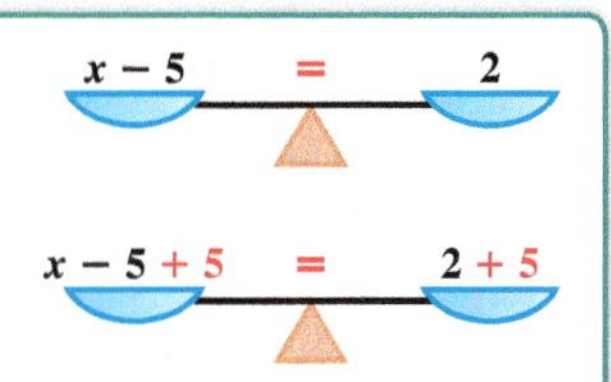

Figure 1

EXAMPLE 1 Applying the Addition Property of Equality

Solve $x - 16 = 7$.

Our goal is to get an equivalent equation of the form **x = a number.**

$$x - 16 = 7$$

$$x - 16 + 16 = 7 + 16 \qquad \text{Add 16 to each side.}$$

$$x = 23 \qquad \text{Combine like terms.}$$

CHECK Substitute 23 for x in the *original* equation.

$$x - 16 = 7 \qquad \text{Original equation}$$

$$23 - 16 \stackrel{?}{=} 7 \qquad \text{Let } x = 23.$$

$$7 = 7 \checkmark \qquad \text{True}$$

(7 is *not* the solution.)

A true statement results, so **23** is the solution and **{23}** is the solution set.

Work Problem 1 at the Side. ▶

EXAMPLE 2 Applying the Addition Property of Equality

Solve $x - 2.9 = -6.4$.

(Our goal is to isolate x.)

$$x - 2.9 = -6.4$$

$$x - 2.9 + 2.9 = -6.4 + 2.9 \qquad \text{Add 2.9 to each side.}$$

$$x = -3.5$$

CHECK

$$x - 2.9 = -6.4 \qquad \text{Original equation}$$

$$-3.5 - 2.9 \stackrel{?}{=} -6.4 \qquad \text{Let } x = -3.5.$$

$$-6.4 = -6.4 \checkmark \qquad \text{True}$$

A true statement results, so the solution set is $\{-3.5\}$.

Work Problem 2 at the Side. ▶

1 Solve.

GS **(a)** Fill in the blanks.

$$x - 12 = 9$$

$$x - 12 + \_\_\_\_ = 9 + \_\_\_\_$$

$$x = \_\_\_\_$$

CHECK $x - 12 = 9$

$$\_\_\_\_ - 12 \stackrel{?}{=} 9$$

$$\_\_\_\_ = 9 \quad (\textit{True/False})$$

The solution set is ____.

(b) $x - 25 = -18$

(c) $x - 8 = -13$

2 Solve.

(a) $x - 3.7 = -8.1$

(b) $a - 4.1 = 6.3$

Answers

1. (a) 12; 12; 21; 21; 9; True; $\{21\}$ (b) $\{7\}$ (c) $\{-5\}$

2. (a) $\{-4.4\}$ (b) $\{10.4\}$

3 Solve.

GS **(a)** $-15 = a + 12$

Our goal is to isolate ____.

On each side of the equation, (*add / subtract*) 12.

Now solve.

(b) $22 = -16 + r$

Answers

3. **(a)** a; subtract; $\{-27\}$ **(b)** $\{38\}$

Subtraction was previously defined as addition of the opposite. Thus, we can also use the following rule when solving an equation.

Addition Property of Equality Extended to Subtraction

The same number may be *subtracted* from each side of an equation without changing the solution set.

EXAMPLE 3 **Applying the Addition Property of Equality**

Solve $-7 = x + 22$.

Here, the variable x is on the right side of the equation.

$-7 = x + 22$ — The variable can be isolated on *either* side.

$-7 - 22 = x + 22 - 22$ — Subtract 22 from each side.

$-29 = x$, or $x = -29$ — Rewrite; a number $= x$, or $x =$ a number.

CHECK

$-7 = x + 22$ — Original equation

$-7 \stackrel{?}{=} -29 + 22$ — Let $x = -29$.

$-7 = -7$ ✓ — True

The check confirms that the solution set is $\{-29\}$.

◀ **Work Problem 3 at the Side.**

Note

In **Example 3**, what would happen if we subtract $-7 - 22$ incorrectly, obtaining $x = -15$ (instead of $x = -29$) as the last line of the solution? A check should indicate an error.

CHECK

$-7 = x + 22$ — Original equation from **Example 3**

$-7 \stackrel{?}{=} -15 + 22$ — Let $x = -15$.

$-7 = 7$ — False (The left side does *not* equal the right side.)

The false statement indicates that -15 is *not* a solution of the equation. If this happens, rework the problem.

EXAMPLE 4 **Subtracting a Variable Term**

Solve $6x - 8 = 7x$.

$6x - 8 = 7x$

$6x - 8 - 6x = 7x - 6x$ — Subtract $6x$ from each side.

$-8 = x$ — Combine like terms.

CHECK

$6x - 8 = 7x$ — Original equation

$6(-8) - 8 \stackrel{?}{=} 7(-8)$ — Let $x = -8$. (Use parentheses when substituting to avoid errors.)

$-48 - 8 \stackrel{?}{=} -56$ — Multiply.

$-56 = -56$ ✓ — True

A true statements results, so the solution set is $\{-8\}$.

What happens in **Example 4** if we start by subtracting $7x$ from each side?

$6x - 8 = 7x$	Original equation from **Example 4**
$6x - 8 - 7x = 7x - 7x$	Subtract $7x$ from each side.
$-8 - x = 0$	Combine like terms.
$-8 - x + 8 = 0 + 8$	Add 8 to each side.
$-x = 8$	Combine like terms.

This result gives the value of $-x$, but not of x itself. However, it does say that the additive inverse of x is 8, which means that x must be -8.

$x = -8$ Same result as in **Example 4**

We can make the following generalization.

If a is a number and $-x = a$, then $x = -a$.

Work Problem 4 at the Side. ▶

4 **(a)** Solve $11z - 9 = 12z$.

(b) What is the solution set of $-x = 6$?

(c) What is the solution set of $-x = -12$?

EXAMPLE 5 Subtracting a Variable Term (Fractional Coefficients)

Solve $\frac{3}{5}k + 15 = \frac{8}{5}k$.

$$\frac{3}{5}k + 15 = \frac{8}{5}k$$

We must get the terms with k on the same side of the = symbol.

$\frac{3}{5}k + 15 - \frac{3}{5}k = \frac{8}{5}k - \frac{3}{5}k$	Subtract $\frac{3}{5}k$ from each side.
$15 = 1k$	$\frac{3}{5}k - \frac{3}{5}k = 0$; $\frac{8}{5}k - \frac{3}{5}k = \frac{5}{5}k = 1k$
$15 = k$	Multiplicative identity property

From now on we will skip this step.

Check by substituting 15 in the original equation. The solution set is $\{15\}$.

Work Problem 5 at the Side. ▶

5 Solve.

(a) $\frac{7}{2}m + 1 = \frac{9}{2}m$

(b) $\frac{2}{3}x - 4 = \frac{5}{3}x$

EXAMPLE 6 Applying the Addition Property of Equality Twice

Solve $8 - 6p = -7p + 5$.

$8 - 6p = -7p + 5$	
$8 - 6p + 7p = -7p + 5 + 7p$	Add $7p$ to each side.
$8 + p = 5$	Combine like terms.
$8 + p - 8 = 5 - 8$	Subtract 8 from each side.
$p = -3$	Combine like terms.

CHECK Substitute -3 for p in the original equation.

$8 - 6p = -7p + 5$	Original equation
$8 - 6(-3) \stackrel{?}{=} -7(-3) + 5$	Let $p = -3$.
$8 + 18 \stackrel{?}{=} 21 + 5$	Multiply.
$26 = 26$ ✓	True

The check results in a true statement, so the solution set is $\{-3\}$.

Work Problem 6 at the Side. ▶

6 Solve.

GS **(a)** $10 - a = -2a + 9$

$10 - a +$ ____ $= -2a + 9 + 2a$

$10 + a = 9$

$10 + a -$ ____ $= 9 -$ ____

$a =$ ____

The solution set is ____.

(b) $6x - 8 = 12 + 5x$

Answers

4. **(a)** $\{-9\}$ **(b)** $\{-6\}$ **(c)** $\{12\}$

5. **(a)** $\{1\}$ **(b)** $\{-4\}$

6. **(a)** $2a$; 10; 10; -1; $\{-1\}$ **(b)** $\{20\}$

7 Solve.

(a) $9r + 4r + 6 - 2 = 9r + 4 + 3r$

(b) $4x + 6 + 2x - 3 = 9 + 5x - 4$

8 Solve.

(a) $3(2x - 1) - 5x = 8$

GS (b) $2(5 + 2m) - 3(m - 4) = 29$

$2(5) + 2(2m) - \_\_\_\_ m$

$- \_\_\_\_ (\_\_\_\_) = 29$

$10 + 4m - \_\_\_\_ + \_\_\_\_ = 29$

Now complete the solution.

Answers

7. (a) $\{0\}$ (b) $\{2\}$
8. (a) $\{11\}$ (b) 3; 3; -4; $3m$; 12; $\{7\}$

Note

There are often several correct ways to solve an equation. We could begin by adding $6p$ (instead of $7p$) to each side of the equation

$$8 - 6p = -7p + 5.$$ See **Example 6.**

Combining like terms and subtracting 5 from each side gives $3 = -p$. (Try this.) If $3 = -p$, then $-3 = p$, and the variable has been isolated on the right side of the equation. The same solution results.

OBJECTIVE 3 Simplify, and then use the addition property of equality.

EXAMPLE 7 Combining Like Terms When Solving

Solve $3t - 12 + t + 2 = 5 + 3t - 15$.

$$3t - 12 + t + 2 = 5 + 3t - 15$$

$$4t - 10 = -10 + 3t \quad \text{Combine like terms on each side.}$$

$$4t - 10 - 3t = -10 + 3t - 3t \quad \text{Subtract } 3t \text{ from each side.}$$

$$t - 10 = -10 \quad \text{Combine like terms.}$$

$$t - 10 + 10 = -10 + 10 \quad \text{Add 10 to each side.}$$

$$t = 0 \quad \text{Combine like terms.}$$

The real number 0 can be a solution of an equation.

CHECK $3t - 12 + t + 2 = 5 + 3t - 15$ Original equation

$$3(0) - 12 + 0 + 2 \stackrel{?}{=} 5 + 3(0) - 15 \quad \text{Let } t = 0.$$

$$0 - 12 + 0 + 2 \stackrel{?}{=} 5 + 0 - 15 \quad \text{Multiply.}$$

$$-10 = -10 \checkmark \quad \text{True}$$

The check results in a true statement, so the solution set is $\{0\}$.

◀ Work Problem 7 at the Side.

CAUTION

The final line of the CHECK does not give the solution of the equation. It gives a confirmation that the solution found is correct.

EXAMPLE 8 Using the Distributive Property When Solving

Solve $3(2 + 5x) - (1 + 14x) = 6$.

$$3(2 + 5x) - (1 + 14x) = 6 \quad \text{Begin by clearing (removing) the parentheses.}$$

$$3(2 + 5x) - 1(1 + 14x) = 6 \quad -(1 + 14x) = -1(1 + 14x)$$

$$3(2) + 3(5x) - 1(1) - 1(14x) = 6 \quad \text{Distributive property}$$

$$6 + 15x - 1 - 14x = 6 \quad \text{Multiply.}$$

Be careful here, or a sign error may result.

$$x + 5 = 6 \quad \text{Combine like terms.}$$

$$x + 5 - 5 = 6 - 5 \quad \text{Subtract 5 from each side.}$$

$$x = 1 \quad \text{Combine like terms.}$$

Check by substituting 1 in the original equation. The solution set is $\{1\}$.

◀ Work Problem 8 at the Side.

2.1 Exercises

FOR EXTRA HELP — *Go to MyMathLab for worked-out, step-by-step solutions to exercises enclosed in a square and video solutions to ▶ exercises.*

CONCEPT CHECK *Complete each statement with the correct response. The following terms may be used once, more than once, or not at all.*

linear	expression	solution set	multiplication
equation	addition	equivalent equations	variable

1. A(n) __________ includes an equality symbol, while a(n) __________ does not.

2. A(n) __________ equation in one __________ (here x) can be written in the form $Ax + B\ (= / \neq)\ C$.

3. Equations that have exactly the same solution set are __________.

4. The __________ property of equality states that the same expression may be added to or subtracted from each side of an equation without changing the __________.

5. CONCEPT CHECK Which of the following are *not* linear equations in one variable?

A. $x^2 - 5x + 6 = 0$ **B.** $x^3 = x$

C. $3x - 4 = 0$ **D.** $7x - 6x = 3 + 9x$

6. CONCEPT CHECK Decide whether each is an *equation* or an *expression*. If it is an equation, solve it. If it is an expression, simplify it.

(a) $5x + 8 - 4x + 7$ **(b)** $-6m + 12 + 7m - 5$

(c) $5x + 8 - 4x = 7$ **(d)** $-6m + 12 + 7m = -5$

Solve each equation, and check the solution. ***See Examples 1–6.***

7. $x - 4 = 8$ **8.** $x - 8 = 9$ **9.** $x - 5 = -8$ **10.** $x - 7 = -9$

11. $z - 12 = -8$ **12.** $z - 18 = -7$ **13.** $r + 9 = 13$ **14.** $t + 6 = 10$

15. $x + 26 = 17$ **16.** $x + 45 = 24$ **17.** $x - 8.4 = -2.1$ **18.** $x - 15.5 = -5.1$

19. $t - 12.7 = -19.2$ **20.** $t - 8.6 = -17.3$ **21.** $t + 12.3 = -4.6$ **22.** $x + 21.5 = -13.4$

23. $x + \frac{1}{4} = -\frac{1}{2}$ **24.** $x + \frac{2}{3} = -\frac{1}{6}$ **25.** $7 + r = -3$ **26.** $8 + k = -4$

27. $2 = p + 15$ **28.** $3 = z + 17$ **29.** $-4 = x - 14$ **30.** $-7 = x - 22$

31. $5.2 = z - 4.9$ **32.** $11.8 = z - 3.6$ **33.** $-\frac{1}{3} = x - \frac{3}{5}$ **34.** $-\frac{1}{4} = x - \frac{2}{3}$

35. $8x - 3 = 9x$ **36.** $4x - 1 = 5x$ **37.** $3x = 2x + 7$ **38.** $5x = 4x + 9$

39. $10x + 4 = 9x$ **40.** $8t + 5 = 7t$ **41.** $\frac{1}{3}x + 12 = \frac{4}{3}x$ **42.** $\frac{2}{9}x + 17 = \frac{11}{9}x$

43. $\frac{1}{2}x + 5 = -\frac{1}{2}x$ **44.** $\frac{1}{5}x + 7 = -\frac{4}{5}x$ **45.** $\frac{2}{5}w - 6 = \frac{7}{5}w$ **46.** $\frac{2}{7}r - 3 = \frac{9}{7}r$

47. $5.6x + 2 = 4.6x$

48. $9.1x + 5 = 8.1x$

49. $3p + 6 = 10 + 2p$

50. $8x + 4 = -6 + 7x$

51. $5 - x = -2x - 11$

52. $3 - 8x = -9x - 1$

53. $-4z + 7 = -5z + 9$

54. $-6q + 3 = -7q + 10$

Solve each equation, and check the solution. ***See Examples 7 and 8.***

55. $3x + 6 - 10 = 2x - 2$

56. $8k - 4 + 6 = 7k + 1$

57. $6x + 5 + 7x + 3 = 12x + 4$

58. $4x - 3 - 8x + 1 = -5x + 9$

59. $10x + 5x + 7 - 4 = 12x + 3 + 2x$

60. $7p + 4p + 13 - 7 = 7p + 6 + 3p$

61. $5.2q - 4.6 - 7.1q = -0.9q - 4.6$

62. $-4.0x + 2.7 - 1.6x = -4.6x + 2.7$

63. $\frac{5}{7}x + \frac{1}{3} = \frac{2}{5} - \frac{2}{7}x + \frac{2}{5}$

64. $\frac{6}{7}s - \frac{3}{4} = \frac{4}{5} - \frac{1}{7}s + \frac{1}{6}$

65. $5(3x - 2) - 14x = 3$

66. $2(3x - 4) - 5x = 7$

67. $13p + 3(2 - 4p) = 4$

68. $21w + 4(6 - 5w) = 8$

69. $(5x + 6) - (3 + 4x) = 10$

70. $(8r - 3) - (7r + 1) = -6$

71. $2(p + 5) - (9 + p) = -3$

72. $4(k - 6) - (3k + 2) = -5$

73. $-6(2x - 1) + (13x - 6) = 0$

74. $-5(3w - 3) + (16w - 15) = 0$

75. $10(-2x + 1) = -19(x + 1)$

76. $2(-3r + 2) = -5(r - 3)$

77. $-2(8p + 2) - 3(2 - 7p) = 2(4 + 2p)$

78. $4(3 - z) - 5(1 - 2z) = 7(3 + z)$

Many college students juggle a difficult schedule and multiple responsibilities, including school, work, and family demands.

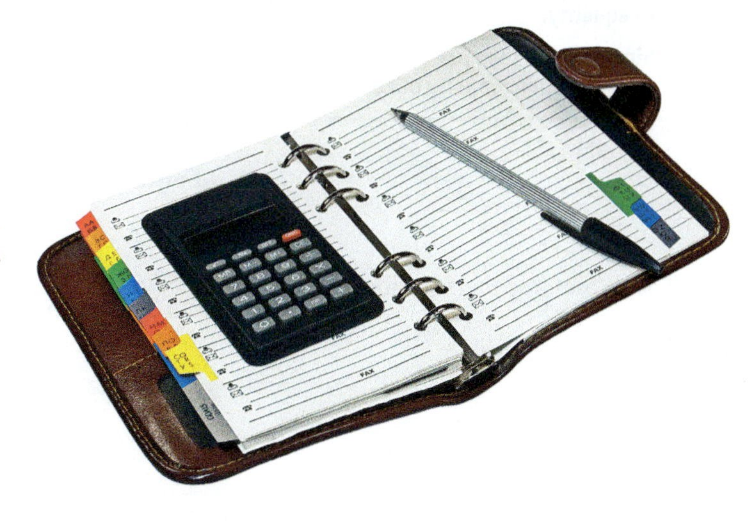

Time Management Tips

- **Read the syllabus for each class.** Understand class policies, such as attendance, late homework, and make-up tests. Find out how you are graded.
- **Make a semester or quarter calendar.** Put test dates and major due dates for *all* your classes on the *same* calendar. Try using a different color pen for each class.
- **Make a weekly schedule.** After you fill in your classes and other regular responsibilities, block off some study periods. Aim for 2 hours of study for each 1 hour in class.
- **Choose a regular study time and place** (such as the campus library). Routine helps.
- **Keep distractions to a minimum.** Get the most out of the time you have set aside for studying by limiting interruptions. Turn off your cell phone. Take a break from social media. Avoid studying in front of the TV.
- **Make "to-do" lists.** Number tasks in order of importance. Cross off tasks as you complete them.
- **Break big assignments into smaller chunks.** Make deadlines for each smaller chunk so that you stay on schedule.
- **Take breaks when studying.** Do not try to study for hours at a time. Take a 10-minute break each hour or so.
- **Ask for help when you need it.** Talk with your instructor during office hours. Make use of the learning center, tutoring center, counseling office, or other resources available at your school.

Now Try This

Think through and answer each question.

1. Evaluate when and where you are currently studying. Are the places you named quiet and comfortable? Are you studying when you are most alert?

   ____

2. How many hours do you have available for studying this week?

   ____

3. Which two or three of the above suggestions will you try this week to improve your time management?

   ____

4. Once the week is over, evaluate how these suggestions worked. What will you do differently next week?

   ____

2.2 The Multiplication Property of Equality

OBJECTIVES

1. Use the multiplication property of equality.
2. Simplify, and then use the multiplication property of equality.

OBJECTIVE 1 Use the multiplication property of equality. The addition property of equality is not sufficient to solve some equations.

$$3x + 2 = 17$$

$$3x + 2 - 2 = 17 - 2 \quad \text{Subtract 2 from each side.}$$

$$3x = 15 \quad \text{Combine like terms.}$$

The coefficient of x here is 3, not 1 as desired. The **multiplication property of equality** is needed to change $3x = 15$ to an equation of the form

$$x = \text{a number.}$$

Since $3x = 15$, both $3x$ and 15 must represent the same number. Multiplying both $3x$ and 15 by the same number will result in an equivalent equation.

Multiplication Property of Equality

If A, B, and C represent real numbers, where $C \neq 0$, then the equations

$$A = B \quad \text{and} \quad AC = BC \quad \text{are equivalent.}$$

In words, each side of an equation may be multiplied by the same nonzero number without changing the solution set.

1 Check that 5 is the solution of $3x = 15$. (GS)

$$3x = 15$$

$$3(\_\_\_\_) \stackrel{?}{=} 15$$

$$\_\_\_\_ = 15 \quad (\textit{True/False})$$

The solution set is ____.

In $3x = 15$, we must change $3x$ to $1x$, or x. To do this, we multiply each side of the equation by $\frac{1}{3}$, the *reciprocal* of 3, because $\frac{1}{3} \cdot 3 = \frac{3}{3} = 1$.

$$3x = 15$$

$$\frac{1}{3}(3x) = \frac{1}{3}(15) \quad \text{Multiply each side by } \tfrac{1}{3}.$$

$$\left(\frac{1}{3} \cdot 3\right) x = \frac{1}{3}(15) \quad \text{Associative property}$$

$$1x = 5 \quad \text{Multiplicative inverse property}$$

$$x = 5 \quad \text{Multiplicative identity property}$$

The product of a number and its reciprocal is 1.

The solution is 5. We can check this result in the original equation.

◀ Work Problem 1 at the Side.

Just as the addition property of equality permits *subtracting* the same number from each side of an equation, the multiplication property of equality permits *dividing* each side of an equation by the same nonzero number.

$$3x = 15$$

$$\frac{3x}{3} = \frac{15}{3} \quad \text{Divide each side by 3.}$$

$$x = 5 \quad \text{Same result as above}$$

Multiplication Property of Equality Extended to Division

We can divide each side of an equation by the same nonzero number without changing the solution. ***Do not, however, divide each side by a variable, because the variable might be equal to 0.***

Answer

1. 5; 15; True; $\{5\}$

Note

It is usually easier to multiply on each side of an equation if the coefficient of the variable is a fraction, and divide on each side if the coefficient is an integer.

To solve $\frac{3}{4}x = 12$, it is easier to multiply by $\frac{4}{3}$ than to divide by $\frac{3}{4}$.

To solve $5x = 20$, it is easier to divide by 5 than to multiply by $\frac{1}{5}$.

EXAMPLE 1 Applying the Multiplication Property of Equality

Solve $5x = 60$.

$$5x = 60$$ Our goal is to isolate x.

$$\frac{5x}{5} = \frac{60}{5}$$ Divide each side by 5, the coefficient of x. (Dividing by 5 is the same as multiplying by $\frac{1}{5}$.)

$$x = 12$$ $\frac{5x}{5} = \frac{5}{5}x = 1x = x$

CHECK Substitute 12 for x in the original equation.

$$5x = 60$$ Original equation

$$5(12) \stackrel{?}{=} 60$$ Let $x = 12$.

$$60 = 60 \checkmark$$ True (60 is *not* the solution.)

A true statement results, so the solution set is $\{12\}$.

Work Problem 2 at the Side. ▶

EXAMPLE 2 Applying the Multiplication Property of Equality

Solve $-25p = 30$.

$$-25p = 30$$

$$\frac{-25p}{-25} = \frac{30}{-25}$$ Divide each side by -25, the coefficient of p.

$$p = -\frac{30}{25}$$ $\frac{a}{-b} = -\frac{a}{b}$

$$p = -\frac{6}{5}$$ Write in lowest terms.

CHECK

$$-25p = 30$$ Original equation

$$\frac{-25}{1}\left(-\frac{6}{5}\right) \stackrel{?}{=} 30$$ Let $p = -\frac{6}{5}$.

$$30 = 30 \checkmark$$ True

The check confirms that the solution set is $\{-\frac{6}{5}\}$.

Work Problem 3 at the Side. ▶

2 Solve.

GS **(a)** Fill in the blanks.

$$7x = 91$$

$$\frac{7x}{\_\_\_\_} = \frac{91}{\_\_\_\_}$$

$$x = \_\_\_\_$$

CHECK $7x = 91$

$7(\_\_\_\_) \stackrel{?}{=} 91$

$\_\_\_\_ = 91$ (*True/False*)

The solution set is ____.

(b) $3r = -12$

(c) $15x = 75$

3 Solve.

(a) $2m = 15$

(b) $-6x = 14$

(c) $10z = -45$

Answers

2. (a) 7; 7; 13; 13; 91; True; $\{13\}$
(b) $\{-4\}$ (c) $\{5\}$

3. (a) $\left\{\frac{15}{2}\right\}$ (b) $\left\{-\frac{7}{3}\right\}$ (c) $\left\{-\frac{9}{2}\right\}$

4 Solve.

(a) $-0.7m = -5.04$

(b) $-63.75 = 12.5k$

5 Solve.

GS (a) $\frac{x}{5} = 5$

What is the coefficient of x here? ____ By what number should we multiply each side? ____

Now solve.

(b) $\frac{p}{4} = -6$

6 Solve.

GS (a) $-\frac{5}{6}t = -15$

By what number should we multiply each side? ____

Now solve.

(b) $\frac{3}{5}k = -21$

Answers

4. (a) $\{7.2\}$ (b) $\{-5.1\}$

5. (a) $\frac{1}{5}$; 5; $\{25\}$ (b) $\{-24\}$

6. (a) $-\frac{6}{5}$; $\{18\}$ (b) $\{-35\}$

EXAMPLE 3 **Solving a Linear Equation (Decimal Coefficient)**

Solve $6.09 = -2.1x$.

$$6.09 = -2.1x$$ Isolate x on the right.

$$\frac{6.09}{-2.1} = \frac{-2.1x}{-2.1}$$ Divide each side by -2.1.

$$-2.9 = x, \quad \text{or} \quad x = -2.9$$

Check by replacing x with -2.9 in the original equation. The solution set is $\{-2.9\}$.

◀ Work Problem 4 at the Side.

EXAMPLE 4 **Solving a Linear Equation (Fractional Coefficient)**

Solve $\frac{x}{4} = 3$.

$$\frac{x}{4} = 3$$

$$\frac{1}{4}x = 3$$ $\frac{x}{4} = \frac{1x}{4} = \frac{1}{4}x$

$$4 \cdot \frac{1}{4}x = 4 \cdot 3$$ Multiply each side by 4, the reciprocal of $\frac{1}{4}$.

$4 \cdot \frac{1}{4}x = 1x = x$ → $$x = 12$$ Multiplicative inverse property; multiplicative identity property

CHECK $$\frac{x}{4} = 3$$ Original equation

$$\frac{12}{4} \stackrel{?}{=} 3$$ Let $x = 12$.

$$3 = 3 \checkmark$$ True

A true statement results, so the solution set is $\{12\}$.

◀ Work Problem 5 at the Side.

EXAMPLE 5 **Solving a Linear Equation (Fractional Coefficient)**

Solve $\frac{3}{4}h = 6$.

$$\frac{3}{4}h = 6$$

It is easier to multiply by $\frac{4}{3}$ than to divide by $\frac{3}{4}$.

$$\frac{4}{3} \cdot \frac{3}{4}h = \frac{4}{3} \cdot 6$$ Multiply each side by $\frac{4}{3}$, the reciprocal of $\frac{3}{4}$.

$$1 \cdot h = \frac{4}{3} \cdot \frac{6}{1}$$ Multiplicative inverse property

$$h = 8$$ Multiplicative identity property; multiply fractions.

Check to confirm that the solution set is $\{8\}$.

◀ Work Problem 6 at the Side.

Note

We can use reasoning to solve an equation such as

$$-k = -15.$$

Because this equation says that the additive inverse (or opposite) of k is -15, k must equal 15. We can also use the multiplication property of equality to obtain the same result. This is done in **Example 6.**

EXAMPLE 6 **Applying the Multiplication Property of Equality When the Coefficient of the Variable Is -1**

Solve $-k = -15$.

$-k = -15$	
$-1 \cdot k = -15$	$-k = -1 \cdot k$
$-1(-1 \cdot k) = -1(-15)$	Multiply each side by -1.
$[-1(-1)] \cdot k = 15$	Associative property; Multiply.
$1 \cdot k = 15$	Multiplicative inverse property
$k = 15$	Multiplicative identity property

CHECK

$-k = -15$	Original equation
$-(15) \stackrel{?}{=} -15$	Let $k = 15$.
$-15 = -15$ ✓	True

A true statement results, so the solution set is $\{15\}$.

Work Problem 7 at the Side. ▶

OBJECTIVE ▶ **2** **Simplify, and then use the multiplication property of equality.**

EXAMPLE 7 **Combining Like Terms When Solving**

Solve $5m + 6m = 33$.

$5m + 6m = 33$	
$11m = 33$	Combine like terms.
$\frac{11m}{11} = \frac{33}{11}$	Divide each side by 11.
$m = 3$	Multiplicative identity property

CHECK

$5m + 6m = 33$	Original equation
$5(3) + 6(3) \stackrel{?}{=} 33$	Let $m = 3$.
$15 + 18 \stackrel{?}{=} 33$	Multiply.
$33 = 33$ ✓	True

A true statement results, so the solution set is $\{3\}$.

Work Problem 8 at the Side. ▶

7 Solve.

GS **(a)** Fill in the blanks.

$$-m = 2$$
$$\_\_\_\_ \cdot m = 2$$
$$\_\_\_\_(-1 \cdot m) = \_\_\_\_ \cdot 2$$
$$m = \_\_\_\_$$

CHECK $-m = 2$

$$-(\_\_\_\_) \stackrel{?}{=} 2$$
$$\_\_\_\_ = 2 \quad (True/False)$$

The solution set is ____.

(b) $-p = -7$

8 Solve.

(a) $7m - 5m = -12$

(b) $4r - 9r = 20$

Answers

7. **(a)** $-1; -1; -1; -2; -2; 2$; True; $\{-2\}$
(b) $\{7\}$

8. **(a)** $\{-6\}$ **(b)** $\{-4\}$

9 Solve.

GS **(a)** $3(2x-2)+6=-24$

$3(2x)+3(\_\_\_\_)+6=-24$

$\_\_\_\_-6+6=-24$

$6x=-24$

$\frac{6x}{\_\_\_\_}=\frac{-24}{\_\_\_\_}$

$x=\_\_\_\_$

Check by substituting ____ for x in the original equation.

The solution set is ____.

(b) $8-2(5x+4)=10$

(c) $5(x+2)-(3x+10)=14$

Answers

9. **(a)** -2; $6x$; 6; 6; -4; -4; $\{-4\}$
(b) $\{-1\}$ **(c)** $\{7\}$

EXAMPLE 8 Using the Distributive Property When Solving

Solve $4(x-1)-2(x-2)=-12$.

Begin by clearing the parentheses.

$4(x-1)-2(x-2)=-12$	
$4x+4(-1)-2x-2(-2)=-12$	Distributive property
$4x-4-2x+4=-12$	Multiply.
$2x=-12$	Combine like terms.
$\frac{2x}{2}=\frac{-12}{2}$	Divide each side by 2.
$x=-6$	

CHECK

$4(x-1)-2(x-2)=-12$	Original equation
$4(-6-1)-2(-6-2)\stackrel{?}{=}-12$	Let $x=-6$.
$4(-7)-2(-8)\stackrel{?}{=}-12$	Subtract within the parentheses.
$-28+16\stackrel{?}{=}-12$	Multiply.
$-12=-12$ ✓	True

The check results in a true statement, so the solution set is $\{-6\}$.

◀ Work Problem 9 at the Side.

2.2 Exercises

FOR EXTRA HELP *Go to MyMathLab for worked-out, step-by-step solutions to exercises enclosed in a square and video solutions to ▶ exercises.*

1. CONCEPT CHECK Indicate whether the addition or multiplication property of equality should be used to solve each equation. *Do not actually solve.*

(a) $3x = 12$ **(b)** $3 + x = 12$

(c) $-x = 4$ **(d)** $-12 = 6 + x$

2. CONCEPT CHECK Which equation does *not* require the use of the multiplication property of equality?

A. $3x - 5x = 6$ **B.** $-\frac{1}{4}x = 12$

C. $5x - 4x = 7$ **D.** $\frac{x}{3} = -2$

CONCEPT CHECK *By what number is it necessary to multiply each side of each equation in order to isolate x on the left side? Do not actually solve.*

3. $\frac{2}{3}x = 8$ **4.** $\frac{4}{5}x = 6$ **5.** $\frac{x}{10} = 3$ **6.** $\frac{x}{100} = 8$

7. $-\frac{9}{2}x = -4$ **8.** $-\frac{8}{3}x = -11$ **9.** $-x = 0.36$ **10.** $-x = 0.29$

CONCEPT CHECK *By what number is it necessary to divide each side of each equation in order to isolate x on the left side? Do not actually solve.*

11. $6x = 5$ **12.** $7x = 10$ **13.** $-4x = 13$ **14.** $-13x = 6$

15. $0.12x = 48$ **16.** $0.21x = 63$ **17.** $-x = 23$ **18.** $-x = 49$

19. CONCEPT CHECK In the solution of a linear equation, the next-to-the-last step reads "$-x = -\frac{3}{4}$." Which of the following would be the solution of this equation?

A. $-\frac{3}{4}$ **B.** $\frac{3}{4}$ **C.** -1 **D.** $\frac{4}{3}$

20. CONCEPT CHECK Which of the following is the solution of the equation

$$-x = -24?$$

A. 24 **B.** -24 **C.** 1 **D.** -1

Solve each equation, and check the solution. ***See Examples 1–8.***

21. $5x = 30$ **22.** $7x = 56$ **23.** $2m = 15$ **24.** $3m = 10$

25. $3a = -15$ ▶ **26.** $5k = -70$ **27.** $10t = -36$ ▶ **28.** $4s = -34$

29. $-6x = -72$ **30.** $-8x = -64$

31. GS
$2r = 0$
$\frac{2r}{\_\_\_\_} = \frac{0}{\_\_\_\_}$
$r = \_\_\_\_$
Solution set: ____

32. GS ▶
$5x = 0$
$\frac{5x}{\_\_\_\_} = \frac{0}{\_\_\_\_}$
$x = \_\_\_\_$
Solution set: ____

33. $-x = 12$

34. $-t = 14$

35. $0.2t = 8$

36. $0.9x = 18$

37. $-2.1m = 25.62$

38. $-3.9a = 31.2$

39. $\frac{1}{4}x = -12$

40. $\frac{1}{5}p = -3$

41. $\frac{z}{6} = 12$

42. $\frac{x}{5} = 15$

43. $\frac{x}{7} = -5$

44. $\frac{k}{8} = -3$

45. $\frac{2}{7}p = 4$

46. $\frac{3}{8}x = 9$

47. $-\frac{7}{9}c = \frac{3}{5}$

48. $-\frac{5}{6}d = \frac{4}{9}$

49. $4x + 3x = 21$

50. $9x + 2x = 121$

51. $3r - 5r = 10$

52. $9p - 13p = 24$

53. $\frac{2}{5}x - \frac{3}{10}x = 2$

54. $\frac{2}{3}x - \frac{5}{9}x = 4$

55. $5m + 6m - 2m = 63$

56. $11r - 5r + 6r = 168$

57. $x + x - 3x = 12$

58. $z - 3z + z = -16$

59. $-6x + 4x - 7x = 0$

60. $-5x + 4x - 8x = 0$

61. $4(3x - 1) + 4 = -36$

62. $5(3x + 2) - 10 = 30$

63. $8 - 4(x + 2) = 16$

64. $18 - 6(4x + 3) = -48$

65. $5(x + 4) - (3x + 20) = -10$

66. $8(x - 2) - (x - 16) = 35$

67. $-3(2p - 8) + 4(p - 6) = -2$

68. $-4(2p + 3) + 2(p + 6) = -18$

Write an equation using the information given in the problem. Use x as the variable. Then solve the equation.

69. When a number is multiplied by -4, the result is 10. Find the number.

70. When a number is multiplied by 4, the result is 6. Find the number.

71. When a number is divided by -5, the result is 2. Find the number.

72. If twice a number is divided by 5, the result is 4. Find the number.

2.3 More on Solving Linear Equations

OBJECTIVES

1. Learn and use the four steps for solving a linear equation.
2. Solve equations that have no solution or infinitely many solutions.
3. Solve equations with fractions or decimals as coefficients.
4. Write expressions for two related unknown quantities.

We now apply *both* properties of equality to solve linear equations.

Work Problem 1 at the Side. ▶

OBJECTIVE 1 Learn and use the four steps for solving a linear equation.

Solving a Linear Equation in One Variable

Step 1 **Simplify each side separately.** Use the distributive property as needed.
- Clear any parentheses.
- Clear any fractions or decimals.
- Combine like terms.

Step 2 **Isolate the variable terms on one side.** Use the addition property of equality so that all terms with variables are on one side of the equation and all constants (numbers) are on the other side.

Step 3 **Isolate the variable.** Use the multiplication property of equality to obtain an equation that has just the variable with coefficient 1 on one side.

Step 4 **Check.** Substitute the value found into the *original* equation. If a true statement results, write the solution set. If not, rework the problem.

Remember that when we solve an equation, our primary goal is to isolate the variable on one side of the equation.

EXAMPLE 1 Solving a Linear Equation

Solve $-6x + 5 = 17$.

Step 1 There are no parentheses, fractions, or decimals, nor are there like terms to combine on one side in this equation. This step is not necessary.

Our goal is to isolate *x*.

$$-6x + 5 = 17$$

Step 2
$$-6x + 5 - 5 = 17 - 5 \quad \text{Subtract 5 from each side.}$$
$$-6x = 12 \quad \text{Combine like terms.}$$

Step 3
$$\frac{-6x}{-6} = \frac{12}{-6} \quad \text{Divide each side by } -6.$$
$$x = -2$$

Step 4 Check by substituting -2 for x in the original equation.

CHECK
$$-6x + 5 = 17 \quad \text{Original equation}$$
$$-6(-2) + 5 \stackrel{?}{=} 17 \quad \text{Let } x = -2.$$
$$12 + 5 \stackrel{?}{=} 17 \quad \text{Multiply.}$$
$$17 = 17 \ ✓ \quad \text{True}$$

17 is *not* the solution.

A true statement results, so the solution set is $\{-2\}$.

Work Problem 2 at the Side. ▶

1 As a review, tell whether the *addition* or *multiplication property of equality* should be used to solve each equation. *Do not actually solve.*

(a) $7 + x = -9$

(b) $-13x = 26$

(c) $-x = \frac{3}{4}$

(d) $-12 = x - 4$

2 Solve.

(a) $-5p + 4 = 19$

(b) $7 + 2m = -3$

Answers

1. **(a)** and **(d)**: addition property of equality
(b) and **(c)**: multiplication property of equality

2. **(a)** $\{-3\}$ **(b)** $\{-5\}$

3 Solve.

GS **(a)** $5 - 8k = 2k - 5$

Begin with Step 2.

$$5 - 8k - 2k = 2k - 5 - \_\_\_\_$$
$$5 - \_\_\_\_ = -5$$
$$5 - 10k - \_\_\_\_ = -5 - \_\_\_\_$$
$$-10k = -10$$
$$\frac{-10k}{\_\_\_\_} = \frac{-10}{\_\_\_\_}$$
$$k = \_\_\_\_$$

Check by substituting ____ for k in the original equation.

The solution set is ____.

(b) $2q + 3 = 4q - 9$

(c) $6x + 7 = -8 + 3x$

Answers

3. **(a)** $2k$; $10k$; 5; 5; -10; -10; 1; 1; $\{1\}$
(b) $\{6\}$ **(c)** $\{-5\}$

EXAMPLE 2 **Solving a Linear Equation**

Solve $3x + 2 = 5x - 8$.

Step 1 There are no parentheses, fractions, or decimals, nor are there like terms to combine on one side in this equation. We begin with Step 2.

$$3x + 2 = 5x - 8$$ Our goal is to isolate x.

Step 2	$3x + 2 - 5x = 5x - 8 - 5x$	Subtract $5x$ from each side.
	$-2x + 2 = -8$	Combine like terms.
	$-2x + 2 - 2 = -8 - 2$	Subtract 2 from each side.
	$-2x = -10$	Combine like terms.
Step 3	$\frac{-2x}{-2} = \frac{-10}{-2}$	Divide each side by -2.
	$x = 5$	

Step 4 Check by substituting 5 for x in the original equation.

CHECK	$3x + 2 = 5x - 8$	Original equation
	$3(5) + 2 \stackrel{?}{=} 5(5) - 8$	Let $x = 5$.
	$15 + 2 \stackrel{?}{=} 25 - 8$	Multiply.
	$17 = 17$ ✓	True

The check confirms that 5 is the solution. The solution set is $\{5\}$.

Note

Remember that a variable can be isolated on either side of an equation. In **Example 2,** x will be isolated on the right if we begin by subtracting $3x$, instead of $5x$, from each side of the equation.

$3x + 2 = 5x - 8$	Equation from **Example 2**
$3x + 2 - 3x = 5x - 8 - 3x$	Subtract $3x$ from each side.
$2 = 2x - 8$	Combine like terms.
$2 + 8 = 2x - 8 + 8$	Add 8 to each side.
$10 = 2x$	Combine like terms.
$\frac{10}{2} = \frac{2x}{2}$	Divide each side by 2.
$5 = x$	The same solution results.

$5 = x$ is equivalent to $x = 5$.

There are often several equally correct ways to solve an equation.

◀ Work Problem **3** at the Side.

EXAMPLE 3 **Solving a Linear Equation**

Solve $4(k-3)-k=k-6$.

Step 1 Clear the parentheses using the distributive property.

$4(k-3)-k=k-6$	
$4(k)+4(-3)-k=k-6$	Distributive property
$4k-12-k=k-6$	Multiply.
$3k-12=k-6$	Combine like terms.
Step 2 $3k-12-k=k-6-k$	Subtract k.
$2k-12=-6$	Combine like terms.
$2k-12+12=-6+12$	Add 12.
$2k=6$	Combine like terms.
Step 3 $\frac{2k}{2}=\frac{6}{2}$	Divide by 2.
$k=3$	

Step 4 Check by substituting 3 for k in the original equation.

CHECK $4(k-3)-k=k-6$	Original equation
$4(3-3)-3\stackrel{?}{=}3-6$	Let $k=3$.
$4(0)-3\stackrel{?}{=}3-6$	Work inside the parentheses.
$-3=-3$ ✓	True

A true statement results, so the solution set is $\{3\}$.

Work Problem 4 at the Side. ▶

EXAMPLE 4 **Solving a Linear Equation**

Solve $8a-(3+2a)=3a+1$.

Step 1 $8a-(3+2a)=3a+1$	
$8a-1(3+2a)=3a+1$	Multiplicative identity property
$8a-3-2a=3a+1$	Distributive property
$6a-3=3a+1$	Combine like terms.
Step 2 $6a-3-3a=3a+1-3a$	Subtract $3a$.
$3a-3=1$	Combine like terms.
$3a-3+3=1+3$	Add 3.
$3a=4$	Combine like terms.
Step 3 $\frac{3a}{3}=\frac{4}{3}$	Divide by 3.
$a=\frac{4}{3}$	

Be careful with signs.

Continued on Next Page

4 Solve.

GS **(a)** $7(p-2)+p=2p+4$

Step 1 Clear the parentheses.

$\_\_\_\_(p)+7(\_\_\_\_)+p=2p+4$

$\_\_\_\_-\_\_\_\_+p=2p+4$

$\_\_\_\_-14=2p+4$

Now complete the solution. Give the solution set.

(b) $11+3(x+1)=5x+16$

Answers

4. **(a)** 7; -2; $7p$; 14; $8p$; The solution set is $\{3\}$.
(b) $\{-1\}$

5 Solve.

(a) $7m - (2m - 9) = 39$

(b) $5x - (x + 9) = x - 4$

6 Solve.

(a) $2(4 + 3r) = 3(r + 1) + 11$

(b) $2 - 3(2 + 6z)$
$= 4(z + 1) - 8$

Answers

5. **(a)** $\{6\}$ **(b)** $\left\{\frac{5}{3}\right\}$

6. **(a)** $\{2\}$ **(b)** $\{0\}$

Step 4 **CHECK**

$8a - (3 + 2a) = 3a + 1$	Original equation
$8\left(\frac{4}{3}\right) - \left[3 + 2\left(\frac{4}{3}\right)\right] \stackrel{?}{=} 3\left(\frac{4}{3}\right) + 1$	Let $a = \frac{4}{3}$.
$\frac{32}{3} - \left[3 + \frac{8}{3}\right] \stackrel{?}{=} 4 + 1$	Multiply.
$\frac{32}{3} - \frac{17}{3} \stackrel{?}{=} 5$	Add.
$5 = 5$ ✓	True

A true statement results, so the solution set is $\left\{\frac{4}{3}\right\}$.

◀ **Work Problem 5 at the Side.**

! CAUTION

In an expression such as $8a - (3 + 2a)$, the $-$ sign acts like a factor of -1 and affects the sign of *every* term within the parentheses.

$$8a - (3 + 2a) \leftarrow \text{Left side of the equation in Example 4}$$
$$= 8a - 1(3 + 2a)$$
$$= 8a + (-1)(3 + 2a)$$
$$= 8a - 3 - 2a$$

Change to $-$ in *both* terms.

EXAMPLE 5 Solving a Linear Equation

Solve $4(4 - 3x) = 32 - 8(x + 2)$.

Do *not* subtract 8 from 32 here.

Be careful with signs.

Step 1	$4(4 - 3x) = 32 - 8(x + 2)$	
	$16 - 12x = 32 - 8x - 16$	Distributive property
	$16 - 12x = 16 - 8x$	Combine like terms.
Step 2	$16 - 12x + 8x = 16 - 8x + 8x$	Add $8x$.
	$16 - 4x = 16$	Combine like terms.
	$16 - 4x - 16 = 16 - 16$	Subtract 16.
	$-4x = 0$	Combine like terms.
Step 3	$\frac{-4x}{-4} = \frac{0}{-4}$	Divide by -4.
	$x = 0$	
Step 4 **CHECK**	$4(4 - 3x) = 32 - 8(x + 2)$	Original equation
	$4[4 - 3(0)] \stackrel{?}{=} 32 - 8(0 + 2)$	Let $x = 0$.
	$4(4) \stackrel{?}{=} 32 - 8(2)$	Work inside the brackets and parentheses.
	$16 = 16$ ✓	True

A true statement results, so the solution set is $\{0\}$. ← $\{0\}$ is a perfectly acceptable solution set.

◀ **Work Problem 6 at the Side.**

OBJECTIVE 2 Solve equations that have no solution or infinitely many solutions. Each equation so far has had exactly one solution. An equation with exactly one solution is a **conditional equation** because it is only true under certain conditions.

EXAMPLE 6 Solving an Equation That Has Infinitely Many Solutions

Solve $5x - 15 = 5(x - 3)$.

$$5x - 15 = 5(x - 3)$$

$$5x - 15 = 5x - 15 \quad \text{Distributive property}$$

$$5x - 15 - 5x = 5x - 15 - 5x \quad \text{Subtract } 5x.$$

$$-15 = -15 \quad \text{Combine like terms.}$$

Notice that the variable "disappeared."

$$-15 + 15 = -15 + 15 \quad \text{Add 15.}$$

$$0 = 0 \quad \text{True}$$

Solution set: {all real numbers}

Because the last statement $(0 = 0)$ is true, *any* real number is a solution. We could have predicted this from the second line in the solution,

$$5x - 15 = 5x - 15 \longleftarrow \text{This is true for } any \text{ value of } x.$$

Try several values for x in the original equation to see that they all satisfy it.

An equation with both sides exactly the same, like $0 = 0$, is an **identity.** An identity is true for all replacements of the variables. As shown above, we write the solution set as **{all real numbers}**.

CAUTION

In **Example 6,** do not write $\{0\}$ as the solution set of the equation. While 0 *is* a solution, there are infinitely many *other* solutions.

For {0} to be the solution set, the last line must include a variable, such as x, and read x = 0 (as in Example 5), not 0 = 0.

EXAMPLE 7 Solving an Equation That Has No Solution

Solve $2x + 3(x + 1) = 5x + 4$.

$$2x + 3(x + 1) = 5x + 4$$

$$2x + 3x + 3 = 5x + 4 \quad \text{Distributive property}$$

$$5x + 3 = 5x + 4 \quad \text{Combine like terms.}$$

$$5x + 3 - 5x = 5x + 4 - 5x \quad \text{Subtract } 5x.$$

$$3 = 4 \quad \text{False}$$

Again, the variable "disappeared."

There is no solution. Solution set: $\varnothing$

A false statement $(3 = 4)$ results. A **contradiction** is an equation that has no solution. Its solution set is the **empty set,** or **null set,** symbolized $\varnothing$.

Work Problem 7 at the Side. ▶

CAUTION

Do not write $\{\varnothing\}$ to represent the empty set.

7 Solve.

(a) $2(x - 6) = 2x - 12$

(b) $3x + 6(x + 1) = 9x - 4$

(c) $8(4 - x) = -8x + 32$

(d) $-4x + 12 = 3 - 4(x - 3)$

Answers

7. (a) {all real numbers} **(b)** $\varnothing$
(c) {all real numbers} **(d)** $\varnothing$

8 Solve.

(a) $\frac{1}{4}x - 4 = \frac{3}{2}x + \frac{3}{4}x$

(b) $\frac{1}{2}x + \frac{5}{8}x = \frac{3}{4}x - 6$

Answers

8. **(a)** $\{-2\}$ **(b)** $\{-16\}$

SOLUTION SETS OF EQUATIONS

Type of Equation	Final Equation in Solution	Number of Solutions	Solution Set
Conditional **(See Examples 1–5.)**	$x =$ a number	One	{a number}
Identity **(See Example 6.)**	A true statement with no variable, such as $0 = 0$	Infinitely many	{all real numbers}
Contradiction **(See Example 7.)**	A false statement with no variable, such as $3 = 4$	None	$\varnothing$

OBJECTIVE 3 Solve equations with fractions or decimals as coefficients. To avoid messy computations, we clear an equation of fractions by multiplying each side by the least common denominator (LCD) of all the fractions in the equation. Doing this will give an equation with only *integer* coefficients.

EXAMPLE 8 Solving a Linear Equation (Fractional Coefficients)

Solve $\frac{2}{3}x - \frac{1}{2}x = -\frac{1}{6}x - 2$.

Step 1

$$\frac{2}{3}x - \frac{1}{2}x = -\frac{1}{6}x - 2$$

Pay particular attention here.

$$6\left(\frac{2}{3}x - \frac{1}{2}x\right) = 6\left(-\frac{1}{6}x - 2\right) \quad \text{Multiply each side by 6, the LCD.}$$

$$6\left(\frac{2}{3}x\right) + 6\left(-\frac{1}{2}x\right) = 6\left(-\frac{1}{6}x\right) + 6(-2) \quad \text{Distributive property; Multiply } each \text{ term inside the parentheses by 6.}$$

The fractions have been cleared.

$$4x - 3x = -x - 12 \quad \text{Multiply.}$$

$$x = -x - 12 \quad \text{Combine like terms.}$$

Step 2

$$x + x = -x - 12 + x \quad \text{Add } x.$$

$$2x = -12 \quad \text{Combine like terms.}$$

Step 3

$$\frac{2x}{2} = \frac{-12}{2} \quad \text{Divide by 2.}$$

$$x = -6$$

Step 4 **CHECK**

$$\frac{2}{3}x - \frac{1}{2}x = -\frac{1}{6}x - 2 \quad \text{Original equation}$$

$$\frac{2}{3}(-6) - \frac{1}{2}(-6) \stackrel{?}{=} -\frac{1}{6}(-6) - 2 \quad \text{Let } x = -6.$$

$$-4 + 3 \stackrel{?}{=} 1 - 2 \quad \text{Multiply.}$$

$$-1 = -1 \checkmark \quad \text{True}$$

The check confirms that the solution set is $\{-6\}$.

◀ Work Problem **8** at the Side.

! CAUTION

When clearing an equation of fractions, be sure to multiply every term on each side of the equation by the LCD.

EXAMPLE 9 Solving a Linear Equation (Fractional Coefficients)

Solve $\frac{1}{3}(x+5) - \frac{3}{5}(x+2) = 1$.

Step 1 We clear the parentheses first. Then we clear the fractions.

$$\frac{1}{3}(x+5) - \frac{3}{5}(x+2) = 1$$

Study Step 1 carefully.

$$\frac{1}{3}(x) + \frac{1}{3}(5) - \frac{3}{5}(x) - \frac{3}{5}(2) = 1 \quad \text{Distributive property}$$

$$\frac{1}{3}x + \frac{5}{3} - \frac{3}{5}x - \frac{6}{5} = 1 \quad \text{Multiply.}$$

Think: $15\left(\frac{1}{3}x\right) = \left(\frac{15}{1} \cdot \frac{1}{3}\right)x = 5x$

$$15\left(\frac{1}{3}x + \frac{5}{3} - \frac{3}{5}x - \frac{6}{5}\right) = 15(1) \quad \text{Multiply each side by 15, the LCD.}$$

$$15\left(\frac{1}{3}x\right) + 15\left(\frac{5}{3}\right) + 15\left(-\frac{3}{5}x\right) + 15\left(-\frac{6}{5}\right) = 15(1) \quad \text{Distributive property}$$

$$5x + 25 - 9x - 18 = 15 \quad \text{Multiply.}$$

$$-4x + 7 = 15 \quad \text{Combine like terms.}$$

Step 2

$$-4x + 7 - 7 = 15 - 7 \quad \text{Subtract 7.}$$

$$-4x = 8 \quad \text{Combine like terms.}$$

Step 3

$$\frac{-4x}{-4} = \frac{8}{-4} \quad \text{Divide by } -4.$$

$$x = -2$$

Step 4 Check to confirm that $\{-2\}$ is the solution set.

Work Problem 9 at the Side. ▶

EXAMPLE 10 Solving a Linear Equation (Decimal Coefficients)

Solve $0.1t + 0.05(20 - t) = 0.09(20)$.

Step 1

$$0.1t + 0.05(20 - t) = 0.09(20)$$

Clear the parentheses first.

$$0.1t + 0.05(20) + 0.05(-t) = 0.09(20) \quad \text{Distributive property}$$

$$0.1t + 1 - 0.05t = 1.8 \quad (*) \text{ Multiply.}$$

Now clear the decimals.

$$100(0.1t + 1 - 0.05t) = 100(1.8) \quad \text{Multiply by 100.}$$

$$100(0.1t) + 100(1) + 100(-0.05t) = 100(1.8) \quad \text{Distributive property}$$

$$10t + 100 - 5t = 180 \quad (**) \text{ Multiply.}$$

$$5t + 100 = 180 \quad \text{Combine like terms.}$$

Step 2

$$5t + 100 - 100 = 180 - 100 \quad \text{Subtract 100.}$$

$$5t = 80 \quad \text{Combine like terms.}$$

Step 3

$$\frac{5t}{5} = \frac{80}{5} \quad \text{Divide by 5.}$$

$$t = 16$$

Step 4 Check to confirm that $\{16\}$ is the solution set.

Work Problem 10 at the Side. ▶

9 Solve.

$$\frac{1}{4}(x+3) - \frac{2}{3}(x+1) = -2$$

10 Solve.

$$0.06(100 - x) + 0.04x = 0.05(92)$$

Answers

9. $\{5\}$

10. $\{70\}$

11 Perform each translation.

(a) Two numbers have a sum of 36. One of the numbers is represented by r. Write an expression for the other number.

(b) Two numbers have a product of 18. One of the numbers is represented by x. Write an expression for the other number.

Answers

11. (a) $36 - r$ (b) $\frac{18}{x}$

Note

In **Example 10,** the decimals are expressed as tenths (0.1 and 1.8) and hundredths (0.05). We chose the least exponent on 10 to eliminate the decimal points, which made all coefficients integers. So, we multiplied by 10^2—that is, 100.

Multiplying by **100** is the same as moving the decimal point **two** places to the right.

$$0.10t + 1.00 - 0.05t = 1.80$$ Equation (*) from **Example 10** with 0s included as placeholders

$$10t + 100 - 5t = 180$$ Multiply by 100. Equation (**) results.

OBJECTIVE 4 **Write expressions for two related unknown quantities.**

Problem-Solving Hint

When we solve applied problems, we must often write *expressions* to relate unknown quantities. We then use these expressions to write the *equation* needed to solve the application. The next example provides preparation for doing this.

EXAMPLE 11 **Translating Phrases into Algebraic Expressions**

Perform each translation.

(a) Two numbers have a sum of 23. If one of the numbers is represented by x, find an expression for the other number.

First, suppose that the sum of two numbers is 23, and one of the numbers is **10**. To find the other number, we would subtract **10** from 23.

$$23 - 10$$ ← This gives 13 as the other number.

Instead of using **10** as one of the numbers, we use x. The other number would be obtained in the same way—by subtracting x from 23.

$$23 - x$$ ← $x - 23$ is *not* correct. Subtraction is *not* commutative.

CHECK We find the sum of the two numbers.

$$x + (23 - x) = 23, \quad \text{as required. ✓}$$

(b) Two numbers have a product of 24. If one of the numbers is represented by x, find an expression for the other number.

Suppose that one of the numbers is **4**. To find the other number, we would divide 24 by **4**.

$$\frac{24}{4}$$ ← This gives 6 as the other number. The product $6 \cdot 4$ is 24.

In the same way, if x is one of the numbers, then we divide 24 by x to find the other number.

$$\frac{24}{x}$$ ← The other number

◀ Work Problem **11** at the Side.

2.3 Exercises

FOR EXTRA HELP — *Go to MyMathLab for worked-out, step-by-step solutions to exercises enclosed in a square and video solutions to ▶ exercises.*

CONCEPT CHECK *Based on the methods of this section, fill in each blank to indicate what we should do first to solve each equation. Do not actually solve.*

1. $7x + 8 = 1$

Use the __________ property of equality to __________ 8 from each side.

2. $7x - 5x + 15 = 8 + x$

On the __________ side, combine __________ terms.

3. $3(2t - 4) = 20 - 2t$

Use the __________ property to clear __________ on the left side of the equation.

4. $\frac{3}{4}z = -15$

Use the __________ property of equality to multiply each side by __________ to obtain z on the left.

5. $\frac{2}{3}x - \frac{1}{6} = \frac{3}{2}x + 1$

Clear __________ by multiplying by __________, the LCD.

6. $0.9x + 0.3x + 3.6 = 6$

Clear __________ by multiplying each side by __________ to obtain $9x$ as the first term on the left.

7. **CONCEPT CHECK** Suppose that when solving three linear equations, we obtain the final results shown in parts (a)–(c). Fill in the blanks in parts (a)–(c), and then match each result with the solution set in choices A–C for the *original* equation.

(a) $6 = 6$ (The original equation is a(n) __________.)

(b) $x = 0$ (The original equation is a(n) __________ equation.)

(c) $-5 = 0$ (The original equation is a(n) __________.)

A. $\{0\}$

B. $\{\text{all real numbers}\}$

C. $\varnothing$

CONCEPT CHECK *Give the letter of the correct choice.*

8. Which linear equation does *not* have all real numbers as solutions?

A. $5x = 4x + x$ **B.** $2(x + 6) = 2x + 12$ **C.** $\frac{1}{2}x = 0.5x$ **D.** $3x = 2x$

9. The expression $12\left(\frac{1}{6}x + \frac{1}{3} - \frac{2}{3}x - \frac{2}{3}\right)$ is equivalent to which of the following?

A. $-6x - 4$ **B.** $2x - 8$ **C.** $-6x + 1$ **D.** $-6x + 4$

10. The expression $100(0.03x - 0.3)$ is equivalent to which of the following?

A. $0.03x - 0.3$ **B.** $3x - 3$ **C.** $3x - 10$ **D.** $3x - 30$

Solve each equation, and check the solution. ***See Examples 1–7.***

11. $3x + 2 = 14$ **12.** $4x + 3 = 27$ **13.** $-5z - 4 = 21$ **14.** $-7w - 4 = 10$

15. $4p - 5 = 2p$

16. $6q - 2 = 3q$

17. $5m + 8 = 7 + 3m$

18. $4r + 2 = r - 6$

19. $10p + 6 = 12p - 4$

20. $-5x + 8 = -3x + 10$

21. $7r - 5r + 2 = 5r - r$

22. $9p - 4p + 6 = 7p - p$

23. $12h - 5 = 11h + 5 - h$

24. $-4x - 1 = -5x + 1 + 3x$

25. $x + 3 = -(2x + 2)$

26. $2x + 1 = -(x + 3)$

27. $3(4x + 2) + 5x = 30 - x$

28. $5(2m + 3) - 4m = 8m + 27$

29. $-2p + 7 = 3 - (5p + 1)$

30. $4x + 9 = 3 - (x - 2)$

31. $6(3w + 5) = 2(10w + 10)$

32. $4(2x - 1) = -6(x + 3)$

33. $-(8x - 2) + 5x - 6 = -4$

34. $-(7x - 5) + 4x - 7 = -2$

35. $24 - 4(7 - 2t) = 4(t - 1)$

36. $8 - 2(2 - x) = 4(x + 1)$

37. $6(3 - x) = -6x + 18$

38. $9(2 - p) = -9p + 18$

39. $3(2x - 4) = 6(x - 2)$

40. $3(6 - 4x) = 2(-6x + 9)$

41. $11x - 5(x + 2) = 6x + 5$

42. $6x - 4(x + 1) = 2x + 4$

43. $6(4x - 1) = 12(2x + 3)$

44. $6(2x + 8) = 4(3x - 6)$

Solve each equation, and check the solution. ***See Examples 8–10.***

45. $\frac{3}{5}t - \frac{1}{10}t = t - \frac{5}{2}$

46. $-\frac{2}{7}r + 2r = \frac{1}{2}r + \frac{17}{2}$

47. $\frac{3}{4}x - \frac{1}{3}x + 5 = \frac{5}{6}x$

48. $\frac{1}{5}x - \frac{2}{3}x - 2 = -\frac{2}{5}x$

49. $\frac{1}{7}(3x + 2) - \frac{1}{5}(x + 4) = 2$

50. $\frac{1}{4}(3x - 1) + \frac{1}{6}(x + 3) = 3$

51. $\frac{1}{9}(x + 18) + \frac{1}{3}(2x + 3) = x + 3$

52. $-\frac{1}{4}(x - 12) + \frac{1}{2}(x + 2) = x + 4$

53. $-\frac{5}{6}q - \left(q - \frac{1}{2}\right) = \frac{1}{4}(q + 1)$

54. $\frac{2}{3}k - \left(k + \frac{1}{4}\right) = \frac{1}{12}(k + 4)$

55. $0.3(30) + 0.15x = 0.2(30) + 0.2x$
(*Hint:* Clear the decimals by multiplying each side of the equation by ______.)

56. $0.2(60) + 0.05x = 0.1(60) + 0.1x$
(*Hint:* Clear the decimals by multiplying each side of the equation by ______.)

57. $0.92x + 0.98(12 - x) = 0.96(12)$

58. $1.00x + 0.05(12 - x) = 0.10(63)$

59. $0.02(5000) + 0.03x = 0.025(5000 + x)$

60. $0.06(10{,}000) + 0.08x = 0.072(10{,}000 + x)$

Solve each equation, and check the solution. ***See Examples 1–10.***

61. $-3(5z + 24) + 2 = 2(3 - 2z) - 4$

62. $-2(2s - 4) - 8 = -3(4s + 4) - 1$

63. $-(6k - 5) - (-5k + 8) = -3$

64. $-(4x + 2) - (-3x - 5) = 3$

65. $8(t - 3) + 4t = 6(2t + 1) - 10$

66. $9(v + 1) - 3v = 2(3v + 1) - 8$

67. $4(x+3)=2(2x+8)-4$

68. $4(x+8)=2(2x+6)+20$

69. $\frac{1}{3}(x+3)+\frac{1}{6}(x-6)=x+3$

70. $\frac{1}{2}(x+2)+\frac{3}{4}(x+4)=x+5$

71. $0.3(x+15)+0.4(x+25)=25$

72. $0.1(x+80)+0.2x=14$

Perform each translation. ***See Example 11.***

73. Two numbers have a sum of 12. One of the numbers is q. What expression represents the other number?

74. Two numbers have a sum of 26. One of the numbers is r. What expression represents the other number?

75. The product of two numbers is 9. One of the numbers is z. What expression represents the other number?

76. The product of two numbers is 13. One of the numbers is k. What expression represents the other number?

77. A football player gained x yards rushing. On the next down, he gained 29 yd. What expression represents the number of yards he gained altogether?

78. A football player gained y yards on a punt return. On the next return, he gained 25 yd. What expression represents the number of yards he gained altogether?

79. Monica is m years old. What expression represents her age 12 yr from now? 2 yr ago?

80. Chandler is b years old. What expression represents his age 3 yr ago? 5 yr from now?

81. Tom has r quarters. Express the value of the quarters in cents.

82. Jean has y dimes. Express the value of the dimes in cents.

83. A bank teller has t dollars, all in \$5 bills. What expression represents the number of \$5 bills the teller has?

84. A store clerk has v dollars, all in \$10 bills. What expression represents the number of \$10 bills the clerk has?

85. A plane ticket costs x dollars for an adult and y dollars for a child. Find an expression that represents the total cost for 3 adults and 2 children.

86. A concert ticket costs p dollars for an adult and q dollars for a child. Find an expression that represents the total cost for 4 adults and 6 children.

Study Skills

USING STUDY CARDS REVISITED

Two additional types of study cards follow. Use challenging problem and practice quiz cards to do the following.

- To help you understand and learn the material
- To quickly review when you have a few minutes
- To review before a quiz or test

Challenging Problem Cards

When you are doing your homework and encounter a "difficult" problem, write the procedure to work the problem on the front of a card in words. Include special notes or tips (like what *not* to do). On the back of the card, work an example. Show all steps, and label what you are doing.

Front of Card

When solving a linear equation, be careful when clearing parentheses if there is a minus sign in front. p. 135

$6x - (x + 3)$

The minus sign acts like −1, so change the sign of every term inside the parentheses.

Back of Card

Solve.

	$6x - (x + 3) = 7$	
	$6x - 1(x + 3) = 7$	
Change both signs.	$6x - x - 3 = 7$	*Distributive property*
	$5x - 3 = 7$	*Combine like terms.*
	$5x - 3 + 3 = 7 + 3$	*Add 3.*
	$5x = 10$	*Combine like terms.*
	$\frac{5x}{5} = \frac{10}{5}$	*Divide by 5.*
Solution set: {2}	$x = 2$	

Practice Quiz Cards

Write a problem with direction words (like *solve, simplify*) on the front of a card, and work the problem on the back. Make one for each type of problem you learn.

Front of Card

Solve $4(3x - 4) = 2(6x - 9) + 2.$ p. 137

Back of Card

	$4(3x - 4) = 2(6x - 9) + 2$	
	$12x - 16 = 12x - 18 + 2$	*Distributive property*
	$12x - 16 = 12x - 16$	*Combine like terms.*
	$12x - 16 + 16 = 12x - 16 + 16$	*Add 16.*
	$12x = 12x$	*Combine like terms.*
	$12x - 12x = 12x - 12x$	*Subtract 12x.*
When both sides of an equation are the same, it is called an identity.	$0 = 0$	*True*

Any real number will work, so the solution set is {all real numbers} (not just {0}).

Now Try This

Make a challenging problem card and a practice quiz card for material you are learning now.

Summary Exercises *Applying Methods for Solving Linear Equations*

CONCEPT CHECK *Decide whether each of the following is an* equation *or an* expression. *If it is an equation, solve it. If it is an expression, simplify it.*

1. $x + 2 = -3$
2. $4p - 6 + 3p - 8$
3. $-(m - 1) - (3 + 2m)$
4. $6q - 9 = 12 + 3q$
5. $5x - 9 = 3(x - 3)$
6. $\frac{1}{2}(x + 10) - \frac{2}{3}x$

Solve each equation, and check the solution.

7. $-6z = -14$
8. $2m + 8 = 16$
9. $12.5x = -63.75$
10. $-x = -12$
11. $\frac{4}{5}x = -20$
12. $7m - 5m = -12$
13. $-x = 6$
14. $\frac{x}{-2} = 8$
15. $4x + 2(3 - 2x) = 6$
16. $x - 16.2 = 7.5$
17. $7m - (2m - 9) = 39$
18. $2 - (m + 4) = 3m - 2$
19. $-3(m - 4) + 2(5 + 2m) = 29$
20. $-0.3x + 2.1(x - 4) = -6.6$
21. $0.08x + 0.06(x + 9) = 1.24$
22. $3(m + 5) - 1 + 2m = 5(m + 2)$
23. $-2t + 5t - 9 = 3(t - 4) - 5$
24. $2.3x + 13.7 = 1.3x + 2.9$
25. $0.2(50) + 0.8r = 0.4(50 + r)$
26. $r + 9 + 7r = 4(3 + 2r) - 3$
27. $2(3 + 7x) - (1 + 15x) = 2$
28. $0.6(100 - x) + 0.4x = 0.5(92)$
29. $\frac{1}{4}x - 4 = \frac{3}{2}x + \frac{3}{4}x$
30. $\frac{3}{4}(z - 2) - \frac{1}{3}(5 - 2z) = -2$

2.4 An Introduction to Applications of Linear Equations

OBJECTIVES

1. Learn the six steps for solving applied problems.
2. Solve problems involving unknown numbers.
3. Solve problems involving sums of quantities.
4. Solve problems involving consecutive integers.
5. Solve problems involving complementary and supplementary angles.

OBJECTIVE 1 Learn the six steps for solving applied problems.

Solving an Applied Problem

Step 1 **Read** the problem, several times if necessary. *What information is given? What is to be found?*

Step 2 **Assign a variable** to represent the unknown value. Use a sketch, diagram, or table, as needed. Express any other unknown values in terms of the variable.

Step 3 **Write an equation** using the variable expression(s).

Step 4 **Solve** the equation.

Step 5 **State the answer.** Label it appropriately. *Does the answer seem reasonable?*

Step 6 **Check** the answer in the words of the *original* problem.

OBJECTIVE 2 Solve problems involving unknown numbers.

EXAMPLE 1 Finding the Value of an Unknown Number

The product of 4, and a number decreased by 7, is 100. What is the number?

Step 1 **Read** the problem carefully. We are asked to find a number.

Step 2 **Assign a variable.** Let $x =$ the number.

Step 3 **Write an equation.**

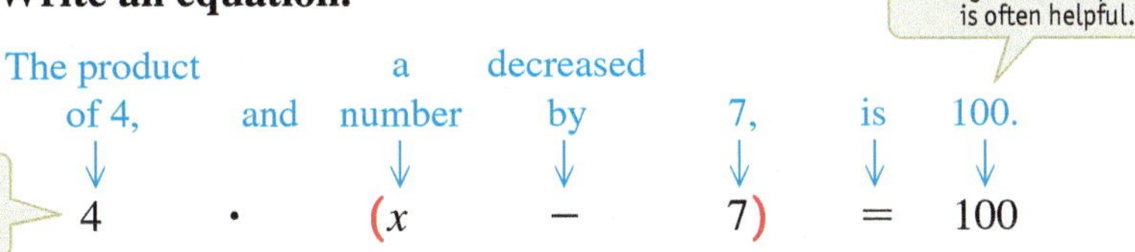

Because of the commas in the given problem, writing the equation as $4x - 7 = 100$ is *incorrect*. The equation $4x - 7 = 100$ corresponds to *"The product of* 4 *and a number, decreased by* 7, *is* 100."

Step 4 **Solve** the equation.

$$4(x-7) = 100 \quad \text{Equation from Step 3}$$

$$4x - 28 = 100 \quad \text{Distributive property}$$

$$4x - 28 + 28 = 100 + 28 \quad \text{Add 28.}$$

$$4x = 128 \quad \text{Combine like terms.}$$

$$\frac{4x}{4} = \frac{128}{4} \quad \text{Divide by 4.}$$

$$x = 32$$

Step 5 **State the answer.** The number is 32.

Step 6 **Check.** The number 32 decreased by 7 is 25. The product of 4 and 25 is 100, as required. The answer, 32, is correct.

Work Problem 1 at the Side. ▶

1 Solve each problem.

GS (a) If 5 is added to a number, the result is 7 less than 3 times the number. Find the number.

Step 1
We must find a ________.

Step 2
Let $x =$ the ________.

Step 3

If 5 is added to a number,	the result is	7 less than 3 times the number.
↓	↓	↓
________	=	________

Complete Steps 4–6 to solve the problem. Give the answer.

(b) The product of 9, and 4 more than twice a number, is 15 less than the number. Find the number.

Answers

1. (a) number; number; $x + 5$ (or $5 + x$); $3x - 7$; The number is 6.
(b) -3

2 Solve each problem.

GS **(a)** The 150-member Iowa legislature includes 12 fewer Democrats than Republicans. (No other parties are represented.) How many Democrats and Republicans are there in the legislature? (Data from www.legis.iowa.gov)

Step 1
We must find the number of Democrats and ________.

Step 2
Let x = the number of Republicans.

Then ________ = the number of ________.

Complete Steps 3–6 to solve the problem. Give the equation and the answer.

(b) At the 2014 Winter Olympics, Germany won 10 more medals than China. The two countries won a total of 28 medals. How many medals did each country win? (Data from *The World Almanac and Book of Facts*.)

Answers

2. **(a)** Republicans; $x - 12$; Democrats; $x + (x - 12) = 150$; Republicans: 81; Democrats: 69
(b) Germany: 19; China: 9

OBJECTIVE 3 Solve problems involving sums of quantities.

Problem-Solving Hint

In general, to solve problems involving sums of quantities, choose a variable to represent one of the unknowns. ***Then represent the other quantity in terms of the same variable.***

EXAMPLE 2 Finding Numbers of Olympic Medals

At the 2014 Winter Olympics in Sochi, Russia, the United States won 13 more medals than France. The two countries won a total of 43 medals. How many medals did each country win? (Data from *The World Almanac and Book of Facts.*)

Step 1 **Read** the problem. We are given the total number of medals and asked to find the number each country won.

Step 2 **Assign a variable.**

Let x = the number of medals France won.

Then $x + 13$ = the number of medals the United States won.

Step 3 **Write an equation.**

The total	is	the number of medals France won	plus	the number of medals the U.S. won.
↓	↓	↓	↓	↓
43	=	x	+	$(x + 13)$

Step 4 **Solve** the equation.

$$43 = 2x + 13 \qquad \text{Combine like terms.}$$

$$43 - 13 = 2x + 13 - 13 \qquad \text{Subtract 13.}$$

$$30 = 2x \qquad \text{Combine like terms.}$$

$$\frac{30}{2} = \frac{2x}{2} \qquad \text{Divide by 2.}$$

$$15 = x, \quad \text{or} \quad x = 15 \leftarrow \text{Medals France won}$$

Step 5 **State the answer.** The variable x represents the number of medals France won, so France won 15 medals.

$$x + 13$$
$$= 15 + 13$$
$$= 28 \leftarrow \text{Medals the United States won}$$

Step 6 **Check.** The United States won 28 medals and France won 15, so the total number of medals was

$$28 + 15 = 43.$$

The United States won 13 more medals than France, so the difference was

$$28 - 15 = 13.$$

All conditions of the problem are satisfied.

◀ Work Problem 2 at the Side.

Problem-Solving Hint

The problem in **Example 2** could also be solved by letting x represent the number of medals the United States won. Then $x - 13$ would represent the number of medals France won. The equation would be different.

$$43 = x + (x - 13) \quad \text{Alternative equation for Example 2}$$

The solution of this equation is 28, which is the number of U.S. medals. The number of French medals would be $28 - 13 = 15$. ***The answers are the same,*** whichever approach is used, even though the equation and its solution are different.

EXAMPLE 3 Analyzing a Gasoline/Oil Mixture

A lawn trimmer uses a mixture of gasoline and oil. The mixture contains 16 oz of gasoline for each 1 oz of oil. If the tank holds 68 oz of the mixture, how many ounces of oil and how many ounces of gasoline does it require when it is full?

Step 1 **Read** the problem. We must find how many ounces of oil and gasoline are needed to fill the tank.

Step 2 **Assign a variable.**

Let x = the number of ounces of oil required.

Then $16x$ = the number of ounces of gasoline required.

A diagram like the following is sometimes helpful.

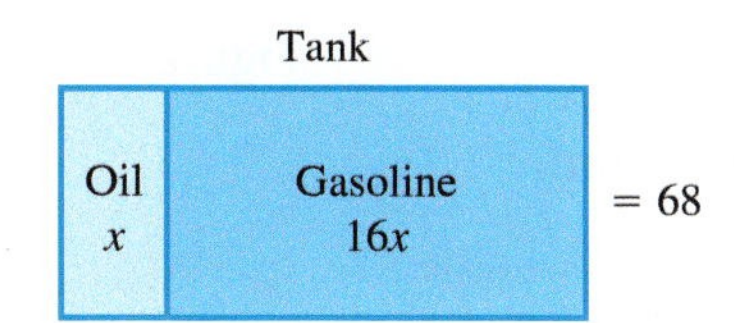

Step 3 **Write an equation.**

Amount of gasoline	plus	amount of oil	is	total amount in tank.
↓	↓	↓	↓	↓
$16x$	$+$	x	$=$	68

Step 4 **Solve.**

$$17x = 68 \quad \text{Combine like terms.}$$

$$\frac{17x}{17} = \frac{68}{17} \quad \text{Divide by 17.}$$

$$x = 4$$

Step 5 **State the answer.** When full, the lawn trimmer requires 4 oz of oil, and

$$16x = 16(4)$$
$$= 64 \text{ oz of gasoline.}$$

Step 6 **Check.** Because $4 + 64 = 68$, and 64 is 16 times 4, the answer is correct.

Work Problem 3 at the Side. ▶

3 Solve each problem.

(a) At a club meeting, each member brought two nonmembers. If a total of 27 people attended, how many were members and how many were nonmembers?

Meeting

Members x	Nonmembers $2x$	$= 27$

Step 1
We must find the number of ________ and nonmembers.

Step 2
Let x = the number of members.

Then ________ = the number of ________.

Step 3
Write an equation.

________ $= 27$

Complete Steps 4–6 to solve the problem. Give the answer.

(b) A fly spray mixture requires 7 oz of water for each 1 oz of essential oil. To fill a quart bottle (32 oz), how many ounces of water and how many ounces of essential oil are required?

Answers

3. **(a)** members; $2x$; nonmembers; $x + 2x$; members: 9; nonmembers: 18
(b) water: 28 oz; essential oil: 4 oz

4 Solve each problem.

GS (a) Over a 6-hr period, a basketball player spent twice as much time lifting weights as practicing free throws and 2 hr longer watching game films than practicing free throws. How many hours did he spend on each task?

Steps 1 and 2
The unknown found in both pairs of comparisons is the time spent ________.

Let $x =$ the time spent practicing free throws.

Then ____ = the time spent lifting weights,

and ____ = the time spent watching game films.

Step 3
Write an equation.

__________________ = 6

Complete Steps 4–6 to solve the problem. Give the answer.

(b) A piece of pipe is 50 in. long. It is cut into three pieces. The longest piece is 10 in. longer than the middle-sized piece, and the shortest piece measures 5 in. less than the middle-sized piece. Find the lengths of the three pieces.

Answers

4. (a) practicing free throws; $2x$; $x + 2$; $x + 2x + (x + 2)$; practicing free throws: 1 hr; lifting weights: 2 hr; watching game films: 3 hr
(b) longest: 25 in.; middle: 15 in.; shortest: 10 in.

Problem-Solving Hint

Sometimes we must find three unknown quantities. ***When the three unknown quantities are compared in pairs, let the variable represent the unknown found in both pairs.***

EXAMPLE 4 Dividing a Board into Pieces

A project calls for three pieces of wood. The longest piece must be twice the length of the middle-sized piece. The shortest piece must be 10 in. shorter than the middle-sized piece. If a board 70 in. long is to be used, how long must each piece be?

Step 1 **Read** the problem. Three lengths must be found.

Step 2 **Assign a variable.** The middle-sized piece appears in both pairs of comparisons, so let x represent the length, in inches, of the middle-sized piece.

Let $x =$ the length of the middle-sized piece.

Then $2x =$ the length of the longest piece,

and $x - 10 =$ the length of the shortest piece.

A sketch is helpful here. See **Figure 2.**

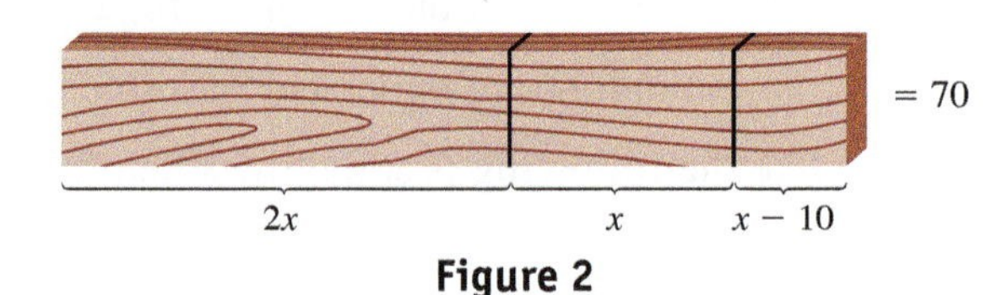

Figure 2

Step 3 **Write an equation.**

Longest	plus	middle-sized	plus	shortest	is	total length.
↓	↓	↓	↓	↓	↓	↓
$2x$	$+$	x	$+$	$(x - 10)$	$=$	70

Step 4 **Solve.**

$4x - 10 = 70$ Combine like terms.

$4x - 10 + 10 = 70 + 10$ Add 10.

$4x = 80$ Combine like terms.

$\frac{4x}{4} = \frac{80}{4}$ Divide by 4.

$x = 20$

Step 5 **State the answer.** The middle-sized piece is 20 in. long, the longest piece is

$$2(20) = 40 \text{ in. long,}$$

and the shortest piece is

$$20 - 10 = 10 \text{ in. long.}$$

Step 6 **Check.** The sum of the lengths is 70 in. All conditions of the problem are satisfied.

◀ Work Problem 4 at the Side.

OBJECTIVE 4 Solve problems involving consecutive integers. Two integers that differ by 1 are **consecutive integers.** For example, 3 and 4, 6 and 7, and -2 and -1 are pairs of consecutive integers. See **Figure 3.**

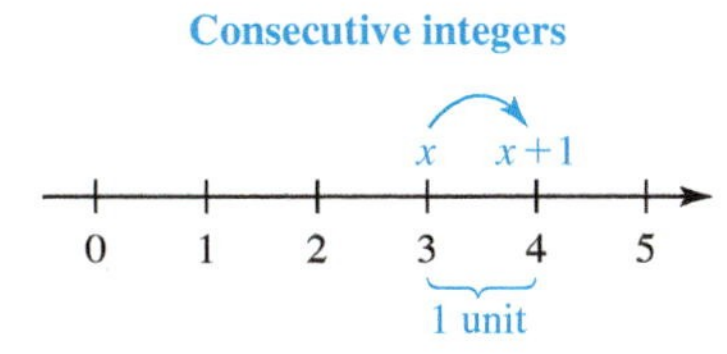

Figure 3

In general, if x represents an integer, then x + 1 represents the next greater consecutive integer.

EXAMPLE 5 Finding Consecutive Integers

Two pages that face each other in this book have 301 as the sum of their page numbers. What are the page numbers?

Step 1 **Read** the problem. Because the two pages face each other, they must have page numbers that are consecutive integers.

Step 2 **Assign a variable.**

Let $x =$ the lesser page number.

Then $x + 1 =$ the greater page number.

Figure 4 illustrates this situation.

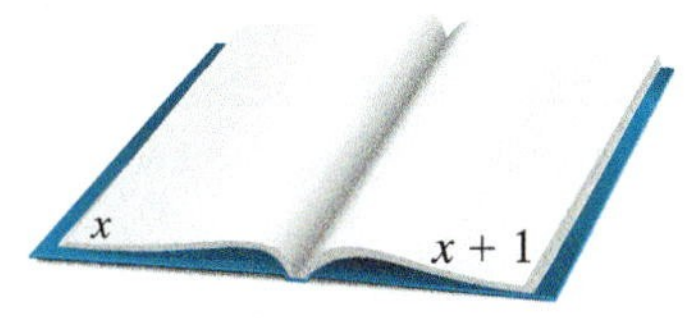

Figure 4

Step 3 **Write an equation.**

Lesser page number	plus	greater page number	is	the sum.
↓	↓	↓	↓	↓
x	$+$	$(x + 1)$	$=$	301

Step 4 **Solve.**

$2x + 1 = 301$ Combine like terms.

$2x + 1 - 1 = 301 - 1$ Subtract 1.

$2x = 300$ Combine like terms.

$\frac{2x}{2} = \frac{300}{2}$ Divide by 2.

$x = 150$

Step 5 **State the answer.** The lesser page number is 150, and the greater is

$150 + 1 = 151.$

(Your text is opened to these two pages.)

Step 6 **Check.** The sum of 150 and 151 is 301. The answer is correct.

Work Problem 5 at the Side. ▶

---

5 Solve the problem.

Two pages that face each other in this book have a sum of 569. What are the page numbers?

Answer

5. 284, 285

Consecutive *even* integers, such as 2 and 4, and 8 and 10, differ by 2. Similarly, **consecutive *odd* integers,** such as 1 and 3, and 9 and 11, also differ by 2. See **Figure 5.**

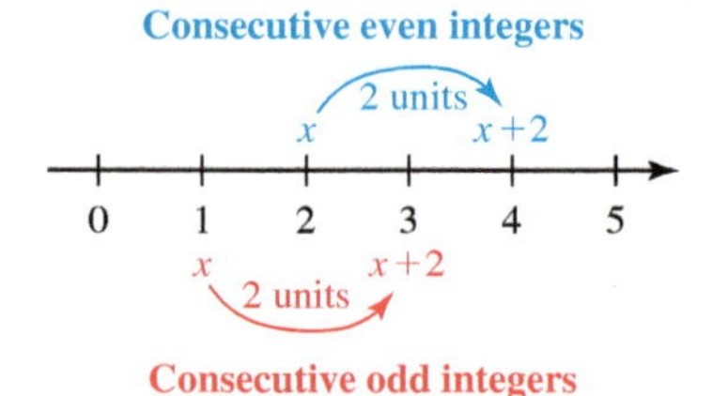

Figure 5

In general, if x represents an even (or odd) integer, then x + 2 represents the next greater consecutive even (or odd) integer, respectively.

Problem-Solving Hint

If $x =$ the lesser (or least) integer in a consecutive integer problem, then the following apply.

- For two consecutive integers, use $\quad x, \quad x + 1.$
- For two consecutive *even* integers, use $\quad x, \quad x + 2.$
- For two consecutive *odd* integers, use $\quad x, \quad x + 2.$

EXAMPLE 6 **Finding Consecutive Odd Integers**

If the lesser of two consecutive odd integers is doubled, the result is 7 more than the greater of the two integers. Find the two integers.

Step 1 **Read** the problem. We must find two consecutive odd integers.

Step 2 **Assign a variable.**

Let $\quad x =$ the lesser consecutive odd integer.

Then $\quad x + 2 =$ the greater consecutive odd integer.

Step 3 **Write an equation.**

If the lesser is doubled,	the result is	7	more than	the greater.
↓	↓	↓	↓	↓
$2x$	$=$	7	$+$	$(x + 2)$

Step 4 **Solve.**

$$2x = 9 + x \qquad \text{Combine like terms.}$$
$$2x - x = 9 + x - x \qquad \text{Subtract } x.$$
$$x = 9 \qquad \text{Combine like terms.}$$

Step 5 **State the answer.** The lesser integer is 9. The greater is

$$9 + 2 = 11.$$

Step 6 **Check.** When 9 is doubled, we obtain 18, which is 7 more than the greater odd integer, 11. The answer is correct.

◀ **Work Problem 6 at the Side.**

6 Solve each problem.

GS **(a)** The sum of two consecutive even integers is 254. Find the integers.

Step 1
We must find two consecutive ______.

Step 2
Let $x =$ the lesser of the two ______ even integers.

Then ______ $=$ the greater of the two consecutive even integers.

Step 3

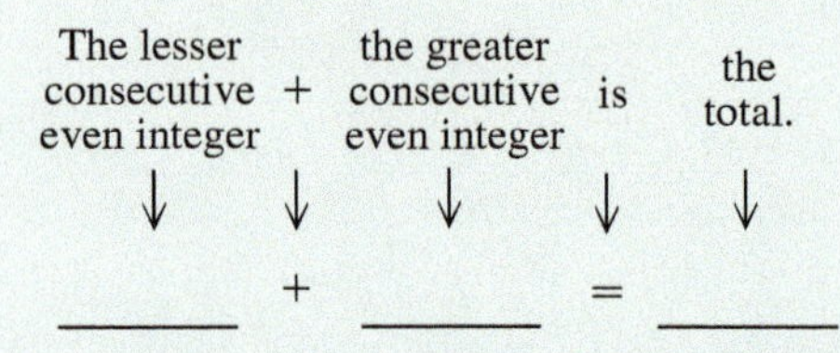

Complete Steps 4–6 to solve the problem. Give the answer.

(b) Find two consecutive odd integers such that the sum of twice the lesser and three times the greater is 191.

Answers

6. (a) even integers; consecutive; $x + 2$; x; $x + 2$; 254; 126 and 128
(b) 37, 39

OBJECTIVE 5 Solve problems involving complementary and supplementary angles. An angle can be measured using a unit called the degree (°), which is $\frac{1}{360}$ of a complete rotation. See **Figure 6.**

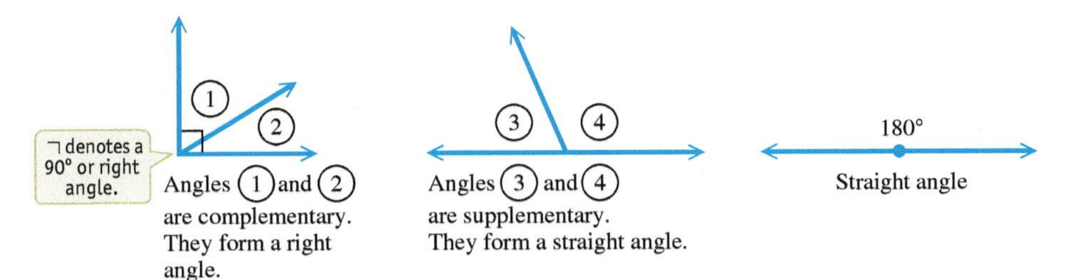

Figure 6

- Two angles whose sum is 90° are **complementary,** or *complements* of each other.
- An angle that measures 90° is a **right angle.**
- Two angles whose sum is 180° are **supplementary,** or *supplements* of each other.
- An angle that measures 180° is a **straight angle.**

Problem-Solving Hint

Let x represent the degree measure of an angle.

90 − x represents the degree measure of its complement.

180 − x represents the degree measure of its supplement.

EXAMPLE 7 Finding the Measure of an Angle

Find the measure of an angle whose complement is five times its measure.

Step 1 **Read** the problem. We must find the measure of an angle.

Step 2 **Assign a variable.**

Let $x =$ the degree measure of the angle.

Then $90 - x =$ the degree measure of its complement.

Step 3 **Write an equation.**

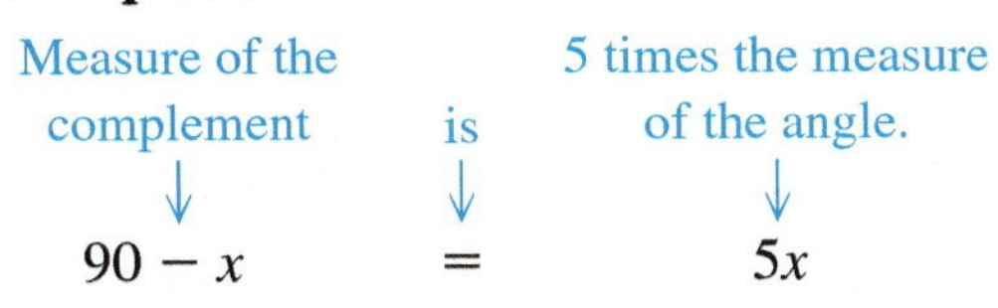

$$90 - x = 5x$$

Step 4 **Solve.**

$$90 - x + x = 5x + x \quad \text{Add } x.$$

$$90 = 6x \quad \text{Combine like terms.}$$

$$\frac{90}{6} = \frac{6x}{6} \quad \text{Divide by 6.}$$

$$15 = x, \quad \text{or} \quad x = 15$$

Step 5 **State the answer.** The measure of the angle is 15°.

Step 6 **Check.** If the angle measures 15°, then its complement measures 90° − 15° = 75°, which is equal to five times 15°, as required.

Work Problems 7 and 8 at the Side. ▶

---

7 Fill in the blank below each figure. Then solve.

(a) Find the complement of an angle that measures 26°.

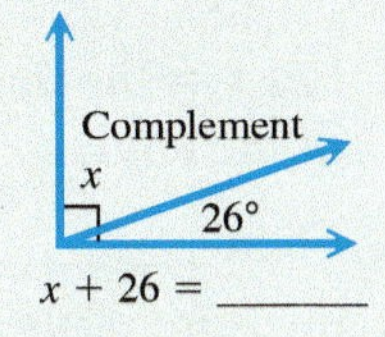

$x + 26 =$ ______

(b) Find the supplement of an angle that measures 92°.

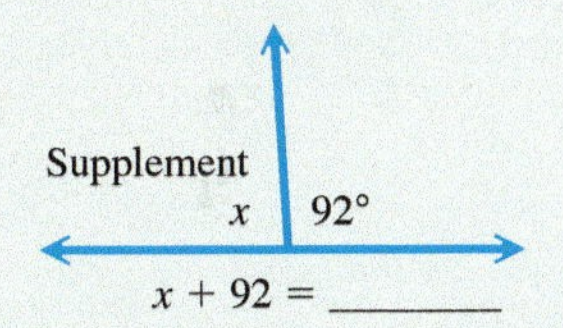

$x + 92 =$ ______

8 Solve each problem. Give the equation and the answer.

Let $x =$ the degree measure of the angle.

(a) Find the measure of an angle whose complement is eight times its measure.

(b) Find the measure of an angle whose supplement is twice its measure.

Answers

7. (a) 90; 64° (b) 180; 88°

8. (a) $90 - x = 8x$; 10°
(b) $180 - x = 2x$; 60°

9 Solve each problem.

GS (a) Find the measure of an angle such that twice its complement is 30° less than its supplement.

Let x = the degree measure of the angle.

Then ________ = the degree measure of its complement,

and ________ = the degree measure of its supplement.

Complete the solution. Give the equation and the answer.

(b) Find the measure of an angle whose supplement is 46° less than three times its complement.

EXAMPLE 8 Finding the Measure of an Angle

Find the measure of an angle whose supplement is 10° more than twice its complement.

Step 1 **Read** the problem. We must find the measure of an angle, given information about its complement and its supplement.

Step 2 **Assign a variable.**

Let x = the degree measure of the angle.

Then $90 - x$ = the degree measure of its complement,

and $180 - x$ = the degree measure of its supplement.

We can visualize this information using a sketch. See **Figure 7.**

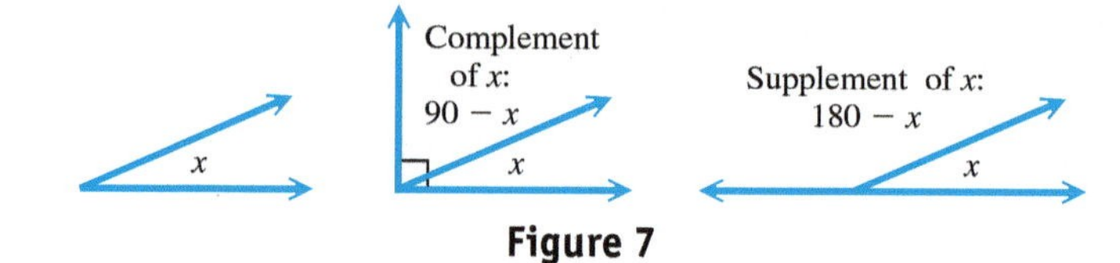

Figure 7

Step 3 **Write an equation.**

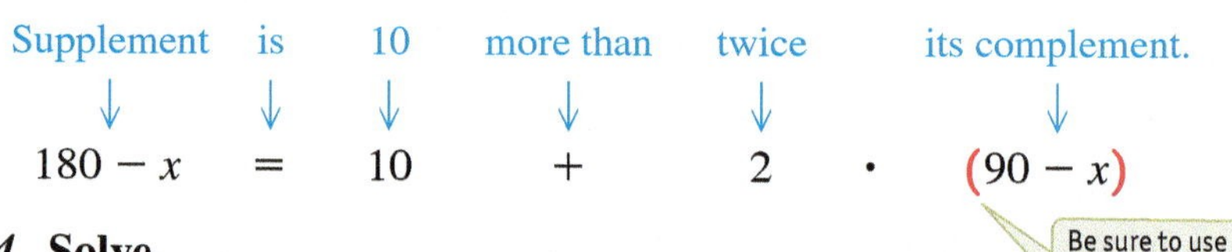

Step 4 **Solve.**

$$180 - x = 10 + 2(90 - x)$$

$$180 - x = 10 + 180 - 2x \quad \text{Distributive property}$$

$$180 - x = 190 - 2x \quad \text{Combine like terms.}$$

$$180 - x + 2x = 190 - 2x + 2x \quad \text{Add } 2x.$$

$$180 + x = 190 \quad \text{Combine like terms.}$$

$$180 + x - 180 = 190 - 180 \quad \text{Subtract 180.}$$

$$x = 10$$

Step 5 **State the answer.** The measure of the angle is 10°.

Step 6 **Check.** The complement of 10° is 80° and the supplement of 10° is 170°. Also, 170° is equal to 10° more than twice 80° (that is, $170 = 10 + 2(80)$ is true). Therefore, the answer is correct.

◀ Work Problem **9** at the Side.

Answers

9. **(a)** $90 - x$; $180 - x$;
$2(90 - x) = (180 - x) - 30$;
30°
(b) 22°

2.4 Exercises

FOR EXTRA HELP — Go to MyMathLab for worked-out, step-by-step solutions to exercises enclosed in a square and video solutions to exercises.

1. Give the six steps introduced in this section for solving application problems.

Step 1: ____________ *Step 4:* ____________

Step 2: ____________ *Step 5:* ____________

Step 3: ____________ *Step 6:* ____________

2. **CONCEPT CHECK** List some of the words that translate as "=" when writing an equation to solve an applied problem.

CONCEPT CHECK *Which choice would* **not** *be a reasonable answer? Justify your response.*

3. A problem requires finding the number of cars on a dealer's lot.

A. 0 **B.** 45 **C.** 1 **D.** $6\frac{1}{2}$

4. A problem requires finding the number of hours a light bulb is on during a day.

A. 0 **B.** 4.5 **C.** 13 **D.** 25

5. A problem requires finding the distance traveled in miles.

A. -10 **B.** 1.8 **C.** $10\frac{1}{2}$ **D.** 50

6. A problem requires finding the time in minutes.

A. 0 **B.** 10.5 **C.** -5 **D.** 90

CONCEPT CHECK *Fill in each blank with the correct response.*

7. Consecutive integers differ by ________, such as 15 and ________, and -8 and ________.

If x represents an integer, then ________ represents the next greater integer.

8. Consecutive odd integers are ________ integers that differ by ________, such as ________ and 13.

Consecutive even integers are ________ integers that differ by ________, such as 12 and ________.

9. Two angles whose measures sum to 90° are ________ angles. Two angles whose measures sum to 180° are ________ angles.

10. A right angle has measure ________. A straight angle has measure ________.

CONCEPT CHECK *Answer each question.*

11. Is there an angle that is equal to its supplement? Is there an angle that is equal to its complement? If the answer is yes to either question, give the measure of the angle.

12. If x represents an even integer, how can we express the next *smaller* consecutive integer in terms of x? The next *smaller* even integer?

Solve each problem. In each case, give the equation using x as the variable, and give the answer. ***See Example 1.***

13. The product of 8, and a number increased by 6, is 104. What is the number?

14. The product of 5, and 3 more than twice a number, is 85. What is the number?

15. If 2 is added to five times a number, the result is equal to 5 more than four times the number. Find the number.

16. If four times a number is added to 8, the result is three times the number, added to 5. Find the number.

17. Two less than three times a number is equal to 14 more than five times the number. What is the number?

18. Nine more than five times a number is equal to 3 less than seven times the number. What is the number?

19. If 2 is subtracted from a number and this difference is tripled, the result is 6 more than the number. Find the number.

20. If 3 is added to a number and this sum is doubled, the result is 2 more than the number. Find the number.

21. The sum of three times a number and 7 more than the number is the same as the difference of -11 and twice the number. What is the number?

22. If 4 is added to twice a number and this sum is multiplied by 2, the result is the same as if the number is multiplied by 3 and 4 is added to the product. What is the number?

GS *Complete the six problem-solving steps to solve each problem.* ***See Example 2.***

23. Pennsylvania and New York were among the states with the most remaining drive-in movie screens in 2015. Pennsylvania had 1 less screen than New York, and there were 55 screens total in the two states. How many drive-in movie screens remained in each state? (Data from www.Drive-Ins.com)

Step 1 **Read.** What are we asked to find?

____

____

Step 2 **Assign a variable.** Let $x =$ the number of screens in New York.

Then $x - 1 =$ ______________________

____

Step 3 **Write an equation.**

____ + ____ = 55

Step 4 **Solve** the equation.

$x =$ ____

Step 5 **State the answer.** New York had ____ screens. Pennsylvania had ____ $- 1 =$ ____ screens.

Step 6 **Check.** The number of screens in Pennsylvania was ____ less than the number of ____________. The total number of screens was 28 + ____ = ____.

24. The total number of television viewers for the final episodes of *M*A*S*H* and *Cheers* was about 92 million, with 8 million more people watching the *M*A*S*H* episode than the *Cheers* episode. How many people watched each episode? (Data from Nielsen Media Research.)

Step 1 **Read.** What are we asked to find?

____

____

Step 2 **Assign a variable.** Let $x =$ the number of people who watched the *Cheers* episode.

Then $x + 8 =$ ______________________

____

Step 3 **Write an equation.**

____ + ____ = ____

Step 4 **Solve** the equation.

$x =$ ____

Step 5 **State the answer.** ____ million people watched the *Cheers* episode, while ____ + 8 = ____ million watched *M*A*S*H*.

Step 6 **Check.** The number of people who watched the *M*A*S*H** episode was ____ million more than the number who watched the *Cheers* episode. The total number of viewers in millions was 42 + ____ = ____.

Solve each problem. **See Example 2.**

25. The total number of Democrats and Republicans in the U.S. House of Representatives during the 114th session (2015–2017) was 435. There were 59 more Republicans than Democrats. How many members of each party were there? (Data from *The World Almanac and Book of Facts.*)

26. During the 114th session, the U.S. Senate had a total of 98 Democrats and Republicans. There were 10 fewer Democrats than Republicans. How many Democrats and Republicans were there in the Senate? (Data from *The World Almanac and Book of Facts.*)

27. Taylor Swift and Kenny Chesney had the two top-grossing North American concert tours in 2015, together generating $315.8 million in ticket sales. If Kenny Chesney took in $83 million less than Taylor Swift, how much did each tour generate? (Data from Pollstar.)

28. In the United States in 2014, Honda Accord sales were 41 thousand less than Toyota Camry sales, and 817 thousand of these two cars were sold. How many of each make of car were sold? (Data from *The World Almanac and Book of Facts.*)

29. In the 2015–2016 NBA regular season, the Golden State Warriors won 1 more than eight times as many games as they lost. The Warriors played 82 games. How many wins and losses did the team have? (Data from www.nba.com)

30. In the 2015 regular MLB season, the Kansas City Royals won 39 fewer than twice as many games as they lost. The Royals played 162 regular season games. How many wins and losses did the team have? (Data from www.mlb.com)

31. A one-cup serving of orange juice contains 3 mg less than four times the amount of vitamin C as a one-cup serving of pineapple juice. Servings of the two juices contain a total of 122 mg of vitamin C. How many milligrams of vitamin C are in a serving of each type of juice? (Data from U.S. Agriculture Department.)

32. A one-cup serving of pineapple juice has 9 more than three times as many calories as a one-cup serving of tomato juice. Servings of the two juices contain a total of 173 calories. How many calories are in a serving of each type of juice? (Data from U.S. Agriculture Department.)

Solve each problem. **See Example 3.**

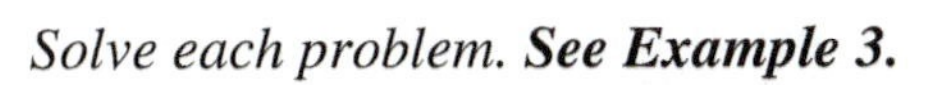

33. A recipe for whole-grain bread calls for 1 oz of rye flour for every 4 oz of whole-wheat flour. How many ounces of each kind of flour should be used to make a loaf of bread weighing 32 oz?

34. U.S. five-cent coins are made from a combination of nickel and copper. For every 1 lb of nickel, 3 lb of copper are used. How many pounds of copper would be needed to make 560 lb of five-cent coins? (Data from The United States Mint.)

35. A medication contains 9 mg of active ingredients for every 1 mg of inert ingredients. How much of each kind of ingredient would be contained in a single 250-mg caplet?

36. A recipe for salad dressing uses 2 oz of olive oil for each 1 oz of red wine vinegar. If 42 oz of salad dressing are needed, how many ounces of each ingredient should be used?

37. The value of a "Mint State-63" (uncirculated) 1950 Jefferson nickel minted at Denver is $\frac{4}{3}$ the value of a similar condition 1944 nickel minted at Philadelphia. Together, the value of the two coins is \$28.00. What is the value of each coin? (Data from Yeoman, R., *A Guide Book of United States Coins*.)

38. In one day, a store sold $\frac{8}{5}$ as many DVDs as Blu-ray discs. The total number of DVDs and Blu-ray discs sold that day was 273. How many DVDs were sold?

39. The world's largest taco contained approximately 1 kg of onion for every 6.6 kg of grilled steak. The total weight of these two ingredients was 617.6 kg. To the nearest tenth of a kilogram, how many kilograms of each ingredient were used? (Data from *Guinness World Records*.)

40. As of 2015, the combined population of China and India was estimated at 2.7 billion. If there were about 93% as many people living in India as China, what was the population of each country, to the nearest tenth of a billion? (Data from United Nations Population Division.)

Solve each problem. ***See Example 4.***

41. A party-length submarine sandwich is 59 in. long. It is to be cut into three pieces so that the middle piece is 5 in. longer than the shortest piece and the shortest piece is 9 in. shorter than the longest piece. How long should the three pieces be?

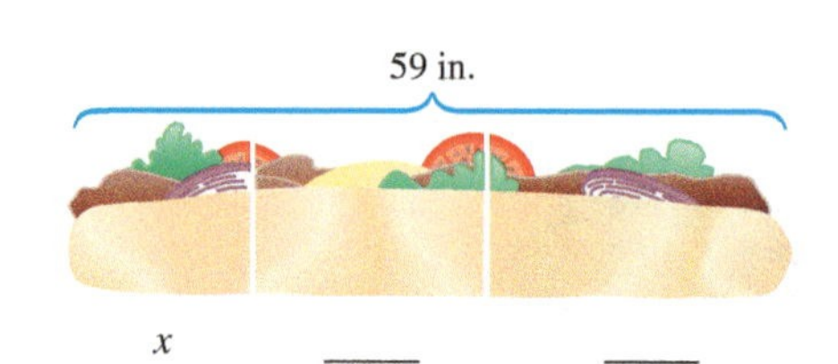

42. A three-foot-long deli sandwich must be split into three pieces so that the middle piece is twice as long as the shortest piece and the shortest piece is 8 in. shorter than the longest piece. How long should the three pieces be? (*Hint:* How many inches are in 3 ft?)

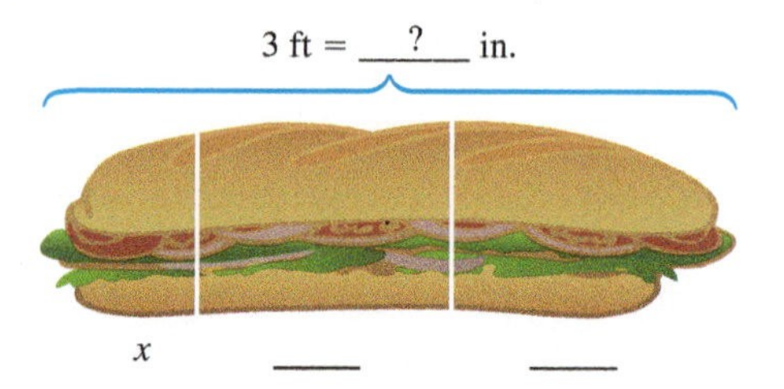

43. An office manager booked 55 airline tickets. He booked 7 more tickets on American Airlines than United Airlines. On Southwest Airlines, he booked 4 more than twice as many tickets as on United. How many tickets did he book on each airline?

44. A mathematics textbook editor spent 7.5 hr making telephone calls, writing e-mails, and attending meetings. She spent twice as much time attending meetings as making telephone calls and 0.5 hr longer writing e-mails than making telephone calls. How many hours did she spend on each task?

45. The United States earned 28 medals at the 2014 Winter Olympics. The number of silver medals earned was 5 less than the number of bronze medals. The number of gold medals was 2 more than the number of silver medals. How many of each kind of medal did the United States earn? (Data from www.espn.go.com)

46. Russia earned 33 medals at the 2014 Winter Olympics. The number of gold medals earned was 4 more than the number of bronze medals. The number of silver medals earned was 2 more than the number of bronze medals. How many of each kind of medal did Russia earn? (Data from www.espn.go.com)

47. Venus is 31.2 million mi farther from the sun than Mercury, while Earth is 57 million mi farther from the sun than Mercury. If the total of the distances from these three planets to the sun is 196.2 million mi, how far away from the sun is Mercury? (All distances given here are mean (*average*) distances.) (Data from *The New York Times Almanac.*)

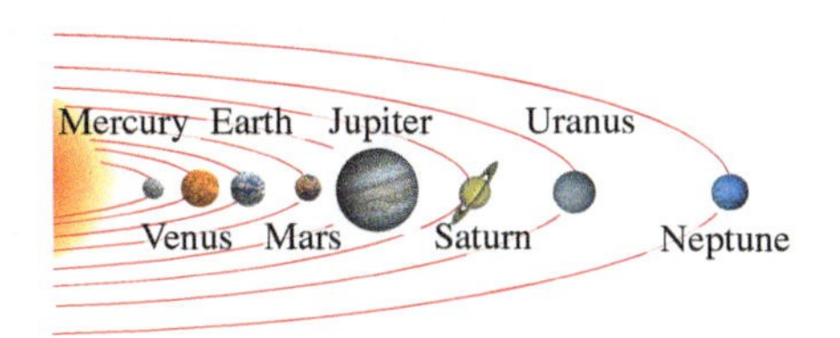

48. Saturn, Jupiter, and Uranus have a total of 156 known satellites (moons). Jupiter has 5 more satellites than Saturn, and Uranus has 35 fewer satellites than Saturn. How many known satellites does Uranus have? (Data from http://solarsystem.nasa.gov)

49. The sum of the measures of the angles of any triangle is 180°. In triangle ABC, angles A and B have the same measure, while the measure of angle C is 60° greater than each of A and B. What are the measures of the three angles?

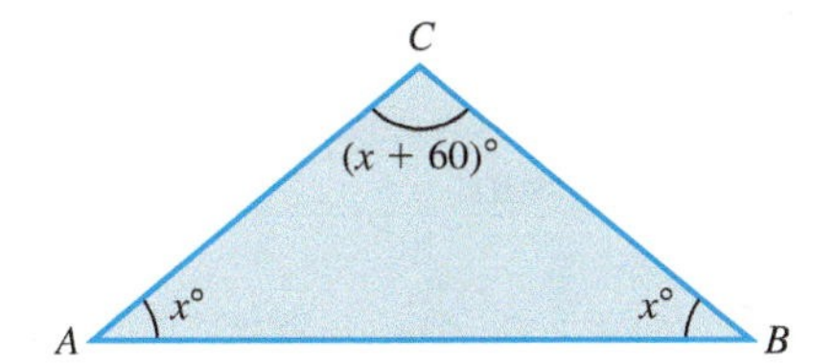

50. The sum of the measures of the angles of any triangle is 180°. In triangle ABC, the measure of angle A is 141° more than the measure of angle B. The measure of angle B is the same as the measure of angle C. Find the measure of each angle.

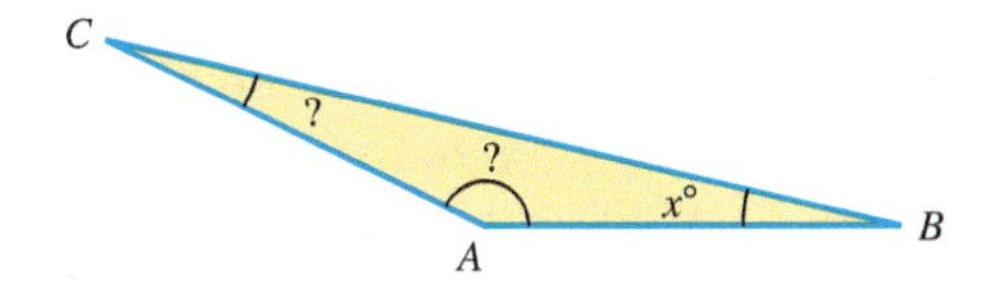

Solve each problem. **See Examples 5 and 6.**

51. The numbers on two consecutively numbered gym lockers have a sum of 137. What are the locker numbers?

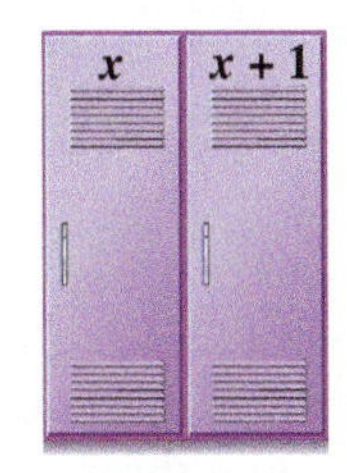

52. The sum of two consecutive check numbers is 357. Find the numbers.

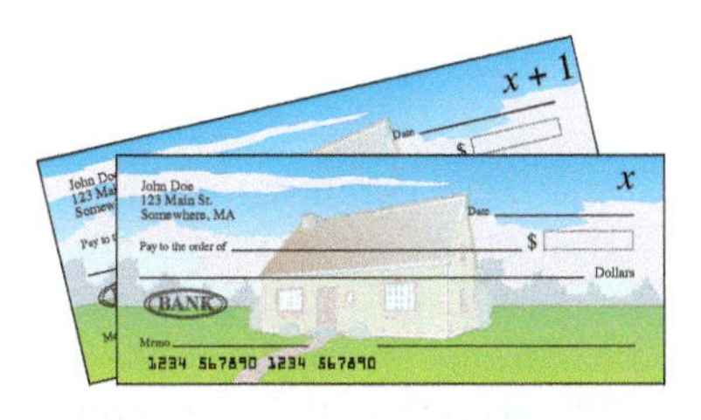

53. Two pages that are back-to-back in this book have 203 as the sum of their page numbers. What are the page numbers?

54. Two hotel rooms have room numbers that are consecutive integers. The sum of the numbers is 515. What are the two room numbers?

55. Find two consecutive even integers such that the lesser added to three times the greater gives a sum of 46.

56. Find two consecutive even integers such that six times the lesser added to the greater gives a sum of 86.

57. Find two consecutive odd integers such that 59 more than the lesser is four times the greater.

58. Find two consecutive odd integers such that twice the greater is 17 more than the lesser.

59. When the lesser of two consecutive integers is added to three times the greater, the result is 43. Find the integers.

60. If five times the lesser of two consecutive integers is added to three times the greater, the result is 59. Find the integers.

61. If the sum of three consecutive even integers is 60, what is the least of the three even integers? (*Hint:* If x and $x + 2$ represent the first two consecutive even integers, how would we represent the largest consecutive even integer?)

62. If the sum of three consecutive odd integers is 69, what is the largest of the three odd integers? (*Hint:* If x and $x + 2$ represent the first two consecutive odd integers, how would we represent the largest consecutive odd integer?)

Solve each problem. ***See Examples 7 and 8.***

63. Find the measure of an angle whose complement is four times its measure. (*Hint:* If x represents the measure of the unknown angle, how would we represent its complement?)

64. Find the measure of an angle whose complement is five times its measure.

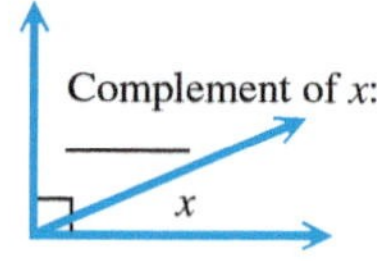

65. Find the measure of an angle whose supplement is eight times its measure. (*Hint:* If x represents the measure of the unknown angle, how would we represent its supplement?)

66. Find the measure of an angle whose supplement is three times its measure.

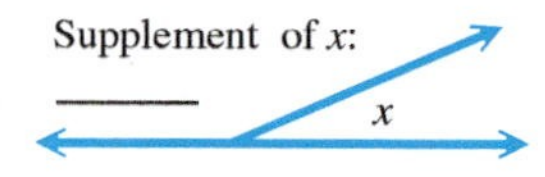

67. Find the measure of an angle whose supplement measures 39° more than twice its complement.

68. Find the measure of an angle whose supplement measures 38° less than three times its complement.

69. Find the measure of an angle such that the difference between the measures of its supplement and three times its complement is 10°.

70. Find the measure of an angle such that the sum of the measures of its complement and its supplement is 160°.

2.5 Formulas and Additional Applications from Geometry

OBJECTIVES

1. Solve a formula for one variable, given the values of the other variables.
2. Use a formula to solve an applied problem.
3. Solve problems involving vertical angles and straight angles.
4. Solve a formula for a specified variable.

A **formula** is an equation in which variables are used to describe a relationship. For example, formulas exist for finding perimeters and areas of geometric figures, calculating money earned on bank savings, and converting among measurements.

$$P = 4s, \quad \mathcal{A}^* = \pi r^2, \quad I = prt, \quad F = \frac{9}{5}C + 32 \qquad \text{Formulas}$$

Many of the formulas used in this text are given at the back of the text.

OBJECTIVE 1 Solve a formula for one variable, given the values of the other variables. The **area** of a plane (two-dimensional) geometric figure is a measure of the surface covered by the figure. Area is measured in square units.

EXAMPLE 1 Using Formulas to Evaluate Variables

Find the value of the remaining variable in each formula.

(a) $\mathcal{A} = LW$; $\quad \mathcal{A} = 64, L = 10$

This formula gives the area of a rectangle. See **Figure 8.**

$$\mathcal{A} = LW \qquad \text{Solve for } W.$$

$$64 = 10W \qquad \text{Let } \mathcal{A} = 64 \text{ and } L = 10.$$

$$\frac{64}{10} = \frac{10W}{10} \qquad \text{Divide by 10.}$$

$$6.4 = W$$

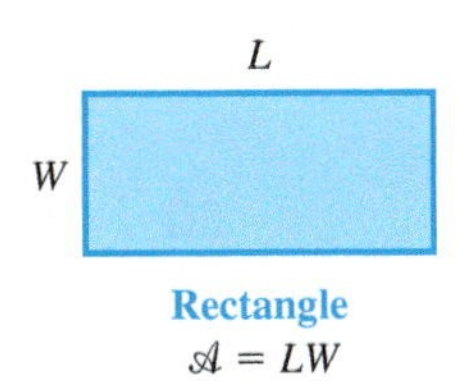

Rectangle
$\mathcal{A} = LW$

Figure 8

The width is 6.4. Since $10(6.4) = 64$, the given area, the answer checks.

(b) $\mathcal{A} = \frac{1}{2}h(b + B)$; $\quad \mathcal{A} = 210, B = 27, h = 10$

This formula gives the area of a trapezoid. See **Figure 9.**

$$\mathcal{A} = \frac{1}{2}h(b + B)$$

$$210 = \frac{1}{2}(10)(b + 27) \qquad \text{Solve for } b. \quad \text{Let } \mathcal{A} = 210,\ h = 10,\ B = 27.$$

$$210 = 5(b + 27) \qquad \text{Multiply } \tfrac{1}{2}(10).$$

$$210 = 5b + 135 \qquad \text{Distributive property}$$

$$210 - 135 = 5b + 135 - 135 \qquad \text{Subtract 135.}$$

$$75 = 5b \qquad \text{Combine like terms.}$$

$$\frac{75}{5} = \frac{5b}{5} \qquad \text{Divide by 5.}$$

$$15 = b$$

b
h
B
Trapezoid
$\mathcal{A} = \frac{1}{2}h(b + B)$

Figure 9

The length of the shorter parallel side, b, is 15. Since $\frac{1}{2}(10)(15 + 27) = 210$, the given area, the answer checks.

Work Problem 1 at the Side. ▶

1 Find the value of the remaining variable in each formula.

(a) $\mathcal{A} = bh$
(area of a parallelogram);
$\mathcal{A} = 96, h = 8$

GS **(b)** $I = prt$ (simple interest);
$I = 246, r = 0.06$ (that is, 6%), $t = 2$

$$I = prt$$

$$\_\_\_\_ = p(\_\_\_\_)(\_\_\_\_)$$

$$246 = \_\_\_\_p$$

$$\frac{246}{\_\_\_\_} = \frac{0.12p}{0.12}$$

$$\_\_\_\_ = p$$

(c) $P = 2L + 2W$
(perimeter of a rectangle);
$P = 126, W = 25$

Answers

1. **(a)** $b = 12$
(b) 246; 0.06; 2; 0.12; 0.12; 2050
(c) $L = 38$

*In this text, we use $\mathcal{A}$ to denote area.

OBJECTIVE 2 Use a formula to solve an applied problem. The **perimeter** of a plane (two-dimensional) geometric figure is the measure of the outer boundary of the figure. For a polygon (such as a rectangle, square, or triangle), it is the sum of the lengths of the sides.

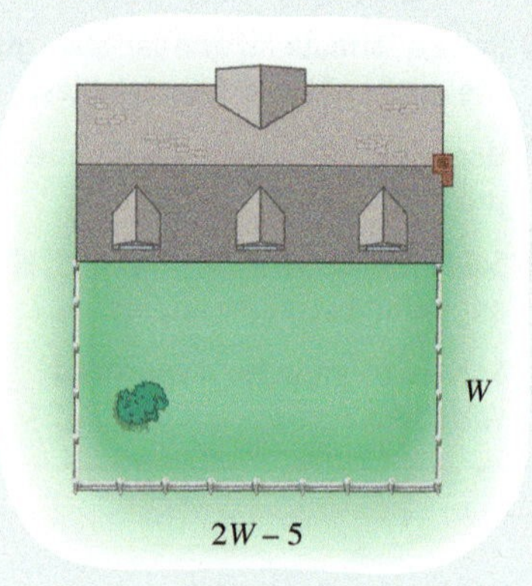

Figure 10

EXAMPLE 2 Finding the Dimensions of a Rectangular Yard

A backyard is in the shape of a rectangle. The length is 5 m less than twice the width, and the perimeter is 80 m. Find the dimensions of the yard.

Step 1 **Read** the problem. We must find the dimensions of the yard.

Step 2 **Assign a variable.** Let $W =$ the width of the lot, in meters. The length is 5 m less than twice the width, so the length is given by $L = 2W - 5$. See **Figure 10.**

Step 3 **Write an equation.** Use the formula for the perimeter of a rectangle.

$$P = 2L + 2W$$

$$\text{Perimeter} = 2 \cdot \text{Length} + 2 \cdot \text{Width}$$

$$80 = 2(2W - 5) + 2W$$ Substitute 80 for perimeter P and $2W - 5$ for length L.

Step 4 **Solve.**

$80 = 4W - 10 + 2W$ Distributive property

$80 = 6W - 10$ Combine like terms.

$80 + 10 = 6W - 10 + 10$ Add 10.

$90 = 6W$ Combine like terms.

$\frac{90}{6} = \frac{6W}{6}$ Divide by 6.

$15 = W$

Step 5 **State the answer.** The width is 15 m. The length is $2(15) - 5 = 25$ m.

Step 6 **Check.** If the width of the yard is 15 m and the length is 25 m, the perimeter is $2(25) + 2(15) = 80$ m, as required.

◀ **Work Problem 2 at the Side.**

2 Solve the problem.
A farmer has 800 m of fencing material to enclose a rectangular field. The width of the field is 175 m. Find the length of the field.

EXAMPLE 3 Finding the Dimensions of a Triangle

The longest side of a triangle is 3 ft longer than the shortest side. The medium side is 1 ft longer than the shortest side. If the perimeter of the triangle is 16 ft, what are the lengths of the three sides?

Step 1 **Read** the problem. We are given the perimeter of a triangle and must find the lengths of the three sides.

Step 2 **Assign a variable.** The shortest side is mentioned in each pair of comparisons in the problem.

Let $s =$ the length of the shortest side, in feet,

$s + 1 =$ the length of the medium side, in feet, and

$s + 3 =$ the length of the longest side, in feet.

It is a good idea to draw a sketch. See **Figure 11.**

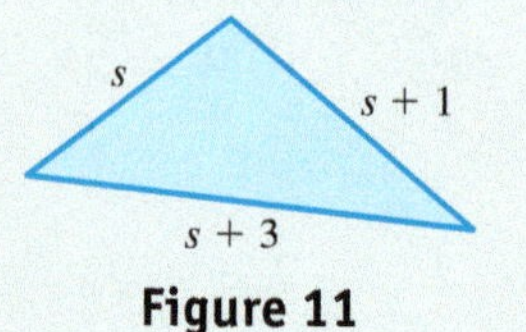

Figure 11

Continued on Next Page

Answer

2. 225 m

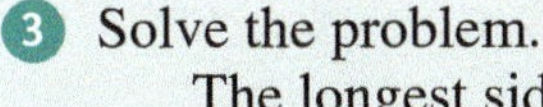

Step 3 **Write an equation.** Use the formula for the perimeter of a triangle.

$$P = a + b + c \quad \text{Perimeter of a triangle}$$

$$16 = s + (s + 1) + (s + 3) \quad \text{Substitute.}$$

Step 4 **Solve.** $16 = 3s + 4$ Combine like terms.

$$16 - 4 = 3s + 4 - 4 \quad \text{Subtract 4.}$$

$$12 = 3s \quad \text{Combine like terms.}$$

$$\frac{12}{3} = \frac{3s}{3} \quad \text{Divide by 3.}$$

$$4 = s$$

Step 5 **State the answer.** The length of the shortest side, s, is **4** ft.

$$s + 1 = 4 + 1 = 5 \text{ ft} \quad \text{Length of the medium side}$$

$$s + 3 = 4 + 3 = 7 \text{ ft} \quad \text{Length of the longest side}$$

Step 6 **Check.** The medium side, 5 ft, is 1 ft longer than the shortest side, and the longest side, 7 ft, is 3 ft longer than the shortest side. The perimeter is

$$4 + 5 + 7 = 16 \text{ ft}, \quad \text{as required.}$$

Work Problem 3 at the Side. ▶

3 Solve the problem.

The longest side of a triangle is 1 in. longer than the medium side. The medium side is 5 in. longer than the shortest side. If the perimeter is 32 in., what are the lengths of the three sides?

EXAMPLE 4 Finding the Height of a Triangular Sail

The area of a triangular sail of a sailboat is 126 ft^2. (Recall that ft^2 means "square feet.") The base of the sail is 12 ft. Find the height of the sail.

Figure 12

Step 1 **Read** the problem. We must find the height of the triangular sail.

Step 2 **Assign a variable.** Let $h =$ the height of the sail, in feet. See **Figure 12.**

Step 3 **Write an equation.** Use the formula for the area of a triangle.

$$\mathcal{A} = \frac{1}{2}bh \quad \mathcal{A} \text{ is the area, } b \text{ is the base, and } h \text{ is the height.}$$

$$126 = \frac{1}{2}(12)h \quad \text{Substitute } \mathcal{A} = 126,\ b = 12.$$

Step 4 **Solve.** $126 = 6h$ Multiply.

$$\frac{126}{6} = \frac{6h}{6} \quad \text{Divide by 6.}$$

$$21 = h$$

Step 5 **State the answer.** The height of the sail is **21** ft.

Step 6 **Check** to see that the values $\mathcal{A} = 126$, $b = 12$, and $h = 21$ satisfy the formula for the area of a triangle.

$$126 = \frac{1}{2}(12)(21) \text{ is true.}$$

Work Problem 4 at the Side. ▶

4 Solve the problem.

The area of a triangle is 120 m^2. The height is 24 m. Find the length of the base of the triangle.

Answers

3. 7 in.; 12 in.; 13 in.

4. 10 m

5 Find the measure of each marked angle.

(a)

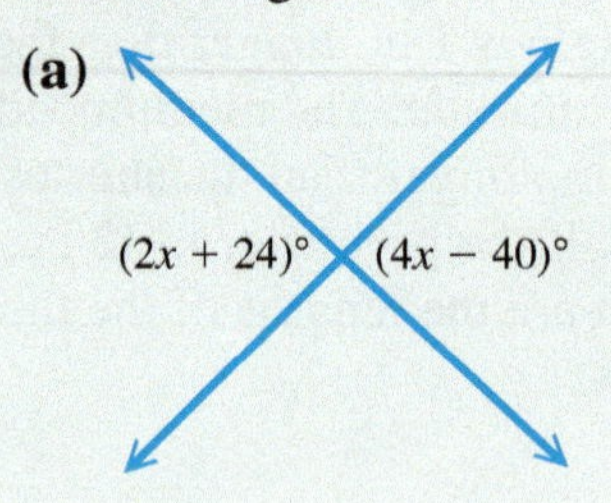

(b)

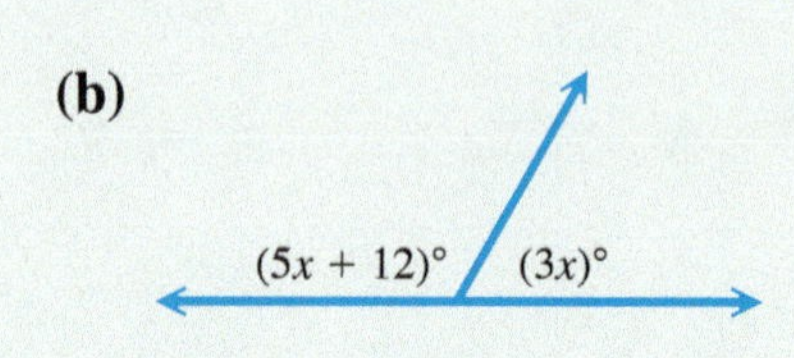

GS **(c)**

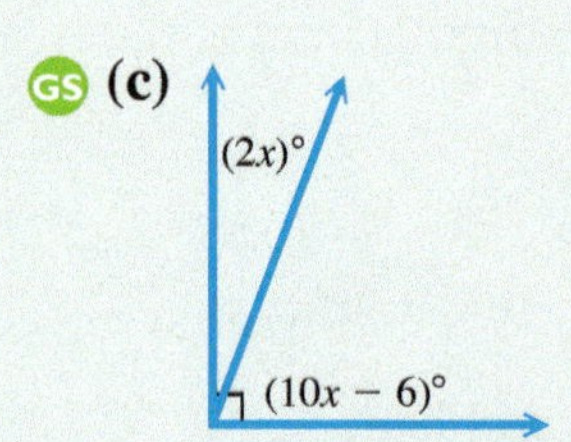

Because of the ⌐ symbol, the marked angles are ________ angles and have a sum of ____. The equation to use is

________ $= 90$.

Solve this equation to find that $x =$ ____. Then substitute to find the measure of each angle.

$2x = 2($____$)$

$=$ ____

$10x - 6 = 10($____$) - 6$

$=$ ____

The two angles measure ________.

Answers

5. **(a)** Both measure 88°. **(b)** 117° and 63° **(c)** complementary; 90°; $2x + (10x - 6)$; 8; 8; 16; 8; 74; 16° and 74°

OBJECTIVE 3 Solve problems involving vertical angles and straight angles. **Figure 13** shows two intersecting lines forming angles that are numbered ①, ②, ③, and ④. Angles ① and ③ lie "opposite" each other. They are **vertical angles.** Another pair of vertical angles is ② and ④. ***Vertical angles have equal measures.***

Consider angles ① and ②. When their measures are added, we obtain 180°, the measure of a **straight angle.** There are three other angle pairs that form straight angles:

② and ③, ③ and ④, and ① and ④.

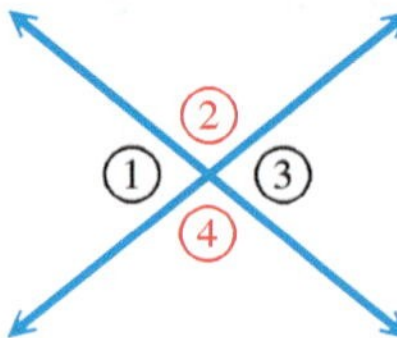

Figure 13

EXAMPLE 5 Finding Angle Measures

Refer to **Figure 14.**

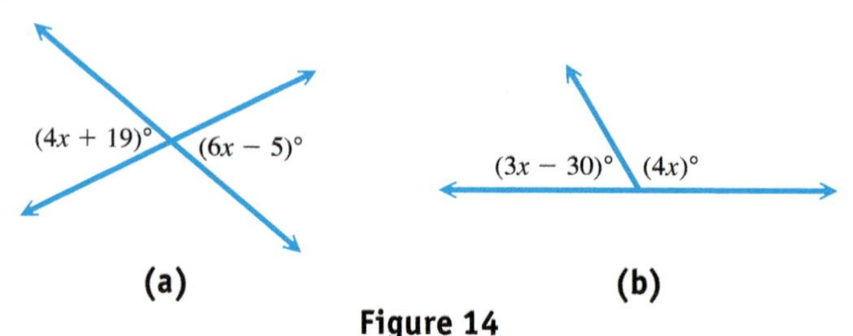

Figure 14

(a) Find the measure of each marked angle in **Figure 14(a).**

The marked angles are vertical angles, so they have equal measures.

$4x + 19 = 6x - 5$	Set $4x + 19$ equal to $6x - 5$.
$19 = 2x - 5$	Subtract $4x$.
$24 = 2x$	Add 5.
$12 = x$	Divide by 2.

This is *not* the angle measure.

Replace x with 12 in the expression for the measure of each angle.

$4x + 19$		$6x - 5$	
$= 4(12) + 19$	Let $x = 12$.	$= 6(12) - 5$	Let $x = 12$.
$= 48 + 19$	Multiply.	$= 72 - 5$	Multiply.
$= 67$	Add.	$= 67$	Subtract.

Each angle measures 67°.

(b) Find the measure of each marked angle in **Figure 14(b).**

The measures of the marked angles must add to 180° because together they form a straight angle. (They are also *supplements* of each other.)

$(3x - 30) + 4x = 180$	Supplementary angles sum to 180°.
$7x - 30 = 180$	Combine like terms.
$7x = 210$	Add 30.
$x = 30$	Divide by 7.

Don't stop here!

Replace x with 30 in the measure of each marked angle.

$3x - 30 = 3(30) - 30 = 90 - 30 = 60$

$4x = 4(30) = 120$

The measures of the angles add to 180°, as required.

The two angle measures are 60° and 120°.

◀ Work Problem **5** at the Side.

OBJECTIVE 4 Solve a formula for a specified variable. Sometimes we want to rewrite a formula in terms of a *different* variable in the formula. For example, consider $\mathcal{A} = LW$, the formula for the area of a rectangle.

How can we rewrite $\mathcal{A} = LW$ in terms of W?

The process whereby we do this involves **solving for a specified variable,** or **solving a literal equation.**

To solve a formula for a specified variable, we use the *same* steps that we used to solve an equation with just one variable. Consider the parallel reasoning to solve each of the following for x.

$3x + 4 = 13$		$ax + b = c$	
$3x + 4 - 4 = 13 - 4$	Subtract 4.	$ax + b - b = c - b$	Subtract b.
$3x = 9$		$ax = c - b$	
$\frac{3x}{3} = \frac{9}{3}$	Divide by 3.	$\frac{ax}{a} = \frac{c - b}{a}$	Divide by a.
$x = 3$		$x = \frac{c - b}{a}$	

When we solve a formula for a specified variable, we treat the specified variable as if it were the ONLY variable in the equation, and treat the other variables as if they were constants (numbers).

EXAMPLE 6 Solving for a Specified Variable

Solve $\mathcal{A} = LW$ for W.

Think of undoing what has been done to W. Since W is multiplied by L, undo the multiplication by dividing each side of $\mathcal{A} = LW$ by L.

$$\mathcal{A} = LW$$ (Our goal is to isolate W.)

$$\frac{\mathcal{A}}{L} = \frac{LW}{L} \quad \text{Divide by } L.$$

$$\frac{\mathcal{A}}{L} = W, \quad \text{or} \quad W = \frac{\mathcal{A}}{L} \qquad \frac{LW}{L} = \frac{L}{L} \cdot W = 1 \cdot W = W$$

Work Problem 6 at the Side. ▶

EXAMPLE 7 Solving for a Specified Variable

Solve $P = 2L + 2W$ for L.

$$P = 2L + 2W$$ (Our goal is to isolate L.)

$$P - 2W = 2L + 2W - 2W \quad \text{Subtract } 2W.$$

$$P - 2W = 2L \quad \text{Combine like terms.}$$

$$\frac{P - 2W}{2} = \frac{2L}{2} \quad \text{Divide by 2.}$$

$$\frac{P - 2W}{2} = L, \quad \text{or} \quad L = \frac{P - 2W}{2} \qquad \frac{2L}{2} = \frac{2}{2} \cdot L = 1 \cdot L = L$$

Work Problem 7 at the Side. ▶

---

6 Solve each formula for the specified variable.

(a) $W = Fd$ for F

GS **(b)** $I = prt$ for t

Our goal is to isolate ____.

$$I = prt$$

$$\frac{I}{\_\_\_\_} = \frac{prt}{\_\_\_\_}$$

$$\_\_\_\_ = t$$

7 Solve for the specified variable.

GS **(a)** $Ax + By = C$ for A

Our goal is to isolate ____. To do this, subtract ____ from each side. Then ____ each side by ____.

Show these steps and write the formula solved for A.

(b) $Ax + By = C$ for B

Answers

6. (a) $F = \frac{W}{d}$ (b) t; pr; pr; $\frac{I}{pr}$

7. (a) A; By; divide; x; $A = \frac{C - By}{x}$

(b) $B = \frac{C - Ax}{y}$

8 Solve each formula for the specified variable.

(a) $x = u + zs$ for z

(b) $A = p(1 + rt)$ for t

9 Solve each equation for y.

(a) $5x + y = 3$

(b) $x - 2y = 8$

Answers

8. **(a)** $z = \frac{x - u}{s}$ **(b)** $t = \frac{A - p}{pr}$

9. **(a)** $y = -5x + 3$ **(b)** $y = \frac{1}{2}x - 4$

EXAMPLE 8 Solving for a Specified Variable

Solve $\mathscr{A} = \frac{1}{2}h(b + B)$ for B.

Our goal is to isolate B.

$$\mathscr{A} = \frac{1}{2}h(b + B)$$

$$\mathscr{A} = \frac{1}{2}hb + \frac{1}{2}hB \quad \text{Clear the parentheses using the distributive property.}$$

$$2 \cdot \mathscr{A} = 2\left(\frac{1}{2}hb + \frac{1}{2}hB\right) \quad \text{Multiply each side by 2 to clear the fractions.}$$

$$2 \cdot \mathscr{A} = 2 \cdot \frac{1}{2}hb + 2 \cdot \frac{1}{2}hB \quad \text{Distributive property}$$

$$2\mathscr{A} = hb + hB \quad \text{Multiply; } 2 \cdot \tfrac{1}{2} = \tfrac{2}{2} = 1$$

$$2\mathscr{A} - hb = hb + hB - hb \quad \text{Subtract } hb.$$

$$2\mathscr{A} - hb = hB \quad \text{Combine like terms.}$$

$$\frac{2\mathscr{A} - hb}{h} = \frac{hB}{h} \quad \text{Divide by } h.$$

$$\frac{2\mathscr{A} - hb}{h} = B, \quad \text{or} \quad B = \frac{2\mathscr{A} - hb}{h}$$

◀ Work Problem **8** at the Side.

EXAMPLE 9 Solving for a Specified Variable

Solve each equation for y.

(a)

$$2x - y = 7$$

Our goal is to isolate y.

$$2x - y - 2x = 7 - 2x \quad \text{Subtract } 2x.$$

$$-y = 7 - 2x \quad \text{Combine like terms.}$$

$$-1(-y) = -1(7 - 2x) \quad \text{Multiply by } -1.$$

$$y = -7 + 2x \quad \text{Multiply; distributive property}$$

$$y = 2x - 7 \quad -a + b = b - a$$

We could have added y and subtracted 7 from each side of the equation to isolate y on the right, giving $2x - 7 = y$, a different form of the same result.

(b)

$$-3x + 2y = 6$$

$$-3x + 2y + 3x = 6 + 3x \quad \text{Add } 3x.$$

$$2y = 3x + 6 \quad \text{Combine like terms; commutative property}$$

$$\frac{2y}{2} = \frac{3x + 6}{2} \quad \text{Divide by 2.}$$

Be careful here.

$$y = \frac{3x}{2} + \frac{6}{2} \quad \frac{a + b}{c} = \frac{a}{c} + \frac{b}{c}$$

$\frac{3x}{2} = \frac{3}{2} \cdot \frac{x}{1} = \frac{3}{2}x$

$$y = \frac{3}{2}x + 3 \quad \text{Simplify.}$$

Although we could have given the answer as $y = \frac{3x + 6}{2}$, we simplified further in preparation for later work.

◀ Work Problem **9** at the Side.

2.5 Exercises

FOR EXTRA HELP Go to MyMathLab for worked-out, step-by-step solutions to exercises enclosed in a square and video solutions to ▶ exercises.

CONCEPT CHECK *Give a one-sentence definition of each term.*

1. Perimeter of a plane geometric figure
2. Area of a plane geometric figure

CONCEPT CHECK *Decide whether* perimeter *or* area *would be used to solve a problem concerning the measure of the quantity.*

3. Sod for a lawn
4. Carpeting for a bedroom
5. Baseboards for a living room
6. Fencing for a yard
7. Fertilizer for a garden
8. Tile for a bathroom
9. Determining the cost of planting rye grass in a lawn for the winter
10. Determining the cost of replacing a linoleum floor with a wood floor

Find the value of the remaining variable in each formula. Use 3.14 *as an approximation for* π *(pi).* ***See Example 1.***

11. $P = 2L + 2W$ (perimeter of a rectangle); $L = 8$, $W = 5$

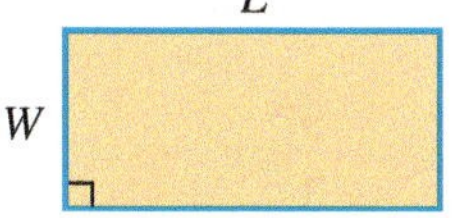

12. $P = 2L + 2W$; $L = 6$, $W = 4$

13. ▶ $\mathcal{A} = \frac{1}{2}bh$ (area of a triangle); $b = 8$, $h = 16$

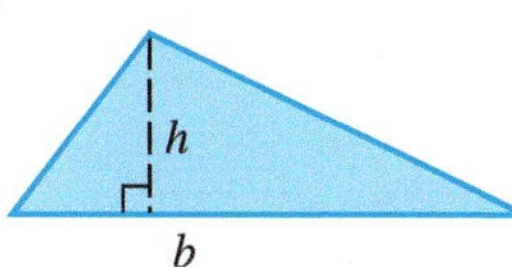

14. $\mathcal{A} = \frac{1}{2}bh$; $b = 10$, $h = 14$

15. $P = a + b + c$ (perimeter of a triangle); $P = 12$, $a = 3$, $c = 5$

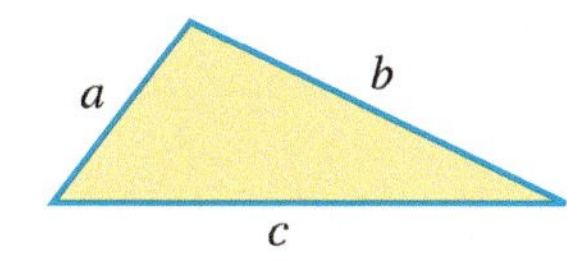

16. $P = a + b + c$; $P = 15$, $a = 3$, $b = 7$

17. ▶ $d = rt$ (distance formula); $d = 252$, $r = 45$

18. $d = rt$; $d = 100$, $t = 2.5$

19. $\mathcal{A} = \frac{1}{2}h(b + B)$ (area of a trapezoid); $\mathcal{A} = 91$, $b = 12$, $B = 14$

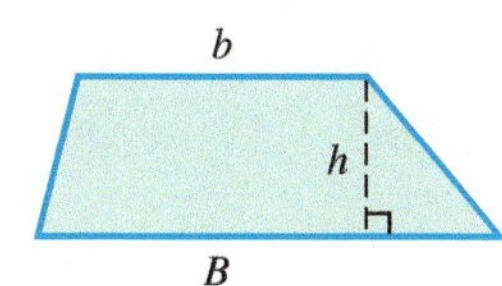

20. $\mathcal{A} = \frac{1}{2}h(b + B)$; $\mathcal{A} = 75$, $b = 19$, $B = 31$

21. $C = 2\pi r$ (circumference of a circle); $C = 16.328$

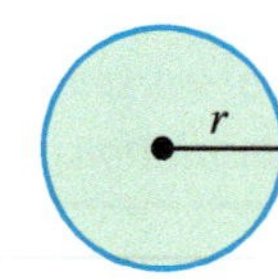

22. $C = 2\pi r$; $C = 8.164$

23. $\mathcal{A} = \pi r^2$ (area of a circle); $r = 4$

24. $\mathcal{A} = \pi r^2$; $r = 12$

*The **volume** of a three-dimensional object is a measure of the space occupied by the object. For example, we would need to know the volume of a gasoline tank in order to know how many gallons of gasoline it would take to completely fill the tank. Volume is measured in cubic units.*

In each exercise, a formula for the volume (V) of a three-dimensional object is given, along with values for the other variables. Evaluate V. (Use 3.14 as an approximation for π.) ***See Example 1.***

25. $V = LWH$ (volume of a rectangular box); $L = 10$, $W = 5$, $H = 3$

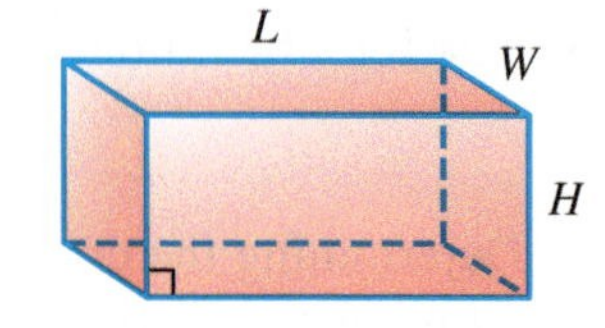

26. $V = LWH$; $L = 12$, $W = 8$, $H = 4$

27. $V = \frac{1}{3}Bh$ (volume of a pyramid); $B = 12$, $h = 13$

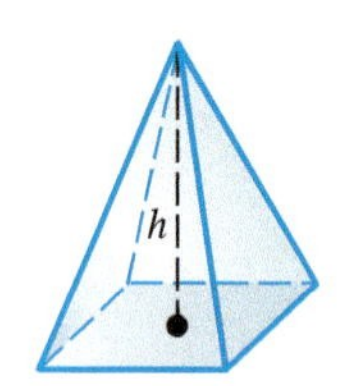

28. $V = \frac{1}{3}Bh$; $B = 36$, $h = 4$

29. $V = \frac{4}{3}\pi r^3$ (volume of a sphere); $r = 12$

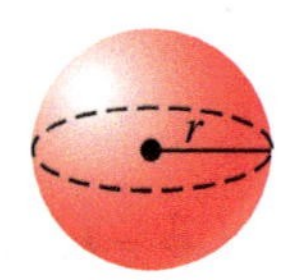

30. $V = \frac{4}{3}\pi r^3$; $r = 6$

Simple interest I in dollars is calculated using the following formula.

$$I = prt \quad \text{Simple interest formula}$$

Here, p represents the principal, or amount, in dollars that is invested or borrowed, r represents the annual interest rate, expressed as a percent, and t represents time, in years.

In each problem, find the value of the remaining variable in the simple interest formula. ***See Example 1.*** *(Hint: Write percents as decimals.)*

31. $p = \$7500$, $r = 4\%$, $t = 2$ yr

32. $p = \$3600$, $r = 3\%$, $t = 4$ yr

33. $I = \$33$, $r = 2\%$, $t = 3$ yr

34. $I = \$270$, $r = 5\%$, $t = 6$ yr

35. $I = \$180$, $p = \$4800$, $r = 2.5\%$
(*Hint:* 2.5% written as a decimal is _____.)

36. $I = \$162$, $p = \$2400$, $r = 1.5\%$
(*Hint:* 1.5% written as a decimal is _____.)

Solve each problem. ***See Examples 2 and 3.***

37. The length of a rectangle is 9 in. more than the width. The perimeter is 54 in. Find the length and the width of the rectangle.

38. The width of a rectangle is 3 ft less than the length. The perimeter is 62 ft. Find the length and the width of the rectangle.

39. The perimeter of a rectangle is 36 m. The length is 2 m more than three times the width. Find the length and the width of the rectangle.

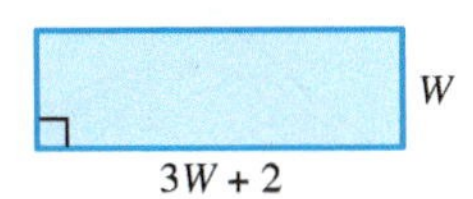

40. The perimeter of a rectangle is 36 yd. The width is 18 yd less than twice the length. Find the length and the width of the rectangle.

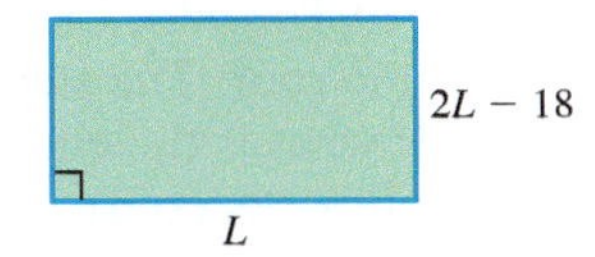

41. The longest side of a triangle is 3 in. longer than the shortest side. The medium side is 2 in. longer than the shortest side. If the perimeter of the triangle is 20 in., what are the lengths of the three sides?

42. The perimeter of a triangle is 28 ft. The medium side is 4 ft longer than the shortest side, while the longest side is twice as long as the shortest side. What are the lengths of the three sides?

43. Two sides of a triangle have the same length. The third side measures 4 m less than twice that length. The perimeter of the triangle is 24 m. Find the lengths of the three sides.

44. A triangle is such that its medium side is twice as long as its shortest side and its longest side is 7 yd less than three times its shortest side. The perimeter of the triangle is 47 yd. Find the lengths of the three sides.

Use a formula to write an equation for each application, and then solve. (Use 3.14 *as an approximation for* π.) ***Formulas are found at the back of this text. See Examples 2–4.***

45. One of the largest fashion catalogues in the world was published in Hamburg, Germany. Each of the 212 pages in the catalogue measured 1.2 m by 1.5 m. What was the perimeter of a page? What was the area? (Data from *Guinness World Records.*)

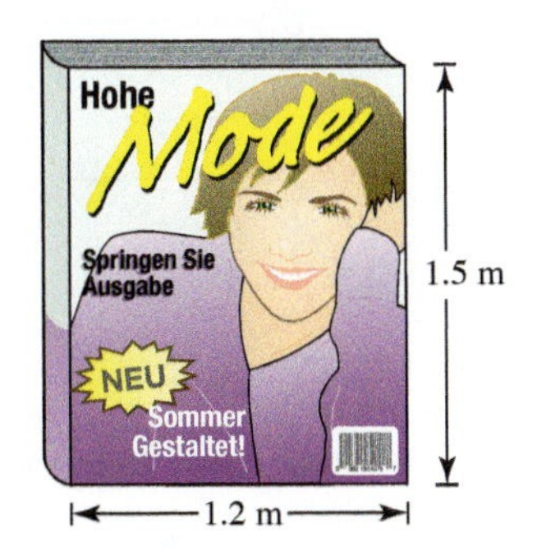

46. One of the world's largest mandalas (sand paintings) measures 12.24 m by 12.24 m. What is the perimeter of the sand painting? To the nearest hundredth of a square meter, what is the area? (Data from *Guinness World Records.*)

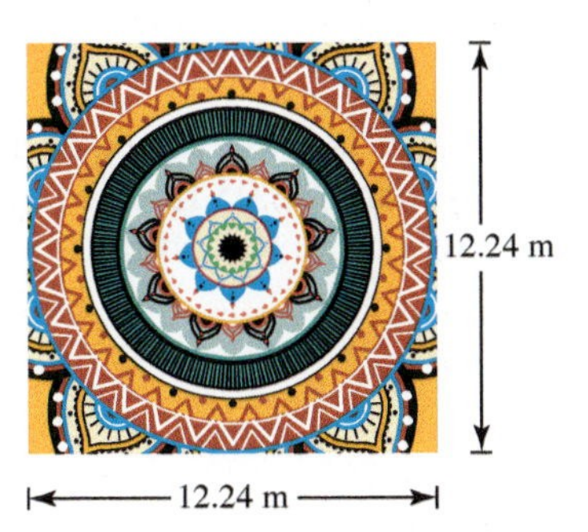

47. The area of a triangular road sign is 70 ft^2. If the base of the sign measures 14 ft, what is the height of the sign?

48. The area of a triangular advertising banner is 96 ft^2. If the height of the banner measures 12 ft, find the measure of the base.

49. A prehistoric ceremonial site dating to about 3000 B.C. was discovered at Stanton Drew in southwestern England. The site, which is larger than Stonehenge, is a nearly perfect circle, consisting of nine concentric rings that probably held upright wooden posts. Around this timber temple is a wide, encircling ditch enclosing an area with a diameter of 443 ft. Find this enclosed area to the nearest thousand square feet. (Data from *Archaeology.*)

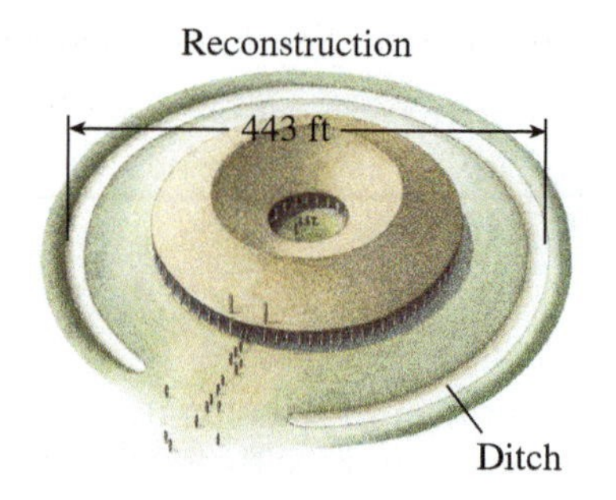

50. The Rogers Centre in Toronto, Canada, is the first stadium with a hard-shell, retractable roof. The steel dome is 630 ft in diameter. To the nearest foot, what is the circumference of this dome? (Data from www.ballparks.com)

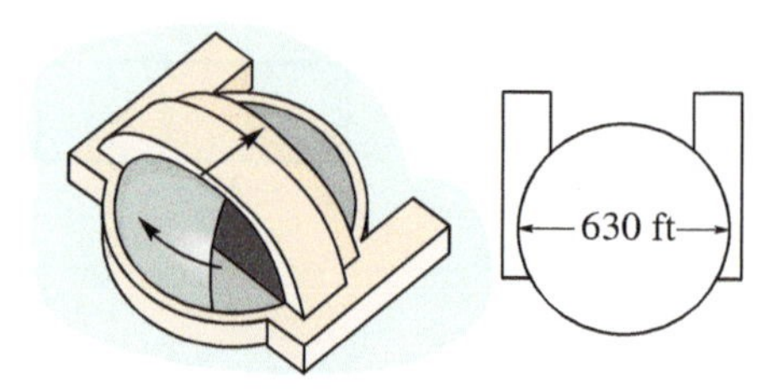

51. One of the largest drums ever constructed was made from Japanese cedar and cowhide, with radius 7.87 ft. What was the area of the circular face of the drum? What was the circumference of the drum? Round answers to the nearest hundredth. (Data from *Guinness World Records.*)

52. A drum played at the Royal Festival Hall in London had radius 6.5 ft. What was the area of the circular face of the drum? What was the circumference of the drum? (Data from *Guinness World Records.*)

53. The survey plat depicted here shows two lots that form a trapezoid. The measures of the parallel sides are 115.80 ft and 171.00 ft. The height of the trapezoid is 165.97 ft. Find the combined area of the two lots. Round the answer to the nearest hundredth of a square foot.

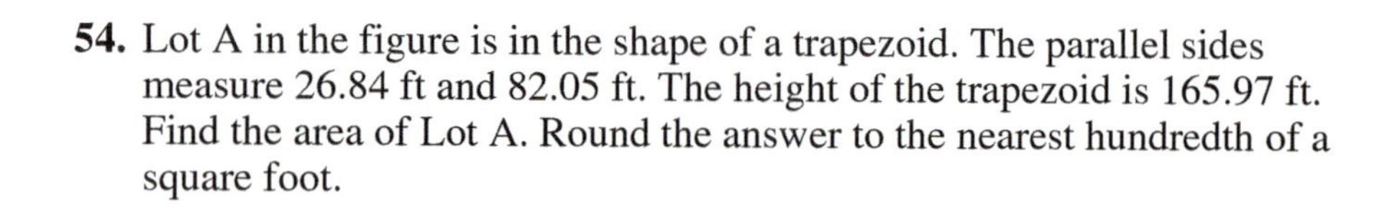

54. Lot A in the figure is in the shape of a trapezoid. The parallel sides measure 26.84 ft and 82.05 ft. The height of the trapezoid is 165.97 ft. Find the area of Lot A. Round the answer to the nearest hundredth of a square foot.

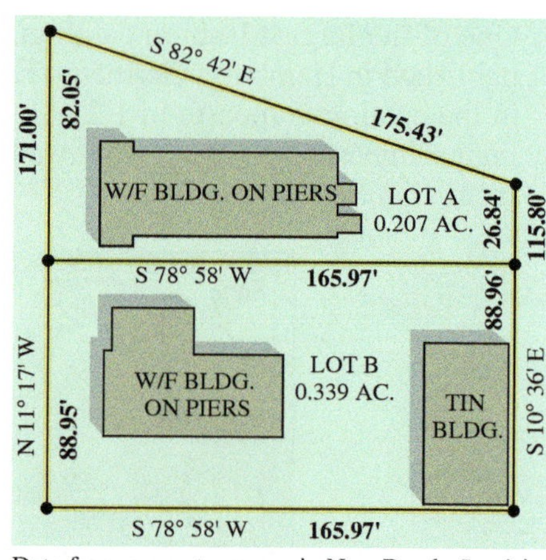

Data from property survey in New Roads, Louisiana.

55. The U.S. Postal Service requires that any box sent by Priority Mail® have length plus girth (distance around) totaling no more than 108 in. The maximum volume that meets this condition is contained by a box with a square end 18 in. on each side. What is the length of the box? What is the maximum volume? (Data from United States Postal Service.)

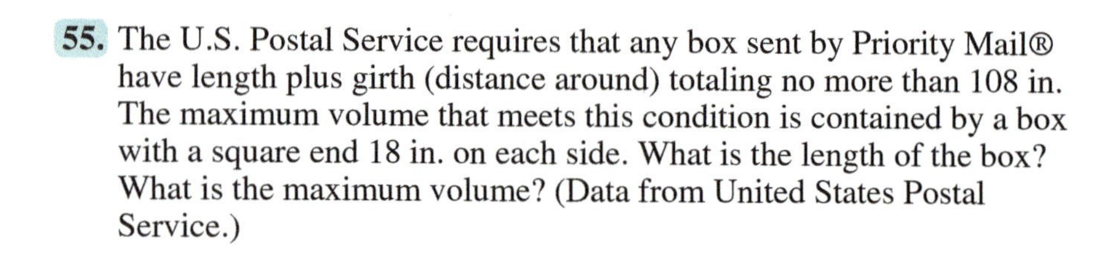

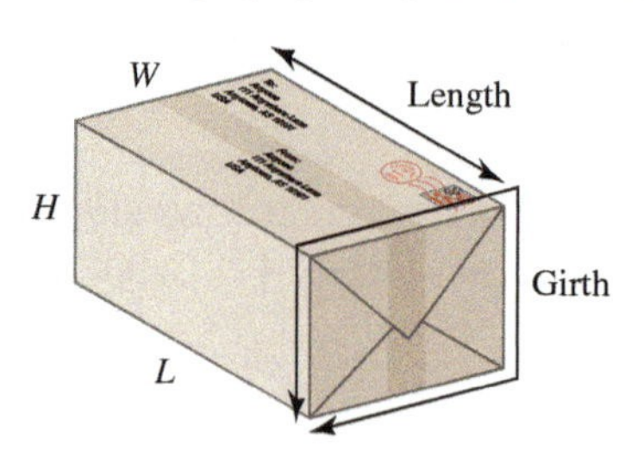

56. One of the world's largest sandwiches, made by Wild Woody's Chill and Grill in Roseville, Michigan, was 12 ft long, 12 ft wide, and $17\frac{1}{2}$ in. ($1\frac{11}{24}$ ft) thick. What was the volume of the sandwich? (Data from *Guinness World Records*.)

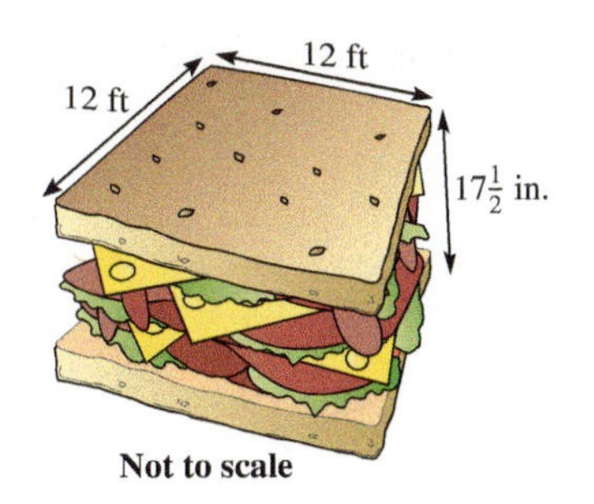

Find the measure of each marked angle. **See Example 5.**

57.

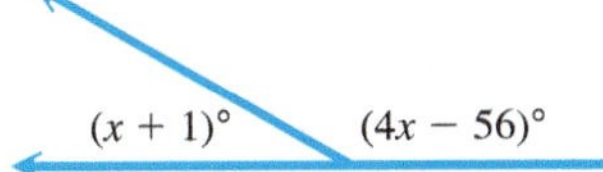

58.

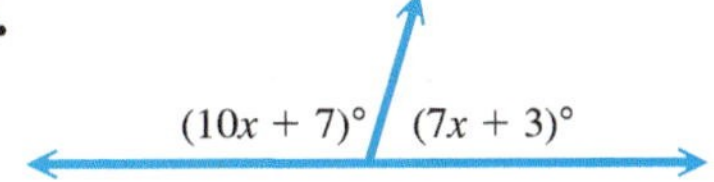

59.

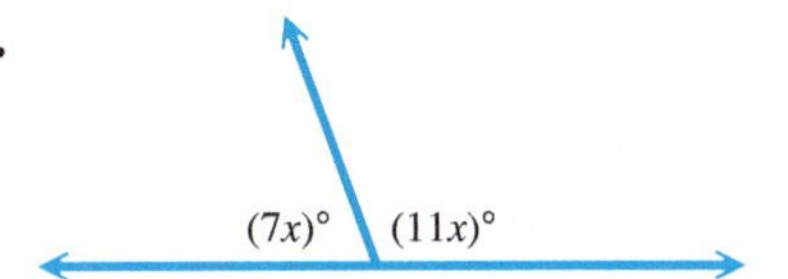

60.

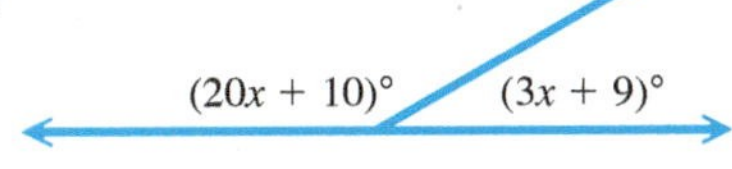

61.

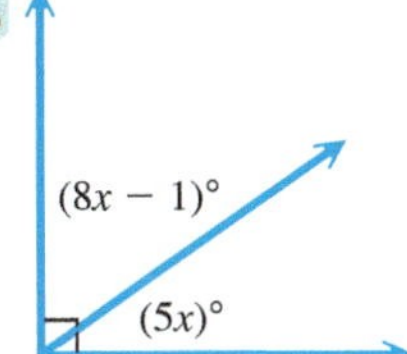

62.

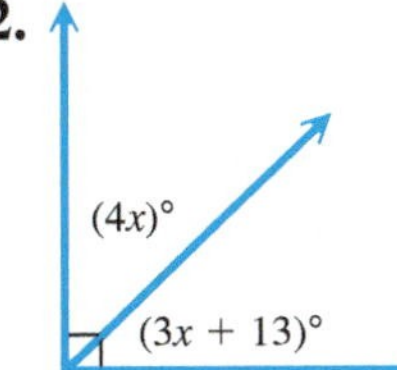

63.

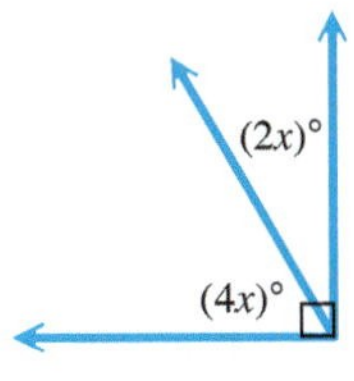

64.

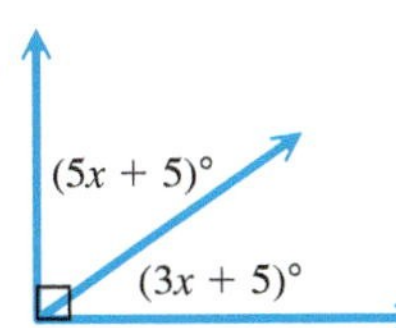

65.

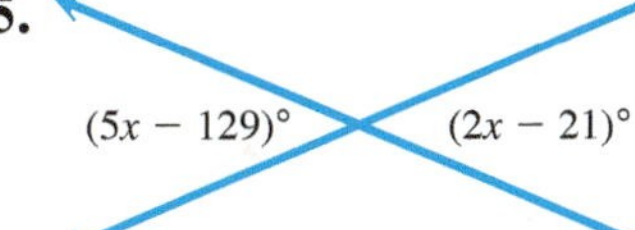

66.

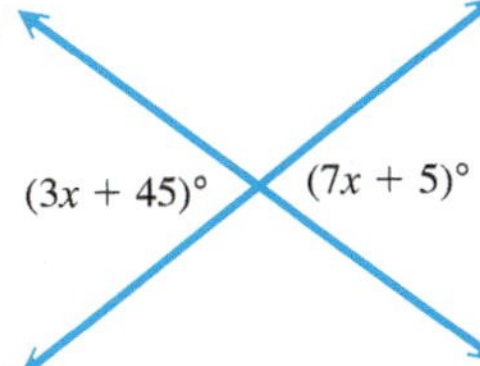

67.

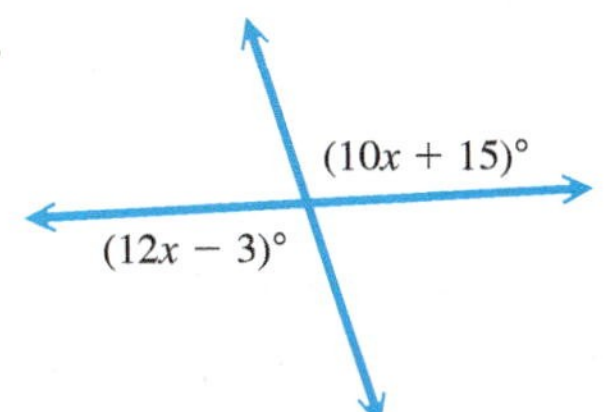

68.

$(11x - 37)°$

$(7x + 27)°$

Solve each formula for the specified variable. **See Examples 6–8.**

69. $d = rt$ for t

70. $d = rt$ for r

71. $V = LWH$ for H

72. $V = LWH$ for L

73. $P = a + b + c$ for b

74. $P = a + b + c$ for c

75. $C = 2\pi r$ for r

76. $C = \pi d$ for d

77. $I = prt$ for r

78. $I = prt$ for p

79. $\mathcal{A} = \frac{1}{2}bh$ for h

80. $\mathcal{A} = \frac{1}{2}bh$ for b

81. $V = \frac{1}{3}\pi r^2 h$ for h

82. $V = \pi r^2 h$ for h

83. $P = 2L + 2W$ for W

84. $A = p + prt$ for r

85. $y = mx + b$ for m

86. $y = mx + b$ for x

87. $Ax + By = C$ for y

88. $Ax + By = C$ for x

89. $M = C(1 + r)$ for r

90. $A = p(1 + rt)$ for r

91. $P = 2(a + b)$ for a

92. $P = 2(a + b)$ for b

93. $f = a(x - h)$ for x

94. $f = a(x - h)$ for h

95. $S = \frac{1}{2}(a + b + c)$ for b

96. $S = \frac{1}{2}(a + b + c)$ for c

97. $C = \frac{5}{9}(F - 32)$ for F

98. $\mathcal{A} = \frac{1}{2}h(b + B)$ for b

Solve each equation for y. ***See Example 9.***

99. $6x + y = 4$

100. $3x + y = 6$

101. $5x - y = 2$

102. $4x - y = 1$

103. $-3x + 5y = -15$

104. $-2x + 3y = -9$

105. $x - 3y = 12$

106. $x - 5y = 10$

Relating Concepts (Exercises 107–110) For Individual or Group Work

The climax of any sports season is the playoffs. Baseball fans eagerly debate predictions of which team will win the pennant for their division. The magic number for each first-place team is often reported in media outlets. The ***magic number*** *(sometimes called the* ***elimination number****) is the combined number of wins by the first-place team and losses by the second-place team that would clinch the title for the first-place team.*

To calculate the magic number, consider the following conditions.

> *The number of wins for the first-place team (W_1) plus the magic number (M) is one more than the sum of the number of wins to date (W_2) and the number of games remaining in the season (N_2) for the second-place team.*

Work Exercises 107–110 in order, *to see the relationships among these concepts.*

107. Use the variable definitions to write an equation involving the magic number.

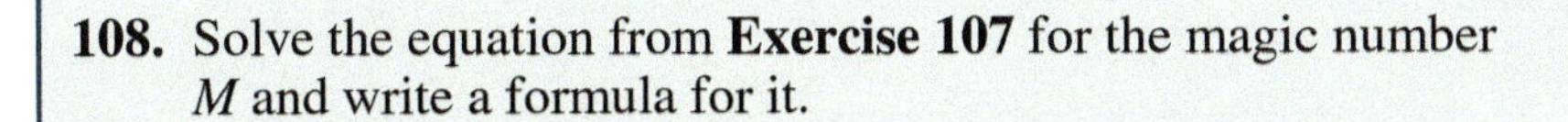

108. Solve the equation from **Exercise 107** for the magic number M and write a formula for it.

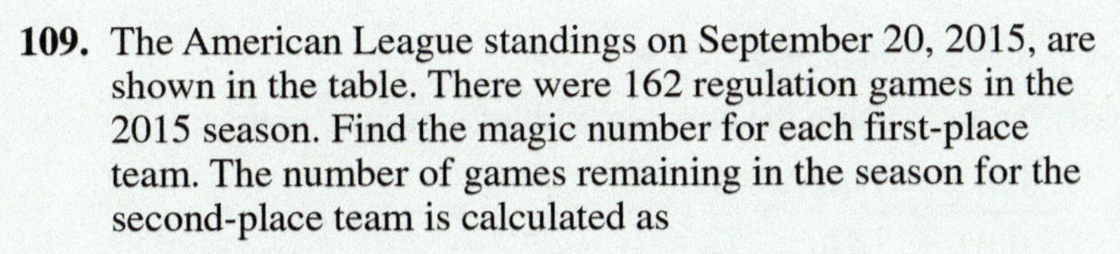

109. The American League standings on September 20, 2015, are shown in the table. There were 162 regulation games in the 2015 season. Find the magic number for each first-place team. The number of games remaining in the season for the second-place team is calculated as

$$N_2 = 162 - (W_2 + L_2),$$

where L_2 represents the number of losses for the second-place team.

(a) AL East: Toronto vs New York

Magic Number ______

(b) AL Central: Kansas City vs Minnesota

Magic Number ______

(c) AL West: Texas vs Houston

Magic Number ______

110. Calculate the magic number for Oakland vs Texas. (Treat Oakland as though it were the second-place team.) How can we interpret the result?

AMERICAN LEAGUE

East	W	L	PCT	GB
Toronto Blue Jays	85	64	.570	—
New York Yankees	82	66	.554	2.5
Baltimore Orioles	73	76	.490	12.0
Tampa Bay Rays	72	77	.483	13.0
Boston Red Sox	71	77	.480	13.5
Central	**W**	**L**	**PCT**	**GB**
Kansas City Royals	87	62	.584	—
Minnesota Twins	76	73	.510	11.0
Cleveland Indians	74	74	.500	12.5
Chicago White Sox	70	78	.473	16.5
Detroit Tigers	69	79	.466	17.5
West	**W**	**L**	**PCT**	**GB**
Texas Rangers	80	69	.537	—
Houston Astros	79	71	.527	1.5
Los Angeles Angels	76	73	.510	4.0
Seattle Mariners	73	77	.487	7.5
Oakland Athletics	64	86	.427	16.5

Data from mlb.com

2.6 Ratio, Proportion, and Percent

OBJECTIVES

1. Write ratios.
2. Solve proportions.
3. Solve applied problems using proportions.
4. Find percents and percentages.

OBJECTIVE 1 Write ratios. A **ratio** is a comparison of two quantities using a quotient.

Ratio

The ratio of the number a to the number b (where $b \neq 0$) is written as follows.

$$a \text{ to } b, \quad a:b, \quad \text{or} \quad \frac{a}{b}$$

Writing a ratio as a quotient $\frac{a}{b}$ is most common in algebra.

Examples: 2 to 3, 2 : 3, $\frac{2}{3}$

EXAMPLE 1 Writing Word Phrases as Ratios

Write a ratio for each word phrase.

(a) 5 hr to 3 hr $\quad \frac{5 \text{ hr}}{3 \text{ hr}} = \frac{5}{3}$

(b) 6 hr to 3 days

First convert 3 days to hours.

$$3 \text{ days} = 3 \cdot 24 = 72 \text{ hr} \qquad 1 \text{ day} = 24 \text{ hr}$$

Now write the ratio using the common unit of measure, hours.

$$\frac{6 \text{ hr}}{3 \text{ days}} = \frac{6 \text{ hr}}{72 \text{ hr}} = \frac{6}{72}, \quad \text{or} \quad \frac{1}{12} \qquad \text{Write in lowest terms.}$$

◀ Work Problem 1 at the Side.

1. Write each ratio.

(a) 9 women to 5 women

GS **(b)** 4 in. to 1 ft

First convert 1 ft to ____ in.

Then write a ratio.

(c) 8 months to 2 yr

An application of ratios is in *unit pricing,* to see which size of an item offered in different sizes produces the best price per unit.

EXAMPLE 2 Finding Price per Unit

A Jewel-Osco supermarket charges the following prices for a jar of extra crunchy peanut butter.

PEANUT BUTTER

Size	Price
18 oz	\$3.49
28 oz	\$4.99
40 oz	\$6.79

Which size is the best buy? That is, which size has the lowest unit price?

Continued on Next Page

Answers

1. (a) $\frac{9}{5}$ (b) 12; $\frac{4}{12}$, or $\frac{1}{3}$ (c) $\frac{8}{24}$, or $\frac{1}{3}$

To find the best buy, write ratios comparing the price for each size jar to the number of units (ounces) per jar.

PEANUT BUTTER

Size	Price	Unit Price (dollars per ounce)
18 oz	\$3.49	$\frac{\$3.49}{18} = \$0.194$
28 oz	\$4.99	$\frac{\$4.99}{28} = \$0.178$
40 oz	\$6.79	$\frac{\$6.79}{40} = \$0.170$ ← Best buy

To find the price per ounce, the number of ounces goes in the denominator.

(Results are rounded to the nearest thousandth.)

Because the 40-oz size produces the lowest unit price, it is the best buy. Buying the largest size does not always provide the best buy, although it often does, as in this case.

Work Problem 2 at the Side. ▶

OBJECTIVE 2 Solve proportions. A ratio is used to compare two numbers or amounts. A **proportion** says that two ratios are equal. For example, the proportion

$$\frac{3}{4} = \frac{15}{20}$$

A proportion is a special type of equation.

says that the ratios $\frac{3}{4}$ and $\frac{15}{20}$ are equal. In the proportion

$$\frac{a}{b} = \frac{c}{d} \quad (\text{where } b, d \neq 0),$$

a, b, c, and d are the **terms** of the proportion. The a and d terms are the **extremes,** and the b and c terms are the **means.** We read the proportion $\frac{a}{b} = \frac{c}{d}$ as **"*a* is to *b* as *c* is to *d*."** Multiplying each side of this proportion by the common denominator, bd, gives the following.

$$bd \cdot \frac{a}{b} = bd \cdot \frac{c}{d} \qquad \text{Multiply each side by } bd.$$

$$\frac{b}{b}(d \cdot a) = \frac{d}{d}(b \cdot c) \qquad \text{Associative and commutative properties}$$

$$ad = bc \qquad \text{Commutative and identity properties}$$

We can also find the products ad and bc by multiplying diagonally.

$$ad = bc$$

$$\frac{a}{b} = \frac{c}{d}$$

For this reason, ad and bc are the **cross products of the proportion.**

Cross Products of a Proportion

If $\frac{a}{b} = \frac{c}{d}$, then the cross products ad and bc of the proportion are equal—that is, ***the product of the extremes equals the product of the means.***

Also, if $ad = bc$, then $\frac{a}{b} = \frac{c}{d}$ (where $b, d \neq 0$).

---

2 Solve each problem.

GS **(a)** A supermarket charges the following prices for a popular brand of pork and beans.

PORK AND BEANS

Size	Price
8 oz	\$0.76
28 oz	\$1.00
53 oz	\$1.99

Calculate the unit price to the nearest thousandth for each size.

8 oz: $\frac{\$0.76}{\_\_\_} = \_\_\_$

28 oz: $\frac{\_\_\_}{28} = \_\_\_$

53 oz: $\frac{\_\_\_}{\_\_\_} = \_\_\_$

Which size is the best buy? What is the unit price for that size?

(b) A supermarket charges the following prices for a certain brand of laundry detergent.

LAUNDRY DETERGENT

Size	Price
75 oz	\$8.94
100 oz	\$13.97
150 oz	\$19.97

Which size is the best buy? What is the unit price to the nearest thousandth for that size?

Answers

2. **(a)** 8; \$0.095; \$1.00; \$0.036; \$1.99; 53; \$0.038; 28 oz; \$0.036 per oz
(b) 75 oz; \$0.119 per oz

3 Solve each proportion.

GS **(a)** $\frac{y}{6} = \frac{35}{42}$

$y \cdot$ ____ $=$ ____ $\cdot 35$

____ $y =$ ____

$y =$ ____

The solution set is ____.

(b) $\frac{a}{24} = \frac{15}{16}$

4 Solve each equation.

(a) $\frac{z}{2} = \frac{z+1}{3}$

(b) $\frac{p+3}{3} = \frac{p-5}{4}$

Note

If $\frac{a}{c} = \frac{b}{d}$, then $ad = cb$, or $ad = bc$. This means that the two proportions are equivalent, and the proportion

$$\frac{a}{b} = \frac{c}{d} \quad \text{can also be written as} \quad \frac{a}{c} = \frac{b}{d} \quad (\text{where } c, d \neq 0).$$

Sometimes one form is more convenient to work with than the other.

Four numbers are used in a proportion. If any three of these numbers are known, the fourth can be found.

EXAMPLE 3 Finding an Unknown in a Proportion

Solve the proportion $\frac{5}{9} = \frac{x}{63}$.

$$\frac{5}{9} = \frac{x}{63}$$ Solve for x.

$5 \cdot 63 = 9 \cdot x$ Cross products must be equal.

$315 = 9x$ Multiply.

$35 = x$ Divide by 9.

Check by substituting 35 for x in the proportion. The solution set is $\{35\}$.

◀ Work Problem 3 at the Side.

CAUTION

The cross-product method cannot be used directly if there is more than one term on either side of the equality symbol.

$$\underbrace{\frac{m-1}{5} = \frac{m+1}{3} - 4}_{\text{2 terms}}, \quad \underbrace{\frac{x}{3} + \frac{5}{4}}_{\text{2 terms}} = \frac{1}{2}$$

Do ***not*** use the cross-product method to solve equations in this form.

EXAMPLE 4 Solving an Equation Using Cross Products

Solve the equation $\frac{m-2}{5} = \frac{m+1}{3}$.

$$\frac{m-2}{5} = \frac{m+1}{3}$$

Be sure to use parentheses.

$3(m-2) = 5(m+1)$ (*) Cross products

$3m - 6 = 5m + 5$ Distributive property

$-2m - 6 = 5$ Subtract $5m$.

$-2m = 11$ Add 6.

$m = -\frac{11}{2}$ Divide by -2.

Check to confirm that the solution set is $\{-\frac{11}{2}\}$.

◀ Work Problem 4 at the Side.

Answers

3. **(a)** 42; 6; 42; 210; 5; $\{5\}$ **(b)** $\left\{\frac{45}{2}\right\}$

4. **(a)** $\{2\}$ **(b)** $\{-27\}$

Note

When we set cross products equal to each other, we are actually multiplying each ratio in the proportion by a common denominator.

$$\frac{m-2}{5} = \frac{m+1}{3} \qquad \text{See Example 4.}$$

$$15\left(\frac{m-2}{5}\right) = 15\left(\frac{m+1}{3}\right) \qquad \text{Multiply each ratio by 15, the LCD.}$$

$$3(m-2) = 5(m+1) \qquad \text{This is equation (*) from Example 4.}$$

$15\left(\frac{m-2}{5}\right) = 15 \cdot \frac{1}{5}(m-2) = 3(m-2)$

OBJECTIVE 3 Solve applied problems using proportions.

EXAMPLE 5 Applying Proportions

After Lee Ann pumped 5.0 gal of gasoline, the display showing the price read \$15.50. When she finished pumping the gasoline, the price display read \$44.95. How many gallons did she pump?

To solve this problem, set up a proportion, with prices in the numerators and gallons in the denominators. Let x = the number of gallons pumped.

$$\frac{\$15.50}{5.0} = \frac{\$44.95}{x} \qquad \begin{matrix}\leftarrow \text{Price} \\ \leftarrow \text{Gallons}\end{matrix}$$

Be sure that numerators represent the *same* quantities and denominators represent the *same* quantities.

$15.50x = 5.0(44.95)$ Cross products

$15.50x = 224.75$ Multiply.

$x = 14.5$ Divide by 15.50.

She pumped a total of 14.5 gal. Check this answer. Notice that the way the proportion was set up uses the fact that the unit price is the same, no matter how many gallons are purchased.

Work Problem 5 at the Side. ▶

OBJECTIVE 4 Find percents and percentages. ***A percent is a ratio where the second number is always 100.***

50% represents the ratio of 50 to 100, that is, $\frac{50}{100}$, or, **0.50**.

27% represents the ratio of 27 to 100, that is, $\frac{27}{100}$, or, **0.27**.

The word **percent** means **"per 100."** One percent means "one per 100."

$$\mathbf{1\% = \frac{1}{100}, \quad \text{or} \quad 1\% = 0.01} \qquad \text{Percent, decimal, and fraction equivalents}$$

We can solve a percent problem involving $x\%$ by writing it as a proportion. The amount, or **percentage,** is compared to the **base** (the whole amount).

$$\frac{\textbf{amount}}{\textbf{base}} = \frac{x}{100}$$

We can also write this proportion as follows.

$$\frac{\text{amount}}{\text{base}} = \text{percent (as a decimal)} \qquad \frac{x}{100} \text{ or } 0.01x \text{ is equivalent to } x \text{ percent.}$$

$$\textbf{amount = percent (as a decimal)} \cdot \textbf{base} \qquad \text{Basic percent equation}$$

5 Solve each problem.

(a) Twelve gallons of diesel fuel costs \$40.80. To the nearest cent, how much would 16.5 gal of the same fuel cost?

(b) Eight quarts of oil cost \$14.00. How much do 5 qt of oil cost?

Answers

5. (a) \$56.10 (b) \$8.75

6 Solve each problem.

(a) What is 20% of 70?

(b) 40% of what number is 130?

(c) 121 is what percent of 484?

7 Solve each problem.

(a) A winter coat is on a clearance sale for \$48. The regular price is \$120. What percent of the regular price is the savings?

GS **(b)** Mark scored 34 points on a test, which was 85% of the possible points. How many possible points were on the test?

Write the percent equation. Give the percent as a decimal.

34	was	85%	of	what number?
↓	↓	↓	↓	↓
34	____	____	____	n

Complete the solution, and give the answer to the problem.

Answers

6. (a) 14 (b) 325 (c) 25%

7. (a) 60% (b) =; 0.85; $\cdot$; 40 points

EXAMPLE 6 Solving Percent Equations

Solve each problem.

(a) What is 15% of 600?

Let n = the number. The word *of* indicates multiplication.

What	is	15%	of	600?
↓	↓	↓	↓	↓
n	$=$	0.15	$\cdot$	600

Translate each word or phrase to write the equation.

Write the percent equation.

Write 15% as a decimal.

$n = 90$ Multiply.

Thus, **90** is 15% of 600.

(b) 32% of what number is 64?

32%	of	what number	is	64?
↓	↓	↓	↓	↓
0.32	$\cdot$	n	$=$	64

Write 32% as a decimal.

Write the percent equation.

$n = \frac{64}{0.32}$ Divide by 0.32.

$n = 200$ Simplify.

32% of **200** is 64.

(c) 90 is what percent of 360?

90	is	what percent	of	360?
↓	↓	↓	↓	↓
90	$=$	p	$\cdot$	360

Write the percent equation.

$\frac{90}{360} = p$ Divide by 360.

$0.25 = p$, or $25\% = p$ Simplify. Write 0.25 as a percent.

Thus, 90 is **25%** of 360.

◀ Work Problem 6 at the Side.

EXAMPLE 7 Solving an Applied Percent Problem

A newspaper ad offered a set of tires at a sale price of \$258. The regular price was \$300. What percent of the regular price was the savings?

The savings amounted to \$300 − \$258 = \$42. We can now restate the problem: *What percent of* 300 *is* 42?

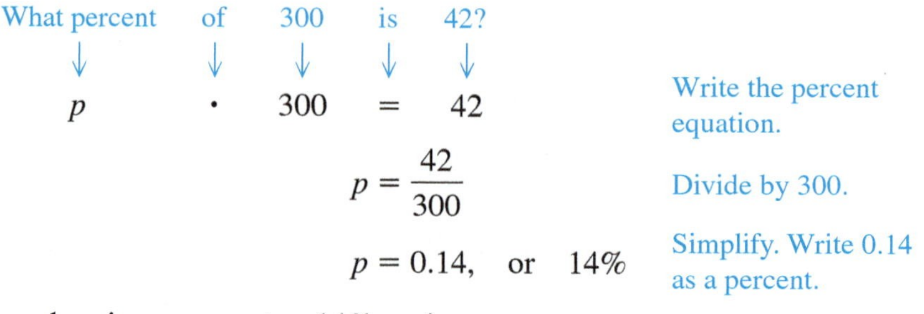

What percent	of	300	is	42?
↓	↓	↓	↓	↓
p	$\cdot$	300	$=$	42

Write the percent equation.

$p = \frac{42}{300}$ Divide by 300.

$p = 0.14$, or 14% Simplify. Write 0.14 as a percent.

The sale price represents a 14% savings.

◀ Work Problem 7 at the Side.

2.6 Exercises

FOR EXTRA HELP

Go to MyMathLab for worked-out, step-by-step solutions to exercises enclosed in a square and video solutions to ▶ exercises.

1. CONCEPT CHECK Ratios are used to _______ two numbers or quantities. Which of the following indicate the ratio of a to b?

A. $\frac{a}{b}$ **B.** $\frac{b}{a}$ **C.** $a \cdot b$ **D.** $a : b$

2. CONCEPT CHECK A proportion says that two _______ are equal. The equation

$$\frac{a}{b} = \frac{c}{d} \quad (\text{where } b, d \neq 0)$$

is a _______, where ad and bc are the _______.

3. CONCEPT CHECK Match each ratio in Column I with the ratio equivalent to it in Column II.

I	II
(a) 75 to 100	**A.** 80 to 100
(b) 5 to 4	**B.** 50 to 100
(c) $\frac{1}{2}$	**C.** 3 to 4
(d) 4 to 5	**D.** 15 to 12

4. CONCEPT CHECK Which of the following represent a ratio of 4 days to 2 weeks?

A. $\frac{4}{2}$ **B.** $\frac{4}{7}$ **C.** $\frac{4}{14}$ **D.** $\frac{2}{1}$

E. $\frac{2}{7}$ **F.** $\frac{1}{2}$ **G.** $\frac{2}{4}$ **H.** $\frac{7}{2}$

Write a ratio for each word phrase. Express fractions in lowest terms. ***See Example 1.***

5. 60 ft to 70 ft

6. 30 mi to 40 mi

7. 72 dollars to 220 dollars ▶

8. 80 people to 120 people

9. 30 in. to 8 ft

10. 8 ft to 20 yd

11. 16 min to 1 hr ▶

12. 24 min to 2 hr

13. 2 yd to 60 in.

14. 3 days to 40 hr

Find the best buy for each item. Give the unit price to the nearest thousandth for that size. ***See Example 2.*** *(Data from Jewel-Osco and HyVee.)*

15. GRANULATED SUGAR

Size	Price
4 lb	$3.29
10 lb	$7.49

16. APPLESAUCE

Size	Price
23 oz	$1.99
48 oz	$3.49

17. ORANGE JUICE

Size	Price
64 oz	$2.99
89 oz	$4.79
128 oz	$6.49

18. SALAD DRESSING

Size	Price
8 oz	$1.69
16 oz	$1.97
36 oz	$5.99

19. MAPLE SYRUP

Size	Price
8.5 oz	$5.79
12.5 oz	$7.99
32 oz	$16.99

20. MOUTHWASH

Size	Price
16.9 oz	$3.39
33.8 oz	$3.49
50.7 oz	$5.29

21. TOMATO KETCHUP

Size	Price
32 oz	$1.79
36 oz	$2.69
40 oz	$2.49
64 oz	$4.38

22. GRAPE JELLY

Size	Price
1 lb	$0.79
2 lb	$1.49
5 lb	$3.59
20 lb	$12.99

Solve each equation. ***See Examples 3 and 4.***

23. $\frac{k}{4} = \frac{175}{20}$ **24.** $\frac{x}{6} = \frac{18}{4}$ **25.** $\frac{49}{56} = \frac{z}{8}$ **26.** $\frac{20}{100} = \frac{z}{80}$

27. $\frac{x}{4} = \frac{12}{30}$ **28.** $\frac{x}{6} = \frac{5}{21}$ **29.** $\frac{8}{12} = \frac{12k}{18}$ **30.** $\frac{14}{10} = \frac{21t}{15}$

31. $\frac{z}{4} = \frac{z+1}{6}$ **32.** $\frac{m}{5} = \frac{m-2}{2}$ **33.** $\frac{3y-2}{5} = \frac{6y-5}{11}$ **34.** $\frac{2r+8}{4} = \frac{3r-9}{3}$

35. $\frac{5k+1}{6} = \frac{3k-2}{3}$ **36.** $\frac{x+4}{6} = \frac{x+10}{8}$ **37.** $\frac{2p+7}{3} = \frac{p-1}{4}$ **38.** $\frac{3m-2}{5} = \frac{4-m}{3}$

Solve each problem. ***See Example 5.***

39. If 16 candy bars cost \$20.00, how much do 24 candy bars cost?

40. If 12 ring tones cost \$30.00, how much do 8 ring tones cost?

41. If 6 gal of premium gasoline cost \$22.74, how much would it cost to completely fill a 15-gal tank?

42. If sales tax on a \$16.00 DVD is \$1.32, how much would the sales tax be on a \$120.00 DVD player?

43. Biologists tagged 500 fish in Grand Bay. At a later date, they found 7 tagged fish in a sample of 700. Estimate the total number of fish in Grand Bay to the nearest hundred.

44. Researchers at West Okoboji Lake tagged 840 fish. A later sample of 1000 fish contained 18 that were tagged. Approximate the fish population in West Okoboji Lake to the nearest hundred.

45. The distance between Kansas City, Missouri, and Denver is 600 mi. On a certain wall map, this is represented by a length of 2.4 ft. On the map, how many feet would there be between Memphis and Philadelphia, two cities that are actually 1000 mi apart?

46. The distance between Singapore and Tokyo is 3300 mi. On a certain wall map, this distance is represented by a length of 11 in. The actual distance between Mexico City and Cairo is 7700 mi. How far apart are they on the same map to the nearest tenth?

47. A wall map of the United States has a distance of 8.0 in. between New Orleans and Chicago, two cities that are actually 912 mi apart. The actual distance between Milwaukee and Seattle is 1940 mi. How far apart on this map are Milkwaukee and Seattle?

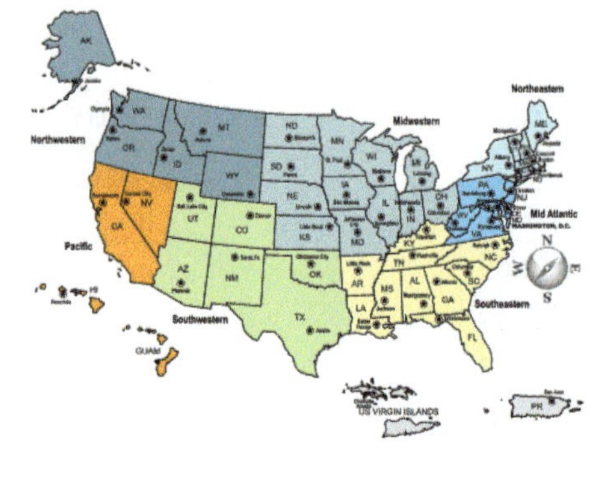

48. On a world globe, the distance between Capetown and Bangkok, two cities that are actually 10,080 km apart, is 12.4 in. The actual distance between Moscow and Berlin is 1610 km. How far apart on this globe are Moscow and Berlin?

49. According to the directions on the label of a bottle of Armstrong® Concentrated Floor Cleaner, for routine cleaning, $\frac{1}{4}$ cup of cleaner should be mixed with 1 gal of warm water. How much cleaner should be mixed with $10\frac{1}{2}$ gal of water?

50. The directions on the bottle mentioned in **Exercise 49** also specify that, for extra-strength cleaning, $\frac{1}{2}$ cup of cleaner should be used for each 1 gal of water. For extra-strength cleaning, how much cleaner should be mixed with $15\frac{1}{2}$ gal of water?

51. On January 18, 2016, the exchange rate between euros and U.S. dollars was 1 euro to \$1.0889. Ashley went to Rome and exchanged her U.S. currency for euros, receiving 300 euros. How much in U.S. dollars did she exchange? (Data from www.exchange-rates.org)

52. If 12 U.S. dollars can be exchanged for 218.64 Mexican pesos, how many pesos can be obtained for \$100?

Two triangles are ***similar*** *if they have the same shape (but not necessarily the same size). Similar triangles have sides that are proportional. The figure shows two similar triangles. Notice that the ratios of the corresponding sides all equal* $\frac{3}{2}$.

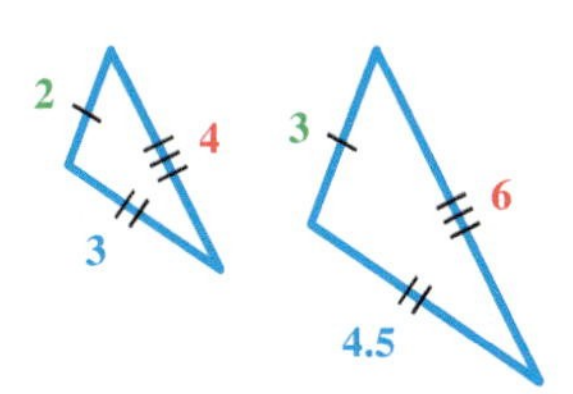

$$\frac{3}{2} = \frac{3}{2} \qquad \frac{4.5}{3} = \frac{3}{2} \qquad \frac{6}{4} = \frac{3}{2}$$

If we know that two triangles are similar, we can set up a proportion to solve for the length of an unknown side.

Find the lengths x and y as needed in each pair of similar triangles.

53. GS

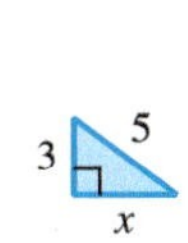

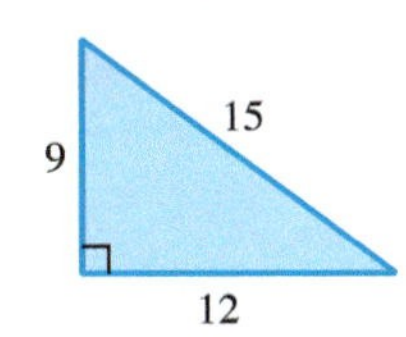

Complete the proportion and then solve for x.

$$\frac{x}{12} = \frac{3}{\_\_\_\_}$$

54. GS

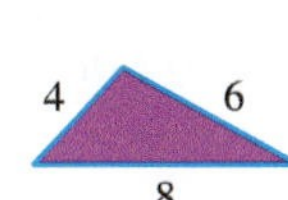

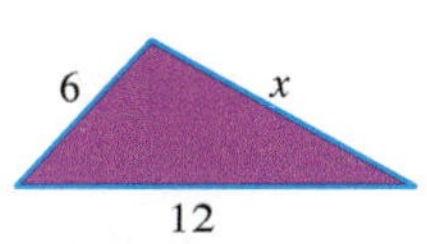

Complete the proportion and then solve for x.

$$\frac{x}{6} = \frac{12}{\_\_\_\_}$$

55.

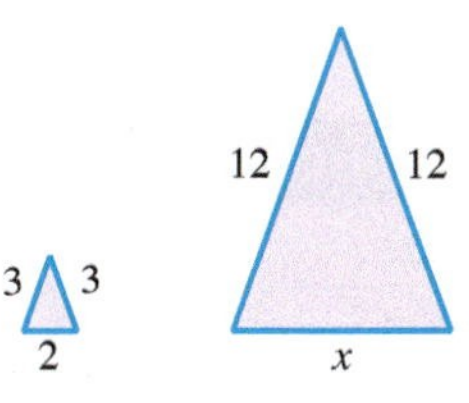

56.

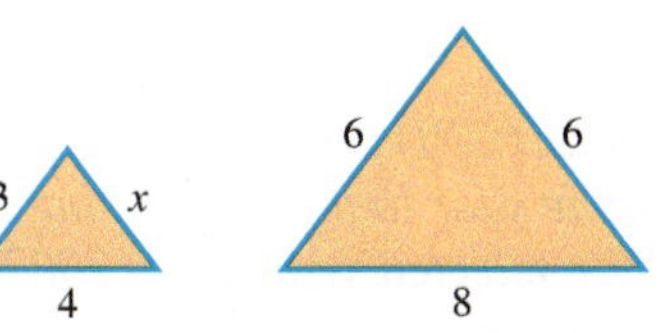

57.

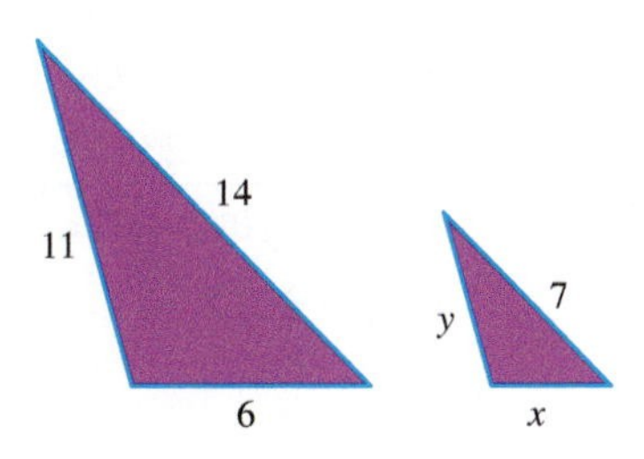

58.

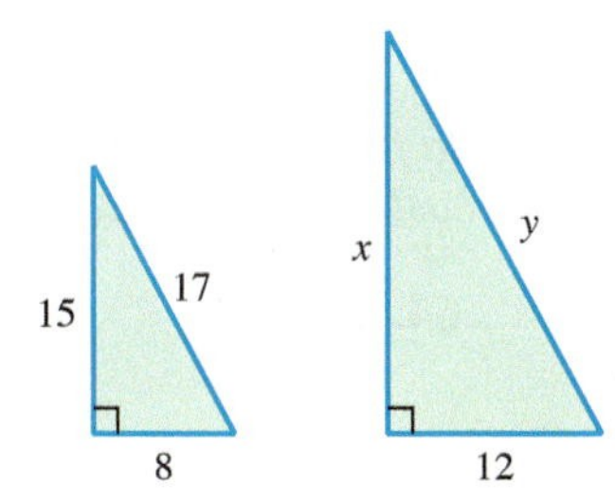

Use the information in each problem to complete the diagram of similar triangles. Then write a proportion and solve the problem.

59. An enlarged version of the chair used by George Washington at the Constitutional Convention casts a shadow 18 ft long at the same time a vertical pole 12 ft high casts a shadow 4 ft long. How tall is the chair?

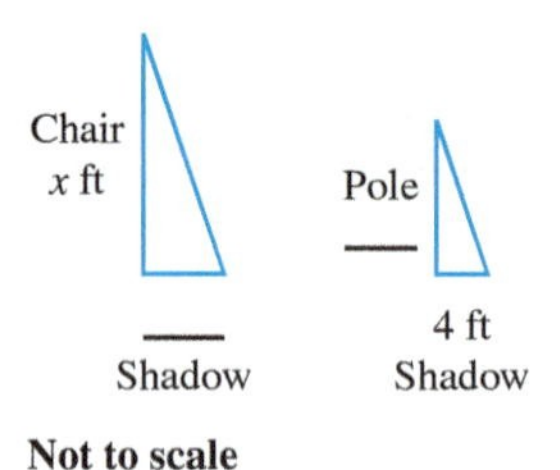

Not to scale

60. One of the tallest candles ever constructed was exhibited at the 1897 Stockholm Exhibition. If it cast a shadow 5 ft long at the same time a vertical pole 32 ft high cast a shadow 2 ft long, how tall was the candle?

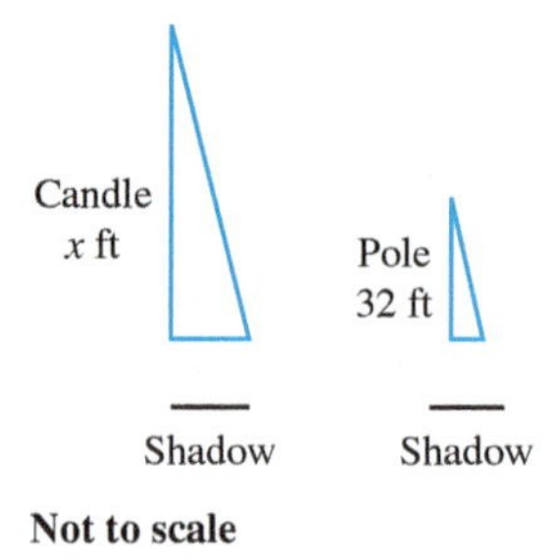

Not to scale

Children are often given antibiotics in liquid form, called an oral suspension. Pharmacists make up these suspensions by mixing the medication in powder form with water. They use proportions to calculate the volume of the suspension for the amount of medication that has been prescribed. For each problem, do each of the following.

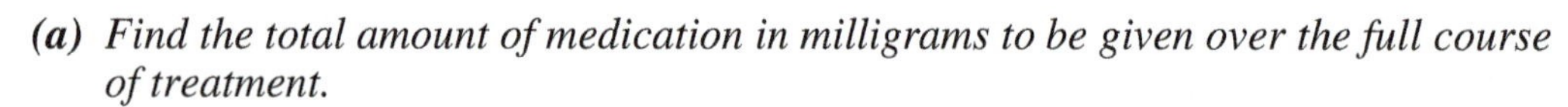

(a) Find the total amount of medication in milligrams to be given over the full course of treatment.

(b) Write a proportion that can be solved to find the total volume of the liquid suspension that the pharmacist will prepare. Use x as the variable.

(c) Solve the proportion to determine the total volume of the oral suspension.

61. Logan's pediatric nurse practitioner has prescribed 375 mg of amoxil a day for 7 days to treat an infection. The pharmacist uses 125 mg of amoxil in each 5 mL of the suspension. (Data from www.drugs.com)

62. An amoxil oral suspension can also be made by using 250 mg for each 5 mL of suspension. Ava's pediatrician prescribed 900 mg a day for 10 days to treat her bronchitis. (Data from www.drugs.com)

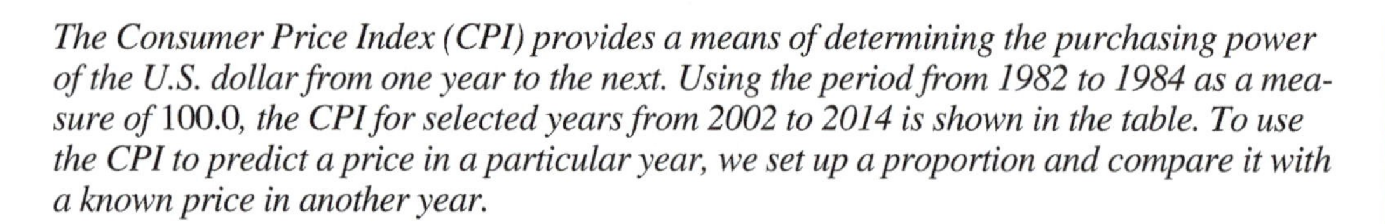

The Consumer Price Index (CPI) provides a means of determining the purchasing power of the U.S. dollar from one year to the next. Using the period from 1982 to 1984 as a measure of 100.0, the CPI for selected years from 2002 to 2014 is shown in the table. To use the CPI to predict a price in a particular year, we set up a proportion and compare it with a known price in another year.

$$\frac{\text{price in year } A}{\text{index in year } A} = \frac{\text{price in year } B}{\text{index in year } B}$$

Year	Consumer Price Index
2002	179.9
2004	188.9
2006	201.6
2008	215.3
2010	218.1
2012	229.6
2014	236.7

Data from U.S. Bureau of Labor Statistics.

Use the CPI figures in the table to find the amount that would be charged for using the same amount of electricity that cost \$225 in 2002. Give answers to the nearest dollar.

63. in 2004

64. in 2006

65. in 2010

66. in 2014

Solve each problem. ***See Examples 6 and 7.***

67. What is 14% of 780?

68. What is 26% of 480?

69. What is 120% of 45?

70. What is 150% of 78?

71. 42% of what number is 294?

72. 18% of what number is 108?

73. 120% of what number is 510?

74. 140% of what number is 315?

75. 4 is what percent of 50?

76. 8 is what percent of 64?

77. What percent of 30 is 36?

78. What percent of 48 is 96?

79. Clayton earned 48 points on a 60-point geometry project. What percent of the total points did he earn?

80. On a 75-point algebra test, Grady scored 63 points. What percent of the total points did he score?

81. A laptop computer that has a regular price of $700 is on sale for $504. What percent of the regular price is the savings?

82. An all-in-one desktop computer that has a regular price of $980 is on sale for $833. What percent of the regular price is the savings?

83. Tyler has a monthly income of $1500. His rent is $480 per month. What percent of his monthly income is his rent?

84. Lily has a monthly income of $2200. She has budgeted $154 per month for entertainment. What percent of her monthly income did she budget for entertainment?

85. Anna saved $1950, which was 65% of the total amount she needed for a used car. What was the total amount she needed for the car?

86. Bryn had $525, which was 70% of the total amount she needed for a deposit on an apartment. What was the total deposit she needed?

Work each percent problem. Round all money amounts to the nearest dollar and percents to the nearest tenth, as needed. ***See Examples 5–7.***

87. A family of four with a monthly income of $3800 plans to spend 8% of this amount on entertainment. How much will be spent on entertainment?

88. George earns $3200 per month. He saves 12% of this amount. How much does he save?

89. In 2014, the U.S. civilian labor force consisted of 155,922,000 persons. Of this total, 9,617,000 were unemployed. What percent of the U.S. civilian labor force was unemployed? (Data from U.S. Bureau of Labor Statistics.)

The number unemployed	was	what percent	of	the total civilian labor force?
↓	↓	↓	↓	↓
______	=	p	___	______

90. In 2014, the U.S. civilian labor force consisted of 155,922,000 persons. Of this total, 2,237,000 were employed in agricultural industries. What percent were employed in agricultural industries? (Data from U.S. Bureau of Labor Statistics.)

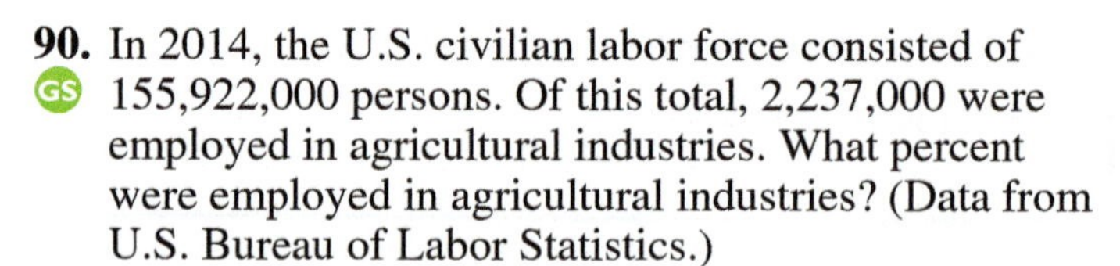

What percent	of	the total U.S. labor force	was	employed in agriculture?
↓	↓	↓	↓	↓
p	___	______	=	______

91. In 2015, U.S. households owned 312,100,000 pets. Of these, 77,800,000 were dogs. What percent of the pets were *not* dogs? (Data from American Pet Product Manufacturers Association.)

92. Of the 1.79 million bachelor's degrees earned in 2011–12, 20.4% were earned in business. How many degrees earned were *not* in business? (Data from National Center for Education Statistics.)

93. The 1916 dime minted in Denver is quite rare. The 1979 edition of *A Guide Book of United States Coins* listed its value in Extremely Fine condition as \$625. The 2015 value had increased to \$6000. What was the percent increase in the value of this coin? (*Hint:* First subtract to find the increase in value. Then write a percent equation that uses this increase.)

94. Here is a common business problem:

If the sales tax rate is 6.5% and we have collected \$3400 in sales tax, how much were sales?

(*Hint:* To solve using a percent equation, ask "6.5% *of what number is* \$3400?" Write 6.5% as a decimal.)

Relating Concepts (Exercises 95–98) For Individual or Group Work

Work Exercises 95–98 in order. *The steps justify the method of solving a proportion using cross products.*

95. What is the LCD of the fractions in the following equation?

$$\frac{x}{6} = \frac{2}{5}$$

96. Solve the equation in **Exercise 95** as follows.

(a) Multiply each side by the LCD. What equation results?

(b) Solve the equation from part (a) by dividing each side by the coefficient of x.

97. Solve the equation in **Exercise 95** using cross products.

98. Compare the answers from **Exercises 96(b) and 97.** What do you notice?

Summary Exercises *Applying Problem-Solving Techniques*

The following problems are of the various types discussed in this chapter. Solve each problem.

1. On an algebra test, the highest grade was 42 points more than the lowest grade. The sum of the two grades was 138. Find the lowest grade.

2. Find the measure of an angle whose supplement is 35° more than twice its complement.

3. If 2 is added to five times a number, the result is equal to 5 more than four times the number. Find the number.

4. Find two consecutive even integers such that four times the greater added to the lesser is 98.

5. Find the measures of the marked angles.

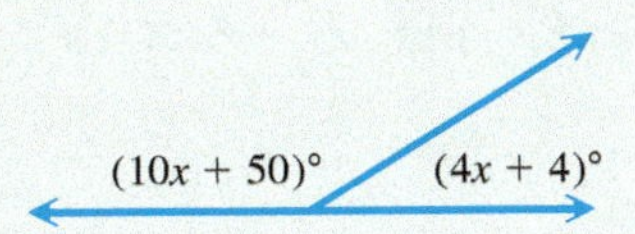

6. Find the measures of the marked angles.

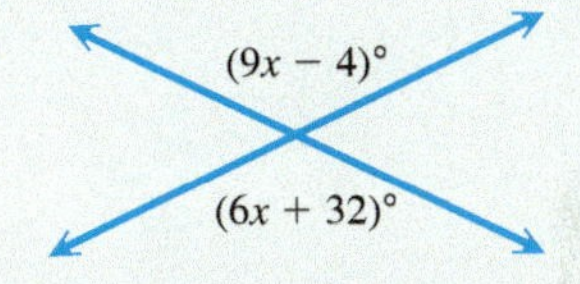

7. The perimeter of a certain square is seven times the length of a side, decreased by 12. Find the length of a side.

8. A store has 39 qt of milk, some in pint cartons and some in quart cartons. There are six times as many quart cartons as pint cartons. How many quart cartons are there? (*Hint:* 1 qt = 2 pt)

9. A music player that normally sells for \$90 is on sale for \$75. What is the percent discount on the player?

10. Two slices of bacon contain 85 calories. How many calories are there in twelve slices of bacon?

11. Athletes in vigorous training programs can eat 50 calories per day for every 2.2 lb of body weight. To the nearest hundred, how many calories can a 175 lb athlete consume per day? (Data from *The Gazette*.)

12. In the 2012 Summer Olympics in London, England, the United States won 16 more medals than China, and Russia won 6 fewer medals than China. The total number of medals won by the United States, China, and Russia was 274. How many medals did each country win? (Data from: www.espn.go.com)

13. Find the best buy (based on price per unit). Give the unit price to the nearest thousandth for that size. (Data from HyVee.)

SPAGHETTI SAUCE

Size	Price
14 oz	\$1.79
24 oz	\$1.77
48 oz	\$3.65

14. A fully inflated professional basketball has a circumference of 78 cm. What is the radius of a circular cross section through the center of the ball? (Use 3.14 as the approximation for π.) Round the answer to the nearest hundredth.

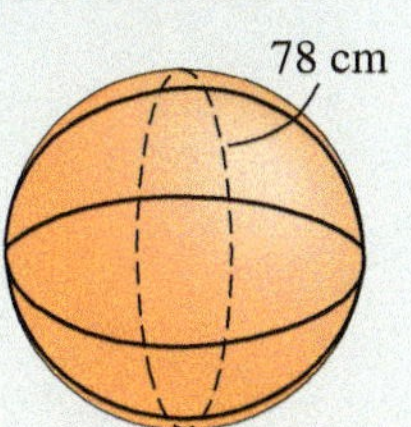

2.7 Solving Linear Inequalities

OBJECTIVES

1. Graph intervals on a number line.
2. Use the addition property of inequality.
3. Use the multiplication property of inequality.
4. Solve linear inequalities.
5. Solve applied problems using inequalities.
6. Solve linear inequalities with three parts.

An **inequality** relates algebraic expressions using the symbols

$<$ "is less than," $\leq$ "is less than or equal to,"

$>$ "is greater than," $\geq$ "is greater than or equal to."

In each case, the interpretation is based on reading the symbol from left to right.

Linear Inequality in One Variable

A **linear inequality in one variable** (here x) can be written in the form

$$Ax + B < C, \quad Ax + B \leq C, \quad Ax + B > C, \quad \text{or} \quad Ax + B \geq C,$$

where A, B, and C represent real numbers and $A \neq 0$.

Examples: $x + 5 < 2$, $z - \frac{3}{4} \geq 5$, and $2k + 5 \leq 10$ Linear inequalities

We solve a linear inequality by finding all of its real number solutions. For example, the set

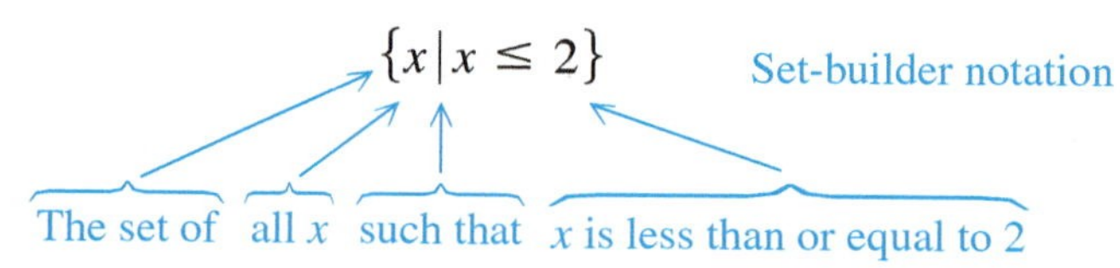

includes *all real numbers* that are less than or equal to 2, not just the *integers* less than or equal to 2.

OBJECTIVE 1 Graph intervals on a number line. Graphing is a good way to show the solution set of an inequality. To graph all real numbers belonging to the set

$$\{x \mid x \leq 2\},$$

we place a square bracket at 2 on a number line and draw an arrow extending from the bracket to the left (because all numbers *less than* 2 are also part of the graph). See **Figure 15.**

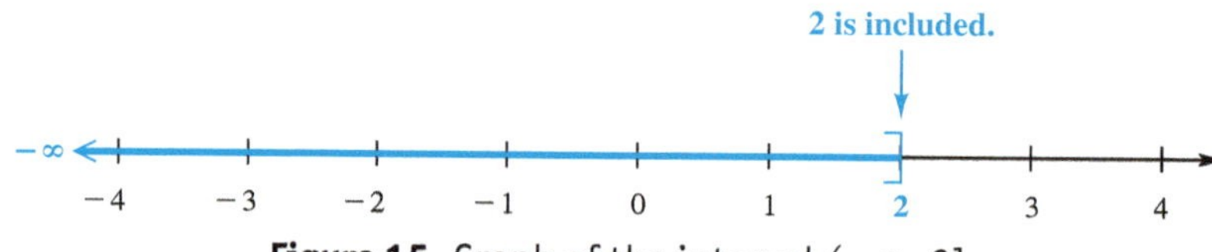

Figure 15 Graph of the interval $(-\infty, 2]$

The set of numbers less than or equal to 2 is an example of an **interval** on a number line. We can write this interval using **interval notation** as follows.

$$(-\infty, 2] \quad \text{Interval notation}$$

The **negative infinity symbol** $-\infty$ here does not indicate a number, but shows that the interval includes *all* real numbers less than 2. Again, the square bracket indicates that 2 is part of the solution. Intervals that continue indefinitely in the positive direction are written with the **positive infinity symbol** ∞.

EXAMPLE 1 Graphing an Interval on a Number Line

Write the inequality $x > -5$ in interval notation, and graph the interval.

Here x can represent any value greater than -5 but *cannot* equal -5, written $(-5, \infty)$. We place a parenthesis at -5 and draw an arrow to the right, as in **Figure 16.** The parenthesis indicates that -5 is *not* part of the graph.

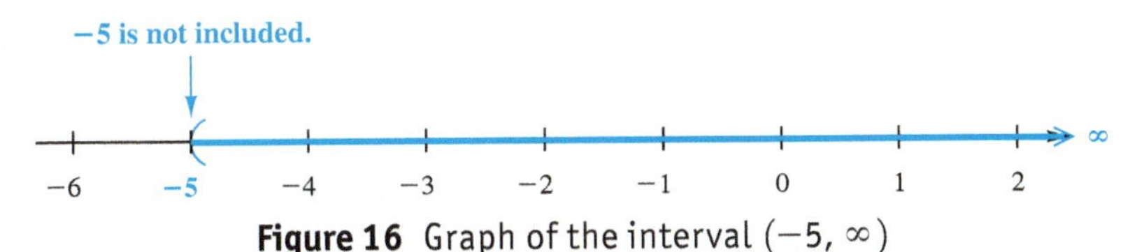

Figure 16 Graph of the interval $(-5, \infty)$

Work Problem **1** at the Side. ▶

Important Concepts Regarding Interval Notation

1. A parenthesis indicates that an endpoint is *not included* in a solution set.
2. A bracket indicates that an endpoint is *included* in a solution set.
3. A parenthesis is *always* used next to an infinity symbol, $-\infty$ or ∞.
4. The set of all real numbers is written in interval notation as $(-\infty, \infty)$.

EXAMPLE 2 Graphing an Interval on a Number Line

Write the inequality $3 > x$ in interval notation, and graph the interval.

The statement $3 > x$ means the same as $x < 3$. ***The inequality symbol continues to point to the lesser value.*** The graph of $x < 3$, written in interval notation as $(-\infty, 3)$, is shown in **Figure 17.**

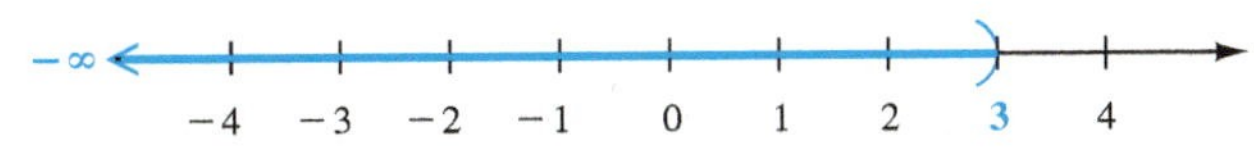

Figure 17 Graph of the interval $(-\infty, 3)$

Work Problem **2** at the Side. ▶

METHODS OF EXPRESSING SOLUTION SETS OF LINEAR INEQUALITIES

Set-Builder Notation	Interval Notation	Graph
$\{x \mid x < a\}$	$(-\infty, a)$	a
$\{x \mid x \leq a\}$	$(-\infty, a]$	a
$\{x \mid x > a\}$	(a, ∞)	a
$\{x \mid x \geq a\}$	$[a, \infty)$	a
$\{x \mid x$ is a real number$\}$	$(-\infty, \infty)$	

OBJECTIVE 2 Use the addition property of inequality. Consider the true inequality $2 < 5$. If 4 is added to each side, the result is also a true statement.

$$2 + 4 < 5 + 4 \quad \text{Add 4.}$$
$$6 < 9 \quad \text{True}$$

This example suggests the **addition property of inequality.**

1 Write each inequality in interval notation, and graph the interval.

(a) $x \leq 3$

(b) $x > -4$

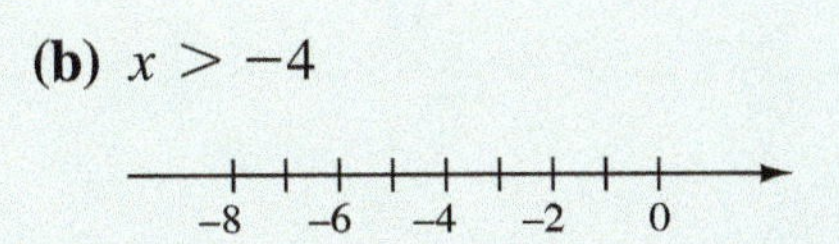

(c) $x \leq -\frac{3}{4}$

2 Write each inequality in interval notation and graph the interval.

(a) $-4 \geq x$

(b) $0 < x$

Answers

1. (a) $(-\infty, 3]$

(b) $(-4, \infty)$

(c) $\left(-\infty, -\frac{3}{4}\right]$

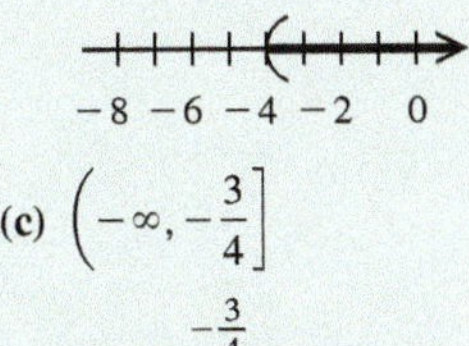

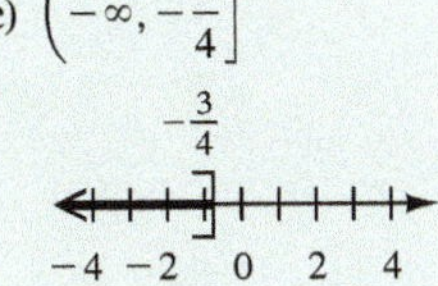

2. (a) $(-\infty, -4]$

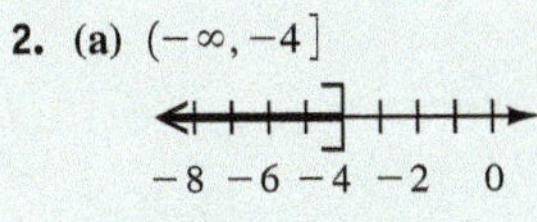

(b) $(0, \infty)$

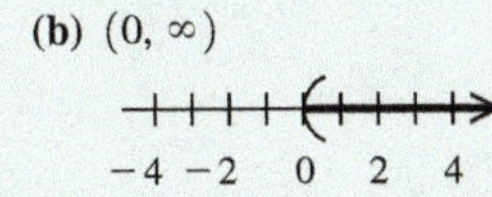

3 Solve each inequality, and graph the solution set.

(a) $-1 + 8r < 7r + 2$

(b) $5 + 5x \geq 4x + 3$

Addition Property of Inequality

If A, B, and C represent real numbers, then the inequalities

$$A < B \quad \text{and} \quad A + C < B + C \quad \text{are equivalent.*}$$

In words, the same number may be added to each side of an inequality without changing the solution set.

*This also applies to $A \leq B$, $A > B$, and $A \geq B$.

As with the addition property of equality, the same number may be subtracted from each side of an inequality.

EXAMPLE 3 **Using the Addition Property of Inequality**

Solve $7 + 3x \geq 2x - 5$, and graph the solution set.

$7 + 3x \geq 2x - 5$ — As with equations, our goal is to isolate x.

$7 + 3x - 2x \geq 2x - 5 - 2x$ — Subtract $2x$.

$7 + x \geq -5$ — Combine like terms.

$7 + x - 7 \geq -5 - 7$ — Subtract 7.

$x \geq -12$ — Combine like terms.

The solution set is $[-12, \infty)$. Its graph is shown in **Figure 18.**

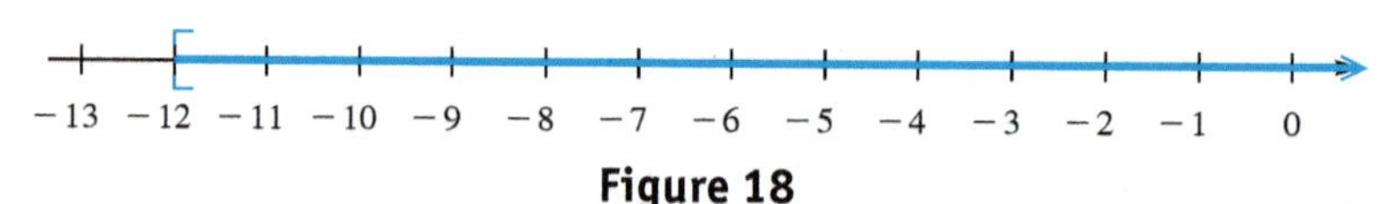

Figure 18

◀ **Work Problem 3 at the Side.**

Note

Because an inequality has many solutions, we cannot check all of them by substitution as we did with the single solution of an equation. To check the solutions in the interval $[-12, \infty)$ in **Example 3,** we first substitute -12 for x in the related *equation*.

CHECK

$7 + 3x = 2x - 5$ — Related equation

$7 + 3(-12) \stackrel{?}{=} 2(-12) - 5$ — Let $x = -12$.

$7 - 36 \stackrel{?}{=} -24 - 5$ — Multiply.

$-29 = -29$ ✓ — True

A true statement results, so -12 is indeed the "boundary" point. Now we test a number other than -12 from the interval $[-12, \infty)$. We choose 0.

CHECK

$7 + 3x \geq 2x - 5$ — Original inequality

$7 + 3(0) \stackrel{?}{\geq} 2(0) - 5$ — Let $x = 0$.

$7 \geq -5$ ✓ — True (0 is easy to substitute.)

Again, a true statement results, so the checks confirm that solutions to the inequality are in the interval $[-12, \infty)$. Any number "outside" the interval $[-12, \infty)$, that is, any number in $(-\infty, -12)$, will give a false statement when tested. (Try this with $x = -13$. A false statement, $-32 \geq -31$, results.)

Answers

3. (a) $(-\infty, 3)$

−4 −2 0 2 3 4

(b) $[-2, \infty)$

−4 −2 0 2 4

OBJECTIVE 3 Use the multiplication property of inequality. Consider the true inequality $3 < 7$. Multiply each side by the positive number 2.

$$3 < 7$$
$$2(3) < 2(7) \quad \text{Multiply by 2.}$$
$$6 < 14 \quad \text{True}$$

The result is a true statement. Now multiply each side of $3 < 7$ by the negative number -5.

$$3 < 7$$
$$-5(3) < -5(7) \quad \text{Multiply by } -5.$$
$$-15 < -35 \quad \text{False}$$

To obtain a true statement when multiplying each side by -5, ***we must reverse the direction of the inequality symbol.***

$$3 < 7$$
$$-5(3) > -5(7) \quad \text{Multiply by } -5. \text{ Reverse the direction of the symbol.}$$
$$-15 > -35 \quad \text{True}$$

Work Problem 4 at the Side. ▶

These examples suggest the **multiplication property of inequality.**

Multiplication Property of Inequality

Let A, B, and C represent real numbers, where $C \neq 0$.

1. If C is ***positive***, then the inequalities

$$A < B \quad \text{and} \quad AC < BC \quad \text{are equivalent.*}$$

2. If C is ***negative***, then the inequalities

$$A < B \quad \text{and} \quad AC > BC \quad \text{are equivalent.*}$$

In words, each side of an inequality may be multiplied by the same positive number without changing the direction of the inequality symbol. ***If the multiplier is negative, we must reverse the direction of the inequality symbol.***

*This also applies to $A \leq B$, $A > B$, and $A \geq B$.

As with the multiplication property of equality, the same nonzero number may be divided into each side.

Note the following differences for positive and negative numbers.

1. When each side of an inequality is multiplied or divided by a ***positive number,*** the direction of the inequality symbol *does not change.*
2. When each side of an inequality is multiplied or divided by a ***negative number, reverse the direction of the inequality symbol.***

4 Work each of the following.

(a) Multiply each side of

$$-3 < 7$$

by 2 and then by -5. Reverse the direction of the inequality symbol if necessary to make a true statement.

(b) Multiply each side of

$$3 > -7$$

by 2 and then by -5. Reverse the direction of the inequality symbol if necessary to make a true statement.

(c) Multiply each side of

$$-7 < -3$$

by 2 and then by -5. Reverse the direction of the inequality symbol if necessary to make a true statement.

Answers

4. **(a)** $-6 < 14$; $15 > -35$
(b) $6 > -14$; $-15 < 35$
(c) $-14 < -6$; $35 > 15$

EXAMPLE 4 **Using the Multiplication Property of Inequality**

Solve each inequality, and graph the solution set.

(a) $3x < -18$

We divide each side by 3, a positive number, so the direction of the inequality symbol *does not* change. ***(It does not matter that the number on the right side of the inequality is negative.)***

$$3x < -18$$

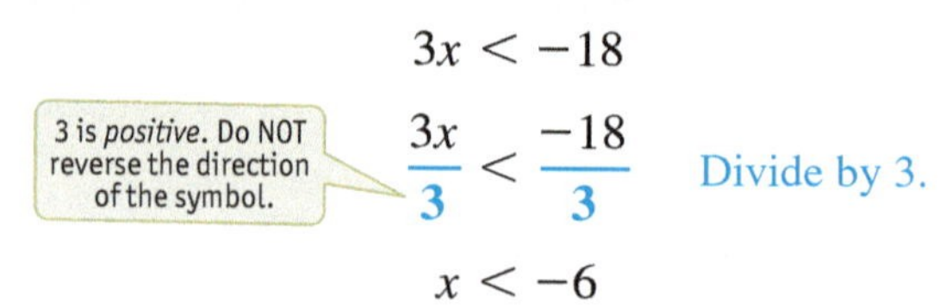

$$\frac{3x}{3} < \frac{-18}{3}$$

$$x < -6$$

The solution set is $(-\infty, -6)$. The graph is shown in **Figure 19.**

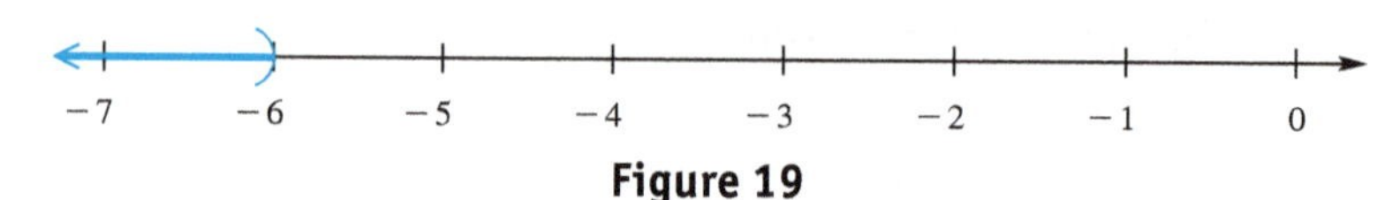

Figure 19

(b) $-4t \geq 8$

Each side of the inequality must be divided by -4, a negative number, which *does* require changing the direction of the inequality symbol.

$$-4t \geq 8$$

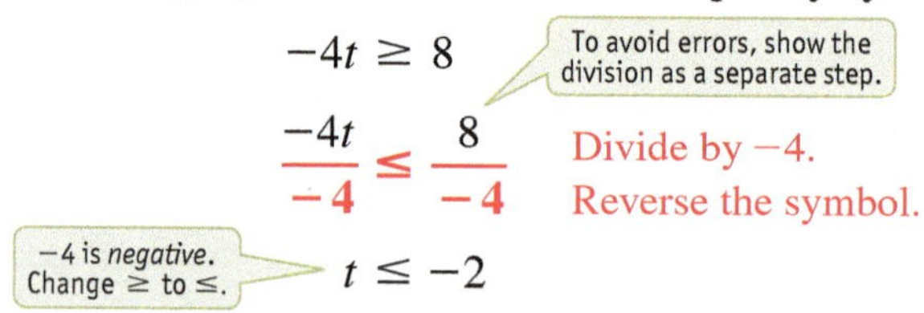

$$\frac{-4t}{-4} \leq \frac{8}{-4}$$

$$t \leq -2$$

The solution set $(-\infty, -2]$ is graphed in **Figure 20.**

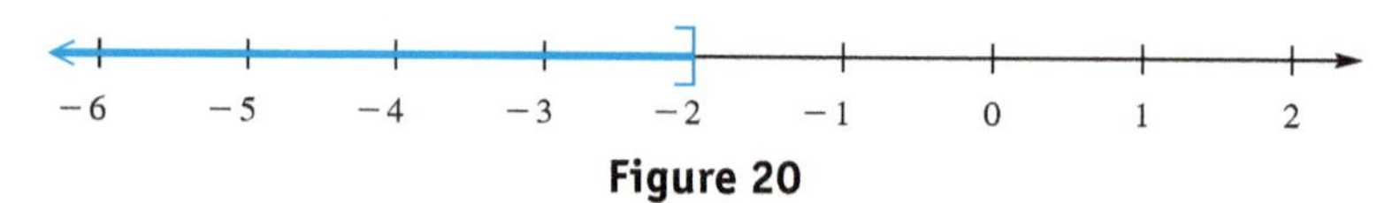

Figure 20

◀ **Work Problem 5 at the Side.**

OBJECTIVE 4 **Solve linear inequalities.**

Solving a Linear Inequality in One Variable

Step 1 **Simplify each side separately.** Use the distributive property as needed.

- Clear any parentheses.
- Clear any fractions or decimals.
- Combine like terms.

Step 2 **Isolate the variable terms on one side.** Use the addition property of inequality so that all terms with variables are on one side of the inequality and all constants (numbers) are on the other side.

Step 3 **Isolate the variable.** Use the multiplication property of inequality to obtain an inequality in one of the following forms, where k is a constant (number).

variable $< k$, variable $\leq k$, variable $> k$, or variable $\geq k$

Remember: Reverse the direction of the inequality symbol only when multiplying or dividing each side of an inequality by a negative number.

---

5 Solve each inequality, and graph the solution set.

GS **(a)** $9x < -18$

$$\frac{9x}{9}\ (<\!/\!>)\ \frac{-18}{\_\_\_\_}$$

$$x\ (<\!/\!>)\ \_\_\_\_$$

The solution set is ____.

GS **(b)** $-2r > -12$

$$\frac{-2r}{\_\_\_\_}\ (<\!/\!>)\ \frac{-12}{\_\_\_\_}$$

$$r\ (<\!/\!>)\ \_\_\_\_$$

The solution set is ____.

(c) $-5p \leq 0$

Answers

5. (a) $<$; 9; $<$; -2; $(-\infty, -2)$

−4 −2 0

(b) -2; $<$; -2; $<$; 6; $(-\infty, 6)$

0 2 4 6

(c) $[0, \infty)$

−3 −1 0 1 3

EXAMPLE 5 **Solving a Linear Inequality**

Solve $3x + 2 - 5 > -x + 7 + 2x$, and graph the solution set.

Step 1 Combine like terms and simplify.

$$3x + 2 - 5 > -x + 7 + 2x$$
$$3x - 3 > x + 7$$

Step 2 Use the addition property of inequality.

$$3x - 3 - x > x + 7 - x \quad \text{Subtract } x.$$
$$2x - 3 > 7 \quad \text{Combine like terms.}$$
$$2x - 3 + 3 > 7 + 3 \quad \text{Add 3.}$$
$$2x > 10 \quad \text{Combine like terms.}$$

Step 3 Use the multiplication property of inequality.

$$\frac{2x}{2} > \frac{10}{2} \quad \text{Divide by 2.}$$

Because 2 is positive, keep the symbol >.

$$x > 5$$

The solution set is $(5, \infty)$. Its graph is shown in **Figure 21.**

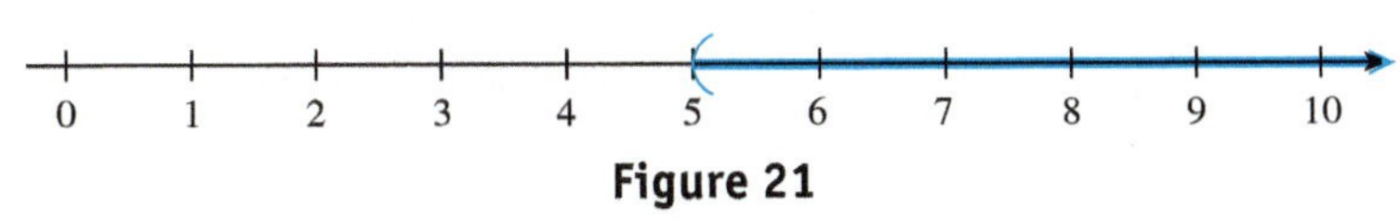

Figure 21

Work Problem **6** at the Side. ▶

EXAMPLE 6 **Solving a Linear Inequality**

Solve $5(k - 3) - 7k \geq 4(k - 3) + 9$, and graph the solution set.

Step 1 $5(k - 3) - 7k \geq 4(k - 3) + 9$ Start by clearing parentheses.

$$5k - 15 - 7k \geq 4k - 12 + 9 \quad \text{Distributive property}$$
$$-2k - 15 \geq 4k - 3 \quad \text{Combine like terms.}$$

Step 2

$$-2k - 15 - 4k \geq 4k - 3 - 4k \quad \text{Subtract } 4k.$$
$$-6k - 15 \geq -3 \quad \text{Combine like terms.}$$
$$-6k - 15 + 15 \geq -3 + 15 \quad \text{Add 15.}$$
$$-6k \geq 12 \quad \text{Combine like terms.}$$

Step 3

$$\frac{-6k}{-6} \leq \frac{12}{-6} \quad \text{Divide by } -6. \text{ Reverse the symbol.}$$

Because −6 is negative, change ≥ to ≤.

$$k \leq -2$$

The solution set is $(-\infty, -2]$. Its graph is shown in **Figure 22.**

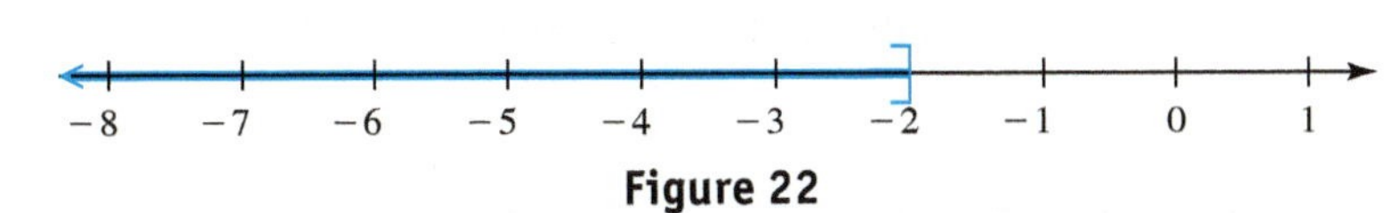

Figure 22

Work Problem **7** at the Side. ▶

6 Solve.

$$7x - 6 + 1 \geq 5x - x + 2$$

Graph the solution set.

7 Solve.

$$-15 - (2x + 1) \geq 4(x - 1) - 3x$$

Graph the solution set.

Answers

6. $\left[\frac{7}{3}, \infty\right)$

7. $(-\infty, -4]$

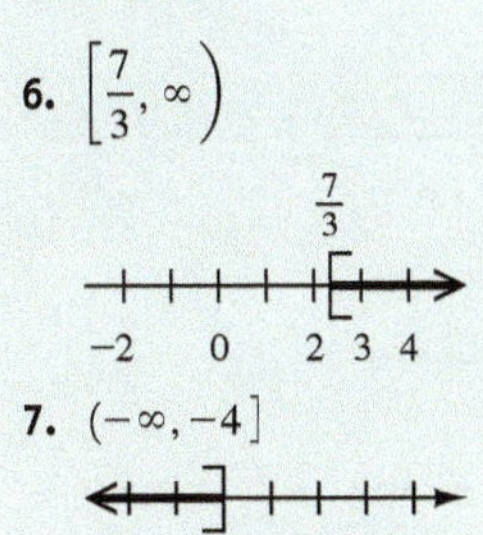

8 Solve each inequality, and graph the solution set.

(a) $\frac{1}{8}(x+4) \geq \frac{1}{6}(2x+8)$

(b) $\frac{1}{2}(3x-1) > \frac{1}{5}(x+4)$

9 Translate each statement into an inequality, using x as the variable.

(a) The total cost is less than \$10.

(b) Chicago received at most 5 in. of snow.

(c) The car's speed exceeded 60 mph.

(d) You must be at least 18 yr old to vote.

Answers

8. (a) $(-\infty, -4]$

−6 −4 −2 0

(b) $(1, \infty)$

−4 −2 0 1 2 4

9. (a) $x < 10$ **(b)** $x \leq 5$ **(c)** $x > 60$ **(d)** $x \geq 18$

EXAMPLE 7 Solving a Linear Inequality with Fractions

Solve $\frac{3}{4}(x-6) < \frac{2}{3}(5x+1)$, and graph the solution set.

Step 1 $\frac{3}{4}(x-6) < \frac{2}{3}(5x+1)$ — Clear the parentheses first. Then clear the fractions.

$\frac{3}{4}x - \frac{9}{2} < \frac{10}{3}x + \frac{2}{3}$ Distributive property

$12\left(\frac{3}{4}x - \frac{9}{2}\right) < 12\left(\frac{10}{3}x + \frac{2}{3}\right)$ Multiply each side by the LCD, 12.

$9x - 54 < 40x + 8$ Distributive property

Step 2 $9x - 54 - 40x < 40x + 8 - 40x$ Subtract $40x$.

$-31x - 54 < 8$ Combine like terms.

$-31x - 54 + 54 < 8 + 54$ Add 54.

$-31x < 62$ Combine like terms.

Step 3 $\frac{-31x}{-31} > \frac{62}{-31}$ Divide by -31. Reverse the symbol.

$x > -2$

The solution set is $(-2, \infty)$. Its graph is shown in **Figure 23.**

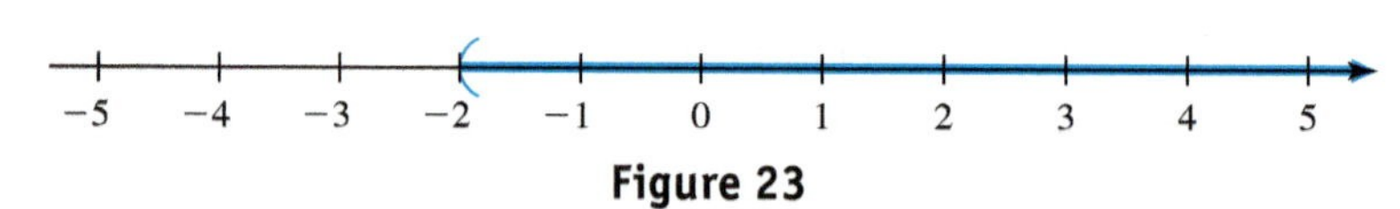

Figure 23

◀ **Work Problem 8 at the Side.**

OBJECTIVE 5 Solve applied problems using inequalities.

WORDS AND PHRASES THAT INDICATE INEQUALITY

Phrase/Word	Example	Inequality
Is greater than	A number *is greater than* 4	$x > 4$
Is less than	A number *is less than* −12	$x < -12$
Exceeds	A number *exceeds* 3.5	$x > 3.5$
Is at least	A number *is at least* 6	$x \geq 6$
Is at most	A number *is at most* 8	$x \leq 8$

◀ **Work Problem 9 at the Side.**

CAUTION

Do not confuse statements such as "5 is more than a number" with the phrase "5 more than a number." The first of these is expressed as $5 > x$, while the second is expressed as $x + 5$, or $5 + x$.

The next example uses the idea of finding the average of a number of scores. ***In general, to find the average of n numbers, add the numbers and divide by n.*** We continue to use the six problem-solving steps, changing Step 3 to "Write an inequality."

EXAMPLE 8 Finding an Average Test Score

Brent has scores of 86, 88, and 78 (out of a possible 100) on each of his first three tests in geometry. If he wants an average of at least 80 after his fourth test, what are the possible scores he can make on that test?

Step 1 **Read** the problem again.

Step 2 **Assign a variable.** Let x = Brent's score on his fourth test.

Step 3 **Write an inequality.**

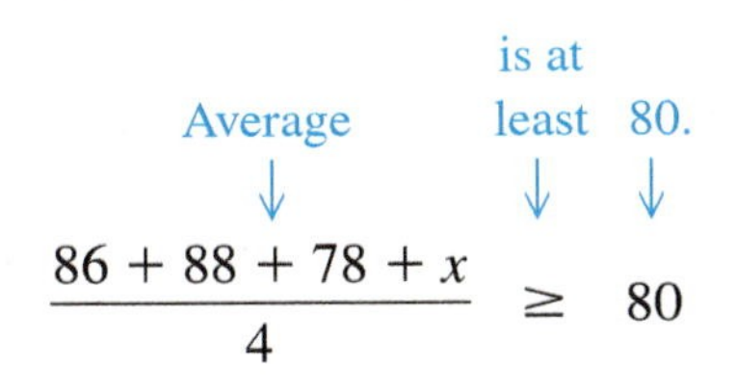

$$\frac{86 + 88 + 78 + x}{4} \geq 80$$ To find his average after four tests, add the test scores and divide by 4.

Step 4 **Solve.**

$$\frac{252 + x}{4} \geq 80$$ Add the known scores.

$$4\left(\frac{252 + x}{4}\right) \geq 4(80)$$ Multiply by 4.

$$252 + x \geq 320$$

$$252 + x - 252 \geq 320 - 252$$ Subtract 252.

$$x \geq 68$$ Combine like terms.

Step 5 **State the answer.** He must score 68 or more on the fourth test to have an average of *at least* 80.

Step 6 **Check.** $\frac{86 + 88 + 78 + 68}{4} = \frac{320}{4} = 80$

To complete the check, also show that any number greater than 68 (but less than or equal to 100) makes the average greater than 80.

Work Problem 10 at the Side. ▶

OBJECTIVE 6 Solve linear inequalities with three parts. An inequality that says that one number is *between* two other numbers is a **three-part inequality.** For example,

$-3 < 5 < 7$ says that 5 is *between* -3 and 7.

EXAMPLE 9 Graphing a Three-Part Inequality

Write the inequality $-3 \leq x < 2$ in interval notation, and graph the interval.

The statement is read "-3 *is less than or equal to x* **and** *x is less than* 2." We want the set of numbers that are *between* -3 and 2, with -3 included and 2 excluded. In interval notation, we write $[-3, 2)$, using a square bracket at -3 because -3 is part of the graph and a parenthesis at 2 because 2 is not part of the graph. See **Figure 24.**

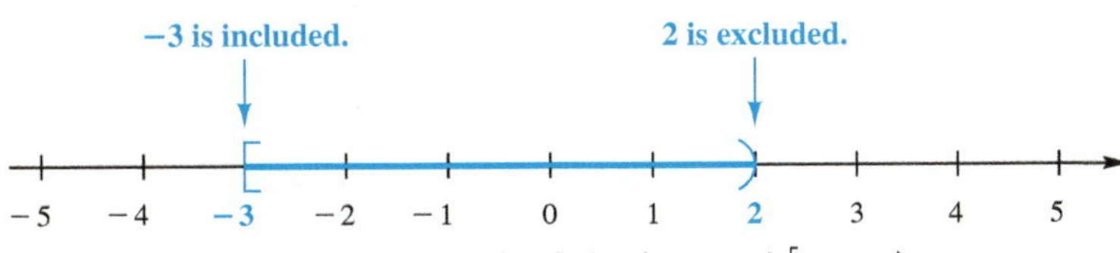

Figure 24 Graph of the interval $[-3, 2)$

Work Problem 11 at the Side. ▶

10 Solve each problem.

(a) Matthew has grades of 98 and 85 on his first two tests in algebra. If he wants an average of at least 92 after his third test, what score must he make on that test?

(b) Maggie has scores of 98, 86, and 88 on her first three tests in algebra. If she wants an average of at least 90 after her fourth test, what score must she make on her fourth test?

11 Write each inequality in interval notation, and graph the interval.

(a) $-7 < x < -2$

(b) $-6 < x \leq -4$

Answers

10. (a) 93 or more (b) 88 or more

11. (a) $(-7, -2)$

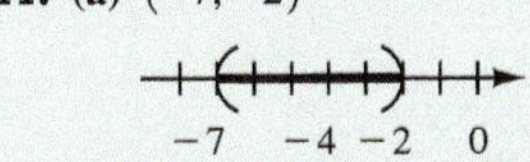

(b) $(-6, -4]$

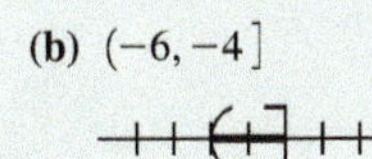

12 Solve each inequality, and graph the solution set.

(a) $2 \le 3x - 1 \le 8$

(b) $-4 < \frac{3}{2}x - 1 < 0$

The three-part inequality $3 < x + 2 < 8$ says that $x + 2$ is between 3 and 8. We solve this inequality as follows.

$$3 - 2 < x + 2 - 2 < 8 - 2 \quad \text{Subtract 2 from } each \text{ part.}$$

$$1 < x < 6$$

The idea is to obtain an inequality in the form

a number $< x <$ **another number.**

CAUTION

Three-part inequalities are written so that the symbols point in the same direction and both point toward the lesser number. Do *not* write $8 < x + 2 < 3$, which would imply that $8 < 3$, a **false** statement.

EXAMPLE 10 Solving a Three-Part Inequality

Solve each inequality, and graph the solution set.

(a)

$$4 < 3x - 5 < 10 \quad \text{Work with all three parts at the same time.}$$

$$4 + 5 < 3x - 5 + 5 < 10 + 5 \quad \text{Add 5 to each part.}$$

$$9 < 3x < 15 \quad \text{Combine like terms.}$$

$$\frac{9}{3} < \frac{3x}{3} < \frac{15}{3} \quad \text{Divide each part by 3.}$$

Remember to divide all *three* parts by 3.

$$3 < x < 5$$

The solution set is $(3, 5)$. Its graph is shown in **Figure 25**.

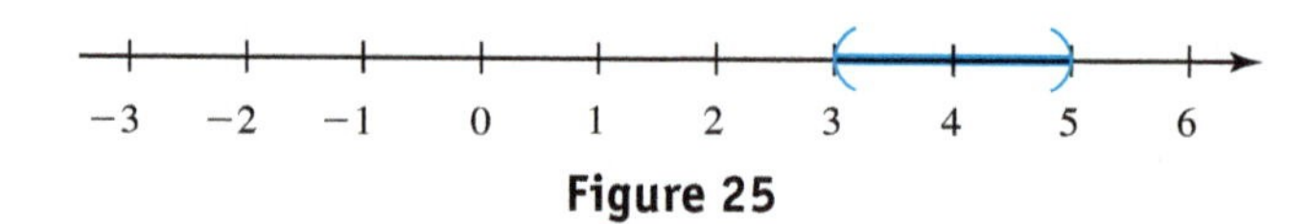

Figure 25

(b)

$$-4 \le \frac{2}{3}m - 1 < 8 \quad \text{Work with all three parts at the same time.}$$

$$3(-4) \le 3\left(\frac{2}{3}m - 1\right) < 3(8) \quad \text{Multiply each part by 3 to clear the fraction.}$$

$$-12 \le 2m - 3 < 24 \quad \text{Multiply. Use the distributive property.}$$

$$-12 + 3 \le 2m - 3 + 3 < 24 + 3 \quad \text{Add 3 to each part.}$$

$$-9 \le 2m < 27 \quad \text{Combine like terms.}$$

$$\frac{-9}{2} \le \frac{2m}{2} < \frac{27}{2} \quad \text{Divide each part by 2.}$$

$$-\frac{9}{2} \le m < \frac{27}{2}$$

The solution set is $\left[-\frac{9}{2}, \frac{27}{2}\right)$. Its graph is shown in **Figure 26**.

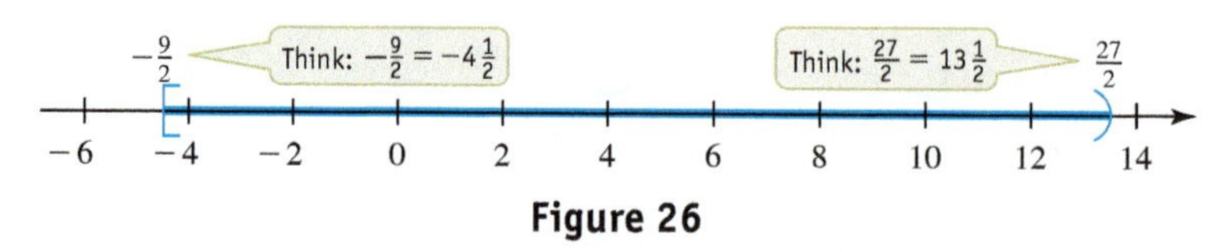

Figure 26

◄ Work Problem 12 at the Side.

Answers

12. (a) $[1, 3]$

(b) $\left(-2, \frac{2}{3}\right)$

Note

The inequality in **Example 10(b)** could also be solved as follows.

$$-4 \le \frac{2}{3}m - 1 < 8 \quad \text{Inequality from Example 10(b)}$$

$$-4 + 1 \le \frac{2}{3}m - 1 + 1 < 8 + 1 \quad \text{Add 1 to each part.}$$

$$-3 \le \frac{2}{3}m < 9$$

$$\frac{3}{2}(-3) \le \frac{3}{2}\left(\frac{2}{3}m\right) < \frac{3}{2}(9) \quad \text{Multiply each part by } \tfrac{3}{2}.$$

$$-\frac{9}{2} \le m < \frac{27}{2}$$

The same solution set $\left[-\frac{9}{2}, \frac{27}{2}\right)$ results.

Be especially careful of whether to use parentheses or square brackets when writing and graphing solution sets of three part inequalities. The following table illustrates the four possibilities that may occur.

METHODS OF EXPRESSING SOLUTION SETS OF THREE-PART INEQUALITIES

Set-Builder Notation	Interval Notation	Graph
$\{x \mid a < x < b\}$	(a, b)	a ( ) b
$\{x \mid a < x \le b\}$	$(a, b]$	a ( ] b
$\{x \mid a \le x < b\}$	$[a, b)$	a [ ) b
$\{x \mid a \le x \le b\}$	$[a, b]$	a [ ] b

2.7 Exercises

FOR EXTRA HELP — *Go to MyMathLab for worked-out, step-by-step solutions to exercises enclosed in a square and video solutions to ▶ exercises.*

CONCEPT CHECK *Work each problem.*

1. When graphing an inequality, use a parenthesis if the inequality symbol is _________ or _________. Use a square bracket if the inequality symbol is _________ or _________.

2. *True* or *false*? In interval notation, a square bracket is sometimes used next to an infinity symbol.

3. In interval notation, the set $\{x \mid x > 0\}$ is written _______.

4. In interval notation, the set of all real numbers is written _______.

CONCEPT CHECK *Write an inequality using the variable x that corresponds to each graph of solutions on a number line.*

5.

−4 −3 −2 −1 0 1 2 3

6.

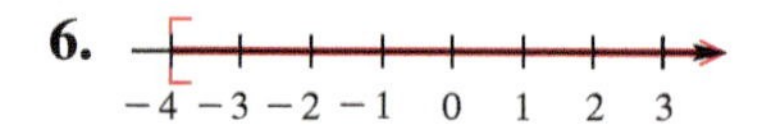

7.

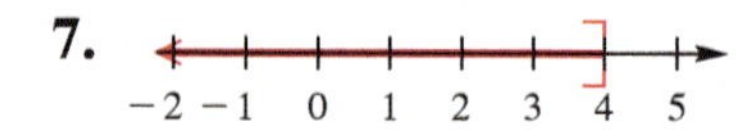

8.

−2 −1 0 1 2 3 4 5

9.

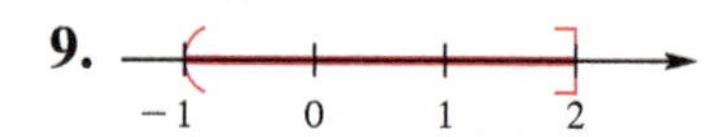

10.

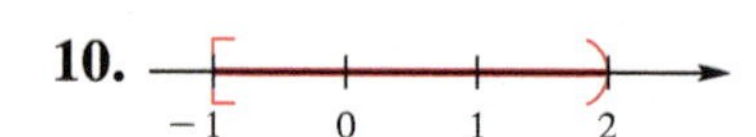

Write each inequality in interval notation, and graph the interval. **See Examples 1, 2, and 9.**

11. $k \leq 4$ ▶

12. $r \leq -10$

13. $x > -3$

14. $x > 3$

15. $8 \leq x \leq 10$

16. $3 \leq x \leq 5$

17. $0 < x \leq 10$ ▶

18. $-3 \leq x < 5$

Solve each inequality. Write the solution set in interval notation, and graph it. **See Example 3.**

19. $z - 8 > -7$

20. $p - 3 > -11$

21. $2k + 3 \geq k + 8$ ▶

22. $3x + 7 \geq 2x + 11$

23. $3n + 5 < 2n - 1$

24. $5x - 2 < 4x - 5$

25. Under what conditions must the inequality symbol be reversed when using the multiplication property of inequality?

26. Explain the steps you would use to solve the inequality $-5x > 20$.

Solve each inequality. Write the solution set in interval notation, and graph it.
See Example 4.

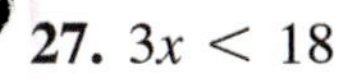

27. $3x < 18$

28. $5x < 35$

29. $2x \geq -20$

30. $6m \geq -24$

31. $-8t > 24$

32. $-7x > 49$

33. $-x \geq 0$

34. $-k < 0$

35. $-\frac{3}{4}r < -15$

36. $-\frac{7}{8}t < -14$

37. $-0.02x \leq 0.06$

38. $-0.03v \geq -0.12$

Solve each inequality. Write the solution set in interval notation, and graph it.
See Examples 3–7.

39. $8x + 9 \leq -15$

40. $6x + 7 \leq -17$

41. $-4x - 3 < 1$

42. $-5x - 4 < 6$

43. $5r + 1 \geq 3r - 9$

44. $6t + 3 < 3t + 12$

45. $6x + 3 + x < 2 + 4x + 4$

46. $-4w + 12 + 9w \geq w + 9 + w$

47. $x - 4 - 7x \geq 2 - 3x - 6$

48. $7x - 6 - 14x < 4 + 5x - 10$

49. $5(t - 1) > 3(t - 2)$

50. $7(m - 2) < 4(m - 4)$

51. $5(x + 3) - 6x \leq 3(2x + 1) - 4x$

52. $2(x - 5) + 3x < 4(x - 6) + 1$

53. $\frac{2}{3}(p + 3) > \frac{5}{6}(p - 4)$

54. $\frac{7}{9}(n - 4) \leq \frac{4}{3}(n + 5)$

55. $\frac{1}{3}(5x - 4) \geq \frac{2}{5}(x + 3)$

56. $\frac{5}{12}(5x - 7) < \frac{5}{6}(x - 5)$

57. $4x - (6x + 1) \leq 8x + 2(x - 3)$

58. $2x - (4x + 3) < 6x + 3(x + 4)$

59. $5(2k + 3) - 2(k - 8) > 3(2k + 4) + k - 2$

60. $2(3z - 5) + 4(z + 6) \geq 2(3z + 2) + 3z - 15$

CONCEPT CHECK *Translate each statement into an inequality. Use x as the variable.*

61. You must be at least 16 yr old to drive.

62. Less than 1 in. of rain fell.

63. Denver received more than 8 in. of snow.

64. A full-time student must take at least 12 credits.

65. Tracy could spend at most \$20 on a gift.

66. The wind speed exceeded 40 mph.

Solve each problem. ***See Example 8.***

67. John has grades of 84 and 98 on his first two history tests. What must he score on his third test so that his average is at least 90?

68. Elizabeth has scores of 74 and 82 on her first two algebra tests. What must she score on her third test so that her average is at least 80?

69. A student has scores of 87, 84, 95, and 79 on four quizzes. What must she score on the fifth quiz to have an average of at least 85?

70. Another student has scores of 82, 93, 94, and 86 on four quizzes. What must he score on the fifth quiz to have an average of at least 90?

71. When 2 is added to the difference of six times a number and 5, the result is greater than 13 added to 5 times the number. Find all such numbers.

72. When 8 is subtracted from the sum of three times a number and 6, the result is less than 4 more than the number. Find all such numbers.

73. The formula for converting Celsius temperature to Fahrenheit is $F = \frac{9}{5}C + 32$. The Fahrenheit temperature of Providence, Rhode Island, has never exceeded 104°. How would you describe this using Celsius temperature?

74. The formula for converting Fahrenheit temperature to Celsius is $C = \frac{5}{9}(F - 32)$. If the Celsius temperature on a certain day in San Diego, California, is never more than 25°, how would you describe the corresponding Fahrenheit temperature?

75. For what values of x would the rectangle have perimeter of at least 400?

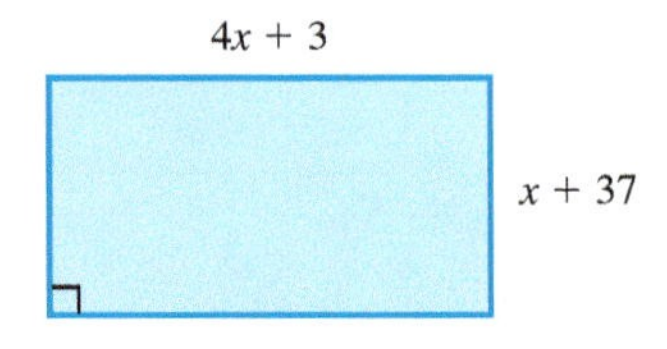

76. For what values of x would the triangle have perimeter of at least 72?

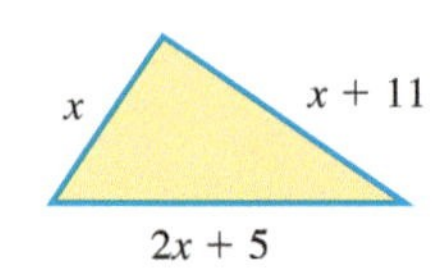

77. An international phone call costs \$2.00, plus \$0.30 per minute or fractional part of a minute. If x represents the number of minutes of the length of the call, then $2 + 0.30x$ represents the cost of the call. If Jorge has \$5.60 to spend on a call, what is the maximum total time he can use the phone?

78. At the Speedy Gas 'n Go, a car wash costs \$4.50, and gasoline is selling for \$3.40 per gal. Terri has \$43.60 to spend, and her car is so dirty that she must have it washed. What is the maximum number of gallons of gasoline that she can purchase?

79. The average monthly precipitation in Houston, Texas, for October, November, and December is 4.6 in. If 5.7 in. falls in October and 4.3 in. falls in November, how many inches must fall in December so that the average monthly precipitation for these months exceeds 4.6 in.? (Data from National Climatic Data Center.)

80. The average monthly precipitation in New Orleans, Louisiana, for June, July, and August in 6.7 in. If 8.1 in. falls in June and 5.7 in. falls in July, how many inches must fall in August so that the average monthly precipitation for these months exceeds 6.7 in.? (Data from National Climatic Data Center.)

Solve each inequality. Write the solution set in interval notation, and graph it. ***See Example 10.***

81. $-5 \leq 2x - 3 \leq 9$

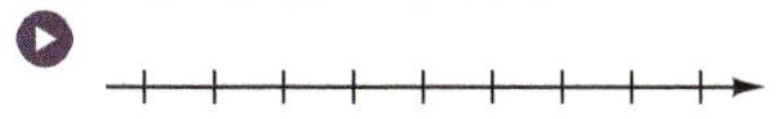

82. $-7 \leq 3x - 4 \leq 8$

83. $10 < 7p + 3 < 24$

84. $-8 \leq 3r - 1 \leq -1$

85. $-12 < -1 + 6m \leq -5$

86. $-14 \leq 1 + 5q < 3$

87. $6 \leq 3(x - 1) < 18$

88. $-4 < 2(x + 1) \leq 6$

89. $-12 \leq \frac{1}{2}z + 1 \leq 4$

90. $-6 \leq 3 + \frac{1}{3}x \leq 5$

91. $1 \leq \frac{2}{3}p + 3 \leq 7$

92. $2 < \frac{3}{4}x + 6 < 12$

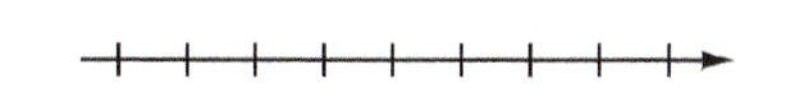

Relating Concepts (Exercises 93–96) For Individual or Group Work

Work Exercises 93–96 in order, *to see the connection between the solution of an equation and the solutions of the corresponding inequalities.*

In Exercises 93–95, solve, write the solution set in interval notation, and graph it.

93. $3x + 2 = 14$

94. $3x + 2 < 14$

95. $3x + 2 > 14$

96. Based on the results from **Exercises 93–95,** if we were to graph the solutions of

$$-4x + 3 = -1, \quad -4x + 3 > -1, \quad \text{and} \quad -4x + 3 < -1$$

on the same number line, describe the graph. Give this solution set using interval notation.

Study Skills

TAKING MATH TESTS

Techniques to Improve Your Test Score	Comments
Come prepared with a pencil, eraser, paper, and calculator, if allowed.	Working in pencil lets you erase, keeping your work neat.
Scan the entire test, note the point values of different problems, and plan your time accordingly.	To do 20 problems in 50 minutes, allow $50 \div 20 = 2.5$ minutes per problem. Spend less time on the easier problems.
Do a "knowledge dump" when you get the test. Write important notes, such as formulas, in a corner of the test.	Writing down tips and information that you've learned at the beginning allows you to relax later.
Read directions carefully, and circle any significant words. When you finish a problem, reread the directions. Did you do what was asked?	Pay attention to any announcements written on the board or made by your instructor. Ask if you don't understand something.
Show all your work. Many teachers give partial credit if some steps are correct, even if the final answer is wrong. ***Write neatly.***	If your teacher can't read your writing, you won't get credit for it. If you need more space to work, ask to use extra paper.
Write down anything that might help solve a problem: a formula, a diagram, etc. If necessary, circle the problem and come back to it later. Do ***not*** erase anything you wrote down.	If you know even a little bit about a problem, write it down. The answer may come to you as you work on it, or you may get partial credit. Don't spend too long on any one problem.
If you can't solve a problem, make a guess. Do not change it unless you find an obvious mistake.	Have a good reason for changing an answer. Your first guess is usually your best bet.
Check that the answer to an application problem is reasonable and makes sense. Reread the problem. Make sure you've answered the question.	Use common sense. Can the father really be seven years old? Would a month's rent be $32,140? Label answers, if needed: $, years, inches, etc.
Check for careless errors. Rework each problem without looking at your previous work. Then compare the two answers.	Reworking a problem from the beginning forces you to rethink it. If possible, use a different method to solve the problem.

Now Try This

Think through and answer each question.

1 **What two or three tips will you try when you take your next math test?**

____

2 **How did the tips you selected work for you when you took your math test?**

____

3 **What will you do differently when taking your next math test?**

____

Chapter 2 Summary

Key Terms

2.1

linear equation A linear equation in one variable (here x) is an equation that can be written in the form $Ax + B = C$, where A, B, and C are real numbers, and $A \neq 0$.

solution set The set of all solutions of an equation is its solution set.

equivalent equations Equations that have exactly the same solution sets are equivalent equations.

2.3

conditional equation A conditional equation is an equation that is true for some values of the variable and false for others.

identity An identity is an equation that is true for all values of the variable.

contradiction A contradiction is an equation that has no solution.

2.4

consecutive integers Two integers that differ by 1 are consecutive integers.

consecutive even (or odd) integers Two even (or odd) integers that differ by 2 are consecutive even (or odd) integers.

complementary angles Two angles whose measures have a sum of 90° are complementary angles.

right angle A right angle measures 90°.

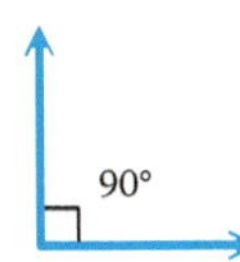

supplementary angles Two angles whose measures have a sum of 180° are supplementary angles.

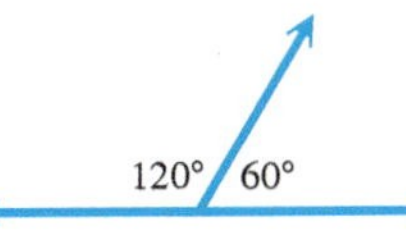

straight angle A straight angle measures 180°.

2.5

formula A formula is an equation in which variables are used to describe a relationship.

area The area of a plane geometric figure is a measure of the surface covered by the figure.

perimeter The perimeter of a plane geometric figure is the measure of the outer boundary of the figure.

vertical angles Vertical angles are angles formed by intersecting lines. They have the same measure. (In the figure, ① and ③ are vertical angles, as are ② and ④.)

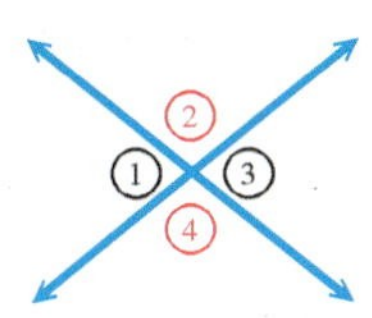

2.6

ratio A ratio is a comparison of two quantities using a quotient.

proportion A proportion is a statement that two ratios are equal.

cross products of a proportion The method of cross products provides a way of determining whether a proportion is true.

$$\frac{a}{b} = \frac{c}{d}$$

ad and *bc* are cross products.

terms In the proportion $\frac{a}{b} = \frac{c}{d}$, the terms are a, b, c, and d. The a and d terms are the **extremes,** and the b and c terms are the **means.**

2.7

inequality An inequality is a statement that relates algebraic expressions using $<$, $\leq$, $>$, or $\geq$.

linear inequality A linear inequality in one variable (here x) can be written in the form $Ax + B < C$, $Ax + B \leq C$, $Ax + B > C$, or $Ax + B \geq C$, where A, B, and C are real numbers, and $A \neq 0$.

interval An interval is a portion of a number line.

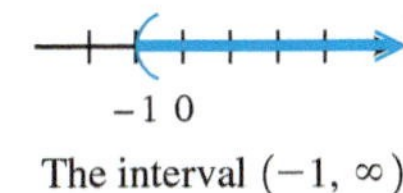

The interval $(-1, \infty)$

interval notation Interval notation is a special notation that uses parentheses ( ) and/or brackets [ ] to describe an interval on a number line.

three-part inequality An inequality that says that one number is between two other numbers is a three-part inequality.

New Symbols

$\varnothing$	empty (null) set	a **to** b, $a:b$, **or** $\frac{a}{b}$	the ratio of a to b	∞	infinity
$1°$	one degree	(a, b)	interval notation for $a < x < b$	$-\infty$	negative infinity
⅂	right angle	$[a, b]$	interval notation for $a \le x \le b$	$(-\infty, \infty)$	set of all real numbers

Test Your Word Power

See how well you have learned the vocabulary in this chapter.

1 A **solution** of an equation is a number that
A. makes an expression undefined
B. makes the equation false
C. makes the equation true
D. makes an expression equal to 0.

2 **Complementary angles** are angles
A. formed by two parallel lines
B. whose sum is 90°
C. whose sum is 180°
D. formed by perpendicular lines.

3 **Supplementary angles** are angles
A. formed by two parallel lines
B. whose sum is 90°
C. whose sum is 180°
D. formed by perpendicular lines.

4 **A ratio**
A. compares two quantities using a quotient
B. says that two quotients are equal
C. is a product of two quantities
D. is a difference of two quantities.

5 **A proportion**
A. compares two quantities using a quotient
B. says that two ratios are equal
C. is a product of two quantities
D. is a difference of two quantities.

6 **An inequality** is
A. a statement that two algebraic expressions are equal
B. a point on a number line
C. an equation with no solutions
D. a statement that relates algebraic expressions using $<$, $\le$, $>$, or $\ge$.

Answers to Test Your Word Power

1. C; *Example:* 8 is the solution of $2x + 5 = 21$.
2. B; *Example:* Angles with measures 35° and 55° are complementary angles.
3. C; *Example:* Angles with measures 112° and 68° are supplementary angles.
4. A; *Example:* $\frac{7 \text{ in.}}{12 \text{ in.}} = \frac{7}{12}$
5. B; *Example:* $\frac{2}{3} = \frac{8}{12}$
6. D; *Examples:* $x < 5$, $7 + 2y \ge 11$, $-5 < 2z - 1 \le 3$

Quick Review

Concepts	Examples
2.1 The Addition Property of Equality The same number may be added to (or subtracted from) each side of an equation without changing the solution set.	Solve. $x - 6 = 12$ $x - 6 + 6 = 12 + 6$ Add 6. $x = 18$ Combine like terms. Solution set: $\{18\}$
2.2 The Multiplication Property of Equality Each side of an equation may be multiplied (or divided) by the same nonzero number without changing the solution set.	Solve. $\frac{3}{4}x = -9$ $\frac{4}{3} \cdot \frac{3}{4}x = \frac{4}{3} \cdot (-9)$ Multiply by $\frac{4}{3}$. $x = -12$ Solution set: $\{-12\}$

Concepts	Examples
2.3 More on Solving Linear Equations **Solving a Linear Equation in One Variable** *Step 1* Simplify each side separately. • Clear any parentheses. • Clear any fractions or decimals. • Combine like terms. *Step 2* Isolate the variable terms on one side. *Step 3* Isolate the variable. *Step 4* Check.	Solve. $2x + 2(x + 1) = 14 + x$ $2x + 2x + 2 = 14 + x$ Distributive property $4x + 2 = 14 + x$ Combine like terms. $4x + 2 - x - 2 = 14 + x - x - 2$ Subtract x. Subtract 2. $3x = 12$ Combine like terms. $\frac{3x}{3} = \frac{12}{3}$ Divide by 3. $x = 4$ CHECK $2(4) + 2(4 + 1) \stackrel{?}{=} 14 + 4$ Let $x = 4$. $18 = 18$ ✓ True Solution set: $\{4\}$
2.4 An Introduction to Applications of Linear Equations **Solving an Applied Problem** *Step 1* Read. *Step 2* Assign a variable. *Step 3* Write an equation. *Step 4* Solve the equation. *Step 5* State the answer. *Step 6* Check.	One number is 5 more than another. Their sum is 21. What are the numbers? We are looking for two numbers. Let x = the lesser number. Then $x + 5$ = the greater number. $x + (x + 5) = 21$ $2x + 5 = 21$ Combine like terms. $2x = 16$ Subtract 5. $x = 8$ Divide by 2. The numbers are 8 and $8 + 5 = 13$. 13 is 5 more than 8, and $8 + 13 = 21$, as required. The answer checks.
2.5 Formulas and Additional Applications from Geometry To find the value of one of the variables in a formula, given values for the others, substitute the known values into the formula.	Find L if $\mathcal{A} = LW$, given that $\mathcal{A} = 24$ and $W = 3$. $\mathcal{A} = LW$ $24 = L \cdot 3$ $\mathcal{A} = 24$, $W = 3$ $\frac{24}{3} = \frac{L \cdot 3}{3}$ Divide by 3. $8 = L$
To solve a formula for one of the variables, isolate that variable by treating the other variables as constants (numbers) and using the steps for solving equations.	Solve $P = 2a + 2b$ for b. $P - 2a = 2a + 2b - 2a$ Subtract $2a$. $P - 2a = 2b$ Combine like terms. $\frac{P - 2a}{2} = \frac{2b}{2}$ Divide by 2. $\frac{P - 2a}{2} = b$, or $b = \frac{P - 2a}{2}$

Concepts | **Examples**

2.6 Ratio, Proportion, and Percent

To write a ratio, express quantities using the same units.

4 ft to 8 in. can be written 48 in. to 8 in., which is the ratio

$$\frac{48}{8}, \quad \text{or} \quad \frac{6}{1}.$$

To solve a proportion, use the method of cross products.

Solve. $\frac{x}{12} = \frac{35}{60}$

$$\begin{aligned} 60x &= 12 \cdot 35 && \text{Cross products} \\ 60x &= 420 && \text{Multiply.} \\ x &= 7 && \text{Divide by 60.} \end{aligned}$$

Solution set: $\{7\}$

To solve a percent problem, use the percent equation.

amount = percent (as a decimal) · base

65 is what percent of 325?

$$65 = p \cdot 325$$

$$\frac{65}{325} = p$$

$$0.2 = p, \quad \text{or} \quad 20\% = p$$

65 is 20% of 325.

2.7 Solving Linear Inequalities

Solving a Linear Inequality in One Variable

Step 1 Simplify each side separately.

- Clear any parentheses.
- Clear any fractions or decimals.
- Combine like terms.

Step 2 Isolate the variable terms on one side.

Step 3 Isolate the variable.

Be sure to reverse the direction of the inequality symbol when multiplying or dividing by a negative number.

Solve and graph the solution set.

$$\begin{aligned} 3(1 - x) + 5 - 2x &> 9 - 6 \\ 3 - 3x + 5 - 2x &> 9 - 6 && \text{Distributive property} \\ 8 - 5x &> 3 && \text{Combine like terms.} \\ 8 - 5x - 8 &> 3 - 8 && \text{Subtract 8.} \\ -5x &> -5 && \text{Combine like terms.} \\ \frac{-5x}{-5} &< \frac{-5}{-5} && \text{Divide by } -5. \text{ Change } > \text{ to } <. \\ x &< 1 \end{aligned}$$

Solution set: $(-\infty, 1)$

To solve a three-part inequality such as

$$4 < 2x + 6 \le 8,$$

work with all three parts at the same time to obtain an inequality in the form

$$\text{a number} < x \le \text{another number}.$$

Solve and graph the solution set.

$$\begin{aligned} 4 &< 2x + 6 \le 8 \\ 4 - 6 &< 2x + 6 - 6 \le 8 - 6 && \text{Subtract 6.} \\ -2 &< 2x \le 2 \\ \frac{-2}{2} &< \frac{2x}{2} \le \frac{2}{2} && \text{Divide by 2.} \\ -1 &< x \le 1 \end{aligned}$$

Solution set: $(-1, 1]$

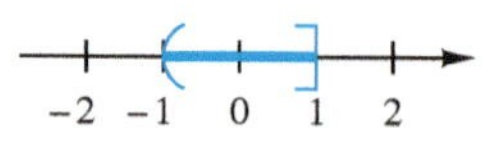

Chapter 2 Review Exercises

2.1–2.3 *Solve each equation. Check the solution.*

1. $x - 7 = 2$

2. $4r - 6 = 10$

3. $5x + 8 = 4x + 2$

4. $8t = 7t + \frac{3}{2}$

5. $4r + 12 - (3r + 12) = 0$

6. $7(2x + 1) = 6(2x - 9)$

7. $-\frac{6}{5}y = -18$

8. $\frac{1}{2}(r - 3) + 2 = \frac{1}{3}(r - 9)$

9. $3x - (-2x + 6) = 4(x - 4) + x$

10. $0.10(x + 80) + 0.20x = 8 + 0.30x$

2.4 *Solve each problem.*

11. If 7 is added to five times a number, the result is equal to three times the number. Find the number.

12. If 4 is subtracted from twice a number, the result is 36. Find the number.

13. The land area of Hawaii is 5213 mi^2 greater than that of Rhode Island. Together, the areas total 7637 mi^2. What is the area of each state?

14. The height of Seven Falls in Colorado is $\frac{5}{2}$ the height (in feet) of Twin Falls in Idaho. The sum of the heights is 420 ft. Find the height of each.

15. The supplement of an angle measures 10 times the measure of its complement. What is the measure of the angle (in degrees)?

16. Find two consecutive odd integers such that when the lesser is added to twice the greater, the result is 24 more than the greater integer.

17. A lawn mower uses a mixture of gasoline and oil. The mixture contains 1 oz of oil for every 32 oz of gasoline. How many ounces of oil and how many ounces of gasoline are required to fill a 132-oz tank completely?

18. A 72-in. board is to be cut into three pieces. The longest piece must be three times as long as the shortest piece. The middle-sized piece must be 7 in. longer than the shortest piece. How long must each piece be?

2.5 *Find the value of the remaining variable in each formula. Use 3.14 as an approximation for π.*

19. $\mathcal{A} = \frac{1}{2}bh$; $\mathcal{A} = 44, b = 8$

20. $\mathcal{A} = \frac{1}{2}h(b + B)$; $b = 3, B = 4, h = 8$

21. $C = 2\pi r$; $C = 29.83$

22. $V = \frac{4}{3}\pi r^3$; $r = 9$

Solve each formula for the specified variable.

23. $\mathcal{A} = bh$ for h

24. $\mathcal{A} = \frac{1}{2}h(b + B)$ for h

Solve each equation for y.

25. $x + y = 11$

26. $3x - 2y = 12$

Find the measure of each marked angle.

27.

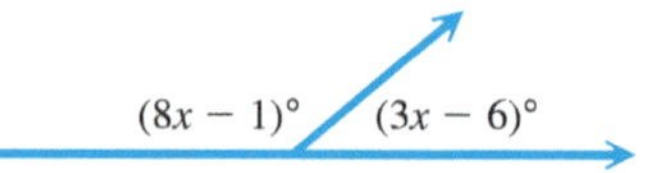

28.

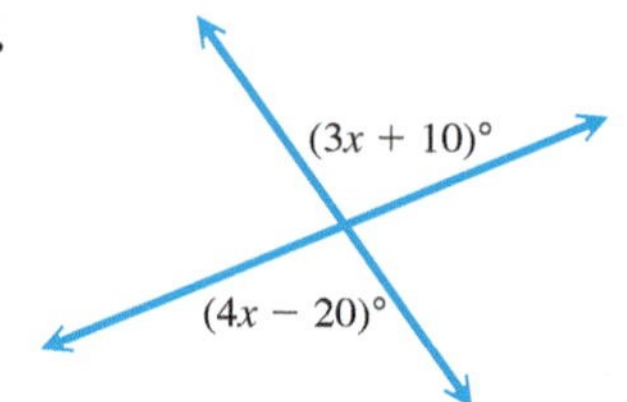

Solve each problem.

29. A cinema screen in Sydney, Australia, has length 97 ft and width 117 ft. What is the perimeter? What is the area? (Data from *Guinness World Records.*)

30. General Sherman, a giant sequoia growing in Sequoia National Park in California, is 271 ft tall and has a circumference of about 85 ft. What is the diameter of the tree? What is its radius? Use 3.14 as an approximation for π. Round answers to the nearest hundredth. (Data from *Guinness World Records.*)

2.6 *Write a ratio for each word phrase. Express fractions in lowest terms.*

31. 60 cm to 40 cm

32. 5 days to 2 weeks

33. 90 in. to 10 ft

Solve each equation.

34. $\frac{p}{21} = \frac{5}{30}$

35. $\frac{5 + x}{3} = \frac{2 - x}{6}$

36. $\frac{y}{5} = \frac{6y - 5}{11}$

Solve each problem.

37. If 2 lb of fertilizer will cover 150 ft^2 of lawn, how many pounds would be needed to cover 500 ft^2?

38. If 8 oz of medicine must be mixed with 20 oz of water, how many ounces of medicine must be mixed with 90 oz of water?

39. The distance between two cities on a road map is 32 cm. The two cities are actually 150 km apart. The distance on the map between two other cities is 80 cm. How far apart are these cities?

40. Find the best buy. Give the unit price to the nearest thousandth for that size. (Data from Jewel-Osco.)

CEREAL

Size	Price
9 oz	\$3.49
14 oz	\$3.99
18 oz	\$4.49

41. What is 8% of 75?

42. What percent of 12 is 21?

43. 6 is what percent of 18?

44. 36% of what number is 900?

45. Nicholas paid \$22,870, including sales tax, for his 2016 Kia Optima. The sales tax rate where he lives is 6%. What was the actual price of the car to the nearest dollar? (Data from www.kia.com)

46. Maureen took the mathematics faculty from a community college out to dinner. The bill was \$304.75. Maureen added a 15% tip and paid for the meal with her corporate credit card. What was the total price she paid to the nearest cent?

47. A laptop computer with a regular price of \$680 is on sale for \$510. What percent of the regular price is the savings?

48. Boyd has a monthly income of \$3200. He has budgeted \$560 per month for auto expenses. What percent of his monthly income did he budget for auto expenses?

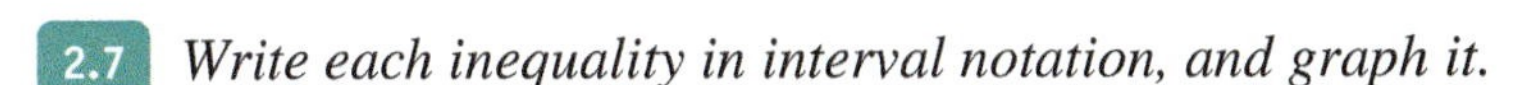

2.7 *Write each inequality in interval notation, and graph it.*

49. $p \geq -4$

50. $x < 7$

51. $-5 \leq k < 6$

52. $r \geq \frac{1}{2}$

Solve each inequality. Write the solution set in interval notation, and graph it.

53. $x + 6 \geq 3$

54. $5t < 4t + 2$

55. $-6x \leq -18$

56. $-8(k - 5) + 2 + 7k \leq -4$

57. $4x - 3x > 10 - 4x + 7x$

58. $3(2w + 5) + 4(8 + 3w) < 5(3w + 2) + 2w$

59. $-3 \leq 2x + 1 < 4$

60. $8 < 3x + 5 \leq 20$

Solve each problem.

61. Justin has grades of 94 and 88 on his first two calculus tests. What possible scores on a third test will give him an average of at least 90?

62. If nine times a number is added to 6, the result is at most 3. Find all such numbers.

Chapter 2 Mixed Review Exercises

Solve.

1. $\frac{x}{7} = \frac{x-5}{2}$

2. $d = 2r$ for r

3. $-2x > -4$

4. $2k - 5 = 4k + 13$

5. $0.05x + 0.02x = 4.9$

6. $2 - 3(t - 5) = 4 + t$

7. $9x - (7x + 2) = 3x + (2 - x)$

8. $\frac{1}{3}s + \frac{1}{2}s + 7 = \frac{5}{6}s + 5 + 2$

9. On a world globe, the distance between Capetown and Bangkok, two cities that are actually 10,080 km apart, is 12.4 in. The actual distance between Moscow and Berlin is 1610 km. How far apart are Moscow and Berlin on this globe, to the nearest inch?

10. In triangle DEF, the measure of angle E is twice the measure of angle D. Angle F has measure 18° less than six times the measure of angle D. Find the measure of each angle.

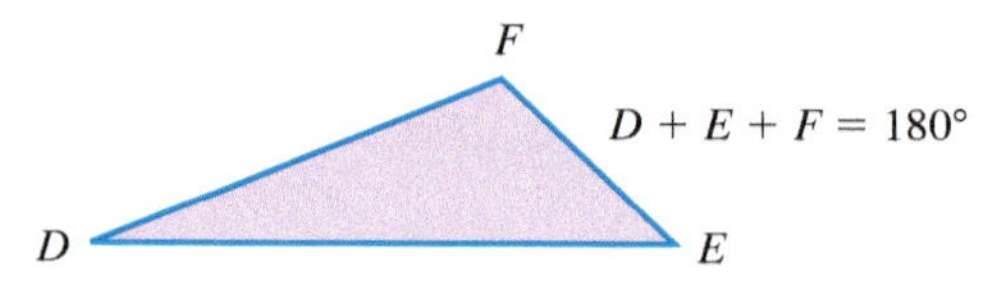

11. The perimeter of a triangle is 96 m. One side is twice as long as another, and the third side is 30 m long. What is the length of the longest side?

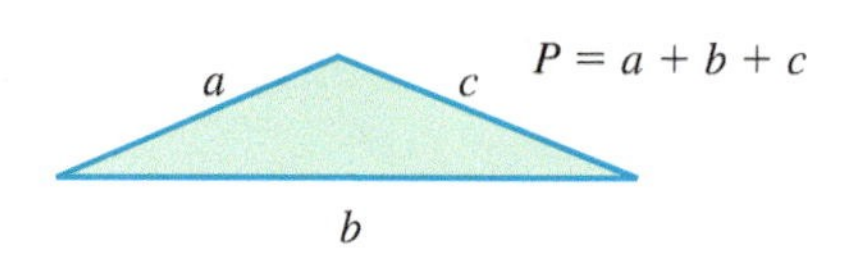

12. The perimeter of a rectangle is 288 ft. The length is 4 ft longer than the width. Find the width.

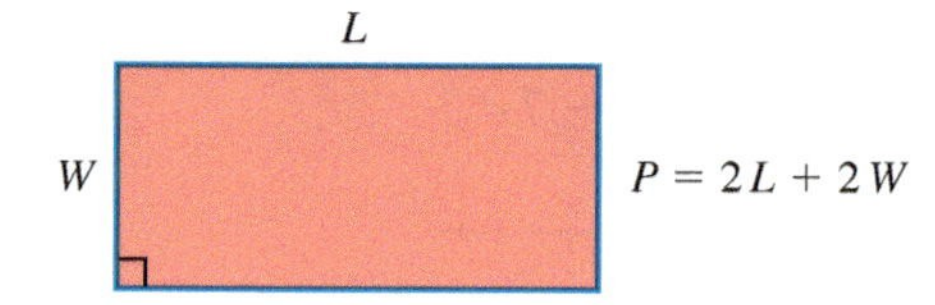

13. Find the best buy. Give the unit price to the nearest thousandth for that size. (Data from Jewel-Osco.)

LAUNDRY DETERGENT

Size	Price
50 oz	\$3.99
100 oz	\$7.29
160 oz	\$9.99

14. Find the measure of each marked angle.

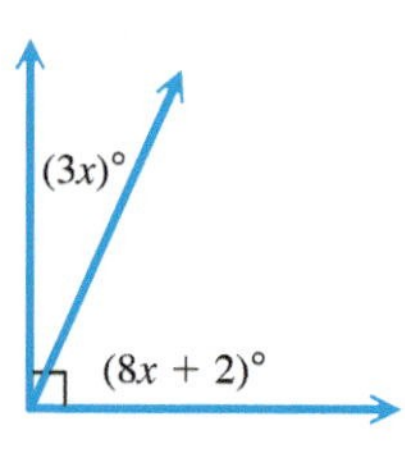

15. Latarsha has grades of 82 and 96 on her first two English tests. What must she make on her third test so that her average will be at least 90?

16. If nine pairs of jeans cost \$355.50, find the cost of five pairs. (Assume all are equally priced.)

Chapter 2 Test

The Chapter Test Prep Videos with step-by-step solutions are available in MyMathLab or on YouTube at http://goo.gl/8aAsWP

Solve each equation, and check the solution.

1. $3x - 7 = 11$

2. $5x + 9 = 7x + 21$

3. $2 - 3(x - 5) = 3 + (x + 1)$

4. $2.3x + 13.7 = 1.3x + 2.9$

5. $-\frac{4}{7}x = -12$

6. $-8(2x + 4) = -4(4x + 8)$

7. $0.06(x + 20) + 0.08(x - 10) = 4.6$

8. $7 - (m - 4) = -3m + 2(m + 1)$

Solve each problem.

9. Wilt Chamberlain and Michael Jordan are the two top NBA all-time point scorers for a single regular season. The total of the points for their best years is 7070. If Wilt Chamberlain scored 2053 points fewer than twice the number of points that Michael Jordan scored, how many points did each player score? (Data from www.landofbasketball.com)

10. Three islands in the Hawaiian island chain are Hawaii (the Big Island), Maui, and Kauai. Together, their areas total 5300 mi^2. The island of Hawaii is 3293 mi^2 larger than the island of Maui, and Maui is 177 mi^2 larger than Kauai. What is the area of each island?

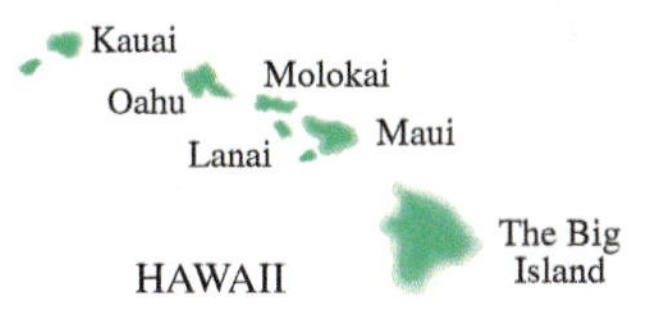

11. If the lesser of two consecutive even integers is tripled, the result is 20 more than twice the greater integer. Find the two integers.

12. Find the measure of an angle if its supplement measures 10° more than three times its complement.

13. The formula for the perimeter of a rectangle is $P = 2L + 2W$.

(a) Solve for W.

(b) If $P = 116$ and $L = 40$, find the value of W.

14. Solve the following equation for y.

$$5x - 4y = 8$$

Find the measure of each marked angle.

15.

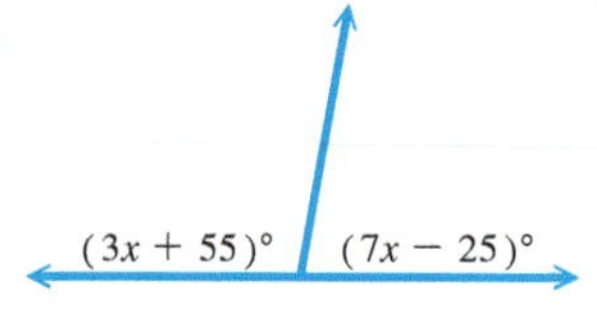

16.

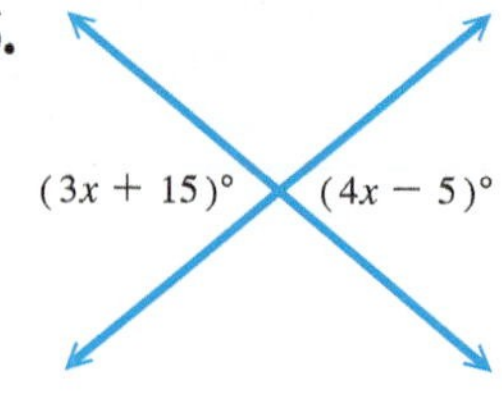

Solve each equation.

17. $\frac{z}{8} = \frac{12}{16}$

18. $\frac{x+5}{3} = \frac{x-3}{4}$

Solve each problem.

19. Find the best buy. Give the unit price to the nearest thousandth for that size. (Data from Jewel-Osco.)

PROCESSED CHEESE SLICES

Size	Price
8 oz	\$2.99
16 oz	\$3.99
48 oz	\$14.69

20. The distance between Milwaukee and Boston is 1050 mi. On a certain map, this distance is represented by 42 in. On the same map, Seattle and Cincinnati are 92 in. apart. What is the actual distance between Seattle and Cincinnati?

21. Dawn has a monthly income of \$2200 and plans to spend 12% of this amount on groceries. How much will be spent on groceries?

22. What percent of 65 is 26?

23. Write an inequality using the variable x that corresponds to each graph of solutions on a number line.

(a) −2 −1 0 1 2 3

(b)

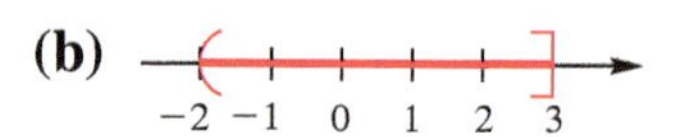

Solve each inequality. Write the solution set in interval notation, and graph it.

24. $-3x > -33$

25. $-0.04x \leq 0.12$

26. $-4x + 2(x - 3) \geq 4x - (3 + 5x) - 7$

27. $-10 < 3x - 4 \leq 14$

28. Shania has scores of 76 and 81 on her first two algebra tests. If she wants an average of at least 80 after her third test, what score must she make on her third test?

Chapters R–2 Cumulative Review Exercises

Write each fraction in lowest terms.

1. $\frac{15}{40}$

2. $\frac{108}{144}$

Perform the indicated operations.

3. $\frac{5}{6} + \frac{1}{4} + \frac{7}{15}$

4. $16\frac{7}{8} - 3\frac{1}{10}$

5. $\frac{9}{8} \cdot \frac{16}{3}$

6. $\frac{3}{4} \div \frac{5}{8}$

7. $4.8 + 12.5 + 16.73$

8. $56.3 - 28.99$

9. $67.8(0.45)$

10. $236.46 \div 4.2$

11. In making dresses, Earth Works uses $\frac{5}{8}$ yd of trim per dress. How many yards of trim would be used to make 56 dresses?

12. A cook wants to increase a recipe for Quaker Quick Grits that serves 4 to make enough for 10 people. The recipe calls for 3 cups of water. How much water will be needed to serve 10?

13. First published in 1953, the digest-sized *TV Guide* changed to a full-sized magazine in 2005. See the figure. The new magazine is 3 in. wider than the old guide. What is the difference in their heights? (Data from *TV Guide*.)

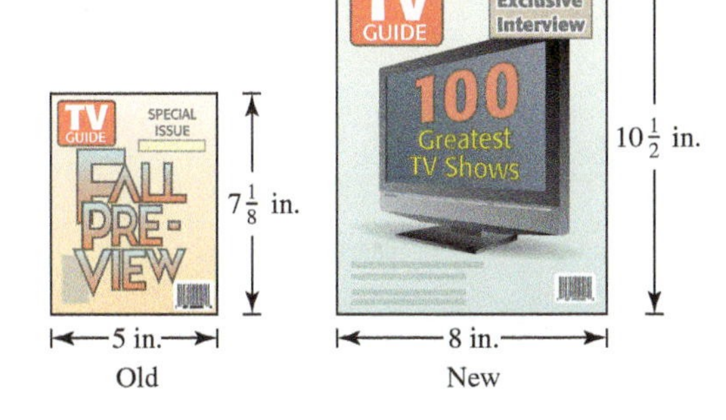

14. A small business owner bought 3 business laptop computers for \$529.99, \$599.99, and \$629.99 and 3 ergonomic office chairs for \$279.99 each. What was the final bill (without tax)? (Data from www.staples.com)

Decide whether each inequality is true *or* false.

15. $\frac{8(7) - 5(6 + 2)}{3 \cdot 5 + 1} \geq 1$

16. $\frac{4(9 + 3) - 4^2}{2 + 3 \cdot 6} \geq 2$

Perform the indicated operations.

17. $-11 + 20 + (-2)$

18. $13 + (-19) + 7$

19. $9 - (-4)$

20. $-2(-5)(-4)$

21. $\frac{4 \cdot 9}{-3}$

22. $\frac{8}{7 - 7}$

23. $(-5 + 8) + (-2 - 7)$

24. $(-7 - 1)(-4) + (-4)$

25. $\frac{-3 - (-5)}{1 - (-1)}$

26. $\frac{6(-4) - (-2)(12)}{3^2 + 7^2}$

27. $\frac{(-3)^2 - (-4)(2^4)}{5 \cdot 2 - (-2)^3}$

28. $\frac{-2(5^3) - 6}{4^2 + 2(-5) + (-2)}$

Find the value of each expression for $x = -2$, $y = -4$, and $z = 3$.

29. $xz^3 - 5y^2$

30. $\dfrac{xz - y^3}{-4z}$

Name the property illustrated by each equation.

31. $7(k + m) = 7k + 7m$

32. $3 + (5 + 2) = 3 + (2 + 5)$

33. $7 + (-7) = 0$

34. $3.5(1) = 3.5$

Simplify each expression.

35. $4p - 6 + 3p - 8$

36. $-4(k + 2) + 3(2k - 1)$

Solve each equation, and check the solution.

37. $2r - 6 = 8$

38. $2(p - 1) = 3p + 2$

39. $4 - 5(a + 2) = 3(a + 1) - 1$

40. $2 - 6(z + 1) = 4(z - 2) + 10$

41. $-(m - 1) = 3 - 2m$

42. $\dfrac{x - 2}{3} = \dfrac{2x + 1}{5}$

43. $\dfrac{2x + 3}{5} = \dfrac{x - 4}{2}$

44. $\dfrac{2}{3}x + \dfrac{3}{4}x = -17$

Solve each formula for the indicated variable.

45. $P = a + b + c + B$ for c

46. $P = 4s$ for s

Solve each inequality. Write the solution set in interval notation, and graph it.

47. $-5z \geq 4z - 18$

48. $6(r - 1) + 2(3r - 5) < -4$

Solve each problem.

49. Abby bought textbooks at the college bookstore for \$276.13, including 6% sales tax. What did the books cost before tax?

50. A used car has a price of \$11,500. For trading in her old car, Shannon will get 25% off. Find the price of the car with the trade-in.

51. The perimeter of a rectangle is 98 cm. The width is 19 cm. Find the length.

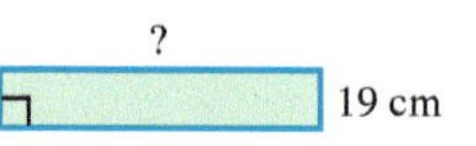

52. The area of a triangle is 104 in.^{2}. The base is 13 in. Find the height.

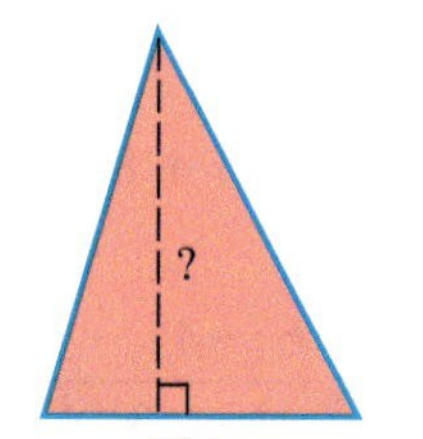

Graphs of Linear Equations and Inequalities in Two Variables

We determine location on a map using *coordinates,* a concept that is based on a *rectangular coordinate system,* one of the topics of this chapter.

3.1 Linear Equations and Rectangular Coordinates

OBJECTIVES

1. Interpret line graphs.
2. Write a solution as an ordered pair.
3. Decide whether a given ordered pair is a solution of a given equation.
4. Complete ordered pairs for a given equation.
5. Complete a table of values.
6. Plot ordered pairs.

OBJECTIVE 1 Interpret line graphs. A **line graph** is used to show changes or trends in data over time. To form a line graph, we connect a series of points representing data with line segments.

EXAMPLE 1 Interpreting a Line Graph

The line graph in **Figure 1** shows average prices of a gallon of regular unleaded gasoline in the United States for the years 2008 through 2015.

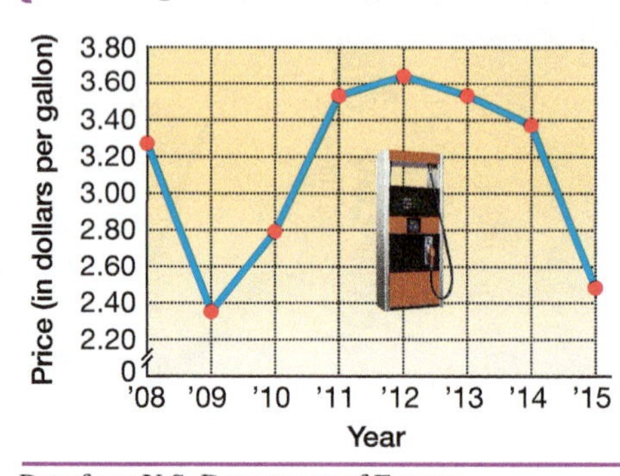

Data from U.S. Department of Energy.

Figure 1

(a) Between which years did the average price of a gallon of gasoline increase?

The line between 2009 and 2010 *rises* from left to right, as do the lines between 2010 and 2011 and between 2011 and 2012. This indicates that the average price of a gallon of gasoline *increased* between 2009 and 2012.

(b) What was the general trend in the average price of a gallon of gasoline from 2012 through 2015?

The line graph *falls* from left to right from 2012 to 2015, so the average price of a gallon of gasoline *decreased* over those years.

(c) Estimate the average price of a gallon of gasoline in 2012 and 2015. About how much did the price decrease between 2012 and 2015?

Move up from 2012 on the horizontal scale to the point plotted for 2012. This point is about one-fourth of the way between the lines on the vertical scale for \$3.60 and \$3.80—that is, about **\$3.65** per gallon.

Locate the point plotted for 2015. Moving across to the vertical scale, this point is about halfway between the lines for \$2.40 and \$2.60—that is, about **\$2.50** per gallon.

Between 2012 and 2015, the average price decreased about

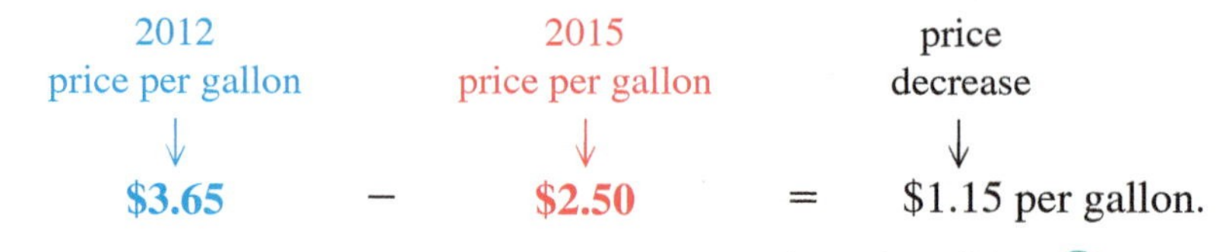

2012
price per gallon
2015
price per gallon
price
decrease
$3.65 − $2.50 = $1.15 per gallon.

◄ Work Problem 1 at the Side.

1 Refer to the line graph in **Figure 1.**

(a) Estimate the average price of a gallon of regular unleaded gasoline in 2009.

(b) About how much did the average price of a gallon of gasoline increase between 2009 and 2012?

Answers

1. (a) \$2.35 **(b)** \$1.30

The line graph in **Figure 1** on the previous page relates years to average prices for a gallon of gasoline. We can also represent these two related quantities using a table of data, as shown in the margin. In table form, we can see more precise data rather than estimating it. Trends in the data are easier to see from the graph, which gives a "picture" of the data.

We can extend these ideas to the subject of this chapter, *linear equations in* ***two*** *variables*. A linear equation in two variables, one for each of the quantities being related, can be used to represent the data in a table or graph.

The graph of a linear equation in two variables is a line.

Year	Average Price (in dollars per gallon)
2008	3.27
2009	2.35
2010	2.79
2011	3.53
2012	3.64
2013	3.53
2014	3.37
2015	2.48

Data from U.S. Department of Energy.

Linear Equation in Two Variables

A **linear equation in two variables** (here x and y) can be written in the form

$$Ax + By = C,$$

where A, B, and C are real numbers and A and B are not both 0. This form is called *standard form.*

Examples: $3x + 4y = 9$, $x - y = 0$, $x + 2y = -8$ — Linear equations in two variables in standard form

Note

Linear equations in two variables that are not written in standard form, such as

$$y = 4x + 5 \quad \text{and} \quad 3x = 7 - 2y,$$

can be algebraically rewritten in this form, as we will discuss later.

OBJECTIVE 2 Write a solution as an ordered pair. Recall that a *solution* of an equation is a number that makes the equation true when it replaces the variable. For example, the linear equation in *one* variable

$$x - 2 = 5 \quad \text{has solution} \quad 7$$

because replacing x with 7 gives a true statement.

A solution of a linear equation in two variables requires two numbers, one for each variable. For example, a true statement results when we replace x with 2 and y with 13 in the equation $y = 4x + 5$.

$$13 = 4(2) + 5 \quad \text{Let } x = 2 \text{ and } y = 13.$$

The pair of numbers $x = 2$ and $y = 13$ gives one solution of the equation $y = 4x + 5$. The phrase "$x = 2$ and $y = 13$" can be abbreviated as a pair of numbers written inside parentheses. ***The x-value is always given first.*** Such a pair of numbers is an **ordered pair.**

x-value → (2, 13) ← y-value
Ordered pair

CAUTION

The ordered pairs (2, 13) and (13, 2) are *not* the same. In the first pair, $x = 2$ and $y = 13$. In the second pair, $x = 13$ and $y = 2$. ***The order in which the numbers are written in an ordered pair is important.***

Work Problem 2 at the Side. ►

2 Write each solution as an ordered pair.

GS **(a)** $x = 5$ and $y = 7$

x-value y-value
(____, ____)

GS **(b)** $y = 6$ and $x = -1$

x-value y-value
(____, ____)

(c) $x = \frac{2}{3}$ and $y = -12$

(d) $y = 1.5$ and $x = -2.4$

(e) $x = 0$ and $y = 0$

Answers

2. **(a)** 5; 7 **(b)** -1; 6
(c) $\left(\frac{2}{3}, -12\right)$ **(d)** $(-2.4, 1.5)$
(e) $(0, 0)$

3 Decide whether each ordered pair is a solution of the equation

$5x + 2y = 20.$

GS **(a)** $(0, 10)$

In this ordered pair, $x =$ ____ and $y =$ ____.

$$5x + 2y = 20$$
$$5(\_\_\_\_) + 2(\_\_\_\_) \stackrel{?}{=} 20$$
$$\_\_\_\_ + 20 \stackrel{?}{=} 20$$
$$\_\_\_\_ = 20$$

Is $(0, 10)$ a solution?

(b) $(2, -5)$

(c) $(-4, 20)$

4 Complete each ordered pair for the equation

$y = 2x - 9.$

GS **(a)** $(5, \_\_\_\_)$

In this ordered pair, $x =$ ____.

We must find the corresponding value of ____.

$$y = 2x - 9$$
$$y = 2(\_\_\_\_) - 9$$
$$y = \_\_\_\_ - 9$$
$$y = \_\_\_\_$$

The ordered pair is ____.

(b) $(2, \_\_\_\_)$

(c) $(\_\_\_\_, 7)$

Answers

3. **(a)** 0; 10; 0; 10; 0; 20; yes **(b)** no **(c)** yes

4. **(a)** 5; y; 5; 10; 1; $(5, 1)$ **(b)** $(2, -5)$ **(c)** $(8, 7)$

OBJECTIVE 3 **Decide whether a given ordered pair is a solution of a given equation.** We substitute the x- and y-values of an ordered pair into a linear equation in two variables to see whether the ordered pair is a solution.

EXAMPLE 2 Deciding Whether Ordered Pairs Are Solutions

Decide whether each ordered pair is a solution of the equation $2x + 3y = 12$.

(a) $(3, 2)$

$$2x + 3y = 12$$
$$2(3) + 3(2) \stackrel{?}{=} 12 \quad \text{Let } x = 3 \text{ and } y = 2.$$
$$6 + 6 \stackrel{?}{=} 12 \quad \text{Multiply.}$$
$$12 = 12 \ \checkmark \quad \text{True}$$

This result is true, so $(3, 2)$ is a solution of $2x + 3y = 12$.

(b) $(-2, -7)$

$$2x + 3y = 12$$
$$2(-2) + 3(-7) \stackrel{?}{=} 12 \quad \text{Let } x = -2 \text{ and } y = -7.$$
$$-4 + (-21) \stackrel{?}{=} 12 \quad \text{Multiply.}$$
$$-25 = 12 \quad \text{False}$$

Use parentheses to avoid errors.

This result is false, so $(-2, -7)$ is *not* a solution of $2x + 3y = 12$.

◀ Work Problem **3** at the Side.

OBJECTIVE 4 **Complete ordered pairs for a given equation.** We substitute a number for one variable to find the value of the other variable.

EXAMPLE 3 Completing Ordered Pairs

Complete each ordered pair for the equation $y = 4x + 5$.

(a) $(7, \_\_\_\_)$ The x-value always comes first.

In this ordered pair, $x = 7$. To find the corresponding value of y, replace x with 7 in the given equation.

$$y = 4x + 5$$
$$y = 4(7) + 5 \quad \text{Let } x = 7.$$
$$y = 28 + 5 \quad \text{Multiply.}$$
$$y = 33 \quad \text{Add.}$$

Solve for the value of y.

The ordered pair is $(7, 33)$.

(b) $(\_\_\_\_, -3)$

In this ordered pair, $y = -3$. Replace y with -3 in the given equation.

$$y = 4x + 5$$
$$-3 = 4x + 5 \quad \text{Let } y = -3.$$
$$-8 = 4x \quad \text{Subtract 5 from each side.}$$
$$-2 = x \quad \text{Divide each side by 4.}$$

Solve for the value of x.

The ordered pair is $(-2, -3)$.

◀ Work Problem **4** at the Side.

OBJECTIVE 5 **Complete a table of values.** Ordered pairs are often displayed in a **table of values.** Although we usually write tables of values vertically, they may be written horizontally.

EXAMPLE 4 **Completing Tables of Values**

Complete each table of values for the given equation. Then write the results as ordered pairs.

(a) $x - 2y = 8$

x	y		Ordered Pairs
2		$\longrightarrow$	(2, ____)
10		$\longrightarrow$	(10, ____)
	0	$\longrightarrow$	(____, 0)
	−2	$\longrightarrow$	(____, −2)

From the first row of the table, let $x = 2$ in the equation. From the second row of the table, let $x = 10$.

	If $x = 2$,		If $x = 10$,
then	$x - 2y = 8$	then	$x - 2y = 8$
becomes	$2 - 2y = 8$	becomes	$10 - 2y = 8$
	$-2y = 6$		$-2y = -2$
	$y = -3$.		$y = 1$.

The first two ordered pairs are $(2, -3)$ and $(10, 1)$. From the third and fourth rows of the table, let $y = 0$ and $y = -2$, respectively.

	If $y = 0$,		If $y = -2$,
then	$x - 2y = 8$	then	$x - 2y = 8$
becomes	$x - 2(0) = 8$	becomes	$x - 2(-2) = 8$
	$x - 0 = 8$		$x + 4 = 8$
	$x = 8$.		$x = 4$.

The last two ordered pairs are $(8, 0)$ and $(4, -2)$. The completed table of values and corresponding ordered pairs follow.

Write y-values in the second column.

Write x-values in the first column.

x	y		Ordered Pairs
2	−3	$\longrightarrow$	(2, −3)
10	1	$\longrightarrow$	(10, 1)
8	0	$\longrightarrow$	(8, 0)
4	−2	$\longrightarrow$	(4, −2)

Each ordered pair is a solution of the given equation $x - 2y = 8$.

(b) $x = 5$ (Using two variables, $x = 5$ could be written $x + 0y = 5$.)

x	y
	−2
	6
	3

The given equation is $x = 5$. No matter which value of y is chosen, the value of x is *always* 5.

x	y		Ordered Pairs
5	−2	$\longrightarrow$	(5, −2)
5	6	$\longrightarrow$	(5, 6)
5	3	$\longrightarrow$	(5, 3)

Continued on Next Page

René Descartes (1596–1650)

The rectangular coordinate system, shown in **Figure 3** on the next page, is also called the **Cartesian coordinate system,** in honor of René Descartes, the French mathematician credited with its invention.

(c) $y = -3$ (Using two variables, $y = -3$ could be written $0x + y = -3$.)

x	y
-5	
0	
2	

The given equation is $y = -3$. No matter which value of x is chosen, the value of y is *always* -3.

x	y	Ordered Pairs
-5	-3	$\longrightarrow (-5, -3)$
0	-3	$\longrightarrow (0, -3)$
2	-3	$\longrightarrow (2, -3)$

◀ Work Problem 5 at the Side.

OBJECTIVE 6 Plot ordered pairs. Recall that a linear equation in *one* variable can have zero, one, or an infinite number of real number solutions. These solutions can be graphed on *one* number line. For example, the linear equation in one variable

$$x - 2 = 5 \quad \text{has solution} \quad 7.$$

The solution 7 is graphed on the number line in **Figure 2.**

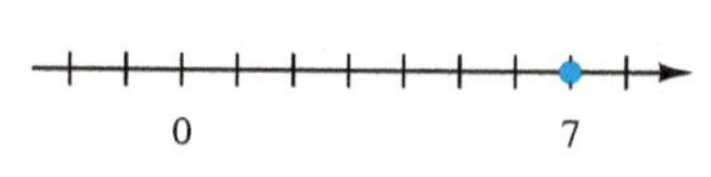

Figure 2

Every linear equation in *two* variables has an infinite number of ordered pairs (x, y) as solutions. To graph these solutions, we need *two* number lines, one for each variable, drawn at right angles as in **Figure 3.** The horizontal number line is the ***x*-axis,** and the vertical line is the ***y*-axis.** The point at which the x-axis and y-axis intersect is the **origin.** Together, the x-axis and y-axis form a **rectangular coordinate system.**

The rectangular coordinate system is divided into four regions, or **quadrants.** These quadrants are numbered counterclockwise. See **Figure 3.**

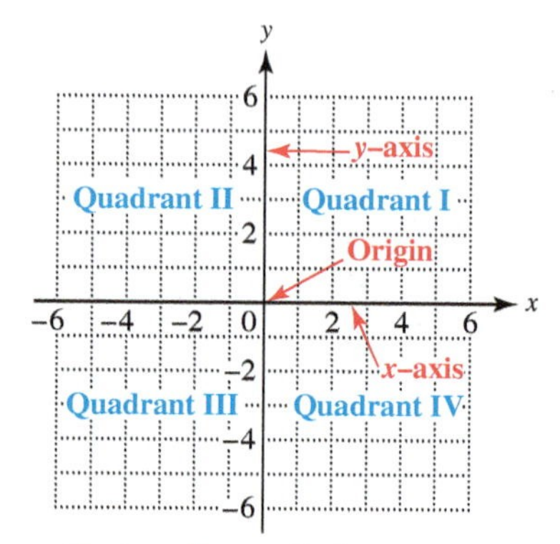

Rectangular coordinate system

Figure 3

The x-axis and y-axis determine a **plane**—a flat surface illustrated by a sheet of paper. By referring to the two axes, every point in the plane can be associated with an ordered pair. The numbers in the ordered pair are the **coordinates** of the point.

Note

In a plane, *both* numbers in the ordered pair are needed to locate a point. The ordered pair is a name for the point.

---

5 Complete each table of values for the given equation. Then write the results as ordered pairs.

GS **(a)** $2x - 3y = 12$

x	y
0	
	0
3	
	-3

From the first row of the table, let $x =$ ____.

$$2x - 3y = 12$$
$$2(\_\_\_\_) - 3y = 12$$
$$\_\_\_\_ - 3y = 12$$
$$-3y = 12$$
$$\frac{-3y}{\_\_\_\_} = \frac{12}{\_\_\_\_}$$
$$y = \_\_\_\_$$

Write ____ for y in the first row of the table. Repeat this process to complete the rest of the table.

(b) $x = -1$

x	y
	-4
	0
-1	2

(c) $y = 4$

x	y
-3	4
2	
5	

Answers

5. (a) 0; 0; 0; -3; -3; -4; -4

x	y
0	-4
6	0
3	-2
$\frac{3}{2}$	-3

$(0, -4)$, $(6, 0)$, $(3, -2)$, $(\frac{3}{2}, -3)$

(b)

x	y
-1	-4
-1	0
-1	-2

$(-1, -4)$, $(-1, 0)$, $(-1, 2)$

(c)

x	y
-3	4
2	4
5	4

$(-3, 4)$, $(2, 4)$, $(5, 4)$

EXAMPLE 5 **Plotting Ordered Pairs**

Plot the given points in a rectangular coordinate system.

(a) $(2, 3)$ **(b)** $(-1, -4)$ **(c)** $(-2, 3)$

(d) $(3, -2)$ **(e)** $\left(\frac{3}{2}, 2\right)$ **(f)** $(4, -3.75)$

(g) $(5, 0)$ **(h)** $(0, -3)$ **(i)** $(0, 0)$

The point $(2, 3)$ from part (a) is **plotted** (graphed) in **Figure 4.** The other points are plotted in **Figure 5.** In each case, we begin at the origin and follow this procedure.

Step 1 Move right or left the number of units that corresponds to the x-coordinate in the ordered pair—***right if the x-coordinate is positive or left if it is negative.***

Step 2 Then turn and move up or down the number of units that corresponds to the y-coordinate in the ordered pair—***up if the y-coordinate is positive or down if it is negative.***

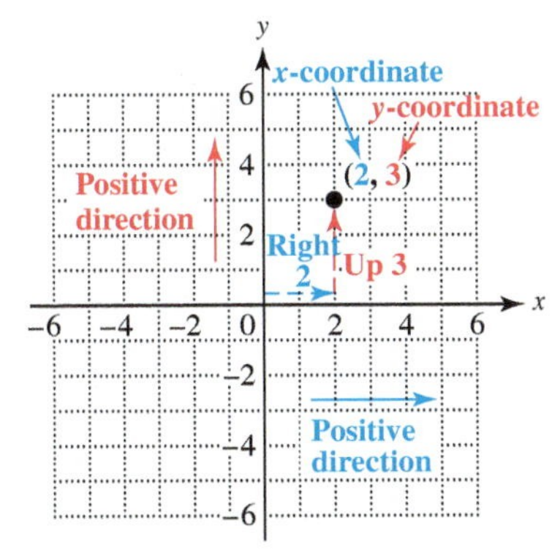

Figure 4

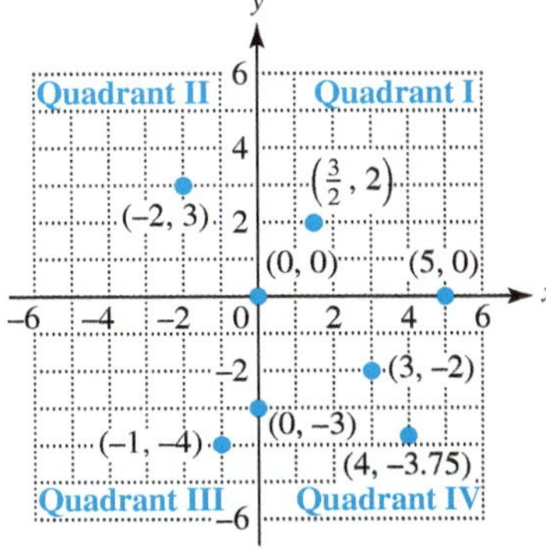

Figure 5

Notice in **Figure 5** that the point $(-2, 3)$ is in quadrant II, whereas the point $(3, -2)$ is in quadrant IV.

The order of the coordinates is important. The x-coordinate is always given first in an ordered pair.

To plot the point $\left(\frac{3}{2}, 2\right)$, think of the improper fraction $\frac{3}{2}$ as the mixed number $1\frac{1}{2}$ and move $\frac{3}{2}$ (or $1\frac{1}{2}$) units to the right along the x-axis. Then turn and go 2 units up, parallel to the y-axis.

The point $(4, -3.75)$ is plotted similarly, by approximating the location of the decimal y-coordinate.

The point $(5, 0)$ lies on the x-axis because the y-coordinate is 0. The point $(0, -3)$ lies on the y-axis because the x-coordinate is 0. The point $(0, 0)$ is at the origin.

Points on the axes themselves are not in any quadrant.

Work Problem 6 at the Side. ▶

We can use a linear equation in two variables to mathematically describe, or *model*, certain real-life situations, as shown in the next example.

6 Plot the given points in a rectangular coordinate system.

(a) $(3, 5)$ **(b)** $(-2, 6)$

(c) $(-4.5, 0)$ **(d)** $(-5, -2)$

(e) $(6, -2)$ **(f)** $(0, -6)$

(g) $(0, 0)$ **(h)** $\left(-3, \frac{5}{2}\right)$

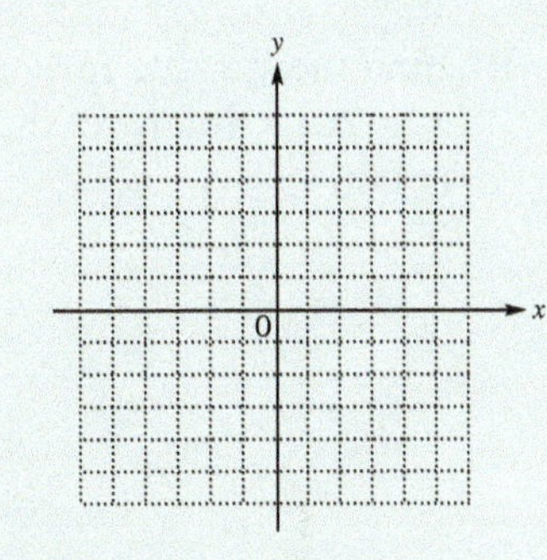

Indicate the points that lie in each quadrant.

quadrant I: ________

quadrant II: ________

quadrant III: ________

quadrant IV: ________

no quadrant: ________

Answers

6.

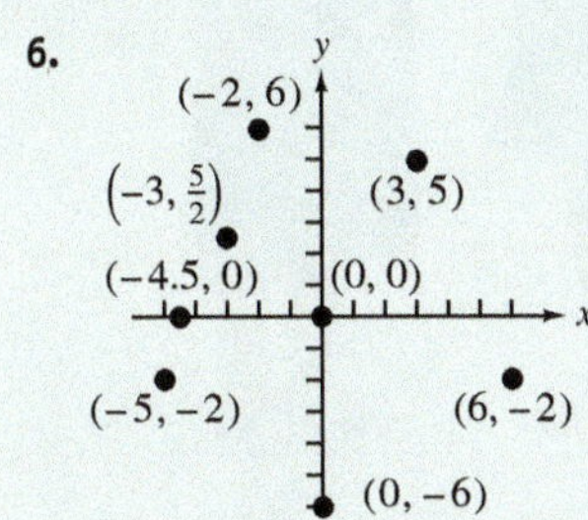

quadrant I: $(3, 5)$

quadrant II: $(-2, 6)$, $\left(-3, \frac{5}{2}\right)$

quadrant III: $(-5, -2)$

quadrant IV: $(6, -2)$

no quadrant: $(-4.5, 0)$, $(0, 0)$, $(0, -6)$

7 Refer to the linear equation in **Example 6.** Round answers to the nearest whole number.

GS **(a)** Find the y-value for $x = 2011$.

$y = -1.571x + 3294$

$y = -1.571(\_\_\_\_) + \_\_\_\_$

$y \approx \_\_\_\_$

(b) Find the y-value for $x = 2013$. Interpret the result.

Answers

7. (a) 2011; 3294; 135
   (b) 132; In 2013, there were about 132 thousand (or 132,000) twin births in the United States.

EXAMPLE 6 Using a Linear Equation to Model Twin Births

The annual number of twin births in the United States from 2008 through 2013 can be approximated by the linear equation

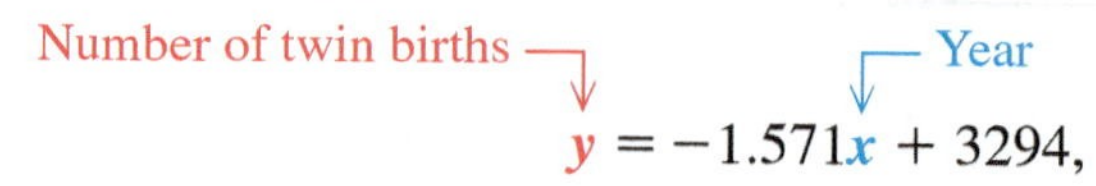

$$y = -1.571x + 3294,$$

which relates x, the year, and y, the number of twin births in thousands. (Data from National Center for Health Statistics.)

(a) Complete the table of values for the given linear equation.

x (Year)	y (Number of Twin Births, in thousands)
2008	
2011	
2013	

To find y when $x = 2008$, we substitute into the equation.

$$y = -1.571x + 3294$$

$$y = -1.571(2008) + 3294 \quad \text{Let } x = 2008.$$

$$y \approx 139 \quad \text{Use a calculator.}$$

≈ means "is approximately equal to."

In 2008, there were about 139 thousand (or 139,000) twin births.

◀ Work Problem **7** at the Side.

Including the results from **Margin Problem 7** gives the completed table.

x (Year)	y (Number of Twin Births, in thousands)		Ordered Pairs (x, y)
2008	139	⟶	(2008, 139)
2011	135	⟶	(2011, 135)
2013	132	⟶	(2013, 132)

Here each year x is paired with a number of twin births y (in thousands).

(b) Graph the ordered pairs found in part (a).

See **Figure 6.** A graph of ordered pairs of data is a **scatter diagram.** A scatter diagram enables us to describe how the two quantities are related. In **Figure 6,** the plotted points could be connected to approximate a straight ***line***, so the variables x (year) and y (number of twin births) have a ***lin**ear* relationship. The decrease in the number of twin births is also reflected.

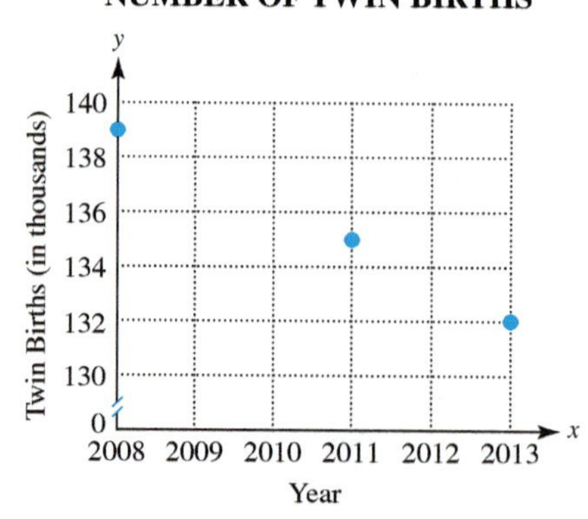

Figure 6

Notice the axis labels and scales. Each square represents 1 unit in the horizontal direction and 2 units in the vertical direction. We show a break in the y-axis, to indicate the jump from 0 to 130.

3.1 Exercises

FOR EXTRA HELP — *Go to MyMathLab for worked-out, step-by-step solutions to exercises enclosed in a square and video solutions to ▶ exercises.*

CONCEPT CHECK *Complete each statement.*

1. The symbol (x, y) (*does* / *does not*) represent an ordered pair, while the symbols $[x, y]$ and $\{x, y\}$ (*do* / *do not*) represent ordered pairs.

2. The origin is represented by the ordered pair ____.

3. In the ordered pair $(4, -1)$, $x =$ ____ and $y =$ ____. This ordered pair (*is* / *is not*) the same as the ordered pair $(-1, 4)$.

4. The point whose graph has coordinates $(-4, 2)$ is in quadrant ____.

5. The point whose graph has coordinates $(0, 5)$ lies on the ____-axis.

6. The ordered pair $(4,$ ____$)$ is a solution of the equation $y = 3$.

7. The ordered pair $($____$, -2)$ is a solution of the equation $x = 6$.

8. The ordered pair $(3, 2)$ is a solution of the equation $2x - 5y =$ ____.

CONCEPT CHECK *Fill in each blank with the word* positive *or the word* negative.

The point with coordinates (x, y) is in

9. quadrant III if x is ______ and y is ______.

10. quadrant II if x is ______ and y is ______.

11. quadrant IV if x is ______ and y is ______.

12. quadrant I if x is ______ and y is ______.

The line graph shows the overall unemployment rate in the U.S. civilian labor force for the years 2009 through 2015. Use the graph to work each problem. ***See Example 1.***

13. Between which years did the unemployment rate increase?

14. What was the general trend in the unemployment rate between 2010 and 2015?

15. Estimate the unemployment rate in 2010 and that in 2015. About how much did the unemployment rate decrease between 2010 and 2015?

16. About how much did the unemployment rate decrease between 2014 and 2015?

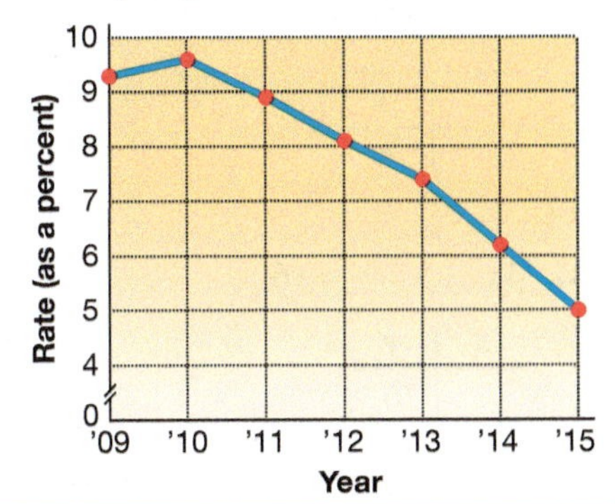

Data from Bureau of Labor Statistics.

The line graph shows the number of new and used passenger cars (in millions) imported into the United States for the years 2009 through 2014. Use the graph to work each problem. **See Example 1.**

Passenger Cars Imported into the U.S.

Data from U.S. Census Bureau.

17. Over which two consecutive years did the number of imported cars increase the most? About how much was this increase?

18. Between which two consecutive years was the number of imported cars about the same? Estimate the number of cars imported during each of these years.

19. Describe what happened to sales of imported cars between 2012 and 2014.

20. During which year(s) were fewer than 6 millions cars imported into the United States?

Decide whether each ordered pair is a solution of the given equation. **See Example 2.**

21. $x + y = 8$; $(0, 8)$

22. $x + y = 9$; $(0, 9)$

23. $2x - y = 6$; $(4, 2)$

24. $2x + y = 5$; $(3, -1)$

25. $5x - 3y = 15$; $(5, 2)$

26. $4x - 3y = 6$; $(2, 1)$

27. $x = -4y$; $(-8, 2)$

28. $y = 3x$; $(2, 6)$

29. $x = -6$; $(5, -6)$

30. $x = -9$; $(8, -9)$

31. $y = 2$; $(2, 4)$

32. $y = 7$; $(7, -2)$

Complete each ordered pair for the equation $y = 2x + 7$. **See Example 3.**

33. $(5, \_\_\_)$

34. $(0, \_\_\_)$

35. $(\_\_\_, 0)$

36. $(\_\_\_, -3)$

Complete each ordered pair for the equation $y = -4x - 4$. **See Example 3.**

37. $(0, \_\_\_)$

38. $(\_\_\_, 0)$

39. $(\_\_\_, 16)$

40. $(\_\_\_, 24)$

Complete each table of values for the given equation. Then write the results as ordered pairs. **See Example 4.**

41. $4x + 3y = 24$

x	y
0	
	0
	4

42. $2x + 3y = 12$

x	y
0	
	0
	8

43. $3x - 5y = -15$

x	y
0	
	0
	−6

44. $4x - 9y = -36$

x	y
	0
0	
	8

45. $x = -9$

x	y
	6
	2
	−3

46. $x = 12$

x	y
	3
	8
	0

47. $y = 6$

x	y
8	
4	
−2	

48. $y = -10$

x	y
4	
0	
−4	

49. $x - 8 = 0$

x	y
	8
	3
	0

50. $x + 4 = 0$

x	y
	4
	0
	−4

51. $y + 2 = 0$

x	y
9	
2	
0	

52. $y - 1 = 0$

x	y
1	
0	
−1	

Give an ordered pair for each point labeled A–H in the figure. (Coordinates of the points shown are integers.) Identify the quadrant in which each point is located. ***See Example 5.***

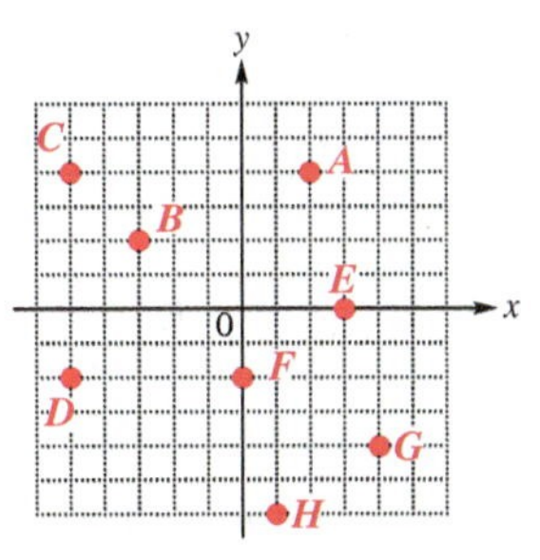

53. *A* **54.** *B* **55.** *C* **56.** *D*

57. *E* **58.** *F* **59.** *G* **60.** *H*

Answer each question.

61. A point (x, y) has the property that $xy < 0$. In which quadrant(s) must the point lie? Explain.

62. A point (x, y) has the property that $xy > 0$. In which quadrant(s) must the point lie? Explain.

Plot each point on the rectangular coordinate system provided. ***See Example 5.***

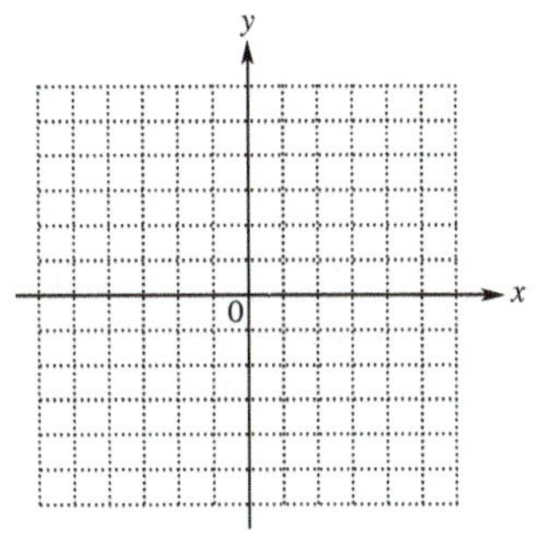

63. $(6, 2)$ **64.** $(5, 3)$ **65.** $(-4, 2)$ **66.** $(-3, 5)$

67. $\left(-\frac{4}{5}, -1\right)$ **68.** $\left(-\frac{3}{2}, -4\right)$ **69.** $(3, -1.75)$ **70.** $(5, -4.25)$

71. $(0, 4)$ **72.** $(0, -3)$ **73.** $(4, 0)$ **74.** $(-3, 0)$

Complete each table of values for the given equation. Then plot the ordered pairs. ***See Examples 4 and 5.***

75. $x - 2y = 6$

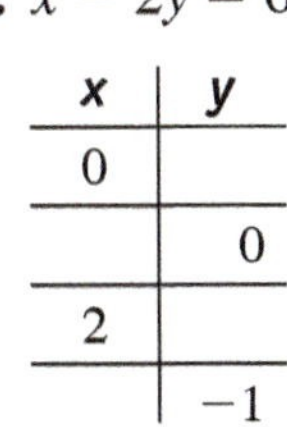

x	y
0	
	0
2	
	−1

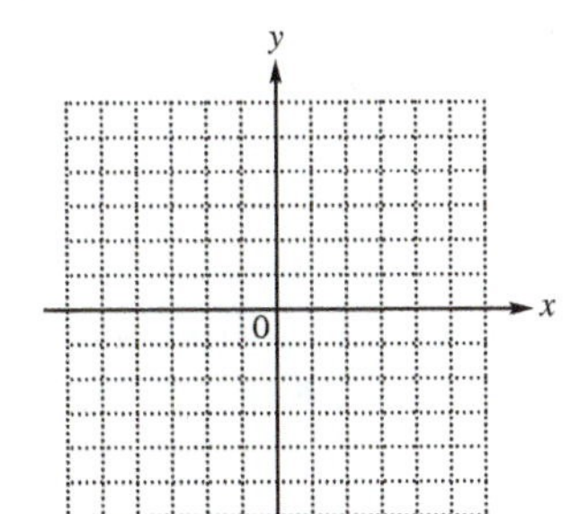

76. $2x - y = 4$

x	y
0	
	0
1	
	−6

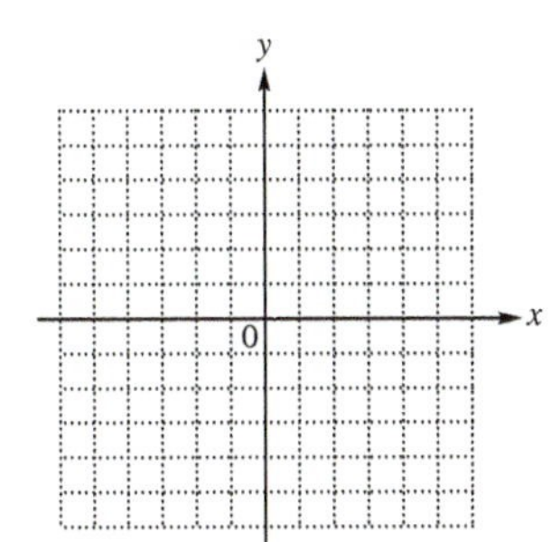

77. $3x - 4y = 12$

x	y
0	
	0
−4	
	−4

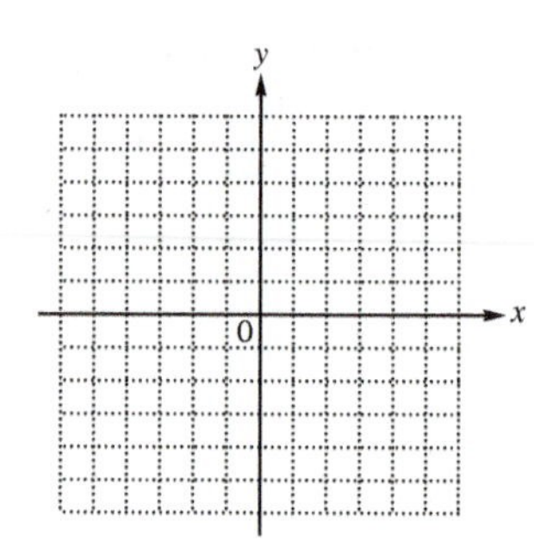

78. $2x - 5y = 10$

x	y
0	
	0
−5	
	−3

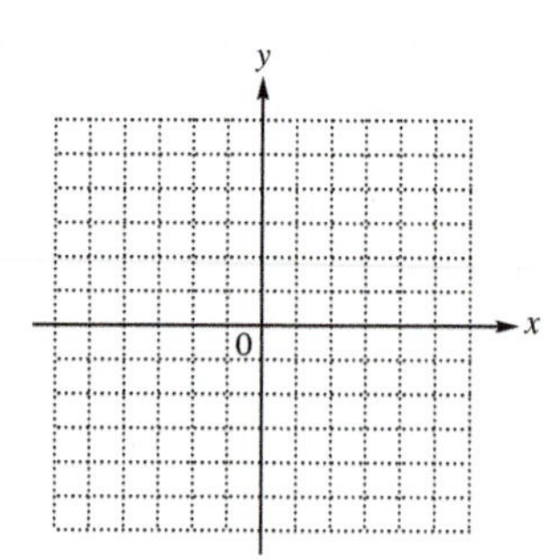

79. $y + 4 = 0$

x	y
0	
5	
−2	
−3	

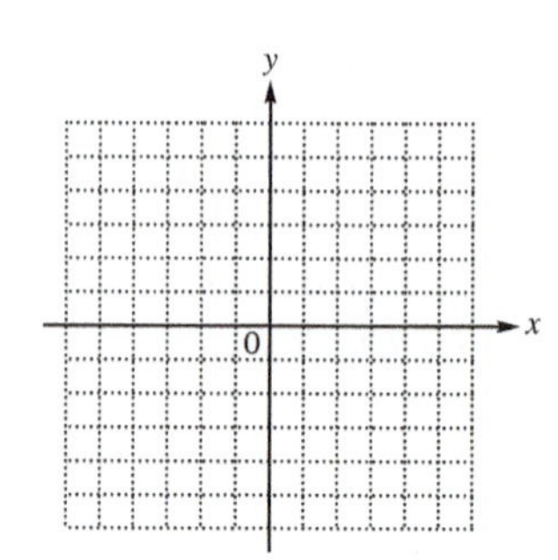

80. $x - 5 = 0$

x	y
	1
	0
	6
	−4

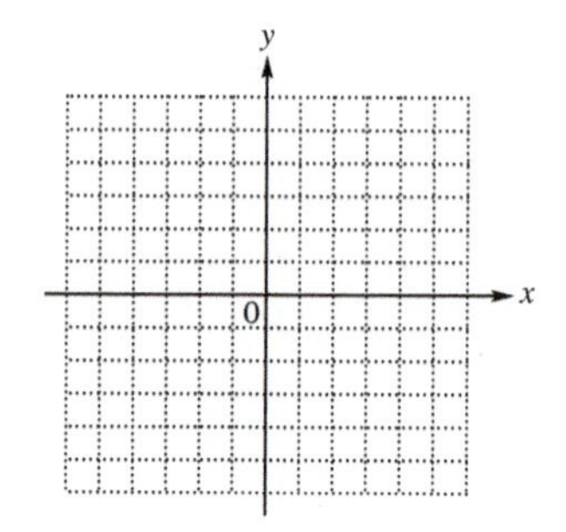

81. Describe the pattern indicated by the plotted points in **Exercises 75–80.**

82. Answer each question.

(a) A line through the plotted points in **Exercise 79** would be (*horizontal / vertical*). What do you notice about the y-coordinates of the ordered pairs?

(b) A line through the plotted points in **Exercise 80** would be (*horizontal / vertical*). What do you notice about the x-coordinates of the ordered pairs?

Work each problem. ***See Example 6.***

83. Suppose that it costs a flat fee of \$20 plus \$5 per day to rent a pressure washer. Therefore, the cost y in dollars to rent the pressure washer for x days is given by

$$y = 5x + 20.$$

Express each of the following using an ordered pair (x, y).

(a) When the washer is rented for 5 days, the cost is \$45. (*Hint:* What does x represent? What does y represent?)

(b) We paid \$50 when we returned the washer, so we must have rented it for 6 days.

84. Suppose that it costs \$5000 to start up a business selling snow cones. Furthermore, it costs \$0.50 per cone in labor, ice, syrup, and overhead. Then the cost y in dollars to make x snow cones is given by

$$y = 0.50x + 5000.$$

Express each of the following using an ordered pair (x, y).

(a) When 100 snow cones are made, the cost is \$5050. (*Hint:* What does x represent? What does y represent?)

(b) When the cost is \$6000, the number of snow cones made is 2000.

85. The table shows the rate (in percent) at which 2-year college students (public) complete a degree within 3 years.

Year	Percent
2009	28.3
2010	28.0
2011	26.9
2012	25.4
2013	22.5
2014	21.9

Data from ACT.

(a) Write the data from the table as ordered pairs (x, y), where x represents the year and y represents the percent.

(b) What would the ordered pair $(2000, 32.4)$ mean in the context of this problem?

(c) Make a scatter diagram of the data using the ordered pairs from part (a).

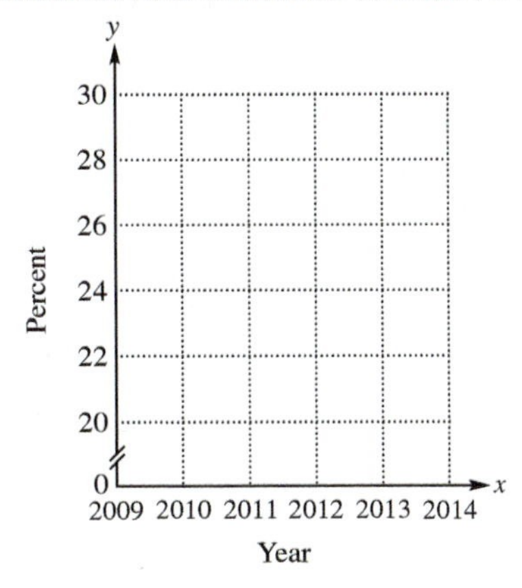

(d) Describe the pattern indicated by the points on the scatter diagram. What happened to the rates at which 2-year college students complete a degree within 3 years?

86. The table shows the number of U.S. students studying abroad (in thousands) for recent academic years.

Academic Year	Number of Students (in thousands)
2008	260
2009	271
2010	274
2011	283
2012	289
2013	304

Data from Institute of International Education.

(a) Write the data from the table as ordered pairs (x, y), where x represents the year and y represents the number of U.S. students studying abroad, in thousands.

(b) What does the ordered pair $(2000, 154)$ mean in the context of this problem?

(c) Make a scatter diagram of the data using the ordered pairs from part (a).

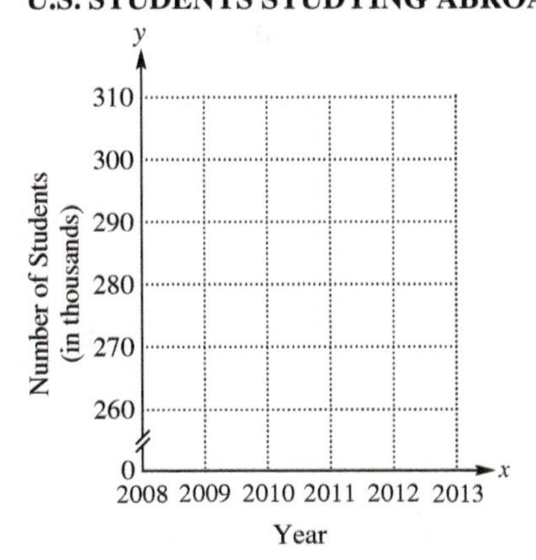

(d) Describe the pattern indicated by the points on the scatter diagram. What was the trend in the number of U.S. students studying abroad during these years?

87. The maximum benefit for the heart from exercising occurs if the heart rate is in the target heart rate zone. The lower limit of this target zone can be approximated by the linear equation

$$y = -0.5x + 108,$$

where x represents age and y represents heartbeats per minute. (Data from www.fitresource.com)

(a) Complete the table of values for this linear equation.

Age	Heartbeats (per minute)
20	
40	
60	
80	

(b) Write the data from the table of values as ordered pairs (x, y).

(c) Make a scatter diagram of the data. Do the points lie in a linear pattern?

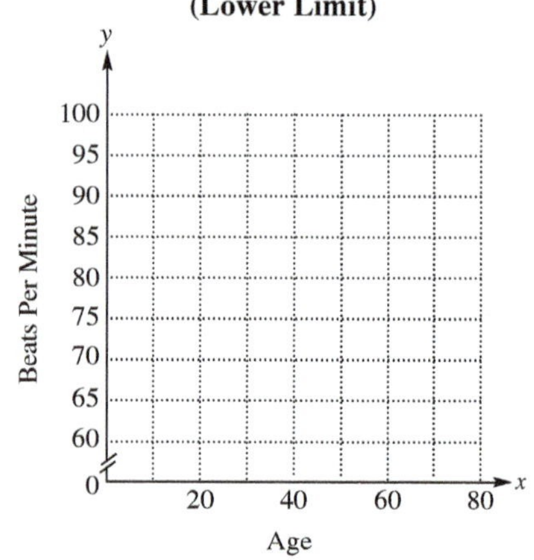

88. (See **Exercise 87.**) The upper limit of the target heart rate zone can be approximated by the linear equation

$$y = -0.8x + 173,$$

where x represents age and y represents heartbeats per minute. (Data from www.fitresource.com)

(a) Complete the table of values for this linear equation.

Age	Heartbeats (per minute)
20	
40	
60	
80	

(b) Write the data from the table of values as ordered pairs (x, y).

(c) Make a scatter diagram of the data. Describe the pattern indicated by the data.

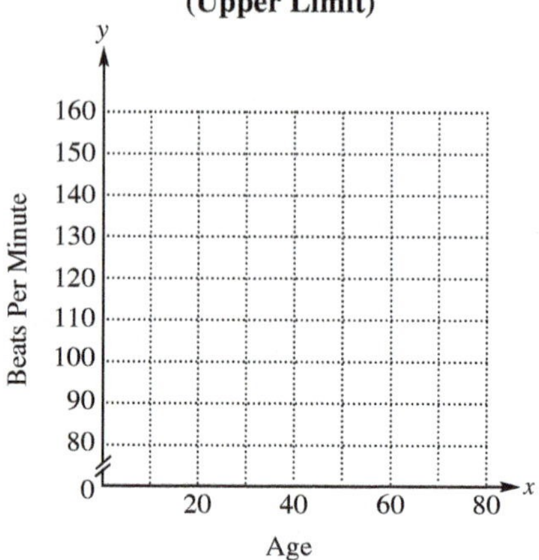

89. See **Exercises 87 and 88.** What is the target heart rate zone for age 20? Age 40?

90. See **Exercises 87 and 88.** What is the target heart rate zone for age 60? Age 80?

A*n exam is a learning opportunity—learn from your mistakes.* After a test is returned, do the following:

- **Note what you got wrong and why you had points deducted.**
- **Figure out how to solve the problems you missed.** Check your text or notes, or ask your instructor. Rework the problems correctly.
- **Keep all quizzes and tests that are returned to you.** Use them to study for future tests and the final exam.

Typical Reasons for Errors on Math Tests

1. You read the directions wrong.
2. You read the question wrong or skipped over something.
3. You made a computation error.
4. You made a careless error. (For example, you incorrectly copied a correct answer onto a separate answer sheet.)
5. Your answer was not complete.
6. You labeled your answer wrong. (For example, you labeled an answer "ft" instead of "ft²".)
7. You didn't show your work.

These are test-taking errors. They are easy to correct if you read carefully, show all your work, proofread, and double-check units and labels.

8. You didn't understand a concept.
9. You were unable to set up the problem (in an application).
10. You were unable to apply a procedure.

These are test preparation errors. Be sure to practice all the kinds of problems that you will see on tests.

Now Try This

1. Below are sample charts for tracking your test-taking progress. Refer to the tests you have taken so far in your course. For each test, check the appropriate box in the charts to indicate that you made an error in a particular category.
2. What test-taking errors did you make? Do you notice any patterns?

   ____

   ____

3. What test preparation errors did you make? Do you notice any patterns?

   ____

   ____

4. What will you do to avoid these kinds of errors on your next test?

   ____

   ____

TEST-TAKING ERRORS

Test	Read directions wrong	Read question wrong	Computation error	Careless error	Not complete	Labeled wrong	Didn't show work
1							
2							
3							

TEST PREPARATION ERRORS

Test	Didn't understand concept	Didn't set up problem correctly	Couldn't apply a procedure to solve
1			
2			
3			

3.2 Graphing Linear Equations in Two Variables

OBJECTIVES

1. Graph linear equations by plotting ordered pairs.
2. Find intercepts.
3. Graph linear equations of the form $Ax + By = 0$.
4. Graph linear equations of the form $y = b$ or $x = a$.
5. Use a linear equation to model data.

OBJECTIVE 1 Graph linear equations by plotting ordered pairs. There are infinitely many ordered pairs that satisfy an equation in two variables. We find these ordered-pair solutions by choosing as many values of x (or y) as we wish and then completing each ordered pair.

For example, consider the equation $x + 2y = 7$. If we choose $x = 1$, then we can substitute to find the corresponding value of y.

$$x + 2y = 7 \quad \text{Given equation}$$
$$1 + 2y = 7 \quad \text{Let } x = 1.$$
$$2y = 6 \quad \text{Subtract 1.}$$
$$y = 3 \quad \text{Divide by 2.}$$

If $x = 1$, then $y = 3$, and the ordered pair $(1, 3)$ is a solution of the equation.

$$1 + 2(3) = 7 \checkmark \quad \text{A true statement results, so } (1, 3) \text{ is a solution.}$$

◄ Work Problem 1 at the Side.

1. Complete each ordered pair to find additional solutions of the equation

$$x + 2y = 7.$$

GS **(a)** $(-1, \_\_\_)$

Substitute ___ for x in the equation and solve for the variable ___.

(b) $(3, \_\_\_)$

(c) $(5, \_\_\_)$

(d) $(7, \_\_\_)$

Figure 7 shows a graph of the ordered-pair solution found above and those in **Margin Problem 1** for $x + 2y = 7$.

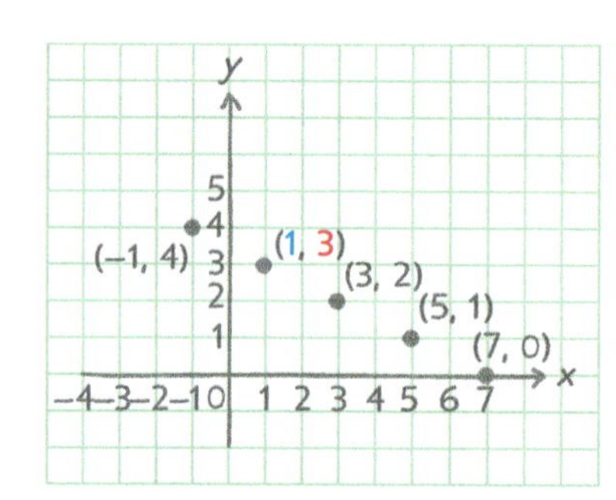

Figure 7

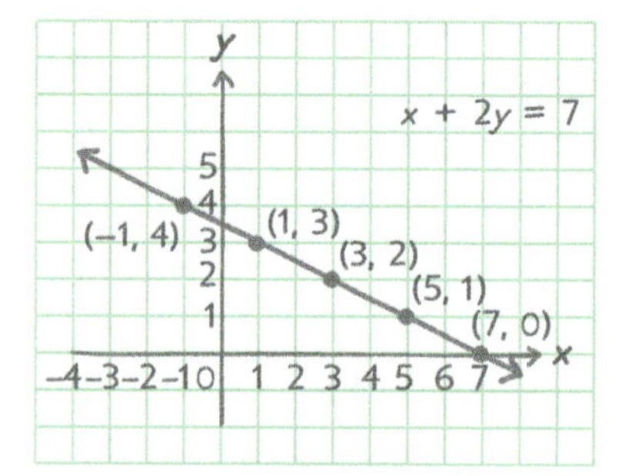

Figure 8

Notice that the points plotted in **Figure 7** all appear to lie on a straight line, as shown in **Figure 8.** In fact, the following is true.

Every point on the line represents a solution of the equation $x + 2y = 7$, and every solution of the equation corresponds to a point on the line.

This line gives a "picture" of all the solutions of the equation

$$x + 2y = 7.$$

The line extends indefinitely in both directions, as suggested by the arrowhead on each end, and is the **graph** of the equation $x + 2y = 7$. **Graphing** is the process of plotting ordered pairs and drawing a line (or curve) through the corresponding points.

Graph of a Linear Equation

The graph of any linear equation in two variables is a straight line.

Notice that the word ***line*** appears in the term "***line***ar equation."

Answers

1. (a) -1; y; $(-1, 4)$
   (b) $(3, 2)$ (c) $(5, 1)$ (d) $(7, 0)$

EXAMPLE 1 **Graphing a Linear Equation**

Graph $x - y = -3$.

At least two different ordered pairs are needed to draw the graph. To find them, we arbitrarily choose values for x or y and substitute them into the equation. We choose $x = 0$ to find one ordered pair and $y = 0$ to find another.

$x - y = -3$
$0 - y = -3$ — 0 is easy to substitute.
$-y = -3$ — Subtract.
$y = 3$ — Multiply by -1.

One ordered pair is $(0, 3)$.

$x - y = -3$
$x - 0 = -3$ — 0 is easy to substitute.
$x = -3$ — Subtract.

One ordered pair is $(-3, 0)$.

We find a third ordered pair (as a check) by choosing some other value for x or y. We let $x = 2$.

$x - y = -3$
$2 - y = -3$ — We arbitrarily let $x = 2$. Other numbers could be used for x, or for y, instead.
$-y = -5$ — Subtract 2.
$y = 5$ — Multiply by -1.

This gives the ordered pair $(2, 5)$. We plot the three ordered-pair solutions and draw a line through them. See **Figure 9.**

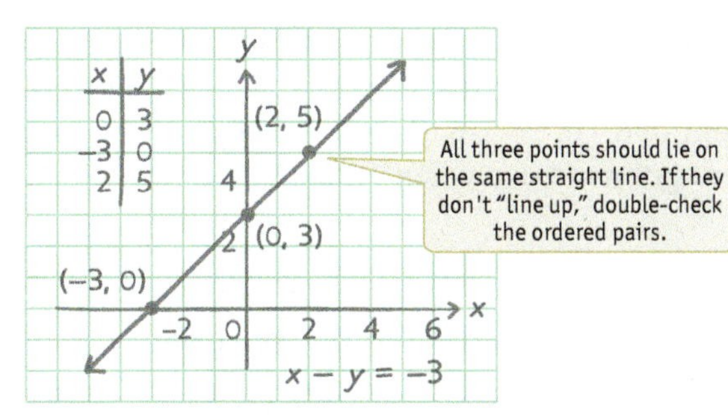

Figure 9

Work Problem **2** at the Side. ▶

EXAMPLE 2 **Graphing a Linear Equation**

Graph $4x - 5y = 20$.

To find three ordered pairs that are solutions of $4x - 5y = 20$, we choose three arbitrary values for x or y that we think will be easy to substitute.

Let $x = 0$.

$4x - 5y = 20$
$4(0) - 5y = 20$
$0 - 5y = 20$
$-5y = 20$
$y = -4$

Ordered pair: $(0, -4)$

Let $y = 0$.

$4x - 5y = 20$
$4x - 5(0) = 20$
$4x - 0 = 20$
$4x = 20$
$x = 5$

Ordered pair: $(5, 0)$

Let $y = 2$.

$4x - 5y = 20$
$4x - 5(2) = 20$
$4x - 10 = 20$
$4x = 30$
$x = 7\frac{1}{2}$ — $\frac{30}{4} = 7\frac{1}{2}$

Ordered pair: $(7\frac{1}{2}, 2)$

Continued on Next Page

2 Graph the linear equation.

GS $x + y = 6$

x	y
0	
	0
4	

From the first row of the table, let $x =$ ____.

$x + y = 6$
____ $+ y = 6$
$y =$ ____

Write ____ for the y-value in the first row of the table.

The first ordered pair is ____.

Repeat this process to complete the table. Then plot the corresponding points, and draw the ____ through them.

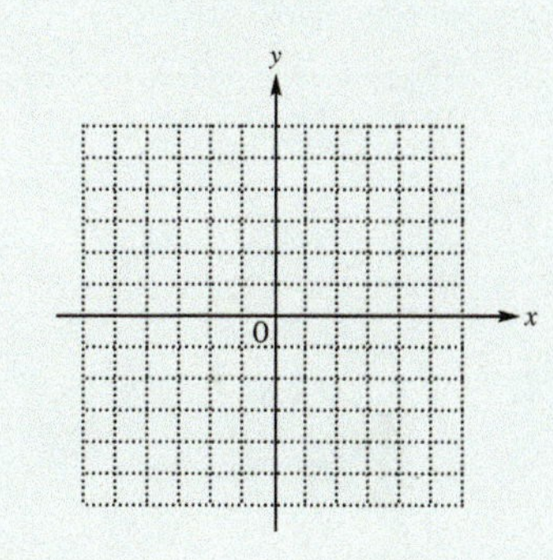

Answer

2. 0; 0; 6; 6; (0, 6); line

x	y
0	6
6	0
4	2

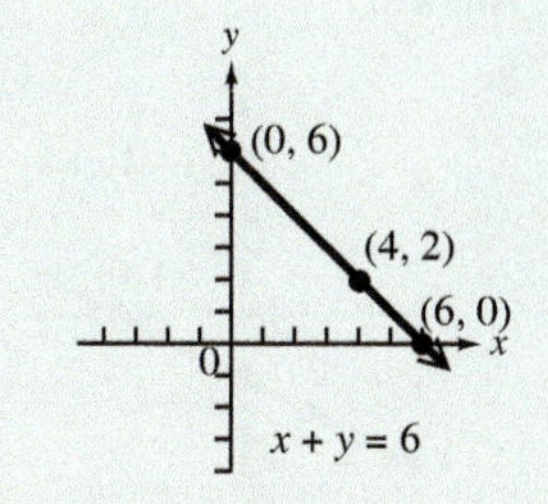

3 Graph $2x - 4y = 8$.

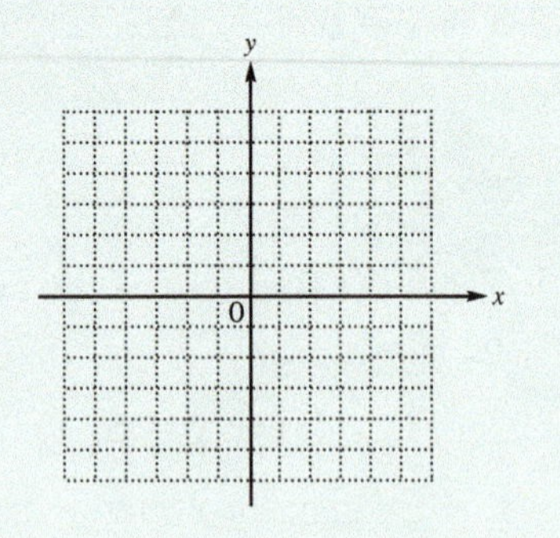

We plot the three ordered-pair solutions and draw a line through them. See **Figure 10.** Two points determine the line, and the third point is used to check that no errors have been made.

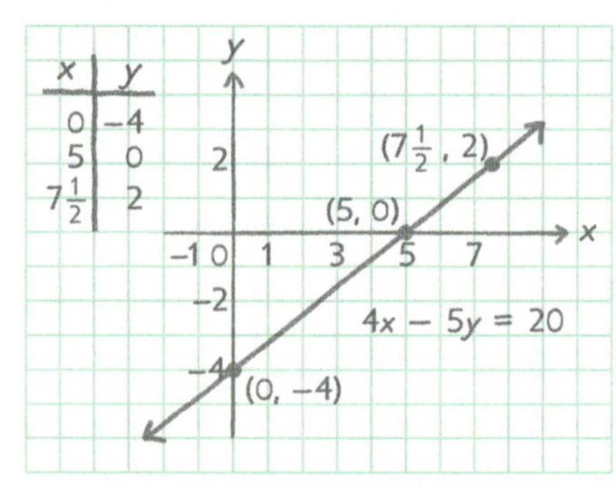

Figure 10

◀ **Work Problem 3 at the Side.**

Note

The ordered pairs that we find and use to graph an equation are *solutions* of the equation. For each value of *x*, there will be a corresponding value of *y*, and for each value of *y*, there will be a corresponding value of *x*. Substituting each ordered pair into the equation should produce a true statement.

OBJECTIVE 2 Find intercepts. In **Figure 10**, the graph intersects (crosses) the x-axis at $(5, 0)$ and the y-axis at $(0, -4)$. For this reason, $(5, 0)$ is the ***x*-intercept** and $(0, -4)$ is the ***y*-intercept** of the graph. The intercepts are convenient points to use when graphing a linear equation.

Finding Intercepts

To find the x-intercept, let $y = 0$ in the given equation and solve for x. Then $(x, 0)$ is the x-intercept.

To find the y-intercept, let $x = 0$ in the given equation and solve for y. Then $(0, y)$ is the y-intercept.

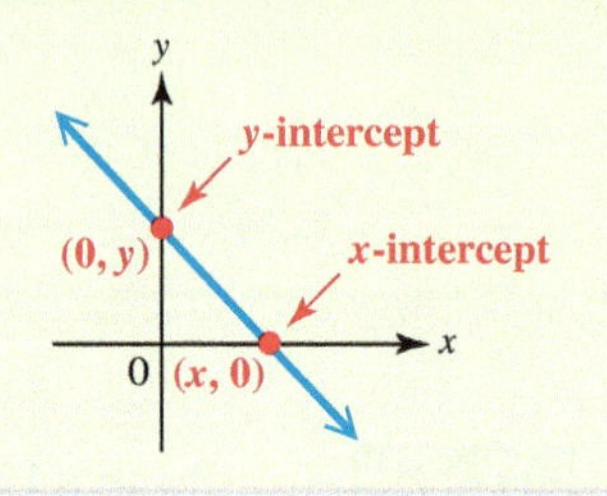

EXAMPLE 3 Graphing a Linear Equation Using Intercepts

Find the intercepts for the graph of $2x + y = 4$. Then draw the graph.

To find the intercepts, we first let $x = 0$ and then let $y = 0$. To find a third point, we arbitrarily let $x = 1$.

Let $x = 0$.	Let $y = 0$.	Let $x = 1$.
$2x + y = 4$	$2x + y = 4$	$2x + y = 4$
$2(0) + y = 4$	$2x + 0 = 4$	$2(1) + y = 4$
$0 + y = 4$	$2x = 4$	$2 + y = 4$
$y = 4$	$x = 2$	$y = 2$
y-intercept: $(0, 4)$	x-intercept: $(2, 0)$	Third point: $(1, 2)$

Continued on Next Page

Answer

3.

$2x - 4y = 8$

(4, 0)

(−2, −3)

(0, −2)

The graph, with the two intercepts in red, is shown in **Figure 11.**

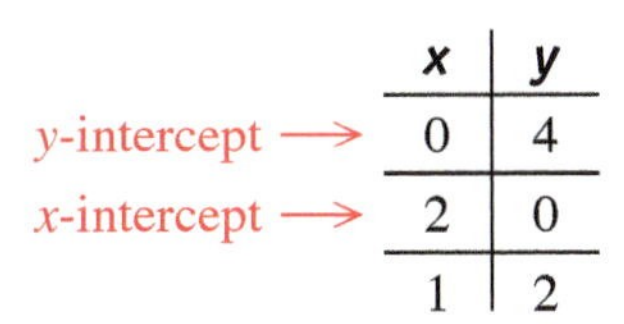

	x	y
y-intercept ⟶	0	4
x-intercept ⟶	2	0
	1	2

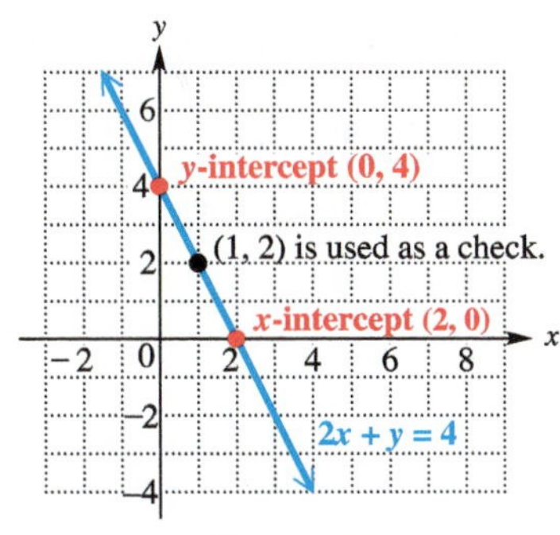

Figure 11

Work Problem 4 at the Side. ▶

EXAMPLE 4 Graphing a Linear Equation Using Intercepts

Graph $y = -\frac{3}{2}x + 3$.

Although this linear equation is not in standard form $Ax + By = C$, it *could* be written in that form. To find the intercepts, we first let $x = 0$ and then let $y = 0$.

$$y = -\frac{3}{2}x + 3$$

$$y = -\frac{3}{2}(0) + 3 \quad \text{Let } x = 0.$$

$$y = 0 + 3 \quad \text{Multiply.}$$

$$y = 3 \quad \text{Add.}$$

y-intercept: $(0, 3)$

$$y = -\frac{3}{2}x + 3$$

$$0 = -\frac{3}{2}x + 3 \quad \text{Let } y = 0.$$

$$\frac{3}{2}x = 3 \quad \text{Add } \tfrac{3}{2}x.$$

$$x = 2 \quad \text{Multiply by } \tfrac{2}{3}.$$

x-intercept: $(2, 0)$

To find a third point, we arbitrarily let $x = -2$.

$$y = -\frac{3}{2}x + 3$$

Choosing a multiple of 2 makes multiplying by $-\frac{3}{2}$ easier.

$$y = -\frac{3}{2}(-2) + 3 \quad \text{Let } x = -2.$$

$$y = 3 + 3 \quad \text{Multiply.}$$

$$y = 6 \quad \text{Add.}$$

Third point: $(-2, 6)$

We plot the three ordered-pair solutions and draw a line through them, as shown in **Figure 12.**

x	y
0	3
2	0
−2	6

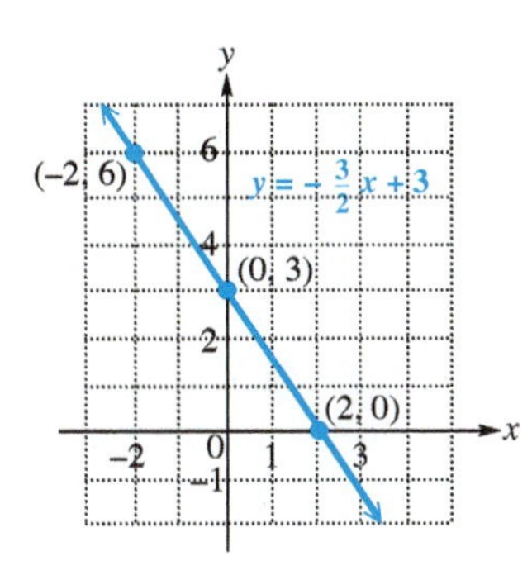

Figure 12

Work Problem 5 at the Side. ▶

4 Find the intercepts for the graph of

$$5x + 2y = 10.$$

Then draw the graph.

Find the y-intercept by letting $x =$ ____ in the equation.
The y-intercept is ____.

Find the x-intercept by letting $y =$ ____ in the equation.
The x-intercept is ____.

Find a third point with $x = 4$ as a check, and graph the equation.

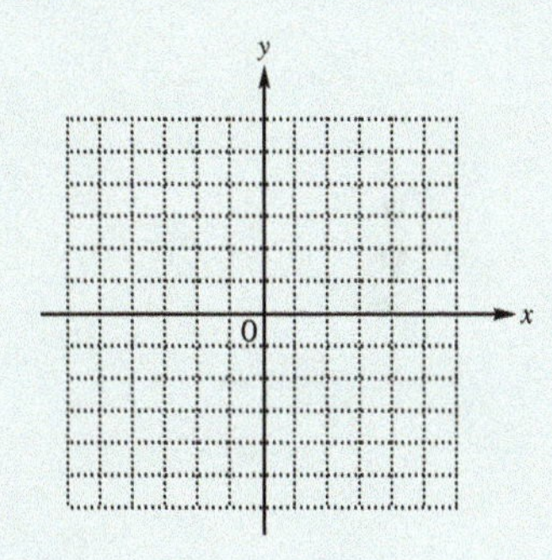

5 Graph $y = \frac{2}{3}x - 2$.

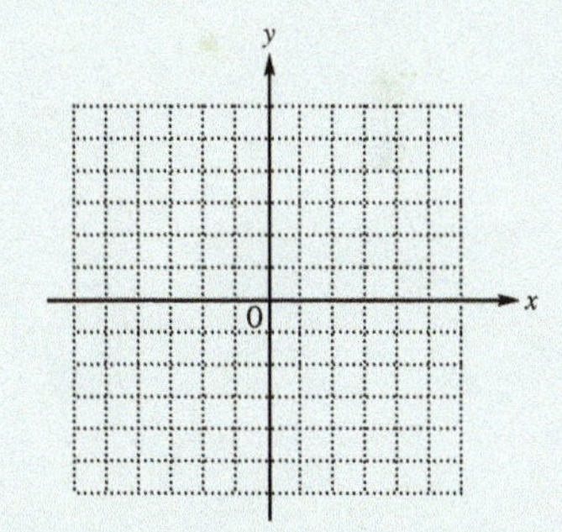

Answers

4. 0; (0, 5); 0; (2, 0)

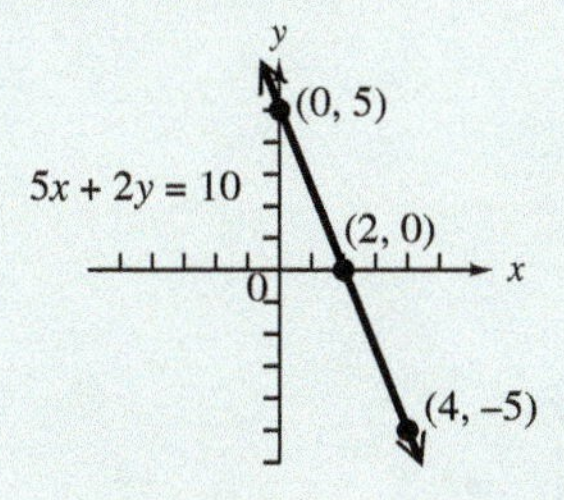

5.

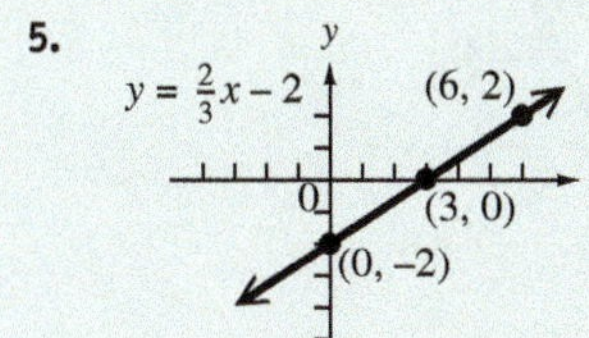

6 Graph $2x - y = 0$.

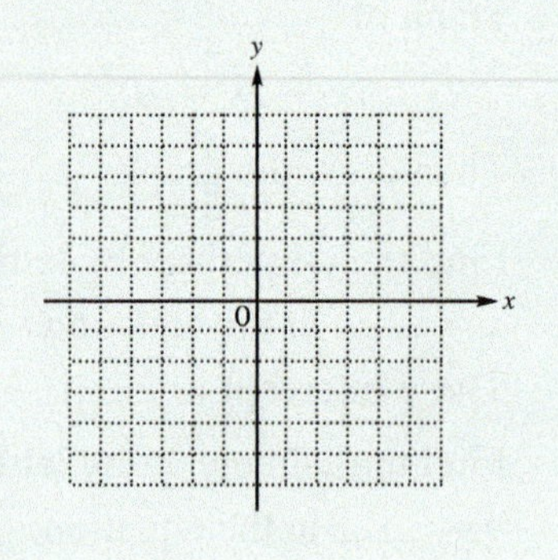

OBJECTIVE 3 **Graph linear equations of the form $Ax + By = 0$.**

EXAMPLE 5 **Graphing an Equation with x- and y-Intercepts (0, 0)**

Graph $x - 3y = 0$.

To find the y-intercept, let $x = 0$.

$$x - 3y = 0$$
$$0 - 3y = 0 \quad \text{Let } x = 0.$$
$$-3y = 0 \quad \text{Subtract.}$$
$$y = 0 \quad \text{Divide by } -3.$$

y-intercept: $(0, 0)$

To find the x-intercept, let $y = 0$.

$$x - 3y = 0$$
$$x - 3(0) = 0 \quad \text{Let } y = 0.$$
$$x - 0 = 0 \quad \text{Multiply.}$$
$$x = 0 \quad \text{Subtract.}$$

x-intercept: $(0, 0)$

The x- and y-intercepts are the *same* point, $(0, 0)$. We select *two other values* for x or y to find two other points. We choose $x = 6$ and $x = -3$.

$$x - 3y = 0$$
$$6 - 3y = 0 \quad \text{Let } x = 6.$$
$$-3y = -6 \quad \text{Subtract 6.}$$
$$y = 2 \quad \text{Divide by } -3.$$

Ordered pair: $(6, 2)$

$$x - 3y = 0$$
$$-3 - 3y = 0 \quad \text{Let } x = -3.$$
$$-3y = 3 \quad \text{Add 3.}$$
$$y = -1 \quad \text{Divide by } -3.$$

Ordered pair: $(-3, -1)$

We use the three ordered-pair solutions to draw the graph in **Figure 13.**

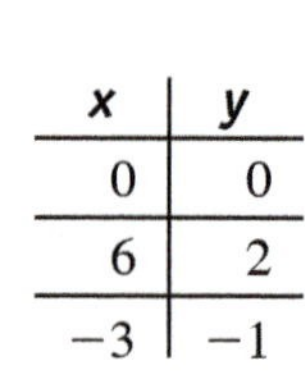

x	y
0	0
6	2
-3	-1

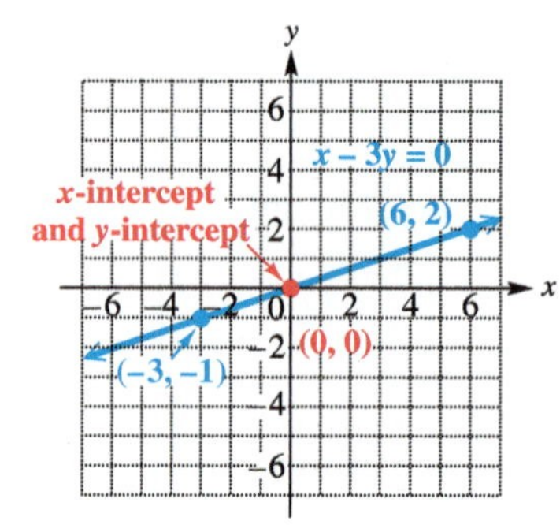

Figure 13

◀ **Work Problem 6 at the Side.**

Line through the Origin

The graph of a linear equation of the form

$$Ax + By = 0,$$

where A and B are nonzero real numbers, passes through the origin $(0, 0)$.

Examples: $2x + 3y = 0$, $x = y$, $-4y = 3x$ The last two can be written in $Ax + By = 0$ form.

OBJECTIVE 4 **Graph linear equations of the form $y = b$ or $x = a$.** Consider the following linear equations.

$$y = -4 \quad \text{can be written as} \quad 0x + y = -4.$$
$$x = 3 \quad \text{can be written as} \quad x + 0y = 3.$$

When the coefficient of x or y is 0, the graph is a horizontal or vertical line.

Answer

6.

y
x
4
2
(2, 4)
(0, 0)
(−2, −4)
$2x - y = 0$

EXAMPLE 6 Graphing a Horizontal Line ($y = b$)

Graph $y = -4$.

For any value of x, the value of y is always -4. Three ordered-pair solutions of the equation are shown in the table of values. Drawing a line through these points gives the **horizontal line** in **Figure 14.**

x can be *any* real number.

x	y
-2	-4
0	-4
3	-4

y must be -4.

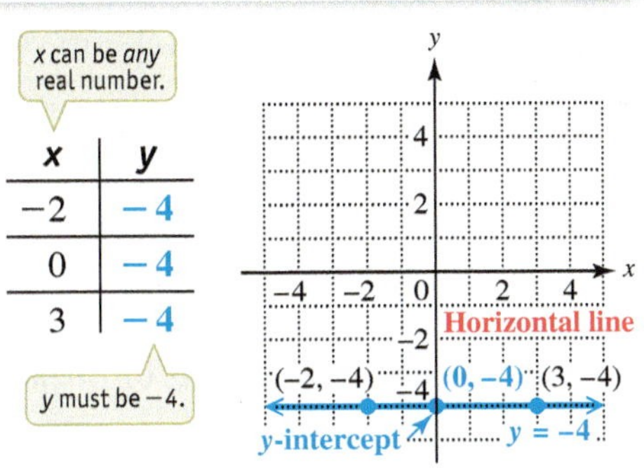

Figure 14

The y-intercept is $(0, -4)$.

There is no x-intercept.

Work Problem 7 at the Side. ▶

EXAMPLE 7 Graphing a Vertical Line ($x = a$)

Graph $x - 3 = 0$.

First we add 3 to each side of the equation $x - 3 = 0$ to obtain $x = 3$. All ordered-pair solutions of this equation have x-coordinate 3. Any number can be used for y. Three ordered pairs that satisfy the equation are given in the table of values. The graph is the **vertical line** in **Figure 15.**

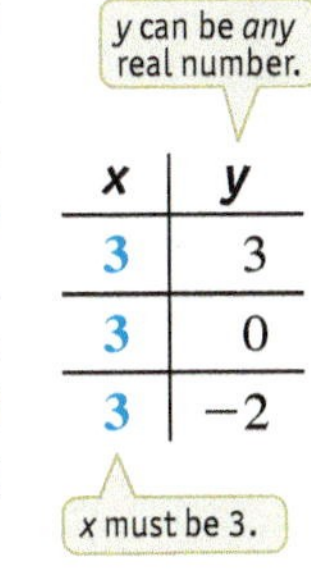
y can be *any* real number.

x	y
3	3
3	0
3	-2

x must be 3.

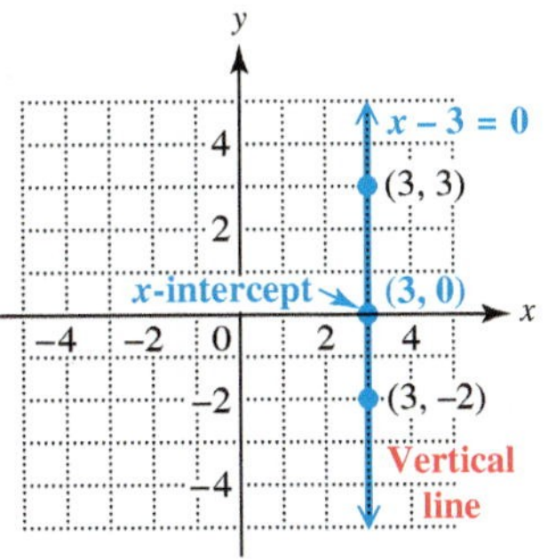

Figure 15

The x-intercept is $(3, 0)$.

There is no y-intercept.

Work Problem 8 at the Side. ▶

From **Examples 6 and 7,** we make the following observations.

Horizontal and Vertical Lines

The graph of $\mathbf{y = b}$, where b is a real number, is a **horizontal line** with y-intercept $(0, b)$ and no x-intercept (unless the horizontal line is the x-axis itself).

Examples: $y = 5$ and $y + 2 = 0$ (which can be written $y = -2$)

The graph of $\mathbf{x = a}$, where a is a real number, is a **vertical line** with x-intercept $(a, 0)$ and no y-intercept (unless the vertical line is the y-axis itself).

Examples: $x = -1$ and $x - 6 = 0$ (which can be written $x = 6$)

Keep the following in mind regarding the x- and y-axes.

- ***The x-axis is the horizontal line given by the equation $y = 0$.***
- ***The y-axis is the vertical line given by the equation $x = 0$.***

7 Graph $y = -5$.

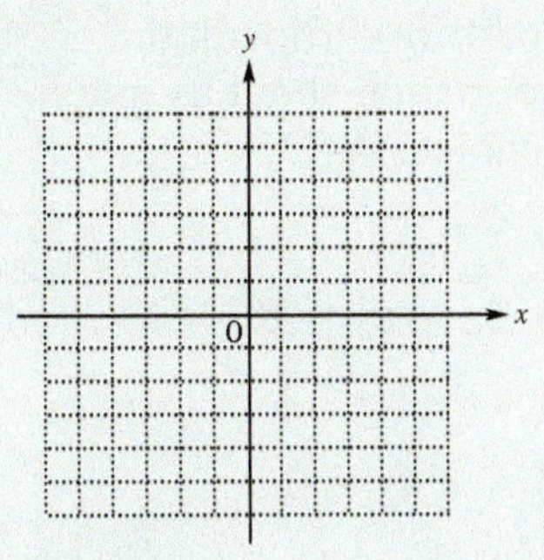

This is the graph of a (*horizontal* / *vertical*) line. There is no ____-intercept.

8 Graph $x + 4 = 6$.

GS First subtract ____ from each side to obtain the equivalent equation $x =$ ____.

Now graph this equation.

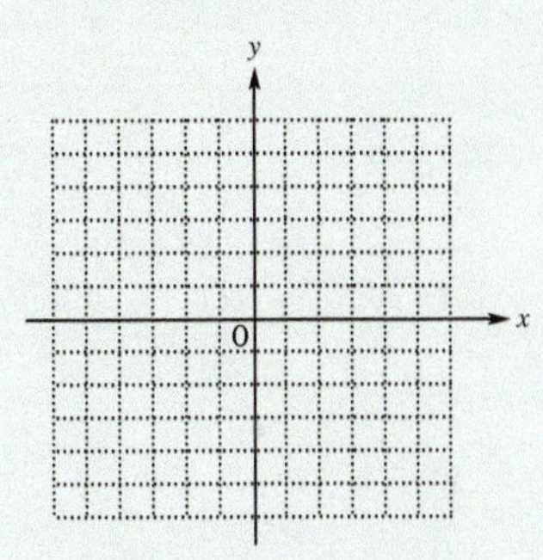

This is the graph of a (*horizontal* / *vertical*) line. There is no ____-intercept.

Answers

7.

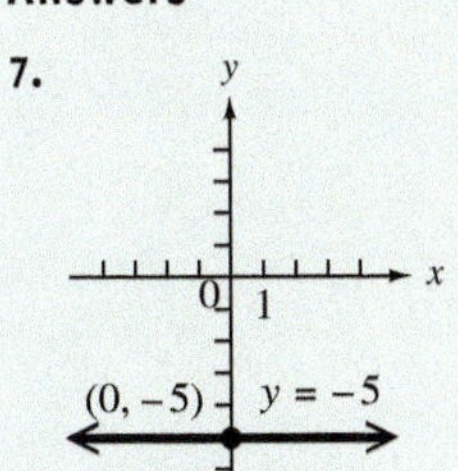

horizontal; x

8. 4; 2

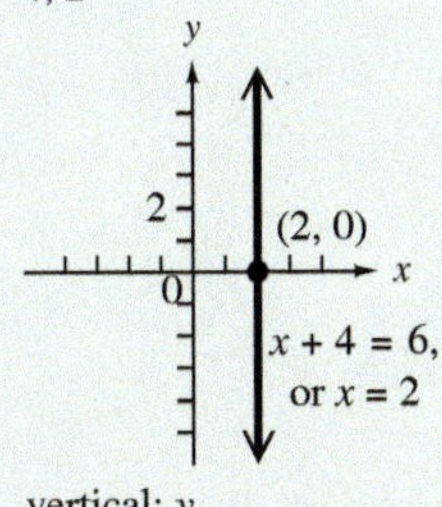

vertical; y

9 Match each linear equation in parts (a)–(d) with the information about its graph in choices A–D.

(a) $x = 5$

(b) $2x - 5y = 8$

(c) $y - 2 = 3$

(d) $x + 4y = 0$

A. The graph of the equation is a horizontal line.

B. The graph of the equation passes through the origin.

C. The graph of the equation is a vertical line.

D. The graph of the equation passes through the point (9, 2).

Answers

9. (a) C (b) D (c) A (d) B

CAUTION

The equations of horizontal and vertical lines are often confused with each other.

- The graph of $y = b$ is parallel to the x-axis (for $b \neq 0$).
- The graph of $x = a$ is parallel to the y-axis (for $a \neq 0$).

A summary of the forms of linear equations from this section follows.

FORMS OF LINEAR EQUATIONS

Equation	To Graph	Example
$Ax + By = C$ (where A, B, and C are real numbers not equal to 0)	Find any two points on the line. A good choice is to find the intercepts. Let $x = 0$, and find the corresponding value of y. Then let $y = 0$, and find x. As a check, find a third point by choosing a value for x or y that has not yet been used.	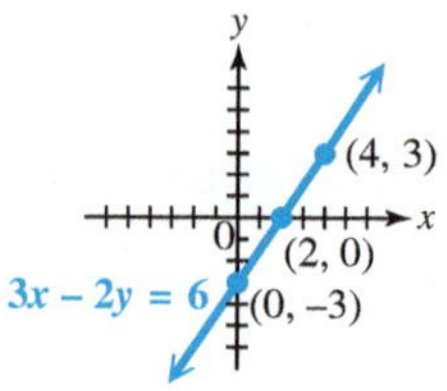 y; x; 0; (4, 3); (2, 0); (0, −3); $3x - 2y = 6$
$Ax + By = 0$	The graph passes through (0, 0). To find additional points that lie on the graph, choose any value for x or y, except 0.	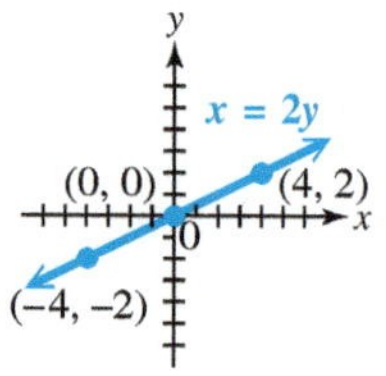 y; x; 0; $x = 2y$; (0, 0); (4, 2); (−4, −2)
$y = b$	Draw a horizontal line, through the point $(0, b)$.	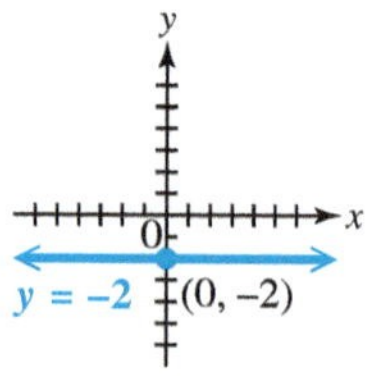 y; x; 0; $y = -2$; (0, −2)
$x = a$	Draw a vertical line, through the point $(a, 0)$.	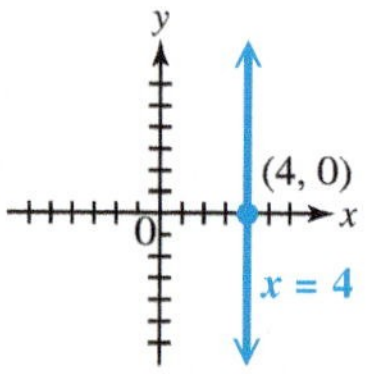 y; x; 0; (4, 0); $x = 4$

◄ Work Problem 9 at the Side.

OBJECTIVE 5 **Use a linear equation to model data.**

EXAMPLE 8 **Using a Linear Equation to Model Credit Card Debt**

The amount of credit card debt y in billions of dollars in the United States from 2010 through 2015 can be modeled by the linear equation

$$y = 179.0x + 2600,$$

where $x = 0$ represents the year 2010, $x = 1$ represents 2011, and so on. (Data from Board of Governors of the Federal Reserve System.)

Continued on Next Page

(a) Use the equation to approximate credit card debt in the years 2010, 2013, and 2015.

Substitute the appropriate value for each year x to find credit card debt in that year.

	$y = 179.0x + 2600$	Given linear equation
For 2010:	$y = 179.0(0) + 2600$	Replace x with 0.
	$y = 2600$ billion dollars	Multiply, and then add.
For 2013:	$y = 179.0(3) + 2600$	$2013 - 2010 = 3$
	$y = 3137$ billion dollars	Replace x with 3.
For 2015:	$y = 179.0(5) + 2600$	$2015 - 2010 = 5$
	$y = 3495$ billion dollars	Replace x with 5.

(b) Write the information from part (a) as three ordered pairs, and use them to graph the given linear equation.

Because x represents the year and y represents the debt, the three ordered pairs are

$$(0, 2600), \quad (3, 3137), \quad \text{and} \quad (5, 3495).$$

See **Figure 16.** (Arrowheads are not included with the graphed line because the data are for the years 2010 to 2015 only—that is, from $x = 0$ to $x = 5$.)

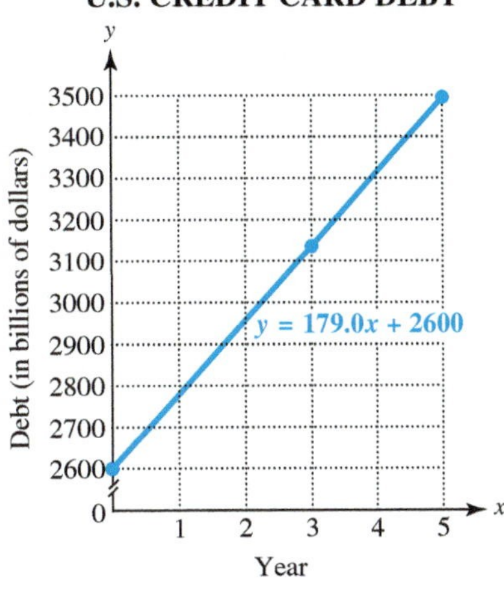

Figure 16

(c) Use the graph and then the equation to determine the year when credit card debt was about \$3300 billion.

Locate 3300 on the vertical axis in **Figure 16.** Move horizontally from 3300 to the graphed line, then down from the line to the value of x on the horizontal scale to find the year, here 4. Because $x = 0$ represents 2010, $x = 4$ represents the year

$$2010 + 4 = \mathbf{2014}. \leftarrow \text{Year when credit card debt was 3300 billion dollars}$$

To use the equation to find the year, substitute 3300 for y and solve for x.

$y = 179.0x + 2600$	Given linear equation
$3300 = 179.0x + 2600$	Let $y = 3300$.
$700 = 179.0x$	Subtract 2600.
$3.9 \approx x$	Divide by 179.0.

Rounding up to the nearest year gives $x = 4$—that is, **2014**—which agrees with the graph.

Work Problem **10** at the Side. ▶

10 Refer to **Example 8.**

(a) Use the equation to approximate credit card debt in the year 2012.

(b) Use the graph and then the equation to determine the year when credit card debt was about \$2800 billion. (Round down for the year.)

Answers

10. **(a)** 2958 billion dollars
(b) 2011; 2011

3.2 Exercises

FOR EXTRA HELP *Go to MyMathLab for worked-out, step-by-step solutions to exercises enclosed in a square and video solutions to ▶ exercises.*

CONCEPT CHECK *Fill in each blank with the correct response.*

1. A linear equation in two variables can be written in the form $Ax +$ ____ $=$ ____, where A, B, and C are real numbers and A and B are not both ____.

2. The graph of any linear equation in two variables is a straight ______. Every point on the line represents a ______ of the equation.

3. CONCEPT CHECK Match the information about each graph in Column I with the correct linear equation in Column II.

I	II
(a) The graph of the equation has y-intercept $(0, -4)$.	**A.** $3x + y = -4$
(b) The graph of the equation has $(0, 0)$ as x-intercept and y-intercept.	**B.** $x - 4 = 0$
(c) The graph of the equation does not have an x-intercept.	**C.** $y = 4x$
(d) The graph of the equation has x-intercept $(4, 0)$.	**D.** $y = 4$

4. CONCEPT CHECK Which of these equations have a graph with only one intercept?

A. $x + 8 = 0$ **B.** $x - y = 3$ **C.** $x + y = 0$ **D.** $y = 4$

CONCEPT CHECK *Identify the intercepts of each graph. (Coordinates of the points shown are integers.)*

5.

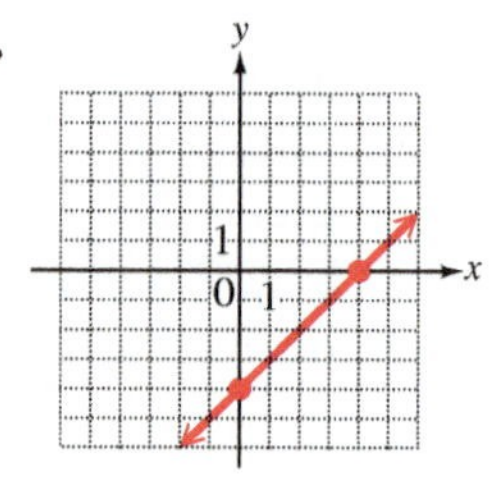

6.

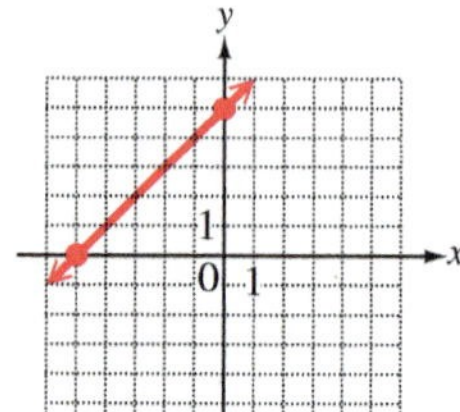

7.

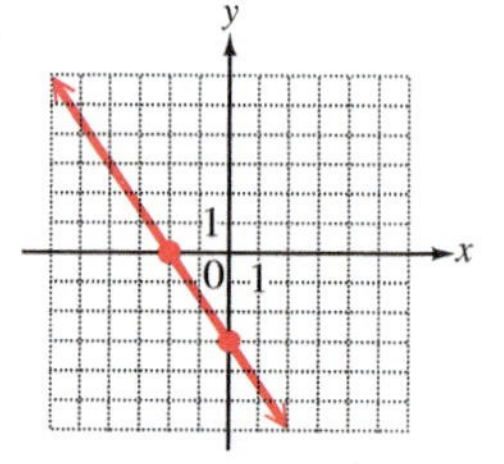

8.

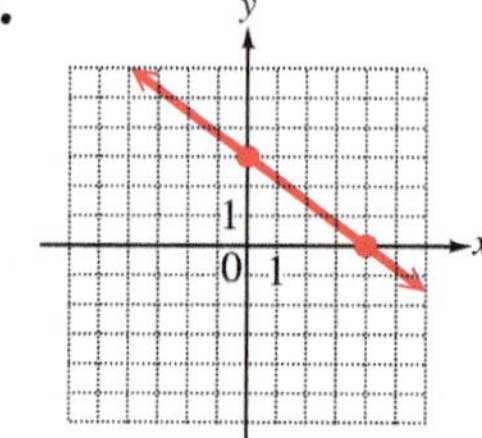

Complete the given ordered-pair solutions of each equation. Then graph the equation by plotting the points and drawing a line through them. **See Examples 1–4.**

9. ▶ $x + y = 5$

(0, ____), (____, 0), (2, ____)

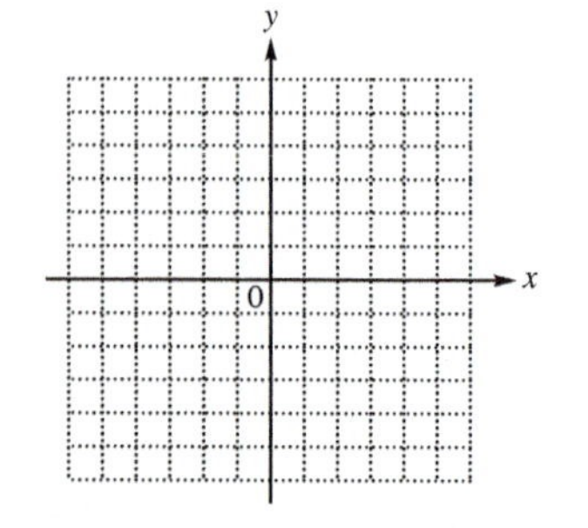

10. $x - y = 2$

(0, ____), (____, 0), (5, ____)

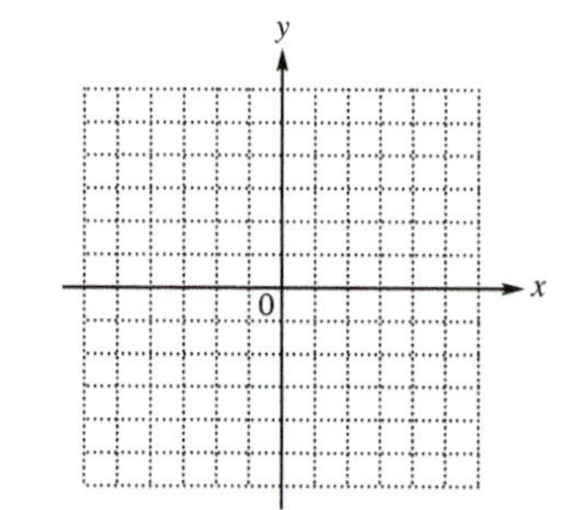

11. ▶ 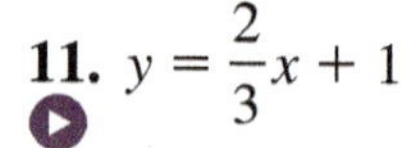 $y = \frac{2}{3}x + 1$

(0, ____), (3, ____), (−3, ____)

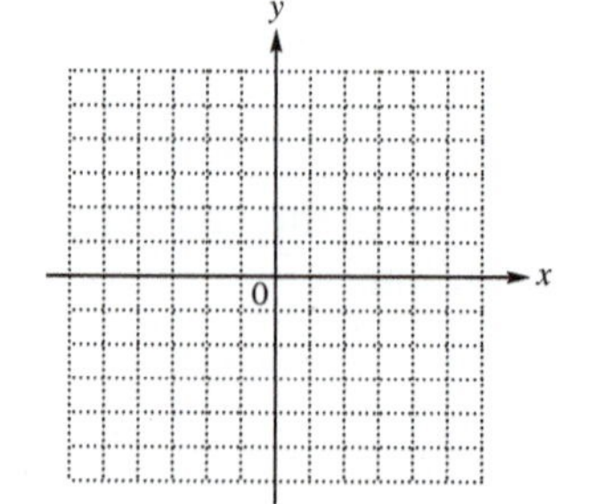

12. $y = -\frac{3}{4}x + 2$

$(0, \_\_\_\_), (4, \_\_\_\_), (-4, \_\_\_\_)$

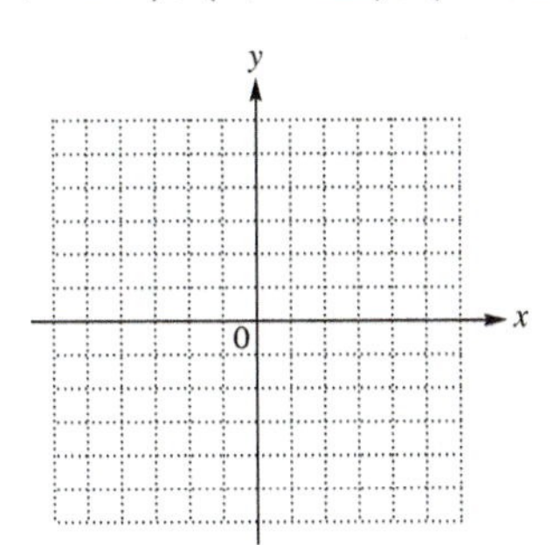

13. $3x = -y - 6$

$(0, \_\_\_\_), (\_\_\_\_, 0), (-\frac{1}{3}, \_\_\_\_)$

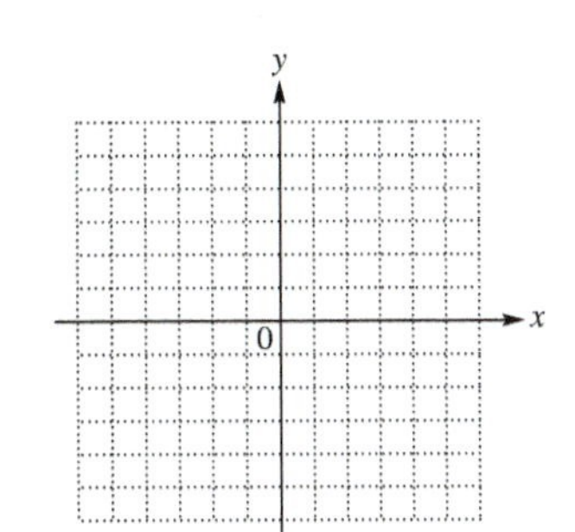

14. $x = 2y + 3$

$(\_\_\_\_, 0), (0, \_\_\_\_), (\_\_\_\_, \frac{1}{2})$

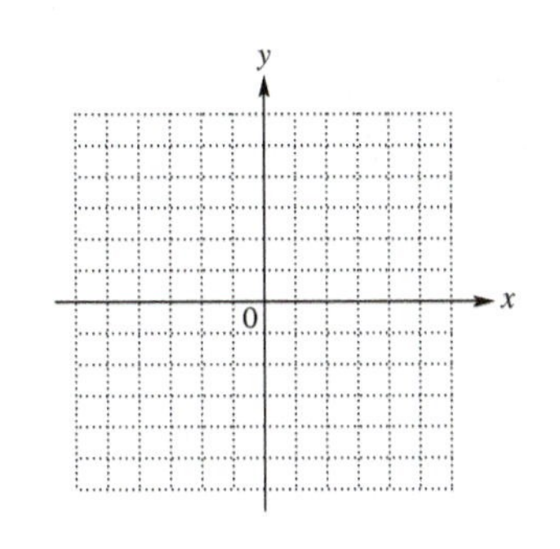

Find the x- and y-intercepts for the graph of each equation. ***See Examples 1–5.***

15. $x - y = 8$

GS To find the y-intercept, let $x =$ ____.
The y-intercept is ____.
To find the x-intercept, let $y =$ ____.
The x-intercept is ____.

16. 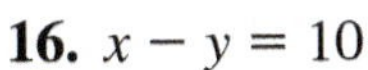 $x - y = 10$

GS To find the y-intercept, let $x =$ ____.
The y-intercept is ____.
To find the x-intercept, let $y =$ ____.
The x-intercept is ____.

17. $2x - 3y = 24$

y-intercept: ____
x-intercept: ____

18. $-3x + 8y = 48$

y-intercept: ____
x-intercept: ____

19. $x + 6y = 0$

y-intercept: ____
x-intercept: ____

20. $3x - y = 0$

y-intercept: ____
x-intercept: ____

Graph each linear equation using intercepts. ***See Examples 1–7.***

21. $x - y = 4$

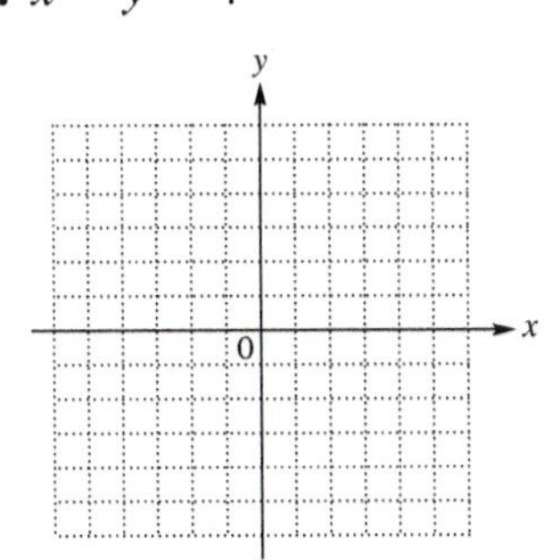

22. 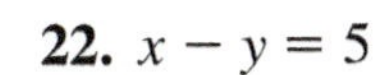 $x - y = 5$

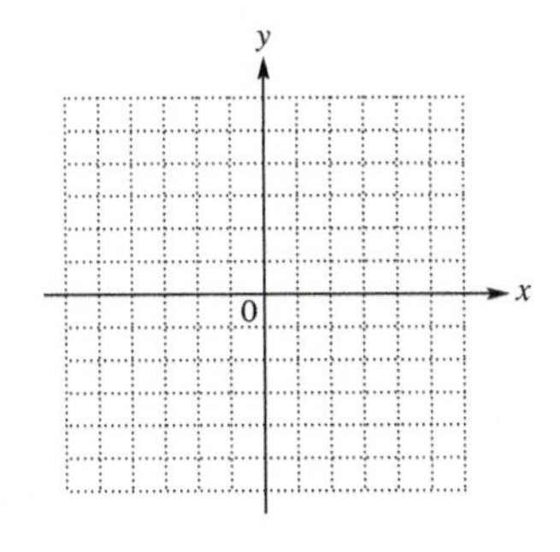

23. $2x + y = 6$

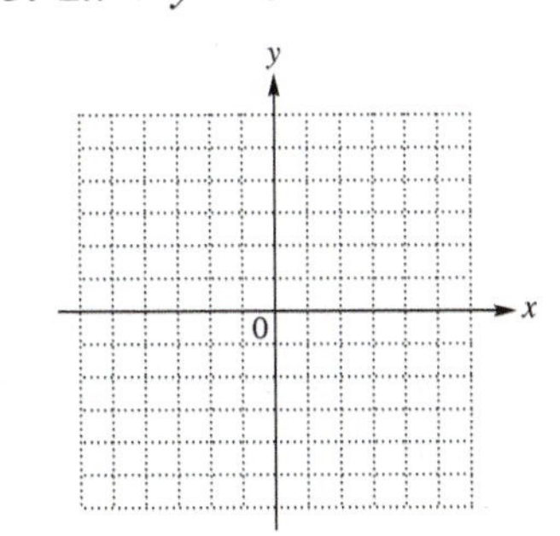

24. $-3x + y = -6$

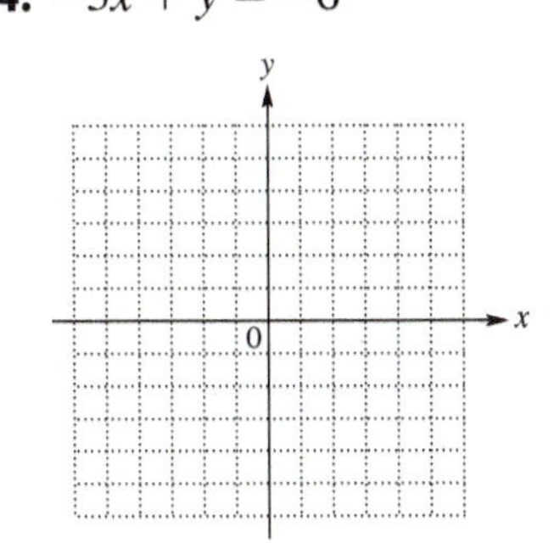

25. $y = x - 2$

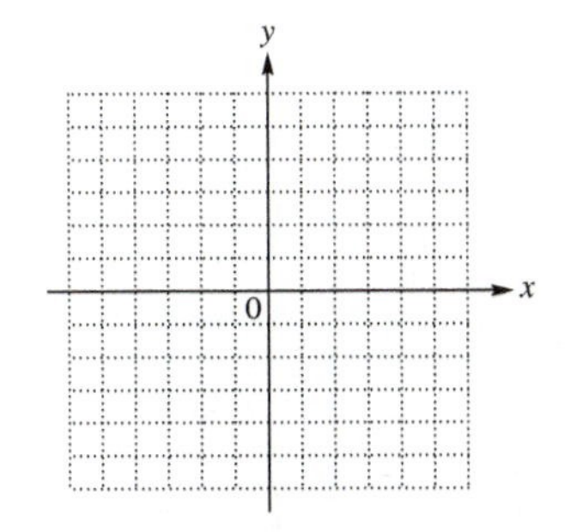

26. 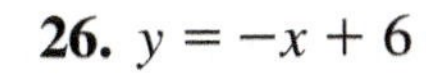 $y = -x + 6$

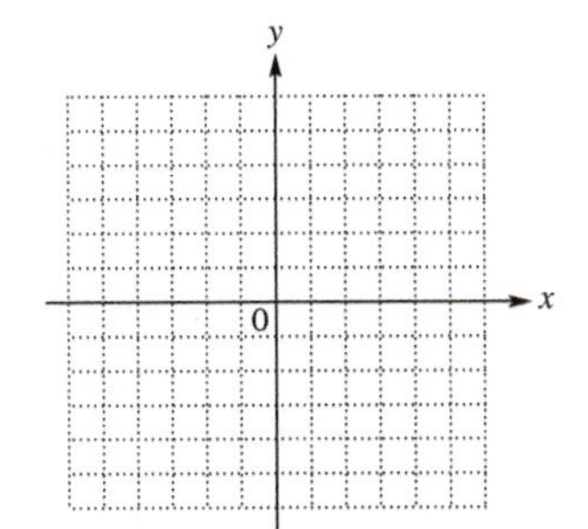

27. 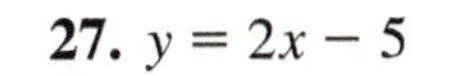 $y = 2x - 5$

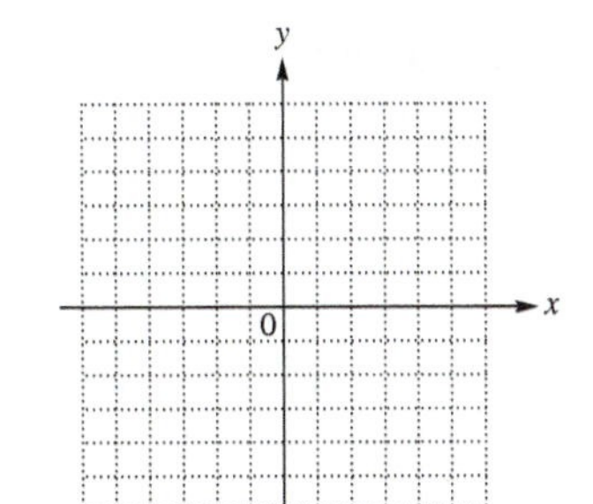

28. 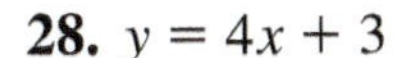 $y = 4x + 3$

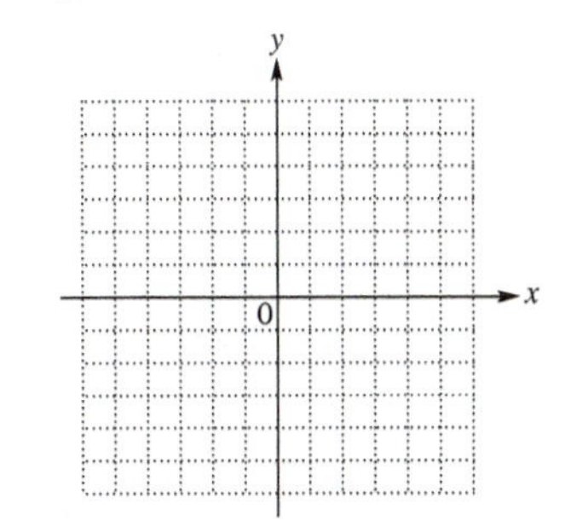

29. $3x + 7y = 14$

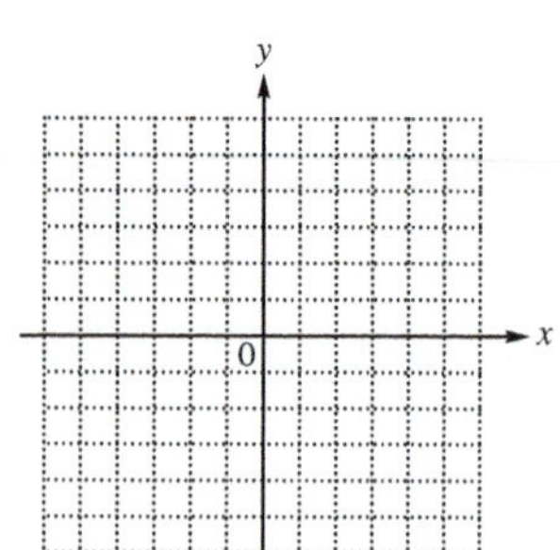

30. $6x - 5y = 18$

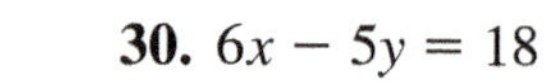

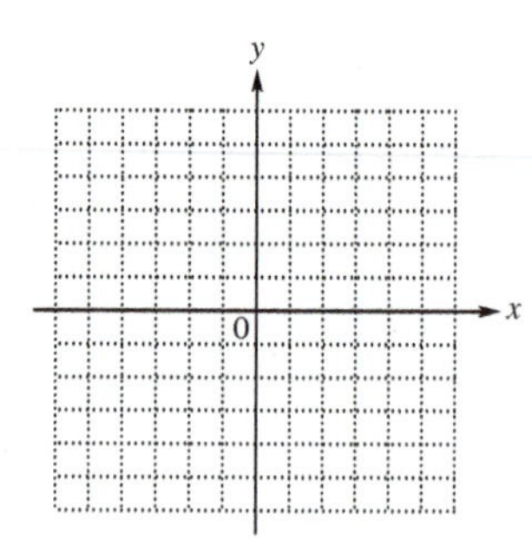

31. $y - 2x = 0$

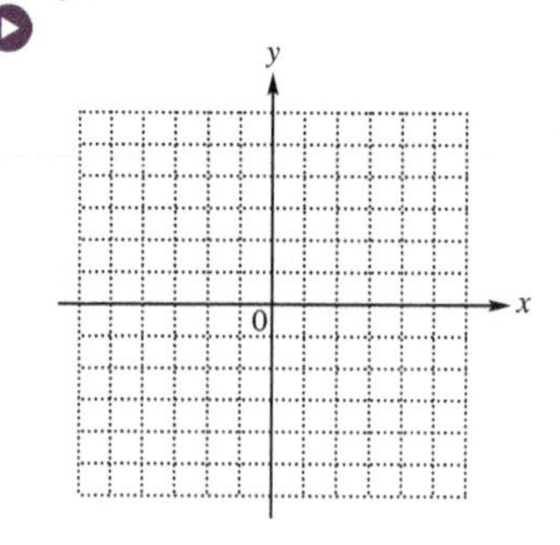

32. $y + 3x = 0$

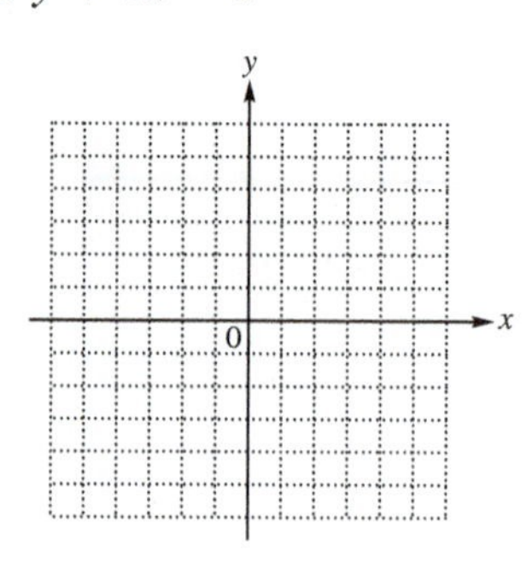

33. $y = -6x$

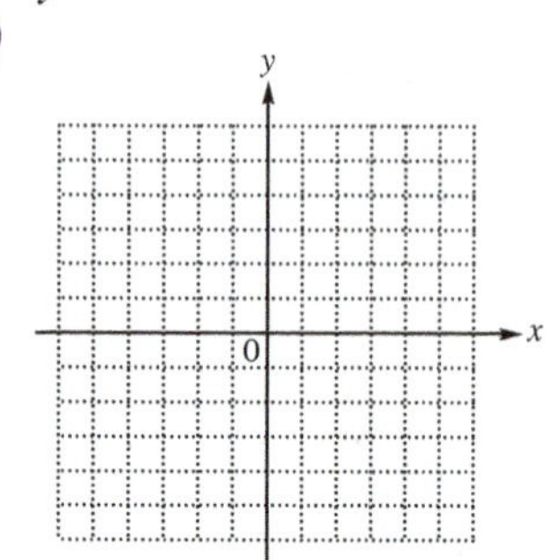

34. $y = 4x$

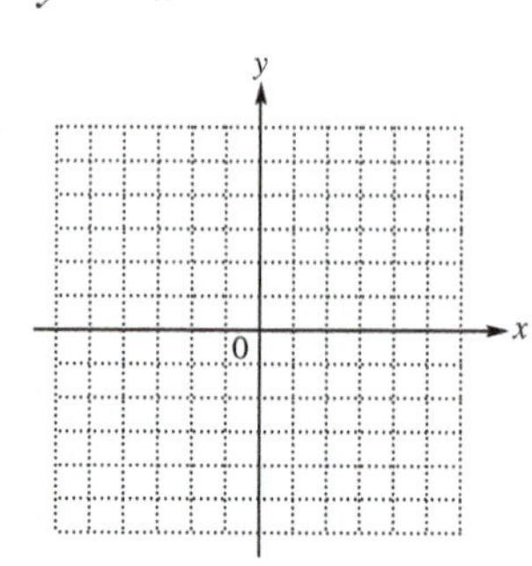

35. $y = -1$

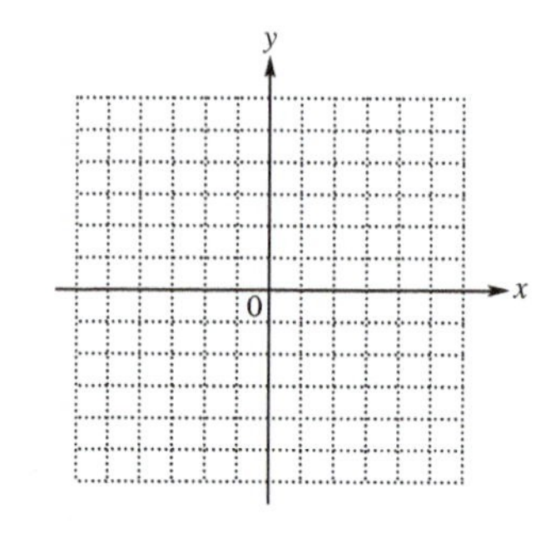

36. $y = 3$

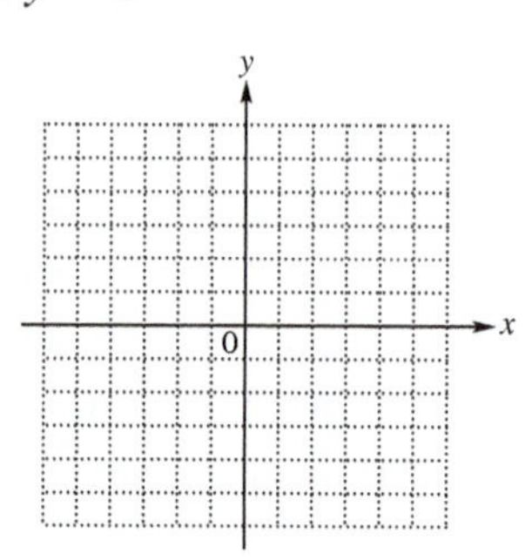

37. $x = -2$

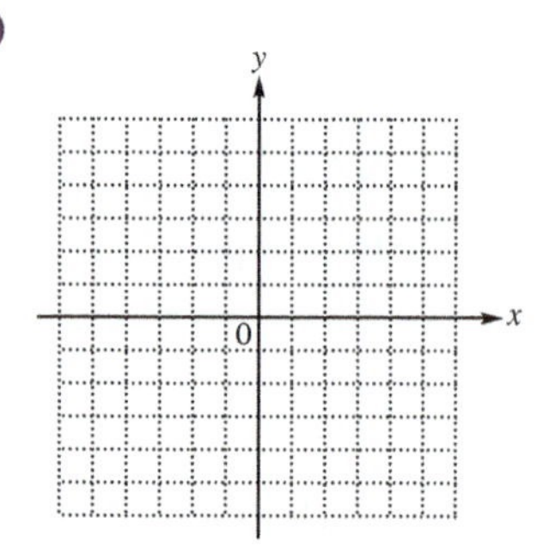

38. $x = 4$

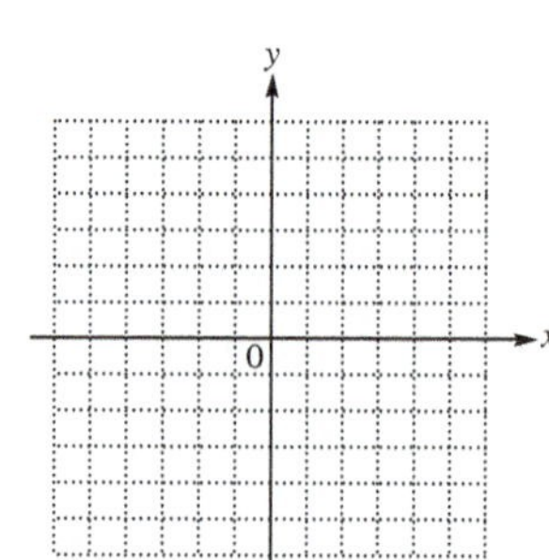

39. $y - 2 = 0$

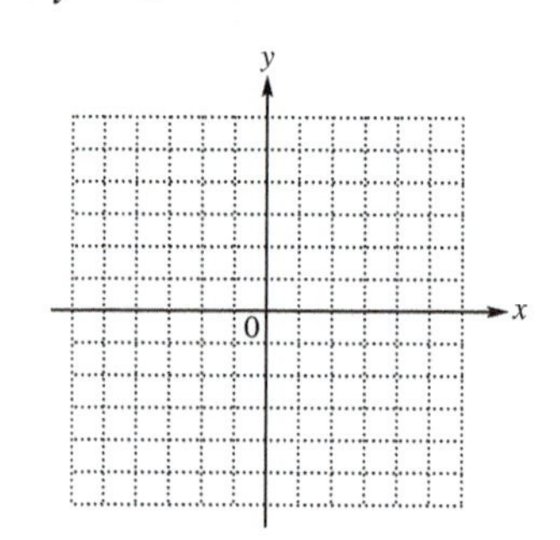

40. $y + 3 = 0$

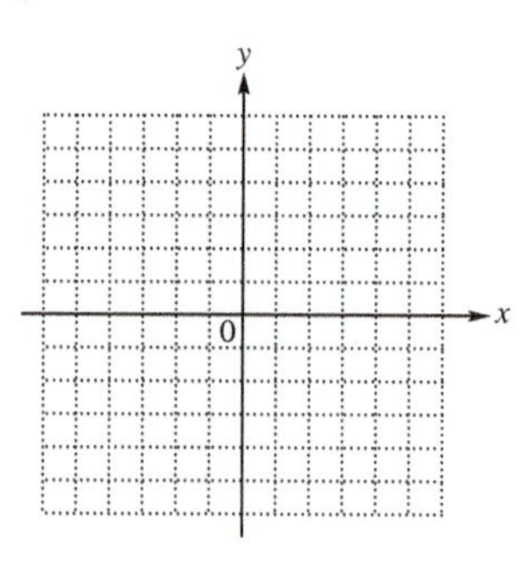

41. $x + 2 = 8$

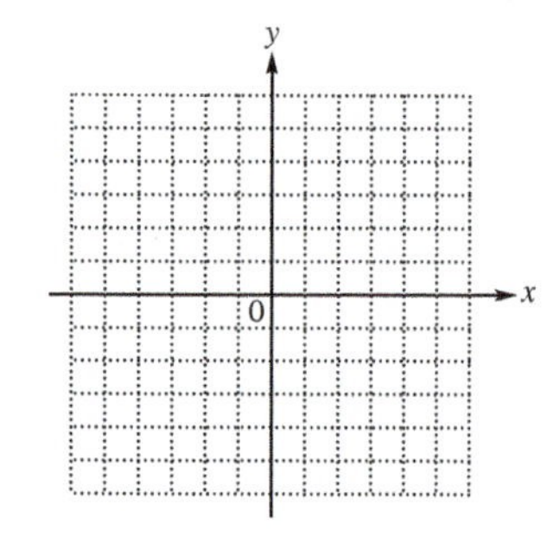

42. $x - 1 = -4$

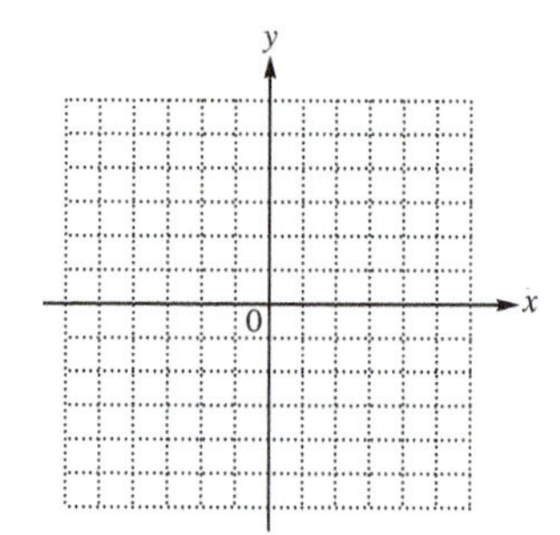

43. $-3y = 15$

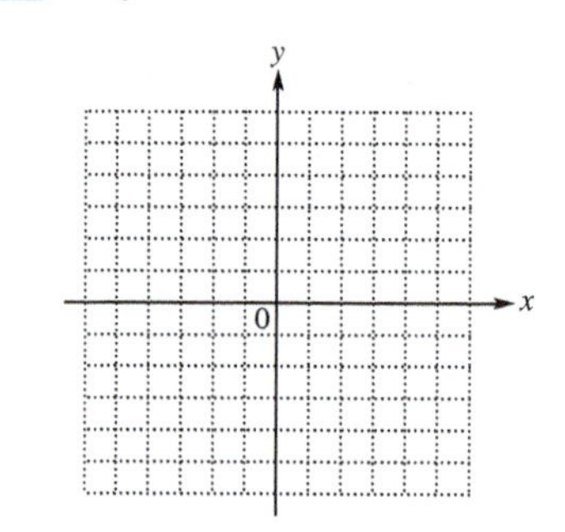

44. $-2y = 12$

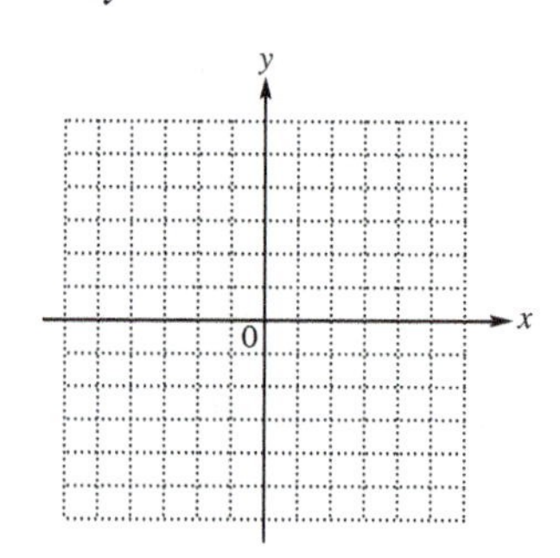

45. CONCEPT CHECK Match each equation in parts (a)–(d) with its graph in choices A–D.

(a) $x = -2$ **(b)** $y = -2$ **(c)** $x = 2$ **(d)** $y = 2$

A.

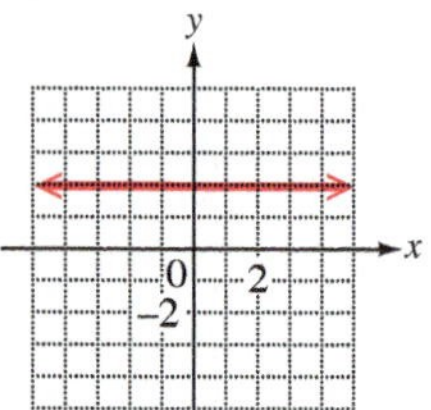

B.

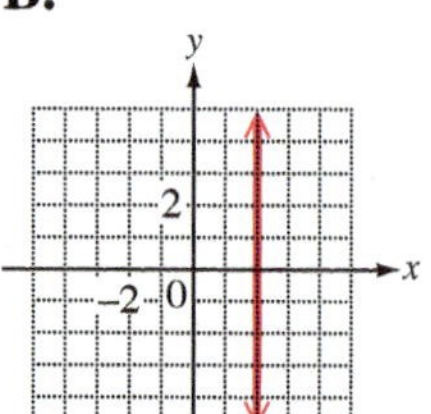

C.

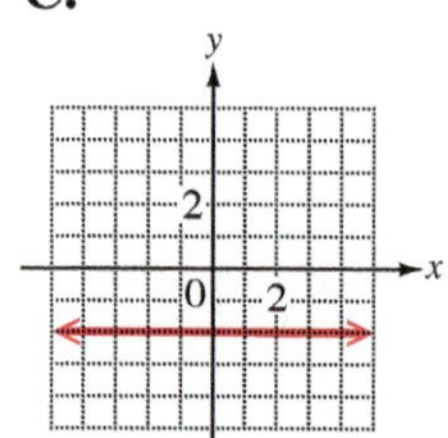

D.

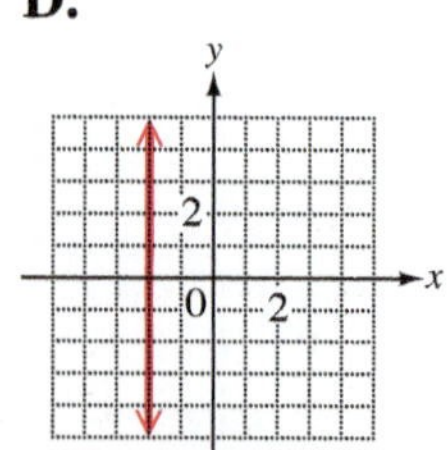

46. **CONCEPT CHECK** What is the equation of the x-axis? The y-axis?

Describe what the graph of each linear equation will look like in the coordinate plane. (Hint: Rewrite the equation if necessary so that it is in a more recognizable form.)

47. $3x = y - 9$

48. $x - 10 = 1$

49. $3y = -6$

50. $2x = 4y$

Solve each problem. ***See Example 8.***

51. The height y (in centimeters) of a woman can be approximated by the linear equation

$$y = 3.9x + 73.5,$$

where x is the length of her radius bone in centimeters.

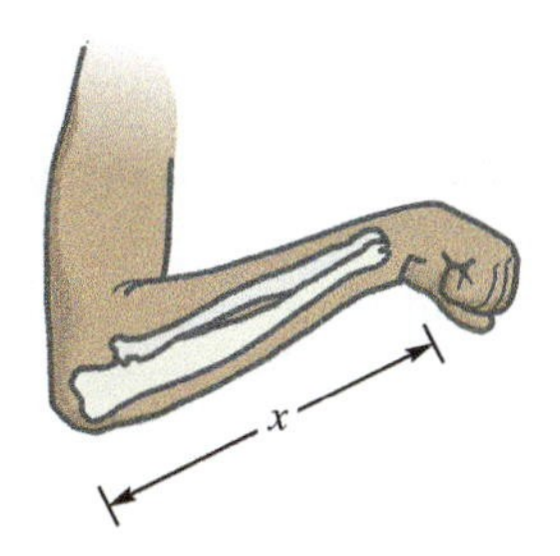

(a) Use the equation to approximate the heights of women with radius bones of lengths 20 cm, 22 cm, and 26 cm.

(b) Write the information from part (a) as three ordered pairs.

(c) Graph the equation for $x \geq 20$, using the data from part (b).

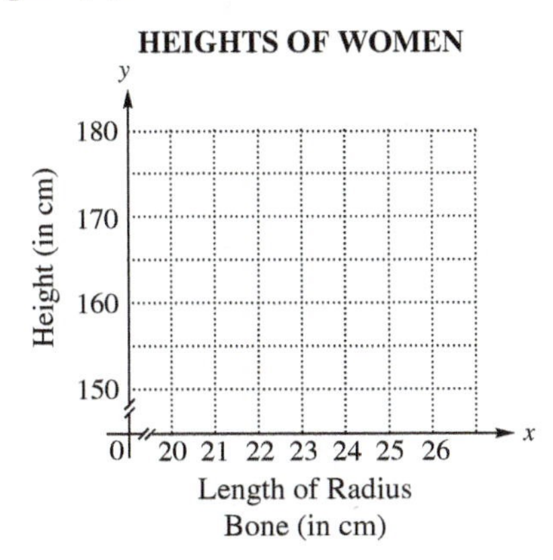

(d) Use the graph to estimate the length of the radius bone in a woman who is 167 cm tall. Then use the equation to find the length of this radius bone to the nearest centimeter. (*Hint:* Substitute for y in the equation.)

52. The weight y (in pounds) of a man taller than 60 in. can be approximated by the linear equation

$$y = 5.5x - 220,$$

where x is the height of the man in inches.

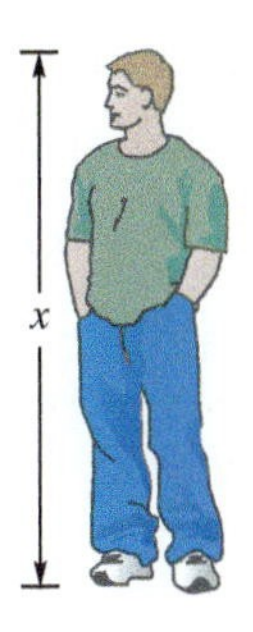

(a) Use the equation to approximate the weights of men whose heights are 62 in., 66 in., and 72 in.

(b) Write the information from part (a) as three ordered pairs.

(c) Graph the equation for $x \geq 62$, using the data from part (b).

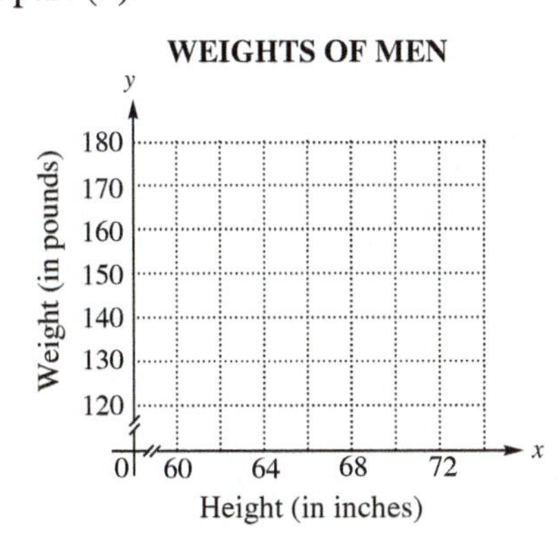

(d) Use the graph to estimate the height of a man who weighs 155 lb. Then use the equation to find the height of this man to the nearest inch. (*Hint:* Substitute for y in the equation.)

53. As a fundraiser, a school club is selling posters. The printer charges a \$25 set-up fee, plus \$0.75 for each poster. The cost y in dollars to print x posters is given by the linear equation

$$y = 0.75x + 25.$$

(a) What is the cost y in dollars to print 50 posters? To print 100 posters?

(b) Find the number of posters x if the printer billed the club for costs of \$175.

(c) Write the information from parts (a) and (b) as three ordered pairs.

(d) Use the data from part (c) to graph the equation for $x \geq 0$.

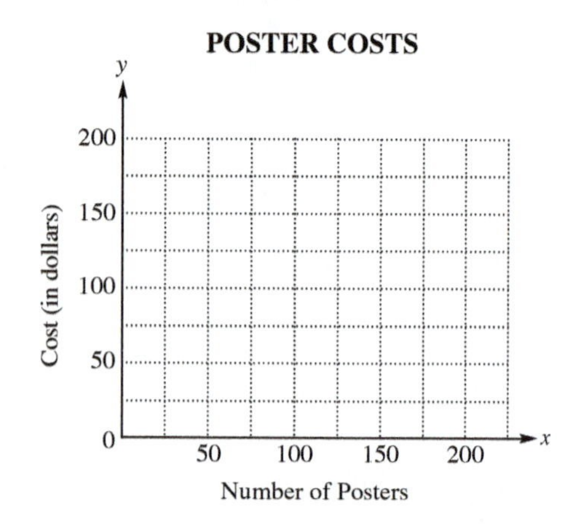

(e) Use the graph in part (d) to estimate the cost to print 125 posters. Then find the actual cost using the given equation.

54. A gas station is selling gasoline for \$3.50 per gal and charges \$7 for a car wash. The cost y in dollars for x gallons of gasoline and a car wash is given by the linear equation

$$y = 3.50x + 7.$$

(a) What is the cost y in dollars for 9 gal of gasoline and a car wash? For 4 gal of gasoline and a car wash?

(b) Find the number of gallons of gasoline x if the cost for the gasoline and a car wash is \$35.

(c) Write the information from parts (a) and (b) as three ordered pairs.

(d) Use the data from part (c) to graph the equation for $x \geq 0$.

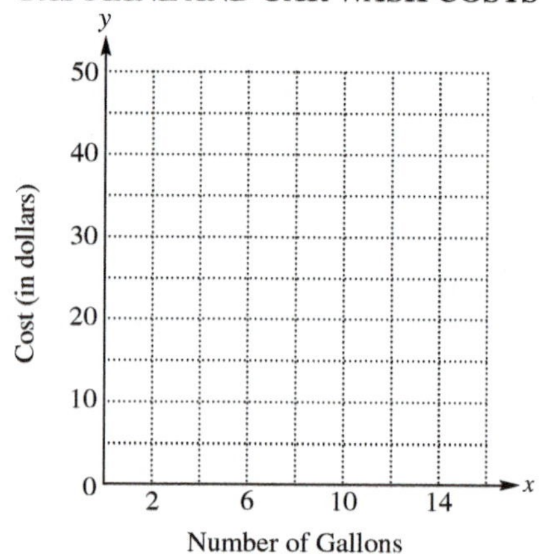

(e) Use the graph in part (d) to estimate the cost of 12 gal of gasoline and a car wash. Then find the actual cost using the given equation.

55. The graph shows the value of a certain sport-utility vehicle over the first 5 yr of ownership. Use the graph to do the following.

(a) Determine the initial value of the SUV.

(b) Find the **depreciation** (loss in value) from the original value after the first 4 yr.

(c) What is the annual or yearly depreciation in each of the first 5 yr?

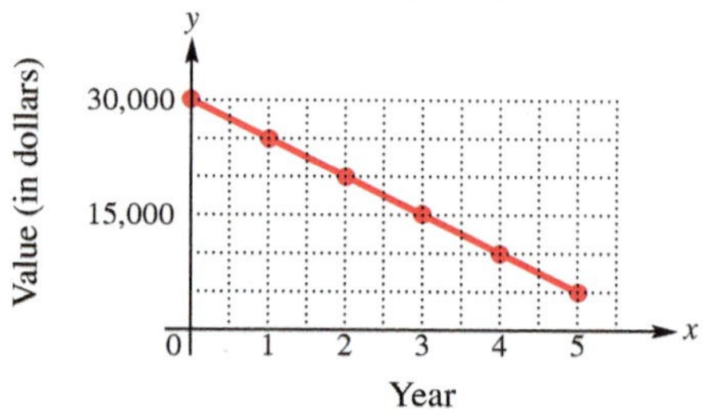

(d) What does the ordered pair (5, 5000) mean in the context of this problem?

56. Demand for an item is often closely related to its price. As price increases, demand decreases, and as price decreases, demand increases. Suppose demand for a video game is 2000 units when the price is \$40, and demand is 2500 units when the price is \$30.

(a) Let x be the price and y be the demand for the game. Graph the two given pairs of prices and demands on the graph to the right.

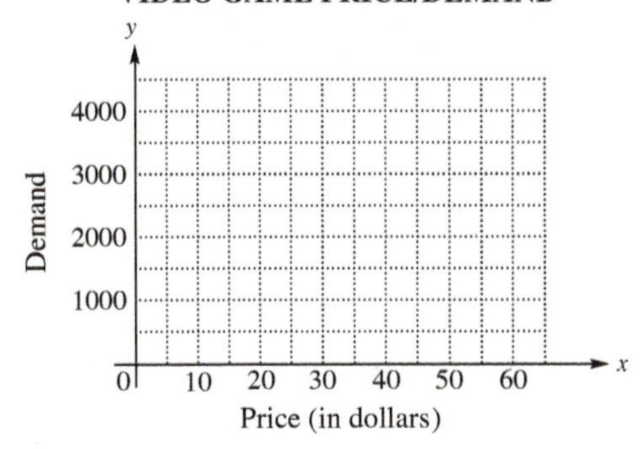

(b) Assume the relationship is linear. Draw a line through the two points from part (a). From the graph, estimate the demand if the price drops to \$20.

(c) Use the graph to estimate the price if the demand is 3500 units.

(d) Write the prices and demands from parts (b) and (c) as ordered pairs.

57. U.S. per capita consumption y of cheese in pounds from 2000 through 2014 is shown in the graph and modeled by the linear equation

$$y = 0.307x + 30.1,$$

where $x = 0$ represents 2000, $x = 2$ represents 2002, and so on.

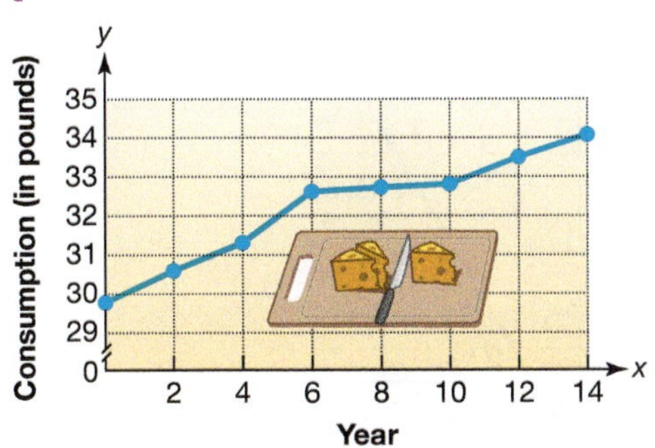

Data from U.S. Department of Agriculture.

(a) Use the equation to approximate cheese consumption (to the nearest tenth) in 2000, 2010, and 2014.

(b) Use the graph to estimate cheese consumption for the same years.

(c) How do the approximations using the equation compare to the estimates from the graph?

(d) The USDA projects that per capita consumption of cheese in 2022 will be 36.8 lb. Use the equation to approximate per capita cheese consumption (to the nearest tenth) in 2022. How does the approximation compare to the USDA projection?

58. The number of U.S. marathon finishers y in thousands from 1990 through 2013 are shown in the graph and modeled by the linear equation

$$y = 13.8x + 224,$$

where $x = 0$ represents 1990, $x = 5$ represents to 1995, and so on.

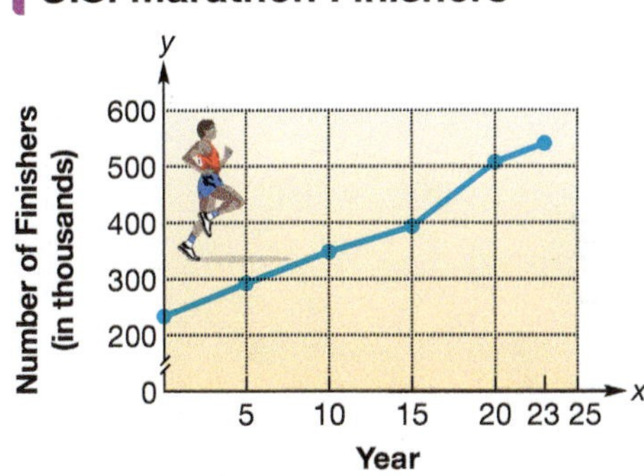

Data from Running U.S.A.

(a) Use the equation to approximate the number of U.S. marathon finishers (to the nearest thousand) in 2000, 2010, and 2013.

(b) Use the graph to estimate the number of U.S. marathon finishers for the same years.

(c) How do the approximations using the equation compare to the estimates from the graph?

(d) Use the graph and then the equation to determine the year when there were about 400 thousand U.S. marathon finishers. Round up to the nearest year when using the equation.

3.3 The Slope of a Line

OBJECTIVES

1. Find the slope of a line given two points.
2. Find the slope from the equation of a line.
3. Use slope to determine whether two lines are parallel, perpendicular, or neither.

An important characteristic of the lines we graphed in the previous section is their slant or "steepness," as viewed from *left to right*. See **Figure 17.**

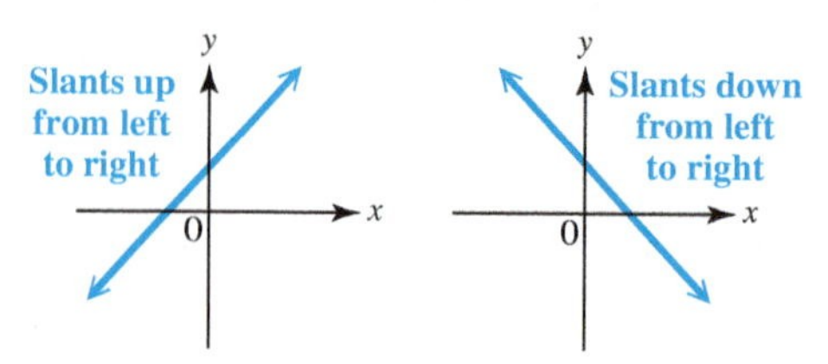

Figure 17

One way to measure the steepness of a line is to compare the vertical change in the line to the horizontal change while moving along the line from one fixed point to another. This measure of steepness is the *slope* of the line.

OBJECTIVE 1 Find the slope of a line given two points. To find the steepness, or slope, of the line in **Figure 18,** we begin at point Q and move to point P. The vertical change, or **rise,** is the change in the y-values, which is the difference

$$6 - 1 = 5 \text{ units.}$$

The horizontal change, or **run,** from Q to P is the change in the x-values, which is the difference

$$5 - 2 = 3 \text{ units.}$$

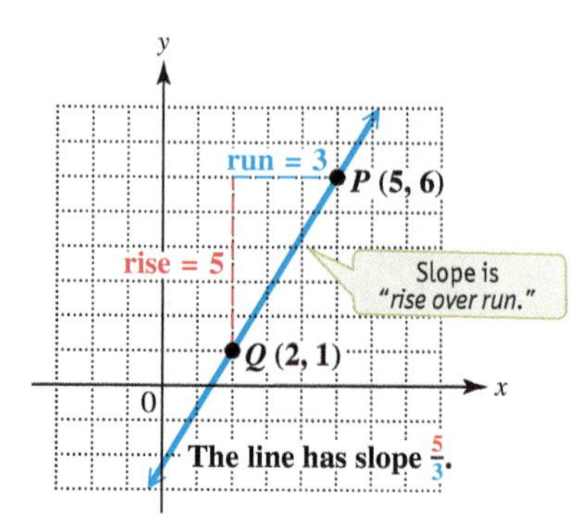

Figure 18

One way to compare two numbers is by using a ratio. **Slope** is the ratio of the vertical change in y to the horizontal change in x. The line in **Figure 18** has

$$\text{slope} = \frac{\text{vertical change in } y \text{ (rise)}}{\text{horizontal change in } x \text{ (run)}} = \frac{5}{3}.$$

To confirm this ratio, we can count grid squares. We start at point Q in **Figure 18** and count *up* 5 grid squares to find the vertical change (rise). To find the horizontal change (run) and arrive at point P, we count to the *right* 3 grid squares. The slope is $\frac{5}{3}$, as found above.

Slope of a Line

Slope is a single number that allows us to determine the direction in which a line is slanting from left to right, as well as how much slant there is to the line.

EXAMPLE 1 Finding the Slope of a Line

Find the slope of the line in **Figure 19.**

We use the coordinates of the two points shown on the line. The vertical change is the difference of the y-values.

$$-1 - 3 = -4$$

The horizontal change is the difference of the x-values.

$$6 - 2 = 4$$

Thus, the line has

$$\text{slope} = \frac{\text{change in } y \text{ (rise)}}{\text{change in } x \text{ (run)}} = \frac{-4}{4}, \text{ or } -1.$$

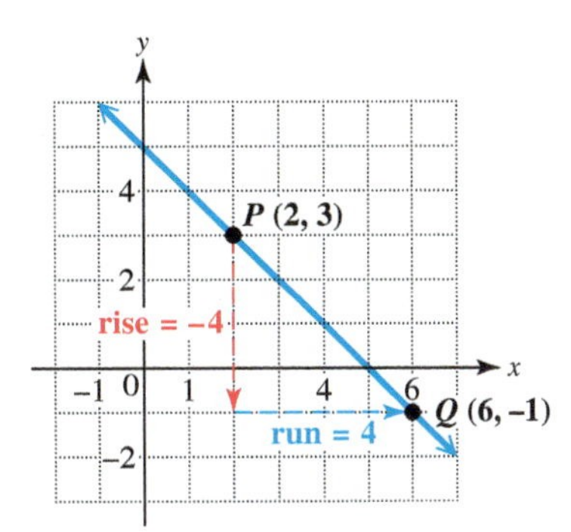

Figure 19

Counting grid squares, we begin at point P and count *down* 4 grid squares. Because we counted down, we write the vertical change as a negative number, -4 here. Then we count to the *right* 4 grid squares to reach point Q. The slope is $\frac{-4}{4}$, or -1.

Work Problem 1 at the Side. ▶

Note

The slope of a line is the same for any two points on the line. In **Figure 19,** locate the points $(3, 2)$ and $(5, 0)$, which also lie on the line. Start at $(3, 2)$ and count ***down* 2 units** and then to the ***right* 2 units** to arrive at the point $(5, 0)$.

$$\text{The slope is } \frac{-2}{2}, \text{ or } -1,$$

the same slope found in **Example 1** using the points $(2, 3)$ and $(6, -1)$.

The concept of slope is used in many everyday situations. See **Figure 20.**

- A highway with a 10%, or $\frac{1}{10}$, grade (or slope) rises 1 m for every 10 m of horizontal run.
- A roof with pitch (or slope) $\frac{5}{12}$ rises 5 ft for every 12 ft that it runs horizontally.
- A stairwell with slope $\frac{8}{12}$ $\left(\text{or } \frac{2}{3}\right)$ indicates a vertical rise of 8 ft for a horizontal run of 12 ft.

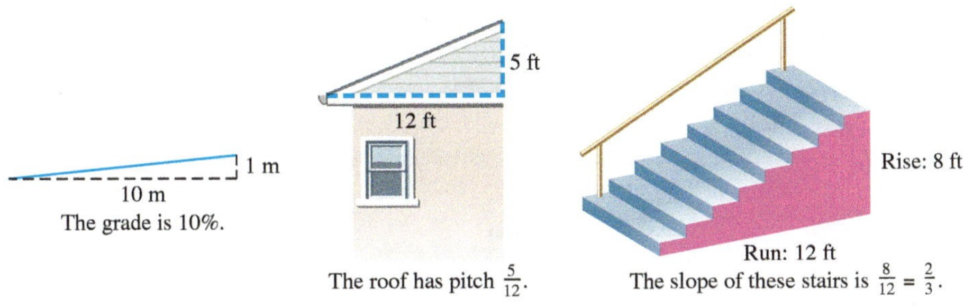

Figure 20

We can generalize the preceding discussion and find the slope of a line through two nonspecific points (x_1, y_1) and (x_2, y_2). This notation is called **subscript notation.** Read x_1 as **"x-sub-one"** and x_2 as **"x-sub-two."**

---

1 Find the slope of each line.

GS **(a)**

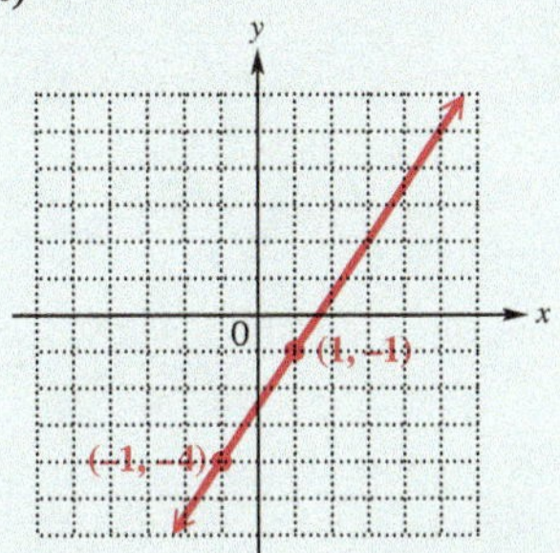

Begin at $(-1, -4)$.
Count (*up* / *down*) ____ units.
Then count to the (*right* / *left*) ____ units.

$$\text{slope} = \frac{\text{vertical change in } y}{\text{horizontal change in } x} = \frac{\_\_\_}{\_\_\_}$$

The slope is ____.

(b)

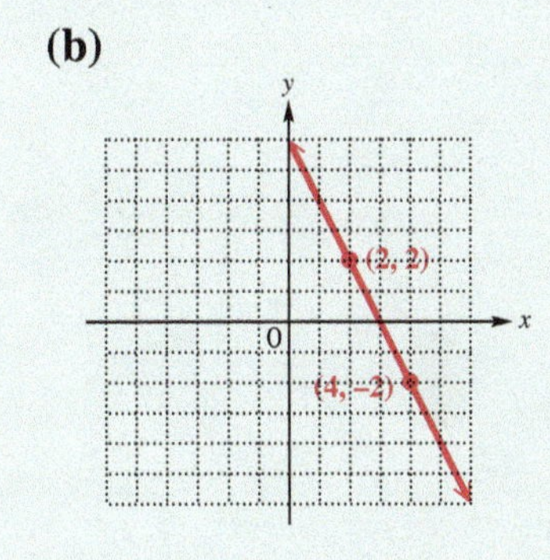

Answers

1. (a) up; 3; right; 2; 3; 2; $\frac{3}{2}$
   (b) -2

Moving along the line from the point (x_1, y_1) in **Figure 21** to the point (x_2, y_2), we see that y changes by $y_2 - y_1$ units. This is the vertical change (rise). Similarly, x changes by $x_2 - x_1$ units, which is the horizontal change (run). The slope of the line is the ratio of $y_2 - y_1$ to $x_2 - x_1$.

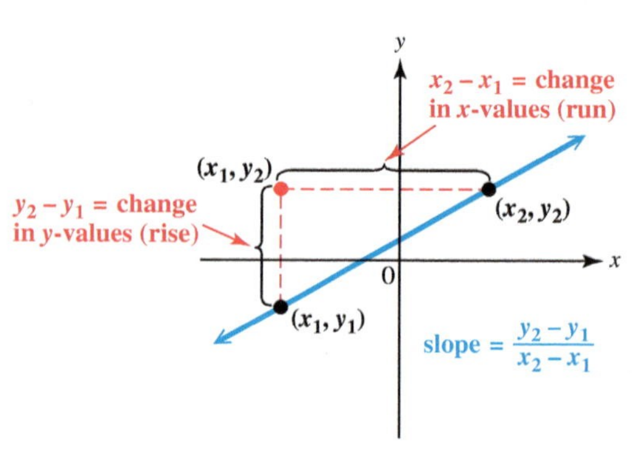

Figure 21

Slope Formula

The **slope** ***m*** of the line passing through the points (x_1, y_1) and (x_2, y_2) is defined as follows. (Traditionally, the letter m represents slope.)

$$m = \frac{\text{change in } y}{\text{change in } x} = \frac{y_2 - y_1}{x_2 - x_1} \quad (\text{where } x_1 \neq x_2)$$

The slope gives the change in y for each unit of change in x.

Note

Subscript notation is used to identify a point. It does *not* indicate an operation. ***Notice the difference between x_2, which represents a nonspecific value, and x^2, which means $x \cdot x$.*** Read x_2 as *"x-sub-two," not "x squared."*

EXAMPLE 2 **Finding Slopes of Lines Given Two Points**

Find the slope of each line.

(a) The line passing through $(-4, 7)$ and $(1, -2)$

Label the given points, and then apply the slope formula.

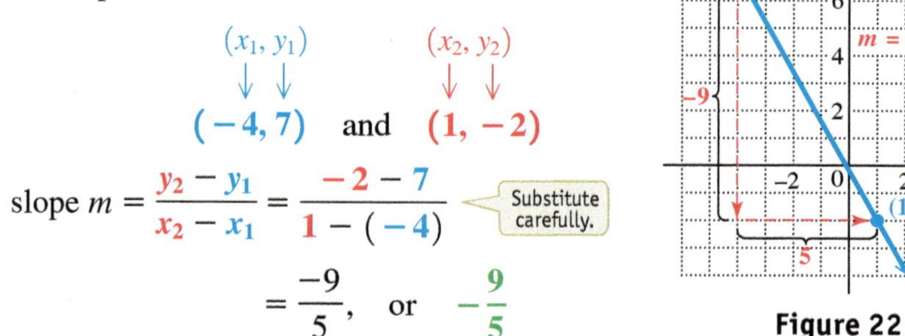

$$\begin{matrix} (x_1, y_1) & & (x_2, y_2) \\ \downarrow\ \downarrow & & \downarrow\ \downarrow \\ (-4, 7) & \text{and} & (1, -2) \end{matrix}$$

$$\text{slope } m = \frac{y_2 - y_1}{x_2 - x_1} = \frac{-2 - 7}{1 - (-4)} \quad \text{Substitute carefully.}$$

$$= \frac{-9}{5}, \quad \text{or} \quad -\frac{9}{5}$$

Figure 22

Begin at $(-4, 7)$ and count grid squares in **Figure 22** to confirm that the slope is $\frac{-9}{5}$, or $-\frac{9}{5}$.

Continued on Next Page

(b) The line passing through $(-9, -2)$ and $(12, 5)$

Label the points, and then apply the slope formula.

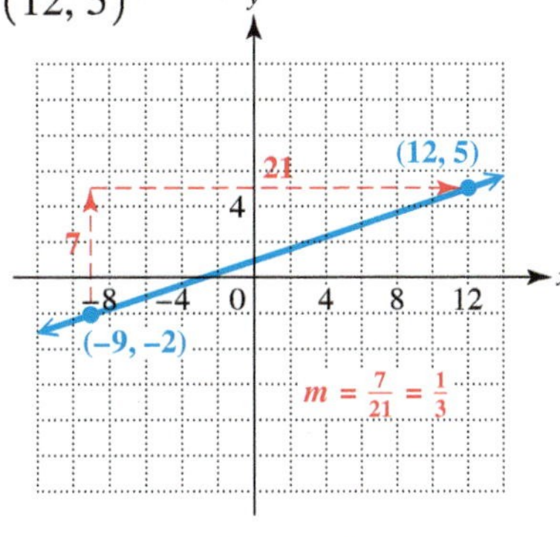

Figure 23

$$\begin{matrix} (x_1, y_1) & & (x_2, y_2) \\ \downarrow\ \downarrow & & \downarrow\ \downarrow \\ (-9, -2) & \text{and} & (12, 5) \end{matrix}$$

$$\text{slope } m = \frac{y_2 - y_1}{x_2 - x_1} = \frac{5 - (-2)}{12 - (-9)}$$

$$= \frac{7}{21} \quad \text{Subtract.}$$

$$= \frac{1}{3} \quad \text{Write in lowest terms.}$$

Confirm this calculation using **Figure 23.** (Note the scale on the axes.)

The same slope is obtained if we label the points in reverse order.

$$\begin{matrix} (x_2, y_2) & & (x_1, y_1) \\ \downarrow\ \downarrow & & \downarrow\ \downarrow \\ (-9, -2) & \text{and} & (12, 5) \end{matrix}$$

It makes no difference which point is identified as (x_1, y_1) or (x_2, y_2).

$$\text{slope } m = \frac{y_2 - y_1}{x_2 - x_1} = \frac{-2 - 5}{-9 - 12} \quad \text{Substitute.}$$

$$= \frac{1}{3} \quad \frac{-7}{-21} \text{ also equals } \frac{1}{3}. \text{ The same slope results.}$$

Start with the values of the *same* point. Subtract the values of the other point.

Work Problem 2 at the Side. ▶

The slopes of the lines in **Figures 22 and 23** suggest the following.

Orientation of Lines with Positive and Negative Slopes

A line with positive slope rises (slants up) from left to right.

A line with negative slope falls (slants down) from left to right.

EXAMPLE 3 Finding the Slope of a Horizontal Line

Find the slope of the line passing through $(-5, 4)$ and $(2, 4)$.

$$m = \frac{y_2 - y_1}{x_2 - x_1} = \frac{4 - 4}{-5 - 2} \quad \begin{matrix}\text{Subtract } y\text{-values.} \\ \text{Subtract } x\text{-values in the } \textit{same} \text{ order.}\end{matrix}$$

$$= \frac{0}{-7}$$

$$= 0 \quad \text{Slope 0}$$

As shown in **Figure 24,** the line passing through these two points is horizontal, with equation $y = 4$.

All horizontal lines have slope 0.

This is because the difference of the y-values for any two points on a horizontal line is always 0, which results in a 0 *numerator* when calculating slope.

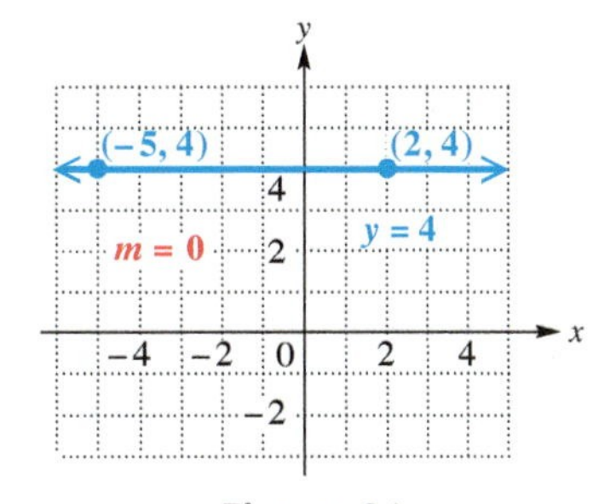

Figure 24

2 Find the slope of each line.

GS **(a)** The line passing through $(6, -2)$ and $(5, 4)$

Label the points.

$$\begin{matrix} (x_1, y_1) & & (\_\_\_, \_\_\_) \\ \downarrow\ \downarrow & & \downarrow\ \downarrow \\ (6, -2) & \text{and} & (5,\ \ 4) \end{matrix}$$

$$\text{slope } m = \frac{y_2 - y_1}{x_2 - x_1}$$

$$= \frac{\_\_\_ - (\_\_\_)}{\_\_\_ - 6}$$

$$= \frac{6}{\_\_\_}$$

$$= -6$$

(b) The line passing through $(-3, 5)$ and $(-4, -7)$

(c) The line passing through $(6, -8)$ and $(-2, 4)$

(Find this slope in two different ways as in **Example 2(b).**)

Answers

2. (a) x_2; y_2; 4; −2; 5; −1

(b) 12 **(c)** $-\frac{3}{2}$; $-\frac{3}{2}$

3 Find the slope of each line.

(a) The line passing through (2, 5) and (−1, 5)

(b) The line passing through (3, 1) and (3, −4)

4 Find the slope of each line.

GS **(a)** The line with equation $y = -1$

This is the equation of a (*horizontal* / *vertical*) line. It has (0 / *undefined*) slope.

(b) The line with equation $x - 4 = 0$

Rewrite this equation in an equivalent form.

$$x - 4 = 0$$

$$x = \_\_\_\_$$

This is the equation of a (*horizontal* / *vertical*) line. It has (0 / *undefined*) slope.

Answers

3. (a) 0 (b) undefined
4. (a) horizontal; 0
(b) 4; vertical; undefined

EXAMPLE 4 Applying the Slope Concept to a Vertical Line

Find the slope of the line passing through (6, 2) and (6, −4).

$$m = \frac{y_2 - y_1}{x_2 - x_1} = \frac{2 - (-4)}{6 - 6}$$ Subtract y-values. Subtract x-values in the *same* order.

$$= \frac{6}{0}$$ Undefined slope

Because division by 0 is undefined, this line has undefined slope. (This is why the slope formula has the restriction $x_1 \neq x_2$.)

The graph in **Figure 25** shows that this line is vertical, with equation $x = 6$.

The slope of any vertical line is undefined.

This is because the difference of the x-values for any two points on a vertical line is always 0, which results in a 0 *denominator* when calculating slope.

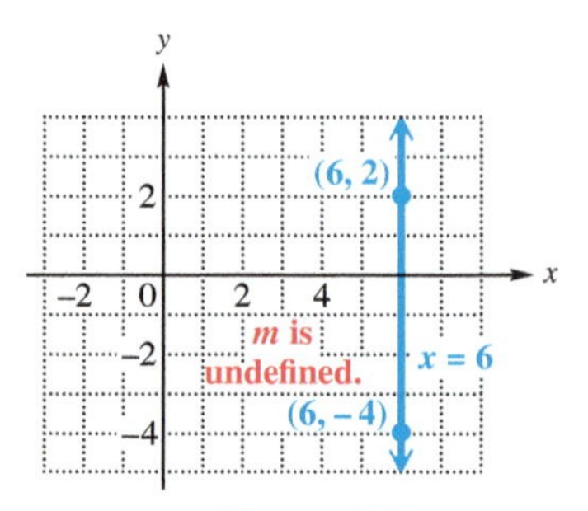

Figure 25

◀ Work Problem **3** at the Side.

Slopes of Horizontal and Vertical Lines

A horizontal line, which has an equation of the form $y = b$ (where b is a constant (number)), has **slope 0.**

A vertical line, which has an equation of the form $x = a$ (where a is a constant (number)), has **undefined slope.**

◀ Work Problem **4** at the Side.

Figure 26 summarizes the four cases for slopes of lines.

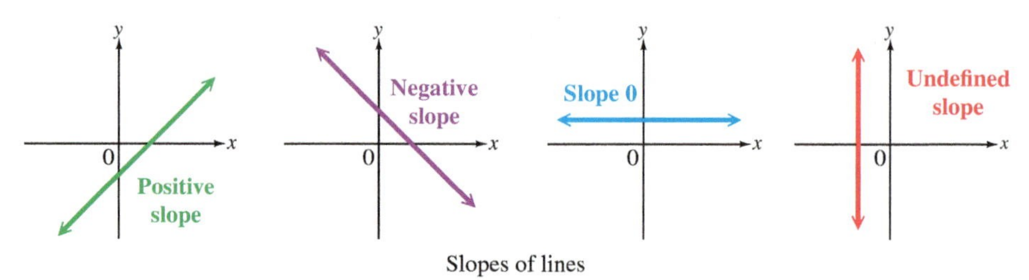

Slopes of lines

Figure 26

OBJECTIVE 2 Find the slope from the equation of a line. Consider the linear equation

$$y = -3x + 5.$$

We can find the slope of this line using any two points on the line. Because the equation is solved for y, it involves less work to choose two different values of x and then find the corresponding values of y. We arbitrarily choose $x = -2$ and $x = 4$.

$y = -3x + 5$

$y = -3(-2) + 5$ Let $x = -2$.

$y = 11$ Multiply. Add.

Ordered pair: (−2, 11)

$y = -3x + 5$

$y = -3(4) + 5$ Let $x = 4$.

$y = -7$ Multiply. Add.

Ordered pair: (4, −7)

Now we apply the slope formula using the two points $(-2, 11)$ and $(4, -7)$.

$$m = \frac{11 - (-7)}{-2 - 4} = \frac{18}{-6} = -3$$

The slope, -3, is the same number as the coefficient of x in the given equation $y = -3x + 5$. It can be shown that this always happens, *as long as the equation is solved for y.* This fact is used to find the slope of a line from its equation.

Finding the Slope of a Line from Its Equation

Step 1 Solve the equation for y.

Step 2 The slope is given by the coefficient of x.

EXAMPLE 5 **Finding Slopes from Equations**

Find the slope of each line.

(a) $2x - 5y = 4$

Step 1 Solve the equation for y.

$$2x - 5y = 4 \quad \text{Isolate } y \text{ on one side.}$$

$$-5y = 4 - 2x \quad \text{Subtract } 2x.$$

$$-5y = -2x + 4 \quad \text{Commutative property}$$

$$\frac{-2x}{-5} = \frac{-2}{-5}x = \frac{2}{5}x$$

$$y = \frac{2}{5}x - \frac{4}{5} \quad \text{Divide } each \text{ term by } -5.$$

↑ Slope

Step 2 The slope is given by the coefficient of x, so the slope is $\frac{2}{5}$.

(b)

$$8x + 4y = 1$$

Solve for y.

$$4y = 1 - 8x \quad \text{Subtract } 8x.$$

$$4y = -8x + 1 \quad \text{Commutative property}$$

$$y = -2x + \frac{1}{4} \quad \text{Divide } each \text{ term by } 4.$$

The slope of this line is given by the coefficient of x, which is -2.

(c)

$$3y + x = -3$$

We omit the step showing the commutative property.

$$3y = -x - 3 \quad \text{Subtract } x.$$

$$y = \frac{-x}{3} - 1 \quad \text{Divide } each \text{ term by } 3.$$

The slope is $-\frac{1}{3}$, *not* $\frac{-x}{3}$ or $-\frac{x}{3}$.

$$y = -\frac{1}{3}x - 1 \quad \frac{-x}{3} = \frac{-1x}{3} = -\frac{1}{3}x$$

The coefficient of x is $-\frac{1}{3}$, so the slope of this line is $-\frac{1}{3}$.

Work Problem 5 at the Side. ▶

5 Find the slope of each line.

GS **(a)** $y = -\frac{7}{2}x + 1$

Because this equation is already solved for y, Step 1 is not needed. The slope, which is given by the ______ of ____, can be read directly from the equation. The slope is ____.

GS **(b)** $4y = 4x - 3$

Solve for y here by dividing each term by ____ to obtain the equation

$y =$ ______.

The coefficient of x is ____, so the slope is ____.

(c) $3x + 2y = 9$

(d) $5y - x = 10$

Answers

5. (a) coefficient; x; $-\frac{7}{2}$

(b) 4; $x - \frac{3}{4}$; 1; 1

(c) $-\frac{3}{2}$ (d) $\frac{1}{5}$

OBJECTIVE 3 Use slope to determine whether two lines are parallel, perpendicular, or neither. Two lines in a plane that never intersect are **parallel.** We use slopes to tell whether two lines are parallel.

Figure 27 shows the graphs of $x + 2y = 4$ and $x + 2y = -6$. These lines appear to be parallel. We solve each equation for y to find the slope.

$$x + 2y = 4$$
$$2y = -x + 4 \quad \text{Subtract } x.$$
$$y = \frac{-x}{2} + 2 \quad \text{Divide by 2.}$$
$$y = -\frac{1}{2}x + 2$$

Slope ($\frac{-x}{2} = \frac{-1x}{2} = -\frac{1}{2}x$)

$$x + 2y = -6$$
$$2y = -x - 6 \quad \text{Subtract } x.$$
$$y = \frac{-x}{2} - 3 \quad \text{Divide by 2.}$$
$$y = -\frac{1}{2}x - 3$$

Slope (The slope is $-\frac{1}{2}$, *not* $-\frac{x}{2}$.)

Each line has slope $-\frac{1}{2}$. ***Nonvertical parallel lines always have equal slopes.***

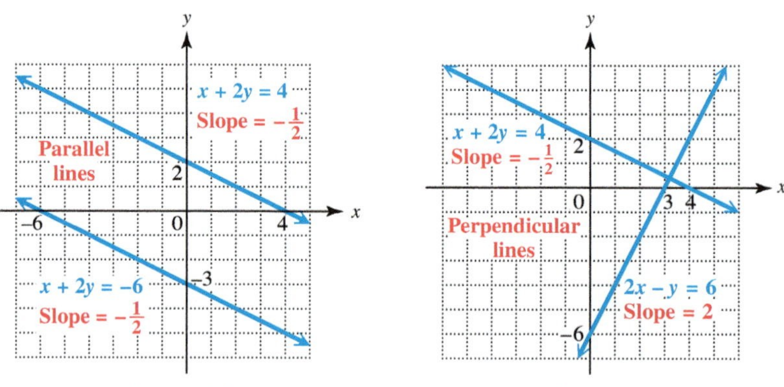

Figure 27 **Figure 28**

Figure 28 shows the graphs of $x + 2y = 4$ and $2x - y = 6$. These lines appear to be **perpendicular** (that is, they intersect at a 90° angle). As shown on the left above, solving $x + 2y = 4$ for y gives $y = -\frac{1}{2}x + 2$, with slope $-\frac{1}{2}$. We solve $2x - y = 6$ for y to find the slope.

$$2x - y = 6$$
$$-y = -2x + 6 \quad \text{Subtract } 2x.$$
$$y = 2x - 6 \quad \text{Multiply by } -1.$$

Slope

The product of the slopes of the two lines is

$$-\frac{1}{2}(2) = -1.$$

This condition is true in general.

The product of the slopes of two perpendicular lines, neither of which is vertical, is always −1.

This means that the slopes of perpendicular lines are negative (or opposite) reciprocals—if one slope is the nonzero number a, then the other is $-\frac{1}{a}$. (This is proved in more advanced courses.) The table in the margin shows several examples.

NEGATIVE RECIPROCALS

Number	Negative Reciprocal
$\frac{3}{4}$	$-\frac{4}{3}$
$\frac{1}{2}$	$-\frac{2}{1}$, or -2
-6, or $-\frac{6}{1}$	$\frac{1}{6}$
-0.4, or $-\frac{4}{10}$	$\frac{10}{4}$, or 2.5

The product of a number and its negative reciprocal is −1.

Slopes of Parallel and Perpendicular Lines

Two lines with the same slope are parallel.

Two lines whose slopes have a product of -1 are perpendicular.

EXAMPLE 6 Deciding Whether Two Lines Are Parallel or Perpendicular

Decide whether each pair of lines is *parallel, perpendicular,* or *neither.*

(a) $x + 3y = 5$ and $-3x + y = 3$

Find the slope of each line by first solving each equation for y.

$x + 3y = 5$	$-3x + y = 3$
$3y = -x + 5$ Subtract x.	$y = 3x + 3$ Add $3x$.
$y = -\frac{1}{3}x + \frac{5}{3}$ Divide by 3.	
The slope is $-\frac{1}{3}$.	The slope is 3.

Because the slopes are not equal, the lines are not parallel. Check the product of the slopes: $-\frac{1}{3}(3) = -1$. The two lines are *perpendicular* because the product of their slopes is -1. See **Figure 29.**

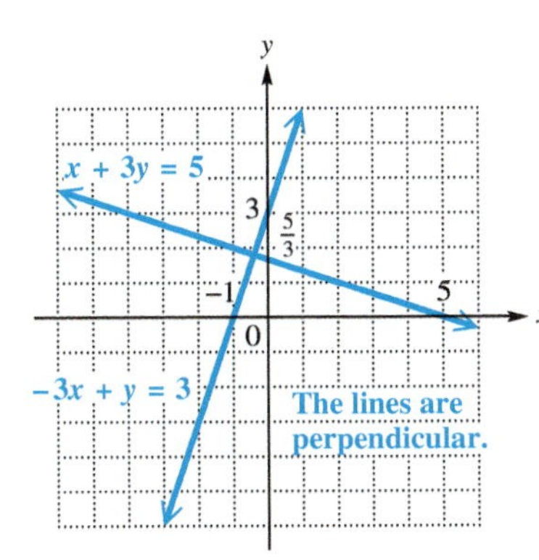

Figure 29

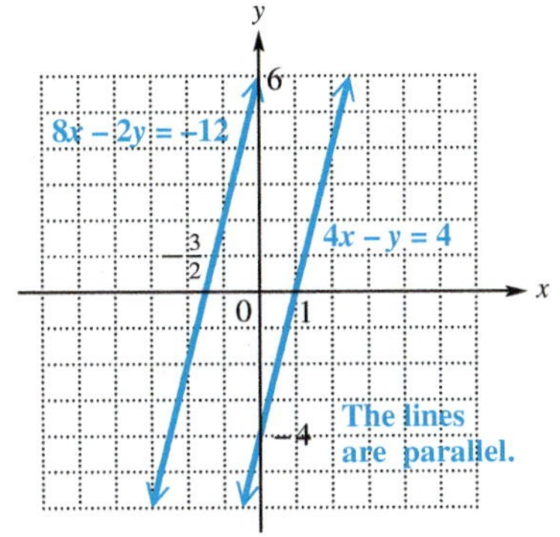

Figure 30

(b) $4x - y = 4$ $\xrightarrow{\text{Solve each equation for } y.}$ $y = 4x - 4$

$8x - 2y = -12$ $\qquad$ $y = 4x + 6$

Both lines have slope 4, so the lines are *parallel.* See **Figure 30.**

(c) $4x + 3y = 6$ $\xrightarrow{\text{Solve each equation for } y.}$ $y = -\frac{4}{3}x + 2$

$2x - y = 5$ $\qquad$ $y = 2x - 5$

The slopes are $-\frac{4}{3}$ and 2. They are not the same $\left(-\frac{4}{3} \neq 2\right)$, nor are they negative reciprocals $\left(-\frac{4}{3}(2) \neq -1\right)$. The two lines are *neither* parallel nor perpendicular.

(d) $5x - y = 1$ $\xrightarrow{\text{Solve each equation for } y.}$ $y = 5x - 1$

$x - 5y = -10$ $\qquad$ $y = \frac{1}{5}x + 2$

$5\left(\frac{1}{5}\right) = 1$, *not* -1.

The slopes, 5 and $\frac{1}{5}$, neither are the same, nor are they negative reciprocals. The two lines are *neither* parallel nor perpendicular.

Work Problem 6 at the Side. ▶

6 Decide whether each pair of lines is *parallel, perpendicular,* or *neither.*

(a) $x + y = 6$
$x + y = 1$

(b) $3x - y = 4$
$x + 3y = 9$

(c) $2x - y = 5$
$2x + y = 3$

(d) $3x - 7y = 35$
$7x - 3y = -6$

Answers

6. (a) parallel **(b)** perpendicular **(c)** neither **(d)** neither

3.3 Exercises

FOR EXTRA HELP *Go to MyMathLab for worked-out, step-by-step solutions to exercises enclosed in a square and video solutions to ▶ exercises.*

CONCEPT CHECK *Work each problem involving slope.*

1. Slope is used to measure the _________ of a line. Slope is the (*horizontal* / *vertical*) change compared to the (*horizontal* / *vertical*) change while moving along the line from one point to another.

2. Slope is the _________ of the vertical change in _________, called the (*rise* / *run*), to the horizontal change in _________, called the (*rise* / *run*).

3. Use at the graph at the right to answer the following.

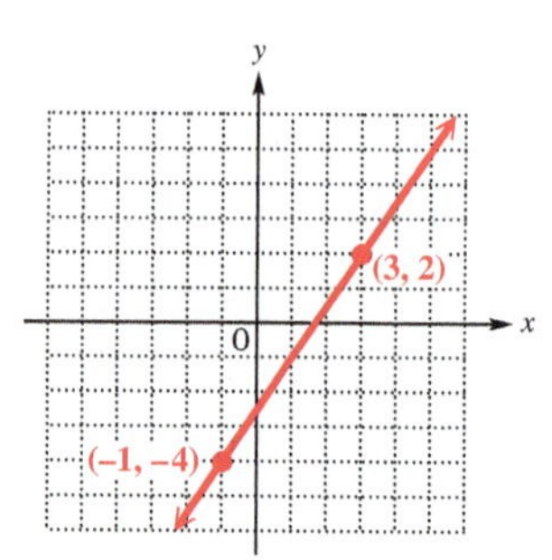

(a) Start at the point $(-1, -4)$ and count vertically up to the horizontal line that goes through the other plotted point. What is this vertical change? (Remember: "up" means positive, "down" means negative.) _______

(b) From this new position, count horizontally to the other plotted point. What is this horizontal change? (Remember: "right" means positive, "left" means negative.) _______

(c) What is the ratio (quotient) of the numbers found in parts (a) and (b)? _______ What do we call this number? ___________

(d) If we were to *start* at the point $(3, 2)$ and *end* at the point $(-1, -4)$, would the answer to part (c) be the same? Explain.

4. Match the graph of each line in parts (a)–(d) with its slope in choices A–D. (Coordinates of the points shown are integers.)

(a)

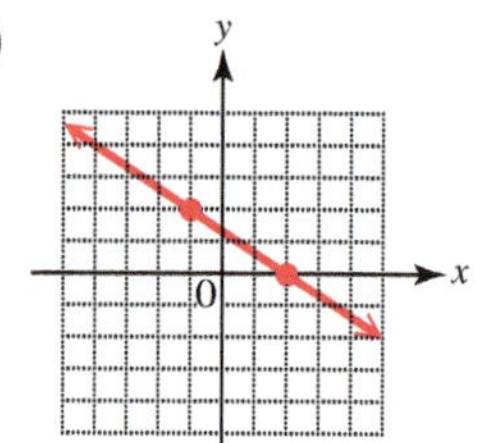

(b)

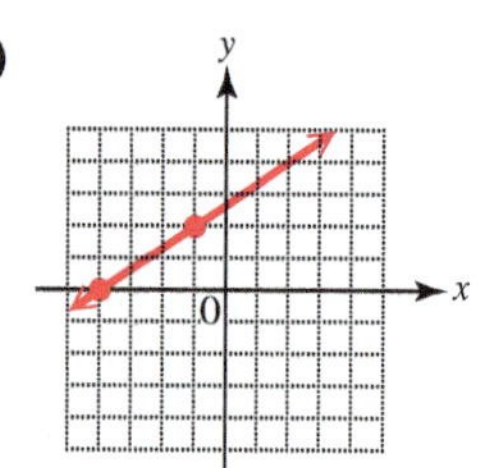

(c)

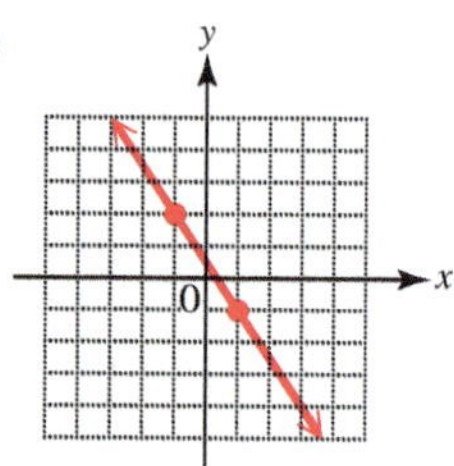

(d)

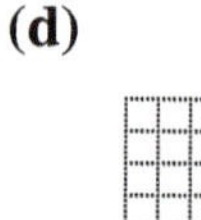

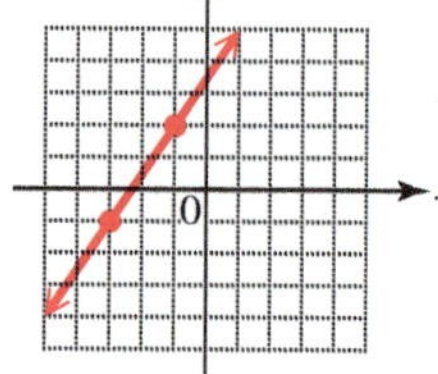

A. $\frac{2}{3}$ **B.** $\frac{3}{2}$ **C.** $-\frac{2}{3}$ **D.** $-\frac{3}{2}$

5. On the given axes, sketch the graph of a straight line having the indicated slope.

(a) Negative **(b)** Positive **(c)** Undefined **(d)** 0

6. Decide whether the line with the given slope m *rises from left to right*, *falls from left to right*, is *horizontal*, or is *vertical*.

(a) $m = -4$ **(b)** $m = 0$ **(c)** m is undefined. **(d)** $m = \frac{3}{7}$

CONCEPT CHECK *The figure at the right shows a line that has a positive slope (because it rises from left to right) and a positive y-value for the y-intercept (because it intersects the y-axis above the origin).*

For each line graphed, decide whether

(a) *the slope is* positive, negative, *or* 0 *and*

(b) *the y-value of the y-intercept is* positive, negative, *or* 0.

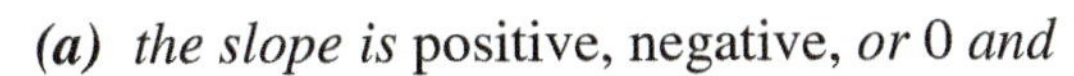

7.

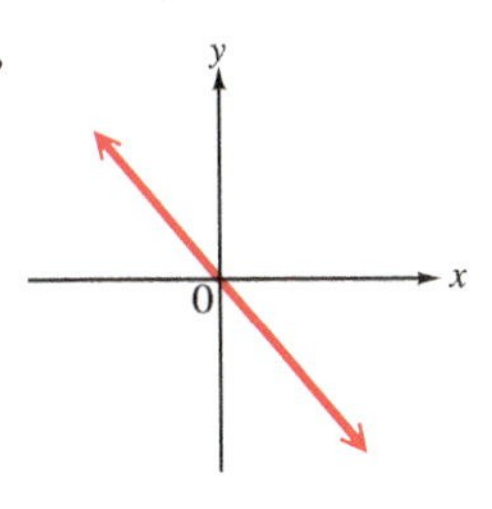

8.

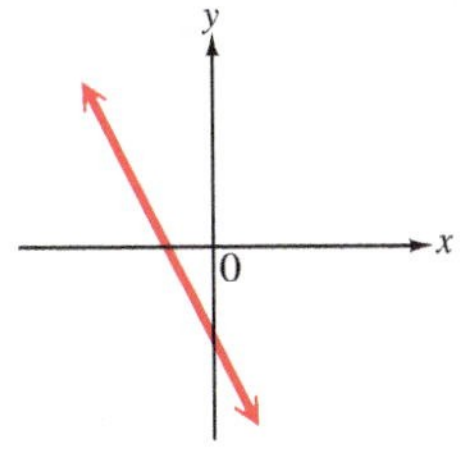

9.

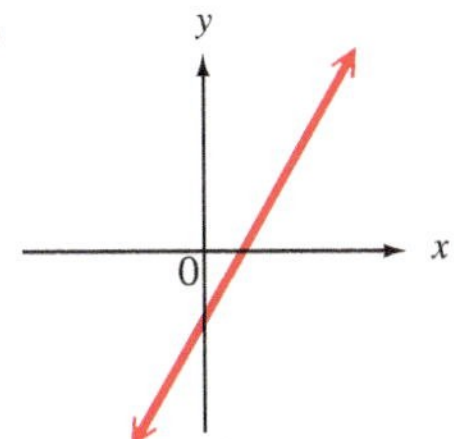

10.

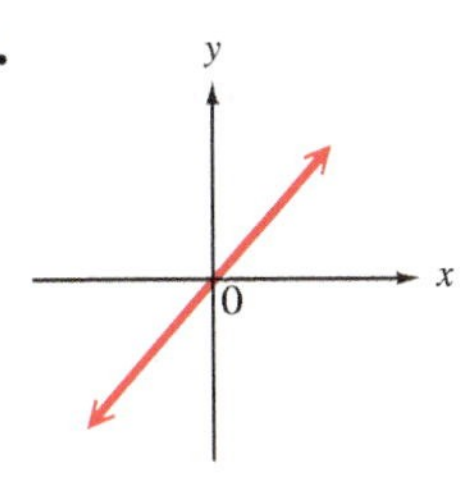

11.

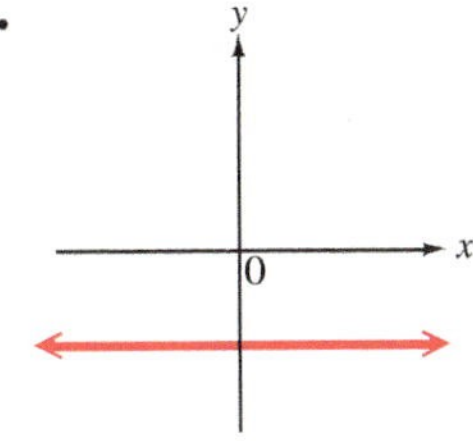

12.

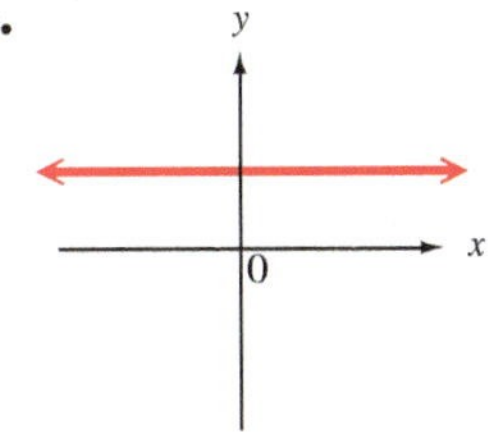

CONCEPT CHECK *Answer each question.*

13. What is the slope (or grade) of this hill?

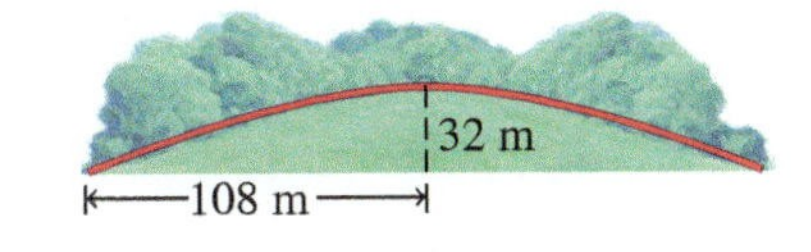

14. What is the slope (or pitch) of this roof?

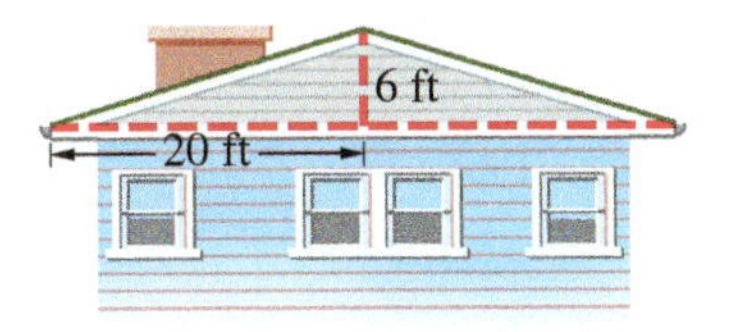

15. What is the slope of the slide? (*Hint:* The slide *drops* 8 ft vertically as it extends 12 ft horizontally.)

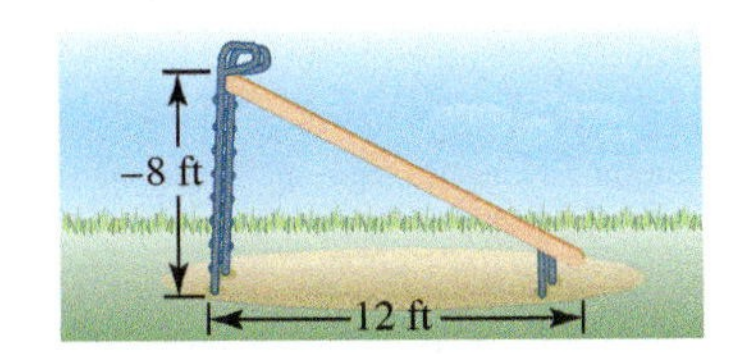

16. What is the slope (or grade) of this ski slope? (*Hint:* The ski slope *drops* 25 ft vertically as it extends 100 ft horizontally.)

17. **CONCEPT CHECK** A student found the slope of the line through the points $(2, 5)$ and $(-1, 3)$ as follows.

$$\frac{3-5}{2-(-1)} = \frac{-2}{3}, \quad \text{or} \quad -\frac{2}{3} \leftarrow \text{His answer}$$

What Went Wrong? Give the correct slope.

18. **CONCEPT CHECK** A student found the slope of the line through the points $(-2, 4)$ and $(6, -1)$ as follows.

$$\frac{-2-6}{4-(-1)} = \frac{-8}{5}, \quad \text{or} \quad -\frac{8}{5} \leftarrow \text{Her answer}$$

What Went Wrong? Give the correct slope.

Use the coordinates of the indicated points to find the slope of each line. (Coordinates of the points shown are integers.) **See Examples 1–4.**

19.

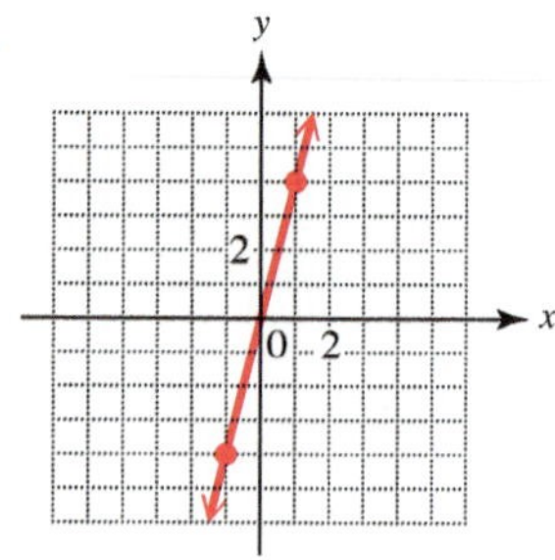

20.

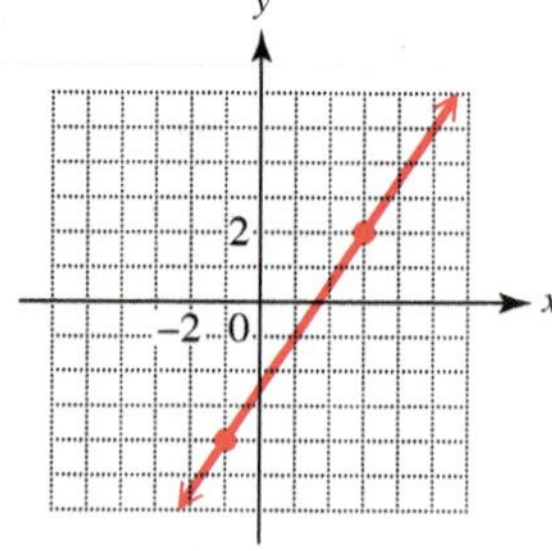

21.

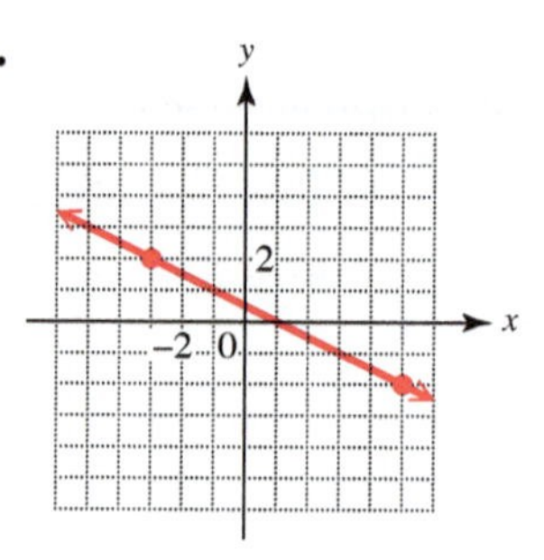

22.

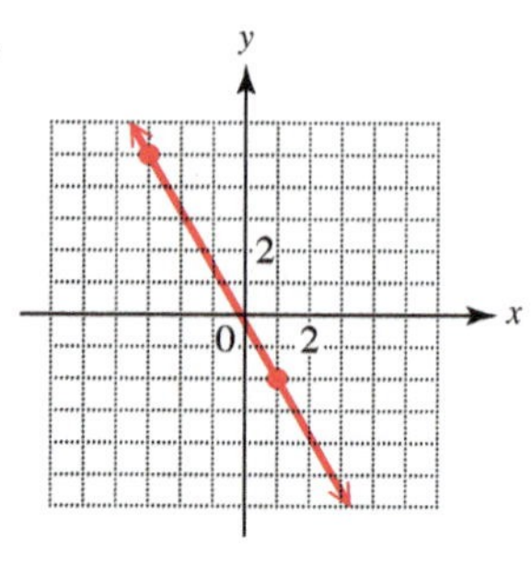

23.

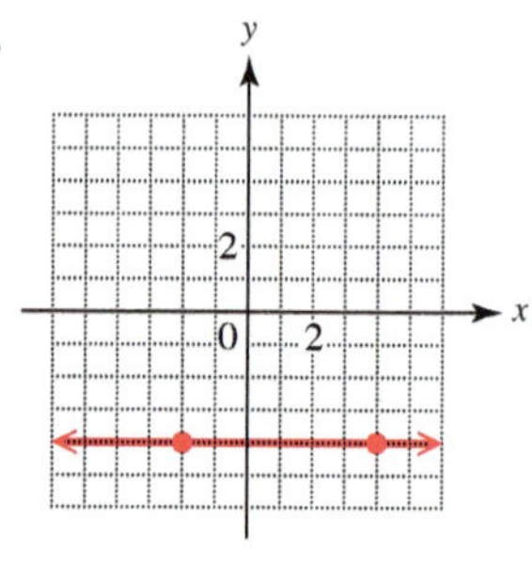

24.

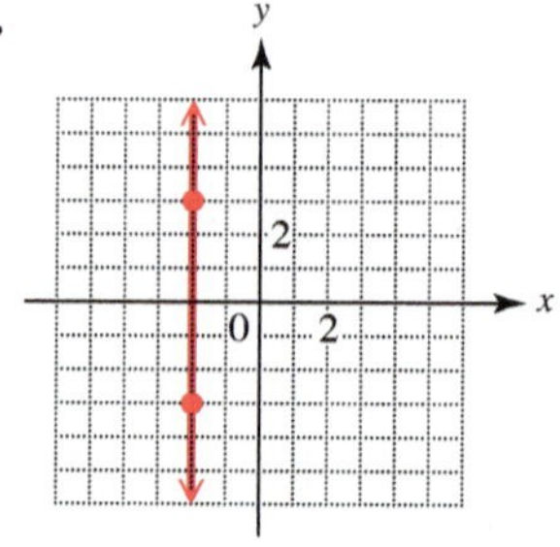

Find the slope of the line passing through each pair of points. **See Examples 2–4.**

25. $(1, -2)$ and $(-3, -7)$

26. $(4, -1)$ and $(-2, -8)$

27. $(0, 3)$ and $(-2, 0)$

28. $(-8, 0)$ and $(0, -5)$

29. $(-2, 4)$ and $(-3, 7)$

30. $(-4, -5)$ and $(-5, -8)$

31. $(4, 3)$ and $(-6, 3)$

32. $(6, -5)$ and $(-12, -5)$

33. $(-12, 3)$ and $(-12, -7)$

34. $(-8, 6)$ and $(-8, -1)$

35. $(4.8, 2.5)$ and $(3.6, 2.2)$

36. $(3.1, 2.6)$ and $(1.6, 2.1)$

37. $\left(-\frac{7}{5}, \frac{3}{10}\right)$ and $\left(\frac{1}{5}, -\frac{1}{2}\right)$

38. $\left(-\frac{4}{3}, \frac{1}{2}\right)$ and $\left(\frac{1}{3}, -\frac{5}{6}\right)$

Find the slope of each line. **See Example 5.**

39. $y = 5x + 12$

40. $y = 2x - 3$

41. $4y = x + 1$

42. $2y = -x + 4$

43. $3x - 2y = 3$

44. $6x - 4y = 4$

45. $-2y - 3x = 5$

46. $-4y - 3x = 2$

47. $y = 6$

48. $y = 4$

49. $x = -2$

50. $x = 5$

51. $x - y = 0$

52. $x + y = 0$

Find the slope of each line in two ways by doing the following.

(a) Give any two points that lie on the line, and use them to determine the slope.

(b) Solve the equation for y, and identify the slope from the equation.

See Objective 2 and Example 5.

53. $2x + y = 10$

54. $-4x + y = -8$

55. $5x - 3y = 15$

56. $3x + 2y = 12$

Each table of values gives several points that lie on a line.

(a) Use any two of the ordered pairs to find the slope of the line.

(b) What is the x-intercept of the line? The y-intercept?

Then graph the line.

57.

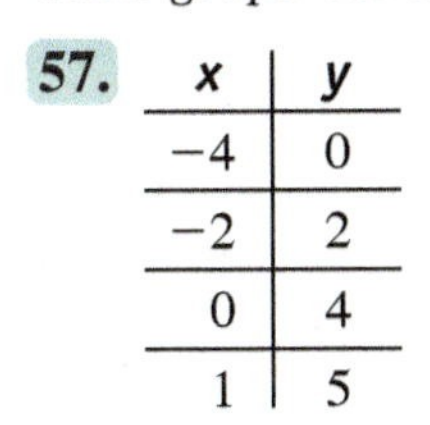

x	y
-4	0
-2	2
0	4
1	5

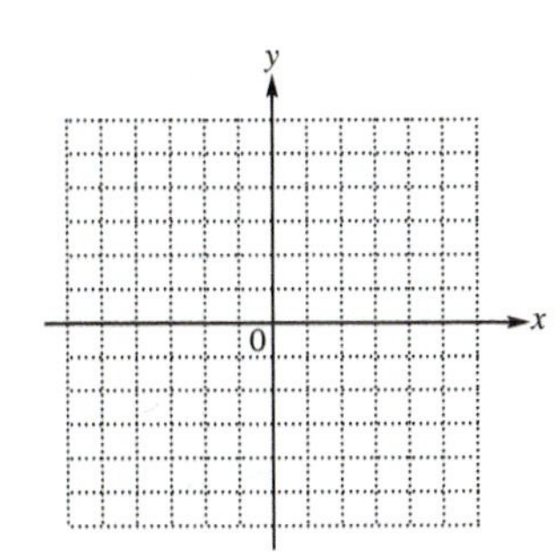

58.

x	y
-4	3
-1	0
0	-1
2	-3

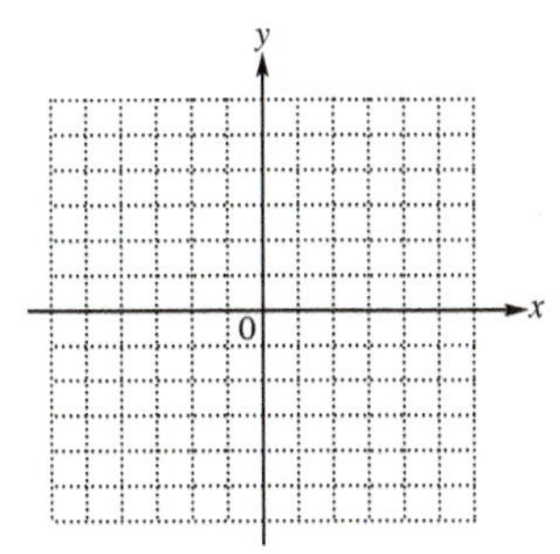

59.

x	y
3	-3
0	-2
-3	-1
-6	0

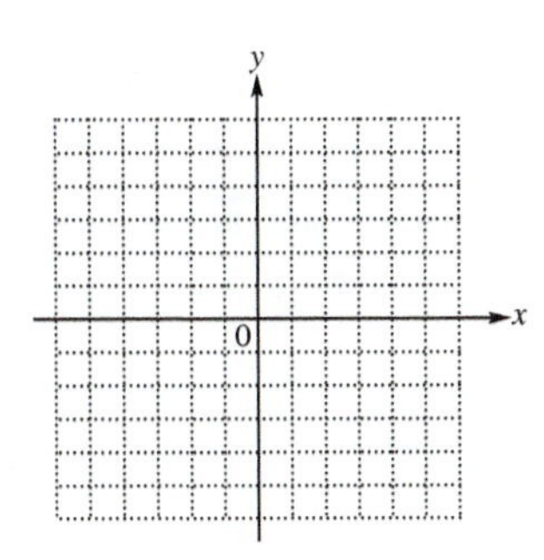

60.

x	y
-1	-6
0	-4
2	0
5	6

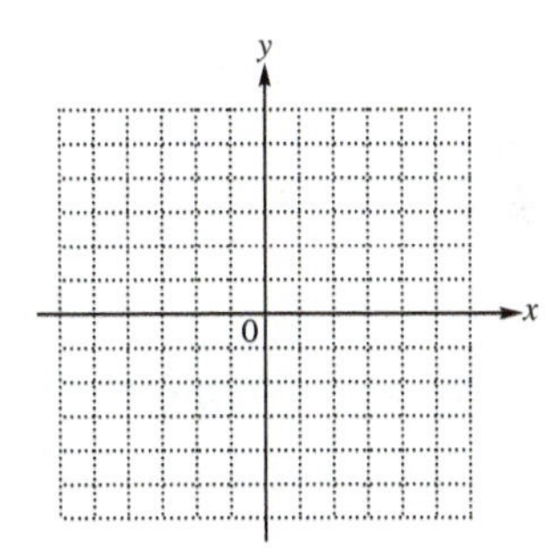

CONCEPT CHECK *Answer each question.*

61. What is the slope of a line whose graph is

(a) parallel to the graph of $3x + y = 7$?

(b) perpendicular to the graph of $3x + y = 7$?

62. What is the slope of a line whose graph is

(a) parallel to the graph of $-5x + y = -3$?

(b) perpendicular to the graph of $-5x + y = -3$?

63. If two lines are both vertical or both horizontal, which of the following are they?

A. Parallel

B. Perpendicular

C. Neither parallel nor perpendicular

64. If a line is vertical, what is true of any line that is perpendicular to it?

For each pair of equations, give the slopes of the lines, and then decide whether the two lines are parallel, perpendicular, *or* neither. ***See Example 6.***

65. $-4x + 3y = 4$
$-8x + 6y = 0$

66. $2x + 5y = 4$
$4x + 10y = 1$

67. $5x - 3y = -2$
$3x - 5y = -8$

68. $8x - 9y = 6$
$8x + 6y = -5$

69. $3x - 5y = -1$
$5x + 3y = 2$

70. $3x - 2y = 6$
$2x + 3y = 3$

71. $6x + y = 1$
$x + 6y = 18$

72. $3x - 4y = 12$
$4x + 3y = 12$

Relating Concepts (Exercises 73–78) For Individual or Group Work

Figure A *gives the percent of first-time full-time freshmen at 4-year colleges and universities who planned to major in the Biological Sciences.* **Figure B** *shows the percent of the same group of students who planned to major in Business.*

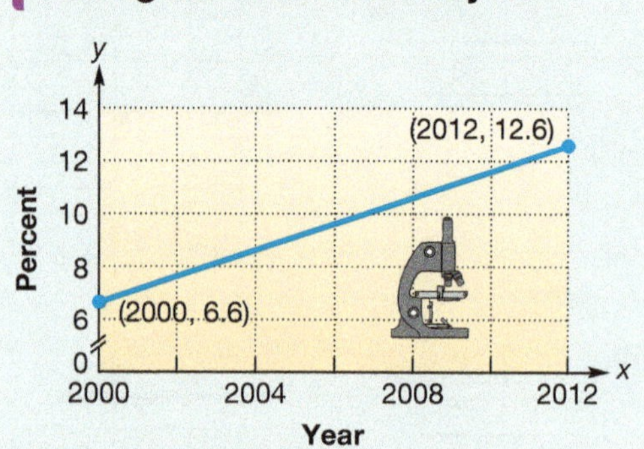

Data from Higher Education Research Institute.

Figure A

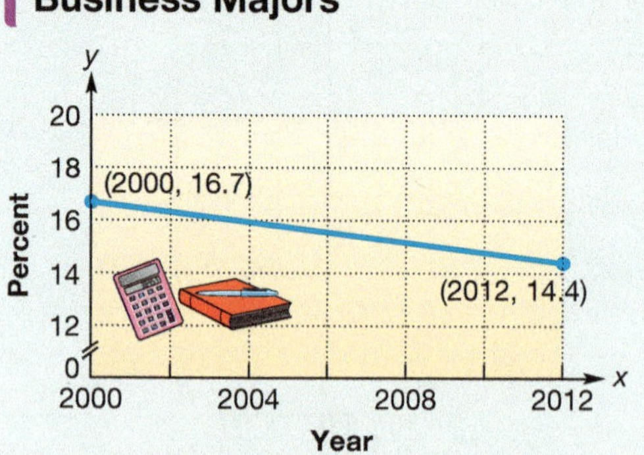

Data from Higher Education Research Institute.

Figure B

Work Exercises 73–78 in order.

73. Use the given ordered pairs to find the slope of the line in **Figure A.**

74. The slope of the line in **Figure A** is (*positive / negative*). This means that during the period represented, the percent of freshmen planning to major in the Biological Sciences (*increased / decreased*).

75. The slope of a line represents the *rate of change.* Based on **Figure A,** what was the increase in the percent of freshmen *per year* who planned to major in the Biological Sciences during the period shown?

76. Use the given ordered pairs to find the slope of the line in **Figure B** to the nearest tenth.

77. The slope of the line in **Figure B** is (*positive / negative*). This means that during the period shown, the percent of freshmen planning to major in Business (*increased / decreased*).

78. Based on **Figure B,** what was the decrease in the percent of freshmen *per year* who planned to major in Business?

Study Skills

PREPARING FOR YOUR MATH FINAL EXAM

Your math final exam is likely to be a comprehensive exam, which means it will cover material from the entire term. **One way to prepare for it now is by working a set of Cumulative Review Exercises** each time your class finishes a chapter. This continual review will help you remember concepts and procedures as you progress through the course.

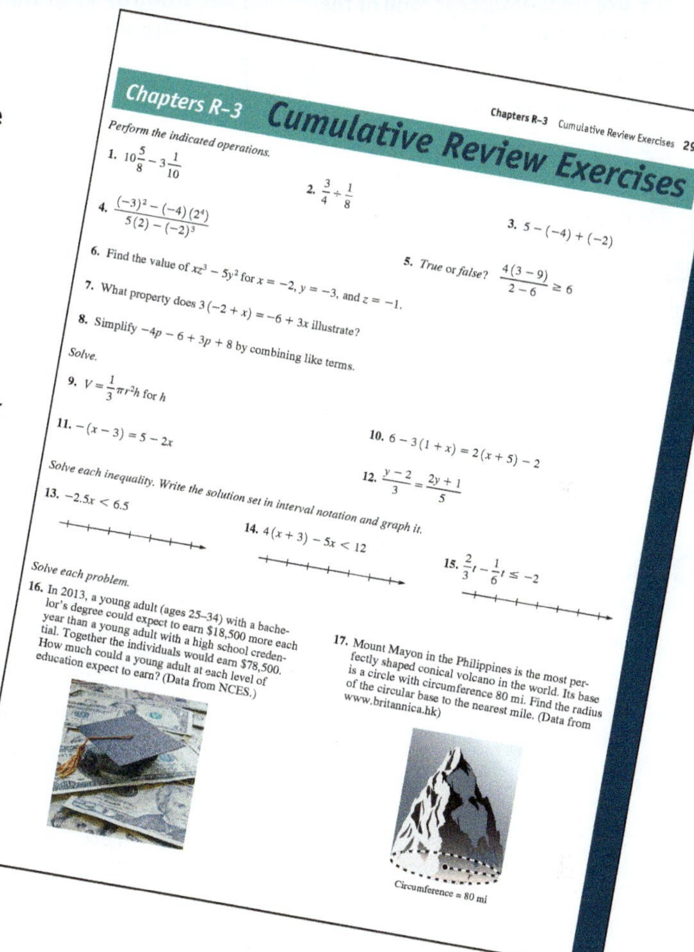
Chapters R–3 Cumulative Review Exercises

Chapters R–3 Cumulative Review Exercises 293

Perform the indicated operations.

1. $10\frac{5}{8} - 3\frac{1}{10}$
2. $\frac{3}{4} \div \frac{1}{8}$
3. $5 - (-4) + (-2)$
4. $\frac{(-3)^2 - (-4)(2^4)}{5(2) - (-2)^3}$
5. *True* or *false*? $\frac{4(3-9)}{2-6} \geq 6$
6. Find the value of $xz^3 - 5y^2$ for $x = -2$, $y = -3$, and $z = -1$.
7. What property does $3(-2 + x) = -6 + 3x$ illustrate?
8. Simplify $-4p - 6 + 3p + 8$ by combining like terms.

Solve.

9. $V = \frac{1}{3}\pi r^2 h$ for h
10. $6 - 3(1 + x) = 2(x + 5) - 2$
11. $-(x - 3) = 5 - 2x$
12. $\frac{y-2}{3} = \frac{2y+1}{5}$

Solve each inequality. Write the solution set in interval notation and graph it.

13. $-2.5x < 6.5$
14. $4(x + 3) - 5x < 12$
15. $\frac{2}{3}t - \frac{1}{6}t \leq -2$

Solve each problem.

16. In 2013, a young adult (ages 25–34) with a bachelor's degree could expect to earn $18,500 more each year than a young adult with a high school credential. Together the individuals would earn $78,500. How much could a young adult at each level of education expect to earn? (Data from NCES.)
17. Mount Mayon in the Philippines is the most perfectly shaped conical volcano in the world. Its base is a circle with circumference 80 mi. Find the radius of the circular base to the nearest mile. (Data from www.britannica.hk)

Circumference = 80 mi

Final Exam Preparation Suggestions

1. **Figure out the grade you need to earn on the final exam to get the course grade you want.** Check your course syllabus for grading policies, or ask your instructor if you are not sure.
2. **Create a final exam week plan.** Set priorities that allow you to spend extra time studying. This may mean making adjustments, in advance, in your work schedule or enlisting extra help with family responsibilities.
3. **Use the following suggestions to guide your studying.**
   - **Begin reviewing several days before the final exam.** DON'T wait until the last minute.
   - **Know exactly which chapters and sections will be covered.**
   - **Divide up the chapters.** Decide how much you will review each day.
   - **Keep returned quizzes and tests. Use them to review.**
   - **Practice all types of problems. Use the Cumulative Review Exercises** at the end of each chapter in your text beginning in Chapter 2. All answers are given in the answer section.
   - **Review or rewrite your notes** to create summaries of key information.
   - **Make study cards for all types of problems.** Carry the cards with you, and review them whenever you have a few minutes.
   - **Take plenty of short breaks as you study to reduce physical and mental stress.** Exercising, listening to music, and enjoying a favorite activity are effective stress busters.

Finally, *DON'T* stay up all night the night before an exam—*get a good night's sleep.*

Now Try This

1. How many points do you need to earn on your math final exam to get the grade you want in your course?

   ____

2. What adjustments to your usual routine or schedule do you need to make for final exam week? List two or three.

   ____

   ____

3. Which of the suggestions for studying will you use as you prepare for your math final exam?

   ____

   ____

3.4 Slope-Intercept Form of a Linear Equation

OBJECTIVES

1. Use slope-intercept form of the equation of a line.
2. Graph a line using its slope and a point on the line.
3. Write an equation of a line using its slope and any point on the line.
4. Graph and write equations of horizontal and vertical lines.

OBJECTIVE 1 Use slope-intercept form of the equation of a line. Recall that we can find the slope (steepness) of a line by solving the equation of the line for y. In that form, the slope is the coefficient of x. For example, the line with equation

$$y = 2x + 3 \quad \text{has slope} \quad 2.$$

What does the number 3 represent? To find out, suppose that a line has slope m and y-intercept $(0, b)$. We can find an equation of this line by choosing another point (x, y) on the line, as shown in **Figure 31.** Then we apply the slope formula.

$$m = \frac{y - b}{x - 0} \quad \begin{matrix}\leftarrow \text{Change in } y\text{-values} \\ \leftarrow \text{Change in } x\text{-values}\end{matrix}$$

$$m = \frac{y - b}{x} \quad \text{Subtract in the denominator.}$$

$$mx = y - b \quad \text{Multiply by } x.$$

$$mx + b = y \quad \text{Add } b.$$

$$y = mx + b \quad \text{Interchange sides.}$$

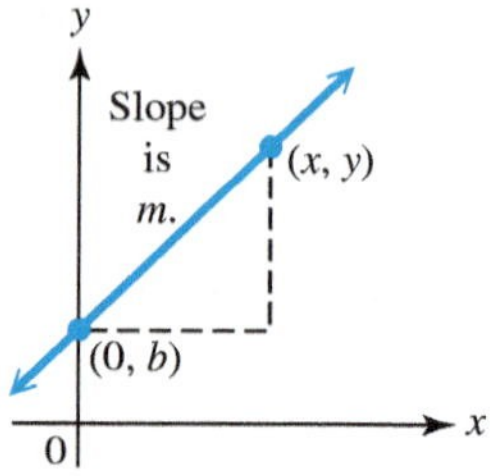

Figure 31

This result is the *slope-intercept form* of the equation of a line. Both the *slope* and the *y-intercept* of the line can be read directly from this form. For the line with equation $y = 2x + 3$, the number 3 gives the y-intercept $(0, 3)$.

Slope-Intercept Form

The **slope-intercept form** of the equation of a line with slope m and y-intercept $(0, b)$ is

$$y = mx + b.$$

Slope ↑ (for m); ↑ $(0, b)$ is the y-intercept.

The intercept given is the y-intercept.

EXAMPLE 1 Identifying Slopes and y-Intercepts

Identify the slope and y-intercept of the line with each equation.

(a) $y = -4x + 1$

Slope (-4); y-intercept $(0, 1)$

(b) $y = x - 8$ can be written as $y = 1x + (-8)$.

Slope (1); y-intercept $(0, -8)$

(c) $y = 6x$ can be written as $y = 6x + 0$.

Slope (6); y-intercept $(0, 0)$

(d) $y = \frac{x}{4} - \frac{3}{4}$ can be written as $y = \frac{1}{4}x + \left(-\frac{3}{4}\right)$.

Slope ($\frac{1}{4}$); y-intercept $\left(0, -\frac{3}{4}\right)$

◀ **Work Problem 1 at the Side.**

1 Identify the slope and y-intercept of the line with each equation.

(a) $y = 2x - 6$

(b) $y = -\frac{x}{3} + \frac{7}{3}$

(c) $y = -x$

Answers

1. (a) slope: 2; y-intercept: $(0, -6)$

(b) slope: $-\frac{1}{3}$; y-intercept: $\left(0, \frac{7}{3}\right)$

(c) slope: -1; y-intercept: $(0, 0)$

Note

Slope-intercept form is an especially useful form for a linear equation because of the information we can determine from it. It is also the form used by graphing calculators and the one that describes a *linear function*.

OBJECTIVE 2 Graph a line using its slope and a point on the line. We can use the slope and y-intercept to graph a line.

Graphing a Line Using Its Slope and y-Intercept

Step 1 Write the equation in slope-intercept form

$$y = mx + b,$$

if necessary, by solving for y.

Step 2 Identify the y-intercept. Plot the point $(0, b)$.

Step 3 Identify the slope m of the line. Use the geometric interpretation of slope ("*rise over run*") to find another point on the graph by counting from the y-intercept.

Step 4 Join the two points with a line to obtain the graph. (If desired, obtain a third point, such as the x-intercept, as a check.)

EXAMPLE 2 Graphing Lines Using Slopes and y-Intercepts

Graph the equation of each line using the slope and y-intercept.

(a) $y = \frac{2}{3}x - 1$

Step 1 The equation is given in slope-intercept form.

$$y = \frac{2}{3}x - 1$$

Slope (↑ under $\frac{2}{3}$) Value of b in y-intercept $(0, b)$ (↑ under -1)

Step 2 The y-intercept is $(0, -1)$. Plot this point. See **Figure 32.**

Step 3 The slope is $\frac{2}{3}$. By definition,

$$\text{slope } m = \frac{\text{change in } y \text{ (rise)}}{\text{change in } x \text{ (run)}} = \frac{2}{3}.$$

From the y-intercept, count up 2 units and to the right 3 units to obtain the point $(3, 1)$.

Step 4 Draw the line through the two points $(0, -1)$ and $(3, 1)$ to obtain the graph in **Figure 32.**

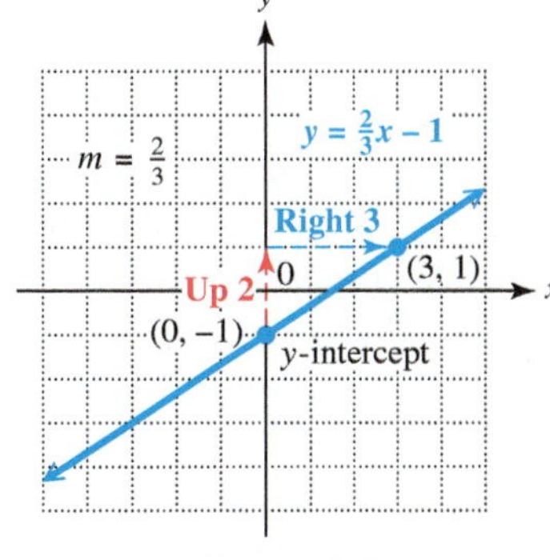

Figure 32

Continued on Next Page

2 Graph $3x - 4y = 8$ using the slope and y-intercept.

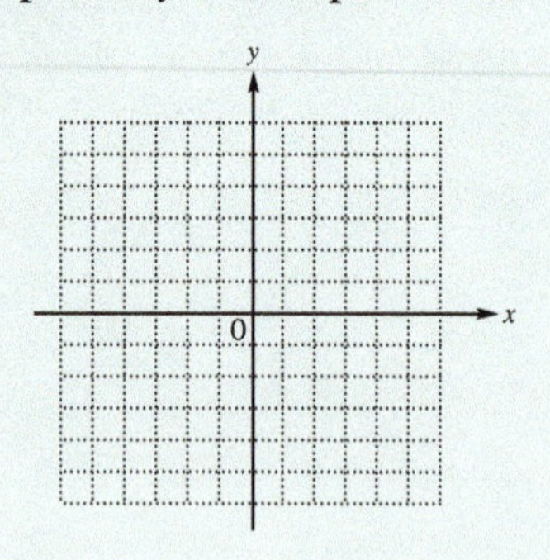

(b) $3x + 4y = 8$

Step 1 Solve for y to write the equation in slope-intercept form.

Isolate y on one side.

$$3x + 4y = 8$$

$$4y = -3x + 8 \quad \text{Subtract } 3x.$$

Slope-intercept form ⟶ $y = -\frac{3}{4}x + 2$ Divide *each* term by 4.

Step 2 The y-intercept is $(0, 2)$. Plot this point. See **Figure 33.**

Step 3 The slope is $-\frac{3}{4}$, which can be written as either $\frac{-3}{4}$ or $\frac{3}{-4}$. We use $\frac{-3}{4}$ here.

$$m = \frac{\text{change in } y \text{ (rise)}}{\text{change in } x \text{ (run)}} = \frac{-3}{4}$$

From the y-intercept, count ***down* 3 units** (because of the negative sign) and to the ***right* 4 units** to obtain the point $(4, -1)$.

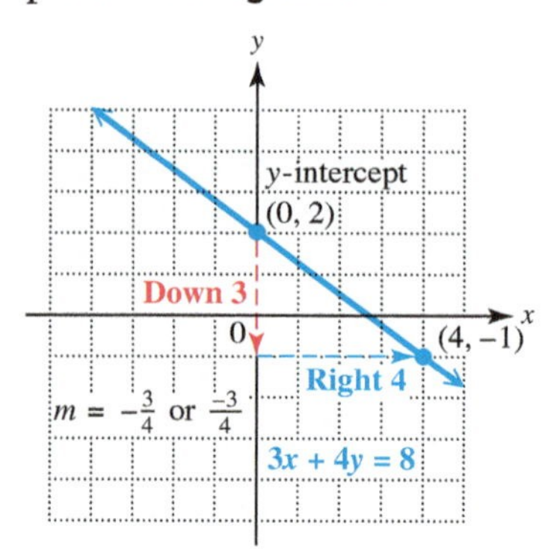

Figure 33

Step 4 Draw the line through the two points $(0, 2)$ and $(4, -1)$ to obtain the graph in **Figure 33.**

◀ **Work Problem 2 at the Side.**

3 Graph the line passing through the point $(2, -3)$, with slope $-\frac{1}{3}$.

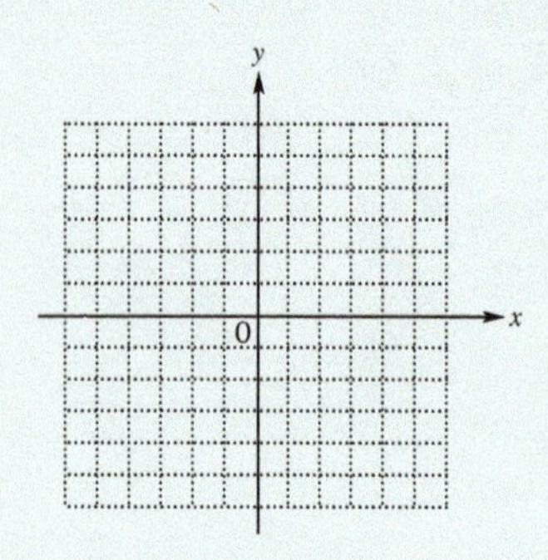

Note

In Step 3 of **Example 2(b),** we could use $\frac{3}{-4}$ for the slope. From the y-intercept $(0, 2)$ in **Figure 33,** count ***up* 3 units** and to the ***left* 4 units** (because of the negative sign) to obtain the point $(-4, 5)$. Verify that this produces the same line.

EXAMPLE 3 Graphing a Line Using Its Slope and a Point

Graph the line passing through the point $(-2, 3)$, with slope -4.

First, plot the point $(-2, 3)$. See **Figure 34.** Then write the slope -4 as

$$\text{slope } m = \frac{\text{change in } y}{\text{change in } x} = -4 = \frac{-4}{1}.$$

Locate another point on the line by counting *down* 4 units from the given point $(-2, 3)$ and then to the *right* 1 unit. Finally, draw the line through this new point $(-1, -1)$ and the given point $(-2, 3)$. See **Figure 34.**

We could have written the slope as $\frac{4}{-1}$ instead. In this case, we would move *up* 4 units from $(-2, 3)$ and then to the *left* 1 unit. Verify that this produces the same line.

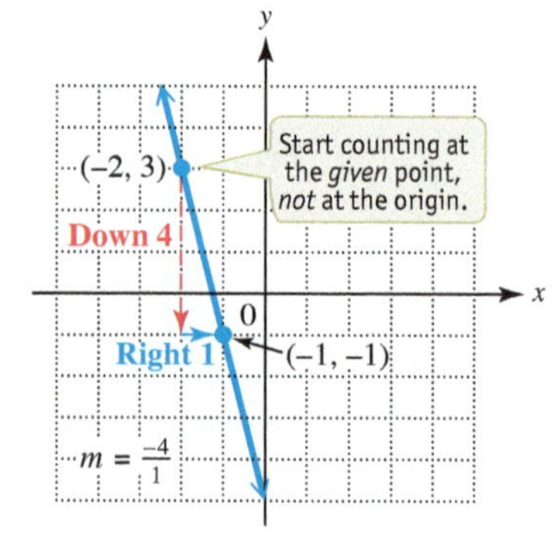

Figure 34

◀ **Work Problem 3 at the Side.**

Answers

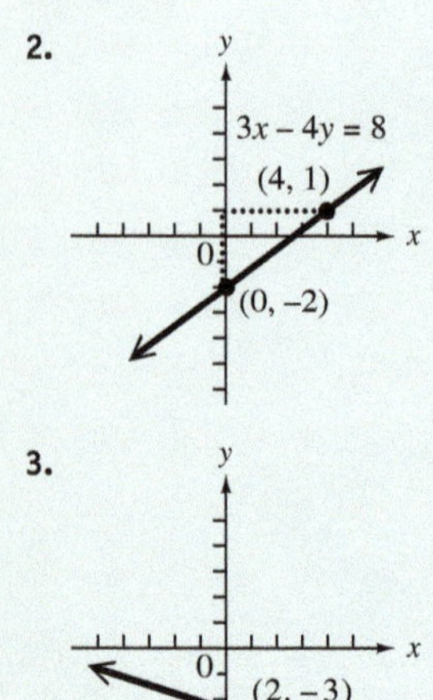

OBJECTIVE 3 Write an equation of a line using its slope and any point on the line. We can use the slope-intercept form to do this.

EXAMPLE 4 Using Slope-Intercept Form to Write Equations of Lines

Write an equation in slope-intercept form of the line passing through the given point and having the given slope.

(a) $(0, -1)$, $m = \frac{2}{3}$

Because the point $(0, -1)$ is the y-intercept, $b = -1$. We can substitute this value for b and the given slope $m = \frac{2}{3}$ directly into slope-intercept form $y = mx + b$ to write an equation.

Slope ↓ ↓ y-intercept is $(0, b)$.

$y = mx + b$ Slope-intercept form

$y = \frac{2}{3}x + (-1)$ Substitute.

$y = \frac{2}{3}x - 1$ Definition of subtraction

(b) $(2, 5)$, $m = 4$

This line passes through the point $(2, 5)$, ***which is not the y-intercept*** because the x-coordinate is 2, ***not* 0. *We cannot substitute for m and b directly as in part (a).***

We can find the y-intercept by substituting $x = 2$ and $y = 5$ from the given point, along with the given slope $m = 4$, into $y = mx + b$ and solving for b.

$y = mx + b$ Slope-intercept form

$5 = 4(2) + b$ Let $y = 5$, $m = 4$, and $x = 2$.

$5 = 8 + b$ Multiply.

$-3 = b$ Subtract 8.

(0, b) is the y-intercept. Don't stop here.

Now substitute the values of m and b into slope-intercept form.

$y = mx + b$ Slope-intercept form

$y = 4x - 3$ Let $m = 4$ and $b = -3$.

Work Problem 4 at the Side. ▶

4 Write an equation in slope-intercept form of the line passing through the given point and having the given slope.

(a) y-intercept $(0, -4)$, slope $\frac{1}{2}$

(b) $(0, 8)$, $m = -1$

(c) $(0, 0)$, $m = 3$

GS **(d)** $(-1, 4)$, $m = -2$

Substitute values and solve for b.

$y = mx + b$

____ = ____(____) + b

$4 =$ ____ $+ b$

____ $= b$

An equation of the line is $y =$ ________.

(e) $(-2, 1)$, $m = 3$

OBJECTIVE 4 Graph and write equations of horizontal and vertical lines.

EXAMPLE 5 Graphing Horizontal and Vertical Lines Using Slope and a Point

Graph each line passing through the given point and having the given slope.

(a) $(4, -2)$, $m = 0$

Recall that a horizontal line has slope 0. To graph this line, plot the point $(4, -2)$ and draw the horizontal line through it. See **Figure 35**.

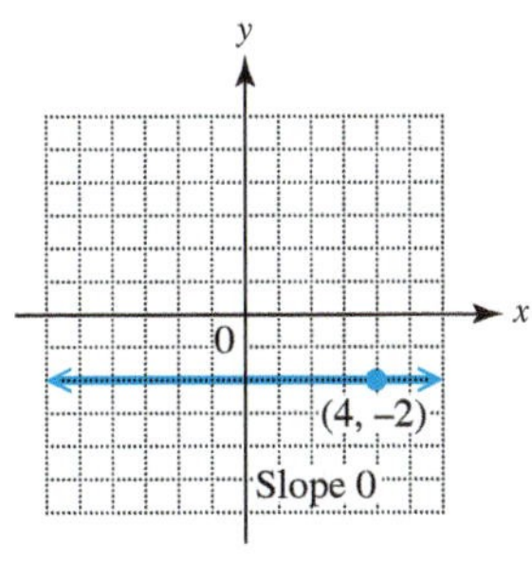

Figure 35

Continued on Next Page

Answers

4. (a) $y = \frac{1}{2}x - 4$
(b) $y = -x + 8$
(c) $y = 3x$
(d) $4; -2; -1; 2; 2; -2x + 2$
(e) $y = 3x + 7$

(b) $(2, -4)$, undefined slope

A vertical line has undefined slope. To graph this line, plot the point $(2, -4)$ and draw the vertical line through it. See **Figure 36.**

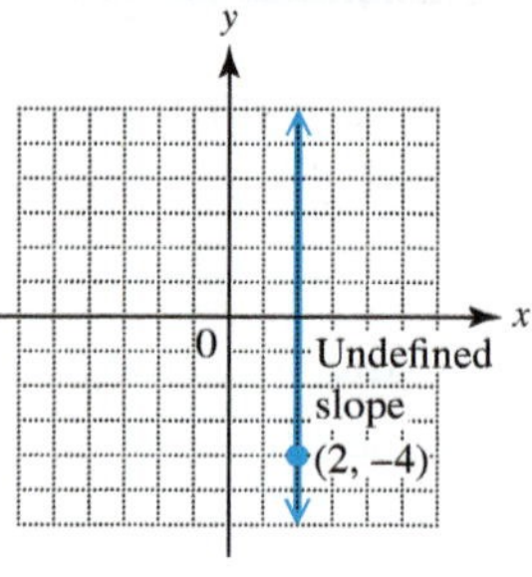

Figure 36

◀ **Work Problem 5 at the Side.**

EXAMPLE 6 **Writing Equations of Horizontal and Vertical Lines**

Write an equation of the line passing through the point $(2, -2)$ and having the given slope.

(a) Slope 0

This line is horizontal because it has slope 0. A horizontal line through the point (a, b) has equation $y = b$. The y-coordinate of the point $(2, -2)$ is -2, so the equation is $y = -2$. See **Figure 37.**

(b) Undefined slope

This line is vertical because it has undefined slope. A vertical line through the point (a, b) has equation $x = a$. The x-coordinate of $(2, -2)$ is 2, so the equation is $x = 2$. See **Figure 37.**

Figure 37

◀ **Work Problem 6 at the Side.**

5 Graph each line passing through the given point and having the given slope.

(a) $(-3, 3)$, undefined slope

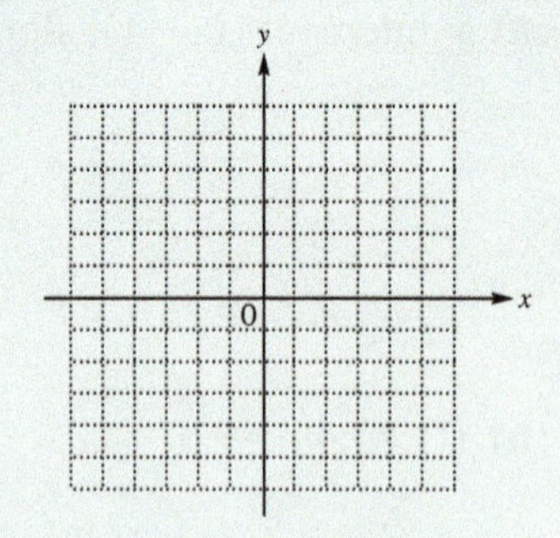

(b) $(3, -3)$, slope 0

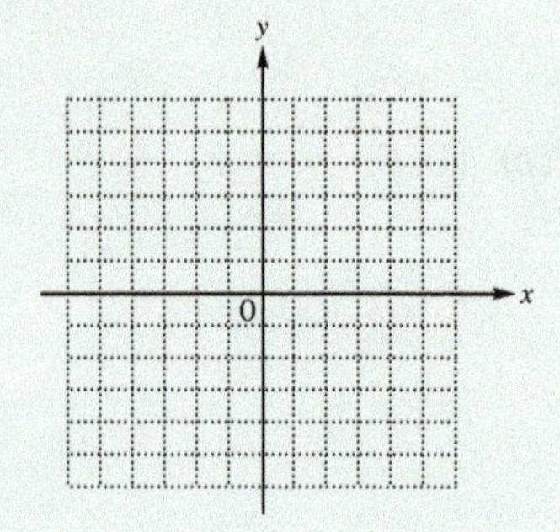

6 Write an equation of the line passing through the point $(-1, 1)$ and having the given slope.

(a) Undefined slope

(b) $m = 0$

Answers

5. (a)

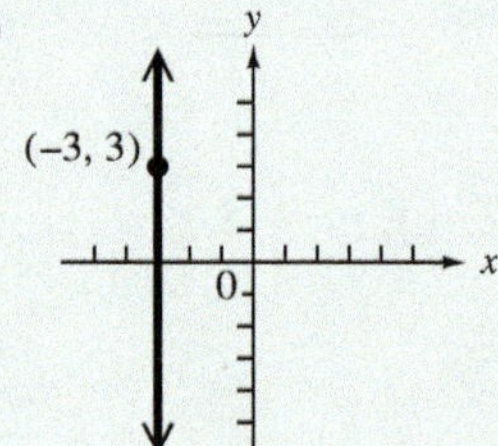

(b)

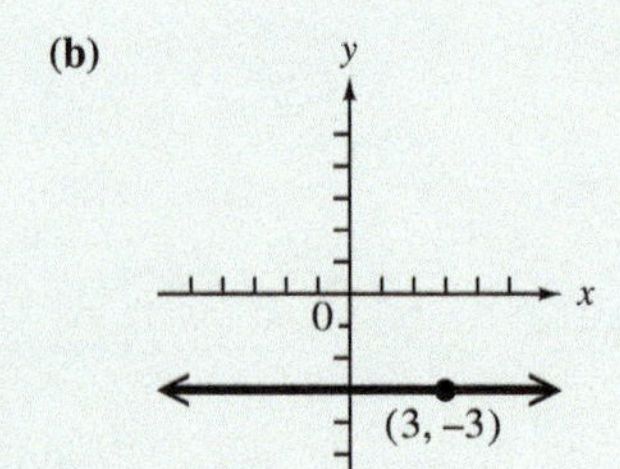

6. (a) $x = -1$ **(b)** $y = 1$

3.4 Exercises

FOR EXTRA HELP

Go to MyMathLab for worked-out, step-by-step solutions to exercises enclosed in a square and video solutions to ▶ exercises.

CONCEPT CHECK *Fill in each blank with the correct response.*

1. In slope-intercept form $y = mx + b$ of the equation of a line, the slope is ______ and the y-intercept is the point ______.

2. The line with equation $y = -\frac{x}{2} - 3$ has slope ____ and y-intercept ______.

CONCEPT CHECK *Work each problem.*

3. Match each equation in parts (a)–(d) with the graph in choices A–D that would most closely resemble its graph.

(a) $y = x + 3$ **(b)** $y = -x + 3$ **(c)** $y = x - 3$ **(d)** $y = -x - 3$

A.

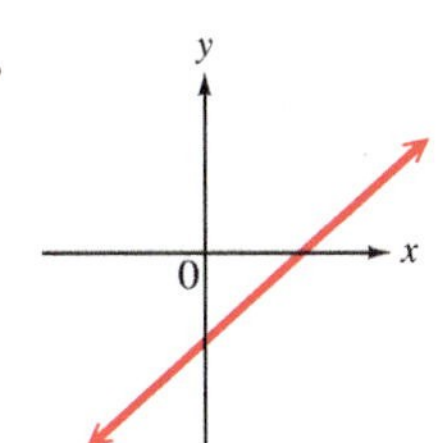

B.

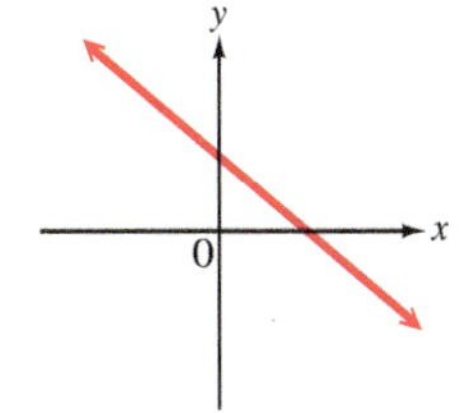

C.

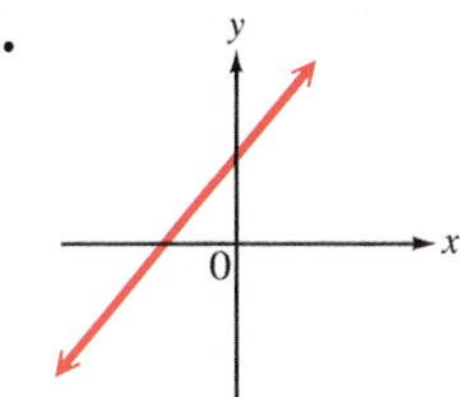

D.

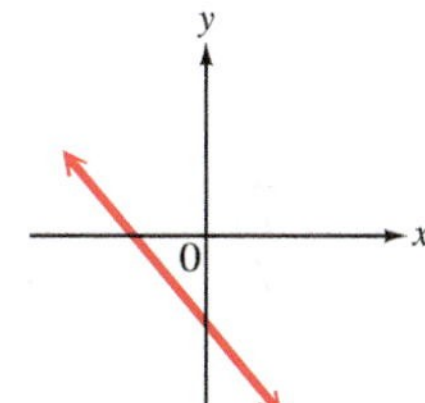

4. Match the description in Column I with the correct equation in Column II.

I	II
(a) Slope -2, passes through the point $(4, 1)$	**A.** $y = 4x$
(b) Slope -2, y-intercept $(0, 1)$	**B.** $y = \frac{1}{4}x$
(c) Passes through the points $(0, 0)$ and $(4, 1)$	**C.** $y = -4x$
(d) Passes through the points $(0, 0)$ and $(1, 4)$	**D.** $y = -2x + 1$
	E. $2x + y = 9$

5. What is the common name given to the vertical line whose x-intercept is the origin?

6. What is the common name given to the line with slope 0 whose y-intercept is the origin?

Identify the slope and y-intercept of the line with each equation. ***See Example 1.***

7. $y = \frac{5}{2}x - 4$

8. $y = \frac{7}{3}x - 6$

9. $y = -x + 9$

10. $y = x + 1$

11. $y = -8x$

12. $y = 2x$

13. $y = \frac{x}{5} - \frac{3}{10}$

14. $y = \frac{x}{7} - \frac{5}{14}$

Graph the equation of each line using the slope and y-intercept. ***See Example 2.***

15. $y = 3x + 2$

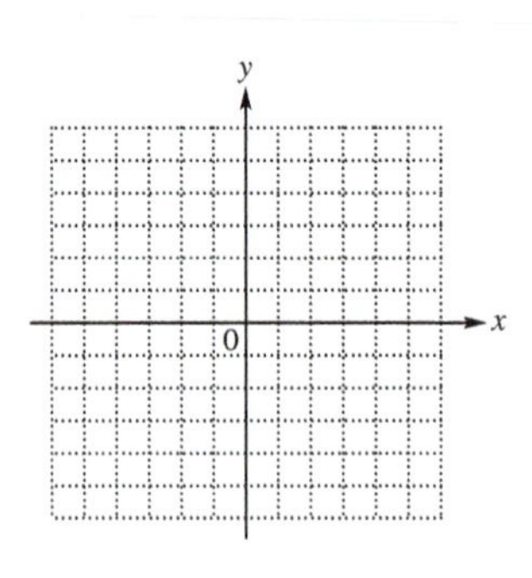

16. $y = 4x - 4$

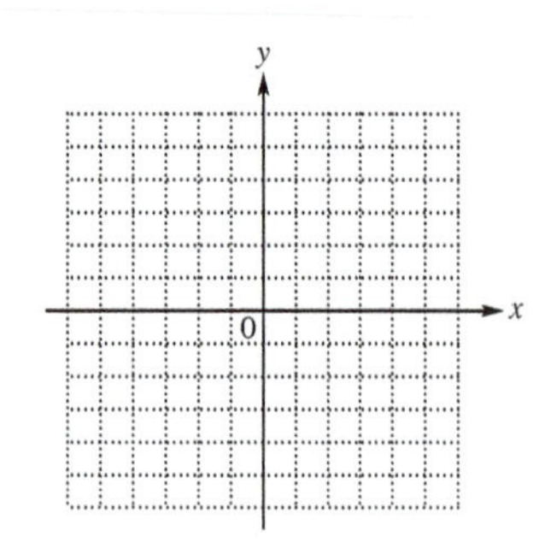

17. $y = \frac{3}{4}x - 1$

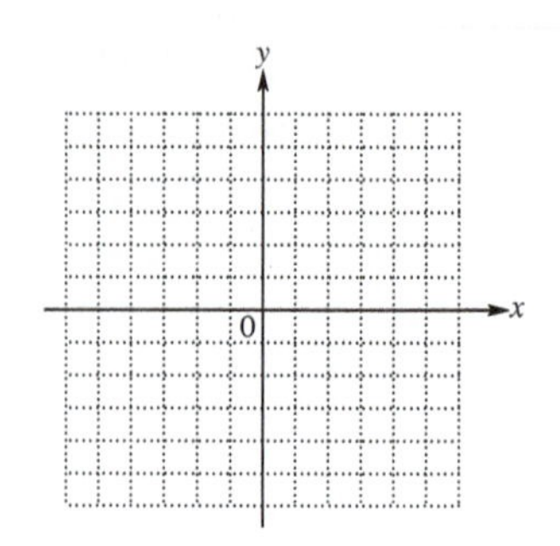

18. $y = \frac{3}{2}x + 2$

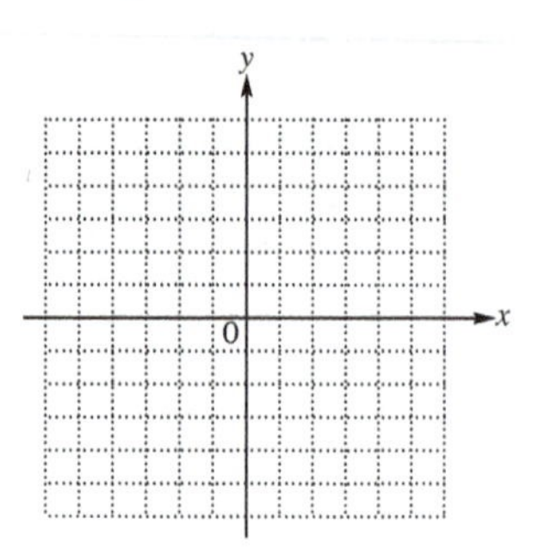

19. $2x + y = -5$

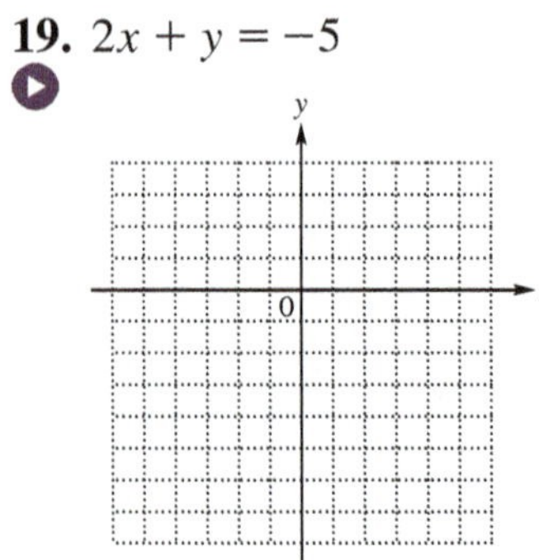

20. $3x + y = -2$

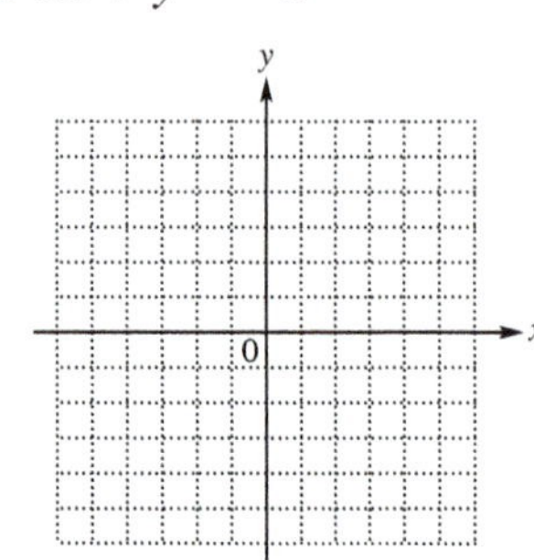

21. $x + 2y = 4$

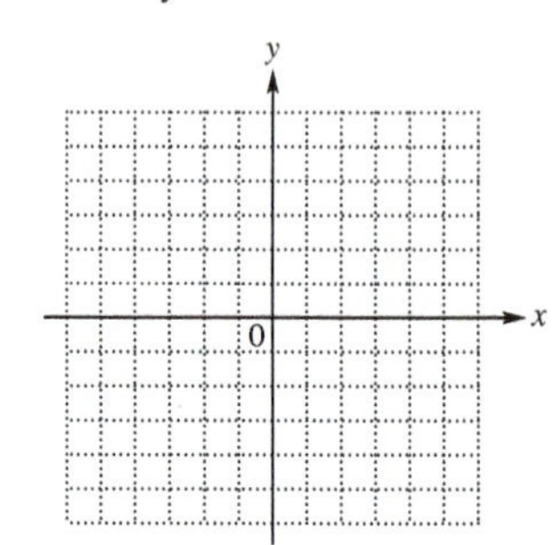

22. $x + 3y = 12$

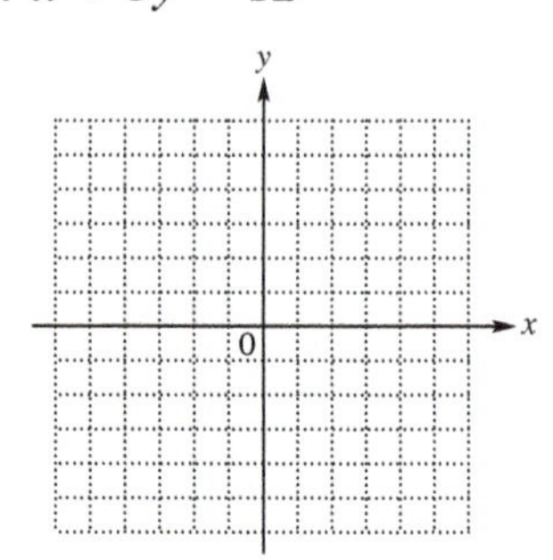

Graph each line passing through the given point and having the given slope.
See Examples 3 and 5.

23. $(0, 1)$, $m = 4$

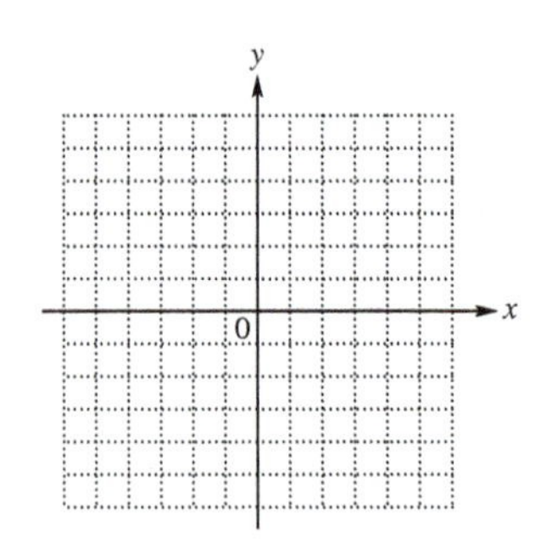

24. $(0, -5)$, $m = -2$

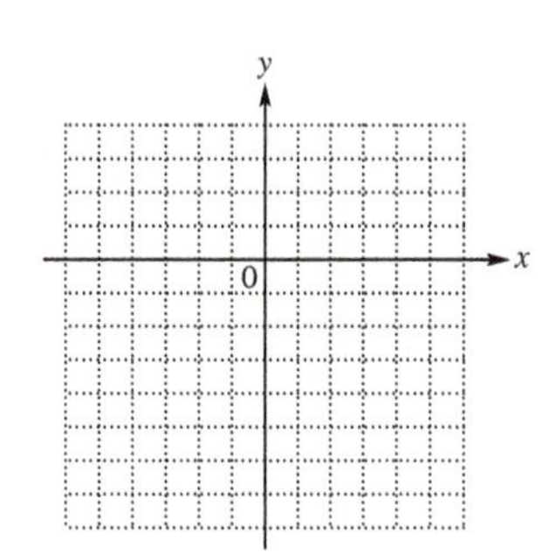

25. $(-2, 3)$, $m = \frac{1}{2}$

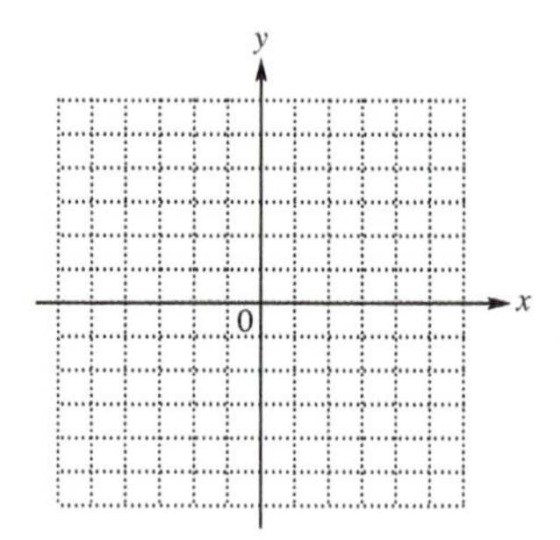

26. $(-4, -1)$, $m = \frac{3}{4}$

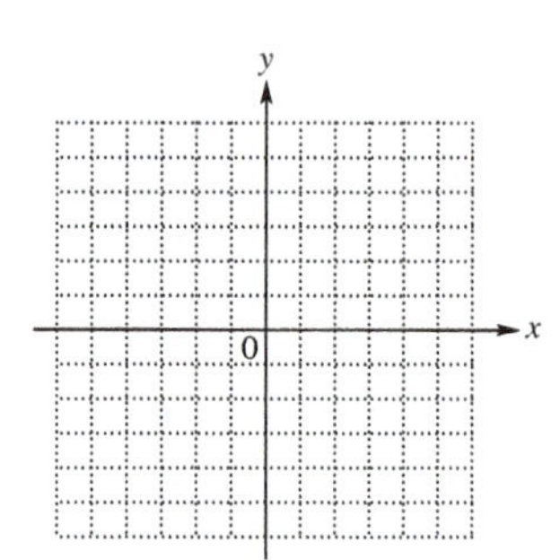

27. $(1, -5)$, $m = -\frac{2}{5}$

28. $(2, -1)$, $m = -\frac{1}{3}$

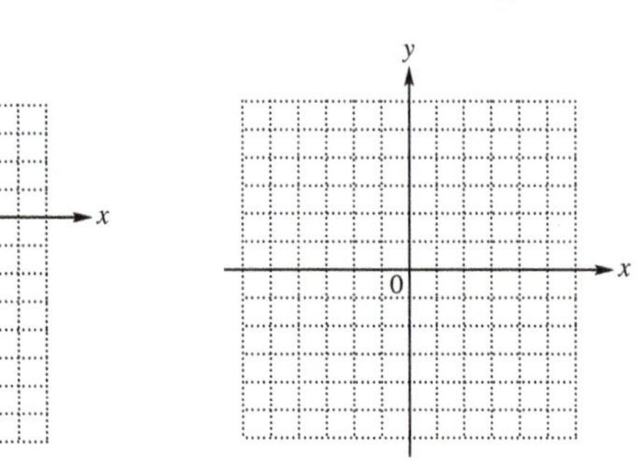

29. $(0, 0)$, $m = \frac{2}{3}$

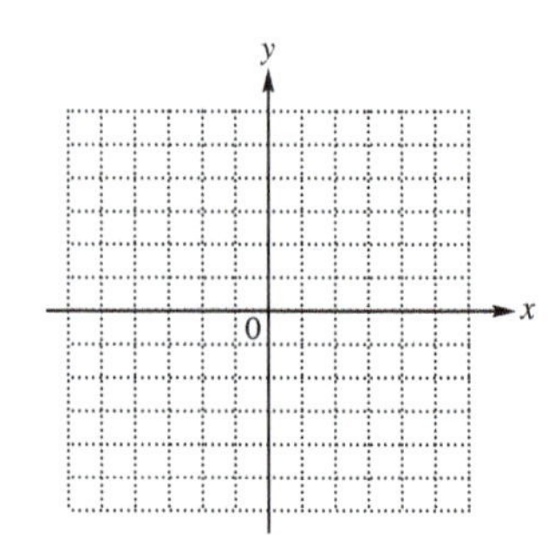

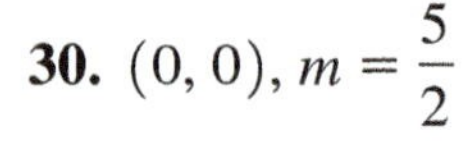

30. $(0, 0)$, $m = \frac{5}{2}$

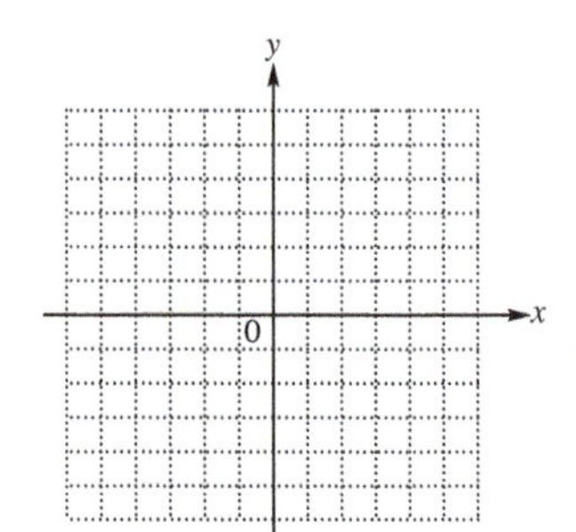

31. $(3, 2)$, $m = 0$

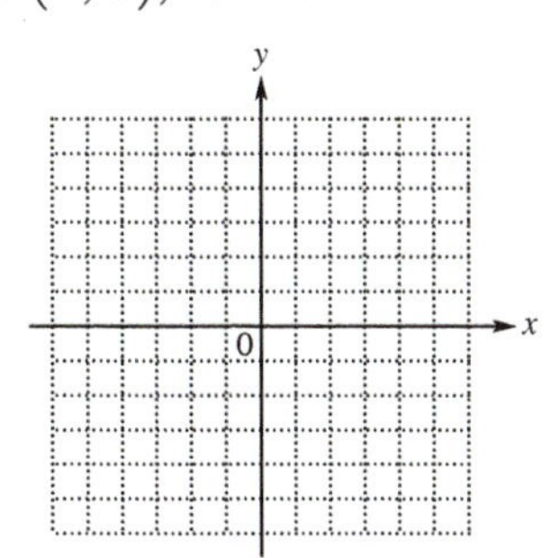

32. $(-2, 3)$, $m = 0$

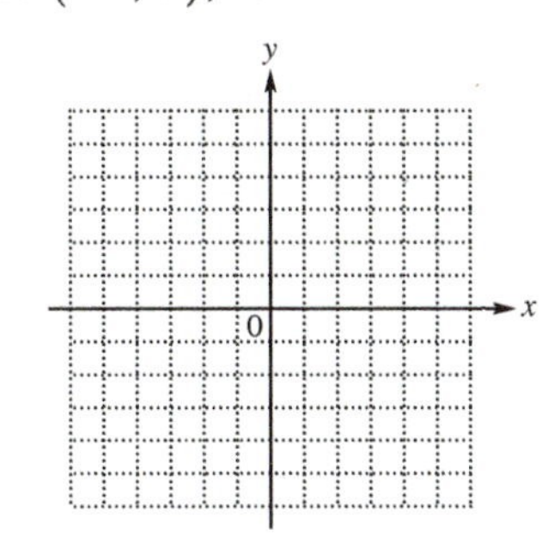

33. $(3, -2)$, undefined slope

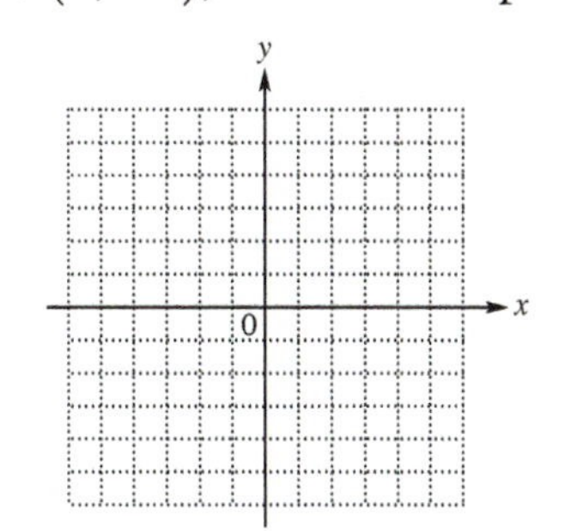

34. $(2, 4)$, undefined slope

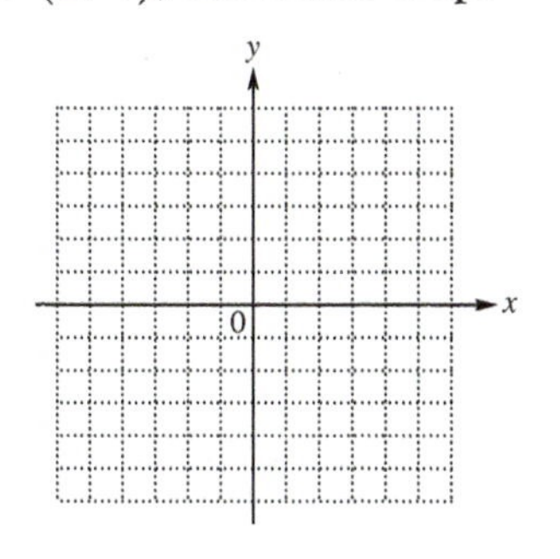

CONCEPT CHECK *Use the geometric interpretation of slope* ("rise over run") *to find the slope of each line. Then, by identifying the y-intercept from the graph, write the slope-intercept form of the equation of the line. (Coordinates of the points shown are integers.)*

35.

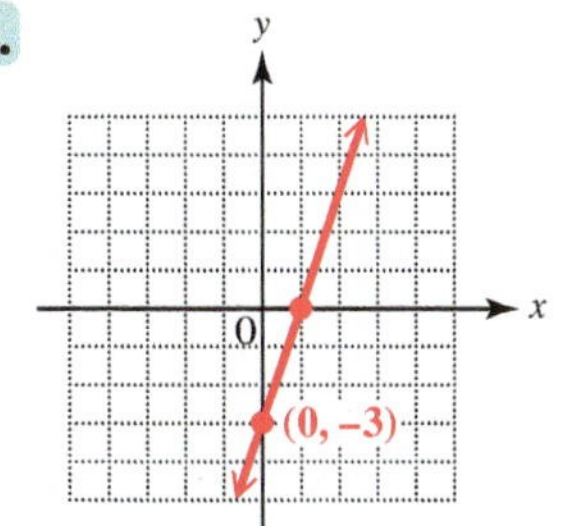

36.

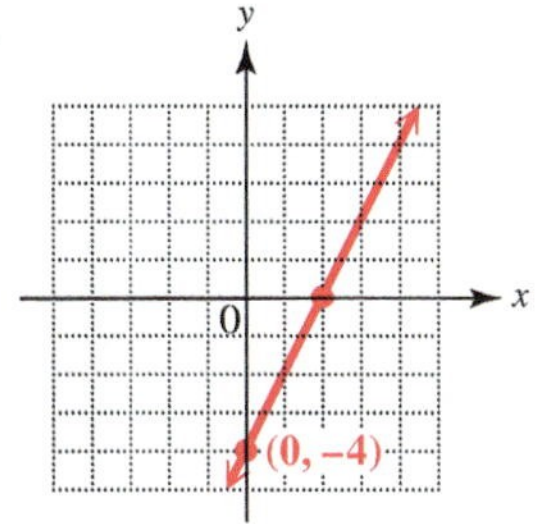

37.

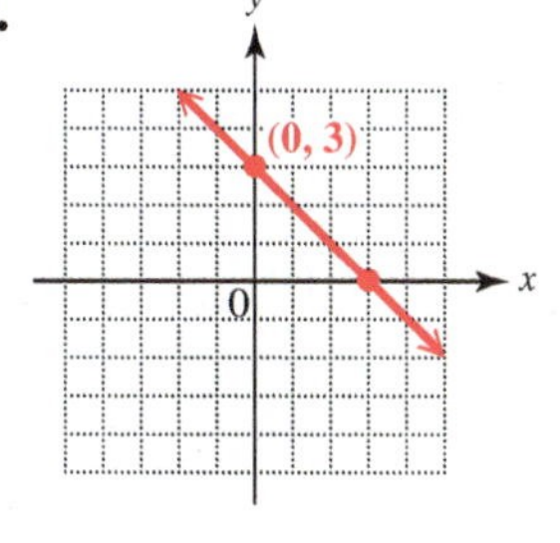

38.

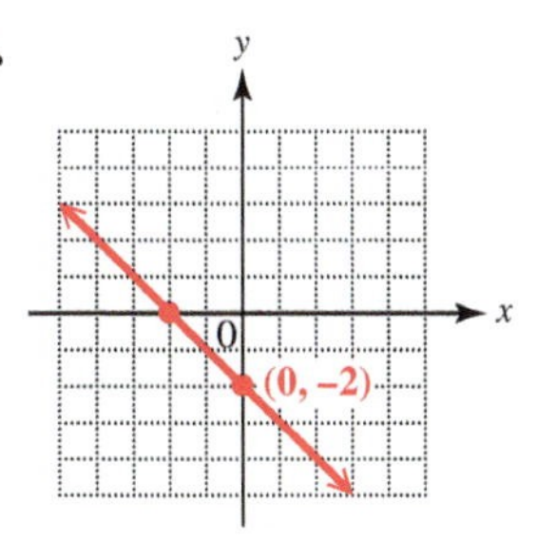

39.

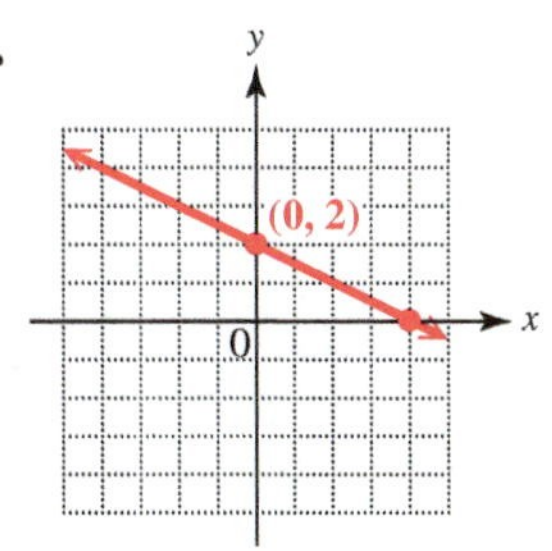

40.

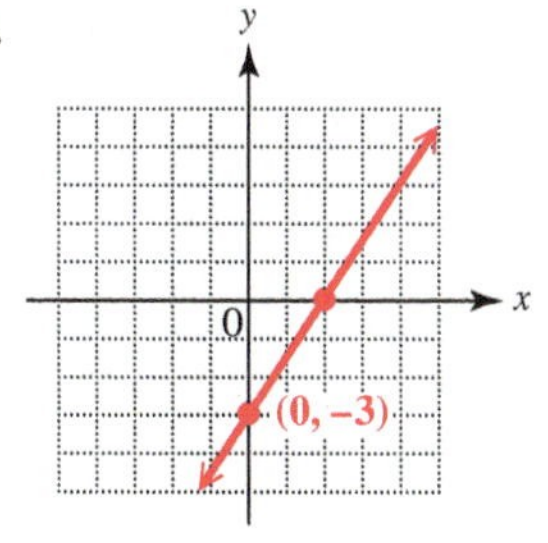

Write an equation in slope-intercept form (if possible) of the line passing through the given point and having the given slope. **See Examples 4 and 6.**

41. slope 4, y-intercept $(0, -3)$

42. slope -5, y-intercept $(0, 6)$

43. $(0, -7)$, $m = -1$

44. $(0, -9)$, $m = 1$

45. $(4, 1)$, $m = 2$

46. $(2, 7)$, $m = 3$

47. $(-1, 3)$, $m = -4$

48. $(-3, 1)$, $m = -2$

49. $(3, -10)$, $m = -2$

50. $(2, -5)$, $m = -4$

51. $(-4, 1)$, $m = \frac{3}{4}$

52. $(2, 1)$, $m = \frac{5}{2}$

53. $(0, 3)$, $m = 0$

54. $(0, -4)$, $m = 0$

55. $(2, -6)$, undefined slope

56. $(-1, 7)$, undefined slope

57. $(6, -6)$, slope 0

58. $(-3, 3)$, slope 0

Each table of values gives several points that lie on a line.

(a) *Use any two of the ordered pairs to find the slope of the line.*

(b) *Identify the y-intercept of the line.*

(c) *Use the slope and y-intercept from parts (a) and (b) to write an equation of the line in slope-intercept form.*

(d) *Then graph the equation.*

59.

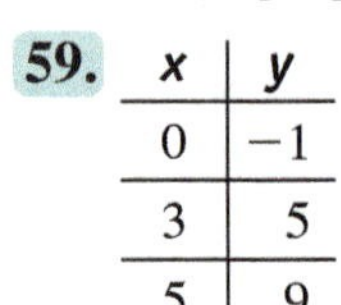

x	y
0	-1
3	5
5	9

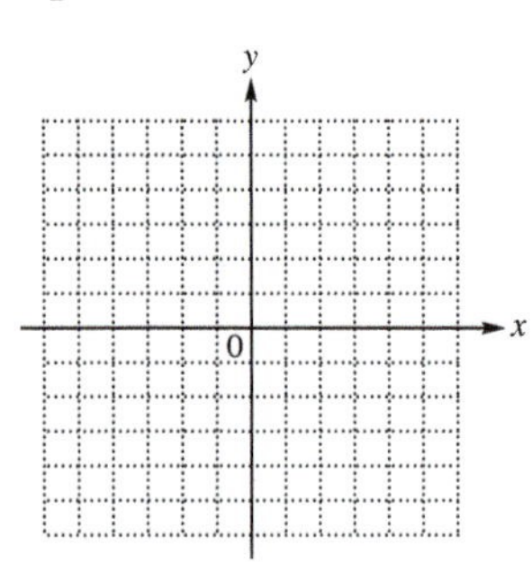

60.

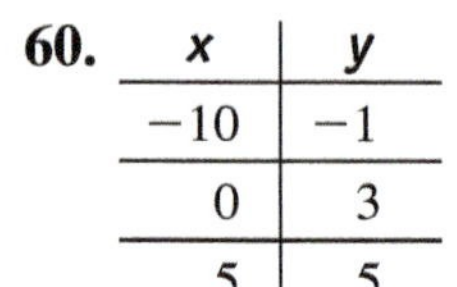

x	y
-10	-1
0	3
5	5

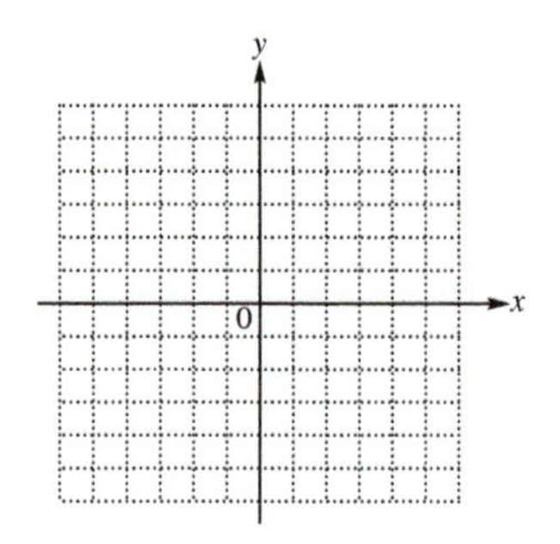

The cost y to produce x items is, in some cases, expressed in the form

$$y = mx + b.$$

The value of b gives the ***fixed cost*** *(the cost that is the same no matter how many items are produced), and the value of m gives the* ***variable cost*** *(the cost to produce an additional item). Use this information to work each problem.*

61. It costs \$400 to start up a business selling campaign buttons. Each button costs \$0.25 to produce.

(a) What is the fixed cost?

(b) What is the variable cost?

(c) Write the cost equation.

(d) What will be the cost of producing 100 campaign buttons, based on the cost equation?

(e) How many campaign buttons will be produced if the total cost is \$775?

62. It costs \$2000 to purchase a copier, and each copy costs \$0.02 to make.

(a) What is the fixed cost?

(b) What is the variable cost?

(c) Write the cost equation.

(d) What will be the cost of producing 10,000 copies, based on the cost equation?

(e) How many copies will be produced if the total cost is \$2600?

Solve each problem.

63. Andrew earns 5% commission on his sales, plus a base salary of \$2000 per month. This is illustrated in the graph and can be modeled by the linear equation

$$y = 0.05x + 2000,$$

where y is his monthly salary in dollars and x is his sales, also in dollars.

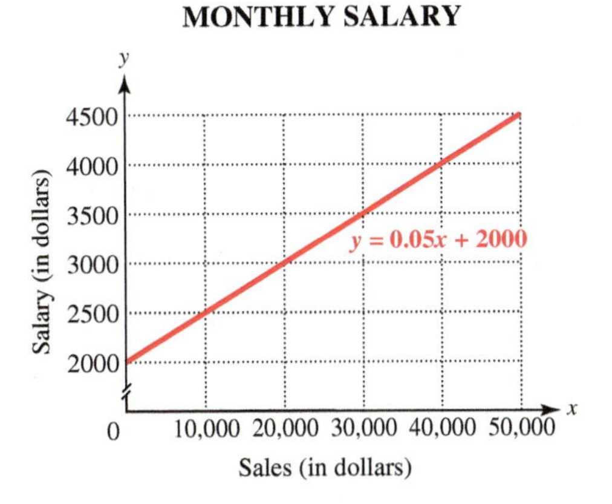

(a) What is the slope? With what does the slope correspond in the problem?

(b) What is the y-intercept? With what does the y-value of the y-intercept correspond in the problem?

(c) Use the equation to determine Andrew's monthly salary if his sales are \$10,000. Confirm this using the graph.

(d) Use the graph to determine his sales if he wants to earn a monthly salary of \$3500. Confirm this using the equation.

64. The cost to rent a moving van is \$0.50 per mile, plus a flat fee of \$100. This is illustrated in the graph and can be modeled by the linear equation

$$y = 0.50x + 100,$$

where y is the total rental cost in dollars and x is the number of miles driven.

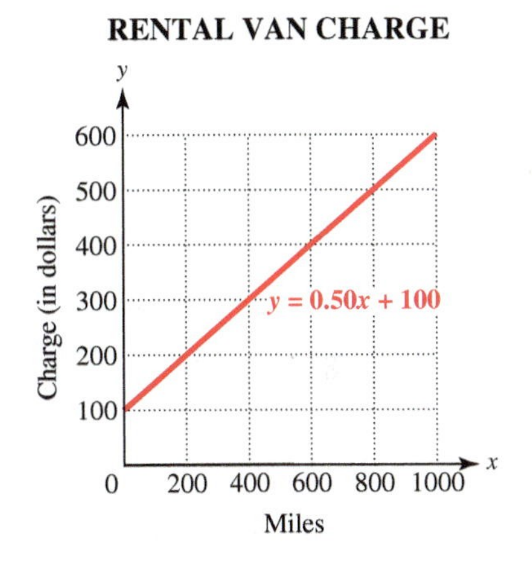

(a) What is the slope? With what does the slope correspond in the problem?

(b) What is the y-intercept? With what does the y-value of the y-intercept correspond in the problem?

(c) Use the equation to determine the total charge if 400 mi are driven. Confirm this using the graph.

(d) Use the graph to determine the number of miles driven if the charge is \$500. Confirm this using the equation.

Relating Concepts (Exercises 65–68) For Individual or Group Work

A line with equation written in slope-intercept form $y = mx + b$ has slope m and y-intercept $(0, b)$. Recall that the standard form of a linear equation in two variables is

$$Ax + By = C, \quad \text{Standard form}$$

where A, B, and C are real numbers and A and B are not both 0. ***Work Exercises 65–68 in order.***

65. Write the standard form of a linear equation in slope-intercept form—that is, solved for y—to show that, in general, the slope is given by $-\frac{A}{B}$ (where $B \neq 0$).

66. Use the fact that $m = -\frac{A}{B}$ to find the slope of the line with each equation.

(a) $2x + 3y = 18$ **(b)** $4x - 2y = -1$ **(c)** $3x - 7y = 21$

67. Refer to the slope-intercept form found in **Exercise 65.** What is the y-intercept?

68. Use the result of **Exercise 67** to find the y-intercept of each line in **Exercise 66.**

3.5 Point-Slope Form of a Linear Equation and Modeling

OBJECTIVES

1. Use point-slope form to write an equation of a line.
2. Write an equation of a line using two points on the line.
3. Write an equation of a line that fits a data set.

OBJECTIVE 1 Use point-slope form to write an equation of a line. There is another form that can be used to write an equation of a line. To develop this form, we let m represent the slope of a line and (x_1, y_1) represent a given point on the line. We let (x, y) represent any other point on the line. See **Figure 38.**

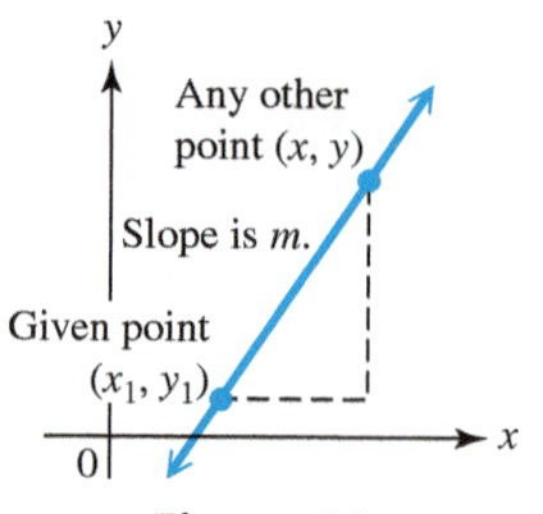

Figure 38

$$m = \frac{y - y_1}{x - x_1} \qquad \text{Definition of slope}$$

$$m(x - x_1) = y - y_1 \qquad \text{Multiply each side by } x - x_1.$$

$$y - y_1 = m(x - x_1) \qquad \text{Interchange sides.}$$

This result is the *point-slope form* of the equation of a line.

Point-Slope Form

The **point-slope form** of the equation of a line with slope m passing through the point (x_1, y_1) is

$$y - y_1 = m(x - x_1).$$

(Slope: m; Given point: (x_1, y_1))

EXAMPLE 1 Using Point-Slope Form to Write Equations

Write an equation of each line. Give the final answer in slope-intercept form.

(a) The line passing through $(-2, 4)$, with slope -3

The given point is $(-2, 4)$ so $x_1 = -2$ and $y_1 = 4$. Also, $m = -3$.

Only y_1, m, and x_1 are replaced with numbers.

$$y - y_1 = m(x - x_1) \qquad \text{Point-slope form}$$

$$y - 4 = -3[x - (-2)] \qquad \text{Let } y_1 = 4,\ m = -3,\ x_1 = -2.$$

$$y - 4 = -3(x + 2) \qquad \text{Definition of subtraction}$$

$$y - 4 = -3x - 6 \qquad \text{Distributive property}$$

$$y = -3x - 2 \qquad \text{Add 4.}$$

The answer is in $y = mx + b$ form as specified.

(b) The line passing through $(4, 2)$, with slope $\frac{3}{5}$

$$y - y_1 = m(x - x_1) \qquad \text{Point-slope form}$$

$$y - 2 = \frac{3}{5}(x - 4) \qquad \text{Let } y_1 = 2,\ m = \tfrac{3}{5},\ x_1 = 4.$$

$$y - 2 = \frac{3}{5}x - \frac{12}{5} \qquad \text{Distributive property}$$

$$y = \frac{3}{5}x - \frac{12}{5} + \frac{10}{5} \qquad \text{Add } 2 = \tfrac{10}{5} \text{ to each side.}$$

$$y = \frac{3}{5}x - \frac{2}{5} \qquad \text{Combine like terms.}$$

Do not clear fractions here because we want the answer in slope-intercept form—that is, solved for y.

◀ Work Problem 1 at the Side.

---

1 Write an equation of each line. Give the final answer in slope-intercept form.

GS **(a)** The line passing through $(-1, 3)$, with slope -2

$$y - y_1 = m(x - x_1)$$

$$y - \_\_\_ = \_\_\_[x - (\_\_\_)]$$

$$y - 3 = -2(x + \_\_\_)$$

$$y - 3 = -2x - \_\_\_$$

$$y = \_\_\_\_\_$$

(b) The line passing through $(5, 2)$, with slope $-\frac{1}{3}$

Answers

1. (a) 3; −2; −1; 1; 2; $-2x + 1$

(b) $y = -\frac{1}{3}x + \frac{11}{3}$

OBJECTIVE 2 Write an equation of a line using two points on the line.

Many of the linear equations we have worked with have been written in the form

$$Ax + By = C,$$

called **standard form,** where A, B, and C are real numbers and A and B are not both 0. In most cases, A, B, and C are rational numbers.

For consistency in this text, we give answers so that A, B, and C are integers with greatest common factor 1 and $A \geq 0$. (If $A = 0$, then we give $B > 0$.)

Note

The definition of standard form is not the same in all texts. A linear equation can be written in many different, yet equally correct, ways. For example,

$$3x + 4y = 12, \quad 6x + 8y = 24, \quad \text{and} \quad -9x - 12y = -36$$

all represent the same set of ordered pairs. When giving answers in standard form, $3x + 4y = 12$ is preferable to the other forms because the greatest common factor of 3, 4, and 12 is 1 and $A \geq 0$.

EXAMPLE 2 Writing an Equation of a Line Using Two Points

Write an equation of the line passing through the points $(3, 4)$ and $(-2, 5)$. Give the final answer in slope-intercept form and then in standard form.

First, find the slope of the line using the given points.

$$\text{slope } m = \frac{4-5}{3-(-2)} \quad \text{Subtract } y\text{-values. Subtract } x\text{-values in the same order.}$$

$$= \frac{-1}{5} \quad \text{Simplify the fraction.}$$

$$= -\frac{1}{5} \quad \frac{-a}{b} = -\frac{a}{b}$$

Now use $m = -\frac{1}{5}$ and either $(-2, 5)$ or $(3, 4)$ as (x_1, y_1) in point-slope form. We choose $(3, 4)$.

$$y - y_1 = m(x - x_1) \quad \text{Point-slope form}$$

$$y - 4 = -\frac{1}{5}(x - 3) \quad \text{Let } y_1 = 4,\ m = -\tfrac{1}{5},\ x_1 = 3.$$

$$y - 4 = -\frac{1}{5}x + \frac{3}{5} \quad \text{Distributive property}$$

$$y = -\frac{1}{5}x + \frac{3}{5} + \frac{20}{5} \quad \text{Add } 4 = \tfrac{20}{5} \text{ to each side.}$$

$$\text{Slope-intercept form} \longrightarrow y = -\frac{1}{5}x + \frac{23}{5} \quad \text{Combine like terms.}$$

$$5y = -x + 23 \quad \text{Multiply by 5 to clear fractions.}$$

$$\text{Standard form} \longrightarrow x + 5y = 23 \quad \text{Add } x.$$

Work Problem 2 at the Side. ▶

2 Write an equation of the line passing through each pair of points. Give the final answer in slope-intercept form and then in standard form.

(a) $(2, 5)$ and $(-1, 6)$

(b) $(-3, 1)$ and $(2, 4)$

Answers

2. (a) $y = -\frac{1}{3}x + \frac{17}{3}$; $x + 3y = 17$

(b) $y = \frac{3}{5}x + \frac{14}{5}$; $3x - 5y = -14$

Note

There is often more than one way to write an equation of a line. Consider **Example 2.**

- The same equation will result using the point $(-2, 5)$ for (x_1, y_1) in point-slope form.
- We could also use slope-intercept form $y = mx + b$ and substitute the slope and either given point, solving for b.

SUMMARY OF THE FORMS OF LINEAR EQUATIONS

Equation	Description	Example
$y = mx + b$	**Slope-intercept form** Slope is m. y-intercept is $(0, b)$.	$y = \frac{3}{2}x - 6$
$y - y_1 = m(x - x_1)$	**Point-slope form** Slope is m. Line passes through (x_1, y_1).	$y + 3 = \frac{3}{2}(x - 2)$
$Ax + By = C$ (where A, B, and C are real numbers and A and B are not both 0.)	**Standard form** Slope is $-\frac{A}{B}$ $(B \neq 0)$. x-intercept is $\left(\frac{C}{A}, 0\right)$ $(A \neq 0)$. y-intercept is $\left(0, \frac{C}{B}\right)$ $(B \neq 0)$.	$3x - 2y = 12$
$x = a$	**Vertical line** Slope is undefined. x-intercept is $(a, 0)$.	$x = 3$
$y = b$	**Horizontal line** Slope is 0. y-intercept is $(0, b)$.	$y = 3$

OBJECTIVE 3 Write an equation of a line that fits a data set. If a given set of data fits a linear pattern—that is, its graph consists of points lying close to a straight line—we can write a linear equation that models the data.

EXAMPLE 3 Writing an Equation of a Line That Models Data

The table lists average annual cost y (in dollars) of tuition and fees for instate students at public 4-year colleges and universities for selected years, where $x = 1$ represents 2001, $x = 3$ represents 2003, and so on.

Year	x	Cost y (in dollars)
2001	1	3735
2003	3	4587
2005	5	5351
2007	7	5943
2009	9	6717
2011	11	7313
2013	13	8312

Data from National Center for Education Statistics.

Continued on Next Page

(a) Plot the data and write an equation that approximates it.

We plot the data as shown in **Figure 39.**

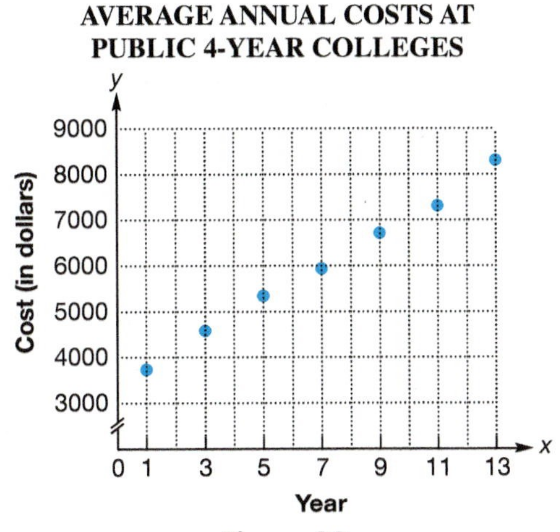

Figure 39

The points appear to lie approximately in a straight line. To find an equation of the line, we choose two ordered pairs $(3, 4587)$ and $(9, 6717)$ from the table and determine the slope of the line through these points.

$$m = \frac{y_2 - y_1}{x_2 - x_1} = \frac{6717 - 4587}{9 - 3} = 355 \qquad \text{Let } (9, 6717) = (x_2, y_2) \text{ and } (3, 4587) = (x_1, y_1).$$

The slope, 355, is positive, indicating that tuition and fees *increased* \$355 each year. Now substitute this slope and the point $(3, 4587)$ in the point-slope form to find an equation of the line.

$$y - y_1 = m(x - x_1) \qquad \text{Point-slope form}$$
$$y - 4587 = 355(x - 3) \qquad \text{Let } (x_1, y_1) = (3, 4587),\ m = 355.$$
$$y - 4587 = 355x - 1065 \qquad \text{Distributive property}$$
$$y = 355x + 3522 \qquad \text{Add 4587.}$$

Thus, the equation $y = 355x + 3522$ can be used to model the data.

(b) Use the equation found in part (a) to determine the cost of tuition and fees in 2013.

We let $x = 13$ (for 2013) in the equation from part (a), and solve for y.

$$y = 355x + 3522 \qquad \text{Equation of the line}$$
$$y = 355(13) + 3522 \qquad \text{Substitute 13 for } x.$$
$$y = 8137 \qquad \text{Multiply, and then add.}$$

Using the equation, tuition and fees in 2013 were \$8137. The corresponding value in the table for $x = 13$ is 8312, so the equation approximates the data reasonably well.

Work Problem 3 at the Side. ▶

Note

In **Example 3,** if we had chosen two different data points, we would have obtained a slightly different equation. See **Margin Problem 3.**

Also, we could have used slope-intercept form $y = mx + b$ (instead of point-slope form) to write an equation that models the data.

3 Refer to **Example 3.**

(a) Use the points $(5, 5351)$ and $(11, 7313)$ to write an equation that approximates the data. Give the final equation in slope-intercept form.

(b) Use the equation from part (a) to determine the cost of tuition and fees in 2009? How well does the equation approximate the 2009 data from the table?

Answers

3. **(a)** $y = 327x + 3716$

(b) \$6659; The corresponding value in the table for $x = 9$ is \$6717, so the equation approximates the data reasonably well.

3.5 Exercises

FOR EXTRA HELP *Go to MyMathLab for worked-out, step-by-step solutions to exercises enclosed in a square and video solutions to ▶ exercises.*

CONCEPT CHECK *Work each problem.*

1. Match each form or description in Column I with the corresponding equation in Column II.

I	II
(a) Point-slope form	**A.** $x = a$
(b) Horizontal line	**B.** $y = mx + b$
(c) Slope-intercept form	**C.** $y = b$
(d) Standard form	**D.** $y - y_1 = m(x - x_1)$
(e) Vertical line	**E.** $Ax + By = C$

2. Write the equation $y + 1 = -2(x - 5)$ first in slope-intercept form and then in standard form.

3. Which equations are equivalent to $2x - 3y = 6$?

A. $y = \frac{2}{3}x - 2$ **B.** $-2x + 3y = -6$ **C.** $y = -\frac{3}{2}x + 3$ **D.** $y - 2 = \frac{2}{3}(x - 6)$

4. In the summary box following **Example 2,** we give the equations

$$y = \frac{3}{2}x - 6 \quad \text{and} \quad y + 3 = \frac{3}{2}(x - 2)$$

as examples of equations in slope-intercept form and point-slope form, respectively. Write each of these equations in standard form. What do you notice?

Write an equation of the line passing through the given point and having the given slope. Give the final answer in slope-intercept form. ***See Example 1.***

5. $(1, 7)$, $m = 5$

6. $(2, 9)$, $m = 6$

7. $(6, -3)$, $m = 1$

8. $(-4, 4)$, $m = 1$

9. $(1, -7)$, $m = -3$

10. $(1, -5)$, $m = -7$

11. $(3, -2)$, $m = -1$

12. $(-5, 4)$, $m = -1$

13. $(-2, 5)$, $m = \frac{2}{3}$

14. $(4, 2)$, $m = -\frac{1}{3}$

15. $(6, -3)$, $m = -\frac{4}{5}$

16. $(7, -2)$, $m = -\frac{7}{2}$

*Write an equation of the line passing through the given pair of points. Give the final answer in **(a)** slope-intercept form and **(b)** standard form.* ***See Example 2.***

17. $(4, 10)$ and $(6, 12)$ **18.** $(8, 5)$ and $(9, 6)$ **19.** $(-1, -7)$ and $(-8, -2)$ **20.** $(-2, -1)$ and $(3, -4)$

21. $(0, -2)$ and $(-3, 0)$ **22.** $(-4, 0)$ and $(0, 2)$ **23.** $\left(\frac{1}{2}, \frac{3}{2}\right)$ and $\left(-\frac{1}{4}, \frac{5}{4}\right)$ **24.** $\left(-\frac{2}{3}, \frac{8}{3}\right)$ and $\left(\frac{1}{3}, \frac{7}{3}\right)$

*Write an equation of the given line through the given points. Give the final answer in **(a)** slope-intercept form and **(b)** standard form.*

25.

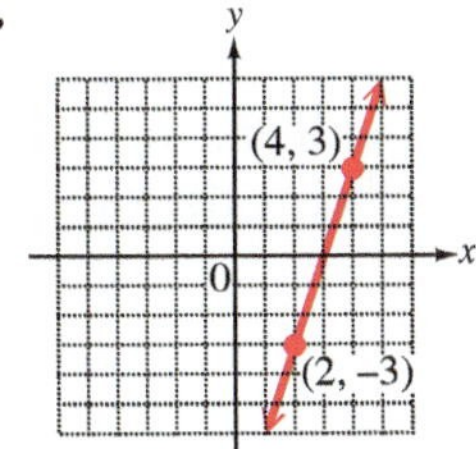

26.

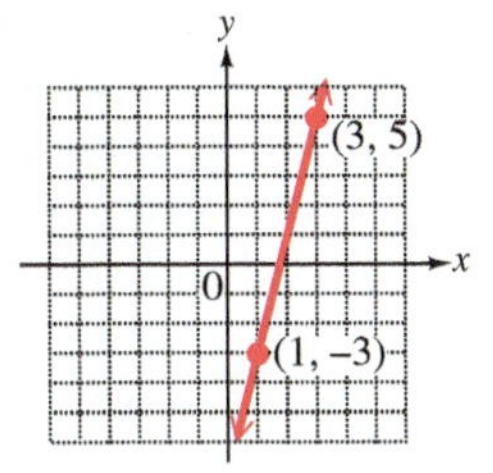

27.

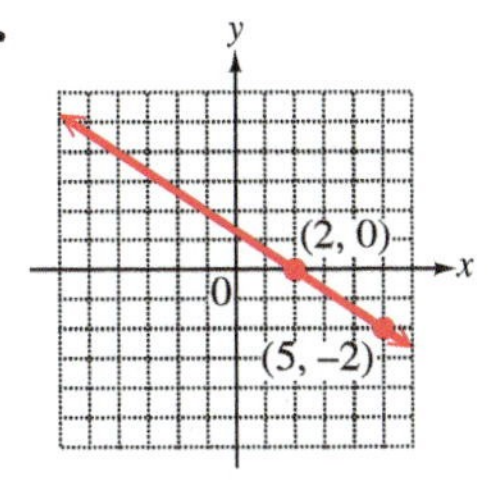

28.

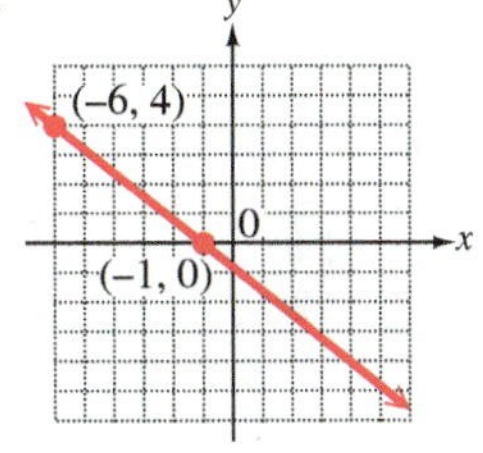

Write an equation of the line satisfying the given conditions. Give the final answer in slope-intercept form. (Hint: Recall the relationships among slopes of parallel and perpendicular lines.)

29. Passing through $(8, -1)$; parallel to a line with slope $\frac{3}{4}$

30. Passing through $(-2, 1)$; perpendicular to a line with slope $\frac{2}{3}$

31. Perpendicular to $x - 2y = 7$; y-intercept $(0, -3)$

32. Parallel to $5x = 2y + 10$; y-intercept $(0, 4)$

33. Passing through $(4, 2)$; perpendicular to $x - 3y = 7$

34. Passing through $(2, 3)$; parallel to $4x - y = -2$

35. Passing through $(2, -3)$; parallel to $3x = 4y + 5$

36. Passing through $(-1, 4)$; perpendicular to $2x + 3y = 8$

Solve each problem. ***See Example 3.***

37. The table lists the average annual cost y (in dollars) of tuition and fees at 2-year public colleges for selected years x, where $x = 1$ represents 2009, $x = 2$ represents 2010, and so on.

Year	x	Cost y (in dollars)
2009	1	2283
2010	2	2441
2011	3	2651
2012	4	2792
2013	5	2882

Data from National Center for Education Statistics.

(a) Write five ordered pairs (x, y) for the data.

(b) Plot the ordered pairs (x, y). Do the points lie approximately in a straight line?

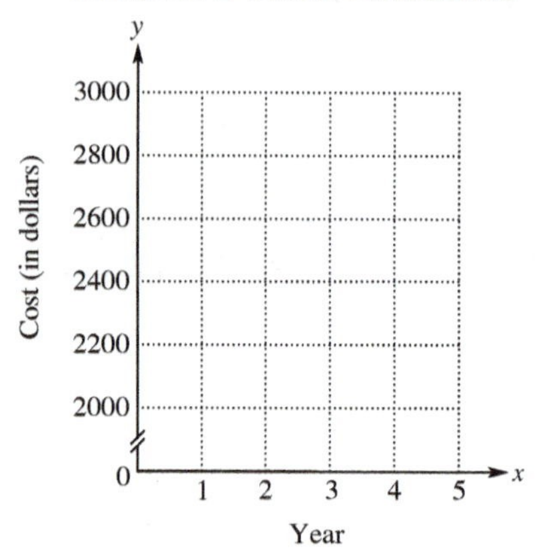

(c) Use the ordered pairs $(2, 2441)$ and $(5, 2882)$ to write the equation of a line that approximates the data. Give the final equation in slope-intercept form.

(d) Use the equation from part (c) to estimate the average annual cost at 2-year colleges in 2015 to the nearest dollar. (*Hint:* What is the value of x for 2015?)

38. The table gives heavy-metal nuclear waste y (in thousands of metric tons) from spent reactor fuel awaiting permanent storage. Here $x = 0$ represents 1995, $x = 5$ represents 2000, and so on.

Year	x	Waste y (in thousands of tons)
1995	0	32
2000	5	42
2010	15	61
2020*	25	76

Data from *Scientific American.*

*Estimate by U.S. Department of Energy.

(a) Write four ordered pairs (x, y) for the data.

(b) Plot the ordered pairs (x, y). Do the points lie approximately in a straight line?

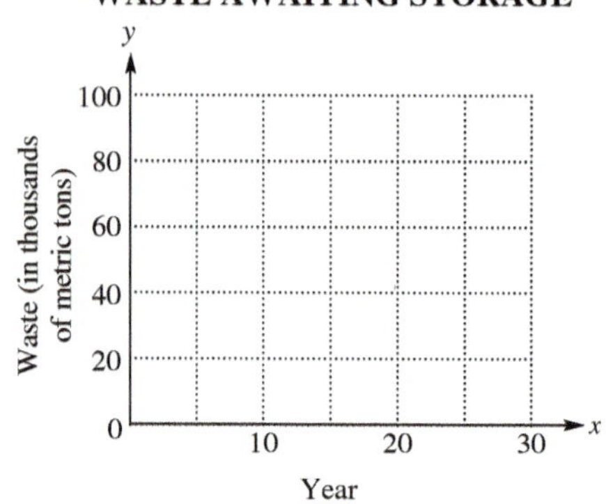

(c) Use the ordered pairs $(0, 32)$ and $(25, 76)$ to write the equation of a line that approximates the data. Give the final equation in slope-intercept form.

(d) Use the equation from part (c) to estimate the amount of nuclear waste in 2015. (*Hint:* What is the value of x for 2015?)

The points on the graph indicate years of life expected at birth y in the United States for selected years x. The graph of a linear equation that models the data is also shown. Here $x = 0$ represents 1930, $x = 10$ represents 1940, and so on.

39. Use the ordered pairs shown on the graph to write an equation of the line that models the data. Give the final equation in slope-intercept form.

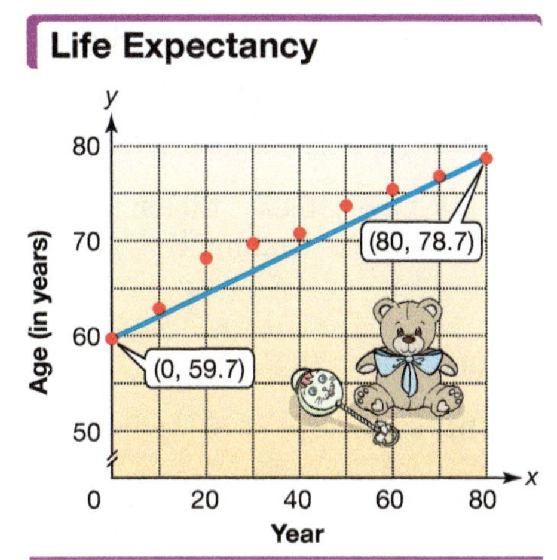

Data from National Center for Health Statistics.

40. Use the equation from **Exercise 39** to do the following.

(a) Find years of life expected at birth in 2000. (*Hint*: What is the value of x for 2000?) Round the answer to the nearest tenth.

(b) How does the answer in part (a) compare to the actual value of 76.8 yr?

(c) Project years of life expected at birth in 2020. (*Hint*: What is the value of x for 2020?) Round the answer to the nearest tenth. Does the answer seem reasonable?

Relating Concepts (Exercises 41–48) For Individual or Group Work

If we think of ordered pairs of the form (C, F), then the two most common methods of measuring temperature, Celsius and Fahrenheit, can be related as follows:

$$\textit{When} \quad C = 0, F = 32, \quad \textit{and when} \quad C = 100, F = 212.$$

Work Exercises 41–48 in order.

41. Write two ordered pairs relating these two temperature scales.

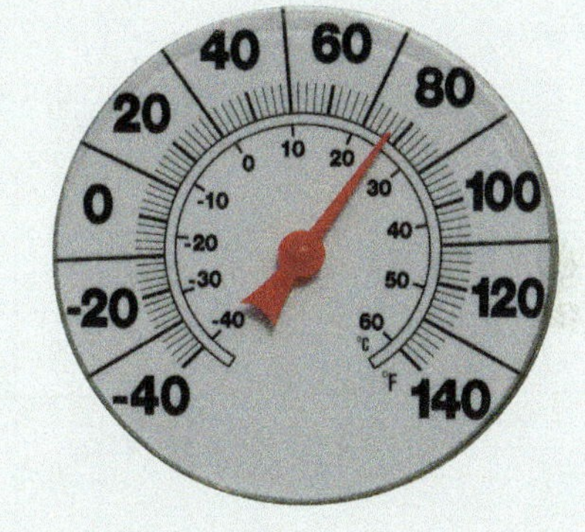

42. Find the slope of the line through the two points.

43. Use the point-slope form to find an equation of the line. (Your variables should be C and F rather than x and y.)

44. Write an equation for F in terms of C.

45. Use the equation from **Exercise 44** to write an equation for C in terms of F.

46. Use the equation from **Exercise 44** to find the Fahrenheit temperature when $C = 30$.

47. Use the equation from **Exercise 45** to find the Celsius temperature when $F = 50$.

48. For what temperature is $F = C$? (Use the thermometer shown above to confirm the answer.)

Summary Exercises *Applying Graphing and Equation-Writing Techniques for Lines*

1. CONCEPT CHECK Match the description of a line in Column I with the correct equation in Column II.

I	II
(a) Slope -0.5, $b = -2$	**A.** $y = -\frac{1}{2}x$
(b) x-intercept $(4, 0)$, y-intercept $(0, 2)$	**B.** $y = -\frac{1}{2}x - 2$
(c) Passes through $(4, -2)$ and $(0, 0)$	**C.** $x - 2y = 2$
(d) $m = \frac{1}{2}$, passes through $(-2, -2)$	**D.** $x + 2y = 4$
	E. $x = 2y$

2. CONCEPT CHECK Which equations are equivalent to $2x + 5y = 20$?

A. $y = -\dfrac{2}{5}x + 4$ **B.** $y - 2 = -\dfrac{2}{5}(x - 5)$ **C.** $y = \dfrac{5}{2}x - 4$ **D.** $2x = 5y - 20$

Graph each line, using the given information or equation.

3. $m = 1$, $b = -2$

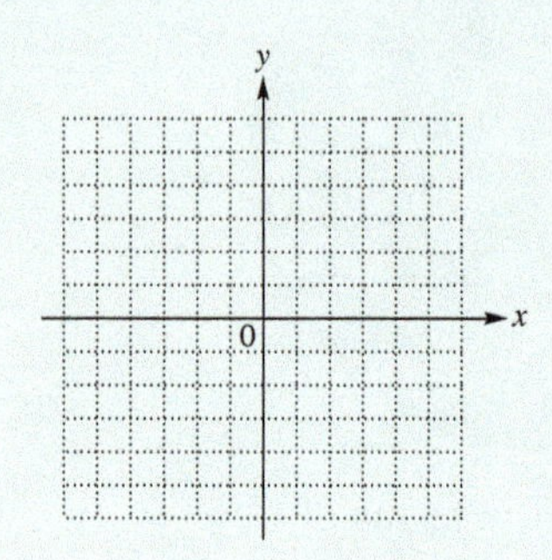

4. $m = 1$, y-intercept $(0, -4)$

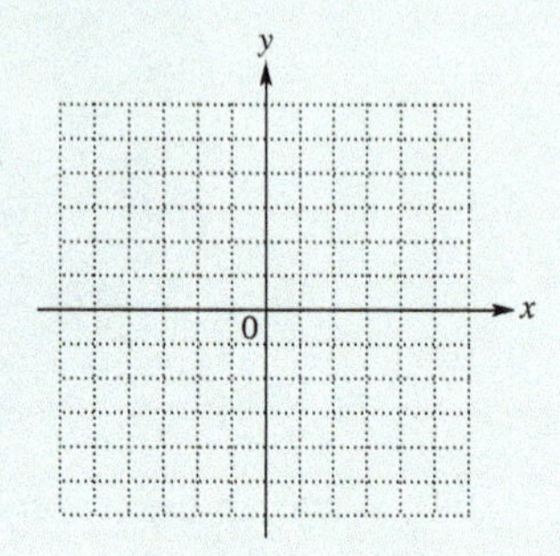

5. $y = -2x + 6$

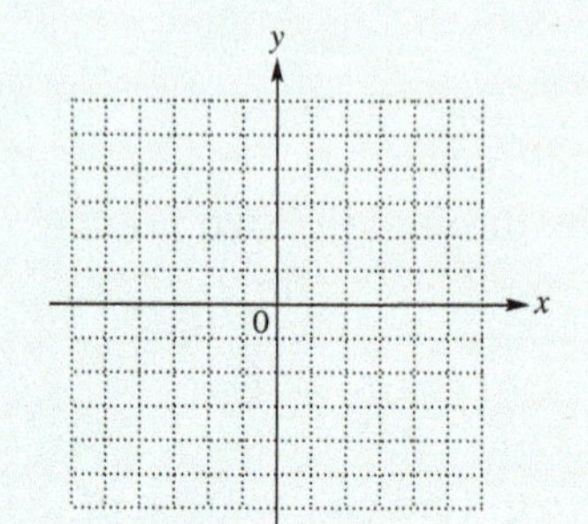

6. $x + 4 = 0$

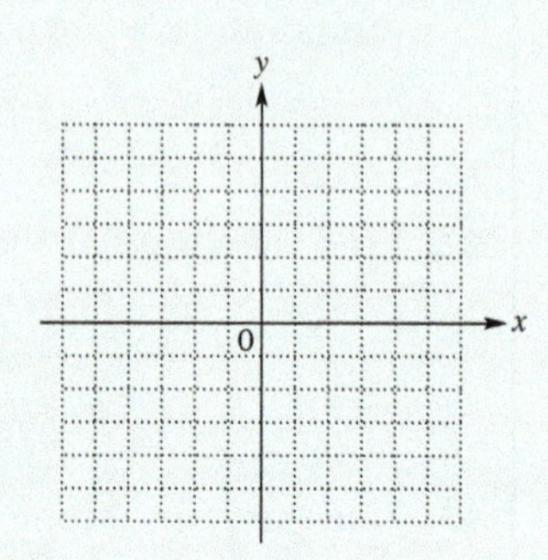

7. $m = -\frac{2}{3}$, passes through $(3, -4)$

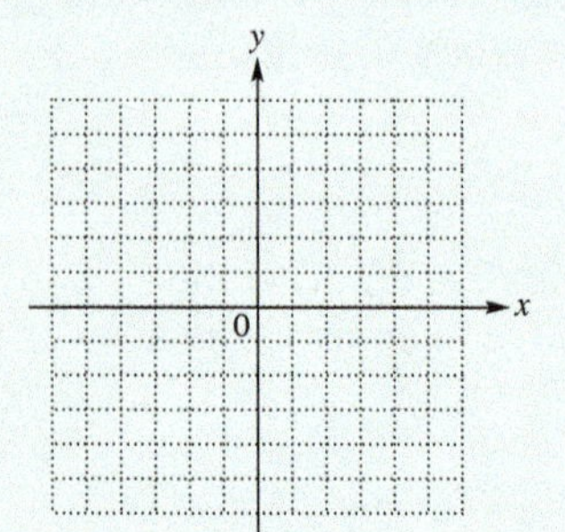

8. $y = -\dfrac{1}{2}x + 2$

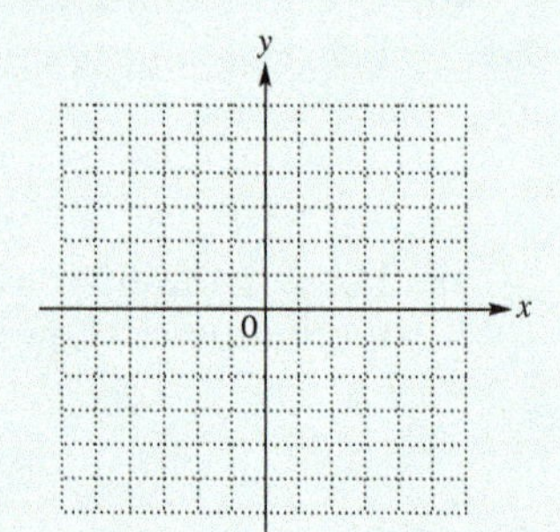

9. $y - 4 = -9$

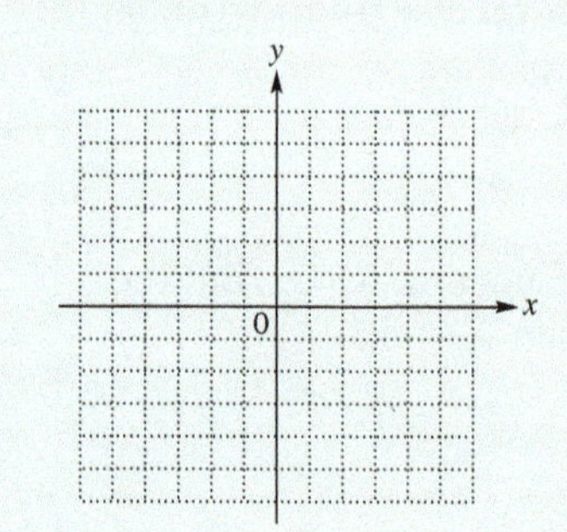

10. $m = -\frac{3}{4}$, passes through $(4, -4)$

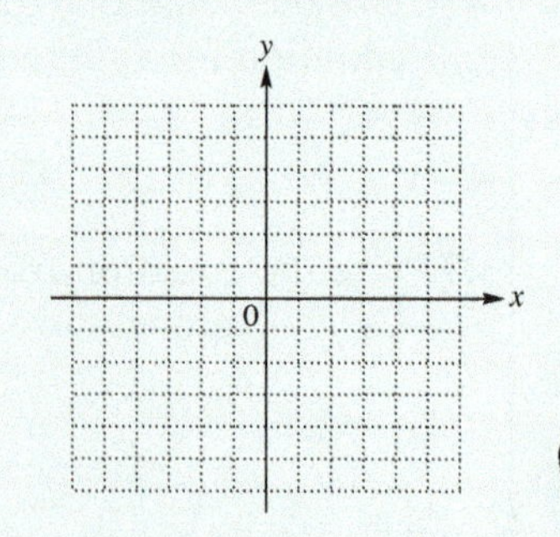

11. Undefined slope, passes through $(3.5, 0)$

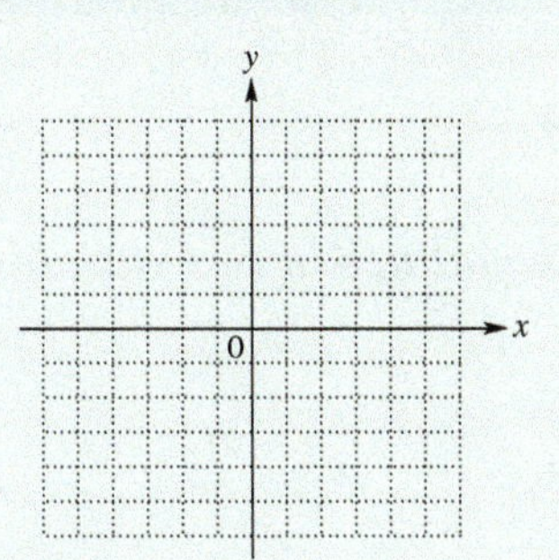

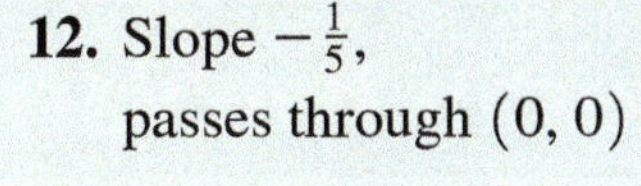
12. Slope $-\frac{1}{5}$, passes through $(0, 0)$

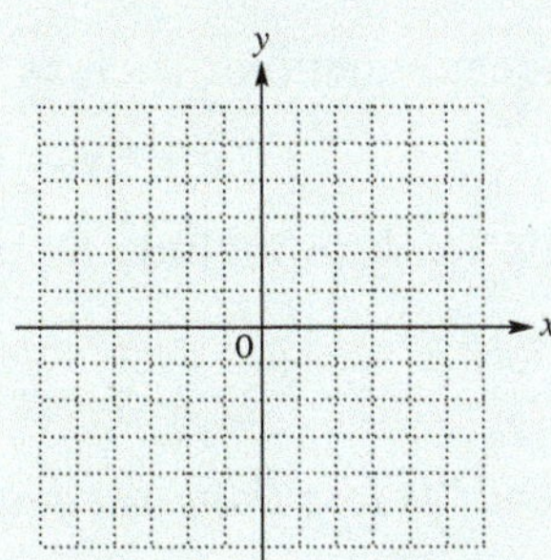

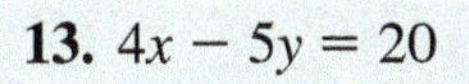
13. $4x - 5y = 20$

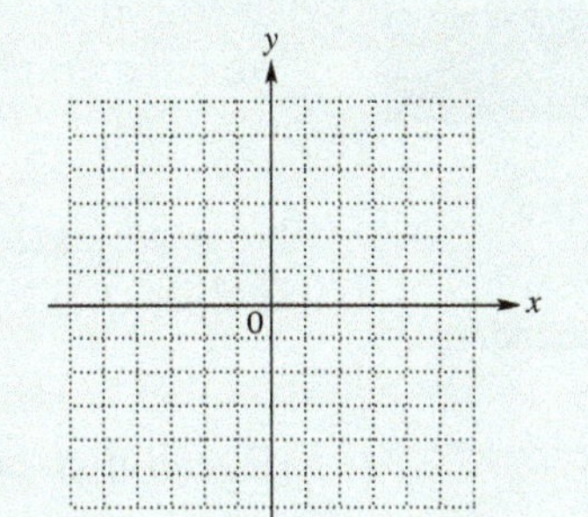

14. $6x - 5y = 30$

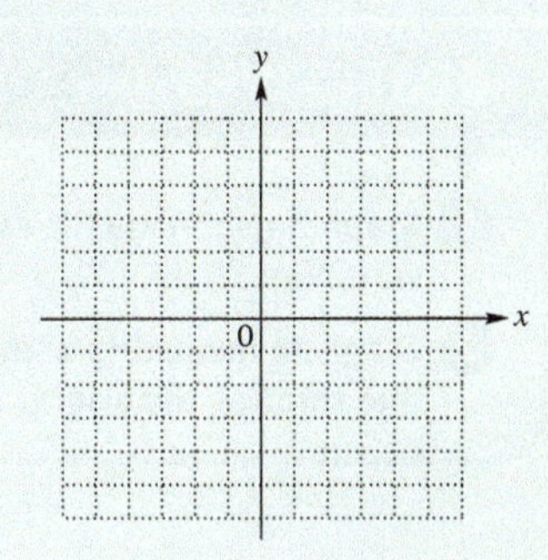

15. $x - 4y = 0$

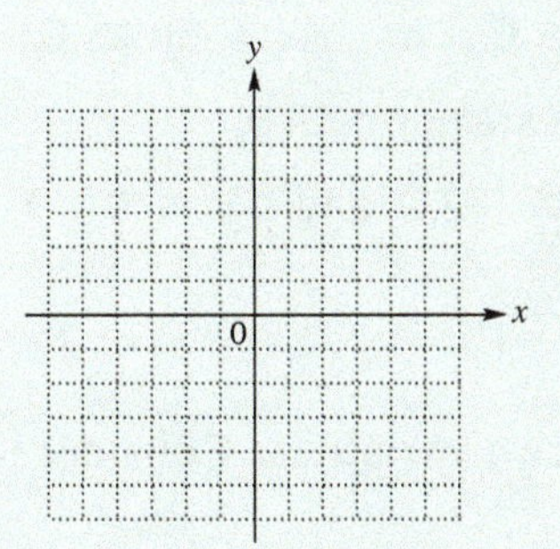

16. $m = 0$, passes through $\left(0, \frac{3}{2}\right)$

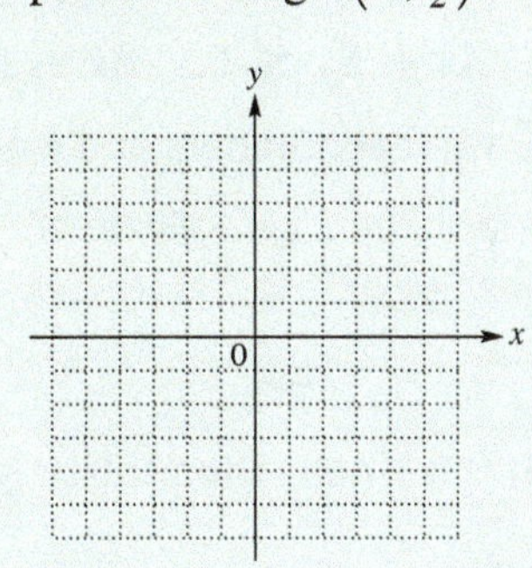

17. $3y = 12 - 2x$

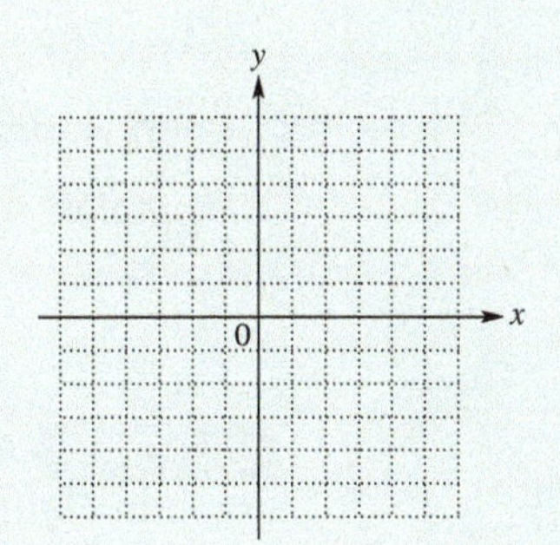

18. $8x = 6y + 24$

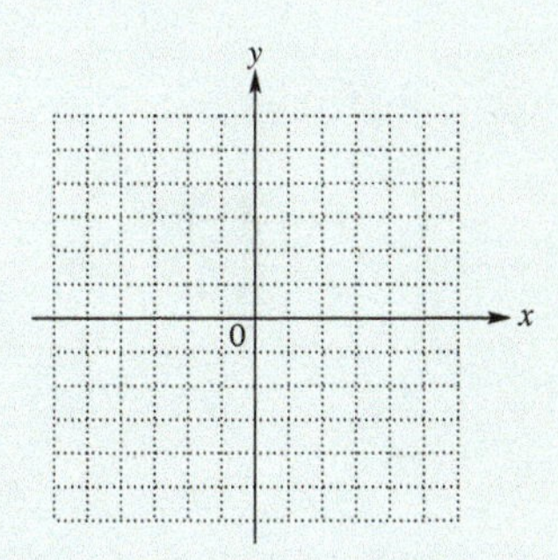

Write an equation in slope-intercept form of each line represented by the table of ordered pairs or the graph.

19.

x	y
3	0
1	4
−1	8

20.

x	y
−6	0
0	8
3	12

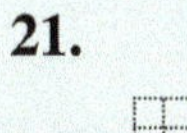
21.

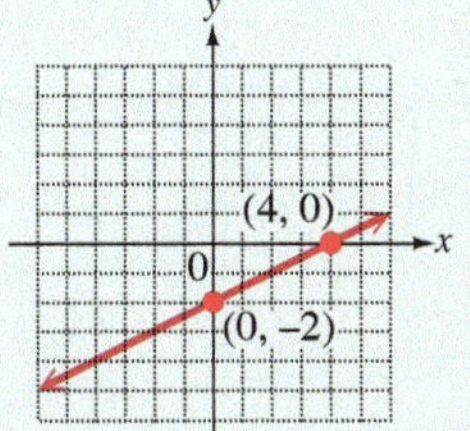

22.

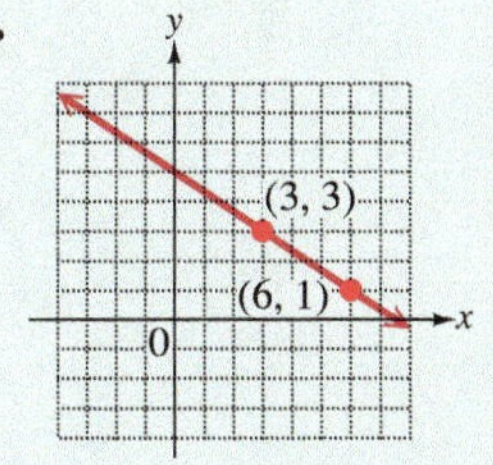

Write an equation of each line. Give the final answer in slope-intercept form if possible.

23. $m = -3$, $b = -6$

24. Passes through $(1, -7)$ and $(-2, 5)$

25. Passes through $(0, 0)$ and $(5, 3)$

26. Passes through $(0, 0)$, undefined slope

27. Passes through $(0, 0)$, $m = 0$

28. $m = \frac{5}{3}$, through $(-3, 0)$

3.6 Graphing Linear Inequalities in Two Variables

OBJECTIVES

1. Graph linear inequalities in two variables.
2. Graph an inequality with a boundary line through the origin.

We have graphed linear equations, such as

$$2x + 3y = 6.$$

Now we extend this work to include *linear inequalities in two variables,* such as

$$2x + 3y \leq 6.$$

(Recall that $\leq$ is read "*is less than or equal to.*")

Linear Inequality in Two Variables

A **linear inequality in two variables** (here x and y) can be written in the form

$$Ax + By < C, \quad Ax + By \leq C, \quad Ax + By > C, \quad \text{or} \quad Ax + By \geq C,$$

where A, B, and C are real numbers and A and B are not both 0.

Examples: $3x - y < 9$, $2x + 5y \geq 0$, $x > -2$, and $y \leq 6$

OBJECTIVE 1 **Graph linear inequalities in two variables.** Consider the graph in **Figure 40.**

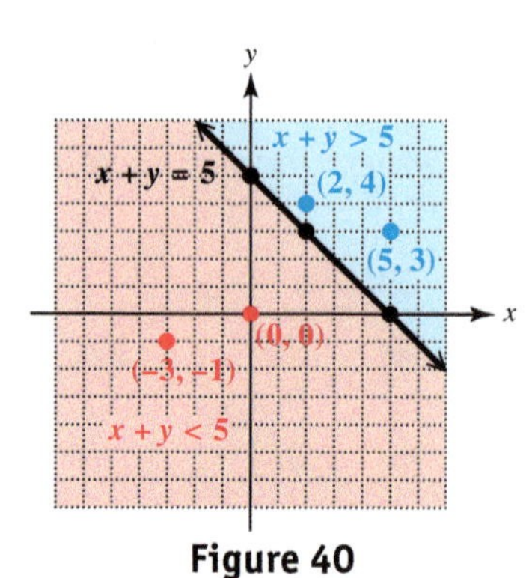

Figure 40

The graph of the line $x + y = 5$ in **Figure 40** divides the points in the rectangular coordinate system into three sets.

1. Those points that lie ***on*** the line itself and satisfy the equation $\boldsymbol{x + y = 5}$ (such as $(0, 5)$, $(2, 3)$, and $(5, 0)$)
2. Those points that lie in the region ***above*** the line and satisfy the inequality $x + y > 5$ (such as $(5, 3)$ and $(2, 4)$)
3. Those points that lie in the region ***below*** the line and satisfy the inequality $x + y < 5$ (such as $(0, 0)$ and $(-3, -1)$)

The graph of the line $\boldsymbol{x + y = 5}$ is the **boundary line** for the inequalities

$$x + y > 5 \quad \text{and} \quad x + y < 5.$$

Graphs of linear inequalities in two variables are regions in the real number plane that may or may not include boundary lines.

EXAMPLE 1 Graphing a Linear Inequality

Graph $2x + 3y \leq 6$.

The inequality $2x + 3y \leq 6$ means that

$$2x + 3y < 6 \quad \text{or} \quad 2x + 3y = 6.$$

We begin by graphing the equation $2x + 3y = 6$, a line with intercepts $(0, 2)$ and $(3, 0)$ as shown in **Figure 41.** This boundary line divides the plane into two regions, one of which satisfies the inequality. To find the correct region, we choose a test point *not* on the boundary line and substitute it into the inequality to see whether the resulting statement is true or false. The point $(0, 0)$ is a convenient choice.

$$2x + 3y < 6 \qquad \text{We are testing the region.}$$

$$2(0) + 3(0) \overset{?}{<} 6 \qquad \text{Let } x = 0 \text{ and } y = 0.$$

$$0 + 0 \overset{?}{<} 6 \qquad \text{Multiply.}$$

$$0 < 6 \qquad \text{True}$$

Use $(0, 0)$ as a test point.

Because a true statement results, we shade the region that includes the test point $(0, 0)$. See **Figure 41.** The shaded region, along with the boundary line because $\leq$ includes equality, is the desired graph.

CHECK To confirm that the correct region is shaded, we select a test point in the region that is ***not*** shaded. We arbitrarily choose $(2, 5)$ and substitute it into the inequality.

$$2x + 3y < 6 \qquad \text{Test the region.}$$

$$2(2) + 3(5) \overset{?}{<} 6 \qquad \text{Let } x = 2 \text{ and } y = 5.$$

$$4 + 15 \overset{?}{<} 6 \qquad \text{Multiply.}$$

$$19 < 6 \qquad \text{False}$$

A false statement results, confirming that we shaded the correct region.

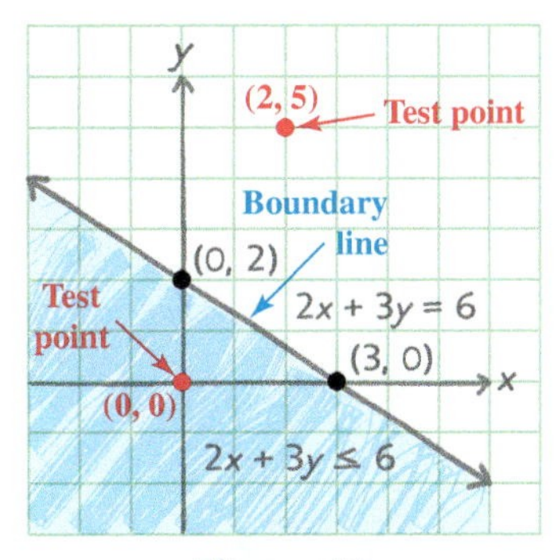

Figure 41

Work Problems 1 and 2 at the Side. ▶

Note

Alternatively in **Example 1,** we can find the required region by solving the given inequality for *y*.

$$2x + 3y \leq 6 \qquad \text{Inequality from Example 1}$$

$$3y \leq -2x + 6 \qquad \text{Subtract } 2x.$$

$$y \leq -\frac{2}{3}x + 2 \qquad \text{Divide each term by 3.}$$

Ordered pairs in which *y* is equal to $-\frac{2}{3}x + 2$ are on the boundary line, so pairs in which *y is less than* $-\frac{2}{3}x + 2$ will be *below* that line. (As we move *down* vertically, the *y*-values *decrease*.) This gives the same region that we shaded in **Figure 41.** (Ordered pairs in which *y is greater than* $-\frac{2}{3}x + 2$ will be *above* the boundary line.)

To solve for *y* in the inequality above, we divided each term by the positive number 3 in the last step. ***Remember to reverse the direction of the inequality symbol when multiplying or dividing an inequality by a negative number.***

1 Use $(0, 0)$ as a test point, and shade the appropriate region for the linear inequality.

$$3x + 4y \leq 12$$

$$3(\_\_\_) + 4(\_\_\_) \overset{?}{\leq} 12$$

$$\_\_\_ \overset{?}{\leq} 12 \quad (\textit{True / False})$$

Shade the region of the graph that (*includes / does not include*) the test point $(0, 0)$.

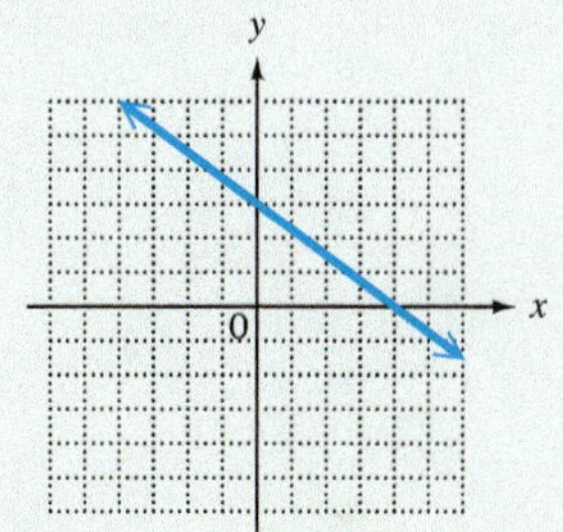

2 Graph $4x - 5y \leq 20$.

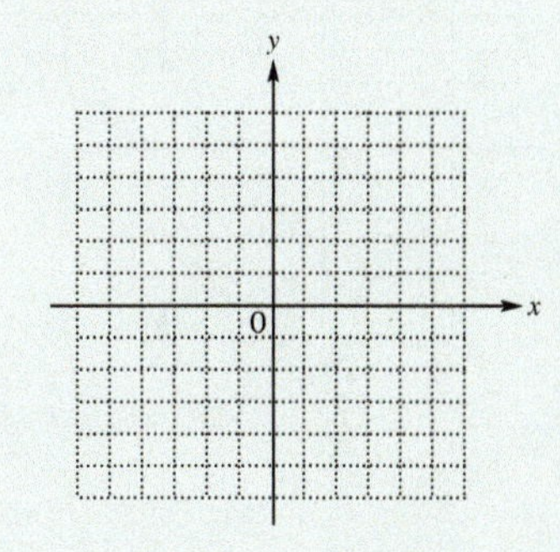

Answers

1. 0; 0; 0; True; includes

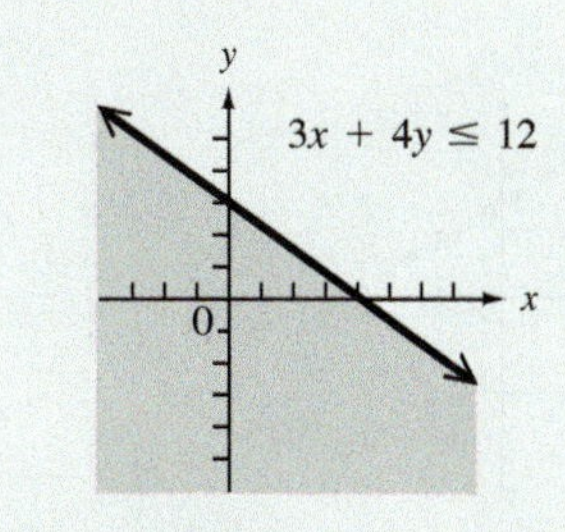

2.

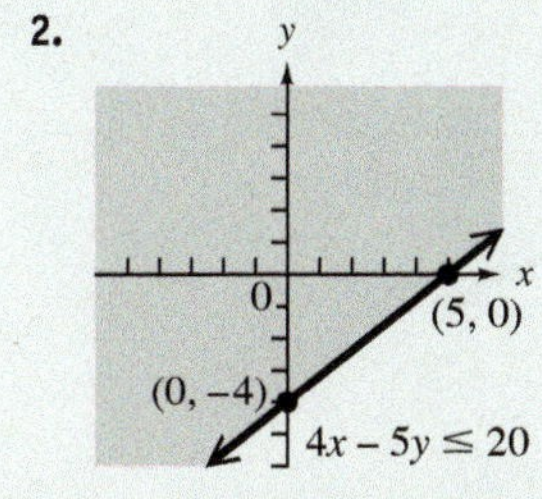

3 Use $(0, 0)$ as a test point and shade the appropriate region for the linear inequality.

$$3x + 5y > 15$$

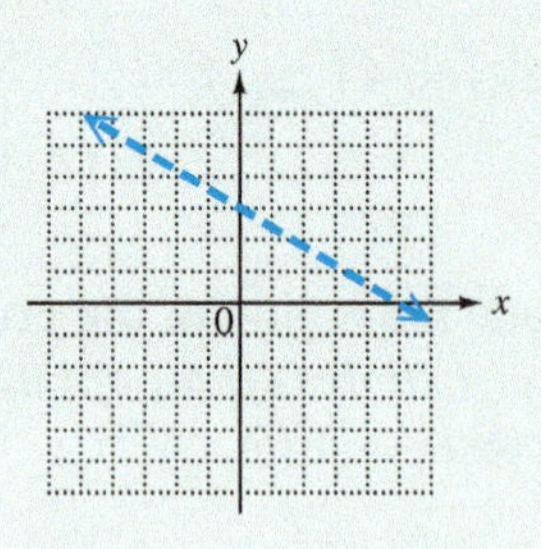

4 Graph $x + 2y > 6$.

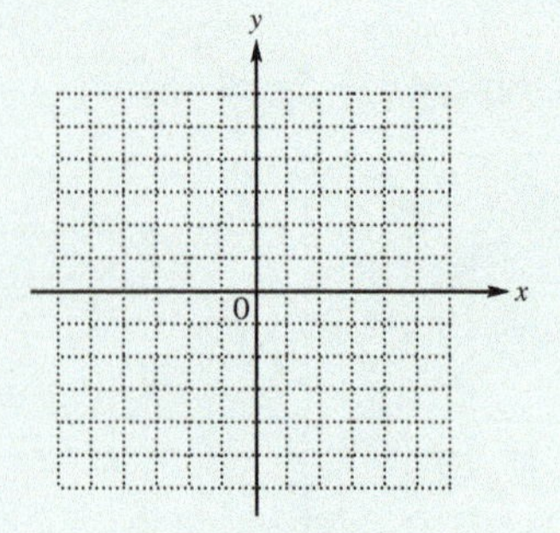

EXAMPLE 2 Graphing a Linear Inequality

Graph $x - y > 5$.

This inequality does *not* involve equality. Therefore, the points on the line $x - y = 5$ do ***not*** belong to the graph. However, the line still serves as a boundary for two regions, one of which satisfies the inequality.

To graph the inequality, first graph the equation $x - y = 5$ using the intercepts $(5, 0)$ and $(0, -5)$. Use a *dashed line* to show that the points on the line are *not* solutions of the inequality $x - y > 5$. See **Figure 42.** Then choose a test point to see which region satisfies the inequality.

$$x - y > 5$$

(0, 0) is a convenient test point.

$$0 - 0 \overset{?}{>} 5 \quad \text{Let } x = 0 \text{ and } y = 0.$$

$$0 > 5 \quad \text{False}$$

Because $0 > 5$ is false, the graph of the inequality is the region that *does not* include $(0, 0)$. Shade the *other* region, as shown in **Figure 42.**

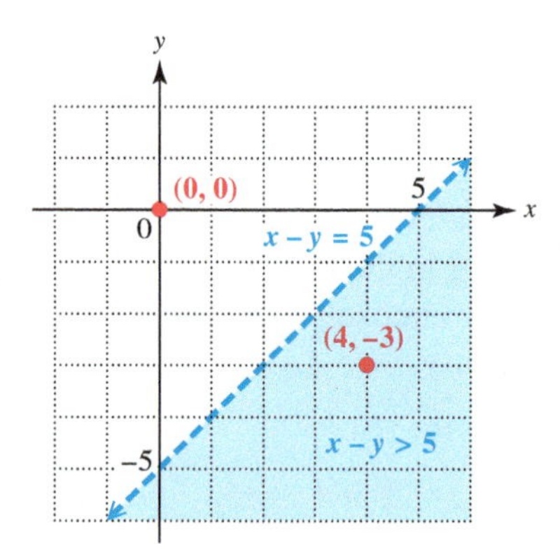

Figure 42

CHECK To confirm that the correct region is shaded, we select a test point in the shaded region. We arbitrarily choose $(4, -3)$.

$$x - y > 5$$

$$4 - (-3) \overset{?}{>} 5 \quad \text{Let } x = 4 \text{ and } y = -3.$$

Use parentheses to avoid errors.

$$7 > 5 \checkmark \quad \text{True}$$

This true statement confirms that the correct region is shaded in **Figure 42.**

◀ **Work Problems 3 and 4 at the Side.**

Graphing a Linear Inequality in Two Variables

Step 1 **Draw the graph of the straight line that is the boundary.**

- Make the line solid if the inequality involves $\leq$ or $\geq$.
- Make the line dashed if the inequality involves $<$ or $>$.

Step 2 **Choose a test point.** Choose any point not on the line, and substitute the coordinates of that point in the inequality.

Step 3 **Shade the appropriate region.** Shade the region that includes the test point if it satisfies the original inequality. Otherwise, shade the region on the other side of the boundary line.

Answers

3.

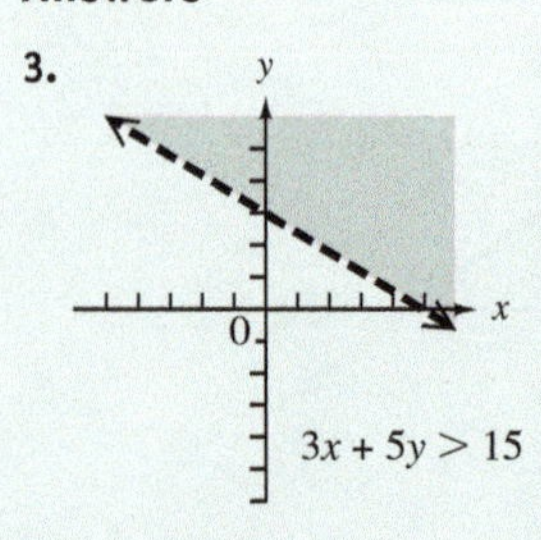

4.

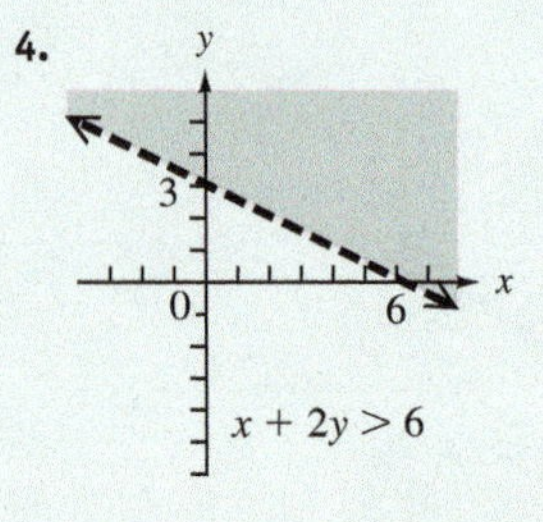

EXAMPLE 3 **Graphing a Linear Inequality**

Graph $x < 3$.

First, we graph $x = 3$, a vertical line passing through the point $(3, 0)$. We use a dashed line because $<$ does not include equality and choose $(0, 0)$ as a test point.

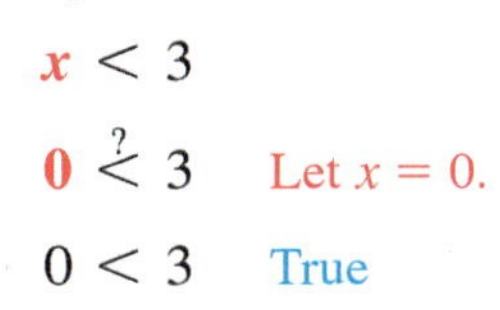

$x < 3$

$0 \overset{?}{<} 3$ Let $x = 0$.

$0 < 3$ True

Because $0 < 3$ is true, we shade the region containing $(0, 0)$, as in **Figure 43.** Intuitively this makes sense—all values of x along the x-axis in the shaded region are indeed less than 3.

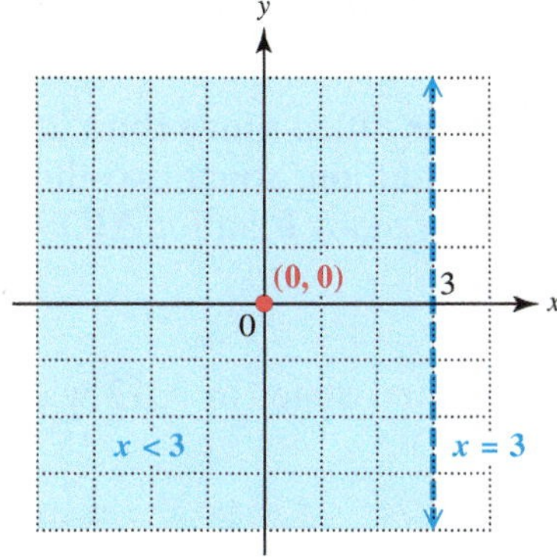

Figure 43

Work Problem 5 at the Side. ▶

OBJECTIVE 2 **Graph an inequality with a boundary line through the origin.** ***If the graph of an inequality has a boundary line through the origin, $(0, 0)$ cannot be used as a test point.***

EXAMPLE 4 **Graphing a Linear Inequality**

Graph $x \leq 2y$.

We graph $x = 2y$ using a solid line by determining several ordered pairs that satisfy the equation.

$x = 2y$	$x = 2y$	$x = 2y$
$0 = 2y$ Let $x = 0$.	$6 = 2y$ Let $x = 6$.	$x = 2(2)$ Let $y = 2$.
$0 = y$	$3 = y$	$x = 4$
Ordered pair: $(0, 0)$	Ordered pair: $(6, 3)$	Ordered pair: $(4, 2)$

The line through these three ordered pairs is shown in **Figure 44.** Because the point $(0, 0)$ is *on* the line $x = 2y$, it cannot be used as a test point. Instead, we choose a test point *off* the line, such as $(1, 3)$.

$x < 2y$ Test the region.

$1 \overset{?}{<} 2(3)$ Let $x = 1$ and $y = 3$.

$1 < 6$ True

A true statement results, so we shade the region containing the test point $(1, 3)$. See **Figure 44.**

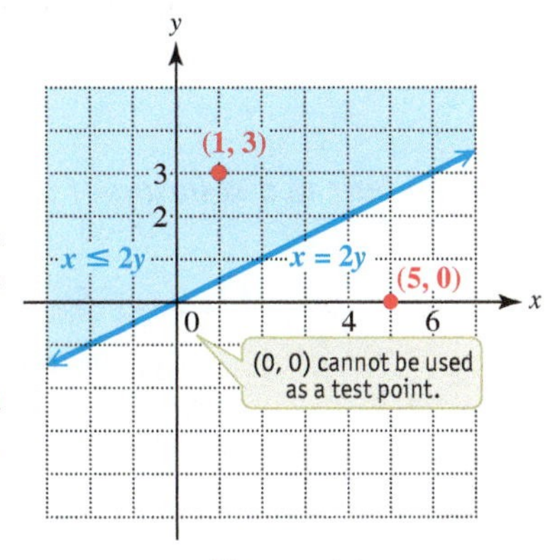

Figure 44

CHECK To confirm that the correct region is shaded, choose a test point, such as $(5, 0)$, in the *other* region.

$x < 2y$ Test the region.

$5 \overset{?}{<} 2(0)$ Let $x = 5$ and $y = 0$.

$5 < 0$ False

A false statement results, confirming that we shaded the correct region.

Work Problem 6 at the Side. ▶

5 Graph $y < 4$.

The graph of this equation has a (*horizontal*/*vertical*) boundary line.

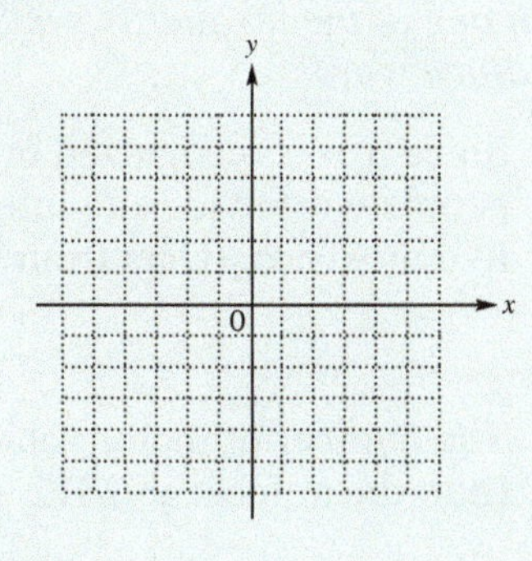

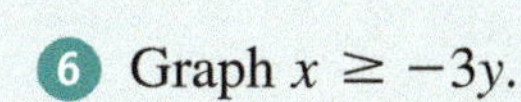

6 Graph $x \geq -3y$.

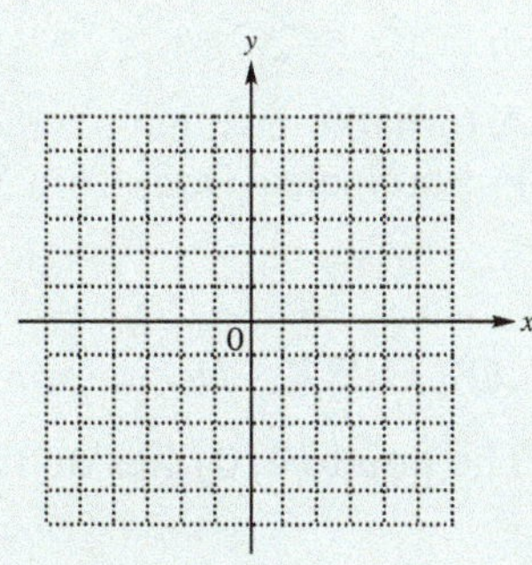

Can $(0, 0)$ be used as a test point here? (*Yes* / *No*)

Answers

5. horizontal

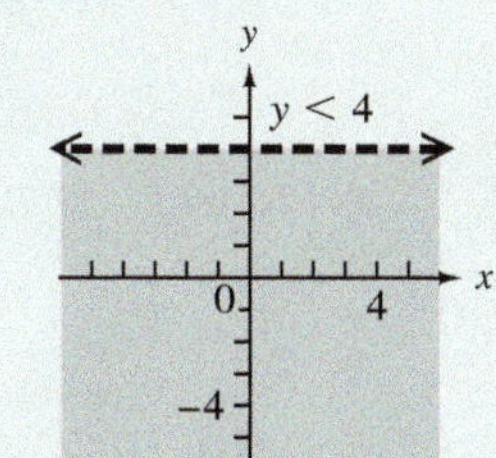

6.

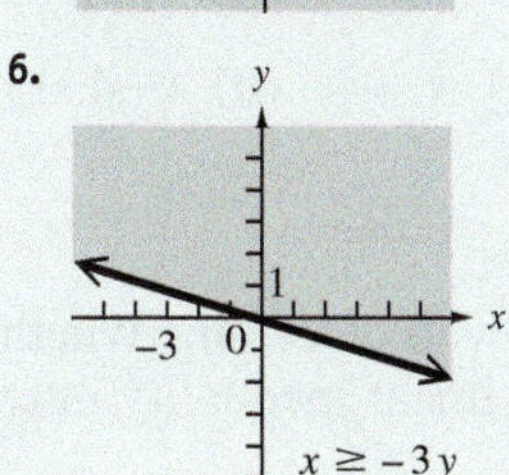

No

3.6 Exercises

FOR EXTRA HELP

Go to MyMathLab for worked-out, step-by-step solutions to exercises enclosed in a square and video solutions to ▶ exercises.

CONCEPT CHECK *Each statement includes one or more phrases that can be symbolized with one of the inequality symbols $<$, $\leq$, $>$, or $\geq$. Give the inequality symbol for the boldface words.*

1. Since it was recognized in 1981, HIV/AIDS has killed **more than** 25 million people worldwide and infected **more than** 60 million, about two-thirds of whom live in Africa. (Data from The President's Emergency Plan for AIDS Relief.)

2. The number of motor vehicle deaths in the United States in 2013 was 1063 **less than** the number in 2012. (Data from NHTSA.)

3. As of January 2016, American Airlines passengers were allowed one carry-on bag, with dimensions (length + width + height) of **at most** 45 in. (Data from www.aa.com)

4. A tornado must have winds of **at least** 65 mph to be rated using the Enhanced Fujita Scale. (Data from National Weather Service.)

CONCEPT CHECK *Decide whether each statement is* true *or* false. *If false, explain why.*

5. The point $(4, 0)$ lies on the graph of $3x - 4y < 12$.

6. The point $(4, 0)$ lies on the graph of $3x - 4y \leq 12$.

7. The point $(0, 0)$ can be used as a test point to determine which region to shade when graphing the linear inequality $x + 4y > 0$.

8. When graphing the linear inequality $3x + 2y \geq 12$, use a dashed line for the boundary line.

9. The points $(4, 1)$ and $(0, 0)$ lie on the graph of $3x - 2y \geq 0$.

10. The graph of $y > x$ does not contain points in quadrant IV.

CONCEPT CHECK *Decide whether the given ordered pair is a solution of the given inequality.*

11. $x - 4y \geq 8$

(a) $(0, 0)$ **(b)** $(0, 2)$ **(c)** $(4, -1)$ **(d)** $(-4, 1)$

12. $2x + 5y < 10$

(a) $(0, 0)$ **(b)** $(5, 0)$ **(c)** $(-5, 2)$ **(d)** $(-2, -3)$

13. Explain how to determine whether to use a dashed line or a solid line when graphing a linear inequality in two variables.

14. Explain why the point $(0, 0)$ is not an appropriate choice for a test point when graphing an inequality whose boundary passes through the origin.

For each inequality, the straight-line boundary has been drawn. Complete each graph by shading the correct region. **See Examples 1–4.**

15. $x + y \geq 4$

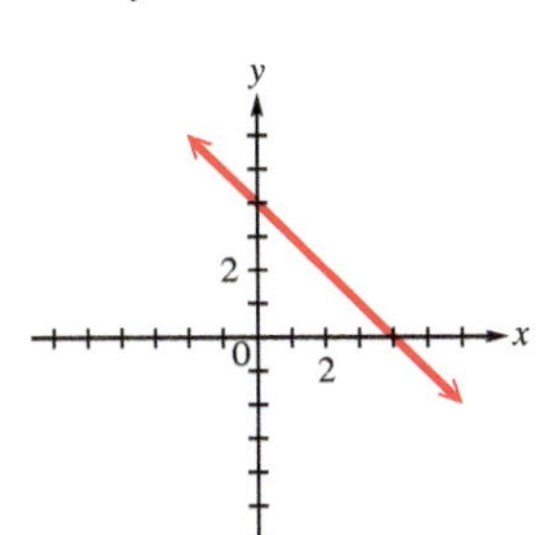

16. $x + y \leq 2$

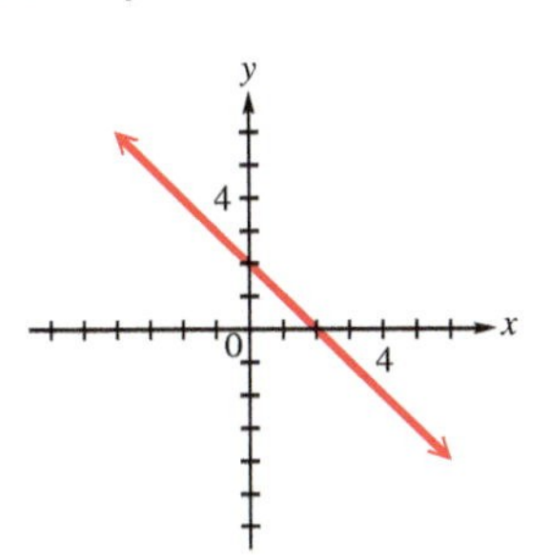

17. $x + 2y \geq 7$

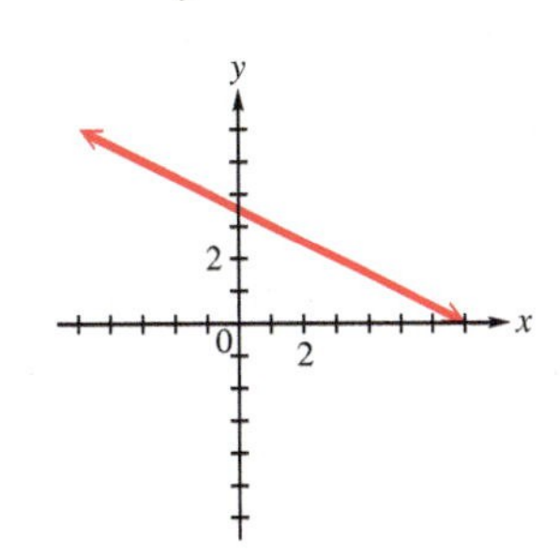

18. $2x + y \geq 5$

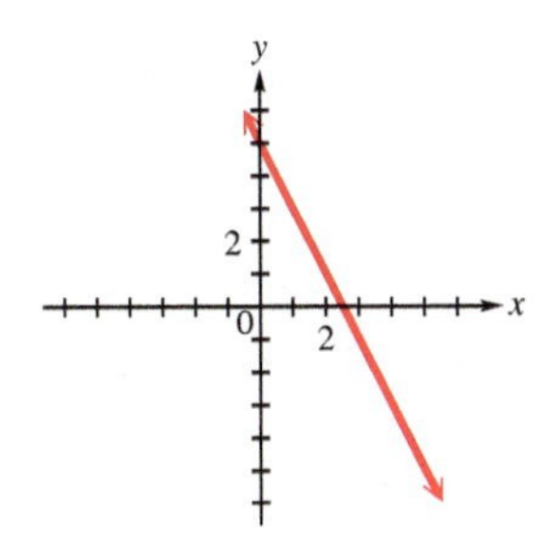

19. $-3x + 4y > 12$

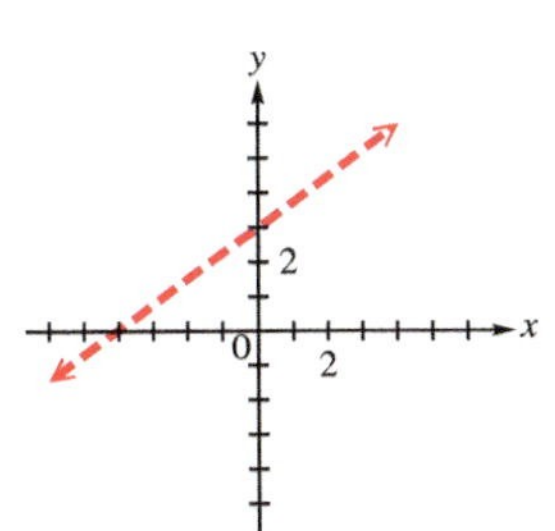

20. $4x - 5y < 20$

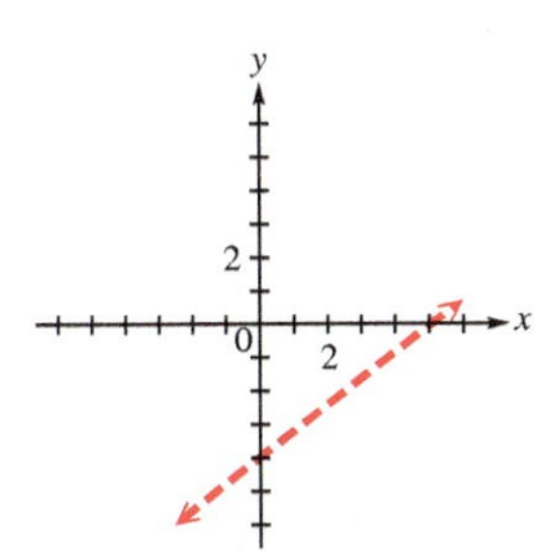

21. $x > 4$

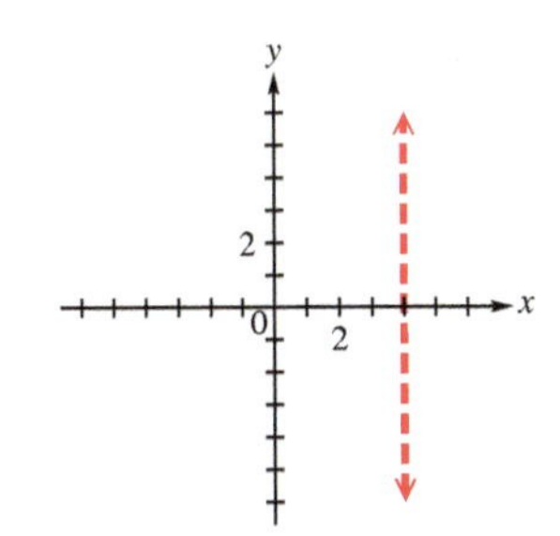

22. $x \leq 0$

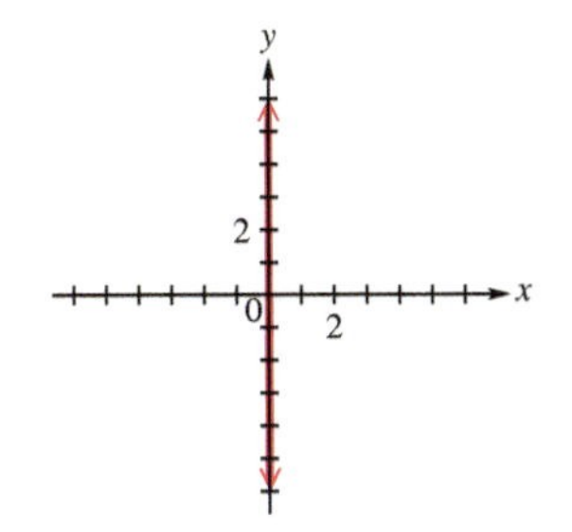

23. $y < 0$

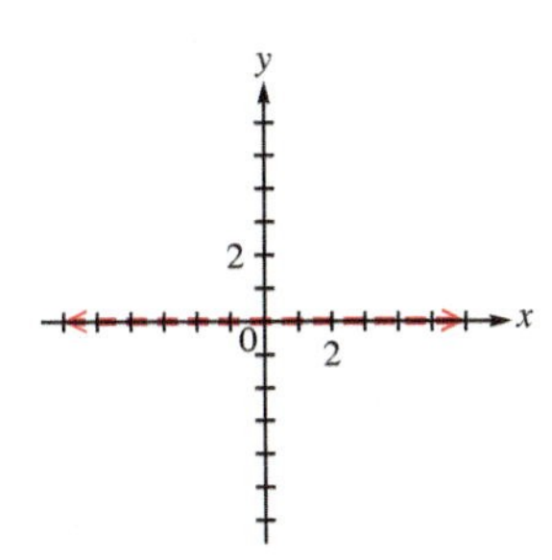

24. $y < -1$

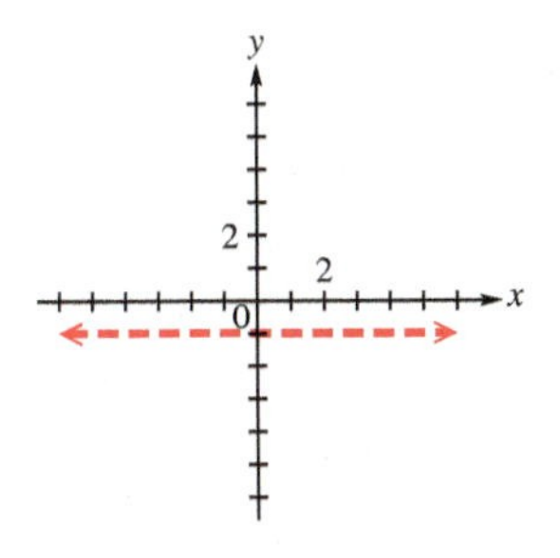

25. $x \geq -y$

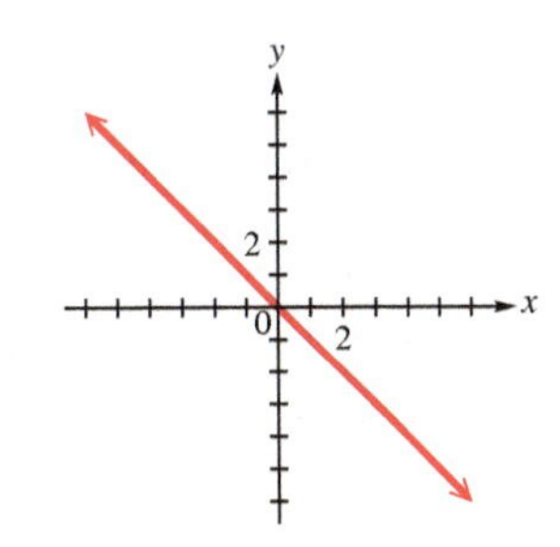

26. $x > 3y$

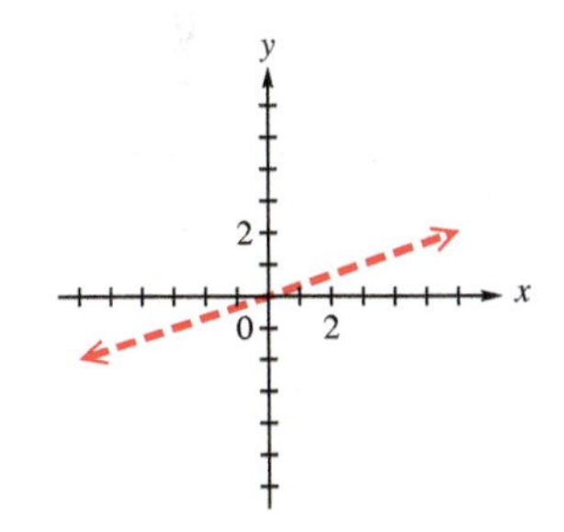

Graph each linear inequality. **See Examples 1–4.**

27. $x + y \leq 5$

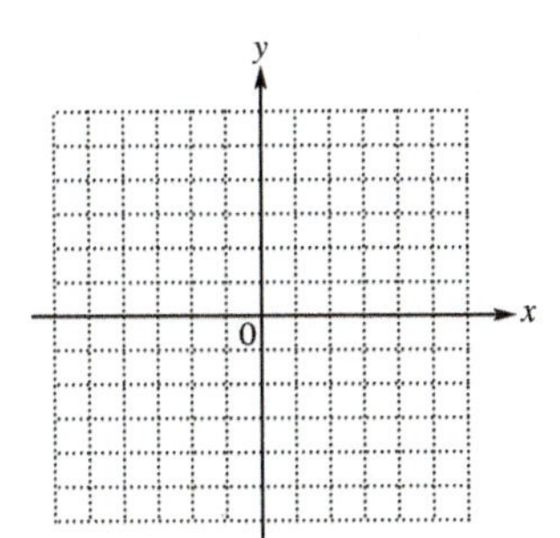

28. $x + y \geq 3$

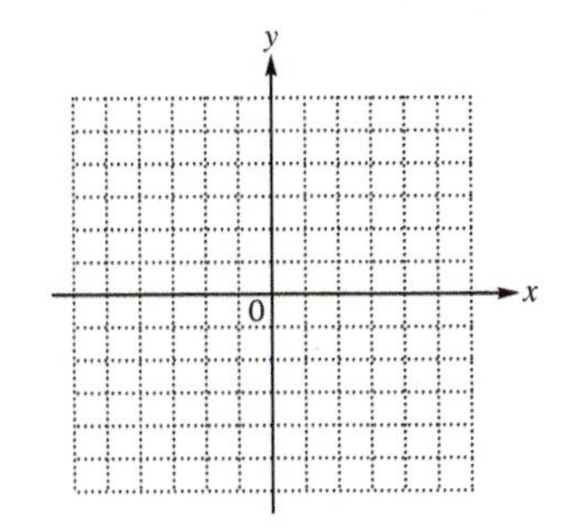

29. $x + 2y < 4$

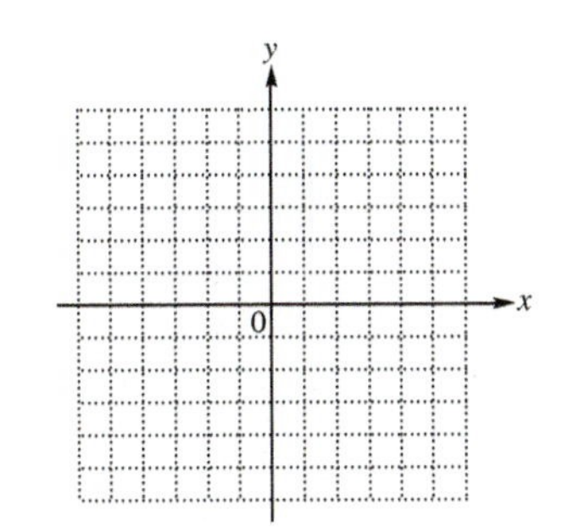

30. $x + 3y > 6$

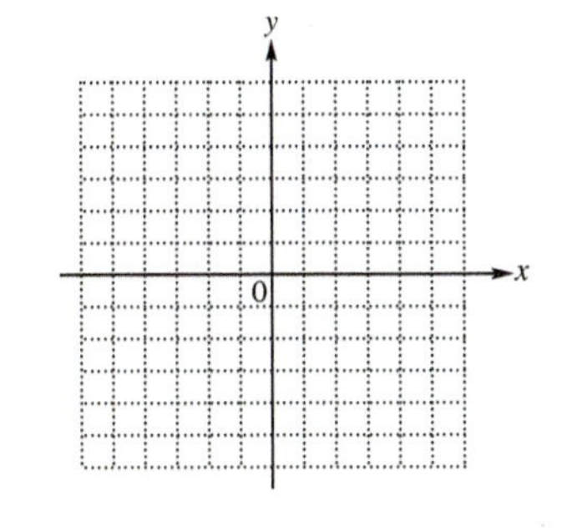

31. $2x + 3y > -6$

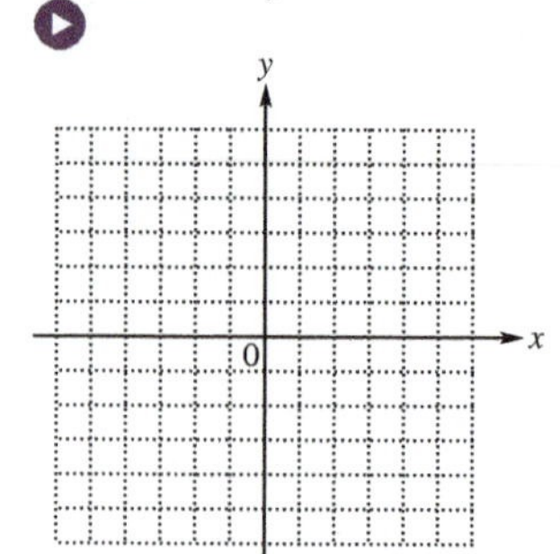

32. $3x + 4y < 12$

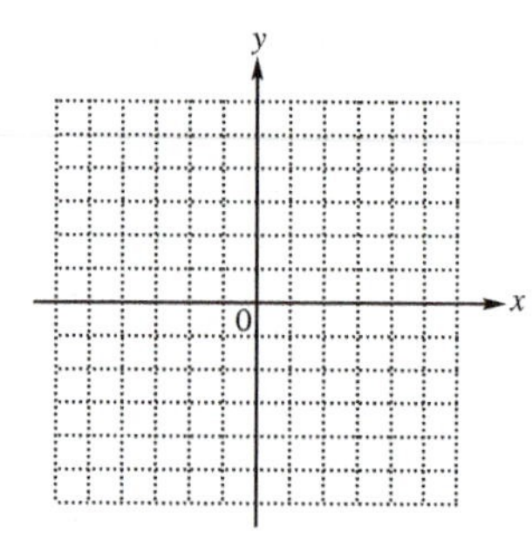

33. $y \geq 2x + 1$

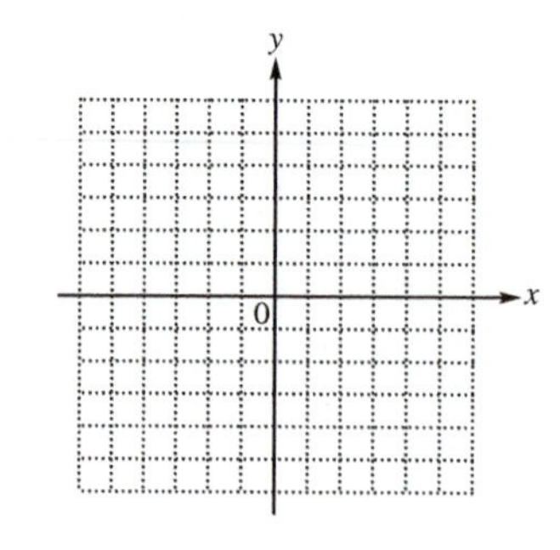

34. $y < -3x + 1$

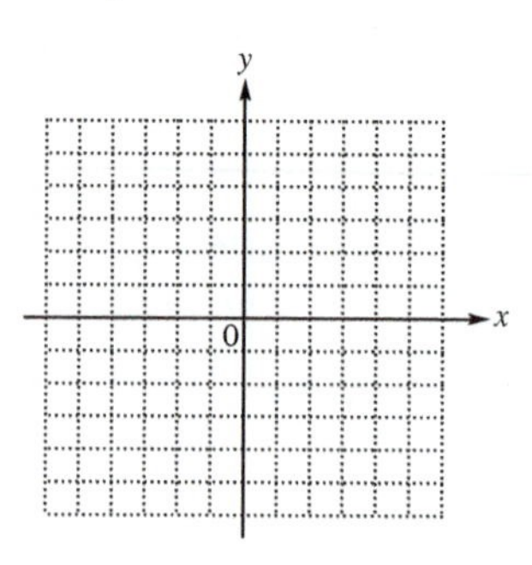

35. $x \leq -2$

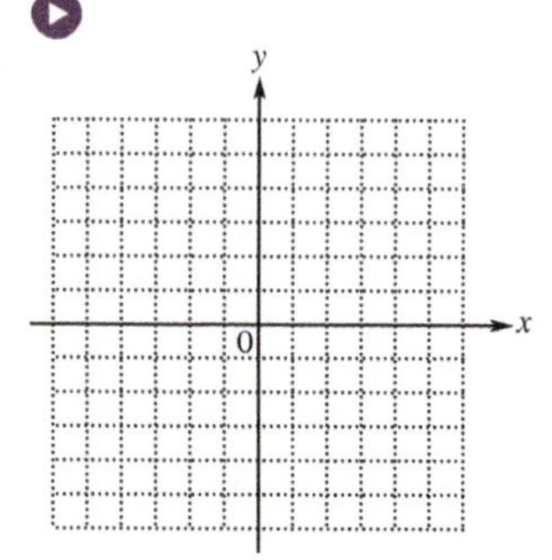

36. $x \geq 1$

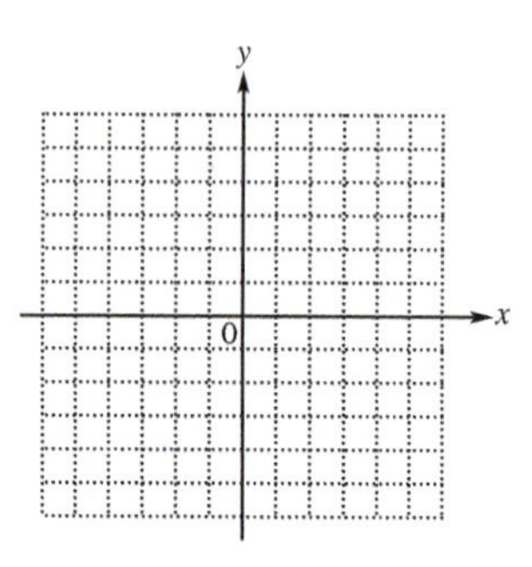

37. $y < 5$

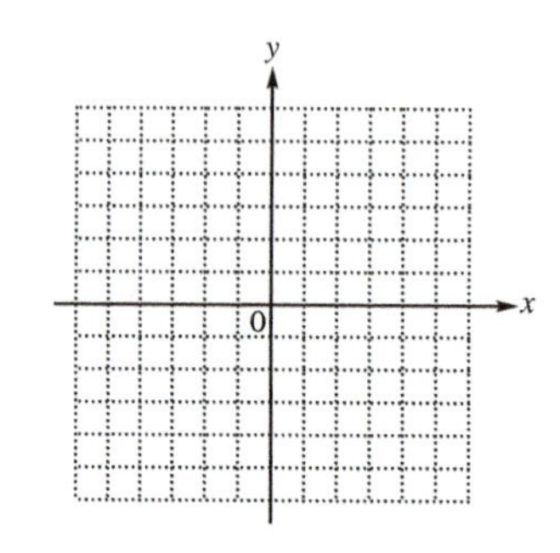

38. $y < -3$

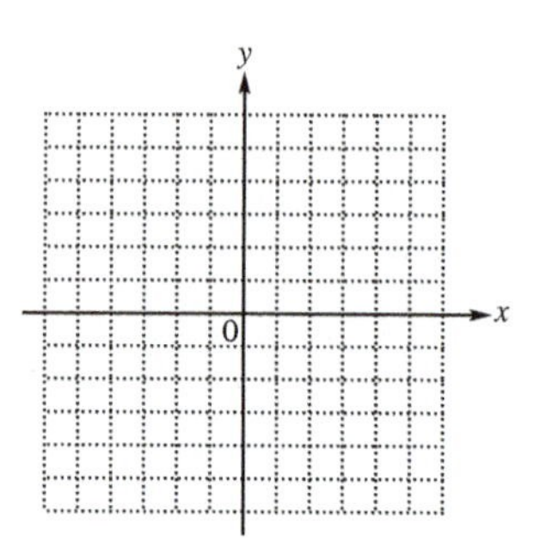

39. $x \geq 0$

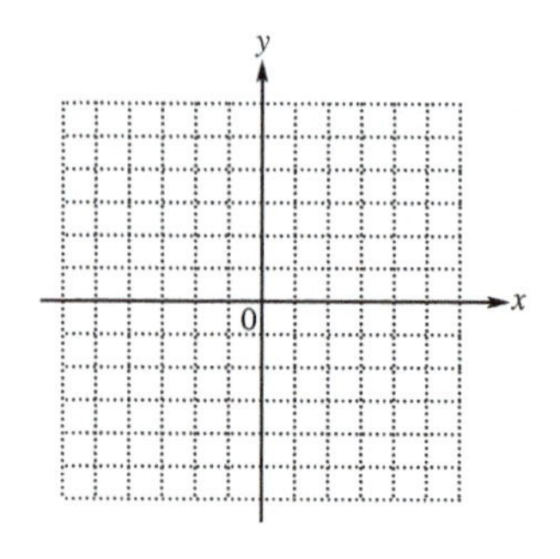

40. $y \geq 0$

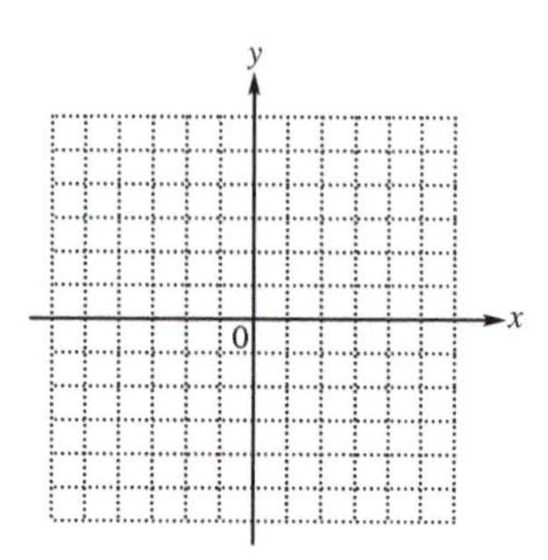

41. $y \geq 4x$

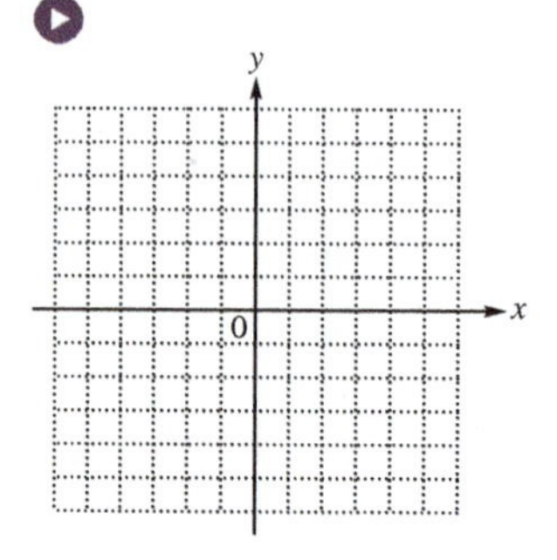

42. $y \leq 2x$

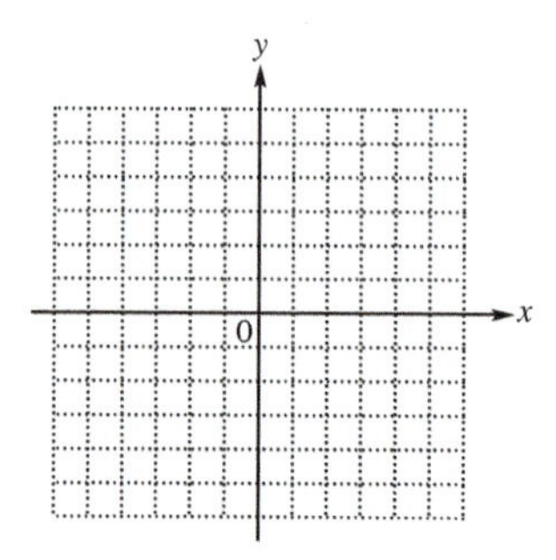

43. $x < -2y$

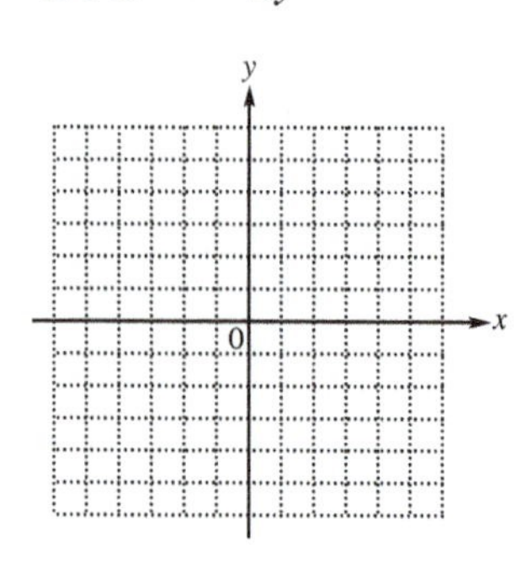

44. $x > -5y$

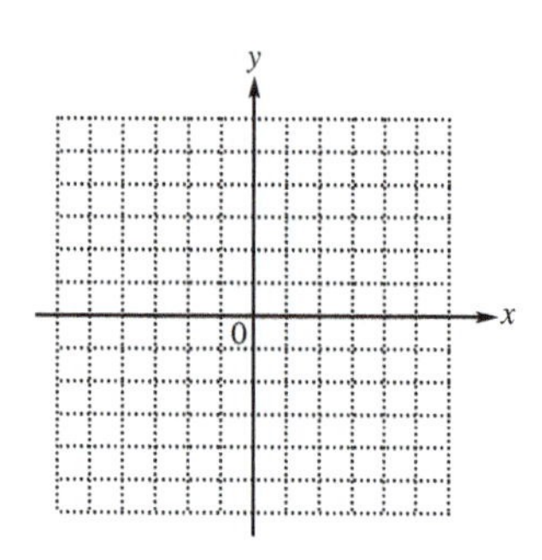

45. $x + y > 0$

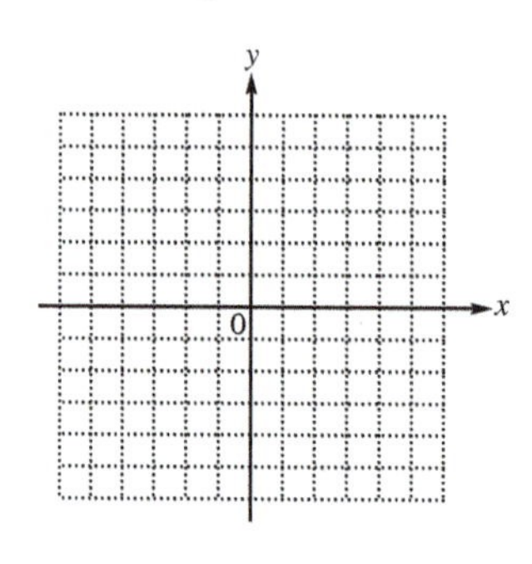

46. $x - 3y < 0$

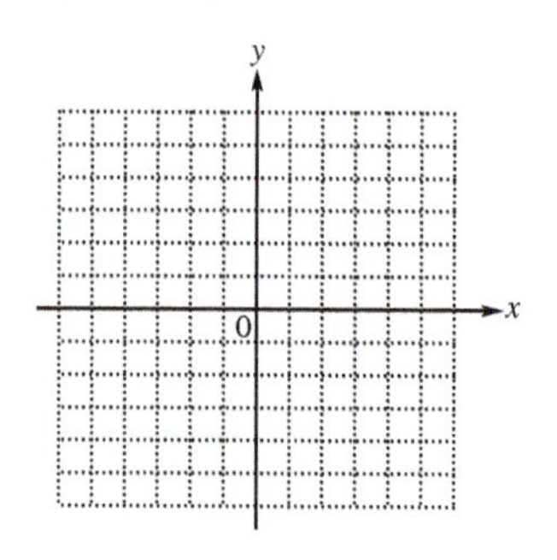

Chapter 3 Summary

Key Terms

3.1

line graph A line graph consists of a series of points that are connected with line segments and is used to show changes or trends in data over time.

linear equation in two variables An equation that can be written in the form $Ax + By = C$ is a linear equation in two variables (here x and y). A, B, and C are real numbers; A and B are not both 0.

ordered pair A pair of numbers written between parentheses in which order is important is an ordered pair.

x	y
0	4 ← Ordered pair (0, 4)
2	0
1	2

Table of values for $2x + y = 4$

table of values A table showing selected ordered pairs of numbers that satisfy an equation is a table of values.

***x*-axis** The horizontal axis in a coordinate system is the x-axis.

***y*-axis** The vertical axis in a coordinate system is the y-axis.

rectangular (Cartesian) coordinate system An x-axis and y-axis drawn at right angles form a coordinate system.

origin The point at which the x-axis and y-axis intersect is the origin.

quadrants A coordinate system divides the plane into four regions, or quadrants.

Rectangular coordinate system

plane A flat surface determined by two intersecting lines is a plane.

coordinates The numbers in an ordered pair are the coordinates of the corresponding point.

plot To plot an ordered pair is to find the corresponding point on a coordinate system.

scatter diagram A graph of ordered pairs of data is a scatter diagram.

3.2

graph The graph of an equation is the set of all points that correspond to the ordered pairs that satisfy the equation.

graphing The process of plotting the ordered pairs that satisfy an equation and drawing a line (or curve) through them is called graphing.

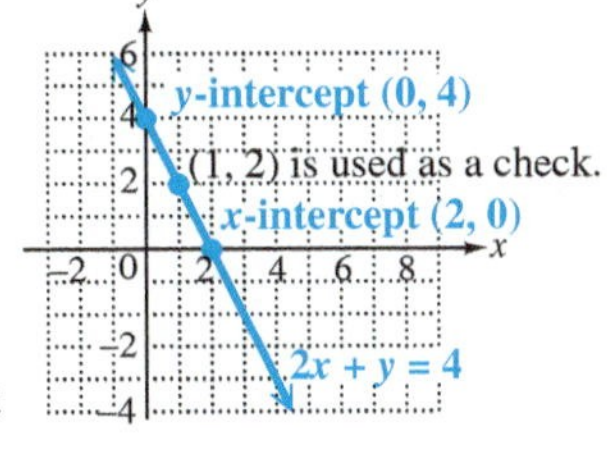

Graph of $2x + y = 4$

***y*-intercept** If a graph intersects the y-axis at b, then the y-intercept is $(0, b)$.

***x*-intercept** If a graph intersects the x-axis at a, then the x-intercept is $(a, 0)$.

3.3

rise Rise is the vertical change between two different points on a line.

run Run is the horizontal change between two different points on a line.

slope The slope of a line is the ratio of the change in y compared to the change in x when moving along the line from one point to another.

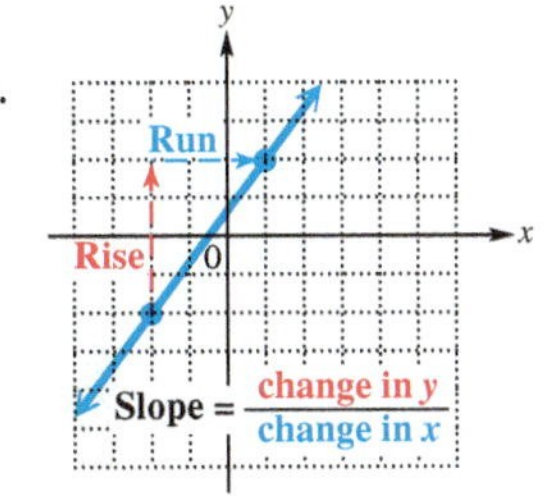

This line has slope $\frac{4}{3}$.

parallel lines Two lines in a plane that never intersect are parallel.

perpendicular lines Perpendicular lines intersect at a 90° angle.

3.6

linear inequality in two variables An inequality that can be written in the form $Ax + By < C$, $Ax + By \leq C$, $Ax + By > C$, or $Ax + By \geq C$ is a linear inequality in two variables (here x and y). A, B, and C are real numbers; A and B are not both 0.

boundary line In the graph of a linear inequality, the boundary line separates the region that satisfies the inequality from the region that does not satisfy the inequality.

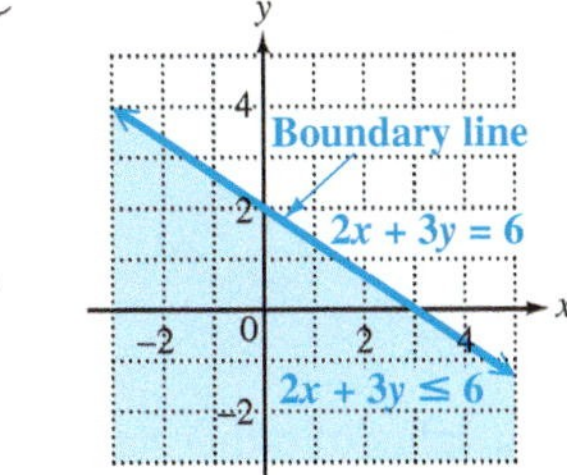

New Symbols

(x, y) ordered pair

(x_1, y_1) subscript notation (read "*x-sub-one, y-sub-one*")

m slope

Test Your Word Power

See how well you have learned the vocabulary in this chapter.

1 An **ordered pair** is a pair of numbers written

A. in numerical order between brackets
B. between parentheses or brackets
C. between parentheses in which order is important
D. between parentheses in which order does not matter.

2 The **coordinates** of a point are

A. the numbers in the corresponding ordered pair
B. the solution of an equation
C. the values of the x- and y-intercepts
D. the graph of the point.

3 An **intercept** is

A. the point where the x-axis and y-axis intersect
B. a pair of numbers written in parentheses in which order is important
C. one of the four regions determined by a rectangular coordinate system
D. the point where a graph intersects the x-axis or the y-axis.

4 The **slope** of a line is

A. the measure of the run over the rise of the line
B. the distance between two points on the line
C. the ratio of the change in y to the change in x along the line
D. the horizontal change compared to the vertical change of two points on the line.

5 Two lines in a plane are **parallel** if

A. they represent the same line
B. they never intersect
C. they intersect at a 90° angle
D. one has a positive slope and one has a negative slope.

6 Two lines in a plane are **perpendicular** if

A. they represent the same line
B. they never intersect
C. they intersect at a 90° angle
D. one has a positive slope and one has a negative slope.

Answers to Test Your Word Power

1. C; *Examples:* $(0, 3)$, $(3, 8)$, $(4, 0)$
2. A; *Example:* The point associated with the ordered pair $(1, 2)$ has x-coordinate 1 and y-coordinate 2.
3. D; *Example:* The graph of the equation $4x - 3y = 12$ has x-intercept $(3, 0)$ and y-intercept $(0, -4)$.
4. C; *Example:* The line through $(3, 6)$ and $(5, 4)$ has slope $\frac{4-6}{5-3} = \frac{-2}{2} = -1$.
5. B; *Example:* See **Figure A**.
6. C; *Example:* See **Figure B**.

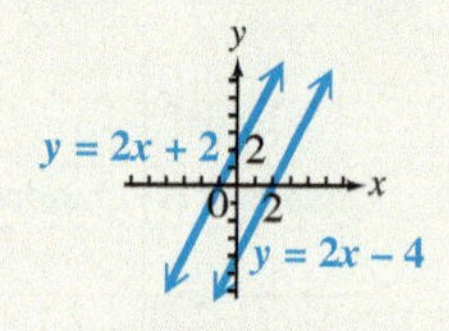

Figure A

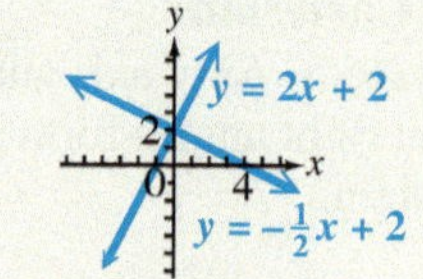

Figure B

Quick Review

Concepts	Examples

3.1 Linear Equations and Rectangular Coordinates

An ordered pair is a solution of an equation if it makes the equation a true statement.

Are $(2, -5)$ and $(0, -6)$ solutions of $4x - 3y = 18$?

$$4(2) - 3(-5) \stackrel{?}{=} 18$$
$$8 + 15 \stackrel{?}{=} 18$$
$$23 = 18 \quad \text{False}$$

$(2, -5)$ is not a solution.

$$4(0) - 3(-6) \stackrel{?}{=} 18$$
$$0 + 18 \stackrel{?}{=} 18$$
$$18 = 18 \checkmark \quad \text{True}$$

$(0, -6)$ is a solution.

If a value of either variable in an equation is given, the value of the other variable can be found by substitution.

Complete the ordered pair $(0, \_\_\_\_)$ for the equation.

$$3x = y + 4$$
$$3(0) = y + 4 \quad \text{Let } x = 0.$$
$$0 = y + 4 \quad \text{Multiply.}$$
$$-4 = y \quad \text{Subtract 4.}$$

The ordered pair is $(0, -4)$.

Concepts	Examples

To plot an ordered pair, begin at the origin.

Step 1 Move right or left the number of units that corresponds to the x-coordinate—right if it is positive or left if it is negative.

Step 2 Then turn and move up or down the number of units that corresponds to the y-coordinate—up if it is positive or down if it is negative.

Plot the ordered pair $(-3, 4)$.

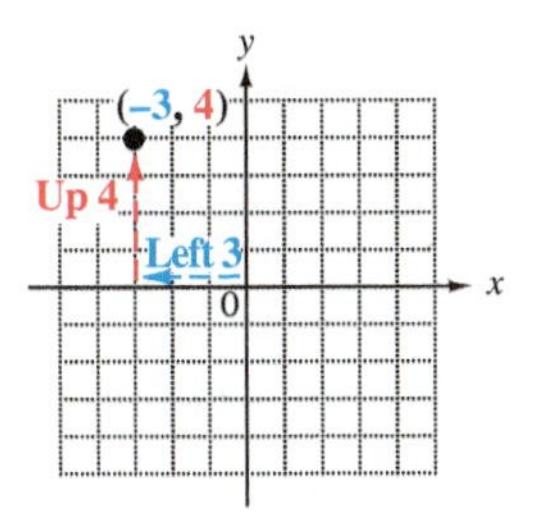

3.2 Graphing Linear Equations in Two Variables

Graphing a Linear Equation in Two Variables

Step 1 Find at least two ordered pairs that are solutions of the equation. (The intercepts are good choices.) It is good practice to find a third ordered pair as a check.

Step 2 Plot the corresponding points.

Step 3 Draw a straight line through the points.

The graph of $\boldsymbol{Ax + By = 0}$ passes through the origin. In this case, find and plot at least one other point that satisfies the equation. Then draw the line through these points.

The graph of $\boldsymbol{y = b}$ is a **horizontal line** through $(0, b)$.

The graph of $\boldsymbol{x = a}$ is a **vertical line** through $(a, 0)$.

Graph $x - 2y = 4$.

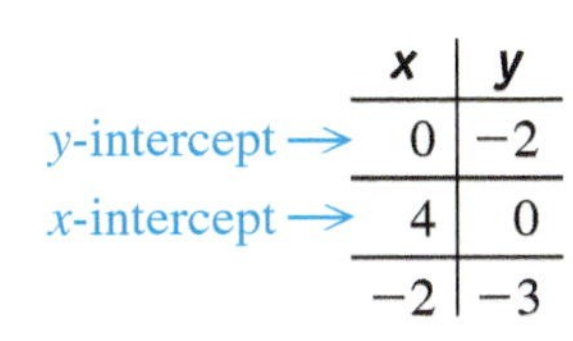

	x	y
y-intercept →	0	−2
x-intercept →	4	0
	−2	−3

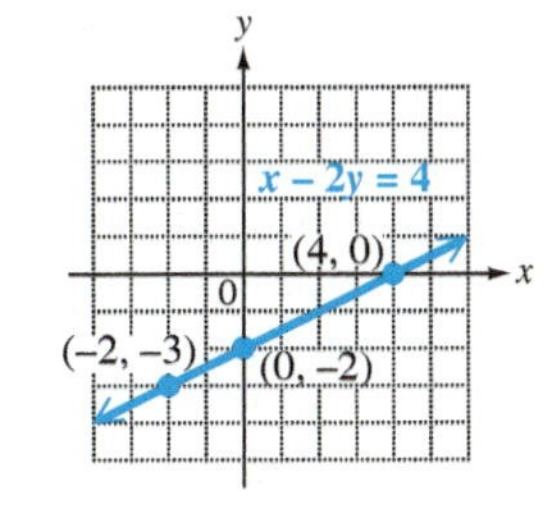

Graph $2x + 3y = 0$.

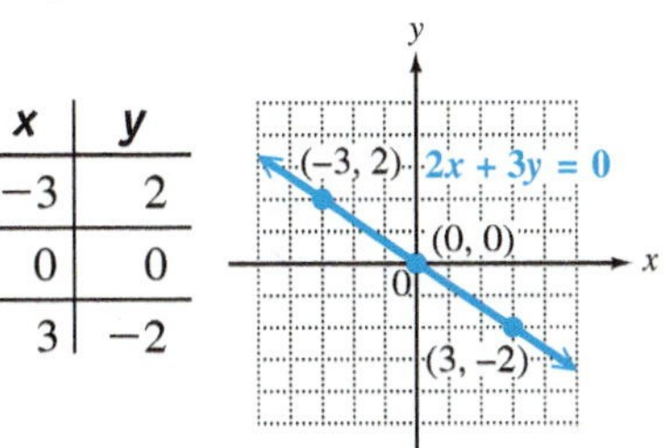

x	y
−3	2
0	0
3	−2

Graph $y = -3$ and $x = -3$.

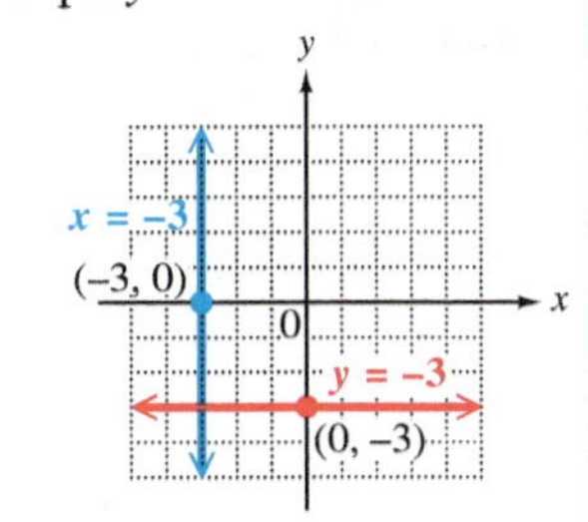

3.3 The Slope of a Line

The slope m of the line passing through the points (x_1, y_1) and (x_2, y_2) is defined as follows.

$$m = \frac{\textbf{change in } y}{\textbf{change in } x} = \frac{y_2 - y_1}{x_2 - x_1} \quad (\textbf{where } x_1 \neq x_2)$$

Horizontal lines have slope 0.

Vertical lines have undefined slope.

Finding the Slope of a Line from Its Equation

Step 1 Solve the equation for y.

Step 2 The slope is given by the coefficient of x.

Parallel lines have the same slope.

Perpendicular lines (neither of which is vertical) have slopes that are negative reciprocals—that is, their product is -1.

Find each slope.

The line passing through $(-2, 3)$ and $(4, -5)$ has slope

$$m = \frac{-5 - 3}{4 - (-2)} = \frac{-8}{6} = -\frac{4}{3}.$$

The line $y = -2$ has slope 0.

The line $x = 4$ has **undefined slope**.

Find the slope of the line with the following equation.

$$3x - 4y = 12$$

$$-4y = -3x + 12 \quad \text{Subtract } 3x.$$

$$y = \frac{3}{4}x - 3 \quad \text{Divide by } -4.$$

↑ Slope (of $\frac{3}{4}$)

The lines $y = 3x - 1$ and $y = 3x + 4$ are parallel because both have slope 3.

The lines $y = -3x - 1$ and $y = \frac{1}{3}x + 4$ are perpendicular because their slopes are -3 and $\frac{1}{3}$, and $-3\left(\frac{1}{3}\right) = -1$.

Concepts	Examples

3.4 Slope-Intercept Form of a Linear Equation

Slope-Intercept Form
$y = mx + b$
m is the slope.
$(0, b)$ is the y-intercept.

Write an equation in slope-intercept form of the line with slope 2 and y-intercept $(0, -5)$.

$y = mx + b$ Slope-intercept form

$y = 2x - 5$

3.5 Point-Slope Form of a Linear Equation and Modeling

Point-Slope Form
$y - y_1 = m(x - x_1)$
m is the slope.
(x_1, y_1) is a point on the line.

Write an equation of the line passing through $(-4, 5)$ with slope $-\frac{1}{2}$.

$y - y_1 = m(x - x_1)$ Point-slope form

$y - 5 = -\frac{1}{2}[x - (-4)]$ Substitute for y_1, m, and x_1.

$y - 5 = -\frac{1}{2}(x + 4)$ Definition of subtraction

$y - 5 = -\frac{1}{2}x - 2$ Distributive property

$y = -\frac{1}{2}x + 3$ Add 5.

Standard Form
$Ax + By = C$
A, B, and C are real numbers and A and B are not both 0.

(In answers, we give A, B, and C as integers with greatest common factor 1 and $A \geq 0$.)

Write the above equation in standard form.

$y = -\frac{1}{2}x + 3$

$2y = -x + 6$ Multiply each term by 2.

$x + 2y = 6$ Add x.

3.6 Graphing Linear Inequalities in Two Variables

Graphing a Linear Inequality in Two Variables

Step 1 Draw the graph of the straight line that is the boundary.

- Make the line solid if the inequality involves $\leq$ or $\geq$.
- Make the line dashed if the inequality involves $<$ or $>$.

Step 2 Choose a test point not on the line, and substitute the coordinates of that point in the inequality.

Step 3 Shade the region that includes the test point if it satisfies the original inequality. Otherwise, shade the region on the other side of the boundary line.

Graph $2x + y \leq 5$.

Graph the boundary line

$$2x + y = 5$$

using the intercepts $(0, 5)$ and $\left(\frac{5}{2}, 0\right)$. Make it solid because the symbol $\leq$ includes equality.

Use $(0, 0)$ as a test point.

$2x + y < 5$

$2(0) + 0 \overset{?}{\leq} 5$

$0 < 5$ True

Shade the region that includes $(0, 0)$.

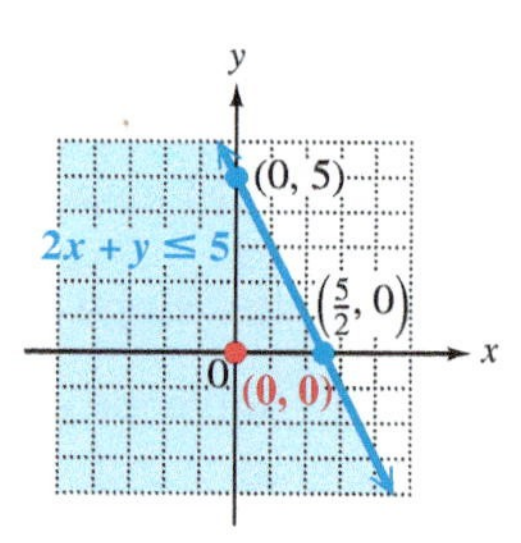

Chapter 3 Review Exercises

3.1 *The line graph shows the percent of 4-year college students at public institutions who earned a degree within 5 years. Use the graph to work each problem.*

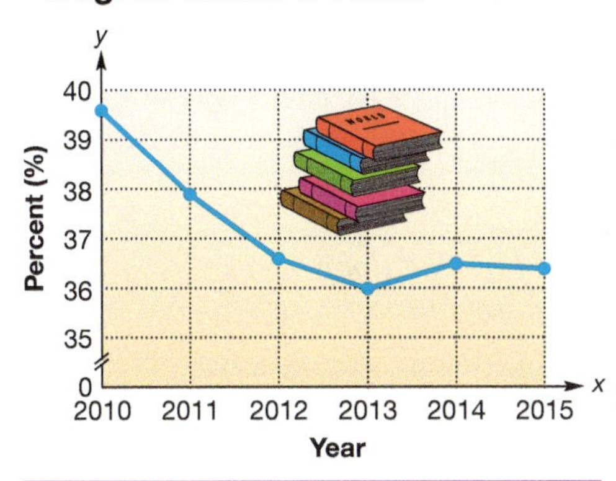

Data from ACT.

1. Between the years 2010 and 2013, what was the general trend in the percent of students who earned a degree within 5 yr?

2. Estimate the percent of students earning a degree within 5 yr in 2010 and 2013.

3. About how much did the percent decrease between 2010 and 2013?

4. What does the ordered pair $(2015, 36.4)$ mean in the context of this problem?

Complete the given ordered pairs for each equation.

5. $y = 3x + 2$; $(-1, \_\_\_)$, $(0, \_\_\_)$, $(\_\_\_, 5)$

6. $4x + 3y = 6$; $(0, \_\_\_)$, $(\_\_\_, 0)$, $(-2, \_\_\_)$

7. $x = 3y$; $(0, \_\_\_)$, $(8, \_\_\_)$, $(\_\_\_, -3)$

8. $x - 7 = 0$; $(\_\_\_, -3)$, $(\_\_\_, 0)$, $(\_\_\_, 5)$

Decide whether each ordered pair is a solution of the given equation.

9. $x + y = 7$; $(2, 5)$

10. $2x + y = 5$; $(-1, 3)$

11. $3x - y = 4$; $\left(\frac{1}{3}, -3\right)$

12. $x = -1$; $(0, -1)$

Identify the quadrant in which each point is located. Then plot the point on the rectangular coordinate system provided.

13. $(2, 3)$

14. $(-4, 2)$

15. $(3, 0)$

16. $(0, -6)$

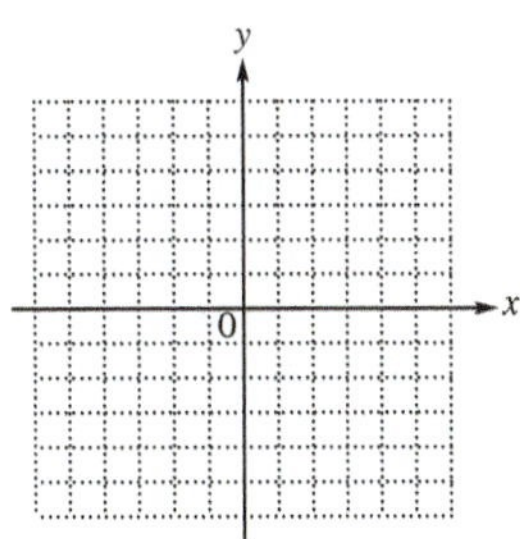

17. If $x > 0$ and $y < 0$, in what quadrant(s) must (x, y) lie? Explain.

18. On what axis does the point $(k, 0)$ lie for any real value of k? The point $(0, k)$? Explain.

3.2 *Graph each linear equation using intercepts.*

19. $2x - y = 3$

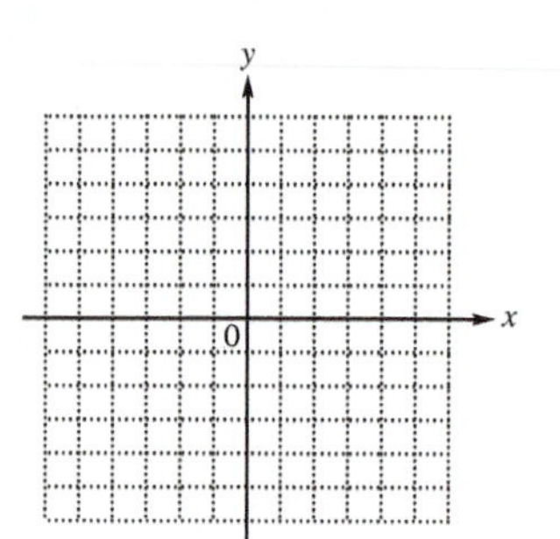

20. $x + 2y = -4$

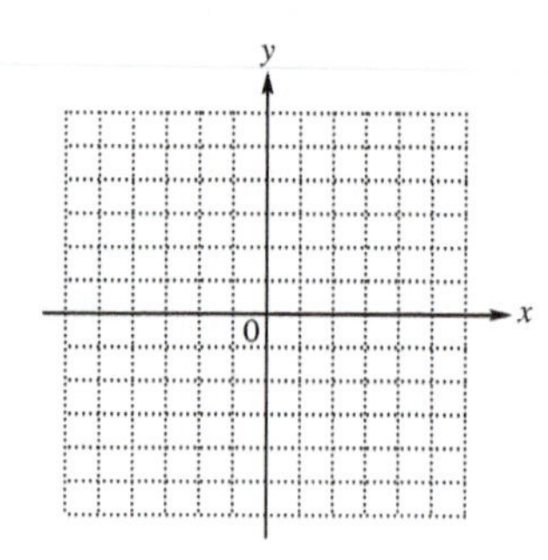

21. $x + y = 0$

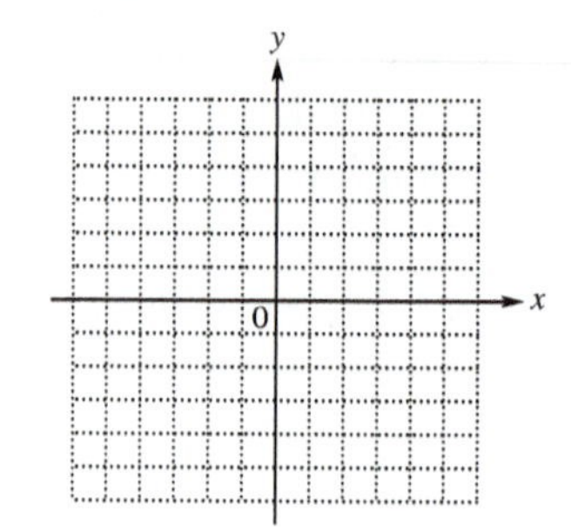

22. $y = 3x + 4$

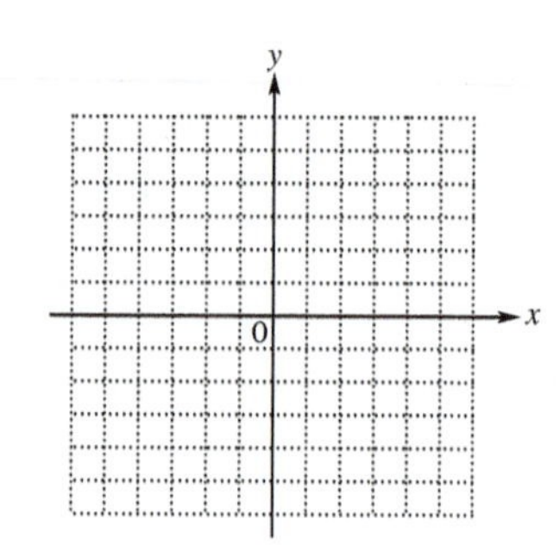

3.3 *Find the slope of each line. (In Exercises 29 and 30, coordinates of the points shown are integers.)*

23. The line passing through $(2, 3)$ and $(-4, 6)$

24. The line passing through $(0, 0)$ and $(-3, 2)$

25. The line passing through $(0, 6)$ and $(1, 6)$

26. The line passing through $(2, 5)$ and $(2, 8)$

27. $y = 3x - 4$

28. $y = \frac{2}{3}x + 1$

29.

30.

31. The line passing through these points

x	y
0	1
2	4
6	10

32. $y = 4$

33. A line parallel to the graph of $y = 2x + 3$

34. A line perpendicular to the graph of $y = -3x + 3$

Decide whether each pair of lines is parallel, perpendicular, *or* neither.

35. $3x + 2y = 6$
$6x + 4y = 8$

36. $x - 3y = 1$
$3x + y = 4$

37. $x - 2y = 8$
$x + 2y = 8$

38. CONCEPT CHECK What is the slope of a line perpendicular to a line with undefined slope?

3.4, 3.5 *Write an equation of each line. Give the final answer in slope-intercept form (if possible).*

39. $m = -1, b = \frac{2}{3}$

40. The line in **Exercise 30**

41. The line passing through $(4, -3)$, $m = 1$

42. The line passing through $(-1, 4)$, $m = \frac{2}{3}$

43. The line passing through $(1, -1)$, $m = -\frac{3}{4}$

44. The line passing through $(2, 1)$ and $(-2, 2)$

45. The line passing through $(-4, 1)$, slope 0

46. The line passing through $\left(\frac{1}{3}, -\frac{3}{4}\right)$, undefined slope

47. Consider the linear equation $x + 3y = 15$.

(a) Write it in the form $y = mx + b$.

(b) What is the slope? The y-intercept?

(c) Use the slope and y-intercept to graph the line. Indicate two points on the graph.

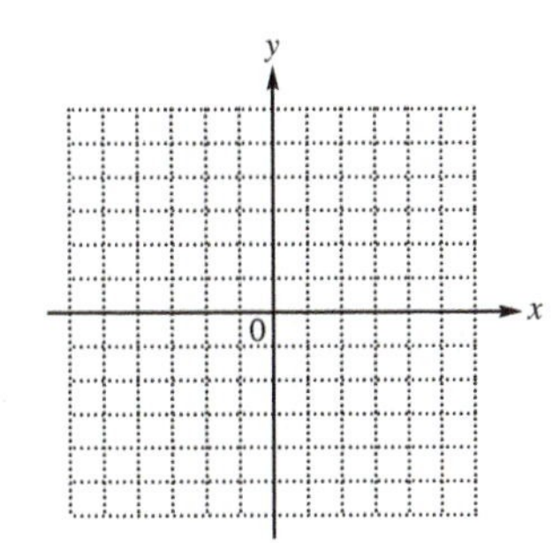

The points on the graph indicate sales y (in millions of dollars) of wearable fitness technology in the United States for selected years x. The graph of a linear equation that models the data is also shown. Here x = 1 represents 2011, x = 2 represents 2012, and so on.

48. Does the line that models the data have positive or negative slope? Explain.

49. Write two ordered pairs (x, y) for the data for 2011 and 2014. Then use these ordered pairs to write an equation of a line that models sales of wearable fitness technology. (Round the slope to the nearest whole number.) Give the final answer in slope-intercept form.

50. Use the equation from **Exercise 49** to approximate sales of wearable fitness technology in 2012, the year data were unavailable. (*Hint:* What is the value of x for 2012?)

U.S. Sales of Wearable Fitness Technology*

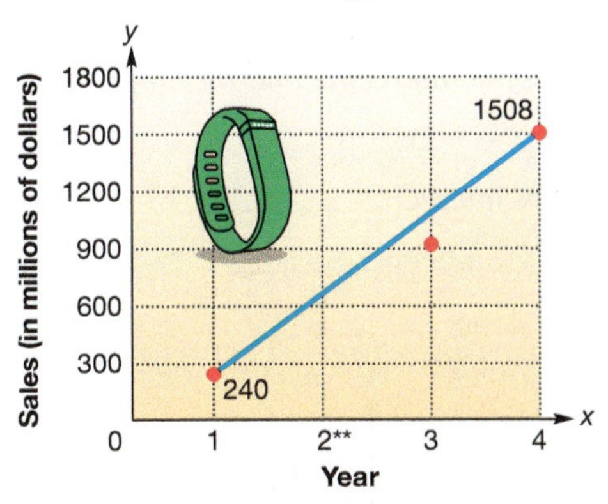

* Includes devices containing pedometers, accelerometers, and heart-rate monitors

** Data for this year are unavailable.

Data from Consumer Electronics Association.

3.6 *Graph each linear inequality.*

51. $3x + 5y > 9$

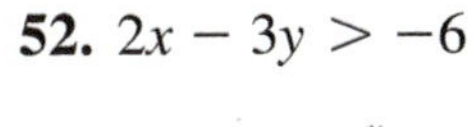

52. $2x - 3y > -6$

53. $x \geq -4$

54. $y \leq -4x$

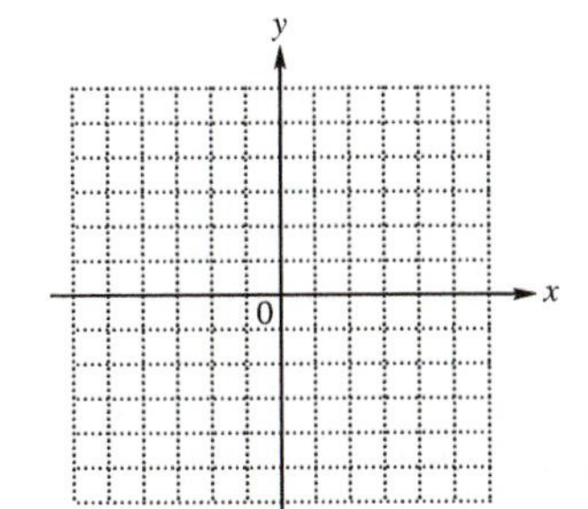

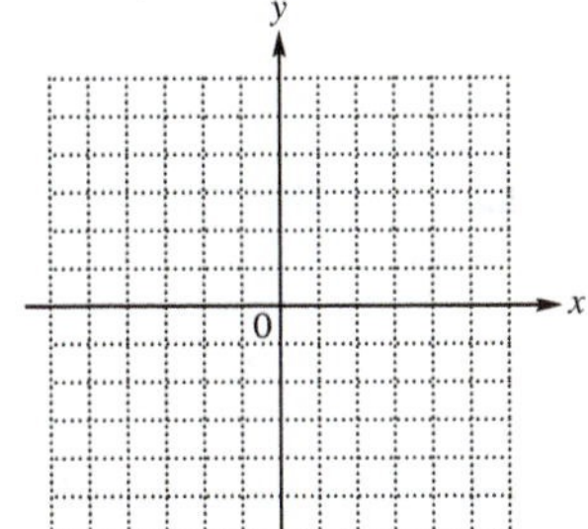

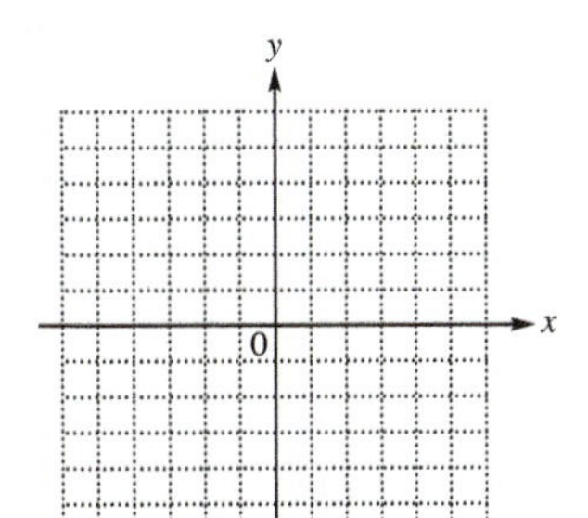

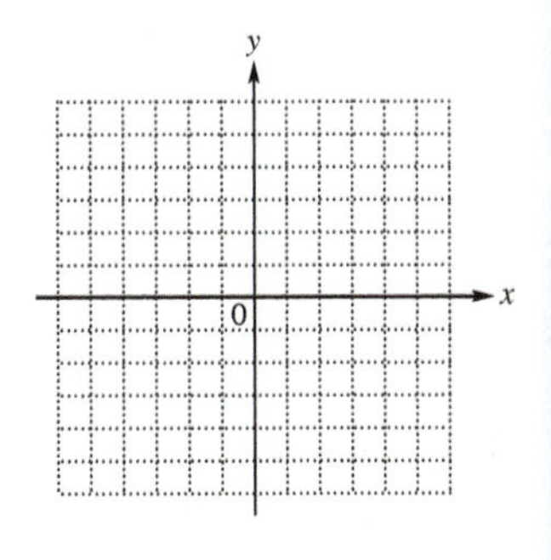

Chapter 3 Mixed Review Exercises

CONCEPT CHECK *Match each statement to the appropriate graph or graphs in choices A–D. Graphs may be used more than once.*

A.

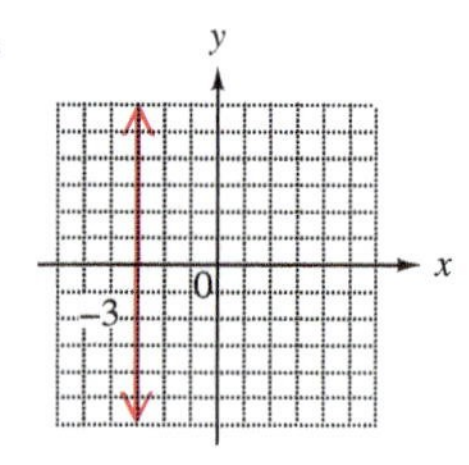

B.

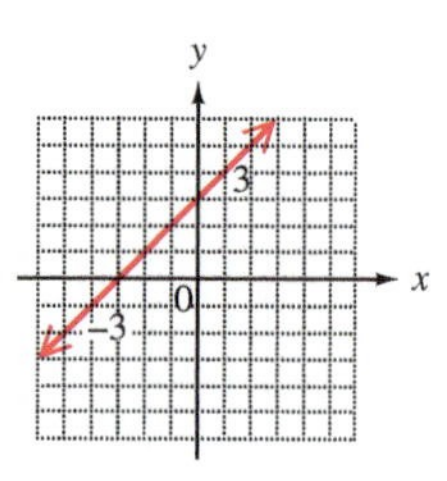

C.

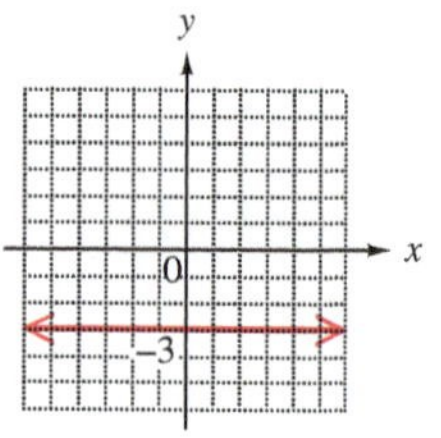
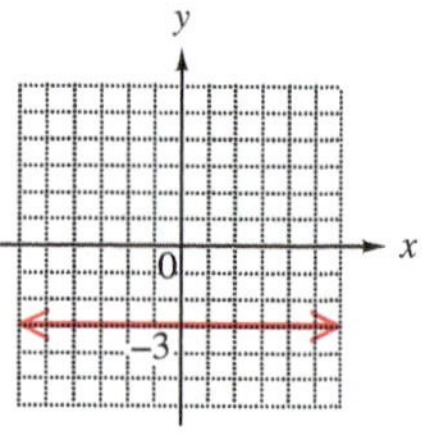

D.

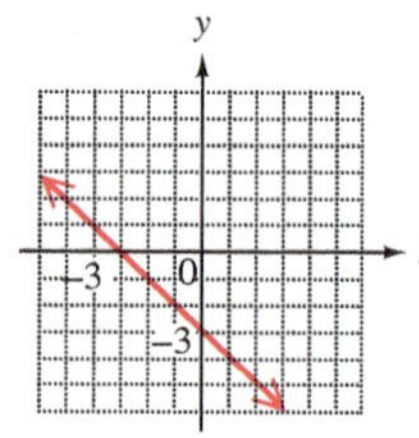
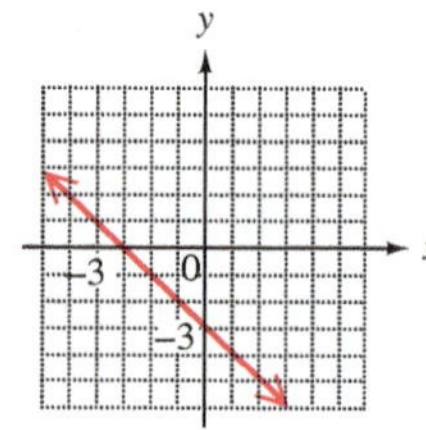

1. The line shown in the graph has undefined slope.

2. The graph of the equation has y-intercept $(0, -3)$.

3. The graph of the equation has x-intercept $(-3, 0)$.

4. The line shown in the graph has negative slope.

5. The graph is that of the equation $y = -3$.

6. The line shown in the graph has slope 1.

Find the intercepts and the slope of each line. Then use them to graph the line.

7. $y = -2x - 5$

y-intercept: ______

x-intercept: ______

slope: ______

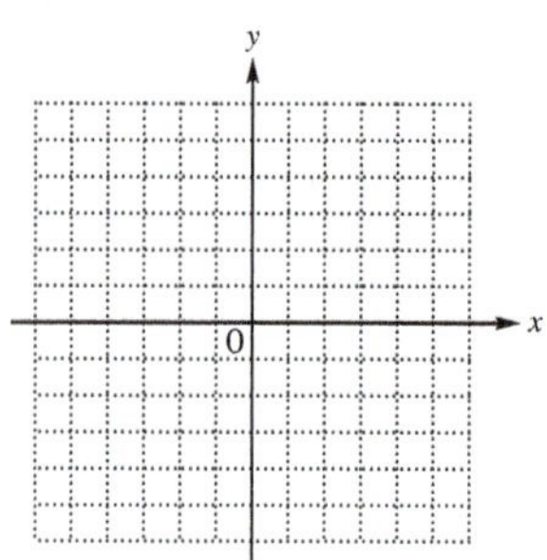

8. $x + 3y = 0$

y-intercept: ______

x-intercept: ______

slope: ______

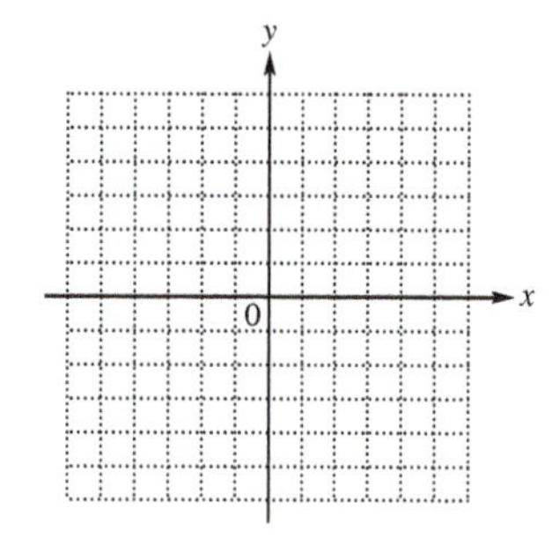

9. $y - 5 = 0$

y-intercept: ______

x-intercept: ______

slope: ______

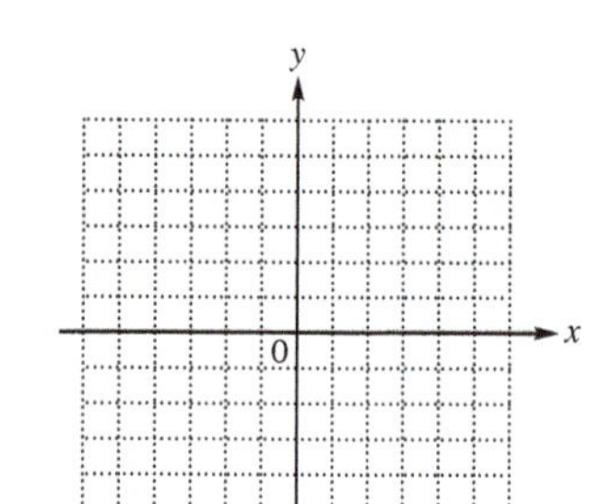

10. $x + 4 = 3$

y-intercept: ______

x-intercept: ______

slope: ______

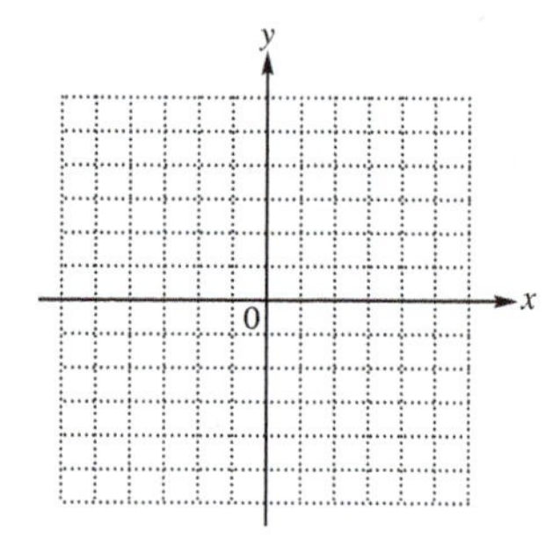

*Write an equation of each line. Give the final answer in **(a)** slope-intercept form and **(b)** standard form.*

11. $m = -\frac{1}{4}, b = -\frac{5}{4}$

12. The line passing through $(8, 6)$, $m = -3$

13. The line passing through $(3, -5)$ and $(-4, -1)$

14. Slope 0, through $(5, -5)$

Chapter 3 Test

The Chapter Test Prep Videos with step-by-step solutions are available in MyMathLab or on YouTube at https://goo.gl/8aAsWP

Work each problem.

1. Complete the ordered pairs (0, ____), (____, 0), and (____, −3) for the equation

$$3x + 5y = -30.$$

2. Is $(4, -1)$ a solution of $4x - 7y = 9$?

3. How do we find the x-intercept of the graph of a linear equation in two variables? How do we find the y-intercept?

4. *True* or *false*: The x-axis is the horizontal line given by the equation $y = 0$.

Graph each linear equation using the intercepts.

5. $3x + y = 6$

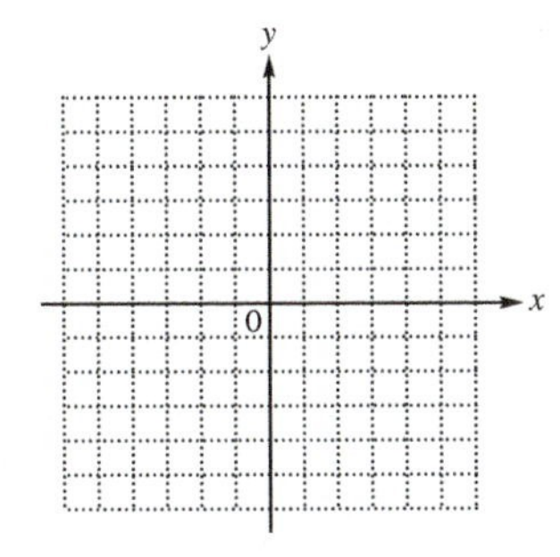

6. $y - 2x = 0$

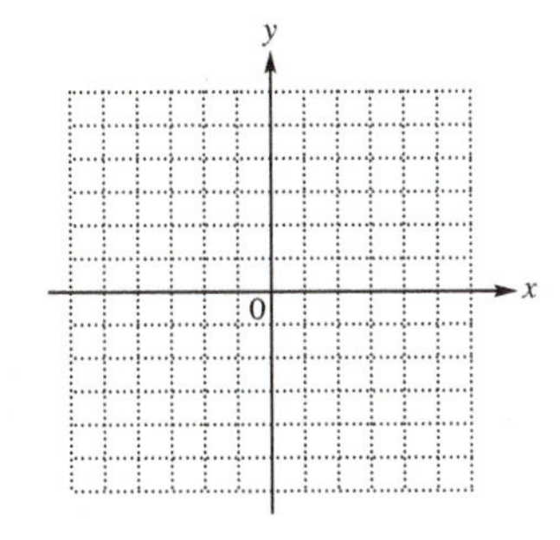

7. $x + 3 = 0$

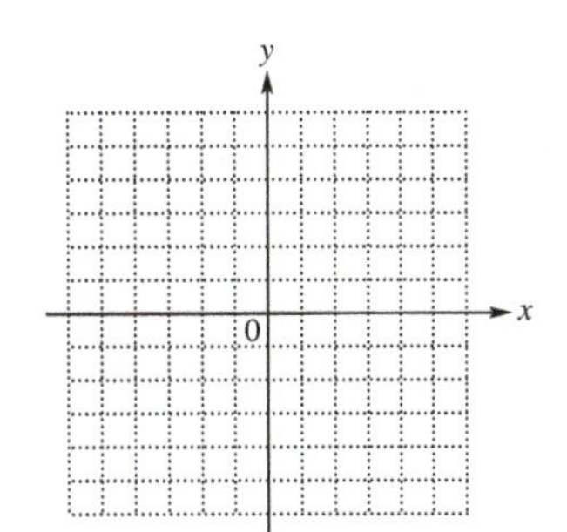

8. $y = 1$

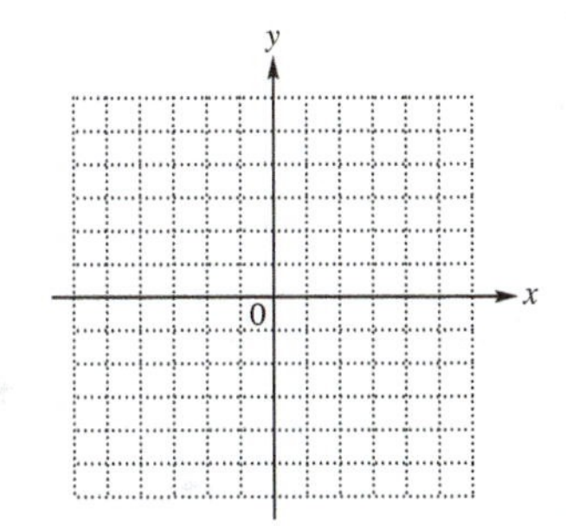

9. Give the slope and y-intercept of the graph of $y = x - 4$. Use them to graph the equation.

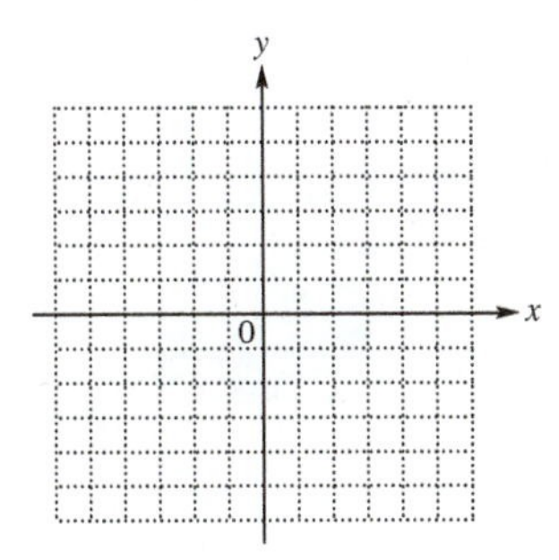

10. Graph the line passing through the point $(-2, 3)$, with slope $-\frac{1}{2}$.

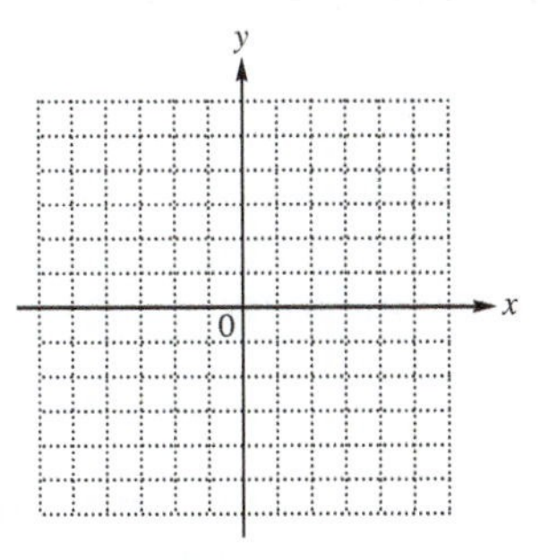

Find the slope of each line. (In Exercise 14, coordinates of the points shown are integers.)

11. The line passing through $(-4, 6)$ and $(-1, -2)$

12. $2x + y = 10$

13. $x + 12 = 0$

14.

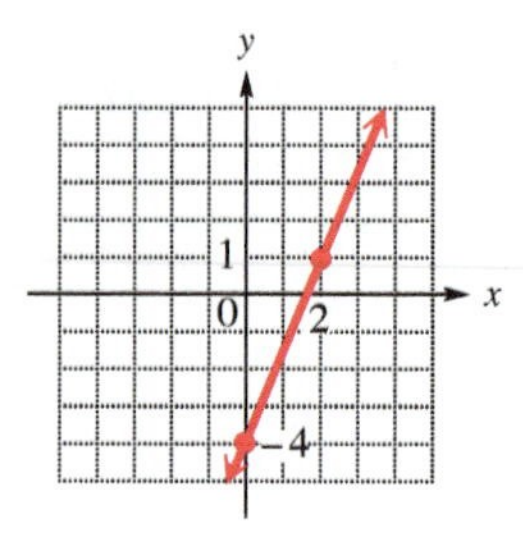

15. A line parallel to the graph of $y - 4 = 6$

16. A line perpendicular to the graph of $y = 4x + 6$

Write an equation for each line. Give the final answer in slope-intercept form.

17. The line passing through $(-1, 4)$, $m = 2$

18. The line in **Exercise 14**

19. The line passing through $(2, -6)$ and $(1, 3)$

20. x-intercept: $(3, 0)$; y-intercept: $\left(0, \frac{9}{2}\right)$

Graph each linear inequality.

21. $x + y \leq 3$

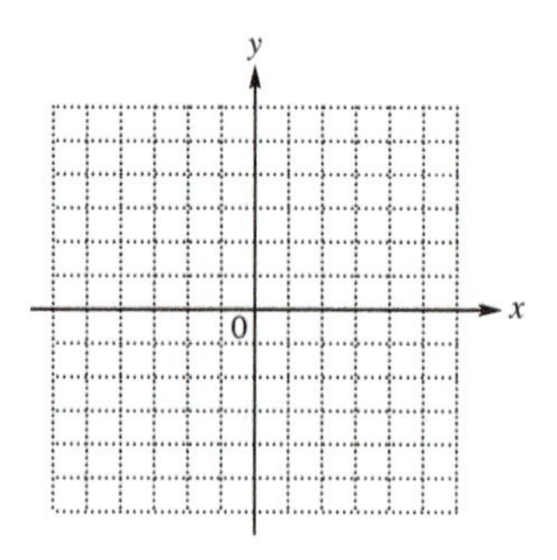

22. $3x - y > 0$

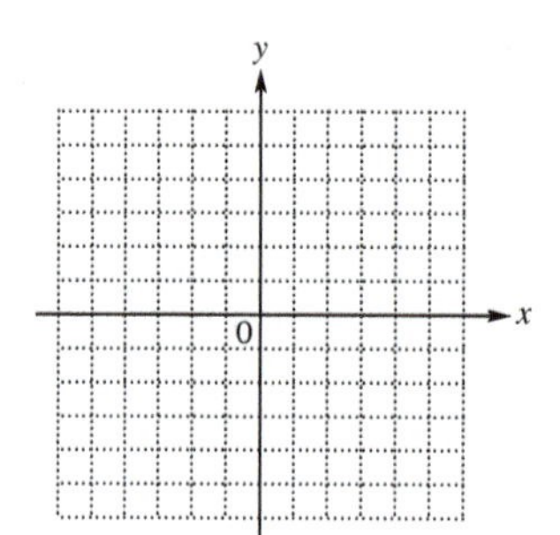

23. $x \leq 0$

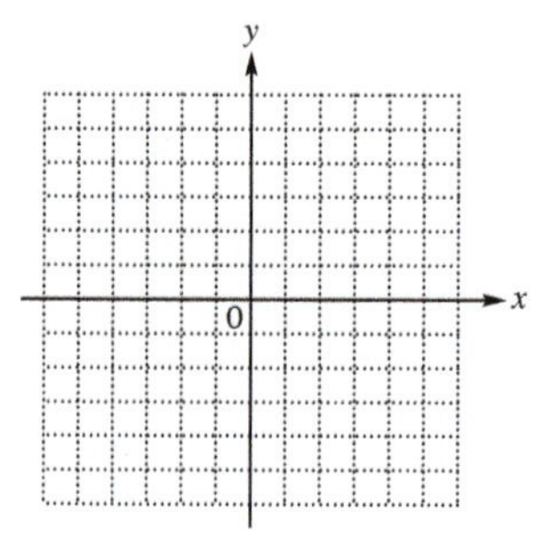

The graph shows worldwide snowmobile sales from 2000 through 2015, where $x = 0$ represents 2000, $x = 1$ represents 2001, and so on. Use the graph to work each problem.

24. Is the slope of the line in the graph positive or negative? Explain.

25. Write two ordered pairs (x, y) for the data points shown in the graph. Use the ordered pairs to find the slope of the line to the nearest tenth.

26. Use the ordered pairs and slope from **Exercise 25** to write an equation of a line that models the data. Give the final equation in slope-intercept form.

Worldwide Snowmobile Sales

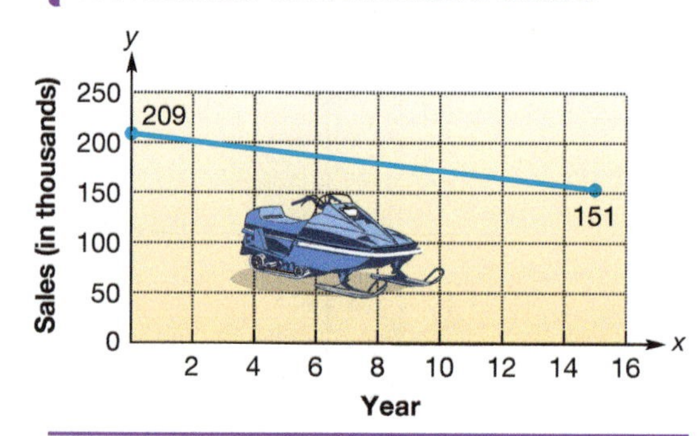

Data from www.snowmobile.org

27. Use the equation from **Exercise 26** to approximate worldwide snowmobile sales in 2010.

28. What does the ordered pair $(15, 151)$ mean in the context of this problem?

Chapters R–3 Cumulative Review Exercises

Perform the indicated operations.

1. $10\frac{5}{8} - 3\frac{1}{10}$

2. $\frac{3}{4} \div \frac{1}{8}$

3. $5 - (-4) + (-2)$

4. $\frac{(-3)^2 - (-4)(2^4)}{5(2) - (-2)^3}$

5. *True* or *false*? $\frac{4(3-9)}{2-6} \geq 6$

6. Find the value of $xz^3 - 5y^2$ for $x = -2$, $y = -3$, and $z = -1$.

7. What property does $3(-2 + x) = -6 + 3x$ illustrate?

8. Simplify $-4p - 6 + 3p + 8$ by combining like terms.

Solve.

9. $V = \frac{1}{3}\pi r^2 h$ for h

10. $6 - 3(1 + x) = 2(x + 5) - 2$

11. $-(x - 3) = 5 - 2x$

12. $\frac{y-2}{3} = \frac{2y+1}{5}$

Solve each inequality. Write the solution set in interval notation and graph it.

13. $-2.5x < 6.5$

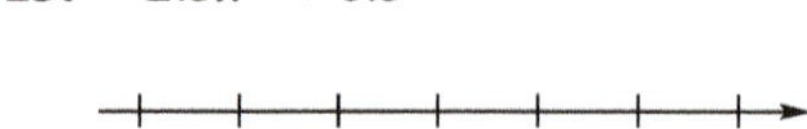

14. $4(x + 3) - 5x < 12$

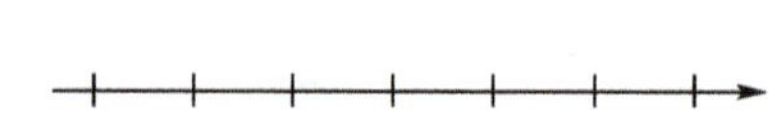

15. $\frac{2}{3}t - \frac{1}{6}t \leq -2$

Solve each problem.

16. In 2013, a young adult (ages 25–34) with a bachelor's degree could expect to earn \$18,500 more each year than a young adult with a high school credential. Together the individuals would earn \$78,500. How much could a young adult at each level of education expect to earn? (Data from NCES.)

17. Mount Mayon in the Philippines is the most perfectly shaped conical volcano in the world. Its base is a circle with circumference 80 mi. Find the radius of the circular base to the nearest mile. (Data from www.britannica.hk)

Circumference = 80 mi

18. The winning times in seconds for the women's 1000 m speed skating event in the Winter Olympics for the years 1960 through 2014 can be closely approximated by the linear equation

$$y = -0.4336x + 94.30,$$

where x is the number of years since 1960. That is, $x = 4$ represents 1964, $x = 8$ represents 1968, and so on. (Data from *The World Almanac and Book of Facts.)*

(a) Use this equation to complete the table of values. Round times to the nearest hundredth of a second.

x	y
20	
38	
42	
	72.62

(b) Write ordered pairs for the data given in the table.

(c) In the context of this problem, what does the ordered pair (54, 70.89) mean?

19. In a recent year, approximately 60 million people from other countries visited the United States. The circle graph shows the distribution of these international visitors by their home countries or regions.

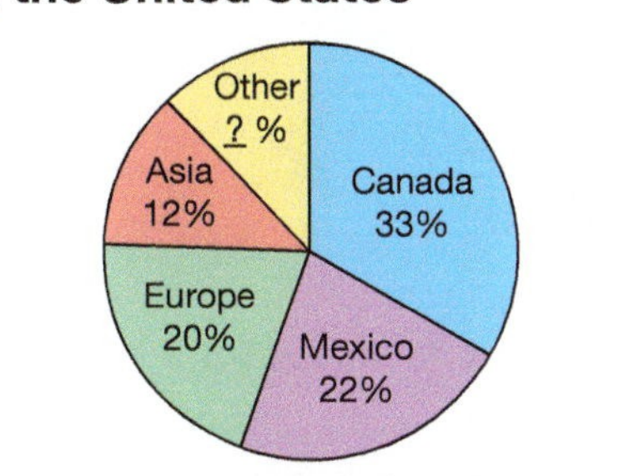

Data from U.S. Department of Commerce.

(a) About how many travelers visited the United States from Canada?

(b) About how many travelers visited the United States from Mexico?

(c) About how many travelers visited the United States from places other than Canada, Mexico, Europe, or Asia?

Consider the linear equation $-3x + 4y = 12$. *Find the following.*

20. The x- and y-intercepts

21. The slope

22. The y-value of the point having x-value 4

23. The graph

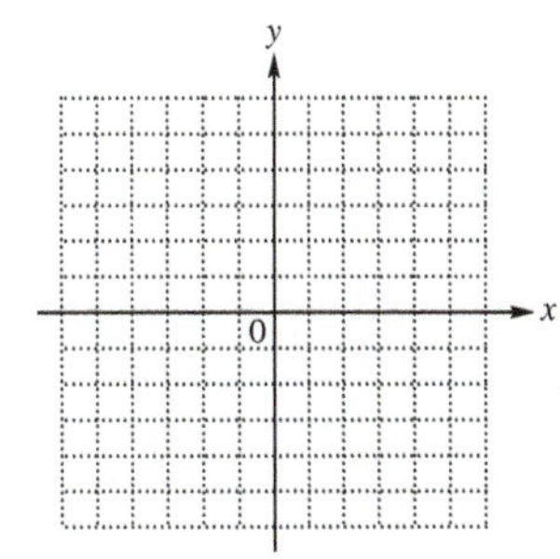

24. Are the lines with equations $x + 5y = -6$ and $y = 5x - 8$ *parallel, perpendicular,* or *neither?*

Write an equation of each line. Give the final answer in slope-intercept form if possible.

25. The line passing through $(2, -5)$, with slope 3

26. The line passing through $(0, 4)$ and $(2, 4)$

Systems of Linear Equations and Inequalities

The point of intersection of two lines can be found using a *system of linear equations*, the subject of this chapter.

4.1 Solving Systems of Linear Equations by Graphing

OBJECTIVES

1. Decide whether a given ordered pair is a solution of a system.
2. Solve linear systems by graphing.
3. Solve special systems by graphing.
4. Identify special systems without graphing.

A **system of linear equations,** or **linear system,** consists of two or more linear equations with the same variables.

$$2x + 3y = 4 \qquad x + 3y = 1 \qquad x - y = 1$$
$$3x - y = -5 \qquad -y = 4 - 2x \qquad y = 3$$

Linear systems

In the system on the right, think of $y = 3$ as an equation in two variables by writing it as $0x + y = 3$.

OBJECTIVE 1 Decide whether a given ordered pair is a solution of a system. A **solution of a system** of linear equations is an ordered pair that makes both equations true at the same time. A solution of an equation is said to *satisfy* the equation.

1 Determine whether the given ordered pair is a solution of the system.

GS **(a)** $(2, 5)$

$$3x - 2y = -4$$
$$5x + y = 15$$

$$3x - 2y = -4$$
$$3(\_\_\_) - 2(\_\_\_) \stackrel{?}{=} -4$$
$$\_\_\_ = -4$$

$$5x + y = 15$$
$$5(\_\_\_) + \_\_\_ \stackrel{?}{=} 15$$
$$\_\_\_ = 15$$

$(2, 5)$ (*is / is not*) a solution.

(b) $(1, -2)$

$$x - 3y = 7$$
$$4x + y = 5$$

$(1, -2)$ (*is / is not*) a solution.

Answers

1. **(a)** 2; 5; −4; 2; 5; 15; is **(b)** is not

EXAMPLE 1 Determining Whether an Ordered Pair Is a Solution

Determine whether the ordered pair $(4, -3)$ is a solution of each system.

(a) $x + 4y = -8$

$3x + 2y = 6$

To decide whether $(4, -3)$ is a solution of the system, substitute 4 for x and -3 for y in each equation.

$x + 4y = -8$		$3x + 2y = 6$	
$4 + 4(-3) \stackrel{?}{=} -8$	Substitute.	$3(4) + 2(-3) \stackrel{?}{=} 6$	Substitute.
$4 + (-12) \stackrel{?}{=} -8$	Multiply.	$12 + (-6) \stackrel{?}{=} 6$	Multiply.
$-8 = -8$ ✓	True	$6 = 6$ ✓	True

Because $(4, -3)$ satisfies *both* equations, it is a solution of the system.

(b) $2x + 5y = -7$

$3x + 4y = 2$

Again, substitute 4 for x and -3 for y in each equation.

$2x + 5y = -7$		$3x + 4y = 2$	
$2(4) + 5(-3) \stackrel{?}{=} -7$	Substitute.	$3(4) + 4(-3) \stackrel{?}{=} 2$	Substitute.
$8 + (-15) \stackrel{?}{=} -7$	Multiply.	$12 + (-12) \stackrel{?}{=} 2$	Multiply.
$-7 = -7$ ✓	True	$0 = 2$	False

The ordered pair $(4, -3)$ is *not* a solution of this system because it does not satisfy the second equation.

◄ Work Problem 1 at the Side.

OBJECTIVE 2 Solve linear systems by graphing. The set of all ordered pairs that are solutions of a system is its **solution set.** One way to find the solution set of a system of two linear equations is to graph both equations on the same axes. Any intersection point would be on both lines and would therefore be a solution of *both* equations. ***Thus, the coordinates of any point where the lines intersect give a solution of the system.***

The graph in **Figure 1** shows that the solution of the system in **Example 1(a)** is the intersection point $(4, -3)$.

The solution (point of intersection) is always written as an ordered pair.

Because *two different* straight lines can intersect at no more than one point, there can never be more than one solution for such a system.

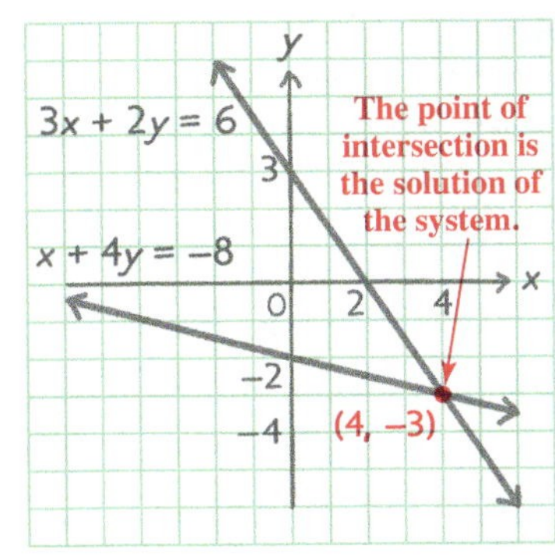

Figure 1

EXAMPLE 2 Solving a System by Graphing

Solve the system of equations by graphing both equations on the same axes.

$$2x + 3y = 4$$
$$3x - y = -5$$

We graph these two equations by plotting several points for each line. Recall that the intercepts are often convenient choices. As review, we show finding the intercepts for $2x + 3y = 4$.

To find the y-intercept, let $x = 0$.

$$2x + 3y = 4$$
$$2(0) + 3y = 4 \quad \text{Let } x = 0.$$
$$3y = 4 \quad \text{Multiply.}$$
$$y = \frac{4}{3} \quad \text{Divide by 3.}$$

To find the x-intercept, let $y = 0$.

$$2x + 3y = 4$$
$$2x + 3(0) = 4 \quad \text{Let } y = 0.$$
$$2x = 4 \quad \text{Multiply.}$$
$$x = 2 \quad \text{Divide by 2.}$$

The tables show the intercepts and a check point for each graph.

$2x + 3y = 4$

x	y	
0	$\frac{4}{3}$	← y-intercept
2	0	← x-intercept
−2	$\frac{8}{3}$	Find a third ordered pair as a check.

$3x - y = -5$

x	y	
0	5	← y-intercept
$-\frac{5}{3}$	0	← x-intercept
−2	−1	

The lines in **Figure 2** suggest that the graphs intersect at the point $(-1, 2)$. We check by substituting -1 for x and 2 for y in *both* equations.

CHECK

$$2x + 3y = 4$$
$$2(-1) + 3(2) \stackrel{?}{=} 4$$
$$4 = 4 \checkmark \quad \text{True}$$
$$3x - y = -5$$
$$3(-1) - 2 \stackrel{?}{=} -5$$
$$-5 = -5 \checkmark \quad \text{True}$$

Figure 2

Because $(-1, 2)$ satisfies both equations, the solution set of this system is $\{(-1, 2)\}$.

Work Problem 2 at the Side. ▶

2 Solve each system of equations by graphing both equations on the same axes.

(a) $5x - 3y = 9$

$x + 2y = 7$

One of the lines is already graphed. Complete the table, and graph the other line.

$5x - 3y = 9$

x	y
0	−3
$\frac{9}{5}$	0
3	2

$x + 2y = 7$

x	y
0	___
___	0
−1	___

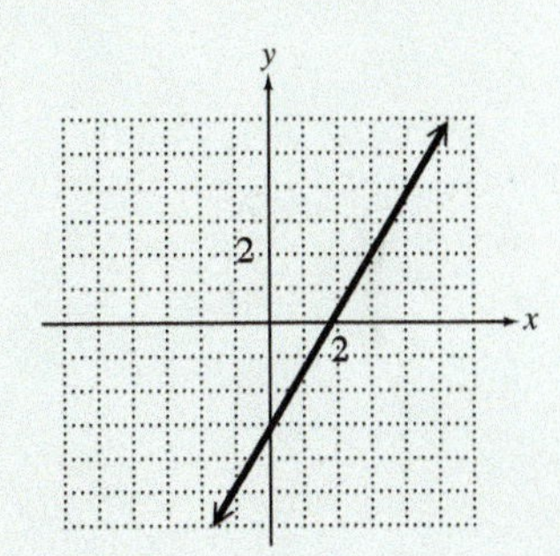

The point of intersection of the lines is the ordered pair ______, so the solution set of the system of equations is ______.

(b) $x + y = 4$

$2x - y = -1$

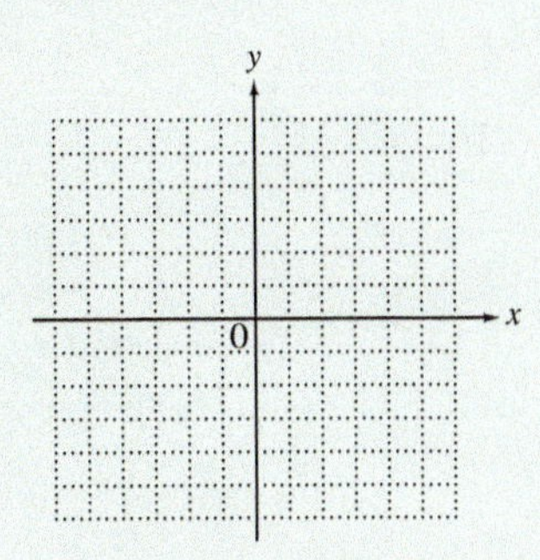

Answers

2. **(a)** $\frac{7}{2}$; 7; 4; (3, 2); $\{(3, 2)\}$ **(b)** $\{(1, 3)\}$

Note

We can also write each equation in a system in slope-intercept form and use the slope and y-intercept to graph each line. For **Example 2,**

$2x + 3y = 4$ becomes $y = -\frac{2}{3}x + \frac{4}{3}$. y-intercept $\left(0, \frac{4}{3}\right)$; slope $-\frac{2}{3}$

$3x - y = -5$ becomes $y = 3x + 5$. y-intercept $(0, 5)$; slope 3, or $\frac{3}{1}$

Confirm that graphing these equations results in the same lines and the same solution shown in **Figure 2** on the preceding page.

Solving a Linear System by Graphing

Step 1 **Graph each equation** of the system on the same coordinate axes.

Step 2 **Find the coordinates of the point of intersection** of the graphs if possible, and write it as an ordered pair.

Step 3 **Check** that the ordered pair is the solution by substituting it in *both* of the original equations. If it satisfies *both* equations, write the solution set.

CAUTION

We recommend using graph paper and a straightedge when solving systems of equations graphically. It may not be possible to determine from the graph the exact coordinates of the point that represents the solution, particularly if those coordinates are not integers. The graphing method does, however, show geometrically how solutions are found and is useful when approximate answers will suffice.

OBJECTIVE 3 Solve special systems by graphing. The graphs of the equations in a system may not intersect at all or may be the same line.

EXAMPLE 3 Solving Special Systems

Solve each system by graphing.

(a) $2x + y = 2$

$2x + y = 8$

The graphs of these lines are shown in **Figure 3.** The two lines are parallel and have no points in common. For such a system, there is no solution. We write the solution set as $\emptyset$.

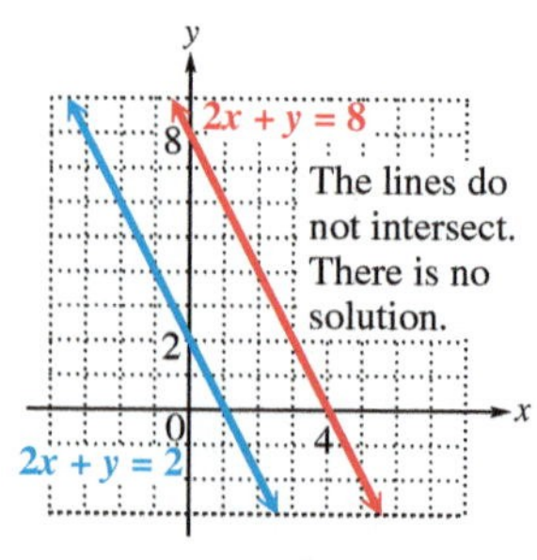

Figure 3

Continued on Next Page

(b) $2x + 5y = 1$

$6x + 15y = 3$

The graphs of these two equations are the same line. See **Figure 4.** We can obtain the second equation by multiplying each side of the first equation by 3. In this case, every point on the line is a solution of the system, and the solution set contains an infinite number of ordered pairs, each of which satisfies both equations of the system.

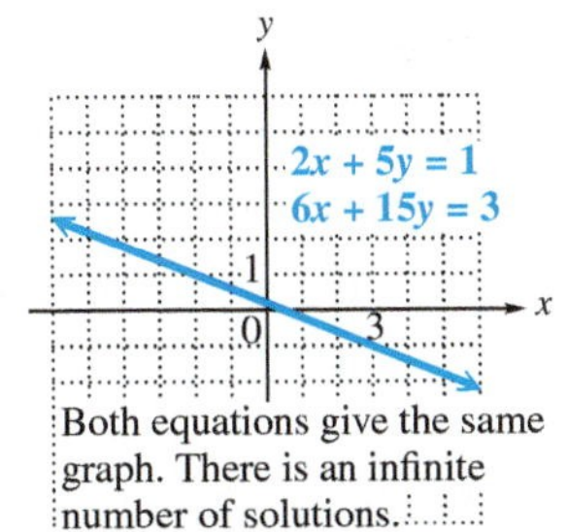

2x + 5y = 1
6x + 15y = 3
Both equations give the same graph. There is an infinite number of solutions.

Figure 4

We write the solution set as

$$\{(x, y) \mid 2x + 5y = 1\},$$

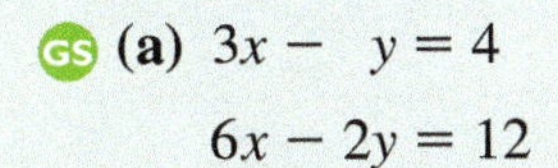

This is the first equation in the system. See the Note on the next page.

read "*the set of ordered pairs* (x, y) *such that* $2x + 5y = 1$." Recall that this notation is called **set-builder notation.**

Work Problem 3 at the Side. ▶

The system in **Example 2** has exactly one solution. A system with at least one solution is a **consistent system.** A system of equations with no solution, such as the one in **Example 3(a),** is an **inconsistent system.**

The equations in **Example 2** are **independent equations** with different graphs. The equations of the system in **Example 3(b)** have the same graph and are equivalent. Because they are different forms of the same equation, these equations are **dependent equations.**

Three Cases for Solutions of Linear Systems with Two Variables

Case 1 The graphs intersect at exactly one point, which gives the (single) ordered-pair solution of the system. The **system is consistent** and the **equations are independent. See Figure 5(a).**

Case 2 The graphs are parallel lines, so there is no solution and the solution set is ∅. The **system is inconsistent** and the **equations are independent. See Figure 5(b).**

Case 3 The graphs are the same line. There is an infinite number of solutions, and the solution set is written in set-builder notation as

$$\{(x, y) \mid \underline{\qquad}\},$$

where one of the equations follows the | symbol. The **system is consistent** and the **equations are dependent. See Figure 5(c).**

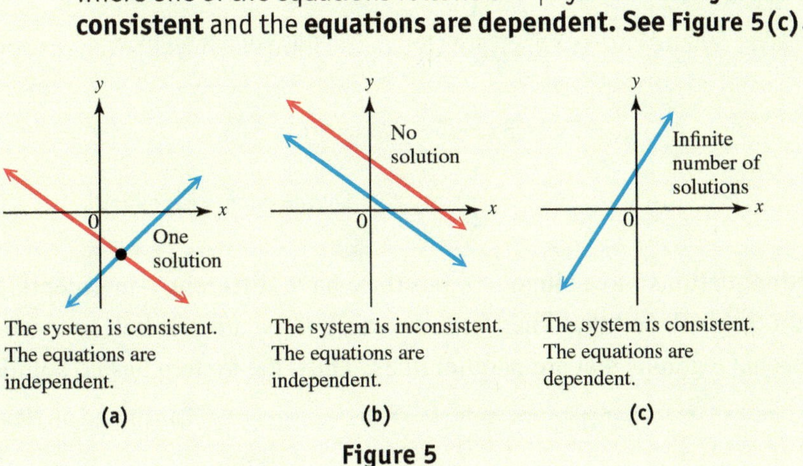

One solution
No solution
Infinite number of solutions

(a) The system is consistent. The equations are independent.

(b) The system is inconsistent. The equations are independent.

(c) The system is consistent. The equations are dependent.

Figure 5

3 Solve each system of equations by graphing.

(a) $3x - y = 4$

$6x - 2y = 12$

One of the lines is already graphed.

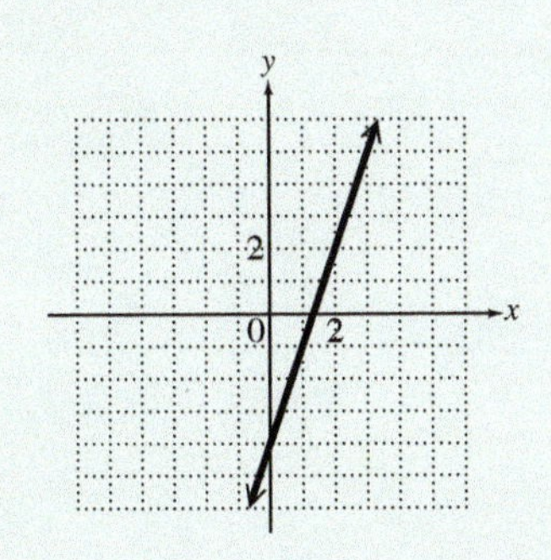

(b) $x - 3y = -2$

$2x - 6y = -4$

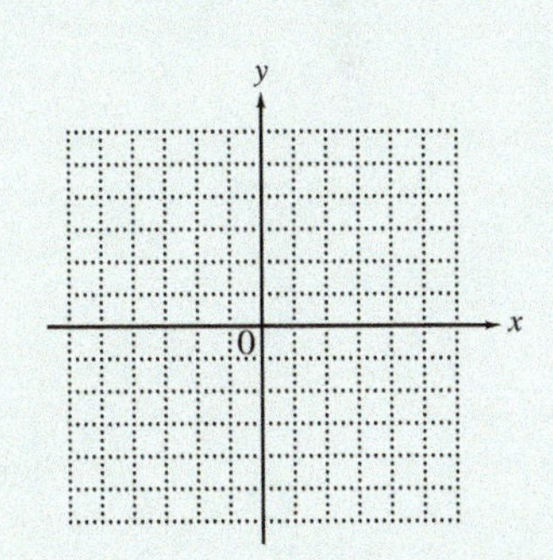

Answers

3. (a) ∅ **(b)** $\{(x, y) \mid x - 3y = -2\}$

Note

When a system has an infinite number of solutions, as in Case 3, either equation of the system can be used to write the solution set.

We prefer to use the equation in standard form with integer coefficients having greatest common factor 1 and positive coefficient of x.

If neither of the given equations is in this form, use an *equivalent* equation that is written this way.

Examples: For the system

$$-6x + 2y = -4$$
$$3x - y = 2,$$

This system has an infinite number of solutions.

we write the solution set using the second equation.

$$\{(x, y) \mid 3x - y = 2\}$$

For the system

$$2x - 4y = 8$$
$$-4x + 8y = -16,$$

This system has an infinite number of solutions.

we divide each term of the first equation by the common factor 2 and write the solution set as follows.

$$\{(x, y) \mid x - 2y = 4\}$$

OBJECTIVE 4 Identify special systems without graphing.

EXAMPLE 4 Identifying the Three Cases by Using Slopes

Describe each system without graphing. State the number of solutions.

(a) $3x + 2y = 6$

$-2y = 3x - 5$

Write each equation in slope-intercept form, $y = mx + b$.

$$3x + 2y = 6 \quad \text{Solve for } y.$$

$$2y = -3x + 6 \quad \text{Subtract } 3x.$$

$$y = -\frac{3}{2}x + 3 \quad \text{Divide each term by 2.}$$

$$-2y = 3x - 5 \quad \text{Solve for } y.$$

$$y = -\frac{3}{2}x + \frac{5}{2} \quad \text{Divide each term by } -2.$$

Both equations have slope $-\frac{3}{2}$ but they have different y-intercepts, $(0, 3)$ and $\left(0, \frac{5}{2}\right)$. Recall that lines with the same slope are parallel, so these equations have graphs that are parallel lines. Thus, the system has no solution.

Continued on Next Page

(b) $2x - y = 4$

$x = \frac{y}{2} + 2$

Again, write each equation in slope-intercept form.

$2x - y = 4$		$x = \frac{y}{2} + 2$	
$-y = -2x + 4$	Subtract $2x$.	$2x = y + 4$	Multiply by 2.
$\mathbf{y = 2x - 4}$	Multiply by -1.	$\mathbf{y = 2x - 4}$	Subtract 4. Interchange sides.

The equations are exactly the same—their graphs are the same line. Any ordered-pair solution of one equation is also a solution of the other equation. Thus, the system has an infinite number of solutions.

(c) $x - 3y = 5$

$2x + y = 8$

Write each equation in slope-intercept form.

$x - 3y = 5$		$2x + y = 8$	
$-3y = -x + 5$	Subtract x.	$y = -2x + 8$ (Slope: -2)	Subtract $2x$.
$y = \frac{1}{3}x - \frac{5}{3}$ (Slope: $\frac{1}{3}$)	Divide by -3.		

The graphs of these equations are neither parallel nor the same line because the slopes are different. The graphs will intersect in one point—thus, the system has exactly one solution.

Work Problem 4 at the Side. ▶

Note

The solution set of the system in **Example 4(a)** is $\emptyset$ because the graphs of the equations of the system are parallel lines. The solution set of the system in **Example 4(b),** written using set-builder notation and the first equation, is

$$\{(x, y) \mid 2x - y = 4\}.$$

If we try to solve the system in **Example 4(c)** by graphing, we will have difficulty identifying the point of intersection of the graphs. We introduce an algebraic method for solving systems like this in the next section.

4 Describe each system without graphing. State the number of solutions.

GS **(a)** $2x - 3y = 5$

$3y = 2x - 7$

Solve the first equation for ____.

$2x - 3y = 5$

$-3y = \_\_\_\_ + 5$

$y = \_\_\_\_ - \frac{5}{3}$

The slope of this line is ____, and the y-intercept is ____.

Now solve the second equation for y, and complete the solution.

(b) $-x + 3y = 2$

$2x - 6y = -4$

(c) $6x + y = 3$

$2x - y = -11$

Answers

4. (a) y; $-2x$; $\frac{2}{3}x$; $\frac{2}{3}$; $\left(0, -\frac{5}{3}\right)$

The equations represent parallel lines. The system has no solution.

(b) The equations represent the same line. The system has an infinite number of solutions.

(c) The equations represent lines that are neither parallel nor the same line. The system has exactly one solution.

4.1 Exercises

FOR EXTRA HELP *Go to MyMathLab for worked-out, step-by-step solutions to exercises enclosed in a square and video solutions to ▶ exercises.*

CONCEPT CHECK *Complete each statement. The following terms may be used once, more than once, or not at all.*

consistent	system of linear equations	inconsistent	solution
ordered pair	independent	linear equation	dependent

1. A(n) ________ consists of two or more linear equations with the (*same* / *different*) variables.

2. A solution of a system of linear equations is a(n) ________ that makes all equations of the system (*true* / *false*) at the same time.

3. The equations of two parallel lines form a(n) ________ system that has (*one* / *no* / *infinitely many*) solution(s). The equations are ________ because their graphs are different.

4. If the graphs of a linear system intersect in one point, the point of intersection is the ________ of the system. The system is ________ and the equations are independent.

5. If two equations of a linear system have the same graph, the equations are ________. The system is ________ and has (*one* / *no* / *infinitely many*) solution(s).

6. If a linear system is inconsistent, the graphs of the two equations are (*intersecting* / *parallel* / *the same*) line(s). The system has no ________.

CONCEPT CHECK *Work each problem.*

7. A student determined that the ordered pair $(1, -2)$ is a solution of the following system. His reasoning was that the ordered pair satisfies the first equation $x + y = -1$ because $1 + (-2) = -1$. **What Went Wrong?**

$$x + y = -1$$
$$2x + y = 4$$

8. The following system has infinitely many solutions. Write its solution set using set-builder notation.

$$6x - 4y = 8$$
$$3x - 2y = 4$$

9. Which ordered pair could be a solution of the system graphed? Why is it the only valid choice?

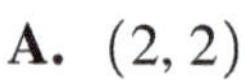

A. $(2, 2)$

B. $(-2, 2)$

C. $(-2, -2)$

D. $(2, -2)$

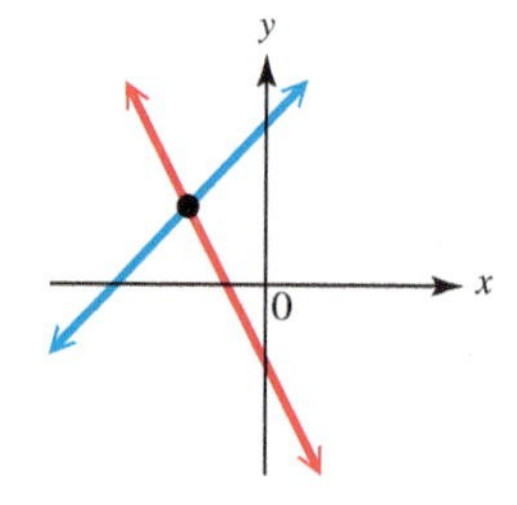

10. Which ordered pair could be a solution of the system graphed? Why is it the only valid choice?

A. $(2, 0)$

B. $(0, 2)$

C. $(-2, 0)$

D. $(0, -2)$

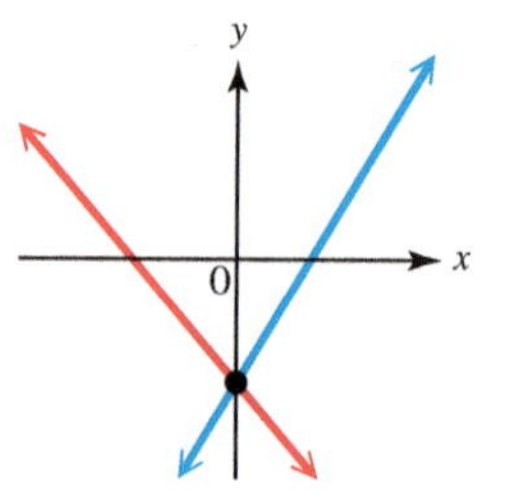

Determine whether the given ordered pair is a solution of the given system. ***See Example 1.***

11. ▶ $\begin{array}{l} x + y = -1 \\ 2x + 5y = 19 \end{array}$; $(2, -3)$

12. $\begin{array}{l} x + 2y = 10 \\ 3x + 5y = 3 \end{array}$; $(4, 3)$

13. ▶ $\begin{array}{l} 3x + 5y = -18 \\ 4x + 2y = -10 \end{array}$; $(-1, -3)$

14. $\begin{array}{l} 2x - 5y = -8 \\ 3x + 6y = -39 \end{array}$; $(-9, -2)$

15. $\begin{aligned} 4x &= 26 - y \\ 3x &= 29 + 4y \end{aligned}$; $(7, -2)$

16. $\begin{aligned} 2x &= 23 - 5y \\ 3x &= 24 + 3y \end{aligned}$; $(9, 1)$

17. $\begin{aligned} -2y &= x + 10 \\ 3y &= 2x + 30 \end{aligned}$; $(6, -8)$

18. $\begin{aligned} 5y &= 3x + 20 \\ 3y &= -2x - 4 \end{aligned}$; $(-5, 2)$

19. $\begin{aligned} 4x + 2y &= 0 \\ x + y &= 0 \end{aligned}$; $(0, 0)$

20. $\begin{aligned} -4x + 4y &= 0 \\ x - y &= 0 \end{aligned}$; $(-1, -1)$

21. $\begin{aligned} y &= \frac{2}{3}x \\ y &= \frac{1}{2}x \end{aligned}$; $(1, 1)$

22. $\begin{aligned} y &= -\frac{3}{2}x \\ y &= -\frac{1}{3}x \end{aligned}$; $(-2, 2)$

Solve each system of equations by graphing. If the system is inconsistent or the equations are dependent, say so. ***See Examples 2 and 3.***

23. $\begin{aligned} x - y &= 2 \\ x + y &= 6 \end{aligned}$

24. $\begin{aligned} x - y &= 3 \\ x + y &= -1 \end{aligned}$

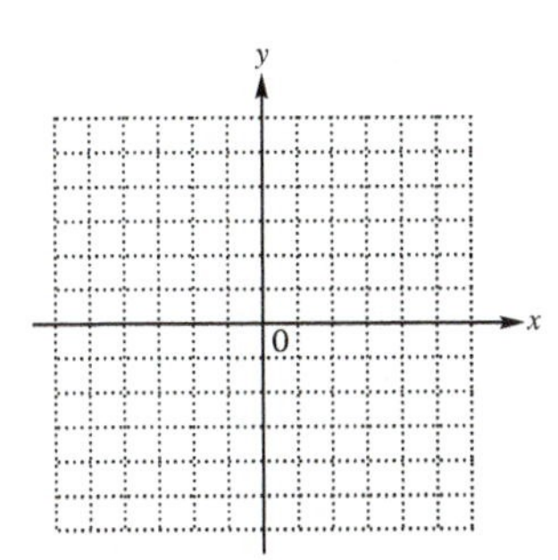

25. $\begin{aligned} x + y &= 4 \\ y - x &= 4 \end{aligned}$

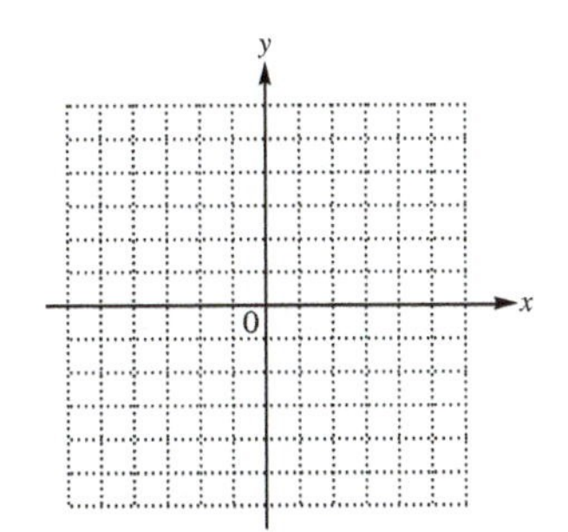

26. $\begin{aligned} x + y &= -5 \\ x - y &= 5 \end{aligned}$

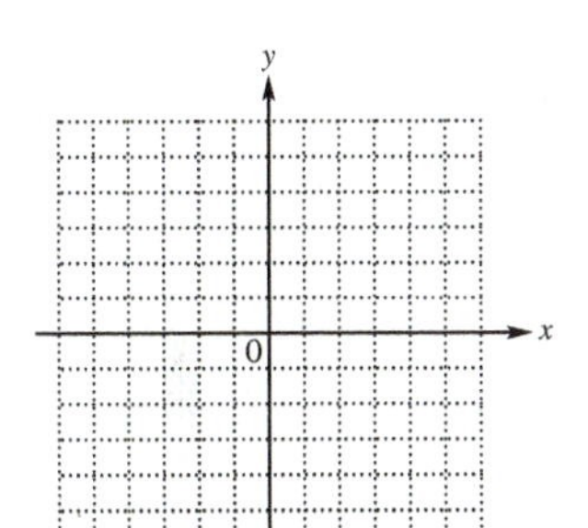

27. $\begin{aligned} x - 2y &= 6 \\ x + 2y &= 2 \end{aligned}$

28. $\begin{aligned} 2x - y &= 4 \\ 4x + y &= 2 \end{aligned}$

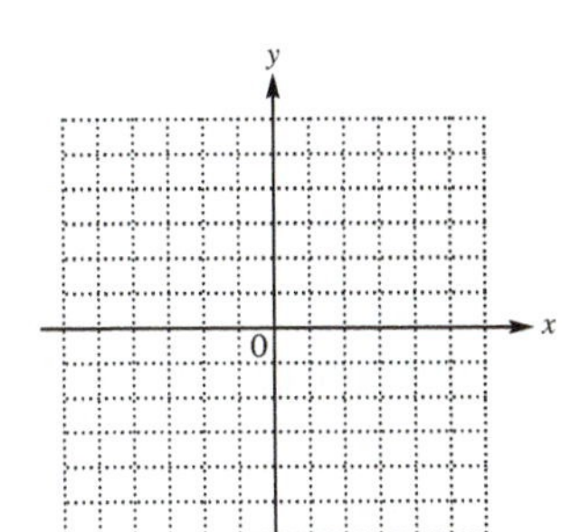

29. $\begin{aligned} 3x - y &= 0 \\ -3x - y &= -6 \end{aligned}$

30. $\begin{aligned} 2x - 3y &= 0 \\ 2x + 3y &= 12 \end{aligned}$

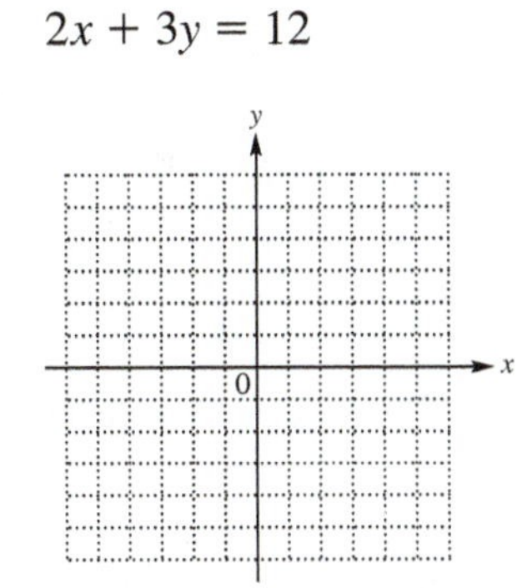

31. $\begin{aligned} 2x - 3y &= -6 \\ y &= -3x + 2 \end{aligned}$

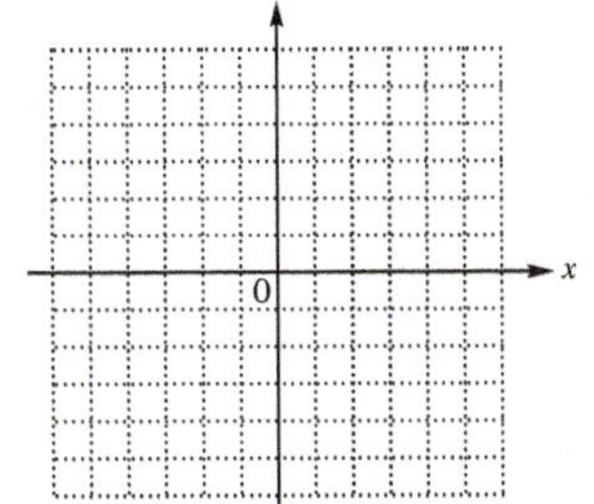

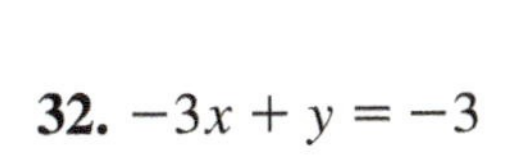

32. $\begin{aligned} -3x + y &= -3 \\ y &= x - 3 \end{aligned}$

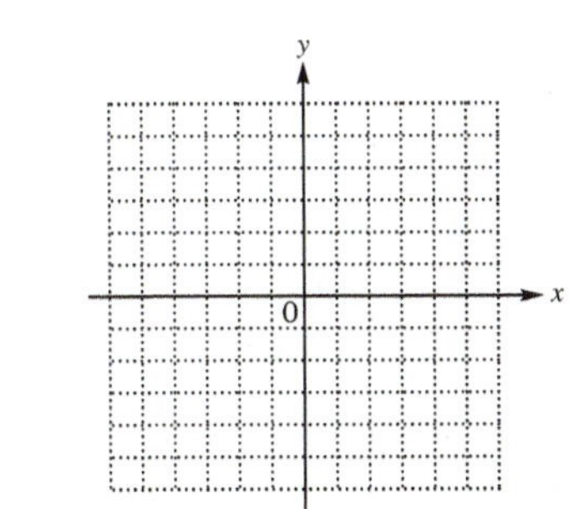

33. $\begin{aligned} x + 2y &= 6 \\ 2x + 4y &= 8 \end{aligned}$

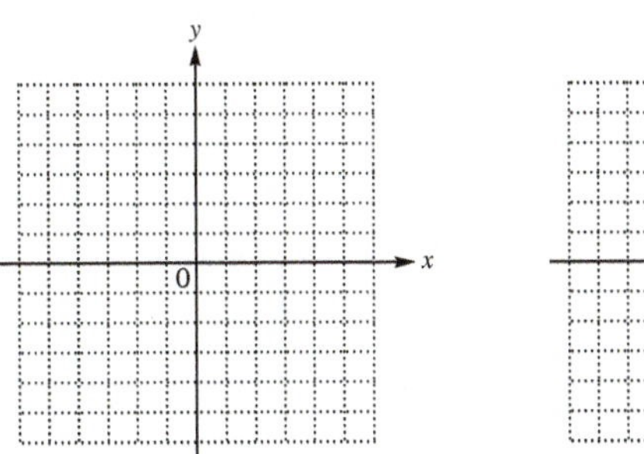

34. $\begin{aligned} 2x - y &= 6 \\ 6x - 3y &= 12 \end{aligned}$

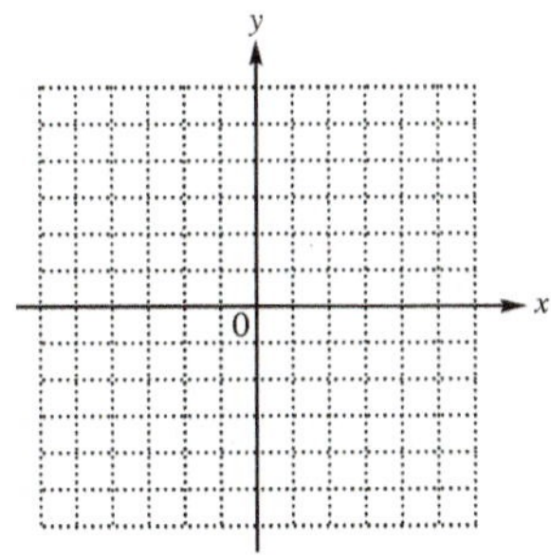

35. $5x - 3y = 2$
$10x - 6y = 4$

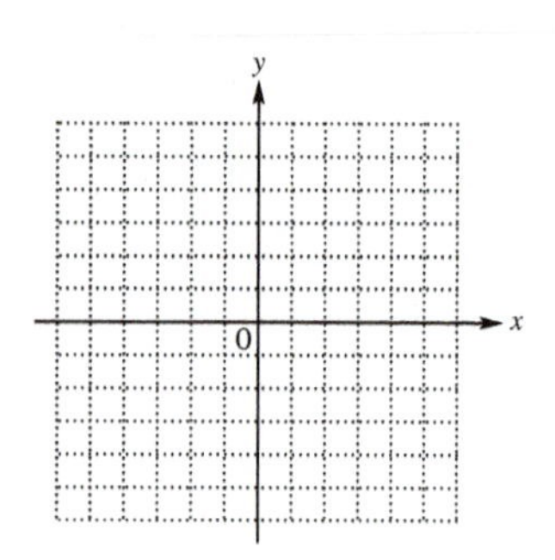

36. $2x - 5y = 8$
$4x - 10y = 16$

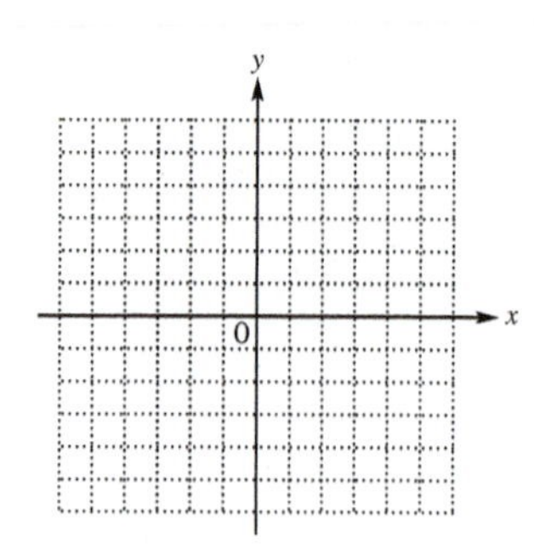

37. $3x - 4y = 24$
$y = -\frac{3}{2}x + 3$

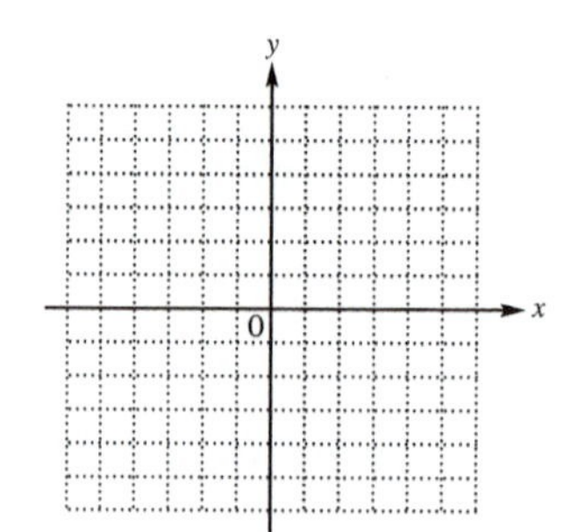

38. $3x - 2y = 12$
$y = -4x + 5$

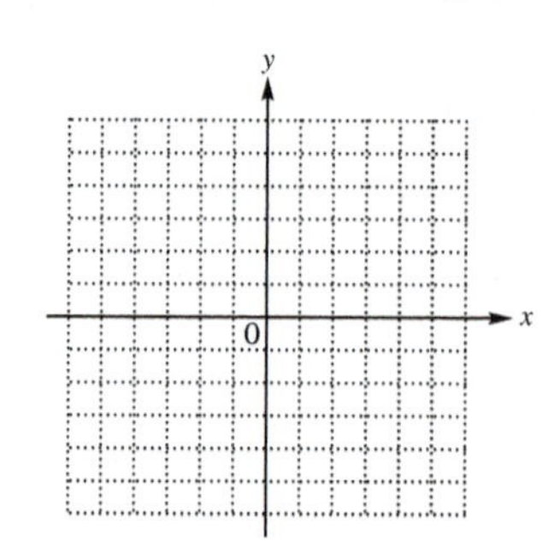

39. $4x - 2y = 8$
$2x = y + 4$

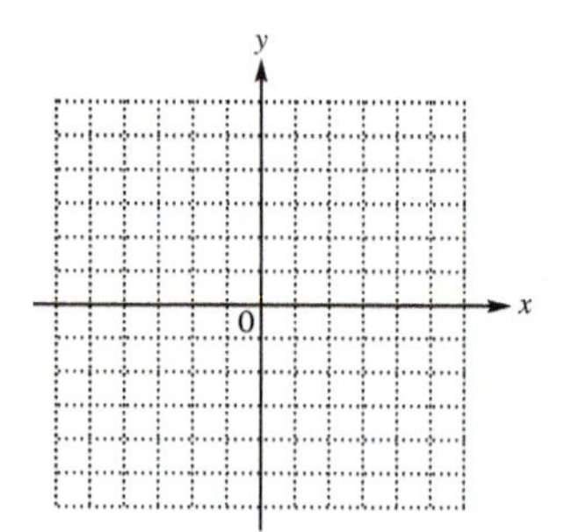

40. $3x = 5 - y$
$6x + 2y = 10$

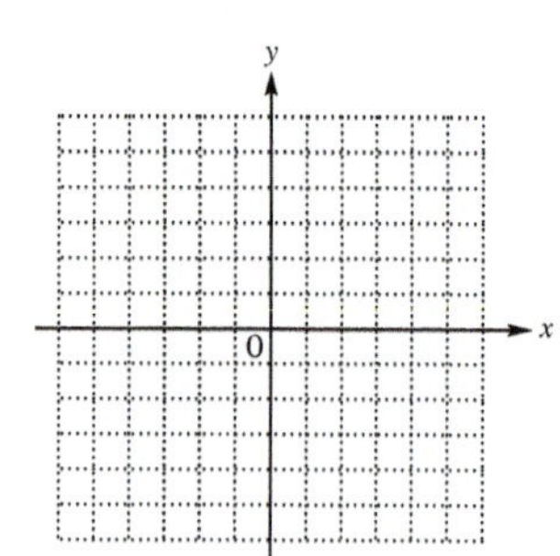

41. $3x = y + 5$
$6x - 5 = 2y$

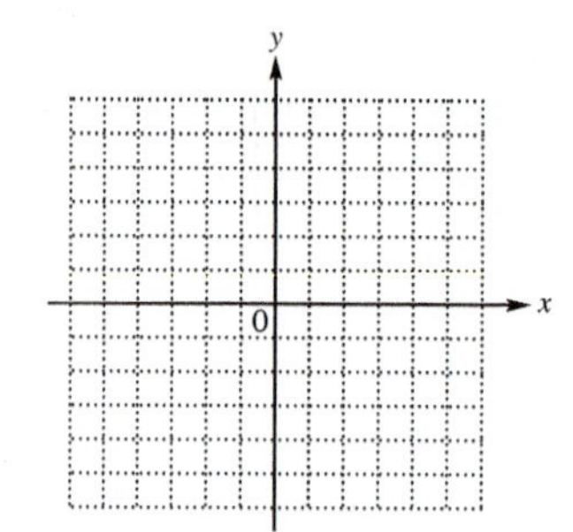

42. $2x = y - 4$
$4x - 2y = -4$

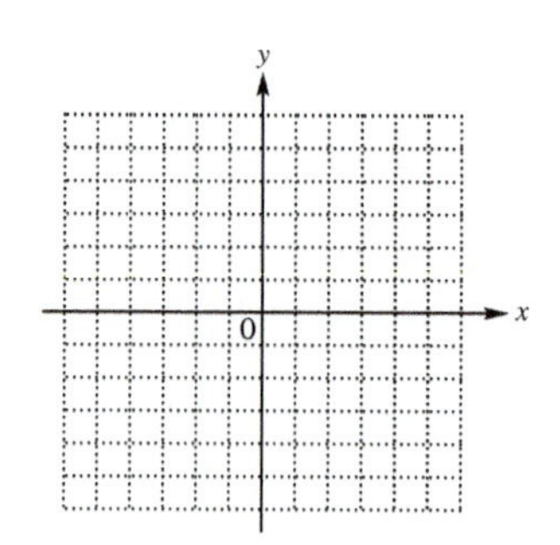

Without graphing, answer the following questions for each linear system. **See Example 4.**

(a) *Is the system inconsistent, are the equations dependent, or neither?*

(b) *Is the graph a pair of intersecting lines, a pair of parallel lines, or one line?*

(c) *Does the system have one solution, no solution, or an infinite number of solutions?*

43. $y - x = -5$
$x + y = 1$

44. $2x + y = 6$
$x - 3y = -4$

45. $x + 2y = 0$
$4y = -2x$

46. $4x - 6y = 10$
$-6x + 9y = -15$

47. $5x + 4y = 7$
$10x + 8y = 4$

48. $y = 3x$
$y + 3 = 3x$

49. $x - 3y = 5$
$2x + y = 8$

50. $2x + 3y = 12$
$2x - y = 4$

51. $5x = 10y$
$\frac{1}{2}x - y = 0$

52. $y = -3x$
$x + \frac{1}{3}y = 0$

53. $3x + 2y = 5$
$-6x - 4y = 10$

54. $2x - y = 6$
$-10x + 5y = -30$

*Economics deals with **supply and demand.** Typically, as the price of an item increases, the demand for the item decreases and the supply increases. If supply and demand can be described by straight-line equations, the point at which the lines intersect determines the **equilibrium supply** and **equilibrium demand.***

The price per unit, p, and the demand, x, for a particular aluminum siding are related by the linear equation $p = 60 - \frac{3}{4}x$, while the price and the supply are related by the linear equation $p = \frac{3}{4}x$. See the figure. Use the graph to work each problem.

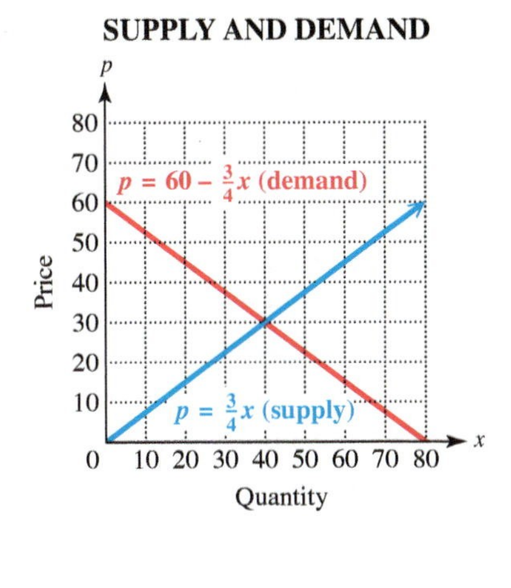

55. At what value of x does supply equal demand? At what value of p does supply equal demand?

56. Express the equilibrium supply and equilibrium demand as an ordered pair of the form (quantity, price).

57. When $x > 40$, does demand exceed supply or does supply exceed demand?

58. When $x < 40$, does demand exceed supply or does supply exceed demand?

The numbers of daily morning and evening newspapers in the United States in selected years over the period 1980–2013 are shown in the graph.

59. For which years were there more evening dailies than morning dailies?

60. Estimate the year in which the number of evening and morning dailies was closest to the same. About how many newspapers of each type were there in that year?

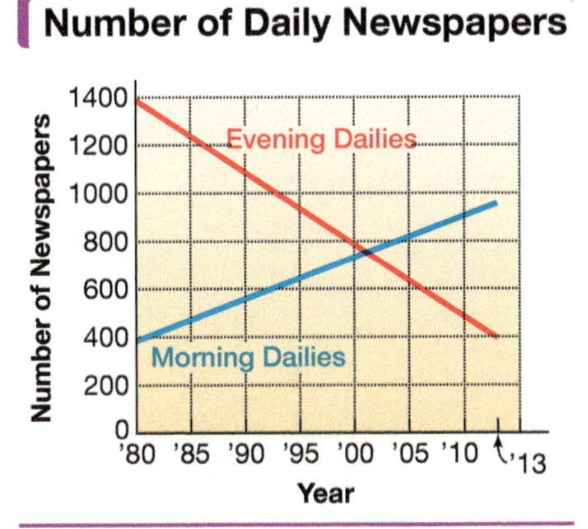

Data from *Editor & Publisher International Year Book.*

Relating Concepts (Exercises 61–64) For Individual or Group Work

*The graph shows sales of music CDs and digital downloads of single songs (in millions) in the United States over selected years. Use the graph to work **Exercises 61–64 in order.***

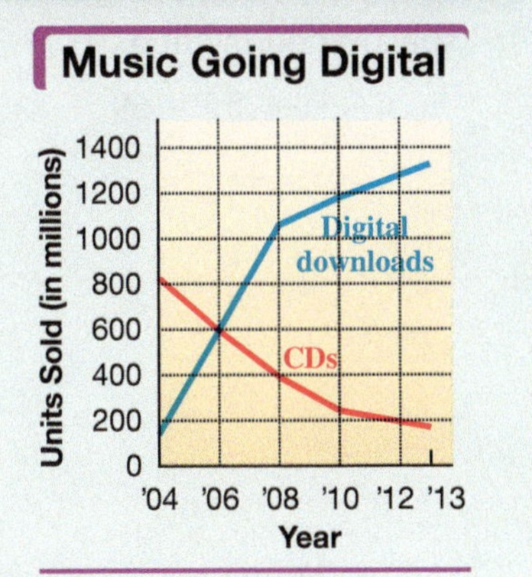

Data from Recording Industry Association of America.

61. In what year did Americans purchase about the same number of CDs as single digital downloads? How many units was this?

62. Express the point of intersection of the two graphs as an ordered pair of the form (year, units in millions).

63. Describe the trend in sales of music CDs over the years 2004 to 2013. If a straight line were used to approximate its graph, would the line have *positive, negative,* or *zero* slope? Explain.

64. If a straight line were used to approximate the graph of sales of digital downloads over the years 2004 to 2013, would the line have *positive, negative,* or *zero* slope? Explain.

4.2 Solving Systems of Linear Equations by Substitution

OBJECTIVES

1. Solve linear systems by substitution.
2. Solve special systems by substitution.
3. Solve linear systems with fractions and decimals.

OBJECTIVE 1 Solve linear systems by substitution.

◀ Work Problem 1 at the Side.

As we see in **Margin Problem 1,** graphing to solve a system of equations has a serious drawback: It is difficult to accurately find a solution such as $\left(\frac{11}{3}, -\frac{4}{9}\right)$ from a graph.

As a result, there are algebraic methods for solving systems of equations. The **substitution method,** which gets its name from the fact that an expression in one variable is *substituted* for the other variable, is one such method.

1 We solved the following system by graphing.

$$2x + 3y = 6$$
$$x - 3y = 5$$

Can we determine the solution? Why or why not?

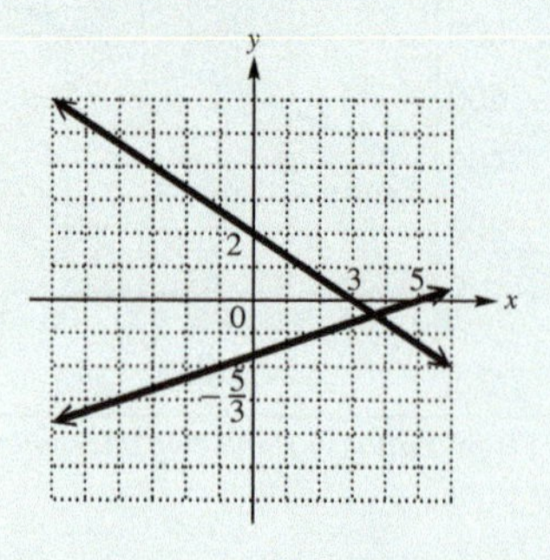

2 Solve the system by the substitution method.

$$3x + 5y = 69$$
$$y = 4x$$

EXAMPLE 1 Using the Substitution Method

Solve the system by the substitution method.

$$3x + 5y = 26 \quad (1)$$
$$y = 2x \quad (2)$$

We number the equations for reference in our discussion.

Equation (2) is already solved for y. This equation says that $y = 2x$, so we substitute $2x$ for y in equation (1).

$3x + 5y = 26$	(1)
$3x + 5(2x) = 26$	Let $y = 2x$.
$3x + 10x = 26$	Multiply.
$13x = 26$	Combine like terms.
$x = 2$	Divide by 13.

Don't stop here.

Now we can find the value of y by substituting 2 for x in either equation. We choose equation (2) because the substitution is easier.

$y = 2x$	(2)
$y = 2(2)$	Let $x = 2$.
$y = 4$	Multiply.

We check that the ordered pair $(2, 4)$ is the solution by substituting **2** for x and **4** for y in *both* equations.

CHECK

$3x + 5y = 26$	(1)	$y = 2x$	(2)
$3(2) + 5(4) \stackrel{?}{=} 26$	Substitute.	$4 \stackrel{?}{=} 2(2)$	Substitute.
$6 + 20 \stackrel{?}{=} 26$	Multiply.	$4 = 4$ ✓	True
$26 = 26$ ✓	True		

Since $(2, 4)$ satisfies both equations, the solution set of the system is $\{(2, 4)\}$.

◀ Work Problem 2 at the Side.

CAUTION

A system is not completely solved until values for both x and y are found. Write the solution set as a set containing an ordered pair.

Answers

1. The solution cannot be determined from the graph because it is not possible to read the exact coordinates.
2. $\{(3, 12)\}$

EXAMPLE 2 Using the Substitution Method

Solve the system by the substitution method.

$$2x + 5y = 7 \quad (1)$$
$$x = -1 - y \quad (2)$$

Equation (2) gives x in terms of y. We substitute $-1 - y$ for x in equation (1).

$$2x + 5y = 7 \quad (1)$$

Be sure to substitute in the *other* equation.

$$2(-1 - y) + 5y = 7 \quad \text{Let } x = -1 - y.$$

Distribute 2 to *both* -1 and $-y$.

$$-2 - 2y + 5y = 7 \quad \text{Distributive property}$$
$$-2 + 3y = 7 \quad \text{Combine like terms.}$$
$$3y = 9 \quad \text{Add 2.}$$
$$y = 3 \quad \text{Divide by 3.}$$

To find x, substitute 3 for y in equation (2).

$$x = -1 - y \quad (2)$$
$$x = -1 - 3 \quad \text{Let } y = 3.$$
$$x = -4 \quad \text{Subtract.}$$

Write the x-coordinate first.

Check that $(-4, 3)$ is the solution.

CHECK

$$2x + 5y = 7 \quad (1)$$
$$2(-4) + 5(3) \stackrel{?}{=} 7 \quad \text{Substitute.}$$
$$-8 + 15 \stackrel{?}{=} 7 \quad \text{Multiply.}$$
$$7 = 7 \checkmark \quad \text{True}$$

$$x = -1 - y \quad (2)$$
$$-4 \stackrel{?}{=} -1 - 3 \quad \text{Substitute.}$$
$$-4 = -4 \checkmark \quad \text{True}$$

Both results are true. The solution set of the system is $\{(-4, 3)\}$.

Work Problem 3 at the Side. ▶

CAUTION

Even though we found y first in **Example 2,** ***the x-coordinate is always written first in the ordered-pair solution of a system.*** The ordered pair $(-4, 3)$ is ***not*** the same as $(3, -4)$.

Solving a Linear System by Substitution

Step 1 **Solve one equation for either variable.** If one of the equations has a variable term with coefficient 1 or -1, choose it because the substitution method is usually easier.

Step 2 **Substitute** for that variable in the other equation. The result should be an equation with just one variable.

Step 3 **Solve** the equation from Step 2.

Step 4 **Find the other value.** Substitute the result from Step 3 into the equation from Step 1 and solve for the other variable.

Step 5 **Check** the values in *both* of the *original* equations. Then write the solution set as a set containing an ordered pair.

3 Solve each system by the substitution method.

(a) $2x + 7y = -12$
$x = 3 - 2y$

(b) $x = y - 3$
$4x + 9y = 1$

Answers

3. **(a)** $\{(15, -6)\}$ **(b)** $\{(-2, 1)\}$

4 Solve each system by the substitution method.

GS (a) Fill in the blanks to solve.

$$x + 4y = -1$$
$$2x - 5y = 11$$

Solve the first equation for x.

$$x = -1 - \_\_\_\_$$

Substitute into the second equation to find y.

$$2(\_\_\_\_) - 5y = 11$$
$$-2 - 8y - 5y = 11$$
$$-2 - \_\_\_\_y = 11$$
$$\_\_\_\_y = 13$$
$$y = -1$$

Find x.

$$x = -1 - 4y$$
$$x = -1 - 4(\_\_\_\_)$$
$$x = \_\_\_\_$$

The solution set is ____.

(b) $2x + 5y = 4$
$x + \ \ y = -1$

Answers

4. (a) $4y$; $-1 - 4y$; 13; -13; -1; 3; $\{(3, -1)\}$
(b) $\{(-3, 2)\}$

EXAMPLE 3 Using the Substitution Method

Solve the system by the substitution method.

$$2x = 4 - y \quad (1)$$
$$5x + 3y = 10 \quad (2)$$

Step 1 Because the coefficient of y in equation (1) is -1, we avoid fractions by choosing this equation and solving it for y.

$$2x = 4 - y \quad (1)$$
$$y + 2x = 4 \quad \text{Add } y.$$
$$y = 4 - 2x \quad \text{Subtract } 2x.$$

Step 2 Now substitute $4 - 2x$ for y in equation (2).

$$5x + 3y = 10 \quad (2)$$
$$5x + 3(4 - 2x) = 10 \quad \text{Let } y = 4 - 2x.$$

Step 3 Solve the equation from Step 2.

$$5x + 12 - 6x = 10 \quad \text{Distributive property}$$

Distribute 3 to *both* 4 and $-2x$.

$$-x + 12 = 10 \quad \text{Combine like terms.}$$
$$-x = -2 \quad \text{Subtract 12.}$$
$$x = 2 \quad \text{Multiply by } -1.$$

Step 4 Equation (1) solved for y is $y = 4 - 2x$. Substitute 2 for x.

$$y = 4 - 2(2) \quad \text{Substitute.}$$
$$y = 0 \quad \text{Multiply, and then subtract.}$$

Step 5 Check that $(2, 0)$ is the solution.

CHECK

$2x = 4 - y$	(1)	$5x + 3y = 10$	(2)
$2(2) \stackrel{?}{=} 4 - 0$	Substitute.	$5(2) + 3(0) \stackrel{?}{=} 10$	Substitute.
$4 = 4$ ✓	True	$10 = 10$ ✓	True

Because both results are true, the solution set of the system is $\{(2, 0)\}$.

◀ Work Problem **4** at the Side.

OBJECTIVE 2 Solve special systems by substitution.

EXAMPLE 4 Solving an Inconsistent System

Solve the system by the substitution method.

$$x = 5 - 2y \quad (1)$$
$$2x + 4y = 6 \quad (2)$$

Equation (1) is solved for x, so we substitute $5 - 2y$ for x in equation (2).

$$2x + 4y = 6 \quad (2)$$
$$2(5 - 2y) + 4y = 6 \quad \text{Let } x = 5 - 2y.$$
$$10 - 4y + 4y = 6 \quad \text{Distributive property}$$
$$10 = 6 \quad \text{False}$$

Continued on Next Page

The false result $10 = 6$ means that the equations in the system have graphs that are parallel lines. The system is inconsistent and has no solution, so the solution set is $\varnothing$.

CHECK We can confirm the solution set by writing each equation in slope-intercept form—that is, solved for y.

$x = 5 - 2y \quad (1)$	$2x + 4y = 6 \quad (2)$
$2y = -x + 5$	$4y = -2x + 6$
$y = -\frac{1}{2}x + \frac{5}{2}$	$y = -\frac{1}{2}x + \frac{3}{2}$

The two lines have the same slope but different y-intercepts. Therefore, they are parallel and do not intersect, confirming that the solution set is $\varnothing$. See **Figure 6.** ✓

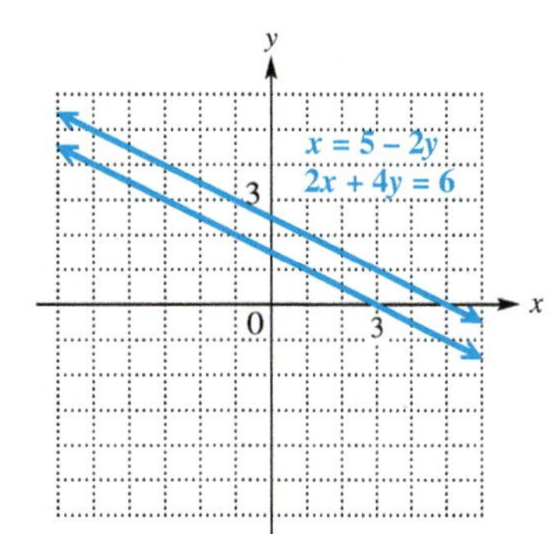

Figure 6

Work Problem 5 at the Side. ▶

5 Solve the system by the substitution method.

$$8x - 2y = 1$$
$$y = 4x - 8$$

EXAMPLE 5 Solving a System with Dependent Equations

Solve the system by the substitution method.

$$3x - y = 4 \qquad (1)$$
$$-9x + 3y = -12 \qquad (2)$$

Begin by solving equation (1) for y to obtain

$$y = 3x - 4. \qquad \text{Equation (1) solved for } y$$

We substitute $3x - 4$ for y in equation (2) and solve the resulting equation.

$$-9x + 3y = -12 \qquad (2)$$
$$-9x + 3(3x - 4) = -12 \qquad \text{Let } y = 3x - 4.$$
$$-9x + 9x - 12 = -12 \qquad \text{Distributive property}$$
$$0 = 0 \qquad \text{Add 12. Combine like terms.}$$

The true result $0 = 0$ means that every solution of one equation is also a solution of the other, so the system has an infinite number of solutions. The solution set written in set-builder notation using equation (1) is

$$\{(x, y) \mid 3x - y = 4\}.$$

CHECK If we multiply equation (1) by -3, we obtain equation (2). Therefore,

$$3x - y = 4 \quad \text{and} \quad -9x + 3y = -12$$

are equivalent equations. They represent the same line. All of the ordered pairs corresponding to points that lie on the common graph are solutions. See **Figure 7.** ✓

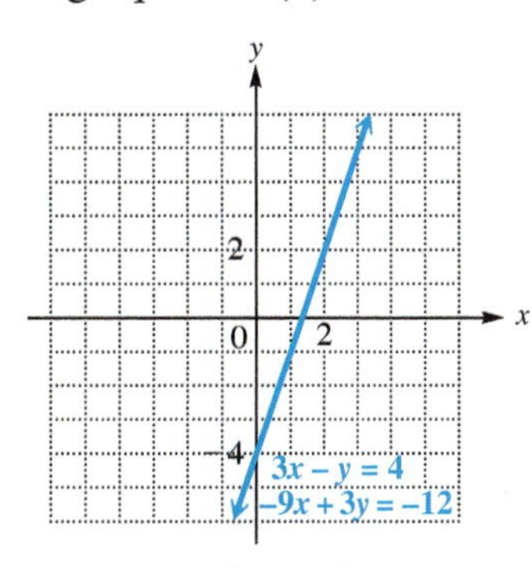

Figure 7

Work Problem 6 at the Side. ▶

6 Solve each system by the substitution method.

(a) $7x - 6y = 10$

$-14x + 20 = -12y$

(b) $8x - y = 4$

$y = 8x + 4$

Answers

5. $\varnothing$

6. **(a)** $\{(x, y) \mid 7x - 6y = 10\}$ **(b)** $\varnothing$

CAUTION

Avoid these common mistakes.

1. Do not give "false" as the solution of an inconsistent system. The correct response is $\varnothing$. **(See Example 4.)**
2. Do not give "true" as the solution of a system of dependent equations. In this text, we write the solution set in set-builder notation using the equation in the system (or an equivalent equation) that is in standard form with integer coefficients having greatest common factor 1 and positive coefficient of x. **(See Example 5.)**

OBJECTIVE 3 **Solve linear systems with fractions and decimals.**

EXAMPLE 6 **Using the Substitution Method (Fractional Coefficients)**

Solve the system by the substitution method.

$$3x + \frac{1}{4}y = 2 \quad (1)$$

$$\frac{1}{2}x + \frac{3}{4}y = -\frac{5}{2} \quad (2)$$

Clear equation (1) of fractions by multiplying each side by 4.

$$4\left(3x + \frac{1}{4}y\right) = 4(2) \quad \text{Multiply by 4, the common denominator.}$$

$$4(3x) + 4\left(\frac{1}{4}y\right) = 4(2) \quad \text{Distributive property}$$

$$12x + y = 8 \quad (3)$$

Now clear equation (2) of fractions by multiplying each side by 4.

$$\frac{1}{2}x + \frac{3}{4}y = -\frac{5}{2} \quad (2)$$

$$4\left(\frac{1}{2}x + \frac{3}{4}y\right) = 4\left(-\frac{5}{2}\right) \quad \text{Multiply by 4, the common denominator.}$$

$$4\left(\frac{1}{2}x\right) + 4\left(\frac{3}{4}y\right) = 4\left(-\frac{5}{2}\right) \quad \text{Distributive property.}$$

$$2x + 3y = -10 \quad (4)$$

The given system of equations has been simplified to an equivalent system.

$$12x + y = 8 \quad (3)$$

$$2x + 3y = -10 \quad (4)$$

To solve this system by substitution, solve equation (3) for y.

$$12x + y = 8 \quad (3)$$

$$y = 8 - 12x \quad \text{Subtract } 12x.$$

Now substitute this result for y in equation (4).

Continued on Next Page

$$2x + 3y = -10 \quad (4)$$

$$2x + 3(8 - 12x) = -10 \quad \text{Let } y = 8 - 12x.$$

$$2x + 24 - 36x = -10 \quad \text{Distributive property}$$

Distribute 3 to both 8 and $-12x$.

$$-34x = -34 \quad \text{Combine like terms. Subtract 24.}$$

$$x = 1 \quad \text{Divide by } -34.$$

Substitute 1 for x in $y = 8 - 12x$ (equation (3) solved for y).

$$y = 8 - 12(1) \quad \text{Let } x = 1.$$

$$y = -4 \quad \text{Multiply, and then subtract.}$$

Check $(1, -4)$ in both of the original equations. The solution set of the system is $\{(1, -4)\}$.

Work Problem 7 at the Side. ▶

EXAMPLE 7 Using the Substitution Method (Decimal Coefficients)

Solve the system by the substitution method.

$$0.5x + 2.4y = 4.2 \quad (1)$$

$$-0.1x + 1.5y = 5.1 \quad (2)$$

Clear each equation of decimals by multiplying by 10.

$$10(0.5x + 2.4y) = 10(4.2) \quad \text{Multiply equation (1) by 10.}$$

$$10(0.5x) + 10(2.4y) = 10(4.2) \quad \text{Distributive property}$$

$$5x + 24y = 42 \quad (3)$$

$$10(-0.1x + 1.5y) = 10(5.1) \quad \text{Multiply equation (2) by 10.}$$

$$10(-0.1x) + 10(1.5y) = 10(5.1) \quad \text{Distributive property}$$

$10(-0.1x) = -1x = -x$

$$-x + 15y = 51 \quad (4)$$

Now solve the equivalent system of equations by substitution.

$$5x + 24y = 42 \quad (3)$$

$$-x + 15y = 51 \quad (4)$$

Equation (4) can be solved for x.

$$x = 15y - 51 \quad \text{Equation (4) solved for } x$$

Substitute this result for x in equation (3).

$$5x + 24y = 42 \quad (3)$$

$$5(15y - 51) + 24y = 42 \quad \text{Let } x = 15y - 51.$$

$$75y - 255 + 24y = 42 \quad \text{Distributive property}$$

$$99y = 297 \quad \text{Combine like terms. Add 255.}$$

$$y = 3 \quad \text{Divide by 99.}$$

Equation (4) solved for x is $x = 15y - 51$. Substitute 3 for y.

$$x = 15(3) - 51 \quad \text{Let } y = 3.$$

$$x = -6 \quad \text{Multiply, and then subtract.}$$

Check $(-6, 3)$ in both of the original equations. The solution set is $\{(-6, 3)\}$.

Work Problem 8 at the Side. ▶

7 Solve each system by the substitution method.

(a) $\frac{2}{3}x + \frac{1}{2}y = 6$

$\frac{1}{2}x - \frac{3}{4}y = 0$

(b) $x + \frac{1}{2}y = \frac{1}{2}$

$\frac{1}{6}x - \frac{1}{3}y = \frac{4}{3}$

8 Solve the system by the substitution method.

$0.2x + 0.3y = 0.5$

$0.3x - 0.1y = 1.3$

Answers

7. **(a)** $\{(6, 4)\}$ **(b)** $\{(2, -3)\}$

8. $\{(4, -1)\}$

4.2 Exercises

FOR EXTRA HELP *Go to MyMathLab for worked-out, step-by-step solutions to exercises enclosed in a square and video solutions to ▶ exercises.*

1. CONCEPT CHECK A student solves the following system and finds that $x = 3$, which is correct. The student gives $\{3\}$ as the solution set.

$$5x - y = 15$$
$$7x + y = 21$$

What Went Wrong? Give the correct solution set.

2. CONCEPT CHECK A student solves the following system and obtains the statement $0 = 0$. The student gives the solution set as $\{(0, 0)\}$.

$$x + y = 4$$
$$2x + 2y = 8$$

What Went Wrong? Give the correct solution set.

CONCEPT CHECK *Answer each question.*

3. When we use the substitution method, how can we tell that a system has no solution?

4. When we use the substitution method, how can we tell that a system has an infinite number of solutions?

Solve each system by the substitution method. Check each solution. ***See Examples 1–5.***

5. $x + y = 12$
$y = 3x$

6. $x + 3y = -28$
$y = -5x$

7. $3x + 2y = 27$
$x = y + 4$

8. $4x + 3y = -5$
$x = y - 3$

9. $3x + 4 = -y$
$2x + y = 0$

10. $2x - 5 = -y$
$x + 3y = 0$

11. $7x + 4y = 13$
$x + y = 1$

12. $3x - 2y = 19$
$x + y = 8$

13. $3x + 5y = 25$
$x - 2y = -10$

14. $5x + 2y = -15$
$2x - y = -6$

15. $y = 6 - x$
$y = 2x + 3$

16. $y = 4x - 4$
$y = -3x - 11$

17. $3x - y = 5$
$y = 3x - 5$

18. $4x - y = -3$
$y = 4x + 3$

19. $2x + 8y = 3$
$x = 8 - 4y$

20. $2x + 10y = 3$
$x = 1 - 5y$

21. $2y = 4x + 24$
$2x - y = -12$

22. $2y = 14 - 6x$
$3x + y = 7$

23. $x + y = 0$
$3x - 3y = 0$

24. $5x + y = 0$
$x - y = 0$

25. $x = y - 4$
$x - y = 1$

26. $x = 2 - y$
$x + y = -5$

27. $6x - 8y = 6$
$2y = -2 + 3x$

28. $3x + 2y = 6$
$6x = 8 + 4y$

Solve each system by the substitution method. Check each solution. ***See Examples 6 and 7.***

29. $\frac{1}{5}x + \frac{2}{3}y = -\frac{8}{5}$
$3x - y = 9$

30. $\frac{1}{3}x - \frac{1}{2}y = -\frac{2}{3}$
$4x + y = 6$

31. $\frac{1}{2}x + \frac{1}{3}y = -\frac{1}{3}$
$\frac{1}{2}x + 2y = -7$

32. $\frac{1}{6}x + \frac{1}{6}y = 1$
$-\frac{1}{2}x - \frac{1}{3}y = -5$

33. $\frac{x}{5} + 2y = \frac{16}{5}$
$\frac{3x}{5} + \frac{y}{2} = -\frac{7}{5}$

34. $\frac{x}{2} + \frac{y}{3} = \frac{7}{6}$
$\frac{x}{4} - \frac{3y}{2} = \frac{9}{4}$

35. $-0.3x + 0.5y = -1.5$
$0.4x + 0.5y = 2$

36. $-0.1x + 0.1y = 0.2$
$0.3x + 0.1y = 0.2$

37. $0.1x + 0.9y = -2$
$0.5x - 0.2y = 4.1$

38. $0.2x - 1.3y = -3.2$
$-0.1x + 2.7y = 9.8$

39. $0.8x - 0.1y = 1.3$
$2.2x + 1.5y = 8.9$

40. $0.3x - 0.1y = 2.1$
$0.6x + 0.3y = -0.3$

Relating Concepts (Exercises 41–44) For Individual or Group Work

A system of linear equations can be used to model the cost and the revenue of a business. ***Work Exercises 41–44 in order.***

41. Suppose that it costs \$5000 to start a business manufacturing and selling bicycles. Each bicycle will cost \$400 to manufacture. Explain why the linear equation

$$y_1 = 400x + 5000 \quad (y_1 \text{ in dollars})$$

gives the *total* cost to manufacture x bicycles.

42. We decide to sell each bicycle for \$600. Write an equation using y_2 (in dollars) to express the revenue when we sell x bicycles.

43. Form a system from the two equations in **Exercises 41 and 42.** Then solve the system, assuming $y_1 = y_2$, that is, cost = revenue.

44. The value of x in **Exercise 43** is the number of bicycles it takes to *break even.* Fill in the blanks:

When ______ bicycles are sold, the break-even point is reached. At that point, we have spent ______ dollars and taken in ______ dollars.

4.3 Solving Systems of Linear Equations by Elimination

OBJECTIVES

1. Solve linear systems by elimination.
2. Multiply when using the elimination method.
3. Use an alternative method to find the second value in a solution.
4. Solve special systems by elimination.

OBJECTIVE 1 Solve linear systems by elimination. Recall that adding the same quantity to each side of an equation results in equal sums.

$$\textbf{If } A = B, \textbf{ then } A + C = B + C.$$

We can take this addition a step further. Adding *equal* quantities, rather than the *same* quantity, to each side of an equation also results in equal sums.

$$\textbf{If } A = B \textbf{ and } C = D, \textbf{ then } A + C = B + D.$$

The **elimination method** uses the addition property of equality to solve systems.

EXAMPLE 1 Using the Elimination Method

Solve the system by the elimination method.

$$x + y = 5 \quad (1)$$
$$x - y = 3 \quad (2)$$

Each equation in this system is a statement of equality, so the sum of the left sides equals the sum of the right sides. Adding vertically in this way gives the following.

The goal is to eliminate a variable.

$$\begin{aligned} x + y &= 5 && (1) \\ x - y &= 3 && (2) \\ \hline 2x \quad &= 8 && \text{Add left sides and add right sides.} \\ x &= 4 && \text{Divide by 2.} \end{aligned}$$

Notice that y has been eliminated. The result, $x = 4$, gives the x-value of the ordered-pair solution of the given system. To find the y-value of the solution, substitute 4 for x in either of the two equations of the system. We choose equation (1).

$$\begin{aligned} x + y &= 5 && (1) \\ 4 + y &= 5 && \text{Let } x = 4. \\ y &= 1 && \text{Subtract 4.} \end{aligned}$$

Check the ordered pair, $(4, 1)$, by substituting 4 for x and 1 for y in both equations of the given system.

CHECK

$x + y = 5$	(1)	$x - y = 3$	(2)
$4 + 1 \stackrel{?}{=} 5$	Substitute.	$4 - 1 \stackrel{?}{=} 3$	Substitute.
$5 = 5$ ✓	True	$3 = 3$ ✓	True

Because *both* results are true, the solution set of the system is $\{(4, 1)\}$.

◀ Work Problem 1 at the Side.

CAUTION

A system is not completely solved until values for both x and y are found. Do not stop after finding the value of only one variable. Remember to write the solution set as a set containing an ordered pair.

---

1 Solve each system by the elimination method.

GS **(a)** Fill in the blanks to solve the following system.

$$\begin{aligned} x + y &= 8 && (1) \\ x - y &= 2 && (2) \\ \hline \_\_\_\_ &= \_\_\_\_ && \text{Add.} \\ x &= \_\_\_\_ \end{aligned}$$

Find y.

$$\begin{aligned} x - y &= 2 && (2) \\ \_\_\_\_ - y &= 2 \\ -y &= \_\_\_\_ \\ y &= \_\_\_\_ \end{aligned}$$

The solution set is ____.

(b) $3x - y = 7$

$2x + y = 3$

Answers

1. (a) $2x$; 10; 5; 5; -3; 3; $\{(5, 3)\}$
(b) $\{(2, -1)\}$

With the elimination method, the idea is to *eliminate* one of the two variables in a system. ***In order for us to do this, one pair of variable terms in the two equations must have coefficients that are opposites.***

Solving a Linear System by Elimination

Step 1 **Write both equations in the form $Ax + By = C$.**

Step 2 **Transform the equations as needed so that the coefficients of one pair of variable terms are opposites.** Multiply one or both equations by appropriate numbers so that the sum of the coefficients of either the x- or y-terms is 0.

Step 3 **Add** the new equations to *eliminate* a variable. The sum should be an equation with just one variable.

Step 4 **Solve** the equation from Step 3 for the remaining variable.

Step 5 **Find the other value.** Substitute the result from Step 4 into either of the original equations, and solve for the other variable.

Step 6 **Check** the values in *both* of the *original* equations. Then write the solution set as a set containing an ordered pair.

It does not matter which variable is eliminated first. Usually we choose the one that is more convenient to work with.

EXAMPLE 2 **Using the Elimination Method**

Solve the system.

$$y + 11 = 2x \quad (1)$$
$$5x = y + 26 \quad (2)$$

Step 1 Write both equations in the form $Ax + By = C$.

$$-2x + y = -11 \quad \text{Subtract } 2x \text{ and } 11 \text{ in equation (1).}$$
$$5x - y = 26 \quad \text{Subtract } y \text{ in equation (2).}$$

Step 2 Because the coefficients of y are 1 and -1, adding will eliminate y. It is not necessary to multiply either equation by a number.

Step 3 Add the two equations.

$$\begin{array}{r} -2x + y = -11 \\ 5x - y = 26 \\ \hline 3x \quad = 15 \end{array} \quad \text{Add in columns.}$$

Step 4 Solve. $\quad x = 5 \quad$ Divide by 3.

Step 5 Find the value of y by substituting 5 for x in either of the original equations.

$$y + 11 = 2x \quad (1)$$
$$y + 11 = 2(5) \quad \text{Let } x = 5.$$
$$y + 11 = 10 \quad \text{Multiply.}$$
$$y = -1 \quad \text{Subtract 11.}$$

Continued on Next Page

Step 6 Check the ordered pair $(5, -1)$ by substituting $x = 5$ and $y = -1$ in both of the original equations.

CHECK

$y + 11 = 2x$	(1)	$5x = y + 26$	(2)
$-1 + 11 \stackrel{?}{=} 2(5)$	Substitute.	$5(5) \stackrel{?}{=} -1 + 26$	Substitute.
$10 = 10$ ✓	True	$25 = 25$ ✓	True

Because both results are true, the solution set is $\{(5, -1)\}$.

◀ Work Problem 2 at the Side.

2 Solve each system by the elimination method.

(a) $2x = y + 2$

$4x + y = 10$

(b) $-5y = -8x + 32$

$4 = 4x + 5y$

OBJECTIVE 2 Multiply when using the elimination method. Sometimes we need to multiply each side of one or both equations in a system by some number so that adding the equations will eliminate a variable.

EXAMPLE 3 Using the Elimination Method

Solve the system.

$$3x - 2y = 10 \quad (1)$$
$$x + 5y = -8 \quad (2)$$

Step 1 The equations are already written in $Ax + By = C$ form.

Step 2 Adding the two equations gives $4x + 3y = 2$, which does not eliminate either variable. However, multiplying equation (2) by -3 and then adding will eliminate the variable x.

Step 3 Add the two equations.

$$3x - 2y = 10 \quad (1)$$
$$-3x - 15y = 24 \quad \text{Multiply equation (2) by } -3.$$
$$-17y = 34 \quad \text{Add.}$$

Step 4 Solve. $y = -2$ Divide by -17.

Step 5 Find the value of x by substituting -2 for y in either of the original equations.

$$x + 5y = -8 \quad (2)$$
$$x + 5(-2) = -8 \quad \text{Let } y = -2.$$
$$x - 10 = -8 \quad \text{Multiply.}$$
$$x = 2 \quad \text{Add 10.}$$

Step 6 Check the ordered pair $(2, -2)$ by substituting $x = 2$ and $y = -2$ in both of the original equations.

CHECK

$3x - 2y = 10$	(1)	$x + 5y = -8$	(2)
$3(2) - 2(-2) \stackrel{?}{=} 10$	Substitute.	$2 + 5(-2) \stackrel{?}{=} -8$	Substitute.
$10 = 10$ ✓	True	$-8 = -8$ ✓	True

Because both results are true, the solution set is $\{(2, -2)\}$.

◀ Work Problem 3 at the Side.

3 Solve each system by the elimination method.

GS **(a)** Fill in the blanks to solve the following system.

$$4x - 3y = 6 \quad (1)$$
$$2x - 7y = -8 \quad (2)$$

Multiply equation (2) by _____ and then add to eliminate the variable x.

$$4x - 3y = 6 \quad (1)$$
$$\_\_\_\_x + 14y = \_\_\_\_$$
$$11y = \_\_\_\_ \quad \text{Add.}$$
$$y = 2$$

Find the value of x by substituting _____ for y in either of the original equations. The value of x is _____.

The solution set is _____.

(b) $5x - y = 7$

$3x + 4y = -5$

> **CAUTION**
> When using the elimination method, remember to *multiply both sides* of an equation by the same nonzero number.

Answers

2. (a) $\{(2, 2)\}$ (b) $\left\{\left(3, -\frac{8}{5}\right)\right\}$

3. (a) -2; -4; 16; 22; 2; 3; $\{(3, 2)\}$
(b) $\{(1, -2)\}$

EXAMPLE 4 **Using the Elimination Method**

Solve the system.

$$2x + 3y = -15 \quad (1)$$
$$5x + 2y = 1 \quad (2)$$

Adding the two equations gives $7x + 5y = -14$, which does not eliminate either variable. However, we can multiply each equation by a suitable number so that the coefficients of one of the two variables are opposites. For example, to eliminate x, multiply each side of equation (1) by 5, and each side of equation (2) by -2.

$$10x + 15y = -75 \quad \text{Multiply both sides of equation (1) by 5.}$$
$$-10x - 4y = -2 \quad \text{Multiply both sides of equation (2) by } -2.$$
$$11y = -77 \quad \text{Add.}$$
$$y = -7 \quad \text{Divide by 11.}$$

The coefficients of x are opposites.

Find the value of x by substituting -7 for y in either equation (1) or (2).

$$5x + 2y = 1 \quad (2)$$
$$5x + 2(-7) = 1 \quad \text{Let } y = -7.$$
$$5x - 14 = 1 \quad \text{Multiply.}$$
$$5x = 15 \quad \text{Add 14.}$$
$$x = 3 \quad \text{Divide by 5.}$$

CHECK

$2x + 3y = -15$ (1)	$5x + 2y = 1$ (2)
$2(3) + 3(-7) \stackrel{?}{=} -15$ Substitute.	$5(3) + 2(-7) \stackrel{?}{=} 1$ Substitute.
$-15 = -15$ ✓ True	$1 = 1$ ✓ True

The solution set is $\{(3, -7)\}$. Write the x-value first.

Work Problem 4 at the Side. ▶

OBJECTIVE ▶ **3 Use an alternative method to find the second value in a solution.** Sometimes it is easier to find the value of the second variable in a solution by using the elimination method twice.

EXAMPLE 5 **Finding the Second Value (Alternative Method)**

Solve the system.

$$4x = 9 - 3y \quad (1)$$
$$5x - 2y = 8 \quad (2)$$

Write equation (1) in $Ax + By = C$ form by adding $3y$ to each side.

$$4x + 3y = 9 \quad (3)$$
$$5x - 2y = 8 \quad (2)$$

One way to proceed is to eliminate y. We multiply each side of equation (3) by 2 and each side of equation (2) by 3, and then add.

Continued on Next Page

4 **(a)** Multiply each equation in the system in **Example 4** by a suitable number so that the variable y is eliminated.

$$2x + 3y = -15$$
$$5x + 2y = 1$$

Complete the solution, and give the solution set. Is it the same as in **Example 4?** (*Yes* / *No*)

(b) Solve the system.

$$6x + 7y = 4$$
$$5x + 8y = -1$$

Answers

4. (a) $\{(3, -7)\}$; Yes **(b)** $\{(3, -2)\}$

5 Solve each system of equations.

(a) $4x + 9y = 3$

$5y = 6 - 3x$

(b) $3y = 8 + 4x$

$6x = 9 - 2y$

6 Solve each system by the elimination method.

(a) $4x + 3y = 10$

$2x + \frac{3}{2}y = 12$

(b) $4x - 6y = 10$

$-10x + 15y = -25$

Answers

5. **(a)** $\left\{\left(\frac{39}{7}, -\frac{15}{7}\right)\right\}$ **(b)** $\left\{\left(\frac{11}{26}, \frac{42}{13}\right)\right\}$

6. **(a)** $\emptyset$
(b) $\{(x, y) \mid 2x - 3y = 5\}$
(*Note:* To write the solution set, we divided each term of the equation $4x - 6y = 10$ by 2 so that the coefficients would have greatest common factor 1.)

$$8x + 6y = 18 \quad \text{Multiply equation (3), } 4x + 3y = 9, \text{ by 2.}$$
$$15x - 6y = 24 \quad \text{Multiply equation (2), } 5x - 2y = 8, \text{ by 3.}$$
$$23x = 42 \quad \text{Add.}$$
$$x = \frac{42}{23} \quad \text{Divide by 23.}$$

The coefficients of y are opposites.

Substituting $\frac{42}{23}$ for x in one of the given equations would give y, but the arithmetic would be complicated. Instead, solve for y by starting again with the original equations in $Ax + By = C$ form (equations (3) and (2)) and eliminating x.

$$20x + 15y = 45 \quad \text{Multiply equation (3), } 4x + 3y = 9, \text{ by 5.}$$
$$-20x + 8y = -32 \quad \text{Multiply equation (2), } 5x - 2y = 8, \text{ by } -4.$$
$$23y = 13 \quad \text{Add.}$$
$$y = \frac{13}{23} \quad \text{Divide by 23.}$$

The coefficients of x are opposites.

Check that the solution set is $\left\{\left(\frac{42}{23}, \frac{13}{23}\right)\right\}$.

◀ **Work Problem 5 at the Side.**

Note

When the value of the first variable is a fraction, the method used in **Example 5** helps avoid arithmetic errors. This method could be used to solve any system.

OBJECTIVE 4 Solve special systems by elimination.

EXAMPLE 6 Solve Special Systems Using the Elimination Method

Solve each system by the elimination method.

(a)
$$2x + 4y = 5 \quad (1)$$
$$4x + 8y = -9 \quad (2)$$

Multiply each side of equation (1) by -2. Then add the two equations.

$$-4x - 8y = -10 \quad \text{Multiply equation (1) by } -2.$$
$$4x + 8y = -9 \quad (2)$$
$$0 = -19 \quad \text{False}$$

The false statement $0 = -19$ indicates that the system has solution set $\emptyset$.

(b)
$$3x - y = 4 \quad (1)$$
$$-9x + 3y = -12 \quad (2)$$

Multiply each side of equation (1) by 3. Then add the two equations.

$$9x - 3y = 12 \quad \text{Multiply equation (1) by 3.}$$
$$-9x + 3y = -12 \quad (2)$$
$$0 = 0 \quad \text{True}$$

A true statement occurs when the equations are equivalent. This indicates that every solution of one equation is also a solution of the other. The solution set is $\{(x, y) \mid 3x - y = 4\}$.

◀ **Work Problem 6 at the Side.**

4.3 Exercises

FOR EXTRA HELP *Go to MyMathLab for worked-out, step-by-step solutions to exercises enclosed in a square and video solutions to ▶ exercises.*

CONCEPT CHECK *In Exercises 1 and 2, answer* true *or* false. *If false, tell why.*

1. To eliminate the x-terms in the following system, we should multiply equation (1) by -2 and then add the result to equation (2).

$$3x + 5y = 7 \quad (1)$$
$$6x + 3y = -10 \quad (2)$$

2. To eliminate the y-terms in the following system, we should multiply equation (2) by 3 and then add the result to equation (1).

$$2x + 12y = 7 \quad (1)$$
$$3x + 4y = 1 \quad (2)$$

3. CONCEPT CHECK When solving the system

$$x + y = 1 \quad (1)$$
$$-x - y = 2 \quad (2)$$

by elimination, a student obtained the false statement

$$0 = 3.$$

He then concluded that the solution set was $\{(0, 3)\}$. ***What Went Wrong?*** Give the correct solution set.

4. CONCEPT CHECK To eliminate the y-terms in the system

$$2x - y = 5 \quad (1)$$
$$-6x + 3y = -15, \quad (2)$$

a student multiplied equation (1) by 3 to obtain

$$6x - 3y = 5.$$

When she added this result to equation (2), she concluded that the solution set was $\emptyset$. ***What Went Wrong?*** Give the correct solution set.

Solve each system by the elimination method. Check each solution. ***See Examples 1 and 2.***

5. ▶ $x + y = 2$; $2x - y = -5$

6. $3x - y = -12$; $x + y = 4$

7. $2x + y = -5$; $x - y = 2$

8. $2x + y = -15$; $-x - y = 10$

9. $x + 2y = 11$; $-x + 3y = 4$

10. $-x - 4y = 10$; $x - 4y = 14$

11. ▶ $2y = -3x$; $-3x - y = 3$

12. $5x = y + 5$; $2y = 5x$

13. $6x - y = -1$; $5y = 17 + 6x$

14. $y = 9 - 6x$; $-6x + 3y = 15$

15. $2x - 6 = -3y$; $5x - 3y = -27$

16. $x - 2 = -y$; $2x = y + 10$

Solve each system by the elimination method. Check each solution. ***See Examples 3–5.***

17. ▶ $2x - y = 12$; $3x + 2y = -3$

18. $x + y = 3$; $-3x + 2y = -19$

19. $x + 3y = 19$; $2x - y = 10$

20. $4x - 3y = -19$; $2x + y = 13$

21. $x + 4y = 16$
$3x + 5y = 20$

22. $2x + y = 8$
$5x - 2y = -16$

23. $5x - 3y = -20$
$-3x + 6y = 12$

24. $4x + 3y = -28$
$5x - 6y = -35$

25. $2x - 8y = 0$
$4x + 5y = 0$

26. $3x - 15y = 0$
$6x + 10y = 0$

27. $x + y = 7$
$x + y = -3$

28. $x - y = 4$
$x - y = -3$

29. $-x + 3y = 4$
$-2x + 6y = 8$

30. $6x - 2y = 24$
$-3x + y = -12$

31. $2x + 3y = 21$
$5x - 2y = -14$

32. $5x + 4y = 12$
$3x + 5y = 15$

33. $3x - 7 = -5y$
$5x + 4y = -10$

34. $2x + 3y = 13$
$6 + 2y = -5x$

35. $4x - 3y = -2$
$5x + 3 = 2y$

36. $2x + 3y = 0$
$4x + 12 = 9y$

37. $24x + 12y = -7$
$16x - 17 = 18y$

38. $9x + 4y = -3$
$6x + 7 = -6y$

39. $5x - 2y = 3$
$10x - 4y = 5$

40. $3x - 5y = 1$
$6x - 10y = 4$

41. $6x + 3y = 0$
$-18x - 9y = 0$

42. $3x - 5y = 0$
$9x - 15y = 0$

43. $3x = 3 + 2y$
$-\frac{4}{3}x + y = \frac{1}{3}$

44. $3x = 27 + 2y$
$x - \frac{7}{2}y = -25$

45. $\frac{1}{5}x + y = \frac{6}{5}$

$\frac{1}{10}x + \frac{1}{3}y = \frac{5}{6}$

46. $\frac{1}{3}x + \frac{1}{2}y = \frac{13}{6}$

$\frac{1}{2}x - \frac{1}{4}y = -\frac{3}{4}$

47. $2.4x + 1.7y = 7.6$

$1.2x - 0.5y = 9.2$

48. $0.5x + 3.4y = 13$

$1.5x - 2.6y = -25$

Relating Concepts (Exercises 49–52) For Individual or Group Work

The graph shows average U.S. movie theater ticket prices from 2004 through 2014. In 2004 the average ticket price was \$6.21, *as represented on the graph by the point* $P(2004, 6.21)$. *In 2014, the average ticket price was* \$8.17, *as represented on the graph by the point* $Q(2014, 8.17)$. ***Work Exercises 49–52 in order.***

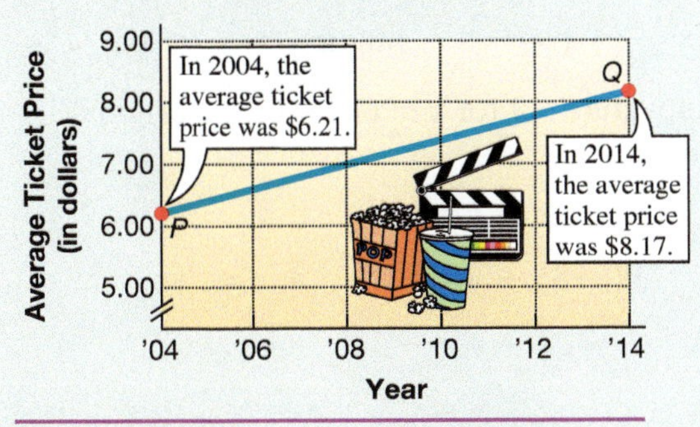

Data from Motion Picture Association of America.

49. The line segment has an equation that can be written in the form

$$y = ax + b.$$

Using the coordinates of point P with $x = 2004$ and $y = 6.21$, write an equation in the variables a and b.

50. Using the coordinates of point Q with $x = 2014$ and $y = 8.17$, write a second equation in the variables a and b.

51. Write the system of equations formed from the two equations in **Exercises 49 and 50.** Solve the system, giving the values of a and b to three decimal places. (*Hint*: Eliminate b using the elimination method.)

52. Answer each question.

(a) What is the equation of the segment PQ?

(b) Let $x = 2013$ in the equation from part (a), and solve for y (to two decimal places). How does the result compare with the actual figure of \$8.13?

Summary Exercises *Applying Techniques for Solving Systems of Linear Equations*

EXAMPLE **Comparing the Substitution and Elimination Methods**

Consider the following system.

$$3x + y = -2 \quad (1)$$
$$5x + 2y = 4 \quad (2)$$

(a) Solve by the substitution method.

Solve equation (1) for y.

$$y = -2 - 3x \quad \text{Subtract } 3x.$$

Substitute $-2 - 3x$ for y in equation (2).

$$5x + 2y = 4 \quad (2)$$
$$5x + 2(-2 - 3x) = 4 \quad \text{Let } y = -2 - 3x.$$
$$5x - 4 - 6x = 4$$
$$-x = 8 \quad \text{Combine like terms. Add 4.}$$
$$x = -8 \quad \text{Multiply by } -1.$$

Substitute -8 for x in $y = -2 - 3x$ (that is, equation (1) solved for y.)

$$y = -2 - 3(-8) \quad \text{Let } x = -8.$$
$$y = 22 \quad \text{Multiply. Subtract.}$$

Check that the solution set is $\{(-8, 22)\}$.

(b) Solve by the elimination method.

Multiply equation (1) by -2 and add the result to equation (2) to eliminate the y-terms.

$$-6x - 2y = 4 \quad \text{Multiply equation (1) by } -2.$$
$$5x + 2y = 4 \quad (2)$$
$$-x = 8 \quad \text{Add.}$$
$$x = -8 \quad \text{Multiply by } -1.$$

Substitute -8 for x in either of the original equations.

$$3x + y = -2 \quad (1)$$
$$3(-8) + y = -2 \quad \text{Let } x = -8.$$
$$-24 + y = -2 \quad \text{Multiply.}$$
$$y = 22 \quad \text{Add 24.}$$

Check that the solution set is $\{(-8, 22)\}$.

Some systems are more easily solved using one method than using the other.

Guidelines for Choosing a Method to Solve a System of Linear Equations

1. If one of the equations of the system is already solved for one of the variables, the substitution method is the better choice.

$$3x + 4y = 9 \qquad \qquad x = 3y - 7$$
$$y = 2x - 6 \quad \text{and} \quad -5x + 3y = 9$$

2. If both equations are in $Ax + By = C$ form and none of the variables has coefficient -1 or 1, the elimination method is the better choice.

$$4x - 11y = 3$$
$$-2x + 3y = 4$$

3. If one or both of the equations are in $Ax + By = C$ form and the coefficient of one of the variables is -1 or 1, either method is appropriate.

$$3x + y = -2$$
$$5x + 2y = 4$$

This system is solved by both methods in the example above for comparison.

CONCEPT CHECK *Use the preceding guidelines to solve each problem.*

1. To minimize the amount of work required, determine whether you would use the substitution or the elimination method to solve each system, and why. *Do not actually solve.*

(a) $3x + 2y = 18$
$y = 3x$

(b) $3x + y = -7$
$x - y = -5$

(c) $3x - 2y = 0$
$9x + 8y = 7$

2. Which system would be easier to solve using the substitution method? Why?

System A: $5x - 3y = 7$
$2x + 8y = 3$

System B: $7x + 2y = 4$
$y = -3x + 1$

In each problem, **(a)** *solve the system by the substitution method,* **(b)** *solve the system by the elimination method, and* **(c)** *tell which method you prefer for that particular system and why.*

3. $4x - 3y = -8$
$x + 3y = 13$

4. $2x + 5y = 0$
$x = -3y + 1$

Solve each system by the method of your choice. (For Exercises 5–7, see your answers for ***Exercise 1.****)*

5. $3x + 2y = 18$
$y = 3x$

6. $3x + y = -7$
$x - y = -5$

7. $3x - 2y = 0$
$9x + 8y = 7$

8. $x + y = 7$
$x = -3 - y$

9. $5x - 4y = 15$
$-3x + 6y = -9$

10. $4x + 2y = 3$
$y = -x$

11. $3x = 7 - y$
$2y = 14 - 6x$

12. $3x - 5y = 7$
$2x + 3y = 30$

13. $5x = 7 + 2y$
$5y = 5 - 3x$

14. $4x + 3y = 1$
$3x + 2y = 2$

15. $2x - 3y = 7$
$-4x + 6y = 14$

16. $7x - 4y = 0$
$3x = 2y$

17. $6x + 5y = 13$
$3x + 3y = 4$

18. $x - 3y = 7$
$4x + y = 5$

Solve each system by any method. First clear all fractions or decimals.

19. $\frac{1}{4}x - \frac{1}{5}y = 9$
$y = 5x$

20. $\frac{1}{5}x + \frac{2}{3}y = -\frac{8}{5}$
$x - 5y = 17$

21. $\frac{1}{6}x + \frac{1}{6}y = 2$
$-\frac{1}{2}x - \frac{1}{3}y = -8$

22. $\frac{x}{5} + 2y = \frac{8}{5}$
$\frac{3x}{5} + \frac{y}{2} = -\frac{7}{10}$

23. $\frac{x}{5} + y = 6$
$\frac{x}{10} + \frac{y}{3} = \frac{5}{6}$

24. $\frac{2}{5}x + \frac{4}{3}y = -8$
$\frac{7}{10}x - \frac{2}{9}y = 9$

25. $0.2x + 0.3y = 1.0$
$-0.3x + 0.1y = 1.8$

26. $0.3x - 0.2y = 0.9$
$0.2x - 0.3y = 0.1$

4.4 Applications of Linear Systems

OBJECTIVES

1. Solve problems about unknown numbers.
2. Solve problems about quantities and their costs.
3. Solve problems about mixtures.
4. Solve problems about distance, rate (or speed), and time.

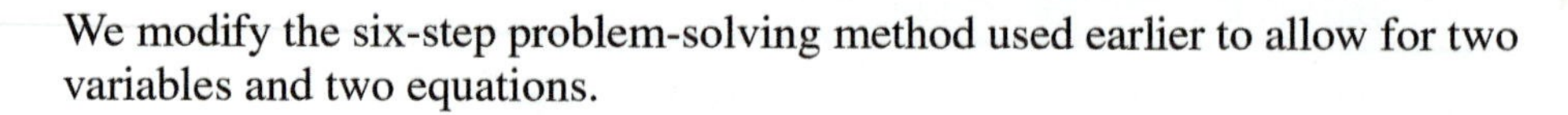

We modify the six-step problem-solving method used earlier to allow for two variables and two equations.

Solving an Applied Problem Using a System of Equations

Step 1 **Read** the problem carefully. *What information is given? What is to be found?*

Step 2 **Assign variables** to represent the unknown values. Make a sketch, diagram, or table, as needed. Write down what each variable represents.

Step 3 **Write two equations** using both variables.

Step 4 **Solve** the system of two equations.

Step 5 **State the answer.** Label it appropriately. *Does it seem reasonable?*

Step 6 **Check** the answer in the words of the *original* problem.

OBJECTIVE 1 Solve problems about unknown numbers.

EXAMPLE 1 Solving a Problem about Two Unknown Numbers

In a recent year, consumer sales of sports equipment were \$307 million more for snow skiing than for snowboarding. Together, total equipment sales for these two sports were \$931 million. What were equipment sales for each sport? (Data from National Sporting Goods Association.)

Step 1 **Read** the problem carefully. We must find equipment sales (in millions of dollars) for snow skiing and for snowboarding. We know how much more equipment sales were for snow skiing than for snowboarding. Also, we know the total sales.

Step 2 **Assign variables.**

Let x = equipment sales for skiing (in millions of dollars),

and y = equipment sales for snowboarding (in millions of dollars).

Step 3 **Write two equations.**

$x = 307 + y$ Equipment sales for skiing were \$307 million more than equipment sales for snowboarding. (1)

$x + y = 931$ Total sales were \$931 million. (2)

Step 4 **Solve** the system of equations from Step 3. We use the substitution method because the first equation is already solved for x.

$x + y = 931$ (2)

$(307 + y) + y = 931$ Let $x = 307 + y$.

$307 + 2y = 931$ Combine like terms.

$2y = 624$ Subtract 307.

Don't stop here. $y = 312$ Divide by 2.

Continued on Next Page

To find the value of x, we substitute 312 for y in equation (1) or (2).

$$x = 307 + y \quad (1)$$

$$x = 307 + 312 \quad \text{Let } y = 312 \text{ in equation (1).}$$

$$x = 619 \quad \text{Add.}$$

Step 5 **State the answer.** Equipment sales for skiing were \$619 million, and equipment sales for snowboarding were \$312 million.

Step 6 **Check** the answer in the original problem. Because

$$619 = 307 + 312 \quad \text{and} \quad 619 + 312 = 931$$

are both true, the answer satisfies the information given.

Work Problem 1 at the Side. ▶

CAUTION

If an applied problem asks for *two* values as in **Example 1,** be sure to give both of them in the answer.

OBJECTIVE 2 Solve problems about quantities and their costs.

EXAMPLE 2 Solving a Problem about Quantities and Costs

For a production of *Jersey Boys,* main floor tickets cost \$127 and rear mezzanine tickets cost \$97. Club members spent a total of \$3150 for 30 tickets. How many tickets of each kind did they buy? (Data from www.broadway.com)

Step 1 **Read** the problem, several times if needed.

Step 2 **Assign variables.**

Let x = the number of main floor tickets,

and y = the number of rear mezzanine tickets.

Summarize the information given in the problem in a table.

	Number of Tickets	Price per Ticket (in dollars)	Total Value (in dollars)
Main Floor	x	127	$127x$
Mezzanine	y	97	$97y$
Total	30	XXXXXX	3150

Multiply Number of Tickets by Price per Ticket to find Total Value.

Step 3 **Write two equations.**

$$x + y = 30 \quad \text{Total number of tickets was 30.} \quad (1)$$

$$127x + 97y = 3150 \quad \text{Total value of tickets was \$3150.} \quad (2)$$

Step 4 **Solve** the system formed in Step 3. We use the elimination method, although the substitution method could be used if desired.

$$x + y = 30 \quad (1)$$

$$127x + 97y = 3150 \quad (2)$$

Continued on Next Page

1 Solve each problem.

(a) In a recent year, consumer sales of sports equipment were \$26 million less for tennis than for archery. Together, total equipment sales for these two sports were \$876 million. What were equipment sales for each sport? (Data from National Sporting Goods Association.)

Step 1
We must find ______.

Step 2
Let x = equipment sales for tennis (in millions of dollars), and y = equipment sales for ______ (in millions of dollars).

Step 3
Write two equations.

$x =$ ______ (1)

(See the first sentence in the problem.)

______ (2)

(See the second sentence in the problem.)

Complete Steps 4–6 to solve the problem. Give the answer.

(b) Two of the most popular movies of 2015 were *Star Wars Episode VII* and *Jurassic World*. Together, their domestic gross was about \$1394 million. *Jurassic World* grossed \$90 million less than *Star Wars*. How much did each movie gross? (Data from www.the-numbers.com)

Answers

1. **(a)** equipment sales for each sport; archery; $y - 26$; $x + y = 876$; equipment sales for tennis: \$425 million; equipment sales for archery: \$451 million
(b) *Star Wars Episode VII*: \$742 million; *Jurassic World*: \$652 million

$$
\begin{aligned}
-97x - 97y &= -2910 && \text{Multiply equation (1) by } -97.\\
127x + 97y &= 3150 && (2)\\
\hline
30x &= 240 && \text{Add.}\\
\text{Main floor tickets} \rightarrow x &= 8 && \text{Divide by 30.}
\end{aligned}
$$

Substitute 8 for x in equation (1).

$$
\begin{aligned}
x + y &= 30 && (1)\\
8 + y &= 30 && \text{Let } x = 8.\\
\text{Mezzanine tickets} \rightarrow y &= 22 && \text{Subtract 8.}
\end{aligned}
$$

Step 5 **State the answer.** Club members bought 8 main floor tickets and 22 rear mezzanine tickets.

Step 6 **Check.** The sum of 8 and 22 is 30, so the total number of tickets is correct. Because 8 tickets were purchased at \$127 each and 22 at \$97 each, the amount spent on all the tickets is

$$\$127(8) + \$97(22) = \$3150, \quad \text{as required.}$$

◀ Work Problem 2 at the Side.

2 For a production of *The Lion King,* orchestra tickets cost \$209 and rear mezzanine tickets cost \$191. If a group of 18 people attended the show and spent \$3528 for their tickets, how many of each kind of ticket did they buy? (Data from www.broadway.com)

(a) Complete the table.

	Number of Tickets	Price per Ticket (dollars)	Total Value (dollars)
Orchestra	x	____	____
Mezzanine	y	____	____
Total	____	XXX	____

(b) Write a system of equations.

(c) Use the system of equations to solve the problem. Check the answer in the words of the original problem.

OBJECTIVE 3 Solve problems about mixtures. Previously, we solved percent problems using one variable. Mixture problems that involve percent can be solved using a system of two equations in two variables.

EXAMPLE 3 Solving a Mixture Problem Involving Percent

A pharmacist needs 100 L of 50% alcohol solution. She has a 30% alcohol solution and an 80% alcohol solution, which she can mix. How many liters of each will be required to make the 100 L of a 50% alcohol solution?

Step 1 **Read** the problem. Note the percent of each solution and of the mixture.

Step 2 **Assign variables.**

Let x = the number of liters of 30% alcohol needed,

and y = the number of liters of 80% alcohol needed.

Liters of Mixture	Percent (as a decimal)	Liters of Pure Alcohol
x	0.30	$0.30x$
y	0.80	$0.80y$
100	0.50	$0.50(100)$

Summarize the information in a table. Percents are written as decimals.

Figure 8 gives an idea of what is happening in this problem.

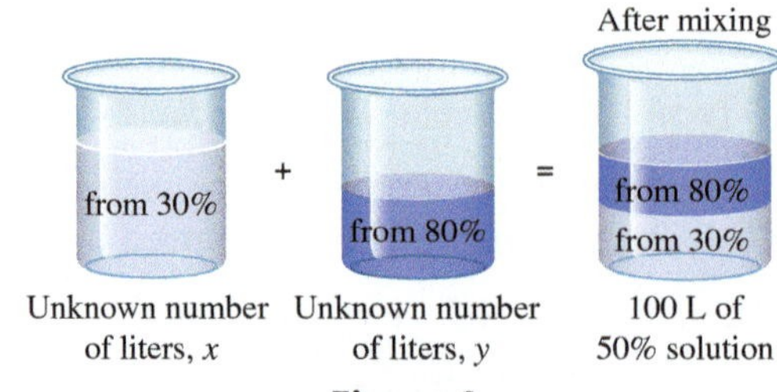

Figure 8

Continued on Next Page

Answers

2. (a)

	Number of Tickets	Price per Ticket (dollars)	Total Value (dollars)
Orchestra	x	209	$209x$
Mezzanine	y	191	$191y$
Total	18	XXX	3528

(b) $x + y = 18$
$209x + 191y = 3528$

(c) orchestra: 5; mezzanine: 13

Step 3 **Write two equations.** The total number of liters in the final mixture will be 100, which gives one equation.

$$x + y = 100$$ Refer to the first column of the table.

To find the amount of pure alcohol in each mixture, multiply the number of liters by the concentration. (Refer to the table.) The amount of pure alcohol in the 30% solution added to the amount of pure alcohol in the 80% solution will equal the amount of pure alcohol in the final 50% solution. This gives a second equation.

$$0.30x + 0.80y = 0.50(100)$$ Refer to the last column of the table.

$$0.30x + 0.80y = 50$$ $0.50(100) = 50$

This is a simpler, yet equivalent, equation. $$3x + 8y = 500$$ Multiply each side by 10 to clear the decimals.

These two equations form a system.

Be sure to write *two* equations.

$$x + y = 100 \quad (1)$$
$$3x + 8y = 500 \quad (2)$$

Step 4 **Solve** the system. We use the substitution method, although the elimination method could also be used. Solving equation (1) for x gives

$$x = 100 - y.$$

Substitute $100 - y$ for x in equation (2).

$$3x + 8y = 500 \quad (2)$$
$$3(100 - y) + 8y = 500 \quad \text{Let } x = 100 - y.$$
$$300 - 3y + 8y = 500 \quad \text{Distributive property}$$
$$300 + 5y = 500 \quad \text{Combine like terms.}$$
$$5y = 200 \quad \text{Subtract 300.}$$
Liters of 80% solution → $$y = 40 \quad \text{Divide by 5.}$$

To find x, substitute 40 for y in $x = 100 - y$ (equation (1) solved for x).

$$x = 100 - 40 \quad \text{Let } y = 40.$$
Liters of 30% solution → $$x = 60 \quad \text{Subtract.}$$

Step 5 **State the answer.** The pharmacist should use 60 L of the 30% solution and 40 L of the 80% solution.

Step 6 **Check** the answer in the original problem. Because

Use the original equations written from the table.

$$60 + 40 = 100 \quad \text{and} \quad 0.30(60) + 0.80(40) = 0.50(100)$$

are both true, this mixture will give the 100 L of 50% solution.

Work Problems 3 and 4 at the Side. ▶

Note

Whether to use the substitution or the elimination method to solve a system is often personal choice. In **Examples 2 and 3,** either method can be used efficiently because equation (1) has coefficients of 1 for both variables x and y.

---

3 (GS) How many liters of a 25% alcohol solution must be mixed with a 12% solution to obtain 13 L of a 15% solution?

(a) Complete the table.

Liters	Percent (as a decimal)	Liters of Pure Alcohol
x	0.25	$0.25x$
y	0.12	______
13	0.15	______

(b) Write a system of equations, and use it to solve the problem.

4 Solve the problem.

Joe needs 60 milliliters (mL) of 20% acid solution for a chemistry experiment. The lab has on hand only 10% and 25% solutions. How much of each should he mix to make the desired amount of 20% solution?

Answers

3. (a)

Liters	Percent (as a decimal)	Liters of Pure Alcohol
x	0.25	$0.25x$
y	0.12	$0.12y$
13	0.15	$0.15(13)$

(b) $x + y = 13$
$0.25x + 0.12y = 0.15(13)$;
3 L of 25%; 10 L of 12%

4. 20 mL of 10%; 40 mL of 25%

OBJECTIVE 4 Solve problems about distance, rate (or speed), and time.

If an automobile travels at an average rate of 50 mph for 2 hr, then it travels

$$50 \times 2 = 100 \text{ mi.}$$

This is an example of the basic relationship between distance, rate, and time,

distance = rate × time, given by the formula $d = rt$.

◀ Work Problem 5 at the Side.

5 Solve using the distance formula $d = rt$.

A small plane traveled from Stockholm, Sweden, to Oslo, Norway, averaging 244 km per hr. The trip took 1.7 hr. To the nearest kilometer, what is the distance between the two cities?

EXAMPLE 4 Solving a Problem about Distance, Rate, and Time

Two executives in cities 400 mi apart drive to a business meeting at a location on the line between their cities. They meet after 4 hr. Find the rate (speed) of each car if one car travels 20 mph faster than the other.

Steps 1 and 2 **Read** the problem carefully. **Assign variables.**

Let x = the rate of the faster car, and y = the rate of the slower car.

Make a table using the formula $d = rt$, and draw a sketch. See **Figure 9.**

	r	t	d
Faster Car	x	4	$x \cdot 4$, or $4x$
Slower Car	y	4	$y \cdot 4$, or $4y$

Because each car travels for 4 hr, the time t for each car is 4. Find d, using $d = rt$ (or $rt = d$).

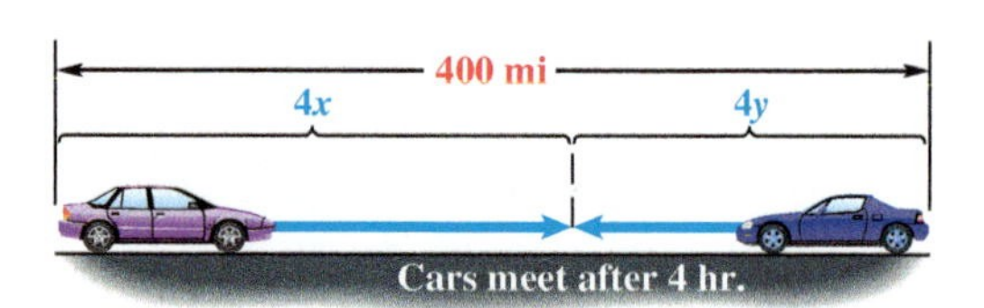

Figure 9

Step 3 **Write two equations.** The total distance traveled by both cars is 400 mi, which gives equation (1). The faster car travels 20 mph faster than the slower car, which gives equation (2).

$$4x + 4y = 400 \quad (1)$$
$$x = 20 + y \quad (2)$$

Step 4 **Solve** the system by substitution since equation (2) is already solved for x. Replace x with $20 + y$ in equation (1).

$$4x + 4y = 400 \quad (1)$$
$$4(20 + y) + 4y = 400 \quad \text{Let } x = 20 + y.$$
$$80 + 4y + 4y = 400 \quad \text{Distributive property}$$
$$80 + 8y = 400 \quad \text{Combine like terms.}$$
$$\text{Slower car} \rightarrow y = 40 \quad \text{Subtract 80. Divide by 8.}$$

To find x, substitute 40 for y in equation (2), $x = 20 + y$.

$$x = 20 + 40 \quad \text{Let } y = 40.$$
$$\text{Faster car} \rightarrow x = 60 \quad \text{Add.}$$

Step 5 **State the answer.** The rates of the cars are 40 mph and 60 mph.

Step 6 **Check** the answer. Each car travels for 4 hr, so total distance is

$$4(60) + 4(40) = 240 + 160 = 400 \text{ mi}, \quad \text{as required.}$$

◀ Work Problems 6 and 7 at the Side.

6 (GS) Two cars that were 450 mi apart traveled toward each other. They met after 5 hr. If one car traveled twice as fast as the other, what were their rates?

(a) Complete this table.

	r	t	d
Faster Car	x	5	___
Slower Car	y	5	___

(b) Write a system of equations, and use it to solve the problem.

7 Solve the problem.

From a truck stop, two trucks travel in opposite directions on a straight highway. In 3 hr they are 405 mi apart. Find the rate of each truck if one travels 5 mph faster than the other.

Answers

5. 415 km

6. (a)

	r	t	d
Faster Car	x	5	$5x$
Slower Car	y	5	$5y$

(b) $5x + 5y = 450$
$x = 2y$;
faster car: 60 mph; slower car: 30 mph

7. faster truck: 70 mph; slower truck: 65 mph

EXAMPLE 5 Solving a Problem about Distance, Rate, and Time

A plane flies 560 mi in 1.75 hr traveling with the wind. The return trip against the same wind takes the plane 2 hr. Find the rate (speed) of the plane and the wind speed.

Steps 1 and 2 **Read** the problem several times. **Assign variables.**

Let $x =$ the rate of the plane in still air,

and $y =$ the wind speed.

When the plane is traveling *with* the wind, the wind "pushes" the plane. In this case, the rate (speed) of the plane is the *sum* of the rate of the plane and the wind speed, $(x + y)$ mph. See **Figure 10.**

When the plane is traveling *against* the wind, the wind "slows" the plane down. In this case, the rate (speed) of the plane is the *difference* between the rate of the plane and the wind speed, $(x - y)$ mph. Again, see **Figure 10.**

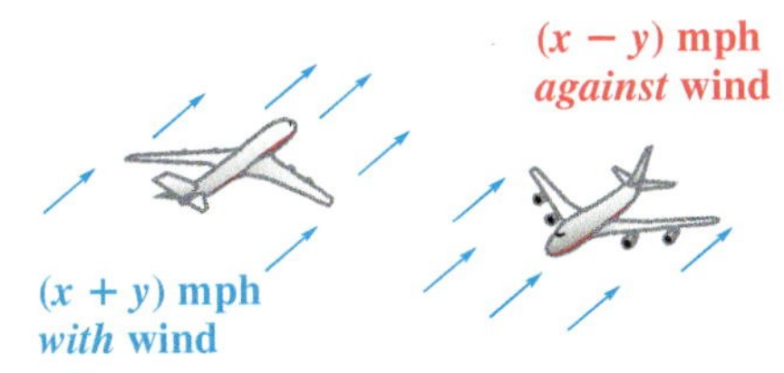

Figure 10

	r	t	d
With Wind	$x + y$	1.75	560
Against Wind	$x - y$	2	560

Summarize this information in a table. The distance is the same both ways.

Step 3 **Write two equations.** Refer to the table and use the formula $d = rt$ (or $rt = d$).

$$(x + y)1.75 = 560 \xrightarrow{\text{Divide by 1.75.}} x + y = 320 \quad (1)$$

$$(x - y)2 = 560 \xrightarrow{\text{Divide by 2.}} x - y = 280 \quad (2)$$

Step 4 **Solve** the system of equations using the elimination method.

$$x + y = 320 \quad (1)$$
$$x - y = 280 \quad (2)$$
$$2x \quad = 600 \quad \text{Add.}$$
$$x = 300 \quad \text{Divide by 2.}$$

Because $x + y = 320$ and $x = 300$, it follows that $y = 20$.

Step 5 **State the answer.** The rate of the plane is 300 mph, and the wind speed is 20 mph.

Step 6 **Check.** Using equations (1) and (2), we see that $300 + 20 = 320$ and $300 - 20 = 280$ are both true. The answer is correct.

Work Problem 8 at the Side. ▶

! CAUTION

Be careful. ***When we use two variables to solve a problem, we must write two equations.***

8 Solve each problem.

GS **(a)** In 1 hr, Gigi can row 2 mi against the current or 10 mi with the current. Find the rate of the current and Gigi's rate in still water.

Steps 1 and 2
Let $x =$ the rate of the current, and $y =$ ______ in still water.

Then her rate *against* the current is (______) mph, and her rate with the current is (______) mph.

Complete the table.

	r	t	d
Against Current	___	___	___
With Current	___	___	___

Step 3
Write two equations.

$(y - x) \cdot 1 =$ ______ (1)

(See the first row of the table.)

____________ (2)

(See the second row of the table.)

Complete Steps 4–6 to solve the problem. Give the answer.

(b) In 1 hr, a boat travels 15 mi with the current and 9 mi against the current. Find the rate of the boat and the rate of the current.

Answers

8. (a) Gigi's rate; $y - x$; $y + x$

	r	t	d
Against Current	$y - x$	1	2
With Current	$y + x$	1	10

2; $(y + x) \cdot 1 = 10$;
rate of the current: 4 mph;
Gigi's rate: 6 mph

(b) rate of the boat: 12 mph;
rate of the current: 3 mph

4.4 Exercises

FOR EXTRA HELP *Go to MyMathLab for worked-out, step-by-step solutions to exercises enclosed in a square and video solutions to ▶ exercises.*

CONCEPT CHECK *Choose the correct response.*

1. Which expression represents the monetary value of x 20-dollar bills?

A. $\frac{x}{20}$ dollars **B.** $\frac{20}{x}$ dollars **C.** $(20 + x)$ dollars **D.** $20x$ dollars

2. Which expression represents the cost of t pounds of candy that sells for \$1.95 per lb?

A. $\$1.95t$ **B.** $\frac{\$1.95}{t}$ **C.** $\frac{t}{\$1.95}$ **D.** $\$1.95 + t$

3. Which expression represents the amount of interest earned on d dollars invested at an interest rate of 2% for 1 yr?

A. $2d$ dollars **B.** $0.02d$ dollars **C.** $0.2d$ dollars **D.** $200d$ dollars

4. Which expression represents the amount of pure alcohol in x liters of a 25% alcohol solution?

A. $25x$ liters **B.** $(25 + x)$ liters **C.** $0.25x$ liters **D.** $(0.25 + x)$ liters

5. According to *Natural History* magazine, the speed of a cheetah is 70 mph. If a cheetah runs for x hours, how many miles does the cheetah cover?

A. $(70 + x)$ miles **B.** $(70 - x)$ miles **C.** $\frac{70}{x}$ miles **D.** $70x$ miles

6. How far does a car travel in 2.5 hr if it travels at an average rate of x miles per hour?

A. $(x + 2.5)$ miles **B.** $\frac{2.5}{x}$ miles **C.** $\frac{x}{2.5}$ miles **D.** $2.5x$ miles

7. What is the rate of a plane that travels at 650 mph *with* a wind of r mph?

▶ **A.** $\frac{r}{650}$ mph **B.** $(650 - r)$ mph **C.** $(650 + r)$ mph **D.** $(r - 650)$ mph

8. What is the rate of a plane that travels at 650 mph *against* a wind of r mph?

A. $(650 + r)$ mph **B.** $\frac{650}{r}$ mph **C.** $(650 - r)$ mph **D.** $(r - 650)$ mph

9. Suppose that x liters of a 40% acid solution are mixed with y liters of a 35% solution to obtain 100 L of a 38% solution. One equation in a system for solving this problem is

$$x + y = 100.$$

Which one of the following is the other equation?

A. $0.35x + 0.40y = 0.38(100)$ **B.** $0.40x + 0.35y = 0.38(100)$

C. $35x + 40y = 38$ **D.** $40x + 35y = 0.38(100)$

10. Suppose that two trucks leave a rest stop traveling in opposite directions on a straight highway. One truck travels 15 mph faster than the other, and in 8 hr they are 840 mi apart. If x and y represent the rates of the trucks, one equation in a system for solving this problem is

$$x = 15 + y.$$

Which one of the following is the other equation?

A. $8x + 8y = 840$ **B.** $8x + y = 840$

C. $x + y = 840$ **D.** $8x - 8y = 840$

GS *Refer to the six-step problem-solving method. Fill in the blanks for Steps 2 and 3 for each problem, and then complete the solution by applying Steps 4–6.*

11. The sum of two numbers is 98 and the difference between them is 48. Find the two numbers.

Step 1 **Read** the problem carefully.

Step 2 **Assign variables.**

Let $x =$ the first number

and $y =$ ____________.

Step 3 **Write two equations.**

First equation: $x + y = 98$

Second equation: ________

12. The sum of two numbers is 201 and the difference between them is 11. Find the two numbers.

Step 1 **Read** the problem carefully.

Step 2 **Assign variables.**

Let $x =$ __________

and $y =$ the second number.

Step 3 **Write two equations.**

First equation: ________

Second equation: $x - y = 11$

Solve each problem using a system of equations. ***See Example 1.***

13. Two of the longest-running shows on Broadway are *The Phantom of the Opera* and *The Lion King*. As of February 10, 2016, there had been a total of 19,272 performances of the two shows, with 4066 more performances of *The Phantom of the Opera* than *The Lion King*. How many performances were there of each show? (Data from www.playbill.com)

14. During Broadway runs of *A Chorus Line* and *Beauty and the Beast*, there have been 676 fewer performances of *Beauty and the Beast* than of *A Chorus Line*. There were a total of 11,598 performances of the two shows. How many performances were there of each show? (Data from The Broadway League.)

15. Two domestic top-grossing movies of 2015 were *Furious 7* and *Minions*. *Minions* grossed $17 million less than *Furious 7*, and together the two films took in $689 million. How much did each of these movies earn? (Data from www.boxofficemojo.com)

16. During their opening weekends, the movies *Furious 7* and *Minions* grossed a total of $263 million, with *Furious 7* grossing $31 million more than *Minions*. How much did each of these movies earn during their opening weekends? (Data from www.boxofficemojo.com)

17. The Terminal Tower in Cleveland, Ohio, is 239 ft shorter than the Key Tower, also in Cleveland. The total of the heights of the two buildings is 1655 ft. Find the heights of the buildings. (Data from *The World Almanac and Book of Facts.*)

18. The total of the heights of the Chase Tower and the One America Tower, both in Indianapolis, Indiana, is 1234 ft. The Chase Tower is 168 ft taller than the One America Tower. Find the heights of the two buildings. (Data from *The World Almanac and Book of Facts.*)

19. GS An official playing field (including end zones) for the Indoor Football League has length 38 yd longer than its width. The perimeter of the rectangular field is 188 yd. Find the length and width of the field. (Data from Indoor Football League.)

Steps 1 and 2 **Read** carefully, and **assign** ______.

Let x = the length (in yards),

and y = the ______ (in yards).

Step 3 **Write two equations.**

Equation (1): See the first sentence in the problem. Express the length in terms of the width.

$x =$ ______ (1)

Equation (2): See the second sentence in the problem. Perimeter of a rectangle equals twice the ______ plus ______ the width.

$2x +$ ______ $=$ ______ (2)

Now complete the solution.

20. GS Pickleball is a combination of badminton, tennis, and ping pong. The perimeter of the rectangular-shaped court is 128 ft. The width is 24 ft shorter than the length. Find the length and width of the court. (Data from www.sportsknowhow.com)

Steps 1 and 2 **Read** carefully, and **assign** ______.

Let x = the ______ (in feet),

and y = the width (in feet).

Step 3 **Write two equations.**

Equation (1): Perimeter of a rectangle equals ______ the length plus twice the ______.

______ $+ 2y =$ ______ (1)

Equation (2): See the third sentence in the problem. Express the width in terms of the length.

$y =$ ______ (2)

Now complete the solution.

Suppose that x units of a product cost C dollars to manufacture and earn revenue of R dollars. The value of x, where the expressions for C and R are equal, is the ***break-even quantity,*** *the number of units that produce 0 profit.*

In each problem, ***(a)*** *find the break-even quantity, and* ***(b)*** *decide whether the product should be produced, based on whether it will earn a profit. (Profit equals revenue minus cost.)*

21. $C = 85x + 900$; $R = 105x$;
No more than 38 units can be sold.

22. $C = 105x + 6000$; $R = 255x$;
No more than 400 units can be sold.

For each problem, complete any tables. Then solve the problem using a system of equations. ***See Example 2.***

23. A motel clerk counts his \$1 and \$10 bills at the end of a day. He finds that he has a total of 74 bills having a combined monetary value of \$326. Find the number of bills of each denomination that he has.

Number of Bills	Denomination of Bill (in dollars)	Total Value (in dollars)
x	1	______
y	10	______
74	XXXXXX	326

24. Carly is a bank teller. At the end of a day, she has a total of 69 \$5 and \$10 bills. The total value of the money is \$590. How many of each denomination does she have?

Number of Bills	Denomination of Bill (in dollars)	Total Value (in dollars)
x	5	$5x$
y	10	______
______	XXXXXX	______

25. Tracy bought each of her seven nephews a DVD of *Ant-Man* or a Blu-ray disc of *The Martian*. Each DVD cost $16.99. Each Blu-ray disc cost $22.42. Tracy spent a total of $129.79. How many of each did she buy?

26. Terry bought each of his five nieces a DVD of *The Peanuts Movie* or a Blu-ray disc of *Frozen*. Each DVD cost $14.99. Each Blu-ray disc cost $24.99. Terry spent a total of $84.95. How many of each did he buy?

27. GS Maria has twice as much money invested at 5% simple annual interest as she does at 4%. If her yearly income from these two investments is $350, how much does she have invested at each rate?

Amount Invested (in dollars)	Rate of Interest	Interest for One Year (in dollars)
x	5%, or 0.05	$0.05x$
y	______	______
XXXXX	XXXXXX	350

Equation (1): $x =$ ______

Equation (2): $0.05x +$ ______ $=$ ______

28. GS Charles invested in two accounts, one paying 3% simple annual interest and the other paying 2%. He earned a total of $880 interest. If he invested three times as much in the 3% account as in the 2% account, how much did he invest at each rate?

Amount Invested (in dollars)	Rate of Interest	Interest for One Year (in dollars)
x	______	______
y	______	______
XXXXXX	XXXXXX	______

Equation (1): ______ $+$ ______ $= 880$

Equation (2): $x =$ ______

29. The two top-grossing North American concert tours in 2015 were Taylor Swift and Kenny Chesney. Based on average ticket prices, it cost a total of $1097 to buy six tickets for Taylor Swift and five tickets for Kenny Chesney. Three tickets for Taylor Swift and four tickets for Kenny Chesney cost $676. How much did an average ticket cost for each tour? (Data from Pollstar.)

30. Two other popular North American concert tours in 2015 were the Rolling Stones and Garth Brooks. Based on average ticket prices, it cost a total of $1592 to buy eight tickets for the Rolling Stones and three tickets for Garth Brooks. Four tickets for the Rolling Stones and five tickets for Garth Brooks cost $1020. How much did an average ticket cost for each tour? (Data from Pollstar.)

For each problem, complete any tables. Then solve the problem using a system of equations. ***See Example 3.***

31. A 40% dye solution is to be mixed with a 70% dye solution to make 120 L of a 50% solution. How many liters of the 40% and 70% solutions will be needed?

Liters of Solution	Percent (as a decimal)	Liters of Pure Dye
x	0.40	______
y	0.70	______
120	0.50	______

32. A 90% antifreeze solution is to be mixed with a 75% solution to make 120 L of a 78% solution. How many liters of the 90% and 75% solutions will be used?

Liters of Solution	Percent (as a decimal)	Liters of Pure Antifreeze
x	0.90	______
y	0.75	______
120	0.78	______

33. Ahmad wishes to mix coffee worth \$6 per lb with coffee worth \$3 per lb to obtain 90 lb of a mixture worth \$4 per lb. How many pounds of the \$6 and the \$3 coffees will be needed?

Number of Pounds	Dollars per Pound	Cost (in dollars)
x	6	______
y	______	______
90	______	______

34. Mariana wishes to blend candy selling for \$1.20 per lb with candy selling for \$1.80 per lb to obtain 45 lb of a mixture that will be sold for \$1.40 per lb. How many pounds of the \$1.20 and the \$1.80 candies should be used to make the mixture?

Number of Pounds	Dollars per Pound	Cost (in dollars)
x	______	______
y	1.80	______
45	______	______

35. How many pounds of nuts selling for \$6 per lb and raisins selling for \$3 per lb should Kelli combine to obtain 60 lb of a trail mix selling for \$5 per lb?

36. Callie is preparing cheese trays. She uses some cheeses that sell for \$8 per lb and others that sell for \$12 per lb. How many pounds of each cheese should she use in order for the mixed cheeses on the trays to weigh a total of 56 lb and sell for \$10.50 per lb?

For each problem, complete any tables. Then solve the problem using a system of equations. ***See Examples 4 and 5.***

37. Two trains start from towns 495 mi apart and travel toward each other on parallel tracks. They pass each other 4.5 hr later. If one train travels 10 mph faster than the other, find the rate of each train.

	r	t	d
Train 1	x	______	______
Train 2	y	______	______

Equation (1): $4.5x +$ ______ $=$ ______

Equation (2): $x =$ ______ $+$ ______

38. Two trains that are 495 mi apart travel toward each other. They pass each other 5 hr later. If one train travels half as fast as the other, find the rate of each train.

	r	t	d
Train 1	x	______	______
Train 2	______	______	$5y$

Equation (1): ______ $+ 5y =$ ______

Equation (2): $x =$ ______ y, or ______ $= y$

39. **RAGBRAI**®, the Des Moines **R**egister's **A**nnual **G**reat **B**icycle **R**ide **A**cross **I**owa, is the longest and oldest touring bicycle ride in the world. Suppose a cyclist began the 420-mi ride on July 23, 2016, in western Iowa at the same time that a car traveling toward it left eastern Iowa. If the bicycle and the car met after 7 hr and the car traveled 33 mph faster than the bicycle, find the average rate of each. (Data from www.ragbrai.com)

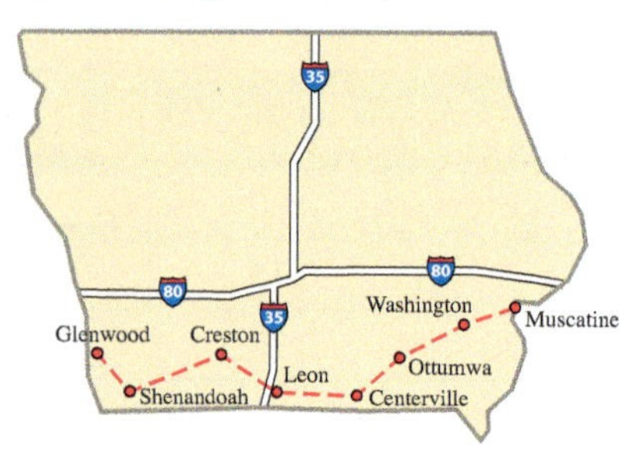

40. Suppose two planes leave Atlanta's Hartsfield Airport at the same time, one traveling east and the other traveling west. If the planes are 2100 mi apart after 2 hr and one plane travels 50 mph faster than the other, find the rate of each plane.

41. Toledo and Cincinnati are 200 mi apart. A car leaves Toledo traveling toward Cincinnati, and another car leaves Cincinnati at the same time, traveling toward Toledo. The car leaving Toledo averages 15 mph faster than the other, and they meet after 1 hr, 36 min. What are the rates of the cars?

42. Kansas City and Denver are 600 mi apart. Two cars start from these cities, traveling toward each other. They meet after 6 hr. Find the rate of each car if one travels 30 mph slower than the other.

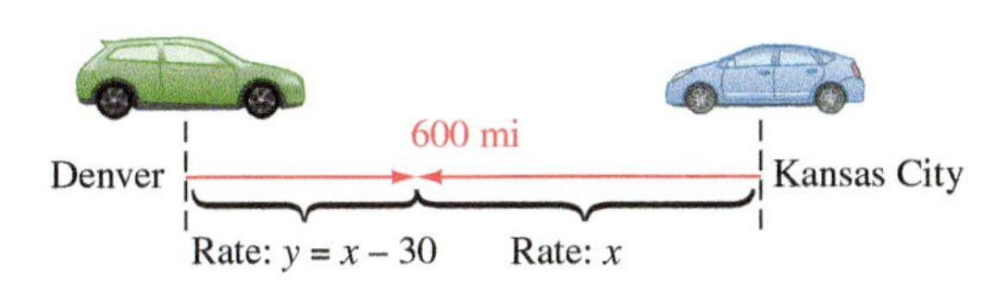

43. A boat takes 3 hr to go 24 mi upstream. It can go 36 mi downstream in the same time. Find the rate of the current and the rate of the boat in still water.

Let $x =$ the rate of the boat in still water and $y =$ the rate of the current.

	r	t	d
Downstream	$x + y$	____	36
Upstream	$x - y$	____	____

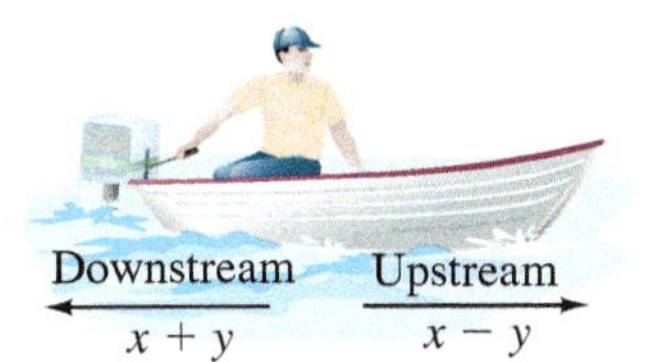

44. It takes a boat $1\frac{1}{2}$ hr to go 12 mi downstream, and 6 hr to return. Find the rate of the boat in still water and the rate of the current.

Let $x =$ the rate of the boat in still water and $y =$ the rate of the current.

	r	t	d
Downstream	$x + y$	$\frac{3}{2}$	____
Upstream	____	____	____

45. If a plane can travel 440 mph against the wind and 500 mph with the wind, find the wind speed and the rate of the plane in still air.

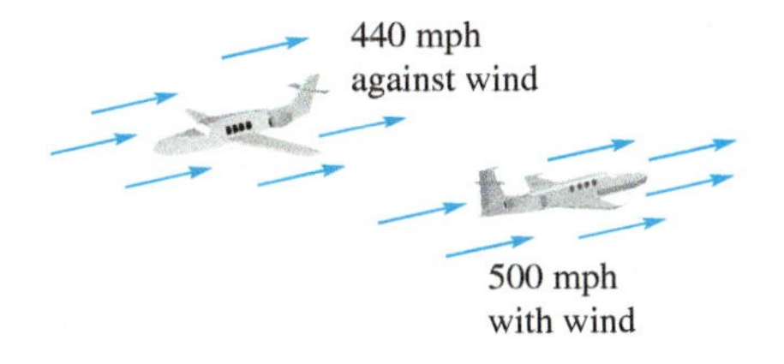

46. A small plane travels 200 mph with the wind and 120 mph against it. Find the wind speed and the rate of the plane in still air.

47. At the beginning of a bicycle ride for charity, Roberto and Juana are 30 mi apart. If they leave at the same time and ride in the same direction, Roberto overtakes Juana in 6 hr. If they ride toward each other, they meet in 1 hr. What are their rates?

48. Mr. Abbot left Farmersville in a plane at noon to travel to Exeter. Mr. Baker left Exeter in his automobile at 2 P.M. to travel to Farmersville. It is 400 mi from Exeter to Farmersville. If the sum of their rates was 120 mph, and if they crossed paths at 4 P.M., find the rate of each.

4.5 Solving Systems of Linear Inequalities

OBJECTIVE

1 Solve systems of linear inequalities by graphing.

Recall that to graph the solutions of $x + 3y > 12$, we first graph the boundary line $x + 3y = 12$ by finding and plotting a few ordered pairs that satisfy the equation. (The x- and y-intercepts are good choices.) Because the $>$ symbol does not include equality, the points on the line do *not* satisfy the inequality, and we graph it using a dashed line.

To decide which region includes the points that are solutions, we choose a test point not on the line.

$$x + 3y > 12 \quad \text{Original inequality}$$

$$0 + 3(0) \overset{?}{>} 12 \quad \text{Let } x = 0 \text{ and } y = 0.$$

$$0 > 12 \quad \text{False}$$

(0, 0) is a convenient test point.

This false result indicates that the solutions are those points on the side of the line that does *not* include $(0, 0)$, as shown in **Figure 11.**

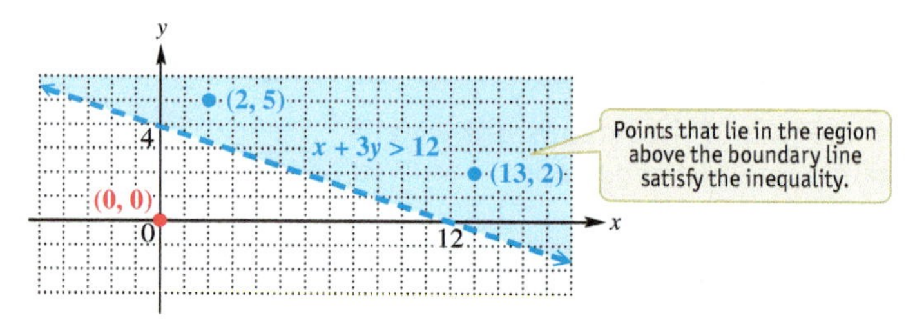

Figure 11

OBJECTIVE 1 Solve systems of linear inequalities by graphing. A **system of linear inequalities** consists of two or more linear inequalities. The **solution set of a system of linear inequalities** includes all points that make all inequalities of the system true at the same time.

Solving a System of Linear Inequalities

Step 1 **Graph each linear inequality on the same axes.**

Step 2 **Choose the intersection.** Indicate the solution set of the system by shading the intersection of the graphs—that is, the region where the graphs overlap.

EXAMPLE 1 Solving a System of Linear Inequalities

Graph the solution set of the system.

$$3x + 2y \leq 6$$

$$2x - 5y > 10$$

Step 1 Graph $3x + 2y \leq 6$ with the solid boundary line $3x + 2y = 6$ using the intercepts $(0, 3)$ and $(2, 0)$. Determine the region to shade.

$$3x + 2y < 6 \quad \text{Test the region.}$$

$$3(0) + 2(0) \overset{?}{<} 6 \quad \text{Use } (0, 0) \text{ as a test point.}$$

$$0 < 6 \quad \text{True}$$

Shade the region containing $(0, 0)$. See **Figure 12(a)** on the next page.

Continued on Next Page

Now graph the inequality $2x - 5y > 10$ with dashed boundary line $2x - 5y = 10$ using the intercepts $(0, -2)$ and $(5, 0)$. Determine the region to shade.

$$2x - 5y > 10 \quad \text{Test the region.}$$
$$2(0) - 5(0) \overset{?}{>} 10 \quad \text{Use } (0, 0) \text{ as a test point.}$$
$$0 > 10 \quad \text{False}$$

Shade the region that does *not* contain $(0, 0)$. See **Figure 12(b).**

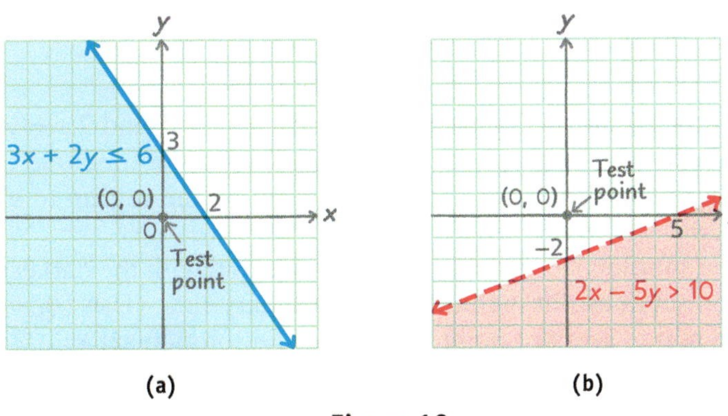

Figure 12

Step 2 The solution set of this system includes all points in the intersection (overlap) of the graphs of the two inequalities. As shown in **Figure 13,** this intersection is the purple shaded region and portion of the boundary line $3x + 2y = 6$ that surrounds it.

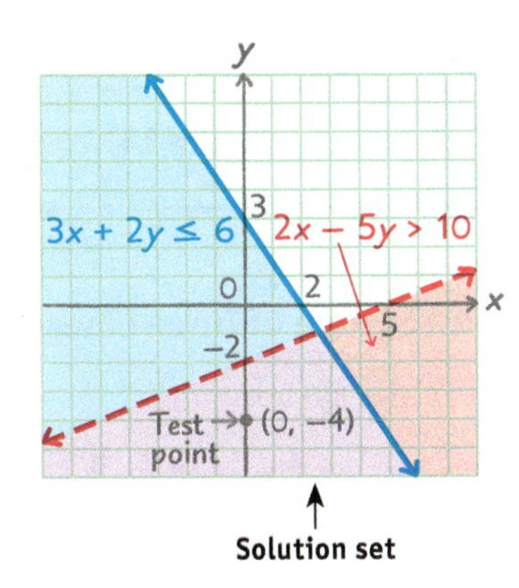

Figure 13

CHECK To confirm the solution set in **Figure 13,** select a test point in the gray shaded region, such as $(0, -4)$, and substitute it into *both* inequalities to make sure that true statements result. (Using an ordered pair that has one coordinate 0 makes the substitution easier.)

$3x + 2y < 6$	$2x - 5y > 10$
$3(0) + 2(-4) \overset{?}{<} 6$ Test $(0, -4)$.	$2(0) - 5(-4) \overset{?}{>} 10$ Test $(0, -4)$.
$-8 < 6$ True	$20 > 10$ True

True statements result, so we have shaded the correct region in **Figure 13.** Test points selected in the other three regions will satisfy only one of the inequalities or neither of them. (Verify this.) ✓

Work Problem 1 at the Side. ▶

1 Graph the solution set of each system.

GS **(a)** $x - 2y \le 8$

$3x + y > 6$

(The graphs of $x - 2y = 8$ and $3x + y = 6$ are shown.)

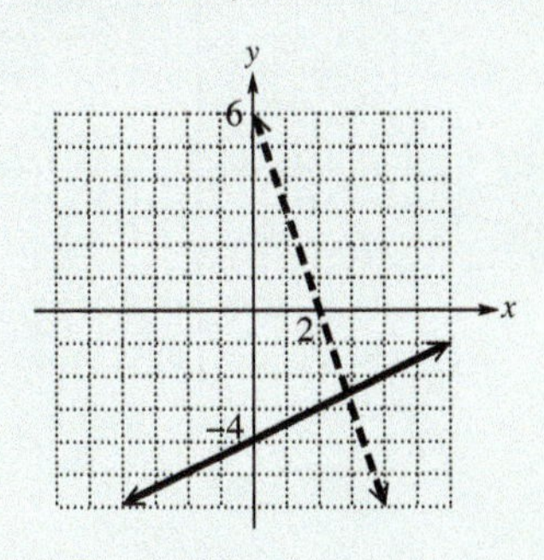

(b) $4x - 2y \le 8$

$x + 3y \ge 3$

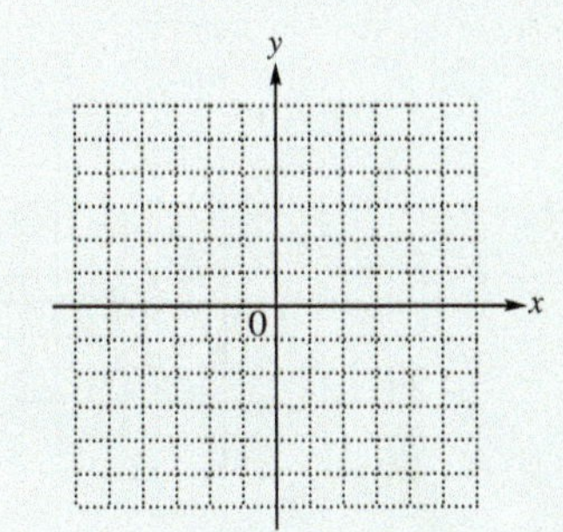

Answers

1. (a)

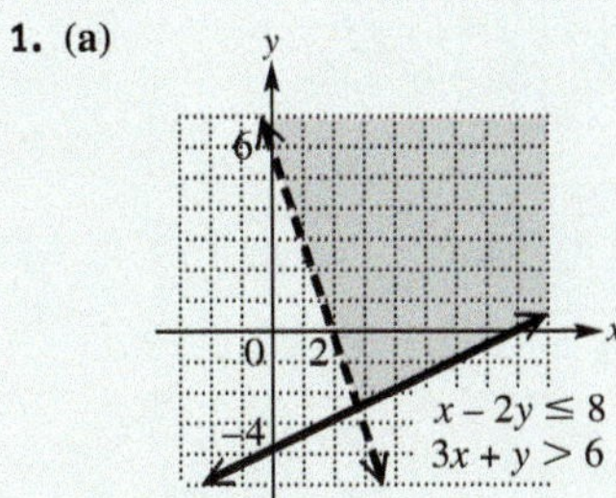

(b)

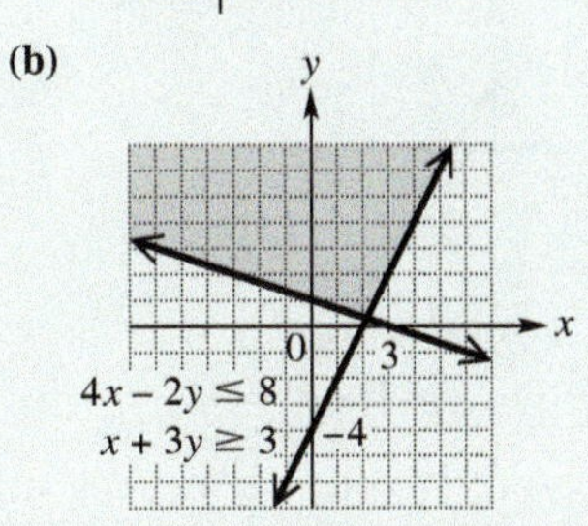

2 Graph the solution set of the system.

$$x + 2y < 0$$
$$3x - 4y < 12$$

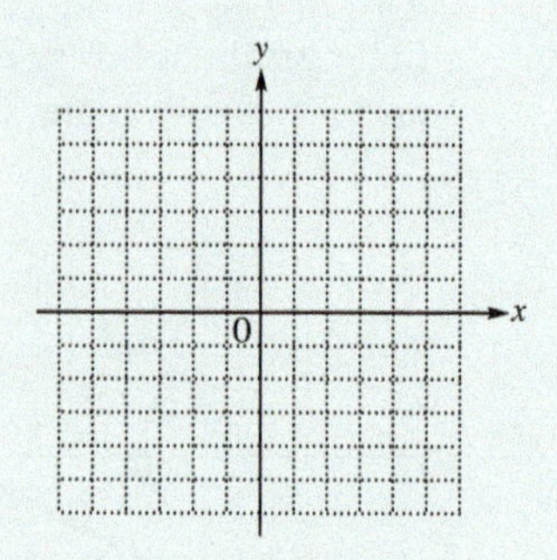

3 Graph the solution set of the system.

$$3x + 2y \leq 12$$
$$x \leq 2$$
$$y \leq 4$$

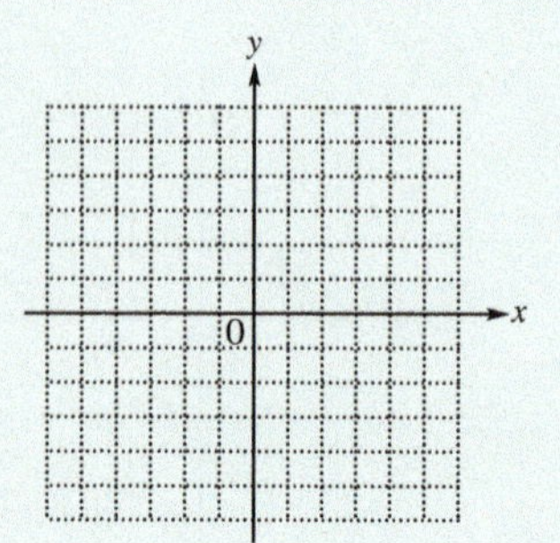

Answers

2.

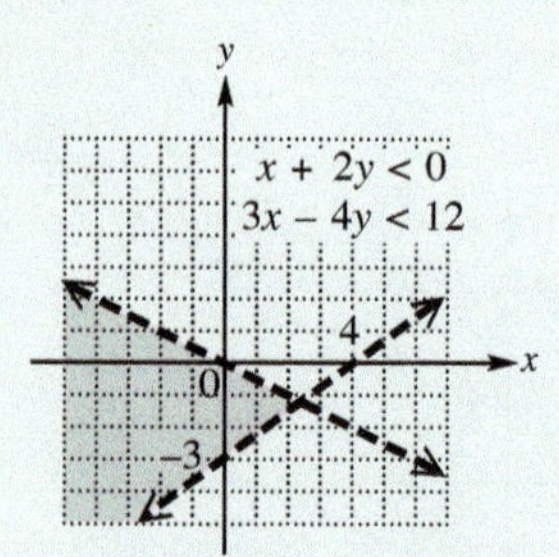

3.

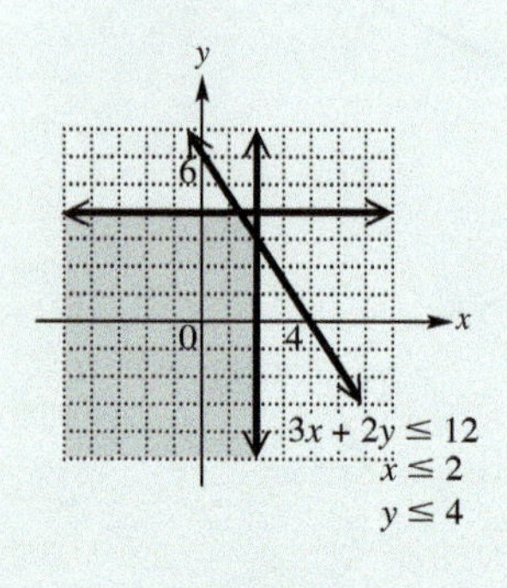

Note

We usually do all the work on one set of axes. In the remaining examples, only one graph is shown. Be sure that the region of the final solution set is clearly indicated.

EXAMPLE 2 Solving a System of Linear Inequalities

Graph the solution set of the system.

$$x - y > 5$$
$$2x + y < 2$$

Figure 14 shows the graphs of both $x - y > 5$ and $2x + y < 2$. Dashed lines show that the graphs of the inequalities do not include their boundary lines. Use $(0, 0)$ as a test point to determine the region to shade for each inequality.

The solution set of the system is the region with the darkest shading. The solution set does not include either boundary line. (Use $(0, -6)$ in the gray shaded region as a test point to confirm the solution set.)

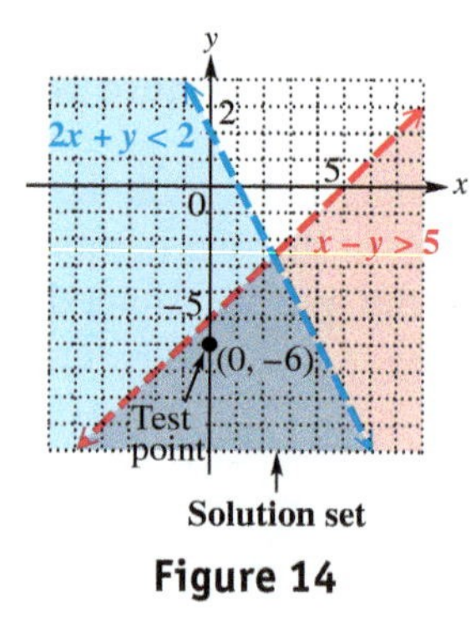

Figure 14

◀ Work Problem 2 at the Side.

EXAMPLE 3 Solving a System of Three Linear Inequalities

Graph the solution set of the system.

$$4x - 3y \leq 8$$
$$x \geq 2$$
$$y \leq 4$$

Graph the solid boundary line $4x - 3y = 8$ through the intercepts $(2, 0)$ and $\left(0, -\frac{8}{3}\right)$. (Because the y-intercept does not have integer coordinates, we also use the point $(-1, -4)$ to help draw an accurate line.) Recall that $x = 2$ is a vertical line through the point $(2, 0)$, and $y = 4$ is a horizontal line through the point $(0, 4)$. Use $(0, 0)$ as a test point to determine the region to shade for each inequality.

The graph of the solution set is the shaded region in **Figure 15,** including all boundary lines. (Here, use $(3, 2)$ as a test point to confirm that the correct region is shaded.)

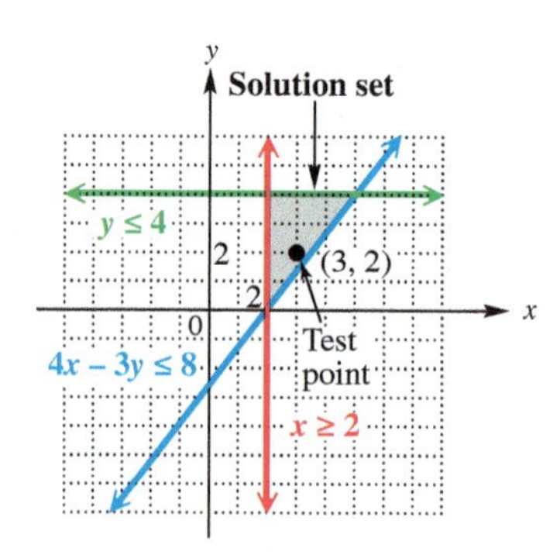

Figure 15

◀ Work Problem 3 at the Side.

4.5 Exercises

FOR EXTRA HELP — Go to MyMathLab for worked-out, step-by-step solutions to exercises enclosed in a square and video solutions to ▶ exercises.

CONCEPT CHECK *Match each system of inequalities with the correct graph from choices A–D.*

1. $x \geq 5$
$y \leq -3$

2. $x \leq 5$
$y \geq -3$

3. $x > 5$
$y < -3$

4. $x < 5$
$y > -3$

A.

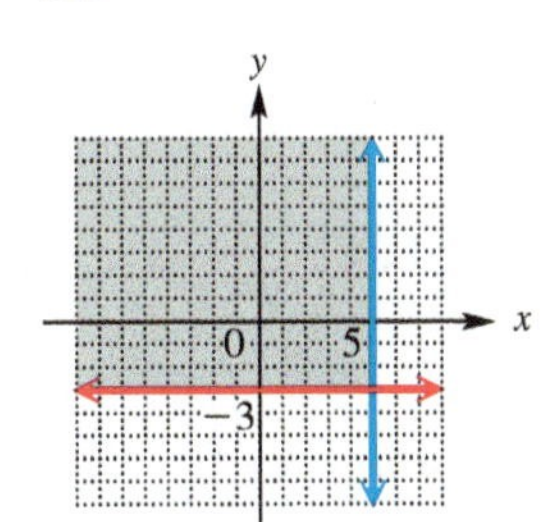

B.

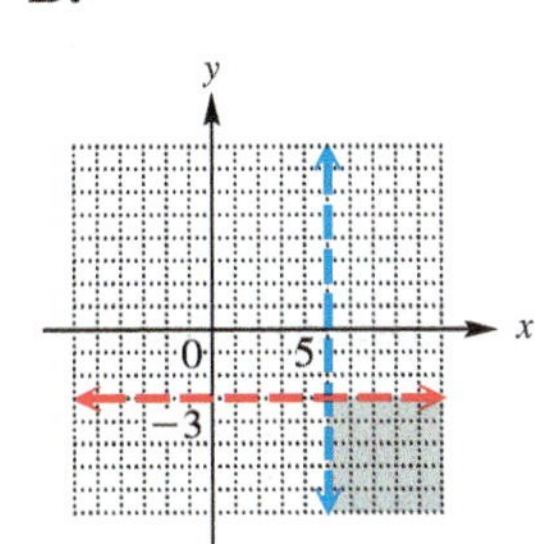

C.

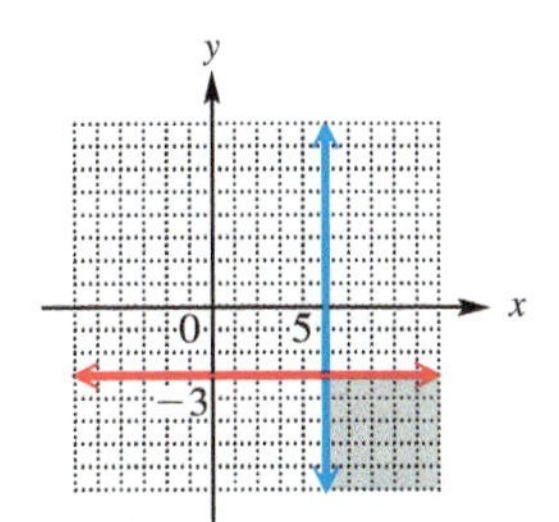

D.

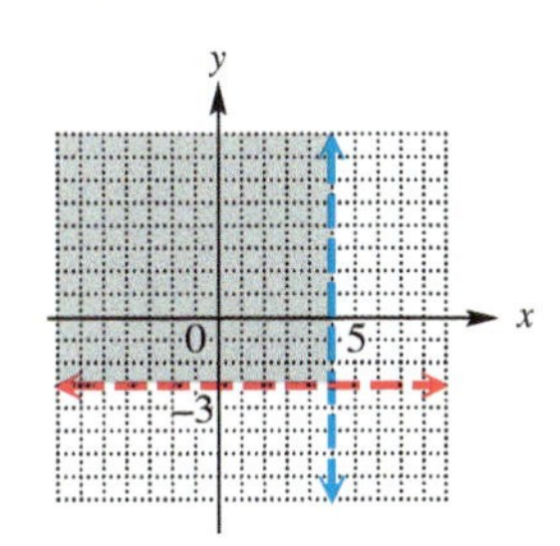

CONCEPT CHECK *Decide whether each ordered pair is a solution of the given system of inequalities. Then shade the solution set of each system. Boundary lines are already graphed.*

5. $x - 3y \leq 6$
$x \geq -4$
(a) $(5, -4)$ **(b)** $(0, 0)$

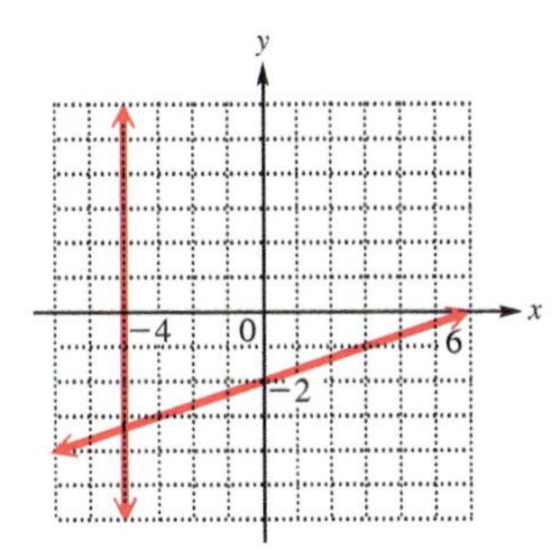

6. $x - 2y \geq 4$
$x \leq -2$
(a) $(0, 0)$ **(b)** $(-4, -5)$

7. $x + y > 4$
$5x - 3y < 15$
(a) $(3, 3)$ **(b)** $(5, 0)$

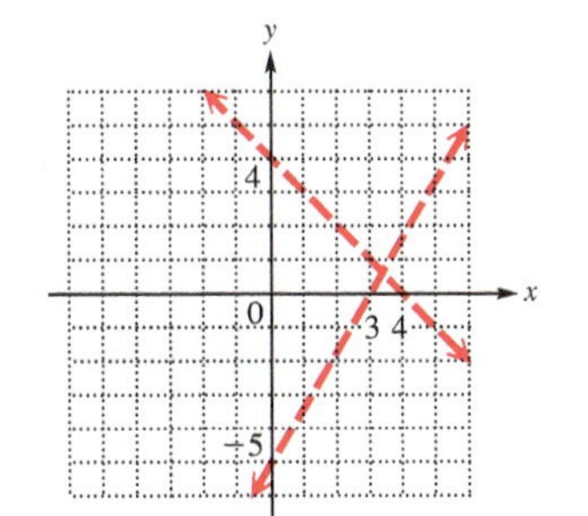

8. $3x - 2y > 12$
$4x + 3y < 12$
(a) $(3, -3)$ **(b)** $(6, 0)$

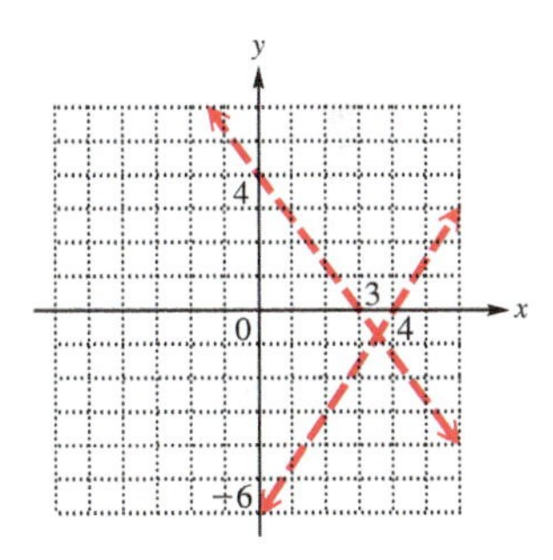

Graph the solution set of each system of linear inequalities. **See Examples 1–3.**

9. ▶ $x + y \leq 6$
$x - y \geq 1$

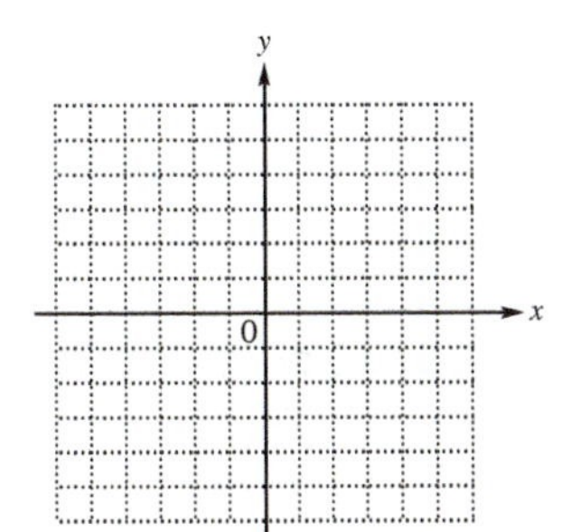

10. $x + y \leq 2$
$x - y \geq 3$

11. $4x + 5y \geq 20$
$x - 2y \leq 5$

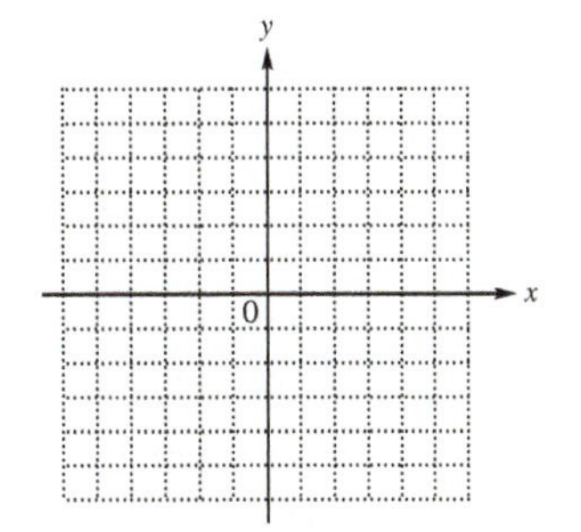

12. $x + 4y \leq 8$
$2x - y \geq 4$

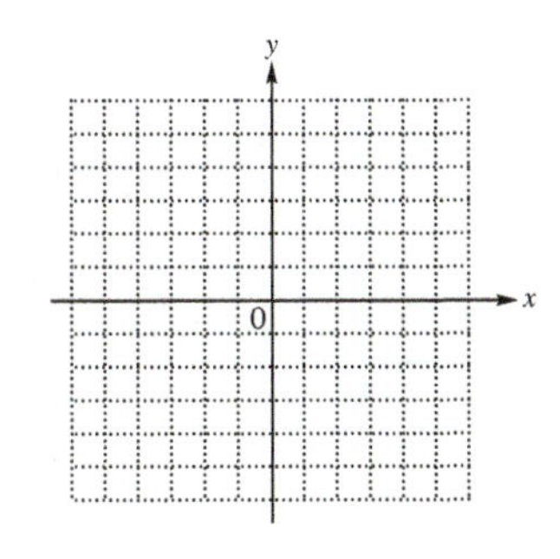

13. $2x + 3y < 6$
$x - y < 5$

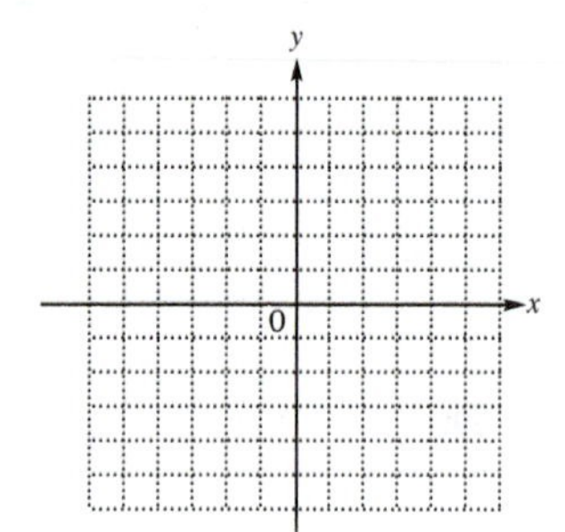

14. $x + 2y < 4$
$x - y < -1$

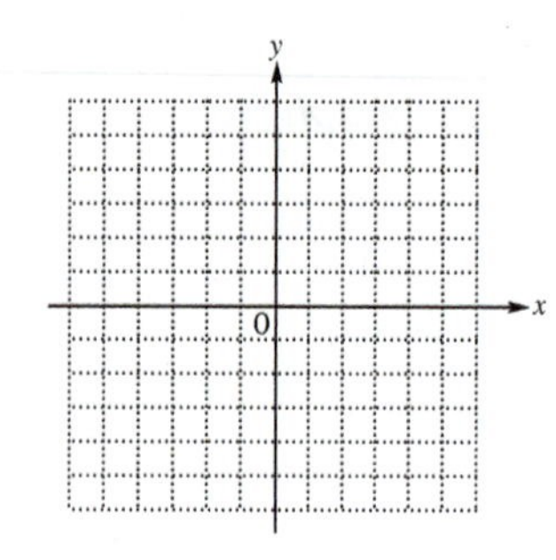

15. $y \le 2x - 5$
$x < 3y + 2$

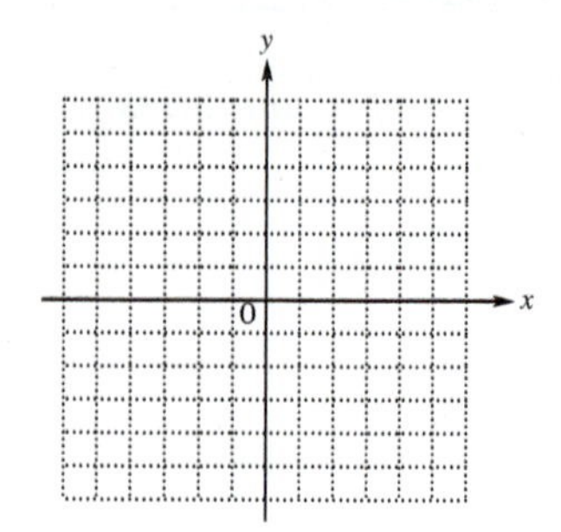

16. $x \ge 2y + 6$
$y > -2x + 4$

17. $4x + 3y < 6$
$x - 2y > 4$

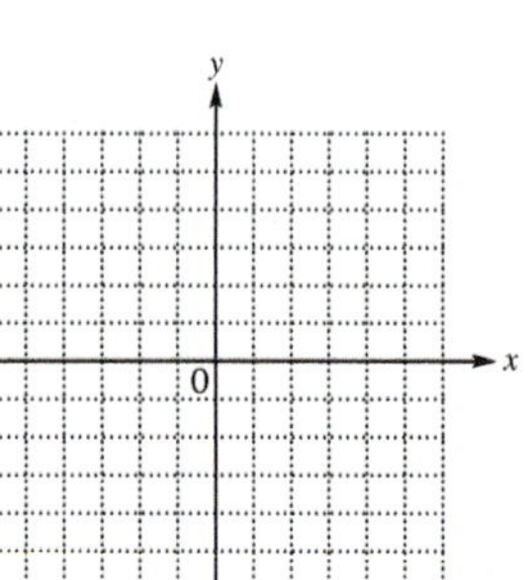

18. $3x + y > 4$
$x + 2y < 2$

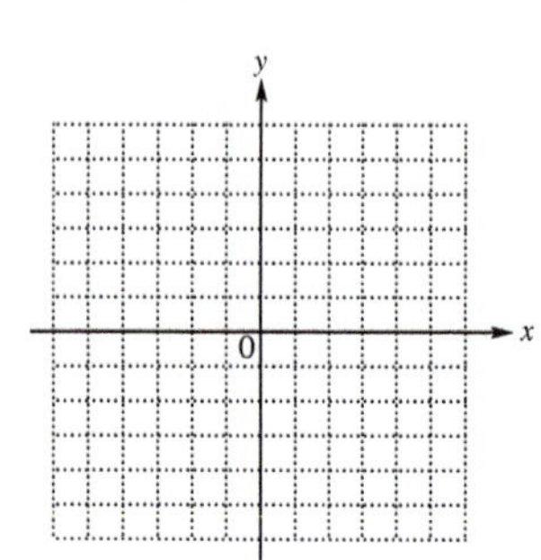

19. $x \le 2y + 3$
$x + y < 0$

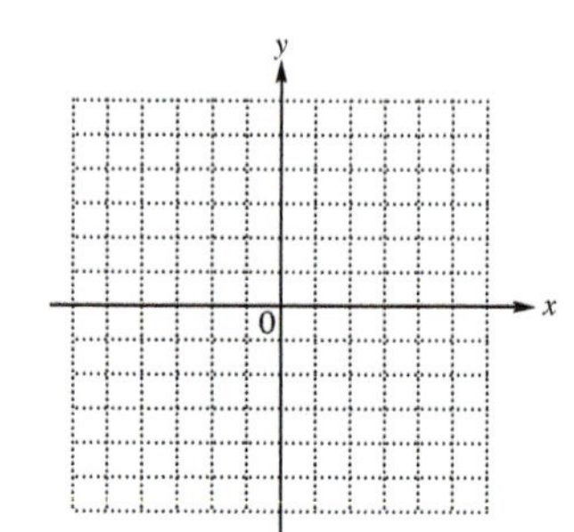

20. $x \le 4y + 3$
$x + y > 0$

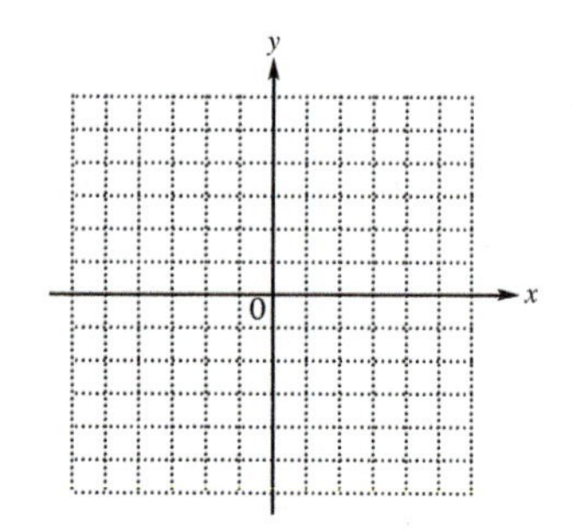

21. $x - 3y \le 6$
$x \ge -5$

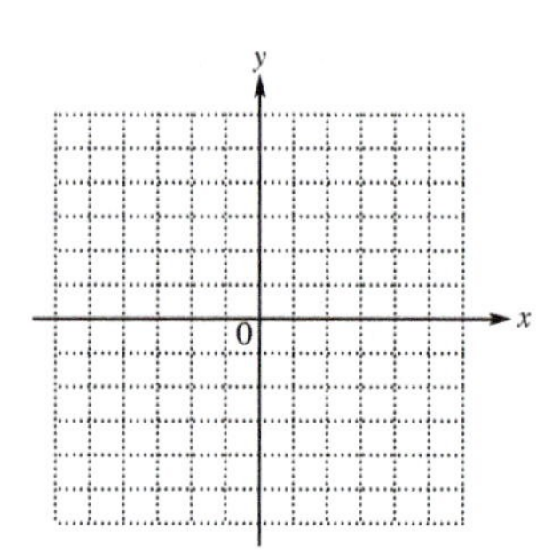

22. $x - 2y \ge 2$
$x \le -3$

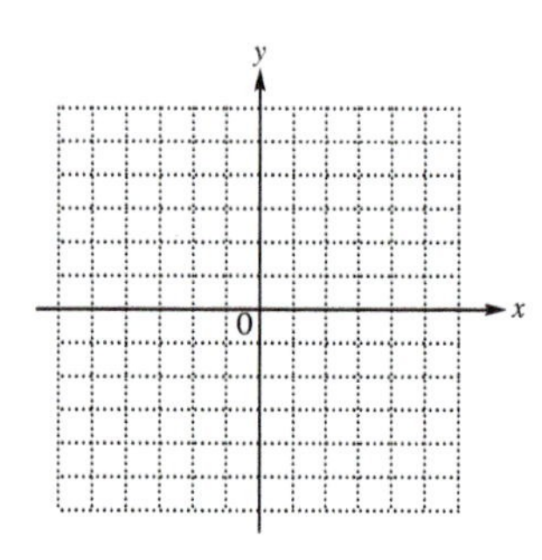

23. $4x + 5y < 8$
$y > -2$
$x > -4$

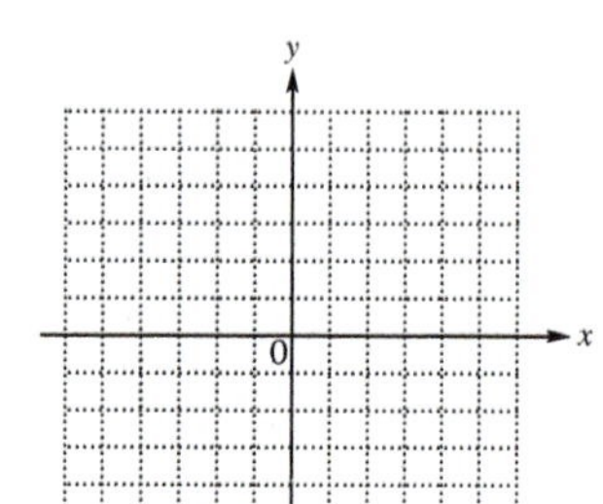

24. $x - 2y \ge -2$
$y \ge -2$
$x \le 3$

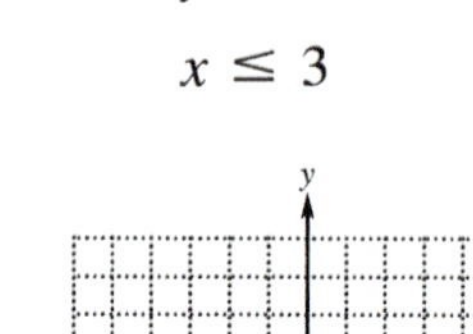

25. $x + y \ge -3$
$x - y \le 3$
$y \le 3$

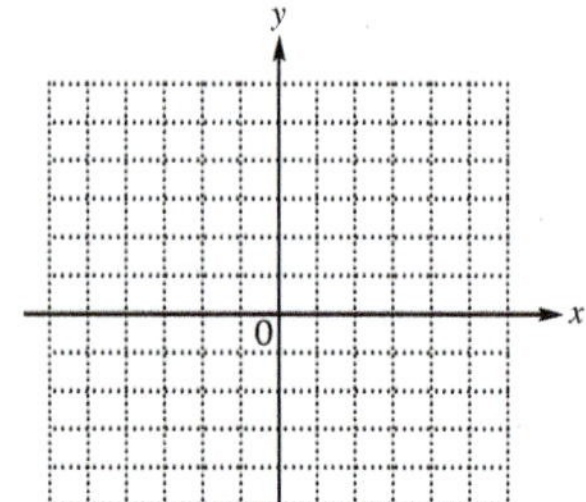

26. $x + y < 4$
$x - y > -4$
$y > -1$

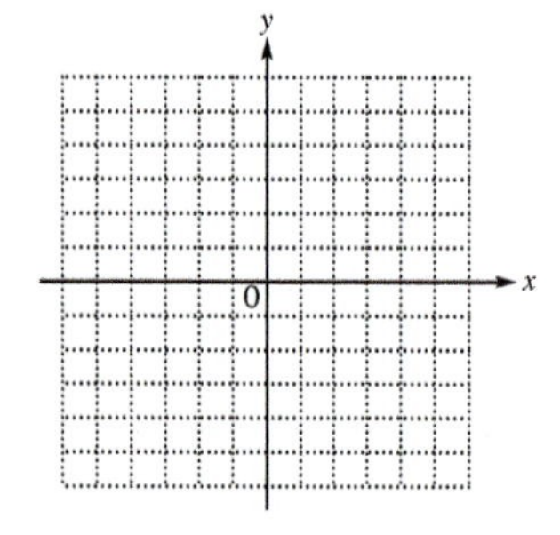

27. $3x - 2y \ge 6$
$x + y \le 4$
$x \ge 0$
$y \ge -4$

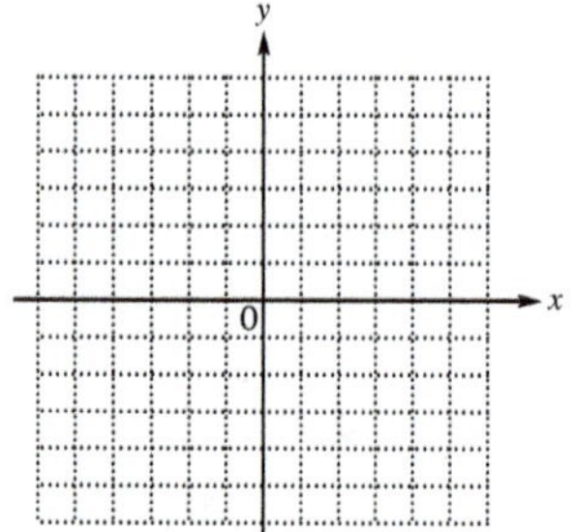

28. $2x - 3y < 6$
$x + y > 3$
$x < 4$
$y < 4$

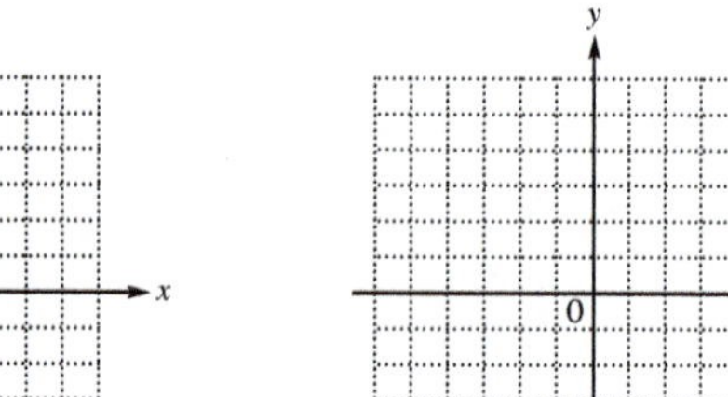

Chapter 4 Summary

Key Terms

4.1

system of linear equations A system of linear equations (or **linear system**) consists of two or more linear equations with the same variables.

solution of a system A solution of a system of linear equations is an ordered pair that makes all the equations of the system true at the same time.

solution set of a system The set of all ordered pairs that are solutions of a system is its solution set.

consistent system A system of equations with at least one solution is a consistent system.

inconsistent system An inconsistent system is a system of equations with no solution.

independent equations Equations of a system that have different graphs are independent equations.

dependent equations Equations of a system that have the same graph (because they are different forms of the same equation) are dependent equations.

4.5

system of linear inequalities A system of linear inequalities consists of two or more linear inequalities.

solution set of a system of linear inequalities The solution set of a system of linear inequalities includes all ordered pairs that make all inequalities of the system true at the same time.

Test Your Word Power

See how well you have learned the vocabulary in this chapter.

1 A **system of linear equations** consists of
- **A.** at least two linear equations with different variables
- **B.** two or more linear equations that have an infinite number of solutions
- **C.** two or more linear equations with the same variables
- **D.** two or more linear inequalities.

2 A **solution of a system** of linear equations is
- **A.** an ordered pair that makes one equation of the system true
- **B.** an ordered pair that makes all the equations of the system true at the same time
- **C.** any ordered pair that makes one or the other or both equations of the system true
- **D.** the set of values that make all the equations of the system false.

3 A **consistent system** is a system of equations
- **A.** with at least one solution
- **B.** with no solution
- **C.** with graphs that do not intersect
- **D.** with solution set $\emptyset$.

4 An **inconsistent system** is a system of equations
- **A.** with one solution
- **B.** with no solution
- **C.** with an infinite number of solutions
- **D.** that have the same graph.

5 **Dependent equations**
- **A.** have different graphs
- **B.** have no solution
- **C.** have one solution
- **D.** are different forms of the same equation.

Answers to Test Your Word Power

1. C; *Example:* $2x + y = 7$
$3x - y = 3$

2. B; *Example:* The ordered pair $(2, 3)$ satisfies both equations of the system in the Answer 1 example, so it is a solution of the system.

3. A; *Example:* The system in the Answer 1 example is consistent. The graphs of the equations intersect at exactly one point, in this case the solution $(2, 3)$.

4. B; *Example:* The equations of two parallel lines make up an inconsistent system. Their graphs never intersect, so there is no solution of the system.

5. D; *Example:* The equations $4x - y = 8$ and $8x - 2y = 16$ are dependent because their graphs are the same line. Multiplying the first equation by 2 gives a form that is equivalent to the second equation.

Quick Review

Concepts	Examples

4.1 Solving Systems of Linear Equations by Graphing

An ordered pair is a solution of a system if it makes all equations of the system true at the same time.

Is $(4, -1)$ a solution of the system $\begin{matrix} x + y = 3 \\ 2x - y = 9 \end{matrix}$?

Because $4 + (-1) = 3$ and $2(4) - (-1) = 9$ are both true, $(4, -1)$ is a solution.

To solve a linear system by graphing, follow these steps.

Step 1 Graph each equation of the system on the same axes.

Step 2 Find the coordinates of the point of intersection.

Step 3 Check. Write the solution set.

Solve the system by graphing.

$$x + y = 5$$

$$2x - y = 4$$

The ordered pair $(3, 2)$ satisfies *both* equations, so $\{(3, 2)\}$ is the solution set.

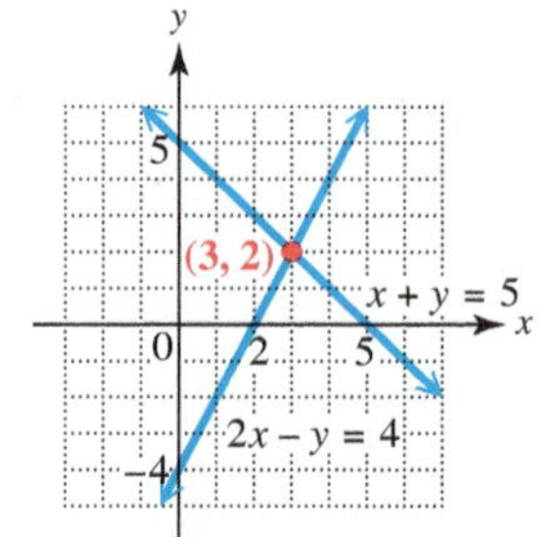

If the graphs of the equations do not intersect (that is, the lines are parallel), then the system has *no solution* and the solution set is $\emptyset$.

If the graphs of the equations are the same line, then the system has an *infinite number of solutions*. Use set-builder notation to write the solution set as

$$\{(x, y) \mid \_\_\_\_\_\_\_\_\},$$

where a form of the equation is written on the blank.

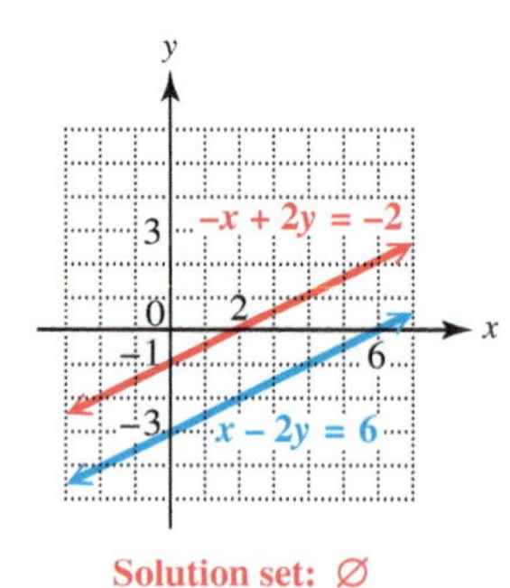

Solution set: $\emptyset$

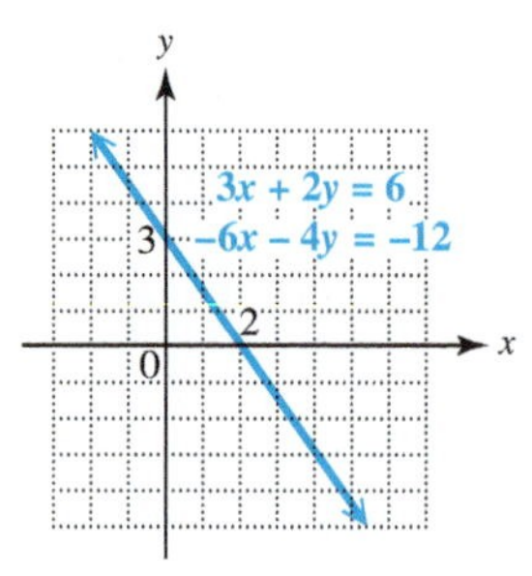

Solution set: $\{(x, y) \mid 3x + 2y = 6\}$

4.2 Solving Systems of Linear Equations by Substitution

Step 1 Solve one equation for either variable.

Step 2 Substitute for that variable in the other equation to obtain an equation in one variable.

Step 3 Solve the equation from Step 2.

Step 4 Find the other value by substituting the result from Step 3 into the equation from Step 1 and solving for the remaining variable.

Step 5 Check. Write the solution set.

Solve the system by substitution.

$$x + 2y = -5 \quad (1)$$

$$y = -2x - 1 \quad (2)$$

Equation (2) is already solved for y.

Substitute $-2x - 1$ for y in equation (1).

$$\begin{aligned} x + 2y &= -5 && (1) \\ x + 2(-2x - 1) &= -5 && \text{Let } y = -2x - 1. \\ x - 4x - 2 &= -5 && \text{Distributive property} \\ -3x - 2 &= -5 && \text{Combine like terms.} \\ -3x &= -3 && \text{Add 2.} \\ x &= 1 && \text{Divide by } -3. \end{aligned}$$

To find y, let $x = 1$ in equation (2).

$$\begin{aligned} y &= -2x - 1 && (2) \\ y &= -2(1) - 1 && \text{Let } x = 1. \\ y &= -3 && \text{Multiply, and then subtract.} \end{aligned}$$

A check confirms that $\{(1, -3)\}$ is the solution set.

Concepts	Examples

4.3 Solving Systems of Linear Equations by Elimination

Step 1 Write both equations in standard form $Ax + By = C$.

Step 2 Multiply to transform the equations so that the coefficients of one pair of variable terms are opposites.

Step 3 Add the equations to *eliminate* a variable.

Step 4 Solve the equation from Step 3.

Step 5 Find the other value by substituting the result from Step 4 into either of the original equations and solving for the remaining variable.

Step 6 Check. Write the solution set.

Solve the system by elimination.

$$x + 3y = 7 \quad (1)$$
$$3x - y = 1 \quad (2)$$

Multiply equation (1) by -3 to eliminate the x-terms.

$$-3x - 9y = -21 \quad \text{Multiply equation (1) by } -3.$$
$$3x - y = 1 \quad (2)$$
$$-10y = -20 \quad \text{Add.}$$
$$y = 2 \quad \text{Divide by } -10.$$

Substitute to find the value of x.

$$x + 3y = 7 \quad (1)$$
$$x + 3(2) = 7 \quad \text{Let } y = 2.$$
$$x + 6 = 7 \quad \text{Multiply.}$$
$$x = 1 \quad \text{Subtract 6.}$$

Because $1 + 3(2) = 7$ and $3(1) - 2 = 1$, the solution $(1, 2)$ checks. The solution set is $\{(1, 2)\}$.

4.4 Applications of Linear Systems

Use the modified six-step method.

Step 1 Read the problem carefully.

Step 2 Assign variables to represent the unknown values. Make a sketch, diagram, or table, as needed.

Step 3 Write two equations using both variables.

Step 4 Solve the system of two equations.

Step 5 State the answer.

Step 6 Check the answer in the words of the *original* problem.

The sum of two numbers is 30. Their difference is 6. Find the numbers.

Let x = one number, and let y = the other number.

$$x + y = 30 \quad (1)$$
$$x - y = 6 \quad (2)$$
$$2x = 36 \quad \text{Add.}$$
$$x = 18 \quad \text{Divide by 2.}$$

Let $x = 18$ in equation (1), $x + y = 30$.

$$18 + y = 30 \quad \text{Let } x = 18.$$
$$y = 12 \quad \text{Subtract 18.}$$

The two numbers are 18 and 12.

The sum of 18 and 12 is 30, and the difference of 18 and 12 is 6, so the answer checks.

4.5 Solving Systems of Linear Inequalities

To solve a system of linear inequalities, follow these steps.

Step 1 Graph each inequality on the same axes.

Step 2 Choose the intersection. The solution set of the system is formed by the overlap of the regions of the two graphs.

Graph the solution set of the system.

$$2x + 4y \geq 5$$
$$x \geq 1$$

First graph the solid boundary lines $2x + 4y = 5$ and $x = 1$. Then use a test point, such as $(0, 0)$, to determine the region to shade for each inequality. The intersection, the gray shaded region, is the solution set of the system.

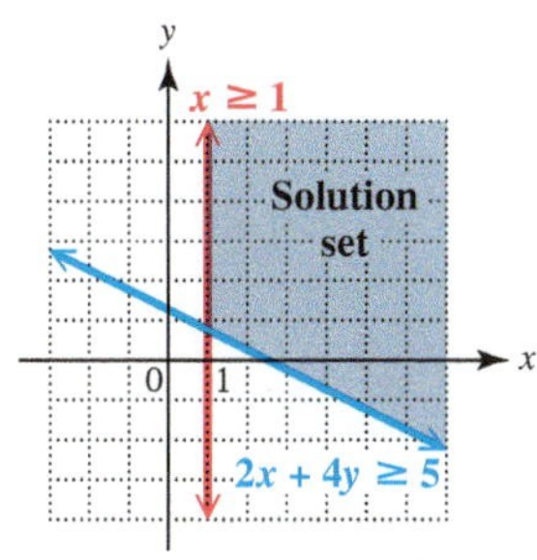

Chapter 4 Review Exercises

4.1 *Determine whether the given ordered pair is a solution of the given system.*

1. $\begin{aligned} 4x - 2y &= 4 \\ 5x + y &= 19 \end{aligned}$; $(3, 4)$

2. $\begin{aligned} x - 4y &= -13 \\ 2x + 3y &= 4 \end{aligned}$; $(-5, 2)$

Solve each system by graphing.

3. $x + y = 4$
$2x - y = 5$

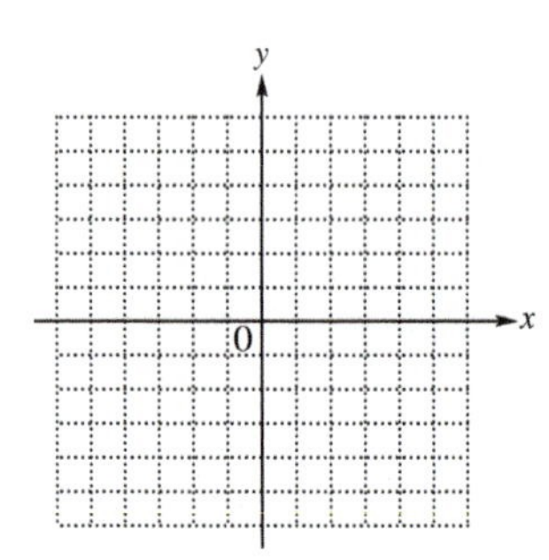

4. $x - 2y = 4$
$2x + y = -2$

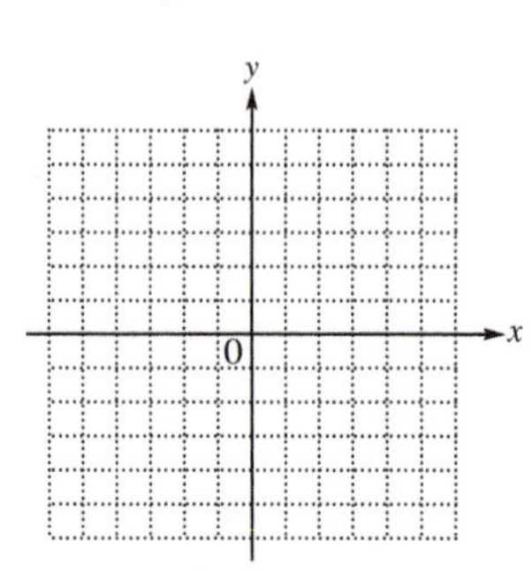

5. $x - 2 = 2y$
$2x - 4y = 4$

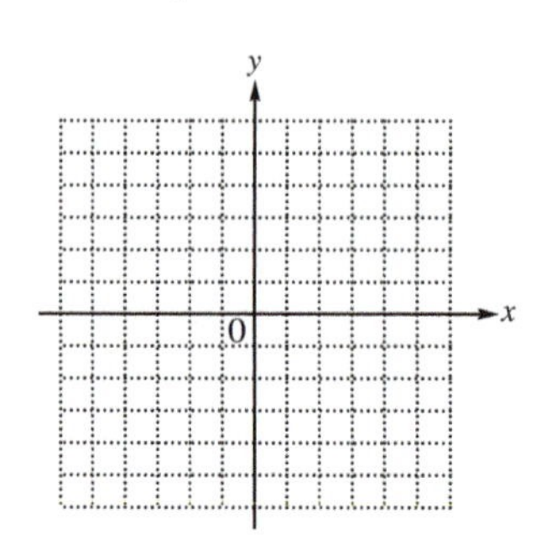

6. $2x + 4 = 2y$
$y - x = -3$

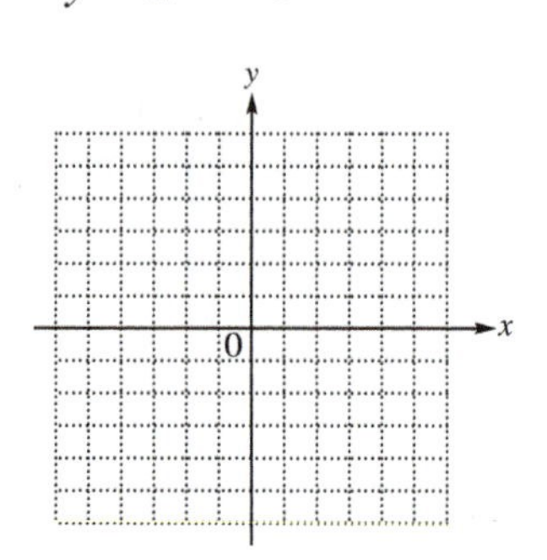

4.2 *Solve each system by the substitution method.*

7. $3x + y = 7$
$x = 2y$

8. $2x - 5y = -19$
$y = x + 2$

9. $4x + 5y = 44$
$x + 2 = 2y$

10. $5x + 15y = 3$
$x + 3y = 2$

4.3 *Solve each system by the elimination method.*

11. $2x - y = 13$
$x + y = 8$

12. $3x - y = -13$
$x - 2y = -1$

13. $-4x + 3y = 25$
$6x - 5y = -39$

14. $3x - 4y = 9$
$6x - 8y = 18$

4.1–4.3 *Solve each system by any method.*

15. $x - 2y = 5$
$y = x - 7$

16. $5x - 3y = 11$
$2y = x - 4$

17. $5x - 4y = 0$
$-3x + 2y = 0$

18. $6y = 10x - 15$
$14x - 8y = 21$

19. $\frac{x}{2} + \frac{y}{3} = 7$
$\frac{x}{4} + \frac{2y}{3} = 8$

20. $\frac{3x}{4} - \frac{y}{3} = \frac{7}{6}$
$\frac{x}{2} + \frac{2y}{3} = \frac{5}{3}$

21. $0.2x + 1.2y = -1$
$0.1x + 0.3y = 0.1$

22. $0.1x + y = 1.6$
$0.6x + 0.5y = -1.4$

4.4 *For each problem, complete any tables. Then solve the problem using a system of equations.*

23. The two leading pizza chains in the United States are Pizza Hut and Domino's. In 2015, Pizza Hut had 3976 more locations than Domino's, and together the two chains had 27,234 locations. How many locations did each chain have? (Data from www.pizzatoday.com)

24. Together, the average paid circulation for *Reader's Digest* and *People* magazines in 2014 was 6.2 million. The circulation for *People* was 0.8 million more than that of *Reader's Digest*. What were the circulation figures for each magazine? (Data from Audit Bureau of Circulations.)

25. Candy that sells for \$1.30 per lb is to be mixed with candy selling for \$0.90 per lb to make 100 lb of a mix that will sell for \$1 per lb. How much of each type candy should be used?

Number of Pounds	Cost per Pound (in dollars)	Total Value (in dollars)
______	1.30	$1.30x$
y	______	______
100	1.00	______

26. A cashier has 20 bills, all of which are \$10 or \$20 bills. The total value of the money is \$330. How many of each type of bill does the cashier have?

Number of Bills	Denomination of Bills (in dollars)	Total Value (in dollars)
x	10	______
______	______	$20y$
______	XXXXXXXX	330

27. The perimeter of a rectangle is 90 m. Its length is $1\frac{1}{2}$ times its width. Find the length and width of the rectangle.

28. A certain plane flying with the wind travels 540 mi in 2 hr. Later, flying against the same wind, the plane travels 690 mi in 3 hr. Find the rate of the plane in still air and the wind speed.

29. After taxes, Ms. Cesar's game show winnings were \$18,000. She invested part of it at 3% annual simple interest and the rest at 4%. Her interest income for the first year was \$650. How much did she invest at each rate?

Amount Invested (in dollars)	Percent (as a decimal)	Interest (in dollars)
x	0.03	______
y	______	______
18,000	XXXXXX	______

30. A 40% antifreeze solution is to be mixed with a 70% solution to make 90 L of a 50% solution. How many liters of the 40% and 70% solutions will be needed?

Number of Liters	Percent (as a decimal)	Amount of Pure Antifreeze
x	0.40	______
y	______	______
90	0.50	______

Graph the solution set of each system of linear inequalities.

31. $x + y \geq 2$
$x - y \leq 4$

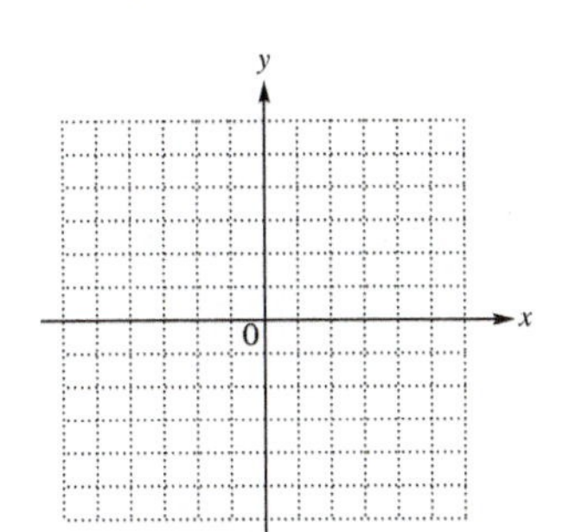

32. $y \geq 2x$
$2x + 3y \leq 6$

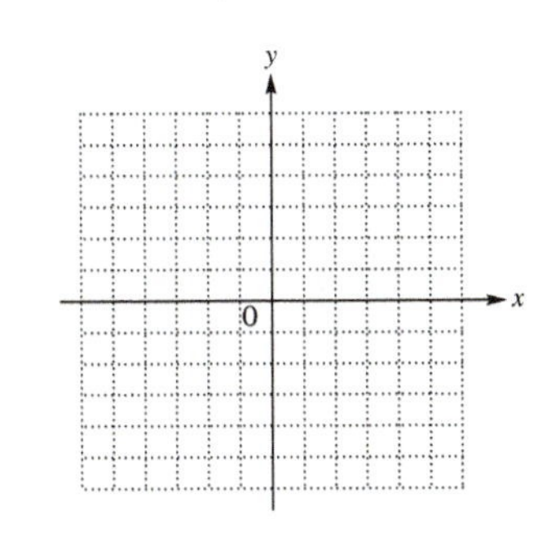

33. $x + y < 3$
$2x > y$

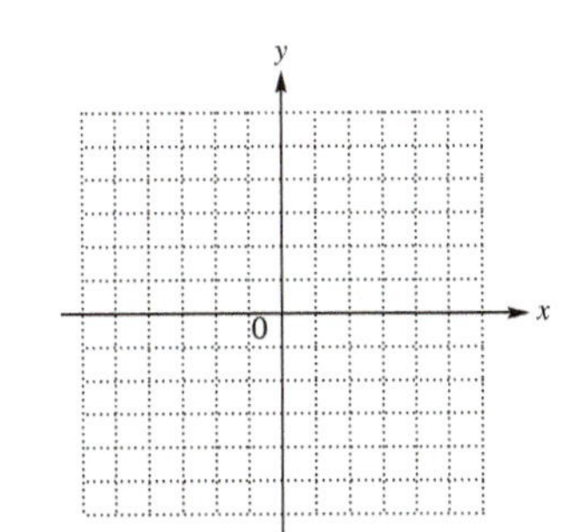

34.
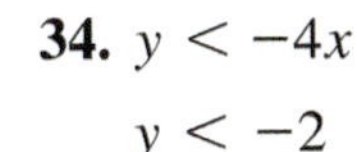
$y < -4x$
$y < -2$

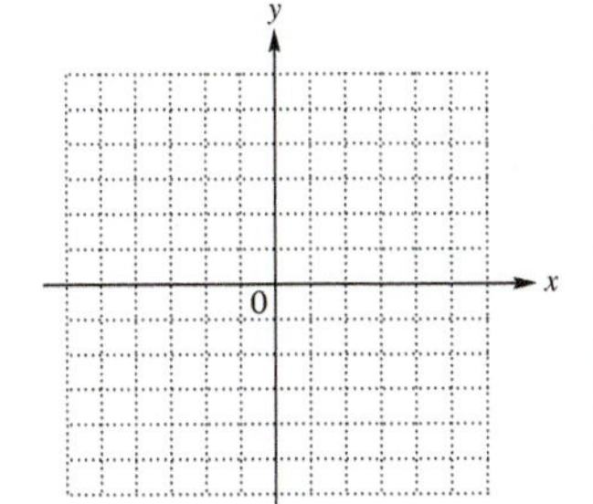

Chapter 4 Mixed Review Exercises

Solve.

1. $3x + 4y = 6$
   $4x - 5y = 8$

2. $\frac{3x}{2} + \frac{y}{5} = -3$
   $4x + \frac{y}{3} = -11$

3. $x + 6y = 3$
   $2x + 12y = 2$

4. $x + y < 5$
   $x - y \geq 2$

5. $y \leq 2x$
   $x + 2y > 4$

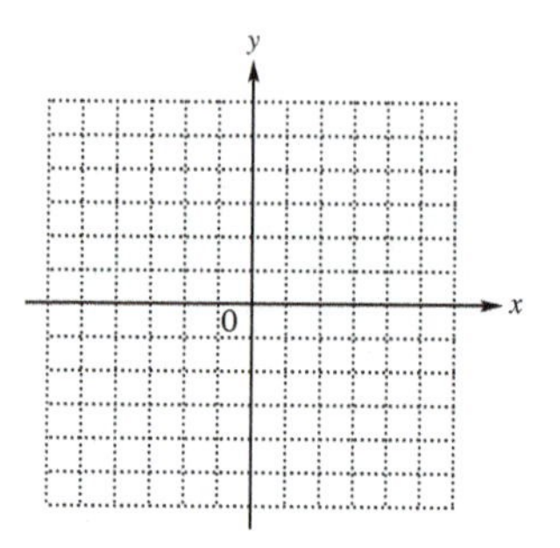

6. The perimeter of an isosceles triangle is 29 in. One side of the triangle is 5 in. longer than each of the two equal sides. Find the lengths of the sides of the triangle.

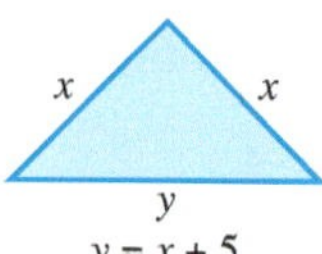

7. In Super Bowl 50, the Denver Broncos beat the Carolina Panthers by 14 points, and the winning score was 4 points more than twice the losing score. What was the final score of the game? (Data from NFL.)

8. Eboni compared the monthly payments she would incur for two types of mortgages: fixed-rate and variable-rate. Her observations led to the graph shown.
   (a) For which years would the monthly payment be more for the fixed-rate mortgage than for the variable-rate mortgage?
   (b) In what year would the payments be the same, and what would those payments be?

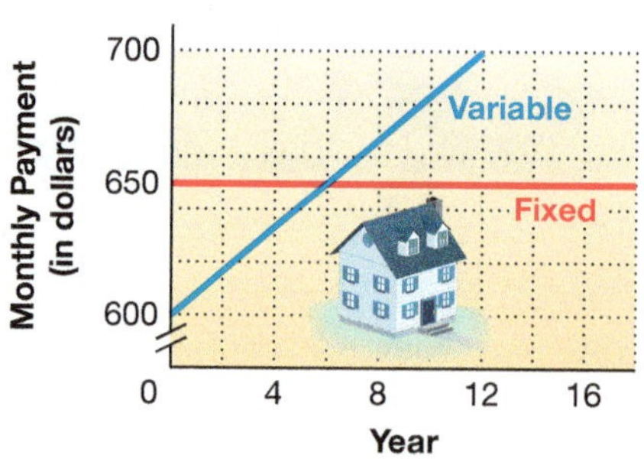

CONCEPT CHECK *Answer each question.*

9. Which system of linear inequalities is graphed in the figure?
   A. $x \leq 3$, $y \leq 1$
   B. $x \leq 3$, $y \geq 1$
   C. $x \geq 3$, $y \leq 1$
   D. $x \geq 3$, $y \geq 1$

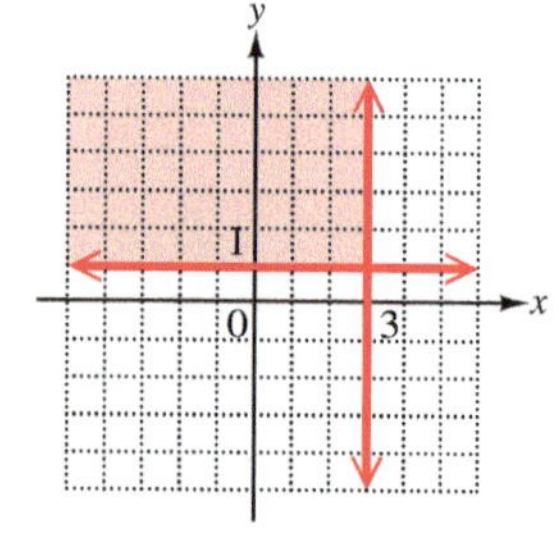

10. Which system of linear inequalities has no solution? (Do not graph.)
   A. $x \geq 4$, $y \leq 3$
   B. $x + y > 4$, $x + y < 3$
   C. $x > 2$, $y < 1$
   D. $x + y < 4$, $x - y < 3$

Chapter 4 Test

The Chapter Test Prep Videos with step-by-step solutions are available in MyMathLab or on YouTube at https://goo.gl/8aAsWP

1. Decide whether each ordered pair is a solution of the system.

$$2x + y = -3$$
$$x - y = -9$$

(a) $(1, -5)$ **(b)** $(1, 10)$ **(c)** $(-4, 5)$

2. Solve the system by graphing.

$$2x + y = 1$$
$$3x - y = 9$$

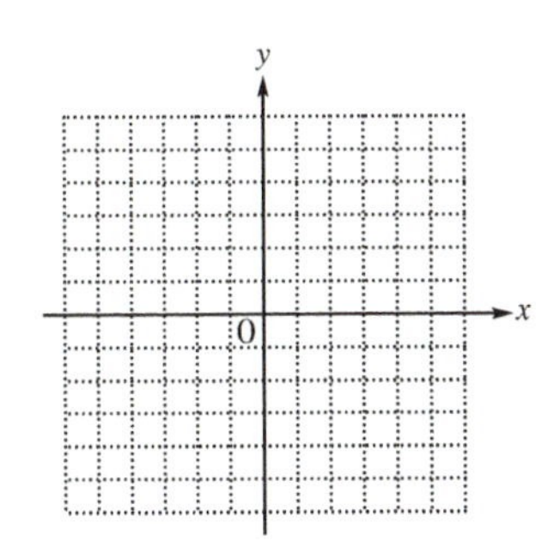

3. Suppose that the graph of a system of two linear equations consists of lines that have the same slope but different y-intercepts. How many solutions does the system have?

Solve each system by the substitution method.

4. $2x + y = -4$
$x = y + 7$

5. $4x + 3y = -35$
$x + y = 0$

6. $y = 6x - 8$
$y = -3x - 11$

Solve each system by the elimination method.

7. $2x - y = 4$
$3x + y = 21$

8. $4x + 2y = 2$
$5x + 4y = 7$

9. $3x + 4y = 9$
$2x + 5y = 13$

10. $6x - 5y = 0$
$-2x + 3y = 0$

11. $4x + 5y = 2$
$-8x - 10y = 6$

12. $-2x + 5y = 14$
$7x + 6y = -2$

Solve each system by any method.

13. $3x = 6 + y$

$6x - 2y = 12$

14. $\frac{6}{5}x - \frac{1}{3}y = -20$

$-\frac{2}{3}x + \frac{1}{6}y = 11$

Solve each problem.

15. The distance between Memphis and Atlanta is 782 mi less than the distance between Minneapolis and Houston. Together, the two distances total 1570 mi. How far is it between Memphis and Atlanta? How far is it between Minneapolis and Houston? (Data from *Rand McNally Road Atlas.*)

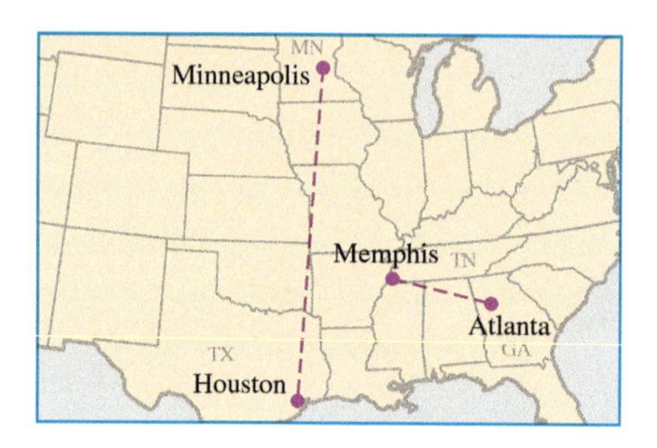

16. In 2015, a total of 6.7 million people visited the Statue of Liberty and the Mount Rushmore National Memorial. The Statue of Liberty had 1.9 million more visitors than the Mount Rushmore National Memorial. How many visitors did each of these attractions have? (Data from National Park Service, Department of the Interior.)

17. A 15% solution of alcohol is to be mixed with a 40% solution to make 50 L of a final mixture that is 30% alcohol. How much of each of the original solutions should be used?

Liters of Solution	Percent (as a decimal)	Liters of Pure Alcohol

18. Two cars leave from Perham, Minnesota, and travel in the same direction. One car travels $1\frac{1}{3}$ times as fast as the other. After 3 hr they are 45 mi apart. What are the rates of the cars?

	r	t	d
Faster Car			
Slower Car			

Graph the solution set of each system of linear inequalities.

19. $2x + 7y \leq 14$

$x - y \geq 1$

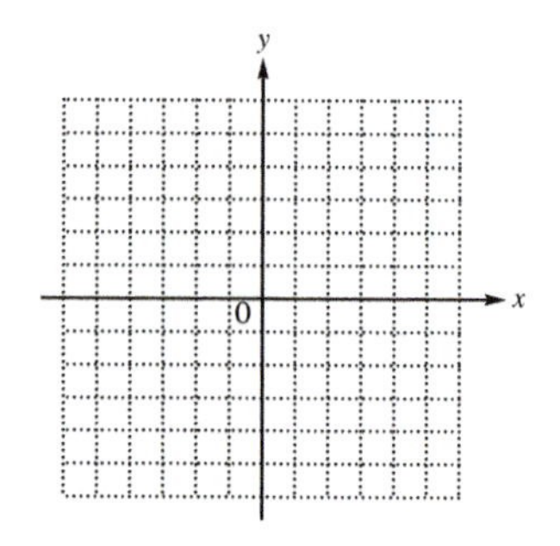

20. $2x - y > 6$

$4y + 12 \geq -3x$

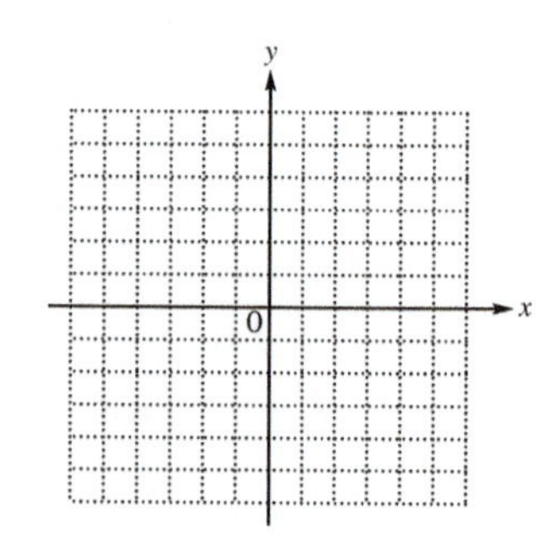

Chapters R–4 Cumulative Review Exercises

1. List all integer factors of 40.

2. Find the value of the expression for $x = 1$ and $y = 5$.

$$\frac{3x^2 + 2y^2}{10y + 3}$$

Name the property that justifies each statement.

3. $5 + (-4) = (-4) + 5$

4. $r(s - k) = rs - rk$

5. $-\frac{2}{3} + \frac{2}{3} = 0$

6. Evaluate $-2 + 6[3 - (4 - 9)]$.

7. Solve the formula $P = \frac{kT}{V}$ for T.

Solve each linear equation.

8. $2 - 3(6x + 2) = 4(x + 1) + 18$

9. $\frac{3}{2}\left(\frac{1}{3}x + 4\right) = 6\left(\frac{1}{4} + x\right)$

Solve each linear inequality. Write the solution set in interval notation.

10. $-\frac{5}{6}x < 15$

11. $-8 < 2x + 3$

12. The iPad Pro tablet computer has a perimeter of 41.4 in., and its width is 3.3 in. less than its length. What are its dimensions? (Data from www.apple.com)

Graph each linear equation.

13. $x - y = 4$

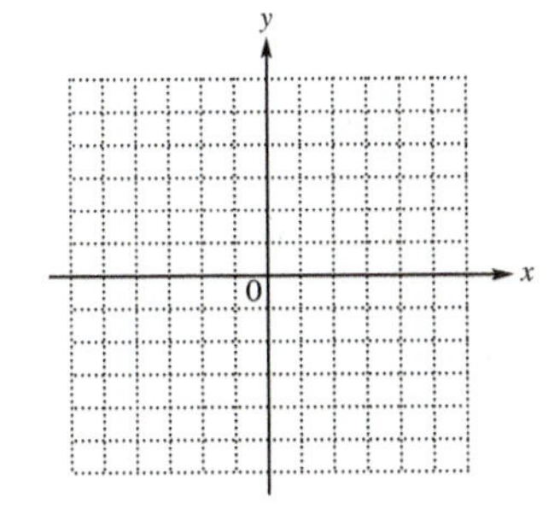

14. $3x + y = 6$

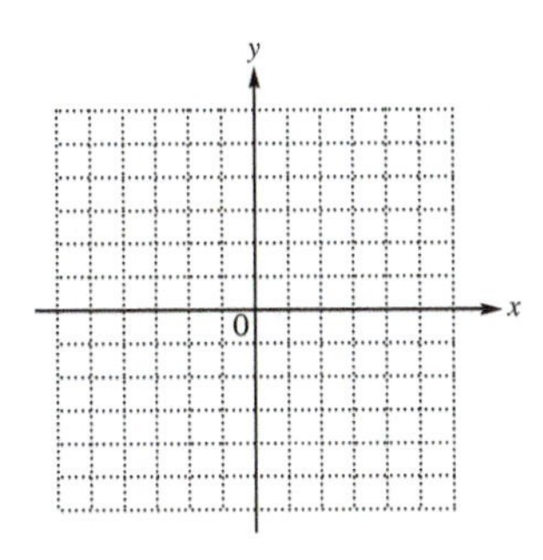

Find the slope of each line.

15. Through $(-5, 6)$ and $(1, -2)$

16. Perpendicular to the line $y = 4x - 3$

Write an equation in slope-intercept form for each line.

17. Through $(-4, 1)$ with slope $\frac{1}{2}$

18. Through the points $(1, 3)$ and $(-2, -3)$

19. Write an equation for each line described.

(a) Vertical, through $(9, -2)$

(b) Horizontal, through $(4, -1)$

Solve each system by any method.

20. $2x - y = -8$
$x + 2y = 11$

21. $4x + 5y = -8$
$3x + 4y = -7$

22. $3x + 4y = 2$
$6x + 8y = 1$

Solve each problem using a system of equations.

23. Admission prices at a high school football game were \$6 for adults and \$2 for children. The total value of the tickets sold was \$2528, and 454 tickets were sold. How many adults and how many children attended the game?

Kind of Ticket	Number Sold	Cost of Each (in dollars)	Total Value (in dollars)
Adult	x	6	$6x$
Child	y	____	____
Total	454	XXXX	____

24. The perimeter of a triangle is 53 in. If two sides are of equal length, and the third side measures 4 in. less than each of the equal sides, what are the lengths of the three sides?

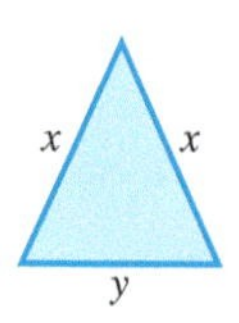

25. Graph the solution set of the system.

$x + 2y \leq 12$
$2x - y \leq 8$

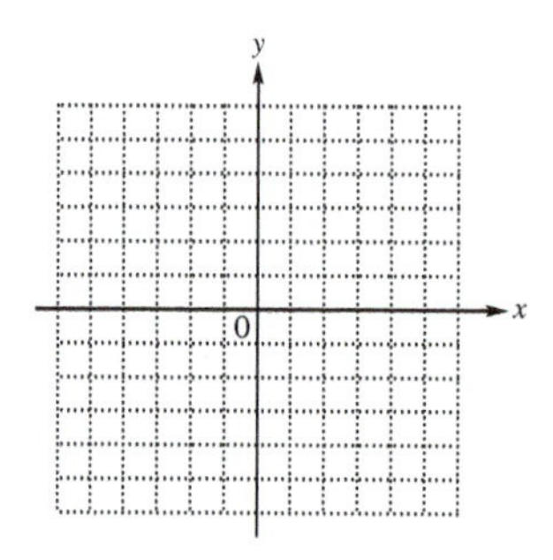

Exponents and Polynomials

Exponents and *scientific notation*, two of the topics of this chapter, are often used to express very large or very small numbers. Using this notation, one light-year, which is about 6 trillion miles, is written

$$6 \times 10^{12}.$$

5.1 The Product Rule and Power Rules for Exponents

OBJECTIVES

1. Use exponents.
2. Use the product rule for exponents.
3. Use the rule $(a^m)^n = a^{mn}$.
4. Use the rule $(ab)^m = a^m b^m$.
5. Use the rule $\left(\frac{a}{b}\right)^m = \frac{a^m}{b^m}$.
6. Use combinations of the rules for exponents.
7. Use the rules for exponents in a geometry problem.

OBJECTIVE 1 Use exponents. Recall that we can use exponents to write repeated products. In the expression 5^2, the number 5 is the **base** and 2 is the **exponent,** or **power.** The expression 5^2 is an **exponential expression.** Although we do not usually write the exponent when it is 1, in general,

$$a^1 = a, \quad \text{for any quantity } a.$$

EXAMPLE 1 Using Exponents

Write $3 \cdot 3 \cdot 3 \cdot 3 \cdot 3$ in exponential form and evaluate.

Since 3 occurs as a factor five times, the base is 3 and the exponent is 5. The exponential expression is 3^5, read "3 *to the fifth power,*" or simply "3 *to the fifth.*"

$$\underbrace{3 \cdot 3 \cdot 3 \cdot 3 \cdot 3}_{5 \text{ factors of } 3} \quad \text{means} \quad 3^5, \quad \text{which equals} \quad 243.$$

◀ Work Problem 1 at the Side.

1 Write each product in exponential form and evaluate.

(a) $2 \cdot 2 \cdot 2 \cdot 2$

(b) $(-3)(-3)(-3)$

EXAMPLE 2 Evaluating Exponential Expressions

Identify the base and the exponent of each expression. Then evaluate.

Expression	Base	Exponent	Value
(a) 5^4	5	4	$5 \cdot 5 \cdot 5 \cdot 5$, which equals 625
(b) -5^4	5	4	$-1 \cdot (5 \cdot 5 \cdot 5 \cdot 5)$, which equals -625
(c) $(-5)^4$	-5	4	$(-5)(-5)(-5)(-5)$, which equals 625

◀ Work Problem 2 at the Side.

2 Identify the base and the exponent of each expression. Then evaluate.

(a) 2^5 **(b)** -2^5

(c) $(-2)^5$ **(d)** 4^2

(e) -4^2 **(f)** $(-4)^2$

! CAUTION

Compare **Examples 2(b) and 2(c).** In -5^4, the absence of parentheses means that the exponent 4 applies only to the base 5, not -5. In $(-5)^4$, the parentheses mean that the exponent 4 applies to the base -5.

In summary, $-a^n$ and $(-a)^n$ are not necessarily the same.

Expression	Base	Exponent	Example
$-a^n$	a	n	$-3^2 = -(3 \cdot 3) = -9$
$(-a)^n$	$-a$	n	$(-3)^2 = (-3)(-3) = 9$

OBJECTIVE 2 Use the product rule for exponents. To develop the product rule, we use the definition of an exponent. Consider the following.

$$2^4 \cdot 2^3$$

$$= \underbrace{(2 \cdot 2 \cdot 2 \cdot 2)}_{4 \text{ factors}}\underbrace{(2 \cdot 2 \cdot 2)}_{3 \text{ factors}}$$

$$= \underbrace{2 \cdot 2 \cdot 2 \cdot 2 \cdot 2 \cdot 2 \cdot 2}_{4 + 3 = 7 \text{ factors}}$$

$$= 2^7$$

Answers

1. (a) 2^4; 16 (b) $(-3)^3$; -27
2. (a) 2; 5; 32 (b) 2; 5; -32 (c) -2; 5; -32 (d) 4; 2; 16 (e) 4; 2; -16 (f) -4; 2; 16

Also, consider that

$$
\begin{aligned}
&6^2 \cdot 6^3 \\
&= (6 \cdot 6)(6 \cdot 6 \cdot 6) \\
&= 6 \cdot 6 \cdot 6 \cdot 6 \cdot 6 \\
&= 6^5.
\end{aligned}
$$

Generalizing from these examples, we have the following.

$2^4 \cdot 2^3$ is equal to 2^{4+3}, which equals 2^7.

$6^2 \cdot 6^3$ is equal to 6^{2+3}, which equals 6^5.

In each case, adding the exponents gives the exponent of the product, suggesting the **product rule for exponents.**

Product Rule for Exponents

For any positive integers m and n, $a^m \cdot a^n = a^{m+n}$.
(Keep the same base and add the exponents.)

Example: $6^2 \cdot 6^5 = 6^{2+5} = 6^7$

CAUTION

When using the product rule, keep the same base and add the exponents. Do ***not*** multiply the bases.

$6^2 \cdot 6^5$ is equal to 6^7, ***not*** 36^7.

EXAMPLE 3 Using the Product Rule

Use the product rule for exponents to simplify each expression, if possible.

(a) $6^3 \cdot 6^5$

$= 6^{3+5}$ Product rule

$= 6^8$ Add the exponents.

(b) $(-4)^7(-4)^2$

$= (-4)^{7+2}$ Product rule

$= (-4)^9$ Add the exponents.

(c) $x^2 \cdot x$

$= x^2 \cdot x^1$ $a = a^1$, for all a.

$= x^{2+1}$ Product rule

$= x^3$ Add the exponents.

(d) $m^4 m^3 m^5$

$= m^{4+3+5}$ Product rule

$= m^{12}$ Add the exponents.

(e) $2^3 \cdot 3^2$ ***The bases are different.***

Think: 2^3 means $2 \cdot 2 \cdot 2$.

Think: 3^2 means $3 \cdot 3$.

$= 8 \cdot 9$ Evaluate 2^3 and 3^2.

$= 72$ Multiply.

The product rule does not apply.

(f) $2^3 + 2^4$ ***This is a sum, not a product.***

$= 8 + 16$ Evaluate 2^3 and 2^4.

$= 24$ Add.

The product rule does not apply.

Work Problem 3 at the Side. ▶

3 Use the product rule for exponents to simplify each expression, if possible.

GS **(a)** $8^2 \cdot 8^5$

$= 8^{\_\_+\_\_}$

$= \_\_\_\_$

(b) $(-7)^5(-7)^3$

GS **(c)** $y^3 \cdot y$

$= y^3 \cdot y^{\_\_}$

$= y^{\_\_+\_\_}$

$= \_\_\_\_$

(d) $z^2 z^5 z^6$

(e) $4^2 \cdot 3^5$

(f) $6^4 + 6^2$

Answers

3. **(a)** 2; 5; 8^7 **(b)** $(-7)^8$
(c) 1; 3; 1; y^4 **(d)** z^{13}
(e) The product rule does not apply. (The product is 3888.)
(f) The product rule does not apply. (The sum is 1332.)

4 Multiply.

GS (a) $5m^2 \cdot 2m^6$

$= (5 \cdot \_\_\_\_) \cdot (m^{—} \cdot m^{—})$

$= \_\_\_\_ m^{— + —}$

$= \_\_\_\_$

(b) $3p^5 \cdot 9p^4$

(c) $-7p^5(3p^8)$

5 Use power rule (a) for exponents to simplify each expression.

GS (a) $(5^3)^4$

$= 5^{— \cdot —}$

$= \_\_\_\_$

(b) $(6^2)^5$

(c) $(a^6)^5$

Answers

4. (a) 2; 2; 6; 10; 2; 6; $10m^8$
(b) $27p^9$ (c) $-21p^{13}$

5. (a) 3; 4; 5^{12} (b) 6^{10} (c) a^{30}

EXAMPLE 4 Using the Product Rule

Multiply $2x^3$ and $3x^7$.

$2x^3 \cdot 3x^7$ ($2x^3 = 2 \cdot x^3$; $3x^7 = 3 \cdot x^7$)

$= (2 \cdot 3) \cdot (x^3 \cdot x^7)$ Commutative and associative properties

$= 6x^{3+7}$ Multiply; product rule

$= 6x^{10}$ Add the exponents.

◀ Work Problem 4 at the Side.

CAUTION

Note the important difference between *adding* and *multiplying* exponential expressions.

$8x^3 + 5x^3$ means $(8 + 5)x^3$, which equals $13x^3$.

$(8x^3)(5x^3)$ means $(8 \cdot 5)x^{3+3}$, which equals $40x^6$.

OBJECTIVE 3 Use the rule $(a^m)^n = a^{mn}$. We can simplify an expression such as $(5^2)^4$ with the product rule for exponents, as follows.

$(5^2)^4$

$= 5^2 \cdot 5^2 \cdot 5^2 \cdot 5^2$ Definition of exponent

$= 5^{2+2+2+2}$ Product rule

$= 5^8$ Add.

Observe that $2 \cdot 4 = 8$. This example suggests **power rule (a) for exponents.**

Power Rule (a) for Exponents

For any positive integers m and n, $(a^m)^n = a^{mn}$.

(Raise a power to a power by multiplying exponents.)

Example: $(3^4)^2 = 3^{4 \cdot 2} = 3^8$

EXAMPLE 5 Using Power Rule (a)

Use power rule (a) for exponents to simplify each expression.

(a) $(2^5)^3$
$= 2^{5 \cdot 3}$
$= 2^{15}$

(b) $(5^7)^2$
$= 5^{7 \cdot 2}$
$= 5^{14}$

(c) $(x^2)^5$
$= x^{2 \cdot 5}$ Power rule (a)
$= x^{10}$ Multiply.

◀ Work Problem 5 at the Side.

OBJECTIVE 4 Use the rule $(ab)^m = a^m b^m$. Consider the following.

$(4x)^3$

$= (4x)(4x)(4x)$ Definition of exponent

$= 4 \cdot 4 \cdot 4 \cdot x \cdot x \cdot x$ Commutative and associative properties

$= 4^3x^3$ Definition of exponent

The example $(4x)^3 = 4^3x^3$ suggests **power rule (b) for exponents.**

Power Rule (b) for Exponents

For any positive integer m, $(ab)^m = a^m b^m$.
(Raise a product to a power by raising each factor to the power.)

Example: $(2p)^5 = 2^5 p^5$

EXAMPLE 6 Using Power Rule (b)

Use power rule (b) for exponents to simplify each expression.

(a) $(3xy)^2$

$= 3^2x^2y^2$ Power rule (b)

$= 9x^2y^2$ $3^2 = 3 \cdot 3 = 9$

(b) $3(xy)^2$

$= 3(x^2y^2)$ Power rule (b)

$= 3x^2y^2$ Multiply.

Compare parts (a) and (b). Pay attention to the use of parentheses.

(c) $(2m^2p^3)^4$

$= 2^4(m^2)^4(p^3)^4$ Power rule (b)

$= 2^4m^8p^{12}$ Power rule (a)

$= 16m^8p^{12}$ $2^4 = 2 \cdot 2 \cdot 2 \cdot 2 = 16$

(d) $(-5^6)^3$

$= (-1 \cdot 5^6)^3$ $-a = -1 \cdot a$

$= (-1)^3(5^6)^3$ Power rule (b)

Raise −1 to the designated power.

$= -1 \cdot 5^{18}$ Power rule (a)

$= -5^{18}$ Multiply.

Work Problem 6 at the Side. ▶

CAUTION

Power rule (b) does not apply to a sum.

$(4x)^2 = 4^2x^2$, *but* $(4 + x)^2 \neq 4^2 + x^2$.

OBJECTIVE 5 Use the rule $\left(\frac{a}{b}\right)^m = \frac{a^m}{b^m}$. Because the quotient $\frac{a}{b}$ can be written as $a \cdot \frac{1}{b}$, we use this fact, power rule (b), and properties of real numbers to obtain **power rule (c) for exponents.**

Power Rule (c) for Exponents

For any positive integer m, $\left(\frac{a}{b}\right)^m = \frac{a^m}{b^m}$ (where $b \neq 0$).

(Raise a quotient to a power by raising both numerator and denominator to that power.)

Example: $\left(\frac{5}{3}\right)^2 = \frac{5^2}{3^2}$

---

6 Use power rule (b) for exponents to simplify each expression.

GS **(a)** $(2ab)^4$

$= 2^{—}a^{—}b^{—}$

$=$ ____

(b) $5(mn)^3$

(c) $(3a^2b^4)^5$

(d) $(-5m^2)^3$

Answers

6. **(a)** 4; 4; 4; $16a^4b^4$
(b) $5m^3n^3$ **(c)** $243a^{10}b^{20}$ **(d)** $-125m^6$

7 Use power rule (c) for exponents to simplify each expression. Assume that all variables represent nonzero real numbers.

GS **(a)** $\left(\frac{5}{2}\right)^4$

$= \frac{5^{—}}{2^{—}}$

$=$ ____

(b) $\left(\frac{1}{3}\right)^5$

(c) $\left(\frac{p}{q}\right)^2$

(d) $\left(\frac{r}{t}\right)^3$

(e) $\left(\frac{1}{x}\right)^{10}$

Answers

7. (a) 4; 4; $\frac{625}{16}$ (b) $\frac{1}{243}$ (c) $\frac{p^2}{q^2}$ (d) $\frac{r^3}{t^3}$ (e) $\frac{1}{x^{10}}$

EXAMPLE 7 Using Power Rule (c)

Use power rule (c) for exponents to simplify each expression.

(a) $\left(\frac{2}{3}\right)^5$

$= \frac{2^5}{3^5}$ Power rule (c)

$= \frac{32}{243}$ Simplify.

(b) $\left(\frac{1}{5}\right)^4$

$= \frac{1^4}{5^4}$

$= \frac{1}{625}$

(c) $\left(\frac{m}{n}\right)^4$

$= \frac{m^4}{n^4}$

(where $n \neq 0$)

◀ Work Problem 7 at the Side.

Note

In **Example 7(b),** we used the fact that $1^4 = 1$ because $1 \cdot 1 \cdot 1 \cdot 1 = 1$.

In general, $\quad 1^n = 1, \quad$ ***for any integer n.***

Rules for Exponents

For positive integers m and n, the following hold true.

		Examples
Product rule	$a^m \cdot a^n = a^{m+n}$	$6^2 \cdot 6^5 = 6^{2+5} = 6^7$
Power rules (a)	$(a^m)^n = a^{mn}$	$(3^4)^2 = 3^{4 \cdot 2} = 3^8$
(b)	$(ab)^m = a^m b^m$	$(2p)^5 = 2^5 p^5$
(c)	$\left(\frac{a}{b}\right)^m = \frac{a^m}{b^m} \quad (b \neq 0)$	$\left(\frac{5}{3}\right)^2 = \frac{5^2}{3^2}$

OBJECTIVE 6 **Use combinations of the rules for exponents.**

EXAMPLE 8 Using Combinations of the Rules

Simplify each expression.

(a) $\left(\frac{2}{3}\right)^2 \cdot 2^3$

$= \frac{2^2}{3^2} \cdot \frac{2^3}{1}$ Power rule (c)

$= \frac{2^2 \cdot 2^3}{3^2 \cdot 1}$ Multiply fractions.

$= \frac{2^{2+3}}{3^2}$ Product rule; Multiply.

$= \frac{2^5}{3^2}$ Add.

$= \frac{32}{9}$ Apply the exponents.

(b) $(5x)^3(5x)^4$

$= (5x)^7$ Product rule

$= 5^7x^7$ Power rule (b)

An equally correct way to simplify this expression follows.

$(5x)^3(5x)^4$ **Alternative solution**

$= 5^3x^35^4x^4$ Power rule (b)

$= 5^3 \cdot 5^4 \cdot x^3x^4$ Commutative property

$= 5^{3+4}x^{3+4}$ Product rule

$= 5^7x^7$ Add the exponents.

Continued on Next Page

(c) $(2x^2y^3)^4(3xy^2)^3$

$= 2^4(x^2)^4(y^3)^4 \cdot 3^3x^3(y^2)^3$ Power rule (b)

$= 2^4x^8y^{12} \cdot 3^3x^3y^6$ Power rule (a)

$= 2^4 \cdot 3^3 \cdot x^8x^3y^{12}y^6$ Commutative and associative properties

$= 16 \cdot 27 \cdot x^{11}y^{18}$ Apply the exponents; product rule

$= 432x^{11}y^{18}$ Multiply $16 \cdot 27$.

Notice that $(2x^2y^3)^4$ means $2^4x^{2 \cdot 4}y^{3 \cdot 4}$, ***not*** $(2 \cdot 4)\,x^{2 \cdot 4}y^{3 \cdot 4}$.

(d) $(-x^3y)^2(-x^5y^4)^3$

$= (-1x^3y)^2 \cdot (-1x^5y^4)^3$ $\quad -a = -1 \cdot a$

Think of the negative sign in each factor as -1.

$= (-1)^2(x^3)^2(y^2) \cdot (-1)^3(x^5)^3(y^4)^3$ Power rule (b)

$= (-1)^2x^6y^2 \cdot (-1)^3x^{15}y^{12}$ Power rule (a)

$= (-1)^5x^{21}y^{14}$ Product rule

$= -x^{21}y^{14}$ Simplify; $(-1)^5 = -1$

Work Problem 8 at the Side. ▶

OBJECTIVE 7 **Use the rules for exponents in a geometry problem.**

EXAMPLE 9 **Using Area Formulas**

Find an expression that represents, in appropriate units, the area of each figure.

(a)

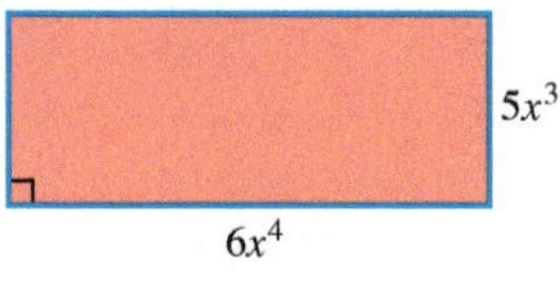

Figure 1

(b)

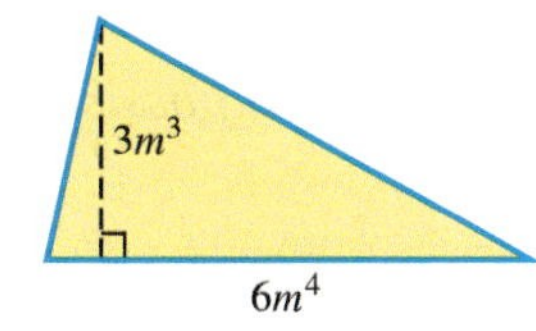

Figure 2

For **Figure 1,** use the formula for the area of a rectangle.

$\mathcal{A} = LW$ Area formula

$\mathcal{A} = (6x^4)(5x^3)$ Substitute.

$\mathcal{A} = 6 \cdot 5 \cdot x^{4+3}$ Commutative property; product rule

$\mathcal{A} = 30x^7$ Multiply. Add the exponents.

Figure 2 is a triangle with base $6m^4$ and height $3m^3$.

$\mathcal{A} = \frac{1}{2}bh$ Area formula

$\mathcal{A} = \frac{1}{2}(6m^4)(3m^3)$ Substitute.

$\mathcal{A} = \frac{1}{2}(6 \cdot 3 \cdot m^{4+3})$ Properties of real numbers; product rule

$\mathcal{A} = 9m^7$ Multiply. Add the exponents.

Work Problem 9 at the Side. ▶

8 Simplify each expression.

GS **(a)** $(2m)^5(2m)^3$

$= (2m)^{—}$

$= \_\_\_\_$

(b) $\left(\frac{5k^3}{3}\right)^2$

(c) $\left(\frac{1}{5}\right)^4(2x)^2$

(d) $(-3xy^2)^3(x^2y)^4$

9 Find an expression that represents, in appropriate units, the area of each figure.

(a)

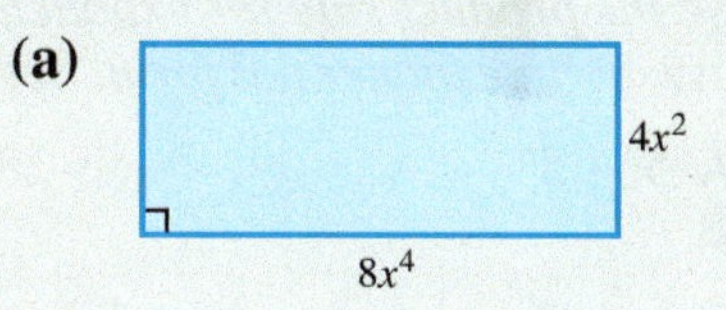

(b)

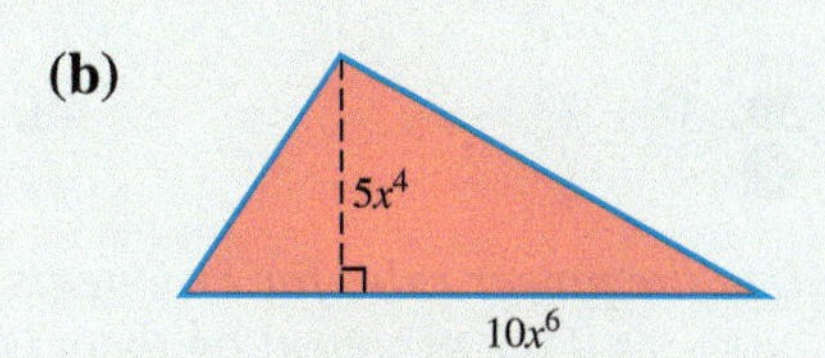

Answers

8. **(a)** 8; 2^8m^8, or $256m^8$

(b) $\frac{25k^6}{9}$ **(c)** $\frac{4x^2}{625}$ **(d)** $-27x^{11}y^{10}$

9. **(a)** $32x^6$ **(b)** $25x^{10}$

5.1 Exercises

FOR EXTRA HELP — Go to MyMathLab for worked-out, step-by-step solutions to exercises enclosed in a square and video solutions to ▶ exercises.

CONCEPT CHECK *Decide whether each statement is* true *or* false. *If false, correct the right-hand side of the statement.*

1. $3^3 = 9$

2. $(-2)^4 = 2^4$

3. $(a^2)^3 = a^5$

4. $\left(\frac{1}{4}\right)^2 = \frac{1}{4^2}$

5. $-2^2 = 4$

6. $2^3 \cdot 2^4 = 4^7$

7. $(3x)^2 = 6x^2$

8. $(-x)^2 = x^2$

Write each expression in exponential form. **See Example 1.**

9. $t \cdot t \cdot t \cdot t \cdot t \cdot t \cdot t$

10. $w \cdot w \cdot w \cdot w \cdot w \cdot w$

11. $\left(\frac{1}{2}\right)\left(\frac{1}{2}\right)\left(\frac{1}{2}\right)\left(\frac{1}{2}\right)\left(\frac{1}{2}\right)$

12. $\left(-\frac{1}{4}\right)\left(-\frac{1}{4}\right)\left(-\frac{1}{4}\right)\left(-\frac{1}{4}\right)$

13. $(-8p)(-8p)$

14. $(-7x)(-7x)(-7x)$

Identify the base and the exponent of each expression. Then evaluate, if possible. **See Example 2.**

15. ▶ 3^5

16. 2^7

17. ▶ $(-3)^5$

18. $(-2)^7$

19. $(-6)^2$

20. $(-9)^2$

21. -6^2

22. -9^2

23. $(-2x)^4$

24. $(-4x)^4$

25. $-2x^4$

26. $-4x^4$

Use the product rule for exponents to simplify each expression, if possible. Write answers in exponential form. **See Examples 3 and 4.**

27. ▶ $5^2 \cdot 5^6$

28. $3^6 \cdot 3^7$

29. $4^2 \cdot 4^7 \cdot 4^3$

30. $5^3 \cdot 5^8 \cdot 5^2$

31. $(-7)^3(-7)^6$

32. $(-9)^8(-9)^5$

33. ▶ $t^3t^8t^{13}$

34. $n^5n^6n^9$

35. $-8r^4(7r^3)$

36. $10a^7(-4a^3)$

37. ▶ $(-6p^5)(-7p^5)$

38. $(-5w^8)(-9w^8)$

39. ▶ $3^8 + 3^9$

40. $4^{12} + 4^5$

41. $5^8 \cdot 3^8$

42. $6^3 \cdot 8^3$

Use the power rules for exponents to simplify each expression. Evaluate coefficients in answers if the exponent on them is 4 or less. **See Examples 5–7.**

43. ▶ $(4^3)^2$

44. $(8^3)^6$

45. ▶ $(t^4)^5$

46. $(y^6)^5$

47. $(7r)^3$

48. $(5x)^4$

49. $(-5^2)^6$

50. $(-9^4)^8$

51. $(-8^3)^5$

52. $(-7^5)^7$

53. ▶ $(5xy)^5$

54. $(9pq)^6$

55. $8(qr)^3$

56. $4(vw)^5$

57. $\left(\frac{9}{5}\right)^8$

58. $\left(\frac{12}{7}\right)^6$

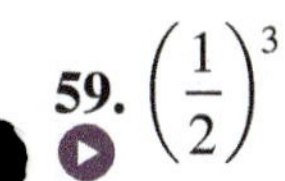

59. $\left(\frac{1}{2}\right)^3$ **60.** $\left(\frac{1}{3}\right)^3$ **61.** $\left(\frac{a}{b}\right)^3 \quad (b \neq 0)$ **62.** $\left(\frac{r}{t}\right)^4 \quad (t \neq 0)$

63. $(-2x^2y)^3$ **64.** $(-5m^4p^2)^3$ **65.** $(3a^3b^2)^2$ **66.** $(4x^3y^5)^4$

Simplify each expression. (Do not evaluate the exponents on coefficients in answers in Exercises 67–76.) ***See Example 8.***

67. $\left(\frac{5}{2}\right)^3 \cdot \left(\frac{5}{2}\right)^2$ **68.** $\left(\frac{3}{4}\right)^5 \cdot \left(\frac{3}{4}\right)^6$ **69.** $\left(\frac{9}{8}\right)^3 \cdot 9^2$ **70.** $\left(\frac{8}{5}\right)^4 \cdot 8^3$

71. $(2x)^9(2x)^3$ **72.** $(6y)^5(6y)^8$ **73.** $(-6p)^4(-6p)$ **74.** $(-13q)^6(-13q)$

75. $(6x^2y^3)^5$ **76.** $(5r^5t^6)^7$ **77.** $(x^2)^3(x^3)^5$ **78.** $(y^4)^5(y^3)^5$

79. $(2w^2x^3y)^2(x^4y)^5$ **80.** $(3x^4y^2z)^3(yz^4)^5$ **81.** $(-r^4s)^2(-r^2s^3)^5$ **82.** $(-ts^6)^4(-t^3s^5)^3$

83. $\left(\frac{4x^2}{5}\right)^3$ **84.** $\left(\frac{3z^5}{4}\right)^2$ **85.** $\left(\frac{5a^2b^5}{c^6}\right)^3 \quad (c \neq 0)$ **86.** $\left(\frac{6x^3y^9}{z^5}\right)^4 \quad (z \neq 0)$

87. $(-5m^3p^4q)^2(p^2q)^3$ **88.** $(-a^4b^5)(-6a^3b^3)^2$ **89.** $(2x^2y^3z)^4(xy^2z^3)^2$ **90.** $(4q^2r^3s^5)^3(qr^2s^3)^4$

91. CONCEPT CHECK A student simplified $(10^2)^3$ as 1000^6. ***What Went Wrong?*** Simplify correctly.

92. CONCEPT CHECK A student simplified $(3x^2y^3)^4$ as $12x^8y^{12}$. ***What Went Wrong?*** Simplify correctly.

Find an expression that represents, in appropriate units, the area of each figure. In Exercise 96, leave π in the answer. ***See Example 9.*** *(If necessary, refer to the formulas at the back of this text. The ⌝ in the figures indicates 90° (right) angles.)*

93.

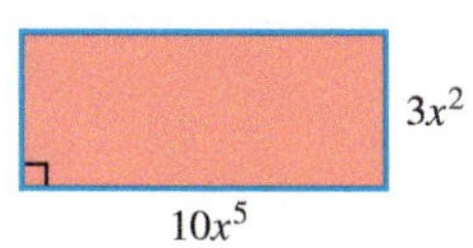

94.

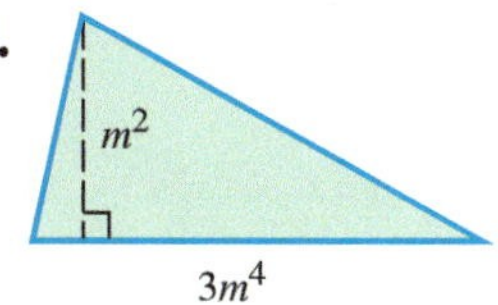

95.

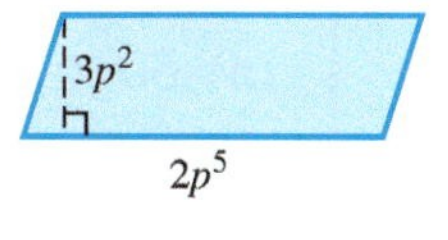

96.

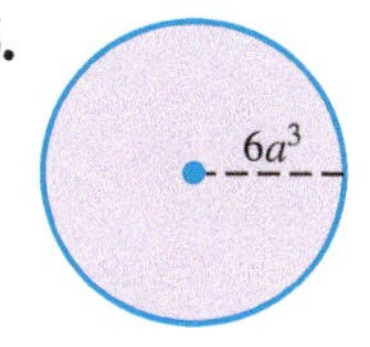

Find an expression that represents the volume of each figure. (If necessary, refer to the formulas at the back of this text.)

97.

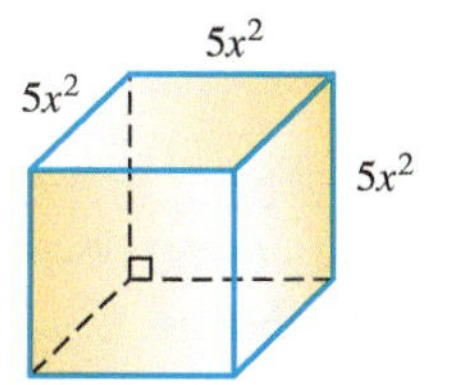

98.

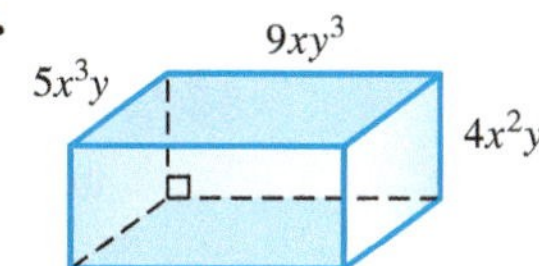

5.2 Integer Exponents and the Quotient Rule

OBJECTIVES

1. Use 0 as an exponent.
2. Use negative numbers as exponents.
3. Use the quotient rule for exponents.
4. Use combinations of the rules for exponents.

Consider the following list.

$$2^4 = 16$$
$$2^3 = 8$$
$$2^2 = 4$$

As exponents decrease by 1, the results are divided by 2 each time.

Each time we decrease the exponent by 1, the value is divided by 2 (the base). We can continue the list to lesser and lesser integer exponents.

$$2^1 = 2$$
$$2^0 = 1$$
$$2^{-1} = \frac{1}{2}$$

We continue the pattern here.

◀ Work Problem 1 at the Side.

1 Continue the list of exponentials using −2, −3, and −4 as exponents.

$2^{-2} = \_\_\_\_$

$2^{-3} = \_\_\_\_$

$2^{-4} = \_\_\_\_$

From the preceding list and the answers to **Margin Problem 1,** it appears that we should define 2^0 as 1 and bases raised to negative exponents as reciprocals of those bases.

OBJECTIVE 1 Use 0 as an exponent. The definitions of 0 and negative exponents must be consistent with the rules for exponents developed earlier. For example, if we define 6^0 to be 1, then

$$6^0 \cdot 6^2 = 1 \cdot 6^2 = 6^2 \quad \text{and} \quad 6^0 \cdot 6^2 = 6^{0+2} = 6^2.$$

The product rule is satisfied. The power rules are also valid for a 0 exponent. Thus, we define a 0 exponent as follows.

Zero Exponent

For any nonzero real number a, $\quad a^0 = 1.$

Example: $17^0 = 1$

2 Evaluate. Assume that all variables represent nonzero real numbers.

(a) 28^0 **(b)** $(-16)^0$

(c) -7^0 **(d)** m^0

(e) $-2p^0$ **(f)** $(5r)^0$

(g) $13^0 + 2^0$ **(h)** $14^0 - 12^0$

EXAMPLE 1 Using Zero Exponents

Evaluate.

(a) $60^0 = 1$

(b) $(-60)^0 = 1$

(c) $-60^0 = -(1) = -1$

(d) $y^0 = 1 \quad (y \neq 0)$

(e) $6y^0 = 6(1) = 6 \quad (y \neq 0)$

(f) $(6y)^0 = 1 \quad (y \neq 0)$

(g) $8^0 + 11^0 = 1 + 1 = 2$

(h) $-8^0 - 11^0 = -1 - 1 = -2$

◀ Work Problem 2 at the Side.

! CAUTION

Look again at **Examples 1(b) and 1(c).** In $(-60)^0$, the base is -60, and since any nonzero base raised to the 0 exponent is 1, $(-60)^0 = 1$. In -60^0, which can be written $-(60)^0$, the base is 60, so $-60^0 = -1$.

Answers

1. $2^{-2} = \frac{1}{4}$; $2^{-3} = \frac{1}{8}$; $2^{-4} = \frac{1}{16}$

2. (a) 1 (b) 1 (c) −1 (d) 1 (e) −2 (f) 1 (g) 2 (h) 0

$2^4 = 16$

$2^3 = 8$

$2^2 = 4$

$2^1 = 2$

$2^0 = 1$

$2^{-1} = \frac{1}{2}$

$2^{-2} = \frac{1}{4}$

$2^{-3} = \frac{1}{8}$

$2^{-4} = \frac{1}{16}$

OBJECTIVE 2 Use negative numbers as exponents. Review the list of exponentials in the margin. Because $2^{-2} = \frac{1}{4}$ and $2^{-3} = \frac{1}{8}$, we can make a conjecture that 2^{-n} should equal $\frac{1}{2^n}$. *Is the product rule valid in such cases?* For example, if we multiply 6^{-2} by 6^2, we obtain

$$6^{-2} \cdot 6^2 = 6^{-2+2} = 6^0 = 1.$$

The expression 6^{-2} behaves as if it were the reciprocal of 6^2—that is, their product is 1. The reciprocal of 6^2 may be written $\frac{1}{6^2}$, leading us to define 6^{-2} as $\frac{1}{6^2}$, and generalize accordingly.

Negative Exponents

For any nonzero real number a and any integer n, $a^{-n} = \frac{1}{a^n}$.

Example: $3^{-2} = \frac{1}{3^2}$

By definition, a^{-n} and a^n are reciprocals.

$$a^n \cdot a^{-n} = a^n \cdot \frac{1}{a^n} = 1$$

Because $1^n = 1$, the definition of a^{-n} can also be written as follows.

$$a^{-n} = \frac{1}{a^n} = \frac{1^n}{a^n} = \left(\frac{1}{a}\right)^n$$

For example, $\quad 6^{-3} = \left(\frac{1}{6}\right)^3$ and $\left(\frac{1}{3}\right)^{-2} = 3^2$.

EXAMPLE 2 Using Negative Exponents

Write with positive exponents and simplify. Assume that all variables represent nonzero real numbers.

(a) $3^{-2} = \frac{1}{3^2} = \frac{1}{9}$ $\quad a^{-n} = \frac{1}{a^n}$

(b) $5^{-3} = \frac{1}{5^3} = \frac{1}{125}$ $\quad a^{-n} = \frac{1}{a^n}$

(c) $\left(\frac{1}{2}\right)^{-3} = 2^3 = 8$ $\quad \frac{1}{2}$ and 2 are reciprocals. (Reciprocals have a product of 1.)

Notice that we can change the base to its reciprocal if we also change the sign of the exponent.

(d) $\left(\frac{2}{5}\right)^{-4}$

$= \left(\frac{5}{2}\right)^4$ $\quad \frac{2}{5}$ and $\frac{5}{2}$ are reciprocals.

$= \frac{5^4}{2^4}$ $\quad$ Power rule (c)

$= \frac{625}{16}$ $\quad$ Apply the exponents.

(e) $\left(\frac{4}{3}\right)^{-5}$

$= \left(\frac{3}{4}\right)^5$ $\quad \frac{4}{3}$ and $\frac{3}{4}$ are reciprocals.

$= \frac{3^5}{4^5}$ $\quad$ Power rule (c)

$= \frac{243}{1024}$ $\quad$ Apply the exponents.

Continued on Next Page

3 Write with positive exponents and simplify. Assume that all variables represent nonzero real numbers.

(a) 4^{-3}

(b) 6^{-2}

(c) $\left(\frac{1}{4}\right)^{-2}$

(d) $\left(\frac{2}{3}\right)^{-2}$

(e) $2^{-1} + 5^{-1}$

(f) $7m^{-5}$

(g) $\frac{1}{z^{-6}}$

(h) p^2q^{-5}

Answers

3. (a) $\frac{1}{64}$ (b) $\frac{1}{36}$ (c) 16 (d) $\frac{9}{4}$ (e) $\frac{7}{10}$ (f) $\frac{7}{m^5}$ (g) z^6 (h) $\frac{p^2}{q^5}$

(f) $4^{-1} - 2^{-1}$

$= \frac{1}{4} - \frac{1}{2}$ Apply the exponents.

$= \frac{1}{4} - \frac{2}{4}$ Find a common denominator.

$= -\frac{1}{4}$ Subtract.

(g) $3p^{-2}$

$= \frac{3}{1} \cdot \frac{1}{p^2}$ $a^{-n} = \frac{1}{a^n}$

$= \frac{3}{p^2}$ Multiply.

(h) $\frac{1}{x^{-4}}$

$= \frac{1^{-4}}{x^{-4}}$ $1^n = 1$, for any integer n

$= \left(\frac{1}{x}\right)^{-4}$ Power rule (c)

$= x^4$ $\frac{1}{x}$ and x are reciprocals.

(i) x^3y^{-4}

$= \frac{x^3}{1} \cdot \frac{1}{y^4}$ $a^{-n} = \frac{1}{a^n}$

$= \frac{x^3}{y^4}$ Multiply.

In general, $\frac{1}{a^{-n}} = a^n$.

◀ Work Problem **3** at the Side.

CAUTION

A negative exponent does not indicate a negative number. Negative exponents lead to reciprocals.

Expression	Example	
a^{-n}	$3^{-2} = \frac{1}{3^2} = \frac{1}{9}$	Not negative
$-a^{-n}$	$-3^{-2} = -\frac{1}{3^2} = -\frac{1}{9}$	Negative

Consider the following.

$$\frac{2^{-3}}{3^{-4}}$$

$$= \frac{\frac{1}{2^3}}{\frac{1}{3^4}}$$ Definition of negative exponent

$$= \frac{1}{2^3} \div \frac{1}{3^4}$$ $\frac{a}{b}$ means $a \div b$.

$$= \frac{1}{2^3} \cdot \frac{3^4}{1}$$ To divide, multiply by the reciprocal of the divisor.

$$= \frac{3^4}{2^3}$$ Multiply.

Therefore, $\frac{2^{-3}}{3^{-4}} = \frac{3^4}{2^3}$.

Negative-to-Positive Rules for Exponents

For any nonzero real numbers a and b and any integers m and n,

$$\frac{a^{-m}}{b^{-n}} = \frac{b^n}{a^m} \quad \text{and} \quad \left(\frac{a}{b}\right)^{-m} = \left(\frac{b}{a}\right)^m.$$

Examples: $\frac{3^{-5}}{2^{-4}} = \frac{2^4}{3^5}$ and $\left(\frac{4}{5}\right)^{-3} = \left(\frac{5}{4}\right)^3$

EXAMPLE 3 Changing from Negative to Positive Exponents

Write with positive exponents and simplify. Assume that all variables represent nonzero real numbers.

(a) $\frac{4^{-2}}{5^{-3}} = \frac{5^3}{4^2} = \frac{125}{16}$

(b) $\frac{m^{-5}}{p^{-1}} = \frac{p^1}{m^5} = \frac{p}{m^5}$

(c) $\frac{a^{-2}b}{3d^{-3}} = \frac{bd^3}{3a^2}$ Notice that b in the numerator and the coefficient 3 in the denominator are not affected.

(d) $\left(\frac{x}{2y}\right)^{-4}$

$= \left(\frac{2y}{x}\right)^4$ Negative-to-positive rule

$= \frac{2^4y^4}{x^4}$ Power rules (b) and (c)

$= \frac{16y^4}{x^4}$ Apply the exponent.

Work Problem 4 at the Side. ▶

! CAUTION

We cannot use the rule $\frac{a^{-m}}{b^{-n}} = \frac{b^n}{a^m}$ to change negative exponents to positive exponents if the exponents occur in a *sum* or *difference* of terms.

Example: $\frac{5^{-2} + 3^{-1}}{7 - 2^{-3}}$ written with positive exponents is $\frac{\frac{1}{5^2} + \frac{1}{3}}{7 - \frac{1}{2^3}}$.

OBJECTIVE 3 Use the quotient rule for exponents. Consider a quotient of two exponential expressions with the same base.

$$\frac{6^5}{6^3} = \frac{6 \cdot 6 \cdot 6 \cdot 6 \cdot 6}{6 \cdot 6 \cdot 6} = 6^2 \quad \text{Divide out the common factors.}$$

The difference of the exponents, $5 - 3 = 2$, is the exponent in the quotient. Also,

$$\frac{6^2}{6^4} = \frac{6 \cdot 6}{6 \cdot 6 \cdot 6 \cdot 6} = \frac{1}{6^2} = 6^{-2}. \quad \text{Divide out the common factors.}$$

Here, $2 - 4 = -2$. These examples suggest the **quotient rule for exponents.**

---

4 Write with positive exponents and simplify. Assume that all variables represent nonzero real numbers.

GS (a) $\frac{7^{-1}}{5^{-4}}$

$= \frac{5^{—}}{7^{—}}$

$=$ ____

(b) $\frac{x^{-3}}{y^{-2}}$

(c) $\frac{4h^{-5}}{m^{-2}k}$

GS (d) $\left(\frac{3m}{p}\right)^{-2}$

$= \left(\frac{p}{3m}\right)^{—}$

$= \frac{p^{—}}{3^{—}m^{—}}$

$=$ ____

Answers

4. (a) 4; 1; $\frac{625}{7}$
(b) $\frac{y^2}{x^3}$ (c) $\frac{4m^2}{h^5k}$
(d) 2; 2; 2; 2; $\frac{p^2}{9m^2}$

5 Simplify. Assume that all variables represent nonzero real numbers.

GS (a) $\frac{5^{11}}{5^8}$

$= 5^{\_\_-\_\_}$

$= 5^{\_\_}$

$= \_\_\_\_$

(b) $\frac{4^7}{4^{10}}$

GS (c) $\frac{6^{-5}}{6^{-2}}$

$= 6^{\_\_-(\_\_)}$

$= 6^{\_\_}$

$= \frac{1}{6^{\_\_}}$

$= \_\_\_\_$

(d) $\frac{8^4 m^9}{8^5 m^{10}}$

(e) $\frac{(x+y)^{-3}}{(x+y)^{-4}} \quad (x \neq -y)$

Answers

5. (a) 11; 8; 3; 125 (b) $\frac{1}{64}$

(c) -5; -2; -3; 3; $\frac{1}{216}$

(d) $\frac{1}{8m}$ (e) $x + y$

Quotient Rule for Exponents

For any nonzero real number a and any integers m and n,

$$\frac{a^m}{a^n} = a^{m-n}.$$

(Keep the same base and subtract the exponents.)

Example: $\frac{5^8}{5^4} = 5^{8-4} = 5^4$

CAUTION

A common **error** is to write $\frac{5^8}{5^4} = 1^{8-4} = 1^4$. ***This is incorrect.***

$\frac{5^8}{5^4} = 5^{8-4} = 5^4$ The quotient must have the *same base*, which is 5 here.

We can confirm this by writing out the factors.

$$\frac{5^8}{5^4} = \frac{5 \cdot 5 \cdot 5 \cdot 5 \cdot 5 \cdot 5 \cdot 5 \cdot 5}{5 \cdot 5 \cdot 5 \cdot 5} = 5^4$$

EXAMPLE 4 Using the Quotient Rule

Simplify. Assume that all variables represent nonzero real numbers.

(a) $\frac{5^8}{5^6} = 5^{8-6} = 5^2 = 25$ Keep the same base.

(b) $\frac{4^2}{4^9} = 4^{2-9} = 4^{-7} = \frac{1}{4^7}$

(c) $\frac{5^{-3}}{5^{-7}} = 5^{-3-(-7)} = 5^4 = 625$ Be careful with signs.

(d) $\frac{q^5}{q^{-3}} = q^{5-(-3)} = q^8$

(e) $\frac{3^2 x^5}{3^4 x^3}$

$= \frac{3^2}{3^4} \cdot \frac{x^5}{x^3}$

$= 3^{2-4} \cdot x^{5-3}$ Quotient rule

$= 3^{-2}x^2$ Subtract.

$= \frac{x^2}{3^2}$ Definition of negative exponent

$= \frac{x^2}{9}$ Apply the exponent.

(f) $\frac{(m+n)^{-2}}{(m+n)^{-4}}$

$= (m+n)^{-2-(-4)}$

$= (m+n)^{-2+4}$

$= (m+n)^2 \quad (m \neq -n)$

The restriction $m \neq -n$ is necessary to prevent a denominator of 0 in the original expression. Division by 0 is undefined.

(g) $\frac{7x^{-3}y^2}{2^{-1}x^2y^{-5}}$

$= \frac{7 \cdot 2^1 \cdot y^2 y^5}{x^2 x^3}$ Negative-to-positive rule

$= \frac{14y^7}{x^5}$ Multiply; product rule

◀ Work Problem 5 at the Side.

Summary of Definitions and Rules for Exponents

For all integers m and n and all real numbers a and b for which the following are defined, these definitions and rules hold true.

		Examples
Product rule	$a^m \cdot a^n = a^{m+n}$	$7^4 \cdot 7^5 = 7^{4+5} = 7^9$
Zero exponent	$a^0 = 1$	$(-3)^0 = 1$
Negative exponent	$a^{-n} = \frac{1}{a^n}$	$5^{-3} = \frac{1}{5^3}$
Quotient rule	$\frac{a^m}{a^n} = a^{m-n}$	$\frac{2^2}{2^5} = 2^{2-5} = 2^{-3} = \frac{1}{2^3}$
Power rules (a)	$(a^m)^n = a^{mn}$	$(4^2)^3 = 4^{2 \cdot 3} = 4^6$
(b)	$(ab)^m = a^m b^m$	$(3k)^4 = 3^4 k^4$
(c)	$\left(\frac{a}{b}\right)^m = \frac{a^m}{b^m}$	$\left(\frac{2}{3}\right)^2 = \frac{2^2}{3^2}$
Negative-to-positive rules	$\frac{a^{-m}}{b^{-n}} = \frac{b^n}{a^m}$	$\frac{2^{-4}}{5^{-3}} = \frac{5^3}{2^4}$
	$\left(\frac{a}{b}\right)^{-m} = \left(\frac{b}{a}\right)^m$	$\left(\frac{4}{7}\right)^{-2} = \left(\frac{7}{4}\right)^2$

OBJECTIVE 4 **Use combinations of the rules for exponents.**

EXAMPLE 5 **Using a Combination of the Rules**

Simplify. Assume that all variables represent nonzero real numbers.

(a) $\frac{(4^2)^3}{4^5}$

$= \frac{4^6}{4^5}$ Power rule (a)

$= 4^{6-5}$ Quotient rule

$= 4^1$ Subtract.

$= 4$ $a^1 = a$, for all a.

(b) $x^2 \cdot x^{-6} \cdot x^{-1}$

$= x^{2+(-6)+(-1)}$ Product rule

$= x^{-5}$ Add.

$= \frac{1}{x^5}$ Definition of negative exponent

(c) $(2x)^3(2x)^2$

$= (2x)^5$ Product rule

$= 2^5x^5$ Power rule (b)

$= 32x^5$ $2^5 = 32$

$(2x)^3(2x)^2$ **Alternative solution**

$= 2^3x^3 \cdot 2^2x^2$ Power rule (b)

$= 2^{3+2}x^{3+2}$ Product rule

$= 2^5x^5$ Add exponents.

$= 32x^5$ $2^5 = 32$

Continued on Next Page

6 Simplify. Assume that all variables represent nonzero real numbers.

(a) $\dfrac{(3^4)^2}{3^3}$

(b) $z^{-2} \cdot z^{-1} \cdot z^4$

(c) $(4t)^5(4t)^{-3}$

(d) $\left(\dfrac{5}{2z^4}\right)^{-3}$

(e) $\left(\dfrac{6y^{-4}}{7^{-1}z^5}\right)^{-2}$

(f) $\dfrac{(6x)^{-1}}{(3x^2)^{-2}}$

Answers

6. (a) 243 (b) z (c) $16t^2$

(d) $\dfrac{8z^{12}}{125}$ (e) $\dfrac{y^8z^{10}}{1764}$ (f) $\dfrac{3x^3}{2}$

(d) $\left(\dfrac{2x^3}{5}\right)^{-4}$

$= \left(\dfrac{5}{2x^3}\right)^4$ Negative-to-positive rule

$= \dfrac{5^4}{2^4x^{12}}$ Power rules (a)–(c)

$= \dfrac{625}{16x^{12}}$ Apply the exponents.

(e) $\left(\dfrac{3x^{-2}}{4^{-1}y^3}\right)^{-3}$

$= \dfrac{3^{-3}x^6}{4^3y^{-9}}$ Power rules (a)–(c)

$= \dfrac{x^6y^9}{4^3 \cdot 3^3}$ Negative-to-positive rule

$= \dfrac{x^6y^9}{1728}$ $4^3 \cdot 3^3 = 64 \cdot 27 = 1728$

(f) $\dfrac{(4m)^{-3}}{(3m)^{-4}}$

$= \dfrac{4^{-3}m^{-3}}{3^{-4}m^{-4}}$ Power rule (b)

$= \dfrac{3^4m^4}{4^3m^3}$ Negative-to-positive rule

$= \dfrac{3^4m^{4-3}}{4^3}$ Quotient rule

$= \dfrac{3^4m}{4^3}$ Subtract.

$= \dfrac{81m}{64}$ Apply the exponents.

$\dfrac{(4m)^{-3}}{(3m)^{-4}}$ **Alternative solution**

$= \dfrac{(3m)^4}{(4m)^3}$

$= \dfrac{3^4m^4}{4^3m^3}$

Use negative-to-positive rule first, followed by power rule (b).

The rest of the solution can be done as shown at the left.

◀ **Work Problem 6 at the Side.**

Note

Because steps can be done in several different orders, there are many correct ways to simplify expressions like those in **Example 5.** See the alternative solutions shown.

5.2 Exercises

FOR EXTRA HELP Go to MyMathLab for worked-out, step-by-step solutions to exercises enclosed in a square and video solutions to ▶ exercises.

CONCEPT CHECK *Decide whether each expression is* positive, negative, *or* 0.

1. $(-2)^{-3}$ **2.** $(-3)^{-2}$ **3.** -2^4 **4.** -3^6

5. $\left(\frac{1}{4}\right)^{-2}$ **6.** $\left(\frac{1}{5}\right)^{-2}$ **7.** $1 - 5^0$ **8.** $1 - 7^0$

CONCEPT CHECK *Match each expression in Column I with the equivalent expression in Column II. Choices in Column II may be used once, more than once, or not at all. (In Exercise 9, $x \neq 0$.)*

I	II	I	II
9. (a) x^0	**A.** 0	**10. (a)** -2^{-4}	**A.** 8
(b) $-x^0$	**B.** 1	**(b)** $(-2)^{-4}$	**B.** 16
(c) $7x^0$	**C.** -1	**(c)** 2^{-4}	**C.** $-\frac{1}{16}$
(d) $(7x)^0$	**D.** 7	**(d)** $\frac{1}{2^{-4}}$	**D.** -8
(e) $-7x^0$	**E.** -7	**(e)** $\frac{1}{-2^{-4}}$	**E.** -16
(f) $(-7x)^0$	**F.** $\frac{1}{7}$	**(f)** $\frac{1}{(-2)^{-4}}$	**F.** $\frac{1}{16}$

Decide whether each expression is equal to 0, 1, *or* -1. ***See Example 1.***

11. 9^0 ▶ **12.** 5^0 **13.** $(-4)^0$ **14.** $(-10)^0$ **15.** -9^0

16. -5^0 **17.** $(-2)^0 - 2^0$ **18.** $(-8)^0 - 8^0$ **19.** $\frac{0^{10}}{10^0}$ **20.** $\frac{0^5}{5^0}$

Evaluate each expression. ***See Examples 1 and 2.***

21. 4^{-3} ▶ **22.** 5^{-4} **23.** $\left(\frac{1}{2}\right)^{-4}$ **24.** $\left(\frac{1}{3}\right)^{-3}$ **25.** $\left(\frac{6}{7}\right)^{-2}$ ▶

26. $\left(\frac{2}{3}\right)^{-3}$ **27.** $(-3)^{-4}$ **28.** $(-4)^{-3}$ **29.** $3x^0 \quad (x \neq 0)$ **30.** $-5t^0 \quad (t \neq 0)$

31. $(3x)^0 \quad (x \neq 0)$ **32.** $(-5t)^0 \quad (t \neq 0)$ **33.** $7^0 + 9^0$ **34.** $8^0 + 6^0$

35. $5^{-1} + 3^{-1}$ **36.** $6^{-1} + 2^{-1}$ **37.** $-2^{-1} + 3^{-2}$ **38.** $(-3)^{-2} + (-4)^{-1}$

Simplify each expression. Evaluate coefficients if the exponent is 4 or less. Assume that all variables represent nonzero real numbers and no denominators are equal to 0. **See Examples 2–4.**

39. $\frac{5^8}{5^5}$ **40.** $\frac{11^6}{11^4}$ **41.** $\frac{9^4}{9^5}$ **42.** $\frac{7^3}{7^4}$ **43.** $\frac{6^{-3}}{6^2}$

44. $\frac{4^{-2}}{4^3}$ **45.** $\frac{3^{-4}}{3^{-7}}$ **46.** $\frac{2^{-5}}{2^{-9}}$ **47.** $\frac{1}{6^{-3}}$ **48.** $\frac{1}{5^{-2}}$

49. $\frac{2}{r^{-4}}$ **50.** $\frac{3}{s^{-8}}$ **51.** $\frac{3^{-2}}{5^{-3}}$ **52.** $\frac{6^{-2}}{5^{-4}}$ **53.** $-4x^{-3}$

54. $-8z^{-5}$ **55.** p^5q^{-8} **56.** $x^{-8}y^4$ **57.** $\frac{r^5}{r^{-4}}$ **58.** $\frac{a^6}{a^{-4}}$

59. $\frac{6^4x^8}{6^5x^3}$ **60.** $\frac{3^8y^5}{3^{10}y^2}$ **61.** $\frac{6y^3}{2y}$ **62.** $\frac{15m^2}{3m}$ **63.** $\frac{3x^5}{3x^2}$

64. $\frac{10p^8}{10p^4}$ **65.** $\frac{x^{-3}y}{4z^{-2}}$ **66.** $\frac{p^{-5}q^4}{9r^{-3}}$ **67.** $\frac{(a+b)^{-3}}{(a+b)^{-4}}$ **68.** $\frac{(x+y)^{-8}}{(x+y)^{-9}}$

Simplify each expression. Evaluate coefficients if the exponent is 4 or less. Assume that all variables represent nonzero real numbers. **See Example 5.**

69. $\frac{(7^4)^3}{7^9}$ **70.** $\frac{(5^3)^2}{5^2}$ **71.** $x^{-3} \cdot x^5 \cdot x^{-4}$ **72.** $y^{-8} \cdot y^5 \cdot y^{-2}$

73. $\frac{(3x)^{-2}}{(4x)^{-3}}$ **74.** $\frac{(2y)^{-3}}{(5y)^{-4}}$ **75.** $\left(\frac{x^{-1}y}{z^2}\right)^{-2}$ **76.** $\left(\frac{p^{-4}q}{r^{-3}}\right)^{-3}$

77. $(6x)^4(6x)^{-3}$ **78.** $(10y)^9(10y)^{-8}$ **79.** $\frac{(m^7n)^{-2}}{m^{-4}n^3}$ **80.** $\frac{(m^2n^4)^{-3}}{m^{-2}n^5}$

81. $\frac{5x^{-3}}{(4x)^2}$ **82.** $\frac{-3k^{-5}}{(2k)^2}$ **83.** $\left(\frac{2p^{-1}q}{3^{-1}m^2}\right)^2$ **84.** $\left(\frac{3xy^{-2}}{2^{-1}y}\right)^2$

85. CONCEPT CHECK A student simplified $\frac{16^3}{2^2}$ as shown.

$$\frac{16^3}{2^2} = \left(\frac{16}{2}\right)^{3-2} = 8^1 = 8 \quad \text{Incorrect}$$

What Went Wrong? Give the correct answer.

86. CONCEPT CHECK A student simplified 5^{-4} as shown.

$$5^{-4} = -5^4 = -625 \quad \text{Incorrect}$$

What Went Wrong? Give the correct answer.

Summary Exercises *Applying the Rules for Exponents*

CONCEPT CHECK *Decide whether each expression is* positive *or* negative.

1. $(-5)^4$

2. -5^4

3. $(-2)^5$

4. -2^5

5. $\left(-\frac{3}{7}\right)^2$

6. $\left(-\frac{3}{7}\right)^3$

7. $\left(-\frac{3}{7}\right)^{-2}$

8. $\left(-\frac{3}{7}\right)^{-5}$

Simplify each expression. Evaluate coefficients if the exponent is 6 *or less. Assume that all variables represent nonzero real numbers.*

9. $\left(\frac{6x^2}{5}\right)^{12}$

10. $\left(\frac{rs^2t^3}{3t^4}\right)^6$

11. $(10x^2y^4)^2(10xy^2)^3$

12. $(-2ab^3c)^4(-2a^2b)^2$

13. $\left(\frac{9wx^3}{y^4}\right)^3$

14. $(4x^{-2}y^{-3})^{-2}$

15. $\frac{c^{11}(c^2)^4}{(c^3)^3(c^2)^{-6}}$

16. $\left(\frac{k^4t^2}{k^2t^{-4}}\right)^{-2}$

17. $5^{-1}+6^{-1}$

18. $\frac{(3y^{-1}z^3)^{-1}(3y^2)}{(y^3z^2)^{-3}}$

19. $\frac{(2xy^{-1})^3}{2^3x^{-3}y^2}$

20. $-8^0+(-8)^0$

21. $(z^4)^{-3}(z^{-2})^{-5}$

22. $\left(\frac{r^2st^5}{3r}\right)^{-2}$

23. $\frac{(3^{-1}x^{-3}y)^{-1}(2x^2y^{-3})^2}{(5x^{-2}y^2)^{-2}}$

24. $\left(\frac{5x^2}{3x^{-4}}\right)^{-1}$

25. $\left(\frac{-2x^{-2}}{2x^2}\right)^{-2}$

26. $\frac{(x^{-4}y^2)^3(x^2y)^{-1}}{(xy^2)^{-3}}$

27. $\dfrac{(a^{-2}b^{3})^{-4}}{(a^{-3}b^{2})^{-2}(ab)^{-4}}$

28. $(2a^{-30}b^{-29})(3a^{31}b^{30})$

29. $5^{-2} + 6^{-2}$

30. $\left(\dfrac{(x^{47}y^{23})^{2}}{x^{-26}y^{-42}}\right)^{0}$

31. $\left(\dfrac{7a^{2}b^{3}}{2}\right)^{3}$

32. $-(-12^{0})$

33. $-(-12)^{0}$

34. $\dfrac{0^{12}}{12^{0}}$

35. $\dfrac{(2xy^{-3})^{-2}}{(3x^{-2}y^{4})^{-3}}$

36. $\left(\dfrac{a^{2}b^{3}c^{4}}{a^{-2}b^{-3}c^{-4}}\right)^{-2}$

37. $(6x^{-5}z^{3})^{-3}$

38. $(2p^{-2}qr^{-3})(2p)^{-4}$

39. $\dfrac{(xy)^{-3}(xy)^{5}}{(xy)^{-4}}$

40. $42^{0} - (-12)^{0}$

41. $\dfrac{(7^{-1}x^{-3})^{-2}(x^{4})^{-6}}{7^{-1}x^{-3}}$

42. $\left(\dfrac{3^{-4}x^{-3}}{3^{-3}x^{-6}}\right)^{-2}$

43. $(5p^{-2}q)^{-3}(5pq^{3})^{4}$

44. $8^{-1} + 6^{-1}$

45. $\left(\dfrac{4r^{-6}s^{-2}t}{2r^{8}s^{-4}t^{2}}\right)^{-1}$

46. $(13x^{-6}y)(13x^{-6}y)^{-1}$

47. $\dfrac{(8pq^{-2})^{4}}{(8p^{-2}q^{-3})^{3}}$

48. $\left(\dfrac{mn^{-2}p}{m^{2}np^{4}}\right)^{-2}\left(\dfrac{mn^{-2}p}{m^{2}np^{4}}\right)^{3}$

49. $-(-3^{0})^{0}$

50. $5^{-1} - 8^{-1}$

5.3 An Application of Exponents: Scientific Notation

OBJECTIVES

1. Express numbers in scientific notation.
2. Convert numbers in scientific notation to standard notation.
3. Use scientific notation in calculations.

OBJECTIVE 1 Express numbers in scientific notation. Numbers occurring in science are often extremely large (such as the distance from Earth to the sun, 93,000,000 mi) or extremely small (the wavelength of blue light, approximately 0.000000475 m). Because of the difficulty of working with many zeros, scientists often express such numbers with exponents using *scientific notation.*

Scientific Notation

A number is written in **scientific notation** when it is expressed in the form

$$a \times 10^n,$$

where $1 \leq |a| < 10$ and n is an integer.

In **scientific notation,** there is *always* one nonzero digit to the left of the decimal point.

Scientific notation		
3.19×10^1	$= 3.19 \times 10 = 31.9$	Decimal point moves 1 place to the right.
3.19×10^2	$= 3.19 \times 100 = 319.$	Decimal point moves 2 places to the right.
3.19×10^3	$= 3.19 \times 1000 = 3190.$	Decimal point moves 3 places to the right.
3.19×10^{-1}	$= 3.19 \times 0.1 = 0.319$	Decimal point moves 1 place to the left.
3.19×10^{-2}	$= 3.19 \times 0.01 = 0.0319$	Decimal point moves 2 places to the left.
3.19×10^{-3}	$= 3.19 \times 0.001 = 0.00319$	Decimal point moves 3 places to the left.

Note

In scientific notation, a multiplication cross $\times$ is commonly used.

A number in scientific notation is always written with the decimal point after the first nonzero digit and then multiplied by the appropriate power of 10.

Example: 56,200 is written 5.62×10^4 because

$$56{,}200 = 5.62 \times \mathbf{10{,}000} = 5.62 \times \mathbf{10^4}.$$

Additional examples:

42,000,000	is written	4.2×10^7.
0.000586	is written	5.86×10^{-4}.
2,000,000,000	is written	2×10^9.

It is not necessary to write 2.0.

To write a positive number in scientific notation, follow the steps given on the next page. (For a negative number, follow these steps using the *absolute value* of the number. Then make the result negative.)

1 Write each number in scientific notation.

GS (a) 7500

The first nonzero digit is ____.

The decimal point should be moved ____ places.

$7500 = \_\_\_\_ \times 10^{—}$

(b) 5,870,000

GS (c) 0.057102

The first nonzero digit is ____.

The decimal point should be moved ____ places.

$0.057102 = \_\_\_\_ \times 10^{—}$

(d) -0.00062

Answers

1. (a) 7; 3; 7.5; 3 (b) 5.87×10^6
(c) 5; 2; 5.7102; -2
(d) -6.2×10^{-4}

Converting a Positive Number to Scientific Notation

Step 1 **Position the decimal point.** Place a caret ^ to the right of the first nonzero digit, where the decimal point will be placed.

Step 2 **Determine the numeral for the exponent.** Count the number of digits from the decimal point to the caret. This number gives the absolute value of the exponent on 10.

Step 3 **Determine the sign for the exponent.** Decide whether multiplying by 10^n should make the result of Step 1 greater or less.

- The exponent should be positive to make the result greater.
- The exponent should be negative to make the result less.

EXAMPLE 1 Using Scientific Notation

Write each number in scientific notation.

(a) 93,000,000

Step 1 Place a caret to the right of the 9 (the first nonzero digit) to mark the new location of the decimal point.

$$9_{\wedge}3{,}000{,}000$$

Step 2 Count from the decimal point, which is understood to be after the last 0, to the caret.

$$9.3{,}000{,}000. \leftarrow \text{Decimal point}$$

Count 7 places.

Step 3 Here 9.3 is to be made greater, so the exponent on 10 is positive.

$$93{,}000{,}000 = 9.3 \times 10^7$$

(b) $63{,}200{,}000{,}000 = 6.3200000000 = 6.32 \times 10^{10}$

10 places

(c) 0.00462

Move the decimal point to the right of the first nonzero digit and count the number of places the decimal point was moved.

$$0.00462 \quad \text{3 places}$$

Because 0.00462 is *less* than 4.62, the exponent must be *negative*.

$$0.00462 = 4.62 \times 10^{-3}$$

(d) $-0.0000762 = -7.62 \times 10^{-5}$

5 places

Remember the *negative* sign.

◀ Work Problem 1 at the Side.

Note

When writing a positive number in scientific notation, think as follows.

- If the original number is "large," like 93,000,000, use a *positive* exponent on 10 because positive is greater than negative.
- If the original number is "small," like 0.00462, use a *negative* exponent on 10 because negative is less than positive.

OBJECTIVE 2 Convert numbers in scientific notation to standard notation. To convert a number written in scientific notation to a number without exponents, work in reverse.

Multiplying a positive number by a positive power of 10 will make the number greater. Multiplying by a negative power of 10 will make the number less.

We refer to a number such as 475 as the **standard notation** of 4.75×10^2.

EXAMPLE 2 Writing Numbers in Standard Notation

Write each number in standard notation.

(a) 6.2×10^3

Because the exponent is positive, make 6.2 greater by moving the decimal point 3 places to the right. It is necessary to attach two 0s.

$$6.2 \times 10^3 = 6.200 = 6200$$

(b) $4.283 \times 10^5 = 4.28300 = 428{,}300$ Move 5 places to the right. Attach 0s as necessary.

(c) $5.41 \times 10^0 = 5.41$ $10^0 = 1$

(d) $-9.73 \times 10^{-2} = -09.73 = -0.0973$ Move 2 places to the left.

The exponent tells the number of places and the direction in which the decimal point is moved.

Work Problem 2 at the Side. ▶

OBJECTIVE 3 Use scientific notation in calculations.

EXAMPLE 3 Multiplying and Dividing with Scientific Notation

Perform each calculation. Write answers in both scientific notation and standard notation.

(a) $(7 \times 10^3)(5 \times 10^4)$

$= (7 \times 5)(10^3 \times 10^4)$ Commutative and associative properties

$= 35 \times 10^7$ Multiply. Use the product rule.

$= (3.5 \times 10^1) \times 10^7$ Write 35 in scientific notation.

$= 3.5 \times (10^1 \times 10^7)$ Associative property

$= 3.5 \times 10^8$ Answer in scientific notation

$= 350{,}000{,}000$ Answer in standard notation

Don't stop. This number is *not* in scientific notation because 35 is not between 1 and 10.

(b) $\dfrac{4 \times 10^{-5}}{2 \times 10^3}$

$= \dfrac{4}{2} \times \dfrac{10^{-5}}{10^3}$

$= 2 \times 10^{-8}$ Answer in scientific notation

$= 0.00000002$ Answer in standard notation

Work Problem 3 at the Side. ▶

---

2 Write each number in standard notation.

GS **(a)** 1.2×10^4

Move the decimal point ____ places to the ____.

$1.2 \times 10^4 =$ ______

(b) 8.7×10^5

(c) 7.004×10^0

GS **(d)** 5.49×10^{-3}

Move the decimal point ____ places to the ____.

$5.49 \times 10^{-3} =$ ______

(e) -5.27×10^{-1}

3 Perform each calculation. Write answers in both scientific notation and standard notation.

(a) $(2.6 \times 10^4)(2 \times 10^{-6})$

(b) $(3 \times 10^5)(5 \times 10^{-2})$

(c) $\dfrac{4.8 \times 10^2}{2.4 \times 10^{-3}}$

Answers

2. **(a)** 4; right; 12,000
(b) 870,000 **(c)** 7.004
(d) 3; left; 0.00549 **(e)** -0.527

3. **(a)** 5.2×10^{-2}; 0.052
(b) 1.5×10^4; 15,000
(c) 2×10^5; 200,000

4 Solve the problem.

See **Example 4.** About how much would 8,000,000 nanometers measure in inches?

5 Solve the problem.

If the speed of light is approximately 3.0×10^5 km per sec, how many seconds does it take light to travel approximately 1.5×10^8 km from the sun to Earth? (Data from *The World Almanac and Book of Facts.*)

Answers

4. 3.1496×10^{-1} in., or 0.31496 in.

5. 5×10^2 sec, or 500 sec

Note

Multiplying or dividing numbers written in scientific notation may produce an answer in the form $a \times 10^0$. Because $10^0 = 1$, $a \times 10^0 = a$.

Example: $(8 \times 10^{-4})(5 \times 10^4) = 40 \times 10^0 = 40$ $10^0 = 1$

Also, if $a = 1$, then $a \times 10^n = 10^n$.

Example: 1,000,000 could be written as 10^6 instead of 1×10^6.

EXAMPLE 4 Using Scientific Notation to Solve an Application

A *nanometer* is a very small unit of measure that is equivalent to about 0.00000003937 in. About how much would 700,000 nanometers measure in inches? (Data from *The World Almanac and Book of Facts.*)

Write each number in scientific notation, and then multiply.

$700{,}000(0.00000003937)$

$= (7 \times 10^5)(3.937 \times 10^{-8})$ Write in scientific notation.

$= (7 \times 3.937)(10^5 \times 10^{-8})$ Properties of real numbers

$= 27.559 \times 10^{-3}$ Multiply; product rule

Don't stop here.

$= (2.7559 \times 10^1) \times 10^{-3}$ Write 27.559 in scientific notation.

$= 2.7559 \times 10^{-2}$ Product rule

$= 0.027559$ Write in standard notation.

Thus, 700,000 nanometers would measure

2.7559×10^{-2} in., or 0.027559 in.

◀ Work Problem **4** at the Side.

EXAMPLE 5 Using Scientific Notation to Solve an Application

In 2016, outstanding public debt was $\$1.9008 \times 10^{13}$ (which is more than \$19 trillion). The population of the United States was approximately 323 million that year. What was each citizen's share of this debt? (Data from U.S. Department of the Treasury; U.S. Census Bureau.)

Divide to obtain each citizen's share.

$\frac{1.9008 \times 10^{13}}{323{,}000{,}000}$

$= \frac{1.9008 \times 10^{13}}{3.23 \times 10^8}$ Write 323 million in scientific notation.

$= \frac{1.9008}{3.23} \times 10^5$ Quotient rule

$\approx 0.5885 \times 10^5$ Divide. Round to 4 decimal places.

$= 58{,}850$ Write in standard notation.

Each citizen would have to pay about \$58,850.

◀ Work Problem **5** at the Side.

5.3 Exercises

FOR EXTRA HELP — Go to MyMathLab for worked-out, step-by-step solutions to exercises enclosed in a square and video solutions to ▶ exercises.

CONCEPT CHECK *Match each number written in scientific notation in Column I with the correct choice from Column II. Not all choices in Column II will be used.*

I	II	I	II
1. (a) 4.6×10^{-4}	**A.** 46,000	**2. (a)** 1×10^{9}	**A.** 1 billion
(b) 4.6×10^{4}	**B.** 460,000	**(b)** 1×10^{6}	**B.** 100 million
(c) 4.6×10^{5}	**C.** 0.00046	**(c)** 1×10^{8}	**C.** 1 million
(d) 4.6×10^{-5}	**D.** 0.000046	**(d)** 1×10^{10}	**D.** 10 billion
	E. 4600		**E.** 100 billion

CONCEPT CHECK *Determine whether or not the given number is written in scientific notation as defined in* ***Objective 1.*** *If it is not, write it as such.*

3. 4.56×10^{3} **4.** 7.34×10^{5} **5.** 5,600,000 **6.** 34,000

7. 0.004 **8.** 0.0007 **9.** 0.8×10^{2} **10.** 0.9×10^{3}

GS *Complete each of the following.*

11. Write each number in scientific notation.

(a) 63,000

The first nonzero digit is _____. The decimal point should be moved _____ places.

$63{,}000 = \_\_\_\_\_ \times 10^{—}$

(b) 0.0571

The first nonzero digit is _____. The decimal point should be moved _____ places.

$0.0571 = \_\_\_\_\_ \times 10^{—}$

12. Write each number in standard notation.

(a) 4.2×10^{3}

Move the decimal point _____ places to the (*right*/*left*).

$4.2 \times 10^{3} = \_\_\_\_\_\_$

(b) 6.42×10^{-3}

Move the decimal point _____ places to the (*right*/*left*).

$6.42 \times 10^{-3} = \_\_\_\_\_\_$

Write each number in scientific notation. ***See Example 1.***

13. ▶ 5,876,000,000 **14.** 9,994,000,000 **15.** 82,350 **16.** 78,330

17. ▶ 0.000007 **18.** 0.0000004 **19.** −0.00203 **20.** −0.0000578

Write each number in standard notation. ***See Example 2.***

21. ▶ 7.5×10^{5} **22.** 8.8×10^{6} **23.** 5.677×10^{12} **24.** 8.766×10^{9}

25. 1×10^{12} **26.** 1×10^{7} **27.** -6.21×10^{0} **28.** -8.56×10^{0}

29. ▶ 7.8×10^{-4} **30.** 8.9×10^{-5} **31.** 5.134×10^{-9} **32.** 7.123×10^{-10}

*Each statement contains a number in **boldface italic** type. If the number is in scientific notation, write it in standard notation. If the number is not in scientific notation, write it as such.* ***See Examples 1 and 2.***

33. A *muon* is an atomic particle related to an electron. The half-life of a muon is about 2 millionths $(\boldsymbol{2 \times 10^{-6}})$ of a second. (Data from www.schoolphysics.co.uk)

34. There are 13 red balls and 39 black balls in a box. Mix them up and draw 13 out one at a time without returning any ball . . . the probability that the 13 drawings each will produce a red ball is . . . $\boldsymbol{1.6 \times 10^{-12}}$.

35. An electron and a positron attract each other in two ways: the electromagnetic attraction of their opposite electric charges, and the gravitational attraction of their two masses. The electromagnetic attraction is

4,200,000,000,000,000,000,000,000,000,000,000,000,000,000,000

times as strong as the gravitational. (Data from Asimov, I., *Isaac Asimov's Book of Facts.*)

36. The name "googol" applies to the number

***10,000,000,000,000,000,000,000,000,000,000,000,000,000,000,000,000,000,000,000,000,
000,000,000,000,000,000,000,000,000,000,000,000,000,000,000,000.***

The Web search engine Google honors this number. Sergey Brin, president and cofounder of Google, Inc., was a mathematics major. He chose the name Google to describe the vast reach of this search engine. (Data from *The Gazette.*)

*Perform the indicated operations. Write each answer in **(a)** scientific notation and **(b)** standard notation.* ***See Example 3.***

37. $(2 \times 10^8)(3 \times 10^3)$

38. $(3 \times 10^7)(3 \times 10^3)$

39. $(5 \times 10^4)(3 \times 10^2)$

40. $(8 \times 10^5)(2 \times 10^3)$

41. $(4 \times 10^{-6})(2 \times 10^3)$

42. $(3 \times 10^{-7})(2 \times 10^2)$

43. $(6 \times 10^3)(4 \times 10^{-2})$

44. $(7 \times 10^5)(3 \times 10^{-4})$

45. $(3 \times 10^{-4})(-2 \times 10^8)$

46. $(4 \times 10^{-3})(-2 \times 10^7)$

47. $(9 \times 10^4)(7 \times 10^{-7})$

48. $(6 \times 10^4)(8 \times 10^{-8})$

49. $\dfrac{9 \times 10^{-5}}{3 \times 10^{-1}}$

50. $\dfrac{12 \times 10^{-4}}{4 \times 10^{-3}}$

51. $\dfrac{8 \times 10^3}{2 \times 10^2}$

52. $\dfrac{15 \times 10^4}{3 \times 10^3}$

53. $\dfrac{2.6 \times 10^{-3}}{2 \times 10^2}$

54. $\dfrac{9.5 \times 10^{-1}}{5 \times 10^3}$

55. $\dfrac{4 \times 10^5}{8 \times 10^2}$

56. $\dfrac{3 \times 10^9}{6 \times 10^5}$

57. $\dfrac{15 \times 10^{-4} \times 12 \times 10^5}{5 \times 10^3 \times 4 \times 10^{-8}}$

58. $\dfrac{24 \times 10^{-3} \times 18 \times 10^{4}}{6 \times 10^{6} \times 9 \times 10^{-2}}$

59. $\dfrac{2.6 \times 10^{-3} \times 7.0 \times 10^{-1}}{2 \times 10^{2} \times 3.5 \times 10^{-3}}$

60. $\dfrac{9.5 \times 10^{-1} \times 2.4 \times 10^{4}}{5 \times 10^{3} \times 1.2 \times 10^{-2}}$

Use scientific notation to calculate the answer to each problem. ***See Examples 4 and 5.***

61. The Double Helix Nebula, a conglomeration of dust and gas stretching across the center of the Milky Way galaxy, is 25,000 light-years from Earth. If one light-year is about 6,000,000,000,000 mi, about how many miles is the Double Helix Nebula from Earth? (Data from www.spitzer.caltech.edu)

62. Pollux, one of the brightest stars in the night sky, is 33.7 light-years from Earth. If one light-year is about 6,000,000,000,000 mi, about how many miles is Pollux from Earth? (Data from *The World Almanac and Book of Facts.*)

63. In 2016, the population of the United States was about 322.9 million. To the nearest dollar, calculate how much each person in the United States would have had to contribute in order to make one person a trillionaire (that is, to give that person $1,000,000,000,000). (Data from U.S. Census Bureau.)

64. In 2015, Congress raised the debt limit to $\$1.81 \times 10^{13}$. To the nearest dollar, about how much was this for every man, woman, and child in the country? Use 321 million as the population of the United States. (Data from U.S. Census Bureau.)

65. In 2014, the U.S. government collected about $4372 per person in individual income taxes. If the population at that time was 319,000,000, how much did the government collect in taxes for 2014? (Data from *The World Almanac and Book of Facts.*)

66. In 2014, the state of Iowa had about 8.8×10^{4} farms with an average of 3.47×10^{2} acres per farm. What was the total number of acres devoted to farmland in Iowa that year? (Data from U.S. Department of Agriculture.)

67. Light travels at a speed of 1.86×10^{5} mi per sec. When Venus is 6.68×10^{7} mi from the sun, how long does it take light to travel from the sun to Venus? Round to two decimal places. (Data from *The World Almanac and Book of Facts.*)

68. The distance to Earth from Pluto is 4.58×10^{9} km. *Pioneer 10* transmitted radio signals from Pluto to Earth at the speed of light, 3.00×10^{5} km per sec. How long (in seconds) did it take for the signals to reach Earth? Round to two decimal places.

69. One of the world's fastest computers can perform 10,000,000,000,000,000 calculations per second. How many can it perform per minute? Per hour? (Data from www.japantimes.co.jp)

70. One of the world's fastest computers can perform 33.86 quadrillion calculations per second. (*Hint:* 1 quadrillion $= 1 \times 10^{15}$) How many can it perform per minute? Per hour? (Data from www.top500.org)

Calculators can express numbers in scientific notation using notation such as

$$5.4\text{E}3 \quad \textit{to represent} \quad 5.4 \times 10^3.$$

Similarly, $5.4\text{E}^{-}3$ *represents* 5.4×10^{-3}. *Predict the display the calculator would give for the expression shown in each screen.*

71. `.00000047`

72. `.000021`

73. `(8E5)/(4E⁻2)`

74. `(9E⁻4)/(3E3)`

Relating Concepts (Exercises 75–78) For Individual or Group Work

In 1935, Charles F. Richter devised a scale to compare the intensities, or relative powers, of earthquakes. The **intensity** *of an earthquake is measured relative to the intensity of a standard* **zero-level** *earthquake of intensity* I_0. *The relationship is equivalent to*

$$I = I_0 \times 10^R, \quad \text{where } R \text{ is the } \textbf{Richter scale} \text{ measure.}$$

For example, if an earthquake has magnitude 5.0 *on the Richter scale, then its intensity is calculated as*

$$I = I_0 \times 10^{5.0} = I_0 \times 100{,}000,$$

which is 100,000 *times as intense as a zero-level earthquake.*

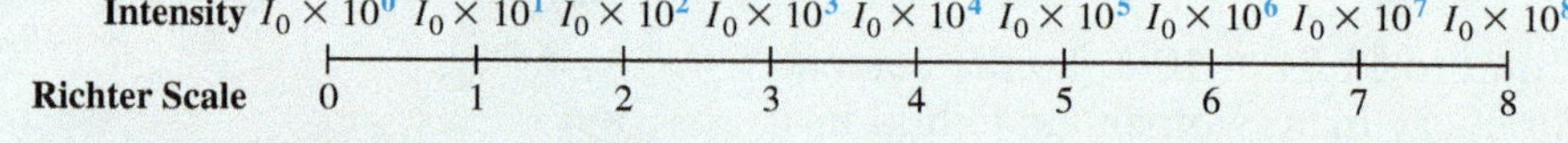

To compare two earthquakes, such as one that measures 8.0 *to one that measures* 5.0, *calculate the ratio of their intensities.*

$$\frac{\text{intensity } 8.0}{\text{intensity } 5.0} = \frac{I_0 \times 10^{8.0}}{I_0 \times 10^{5.0}} = \frac{10^8}{10^5} = 10^{8-5} = 10^3 = 1000$$

An earthquake that measures 8.0 *is* 1000 *times as intense as one that measures* 5.0.

Use the information in the table to ***work Exercises 75–78 in order.***

Year	Earthquake Location	Richter Scale Measurement
1960	Chile	9.5
2011	NE Japan	9.0
2007	Southern Sumatra, Indonesia	8.5
2013	Solomon Islands region	8.0
2013	Falkland Islands region	7.0

Data from earthquake.usgs.gov

75. Compare the intensity of the 1960 Chile earthquake with the 2007 Southern Sumatra earthquake.

76. Compare the intensity of the 2013 Solomon Islands earthquake with the 2013 Falkland Islands earthquake.

77. Compare the intensity of the 2011 NE Japan earthquake with the 2013 Falkland Islands earthquake.

78. Suppose an earthquake measures 5.5 on the Richter scale. How would the intensity of the 1960 Chile earthquake compare to it?

5.4 Adding and Subtracting Polynomials

OBJECTIVES

1. Identify terms and coefficients.
2. Combine like terms.
3. Know the vocabulary for polynomials.
4. Evaluate polynomials.
5. Add polynomials.
6. Subtract polynomials.
7. Add and subtract polynomials with more than one variable.

OBJECTIVE 1 Identify terms and coefficients. Recall that in an algebraic expression such as

$$4x^3 + 6x^2 + 5x + 8,$$

the quantities that are added, $4x^3$, $6x^2$, $5x$, and 8, are **terms.** In the term $4x^3$, the number 4 is the **numerical coefficient,** or simply the **coefficient,** of x^3. In the same way, 6 is the coefficient of x^2 in the term $6x^2$, and 5 is the coefficient of x in the term $5x$. The constant term 8 can be thought of as

$$8 \cdot 1 = 8x^0, \quad \text{By definition, } x^0 = 1.$$

so 8 is the coefficient in the term 8. Other examples are given in the table.

TERMS AND THEIR COEFFICIENTS

Term	Numerical Coefficient
8	8
$-7y$	-7
$34r^3$	34
$-26x^5yz^4$	-26
$-k = -1k$	-1
$r = 1r$	1
$\frac{3x}{8} = \frac{3}{8}x$	$\frac{3}{8}$
$\frac{x}{3} = \frac{1x}{3} = \frac{1}{3}x$	$\frac{1}{3}$

EXAMPLE 1 Identifying Coefficients

Identify the coefficient of each term in the expression. Then give the number of terms.

(a) $-6x^4 + x - 3$ can be written as $-6x^4 + 1x + (-3x^0)$.

The coefficients are -6, 1, and -3.

There are three terms:

$-6x^4$, x, and -3.

(b) $5 - v^3$ can be written as $5v^0 + (-1v^3)$.

There are two terms. The coefficients are 5 and -1.

Work Problem 1 at the Side. ▶

1 Identify the coefficient of each term in the expression. Then give the number of terms.

$$2x^3 - x + 10$$

OBJECTIVE 2 Combine like terms. Recall that **like terms** have exactly the same combination of variables, with the same exponents on the variables. ***Only the coefficients may differ.***

Examples of like terms			Examples of unlike terms		
$19m^5$	and	$14m^5$	$7x$	and	$7y$
$-37y^9$	and	y^9	z^4	and	z
$3pq$	and	$-2pq$	$2pq$	and	$2p$
$2xy^2$	and	$-xy^2$	$-4xy^2$	and	$5x^2y$

Using the distributive property, we combine like terms by adding or subtracting their coefficients.

Answer

1. 2; -1; 10; three terms

2 Simplify each expression by combining like terms, if possible.

(a) $5x^4 + 7x^4$

(b) $9pq + 3pq - 2pq$

(c) $r^2 + 3r + 5r^2$

(d) $x + \frac{1}{2}x$

(e) $8t + 6w$

(f) $3x^4 - 3x^2$

Answers

2. **(a)** $12x^4$ **(b)** $10pq$ **(c)** $6r^2 + 3r$ **(d)** $\frac{3}{2}x$
(e) These are unlike terms. They cannot be combined.
(f) These are unlike terms. They cannot be combined.

EXAMPLE 2 Combining Like Terms

Simplify each expression by combining like terms, if possible.

(a) $-4x^3 + 6x^3$

$= (-4 + 6)x^3$ — $ac + bc = (a + b)c$

$= 2x^3$

(b) $9x^6 - 14x^6 + x^6$

$= (9 - 14 + 1)x^6$ — $x^6 = 1x^6$

$= -4x^6$

(c) $y + \frac{2}{3}y$

$= 1y + \frac{2}{3}y$ — $y = 1y$

$= \left(\frac{3}{3} + \frac{2}{3}\right)y$ — $1 = \frac{3}{3}$; Distributive property

$= \frac{5}{3}y$ — Add the fractions.

(d) $8rs - 13rs + 9rs$

$= (8 - 13 + 9)rs$

$= 4rs$

(e) $12m^2 + 5m + 4m^2$

$= (12 + 4)m^2 + 5m$

$= 16m^2 + 5m$ — Stop here. These are unlike terms.

(f) $5u + 11v$

These are unlike terms. They cannot be combined.

◀ **Work Problem 2 at the Side.**

CAUTION

In **Example 2(e),** we cannot combine $16m^2$ and $5m$ because the exponents on the variables are different. ***Unlike terms have different variables or different exponents on the same variables.***

OBJECTIVE 3 Know the vocabulary for polynomials.

Polynomial in *x*

A **polynomial in *x*** is a term or the sum of a finite number of terms of the form

ax^n, for any real number a and any whole number n.

For example, the expression

$$\underbrace{16x^8 - 7x^6 + 5x^4 - 3x^2 + 4}_{\text{Descending powers of } x}$$ Polynomial in x (The 4 can be written as $4x^0$.)

is a polynomial in x. It is written in **descending powers,** because the exponents on x decrease from left to right. By contrast, the expression

$$2x^3 - x^2 + \frac{4}{x}, \quad \text{or} \quad 2x^3 - x^2 + 4x^{-1}, \qquad \text{Not a polynomial}$$

is *not* a polynomial in x. A variable appears in a denominator, which can be written as a factor to a negative power.

Note

A polynomial can use any variable, not just x (see **Example 2(c)**), and have terms with more than one variable (see **Example 2(d)**).

The **degree of a term** is the sum of the exponents on the variables. The **degree of a polynomial** is the greatest degree of any nonzero term of the polynomial. The table gives several examples.

DEGREES OF TERMS AND POLYNOMIALS

Term	Degree	Polynomial	Degree
$3x^4$	4	$3x^4 - 5x^2 + 6$	4
$5x$, or $5x^1$	1	$5x + 7$	1
-7, or $-7x^0$	0	$x^5 + 3x^6 - 7$	6
$2x^2y$, or $2x^2y^1$	$2 + 1 = 3$	$xy - 5y^2 + 2x^2y$	3

Some polynomials with a specific number of terms have special names.

- A polynomial with exactly one term is a **monomial.** (*Mono-* means "one," as in *mono*rail.)

$$9m, \quad -6y^5, \quad x^2, \quad \text{and} \quad 6 \qquad \text{Monomials}$$

- A polynomial with exactly two terms is a **binomial.** (*Bi-* means "two," as in *bi*cycle.)

$$-9x^4 + 9x^3, \quad 8m^2 + 6m, \quad \text{and} \quad 3t - 10 \qquad \text{Binomials}$$

- A polynomial with exactly three terms is a **trinomial.** (*Tri-* means "three," as in *tri*angle.)

$$9m^3 - 4m^2 + 6, \quad \frac{19}{3}y^2 + \frac{8}{3}y + 5, \quad \text{and} \quad -3z^5 - z^2 + z \qquad \text{Trinomials}$$

EXAMPLE 3 Classifying Polynomials

For each polynomial, first simplify, if possible. Then give the degree and tell whether the simplified polynomial is a *monomial*, a *binomial*, a *trinomial*, or *none of these*.

(a) $2x^3 + 5$

The polynomial cannot be simplified. It is a *binomial* of degree 3.

(b) $6x - 8x + 13x$

$= 11x$ Combine like terms to simplify.

The degree is 1 (here $x = x^1$). The simplified polynomial is a *monomial*.

(c) $4xy - 5xy + 2xy$

$= xy$ Combine like terms.

The degree is 2 (because $xy = x^1y^1$, and $1 + 1 = 2$). The simplified polynomial is a *monomial*.

(d) $2x^2 - 3x + 8x - 12$

$= 2x^2 + 5x - 12$ Combine like terms.

The degree is 2. The simplified polynomial is a *trinomial*.

Work Problem 4 at the Side. ▶

3 Choose all descriptions that apply for each of the expressions in parts (a)–(d).

A. Polynomial

B. Polynomial written in descending powers

C. Not a polynomial

(a) $3m^5 + 5m^2 - 2m + 1$

(b) $2p^4 + p^6$

(c) $\frac{1}{x} + 2x^2 + 3$

(d) $x - 3$

4 For each polynomial, first simplify, if possible. Then give the degree and tell whether the simplified polynomial is a *monomial,* a *binomial,* a *trinomial,* or *none of these.*

(a) $3x^2 + 2x - 4$

(b) $x^3 + 4x^3$

(c) $x^8 - x^7 + 2x^8$

(d) $6ab - 7ab + 3ab$

Answers

3. **(a)** A, B **(b)** A **(c)** C **(d)** A, B

4. **(a)** The polynomial cannot be simplified; degree 2; trinomial
(b) $5x^3$; degree 3; monomial
(c) $3x^8 - x^7$; degree 8; binomial
(d) $2ab$; degree 2; monomial

5 Find the value of $2x^3 + 8x - 6$ in each case.

GS **(a)** For $x = -1$

$$2x^3 + 8x - 6$$
$$= 2(\_\_)^3 + 8(\_\_) - 6$$
$$= 2(\_\_) + 8(-1) - 6$$
$$= \_\_\_ - \_\_\_ - 6$$
$$= \_\_\_$$

(b) For $x = 4$

(c) For $x = -2$

6 Find each sum.

(a) Add $4x^3 - 3x^2 + 2x$ and $6x^3 + 2x^2 - 3x$.

(b) Add $x^2 - 2x + 5$ and $4x^2 - 2$.

Answers

5. **(a)** $-1; -1; -1; -2; 8; -16$
(b) 154 **(c)** -38

6. **(a)** $10x^3 - x^2 - x$
(b) $5x^2 - 2x + 3$

OBJECTIVE 4 **Evaluate polynomials.** When we *evaluate* an expression, we find its *value*. A polynomial usually represents different numbers for different values of the variable.

EXAMPLE 4 Evaluating a Polynomial

Find the value of $3x^4 + 5x^3 - 4x - 4$ for **(a)** $x = -2$ and for **(b)** $x = 3$.

(a) $3x^4 + 5x^3 - 4x - 4$

$= 3(-2)^4 + 5(-2)^3 - 4(-2) - 4$ — Substitute -2 for x.

$= 3(16) + 5(-8) - 4(-2) - 4$ — Apply the exponents.

$= 48 - 40 + 8 - 4$ — Multiply.

$= 12$ — Add and subtract.

Use parentheses to avoid errors.

(b) $3x^4 + 5x^3 - 4x - 4$

$= 3(3)^4 + 5(3)^3 - 4(3) - 4$ — Let $x = 3$.

$= 3(81) + 5(27) - 4(3) - 4$ — Apply the exponents.

$= 243 + 135 - 12 - 4$ — Multiply.

$= 362$ — Add and subtract.

◀ Work Problem **5** at the Side.

OBJECTIVE 5 **Add polynomials.**

Adding Polynomials

To add two polynomials, combine (add) like terms.

EXAMPLE 5 Adding Polynomials Vertically

Find each sum.

(a) Add $6x^3 - 4x^2 + 3$ and $-2x^3 + 7x^2 - 5$.

$$\begin{array}{r} 6x^3 - 4x^2 + 3 \\ +\ \underline{(-2x^3 + 7x^2 - 5)} \end{array}$$

Write like terms in columns.

Now add, column by column.

$$\begin{array}{rrr} 6x^3 & -4x^2 & 3 \\ \underline{-2x^3} & \underline{7x^2} & \underline{-5} \\ 4x^3 & 3x^2 & -2 \end{array}$$

Add the coefficients only. Do *not* add the exponents.

Add the three sums together to obtain the answer.

$$4x^3 + 3x^2 + (-2) = 4x^3 + 3x^2 - 2$$

(b) Add $2x^2 - 4x + 3$ and $x^3 + 5x$.

Write like terms in columns and add column by column.

$$\begin{array}{r} 2x^2 - 4x + 3 \\ +\ \underline{(x^3 \qquad\quad + 5x \qquad)} \\ x^3 + 2x^2 + \ x + 3 \end{array}$$

Leave spaces for missing terms.

◀ Work Problem **6** at the Side.

The polynomials in **Example 5** also could be added horizontally.

EXAMPLE 6 Adding Polynomials Horizontally

Find each sum.

(a) Add $6x^3 - 4x^2 + 3$ and $-2x^3 + 7x^2 - 5$.

$$(6x^3 - 4x^2 + 3) + (-2x^3 + 7x^2 - 5) = 4x^3 + 3x^2 - 2$$

Combine like terms. Same answer as found in **Example 5(a)**

(b) Add $2x^2 - 4x + 3$ and $x^3 + 5x$.

$(2x^2 - 4x + 3) + (x^3 + 5x)$

$= x^3 + 2x^2 - 4x + 5x + 3$ Commutative property

$= x^3 + 2x^2 + x + 3$ See **Example 5(b)**.

Work Problem 7 at the Side. ▶

OBJECTIVE 6 Subtract polynomials. Recall that the difference $x - y$ is defined as $x + (-y)$. (We find the difference $x - y$ by adding x and the *opposite* of y.)

$7 - 2$ is equivalent to $7 + (-2)$, which equals 5.

$-8 - (-2)$ is equivalent to $-8 + 2$, which equals -6.

A similar method is used to subtract polynomials.

Subtracting Polynomials

To subtract two polynomials, change the sign of each term in the subtrahend (second polynomial) and add the result to the minuend (first polynomial)—that is, add the *opposite* of each term of the second polynomial to the first polynomial.

EXAMPLE 7 Subtracting Polynomials Horizontally

Perform each subtraction.

(a) $(5x - 2) - (3x - 8)$

$= (5x - 2) + [-(3x - 8)]$ Definition of subtraction

$= (5x - 2) + [-1(3x - 8)]$ $-a = -1a$

$= (5x - 2) + (-3x + 8)$ Distributive property

$= 2x + 6$ Combine like terms.

CHECK To check a subtraction problem, use the fact that

if $a - b = c$, then $a = b + c$.

$(3x - 8) + (2x + 6) = 5x - 2$ ✓

(b) Subtract $6x^3 - 4x^2 + 2$ from $11x^3 + 2x^2 - 8$.

$(11x^3 + 2x^2 - 8) - (6x^3 - 4x^2 + 2)$ Be careful to write the problem in the correct order.

$= (11x^3 + 2x^2 - 8) + (-6x^3 + 4x^2 - 2)$

$= 5x^3 + 6x^2 - 10$ Combine like terms. Check as above.

Work Problem 8 at the Side. ▶

7 Find each sum.

(a) Add $2x^4 - 6x^2 + 7$ and $-3x^4 + 5x^2 + 2$.

(b) Add $3x^2 + 4x + 2$ and $6x^3 - 5x - 7$.

8 Perform each subtraction.

(a) $(3x - 8) - (5x - 9)$

(b) $(14y^3 - 6y^2 + 2y - 5) - (2y^3 - 7y^2 - 4y + 6)$

(c) Subtract $-3y^2 + 4y + 6$ from $7y^2 - 11y + 8$.

Answers

7. **(a)** $-x^4 - x^2 + 9$
(b) $6x^3 + 3x^2 - x - 5$

8. **(a)** $-2x + 1$
(b) $12y^3 + y^2 + 6y - 11$
(c) $10y^2 - 15y + 2$

Subtraction also can be done in columns. We will use vertical subtraction when we study polynomial division.

EXAMPLE 8 Subtracting Polynomials Vertically

Subtract by columns. $(14y^3 - 6y^2 + 2y - 5) - (2y^3 - 7y^2 - 4y + 6)$

$$\begin{array}{r} 14y^3 - 6y^2 + 2y - 5 \\ -\ (2y^3 - 7y^2 - 4y + 6) \\ \hline \end{array}$$ Arrange like terms in columns.

Change all signs in the second polynomial (the subtrahend), and then add.

$$\begin{array}{r} 14y^3 - 6y^2 + 2y - 5 \\ +\ (-2y^3 + 7y^2 + 4y - 6) \\ \hline 12y^3 + y^2 + 6y - 11 \end{array}$$

Change each sign.

Add.

◀ Work Problem 9 at the Side.

EXAMPLE 9 Adding and Subtracting More Than Two Polynomials

Perform the indicated operations.

$$(4 - x + 3x^2) - (2 - 3x + 5x^2) + (8 + 2x - 4x^2)$$

Rewrite, changing subtraction to adding the opposite.

$(4 - x + 3x^2) - (2 - 3x + 5x^2) + (8 + 2x - 4x^2)$

$= (4 - x + 3x^2) + (-2 + 3x - 5x^2) + (8 + 2x - 4x^2)$

$= (2 + 2x - 2x^2) + (8 + 2x - 4x^2)$ Combine like terms.

$= 10 + 4x - 6x^2$ Combine like terms.

◀ Work Problem 10 at the Side.

OBJECTIVE 7 Add and subtract polynomials with more than one variable. Polynomials in more than one variable are added and subtracted by combining like terms, just as with single-variable polynomials.

EXAMPLE 10 Adding and Subtracting Multivariable Polynomials

Add or subtract as indicated.

(a) $(4a + 2ab - b) + (3a - ab + b)$

$= 4a + 2ab - b + 3a - ab + b$

$= 7a + ab$ Combine like terms.

(b) $(2x^2y + 3xy + y^2) - (3x^2y - xy - 2y^2)$

$= 2x^2y + 3xy + y^2 - 3x^2y + xy + 2y^2$

$= -x^2y + 4xy + 3y^2$

Be careful with signs. The coefficient of xy is 1.

◀ Work Problem 11 at the Side.

---

9 Subtract by columns.

$(4y^3 - 16y^2 + 2y + 1)$
$- (12y^3 - 9y^2 - 3y + 16)$

10 Perform the indicated operations.

$(6p^4 - 8p^3 + 2p - 1)$
$- (-7p^4 + 6p^2 - 12)$
$+ (p^4 - 3p + 8)$

11 Add or subtract as indicated.

(a) $(3mn + 2m - 4n)$
$+ (-mn + 4m + n)$

(b) $(5p^2q^2 - 4p^2 + 2q)$
$- (2p^2q^2 - p^2 - 3q)$

Answers

9. $-8y^3 - 7y^2 + 5y - 15$
10. $14p^4 - 8p^3 - 6p^2 - p + 19$
11. **(a)** $2mn + 6m - 3n$
(b) $3p^2q^2 - 3p^2 + 5q$

5.4 Exercises

Go to MyMathLab for worked-out, step-by-step solutions to exercises enclosed in a square and video solutions to ▶ exercises.

CONCEPT CHECK *Complete each statement.*

1. In the term $7x^5$, the coefficient is ______ and the exponent is ______.

2. The expression $5x^3 - 4x^2$ has (*one / two / three*) term(s).

3. The degree of the term $-4x^8$ is ______.

4. The polynomial $4x^2 - y^2$ (*is / is not*) an example of a trinomial.

5. When $x^2 + 10$ is evaluated for $x = 4$, the result is ______.

6. ______ is an example of a monomial with coefficient 5, in the variable x, having degree 9.

Identify the coefficient of each term in the expression. Then give the number of terms. **See Example 1.**

7. $6x^4$ ▶

8. $-9y^5$

9. t^4

10. s^7

11. $\frac{x}{5}$

12. $\frac{z}{8}$

13. $-19r^2 - r$

14. $2y^3 - y$

15. $x - 8x^2 + \frac{2}{3}x^3$

16. $v - 2v^3 + \frac{3}{4}v^2$

Simplify each expression by combining like terms, if possible. Write results with more than one term in descending powers of the variable. **See Example 2 and Objective 3.**

17. $-3m^5 + 5m^5$ ▶

18. $-4y^3 + 3y^3$

19. $2r^5 + (-3r^5)$

20. $-19y^2 + 9y^2$

21. $\frac{1}{2}x^4 + \frac{1}{6}x^4$

22. $\frac{3}{10}x^6 + \frac{1}{5}x^6$

23. $-0.5m^2 + 0.2m^5$

24. $-0.9y + 0.9y^2$

25. $-3x^5 + 2x^5 - 4x^5$

26. $6x^3 - 8x^3 + 9x^3$

27. $-4p^7 + 8p^7 + 5p^9$ ▶

28. $-3a^8 + 4a^8 - 3a^2$

29. $-4y^2 + 3y^2 - 2y^2 + y^2$ ▶

30. $3r^5 - 8r^5 + r^5 + 2r^5$

31. $3xy + 5xy - 12xy$

32. $2ab - 4ab - 7ab$

For each polynomial, first simplify if possible, and write the result in descending powers of the variable. Then give the degree, and tell whether the simplified polynomial is a monomial, *a* binomial, *a* trinomial, *or* none of these. **See Example 3.**

33. $6x^4 - 9x$ ▶

34. $7t^3 - 3t$

35. $x^2 + 3x + 6x - 3$

36. $y^2 - 2y - 6y + 5$

37. $5xy + 13xy - 12xy$

38. $-11ab + 2ab - 4ab$

39. $5m^4 - 3m^2 + 6m^5 - 7m^3$

40. $6p^5 + 4p^3 - 8p^4 + 10p^2$

41. $\frac{5}{3}x^4 - \frac{2}{3}x^4$

42. $\frac{4}{5}r^6 + \frac{1}{5}r^6$

43. $0.8x^4 - 0.3x^4 - 0.5x^4 + 7$

44. $1.2t^3 - 0.9t^3 - 0.3t^3 + 9$

Find the value of each polynomial for **(a)** $x = 2$ *and for* **(b)** $x = -1$. ***See Example 4.***

45. $-2x + 3$

46. $5x - 4$

47. $2x^2 + 5x + 1$

48. $-3x^2 + 14x - 2$

49. $2x^5 - 4x^4 + 5x^3 - x^2$

50. $x^4 - 6x^3 + x^2 + 1$

51. $-4x^5 + x^2$

52. $2x^6 - 4x$

Add. ***See Example 5.***

53. $\begin{array}{r} 2x^2 - 4x \\ +\ \underline{(3x^2 + 2x)} \end{array}$

54. $\begin{array}{r} -5y^3 + 3y \\ +\ \underline{(\ 8y^3 - 4y)} \end{array}$

55. $\begin{array}{r} 3m^2 + 5m + 6 \\ +\ \underline{(2m^2 - 2m - 4)} \end{array}$

56. $\begin{array}{r} 4a^3 - 4a^2 - 4 \\ +\ \underline{(6a^3 + 5a^2 - 8)} \end{array}$

57. $\begin{array}{r} \frac{2}{3}x^2 + \frac{1}{5}x + \frac{1}{6} \\ +\ \underline{\left(\frac{1}{2}x^2 - \frac{1}{3}x + \frac{2}{3}\right)} \end{array}$

58. $\begin{array}{r} \frac{4}{7}y^2 - \frac{1}{5}y + \frac{7}{9} \\ +\ \underline{\left(\frac{1}{3}y^2 - \frac{1}{3}y + \frac{2}{5}\right)} \end{array}$

Subtract. ***See Example 8.***

59. $\begin{array}{r} 5y^3 - 3y^2 \\ -\ \underline{(2y^3 + 8y^2)} \end{array}$

60. $\begin{array}{r} -6t^3 + 4t^2 \\ -\ \underline{(\ 8t^3 - 6t^2)} \end{array}$

61. $\begin{array}{r} 12x^4 - x^2 + x \\ -\ \underline{(\ 8x^4 + 3x^2 - 3x)} \end{array}$

62. $\begin{array}{r} 13y^5 - y^3 - 8y^2 \\ -\ \underline{(7y^5 + 5y^3 + y^2)} \end{array}$

63. $\begin{array}{r} 12m^3 - 8m^2 + 6m + 7 \\ -\ \underline{(-3m^3 + 5m^2 - 2m - 4)} \end{array}$

64. $\begin{array}{r} 5a^4 - 3a^3 + 2a^2 - a \\ -\ \underline{(-6a^4 + a^3 - a^2 + a)} \end{array}$

Perform the indicated operations. ***See Examples 6, 7, and 9.***

65. $(2r^2 + 3r - 12) + (6r^2 + 2r)$

66. $(3r^2 + 5r - 6) + (2r - 5r^2)$

67. $(8m^2 - 7m) - (3m^2 + 7m - 6)$

68. $(x^2 + x) - (3x^2 + 2x - 1)$

69. $(16x^3 - x^2 + 3x) + (-12x^3 + 3x^2 + 2x)$

70. $(-2b^6 + 3b^4 - b^2) + (b^6 + 2b^4 + 2b^2)$

71. $(7y^4 + 3y^2 + 2y) - (18y^5 - 5y^3 + y)$

72. $(8t^5 + 3t^3 + 5t) - (19t^4 - 6t^2 + t)$

73. $(-4x^2 + 2x - 3) - (-2x^2 + x - 1) + (-8x^2 + 3x - 4)$

74. $(-8x^2 - 3x + 2) + (4x^2 - 3x + 8) - (-2x^2 - x + 7)$

75. $[(8m^2 + 4m - 7) - (2m^3 - 5m + 2)] - (m^2 + m)$

76. $[(9b^3 - 4b^2 + 3b + 2) - (-2b^3 + b)] - (8b^3 + 6b + 4)$

77. Add $9m^3 - 5m^2 + 4m - 8$ and $-3m^3 + 6m^2 - 6$.

78. Add $12r^5 + 11r^4 - 2r^2$ and $-8r^5 + 3r^3 + 2r^2$.

79. Subtract $9x^2 - 3x + 7$ from $-2x^2 - 6x + 4$.

80. Subtract $-5w^3 + 5w^2 - 7$ from $6w^3 + 8w + 5$.

Find a polynomial that represents, in appropriate units, the perimeter of each square, rectangle, or triangle.

81.

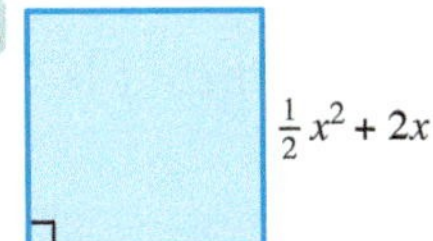

82.

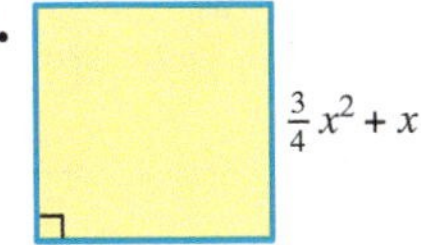

83.

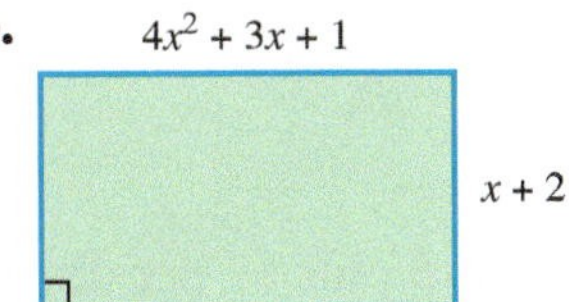

84.

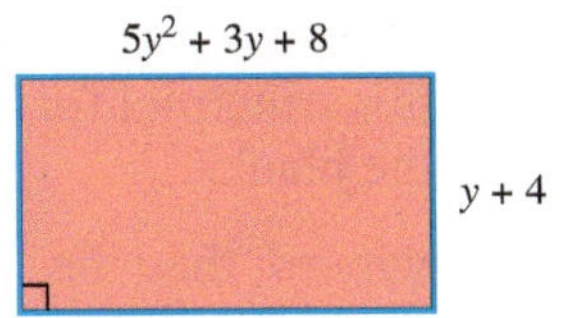

85.

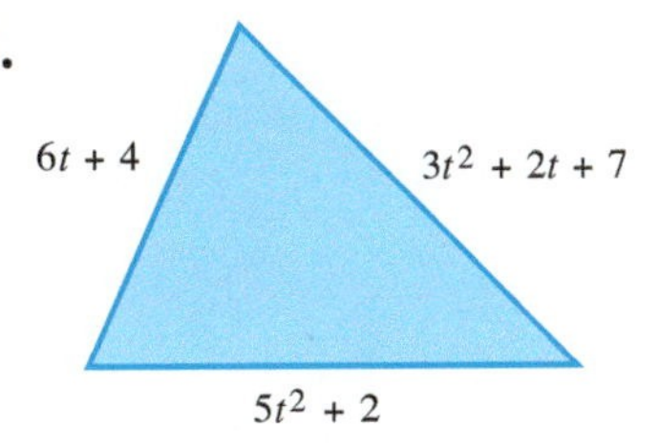

86.

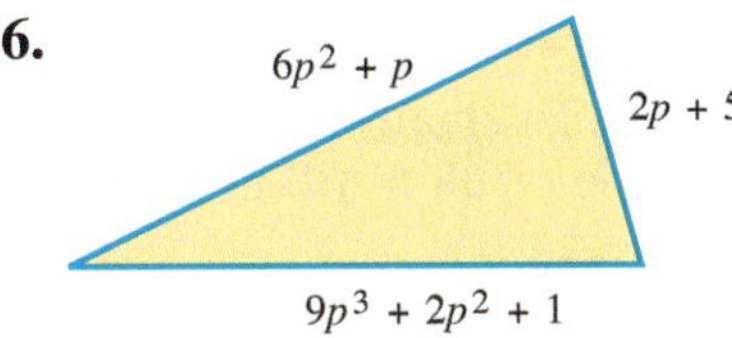

Add or subtract as indicated. ***See Example 10.***

87. $(4x + 2xy - 3) - (-2x + 3xy + 4)$

88. $(8ab + 2a - 3b) - (6ab - 2a + 3b)$

89. $(9a^2b - 3a^2 + 2b) + (4a^2b - 4a^2 - 3b)$

90. $(4xy^3 - 3x + y) + (5xy^3 + 13x - 4y)$

91. $(2c^4d + 3c^2d^2 - 4d^2) - (c^4d + 8c^2d^2 - 5d^2)$

92. $(3k^2h^3 + 5kh + 6k^3h^2) - (2k^2h^3 - 9kh + k^3h^2)$

Find **(a)** *a polynomial that represents the perimeter of each triangle and* **(b)** *the measures of the angles of the triangle. (Hint: The sum of the measures of the angles of any triangle is* 180°.)

93.

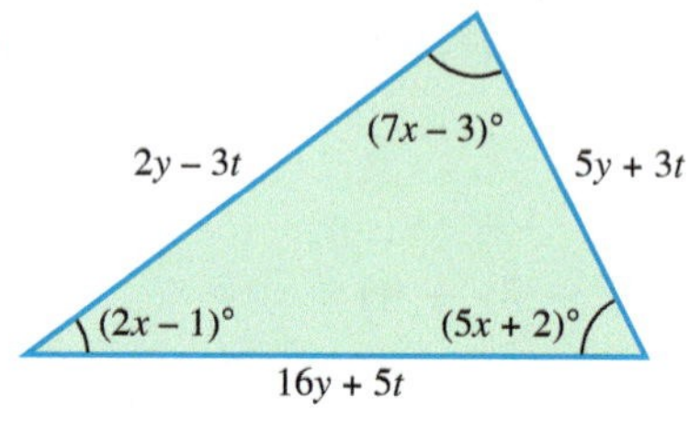

94.

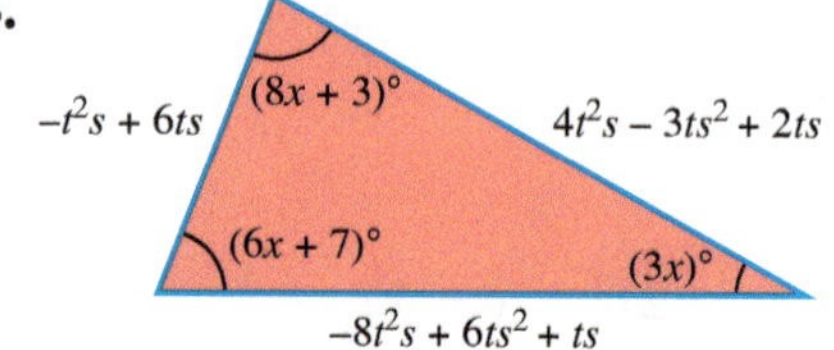

Relating Concepts (Exercises 95–98) For Individual or Group Work

The following binomial models the distance in feet that a car going approximately 68 mph *will skid in t seconds.*

$$100t - 13t^2$$

When we evaluate this binomial for a value of t, we obtain a value for distance. This illustrates the concept of a ***function****—for each input of a time, we obtain one and only one output for distance.*

Exercises 95–98 further illustrate the function concept with polynomials. ***Work them in order.***

95. Evaluate the given binomial

$$100t - 13t^2$$

for $t = 5$. Use the result to fill in the blanks:

In _______ seconds, the car will skid _______ feet.

96. If one "dog" year is estimated to be about seven "human" years, the monomial

$$7x$$

gives the dog's age in human years for x dog years. Evaluate this monomial for $x = 9$. Use the result to fill in the blanks:

If a dog is _______ in dog years, then it is _______ in human years.

97. If it costs \$15 plus \$2 per day to rent a chain saw, the binomial

$$2x + 15$$

gives the cost in dollars to rent the chain saw for x days. Evaluate this binomial for $x = 6$. Use the result to fill in the blanks:

If the saw is rented for _______ days, then the cost is _______ dollars.

98. If an object is projected upward under certain conditions, its height in feet is given by the trinomial

$$-16t^2 + 60t + 80,$$

where t is in seconds. Evaluate this trinomial for $t = 2.5$. Use the result to fill in the blanks:

If _______ seconds have elapsed, then the height of the object is _______ feet.

5.5 Multiplying Polynomials

OBJECTIVES

1. Multiply monomials.
2. Multiply a monomial and a polynomial.
3. Multiply two polynomials.
4. Multiply binomials using the FOIL method.

OBJECTIVE 1 Multiply monomials. Recall that we multiply monomials using the product rule for exponents.

EXAMPLE 1 Multiplying Monomials

Find each product.

(a) $8m^2(-9m)$

$= 8(-9) \cdot m^2m^1$ Commutative and associative properties

$= -72m^{2+1}$ Multiply; product rule

$= -72m^3$ Add.

(b) $4x^3y^2(2x^2y)$

$= 4(2) \cdot x^3x^2 \cdot y^2y^1$ Commutative and associative properties

$= 8x^{3+2}y^{2+1}$ Multiply; product rule

$= 8x^5y^3$ Add.

Work Problem 1 at the Side. ▶

OBJECTIVE 2 Multiply a monomial and a polynomial. We use the distributive property and multiplication of monomials.

EXAMPLE 2 Multiplying Monomials and Polynomials

Find each product.

(a) $4x^2(3x + 5)$ $\quad a(b+c) = ab + ac$

$= 4x^2(3x) + 4x^2(5)$ Distributive property

$= 12x^3 + 20x^2$ Multiply monomials.

(b) $-8m^3(4m^3 + 3m^2 + 2m - 1)$

$= -8m^3(4m^3) + (-8m^3)(3m^2)$
$+ (-8m^3)(2m) + (-8m^3)(-1)$ Distributive property

$= -32m^6 - 24m^5 - 16m^4 + 8m^3$ Multiply monomials.

Work Problem 2 at the Side. ▶

OBJECTIVE 3 Multiply two polynomials. To find the product of the polynomials $x^2 + 3x + 5$ and $x - 4$, think of $x - 4$ as a single quantity and use the distributive property as follows.

$(x^2 + 3x + 5)(x - 4)$

$= x^2(x - 4) + 3x(x - 4) + 5(x - 4)$ Distributive property

$= x^2(x) + x^2(-4) + 3x(x) + 3x(-4) + 5(x) + 5(-4)$ Distributive property again

$= x^3 - 4x^2 + 3x^2 - 12x + 5x - 20$ Multiply monomials.

$= x^3 - x^2 - 7x - 20$ Combine like terms.

Multiplying Polynomials

To multiply two polynomials, multiply each term of the first polynomial by each term of the second polynomial. Then combine like terms.

---

1 Find each product.

(a) $2x(3x)$

(b) $-6x(5x^3)$

(c) $10x^2(-8xy^2)$

(d) $6mn^3(12mn)$

2 Find each product.

GS **(a)** $5m^3(2m + 7)$

$= 5m^3(\_\_\_) + 5m^3(\_\_\_)$

$= \_\_\_\_\_\_$

(b) $2x^4(3x^2 + 2x - 5)$

(c) $-4y^2(3y^3 + 2y^2 - 4y + 8)$

Answers

1. (a) $6x^2$ (b) $-30x^4$
(c) $-80x^3y^2$ (d) $72m^2n^4$
2. (a) $2m$; 7; $10m^4 + 35m^3$
(b) $6x^6 + 4x^5 - 10x^4$
(c) $-12y^5 - 8y^4 + 16y^3 - 32y^2$

EXAMPLE 3 **Multiplying Two Polynomials**

Multiply $(m^2+5)(4m^3-2m^2+4m)$.

$(m^2+5)(4m^3-2m^2+4m)$ — Multiply each term of the first polynomial by each term of the second.

$= m^2(4m^3)+m^2(-2m^2)+m^2(4m)+5(4m^3)+5(-2m^2)+5(4m)$ — Distributive property

$= 4m^5-2m^4+4m^3+20m^3-10m^2+20m$ — Distributive property again

$= 4m^5-2m^4+24m^3-10m^2+20m$ — Combine like terms.

◀ **Work Problem 3 at the Side.**

EXAMPLE 4 **Multiplying Polynomials Vertically**

Multiply $(x^3+2x^2+4x+1)(3x+5)$ vertically.

$$\begin{array}{r} x^3+2x^2+4x+1 \\ \underline{3x+5} \end{array}$$

Write the polynomials vertically.

Begin by multiplying each of the terms in the top row by 5.

$$\begin{array}{r} x^3+2x^2+4x+1 \\ \underline{3x+5} \\ 5x^3+10x^2+20x+5 \end{array}$$

$5(x^3+2x^2+4x+1)$

Now multiply each term in the top row by $3x$. Then add like terms.

$$\begin{array}{r} x^3+2x^2+4x+1 \\ \underline{3x+5} \\ 5x^3+10x^2+20x+5 \\ \underline{3x^4+6x^3+12x^2+3x} \\ 3x^4+11x^3+22x^2+23x+5 \end{array}$$

This process is similar to multiplication of whole numbers.

Place *like* terms in columns so they can be added.

$3x(x^3+2x^2+4x+1)$

Add in columns.

◀ **Work Problem 4 at the Side.**

We can use a rectangle to model polynomial multiplication. For example, to find the product

$$(2x+1)(3x+2),$$

we label a rectangle with each term as shown below on the left. Then we write the product of each pair of monomials in the appropriate box as shown on the right.

	$3x$	2
$2x$		
1		

	$3x$	2
$2x$	$6x^2$	$4x$
1	$3x$	2

The product of the binomials is the sum of these four monomial products.

$$(2x+1)(3x+2)$$
$$= 6x^2+4x+3x+2$$
$$= 6x^2+7x+2 \quad \text{Combine like terms.}$$

◀ **Work Problem 5 at the Side.**

3 Multiply.

(a) $(m+3)(m^2-2m+1)$

(b) $(6p^2+2p-4)(3p^2-5)$

4 Multiply vertically.

$$\begin{array}{r} 3x^2+4x-5 \\ \underline{x+4} \end{array}$$

5 Use the rectangle method to find each product.

GS **(a)** $(4x+3)(x+2)$

	x	2
$4x$	$4x^2$	
3		

(b) $(x+5)(x^2+3x+1)$

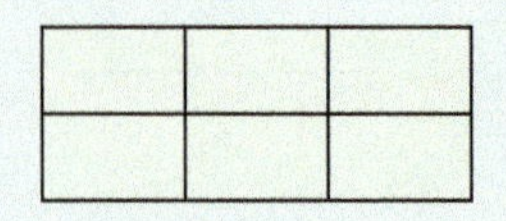

Answers

3. (a) m^3+m^2-5m+3
(b) $18p^4+6p^3-42p^2-10p+20$

4. $3x^3+16x^2+11x-20$

5. (a)

	x	2
$4x$	$4x^2$	$8x$
3	$3x$	6

$4x^2+11x+6$

(b) $x^3+8x^2+16x+5$

OBJECTIVE 4 Multiply binomials using the FOIL method. When multiplying binomials, the **FOIL method** reduces the rectangle method to a systematic approach without the rectangle. Consider this example.

$$(x + 3)(x + 5)$$

$= x(x + 5) + 3(x + 5)$ Distributive property

$= x(x) + x(5) + 3(x) + 3(5)$ Distributive property again

$= x^2 + 5x + 3x + 15$ Multiply.

$= x^2 + 8x + 15$ Combine like terms.

The letters of the word FOIL refer to the positions of the terms.

$(x + 3)(x + 5)$ Multiply the **First terms:** $x(x)$. **F**

$(x + 3)(x + 5)$ Multiply the **Outer terms:** $x(5)$. **O**
This is the **outer product.**

$(x + 3)(x + 5)$ Multiply the **Inner terms:** $3(x)$. **I**
This is the **inner product.**

$(x + 3)(x + 5)$ Multiply the **Last terms:** $3(5)$. **L**

We add the outer product, $5x$, and the inner product, $3x$, mentally so that the three terms of the answer can be written without extra steps.

$$(x + 3)(x + 5)$$
$$= x^2 + 8x + 15$$

FOIL Method for Multiplying Binomials

Step 1 Multiply the two **F**irst terms of the binomials to obtain the first term of the product.

Step 2 Find the **O**uter product and the **I**nner product and combine them (when possible) to obtain the middle term of the product.

Step 3 Multiply the two **L**ast terms of the binomials to obtain the last term of the product.

Step 4 Add the terms found in Steps 1–3.

Example:

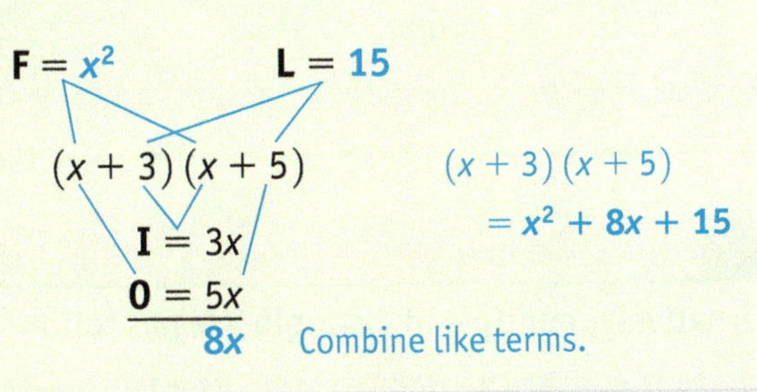

$(x + 3)(x + 5)$
$= x^2 + 8x + 15$

Work Problem 6 at the Side. ▶

6 For the product

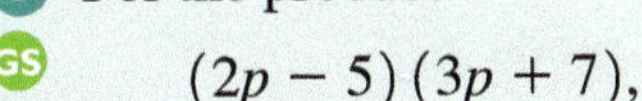

$$(2p - 5)(3p + 7),$$

find and simplify the following.

(a) Product of first terms

____(____)
= ____

(b) Outer product

____(____)
= ____

(c) Inner product

____(____)
= ____

(d) Product of last terms

____(____)
= ____

(e) Complete product in simplified form

____

Answers

6. (a) $2p$; $3p$; $6p^2$ **(b)** $2p$; 7; $14p$
(c) -5; $3p$; $-15p$ **(d)** -5; 7; -35
(e) $6p^2 - p - 35$

7 Use the FOIL method to find each product.

GS **(a)** $(m + 4)(m - 3)$

$= m(\_\_\_\_) + m(\_\_\_\_)$
$+ 4(\_\_\_\_) + 4(\_\_\_\_)$
$= \_\_\_\_\_\_\_\_\_\_\_\_$

(b) $(y + 7)(y + 2)$

(c) $(r - 8)(r - 5)$

8 Multiply.

$(4x - 3)(2y + 5)$

9 Find each product.

GS **(a)** $(6m + 5)(m - 4)$

$= 6m(\_\_\_\_) + 6m(\_\_\_\_)$
$+ 5(\_\_\_\_) + 5(\_\_\_\_)$
$= \_\_\_\_\_\_\_\_\_\_\_\_$

(b) $(3r + 2t)(3r + 4t)$

GS **(c)** $y^2(8y + 3)(2y + 1)$

$= y^2(\_\_\_\_\_\_\_\_\_\_)$
$= \_\_\_\_\_\_\_\_\_\_\_\_$

Answers

7. **(a)** m; -3; m; -3; $m^2 + m - 12$
(b) $y^2 + 9y + 14$
(c) $r^2 - 13r + 40$

8. $8xy + 20x - 6y - 15$

9. **(a)** m; -4; m; -4; $6m^2 - 19m - 20$
(b) $9r^2 + 18rt + 8t^2$
(c) $16y^2 + 14y + 3$; $16y^4 + 14y^3 + 3y^2$

EXAMPLE 5 Using the FOIL Method

Use the FOIL method to find the product $(x + 8)(x - 6)$.

Step 1 **F** Multiply the **first** terms: $x(x) = x^2$.

Step 2 **O** Find the **outer** product: $x(-6) = -6x$.
I Find the **inner** product: $8(x) = 8x$.
Combine. $-6x + 8x = 2x$

Step 3 **L** Multiply the **last** terms: $8(-6) = -48$.

Step 4 The product $(x + 8)(x - 6)$ is $x^2 + 2x - 48$. Add the terms found in Steps 1–3.

Shortcut:

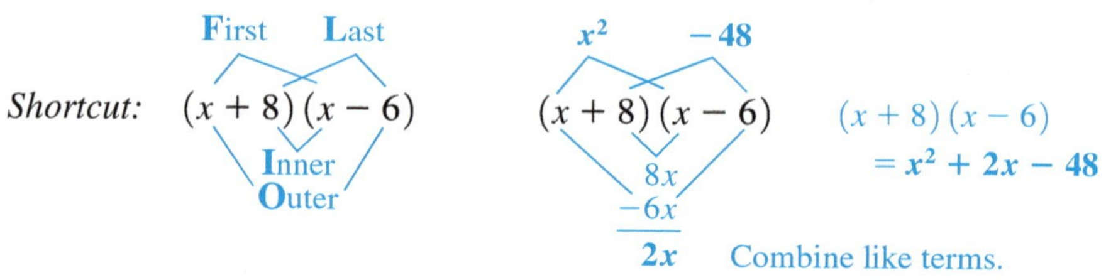

$(x + 8)(x - 6)$
$= x^2 + 2x - 48$

Combine like terms.

◀ Work Problem **7** at the Side.

EXAMPLE 6 Using the FOIL Method

Multiply $(9x - 2)(3y + 1)$.

First	$(9x - 2)(3y + 1)$	$27xy$
Outer	$(9x - 2)(3y + 1)$	$9x$
Inner	$(9x - 2)(3y + 1)$	$-6y$
Last	$(9x - 2)(3y + 1)$	-2

These unlike terms *cannot* be combined.

The product $(9x - 2)(3y + 1)$ is $27xy + 9x - 6y - 2$.

◀ Work Problem **8** at the Side.

EXAMPLE 7 Using the FOIL Method

Find each product.

(a) $(2k + 5y)(k + 3y)$

$= 2k(k) + 2k(3y) + 5y(k) + 5y(3y)$ FOIL method
$= 2k^2 + 6ky + 5ky + 15y^2$ Multiply.
$= 2k^2 + 11ky + 15y^2$ Combine like terms.

(b) $(7p + 2q)(3p - q)$

$= 21p^2 - 7pq + 6pq - 2q^2$
$= 21p^2 - pq - 2q^2$

(c) $2x^2(x - 3)(3x + 4)$

$= 2x^2(3x^2 - 5x - 12)$
$= 6x^4 - 10x^3 - 24x^2$

◀ Work Problem **9** at the Side.

Note

Alternatively, multiply in **Example 7(c)** as follows.

$2x^2(x - 3)(3x + 4)$ Multiply $2x^2$ and $x - 3$ first.
$= (2x^3 - 6x^2)(3x + 4)$ Multiply that product and $3x + 4$.
$= 6x^4 - 10x^3 - 24x^2$ The same answer results.

5.5 Exercises

FOR EXTRA HELP — *Go to MyMathLab for worked-out, step-by-step solutions to exercises enclosed in a square and video solutions to exercises.*

CONCEPT CHECK *Match each product in Column I with the correct polynomial in Column II.*

	I	II
1.	**(a)** $5x^3(6x^7)$	**A.** $125x^{21}$
	(b) $-5x^7(6x^3)$	**B.** $30x^{10}$
	(c) $(5x^7)^3$	**C.** $-216x^9$
	(d) $(-6x^3)^3$	**D.** $-30x^{10}$

	I	II
2.	**(a)** $(x-5)(x+4)$	**A.** $x^2+9x+20$
	(b) $(x+5)(x+4)$	**B.** $x^2-9x+20$
	(c) $(x-5)(x-4)$	**C.** x^2-x-20
	(d) $(x+5)(x-4)$	**D.** x^2+x-20

CONCEPT CHECK *Complete each statement.*

3. In multiplying a monomial by a polynomial, such as in $4x(3x^2+7x^3)=4x(3x^2)+4x(7x^3)$, the first property that is used is the ______________ property.

4. The FOIL method can only be used to multiply two polynomials when both polynomials are (*monomials / binomials / trinomials*).

Find each product. ***See Example 1.***

5. $4x(3x^2)$

6. $6x(7x^4)$

7. $5y^4(-3y^7)$

8. $10p^2(-5p^3)$

9. $-15a^4(-2a^5)$

10. $-3m^6(-5m^4)$

11. $-6m^3(3n^2)$

12. $-2s^2(9r^3)$

13. $4a(5ab)$

14. $7y(3xy)$

15. $2m^2n(-6mn^3)$

16. $8xy^3(-9xy^2)$

Find each product. ***See Example 2.***

17. $2m(3m+2)$

18. $4x(5x+3)$

19. $2y^5(5y^4+2y+3)$

20. $2m^4(3m^2+5m+6)$

21. $2y^3(3y^3+2y+1)$

22. $2m^4(3m^2+5m+1)$

23. $\frac{3}{4}p(8-6p+12p^3)$

24. $\frac{4}{3}x(3+2x+5x^3)$

25. $-4r^3(-7r^2+8r-9)$

26. $-9a^5(-3a^6-2a^4+8a^2)$

27. $3a^2(2a^2-4ab+5b^2)$

28. $4z^3(8z^2+5zy-3y^2)$

29. $7m^3n^2(3m^2+2mn-n^3)$
GS $= 7m^3n^2(\_\_\_\_) + 7m^3n^2(\_\_\_\_) + 7m^3n^2(\_\_\_\_)$
$= \_\_\_\_\_\_\_\_\_\_\_\_$

30. $2p^2q(3p^2q^2-5p+2q^2)$
GS $= \_\_\_\_(3p^2q^2) + \_\_\_\_(-5p) + \_\_\_\_(2q^2)$
$= \_\_\_\_\_\_\_\_\_\_\_\_$

Find each product. **See Examples 3 and 4.**

31. $(6x + 1)(2x^2 + 4x + 1)$

32. $(9a + 2)(9a^2 + a + 1)$

33. $(2r - 1)(3r^2 + 4r - 4)$

34. $(9y - 2)(8y^2 - 6y + 1)$

35. $(4m + 3)(5m^3 - 4m^2 + m - 5)$

36. $(y + 4)(3y^3 - 2y^2 + y + 3)$

37. $(5x^2 + 2x + 1)(x^2 - 3x + 5)$

38. $(2m^2 + m - 3)(m^2 - 4m + 5)$

39. $(6x^4 - 4x^2 + 8x)\left(\frac{1}{2}x + 3\right)$

40. $(8y^6 + 4y^4 - 12y^2)\left(\frac{3}{4}y^2 + 2\right)$

Find each product using the rectangle method shown in the text. Determine the individual terms that should appear on the blanks or in the rectangles, and then give the final product.

41. $(x + 3)(x + 4)$

	x	4
x		
3		

Product: ______________

42. $(x + 5)(x + 2)$

	___	___
x		
5		

Product: ______________

43. $(2x + 1)(x^2 + 3x + 2)$

	x^2	$3x$	2
$2x$			
1			

Product: ______________

44. $(x + 4)(3x^2 + 2x + 1)$

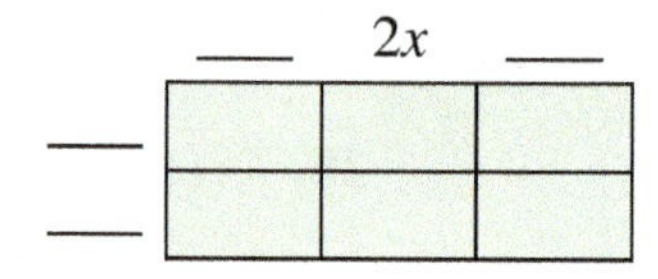

Product: ______________

Find each product. **See Examples 5–7.**

45. $(m + 7)(m + 5)$

46. $(x + 4)(x + 7)$

47. $(n - 2)(n + 3)$

48. $(r - 6)(r + 8)$

49. $(4r + 1)(2r - 3)$

50. $(5x + 2)(2x - 7)$

51. $(3x + 2)(3x - 2)$

52. $(7x + 3)(7x - 3)$

53. $(3q + 1)(3q + 1)$

54. $(4w + 7)(4w + 7)$

55. $(4x + 3)(2y - 1)$

56. $(5x + 7)(3y - 8)$

57. $(3x + 2y)(5x - 3y)$

58. $(5a + 3b)(5a - 4b)$

59. $(3t + 4s)(2t + 5s)$

60. $(8v + 5w)(2v + 3w)$

61. $(-0.3t + 0.4)(t + 0.6)$

62. $(-0.5x + 0.9)(x - 0.2)$

63. $\left(x - \frac{2}{3}\right)\left(x + \frac{1}{4}\right)$

64. $\left(y + \frac{3}{5}\right)\left(y - \frac{1}{2}\right)$

65. $\left(-\frac{5}{4} + 2r\right)\left(-\frac{3}{4} - r\right)$

66. $\left(-\frac{8}{3} + 3k\right)\left(-\frac{2}{3} - k\right)$

67. $x(2x - 5)(x + 3)$

68. $m(4m - 1)(2m + 3)$

69. $3y^3(2y + 3)(y - 5)$

70. $5t^4(t + 3)(3t - 1)$

71. $-8r^3(5r^2 + 2)(5r^2 - 2)$

72. $-5t^4(2t^4 + 1)(2t^4 - 1)$

Find polynomials that represent, in appropriate units, ***(a)*** *the area and* ***(b)*** *the perimeter of each square or rectangle. (If necessary, refer to the formulas at the back of this text.)*

73.

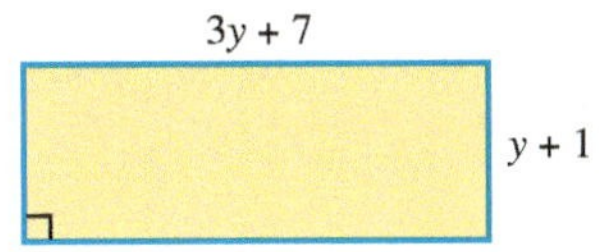

74.

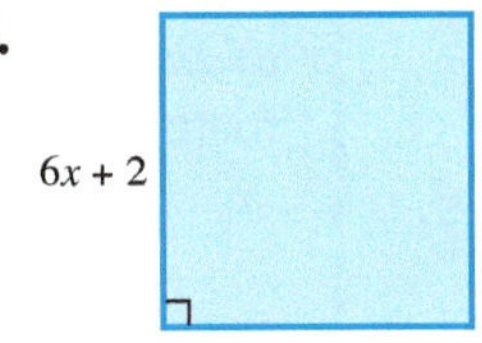

Find a polynomial that represents, in appropriate units, the area of each shaded region. (If necessary, refer to the formulas at the back of this text.)

75.

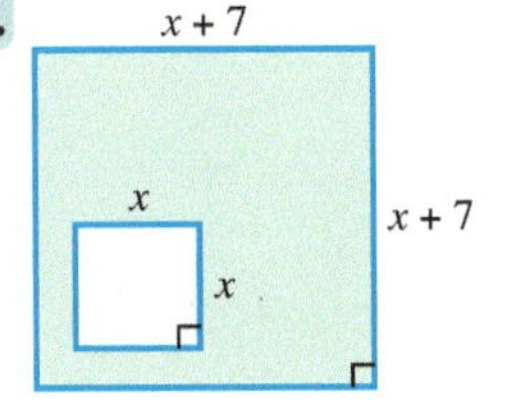

76.

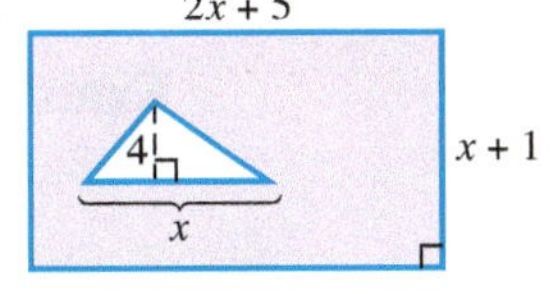

Relating Concepts (Exercises 77–82) For Individual or Group Work

Work Exercises 77–82 in order. *(All units are in feet.)*

77. Find a polynomial that represents the area, in square feet, of the rectangle.

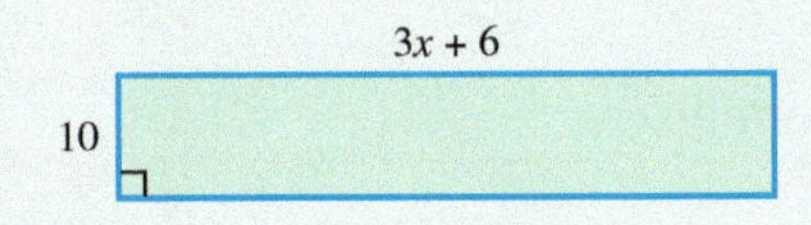

78. Suppose we know that the area of the rectangle is 600 ft^2. Use this information and the polynomial from **Exercise 77** to write an equation in x, and solve it.

79. Refer to **Exercise 78.** What are the dimensions of the rectangle?

80. Use the result of **Exercise 79** to find the perimeter of the rectangle.

81. Suppose the rectangle represents a lawn and it costs \$1.50 per square foot for sod. How much will it cost to sod the entire lawn?

82. Again, suppose the rectangle represents a lawn and it costs \$20.50 per linear foot for fencing. How much will it cost to fence the entire lawn?

5.6 Special Products

OBJECTIVES

1. Square binomials.
2. Find the product of the sum and difference of two terms.
3. Find greater powers of binomials.

OBJECTIVE 1 Square binomials.

EXAMPLE 1 Squaring a Binomial

Find $(m + 3)^2$.

$(m+3)^2$ means $(m+3)(m+3)$.

$$(m + 3)(m + 3)$$
$$= m^2 + m(3) + 3(m) + 9 \quad \text{FOIL method}$$
$$= m^2 + 6m + 9 \quad \text{Combine like terms.}$$

Compare the result, $m^2 + 6m + 9$, to the binomial $m + 3$. The trinomial result includes the squares of the first and last terms of the binomial.

$$m^2 = m^2 \quad \text{and} \quad 3^2 = 9$$

The middle term of the trinomial, 6m, is twice the product of the two terms of the binomial, m and 3. This is true because when we used the FOIL method above, the outer and inner products were $m(3)$ and $3(m)$, and

$$m(3) + 3(m) \quad \text{equals} \quad 2(m)(3).$$

Thus, $(m + 3)^2 = m^2 + 6m + 9.$

(m^2: Square m.; $6m$: $2(m)(3)$; 9: Square 3.)

◀ Work Problem 1 at the Side.

1 Consider the binomial $x + 4$.

GS (a) What is the first term of the binomial? ____

Square it. ____

(b) What is the last term of the binomial? ____

Square it. ____

(c) Find twice the product of the two terms of the binomial.

$2(\_\_\_)(\_\_\_) = \_\_\_$

(d) Use the results of parts (a)–(c) to find $(x + 4)^2$.

____

Example 1 suggests the following rules.

Square of a Binomial

The square of a binomial is a *trinomial* consisting of

the square of the first term + twice the product of the two terms + the square of the last term.

For x and y, the following hold true.

$$(x + y)^2 = x^2 + 2xy + y^2$$
$$(x - y)^2 = x^2 - 2xy + y^2$$

EXAMPLE 2 Squaring Binomials

Square each binomial.

$$(x - y)^2 = x^2 - 2 \cdot x \cdot y + y^2$$

(a) $(t - 8)^2 = t^2 - 2(t)(8) + 8^2$

$$= t^2 - 16t + 64$$

Continued on Next Page

Answers

1. (a) x; x^2 (b) 4; 16 (c) x; 4; $8x$
(d) $x^2 + 8x + 16$

(b) $(5z - 1)^2$

$= (5z)^2 - 2(5z)(1) + 1^2$ $\quad (x - y)^2 = x^2 - 2xy + y^2$

$= 5^2z^2 - 10z + 1$ $\quad (5z)^2 = 5^2z^2 = 25z^2$ by power rule (b).

$= 25z^2 - 10z + 1$

Be careful to square $5z$ correctly.

(c) $(3b + 5r)^2$

$= (3b)^2 + 2(3b)(5r) + (5r)^2$ $\quad (x + y)^2 = x^2 + 2xy + y^2$

$= 9b^2 + 30br + 25r^2$ $\quad (3b)^2 = 3^2b^2 = 9b^2$ and $(5r)^2 = 5^2r^2 = 25r^2$

(d) $(2a - 9x)^2$

$= (2a)^2 - 2(2a)(9x) + (9x)^2$ $\quad (x - y)^2 = x^2 - 2xy + y^2$

$= 4a^2 - 36ax + 81x^2$ $\quad (2a)^2 = 2^2a^2 = 4a^2$ and $(9x)^2 = 9^2x^2 = 81x^2$

(e) $\left(4m + \frac{1}{2}\right)^2$

$= (4m)^2 + 2(4m)\left(\frac{1}{2}\right) + \left(\frac{1}{2}\right)^2$ $\quad (x + y)^2 = x^2 + 2xy + y^2$

$= 16m^2 + 4m + \frac{1}{4}$ $\quad (4m)^2 = 4^2m^2 = 16m^2$ and $\left(\frac{1}{2}\right)^2 = \frac{1}{4}$

(f) $x(4x - 3)^2$

Remember the middle term, $2(4x)(3)$.

$= x(16x^2 - 24x + 9)$ $\quad$ Square the binomial.

$= 16x^3 - 24x^2 + 9x$ $\quad$ Distributive property

Work Problem 2 at the Side. ▶

In the square of a sum, all of the terms are positive. (See Examples 2(c) and (e).) In the square of a difference, the middle term is negative. (See Examples 2(a), (b), and (d).)

! CAUTION

A common error in squaring a binomial is to forget the middle term of the product. In general, remember the following.

$$(x + y)^2 = x^2 + 2xy + y^2, \quad \textit{not} \quad x^2 + y^2.$$

$$(x - y)^2 = x^2 - 2xy + y^2, \quad \textit{not} \quad x^2 - y^2.$$

OBJECTIVE 2 Find the product of the sum and difference of two terms. In binomial products of the form $(x + y)(x - y)$, one binomial is the sum of two terms, and the other is the difference of the *same* two terms. Consider the following.

$(x + 2)(x - 2)$

$= x^2 - 2x + 2x - 4$ $\quad$ FOIL method

$= x^2 - 4$ $\quad$ Combine like terms.

Thus, the product of $x + y$ and $x - y$ is a **difference of two squares.**

---

2 Square each binomial.

GS **(a)** $(t + 6)^2$

$= (\_\_\_)^2 + 2(\_\_\_)(\_\_\_) + (\_\_\_)^2$

$= \_\_\_\_\_\_\_\_$

(b) $(2m - p)^2$

GS **(c)** $(4p + 3q)^2$

$= (\_\_\_)^2 + \_\_\_(\_\_\_)3q + (\_\_\_)^2$

$= \_\_\_\_\_\_\_\_$

(d) $(5r - 6s)^2$

(e) $\left(3k - \frac{1}{2}\right)^2$

(f) $x(2x + 7)^2$

Answers

2. **(a)** t; t; 6; 6; $t^2 + 12t + 36$
(b) $4m^2 - 4mp + p^2$
(c) $4p$; 2; $4p$; $3q$; $16p^2 + 24pq + 9q^2$
(d) $25r^2 - 60rs + 36s^2$
(e) $9k^2 - 3k + \frac{1}{4}$
(f) $4x^3 + 28x^2 + 49x$

3 Find each product.

GS **(a)** $(y+3)(y-3)$

$= (\_\_\_\_)^2 - (\_\_\_\_)^2$

$= \_\_\_\_\_\_\_\_$

(b) $(8-x)(8+x)$

(c) $2x^2(x+5)(x-5)$

Answers

3. **(a)** y; 3; y^2-9
(b) $64-x^2$ **(c)** $2x^4-50x^2$

Product of a Sum and Difference of Two Terms

The product of a sum and difference of two terms is a *binomial* consisting of

the square of the first term − the square of the second term.

For x and y, the following holds true.

$$(x+y)(x-y)=x^2-y^2$$

Note

The expressions $x+y$ and $x-y$, the sum and difference of the *same* two terms, are **conjugates.**

Example: $x+2$ and $x-2$ are conjugates.

EXAMPLE 3 **Finding the Product of a Sum and Difference of Two Terms**

Find each product.

(a) $(x+4)(x-4)$

$= x^2-4^2$ $\quad (x+y)(x-y)=x^2-y^2$

$= x^2-16$ $\quad$ Square 4.

(b) $(10-w)(10+w)$

$= (10+w)(10-w)$ $\quad$ Commutative property

$= 10^2-w^2$ $\quad (x+y)(x-y)=x^2-y^2$

$= 100-w^2$ $\quad$ Square 10.

(c) $x(x+2)(x-2)$

$= x(x^2-4)$ $\quad$ Find the product of the sum and difference of two terms.

$= x^3-4x$ $\quad$ Distributive property

◀ Work Problem 3 at the Side.

EXAMPLE 4 **Finding the Product of a Sum and Difference of Two Terms**

Find each product.

$(x \;\; + y) \;\; (x \;\; - y)$

$\downarrow \quad \downarrow \quad \downarrow \quad \downarrow$

(a) $(5m+3)(5m-3)$

$= (5m)^2-3^2$ $\quad (x+y)(x-y)=x^2-y^2$

$= 25m^2-9$ $\quad$ Apply the exponents.

Be careful to square $5m$ correctly.

(b) $(4x+y)(4x-y)$

$= (4x)^2-y^2$

$= 16x^2-y^2$ $\quad (4x)^2=4^2x^2=16x^2$

Continued on Next Page

(c) $\left(z - \frac{1}{4}\right)\left(z + \frac{1}{4}\right)$

$= z^2 - \left(\frac{1}{4}\right)^2$ $\quad (x - y)(x + y) = (x + y)(x - y) = x^2 - y^2$

$= z^2 - \frac{1}{16}$

(d) $p(2p + 1)(2p - 1)$

$= p(4p^2 - 1)$

$= 4p^3 - p$ $\quad$ Distributive property

(e) $-3(x + y^2)(x - y^2)$

$= -3(x^2 - y^4)$

$= -3x^2 + 3y^4$

Work Problem 4 at the Side. ▶

OBJECTIVE 3 **Find greater powers of binomials.** The methods used in the previous section and this section can be combined to find greater powers of binomials.

EXAMPLE 5 **Finding Greater Powers of Binomials**

Find each product.

(a) $(x + 5)^3$

$= (x + 5)(x + 5)^2$ $\quad a^3 = a \cdot a^2$

$= (x + 5)(x^2 + 10x + 25)$ $\quad$ Square the binomial.

$= x^3 + 10x^2 + 25x + 5x^2 + 50x + 125$ $\quad$ Multiply polynomials.

$= x^3 + 15x^2 + 75x + 125$ $\quad$ Combine like terms.

(b) $(2y - 3)^4$

$= (2y - 3)^2(2y - 3)^2$ $\quad a^4 = a^2 \cdot a^2$

$= (4y^2 - 12y + 9)(4y^2 - 12y + 9)$ $\quad$ Square each binomial.

$= 16y^4 - 48y^3 + 36y^2 - 48y^3 + 144y^2 - 108y + 36y^2 - 108y + 81$ $\quad$ Multiply polynomials.

$= 16y^4 - 96y^3 + 216y^2 - 216y + 81$ $\quad$ Combine like terms.

(c) $-2r(r + 2)^3$

$= -2r(r + 2)(r + 2)^2$ $\quad a^3 = a \cdot a^2$

$= -2r(r + 2)(r^2 + 4r + 4)$ $\quad$ Square the binomial.

$= -2r(r^3 + 4r^2 + 4r + 2r^2 + 8r + 8)$ $\quad$ Multiply polynomials.

$= -2r(r^3 + 6r^2 + 12r + 8)$ $\quad$ Combine like terms.

$= -2r^4 - 12r^3 - 24r^2 - 16r$ $\quad$ Distributive property

Work Problem 5 at the Side. ▶

---

4 Find each product.

(a) $(10m + 7)(10m - 7)$

GS **(b)** $(7p + 2q)(7p - 2q)$

$= (\_\_\_\_)^2 - (\_\_\_\_)^2$

$= \_\_\_\_\_\_\_\_$

(c) $\left(3r - \frac{1}{2}\right)\left(3r + \frac{1}{2}\right)$

(d) $-7(t^2 - q)(t^2 + q)$

5 Find each product.

GS **(a)** $(m + 3)^3$

$= (\_\_\_\_)(\_\_\_\_)^2$

$= (m + 3)(\_\_\_\_\_\_)$

$= \_\_\_\_\_\_\_\_$

(b) $(3k - 2)^4$

(c) $-3x(x - 4)^3$

Answers

4. (a) $100m^2 - 49$ **(b)** $7p$; $2q$; $49p^2 - 4q^2$

(c) $9r^2 - \frac{1}{4}$ **(d)** $-7t^4 + 7q^2$

5. (a) $m + 3$; $m + 3$; $m^2 + 6m + 9$; $m^3 + 9m^2 + 27m + 27$

(b) $81k^4 - 216k^3 + 216k^2 - 96k + 16$

(c) $-3x^4 + 36x^3 - 144x^2 + 192x$

5.6 Exercises

FOR EXTRA HELP — *Go to MyMathLab for worked-out, step-by-step solutions to exercises enclosed in a square and video solutions to ▶ exercises.*

CONCEPT CHECK *Fill in each blank with the correct response.*

1. The square of a binomial is a trinomial consisting of the _______ of the first term + _______ the _______ of the two terms + the _______ of the last term.

2. The product of a sum and difference of two terms is the _______ of the _______ of the two terms.

3. Consider the square of the binomial $2x + 3$:

$$(2x + 3)^2.$$

(a) What is the first term of the binomial? Square it.

(b) Find twice the product of the two terms of the binomial: $2(\_\_\_\_)(\_\_\_\_) = \_\_\_\_$.

(c) What is the last term of the binomial? Square it.

(d) Use the results of parts (a)–(c) to find $(2x + 3)^2$.

4. Repeat **Exercise 3** for the binomial square $(3x - 2)^2$.

Square each binomial. ***See Examples 1 and 2.***

5. $(p + 2)^2$

6. $(r + 5)^2$

7. $(z - 5)^2$

8. $(x - 3)^2$

9. $\left(x - \frac{3}{4}\right)^2$

10. $\left(y + \frac{5}{8}\right)^2$

11. $(v + 0.4)^2$

12. $(w - 0.9)^2$

13. $(4x - 3)^2$

14. $(9y - 4)^2$

15. $(10z + 6)^2$

16. $(5y + 2)^2$

17. $(x + 2y)^2$

18. $(p + 3m)^2$

19. $(2p + 5q)^2$

20. $(8a + 3b)^2$

21. $(4a - 5b)^2$

22. $(9y - 4z)^2$

23. $(0.8t + 0.7s)^2$

24. $(0.7z - 0.3w)^2$

25. $\left(6m - \frac{4}{5}n\right)^2$

26. $\left(5x + \frac{2}{5}y\right)^2$

27. $t(3t - 1)^2$

28. $x(2x + 5)^2$

29. $3t(4t + 1)^2$

30. $2x(7x - 2)^2$

31. $-(4r - 2)^2$

32. $-(3y - 8)^2$

33. Consider the product of the conjugates $(7x + 3y)$ and $(7x - 3y)$:

$$(7x + 3y)(7x - 3y).$$

(a) What is the first term of each binomial factor? Square it.

(b) What is the product of the outer terms? The inner terms? Add them.

(c) What are the last terms of the binomial factors? Multiply them.

(d) Use the results of parts (a)–(c) to find $(7x + 3y)(7x - 3y)$.

34. Repeat **Exercise 33** for the product $(5x + 7y)(5x - 7y)$.

Find each product. ***See Examples 3 and 4.***

35. $(k + 5)(k - 5)$

36. $(x + 8)(x - 8)$

37. $\left(r - \frac{3}{4}\right)\left(r + \frac{3}{4}\right)$

38. $\left(q - \frac{7}{8}\right)\left(q + \frac{7}{8}\right)$

39. $(s + 2.5)(s - 2.5)$

40. $(t + 1.4)(t - 1.4)$

41. $(2w + 5)(2w - 5)$

42. $(3z + 8)(3z - 8)$

43. $(3x + 4y)(3x - 4y)$

44. $(5y + 3x)(5y - 3x)$

45. $(10x + 3y)(10x - 3y)$

46. $(13r + 2z)(13r - 2z)$

47. $\left(7x + \frac{3}{7}\right)\left(7x - \frac{3}{7}\right)$

48. $\left(9y + \frac{2}{3}\right)\left(9y - \frac{2}{3}\right)$

49. $(2x^2 - 5)(2x^2 + 5)$

50. $(9y^2 - 2)(9y^2 + 2)$

51. $q(5q - 1)(5q + 1)$

52. $p(3p + 7)(3p - 7)$

53. $-5(a - b^3)(a + b^3)$

54. $-6(r - s^4)(r + s^4)$

Find each product. ***See Example 5.***

55. $(x + 1)^3$

56. $(y + 2)^3$

57. $(m - 5)^3$

58. $(x - 7)^3$

59. $(2a + 1)^3$

60. $(3m + 1)^3$

61. $(4x - 1)^4$

62. $(2x - 1)^4$

63. $(3r - 2t)^4$

64. $(2z + 5y)^4$

65. $3x^2(x - 3)^3$

66. $4p^3(p + 4)^3$

67. $-8x^2y(x + y)^4$

68. $-5uv^2(u - v)^4$

Find a polynomial that represents, in appropriate units, the area of each figure. (In Exercise 73, leave π in the answer. If necessary, refer to the formulas at the back of this text.)

69.

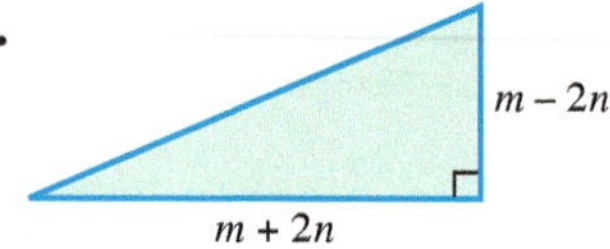

70.

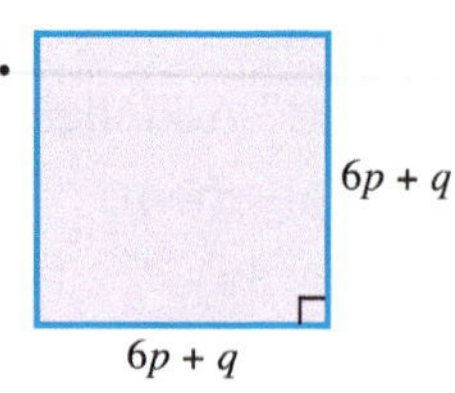

71.

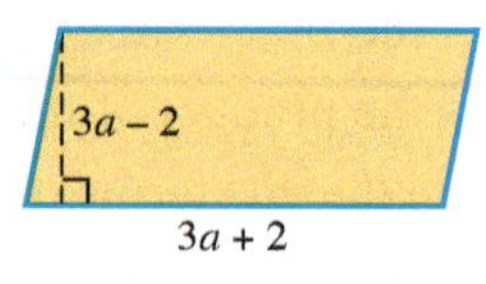

72.

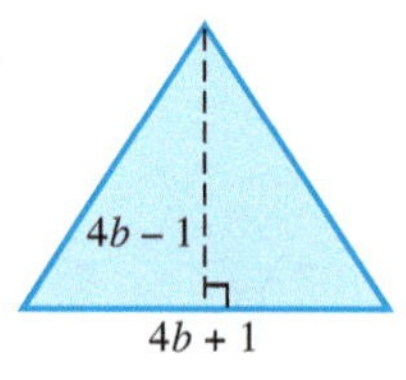

73.

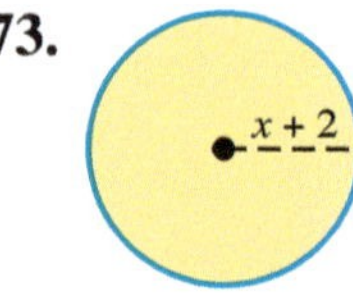

74.

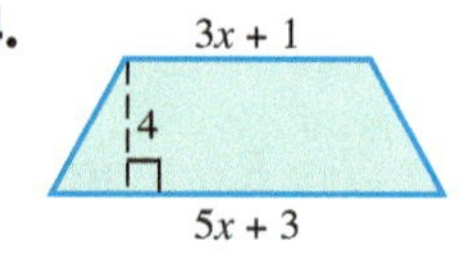

Refer to the figure shown here.

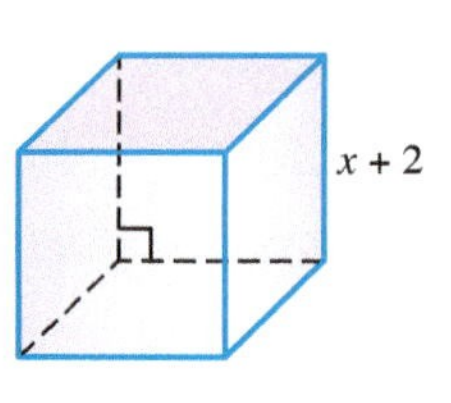

75. Find a polynomial that represents the volume of the cube (in cubic units).

76. If the value of x is 6, what is the volume of the cube (in cubic units)?

Relating Concepts (Exercises 77–86) For Individual or Group Work

*Use the figure and **work Exercises 77–82 in order,** to justify the special product*

$$(x + y)^2 = x^2 + 2xy + y^2.$$

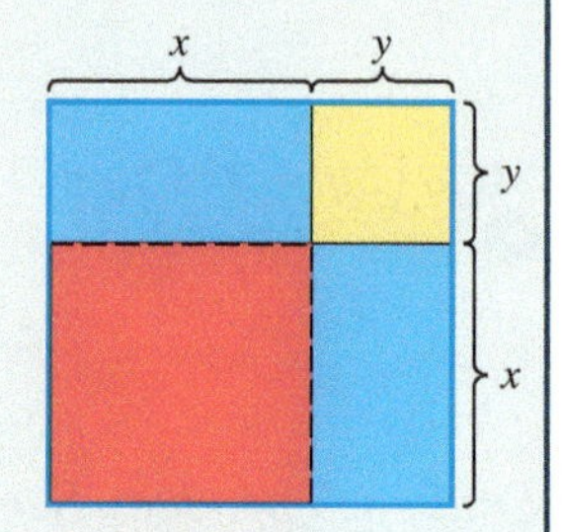

77. Express the area of the large square as the square of a binomial.

78. Give the monomial that represents the area of the red square.

79. Give the monomial that represents the sum of the areas of the blue rectangles.

80. Give the monomial that represents the area of the yellow square.

81. What is the sum of the monomials obtained in **Exercises 78–80?**

82. Explain why the binomial square found in **Exercise 77** must equal the polynomial found in **Exercise 81.**

*To apply the above special product to a purely numerical problem, **work Exercises 83–86 in order.***

83. Using either traditional paper-and-pencil methods or a calculator, evaluate 35^2.

84. The number 35 can be written as $30 + 5$. Therefore, $35^2 = (30 + 5)^2$. Use the special product for squaring a binomial with $x = 30$ and $y = 5$ to write an expression for $(30 + 5)^2$. Do not simplify yet.

85. Use the rules for order of operations to simplify the expression found in **Exercise 84.**

86. Compare the answers to **Exercises 83 and 85.**

5.7 Dividing a Polynomial by a Monomial

OBJECTIVE 1 Divide a polynomial by a monomial. We add two fractions with a common denominator as follows.

$$\frac{a}{c} + \frac{b}{c} = \frac{a + b}{c}$$

In reverse, this statement gives a rule for dividing a polynomial by a monomial.

Dividing a Polynomial by a Monomial

To divide a polynomial by a monomial, divide each term of the polynomial by the monomial.

$$\frac{a + b}{c} = \frac{a}{c} + \frac{b}{c} \quad (\text{where } c \neq 0)$$

Examples: $\frac{2 + 5}{3} = \frac{2}{3} + \frac{5}{3}$ and $\frac{x + 3z}{2y} = \frac{x}{2y} + \frac{3z}{2y} \quad (y \neq 0)$

The parts of a division problem are named as follows.

$$\text{Dividend} \rightarrow \frac{12x^2 + 6x}{6x} = 2x + 1 \leftarrow \text{Quotient}$$

Divisor → $6x$

EXAMPLE 1 Dividing a Polynomial by a Monomial

Divide $5m^5 - 10m^3$ by $5m^2$.

$$\frac{5m^5 - 10m^3}{5m^2}$$ A fraction bar means division.

$$= \frac{5m^5}{5m^2} - \frac{10m^3}{5m^2}$$ Use the preceding rule, with + replaced by −.

$$= m^3 - 2m$$ Quotient rule

CHECK Multiply. $5m^2 \cdot (m^3 - 2m) = 5m^5 - 10m^3$ ✓

Divisor: $5m^2$; Quotient: $m^3 - 2m$; Original polynomial (Dividend): $5m^5 - 10m^3$

Because division by 0 is undefined, the quotient $\frac{5m^5 - 10m^3}{5m^2}$ is undefined if $m = 0$. From now on, we assume that no denominators are 0.

Work Problem 1 at the Side. ▶

EXAMPLE 2 Dividing a Polynomial by a Monomial

Divide.

$$\frac{16a^5 - 12a^4 + 8a^2}{4a^3}$$

Be careful simplifying this expression.

$$= \frac{16a^5}{4a^3} - \frac{12a^4}{4a^3} + \frac{8a^2}{4a^3}$$ Divide each term by $4a^3$.

$$= 4a^2 - 3a + \frac{2}{a}$$ $\frac{8a^2}{4a^3} = \frac{8}{4}a^{2-3} = 2a^{-1} = 2\left(\frac{1}{a}\right) = \frac{2}{a}$

Continued on Next Page

OBJECTIVE

1 Divide a polynomial by a monomial.

1 Divide.

GS (a) $\frac{6p^4 + 18p^7}{3p^2}$

$= \frac{\_\_\_}{3p^2} + \frac{\_\_\_}{3p^2}$

$= \_\_\_\_$

(b) $\frac{12m^6 + 18m^5 + 30m^4}{6m^2}$

(c) $(18r^7 - 9r^2) \div (3r)$

Answers

1. (a) $6p^4$; $18p^7$; $2p^2 + 6p^5$
(b) $2m^4 + 3m^3 + 5m^2$
(c) $6r^6 - 3r$

The quotient $4a^2 - 3a + \frac{2}{a}$ is *not* a polynomial because of the expression $\frac{2}{a}$, which has a variable in the denominator. While the sum, difference, and product of two polynomials are always polynomials, the quotient of two polynomials may not be.

CHECK $4a^3\left(4a^2 - 3a + \frac{2}{a}\right)$ — Divisor × Quotient should equal Dividend.

$= 4a^3(4a^2) + 4a^3(-3a) + 4a^3\left(\frac{2}{a}\right)$ — Distributive property

$= 16a^5 - 12a^4 + 8a^2$ ✓ — Dividend

◀ Work Problem 2 at the Side.

EXAMPLE 3 Dividing a Polynomial by a Monomial

Divide $-7x^3 + 12x^4 - 4x$ by $-4x$.

Write the dividend polynomial in descending powers.

$$\frac{12x^4 - 7x^3 - 4x}{-4x}$$

Write in descending powers before dividing.

$$= \frac{12x^4}{-4x} - \frac{7x^3}{-4x} - \frac{4x}{-4x}$$ Divide each term by $-4x$.

$$= -3x^3 - \frac{7x^2}{-4} - (-1)$$ Quotient rule

$$= -3x^3 + \frac{7x^2}{4} + 1$$

Be careful with signs, and be sure to include 1 in the answer.

CHECK $-4x\left(-3x^3 + \frac{7x^2}{4} + 1\right)$ — Divisor × Quotient should equal Dividend.

$= -4x(-3x^3) - 4x\left(\frac{7x^2}{4}\right) - 4x(1)$ — Distributive property

$= 12x^4 - 7x^3 - 4x$ ✓ — Dividend

◀ Work Problem 3 at the Side.

EXAMPLE 4 Dividing a Polynomial by a Monomial

Divide $180x^4y^{10} - 150x^3y^8 + 120x^2y^6 - 90xy^4 + 100y$ by $-30xy^2$.

Divide each term of the polynomial by $-30xy^2$.

$$\frac{180x^4y^{10} - 150x^3y^8 + 120x^2y^6 - 90xy^4 + 100y}{-30xy^2}$$

$$= \frac{180x^4y^{10}}{-30xy^2} - \frac{150x^3y^8}{-30xy^2} + \frac{120x^2y^6}{-30xy^2} - \frac{90xy^4}{-30xy^2} + \frac{100y}{-30xy^2}$$

$$= -6x^3y^8 + 5x^2y^6 - 4xy^4 + 3y^2 - \frac{10}{3xy}$$

Check by multiplying the divisor by the quotient.

◀ Work Problem 4 at the Side.

---

2 Divide.

GS (a) $\frac{20x^4 - 25x^3 + 5x}{5x^2}$

$= \frac{20x^4}{\_\_\_} - \frac{\_\_\_}{5x^2} + \frac{5x}{\_\_\_}$

$= \_\_\_\_\_\_$

(b) $\frac{50m^4 - 30m^3 + 20m}{10m^3}$

3 Divide.

(a) $\frac{-8p^4 - 6p^3 - 12p^5}{-3p^3}$

(b) $\frac{-9y^6 + 8y^7 - 11y - 4}{y^2}$

4 Divide.

$\frac{45x^4y^3 + 30x^3y^2 - 60x^2y}{15x^2y^2}$

Answers

2. (a) $5x^2$; $25x^3$; $5x^2$; $4x^2 - 5x + \frac{1}{x}$
   (b) $5m - 3 + \frac{2}{m^2}$

3. (a) $4p^2 + \frac{8p}{3} + 2$
   (b) $8y^5 - 9y^4 - \frac{11}{y} - \frac{4}{y^2}$

4. $3x^2y + 2x - \frac{4}{y}$

5.7 Exercises

FOR EXTRA HELP Go to MyMathLab for worked-out, step-by-step solutions to exercises enclosed in a square and video solutions to ▶ exercises.

CONCEPT CHECK *Complete each statement.*

1. In the division $\frac{6x^2+8}{2} = 3x^2+4$, _______ is the dividend, _______ is the divisor, and _______ is the quotient.

2. To check the division shown in **Exercise 1,** multiply _______ by _______ and show that the product is _______.

3. The expression $5x^2 - 3x + 6 + \frac{2}{x}$ (*is / is not*) a polynomial.

4. The expression $\frac{3x+12}{x}$ is undefined if $x =$ _______.

Perform each division. ***See Examples 1–3.***

5. $\frac{12m^4 - 6m^3}{6m^2}$

6. $\frac{35n^5 - 5n^2}{5n}$

7. $\frac{60x^4 - 20x^2 + 10x}{2x}$

8. $\frac{120x^6 - 60x^3 + 80x^2}{2x}$

9. $\frac{20m^5 - 10m^4 + 5m^2}{-5m^2}$

10. $\frac{12t^5 - 6t^3 + 6t^2}{-6t^2}$

11. $\frac{8t^5 - 4t^3 + 4t^2}{2t}$

12. $\frac{8r^4 - 4r^3 + 6r^2}{2r}$

13. $\frac{4a^5 - 4a^2 + 8}{4a}$

14. $\frac{5t^8 + 5t^7 + 15}{5t}$

15. $\frac{12x^5 - 4x^4 + 6x^3}{-6x^2}$

16. $\frac{24x^6 - 14x^5 + 32x^4}{-4x^2}$

17. $\frac{4x^2 + 20x^3 - 36x^4}{4x^2}$

18. $\frac{5x^2 - 30x^4 + 30x^5}{5x^2}$

19. $\frac{-7r^7 + 6r^5 - r^4}{-r^5}$

20. $\frac{-13t^9 + 8t^6 - t^5}{-t^6}$

21. $\frac{-3x^3 - 4x^4 + 2x}{-3x^2}$

22. $\frac{-8x + 6x^3 - 5x^4}{-3x^2}$

23. CONCEPT CHECK What polynomial, when divided by $5x^3$, yields $3x^2 - 7x + 7$ as a quotient?

24. CONCEPT CHECK The quotient of a certain polynomial and $-12y^3$ is $6y^3 - 5y^2 + 2y - 3 + \frac{7}{y}$. What is this polynomial?

Perform each division. ***See Example 1–4.***

25. $\dfrac{27r^4 - 36r^3 - 6r^2 + 3r - 2}{3r}$

$= \dfrac{\_\_\_}{3r} - \dfrac{\_\_\_}{3r} - \dfrac{\_\_\_}{3r} + \dfrac{\_\_\_}{3r} - \dfrac{\_\_\_}{3r}$

$=$ ____________________

26. $\dfrac{8k^4 - 12k^3 - 2k^2 - 2k - 3}{2k}$

$= \dfrac{8k^4}{\_\_\_} - \dfrac{12k^3}{\_\_\_} - \dfrac{2k^2}{\_\_\_} - \dfrac{2k}{\_\_\_} - \dfrac{3}{\_\_\_}$

$=$ ____________________

27. $\dfrac{2m^5 - 6m^4 + 8m^2}{-2m^3}$

28. $\dfrac{6r^5 - 8r^4 + 10r^2}{-2r^3}$

29. $(120x^{11} - 60x^{10} + 140x^9 - 100x^8) \div (10x^{12})$

30. $(120x^{12} - 84x^9 + 60x^8 - 36x^7) \div (12x^9)$

31. $(20a^4b^3 - 15a^5b^2 + 25a^3b) \div (-5a^4b)$

32. $(16y^5z - 8y^2z^2 + 12yz^3) \div (-4y^2z^2)$

33. $(120x^5y^4 - 80x^2y^3 + 40x^2y^4 - 20x^5y^3) \div (20xy^2)$

34. $(200a^5b^6 - 160a^4b^7 - 120a^3b^9 + 40a^2b^2) \div (40a^2b)$

Relating Concepts (Exercises 35–38) For Individual or Group Work

Our system of numeration is a decimal system, based on powers of 10. *Consider the following whole number.*

$$2846$$

Each digit represents the number of powers of 10 *for its place value. The* 2 *represents two thousands* (2×10^3), *the* 8 *represents eight hundreds* (8×10^2), *the* 4 *represents four tens* (4×10^1), *and the* 6 *represents six ones (or units)* (6×10^0). *In expanded form, we write*

$$2846 = (2 \times 10^3) + (8 \times 10^2) + (4 \times 10^1) + (6 \times 10^0).$$

Keeping this information in mind, ***work Exercises 35–38 in order.***

35. Divide 2846 by 2, using paper-and-pencil methods.

$$2\overline{)2846}$$

36. Write the answer from **Exercise 35** in expanded form.

37. Divide the polynomial $2x^3 + 8x^2 + 4x + 6$ by 2.

38. Compare the answers in **Exercises 36 and 37.** For what value of x does the answer in **Exercise 37** equal the answer in **Exercise 36?**

5.8 Dividing a Polynomial by a Polynomial

OBJECTIVES

1. Divide a polynomial by a polynomial.
2. Apply polynomial division to a geometry problem.

OBJECTIVE 1 Divide a polynomial by a polynomial. We use a method of "long division" to do this.

Dividing Whole Numbers	Dividing Polynomials
Step 1	
Divide 6696 by 27. $27\overline{)6696}$	Divide $8x^3 - 4x^2 - 14x + 15$ by $2x + 3$. $2x + 3\overline{)8x^3 - 4x^2 - 14x + 15}$
Step 2	
66 divided by 27 = 2.	$8x^3$ divided by $2x = 4x^2$.
$2 \cdot 27 = 54$	$4x^2(2x + 3) = 8x^3 + 12x^2$
$\begin{array}{r} 2 \\ 27\overline{)6696} \\ 54 \end{array}$	$\begin{array}{r} 4x^2 \\ 2x + 3\overline{)8x^3 - 4x^2 - 14x + 15} \\ 8x^3 + 12x^2 \end{array}$
Step 3	
Subtract.	Subtract.
$\begin{array}{r} 2 \\ 27\overline{)6696} \\ -54 \\ \hline 12 \end{array}$	$\begin{array}{r} 4x^2 \\ 2x + 3\overline{)8x^3 - 4x^2 - 14x + 15} \\ -(8x^3 + 12x^2) \\ \hline -16x^2 \end{array}$
Bring down the next digit.	Bring down the next term.
$\begin{array}{r} 2 \\ 27\overline{)6696} \\ -54 \\ \hline 129 \end{array}$	$\begin{array}{r} 4x^2 \\ 2x + 3\overline{)8x^3 - 4x^2 - 14x + 15} \\ -(8x^3 + 12x^2) \\ \hline -16x^2 - 14x \end{array}$
Step 4	
129 divided by 27 = 4.	$-16x^2$ divided by $2x = -8x$.
$4 \cdot 27 = 108$	$-8x(2x + 3) = -16x^2 - 24x$
$\begin{array}{r} 24 \\ 27\overline{)6696} \\ -54 \\ \hline 129 \\ 108 \end{array}$	$\begin{array}{r} 4x^2 - 8x \\ 2x + 3\overline{)8x^3 - 4x^2 - 14x + 15} \\ -(8x^3 + 12x^2) \\ \hline -16x^2 - 14x \\ -16x^2 - 24x \end{array}$
Step 5	
Subtract. Bring down.	Subtract. Bring down.
$\begin{array}{r} 24 \\ 27\overline{)6696} \\ -54 \\ \hline 129 \\ -108 \\ \hline 216 \end{array}$	$\begin{array}{r} 4x^2 - 8x \\ 2x + 3\overline{)8x^3 - 4x^2 - 14x + 15} \\ -(8x^3 + 12x^2) \\ \hline -16x^2 - 14x \\ -(-16x^2 - 24x) \\ \hline 10x + 15 \end{array}$

Dividing Whole Numbers	Dividing Polynomials
Step 6	
216 divided by 27 = 8.	$10x$ divided by $2x = 5$.
$8 \cdot 27 = 216$	$5(2x+3) = 10x + 15$
$27\overline{)6696}$ quotient 248; -54, 129, -108, 216, -216, Remainder → 0	$2x+3\overline{)8x^3 - 4x^2 - 14x + 15}$ quotient $4x^2 - 8x + 5$; $-(8x^3 + 12x^2)$, $-16x^2 - 14x$, $-(-16x^2 - 24x)$, $10x + 15$, $-(10x + 15)$, Remainder → 0
6696 divided by 27 is 248.	$8x^3 - 4x^2 - 14x + 15$ divided by $2x + 3$ is $4x^2 - 8x + 5$.
Step 7 Multiply to check.	Multiply to check.
CHECK $27 \cdot 248 = 6696$ ✓	**CHECK** $(2x+3)(4x^2 - 8x + 5) = 8x^3 - 4x^2 - 14x + 15$ ✓

EXAMPLE 1 Dividing a Polynomial by a Polynomial

Divide. $\dfrac{3x^2 - 5x - 28}{x - 4}$

$$
\begin{array}{r l}
 & 3x + 7 \quad \leftarrow \text{Quotient} \\
\text{Divisor} \rightarrow x - 4\overline{)3x^2 - 5x - 28} & \leftarrow \text{Dividend} \\
-(3x^2 - 12x) & \leftarrow (3x^2 - 5x) - (3x^2 - 12x) = 7x \\
7x - 28 & \\
-(7x - 28) & \leftarrow (7x - 28) - (7x - 28) = 0 \\
0 &
\end{array}
$$

Step 1 $3x^2$ divided by x is $3x$. $3x(x - 4) = 3x^2 - 12x$

Step 2 Subtract $3x^2 - 12x$ from $3x^2 - 5x$. Bring down -28.

Step 3 $7x$ divided by x is 7. $7(x - 4) = 7x - 28$

Step 4 Subtract $7x - 28$ from $7x - 28$. The remainder is 0.

CHECK Multiply the divisor, $x - 4$, by the quotient, $3x + 7$. The product must be the original dividend, $3x^2 - 5x - 28$.

$$(x - 4)(3x + 7) = 3x^2 + 7x - 12x - 28 \quad \text{FOIL method}$$
$$= 3x^2 - 5x - 28 \checkmark \quad \text{Combine like terms.}$$

(Divisor: $x - 4$; Quotient: $3x + 7$; Dividend: $3x^2 - 5x - 28$)

◀ **Work Problem 1 at the Side.**

1 Divide.

(a) $\dfrac{4x^2 + x - 18}{x - 2}$

(b) $\dfrac{2x^2 + 5x - 25}{x + 5}$

Answers

1. (a) $4x + 9$ **(b)** $2x - 5$

EXAMPLE 2 **Dividing a Polynomial by a Polynomial**

Divide. $\dfrac{5x + 4x^3 - 8 - 4x^2}{2x - 1}$

When we divide two polynomials, we must write both in descending powers of the variable. We must rewrite the dividend polynomial here. Then we divide by $2x - 1$.

$$\begin{array}{r} 2x^2 - x + 2 \\ 2x - 1\overline{)4x^3 - 4x^2 + 5x - 8} \\ -(4x^3 - 2x^2) \\ \hline -2x^2 + 5x \\ -(-2x^2 + x) \\ \hline 4x - 8 \\ -(4x - 2) \\ \hline -6 \leftarrow \text{Remainder} \end{array}$$

Write in descending powers.

In each subtraction, add the opposite.

Step 1 $4x^3$ divided by $2x$ is $2x^2$. $2x^2(2x - 1) = 4x^3 - 2x^2$

Step 2 Subtract $4x^3 - 2x^2$ from $4x^3 - 4x^2$. Bring down the next term, $5x$.

Step 3 $-2x^2$ divided by $2x$ is $-x$. $-x(2x - 1) = -2x^2 + x$

Step 4 Subtract $-2x^2 + x$ from $-2x^2 + 5x$. Bring down the next term, -8.

Step 5 $4x$ divided by $2x$ is 2. $2(2x - 1) = 4x - 2$

Step 6 Subtract $4x - 2$ from $4x - 8$. The remainder is -6. Write the remainder as the numerator of a fraction that has the divisor $2x - 1$ as its denominator. Because there is a nonzero remainder, the answer is not a polynomial.

$$\begin{array}{l} \text{Dividend} \rightarrow \\ \text{Divisor} \rightarrow \end{array} \frac{4x^3 - 4x^2 + 5x - 8}{2x - 1} = \underbrace{2x^2 - x + 2}_{\text{Quotient polynomial}} + \underbrace{\frac{-6}{2x - 1}}_{\text{Fractional part of quotient}} \begin{array}{l} \leftarrow \text{Remainder} \\ \leftarrow \text{Divisor} \end{array}$$

Step 7 Multiply to check.

CHECK $(2x - 1)\left(2x^2 - x + 2 + \dfrac{-6}{2x - 1}\right)$ Multiply Divisor × (Quotient including the Remainder).

$= (2x - 1)(2x^2) + (2x - 1)(-x) + (2x - 1)(2)$

$+ (2x - 1)\left(\dfrac{-6}{2x - 1}\right)$

$= 4x^3 - 2x^2 - 2x^2 + x + 4x - 2 - 6$

$= 4x^3 - 4x^2 + 5x - 8$ ✓ Divisor × Quotient = Dividend

Work Problem 2 at the Side. ▶

CAUTION

Remember to include "$+ \dfrac{\text{remainder}}{\text{divisor}}$" as part of the answer.

2 Divide.

(a) $(x^2 + x^3 + 4x - 6) \div (x - 1)$

(b) $\dfrac{p^3 - 2p^2 + 9 - 5p}{p + 2}$

(c) $\dfrac{6k^3 - 20k - k^2 + 1}{2k - 3}$

Answers

2. (a) $x^2 + 2x + 6$

(b) $p^2 - 4p + 3 + \dfrac{3}{p + 2}$

(c) $3k^2 + 4k - 4 + \dfrac{-11}{2k - 3}$

3 Divide.

(a) $(x^3 - 8) \div (x - 2)$

(b) $\dfrac{r^2 - 5}{r + 4}$

4 Divide.

(a)

$$\frac{2m^5 + m^4 + 6m^3 - 3m^2 - 18}{m^2 + 3}$$

(b) $(2x^4 + 3x^3 - x^2 + 6x + 5) \div (x^2 - 1)$

Answers

3. (a) $x^2 + 2x + 4$
(b) $r - 4 + \dfrac{11}{r + 4}$

4. (a) $2m^3 + m^2 - 6$
(b) $2x^2 + 3x + 1 + \dfrac{9x + 6}{x^2 - 1}$

EXAMPLE 3 Dividing into a Polynomial with Missing Terms

Divide $x^3 - 1$ by $x - 1$.

Here the dividend, $x^3 - 1$, is missing the x^2-term and the x-term. We use **0** as the coefficient for each missing term. Thus, we write

$$x^3 - 1 \quad \text{as} \quad x^3 + 0x^2 + 0x - 1.$$

$$\begin{array}{r} x^2 + x + 1 \\ x - 1\overline{)x^3 + 0x^2 + 0x - 1} \\ -(x^3 - x^2) \qquad\qquad \\ \hline x^2 + 0x \qquad \\ -(x^2 - x) \qquad \\ \hline x - 1 \\ -(x - 1) \\ \hline 0 \end{array}$$

Insert placeholders for the missing terms.

The remainder is **0**. The quotient is $x^2 + x + 1$.

CHECK $(x - 1)(x^2 + x + 1)$

$= x^3 + x^2 + x - x^2 - x - 1$

$= x^3 - 1$ ✓ Divisor × Quotient = Dividend

◀ **Work Problem 3 at the Side.**

EXAMPLE 4 Dividing by a Polynomial with Missing Terms

Divide $x^4 + 2x^3 + 2x^2 - x - 1$ by $x^2 + 1$.

The divisor $x^2 + 1$ has a missing x-term, so we write it as $x^2 + 0x + 1$.

$$\begin{array}{r} x^2 + 2x + 1 \\ x^2 + 0x + 1\overline{)x^4 + 2x^3 + 2x^2 - x - 1} \\ -(x^4 + 0x^3 + x^2) \qquad\qquad \\ \hline 2x^3 + x^2 - x \qquad \\ -(2x^3 + 0x^2 + 2x) \qquad \\ \hline x^2 - 3x - 1 \\ -(x^2 + 0x + 1) \\ \hline -3x - 2 \leftarrow \text{Remainder} \end{array}$$

Insert a placeholder for the missing term.

When the result of subtracting (here $-3x - 2$) is a polynomial of degree *less* than the divisor ($x^2 + 0x + 1$, in this case), that polynomial is the remainder. We write the answer as follows.

$$x^2 + 2x + 1 + \frac{-3x - 2}{x^2 + 1}$$

Remember to include "$+ \dfrac{\text{remainder}}{\text{divisor}}$."

CHECK Show that multiplying $(x^2 + 1)\left(x^2 + 2x + 1 + \dfrac{-3x - 2}{x^2 + 1}\right)$ gives the original dividend, $x^4 + 2x^3 + 2x^2 - x - 1$. ✓

◀ **Work Problem 4 at the Side.**

EXAMPLE 5 Dividing a Polynomial When the Quotient Has Fractional Coefficients

Divide $4x^3 + 2x^2 + 3x + 2$ by $4x - 4$.

$\frac{6x^2}{4x} = \frac{3}{2}x$

$\frac{9x}{4x} = \frac{9}{4}$

$$
\begin{array}{r}
x^2 + \frac{3}{2}x + \frac{9}{4} \\
4x - 4\overline{)4x^3 + 2x^2 + 3x + 2} \\
-\underline{(4x^3 - 4x^2)} \qquad\qquad \\
6x^2 + 3x \qquad \\
-\underline{(6x^2 - 6x)} \qquad \\
9x + 2 \\
-\underline{(9x - 9)} \\
11
\end{array}
$$

The answer is $x^2 + \frac{3}{2}x + \frac{9}{4} + \frac{11}{4x - 4}$.

Work Problem 5 at the Side. ▶

5 Divide $3x^3 + 7x^2 + 7x + 11$ by $3x + 6$.

OBJECTIVE 2 Apply polynomial division to a geometry problem.

EXAMPLE 6 Using an Area Formula

The area of the rectangle in **Figure 3** is given by $(x^3 + 4x^2 + 8x + 8)$ sq. units. The width is given by $(x + 2)$ units. What is its length?

Length = ?

Width = $x + 2$

Area = $x^3 + 4x^2 + 8x + 8$

Figure 3

For a rectangle, $\mathscr{A} = LW$. Solving for L gives $L = \frac{\mathscr{A}}{W}$. Divide the area, $x^3 + 4x^2 + 8x + 8$, by the width, $x + 2$.

$$
\begin{array}{r}
x^2 + 2x + 4 \\
x + 2\overline{)x^3 + 4x^2 + 8x + 8} \\
-\underline{(x^3 + 2x^2)} \qquad\qquad \\
2x^2 + 8x \qquad \\
-\underline{(2x^2 + 4x)} \qquad \\
4x + 8 \\
-\underline{(4x + 8)} \\
0
\end{array}
$$

The quotient $(x^2 + 2x + 4)$ units represents the length of the rectangle.

Work Problem 6 at the Side. ▶

6 The area of a rectangle is given by $(x^3 + 7x^2 + 17x + 20)$ sq. units. The width is given by $(x + 4)$ units. What is its length?

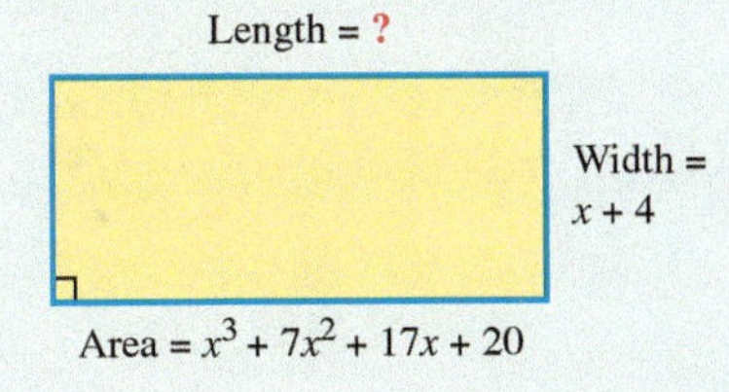

Answers

5. $x^2 + \frac{1}{3}x + \frac{5}{3} + \frac{1}{3x + 6}$

6. $(x^2 + 3x + 5)$ units

5.8 Exercises

FOR EXTRA HELP — Go to MyMathLab for worked-out, step-by-step solutions to exercises enclosed in a square and video solutions to ▶ exercises.

CONCEPT CHECK *Complete the statement or answer the question.*

1. Label the parts of the division problem using the words *quotient*, *divisor*, and *dividend*.

$$(x^3 - 2x^2 - 9) \div (x - 3) = x^2 + x + 3$$

↑ ↑ ↑

______ ______ ______

2. When dividing one polynomial by another, how do we know when to stop dividing?

3. In dividing $12m^2 - 20m + 3$ by $2m - 3$, what is the first step?

4. In the division in **Exercise 3,** what is the second step?

Perform each division. ***See Examples 1–5.***

5. $\dfrac{x^2 - x - 6}{x - 3}$

6. $\dfrac{m^2 - 2m - 24}{m - 6}$

7. $\dfrac{2y^2 + 9y - 35}{y + 7}$

8. $\dfrac{2y^2 + 9y + 7}{y + 1}$

9. $\dfrac{p^2 + 2p + 20}{p + 6}$

10. $\dfrac{x^2 + 11x + 16}{x + 8}$

11. $(r^2 - 8r + 15) \div (r - 3)$

12. $(t^2 + 2t - 35) \div (t - 5)$

13. $\dfrac{4a^2 - 22a + 32}{2a + 3}$

14. $\dfrac{9w^2 + 6w + 10}{3w - 2}$

15. $\dfrac{8x^3 - 10x^2 - x + 3}{2x + 1}$

16. $\dfrac{12t^3 - 11t^2 + 9t + 18}{4t + 3}$

17. $\dfrac{2r^3 - 6r - 5r^2 + 15}{r - 3}$

18. $\dfrac{2y^2 + 5y^3 - y - 8}{y + 1}$

19. $\dfrac{3y^3 + y^2 + 2}{y + 1}$

20. $\dfrac{2r^3 - 6r - 36}{r - 3}$

21. $\dfrac{2x^3 + x + 2}{x + 3}$

22. $\dfrac{3x^3 + x + 5}{x + 1}$

23. $\dfrac{3k^3 - 4k^2 - 6k + 10}{k^2 - 2}$

24. $\dfrac{5z^3 - z^2 + 10z + 2}{z^2 + 2}$

25. $(x^4 - x^2 - 2) \div (x^2 - 2)$

26. $(r^4 + 2r^2 - 3) \div (r^2 - 1)$

27. $\dfrac{x^4 - 1}{x^2 - 1}$

28. $\dfrac{y^3 + 1}{y + 1}$

29. $\dfrac{6p^4 - 15p^3 + 14p^2 - 5p + 10}{3p^2 + 1}$

30. $\dfrac{6r^4 - 10r^3 - r^2 + 15r - 8}{2r^2 - 3}$

31. $(10x^3 + 13x^2 + 4x + 1) \div (5x + 5)$

32. $(6x^3 - 19x^2 - 19x - 4) \div (2x - 8)$

33. $\dfrac{3x^3 + 5x^2 - 9x + 5}{3x - 3}$

34. $\dfrac{5x^3 + 4x^2 + 10x + 20}{5x + 5}$

35. $\dfrac{2x^5 + x^4 + 11x^3 - 8x^2 - 13x + 7}{2x^2 + x - 1}$

36. $\dfrac{4t^5 - 11t^4 - 6t^3 + 5t^2 - t + 3}{4t^2 + t - 3}$

Work each problem. Give answers in units (or as specified). **See Example 6.** *(If necessary, refer to the formulas at the back of this text.)*

37. What expression represents the length of the rectangle?

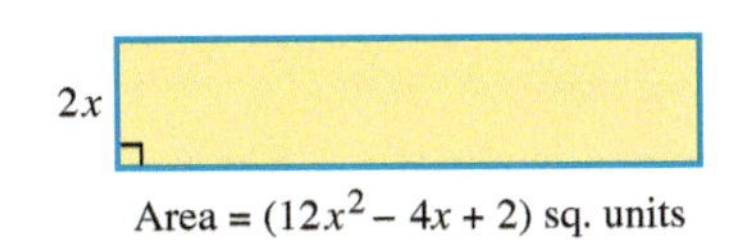

38. What expression represents the length of the base of the triangle?

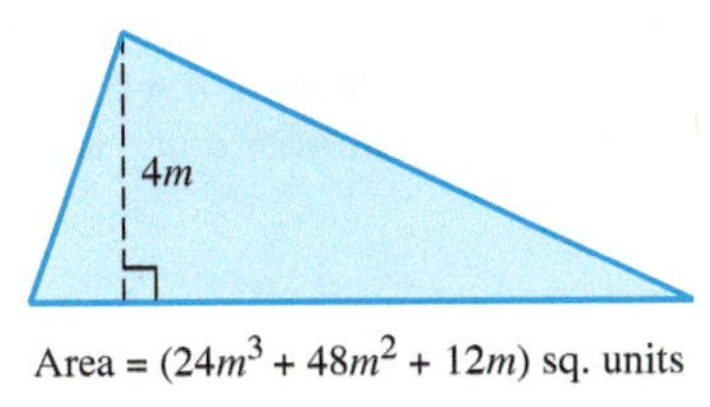

39. Find the measure of the length of the rectangle.

40. Find the measure of the base of the parallelogram.

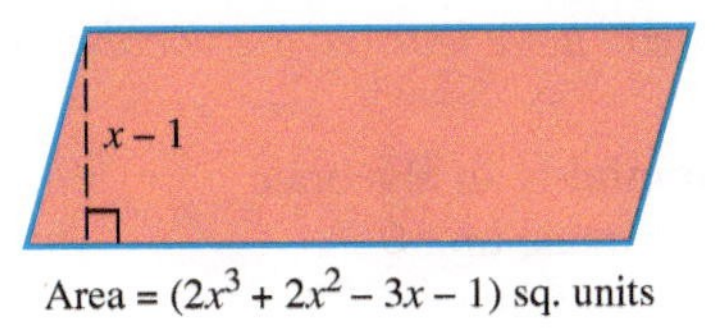

41. If the distance traveled is

$$(5x^3 - 6x^2 + 3x + 14) \text{ miles}$$

and the rate is $(x + 1)$ mph, write an expression, in hours, for the time traveled.

42. If the cost to fertilize a garden is

$$(4x^5 + 3x^4 + 2x^3 + 9x^2 - 29x + 2) \text{ dollars}$$

and fertilizer costs $(x + 2)$ dollars per square yard, write an expression, in square yards, for the area of the garden.

Chapter 5 Summary

Key Terms

5.1

exponential expression A number written with an exponent (or power) is an exponential expression.

3^4 ← Exponent; ↑ Base; } Exponential expression

5.3

scientific notation A number written in the form $a \times 10^n$, where $1 \leq |a| < 10$ and n is an integer, is in scientific notation.

standard notation We refer to a number such as 125 as the standard notation of 1.25×10^2.

5.4

term A term is a number (constant), a variable, or a product or quotient of a number and one or more variables raised to powers.

Numerical coefficient → $-7x^2$ (Term)

like terms Terms with exactly the same variables (including the same exponents) are like terms.

polynomial in *x* A term or the sum of a finite number of terms of the form ax^n, for any real number a and any whole number n, is a polynomial in x.

descending powers A polynomial in x is written in descending powers if the exponents on x in its terms decrease from left to right.

degree of a term The degree of a term is the sum of the exponents on the variables.

degree of a polynomial The degree of a polynomial is the greatest degree of any nonzero term of the polynomial.

monomial A monomial is a polynomial with exactly one term.

binomial A binomial is a polynomial with exactly two terms.

trinomial A trinomial is a polynomial with exactly three terms.

5.5

FOIL method The FOIL method is used to find the product of two binomials. The letters of the word **FOIL** originate as follows: Multiply the **F**irst terms, multiply the **O**uter terms (to obtain the outer product), multiply the **I**nner terms (to obtain the inner product), and multiply the **L**ast terms.

outer product The outer product of $(a + b)(c + d)$ is ad.

inner product The inner product of $(a + b)(c + d)$ is bc.

5.6

conjugate The conjugate of $x + y$ is $x - y$.

New Symbols

x^{-n} x to the negative n power

Test Your Word Power

See how well you have learned the vocabulary in this chapter.

1. A **polynomial** is an algebraic expression made up of
   A. a term or a finite product of terms with positive coefficients and exponents
   B. a term or a finite sum of terms with real coefficients and whole number exponents
   C. the product of two or more terms with positive exponents
   D. the sum of two or more terms with whole number coefficients and exponents.

2. The **degree of a term** is the
   A. number of variables in the term
   B. product of the exponents on the variables
   C. least exponent on the variables
   D. sum of the exponents on the variables.

3. A **trinomial** is a polynomial with
   A. exactly one term
   B. exactly two terms
   C. exactly three terms
   D. more than three terms.

4. A **binomial** is a polynomial with
   A. exactly one term
   B. exactly two terms
   C. exactly three terms
   D. more than three terms.

5. A **monomial** is a polynomial with
   A. exactly one term
   B. exactly two terms
   C. exactly three terms
   D. more than three terms.

Answers to Test Your Word Power

1. B; *Example:* $5x^3 + 2x^2 - 7$
2. D; *Examples:* The term 6 has degree 0, $3x$ has degree 1, $-2x^8$ has degree 8, and $5x^2y^4$ has degree 6.
3. C; *Example:* $2a^2 - 3ab + b^2$
4. B; *Example:* $3t^3 + 5t$
5. A; *Examples:* -5 and $4xy^5$

Quick Review

Concepts	*Examples*
5.1 The Product Rule and Power Rules for Exponents	
For any integers m and n, the following hold true.	Simplify using the rules for exponents.
Product rule $a^m \cdot a^n = a^{m+n}$	$2^4 \cdot 2^5 = 2^{4+5} = 2^9$
Power rules (a) $(a^m)^n = a^{mn}$	$(3^4)^2 = 3^{4 \cdot 2} = 3^8$
(b) $(ab)^m = a^m b^m$	$(6a)^5 = 6^5 a^5$
(c) $\left(\frac{a}{b}\right)^m = \frac{a^m}{b^m}$ (where $b \neq 0$)	$\left(\frac{2}{3}\right)^4 = \frac{2^4}{3^4}$
5.2 Integer Exponents and the Quotient Rule	
For any nonzero real numbers a and b and any integers m and n, the following hold true.	Simplify using the rules for exponents.
Zero exponent $a^0 = 1$	$15^0 = 1$
Negative exponent $a^{-n} = \frac{1}{a^n}$	$5^{-2} = \frac{1}{5^2} = \frac{1}{25}$
Quotient rule $\frac{a^m}{a^n} = a^{m-n}$	$\frac{4^8}{4^3} = 4^{8-3} = 4^5$
Negative-to-positive rules $\frac{a^{-m}}{b^{-n}} = \frac{b^n}{a^m}$ $\left(\frac{a}{b}\right)^{-m} = \left(\frac{b}{a}\right)^m$	$\frac{6^{-2}}{7^{-3}} = \frac{7^3}{6^2}$ $\left(\frac{5}{3}\right)^{-4} = \left(\frac{3}{5}\right)^4$
5.3 An Application of Exponents: Scientific Notation	
To write a positive number in scientific notation $a \times 10^n$, where $1 \leq \lvert a \rvert < 10$, move the decimal point to follow the first nonzero digit.	Write in scientific notation. $247 = 2.47 \times 10^2$ $0.0051 = 5.1 \times 10^{-3}$
1. If moving the decimal point makes the number less, then n is positive. 2. If moving the decimal point makes the number greater, then n is negative. 3. If the decimal point is not moved, then n is 0.	Write in standard notation. $3.25 \times 10^5 = 325{,}000$ $8.44 \times 10^{-6} = 0.00000844$ $-4.8 \times 10^0 = -4.8$
For a negative number, follow these steps using the *absolute value* of the number. Then make the result negative.	

Concepts	Examples

5.4 Adding and Subtracting Polynomials

Adding Polynomials
Combine (add) like terms.

Subtracting Polynomials
Change the sign of each term in the subtrahend (second polynomial) and add the result to the minuend (first polynomial).

Add.

$$\begin{array}{r} 2x^2 + 5x - 3 \\ +\ (5x^2 - 2x + 7) \\ \hline 7x^2 + 3x + 4 \end{array}$$

Subtract.

$$\begin{aligned}&(2x^2 + 5x - 3) - (5x^2 - 2x + 7) \\ &= (2x^2 + 5x - 3) + (-5x^2 + 2x - 7) \\ &= -3x^2 + 7x - 10\end{aligned}$$

5.5 Multiplying Polynomials

Multiply each term of the first polynomial by each term of the second polynomial. Then combine like terms.

FOIL Method for Multiplying Binomials

Step 1 Multiply the two **F**irst terms to obtain the first term of the product.

Step 2 Find the **O**uter product and the **I**nner product and combine them (when possible) to obtain the middle term of the product.

Step 3 Multiply the two **L**ast terms to obtain the last term of the product.

Step 4 Add the terms found in Steps 1–3.

Multiply.

$$\begin{array}{rrrrr} & 3x^3 & -\ 4x^2 & +\ 2x & -\ 7 \\ & & & 4x & +\ 3 \\ \hline & 9x^3 & -\ 12x^2 & +\ 6x & -\ 21 \\ 12x^4 & -\ 16x^3 & +\ 8x^2 & -\ 28x & \\ \hline 12x^4 & -\ 7x^3 & -\ 4x^2 & -\ 22x & -\ 21 \end{array}$$

Multiply. $(2x + 3)(5x - 4)$

$$\begin{aligned} 2x(5x) &= 10x^2 && \text{F} \\ 2x(-4) + 3(5x) &= 7x && \text{O, I} \\ 3(-4) &= -12 && \text{L} \end{aligned}$$

The product is $10x^2 + 7x - 12$.

5.6 Special Products

Square of a Binomial

$$(x + y)^2 = x^2 + 2xy + y^2$$

$$(x - y)^2 = x^2 - 2xy + y^2$$

Product of a Sum and Difference of Two Terms

$$(x + y)(x - y) = x^2 - y^2$$

Multiply.

$$\begin{aligned}&(3x + 1)^2 \\ &= (3x)^2 + 2(3x)(1) + 1^2 \\ &= 9x^2 + 6x + 1\end{aligned}$$

$$\begin{aligned}&(2m - 5n)^2 \\ &= (2m)^2 - 2(2m)(5n) + (5n)^2 \\ &= 4m^2 - 20mn + 25n^2\end{aligned}$$

$$\begin{aligned}&(4a + 3)(4a - 3) \\ &= (4a)^2 - 3^2 \\ &= 16a^2 - 9 \qquad (4a)^2 = 4^2a^2 = 16a^2\end{aligned}$$

5.7 Dividing a Polynomial by a Monomial

Divide each term of the polynomial by the monomial.

$$\frac{a + b}{c} = \frac{a}{c} + \frac{b}{c} \quad \textbf{(where } c \neq 0\textbf{)}$$

Divide. $\dfrac{4x^3 - 2x^2 + 6x - 8}{2x}$

$$= \frac{4x^3}{2x} - \frac{2x^2}{2x} + \frac{6x}{2x} - \frac{8}{2x}$$

Divide each term in the dividend by $2x$, the divisor.

$$= 2x^2 - x + 3 - \frac{4}{x}$$

$\frac{8}{2x} = \frac{8 \cdot 1}{2 \cdot x} = \frac{4}{x}$

5.8 Dividing a Polynomial by a Polynomial

Use "long division."

Divide.

$$\begin{array}{r} 2x - 5 \\ 3x + 4 \overline{)6x^2 - 7x - 21} \\ -(6x^2 + 8x) \\ \hline -15x - 21 \\ -(-15x - 20) \\ \hline -1 \end{array} \leftarrow \text{Remainder}$$

The answer is $2x - 5 + \dfrac{-1}{3x + 4}$.

Chapter 5 Review Exercises

5.1 *Use the product rule, power rules, or both to simplify each expression. Evaluate coefficients if the exponent is 4 or less.*

1. $4^3 \cdot 4^8$

2. $(-5)^6(-5)^5$

3. $-8x^4(9x^3)$

4. $(2x^2)(5x^3)(x^9)$

5. $(19x)^5$

6. $(-4y)^7$

7. $5(pt)^4$

8. $\left(\frac{7}{5}\right)^6$

9. $(3x^2y^3)^3$

10. $(t^4)^8(t^2)^5$

11. $(6x^2z^4)^2(x^3yz^2)^4$

12. $\left(\frac{2m^3n}{p^2}\right)^3$

Solve each problem.

13. Find an expression that represents, in appropriate units, the volume of the figure. (If necessary, refer to the formulas at the back of this text.)

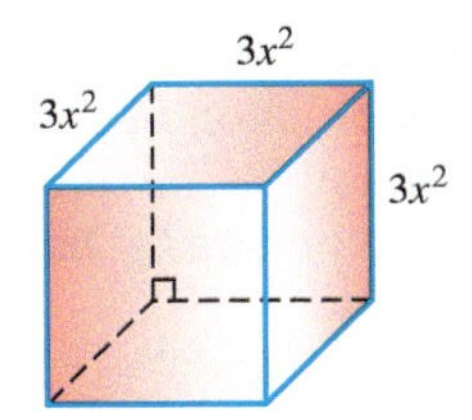

14. CONCEPT CHECK A student incorrectly simplified the expression

$$7^2 + 7^3 \quad \text{as} \quad 7^5.$$

What Went Wrong? Give the correct answer.

5.2 *Evaluate each expression.*

15. -10^0

16. $(-23)^0$

17. $5^0 + 8^0$

18. 2^{-5}

19. $\left(\frac{6}{5}\right)^{-2}$

20. $4^{-2} - 4^{-1}$

Simplify each expression. Assume that all variables represent nonzero real numbers. Evaluate coefficients if the exponent is 4 or less.

21. $\frac{6^{-3}}{6^{-5}}$

22. $\frac{x^{-7}}{x^{-9}}$

23. $\frac{p^{-8}}{p^4}$

24. $\frac{r^{-2}}{r^{-6}}$

25. $(2^4)^2$

26. $(9^3)^{-2}$

27. $(5^{-2})^{-4}$

28. $(8^{-3})^4$

29. $\frac{(m^2)^3}{(m^4)^2}$

30. $\frac{y^4 \cdot y^{-2}}{y^{-5}}$

31. $\frac{r^9 \cdot r^{-5}}{r^{-2} \cdot r^{-7}}$

32. $(-5m^3)^2$

33. $(2y^{-4})^{-3}$

34. $\frac{ab^{-3}}{a^4b^2}$

35. $\frac{(6r^{-1})^2 \cdot (2r^{-4})}{r^{-5}(r^2)^{-3}}$

36. $\frac{(2m^{-5}n^2)^3(3m^2)^{-1}}{m^{-2}n^{-4}(m^{-1})^2}$

5.3 *Write each number in scientific notation.*

37. 48,000,000 **38.** 28,988,000,000 **39.** 0.000065 **40.** 0.0000000824

Write each number in standard notation.

41. 2.4×10^4 **42.** 7.83×10^7 **43.** 8.97×10^{-7} **44.** 9.95×10^{-12}

*Perform the indicated operations. Write each answer in **(a)** scientific notation and **(b)** standard notation.*

45. $(2 \times 10^{-3})(4 \times 10^5)$

46. $\dfrac{8 \times 10^4}{2 \times 10^{-2}}$

47. $\dfrac{12 \times 10^{-5} \times 16 \times 10^4}{4 \times 10^3 \times 8 \times 10^{-2}}$

48. $\dfrac{2.5 \times 10^5 \times 4.8 \times 10^{-4}}{7.5 \times 10^8 \times 1.6 \times 10^{-5}}$

Use scientific notation to calculate the answer to each problem.

49. A computer can perform 466,000,000 calculations per second. How many calculations can it perform per minute? Per hour?

50. In theory, there are 1×10^9 possible Social Security numbers. The population of the United States is about 3×10^8. How many Social Security numbers are available for each person? (Data from U.S. Census Bureau.)

5.4 *For each polynomial, first simplify if possible, and write the result in descending powers of the variable. Then give the degree, and tell whether the simplified polynomial is a* monomial, *a* binomial, *a* trinomial, *or* none of these.

51. $9m^2 + 11m^2 + 2m^2$

52. $-4p + p^3 - p^2 + 8p + 2$

53. $12a^5 - 9a^4 + 8a^3 + 2a^2 - a + 3$

54. $-7y^5 - 8y^4 - y^5 + y^4 + 9y$

Add or subtract as indicated.

55. $\begin{array}{r} -2a^3 + 5a^2 \\ +\ \underline{(-3a^3 - a^2)} \end{array}$

56. $\begin{array}{r} 4r^3 - 8r^2 + 6r \\ +\ \underline{(-2r^3 + 5r^2 + 3r)} \end{array}$

57. $\begin{array}{r} 6y^2 - 8y + 2 \\ -\ \underline{(-5y^2 + 2y - 7)} \end{array}$

58. $\begin{array}{r} -12k^4 - 8k^2 + 7k - 5 \\ -\ \underline{(k^4 + 7k^2 + 11k + 1)} \end{array}$

59. $(2m^3 - 8m^2 + 4) + (8m^3 + 2m^2 - 7)$

60. $(-5y^2 + 3y + 11) + (4y^2 - 7y + 15)$

61. $(6p^2 - p - 8) - (-4p^2 + 2p + 3)$

62. $(12r^4 - 7r^3 + 2r^2) - (5r^4 - 3r^3 + 2r^2 + 1)$

5.5 *Find each product.*

63. $5x(2x + 14)$

64. $-3p^3(2p^2 - 5p)$

65. $(3r - 2)(2r^2 + 4r - 3)$

66. $(2y + 3)(4y^2 - 6y + 9)$

67. $(5p^2 + 3p)(p^3 - p^2 + 5)$

68. $(x + 6)(x - 3)$

69. $(3k - 6)(2k + 1)$

70. $(6p - 3q)(2p - 7q)$

71. $(m^2 + m - 9)(2m^2 + 3m - 1)$

5.6 *Find each product.*

72. $(a + 4)^2$

73. $(3p - 2)^2$

74. $(2r + 5s)^2$

75. $(r + 2)^3$

76. $(2x - 1)^3$

77. $(2z + 7)(2z - 7)$

78. $(6m - 5)(6m + 5)$

79. $(5a + 6b)(5a - 6b)$

80. $3(2x^2 + 5)(2x^2 - 5)$

Work each problem.

81. **CONCEPT CHECK** The square of a binomial leads to a polynomial with how many terms? The product of a sum and difference of two terms leads to a polynomial with how many terms?

82. Explain why $(a + b)^2$ is not equivalent to $a^2 + b^2$.

5.7 *Perform each division.*

83. $\dfrac{-15y^4}{-9y^2}$

84. $\dfrac{-12x^3y^2}{6xy}$

85. $\dfrac{6y^4 - 12y^2 + 18y}{-6y}$

86. $\dfrac{2p^3 - 6p^2 + 5p}{2p^2}$

87. $(5x^{13} - 10x^{12} + 20x^7 - 35x^5) \div (-5x^4)$

88. $(-10m^4n^2 + 5m^3n^3 + 6m^2n^4) \div (5m^2n)$

5.8 *Perform each division.*

89. $(2r^2 + 3r - 14) \div (r - 2)$

90. $\dfrac{12m^2 - 11m - 10}{3m - 5}$

91. $\dfrac{10a^3 + 5a^2 - 14a + 9}{5a^2 - 3}$

92. $\dfrac{2k^4 + 4k^3 + 9k^2 - 8}{2k^2 + 1}$

Chapter 5 Mixed Review Exercises

Perform each indicated operation, or simplify each expression. Evaluate coefficients if the exponent is 4 or less. Assume that all variables represent nonzero real numbers.

1. $19^0 - 3^0$

2. $(3p)^4(3p^{-7})$

3. 7^{-2}

4. $(2k - 7)^2$

5. $\dfrac{2y^3 + 17y^2 + 37y + 7}{2y + 7}$

6. $\left(\dfrac{6r^2s}{5}\right)^4$

7. $-m^5(8m^2 + 10m + 6)$

8. $\left(\dfrac{1}{2}\right)^{-5}$

9. $(25x^2y^3 - 8xy^2 + 15x^3y) \div (5x)$

10. $(6r^{-2})^{-1}$

11. $(2x + y)^3$

12. $2^{-1} + 4^{-1}$

13. $(a + 2)(a^2 - 4a + 1)$

14. $(5y^3 - 8y^2 + 7) - (-3y^3 + y^2 + 2)$

15. $(2r + 5)(5r - 2)$

16. $(12a + 1)(12a - 1)$

Find polynomials that represent, in appropriate units, ***(a)*** *the perimeter and* ***(b)*** *the area of each square or rectangle.*

17.

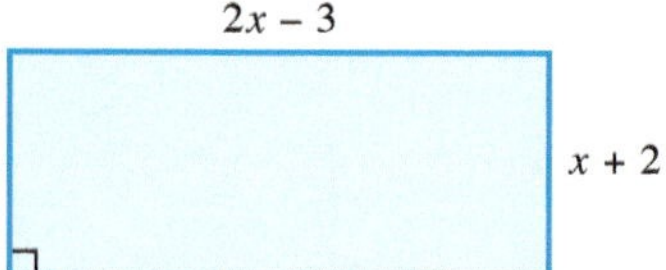

18.

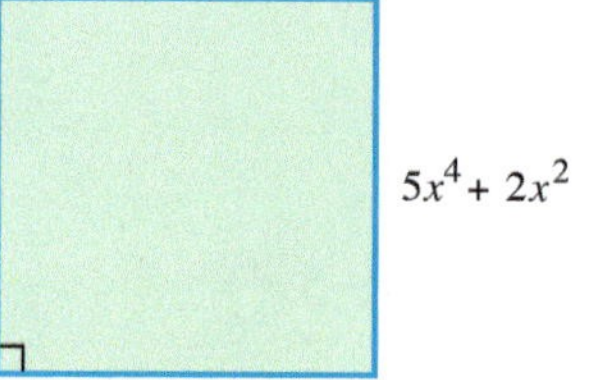

19. CONCEPT CHECK A friend incorrectly simplified

$$\frac{6x^2 - 12x}{6} \quad \text{as} \quad x^2 - 12x.$$

What Went Wrong? Give the correct answer.

20. CONCEPT CHECK What polynomial, when multiplied by $6m^2n$, gives the following product?

$$12m^3n^2 + 18m^6n^3 - 24m^2n^2$$

Chapter 5 Test

The Chapter Test Prep Videos with step-by-step solutions are available in MyMathLab or on YouTube at https://goo.gl/8aAsWP

Evaluate each expression.

1. $(-2)^3(-2)^2$

2. 5^{-4}

3. $(-3)^0 + 4^0$

4. $4^{-1} + 3^{-1}$

Simplify each expression. Evaluate coefficients if the exponent is 4 or less. Assume that all variables represent nonzero real numbers.

5. $\left(\frac{6}{m^2}\right)^3$

6. $\frac{(3x^2y)^2(xy^3)^2}{(xy)^3}$

7. $\frac{8^{-1} \cdot 8^4}{8^{-2}}$

8. $\frac{(x^{-3})^{-2}(x^{-1}y)^2}{(xy^{-2})^2}$

9. Determine whether each expression represents a number that is *positive, negative,* or *zero.*

(a) 3^{-4} **(b)** $(-3)^4$ **(c)** -3^4 **(d)** 3^0 **(e)** $(-3)^0 - 3^0$ **(f)** $(-3)^{-3}$

10. Write each number in scientific notation.

(a) 344,000,000,000 **(b)** 0.00000557

11. Write each number in standard notation.

(a) 2.96×10^7 **(b)** 6.07×10^{-8}

12. A satellite galaxy of the Milky Way, known as the Large Magellanic Cloud, is ***1000*** light-years across. A light-year is equal to ***5,890,000,000,000*** mi. (Data from *USA Today.*)

(a) Write the two boldface italic numbers in scientific notation.

(b) How many miles across is the Large Magellanic Cloud?

For each polynomial, first simplify if possible, and write the result in descending powers of the variable. Then give the degree, and tell whether the simplified polynomial is a monomial, *a* binomial, *a* trinomial, *or* none of these.

13. $5x^2 + 8x - 12x^2$

14. $13n^3 - n^2 + n^4 + 3n^4 - 9n^2$

Perform the indicated operations.

15. $(5t^4 - 3t^2 + 7t + 3) - (t^4 - t^3 + 3t^2 + 8t + 3)$

16. $(2y^2 - 8y + 8) + (-3y^2 + 2y + 3) - (y^2 + 3y - 6)$

17.
$$\begin{array}{r} -6r^5 + 4r^2 - 3 \\ +\ (6r^5 + 12r^2 - 16) \\ \hline \end{array}$$

18.
$$\begin{array}{r} 9t^3 - 4t^2 + 2t + 2 \\ -\ (9t^3 + 8t^2 - 3t - 6) \\ \hline \end{array}$$

19. $3x^2(-9x^3 + 6x^2 - 2x + 1)$

20. $(2r - 3)(r^2 + 2r - 5)$

21. $(t - 8)(t + 3)$

22. $(4x + 3y)(2x - y)$

23. $(5x - 2y)^2$

24. $(10v + 3w)(10v - 3w)$

25. $(x + 1)^3$

26. What polynomial represents, in appropriate units, the perimeter of this square? The area?

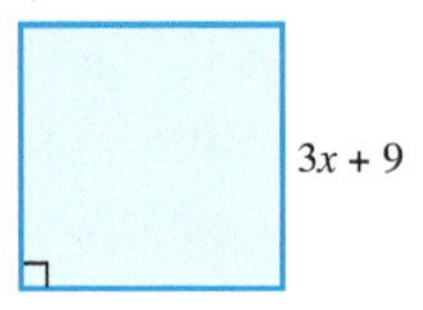

Perform each division.

27. $\dfrac{8y^3 - 6y^2 + 4y + 10}{2y}$

28. $(-9x^2y^3 + 6x^4y^3 + 12xy^3) \div (3xy)$

29. $\dfrac{2x^2 + x - 36}{x - 4}$

30. $(3x^3 - x + 4) \div (x - 2)$

Chapters R–5 Cumulative Review Exercises

Perform each operation.

1. $\frac{2}{3} + \frac{1}{8}$

2. $\frac{7}{4} - \frac{9}{5}$

3. $8.32 - 4.6$

4. 0.07×0.0006

5. A retailer has $34,000 invested in her business. She finds that last year she earned 5.4% on this investment. How much did she earn?

Find the value of each expression for $x = -2$ and $y = 4$.

6. $\frac{4x - 2y}{x + y}$

7. $x^3 - 4xy$

Perform the indicated operations.

8. $\frac{(-13 + 15) - (3 + 2)}{6 - 12}$

9. $-7 - 3[2 + (5 - 8)]$

Decide what property justifies each statement.

10. $(9 + 2) + 3 = 9 + (2 + 3)$

11. $-7 + 7 = 0$

12. $6(4 + 2) = 6(4) + 6(2)$

Solve each equation.

13. $2x - 7x + 8x = 30$

14. $2 - 3(t - 5) = 4 + t$

15. $2(5h + 1) = 10h + 4$

16. $d = rt$ for r

17. $\frac{x}{5} = \frac{x - 2}{7}$

18. $\frac{1}{3}p - \frac{1}{6}p = -2$

19. $0.05x + 0.15(50 - x) = 5.50$

20. $4 - (3x + 12) = -9 - (3x - 1)$

Solve each problem.

21. In any given time period, a 1-oz mouse takes about 16 times as many breaths as a 3-ton elephant. If the two animals take a combined total of 170 breaths per minute, how many breaths does each take during that time? (Data from *Dinosaurs, Spitfires, and Sea Dragons*, McGowan, C., Harvard University Press.)

22. If a number is subtracted from 8 and this difference is tripled, the result is three times the number. Find this number to learn how many times a dolphin rests during a 24-hr period.

Solve each inequality. Write the solution set in interval notation.

23. $-8x \leq -80$

24. $-2(x + 4) > 3x + 6$

25. $-3 \leq 2x + 5 < 9$

Given $2x - 3y = -6$, find the following.

26. The intercepts of the graph

27. The slope of the line

28. The graph of the equation

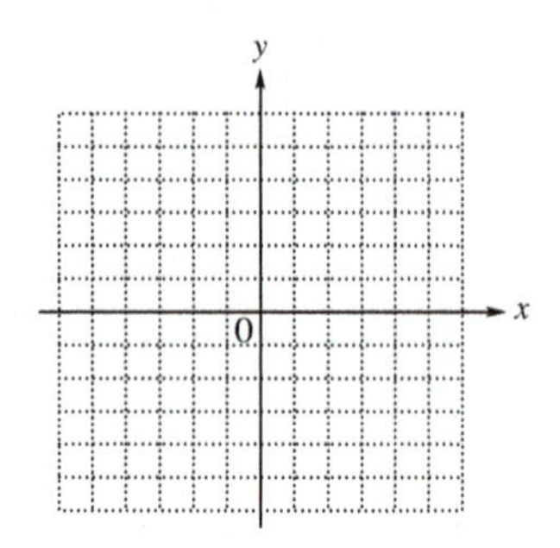

Consider the two points $(-1, 5)$ and $(2, 8)$.

29. Find the slope of the line passing through them.

30. Write, in slope-intercept form, the equation of the line passing through them.

Solve each system of equations using the method indicated.

31. $y = 2x + 5$
$x + y = -4$ (Substitution)

32. $3x + 2y = 2$
$2x + 3y = -7$ (Elimination)

Evaluate each expression.

33. $4^{-1} + 3^0$

34. $2^{-4} \cdot 2^5$

35. $\dfrac{8^{-5} \cdot 8^7}{8^2}$

36. Write $\dfrac{(a^{-3}b^2)^2}{(2a^{-4}b^{-3})^{-1}}$ with positive exponents only.

37. Write 34,500 in scientific notation.

Perform the indicated operations.

38. $(7x^3 - 12x^2 - 3x + 8) + (6x^2 + 4) - (-4x^3 + 8x^2 - 2x - 2)$

39. $6x^5(3x^2 - 9x + 10)$

40. $(7x + 4)(9x + 3)$

41. $(5x + 8)^2$

42. $\dfrac{y^3 - 3y^2 + 8y - 6}{y - 1}$

6

Factoring and Applications

The motion of a freely falling object or of an object that is projected upward can be described using a *quadratic equation*. *Factoring*, a key topic of this chapter, is used when solving some such equations.

6.1 Greatest Common Factors; Factor by Grouping

OBJECTIVES

1. Find the greatest common factor of a list of numbers.
2. Find the greatest common factor of a list of variable terms.
3. Factor out the greatest common factor.
4. Factor by grouping.

To **factor** a number means to write it as the product of two or more numbers.

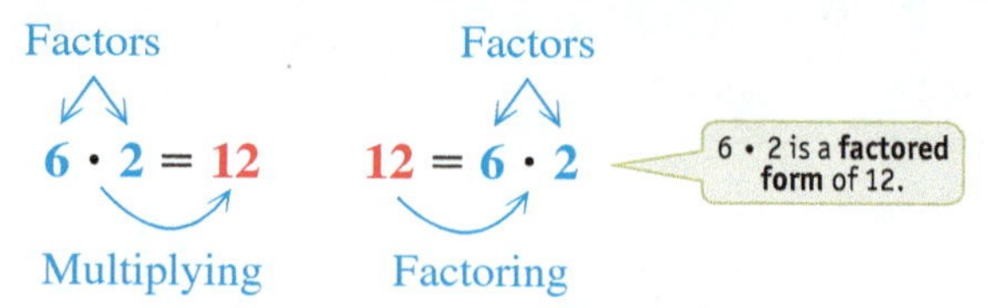

Factoring is a process that "undoes" multiplying. We multiply $6 \cdot 2$ to obtain 12, but we factor 12 by writing it as $6 \cdot 2$.

OBJECTIVE 1 Find the greatest common factor of a list of numbers. An integer that is a factor of two or more integers is a **common factor** of those integers. For example, 6 is a common factor of 18 and 24 because 6 is a factor of both 18 and 24. Other common factors of 18 and 24 are 1, 2, and 3.

The **greatest common factor (GCF)** of a list of integers is the largest common factor of those integers. This means 6 is the greatest common factor of 18 and 24 because it is the largest of their common factors.

Note

Factors of a number are also ***divisors*** of the number. The ***greatest common*** factor is the same as the ***greatest common divisor.*** Here are some divisibility rules for deciding what numbers divide into a given number.

DIVISIBILITY TESTS

A Whole Number Divisible by	Must Have the Following Property
2	Ends in 0, 2, 4, 6, or 8
3	Sum of digits divisible by 3
4	Last two digits form a number divisible by 4
5	Ends in 0 or 5
6	Divisible by both 2 and 3
8	Last three digits form a number divisible by 8
9	Sum of digits divisible by 9
10	Ends in 0

EXAMPLE 1 Finding the Greatest Common Factor for Numbers

Find the greatest common factor for each list of numbers.

(a) 30, 45

$$30 = 2 \cdot 3 \cdot 5$$
$$45 = 3 \cdot 3 \cdot 5$$
Write the prime factored form of each number.

Use each prime the* least *number of times it appears in* all *the factored forms. There is no 2 in the prime factored form of 45, so there will be no 2 in the greatest common factor. The least number of times 3 appears in all the factored forms is 1. The least number of times 5 appears is also 1.

$$\text{GCF} = 3^1 \cdot 5^1 = 15 \qquad 3^1 = 3 \text{ and } 5^1 = 5.$$

Continued on Next Page

(b) 72, 120, 432

$$72 = 2 \cdot 2 \cdot 2 \cdot 3 \cdot 3$$
$$120 = 2 \cdot 2 \cdot 2 \cdot 3 \cdot 5$$
$$432 = 2 \cdot 2 \cdot 2 \cdot 2 \cdot 3 \cdot 3 \cdot 3$$

Write the prime factored form of each number.

The least number of times 2 appears in all the factored forms is 3, and the least number of times 3 appears is 1. There is no 5 in the prime factored form of either 72 or 432.

$$\text{GCF} = 2^3 \cdot 3^1 = 24$$

$2^3 = 8$ and $3^1 = 3$.

We can also align factors vertically and use exponents in the prime factorizations to organize the lists of factors.

$$72 = 2^3 \cdot 3^2$$
$$120 = 2^3 \cdot 3^1 \cdot 5$$
$$432 = 2^4 \cdot 3^3$$
$$\text{GCF} = 2^3 \cdot 3^1 = 24$$

The exponent on a factor in the GCF is the ***least exponent*** *that appears on a factor in* ***all*** *the terms.*

(c) 10, 11, 14

$$10 = 2 \cdot 5$$
$$11 = 11$$
$$14 = 2 \cdot 7$$

Write the prime factored form of each number.

There are no primes common to all three numbers. In such cases,

$$\text{GCF} = 1.$$

Work Problem 1 at the Side. ▶

OBJECTIVE 2 Find the greatest common factor of a list of variable terms. The terms x^4, x^5, x^6, and x^7 have x^4 as the greatest common factor because the least exponent on the variable x in the factored forms is 4.

$$x^4 = 1 \cdot x^4, \quad x^5 = x \cdot x^4, \quad x^6 = x^2 \cdot x^4, \quad x^7 = x^3 \cdot x^4$$
$$\text{GCF} = x^4$$

Finding the Greatest Common Factor (GCF)

Step 1 **Factor.** Write each number in prime factored form.

Step 2 **List common factors.** List each prime number or each variable that is a factor of every term in the list. (If a prime does not appear in one of the prime factored forms, it *cannot* appear in the greatest common factor.)

Step 3 **Choose least exponents.** Use as exponents on the common prime factors the *least* exponents from the prime factored forms.

Step 4 **Multiply** the primes from Step 3. If there are no primes left after Step 3, the greatest common factor is 1.

---

1 Find the greatest common factor for each list of numbers.

GS **(a)** 30, 20, 15

$30 = 2 \cdot 3 \cdot 5$

$20 = 2 \cdot$ ____ $\cdot$ ____

$15 = 3 \cdot$ ____

GCF = ____

(b) 42, 28, 35

(c) 12, 18, 26, 32

(d) 10, 15, 21

Answers

1. **(a)** 2; 5; 5; 5 **(b)** 7 **(c)** 2 **(d)** 1

2 Find the greatest common factor for each list of terms.

(a) $6m^4, 9m^2, 12m^5$

$6m^4 = 2 \cdot \_\_\_\_ \cdot m^4$

$9m^2 = 3 \cdot \_\_\_\_ \cdot \_\_\_\_$

$12m^5 = 2 \cdot 2 \cdot \_\_\_\_ \cdot \_\_\_\_$

GCF = ____

(b) $12p^5, 18q^4$

(c) y^4z^2, y^6z^8, z^9

(d) $12p^{11}, 17q^5$

Answers

2. (a) 3; 3; m^2; 3; m^5; $3m^2$
(b) 6 (c) z^2 (d) 1

EXAMPLE 2 Finding the Greatest Common Factor for Variable Terms

Find the greatest common factor for each list of terms.

(a) $21m^7, \quad 18m^6, \quad 45m^8, \quad 24m^5$

$$21m^7 = 3 \cdot 7 \cdot m^7$$
$$18m^6 = 2 \cdot 3 \cdot 3 \cdot m^6$$
$$45m^8 = 3 \cdot 3 \cdot 5 \cdot m^8$$
$$24m^5 = 2 \cdot 2 \cdot 2 \cdot 3 \cdot m^5$$

Here, 3 is the greatest common factor of the coefficients 21, 18, 45, and 24. The least exponent on m is 5.

$$\text{GCF} = 3m^5$$

(b) $x^4y^2, \quad x^7y^5, \quad x^3y^7, \quad y^{15}$

$$x^4y^2 = x^4 \cdot y^2$$
$$x^7y^5 = x^7 \cdot y^5$$
$$x^3y^7 = x^3 \cdot y^7$$
$$y^{15} = y^{15}$$

There is no x in the last term, y^{15}, so x will not appear in the greatest common factor. There is a y in each term, however, and 2 is the least exponent on y.

$$\text{GCF} = y^2$$

◀ Work Problem **2** at the Side.

OBJECTIVE 3 Factor out the greatest common factor. **Factoring** a polynomial is the process of writing a polynomial sum in factored form as a product. For example, the polynomial

$$3m + 12$$

has two terms, $3m$ and 12. The greatest common factor of these two terms is 3. We can write $3m + 12$ so that each term is a product with 3 as one factor.

$$3m + 12$$
$$= 3 \cdot m + 3 \cdot 4 \quad \text{GCF} = 3$$
$$= 3(m + 4) \quad \text{Distributive property, } a \cdot b + a \cdot c = a(b + c)$$

The factored form of $3m + 12$ is $3(m + 4)$. This process is called **factoring out the greatest common factor.**

! CAUTION

The polynomial $3m + 12$ is *not* in factored form when written as

$$3 \cdot m + 3 \cdot 4. \quad \text{Not in factored form}$$

The terms are factored, but the polynomial is not. The factored form of $3m + 12$ is the ***product***

The factors are 3 and $(m + 4)$. $\quad 3(m + 4)$. In factored form

EXAMPLE 3 Factoring Out the Greatest Common Factor

Write in factored form by factoring out the greatest common factor.

(a) $5y^2 + 10y$

$$= 5y(y) + 5y(2) \quad \text{GCF} = 5y$$
$$= 5y(y + 2) \quad \text{Distributive property}$$

Continued on Next Page

CHECK Multiply the factored form.

$5y(y + 2)$

$= 5y(y) + 5y(2)$ Distributive property, $a(b + c) = ab + ac$

$= 5y^2 + 10y$ ✓ Original polynomial

(b) $20m^5 + 10m^4 - 15m^3$

$= \mathbf{5m^3}(4m^2) + \mathbf{5m^3}(2m) - \mathbf{5m^3}(3)$ GCF $= 5m^3$

$= \mathbf{5m^3}(4m^2 + 2m - 3)$ Factor out $5m^3$.

CHECK $5m^3(4m^2 + 2m - 3)$

$= 5m^3(4m^2) + 5m^3(2m) + 5m^3(-3)$ Distributive property

$= 20m^5 + 10m^4 - 15m^3$ ✓ Original polynomial

(c) $x^5 + x^3$

$= \mathbf{x^3}(x^2) + \mathbf{x^3}(\mathbf{1})$ GCF $= x^3$

$= \mathbf{x^3}(x^2 + \mathbf{1})$ Don't forget the 1.

Check mentally by distributing x^3 over each term inside the parentheses.

(d) $20m^7p^2 - 36m^3p^4$

$= \mathbf{4m^3p^2}(5m^4) - \mathbf{4m^3p^2}(9p^2)$ GCF $= 4m^3p^2$

$= \mathbf{4m^3p^2}(5m^4 - 9p^2)$ Factor out $4m^3p^2$.

Check mentally by distributing $4m^3p^2$ over each term inside the parentheses.

Work Problem 3 at the Side. ▶

EXAMPLE 4 Factoring Out a Negative Common Factor

Write $-8x^4 + 16x^3 - 4x^2$ in factored form.

We can factor out either $4x^2$ or $-4x^2$ here. So that the coefficient of the first term in the trinomial factor will be positive, we factor out $-4x^2$.

$-8x^4 + 16x^3 - 4x^2$ Be careful with signs.

$= \mathbf{-4x^2}(2x^2) \mathbf{- 4x^2}(-4x) \mathbf{- 4x^2}(\mathbf{1})$ $-4x^2$ is a common factor.

$= \mathbf{-4x^2}(2x^2 - 4x + \mathbf{1})$ Factor out $-4x^2$.

CHECK $-4x^2(2x^2 - 4x + 1)$

$= -4x^2(2x^2) - 4x^2(-4x) - 4x^2(1)$ Distributive property

$= -8x^4 + 16x^3 - 4x^2$ ✓ Original polynomial

Work Problem 4 at the Side. ▶

Note

When the coefficient of the first term in a polynomial is negative, we will often factor out the negative common factor, even if it is just -1. It would also be correct to factor out $4x^2$ in **Example 4** to obtain

$$4x^2(-2x^2 + 4x - 1).$$

3 Write in factored form by factoring out the greatest common factor.

GS **(a)** $4x^2 + 6x$

$= 2x(\_\_\_\_) + 2x(\_\_\_\_)$

$= \_\_\_\_(\_\_\_\_ + \_\_\_\_)$

(b) $10y^5 - 8y^4 + 6y^2$

GS **(c)** $m^7 + m^9$

$= m^7(\_\_\_\_) + \_\_\_\_(m^2)$

$= \_\_\_\_(\_\_\_\_ + \_\_\_\_)$

(d) $15x^3 - 10x^2 + 5x$

(e) $8p^5q^2 + 16p^6q^3 - 12p^4q^7$

4 Write

$$-14a^3 - 21a^2 + 7a$$

in factored form by factoring out a negative common factor.

Answers

3. (a) $2x$; 3; $2x$; $2x$; 3
(b) $2y^2(5y^3 - 4y^2 + 3)$
(c) 1; m^7; m^7; 1; m^2
(d) $5x(3x^2 - 2x + 1)$
(e) $4p^4q^2(2p + 4p^2q - 3q^5)$

4. $-7a(2a^2 + 3a - 1)$

5 Write in factored form by factoring out the greatest common factor.

GS **(a)** $r(t-4)+5(t-4)$
$= (t-4)(\_\_\_\_\_\_)$

(b) $x(x+2)+7(x+2)$

(c) $y^2(y+2)-3(y+2)$

(d) $x(x-1)-5(x-1)$

Answers

5. **(a)** $r+5$ **(b)** $(x+2)(x+7)$
(c) $(y+2)(y^2-3)$
(d) $(x-1)(x-5)$

EXAMPLE 5 Factoring Out a Common Binomial Factor

Write in factored form by factoring out the greatest common factor.

Same

(a) $a(a+3)+4(a+3)$ The binomial $a+3$ is the greatest common factor.
$= (a+3)(a+4)$ Factor out $a+3$.

(b) $x^2(x+1)-5(x+1)$
$= (x+1)(x^2-5)$ Factor out $x+1$.

◀ Work Problem **5** at the Side.

Note

In factored forms like those in **Example 5,** the order of the factors does not matter because of the commutative property of multiplication, $ab = ba$.

$(a+3)(a+4)$ can also be written $(a+4)(a+3)$.

OBJECTIVE 4 **Factor by grouping.** ***When a polynomial has four terms, common factors can sometimes be used to factor by grouping.***

EXAMPLE 6 Factoring by Grouping

Factor by grouping.

(a) $2x+6+ax+3a$

Group the first two terms and the last two terms because the first two terms have a common factor of 2 and the last two terms have a common factor of a.

$2x+6+ax+3a$
$= (2x+6)+(ax+3a)$ Group the terms.
$= 2(x+3)+a(x+3)$ Factor each group.

The expression is still not in factored form because it is the *sum* of two terms. Now, however, $x+3$ is a common factor and can be factored out.

$= 2(x+3)+a(x+3)$ $x+3$ is a common factor.
$= (x+3)(2+a)$ Factor out $x+3$.

$(2+a)(x+3)$ is also correct.

The final result is in factored form because it is a ***product.***

CHECK $(x+3)(2+a)$

$= x(2)+x(a)+3(2)+3(a)$ Multiply using the FOIL method.
$= 2x+ax+6+3a$ Simplify.
$= 2x+6+ax+3a$ ✓ Rearrange terms to obtain the original polynomial.

Continued on Next Page

(b) $6ax + 24x + a + 4$

$= (6ax + 24x) + (a + 4)$ Group the terms.

$= 6x(a + 4) + 1(a + 4)$ Factor each group.

Remember the 1.

$= (a + 4)(6x + 1)$ Factor out $a + 4$.

CHECK $(a + 4)(6x + 1)$

$= 6ax + a + 24x + 4$ FOIL method

$= 6ax + 24x + a + 4$ ✓ Rearrange terms to obtain the original polynomial.

(c) $2x^2 - 10x + 3xy - 15y$

$= (2x^2 - 10x) + (3xy - 15y)$ Group the terms.

$= 2x(x - 5) + 3y(x - 5)$ Factor each group.

$= (x - 5)(2x + 3y)$ Factor out the common factor, $x - 5$.

CHECK $(x - 5)(2x + 3y)$

$= 2x^2 + 3xy - 10x - 15y$ FOIL method

$= 2x^2 - 10x + 3xy - 15y$ ✓ Original polynomial

(d) $t^3 + 2t^2 - 3t - 6$

Write a + sign between the groups.

$= (t^3 + 2t^2) + (-3t - 6)$ Group the terms.

$= t^2(t + 2) - 3(t + 2)$ Factor out -3 so there is a common factor, $t + 2$. ***Check:*** $-3(t + 2) = -3t - 6$

Be careful with signs.

$= (t + 2)(t^2 - 3)$ Factor out $t + 2$.

Check by multiplying using the FOIL method.

Work Problem 6 at the Side. ▶

CAUTION

Be careful with signs when grouping in a problem like **Example 6(d).** It is wise to check the factoring in the second step, as shown in the example side comment, before continuing.

Factoring a Polynomial with Four Terms by Grouping

Step 1 **Group terms.** Collect the terms into two groups so that each group has a common factor.

Step 2 **Factor within groups.** Factor out the greatest common factor from each group.

Step 3 **If possible, factor the entire polynomial.** Factor out a common binomial factor from the results of Step 2.

Step 4 **If necessary, rearrange terms.** If Step 2 does not result in a common binomial factor, try a different grouping.

Always check the factored form by multiplying.

6 Factor by grouping.

GS **(a)** $pq + 5q + 2p + 10$

$= (\_\_ + 5q) + (\_\_ + 10)$

$= \_\_(p + 5) + \_\_(p + 5)$

$= (p + 5)(\_\_\_\_)$

(b) $2xy + 3y + 2x + 3$

(c) $2a^2 - 4a + 3ab - 6b$

(d) $x^3 + 3x^2 - 5x - 15$

Answers

6. **(a)** pq; $2p$; q; 2; $q + 2$
(b) $(2x + 3)(y + 1)$
(c) $(a - 2)(2a + 3b)$
(d) $(x + 3)(x^2 - 5)$

7 Factor by grouping.

(a) $6y^2 - 20w + 15y - 8yw$

(b) $9mn - 4 + 12m - 3n$

(c) $12p^2 - 28q - 16pq + 21p$

EXAMPLE 7 Rearranging Terms before Factoring by Grouping

Factor by grouping.

(a) $10x^2 - 12y + 15x - 8xy$

Factoring out the common factor 2 from the first two terms and the common factor x from the last two terms gives the following.

$$(10x^2 - 12y) + (15x - 8xy) \quad \text{Group the terms.}$$
$$= 2(5x^2 - 6y) + x(15 - 8y) \quad \text{Factor each group.}$$

This does not lead to a common factor, so we try rearranging the terms. There is usually more than one way to do this. We try the following.

$$10x^2 - 12y + 15x - 8xy$$
$$= 10x^2 - 8xy - 12y + 15x \quad \text{Commutative property}$$
$$= (10x^2 - 8xy) + (-12y + 15x) \quad \text{Group the terms.}$$
$$= 2x(5x - 4y) + 3(-4y + 5x) \quad \text{Factor each group.}$$
$$= 2x(5x - 4y) + 3(5x - 4y) \quad \text{Rewrite } -4y + 5x.$$
$$= (5x - 4y)(2x + 3) \quad \text{Factor out } 5x - 4y.$$

CHECK $(5x - 4y)(2x + 3)$

$$= 10x^2 + 15x - 8xy - 12y \quad \text{FOIL method}$$
$$= 10x^2 - 12y + 15x - 8xy \checkmark \quad \text{Original polynomial}$$

(b) $2xy + 12 - 3y - 8x$

We must rearrange the terms to obtain two groups that each have a common factor. Trial and error suggests the following grouping.

$$2xy + 12 - 3y - 8x$$

Write a + sign between the two groups.

$$= (2xy - 3y) + (-8x + 12) \quad \text{Group the terms.}$$
$$= y(2x - 3) - 4(2x - 3) \quad \text{Factor each group;}$$

Check: $-4(2x - 3) = -8x + 12$

Be careful with signs.

$$= (2x - 3)(y - 4) \quad \text{Factor out } 2x - 3.$$

Because the quantities in parentheses in the second step must be the same, we factored out -4 rather than 4.

CHECK $(2x - 3)(y - 4)$

$$= 2xy - 8x - 3y + 12 \quad \text{FOIL method}$$
$$= 2xy + 12 - 3y - 8x \checkmark \quad \text{Original polynomial}$$

◀ **Work Problem 7 at the Side.**

CAUTION

Use negative signs carefully when grouping, as in **Example 7(b)**, or a sign error will occur. ***Always check by multiplying.***

Answers

7. (a) $(2y + 5)(3y - 4w)$
(b) $(3m - 1)(3n + 4)$
(c) $(3p - 4q)(4p + 7)$

6.1 Exercises

FOR EXTRA HELP — *Go to MyMathLab for worked-out, step-by-step solutions to exercises enclosed in a square and video solutions to ▶ exercises.*

CONCEPT CHECK *Complete each statement.*

1. To factor a number or quantity means to write it as a ________. Factoring is the opposite, or inverse, process of ________.

2. An integer or variable expression that is a factor of two or more terms is a ________ ________. For example, 12 (*is* / *is not*) a common factor of both 36 and 72 since it ________ evenly into both integers.

Find the greatest common factor for each list of numbers. ***See Example 1.***

3. 12, 16 **4.** 18, 24 **5.** 40, 20, 4 **6.** 50, 30, 5

7. 18, 24, 36, 48 **8.** 15, 30, 45, 75 **9.** 4, 9, 12 **10.** 9, 16, 24

Find the greatest common factor for each list of terms. ***See Example 2.***

11. $16y, 24$ **12.** $18w, 27$ **13.** $30x^3, 40x^6, 50x^7$

14. $60z^4, 70z^8, 90z^9$ **15.** $15m^2, 30m^4, 60m^2$ **16.** $12y^5, 36y^5, 72y^7$

17. x^4y^3, xy^2 **18.** a^4b^5, a^3b **19.** $42ab^3, 36a, 90b, 48ab$

20. $45c^3d, 75c, 90d, 105cd$ **21.** $12m^3n^2, 18m^5n^4, 36m^8n^3$ **22.** $25p^5r^7, 30p^7r^8, 50p^5r^3$

CONCEPT CHECK *An expression is factored when it is written as a product, not a sum. Determine whether each expression is* factored *or* not factored.

23. $2k^2(5k)$ **24.** $2k^2(5k+1)$ **25.** $2k^2+(5k+1)$ **26.** $(2k^2+5k)+1$

27. CONCEPT CHECK A student incorrectly factored

$$18x^3y^2 + 9xy \quad \text{as} \quad 9xy(2x^2y).$$

What Went Wrong? Factor correctly.

28. CONCEPT CHECK When asked to factor completely, a student incorrectly factored

$$12x^2y - 24xy \quad \text{as} \quad 3xy(4x-8).$$

What Went Wrong? Factor correctly.

GS *Complete each factoring by writing each polynomial as the product of two factors.* ***See Example 3.***

29. $9m^4 = 3m^2(\_\_\_\_)$ **30.** $12p^5 = 6p^3(\_\_\_\_)$ **31.** $-8z^9 = -4z^5(\_\_\_\_)$

32. $-15k^{11} = -5k^8(\_\_\_\_)$ **33.** $6m^4n^5 = 3m^3n(\_\_\_\_)$ **34.** $27a^3b^2 = 9a^2b(\_\_\_\_)$

35. $12y + 24 = 12(\_\_\_\_)$ **36.** $18p + 36 = 18(\_\_\_\_)$ **37.** $10a^2 - 20a = 10a(\_\_\_\_)$

38. $15x^2 - 30x = 15x(\_\_\_\_)$ **39.** $8x^2y + 12x^3y^2 = 4x^2y(\_\_\_\_)$ **40.** $18s^3t^2 + 10st = 2st(\_\_\_\_)$

Write in factored form by factoring out the greatest common factor (or a negative common factor if the coefficient of the term of greatest degree is negative). ***See Examples 3–5.***

41. $x^2 - 4x$

42. $m^2 - 7m$

43. $6t^2 + 15t$

44. $8x^2 + 6x$

45. $m^3 - m^2$

46. $p^3 - p^2$

47. $-12x^3 - 6x^2$

48. $-21b^3 - 7b^2$

49. $16z^4 + 24z^2$

50. $100a^5 + 16a^3$

51. $11w^3 - 100$

52. $13z^5 - 80$

53. $8mn^3 + 24m^2n^3$

54. $19p^2y + 38p^2y^3$

55. $-4x^3 + 10x^2 - 6x$

56. $-9z^3 + 6z^2 - 12z$

57. $13y^8 + 26y^4 - 39y^2$

58. $5x^5 + 25x^4 - 20x^3$

59. $45q^4p^5 - 36qp^6 + 81q^2p^3$

60. $125a^3z^5 + 60a^4z^4 - 85a^5z^2$

61. $c(x + 2) + d(x + 2)$

62. $r(5 - x) + t(5 - x)$

63. $a^2(2a + b) - b(2a + b)$

64. $3x(x^2 + 5) - y(x^2 + 5)$

65. $q(p + 4) - 1(p + 4)$

66. $y^2(x - 4) + 1(x - 4)$

CONCEPT CHECK *Students often have difficulty when factoring by grouping because they are not able to tell when the polynomial is completely factored. For example,*

$$5y(2x - 3) + 8t(2x - 3) \quad \text{Not in factored form}$$

is not in factored form because it is the ***sum*** *of two terms:* $5y(2x - 3)$ *and* $8t(2x - 3)$*. However, because* $2x - 3$ *is a common factor of these two terms, the expression can now be factored.*

$$(2x - 3)(5y + 8t) \quad \text{In factored form}$$

The factored form is a ***product*** *of the two factors* $2x - 3$ *and* $5y + 8t$*.*

Determine whether each expression is in factored form *or is* not in factored form*. If it is not in factored form, factor it if possible.*

67. $8(7t + 4) + x(7t + 4)$

68. $3r(5x - 1) + 7(5x - 1)$

69. $(8 + x)(7t + 4)$

70. $(3r + 7)(5x - 1)$

71. $18x^2(y + 4) + 7(y - 4)$

72. $12k^3(s - 3) + 7(s + 3)$

Factor by grouping. ***See Examples 6 and 7.***

73. $5m + mn + 20 + 4n$

74. $ts + 5t + 2s + 10$

75. $6xy - 21x + 8y - 28$

76. $2mn - 8n + 3m - 12$

77. $a^2 - 2a + ab - 2b$

78. $y^2 - 6y + yw - 6w$

79. $7z^2 + 14z - az - 2a$

80. $2b^2 + 3b - 8ab - 12a$

81. $18r^2 + 12ry - 3xr - 2xy$

82. $5m^2 + 15mp - 2mr - 6pr$

83. $w^3 + w^2 + 9w + 9$

84. $y^3 + y^2 + 6y + 6$

85. $3a^3 + 6a^2 - 2a - 4$

86. $10x^3 + 15x^2 - 8x - 12$

87. $16m^3 - 4m^2p^2 - 4mp + p^3$

88. $10t^3 - 2t^2s^2 - 5ts + s^3$

89. $y^2 + 3x + 3y + xy$

90. $m^2 + 14p + 7m + 2mp$

91. $2z^2 + 6w - 4z - 3wz$

92. $2a^2 + 20b - 8a - 5ab$

93. $5m - 6p - 2mp + 15$

94. $7y - 9x - 3xy + 21$

95. $18r^2 - 2ty + 12ry - 3rt$

96. $12a^2 - 4bc + 16ac - 3ab$

Relating Concepts (Exercises 97–100) For Individual or Group Work

In many cases, the choice of which pairs of terms to group when factoring by grouping can be made in different ways. To see this for ***Example 7(b), work Exercises 97–100 in order.***

97. Start with the polynomial from **Example 7(b),** $2xy + 12 - 3y - 8x$, and rearrange the terms as follows: $2xy - 8x - 3y + 12$. What property allows this?

98. Group the first two terms and the last two terms of the rearranged polynomial in **Exercise 97.** Then factor each group.

99. Is the result from **Exercise 98** in factored form? Explain.

100. If the answer to **Exercise 99** is *no,* factor the polynomial. Is the result the same as the one shown for **Example 7(b)?**

6.2 Factoring Trinomials

OBJECTIVES

1. Factor trinomials with coefficient 1 for the second-degree term.
2. Factor such trinomials after factoring out the greatest common factor.

Using the FOIL method, we can find the product of the binomials $k - 3$ and $k + 1$.

$$(k - 3)(k + 1) = k^2 - 2k - 3 \quad \text{Multiplying}$$

Suppose instead that we are given the polynomial $k^2 - 2k - 3$ and want to rewrite it as the product $(k - 3)(k + 1)$.

$$k^2 - 2k - 3 = (k - 3)(k + 1) \quad \text{Factoring}$$

Recall that this process is called *factoring* the polynomial. Factoring reverses, or "undoes," multiplying.

OBJECTIVE 1 Factor trinomials with coefficient 1 for the second-degree term. When factoring polynomials with integer coefficients, we use only integers in the factors. For example, we can factor $x^2 + 5x + 6$ by finding two integers m and n such that

$$x^2 + 5x + 6 \quad \text{is written as} \quad (x + m)(x + n).$$

To find these integers m and n, we multiply the two binomials on the right.

$$(x + m)(x + n)$$
$$= x^2 + nx + mx + mn \quad \text{FOIL method}$$
$$= x^2 + (n + m)x + mn \quad \text{Distributive property}$$

Comparing this result with $x^2 + 5x + 6$ shows that we must find integers m and n having a sum of 5 and a product of 6.

Product of m and n is 6.

$$x^2 + 5x + 6 = x^2 + (n + m)x + mn$$

Sum of m and n is 5.

Because many pairs of integers have a sum of 5, it is best to begin by listing those pairs of integers whose product is 6. Both 5 and 6 are positive, so we consider only pairs in which both integers are positive.

◀ Work Problem 1 at the Side.

1 (a) Complete the table to find pairs of positive integers whose product is 6. Then find the sum of each pair.

Factors of 6	Sums of Factors
6, ___	6 + ___ = ___
___, 2	___ + 2 = ___

(b) Which pair of factors from the table in part (a) has a sum of 5?

From **Margin Problem 1,** we see that the numbers 6 and 1 and the numbers 3 and 2 both have a product of 6, but only the pair 3 and 2 has a sum of 5. So 3 and 2 are the required integers.

$$x^2 + 5x + 6 \quad \text{is factored as} \quad (x + 3)(x + 2).$$

Check by using the FOIL method to multiply the binomials. ***Make sure that the sum of the outer and inner products produces the correct middle term.***

CHECK $\quad (x + 3)(x + 2) = x^2 + 5x + 6$ ✓ Correct

$3x$
$\underline{2x}$
$5x$ Add.

This method of factoring can be used only for trinomials that have 1 as the coefficient of the second-degree (squared) term.

Answers

1. (a) 1; 1; 7; 3; 3; 5 (b) 3, 2

Note

Consider all possible sign combinations for multiplying two bionomials to understand the resulting signs of the terms of a trinomial.

$(x + 3)(x + 2) = x^2 + 5x + 6$ — All signs in the trinomial product are positive.
Both signs positive

$(x - 3)(x - 2) = x^2 - 5x + 6$ — The last term of the trinomial product is positive and the middle term is negative.
Both signs negative

$(x + 3)(x - 2) = x^2 + x - 6$
$(x - 3)(x + 2) = x^2 - x - 6$
Different signs — The last term of the trinomial product is negative and the middle term has the same sign as the number in the binomials with the greater absolute value.

EXAMPLE 1 Factoring a Trinomial (All Positive Terms)

Factor $m^2 + 9m + 14$.

Look for two integers whose product is 14 and whose sum is 9. List the pairs of integers whose products are 14, and examine the sums. Only positive integers are needed because all signs in $m^2 + 9m + 14$ are positive.

Factors of 14	Sums of Factors
14, 1	$14 + 1 = 15$
7, 2	$7 + 2 = 9$

Sum is 9.

The required integers are 7 and 2.

$m^2 + 9m + 14$ factors as $(m + 7)(m + 2)$. — $(m + 2)(m + 7)$ is also correct.

CHECK $(m + 7)(m + 2)$

$= m^2 + 2m + 7m + 14$ FOIL method

$= m^2 + 9m + 14$ ✓ Original polynomial

Work Problem 2 at the Side. ▶

EXAMPLE 2 Factoring a Trinomial (Negative Middle Term)

Factor $x^2 - 9x + 20$.

Find two integers whose product is 20 and whose sum is -9. Because the numbers we are looking for have a *positive product* and a *negative sum*, we consider only pairs of negative integers.

Factors of 20	Sums of Factors
$-20, -1$	$-20 + (-1) = -21$
$-10, -2$	$-10 + (-2) = -12$
$-5, -4$	$-5 + (-4) = -9$

Sum is -9.

The required integers are -5 and -4.

$x^2 - 9x + 20$ factors as $(x - 5)(x - 4)$. — The order of the factors does not matter.

CHECK $(x - 5)(x - 4)$

$= x^2 - 4x - 5x + 20$ FOIL method

$= x^2 - 9x + 20$ ✓ Original polynomial

Work Problem 3 at the Side. ▶

2 Factor each trinomial.

GS **(a)** $y^2 + 12y + 20$

Find two integers whose product is ____ and whose sum is ____. Complete the table.

Factors of 20	Sums of Factors
20, 1	$20 + 1 = 21$
10, ___	$10 +$ ___ $=$ ___
5, ___	$5 +$ ___ $=$ ___

Which pair of factors has the required sum? ______
Now factor the trinomial.

(b) $x^2 + 9x + 18$

3 Factor each trinomial.

GS **(a)** $t^2 - 12t + 32$

Find two integers whose product is ____ and whose sum is ____. Complete the table.

Factors of 32	Sums of Factors
$-32, -1$	$-32 + (-1) = -33$
$-16,$ ___	$-16 + ($___$) =$ ___
$-8,$ ___	$-8 + ($___$) =$ ___

Which pair of factors has the required sum? ______
Now factor the trinomial.

(b) $y^2 - 10y + 24$

Answers

2. **(a)** 20; 12; 2; 2; 12; 4; 4; 9; 10 and 2; $(y + 10)(y + 2)$
(b) $(x + 3)(x + 6)$

3. **(a)** 32; -12; -2; -2; -18; -4; -4; -12; -8 and -4; $(t - 8)(t - 4)$
(b) $(y - 6)(y - 4)$

4 Factor each trinomial.

(a) $z^2 + z - 30$

Factors of -30	Sums of Factors

(b) $x^2 + x - 42$

EXAMPLE 3 Factoring a Trinomial (Negative Last (Constant) Term)

Factor $x^2 + x - 6$.

We must find two integers whose product is -6 and whose sum is 1 (since the coefficient of x, or $1x$, is 1). To obtain a *negative product,* the pairs of integers must have different signs.

Once we find the required pair, we can stop listing factors.

Factors of -6	Sums of Factors	
$6, -1$	$6 + (-1) = 5$	
$-6, 1$	$-6 + 1 = -5$	
$3, -2$	$3 + (-2) = 1$	Sum is 1.

The required integers are 3 and -2.

$$x^2 + x - 6 \quad \text{factors as} \quad (x + 3)(x - 2).$$

CHECK $(x + 3)(x - 2)$

$= x^2 - 2x + 3x - 6$ FOIL method

$= x^2 + x - 6$ ✓ Original polynomial

◀ Work Problem 4 at the Side.

Note

Remember that because of the commutative property of multiplication, the order of the factors does not matter. ***Always check by multiplying.***

5 Factor each trinomial.

(a) $a^2 - 9a - 22$

Factors of -22	Sums of Factors

(b) $r^2 - 6r - 16$

EXAMPLE 4 Factoring a Trinomial (Two Negative Terms)

Factor $p^2 - 2p - 15$.

Find two integers whose product is -15 and whose sum is -2. Because the constant term, -15, is negative, we need pairs of integers with different signs.

Factors of -15	Sums of Factors	
$15, -1$	$15 + (-1) = 14$	
$-15, 1$	$-15 + 1 = -14$	
$5, -3$	$5 + (-3) = 2$	
$-5, 3$	$-5 + 3 = -2$	Sum is -2.

The required integers are -5 and 3.

$$p^2 - 2p - 15 \quad \text{factors as} \quad (p - 5)(p + 3).$$

CHECK Multiply $(p - 5)(p + 3)$ to obtain $p^2 - 2p - 15$. ✓

◀ Work Problem 5 at the Side.

Note

In **Examples 1–4,** we listed factors in descending order (disregarding their signs) when we were looking for the required pair of integers. This helps avoid skipping the correct combination.

Answers

4. (a) $(z + 6)(z - 5)$ **(b)** $(x + 7)(x - 6)$

5. (a) $(a - 11)(a + 2)$ **(b)** $(r - 8)(r + 2)$

Guidelines for Factoring $x^2 + bx + c$

Find two integers whose product is c and whose sum is b.

1. Both integers must be positive if b and c are positive. (See **Example 1.**)
2. Both integers must be negative if c is positive and b is negative. (See **Example 2.**)
3. One integer must be positive and one must be negative if c is negative. (See **Examples 3 and 4.**)

Some trinomials cannot be factored using only integers. Such trinomials are **prime polynomials.**

EXAMPLE 5 Deciding Whether Polynomials Are Prime

Factor each trinomial if possible.

(a) $x^2 - 5x + 12$

As in **Example 2,** both factors must be negative to give a positive product, 12, and a negative sum, -5. List pairs of negative integers whose product is 12, and examine the sums.

Factors of 12	Sums of Factors
$-12, -1$	$-12 + (-1) = -13$
$-6, -2$	$-6 + (-2) = -8$
$-4, -3$	$-4 + (-3) = -7$

No sum is -5.

None of the pairs of integers has a sum of -5. Therefore, the trinomial $x^2 - 5x + 12$ *cannot be factored using only integers.* It is a prime polynomial.

(b) $k^2 - 8k + 11$

There is no pair of integers whose product is 11 and whose sum is -8, so $k^2 - 8k + 11$ is a prime polynomial.

Work Problem 6 at the Side. ▶

EXAMPLE 6 Factoring a Trinomial with Two Variables

Factor $z^2 - 2bz - 3b^2$.

Here, the coefficient of z in the middle term is $-2b$, so we need to find two expressions whose product is $-3b^2$ and whose sum is $-2b$.

Factors of $-3b^2$	Sums of Factors
$3b, -b$	$3b + (-b) = 2b$
$-3b, b$	$-3b + b = -2b$

Sum is $-2b$.

$$z^2 - 2bz - 3b^2 \quad \text{factors as} \quad (z - 3b)(z + b).$$

CHECK $(z - 3b)(z + b)$

$= z^2 + zb - 3bz - 3b^2$ FOIL method

$= z^2 + 1bz - 3bz - 3b^2$ Identity and commutative properties

$= z^2 - 2bz - 3b^2$ ✓ Combine like terms.

Work Problem 7 at the Side. ▶

---

6 Factor each trinomial if possible.

(a) $x^2 + 5x + 8$

(b) $r^2 - 3r - 4$

(c) $m^2 - 2m + 5$

7 Factor each trinomial.

GS **(a)** $b^2 - 3ab - 4a^2$

We need two expressions whose product is ____ and whose sum is ____.

Complete the table.

Factors of $-4a^2$	Sums of Factors
$4a$, ___	$4a + ($___$) =$ ___
$-4a$, ___	$-4a +$ ___ $=$ ___
$2a, -2a$	$2a + (-2a) = 0$

Which pair of factors has the required sum? ________

Now factor the trinomial.

(b) $r^2 - 6rs + 8s^2$

Answers

6. **(a)** prime
(b) $(r - 4)(r + 1)$
(c) prime

7. **(a)** $-4a^2$; $-3a$; $-a$; $-a$; $3a$; a; a; $-3a$; $-4a$ and a; $(b - 4a)(b + a)$
(b) $(r - 4s)(r - 2s)$

OBJECTIVE 2 **Factor such trinomials after factoring out the greatest common factor.** If the terms of a trinomial have a common factor, first factor it out. Then factor the remaining trinomial as in **Examples 1–6.**

EXAMPLE 7 Factoring a Trinomial with a Common Factor

Factor completely.

(a) $4x^5 - 28x^4 + 40x^3$

There is no second-degree term. Look for a common factor.

$4x^5 - 28x^4 + 40x^3$

$= \mathbf{4x^3(x^2 - 7x + 10)}$ Factor out the greatest common factor, $4x^3$.

Now factor $x^2 - 7x + 10$. The integers $\mathbf{-5}$ and $\mathbf{-2}$ have a product of 10 and a sum of -7.

Include $4x^3$. $= \mathbf{4x^3(x - 5)(x - 2)}$ Completely factored form

CHECK $4x^3(x - 5)(x - 2)$

$= 4x^3(x^2 - 2x - 5x + 10)$ FOIL method

$= 4x^3(x^2 - 7x + 10)$ Combine like terms.

$= 4x^5 - 28x^4 + 40x^3$ ✓ Distributive property

(b) $-3y^8 - 18y^7 + 21y^6$

The coefficient of the first term is negative, so we factor out $\mathbf{-3y^6}$.

$-3y^8 - 18y^7 + 21y^6$

$= \mathbf{-3y^6(y^2 + 6y - 7)}$ Factor out the greatest common factor, $-3y^6$.

Now factor $y^2 + 6y - 7$. The integers $\mathbf{7}$ and $\mathbf{-1}$ have a product of -7 and a sum of 6.

$= \mathbf{-3y^6(y + 7)(y - 1)}$ Completely factored form

CHECK $-3y^6(y + 7)(y - 1)$

$= -3y^6(y^2 - y + 7y - 7)$ FOIL method

$= -3y^6(y^2 + 6y - 7)$ Combine like terms.

$= -3y^8 - 18y^7 + 21y^6$ ✓ Distributive property

◀ **Work Problem 8 at the Side.**

CAUTION

When factoring, always look for a common factor first. If the coefficient of the leading term is negative, we will factor out the negative common factor.

Remember to include the common factor as part of the answer, and check by multiplying out the completely factored form.

8 Factor each trinomial completely.

GS **(a)** $2p^3 + 6p^2 - 8p$

$= \_\_\_(\_\_\_ + 3p - \_\_\_)$

$= \_\_\_(\_\_\_\_\_)(p - 1)$

(b) $3y^4 - 27y^3 + 60y^2$

(c) $-3x^4 + 15x^3 - 18x^2$

(d) $-5y^5 - 15y^4 + 90y^3$

Answers

8. **(a)** $2p$; p^2; 4; $2p$; $p + 4$

(b) $3y^2(y - 5)(y - 4)$

(c) $-3x^2(x - 3)(x - 2)$

(d) $-5y^3(y + 6)(y - 3)$

6.2 Exercises

FOR EXTRA HELP *Go to MyMathLab for worked-out, step-by-step solutions to exercises enclosed in a square and video solutions to ▶ exercises.*

CONCEPT CHECK *Answer each question.*

1. When factoring a trinomial in x as $(x + a)(x + b)$, what must be true of a and b, if the coefficient of the constant term of the trinomial is negative?

2. When factoring a trinomial in x as $(x + a)(x + b)$ what must be true of a and b if the coefficient of the constant term is positive?

3. Which one of the following is the correct factored form of $x^2 - 12x + 32$?

A. $(x - 8)(x + 4)$ **B.** $(x + 8)(x - 4)$

C. $(x - 8)(x - 4)$ **D.** $(x + 8)(x + 4)$

4. What would be the first step in factoring

$$2x^3 + 8x^2 - 10x?$$

(See Example 7.)

5. What polynomial can be factored as

$$(a + 9)(a + 4)?$$

6. What polynomial can be factored as

$$(y - 7)(y + 3)?$$

7. CONCEPT CHECK A student factored as follows.

$$x^3 + 3x^2 - 28x$$
$$= x(x^2 + 3x - 28x)$$
$$= (x + 7)(x - 4) \quad \text{Incorrect}$$

What Went Wrong? Factor correctly.

8. CONCEPT CHECK A student incorrectly factored

$$x^2 + x + 6 \quad \text{as} \quad (x + 6)(x - 1).$$

What Went Wrong? Factor correctly if possible.

List all pairs of integers with the given product. Then find the pair whose sum is given. ***See the tables in Examples 1–4.***

9. Product: 12; Sum: 7

10. Product: 18; Sum: 9

11. Product: −24; Sum: −5

12. Product: −36; Sum: −16

GS *Complete each factoring.* ***See Examples 1–4.***

13. $p^2 + 11p + 30$
$= (p + 5)(\_\_\_\_\_)$

14. $x^2 + 10x + 21$
$= (x + 7)(\_\_\_\_\_)$

15. $x^2 + 15x + 44$
$= (x + 4)(\_\_\_\_\_)$

16. $r^2 + 15r + 56$
$= (r + 7)(\_\_\_\_\_)$

17. $x^2 - 9x + 8$
$= (x - 1)(\_\_\_\_\_)$

18. $t^2 - 14t + 24$
$= (t - 2)(\_\_\_\_\_)$

19. $y^2 - 2y - 15$
$= (y + 3)(\_\_\_\_\_)$

20. $t^2 - t - 42$
$= (t + 6)(\_\_\_\_\_)$

21. $x^2 + 9x - 22$
$= (x - 2)(\_\_\_\_\_)$

22. $x^2 + 6x - 27$
$= (x - 3)(\_\_\_\_\_)$

23. $y^2 - 7y - 18$
$= (y + 2)(\_\_\_\_\_)$

24. $y^2 - 2y - 24$
$= (y + 4)(\_\_\_\_\_)$

Factor completely. If a polynomial cannot be factored, write prime. ***See Examples 1–5.***

25. $y^2 + 9y + 8$

26. $a^2 + 9a + 20$

27. $b^2 + 8b + 15$

28. $x^2 + 6x + 8$

29. $m^2 + m - 20$

30. $p^2 + 4p - 5$

31. $x^2 + 3x - 40$

32. $d^2 + 4d - 45$

33. $x^2 + 4x + 5$

34. $t^2 + 11t + 12$

35. $y^2 - 8y + 15$

36. $y^2 - 6y + 8$

37. $z^2 - 15z + 56$

38. $x^2 - 13x + 36$

39. $r^2 - r - 30$

40. $q^2 - q - 42$

41. $a^2 - 8a - 48$

42. $m^2 - 10m - 24$

Factor completely. ***See Example 6.***

43. $r^2 + 3ra + 2a^2$

44. $x^2 + 5xa + 4a^2$

45. $x^2 + 4xy + 3y^2$

46. $p^2 + 9pq + 8q^2$

47. $t^2 - tz - 6z^2$

48. $a^2 - ab - 12b^2$

49. $v^2 - 11vw + 30w^2$

50. $v^2 - 11vx + 24x^2$

51. $a^2 + 2ab - 15b^2$

52. $m^2 + 4mn - 12n^2$

53. $a^2 - 9ab + 18b^2$

54. $h^2 - 11hk + 28k^2$

Factor completely. ***See Example 7.***

55. $4x^2 + 12x - 40$

56. $5y^2 - 5y - 30$

57. $2t^3 + 8t^2 + 6t$

58. $3t^3 + 27t^2 + 24t$

59. $-2x^6 - 8x^5 + 42x^4$

60. $-4y^5 - 12y^4 + 40y^3$

61. $-a^5 - 3a^4b + 4a^3b^2$

62. $-z^{10} + 4z^9y + 21z^8y^2$

63. $5m^5 + 25m^4 - 40m^2$

64. $12k^5 - 6k^3 + 10k^2$

65. $m^3n - 10m^2n^2 + 24mn^3$

66. $y^3z + 3y^2z^2 - 54yz^3$

6.3 Factoring Trinomials by Grouping

OBJECTIVE 1 Factor trinomials by grouping when the coefficient of the second-degree term is not 1. We now extend our work to factor a trinomial such as $2x^2 + 7x + 6$, where the coefficient of x^2 is *not* 1.

EXAMPLE 1 Factoring a Trinomial by Grouping (Coefficient of the Second-Degree Term Not 1)

Factor $2x^2 + 7x + 6$.

To factor this trinomial, we look for two positive integers whose product is $2 \cdot 6 = 12$ and whose sum is 7.

$$2x^2 + 7x + 6$$

Sum is 7. Product is $2 \cdot 6 = 12$.

The required integers are 3 and 4, since $3 \cdot 4 = 12$ and $3 + 4 = 7$. We use these integers to write the middle term $7x$ as $3x + 4x$.

$$2x^2 + 7x + 6$$
$$= 2x^2 + \underbrace{3x + 4x}_{7x} + 6$$
$$= (2x^2 + 3x) + (4x + 6) \quad \text{Group the terms.}$$
$$= x(2x + 3) + 2(2x + 3) \quad \text{Factor each group.}$$

Must be the same factor

$$= (2x + 3)(x + 2) \quad \text{Factor out } 2x + 3.$$

CHECK Multiply $(2x + 3)(x + 2)$ to obtain $2x^2 + 7x + 6$. ✓

Work Problem 1 at the Side. ▶

EXAMPLE 2 Factoring Trinomials by Grouping

Factor each trinomial.

(a) $6r^2 + r - 1$

We must find two integers with a product of $6(-1) = -6$ and a sum of 1 (because the coefficient of r, or $1r$, is 1). The integers are -2 and 3. We write the middle term r as $-2r + 3r$.

$$6r^2 + r - 1$$
$$= 6r^2 - 2r + 3r - 1 \quad r = -2r + 3r$$
$$= (6r^2 - 2r) + (3r - 1) \quad \text{Group the terms.}$$
$$= 2r(3r - 1) + 1(3r - 1) \quad \text{The binomials must be the same. Remember the 1.}$$
$$= (3r - 1)(2r + 1) \quad \text{Factor out } 3r - 1.$$

CHECK Multiply $(3r - 1)(2r + 1)$ to obtain $6r^2 + r - 1$. ✓

Continued on Next Page

OBJECTIVE

1 Factor trinomials by grouping when the coefficient of the second-degree term is not 1.

1 **(a)** In **Example 1,** we factored

$$2x^2 + 7x + 6$$

by writing $7x$ as $3x + 4x$. This trinomial can also be factored by writing $7x$ as $4x + 3x$.

Complete the following.

$$2x^2 + \mathbf{7x} + 6$$
$$= 2x^2 + \mathbf{4x} + \mathbf{3x} + 6$$
$$= (2x^2 + \_\_\_\_) + (3x + \_\_\_\_)$$
$$= 2x(x + \_\_\_\_) + 3(x + \_\_\_\_)$$
$$= (\_\_\_\_\_)(2x + 3)$$

(b) Is the answer in part (a) the same as in **Example 1?** (Remember that the order of the factors does not matter.)

(c) Factor $2z^2 + 5z + 3$ by grouping.

Answers

1. **(a)** $4x$; 6; 2; 2; $x + 2$ **(b)** yes
(c) $(2z + 3)(z + 1)$

2 Factor each trinomial by grouping.

(a) $2m^2 + 7m + 3$

(b) $5p^2 - 2p - 3$

(c) $15k^2 - km - 2m^2$

3 Factor the trinomial completely.

GS $4x^2 - 2x - 30$

$= \_\_(2x^2 - x - 15)$

$= \_\_(2x^2 - \_\_ + \_\_ - 15)$

$= 2[(2x^2 - \_\_) + (5x - 15)]$

$= 2[2x(\_\_\_) + 5(x - 3)]$

$= \_\_\_\_\_\_$

4 Factor each trinomial completely.

(a) $18p^4 + 63p^3 + 27p^2$

(b) $6a^2 + 3ab - 18b^2$

Answers

2. (a) $(2m + 1)(m + 3)$
   (b) $(5p + 3)(p - 1)$
   (c) $(5k - 2m)(3k + m)$

3. 2; 2; $6x$; $5x$; $6x$; $x - 3$; $2(x - 3)(2x + 5)$

4. (a) $9p^2(2p + 1)(p + 3)$
   (b) $3(2a - 3b)(a + 2b)$

(b) $12z^2 - 5z - 2$

Look for two integers whose product is $12(-2) = -24$ and whose sum is -5. The required integers are 3 and -8.

$$12z^2 - 5z - 2$$
$$= 12z^2 + 3z - 8z - 2 \quad -5z = 3z - 8z$$
$$= (12z^2 + 3z) + (-8z - 2) \quad \text{Group the terms.}$$
$$= 3z(4z + 1) - 2(4z + 1) \quad \text{Factor each group.}$$
$$= (4z + 1)(3z - 2) \quad \text{Factor out } 4z + 1.$$

Be careful with signs.

CHECK Multiply $(4z + 1)(3z - 2)$ to obtain $12z^2 - 5z - 2$. ✓

(c) $10m^2 + mn - 3n^2$

Two integers whose product is $10(-3) = -30$ and whose sum is 1 are -5 and 6. Rewrite the trinomial with four terms.

$$10m^2 + mn - 3n^2$$
$$= 10m^2 - 5mn + 6mn - 3n^2 \quad mn = -5mn + 6mn$$
$$= (10m^2 - 5mn) + (6mn - 3n^2) \quad \text{Group the terms.}$$
$$= 5m(2m - n) + 3n(2m - n) \quad \text{Factor each group.}$$
$$= (2m - n)(5m + 3n) \quad \text{Factor out } 2m - n.$$

CHECK Multiply $(2m - n)(5m + 3n)$ to obtain $10m^2 + mn - 3n^2$. ✓

◀ Work Problem 2 at the Side.

EXAMPLE 3 Factoring a Trinomial with a Common Factor by Grouping

Factor $28x^5 - 58x^4 - 30x^3$.

$$28x^5 - 58x^4 - 30x^3$$
$$= 2x^3(14x^2 - 29x - 15) \quad \text{Factor out the greatest common factor, } 2x^3.$$

To factor $14x^2 - 29x - 15$, find two integers whose product is $14(-15) = -210$ and whose sum is -29. Factoring 210 into prime factors helps find these integers.

$$210 = 2 \cdot 3 \cdot 5 \cdot 7$$

Combine the prime factors in pairs in different ways, using one positive factor and one negative factor to obtain -210. The factors 6 and -35 have the correct sum, -29.

$$28x^5 - 58x^4 - 30x^3$$
$$= 2x^3(14x^2 - 29x - 15) \quad \text{From above}$$
$$= 2x^3(14x^2 + 6x - 35x - 15) \quad -29x = 6x - 35x$$
$$= 2x^3[(14x^2 + 6x) + (-35x - 15)] \quad \text{Group the terms.}$$
$$= 2x^3[2x(7x + 3) - 5(7x + 3)] \quad \text{Factor each group.}$$
$$= 2x^3[(7x + 3)(2x - 5)] \quad \text{Factor out } 7x + 3.$$
$$= 2x^3(7x + 3)(2x - 5)$$

Remember the common factor.

Check by multiplying.

◀ Work Problem 3 and 4 at the Side.

6.3 Exercises

FOR EXTRA HELP *Go to MyMathLab for worked-out, step-by-step solutions to exercises enclosed in a square and video solutions to ▶ exercises.*

1. **CONCEPT CHECK** Which pair of integers would be used to rewrite the middle term when factoring $12y^2 + 5y - 2$ by grouping?

   A. $-8, 3$ **B.** $8, -3$ **C.** $-6, 4$ **D.** $6, -4$

2. **CONCEPT CHECK** Which pair of integers would be used to rewrite the middle term when factoring $20b^2 - 13b + 2$ by grouping?

   A. $10, 3$ **B.** $-10, -3$ **C.** $8, 5$ **D.** $-8, -5$

GS *The middle term of each trinomial has been rewritten. Now factor by grouping.* ***See Examples 1 and 2.***

3. $m^2 + 8m + 12$
   $= m^2 + 6m + 2m + 12$

4. $x^2 + 9x + 14$
   $= x^2 + 7x + 2x + 14$

5. $a^2 + 3a - 10$
   $= a^2 + 5a - 2a - 10$

6. $y^2 + 2y - 24$
   $= y^2 - 4y + 6y - 24$

7. $10t^2 + 9t + 2$
   $= 10t^2 + 5t + 4t + 2$

8. $6x^2 + 13x + 6$
   $= 6x^2 + 9x + 4x + 6$

9. $15z^2 - 19z + 6$
   $= 15z^2 - 10z - 9z + 6$

10. $12p^2 - 17p + 6$
    $= 12p^2 - 9p - 8p + 6$

11. $8s^2 - 2st - 3t^2$
    $= 8s^2 + 4st - 6st - 3t^2$

12. $3x^2 - xy - 14y^2$
    $= 3x^2 - 7xy + 6xy - 14y^2$

13. $15a^2 + 22ab + 8b^2$
    $= 15a^2 + 10ab + 12ab + 8b^2$

14. $25m^2 + 25mn + 6n^2$
    $= 25m^2 + 15mn + 10mn + 6n^2$

GS *Complete the steps to factor each trinomial by grouping.* ***See Examples 1 and 2.***

15. $2m^2 + 11m + 12$

    (a) Find two integers whose product is ____ • ____ = ____ and whose sum is ____.

    (b) The required integers are ____ and ____.

    (c) Write the middle term $11m$ as ____ + ____.

    (d) Rewrite the given trinomial using four terms.

    (e) Factor the polynomial in part (d) by grouping.

    (f) Check by multiplying.

16. $6y^2 - 19y + 10$

    (a) Find two integers whose product is ____ • ____ = ____ and whose sum is ____.

    (b) The required integers are ____ and ____.

    (c) Write the middle term $-19y$ as ____ + ______.

    (d) Rewrite the given trinomial using four terms.

    (e) Factor the polynomial in part (d) by grouping.

    (f) Check by multiplying.

Factor each trinomial completely. ***See Examples 1–3.***

17. $2x^2 + 7x + 3$

18. $3y^2 + 13y + 4$

19. $4r^2 + r - 3$

20. $4r^2 + 3r - 10$

21. $8m^2 - 10m - 3$

22. $20x^2 - 28x - 3$

23. $21m^2 + 13m + 2$

24. $38x^2 + 23x + 2$

25. $3a^2 + 10a + 7$

26. $6w^2 + 19w + 10$

27. $12y^2 - 13y + 3$

28. $15a^2 - 16a + 4$

29. $16 + 16x + 3x^2$

30. $18 + 65x + 7x^2$

31. $24x^2 - 42x + 9$

32. $48b^2 - 86b + 10$

33. $2m^3 + 2m^2 - 40m$

34. $3x^3 + 12x^2 - 36x$

35. $-32z^5 + 20z^4 + 12z^3$

36. $-18x^5 - 15x^4 + 75x^3$

37. $12p^2 + 7pq - 12q^2$

38. $6m^2 + 5mn - 6n^2$

39. $6a^2 - 7ab - 5b^2$

40. $25g^2 - 5gh - 2h^2$

41. CONCEPT CHECK On a quiz, a student factored $16x^2 - 24x + 5$ by grouping as follows.

$$16x^2 - 24x + 5$$
$$= 16x^2 - 4x - 20x + 5$$
$$= 4x(4x - 1) - 5(4x - 1) \quad \text{His answer}$$

He did not receive credit for this answer.
What Went Wrong? What is the correct factored form?

42. CONCEPT CHECK On a quiz, a student factored $3k^3 - 12k^2 - 15k$ by first factoring out the common factor $3k$ to obtain $3k(k^2 - 4k - 5)$. Then she wrote the following.

$$k^2 - 4k - 5$$
$$= k^2 - 5k + k - 5$$
$$= k(k - 5) + 1(k - 5)$$
$$= (k - 5)(k + 1) \quad \text{Her answer}$$

What Went Wrong? What is the correct factored form?

6.4 Factoring Trinomials Using the FOIL Method

OBJECTIVE

1. **Factor trinomials using the FOIL method.**

OBJECTIVE 1 Factor trinomials using the FOIL method. This section shows an alternative method of factoring trinomials that uses trial and error.

EXAMPLE 1 Factoring a Trinomial Using FOIL (Coefficient of the Second-Degree Term Not 1)

Factor $2x^2 + 7x + 6$.

We want to write $2x^2 + 7x + 6$ as the product of two binomials.

$$2x^2 + 7x + 6$$
$$= (\_\_\_\_\_)(\_\_\_\_\_)$$

We use the FOIL method in reverse.

The product of the two first terms of the binomials is $2x^2$. The possible factors of $2x^2$ are $2x$ and x, or $-2x$ and $-x$. Because all terms of the trinomial are positive, we consider only positive factors. Thus, we have the following.

$$2x^2 + 7x + 6$$
$$= (2x\_\_\_\_)(x\_\_\_\_)$$

The product of the two last terms of the binomials must be 6. It can be factored as $1 \cdot 6$, $6 \cdot 1$, $2 \cdot 3$, or $3 \cdot 2$. We try each pair of factors in $(2x\_\_\_\_)(x\_\_\_\_)$ to find the pair that gives the correct middle term, $7x$. Begin with 1 and 6 and then try 6 and 1.

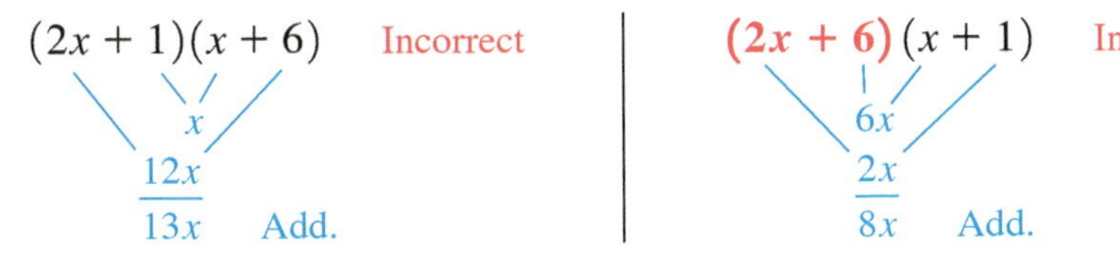

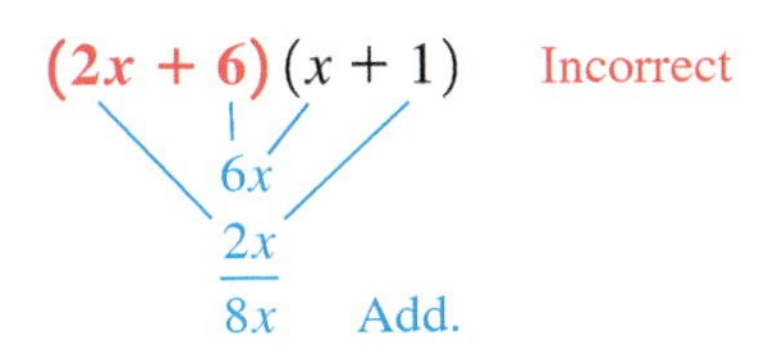

On the right above, $2x + 6 = 2(x + 3)$. The terms of the binomial $2x + 6$ have a common factor of 2, while the terms of $2x^2 + 7x + 6$ have no common factor other than 1. The product $(2x + 6)(x + 1)$ cannot be correct.

If the terms of the original polynomial have greatest common factor 1, then each of its factors will also have terms with GCF 1.

We try the pair 2 and 3 in $(2x\_\_\_\_)(x\_\_\_\_)$. Because of the common factor 2 in the terms of $2x + 2$, the product $(2x + 2)(x + 3)$ will not work. Finally, we try the pair 3 and 2 in $(2x\_\_\_\_)(x\_\_\_\_)$.

$$(2x + 3)(x + 2) = 2x^2 + 7x + 6 \quad \text{Correct}$$

$3x$
$4x$
$7x$ Add.

Thus, $2x^2 + 7x + 6$ factors as $(2x + 3)(x + 2)$.

CHECK Multiply $(2x + 3)(x + 2)$ to obtain $2x^2 + 7x + 6$. ✓

Work Problem 1 at the Side. ▶

1 Factor each trinomial.

GS (a) $2p^2 + 9p + 9$

Try various combinations of factors of $2p^2$ with factors of 9 to find the pair of factors that gives the correct middle term, ____.

Possible Pairs of Factors	Middle Term	Correct?
$(2p + 9)(p + 1)$	$11p$	(Yes/No)
$(2p + 1)(p + 9)$	____	(Yes/No)
$(2p + 3)(p + 3)$	____	(Yes/No)

$2p^2 + 9p + 9$ factors as ________.

(b) $8y^2 + 22y + 5$

Answers

1. (a) $9p$; No; $19p$; No; $9p$; Yes; $(2p + 3)(p + 3)$
   (b) $(4y + 1)(2y + 5)$

2 Factor each trinomial.

(a) $8x^2 + 10x + 3$

(b) $6p^2 + 19p + 10$

3 Factor each trinomial.

(a) $4y^2 - 11y + 7$

(b) $9x^2 - 21x + 10$

Answers

2. (a) $(4x + 3)(2x + 1)$
   (b) $(3p + 2)(2p + 5)$
3. (a) $(4y - 7)(y - 1)$
   (b) $(3x - 5)(3x - 2)$

EXAMPLE 2 Factoring a Trinomial Using FOIL (All Positive Terms)

Factor $8p^2 + 14p + 5$.

The number 8 has several possible pairs of factors, but 5 has only 1 and 5 or -1 and -5, so we begin by considering the factors of 5. We ignore the negative factors because all coefficients in the trinomial are positive. If $8p^2 + 14p + 5$ can be factored, the factors will have this form.

$$(\_\_\_\_ + 5)(\_\_\_\_ + 1)$$

The possible pairs of factors of $8p^2$ are $8p$ and p, or $4p$ and $2p$. We try various combinations, checking to see if the middle term is $14p$.

$(8p + 5)(p + 1)$ Incorrect: $5p + 8p = 13p$ Add.

$(p + 5)(8p + 1)$ Incorrect: $40p + p = 41p$ Add.

$(4p + 5)(2p + 1)$ Correct: $10p + 4p = 14p$ Add.

This last combination produces $14p$, the correct middle term.

$$8p^2 + 14p + 5 \quad \text{factors as} \quad (4p + 5)(2p + 1).$$

CHECK Multiply $(4p + 5)(2p + 1)$ to obtain $8p^2 + 14p + 5$. ✓

◀ Work Problem 2 at the Side.

EXAMPLE 3 Factoring a Trinomial Using FOIL (Negative Middle Term)

Factor $6x^2 - 11x + 3$.

Because 3 has only 1 and 3 or -1 and -3 as factors, it is better here to begin by considering the factors of 3. The last term of the trinomial $6x^2 - 11x + 3$ is positive and the middle term has a negative coefficient, so we consider only negative factors. We need two negative factors because the *product* of two negative factors is positive and their *sum* is negative, as required.

We try -3 and -1 as factors of 3.

$$(\_\_\_\_ - 3)(\_\_\_\_ - 1)$$

The factors of $6x^2$ may be either $6x$ and x, or $2x$ and $3x$.

$(6x - 3)(x - 1)$ Incorrect: $-3x + (-6x) = -9x$ Add.

$(2x - 3)(3x - 1)$ Correct: $-9x + (-2x) = -11x$ Add.

The factors $2x$ and $3x$ produce $-11x$, the correct middle term.

$$6x^2 - 11x + 3 \quad \text{factors as} \quad (2x - 3)(3x - 1).$$

CHECK Multiply $(2x - 3)(3x - 1)$ to obtain $6x^2 - 11x + 3$. ✓

◀ Work Problem 3 at the Side.

Note

Our initial attempt to factor $6x^2 - 11x + 3$ as $(6x - 3)(x - 1)$ in **Example 3** *cannot* be correct because the terms of $6x - 3$ have a common factor of 3, while those of the original polynomial do not.

EXAMPLE 4 Factoring a Trinomial Using FOIL (Negative Constant Term)

Factor $8x^2 + 6x - 9$.

The integer 8 has several possible pairs of factors, as does -9. Because the constant term is negative, one positive factor and one negative factor of -9 are needed. The coefficient of the middle term is relatively small, so we avoid large factors, such as 8 or 9. We try $4x$ and $2x$ as factors of $8x^2$, and 3 and -3 as factors of -9, and check the middle term.

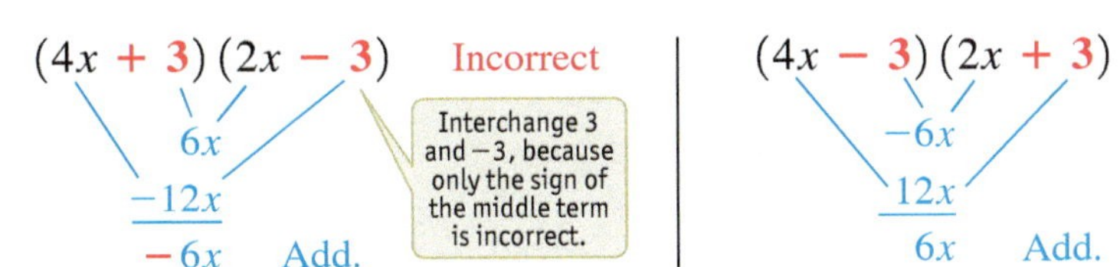

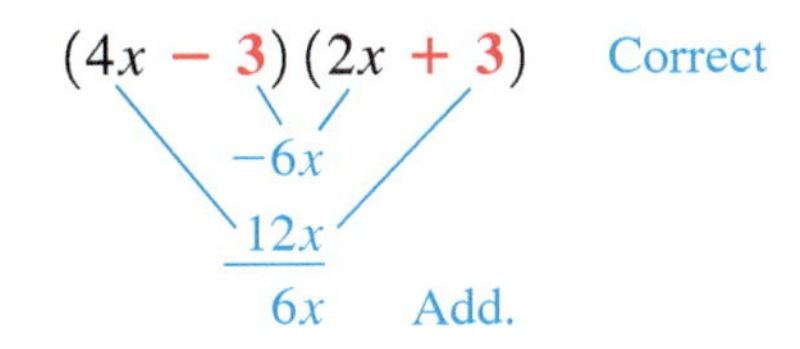

The combination on the right produces $6x$, the correct middle term.

$8x^2 + 6x - 9$ factors as $(4x - 3)(2x + 3)$. *Check by multiplying.*

Work Problem 4 at the Side. ▶

EXAMPLE 5 Factoring a Trinomial with Two Variables

Factor $12a^2 - ab - 20b^2$.

There are several pairs of factors of $12a^2$, including

$$12a \text{ and } a, \quad 6a \text{ and } 2a, \quad \text{and} \quad 4a \text{ and } 3a.$$

There are also many possible pairs of factors of $-20b^2$, including

$$20b \text{ and } -b, \quad -20b \text{ and } b, \quad 10b \text{ and } -2b,$$
$$-10b \text{ and } 2b, \quad 4b \text{ and } -5b, \quad \text{and} \quad -4b \text{ and } 5b.$$

Once again, because the coefficient of the middle term is relatively small, avoid the larger factors. Try the factors $3a$ and $4a$, and $4b$ and $-5b$.

$(3a - 5b)(4a + 4b)$ Incorrect

This cannot be correct because the terms of $(4a + 4b)$ have 4 as a common factor, while the terms of the given trinomial do not.

$(3a + 4b)(4a - 5b)$ Incorrect

$16ab$

$-15ab$

ab Add.

In the factorization attempt on the right above, the middle term is ab, rather than $-ab$. We interchange the signs of the last two terms in the factors to obtain the correct result.

$$(3a - 4b)(4a + 5b)$$
$$= 12a^2 - ab - 20b^2 \quad \text{Correct}$$

Thus, $12a^2 - ab - 20b^2$ factors as $(3a - 4b)(4a + 5b)$. *Check by multiplying.*

Work Problem 5 at the Side. ▶

4 Factor each trinomial if possible.

GS **(a)** $6x^2 + 5x - 4$
$= (3x + \_\_\_\_)(2x - \_\_\_\_)$

(b) $6m^2 - 11m - 10$

(c) $4x^2 - 3x - 7$

(d) $3y^2 + 8y - 6$

5 Factor each trinomial.

GS **(a)** $2x^2 - 5xy - 3y^2$
$= (2x + \_\_\_\_)(x - \_\_\_\_)$

(b) $8a^2 + 2ab - 3b^2$

Answers

4. **(a)** 4; 1
(b) $(2m - 5)(3m + 2)$
(c) $(4x - 7)(x + 1)$
(d) prime

5. **(a)** y; $3y$
(b) $(4a + 3b)(2a - b)$

6 Factor each trinomial.

GS (a) $36z^3 + 102z^2 + 72z$

$= \_\_\_(\_\_\_ + \_\_\_ + 12)$

$= 6z(3z + 4)(\_\_\_\_)$

(b) $10x^3 + 45x^2 - 90x$

(c) $-24x^3 + 32x^2 + 6x$

EXAMPLE 6 Factoring Trinomials with Common Factors

Factor each trinomial.

(a) $15y^3 + 55y^2 + 30y$

$$15y^3 + 55y^2 + 30y$$
$$= 5y(3y^2 + 11y + 6) \quad \text{Factor out the greatest common factor, } 5y.$$

Now factor $3y^2 + 11y + 6$. Try $3y$ and y as factors of $3y^2$ and 2 and 3 as factors of 6. We know that

$$(3y + 3)(y + 2) \text{ is incorrect}$$

because the terms of $(3y + 3)$ have a common factor of 3. So we switch the 3 and 2 and try the following factors.

$$(3y + 2)(y + 3)$$
$$= 3y^2 + 11y + 6 \quad \text{Correct}$$

This leads to the completely factored form.

$$15y^3 + 55y^2 + 30y$$
$$= 5y(3y + 2)(y + 3)$$

Remember the common factor.

CHECK $5y(3y + 2)(y + 3)$

$$= 5y(3y^2 + 9y + 2y + 6) \quad \text{FOIL method}$$
$$= 5y(3y^2 + 11y + 6) \quad \text{Combine like terms.}$$
$$= 15y^3 + 55y^2 + 30y \; ✓ \quad \text{Distributive property}$$

(b) $-24a^3 - 42a^2 + 45a$

The common factor could be $3a$ or $-3a$. If we factor out $-3a$, the first term of the trinomial will be positive, which makes it easier to factor the remaining trinomial.

$$-24a^3 - 42a^2 + 45a$$
$$= -3a(8a^2 + 14a - 15) \quad \text{Factor out } -3a.$$
$$= -3a(4a - 3)(2a + 5) \quad \text{Use trial and error.}$$

It is easier here to factor the trinomial if the first term is $8a^2$ rather than $-8a^2$.

CHECK $-3a(4a - 3)(2a + 5)$

$$= -3a(8a^2 + 20a - 6a - 15) \quad \text{FOIL method}$$
$$= -3a(8a^2 + 14a - 15) \quad \text{Combine like terms.}$$
$$= -24a^3 - 42a^2 + 45a \; ✓ \quad \text{Distributive property}$$

◀ Work Problem 6 at the Side.

CAUTION

Remember to include the common factor in the final factored form.

Answers

6. (a) $6z$; $6z^2$; $17z$; $2z + 3$
 (b) $5x(2x - 3)(x + 6)$
 (c) $-2x(6x + 1)(2x - 3)$

6.4 Exercises

FOR EXTRA HELP — Go to MyMathLab for worked-out, step-by-step solutions to exercises enclosed in a square and video solutions to ▶ exercises.

CONCEPT CHECK *Decide which is the correct factored form of the given polynomial.*

1. $2x^2 - x - 1$
A. $(2x - 1)(x + 1)$ **B.** $(2x + 1)(x - 1)$

2. $3a^2 - 5a - 2$
A. $(3a + 1)(a - 2)$ **B.** $(3a - 1)(a + 2)$

3. $4y^2 + 17y - 15$
A. $(y + 5)(4y - 3)$ **B.** $(2y - 5)(2y + 3)$

4. $12c^2 - 7c - 12$
A. $(6c - 2)(2c + 6)$ **B.** $(4c + 3)(3c - 4)$

5. $4k^2 + 13mk + 3m^2$
A. $(4k + m)(k + 3m)$ **B.** $(4k + 3m)(k + m)$

6. $2x^2 + 11x + 12$
A. $(2x + 3)(x + 4)$ **B.** $(2x + 4)(x + 3)$

GS *Complete each factoring.* ***See Examples 1–6.***

7. ▶ $6a^2 + 7ab - 20b^2$
$= (3a - 4b)(\_\_\_\_\_\_)$

8. $9m^2 - 3mn - 2n^2$
$= (3m + n)(\_\_\_\_\_\_)$

9. $2x^2 + 6x - 8$
$= 2(\_\_\_\_\_\_\_\_\_\_)$
$= 2(\_\_\_\_\_\_)(\_\_\_\_\_\_)$

10. $3x^2 - 9x - 30$
$= 3(\_\_\_\_\_\_\_\_\_\_)$
$= 3(\_\_\_\_\_\_)(\_\_\_\_\_\_)$

11. $-4z^3 + 10z^2 + 6z$
$= -2z(\_\_\_\_\_\_\_\_\_\_)$
$= -2z(\_\_\_\_\_\_)(\_\_\_\_\_\_)$

12. $-15r^3 - 39r^2 + 18r$
$= -3r(\_\_\_\_\_\_\_\_\_\_)$
$= -3r(\_\_\_\_\_\_)(\_\_\_\_\_\_)$

13. CONCEPT CHECK A student factoring the trinomial
$$12x^2 + 7x - 12$$
wrote $(4x + 4)$ as one binomial factor. ***What Went Wrong?*** Factor the trinomial correctly.

14. CONCEPT CHECK A student factoring the trinomial
$$4x^2 + 10x - 6$$
wrote $(4x - 2)(x + 3)$. ***What Went Wrong?*** Factor the trinomial correctly.

Factor each trinomial completely. ***See Examples 1–6.***

15. ▶ $3a^2 + 10a + 7$

16. $7r^2 + 8r + 1$

17. $2y^2 + 7y + 6$

18. $5z^2 + 12z + 4$

19. $15m^2 + m - 2$

20. $6x^2 + x - 1$

21. $12s^2 + 11s - 5$

22. $20x^2 + 11x - 3$

23. ▶ $10m^2 - 23m + 12$

24. $6x^2 - 17x + 12$

25. $8w^2 - 14w + 3$

26. $9p^2 - 18p + 8$

27. $20y^2 - 39y - 11$

28. $10x^2 - 11x - 6$

29. $3x^2 - 15x + 16$

30. $2t^2 + 13t - 18$

31. $20x^2 + 22x + 6$

32. $36y^2 + 81y + 45$

33. $-40m^2q - mq + 6q$

34. $-15a^2b - 22ab - 8b$

35. $15n^4 - 39n^3 + 18n^2$

36. $24a^4 + 10a^3 - 4a^2$

37. $-15x^2y^2 + 7xy^2 + 4y^2$

38. $-14a^2b^3 - 15ab^3 + 9b^3$

39. $5a^2 - 7ab - 6b^2$

40. $6x^2 - 5xy - y^2$

41. $12s^2 + 11st - 5t^2$

42. $25a^2 + 25ab + 6b^2$

43. $6m^6n + 7m^5n^2 + 2m^4n^3$

44. $12k^3q^4 - 4k^2q^5 - kq^6$

If a trinomial has a negative coefficient for the second-degree term, such as $-2x^2 + 11x - 12$, *it may be easier to factor by first factoring out* -1.

$$-2x^2 + 11x - 12$$
$$= -1(2x^2 - 11x + 12) \quad \text{Factor out } -1.$$
$$= -1(2x - 3)(x - 4) \quad \text{Factor the trinomial.}$$

Use this method to factor each trinomial.

45. $-x^2 - 4x + 21$

46. $-x^2 + x + 72$

47. $-3x^2 - x + 4$

48. $-5x^2 + 2x + 16$

49. $-2a^2 - 5ab - 2b^2$

50. $-3p^2 + 13pq - 4q^2$

Relating Concepts (Exercises 51–56) For Individual or Group Work

Often there are several different equivalent forms of an answer that are all correct. ***Work Exercises 51–56 in order,*** *to see this for factoring problems.*

51. Factor the integer 35 as the product of two prime numbers.

52. Factor the integer 35 as the product of the negatives of two prime numbers.

53. Verify that $6x^2 - 11x + 4$ factors as $(3x - 4)(2x - 1)$.

54. Verify that $6x^2 - 11x + 4$ factors as $(4 - 3x)(1 - 2x)$.

55. Compare the two valid factored forms in **Exercises 53 and 54.** How do the factors in each case compare?

56. Suppose we know that the correct factored form of a particular trinomial is $(7t - 3)(2t - 5)$. From **Exercises 51–55,** what is another valid factored form?

6.5 Special Factoring Techniques

OBJECTIVES

1. Factor a difference of squares.
2. Factor a perfect square trinomial.

OBJECTIVE 1 Factor a difference of squares. The rule for finding the product of the sum and difference of the same two terms is

$$(x + y)(x - y) = x^2 - y^2.$$

Reversing this rule leads to the following special factoring rule.

Factoring a Difference of Squares

$$x^2 - y^2 = (x + y)(x - y)$$

For example, $m^2 - 4$

$= m^2 - 2^2$

$= (m + 2)(m - 2).$

Two conditions must be true for a binomial to be a difference of squares.

1. Both terms of the binomial must be **perfect squares,** such as

$$x^2, \quad 9y^2 = (3y)^2, \quad m^4 = (m^2)^2, \quad 25 = 5^2, \quad 1 = 1^2.$$

2. The binomial terms must have different signs (one positive, one negative).

EXAMPLE 1 Factoring Differences of Squares

Factor each binomial if possible.

$$x^2 - y^2 = (x + y)(x - y)$$

(a) $p^2 - 16 = p^2 - 4^2 = (p + 4)(p - 4)$

(b) $x^2 - 8$

Because 8 is not the square of an integer, the binomial $x^2 - 8$ does not satisfy Condition 1 above. It cannot be factored using integers, so it is a prime polynomial.

(c) $p^2 + 16$

The binomial $p^2 + 16$ does not satisfy Condition 2 above. It is a *sum* of squares—it is *not* equal to $(p + 4)(p - 4)$. (See part (a).) We can use the FOIL method and try the following.

$(p - 4)(p - 4)$	$(p + 4)(p + 4)$
$= p^2 - 8p + 16$, not $p^2 + 16$.	$= p^2 + 8p + 16$, not $p^2 + 16$.

Thus, $p^2 + 16$ is a prime polynomial.

Work Problem 1 at the Side. ▶

Sum of Squares

If x and y have no common factors (except 1), the following holds true.

A sum of squares $x^2 + y^2$ cannot be factored using real numbers.

That is, $x^2 + y^2$ is prime. (See **Example 1(c).**)

1 Factor each binomial, if possible.

(a) $x^2 - 81$

$= (x + \_\_\_\_)(x - \_\_\_\_)$

(b) $p^2 - 100$

(c) $t^2 - s^2$

(d) $y^2 - 10$

(e) $x^2 + 36$

(f) $4x^2 + 16$

Answers

1. (a) 9; 9
(b) $(p + 10)(p - 10)$
(c) $(t + s)(t - s)$
(d) prime **(e)** prime **(f)** $4(x^2 + 4)$

2 Factor each binomial.

GS **(a)** $9m^2 - 49$

$= (\_\_\_)^2 - \_\_\_^2$

$= (\_\_\_\_)(\_\_\_\_)$

(b) $64a^2 - 25$

(c) $25a^2 - 64b^2$

3 Factor each binomial completely.

GS **(a)** $50r^2 - 32$

$= 2(\_\_\_\_)$

$= 2[(\_\_\_)^2 - 4^2]$

$= \_\_\_(\_\_\_\_)(\_\_\_\_)$

(b) $27y^2 - 75$

(c) $k^4 - 49$

(d) $81r^4 - 16$

Answers

2. (a) $3m$; 7; $3m + 7$; $3m - 7$
   (b) $(8a + 5)(8a - 5)$
   (c) $(5a + 8b)(5a - 8b)$
3. (a) $25r^2 - 16$; $5r$; 2; $5r + 4$; $5r - 4$
   (b) $3(3y + 5)(3y - 5)$
   (c) $(k^2 + 7)(k^2 - 7)$
   (d) $(9r^2 + 4)(3r + 2)(3r - 2)$

EXAMPLE 2 Factoring Differences of Squares

Factor each binomial.

$$x^2 - y^2 = (x + y)(x - y)$$

(a) $25m^2 - 4 = (5m)^2 - 2^2 = (5m + 2)(5m - 2)$

(b) $49z^2 - 64$

$= (7z)^2 - 8^2$ Write each term as a square.

$= (7z + 8)(7z - 8)$ Factor the difference of squares.

(c) $9x^2 - 4z^2$

$= (3x)^2 - (2z)^2$ Write each term as a square.

$= (3x + 2z)(3x - 2z)$ Factor the difference of squares.

CHECK $(3x + 2z)(3x - 2z)$

$= 9x^2 - 6xz + 6zx - 4z^2$ FOIL method

$= 9x^2 - 4z^2$ ✓ Commutative property; Combine like terms.

◀ Work Problem **2** at the Side.

Note

It is a good idea to check a factored form by multiplying.

EXAMPLE 3 Factoring More Complex Differences of Squares

Factor each binomial completely.

(a) $81y^2 - 36$

$= 9(9y^2 - 4)$ Factor out the GCF, 9.

$= 9[(3y)^2 - 2^2]$ Write each term as a square.

$= 9(3y + 2)(3y - 2)$ Factor the difference of squares.

(b) $p^4 - 36$

$= (p^2)^2 - 6^2$ Write each term as a square.

$= (p^2 + 6)(p^2 - 6)$ Factor the difference of squares.

Neither binomial can be factored further.

(c) $m^4 - 16$

$= (m^2)^2 - 4^2$ Write each term as a square.

$= (m^2 + 4)(m^2 - 4)$ Factor the difference of squares.

Don't stop here.

$= (m^2 + 4)(m + 2)(m - 2)$ Factor the difference of squares again.

◀ Work Problem **3** at the Side.

CAUTION

Factor again when any of the factors is a difference of squares, as in **Example 3(c).** Check by multiplying.

OBJECTIVE 2 Factor a perfect square trinomial. Recall the rules for squaring binomials.

$(x + y)^2$	Squared binomial	$(x - y)^2$	Squared binomial
$= (x + y)(x + y)$		$= (x - y)(x - y)$	
$= \mathbf{x^2 + 2xy + y^2}$	Perfect square trinomial	$= \mathbf{x^2 - 2xy + y^2}$	Perfect square trinomial

A **perfect square trinomial** is a trinomial that is the square of a binomial. For example, $x^2 + 8x + 16$ is a perfect square trinomial because it is the square of the binomial $x + 4$.

$$(x + 4)^2 \quad \text{Squared binomial}$$
$$= (x + 4)(x + 4) \quad \text{Apply the exponent.}$$
$$= x^2 + 8x + 16 \quad \text{Perfect square trinomial}$$

Two conditions must be true for a trinomial to be a perfect square trinomial.

1. Two of its terms must be perfect squares. In the perfect square trinomial $x^2 + 8x + 16$, the terms x^2 and $16 = 4^2$ are perfect squares.
2. *The remaining (middle) term of a perfect square trinomial is always twice the product of the two terms in the squared binomial.*

$$x^2 + 8x + 16$$
$$= x^2 + 2(x)(4) + 4^2 \quad 8x = 2(x)(4)$$
$$= (x + 4)^2 \quad \text{Factor.}$$

The following are *not* perfect square trinomials.

$16x^2 + 4x + 15$ — Violates Condition 1 (Only $16x^2 = (4x)^2$ is a perfect square; 15 is not.)

$x^2 + 6x + 36$ — Violates Condition 2 (x^2 and $36 = 6^2$ are perfect squares, but $2(x)(6) = 12x$, *not* $6x$.)

Reversing the rules for squaring binomials leads to the following.

Factoring Perfect Square Trinomials

$$x^2 + 2xy + y^2 = (x + y)^2$$
$$x^2 - 2xy + y^2 = (x - y)^2$$

EXAMPLE 4 Factoring a Perfect Square Trinomial

Factor $x^2 + 10x + 25$.

The x^2-term is a perfect square, and so is 25, which equals 5^2.

Try to factor the trinomial $x^2 + 10x + 25$ as $(x + 5)^2$.

To check, take twice the product of the two terms in the squared binomial.

$$2 \cdot x \cdot 5 = 10x \leftarrow \text{Middle term of } x^2 + 10x + 25$$

Twice; First term of binomial; Last term of binomial

Because $10x$ is the middle term of the given trinomial $x^2 + 10x + 25$, the trinomial is a perfect square and factors as $(x + 5)^2$.

Work Problem 4 at the Side. ▶

---

4 Factor each trinomial.

GS **(a)** $p^2 + 14p + 49$

The term p^2 is a perfect square. Since $49 = \_\_\_\_^2$, 49 (*is / is not*) a perfect square. Try to factor the trinomial

$$p^2 + 14p + 49$$

as $(p + \_\_\_\_)^2$.

Check by taking twice the product of the two terms of the squared binomial.

$$2 \cdot p \cdot \_\_\_\_ = \_\_\_\_$$

The result (*is / is not*) the middle term of the given trinomial. The trinomial is a perfect square and factors as ______.

(b) $x^2 + 6x + 9$

(c) $y^2 + 22y + 121$

Answers

4. (a) 7; is; 7; 7; $14p$; is; $(p + 7)^2$
(b) $(x + 3)^2$ **(c)** $(y + 11)^2$

5 Factor each trinomial.

(a) $p^2 - 18p + 81$

(b) $16a^2 - 56a + 49$

(c) $121z^2 + 110z + 100$

(d) $64x^2 - 48x + 9$

(e) $27y^3 + 72y^2 + 48y$

EXAMPLE 5 Factoring Perfect Square Trinomials

Factor each trinomial.

(a) $x^2 - 22x + 121$

The first and last terms are perfect squares $(121 = 11^2$ or $(-11)^2)$. Check to see whether the middle term of $x^2 - 22x + 121$ is twice the product of the first and last terms of the binomial $x - 11$.

$$2(x)(-11) = -22x \leftarrow \text{Middle term of } x^2 - 22x + 121$$

Twice ↑ First term ↑ Last term ↑

Thus, $x^2 - 22x + 121$ is a perfect square trinomial.

$$x^2 - 22x + 121 \quad \text{factors as} \quad (x - 11)^2.$$

Same sign

Check by squaring the binomial.

Notice that the sign of the second term in the squared binomial is the same as the sign of the middle term in the trinomial.

(b) $9m^2 - 24m + 16 = (3m)^2 + 2(3m)(-4) + (-4)^2 = (3m - 4)^2$

Perfect Squares; Twice; First term; Last Term

(c) $25y^2 + 20y + 16$

The first and last terms are perfect squares.

$$25y^2 = (5y)^2 \quad \text{and} \quad 16 = 4^2$$

Twice the product of the first and last terms of the binomial $5y + 4$ is

$$2(5y)4 = 40y,$$

which is *not* the middle term of

$$25y^2 + 20y + 16.$$

This trinomial is not a perfect square. In fact, the trinomial cannot be factored even with the methods of the previous sections. It is a prime polynomial.

(d) $12z^3 + 60z^2 + 75z$

$= 3z(4z^2 + 20z + 25)$ Factor out the common factor, $3z$.

$= 3z[(2z)^2 + 2(2z)(5) + 5^2]$ $4z^2 + 20z + 25$ is a perfect square trinomial.

$= 3z(2z + 5)^2$ Factor.

◀ Work Problem **5** at the Side.

Note

1. The sign of the second term in a squared binomial is always the same as the sign of the middle term in the trinomial.
2. The first and last terms of a perfect square trinomial must be *positive* because they are squares. For example, the polynomial $x^2 - 2x - 1$ cannot be a perfect square because the last term is negative.
3. Perfect square trinomials can also be factored using grouping or the FOIL method. Using the method of this section is often easier.

Answers

5. **(a)** $(p - 9)^2$ **(b)** $(4a - 7)^2$ **(c)** prime **(d)** $(8x - 3)^2$ **(e)** $3y(3y + 4)^2$

6.5 Exercises

FOR EXTRA HELP Go to MyMathLab for worked-out, step-by-step solutions to exercises enclosed in a square and video solutions to ▶ exercises.

CONCEPT CHECK *Work each problem.*

1. To help factor a difference of squares, complete the following list of squares.

$1^2 =$ ______ $2^2 =$ ______ $3^2 =$ ______ $4^2 =$ ______ $5^2 =$ ______

$6^2 =$ ______ $7^2 =$ ______ $8^2 =$ ______ $9^2 =$ ______ $10^2 =$ ______

$11^2 =$ ______ $12^2 =$ ______ $13^2 =$ ______ $14^2 =$ ______ $15^2 =$ ______

$16^2 =$ ______ $17^2 =$ ______ $18^2 =$ ______ $19^2 =$ ______ $20^2 =$ ______

2. To use the factoring techniques described in this section, it is helpful to recognize fourth powers of integers. Complete the following list of fourth powers.

$1^4 =$ ______ $2^4 =$ ______ $3^4 =$ ______ $4^4 =$ ______ $5^4 =$ ______

3. Which of the following binomials are differences of squares?

A. $x^2 - 4$ **B.** $y^2 + 9$

C. $2a^2 - 25$ **D.** $9m^2 - 1$

4. Which of the following binomial sums can be factored?

A. $x^2 + 36$ **B.** $x^3 + x$

C. $3x^2 + 12$ **D.** $25x^2 + 49$

5. CONCEPT CHECK On a quiz, a student indicated *prime* when asked to factor the binomial

$$4x^2 + 16,$$

because she said that a sum of squares cannot be factored. ***What Went Wrong?*** Give the correct answer.

6. CONCEPT CHECK When directed to factor $k^4 - 81$ completely, a student did not earn full credit.

$$(k^2 + 9)(k^2 - 9) \quad \text{His answer}$$

The student argued that because his answer does indeed give $k^4 - 81$ when multiplied out, he should be given full credit. ***What Went Wrong?*** Give the correct factored form.

Factor each binomial completely. ***See Examples 1–3.***

7. $y^2 - 25$ ▶

8. $t^2 - 36$

9. $x^2 - 144$

10. $y^2 - 400$

11. $m^2 - 12$ ▶

12. $k^2 - 18$

13. $m^2 + 64$

14. $k^2 + 49$

15. $9r^2 - 4$ ▶

16. $4x^2 - 9$

17. $36x^2 - 16$ ▶

18. $32a^2 - 8$

19. $196p^2 - 225$

20. $361q^2 - 400$

21. $16r^2 - 25a^2$

22. $49m^2 - 100p^2$

23. $16m^2 + 64$

24. $9x^2 + 81$

25. $p^4 - 49$

26. $r^4 - 25$

27. $x^4 - 1$

28. $y^4 - 10{,}000$

29. $p^4 - 256$ ▶

30. $x^4 - 625$

31. CONCEPT CHECK Which of the following are perfect square trinomials?

A. $y^2 - 13y + 36$ **B.** $x^2 + 6x + 9$ **C.** $4z^2 - 4z + 1$ **D.** $16m^2 + 8m - 1$

32. In the polynomial $9y^2 + 14y + 25$, the first and last terms are perfect squares. Can the polynomial be factored? If it can, factor it. If it cannot, explain why it is not a perfect square trinomial.

Factor each trinomial completely. **See Examples 4 and 5.**

33. $w^2 + 2w + 1$

34. $p^2 + 4p + 4$

35. $x^2 - 8x + 16$

36. $x^2 - 10x + 25$

37. $x^2 - 10x + 100$

38. $x^2 - 18x + 36$

39. $2x^2 + 24x + 72$

40. $3y^2 + 48y + 192$

41. $4x^2 + 12x + 9$

42. $25x^2 + 10x + 1$

43. $16x^3 - 40x^2 + 25x$

44. $36y^3 - 60y^2 + 25y$

45. $49x^2 - 28xy + 4y^2$

46. $4z^2 - 12zw + 9w^2$

47. $64x^2 + 48xy + 9y^2$

48. $9t^2 + 24tr + 16r^2$

49. $-50h^3 + 40h^2y - 8hy^2$

50. $-18x^3 - 48x^2y - 32xy^2$

Although we usually factor polynomials using integers, we can apply the same concepts to factoring using fractions and decimals.

$$z^2 - \frac{9}{16}$$

$$= z^2 - \left(\frac{3}{4}\right)^2 \qquad \frac{9}{16} = \left(\frac{3}{4}\right)^2$$

$$= \left(z + \frac{3}{4}\right)\left(z - \frac{3}{4}\right)$$ Factor the difference of squares.

$$x^2 + \frac{2}{5}x + \frac{1}{25}$$

$$= x^2 + 2(x)\left(\frac{1}{5}\right) + \left(\frac{1}{5}\right)^2 \qquad \frac{1}{25} = \left(\frac{1}{5}\right)^2$$

$$= \left(x + \frac{1}{5}\right)^2$$ Factor the perfect square trinomial.

Factor each binomial or trinomial.

51. $p^2 - \frac{1}{9}$

52. $q^2 - \frac{1}{4}$

53. $4m^2 - \frac{9}{25}$

54. $100b^2 - \frac{4}{49}$

55. $x^2 - 0.64$

56. $y^2 - 0.36$

57. $t^2 + t + \frac{1}{4}$

58. $m^2 + \frac{2}{3}m + \frac{1}{9}$

59. $a^2 - \frac{4}{7}a + \frac{4}{49}$

60. $b^2 - \frac{10}{9}b + \frac{25}{81}$

61. $x^2 - 1.0x + 0.25$

62. $y^2 - 1.4y + 0.49$

Summary Exercises *Recognizing and Applying Factoring Strategies*

When factoring, ask these questions to decide on a suitable factoring technique.

Factoring a Polynomial

Question 1 **Is there a common factor other than 1?** If so, factor it out.

Question 2 **How many terms are there in the polynomial?**

Two terms: Is it a difference of squares? If so, factor as in **Section 6.5.**

Three terms: Is it a perfect square trinomial? In this case, factor as in **Section 6.5.**

If the trinomial is not a perfect square trinomial, what is the coefficient of the second-degree term?

- If it is 1, use the factoring method of **Section 6.2.**
- If it is not 1, use the general factoring methods of **Sections 6.3 and 6.4.**

Four terms: Try to factor by grouping, as in **Section 6.1.**

Question 3 **Can any factors be factored further?** If so, factor them.

CAUTION

Be careful when checking the answer to a factoring problem.

1. *Check* that the product of all the factors does indeed yield the original polynomial.
2. *Check* that the original polynomial has been factored ***completely.***

EXAMPLE 1 **Applying Factoring Strategies**

Factor $144x^2 - 169$ completely.

Question 1 **Is there a common factor other than 1?**

There is no common factor other than 1. Proceed to Question 2.

Question 2 **How many terms are there in the polynomial?**

The polynomial $144x^2 - 169$ has two terms. It is a difference of squares because $144x^2 = (12x)^2$ and $169 = 13^2$.

$$144x^2 - 169$$

$$= (12x)^2 - 13^2 \quad \text{Write each term as a square.}$$

$$= (12x + 13)(12x - 13) \quad \text{Factor the difference of squares.}$$

Question 3 **Can any factors be factored further?**

No. The original polynomial has been factored completely.

EXAMPLE 2 Applying Factoring Strategies

Factor $12x^2 + 26xy + 12y^2$ completely.

Question 1 **Is there a common factor other than 1?**

Yes, 2 is a common factor, so factor it out.

$$12x^2 + 26xy + 12y^2$$
$$= 2(6x^2 + 13xy + 6y^2)$$

Question 2 **How many terms are there in the polynomial?**

The polynomial $6x^2 + 13xy + 6y^2$ has three terms, but it is not a perfect square. To factor the trinomial by grouping, begin by finding two integers with a product of $6 \cdot 6$, or 36, and a sum of 13. These integers are 4 and 9.

$12x^2 + 26xy + 12y^2$

$= 2(6x^2 + 13xy + 6y^2)$ Factor out the GCF, 2.

$= 2(6x^2 + 4xy + 9xy + 6y^2)$ $4 \cdot 9 = 36;\ 4 + 9 = 13$

$= 2[(6x^2 + 4xy) + (9xy + 6y^2)]$ Group the terms.

$= 2[2x(3x + 2y) + 3y(3x + 2y)]$ Factor each group.

$= 2(3x + 2y)(2x + 3y)$ Factor out the common factor, $3x + 2y$.

The trinomial $6x^2 + 13xy + 6y^2$ could also be factored by trial and error, using the FOIL method in reverse.

Question 3 **Can any factors be factored further?**

No. The original polynomial has been factored completely.

CONCEPT CHECK *Match each polynomial in Column I with the best choice for factoring it in Column II. The choices in Column II may be used once, more than once, or not at all.*

I	II
1. $12x^2 + 20x + 8$	**A.** Factor out the GCF. No further factoring is possible.
2. $x^2 - 17x + 72$	**B.** Factor a difference of squares once.
3. $-16m^2n + 24mn - 40mn^2$	**C.** Factor a difference of squares twice.
4. $64a^2 - 121b^2$	**D.** Factor a perfect square trinomial.
5. $36p^2 - 60pq + 25q^2$	**E.** Factor by grouping.
6. $z^2 - 4z + 6$	**F.** Factor out the GCF. Then factor a trinomial by grouping or trial and error.
7. $625 - r^4$	**G.** Factor into two binomials by finding two integers whose product is the constant in the trinomial and whose sum is the coefficient of the middle term.
8. $x^6 + 4x^4 - 3x^2 - 12$	**H.** The polynomial is prime.
9. $4w^2 + 49$	
10. $144 - 24z + z^2$	

Factor each polynomial completely. Check by multiplying.

11. $32m^9 + 16m^5 + 24m^3$

12. $2m^2 - 10m - 48$

13. $14k^3 + 7k^2 - 70k$

14. $9z^2 + 64$

15. $6z^2 + 31z + 5$

16. $m^2 - 3mn - 4n^2$

17. $49z^2 - 16y^2$

18. $100n^2r^2 + 30nr^3 - 50n^2r$

19. $16x^3 + 100x$

20. $20 + 5m + 12n + 3mn$

21. $10y^2 - 7yz - 6z^2$

22. $y^4 - 81$

23. $m^2 + 2m - 15$

24. $6y^2 - 5y - 4$

25. $32z^3 + 56z^2 - 16z$

26. $p^2 - 24p + 144$

27. $z^2 - 12z + 36$

28. $9m^2 - 64$

29. $y^2 - 4yk - 12k^2$

30. $16z^2 - 8z + 1$

31. $6y^2 - 6y - 12$

32. $x^2 + 2x + 16$

33. $p^2 - 17p + 66$

34. $a^2 + 17a + 72$

35. $k^2 + 100$

36. $108m^2 - 36m + 3$

37. $z^2 - 3za - 10a^2$

38. $2a^3 + a^2 - 14a - 7$

39. $4k^2 - 12k + 9$

40. $a^2 - 3ab - 28b^2$

41. $16r^2 + 24rm + 9m^2$

42. $3k^2 + 4k - 4$

43. $n^2 - 12n - 35$

44. $a^4 - 625$

45. $16k^2 - 48k + 36$

46. $8k^2 - 10k - 3$

47. $36y^6 - 42y^5 - 120y^4$

48. $5z^3 - 45z^2 + 70z$

49. $8p^2 + 23p - 3$

50. $8k^2 - 2kh - 3h^2$

51. $54m^2 - 24z^2$

52. $4k^2 - 20kz + 25z^2$

53. $6a^2 + 10a - 4$

54. $15h^2 + 11hg - 14g^2$

55. $28a^2 - 63b^2$

56. $10z^2 - 7z - 6$

57. $125m^4 - 400m^3n + 195m^2n^2$

58. $9y^2 + 12y - 5$

59. $9u^2 + 66uv + 121v^2$

60. $36x^2 + 32x + 9$

61. $27p^{10} - 45p^9 - 252p^8$

62. $10m^2 + 25m - 60$

63. $4 - 2q - 6p + 3pq$

64. $k^2 - 121$

65. $64p^2 - 100m^2$

66. $m^3 + 4m^2 - 6m - 24$

67. $100a^2 - 81y^2$

68. $8a^2 + 23ab - 3b^2$

69. $a^2 + 8a + 16$

70. $4y^2 - 25$

71. $2x^2 + 5x + 6$

72. $-3x^3 + 12xy^2$

73. $25a^2 - 70ab + 49b^2$

74. $8t^4 - 8$

75. $-4x^2 + 24xy - 36y^2$

76. $100a^2 - 25b^2$

77. $-2x^2 + 26x - 72$

78. $2m^2 - 15n - 5mn + 6m$

79. $12x^2 + 22x - 20$

80. $y^6 + 5y^4 - 3y^2 - 15$

81. $y^2 - 64$

82. $12p^3 - 54p^2 - 30p$

6.6 Solving Quadratic Equations Using the Zero-Factor Property

OBJECTIVES

1. Solve quadratic equations using the zero-factor property.
2. Solve other equations using the zero-factor property.

Galileo Galilei developed theories to explain physical phenomena. According to legend, Galileo dropped objects of different weights from the Leaning Tower of Pisa to disprove the belief that heavier objects fall faster than lighter objects. He developed a formula for freely falling objects described by

$$d = 16t^2,$$

where d is the distance in feet that an object falls (disregarding air resistance) in t seconds, regardless of weight.

The equation $d = 16t^2$ is a *quadratic equation.*

Galileo Galilei (1564–1642)

Quadratic Equation

A **quadratic equation** (in x here) can be written in the form

$$ax^2 + bx + c = 0,$$

where a, b, and c are real numbers and $a \neq 0$. The given form is called **standard form.**

Examples: $x^2 + 5x + 6 = 0$, $2x^2 - 5x = 3$, $x^2 = 4$ — Quadratic equations

A quadratic equation has a second-degree (squared) term and no terms of greater degree. Of the above examples, only $x^2 + 5x + 6 = 0$ is in standard form.

Work Problems 1 and 2 at the Side. ▶

We have factored many quadratic *expressions* of the form

$$ax^2 + bx + c.$$

In this section, we use factored quadratic expressions to solve quadratic *equations*.

OBJECTIVE 1 Solve quadratic equations using the zero-factor property. We can use the following property to solve some quadratic equations.

Zero-Factor Property

If a and b are real numbers and $ab = 0$, then $a = 0$ or $b = 0$.

In words, if the product of two numbers is 0, then at least one of the numbers must be 0. One number *must* be 0, but both *may* be 0.

1 Which of the following equations are quadratic equations?

A. $y^2 - 4y - 5 = 0$

B. $x^3 - x^2 + 16 = 0$

C. $2z^2 + 7z = -3$

D. $x + 2y = -4$

2 Write each quadratic equation in standard form.

(a) $x^2 - 3x = 4$

(b) $y^2 = 9y - 8$

Answers

1. A, C

2. **(a)** $x^2 - 3x - 4 = 0$
(b) $y^2 - 9y + 8 = 0$

3 Solve each equation.

GS **(a)** $(x+2)(x-5)=0$

To solve, use the zero-factor property.

$\_\_\_\_ = 0$ or $\_\_\_\_ = 0$

Solve these two equations, obtaining

$x = \_\_\_\_$ or $x = \_\_\_\_$.

Check each value in the original equation.

The solution set is $\{\_\_, \_\_\}$.

(b) $(3x-2)(x+6)=0$

(c) $z(2z+5)=0$

Answers

3. **(a)** $x+2$; $x-5$; -2; 5; -2; 5

(b) $\left\{-6, \frac{2}{3}\right\}$ **(c)** $\left\{-\frac{5}{2}, 0\right\}$

EXAMPLE 1 Using the Zero-Factor Property

Solve each equation.

(a) $(x+3)(2x-1)=0$

The product $(x+3)(2x-1)$ is equal to 0. By the zero-factor property, the only way that the product of these two factors can be 0 is if at least one of the factors equals 0. Therefore, $x+3=0$ or $2x-1=0$.

$$x+3=0 \quad \text{or} \quad 2x-1=0 \qquad \text{Zero-factor property}$$

$$x=-3 \quad \text{or} \quad 2x=1 \qquad \text{Solve each equation.}$$

$$x=\frac{1}{2} \qquad \text{Divide each side by 2.}$$

Check these two values by substituting -3 for x in the original equation. ***Then start over*** and substitute $\frac{1}{2}$ for x.

CHECK Let $x=-3$.

$$(x+3)(2x-1)=0$$

$$(-3+3)[2(-3)-1] \stackrel{?}{=} 0$$

$$0(-7) \stackrel{?}{=} 0$$

$$0=0 \ \checkmark \ \text{True}$$

Let $x=\frac{1}{2}$.

$$(x+3)(2x-1)=0$$

$$\left(\frac{1}{2}+3\right)\left(2\cdot\frac{1}{2}-1\right) \stackrel{?}{=} 0$$

$$\frac{7}{2}(1-1) \stackrel{?}{=} 0$$

$$0=0 \ \checkmark \ \text{True}$$

Because true statements result, the solution set is $\{-3, \frac{1}{2}\}$.

(b)

$$y(3y-4)=0$$

$$y=0 \quad \text{or} \quad 3y-4=0 \qquad \text{Zero-factor property}$$

Don't forget that 0 is a solution.

$$3y=4 \qquad \text{Add 4.}$$

$$y=\frac{4}{3} \qquad \text{Divide by 3.}$$

Check each value in the original equation. The solution set is $\{0, \frac{4}{3}\}$.

◀ Work Problem **3** at the Side.

EXAMPLE 2 Solving Quadratic Equations

Solve each equation.

(a) $x^2-5x=-6$

First, write the equation in standard form by adding 6 to each side.

Don't factor x out at this step.

$$x^2-5x=-6$$

$$x^2-5x+6=0 \qquad \text{Add 6.}$$

Now factor x^2-5x+6. Find two numbers whose product is 6 and whose sum is -5. These two numbers are -2 and -3, so we factor as follows.

$$(x-2)(x-3)=0 \qquad \text{Factor.}$$

$$x-2=0 \quad \text{or} \quad x-3=0 \qquad \text{Zero-factor property}$$

$$x=2 \quad \text{or} \quad x=3 \qquad \text{Solve each equation.}$$

Continued on Next Page

CHECK Let $x = 2$.

$$x^2 - 5x = -6$$
$$2^2 - 5(2) \stackrel{?}{=} -6$$
$$4 - 10 \stackrel{?}{=} -6$$
$$-6 = -6 \checkmark \text{ True}$$

Let $x = 3$.

$$x^2 - 5x = -6$$
$$3^2 - 5(3) \stackrel{?}{=} -6$$
$$9 - 15 \stackrel{?}{=} -6$$
$$-6 = -6 \checkmark \text{ True}$$

Both values check, so the solution set is $\{2, 3\}$.

(b) $y^2 = y + 20$ — Write this equation in standard form.

Standard form ⟶	$y^2 - y - 20 = 0$	Subtract y and 20.
	$(y - 5)(y + 4) = 0$	Factor.
	$y - 5 = 0$ or $y + 4 = 0$	Zero-factor property
	$y = 5$ or $y = -4$	Solve each equation.

Check each value in the original equation. The solution set is $\{-4, 5\}$.

Work Problem 4 at the Side. ▶

Solving a Quadratic Equation Using the Zero-Factor Property

Step 1 **Write the equation in standard form**—that is, with all terms on one side of the equality symbol in descending powers of the variable and 0 on the other side.

Step 2 **Factor** completely.

Step 3 **Apply the zero-factor property.** Set each factor with a variable equal to 0.

Step 4 **Solve** the resulting equations.

Step 5 **Check** each value in the original equation. Write the solution set.

EXAMPLE 3 Solving a Quadratic Equation (Common Factor)

Solve $4p^2 + 40 = 26p$.

$4p^2 + 40 = 26p$ — Write this equation in the form $ax^2 + bx + c = 0$.

Step 1	$4p^2 - 26p + 40 = 0$	Subtract $26p$.
This 2 is *not* a solution of the equation.	$2(2p^2 - 13p + 20) = 0$	Factor out 2.
	$2p^2 - 13p + 20 = 0$	Divide each side by 2.
Step 2	$(2p - 5)(p - 4) = 0$	Factor.
Step 3	$2p - 5 = 0$ or $p - 4 = 0$	Zero-factor property
Step 4	$p = \frac{5}{2}$ or $p = 4$	Solve each equation.

Step 5 Substitute to check $\frac{5}{2}$ and 4. The solution set is $\left\{\frac{5}{2}, 4\right\}$.

Work Problem 5 at the Side. ▶

CAUTION

A common error is to include the common factor 2 as a solution in **Example 3.** ***Only factors containing variables lead to solutions.***

---

4. Solve each equation.

(a) $m^2 - 3m - 10 = 0$

(b) $r^2 + 2r = 8$

5. Solve each equation.

GS (a) $10x^2 - 5x - 15 = 0$

$\_\_\_(\_\_\_ - x - 3) = 0$

Divide each side by ___.

$2x^2 - x - 3 = 0$

$(2x - 3)(\_\_\_) = 0$

$2x - 3 = 0$ or $\_\_\_ = 0$

$x = \_\_\_$ or $x = \_\_\_$

Check by substituting in the original equation.

The solution set is ___.

(b) $4x^2 - 2x = 42$

Answers

4. (a) $\{-2, 5\}$ (b) $\{-4, 2\}$

5. (a) $5; 2x^2; 5; x + 1; x + 1; \frac{3}{2}; -1;$ $\left\{-1, \frac{3}{2}\right\}$

(b) $\left\{-3, \frac{7}{2}\right\}$

6 Solve each equation.

(a) $49x^2 - 9 = 0$

(b) $m^2 = 3m$

(c) $p(4p + 7) = 2$

EXAMPLE 4 Solving Quadratic Equations

Solve each equation.

(a)

$$16m^2 - 25 = 0$$

This equation is in standard form $ax^2 + bx + c = 0$. There is no first-degree term because $b = 0$.

$$(4m + 5)(4m - 5) = 0 \quad \text{Factor the difference of squares.}$$

$$4m + 5 = 0 \quad \text{or} \quad 4m - 5 = 0 \quad \text{Zero-factor property}$$

$$4m = -5 \quad \text{or} \quad 4m = 5 \quad \text{Solve each equation.}$$

$$m = -\frac{5}{4} \quad \text{or} \quad m = \frac{5}{4} \quad \text{Divide by 4.}$$

Check $-\frac{5}{4}$ and $\frac{5}{4}$ in the original equation. The solution set is $\left\{-\frac{5}{4}, \frac{5}{4}\right\}$.

(b)

$$y^2 = 2y$$

$$y^2 - 2y = 0$$

This equation is in the form $ax^2 + bx + c = 0$. Here, $c = 0$.

$$y(y - 2) = 0 \quad \text{Factor.}$$

Don't forget to set the variable factor y equal to 0.

$$y = 0 \quad \text{or} \quad y - 2 = 0 \quad \text{Zero-factor property}$$

$$y = 2 \quad \text{Add 2.}$$

A check confirms that the solution set is $\{0, 2\}$.

(c)

$$k(2k + 5) = 3$$

To be in standard form, 0 must be on one side.

$$2k^2 + 5k = 3 \quad \text{Multiply.}$$

$$\text{Standard form} \rightarrow 2k^2 + 5k - 3 = 0 \quad \text{Subtract 3.}$$

$$(2k - 1)(k + 3) = 0 \quad \text{Factor.}$$

$$2k - 1 = 0 \quad \text{or} \quad k + 3 = 0 \quad \text{Zero-factor property}$$

$$2k = 1 \quad \text{or} \quad k = -3 \quad \text{Solve each equation.}$$

$$k = \frac{1}{2}$$

A check confirms that the solution set is $\left\{-3, \frac{1}{2}\right\}$.

◄ Work Problem **6** at the Side.

CAUTION

In **Example 4(b)**, it is tempting to begin by dividing both sides of

$$y^2 = 2y$$

by y to obtain $y = 2$. We do not find the solution 0 if we divide by a variable. (We *may* divide each side of an equation by a *nonzero* real number, however. In **Example 3** we divided each side by 2.)

In **Example 4(c)**, we cannot use the zero-factor property to solve

$$k(2k + 5) = 3$$

in its given form because of the 3 on the right side of the equation. ***The zero-factor property applies only to a product that equals 0.***

Answers

6. (a) $\left\{-\frac{3}{7}, \frac{3}{7}\right\}$ (b) $\{0, 3\}$ (c) $\left\{-2, \frac{1}{4}\right\}$

EXAMPLE 5 Solving Quadratic Equations (Double Solutions)

Solve each equation.

(a) $z^2 - 22z + 121 = 0$ This is a perfect square trinomial.

$(z - 11)^2 = 0$ Factor.

$(z - 11)(z - 11) = 0$ $a^2 = a \cdot a$

$z - 11 = 0$ or $z - 11 = 0$ Zero-factor property

$z = 11$ Add 11.

The *same* factor appears twice, which leads to the *same* solution, called a **double solution.**

CHECK $z^2 - 22z + 121 = 0$ Original equation

$11^2 - 22(11) + 121 \stackrel{?}{=} 0$ Let $z = 11$.

$121 - 242 + 121 \stackrel{?}{=} 0$ Apply the exponent. Multiply.

$0 = 0$ ✓ True

The solution set is $\{11\}$.

(b) $9t^2 + 30t = -25$

$9t^2 + 30t + 25 = 0$ Standard form

$(3t + 5)^2 = 0$ Factor the perfect square trinomial.

$3t + 5 = 0$ or $3t + 5 = 0$ Zero-factor property

$3t = -5$ Subtract 5.

$t = -\frac{5}{3}$ $-\frac{5}{3}$ is a double solution.

CHECK $9t^2 + 30t = -25$ Original equation

$9\left(-\frac{5}{3}\right)^2 + 30\left(-\frac{5}{3}\right) \stackrel{?}{=} -25$ Let $t = -\frac{5}{3}$.

$9\left(\frac{25}{9}\right) + 30\left(-\frac{5}{3}\right) \stackrel{?}{=} -25$ Apply the exponent.

$25 - 50 \stackrel{?}{=} -25$ Multiply.

$-25 = -25$ ✓ True

The solution set is $\left\{-\frac{5}{3}\right\}$.

Work Problem 7 at the Side. ▶

CAUTION

Each equation in **Example 5** has only *one* distinct solution. ***We write a double solution only once in a solution set.***

7 Solve each equation.

GS **(a)** $x^2 + 16x = -64$

$x^2 + 16x + \_\_\_\_ = 0$

$(\_\_\_\_)^2 = 0$

$\_\_\_\_ = 0$ or $\_\_\_\_ = 0$

Solve to obtain $x = \_\_\_\_$.

Check by substituting in the original equation. -8 is a ____ solution.

The solution set is ____.

(b) $4x^2 - 4x + 1 = 0$

(c) $4z^2 + 20z = -25$

Answers

7. (a) 64; $x + 8$; $x + 8$; $x + 8$; -8; double; $\{-8\}$

(b) $\left\{\frac{1}{2}\right\}$ (c) $\left\{-\frac{5}{2}\right\}$

8 Solve each equation.

(a) $2r^3 - 32r = 0$

(b) $x^3 - 3x^2 - 18x = 0$

9 Solve each equation.

(a) $(m + 3)(m^2 - 11m + 10) = 0$

(b) $(2x + 5)(4x^2 - 9) = 0$

Answers

8. (a) $\{-4, 0, 4\}$ **(b)** $\{-3, 0, 6\}$

9. (a) $\{-3, 1, 10\}$ **(b)** $\left\{-\frac{5}{2}, -\frac{3}{2}, \frac{3}{2}\right\}$

Note

Not all quadratic equations can be solved by factoring. A more general method for solving such equations is given later in the text.

OBJECTIVE 2 Solve other equations using the zero-factor property. We can extend the zero-factor property to solve equations that involve more than two factors with variables. (These equations will have at least one term greater than second degree. They are *not* quadratic equations.)

EXAMPLE 6 Solving an Equation with More Than Two Variable Factors

Solve $6z^3 - 6z = 0$.

$$6z^3 - 6z = 0 \quad \text{This is not a quadratic equation because of the degree 3 term.}$$

$$6z(z^2 - 1) = 0 \quad \text{Factor out } 6z.$$

$$6z(z + 1)(z - 1) = 0 \quad \text{Factor } z^2 - 1.$$

By an extension of the zero-factor property, this product can equal 0 only if at least one of the factors equals 0. Write and solve three equations, one for each factor with a variable.

$$6z = 0 \quad \text{or} \quad z + 1 = 0 \quad \text{or} \quad z - 1 = 0 \quad \text{Zero-factor property}$$

$$z = 0 \quad \text{or} \quad z = -1 \quad \text{or} \quad z = 1 \quad \text{Solve each equation.}$$

Check by substituting, in turn, 0, −1, and 1 into the original equation. The solution set is $\{-1, 0, 1\}$.

◀ Work Problem **8** at the Side.

EXAMPLE 7 Solving an Equation with a Quadratic Factor

Solve $(2x - 1)(x^2 - 9x + 20) = 0$.

$$(2x - 1)(x^2 - 9x + 20) = 0 \quad \text{The product of the factors is 0, as required. Do } \textit{not} \text{ multiply.}$$

$$(2x - 1)(x - 5)(x - 4) = 0 \quad \text{Factor } x^2 - 9x + 20.$$

$$2x - 1 = 0 \quad \text{or} \quad x - 5 = 0 \quad \text{or} \quad x - 4 = 0 \quad \text{Zero-factor property}$$

$$x = \frac{1}{2} \quad \text{or} \quad x = 5 \quad \text{or} \quad x = 4 \quad \text{Solve each equation.}$$

Check to verify that the solution set is $\left\{\frac{1}{2}, 4, 5\right\}$.

◀ Work Problem **9** at the Side.

CAUTION

In **Example 7,** it would be unproductive to begin by multiplying the two factors together. The zero-factor property requires the *product* of two or more factors to equal 0. ***Always consider first whether an equation is given in the appropriate form to apply the zero-factor property.***

6.6 Exercises

FOR EXTRA HELP — *Go to MyMathLab for worked-out, step-by-step solutions to exercises enclosed in a square and video solutions to ▶ exercises.*

CONCEPT CHECK *Fill in each blank with the correct response.*

1. A quadratic equation in x is an equation that can be written in the form ___________ $= 0$.

2. The form $ax^2 + bx + c = 0$ is called ___________ form.

3. If a quadratic equation is in standard form, to solve the equation we should begin by attempting to ___________ the polynomial.

4. If the product of two numbers is 0, then at least one of the numbers is ___________. This is the ___________-___________ property.

CONCEPT CHECK *Work each problem.*

5. Identify each equation as *linear* or *quadratic*.

(a) $2x - 5 = 6$ **(b)** $x^2 - 5 = -4$

(c) $x^2 + 2x - 1 = 2x^2$ **(d)** $2^2 + 5x = 0$

6. The number 9 is a *double solution* of the equation

$$(x - 9)^2 = 0.$$

Why is this so?

7. CONCEPT CHECK Look at this "solution."

$$2x(3x - 4) = 0$$

$$x = 2 \quad \text{or} \quad x = 0 \quad \text{or} \quad 3x - 4 = 0$$

$$x = \frac{4}{3}$$

The solution set is $\{2, 0, \frac{4}{3}\}$.

What Went Wrong? Solve the equation correctly.

8. CONCEPT CHECK Look at this "solution."

$$7x^2 - x = 0$$

$$7x - 1 = 0 \quad \text{Divide by } x.$$

$$x = \frac{1}{7} \quad \text{Solve the equation.}$$

The solution set is $\{\frac{1}{7}\}$.

What Went Wrong? Solve the equation correctly.

Solve each equation, and check the solutions. ***See Example 1.***

9. $(x + 5)(x - 2) = 0$

10. $(x - 1)(x + 8) = 0$

11. $(2m - 7)(m - 3) = 0$ ▶

12. $(6k + 5)(k + 4) = 0$

13. $(2x + 1)(6x - 1) = 0$

14. $(3x - 2)(10x + 1) = 0$

15. $t(6t + 5) = 0$

16. $w(4w + 1) = 0$

17. $2x(3x - 4) = 0$

18. $6x(4x - 9) = 0$

19. $(x - 9)(x - 9) = 0$

20. $(y + 1)(y + 1) = 0$

Solve each equation, and check the solutions. ***See Examples 2–7.***

21. $y^2 + 5y + 4 = 0$

22. $p^2 + 8p + 7 = 0$

23. $y^2 - 3y + 2 = 0$

24. $r^2 - 4r + 3 = 0$

25. $x^2 = 24 - 5x$

26. $t^2 = 2t + 15$

27. $x^2 = 3 + 2x$

28. $m^2 = 4 + 3m$

29. $z^2 + 3z = -2$

30. $p^2 - 2p = 3$

31. $m^2 + 8m + 16 = 0$

32. $x^2 + 6x + 9 = 0$

33. $16x^2 = 8x - 1$

34. $25y^2 = 10y - 1$

35. $3x^2 + 5x - 2 = 0$

36. $6r^2 - r - 2 = 0$

37. $12p^2 = 8 - 10p$

38. $18x^2 = 12 + 15x$

39. $9k^2 + 12k = -4$

40. $36x^2 - 60x = -25$

41. $y^2 - 9 = 0$

42. $m^2 - 100 = 0$

43. $16k^2 - 49 = 0$

44. $4w^2 - 9 = 0$

45. $n^2 = 169$

46. $x^2 = 400$

47. $x^2 = 7x$

48. $t^2 = 9t$

49. $6r^2 = 3r$

50. $10y^2 = -5y$

51. $g(g - 7) = -10$

52. $r(r - 5) = -6$

53. $z(2z + 7) = 4$

54. $b(2b+3)=9$

55. $2(y^2-66)=-13y$

56. $3(t^2+4)=20t$

57. $3z(2z+7)=12$

58. $4x(2x+3)=36$

59. $5x^3-20x=0$

60. $3x^3-48x=0$

61. $9y^3-49y=0$

62. $16r^3-9r=0$

63. $(2r+5)(3r^2-16r+5)=0$

64. $(3m+4)(6m^2+m-2)=0$

65. $(2x+7)(x^2+2x-3)=0$

66. $(x+1)(6x^2+x-12)=0$

67. $x^3+x^2-20x=0$

68. $y^3-6y^2+8y=0$

69. $r^4=2r^3+15r^2$

70. $x^4=3x^2+2x^3$

71. $(x-8)(x+6)=6x$

72. $(x-2)(x+9)=4x$

73. $3x(x+1)=(2x+3)(x+1)$

74. $2x(x+3)=(3x+1)(x+3)$

Galileo's formula describing the motion of freely falling objects is

$$d=16t^2.$$

The distance d in feet an object falls depends on the time t elapsed, in seconds. (This is an example of an important mathematical concept, a ***function.****)*

75. (a) Use Galileo's formula and complete the following table. (*Hint:* Substitute each given value into the formula and solve for the unknown value.)

t in seconds	0	1	2	3	____	____
d in feet	0	16	____	____	256	576

(b) When $t=0$, $d=0$. Explain this in the context of the problem.

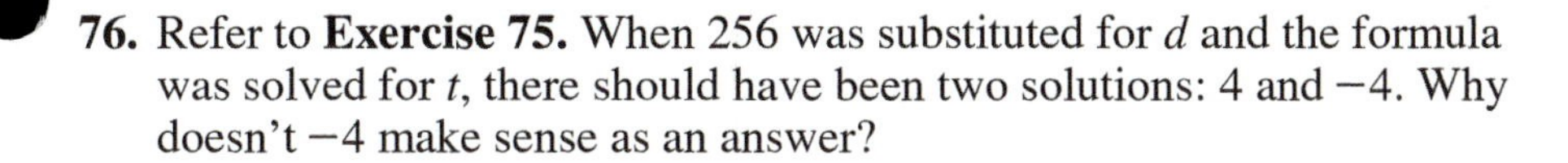

76. Refer to **Exercise 75.** When 256 was substituted for d and the formula was solved for t, there should have been two solutions: 4 and -4. Why doesn't -4 make sense as an answer?

6.7 Applications of Quadratic Equations

OBJECTIVES

1. Solve problems involving geometric figures.
2. Solve problems involving consecutive integers.
3. Solve problems by applying the Pythagorean theorem.
4. Solve problems using given quadratic models.

Solving an Applied Problem

Step 1 **Read** the problem carefully. *What information is given? What is to be found?*

Step 2 **Assign a variable** to represent the unknown value. Use a sketch, diagram, or table, as needed. Express any other unknown values in terms of the variable.

Step 3 **Write an equation** using the variable expression(s).

Step 4 **Solve** the equation.

Step 5 **State the answer.** Label it appropriately. *Does it seem reasonable?*

Step 6 **Check** the answer in the words of the *original* problem.

Problem-Solving Hint

Refer to the formulas at the back of the text as needed when solving application problems.

OBJECTIVE 1 Solve problems involving geometric figures.

EXAMPLE 1 Solving an Area Problem

The Monroes want to plant a rectangular garden in their yard. The width of the garden will be 4 ft less than its length, and they want it to have an area of 96 ft^2. (ft^2 means square feet.) Find the length and width of the garden.

Step 1 **Read** the problem carefully. We need to find the dimensions of a garden with area 96 ft^2.

Step 2 **Assign a variable.** See **Figure 1.**

Let $\quad x =$ the length of the garden.

Then $x - 4 =$ the width. (The width is 4 ft less than the length.)

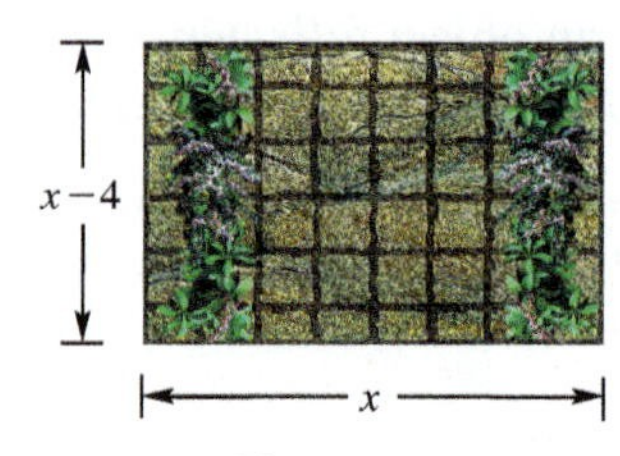

Figure 1

Step 3 **Write an equation.** The area of a rectangle is given by

$$\text{Area} = \text{Length} \cdot \text{Width}. \quad \text{Area formula}$$

Substitute 96 for area, x for length, and $x - 4$ for width.

$$\mathcal{A} = L \cdot W$$

$$\downarrow \quad \downarrow \quad \downarrow$$

$$96 = x(x-4) \quad \text{Let } \mathcal{A} = 96,\ L = x,\ W = x - 4.$$

Continued on Next Page

Step 4 **Solve.**

$$96 = x(x - 4) \quad \text{Equation from Step 3}$$
$$96 = x^2 - 4x \quad \text{Distributive property}$$
$$x^2 - 4x - 96 = 0 \quad \text{Standard form}$$
$$(x - 12)(x + 8) = 0 \quad \text{Factor.}$$
$$x - 12 = 0 \quad \text{or} \quad x + 8 = 0 \quad \text{Zero-factor property}$$
$$x = 12 \quad \text{or} \quad x = -8 \quad \text{Solve each equation.}$$

Step 5 **State the answer.** The solutions are 12 and −8. A rectangle cannot have a side of negative length, so we discard −8. The length of the garden will be 12 ft. The width will be

$$12 - 4 = 8 \text{ ft.}$$

Step 6 **Check.** The width of the garden is 4 ft less than the length, and the area is

$$12 \cdot 8 = 96 \text{ ft}^2, \quad \text{as required.}$$

Work Problem 1 at the Side. ▶

1 Solve each problem.

(a) The length of a rectangular room is 2 m more than the width. The area of the floor is 48 m^2. Find the length and width of the room.

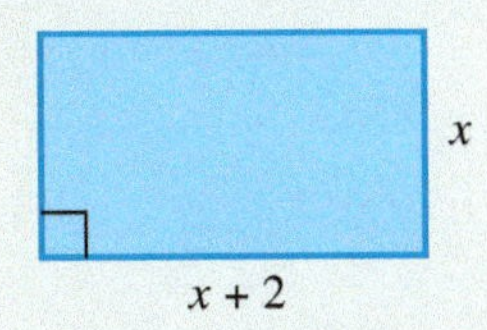

Give the equation using x as the variable, and give the answer.

(b) The length of each side of a square is increased by 4 in. The sum of the areas of the original square and the larger square is 106 in^2. What is the length of a side of the original square?

CAUTION

When solving applied problems, always check solutions against physical facts and discard any answers that are not appropriate.

OBJECTIVE 2 Solve problems involving consecutive integers. Recall that **consecutive integers** (see **Figure 2**) are integers that are next to each other on a number line, such as

$$1 \text{ and } 2, \quad \text{or} \quad -11 \text{ and } -10.$$

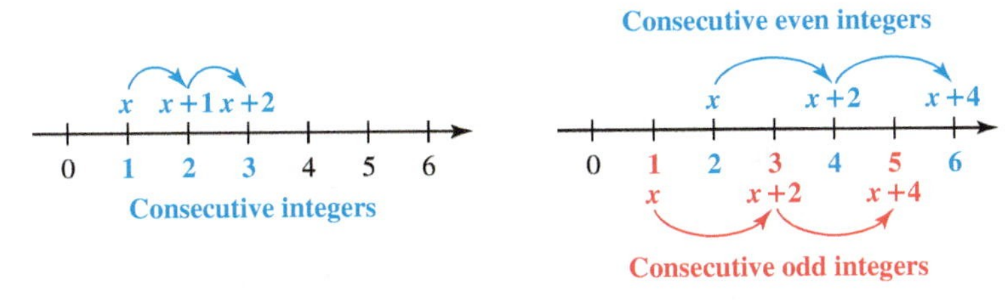

Figure 2 **Figure 3**

Consecutive even integers are *even* integers that are next to each other on a number line, such as

$$4 \text{ and } 6, \quad \text{or} \quad -10 \text{ and } -8.$$

Consecutive odd integers are defined similarly—for example, 3 and 5 are consecutive *odd* integers, as are −13 and −11. See **Figure 3.**

Problem-Solving Hint

If x = the lesser (least) integer in a consecutive integer problem, then the following apply.

- For two consecutive integers, use $x, \; x + 1.$
- For three consecutive integers, use $x, \; x + 1, \; x + 2.$
- For two consecutive even or odd integers, use $x, \; x + 2.$
- For three consecutive even or odd integers, use $x, \; x + 2, \; x + 4.$

In this text, we list consecutive integers in increasing order.

Answers

1. (a) $48 = (x + 2)x$; length: 8 m; width: 6 m
   (b) 5 in.

2 Solve the problem.

The product of the numbers on two consecutively-numbered lockers at a health club is 132. Find the locker numbers.

EXAMPLE 2 Solving a Consecutive Integer Problem

The product of the numbers on two consecutively-numbered post-office boxes is 210. Find the box numbers.

Step 1 **Read** the problem. Note that the boxes are numbered consecutively.

Step 2 **Assign a variable.** See **Figure 4.**

Let $x =$ the first box number.

Then $x + 1 =$ the next consecutive box number.

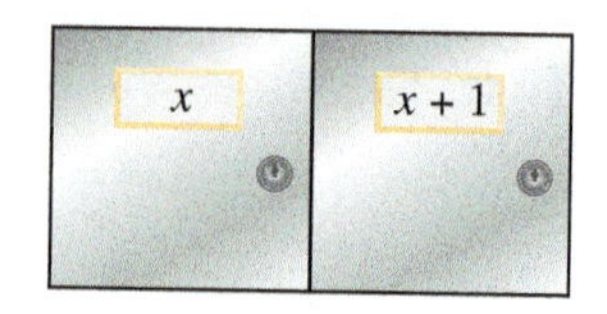

Figure 4

Step 3 **Write an equation.** The product of the box numbers is 210.

The product is 210.

$$x(x + 1) = 210$$

Step 4 **Solve.**

$$x^2 + x = 210 \quad \text{Distributive property}$$

$$x^2 + x - 210 = 0 \quad \text{Standard form}$$

$$(x + 15)(x - 14) = 0 \quad \text{Factor.}$$

$$x + 15 = 0 \quad \text{or} \quad x - 14 = 0 \quad \text{Zero-factor property}$$

$$x = -15 \quad \text{or} \quad x = 14 \quad \text{Solve each equation.}$$

Step 5 **State the answer.** The solutions are -15 and 14. We discard -15 because a box number cannot be negative. If $x = 14$, then

$$x + 1 = 15,$$

so the boxes have numbers 14 and 15.

Step 6 **Check.** The numbers 14 and 15 are consecutive and their product is

$$14 \cdot 15 = 210, \quad \text{as required.}$$

◀ Work Problem **2** at the Side.

EXAMPLE 3 Solving a Consecutive Integer Problem

The product of two consecutive odd integers is 1 less than five times their sum. Find the integers.

Step 1 **Read** carefully. This problem is a little more complicated.

Step 2 **Assign a variable.** We must find two consecutive *odd* integers.

Let $x =$ the lesser integer.

Then $x + 2 =$ the next greater odd integer.

Step 3 **Write an equation.**

The product is five times the sum less 1.

$$x(x + 2) = 5(x + x + 2) - 1$$

Continued on Next Page

Answer

2. 11 and 12

Step 4 **Solve.**

$$x(x+2) = 5(x+x+2) - 1 \quad \text{Equation from Step 3}$$
$$x^2 + 2x = 5x + 5x + 10 - 1 \quad \text{Distributive property}$$
$$x^2 + 2x = 10x + 9 \quad \text{Combine like terms.}$$
$$x^2 - 8x - 9 = 0 \quad \text{Standard form}$$
$$(x-9)(x+1) = 0 \quad \text{Factor.}$$
$$x - 9 = 0 \quad \text{or} \quad x + 1 = 0 \quad \text{Zero-factor property}$$
$$x = 9 \quad \text{or} \quad x = -1 \quad \text{Solve each equation.}$$

Step 5 **State the answer.** We need to find two consecutive odd integers.

If $x = 9$ is the lesser, then $x + 2 = 9 + 2 = 11$ is the greater.

If $x = -1$ is the lesser, then $x + 2 = -1 + 2 = 1$ is the greater.

Do not discard the solution -1 here. There are two sets of answers since integers can be positive *or* negative.

Step 6 **Check.** The product of the first pair of integers is

$$9 \cdot 11 = 99.$$

One less than five times their sum is

$$5(9 + 11) - 1 = 99.$$

Thus, 9 and 11 satisfy the problem. Repeat the check with -1 and 1.

Work Problem 3 at the Side. ▶

3 Solve each problem. Give the equation using x to represent the least integer, and give the answer.

(a) The product of two consecutive even integers is 4 more than two times their sum. Find the integers.

(b) Find three consecutive odd integers such that the product of the least and greatest is 16 more than the middle integer.

CAUTION

Do *not* use $x, x+1, x+3$ to represent consecutive odd integers. To see why, let $x = 3$.

Then $x + 1 = 3 + 1$ or 4, and $x + 3 = 3 + 3$ or 6.

The numbers 3, 4, and 6 are not consecutive odd integers.

OBJECTIVE 3 Solve problems by applying the Pythagorean theorem. Although there is evidence of the discovery of the Pythagorean relationship in earlier civilizations, the Greek mathematician and philosopher Pythagoras generally receives credit for being the first to prove it. This famous theorem from geometry relates the lengths of the sides of a right triangle.

Pythagoras (c. 580–500 B.C.)

Pythagorean Theorem

If a and b are the lengths of the two shorter sides of a right triangle (a triangle with a 90° angle) and c is the length of the longest side, then

$$a^2 + b^2 = c^2.$$

The two shorter sides are the **legs** of the triangle, and the longest side, opposite the right angle, is the **hypotenuse.**

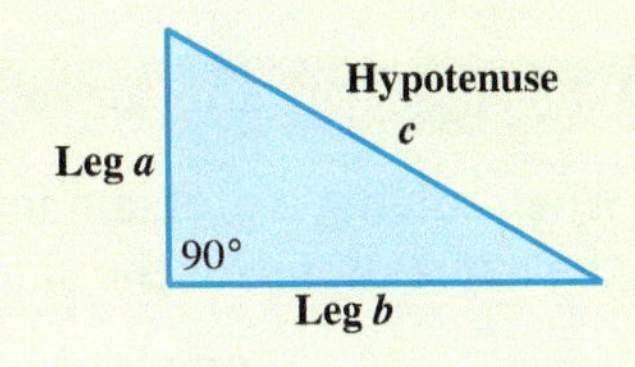

Answers

3. (a) $x(x+2) = 4 + 2(x + x + 2)$; 4 and 6 or -2 and 0

(b) $x(x+4) = 16 + (x+2)$; 3, 5, 7

4 Solve the problem.

The hypotenuse of a right triangle is 3 in. longer than the longer leg. The shorter leg is 3 in. shorter than the longer leg. Find the lengths of the sides of the triangle.

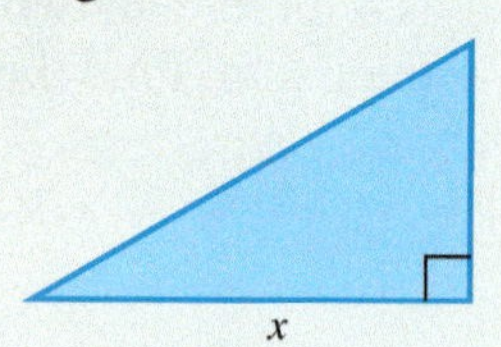

Answer

4. 9 in., 12 in., 15 in.

EXAMPLE 4 Applying the Pythagorean Theorem

Amy and Kevin leave their office, with Amy traveling north and Kevin traveling east. When Kevin is 1 mi farther than Amy from the office, the distance between them is 2 mi more than Amy's distance from the office. Find their distances from the office and the distance between them.

Step 1 **Read** the problem again. We must find three distances.

Step 2 **Assign a variable.**

Let $x =$ Amy's distance from the office.

Then $x + 1 =$ Kevin's distance from the office,

and $x + 2 =$ the distance between them.

Label a right triangle with these expressions, as in **Figure 5.**

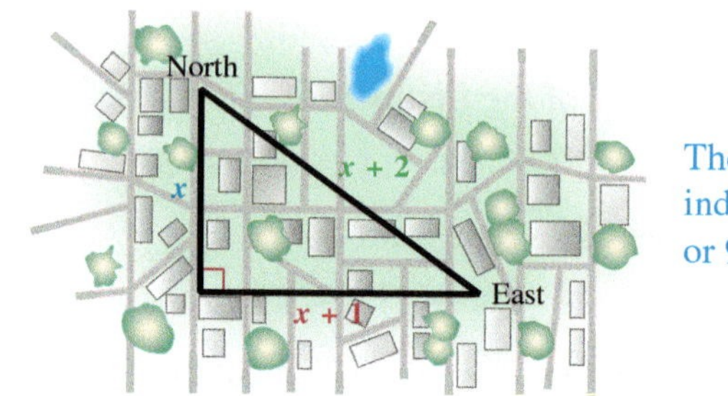

Figure 5

The symbol ⌊ indicates a right, or 90°, angle.

Step 3 **Write an equation.** Substitute into the Pythagorean theorem.

$$a^2 + b^2 = c^2$$

$$x^2 + (x + 1)^2 = (x + 2)^2$$ Be careful to substitute properly.

Step 4 **Solve.**

$$x^2 + x^2 + 2x + 1 = x^2 + 4x + 4$$ Square each binomial.

$$x^2 - 2x - 3 = 0$$ Standard form

$$(x - 3)(x + 1) = 0$$ Factor.

$$x - 3 = 0 \quad \text{or} \quad x + 1 = 0$$ Zero-factor property

$$x = 3 \quad \text{or} \quad x = -1$$ Solve each equation.

Step 5 **State the answer.** Because -1 cannot represent a distance, 3 is the only possible answer. Amy's distance is 3 mi, Kevin's distance is

$$3 + 1 = 4 \text{ mi},$$

and the distance between them is

$$3 + 2 = 5 \text{ mi}.$$

Step 6 **Check.** Because $3^2 + 4^2 = 5^2$ is true, the answer is correct.

◀ Work Problem 4 at the Side.

Problem-Solving Hint

When solving a problem involving the Pythagorean theorem, be sure that the expressions for the sides of the triangle are properly placed.

$$(\text{one leg})^2 + (\text{other leg})^2 = \text{hypotenuse}^2$$

OBJECTIVE 4 Solve problems using given quadratic models. In **Examples 1–4,** we wrote quadratic equations to model, or mathematically describe, various situations and then solved the equations. Now we are given the quadratic models and must use them to determine data.

EXAMPLE 5 Finding the Height of a Ball

A tennis player can hit a ball 180 ft per sec (123 mph). If she hits a ball directly upward, the height h of the ball in feet at time t in seconds is modeled by the quadratic equation

$$h = -16t^2 + 180t + 6.$$

How long will it take for the ball to reach a height of 206 ft?

A height of 206 ft means $h = 206$, so we substitute 206 for h in the equation and then solve for t.

$$h = -16t^2 + 180t + 6$$

$$206 = -16t^2 + 180t + 6 \quad \text{Let } h = 206.$$

$$-16t^2 + 180t + 6 = 206 \quad \text{Interchange sides.}$$

$$-16t^2 + 180t - 200 = 0 \quad \text{Standard form}$$

$$4t^2 - 45t + 50 = 0 \quad \text{Divide by } -4.$$

$$(4t - 5)(t - 10) = 0 \quad \text{Factor.}$$

$$4t - 5 = 0 \quad \text{or} \quad t - 10 = 0 \quad \text{Zero-factor property}$$

$$t = \frac{5}{4} \quad \text{or} \quad t = 10 \quad \text{Solve each equation.}$$

Both answers are acceptable. The ball will be 206 ft above the ground twice—once on its way up and once on its way down—at $\frac{5}{4}$ sec and at 10 sec after it is hit. See **Figure 6.**

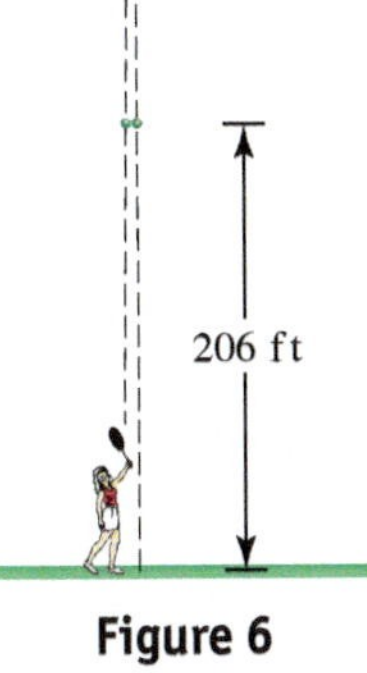

Figure 6

Work Problem 5 at the Side. ▶

5 Solve each problem.

(a) Refer to **Example 5.** How long will it take for the ball to reach a height of 50 ft?

(b) The number of impulses y fired after a nerve has been stimulated is modeled by the quadratic equation

$$y = -x^2 + 2x + 60,$$

where x is in milliseconds (ms) after the stimulation. When will 45 impulses occur? How many solutions do we obtain? Why is only one answer acceptable?

EXAMPLE 6 Modeling Foreign-Born Population of the United States

The foreign-born population of the United States over the years 1930–2010 can be modeled by the quadratic equation

$$y = 0.009665x^2 - 0.4942x + 15.12,$$

where $x = 0$ represents 1930, $x = 10$ represents 1940, and so on, and y is the number of people in millions. (Data from U.S. Census Bureau.)

(a) Use the model to find the foreign-born population in 1980 to the nearest tenth of a million.

Because $x = 0$ represents 1930, $x = 50$ represents 1980. Substitute 50 for x in the equation.

$$y = 0.009665x^2 - 0.4942x + 15.12 \quad \text{Given model}$$

$$y = 0.009665(50)^2 - 0.4942(50) + 15.12 \quad \text{Let } x = 50.$$

$$y \approx 14.6 \quad \text{Round to the nearest tenth.}$$

In 1980, the foreign-born population of the United States was about 14.6 million.

Continued on Next Page

Answers

5. (a) $\frac{1}{4}$ sec and 11 sec

(b) After 5 ms; There are two solutions, -3 and 5; Only one answer makes sense here because a negative answer is not appropriate.

6 Use the model in **Example 6** to find the foreign-born population of the United States in 1990. Give the answer to the nearest tenth of a million. How does it compare to the actual value from the table?

(b) Repeat part (a) for 2010.

$y = 0.009665x^2 - 0.4942x + 15.12$ Given model

$y = 0.009665(80)^2 - 0.4942(80) + 15.12$ For 2010, let $x = 80$.

$y \approx 37.4$ Round to the nearest tenth.

In 2010, the foreign-born population of the United States was about 37.4 million.

(c) The model used in parts (a) and (b) was developed using the data in the table below. How do the results in parts (a) and (b) compare to the actual data from the table?

Year	Foreign-Born Population (millions)
1930	14.2
1940	11.6
1950	10.3
1960	9.7
1970	9.6
1980	**14.1**
1990	19.8
2000	28.4
2010	**37.6**

From the table, the actual value for 1980 is 14.1 million. By comparison, our answer in part (a), 14.6 million, is slightly high.

For 2010, the actual value is 37.6 million. Our answer of 37.4 million in part (b) is slightly low, but a good estimate.

◀ Work Problem 6 at the Side.

Answer

6. 20.3 million; The actual value is 19.8 million, so the answer using the model is slightly high.

6.7 Exercises

FOR EXTRA HELP — Go to MyMathLab for worked-out, step-by-step solutions to exercises enclosed in a square and video solutions to ▶ exercises.

1. **CONCEPT CHECK** Complete each statement to review the six problem-solving steps.

Step 1: ______ the problem carefully.

Step 2: Assign a ______ to represent the unknown value.

Step 3: Write a(n) ______ using the variable expression(s).

Step 4: ______ the equation.

Step 5: State the ______.

Step 6: ______ the answer in the words of the ______ problem.

2. **CONCEPT CHECK** A student solves an applied problem and gets 6 or −3 for the length of the side of a square. Which of these answers is reasonable? Why?

*A geometric figure is given in each exercise. Write the indicated formula. Then, using x as the variable, complete Steps 3–6 for each problem. (Refer to the steps in **Exercise 1** as needed.)*

3.

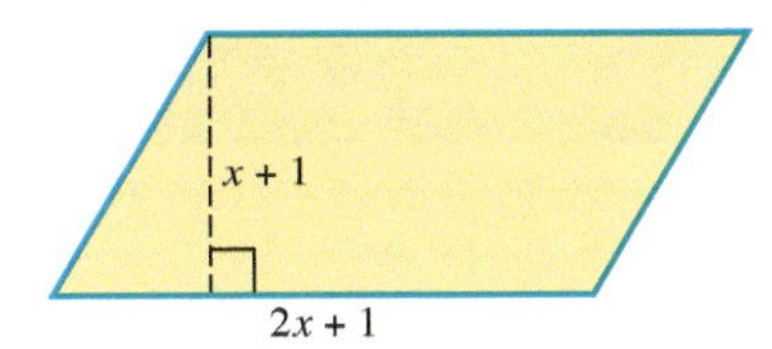

The area of this parallelogram is 45 sq. units. Find its base and height.

Formula for the area of a parallelogram: ______

Step 3: 45 = ______

Step 4: $x =$ ____ or $x =$ ____

Step 5: base: ____ units;

height: ____ units

Step 6: ______ = 45

4.

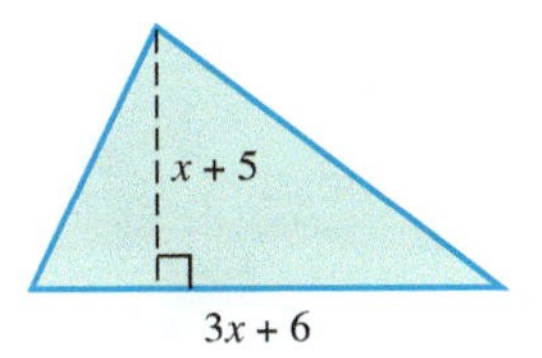

The area of this triangle is 60 sq. units. Find its base and height.

Formula for the area of a triangle: ______

Step 3: 60 = ______

Step 4: $x =$ ____ or $x =$ ____

Step 5: base: ____ units;

height: ____ units

Step 6: ______ = 60

5.

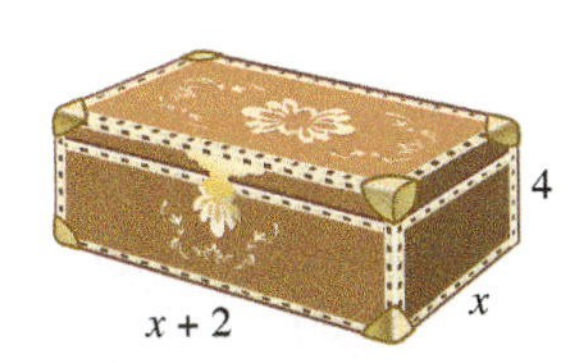

The volume of this box is 192 cu. units. Find its length and width.

Formula for the volume of a rectangular solid: ______

Step 3: ____ = ____$(x + 2)$

Step 4: $x =$ ____ or $x =$ ____

Step 5: length: ____ units;

width: ____ units

Step 6: ______ · 4 = ____

6.

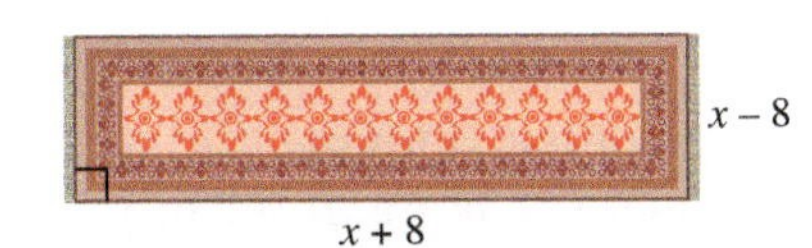

The area of this rug is 80 sq. units. Find its length and width.

Formula for the area of a rectangle: ______

Step 3: ____ = $(x + 8)$ ______

Step 4: $x =$ ____ or $x =$ ____

Step 5: length: ____ units;

width: ____ units

Step 6: ______ = 80

Solve each problem. Check answers to be sure they are reasonable. (If necessary, refer to the formulas at the back of this text.) ***See Example 1.***

7. The length of a standard jewel case is 2 cm more than its width. The area of the rectangular top of the case is 168 cm^2. Find the length and width of the jewel case.

8. A standard DVD case is 6 cm longer than it is wide. The area of the rectangular top of the case is 247 cm^2. Find the length and width of the case.

9. The area of a triangle is 30 in.^{2}. The base of the triangle measures 2 in. more than twice the height of the triangle. Find the measures of the base and the height.

10. A certain triangle has its base equal in measure to its height. The area of the triangle is 72 m^2. Find the equal base and height measure.

11. The dimensions of a rectangular monitor screen are such that its length is 3 in. more than its width. If the length were doubled and if the width were decreased by 1 in., the area would be increased by 150 in.^{2}. What are the length and width of the screen?

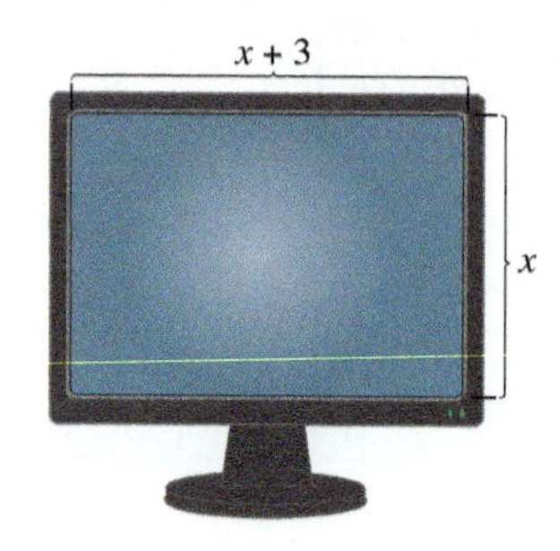

12. A computer keyboard is 11 in. longer than it is wide. If the length were doubled and if 2 in. were added to the width, the area would be increased by 198 in.^{2}. What are the length and width of the keyboard?

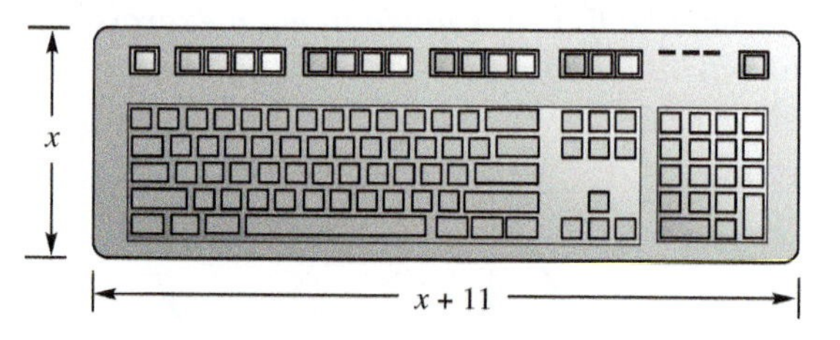

13. A 10-gal aquarium is 3 in. higher than it is wide. Its length is 21 in., and its volume is 2730 in.^{3}. What are the height and width of the aquarium?

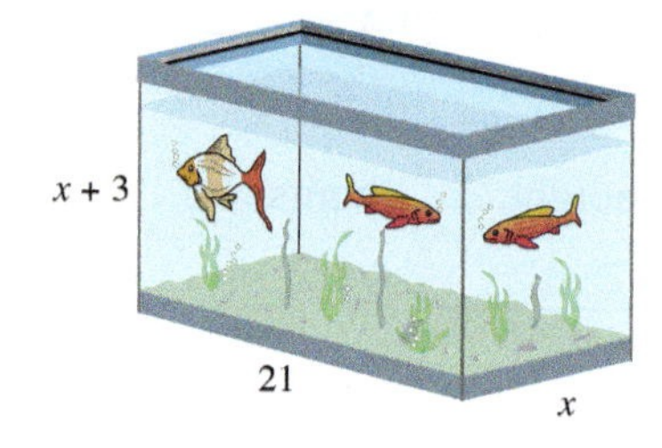

14. A toolbox is 2 ft high, and its width is 3 ft less than its length. If its volume is 80 ft^3, find the length and width of the box.

15. A square mirror has sides measuring 2 ft less than the sides of a square painting. If the difference between their areas is 32 ft^2, find the lengths of the sides of the mirror and the painting.

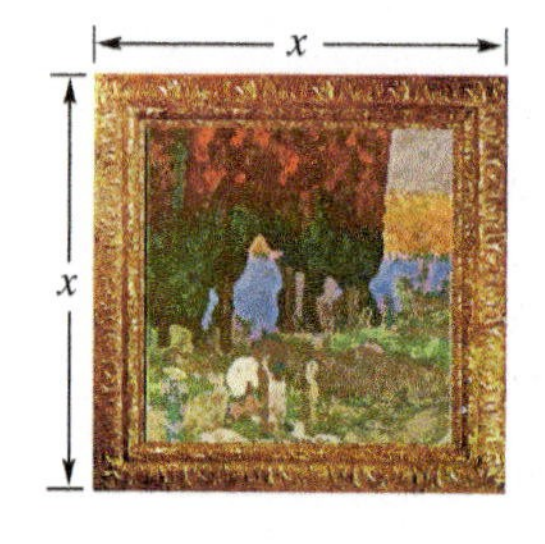

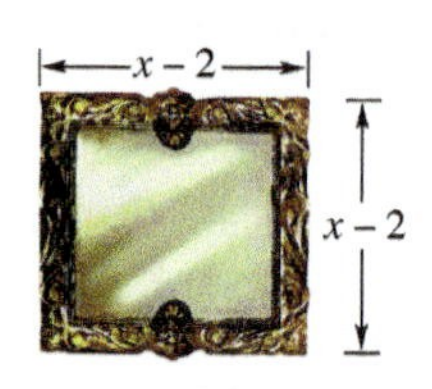

16. The sides of one square have length 3 m more than the sides of a second square. If the area of the larger square is subtracted from 4 times the area of the smaller square, the result is 36 m^2. What are the lengths of the sides of each square?

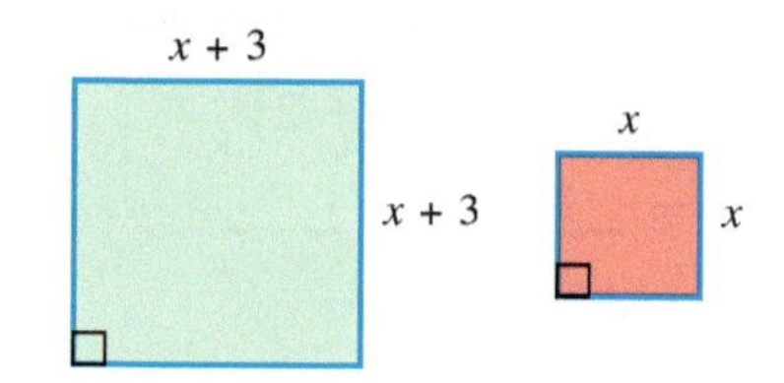

Solve each problem. **See Examples 2 and 3.**

17. The product of the numbers on two consecutive volumes of research data is 420. Find the volume numbers.

18. The product of the page numbers on two facing pages of a book is 600. Find the page numbers.

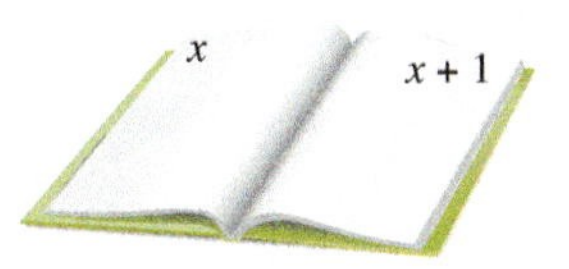

19. The product of two consecutive integers is 11 more than their sum. Find the integers.

20. The product of two consecutive integers is 4 less than four times their sum. Find the integers.

21. Find two consecutive odd integers such that their product is 15 more than three times their sum.

22. Find two consecutive odd integers such that five times their sum is 23 less than their product.

23. Find three consecutive even integers such that the sum of the squares of the lesser two integers is equal to the square of the greatest.

24. Find three consecutive even integers such that the square of the sum of the lesser two integers is equal to twice the greatest.

25. Find three consecutive odd integers such that 3 times the sum of all three is 18 more than the product of the lesser two integers.

26. Find three consecutive odd integers such that the sum of all three is 42 less than the product of the greater two integers.

Solve each problem. **See Example 4.**

27. The hypotenuse of a right triangle is 1 cm longer than the longer leg. The shorter leg is 7 cm shorter than the longer leg. Find the length of the longer leg.

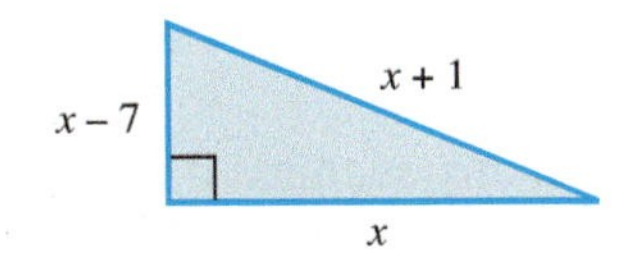

28. The longer leg of a right triangle is 1 m longer than the shorter leg. The hypotenuse is 1 m shorter than twice the shorter leg. Find the length of the shorter leg.

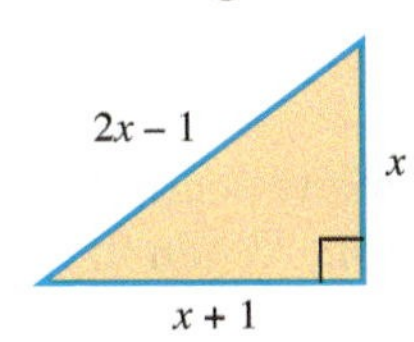

29. Terri works due north of home. Her husband Denny works due east. They leave for work at the same time. By the time Terri is 5 mi from home, the distance between them is 1 mi more than Denny's distance from home. How far from home is Denny?

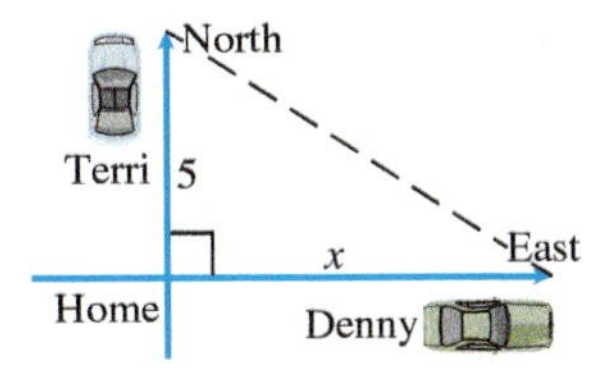

30. Two cars left an intersection at the same time. One traveled north. The other traveled 14 mi farther, but to the east. How far apart were they then, if the distance between them was 4 mi more than the distance traveled east?

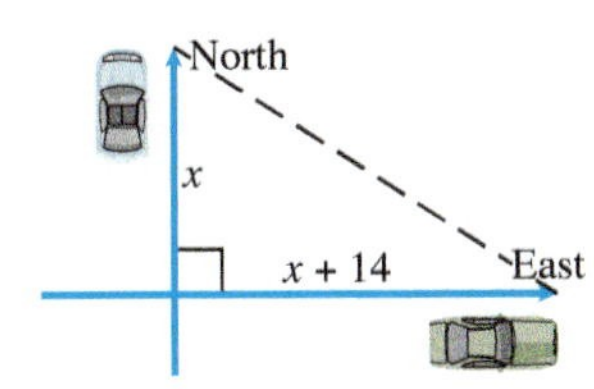

31. The length of a rectangle is 5 in. longer than its width. The diagonal is 5 in. shorter than twice the width. Find the length, width, and diagonal measures.

32. The length of a rectangle is 4 in. longer than its width. The diagonal is 8 in. longer than the width. Find the length, width, and diagonal measures.

33. A ladder is leaning against a building. The distance from the bottom of the ladder to the building is 4 ft less than the length of the ladder. How high up the side of the building is the top of the ladder if that distance is 2 ft less than the length of the ladder?

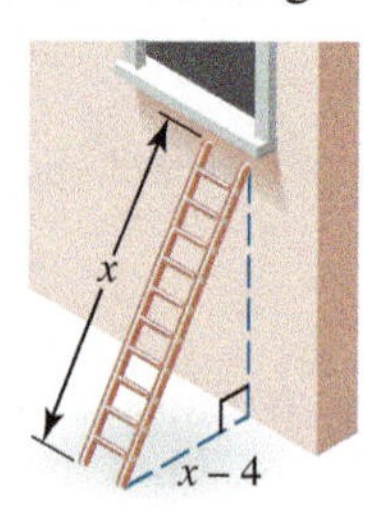

34. A lot has the shape of a right triangle with one leg 2 m longer than the other. The hypotenuse is 2 m less than twice the length of the shorter leg. Find the length of the shorter leg.

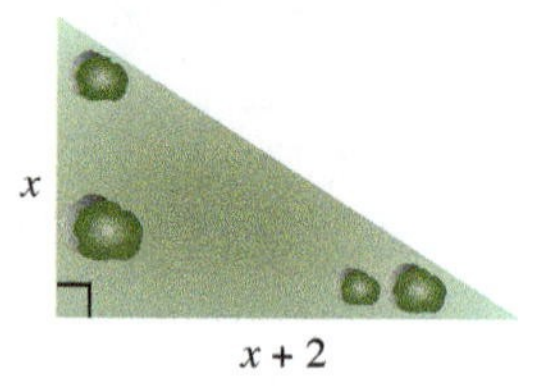

Solve each problem. ***See Example 5.***

35. An object projected from a height of 48 ft with an initial velocity of 32 ft per sec after t seconds has height $h = -16t^2 + 32t + 48$.

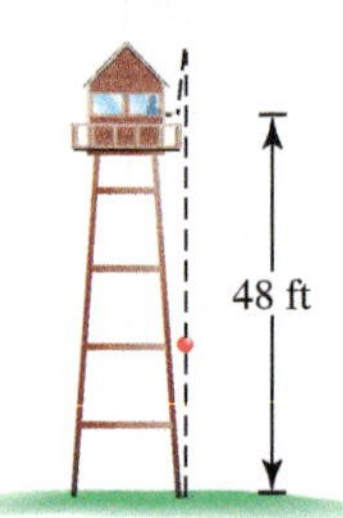

(a) After how many seconds is the height 64 ft? (*Hint:* Let $h = 64$ and solve.)

(b) After how many seconds is the height 60 ft?

(c) After how many seconds does the object hit the ground? (*Hint:* When the object hits the ground, $h = 0$.)

(d) The quadratic equation from part (c) has two solutions, yet only one of them is appropriate for answering the question. Why is this so?

36. If an object is projected upward from ground level with an initial velocity of 64 ft per sec, its height h in feet t seconds later is $h = -16t^2 + 64t$.

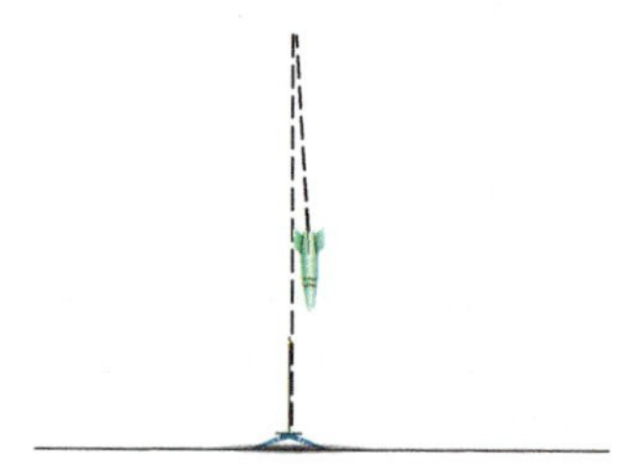

(a) After how many seconds is the height 48 ft? (*Hint:* Let $h = 48$ and solve.)

(b) The object reaches its maximum height 2 sec after it is projected. What is this maximum height?

(c) After how many seconds does the object hit the ground? (*Hint:* When the object hits the ground, $h = 0$.)

(d) The quadratic equation from part (c) has two solutions, yet only one of them is appropriate for answering the question. Why is this so?

If an object is projected upward with an initial velocity of 128 *ft per sec, its height h in feet after t seconds is*

$$h = 128t - 16t^2.$$

Find the height of the object after each period of time. ***See Example 5.***

37. 1 sec

38. 2 sec

39. 4 sec

40. How long does it take the object just described to return to the ground?

Solve each problem. ***See Example 6.***

41. The table shows the number of cellular phone subscribers (in millions) in the United States.

Year	Subscribers (in millions)
2000	109
2002	141
2004	182
2006	233
2008	286
2010	303
2012	326
2014	355

Data from CTIA-The Wireless Association.

We used the data to develop the quadratic equation

$$y = -0.4985x^2 + 25.21x + 100.3,$$

which models the number of cellular phone subscribers y (in millions) in the year x, where $x = 0$ represents 2000, $x = 2$ represents 2002, and so on.

(a) Use the model to find the number of subscribers in 2000 to the nearest million. How does the result compare to the actual data in the table?

(b) What value of x corresponds to 2010?

(c) Use the model to find the number of subscribers in 2010 to the nearest million. How does the result compare to the actual data in the table?

(d) Assuming that the trend in the data continues, use the quadratic equation to estimate the number of subscribers in 2018 to the nearest million.

42. Annual revenue in billions of dollars for eBay is shown in the table.

Year	Revenue (in billions of dollars)
2007	7.67
2008	8.54
2009	8.73
2010	9.16
2011	11.65
2012	14.07
2013	16.05
2014	17.90

Data from Statista—the Statistics Portal.

We used the data to develop the quadratic equation

$$y = 0.185x^2 + 0.223x + 7.70,$$

which models eBay revenues y (in billion of dollars) in the year x, where $x = 0$ represents 2007, $x = 1$ represents 2008, and so on.

(a) Use the model to find eBay revenue in 2011 to the nearest hundredth of a billion. How does the result compare to the actual data in the table?

(b) What value of x corresponds to 2014?

(c) Use the model to find eBay revenue in 2014 to the nearest hundredth of a billion. How does the result compare to the actual revenue in 2014?

(d) Assuming that the trend in the data continues, use the quadratic equation to estimate the number of subscribers in 2016 to the nearest hundredth of a billion.

Relating Concepts (Exercises 43–46) For Individual or Group Work

A proof of the Pythagorean theorem is based on the figures shown. ***Work Exercises 43–46 in order.***

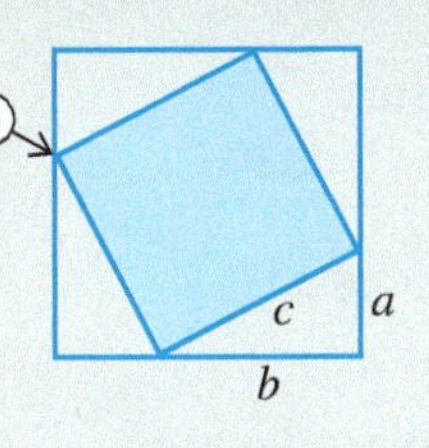

Figure A

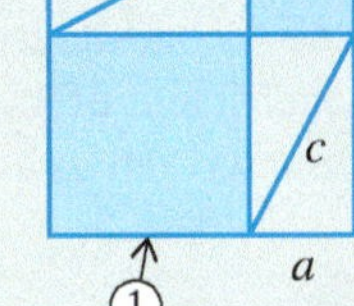

Figure B

43. What is an expression for the area of the dark square labeled ③ in **Figure A?**

44. The five regions in **Figure A** are equal in area to the six regions in **Figure B.** What is an expression for the area of the square labeled ① in **Figure B?**

45. What is an expression for the area of the square labeled ② in **Figure B?**

46. Represent this statement using algebraic expressions: The sum of the areas of the shaded regions in **Figure B** is equal to the area of the shaded region in **Figure A.** What does this equation represent?

Chapter 6 Summary

Key Terms

6.1

factor For integers a and b, if $a \cdot b = c$, then a and b are factors of c.

Factors
$20 = 4 \cdot 5$
Factored form

factored form An expression is in factored form when it is written as a product.

greatest common factor (GCF) The greatest common factor is the largest quantity that is a factor of each of a group of quantities.

factoring The process of writing a polynomial as a product is called factoring.

6.2

prime polynomial A prime polynomial is a polynomial that cannot be factored using only integers.

6.5

perfect square trinomial A perfect square trinomial is a trinomial that can be factored as the square of a binomial.

6.6

quadratic equation A quadratic equation is an equation that can be written in the form $ax^2 + bx + c = 0$, where a, b, and c are real numbers and $a \neq 0$.

standard form The form $ax^2 + bx + c = 0$ is the standard form of a quadratic equation.

6.7

hypotenuse The longest side of a right triangle, opposite the right angle, is the hypotenuse.

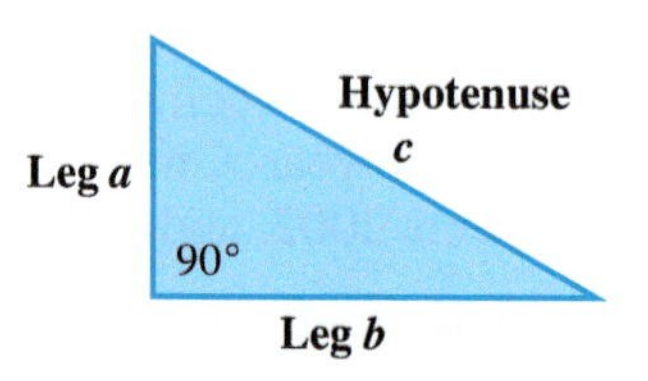

legs The two shorter sides of a right triangle are the legs.

Test Your Word Power

See how well you have learned the vocabulary in this chapter.

1 **Factoring** is
- **A.** a method of multiplying polynomials
- **B.** the process of writing a polynomial as a product
- **C.** the answer in a multiplication problem
- **D.** a way to add the terms of a polynomial.

2 A polynomial is in **factored form** when
- **A.** it is prime
- **B.** it is written as a sum
- **C.** the second-degree term has a coefficient of 1
- **D.** it is written as a product.

3 The greatest common factor of a polynomial is
- **A.** the least integer that divides evenly into all its terms
- **B.** the least expression that is a factor of all its terms
- **C.** the greatest expression that is a factor of all its terms
- **D.** the variable that is common to all its terms.

4 A **perfect square trinomial** is a trinomial
- **A.** that can be factored as the square of a binomial
- **B.** that cannot be factored
- **C.** that is multiplied by a binomial
- **D.** where all terms are perfect squares.

5 A **quadratic equation** is an equation that can be written in the form
- **A.** $y = mx + b$
- **B.** $ax^2 + bx + c = 0 \quad (a \neq 0)$
- **C.** $Ax + By = C$
- **D.** $x = k$.

6 A **hypotenuse** is
- **A.** either of the two shorter sides of a triangle
- **B.** the shortest side of a right triangle
- **C.** the side opposite the right angle in a right triangle
- **D.** the longest side in any triangle.

Answers to Test Your Word Power

1. B; *Example:* $7t^5 - 14t^{10}$ factors as $7t^5(1 - 2t^5)$.

2. D; *Example:* The factored form of $x^2 + 2x - 7x - 14$ is $(x - 7)(x + 2)$.

3. C; *Example:* The greatest common factor of $8x^2$, $22xy$, and $16x^3y^2$ is $2x$.

4. A; *Example:* $a^2 + 2a + 1$ is a perfect square trinomial. its factored form is $(a + 1)^2$.

5. B; *Examples:* $y^2 - 3y + 2 = 0$, $x^2 - 9 = 0$, $2m^2 = 6m + 8$

6. C; *Example:* See the triangle included in the Key Terms above.

Quick Review

Concepts | **Examples**

6.1 Greatest Common Factors; Factor by Grouping

Finding the Greatest Common Factor (GCF)

Step 1 Write each number in prime factored form.

Step 2 List each prime number or each variable that is a factor of every term in the list.

Step 3 Use as exponents on the common prime factors the *least* exponents from the prime factored forms.

Step 4 Multiply the primes from Step 3.

Find the greatest common factor of $4x^2y$, $6x^2y^3$, and $2xy^2$.

$$4x^2y = 2 \cdot 2 \cdot x^2 \cdot y$$
$$6x^2y^3 = 2 \cdot 3 \cdot x^2 \cdot y^3$$
$$2xy^2 = 2 \cdot x \cdot y^2$$
$$\text{GCF} = 2xy$$

Factoring by Grouping

Step 1 Collect the terms into two groups so that each group has a common factor.

Step 2 Factor out the greatest common factor from each group.

Step 3 Factor out a common binomial factor from the results of Step 2.

Step 4 If necessary, rearrange terms and try a different grouping.

Factor by grouping.

$2a^2 + 2ab + a + b$

$= (2a^2 + 2ab) + (a + b)$ Group the terms.

$= 2a(a + b) + 1(a + b)$ Factor each group.

$= (a + b)(2a + 1)$ Factor out $a + b$.

6.2 Factoring Trinomials

To factor $x^2 + bx + c$, find m and n such that $mn = c$ and $m + n = b$.

$mn = c$ (arrow to c)

$$x^2 + bx + c$$

$m + n = b$ (arrow to b)

Then $x^2 + bx + c$ factors as $(x + m)(x + n)$.

Check by multiplying.

Factor $x^2 + 6x + 8$.

$mn = 8$ (arrow to 8)

$$x^2 + 6x + 8$$

$m + n = 6$ (arrow to $6x$)

Find two integers m and n whose product is 8 and whose sum is 6. Here, $m = 2$ and $n = 4$.

$x^2 + 6x + 8$ factors as $(x + 2)(x + 4)$.

CHECK $(x + 2)(x + 4)$

$= x^2 + 4x + 2x + 8$ FOIL method

$= x^2 + 6x + 8$ ✓ Combine like terms.

6.3 Factoring Trinomials by Grouping

To factor $ax^2 + bx + c$, find m and n such that $mn = ac$ and $m + n = b$.

$m + n = b$ (arrow to b)

$$ax^2 + bx + c$$

$mn = ac$ (arrows to a and c)

Then factor $ax^2 + mx + nx + c$ by grouping.

Factor $3x^2 + 14x - 5$. Here, $mn = -15$ and $m + n = 14$.

-15 (from 3 and -5)

Find two integers with a product of $3(-5) = -15$ and a sum of 14. The integers are -1 and 15.

$3x^2 + 14x - 5$

$= 3x^2 - x + 15x - 5$ $14x = -x + 15x$

$= (3x^2 - x) + (15x - 5)$ Group the terms.

$= x(3x - 1) + 5(3x - 1)$ Factor each group.

$= (3x - 1)(x + 5)$ Factor out $3x - 1$.

Concepts	Examples
6.4 Factoring Trinomials Using the FOIL Method	
To factor $ax^2 + bx + c$ using trial and error, apply the FOIL method in reverse.	Factor $3x^2 + 14x - 5$. Because the only positive factors of 3 are 3 and 1, and -5 has possible factors of 1 and -5, or -1 and 5, the only possible factored forms for this trinomial follow. $(3x + 1)(x - 5)$ Incorrect $(3x - 5)(x + 1)$ Incorrect $(3x + 5)(x - 1)$ Incorrect $(3x - 1)(x + 5)$ Correct Multiply to confirm that the final factorization is correct.
6.5 Special Factoring Techniques	
Difference of Squares $x^2 - y^2 = (x + y)(x - y)$	Factor each difference of squares. $4x^2 - 9 = (2x + 3)(2x - 3)$ $100y^4 - 49 = (10y^2 + 7)(10y^2 - 7)$
Perfect Square Trinomials $x^2 + 2xy + y^2 = (x + y)^2$ $x^2 - 2xy + y^2 = (x - y)^2$	Factor each perfect square trinomial. $9x^2 + 6x + 1 = (3x + 1)^2$ $4x^2 - 20x + 25 = (2x - 5)^2$
6.6 Solving Quadratic Equations Using the Zero-Factor Property	
Quadratic Equation A **quadratic equation** (in x here) can be written in the form $ax^2 + bx + c = 0$ Standard form where a, b, and c are real numbers and $a \neq 0$.	The following are examples of quadratic equations. $x^2 - 144 = 0$, $2x^2 + 11x - 21 = 0$ Both are in standard form.
Zero-Factor Property If a and b are real numbers and $ab = 0$, then $a = 0$ or $b = 0$.	If $(x - 2)(x + 3) = 0$, then $x - 2 = 0$ or $x + 3 = 0$, leading to $x = 2$ or $x = -3$.
Solving a Quadratic Equation Using the Zero-Factor Property *Step 1* Write the equation in standard form. *Step 2* Factor. *Step 3* Apply the zero-factor property. *Step 4* Solve the resulting equations. *Step 5* Check. Write the solution set.	Solve $2x^2 = 7x + 15$. $2x^2 - 7x - 15 = 0$ Standard form $(2x + 3)(x - 5) = 0$ Factor. $2x + 3 = 0$ or $x - 5 = 0$ Zero-factor property $2x = -3$ or $x = 5$ Solve each equation. $x = -\frac{3}{2}$ A check confirms that the values $-\frac{3}{2}$ and 5 satisfy the original equation. Solution set: $\{-\frac{3}{2}, 5\}$

Concepts	Examples

6.7 Applications of Quadratic Equations

Step 1 Read the problem.

In a right triangle, one leg measures 2 ft longer than the other. The hypotenuse measures 4 ft longer than the shorter leg. Find the lengths of the three sides of the triangle.

Step 2 Assign a variable.

Let $x =$ the length of the shorter leg.

Then $x + 2 =$ the length of the longer leg,

and $x + 4 =$ the length of the hypotenuse.

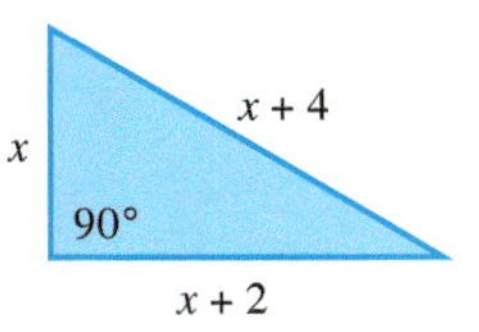

Step 3 Write an equation.

Step 4 Solve the equation.

$$a^2 + b^2 = c^2$$

$$x^2 + (x + 2)^2 = (x + 4)^2 \quad \text{Pythagorean theorem}$$

$$x^2 + x^2 + 4x + 4 = x^2 + 8x + 16 \quad \text{Square each binomial.}$$

$$x^2 - 4x - 12 = 0 \quad \text{Standard form}$$

$$(x - 6)(x + 2) = 0 \quad \text{Factor.}$$

$$x - 6 = 0 \quad \text{or} \quad x + 2 = 0 \quad \text{Zero-factor property}$$

$$x = 6 \quad \text{or} \quad x = -2 \quad \text{Solve each equation.}$$

Step 5 State the answer.

Because -2 cannot represent a length of a triangle, 6 is the only possible answer. The length of the shorter leg is 6 ft, the length of the longer leg is

$$6 + 2 = 8 \text{ ft},$$

and the length of the hypotenuse is

$$6 + 4 = 10 \text{ ft}.$$

Step 6 Check.

Because $6^2 + 8^2 = 10^2$ is true, the answer is correct.

Chapter 6 Review Exercises

6.1 *Factor out the greatest common factor or factor by grouping.*

1. $15t + 45$

2. $60z^3 - 30z$

3. $44x^3 + 55x^2$

4. $100m^2n^3 - 50m^3n^4 + 150m^2n^2$

5. $2xy - 8y + 3x - 12$

6. $6y^2 + 9y + 4xy + 6x$

6.2 *Factor completely.*

7. $x^2 + 10x + 21$

8. $y^2 - 13y + 40$

9. $q^2 + 6q - 27$

10. $r^2 - r - 56$

11. $x^2 + x + 1$

12. $3x^2 + 6x + 6$

13. $r^2 - 4rs - 96s^2$

14. $p^2 + 2pq - 120q^2$

15. $-8p^3 + 24p^2 + 80p$

16. $3x^4 + 30x^3 + 48x^2$

17. $m^2 - 3mn - 18n^2$

18. $y^2 - 8yz + 15z^2$

19. $p^7 - p^6q - 2p^5q^2$

20. $-3r^5 + 6r^4s + 45r^3s^2$

6.3–6.4

21. **CONCEPT CHECK** To begin factoring

$$6r^2 - 5r - 6,$$

what are the possible first terms of the two binomial factors, if we consider only positive integer coefficients?

22. **CONCEPT CHECK** What is the first step we would use to factor the following trinomial?

$$2z^3 + 9z^2 - 5z$$

Factor completely.

23. $2k^2 - 5k + 2$

24. $3r^2 + 11r - 4$

25. $6r^2 - 5r - 6$

26. $10z^2 - 3z - 1$

27. $5t^2 - 11t + 12$

28. $24x^5 - 20x^4 + 4x^3$

29. $-6x^2 + 3x + 30$

30. $10r^3s + 17r^2s^2 + 6rs^3$

31. $-30y^3 - 5y^2 + 10y$

32. $4z^2 - 5z + 7$

33. $-3m^3n + 19m^2n + 40mn$

34. $14a^2 - 27ab - 20b^2$

6.5

35. CONCEPT CHECK Which one of the following is a difference of squares?

A. $32x^2 - 1$ **B.** $4x^2y^2 - 25z^2$

C. $x^2 + 36$ **D.** $25y^3 - 1$

36. CONCEPT CHECK Which one of the following is a perfect square trinomial?

A. $x^2 + x + 1$ **B.** $y^2 - 4y + 9$

C. $4x^2 + 10x + 25$ **D.** $x^2 - 20x + 100$

Factor completely.

37. $n^2 - 64$

38. $25b^2 - 121$

39. $49y^2 - 25w^2$

40. $9m^2 + 81$

41. $144p^2 - 36q^2$

42. $v^4 - 1$

43. $x^2 + 100$

44. $z^2 + 10z + 25$

45. $r^2 - 12r + 36$

46. $9t^2 - 42t + 49$

47. $16m^2 + 40mn + 25n^2$

48. $54x^3 - 72x^2 + 24x$

6.6

Solve each equation, and check the solutions.

49. $(4t + 3)(t - 1) = 0$

50. $(x + 7)(x - 4)(x + 3) = 0$

51. $x(2x - 5) = 0$

52. $z^2 + 4z + 3 = 0$

53. $m^2 - 5m + 4 = 0$

54. $x^2 = -15 + 8x$

55. $3z^2 - 11z - 20 = 0$

56. $81t^2 - 64 = 0$

57. $y^2 = 8y$

58. $n(n - 5) = 6$

59. $t^2 + 14t + 49 = 0$

60. $t^2 = 12(t - 3)$

61. $(5z + 2)(z^2 + 3z + 2) = 0$

62. $x^2 = 9$

63. $(x + 2)(x - 3) = 4x$

64. $(2r + 1)(12r^2 + 5r - 3) = 0$

65. $25w^2 - 90w + 81 = 0$

66. $r(r - 7) = 30$

Solve each problem. (If necessary, refer to the formulas at the back of this text.)

67. The length of a rectangular rug is 6 ft more than the width. The area is 40 ft^2. Find the length and width of the rug.

68. A treasure chest from a sunken galleon has dimensions (in feet) as shown in the figure. Its surface area is 650 ft^2. Find its width.

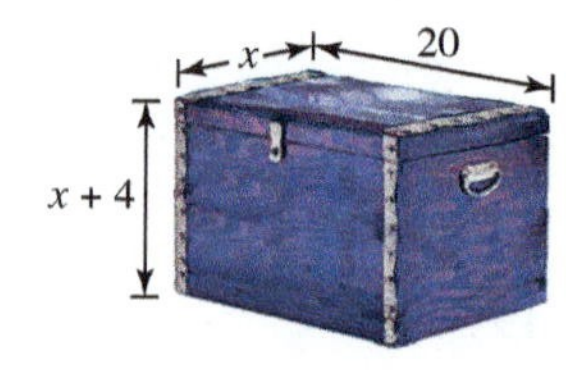

69. The product of two consecutive integers is 29 more than their sum. What are the integers?

70. The product of the lesser two of three consecutive integers is equal to 23 plus the greatest. Find the integers.

71. Two cars left an intersection at the same time. One traveled west, and the other traveled 14 mi less, but to the south. How far apart were they then, if the distance between them was 16 mi more than the distance traveled south?

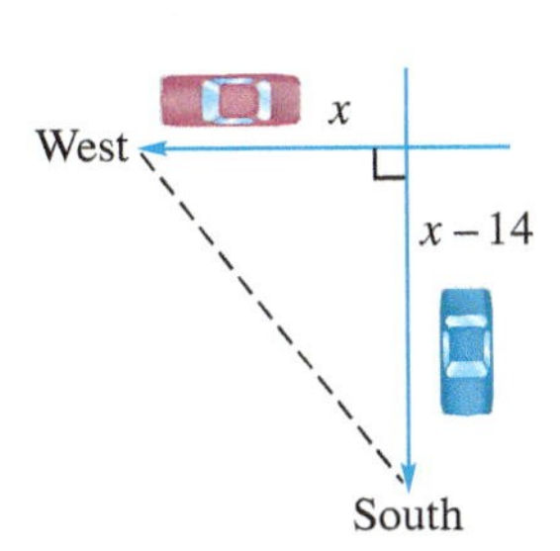

72. The triangular sail of a schooner has an area of 30 m^2. The height of the sail is 4 m more than the base. Find the length of the base of the sail.

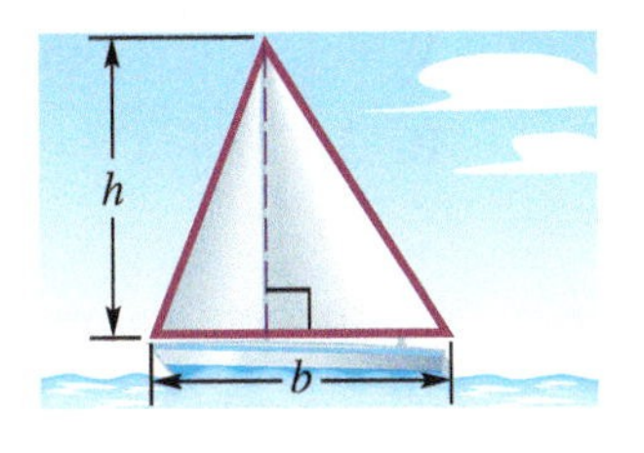

73. The floor plan for a house is a rectangle with length 7 m more than its width. The area is 170 m^2. Find the width and length of the house.

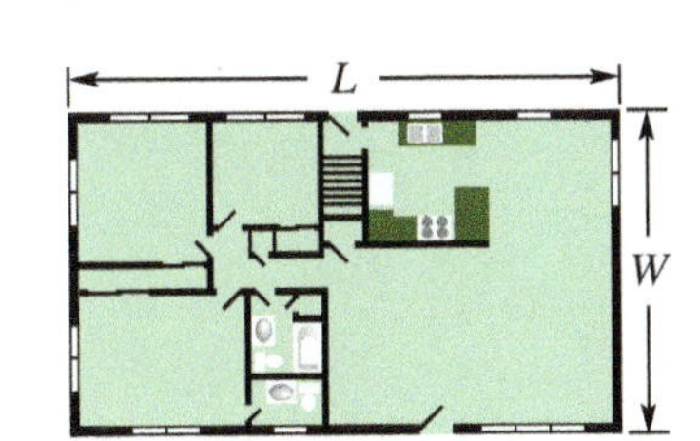

74. If an object is dropped, the distance d in feet it falls in t seconds (disregarding air resistance) is given by the quadratic equation

$$d = 16t^2.$$

Find the distance an object would fall in each of the following times.

(a) 4 sec **(b)** 8 sec

The table shows the number of alternative fuel vehicles, in thousands, in use in the United States.

Year	Number (in thousands)
2001	425
2003	534
2005	592
2007	696
2009	826
2011	1192

Data from U.S. Department of Energy.

We used the data to develop the quadratic equation

$$y = 7.02x^2 - 15.5x + 469,$$

which models the number of vehicles y (in thousands) in year x, where $x = 1$ represents 2001, $x = 2$ represents 2002, and so on.

75. Use the model to find the number of alternative fuel vehicles in 2007, to the nearest thousand. How does the result compare with the actual data in the table?

76. Assuming that the trend in the data continues, use the quadratic equation to estimate the number of alternative fuel vehicles in 2013 to the nearest thousand.

Chapter 6 Mixed Review Exercises

1. **CONCEPT CHECK** Which of the following is *not* factored completely?

   A. $3(7 + t)$ **B.** $3x(7t + 4)$ **C.** $(3 + x)(7t + 4)$ **D.** $3(7t + 4) + x(7t + 4)$

2. **CONCEPT CHECK** A student factored $6x^2 + 16x - 32$ as

   $(2x + 8)(3x - 4)$

   What Went Wrong? Factor the trinomial correctly.

Factor completely.

3. $15m^2 + 20mp - 12m - 16p$
4. $24ab^3 - 56a^2bc^3 + 72a^2b^2$
5. $k^2 + 400$
6. $z^2 - 11zx + 10x^2$
7. $3k^2 + 11k + 10$
8. $y^4 - 625$
9. $6m^3 - 21m^2 - 45m$
10. $25a^2 + 15ab + 9b^2$
11. $2a^5 - 8a^4 - 24a^3$
12. $-12r^2 - 8rq + 15q^2$
13. $100a^2 - 9$
14. $49t^2 + 56t + 16$

Solve each equation, and check the solutions.

15. $t(t - 7) = 0$
16. $x^2 + 3x = 10$
17. $25x^2 + 20x + 4 = 0$
18. $64x^3 - 9x = 0$
19. $6r^2 = 15 - r$
20. $(t + 1)(3t^2 + 19t + 6) = 0$

Solve each problem.

21. A lot is in the shape of a right triangle. The hypotenuse is 3 m longer than the longer leg. The longer leg is 6 m longer than twice the length of the shorter leg. Find the lengths of the sides of the lot.

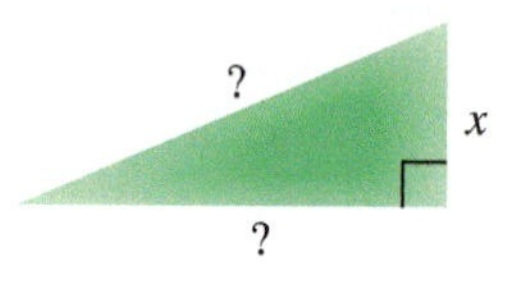

22. A pyramid has a rectangular base with a length that is 2 m more than the width. The height of the pyramid is 6 m, and its volume is 48 m³. Find the length and width of the base.

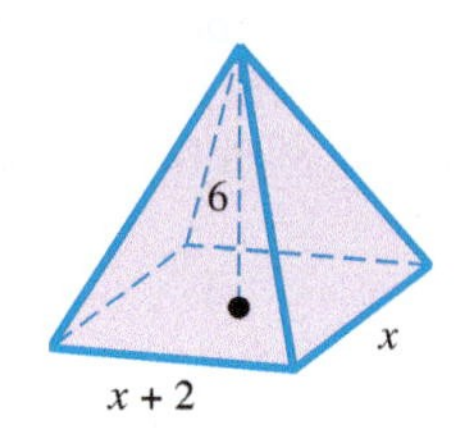

Chapter 6 Test

The Chapter Test Prep Videos with step-by-step solutions are available in MyMathLab or on YouTube at https://goo.gl/8aAsWP

1. Which one of the following is the correct, completely factored form of $2x^2 - 2x - 24$?

A. $(2x + 6)(x - 4)$ **B.** $(x + 3)(2x - 8)$

C. $2(x + 4)(x - 3)$ **D.** $2(x + 3)(x - 4)$

Factor completely.

2. $12x^2 - 30x$

3. $2m^3n^2 + 3m^3n - 5m^2n^2$

4. $2ax - 2bx + ay - by$

5. $x^2 - 9x + 14$

6. $t^2 + 7t + 10$

7. $6x^2 - 19x - 7$

8. $3x^2 - 12x - 15$

9. $10z^2 - 17z + 3$

10. $2x^2 + x - 3$

11. $x^2 + 36$

12. $y^2 - 49$

13. $81a^2 - 121b^2$

14. $x^2 + 16x + 64$

15. $4x^2 - 28xy + 49y^2$

16. $-2x^2 - 4x - 2$

17. $6t^4 + 3t^3 - 108t^2$

18. $4r^2 + 10rt + 25t^2$

19. $4t^3 + 32t^2 + 64t$

20. $x^4 - 81$

Solve each equation.

21. $(x+3)(x-9)=0$

22. $2r^2-13r+6=0$

23. $25x^2-4=0$

24. $x(x-20)=-100$

25. $t^2=3t$

26. $(x+8)(6x^2+13x-5)=0$

Solve each problem.

27. The length of a rectangular flower bed is 3 ft less than twice its width. The area of the bed is 54 ft^2. Find the dimensions of the flower bed.

28. Find two consecutive integers such that the square of the sum of the two integers is 11 more than the lesser integer.

29. A carpenter needs to cut a brace to support a wall stud, as shown in the figure. The brace should be 7 ft less than three times the length of the stud. If the brace will be anchored on the floor 15 ft away from the stud, how long should the brace be?

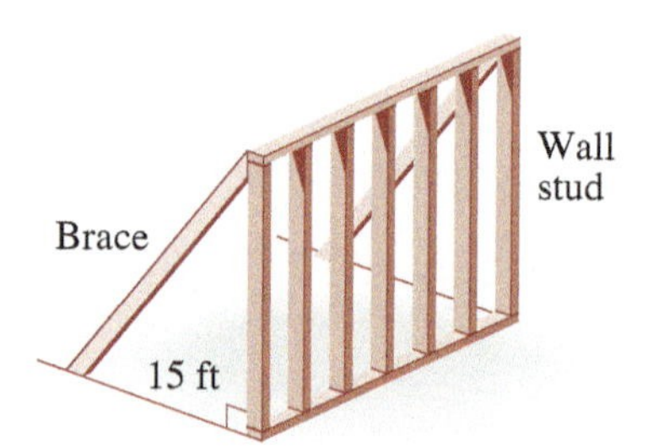

30. The public debt y (in billions of dollars) of the United States from 2000 through 2014 can be approximated by the quadratic equation

$$y=57.53x^2-72.93x+3417,$$

where $x=0$ represents 2000, $x=1$ represents 2001, and so on. Use the model to estimate the public debt, to the nearest billion dollars, in the year 2018. (Data from Bureau of Public Debt.)

Chapters R–6 Cumulative Review Exercises

Solve each equation.

1. $3x + 2(x - 4) = 4(x - 2)$

2. $0.3x + 0.9x = 0.06$

3. $\frac{2}{3}n - \frac{1}{2}(n - 4) = 3$

4. Solve for t: $A = P + Prt$

Solve each problem.

5. From a list of "technology-related items," 500 adults were surveyed as to those items they couldn't live without. Complete the results shown in the table.

Item	Percent That Couldn't Live Without	Number That Couldn't Live Without
Personal computer	46%	____
Cell phone	41%	____
High-speed Internet	____	190
MP3 player	____	60

(Other items included digital cable, HDTV, and electronic gaming console.)
Data from Ipsos for AP.

6. At the 2014 Winter Olympics in Sochi, Russia, the United States won a total of 28 medals. The United States won 2 more gold medals than silver and 5 fewer silver medals than bronze. Find the number of each type of medal won. (Data from *The Gazette*.)

7. In 2014, American women working full time earned, on average, 79 cents for every dollar earned by men working full time. The median annual salary for full-time male workers was $50,383. To the nearest dollar, what was the median annual salary for full-time female workers? (Data from U.S. Bureau of Labor Statistics.)

8. Find the measures of the marked angles.

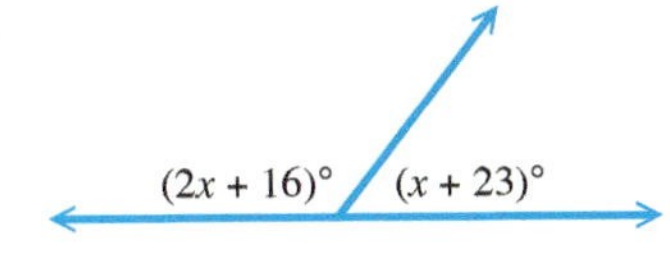

9. Fill in each blank with *positive* or *negative*. The point with coordinates (a, b) is in

(a) quadrant II if a is ________ and b is ________.

(b) quadrant III if a is ________ and b is ________.

Consider the equation $y = 4x + 3$. Find the following.

10. The x- and y-intercepts

11. The slope

12. The graph

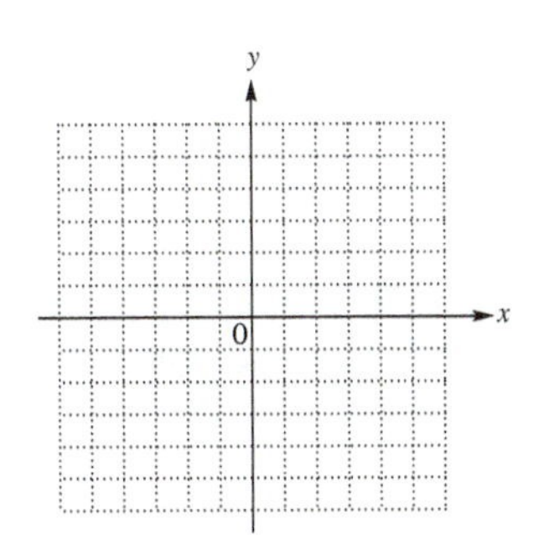

13. The points on the graph show total revenue, in billions of dollars, for Amazon in selected years (2011–2015), along with a graph of a linear equation that models the data.

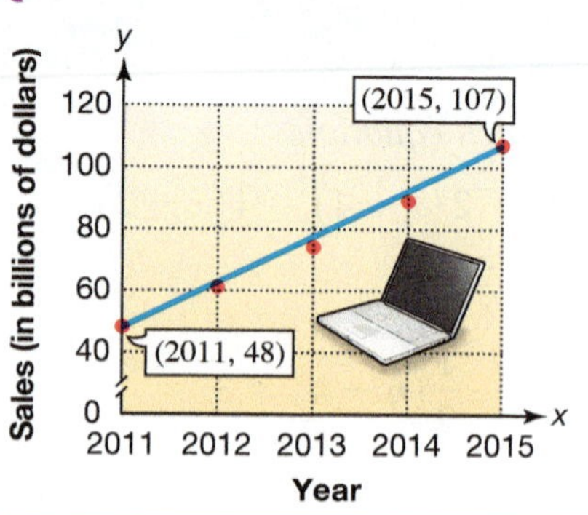

Data from Market Watch.

(a) Use the ordered pairs shown on the graph to find the slope of the line. Interpret this slope.

(b) Use the graph to estimate revenue in the year 2014. Write the answer as an ordered pair of the form (year, sales in billions of dollars).

Solve each system of equations.

14. $4x - y = -6$
$2x + 3y = 4$

15. $5x + 3y = 10$
$2x + \frac{6}{5}y = 5$

Evaluate each expression.

16. $2^{-3} \cdot 2^5$

17. $\left(\frac{3}{4}\right)^{-2}$

18. $\frac{6^5 \cdot 6^{-2}}{6^3}$

19. $\left(\frac{4^{-3} \cdot 4^4}{4^5}\right)^{-1}$

Simplify each expression and write the answer using only positive exponents. Assume no denominators are 0.

20. $\frac{(p^2)^3 p^{-4}}{(p^{-3})^{-1} p}$

21. $\frac{(m^{-2})^3 m}{m^5 m^{-4}}$

Perform the indicated operations.

22. $(2k^2 + 4k) - (5k^2 - 2) - (k^2 + 8k - 6)$

23. $(9x + 6)(5x - 3)$

24. $(3p + 2)^2$

25. $\frac{8x^4 + 12x^3 - 6x^2 + 20x}{2x}$

Factor completely.

26. $2a^2 + 7a - 4$

27. $10m^2 + 19m + 6$

28. $8t^2 + 10tv + 3v^2$

29. $4p^2 - 12p + 9$

30. $25r^2 - 81t^2$

31. $2pq + 6p^3q + 8p^2q$

Solve each equation.

32. $6m^2 + m - 2 = 0$

33. $8x^2 = 64x$

34. $49x^2 - 56x + 16 = 0$

35. The length of the hypotenuse of a right triangle is 3 m more than twice the length of the shorter leg. The longer leg is 7 m longer than the shorter leg. Find the lengths of the sides.

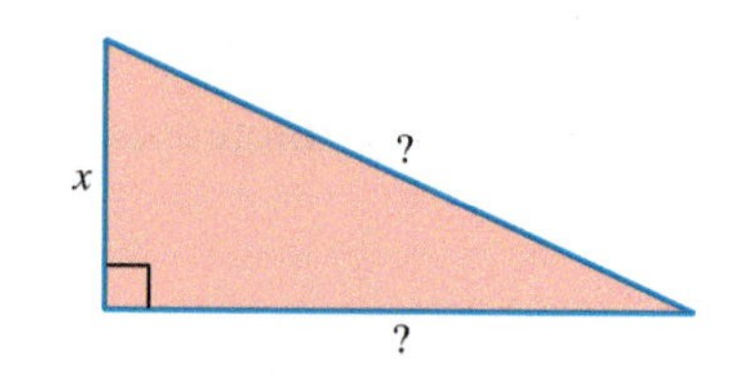

7

Rational Expressions and Applications

The formula $r = \frac{d}{t}$ gives the rate r of a speeding car in terms of its distance d and its time t traveled. This formula involves a *rational expression* (or *algebraic fraction*), the subject of this chapter.

7.1 The Fundamental Property of Rational Expressions

OBJECTIVES

1. Find the numerical value of a rational expression.
2. Find the values of the variable for which a rational expression is undefined.
3. Write rational expressions in lowest terms.
4. Recognize equivalent forms of rational expressions.

The quotient of two integers (with denominator not 0), such as $\frac{2}{3}$ or $-\frac{3}{4}$, is a *rational number.* In the same way, the quotient of two polynomials with denominator not equal to 0 is a *rational expression.*

Rational Expression

A **rational expression** is an expression of the form $\frac{P}{Q}$, where P and Q are polynomials and $Q \neq 0$.

Examples: $\frac{-6x}{x^3+8}$, $\frac{9x}{y+3}$, and $\frac{2m^3}{8}$ Rational expressions

Our work with rational expressions will require much of what we learned on polynomials and factoring, as well as the rules for arithmetic fractions.

OBJECTIVE 1 Find the numerical value of a rational expression. Remember that to *evaluate* an expression means to find its *value.* We use substitution to evaluate a rational expression for a given value of the variable.

EXAMPLE 1 Evaluating Rational Expressions

Find the numerical value of $\frac{3x+6}{2x-4}$ for each value of x.

(a) $x = 1$

$$\frac{3x+6}{2x-4} = \frac{3(1)+6}{2(1)-4} \quad \text{Let } x = 1.$$
$$= \frac{9}{-2}$$
$$= -\frac{9}{2} \quad \frac{a}{-b} = -\frac{a}{b}$$

(b) $x = 0$

$$\frac{3x+6}{2x-4} = \frac{3(0)+6}{2(0)-4} \quad \text{Let } x = 0.$$
$$= \frac{6}{-4}$$
$$= -\frac{3}{2} \quad \text{Lowest terms}$$

(c) $x = 2$

$$\frac{3x+6}{2x-4} = \frac{3(2)+6}{2(2)-4} \quad \text{Let } x = 2.$$
$$= \frac{12}{0}$$

The expression is undefined for $x = 2$.

(d) $x = -2$

$$\frac{3x+6}{2x-4} = \frac{3(-2)+6}{2(-2)-4} \quad \text{Let } x = -2.$$
$$= \frac{0}{-8}$$
$$= 0 \quad \frac{0}{b} = 0$$

◀ Work Problem 1 at the Side.

1 Find the numerical value of each rational expression for $x = -3$, $x = 0$, and $x = 3$.

GS **(a)** $\frac{x}{2x+1}$

For $x = -3$:

$$\frac{x}{2x+1} = \frac{\_\_\_}{2(\_\_\_)+1} = \frac{-3}{\_\_\_+1} = \_\_\_$$

(b) $\frac{2x+6}{x-3}$

Answers

1. (a) For $x = -3$: -3; -3; -6; $\frac{3}{5}$; for $x = 0$, final answer 0; for $x = 3$, final answer $\frac{3}{7}$
(b) 0; -2; undefined

Note

The numerator of a rational expression may be any real number. If the numerator equals 0 and the denominator does not equal 0, then the rational expression equals 0. **See Example 1(d).**

OBJECTIVE 2 Find the values of the variable for which a rational expression is undefined. In the definition of a rational expression $\frac{P}{Q}$, Q cannot equal 0. ***The denominator of a rational expression cannot equal 0 because division by 0 is undefined.*** For example, in the rational expression

$$\frac{8x^2}{x-3}, \quad \leftarrow \text{Denominator cannot equal 0.}$$

the variable x can take on any real number value except 3. When x is 3, the denominator becomes $3 - 3 = 0$, making the expression undefined. Thus, x cannot equal 3. We indicate this restriction by writing $x \neq 3$.

Determining When a Rational Expression Is Undefined

Step 1 Set the denominator of the rational expression equal to 0.

Step 2 Solve this equation.

Step 3 The solutions of the equation are the values that make the rational expression undefined. The variable *cannot* equal these values.

EXAMPLE 2 Finding Values That Make Rational Expressions Undefined

Find values of the variable for which each rational expression is undefined.

(a) $\frac{x+5}{3x+2}$ We must find any value of x that makes the *denominator* equal to 0 because division by 0 is undefined.

Step 1 $3x + 2 = 0$ Set the denominator equal to 0.

Step 2 Solve. $3x = -2$ Subtract 2.

$x = -\frac{2}{3}$ Divide by 3.

Step 3 The given expression is undefined for $-\frac{2}{3}$, so $x \neq -\frac{2}{3}$.

(b) $\frac{9m^2}{m^2 - 5m + 6}$

$m^2 - 5m + 6 = 0$ Set the denominator equal to 0.

$(m-2)(m-3) = 0$ Factor.

$m - 2 = 0$ or $m - 3 = 0$ Zero-factor property

$m = 2$ or $m = 3$ Solve for m.

The given expression is undefined for 2 and 3, so $m \neq 2$, $m \neq 3$.

(c) $\frac{2r}{r^2+1}$ This denominator will not equal 0 for any value of r, because r^2 is always greater than or equal to 0, and adding 1 makes the sum greater than or equal to 1. There are no values for which this expression is undefined.

Work Problem 2 at the Side. ▶

2 Find values of the variable for which each rational expression is undefined. Write answers with the symbol $\neq$.

GS **(a)** $\frac{x+2}{x-5}$

Step 1 ______ $= 0$

Step 2 $x =$ ____

Step 3 The given expression is undefined for ____. Thus, $x\ (= / \neq)\ 5$.

(b) $\frac{3r}{r^2 + 6r + 8}$

(c) $\frac{-5m}{m^2 + 4}$

Answers

2. **(a)** $x - 5$; 5; 5; $\neq$
(b) $r \neq -4$, $r \neq -2$
(c) The expression is never undefined.

3 Write each expression in lowest terms.

(a) $\frac{15}{40}$

(b) $\frac{5x^4}{15x^2}$

(c) $\frac{6p^3}{2p^2}$

Answers

3. (a) $\frac{3}{8}$ (b) $\frac{x^2}{3}$ (c) $3p$

OBJECTIVE 3 Write rational expressions in lowest terms. A common fraction such as $\frac{2}{3}$ is said to be in *lowest terms.*

Lowest Terms

A rational expression $\frac{P}{Q}$ (where $Q \neq 0$) is in **lowest terms** if the greatest common factor of its numerator and denominator is 1.

We use the **fundamental property of rational expressions** to write a rational expression in lowest terms.

Fundamental Property of Rational Expressions

If $\frac{P}{Q}$ (where $Q \neq 0$) is a rational expression and if K represents any polynomial (where $K \neq 0$), then the following holds true.

$$\frac{PK}{QK} = \frac{P}{Q}$$

This property is based on the identity property of multiplication.

$$\frac{PK}{QK} = \frac{P}{Q} \cdot \frac{K}{K} = \frac{P}{Q} \cdot 1 = \frac{P}{Q}$$

EXAMPLE 3 Writing in Lowest Terms

Write each expression in lowest terms.

(a) $\frac{30}{72}$ Rational number

Begin by factoring.

$$= \frac{2 \cdot 3 \cdot 5}{2 \cdot 2 \cdot 2 \cdot 3 \cdot 3}$$

Group any factors common to the numerator and denominator.

$$= \frac{5 \cdot (2 \cdot 3)}{2 \cdot 2 \cdot 3 \cdot (2 \cdot 3)}$$

Use the fundamental property.

$$= \frac{5}{2 \cdot 2 \cdot 3}$$

$$= \frac{5}{12}$$

(b) $\frac{14k^2}{2k^3}$ Rational expression

Write k^2 as $k \cdot k$ and k^3 as $k \cdot k \cdot k$.

$$= \frac{2 \cdot 7 \cdot k \cdot k}{2 \cdot k \cdot k \cdot k}$$

$$= \frac{7(2 \cdot k \cdot k)}{k(2 \cdot k \cdot k)}$$

$$= \frac{7}{k}$$

◀ Work Problem 3 at the Side.

Writing a Rational Expression in Lowest Terms

Step 1 **Factor** the numerator and denominator completely.

Step 2 **Use the fundamental property** to divide out any common factors.

EXAMPLE 4 Writing in Lowest Terms

Write each rational expression in lowest terms.

(a) $\dfrac{3x-12}{5x-20}$

Step 1 $= \dfrac{3(x-4)}{5(x-4)}$ Factor. ($x \neq 4$ because the denominator is 0 for this value.)

Step 2 $= \dfrac{3}{5}$ $\quad \frac{x-4}{x-4} = 1$; Fundamental property

The given expression is equal to $\frac{3}{5}$ for all values of x, where $x \neq 4$ (because the denominator of the original rational expression is 0 when x is 4).

(b) $\dfrac{2y^2-8}{2y+4}$

Step 1 $= \dfrac{2(y^2-4)}{2(y+2)}$ Factor. ($y \neq -2$ because the denominator is 0 for this value.)

$= \dfrac{2(y+2)(y-2)}{2(y+2)}$ Factor the numerator completely. y^2-4 is a difference of squares.

Step 2 $= y-2$ $\quad \frac{2(y+2)}{2(y+2)} = 1$; Fundamental property

(c) $\dfrac{m^2+2m-8}{2m^2-m-6}$

Step 1 $= \dfrac{(m+4)(m-2)}{(2m+3)(m-2)}$ Factor. ($m \neq -\frac{3}{2}, m \neq 2$)

Step 2 $= \dfrac{m+4}{2m+3}$ $\quad \frac{m-2}{m-2} = 1$; Fundamental property

From now on, we write statements of equality of rational expressions with the understanding that they apply only to those real numbers that make neither denominator equal to 0.

(d) $\dfrac{4m+12-mp-3p}{5m+mp+15+3p}$ The numerator and denominator each have four terms, so try to factor by grouping.

Step 1 $= \dfrac{(4m+12)+(-mp-3p)}{(5m+mp)+(15+3p)}$ Group terms in the numerator. Group terms in the denominator.

$= \dfrac{4(m+3)-p(m+3)}{m(5+p)+3(5+p)}$ Factor each group.

$= \dfrac{(m+3)(4-p)}{(5+p)(m+3)}$ Factor out $m+3$. Factor out $5+p$.

Step 2 $= \dfrac{4-p}{5+p}$ $\quad \frac{m+3}{m+3} = 1$; Fundamental property

Work Problem 4 at the Side. ▶

4 Write each rational expression in lowest terms.

GS **(a)** $\dfrac{4y+2}{6y+3}$

$= \dfrac{2(\_\_\_\_)}{3(\_\_\_\_)}$

$= \_\_\_\_$

(b) $\dfrac{8p+8q}{5p+5q}$

(c) $\dfrac{x^2+4x+4}{4x+8}$

(d) $\dfrac{a^2-b^2}{a^2+2ab+b^2}$

(e) $\dfrac{xy+5y-2x-10}{xy+3x+5y+15}$

Answers

4. (a) $2y+1$; $2y+1$; $\frac{2}{3}$ (b) $\frac{8}{5}$ (c) $\frac{x+2}{4}$ (d) $\frac{a-b}{a+b}$ (e) $\frac{y-2}{y+3}$

5 Write $\frac{m-n}{n-m}$ in lowest terms.

CAUTION

Rational expressions cannot be written in lowest terms until after the numerator and denominator have been factored. Only common factors can be divided out, not common terms.

$$\frac{6x+9}{4x+6} = \frac{3(2x+3)}{2(2x+3)} = \frac{3}{2}$$

↑ Divide out the common factor.

$$\frac{6+x}{4x}$$ ← Numerator cannot be factored.

Already in lowest terms

EXAMPLE 5 Writing in Lowest Terms (Factors Are Opposites)

Write $\frac{x-y}{y-x}$ in lowest terms.

To find a common factor, the denominator $y - x$ can be factored as follows.

$y - x$ (We are factoring out -1, **NOT** multiplying by it.)

$= -1(-y + x)$ Factor out -1.

$= -1(x - y)$ Commutative property, $a + b = b + a$

With this result in mind, we simplify as follows.

$\frac{x-y}{y-x}$

$= \frac{1(x-y)}{-1(x-y)}$ $y - x = -1(x - y)$ from above.

$= \frac{1}{-1}$ Fundamental property

$= -1$ Lowest terms

Alternatively, we could factor -1 from the numerator $x - y$.

$\frac{x-y}{y-x}$ Alternative solution

$= \frac{-1(-x+y)}{y-x}$ Factor out -1 in the numerator.

$= \frac{-1(y-x)}{y-x}$ Commutative property

$= -1$ The result is the same.

◀ **Work Problem 5 at the Side.**

CAUTION

Although x and y appear in both the numerator and denominator in **Example 5,** we cannot use the fundamental property right away because they are *terms,* not *factors*. ***Terms are added, while factors are multiplied.***

Answer

5. -1

In **Example 5,** notice that $y - x$ is the **opposite** (or **additive inverse**) of $x - y$. A general rule for this situation follows.

Quotient of Opposites

If the numerator and the denominator of a rational expression are opposites, such as in $\frac{x-y}{y-x}$, then the rational expression is equal to -1.

Based on this result, the following are true.

Numerator and denominator are opposites. $\rightarrow$ $\frac{q-7}{7-q} = -1$ and $\frac{-5a+2b}{5a-2b} = -1$

However, the following expression cannot be simplified further.

$\frac{x-2}{x+2}$ $\leftarrow$ Numerator and denominator are *not* opposites.

EXAMPLE 6 Writing in Lowest Terms (Factors Are Opposites)

Write each rational expression in lowest terms.

(a) $\frac{2-m}{m-2}$

Because $2 - m$ and $m - 2$ (or $-2 + m$) are opposites, this expression equals -1.

(b) $\frac{4x^2-9}{6-4x}$

$= \frac{(2x+3)(2x-3)}{2(3-2x)}$ Factor the numerator and denominator.

$= \frac{(2x+3)(2x-3)}{2(-1)(2x-3)}$ Write $3 - 2x$ in the denominator as $-1(2x - 3)$.

$= \frac{2x+3}{2(-1)}$ Fundamental property

$= \frac{2x+3}{-2}$ Multiply in the denominator.

$= -\frac{2x+3}{2}$ $\quad \frac{a}{-b} = -\frac{a}{b}$

(c) $\frac{3+r}{3-r}$ $\quad 3 - r$ is *not* the opposite of $3 + r$.

The rational expression is already in lowest terms.

Work Problem **6** at the Side. ▶

OBJECTIVE 4 Recognize equivalent forms of rational expressions. It is important to recognize equivalent forms of expressions. For example,

$$0.5, \quad \frac{1}{2}, \quad 50\%, \quad \text{and} \quad \frac{50}{100} \qquad \text{Equivalent expressions}$$

all represent the *same* real number. On a number line, the exact same point would apply to all four of them.

6 Write each rational expression in lowest terms.

(a) $\frac{5-y}{y-5}$

GS **(b)** $\frac{25x^2-16}{12-15x}$

$= \frac{(5x+4)(\_\_\_\_)}{3(\_\_\_\_)}$

$= \frac{(5x+4)(5x-4)}{3(\_\_\_\_)(5x-4)}$

$= \_\_\_\_$

(c) $\frac{9-k}{9+k}$

Answers

6. **(a)** -1

(b) $5x - 4$; $4 - 5x$; -1; $\frac{5x+4}{-3}$, or $-\frac{5x+4}{3}$

(c) It is already in lowest terms.

7 Which rational expressions are equivalent to

$$-\frac{2x-6}{x+3}?$$

A. $\frac{-(2x-6)}{x+3}$

B. $\frac{-2x+6}{x+3}$

C. $\frac{-2x-6}{x+3}$

D. $\frac{2x-6}{-(x+3)}$

E. $\frac{2x-6}{-x-3}$

F. $\frac{2x-6}{x-3}$

A similar situation exists with negative common fractions. The common fraction $-\frac{5}{6}$ can also be written $\frac{-5}{6}$ and $\frac{5}{-6}$, with the negative sign appearing in any of three different positions. All represent the *same* rational number.

Consider the following rational expression.

$$-\frac{2x+3}{2}$$ Final result from **Example 6(b)** in the form $-\frac{a}{b}$

The $-$ sign representing the factor -1 is in front of the expression, aligned with the fraction bar. Although we usually give answers in this form, it is important to be able to recognize other equivalent forms of a rational expression. The factor -1 may instead be placed in the numerator or in the denominator.

Use parentheses.

$$\frac{-(2x+3)}{2} \quad \text{and} \quad \frac{2x+3}{-2}$$

In the first of these two expressions, the distributive property can be applied. Thus,

$$\frac{-(2x+3)}{2} \text{ can also be written } \frac{-2x-3}{2}.$$

Multiply *each* term in the binomial by -1.

CAUTION

$\frac{-2x+3}{2}$ is *not* an equivalent form of $\frac{-(2x+3)}{2}$. ***Be careful to apply the distributive property correctly.***

EXAMPLE 7 Writing Equivalent Forms of a Rational Expression

Write four equivalent forms of the following rational expression.

$$-\frac{3x+2}{x-6}$$

If we apply the negative sign to the numerator, we obtain these equivalent forms.

$$①\rightarrow \frac{-(3x+2)}{x-6}, \quad \text{and, by the distributive property,} \quad \frac{-3x-2}{x-6} \leftarrow ②$$

If we apply the negative sign to the denominator, we obtain two additional forms.

$$③\rightarrow \frac{3x+2}{-(x-6)} \quad \text{and, by distributing once again,} \quad \frac{3x+2}{-x+6} \leftarrow ④$$

◀ Work Problem **7** at the Side.

CAUTION

Recall that $-\frac{5}{6} \neq \frac{-5}{-6}$. Thus, in **Example 7**, it would be incorrect to distribute the negative sign in $-\frac{3x+2}{x-6}$ to *both* the numerator *and* the denominator. (Doing this would actually lead to the *opposite* of the original expression.)

Answer

7. A, B, D, E

7.1 Exercises

FOR EXTRA HELP Go to MyMathLab for worked-out, step-by-step solutions to exercises enclosed in a square and video solutions to ▶ exercises.

CONCEPT CHECK *Fill in each blank with the correct response.*

1. The rational expression $\frac{x-5}{x+3}$ is undefined when x is _______, so $x \neq$ _______. This rational expression is equal to 0 when $x =$ _______.

2. The rational expression $\frac{p-q}{q-p}$ is undefined when the values of p and _______ are the same, so $p \neq$ _______. In all other cases, $\frac{p-q}{q-p}$ is equal to _______.

3. **CONCEPT CHECK** Which rational expressions are equivalent to $-\frac{a}{b}$?

A. $\frac{a}{b}$ **B.** $\frac{a}{-b}$ **C.** $\frac{-a}{b}$ **D.** $\frac{-a}{-b}$ **E.** $-\frac{a}{-b}$ **F.** $-\frac{-a}{b}$

4. **CONCEPT CHECK** Make the correct choice to complete each statement.

(a) $\frac{4-r^2}{4+r^2}$ (*is* / *is not*) equal to -1.

(b) $\frac{5+2x}{3-x}$ and $\frac{-5-2x}{x-3}$ (*are* / *are not*) equivalent rational expressions.

5. Define *rational expression,* and give an example.

6. Why can't the denominator of a rational expression equal 0?

Find the numerical value of each rational expression for **(a)** $x = 2$ *and* **(b)** $x = -3$. ***See Example 1.***

7. ▶ $\frac{3x+1}{5x}$

8. $\frac{5x-2}{4x}$

9. $\frac{x^2-4}{2x+1}$

10. $\frac{2x^2-4x}{3x-1}$

11. $\frac{(-3x)^2}{4x+12}$

12. $\frac{(-2x)^3}{3x+9}$

13. $\frac{5x+2}{2x^2+11x+12}$

14. $\frac{7-3x}{3x^2-7x+2}$

Find any value(s) of the variable for which each rational expression is undefined. Write answers with the symbol $\neq$. ***See Example 2.***

15. $\frac{2}{5y}$

16. $\frac{7}{3z}$

17. $\frac{x+1}{x+6}$

18. $\frac{m+2}{m+5}$

19. ▶ $\frac{4x^2}{3x-5}$

20. $\frac{2x^3}{3x-4}$

21. $\frac{m+2}{m^2+m-6}$

22. $\frac{r-5}{r^2-5r+4}$

23. $\frac{x^2-3x}{4}$

24. $\frac{x^2-4x}{6}$

25. $\frac{3x}{x^2+2}$

26. $\frac{4q}{q^2+9}$

Write each rational number in lowest terms. **See Example 3(a).**

27. $\frac{36}{84}$
28. $\frac{16}{60}$
29. $\frac{54}{198}$
30. $\frac{48}{108}$

Write each rational expression in lowest terms. **See Examples 3(b) and 4.**

31. $\frac{18r^3}{6r}$
32. $\frac{27p^2}{3p}$
33. $\frac{4(y-2)}{10(y-2)}$
34. $\frac{15(m-1)}{9(m-1)}$
35. $\frac{(x+1)(x-1)}{(x+1)^2}$
36. $\frac{(t+5)(t-3)}{(t-1)(t+5)}$
37. $\frac{7m+14}{5m+10}$
38. $\frac{5r+20}{3r+12}$
39. $\frac{16x-8}{14x-7}$
40. $\frac{21x-7}{9x-3}$
41. $\frac{m^2-n^2}{m+n}$
42. $\frac{a^2-b^2}{a-b}$
43. $\frac{12m^2-3}{8m-4}$
44. $\frac{20p^2-45}{6p-9}$
45. $\frac{3m^2-3m}{5m-5}$
46. $\frac{6t^2-6t}{2t-2}$
47. $\frac{9r^2-4s^2}{9r+6s}$
48. $\frac{16x^2-9y^2}{12x-9y}$
49. $\frac{2x^2-3x-5}{2x^2-7x+5}$
50. $\frac{3x^2+8x+4}{3x^2-4x-4}$
51. $\frac{zw+4z-3w-12}{zw+4z+5w+20}$
52. $\frac{km+4k+4m+16}{km+4k+5m+20}$
53. $\frac{ac-ad+bc-bd}{ac-ad-bc+bd}$
54. $\frac{rt-ru-st+su}{rt-ru+st-su}$

Write each rational expression in lowest terms. **See Examples 5 and 6.**

55. $\frac{6-t}{t-6}$
56. $\frac{2-k}{k-2}$
57. $\frac{m^2-1}{1-m}$
58. $\frac{a^2-b^2}{b-a}$
59. $\frac{q^2-4q}{4q-q^2}$
60. $\frac{z^2-5z}{5z-z^2}$
61. $\frac{p+6}{p-6}$
62. $\frac{5-x}{5+x}$

Write four equivalent forms of each rational expression. **See Example 7.**

63. $-\frac{x+4}{x-3}$
64. $-\frac{x+6}{x-1}$
65. $-\frac{2x-3}{x+3}$
66. $-\frac{5x-6}{x+4}$
67. $-\frac{3x-1}{5x-6}$
68. $-\frac{2x-9}{7x-1}$

Solve each problem. (Assume all measures are given in appropriate units.)

69. The area of the rectangle is represented by

$$x^4 + 10x^2 + 21.$$

Find the polynomial that represents the width of the rectangle. (*Hint:* Use $W = \frac{\mathcal{A}}{L}$.)

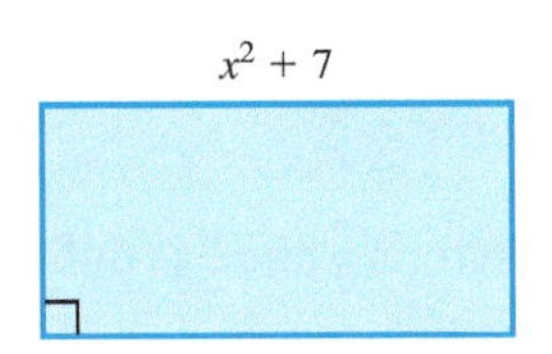

70. The volume of the box is represented by

$$(x^2 + 8x + 15)(x + 4).$$

Find the polynomial that represents the area of the bottom of the box.

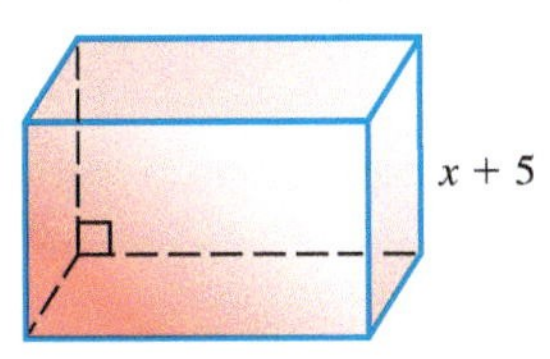

The percent of deaths caused by smoking is modeled by the rational expression

$$\frac{x - 1}{x},$$

where x is the number that tells how many times more likely a smoker is than a nonsmoker to die of lung cancer. This is called the ***incidence rate.*** *(Data from Walker, A.,* Observation and Inference: An Introduction to the Methods of Epidemiology, *Epidemiology Resources Inc.) For example, x* = 10 *means that a smoker is* 10 *times more likely than a nonsmoker to die of lung cancer.*

71. Find the percent of deaths if the incidence rate is the given number.

(a) 5 **(b)** 10 **(c)** 20

72. Can the incidence rate equal 0? Explain.

Relating Concepts (Exercises 73–75) For Individual or Group Work

We have used long division to find a quotient of two polynomials. We obtain the same quotient by expressing a division problem as a rational expression (algebraic fraction) and writing this rational expression in lowest terms, as shown below.

$$\begin{array}{r} x + 4 \\ 2x - 3\overline{)2x^2 + 5x - 12} \\ -(2x^2 - 3x) \\ \hline 8x - 12 \\ -(8x - 12) \\ \hline 0 \end{array}$$

$$\frac{2x^2 + 5x - 12}{2x - 3}$$

$$= \frac{(2x - 3)(x + 4)}{2x - 3} \quad \text{Factor.}$$

$$= x + 4 \quad \text{Fundamental property}$$

Perform the long division. Then simplify the rational expression to show that the result is the same.

73. $4x + 7\overline{)8x^2 + 26x + 21}$ and $\dfrac{8x^2 + 26x + 21}{4x + 7}$

74. $6x + 5\overline{)12x^2 + 16x + 5}$ and $\dfrac{12x^2 + 16x + 5}{6x + 5}$

75. $x + 1\overline{)x^3 + x^2 + x + 1}$ and $\dfrac{x^3 + x^2 + x + 1}{x + 1}$

7.2 Multiplying and Dividing Rational Expressions

OBJECTIVES

1. Multiply rational expressions.
2. Find reciprocals.
3. Divide rational expressions.

OBJECTIVE 1 Multiply rational expressions. The product of two common fractions is found by multiplying the numerators and multiplying the denominators. Rational expressions are multiplied in the same way.

Multiplying Rational Expressions

The product of the rational expressions $\frac{P}{Q}$ and $\frac{R}{S}$ is defined as follows.

$$\frac{P}{Q} \cdot \frac{R}{S} = \frac{PR}{QS}$$

In words: To multiply rational expressions, multiply the numerators and multiply the denominators.

1 Multiply. Write each answer in lowest terms.

(a) $\frac{2}{7} \cdot \frac{5}{10}$

(b) $\frac{3m^2}{2} \cdot \frac{10}{m}$

EXAMPLE 1 Multiplying Rational Expressions

Multiply. Write each answer in lowest terms.

(a) $\frac{3}{10} \cdot \frac{5}{9}$ Rational numbers

(b) $\frac{6}{x} \cdot \frac{x^2}{12}$ Rational expressions

Indicate the product of the numerators and the product of the denominators.

(a) $= \frac{3 \cdot 5}{10 \cdot 9}$

(b) $= \frac{6 \cdot x^2}{x \cdot 12}$

Leave the products in factored form. Factor the numerator and denominator to further identify any common factors. Then use the fundamental property to divide out any common factors and write each product in lowest terms.

(a) $= \frac{3 \cdot 5}{2 \cdot 5 \cdot 3 \cdot 3}$

$= \frac{1}{6}$ Remember to write 1 in the numerator.

(b) $= \frac{6 \cdot x \cdot x}{2 \cdot 6 \cdot x}$

$= \frac{x}{2}$

◀ Work Problem 1 at the Side.

2 Multiply. Write each answer in lowest terms.

(a) $\frac{a+b}{5} \cdot \frac{30}{2(a+b)}$

(b) $\frac{3(p-q)}{q^2} \cdot \frac{q}{2(p-q)^2}$

EXAMPLE 2 Multiplying Rational Expressions

Multiply. Write the answer in lowest terms.

$$\frac{x+y}{2x} \cdot \frac{x^2}{(x+y)^2}$$

$= \frac{(x+y)x^2}{2x(x+y)^2}$ Multiply numerators. Multiply denominators. (Use parentheses here around $x + y$.)

$= \frac{(x+y)x \cdot x}{2x(x+y)(x+y)}$ Factor. Identify the common factors.

$= \frac{x}{2(x+y)}$ $\frac{(x+y)x}{x(x+y)} = 1$; Write in lowest terms.

◀ Work Problem 2 at the Side.

Answers

1. (a) $\frac{1}{7}$ (b) $15m$

2. (a) 3 (b) $\frac{3}{2q(p-q)}$

EXAMPLE 3 **Multiplying Rational Expressions**

Multiply. Write the answer in lowest terms.

$$\frac{x^2 + 3x}{x^2 - 3x - 4} \cdot \frac{x^2 - 5x + 4}{x^2 + 2x - 3}$$

$$= \frac{(x^2 + 3x)(x^2 - 5x + 4)}{(x^2 - 3x - 4)(x^2 + 2x - 3)}$$ Definition of multiplication

$$= \frac{x(x + 3)(x - 4)(x - 1)}{(x - 4)(x + 1)(x + 3)(x - 1)}$$ Factor.

$$= \frac{x}{x + 1}$$ Divide out the common factors. The result is in lowest terms.

The quotients $\frac{x+3}{x+3}$, $\frac{x-4}{x-4}$, and $\frac{x-1}{x-1}$ are all equal to 1, justifying the final product $\frac{x}{x+1}$.

Work Problem **3** at the Side. ▶

OBJECTIVE 2 **Find reciprocals.** If the product of two rational expressions is 1, the rational expressions are **reciprocals** (or **multiplicative inverses**) of each other. The reciprocal of a rational expression is found by interchanging the numerator and the denominator.

$$\frac{2x - 1}{x - 5} \text{ has reciprocal } \frac{x - 5}{2x - 1}.$$

EXAMPLE 4 **Finding Reciprocals of Rational Expressions**

Find the reciprocal of each rational expression.

(a) $\frac{4p^3}{9q}$ has reciprocal $\frac{9q}{4p^3}$. Interchange the numerator and denominator.

(b) $\frac{k^2 - 9}{k^2 - k - 20}$ has reciprocal $\frac{k^2 - k - 20}{k^2 - 9}$. Reciprocals have product 1.

Work Problem **4** at the Side. ▶

OBJECTIVE 3 **Divide rational expressions.** Suppose we have $\frac{7}{8}$ gal of milk and want to find how many quarts we have. Because 1 qt is $\frac{1}{4}$ gal, we ask, "*How many* $\frac{1}{4}$*s are there in* $\frac{7}{8}$?" This would be interpreted as follows.

$$\frac{7}{8} \div \frac{1}{4}, \text{ which can be written } \frac{\frac{7}{8}}{\frac{1}{4}}$$ ← The fraction bar means division.

The fundamental property of rational expressions can be applied to rational number values of P, Q, and K.

$$\frac{P}{Q} = \frac{P \cdot K}{Q \cdot K} = \frac{\frac{7}{8} \cdot \frac{4}{1}}{\frac{1}{4} \cdot \frac{4}{1}} = \frac{\frac{7}{8} \cdot \frac{4}{1}}{1} = \frac{7}{8} \cdot \frac{4}{1}$$ Let $P = \frac{7}{8}$, $Q = \frac{1}{4}$, and $K = \frac{4}{1}$. (K is the reciprocal of Q.)

So, to divide $\frac{7}{8}$ by $\frac{1}{4}$, we multiply $\frac{7}{8}$ by the reciprocal of $\frac{1}{4}$, namely $\frac{4}{1}$ (or 4). Because $\frac{7}{8} \cdot \frac{4}{1} = \frac{7}{2}$, there are $\frac{7}{2}$ qt, or $3\frac{1}{2}$ qt, in $\frac{7}{8}$ gal.

3 Multiply. Write each answer in lowest terms.

(a) $\frac{x^2 + 7x + 10}{3x + 6} \cdot \frac{6x - 6}{x^2 + 2x - 15}$

(b) $\frac{m^2 + 4m - 5}{m + 5} \cdot \frac{m^2 + 8m + 15}{m - 1}$

4 Find the reciprocal of each rational expression.

(a) $\frac{5}{8}$

(b) $\frac{6b^5}{3r^2b}$

(c) $\frac{t^2 - 4t}{t^2 + 2t - 3}$

Answers

3. **(a)** $\frac{2(x - 1)}{x - 3}$ **(b)** $(m + 5)(m + 3)$

4. **(a)** $\frac{8}{5}$ **(b)** $\frac{3r^2b}{6b^5}$ **(c)** $\frac{t^2 + 2t - 3}{t^2 - 4t}$

5 Divide. Write each answer in lowest terms.

(a) $\frac{3}{4} \div \frac{5}{16}$

(b) $\frac{r}{r-1} \div \frac{3r}{r+4}$

GS (c) $\frac{6x-4}{3} \div \frac{15x-10}{9}$

$= \frac{6x-4}{3} \cdot \frac{\_\_\_\_}{\_\_\_\_}$

$= \frac{2(\_\_\_\_)}{3} \cdot \frac{9}{5(\_\_\_\_)}$

$= \frac{2(3x-2) \cdot 3 \cdot 3}{3 \cdot 5(3x-2)}$

$= \_\_\_\_$

6 Divide. Write each answer in lowest terms.

(a) $\frac{5a^2b}{2} \div \frac{10ab^2}{8}$

(b) $\frac{(3t)^2}{w} \div \frac{3t^2}{5w^4}$

Answers

5. (a) $\frac{12}{5}$ (b) $\frac{r+4}{3(r-1)}$

(c) 9; $15x - 10$; $3x - 2$; $3x - 2$; $\frac{6}{5}$

6. (a) $\frac{2a}{b}$ (b) $15w^3$

The preceding discussion illustrates dividing common fractions. Division of rational expressions is defined in the same way.

Dividing Rational Expressions

If $\frac{P}{Q}$ and $\frac{R}{S}$ are any two rational expressions where $\frac{R}{S} \neq 0$, then their quotient is defined as follows.

$$\frac{P}{Q} \div \frac{R}{S} = \frac{P}{Q} \cdot \frac{S}{R} = \frac{PS}{QR}$$

***In words:** To divide one rational expression by another rational expression, multiply the first rational expression (dividend) by the reciprocal of the second rational expression (divisor).*

EXAMPLE 5 Dividing Rational Expressions

Divide. Write each answer in lowest terms.

(a) $\frac{5}{8} \div \frac{7}{16}$ Rational numbers

(b) $\frac{y}{y+3} \div \frac{4y}{y+5}$ Rational expressions

Multiply the dividend by the reciprocal of the divisor.

(a)

$= \frac{5}{8} \cdot \frac{16}{7}$ ← Reciprocal of $\frac{7}{16}$

$= \frac{5 \cdot 16}{8 \cdot 7}$ Multiply.

$= \frac{5 \cdot 8 \cdot 2}{8 \cdot 7}$ Factor 16.

$= \frac{10}{7}$ Lowest terms

(b)

$= \frac{y}{y+3} \cdot \frac{y+5}{4y}$ ← Reciprocal of $\frac{4y}{y+5}$

$= \frac{y(y+5)}{(y+3)(4y)}$ Multiply.

$= \frac{y+5}{4(y+3)}$ Lowest terms

◄ Work Problem 5 at the Side.

EXAMPLE 6 Dividing Rational Expressions

Divide. Write the answer in lowest terms.

$$\frac{(3m)^2}{(2p)^3} \div \frac{6m^3}{16p^2}$$

$= \frac{(3m)^2}{(2p)^3} \cdot \frac{16p^2}{6m^3}$ Multiply by the reciprocal of the divisor.

$= \frac{9m^2}{8p^3} \cdot \frac{16p^2}{6m^3}$ Power rule for exponents, $(ab)^2 = a^2b^2$

$(3m)^2 = 3^2m^2$; $(2p)^3 = 2^3p^3$

$= \frac{9 \cdot 16m^2p^2}{8 \cdot 6p^3m^3}$ Multiply numerators. Multiply denominators.

$= \frac{3 \cdot 3 \cdot 8 \cdot 2 \cdot m^2 \cdot p^2}{8 \cdot 3 \cdot 2 \cdot p^2 \cdot p \cdot m^2 \cdot m}$ Factor.

$= \frac{3}{pm}$, or $\frac{3}{mp}$ Lowest terms; Either form is correct.

◄ Work Problem 6 at the Side.

EXAMPLE 7 **Dividing Rational Expressions**

Divide. Write the answer in lowest terms.

$$\frac{x^2-4}{(x+3)(x-2)} \div \frac{(x+2)(x+3)}{-2x}$$

$$= \frac{x^2-4}{(x+3)(x-2)} \cdot \frac{-2x}{(x+2)(x+3)}$$ Multiply by the reciprocal of the divisor.

$$= \frac{-2x(x^2-4)}{(x+3)(x-2)(x+2)(x+3)}$$ Multiply numerators. Multiply denominators.

$$= \frac{-2x(x+2)(x-2)}{(x+3)(x-2)(x+2)(x+3)}$$ Factor the numerator.

$$= -\frac{2x}{(x+3)^2}$$ Divide out the common factors; $a \cdot a = a^2$; $\frac{-a}{b} = -\frac{a}{b}$

Work Problem 7 at the Side. ▶

EXAMPLE 8 **Dividing Rational Expressions (Factors Are Opposites)**

Divide. Write the answer in lowest terms.

$$\frac{m^2-4}{m^2-1} \div \frac{2m^2+4m}{1-m}$$

$$= \frac{m^2-4}{m^2-1} \cdot \frac{1-m}{2m^2+4m}$$ Multiply by the reciprocal of the divisor.

$$= \frac{(m^2-4)(1-m)}{(m^2-1)(2m^2+4m)}$$ Multiply numerators. Multiply denominators.

$$= \frac{(m+2)(m-2)(1-m)}{(m+1)(m-1)(2m)(m+2)}$$ Factor. $1-m$ and $m-1$ are opposites.

$$= \frac{-1(m-2)}{2m(m+1)}$$ Divide out the common factors. Recall that $\frac{1-m}{m-1} = -1$.

$$= \frac{-m+2}{2m(m+1)}$$ Distribute -1 in the numerator.

$$= \frac{2-m}{2m(m+1)}$$ Rewrite $-m+2$ as $2-m$.

Work Problem 8 at the Side. ▶

Multiplying or Dividing Rational Expressions

Step 1 **Note the operation.** If the operation is division, use the definition of division to rewrite it as multiplication.

Step 2 **Multiply** numerators and multiply denominators.

Step 3 **Factor** all numerators and denominators completely.

Step 4 **Write in lowest terms** using the fundamental property.

Steps 2 and 3 may be interchanged based on personal preference.

7 Divide. Write each answer in lowest terms.

GS **(a)** $\frac{y^2+4y+3}{y+3} \div \frac{y^2-4y-5}{y-3}$

Rewrite as multiplication, and complete the problem.

________ · ________

(b) $\frac{4x(x+3)}{2x+1} \div \frac{-x^2(x+3)}{4x^2-1}$

8 Divide. Write each answer in lowest terms.

(a) $\frac{ab-a^2}{a^2-1} \div \frac{a-b}{a-1}$

(b) $\frac{x^2-9}{2x+6} \div \frac{9-x^2}{4x-12}$

Answers

7. **(a)** $\frac{y^2+4y+3}{y+3} \cdot \frac{y-3}{y^2-4y-5}$; $\frac{y-3}{y-5}$
(b) $-\frac{4(2x-1)}{x}$

8. **(a)** $\frac{-a}{a+1}$ **(b)** $\frac{6-2x}{3+x}$

7.2 Exercises

FOR EXTRA HELP — *Go to MyMathLab for worked-out, step-by-step solutions to exercises enclosed in a square and video solutions to ▶ exercises.*

1. **CONCEPT CHECK** Match each multiplication problem in Column I with the correct product in Column II.

I	II
(a) $\frac{5x^3}{10x^4} \cdot \frac{10x^7}{2x}$	A. $\frac{2}{5x^5}$
(b) $\frac{10x^4}{5x^3} \cdot \frac{10x^7}{2x}$	B. $\frac{5x^5}{2}$
(c) $\frac{5x^3}{10x^4} \cdot \frac{2x}{10x^7}$	C. $\frac{1}{10x^7}$
(d) $\frac{10x^4}{5x^3} \cdot \frac{2x}{10x^7}$	D. $10x^7$

2. **CONCEPT CHECK** Match each division problem in Column I with the correct quotient in Column II.

I	II
(a) $\frac{5x^3}{10x^4} \div \frac{10x^7}{2x}$	A. $\frac{5x^5}{2}$
(b) $\frac{10x^4}{5x^3} \div \frac{10x^7}{2x}$	B. $10x^7$
(c) $\frac{5x^3}{10x^4} \div \frac{2x}{10x^7}$	C. $\frac{2}{5x^5}$
(d) $\frac{10x^4}{5x^3} \div \frac{2x}{10x^7}$	D. $\frac{1}{10x^7}$

Multiply. Write each answer in lowest terms. ***See Examples 1 and 2.***

3. $\frac{4}{9} \cdot \frac{15}{16}$

4. $\frac{10}{21} \cdot \frac{3}{5}$

5. $\frac{15a^2}{14} \cdot \frac{7}{5a}$

6. $\frac{21b^6}{18} \cdot \frac{9}{7b^4}$

7. $\frac{16y^4}{18y^5} \cdot \frac{15y^5}{y^2}$

8. $\frac{20x^5}{-2x^2} \cdot \frac{8x^4}{35x^3}$

9. $\frac{2(c+d)}{3} \cdot \frac{18}{6(c+d)^2}$

10. $\frac{4(y-2)}{x} \cdot \frac{3x}{6(y-2)^2}$

11. $\frac{(x-y)^2}{2} \cdot \frac{24}{3(x-y)}$

12. $\frac{(a+b)^2}{5} \cdot \frac{30}{2(a+b)}$

Find the reciprocal of each rational expression. ***See Example 4.***

13. $\frac{3p^3}{16q}$

14. $\frac{6x^4}{9y^2}$

15. $\frac{r^2+rp}{7}$

16. $\frac{16}{9a^2+36a}$

17. $\frac{z^2+7z+12}{z^2-9}$

18. $\frac{p^2-4p+3}{p^2-3p}$

Divide. Write each answer in lowest terms. ***See Examples 5 and 6.***

19. $\frac{4}{5} \div \frac{13}{20}$

20. $\frac{7}{8} \div \frac{3}{4}$

21. $\frac{9z^4}{3z^5} \div \frac{3z^2}{5z^3}$

22. $\frac{35q^8}{9q^5} \div \frac{25q^6}{10q^5}$

23. $\frac{4t^4}{2t^5} \div \frac{(2t)^3}{-6}$

24. $\frac{-12a^6}{3a^2} \div \frac{(2a)^3}{27a}$

25. $\frac{3}{2y-6} \div \frac{6}{y-3}$

26. $\frac{4m+16}{10} \div \frac{3m+12}{18}$

27. $\frac{(x-3)^2}{6x} \div \frac{x-3}{x^2}$

28. $\frac{2a}{a+4} \div \frac{a^2}{(a+4)^2}$

Multiply or divide. Write each answer in lowest terms. ***See Examples 3, 7, and 8.***

29. $\dfrac{5x-15}{3x+9} \cdot \dfrac{4x+12}{6x-18}$

30. $\dfrac{8r+16}{24r-24} \cdot \dfrac{6r-6}{3r+6}$

31. $\dfrac{2-t}{8} \div \dfrac{t-2}{6}$

32. $\dfrac{4}{m-2} \div \dfrac{16}{2-m}$

33. $\dfrac{5-4x}{5+4x} \cdot \dfrac{4x+5}{4x-5}$

34. $\dfrac{5-x}{5+x} \cdot \dfrac{x+5}{x-5}$

35. $\dfrac{6(m-2)^2}{5(m+4)^2} \cdot \dfrac{15(m+4)}{2(2-m)}$

36. $\dfrac{7(q-1)}{3(q+1)^2} \cdot \dfrac{6(q+1)}{3(1-q)^2}$

37. $\dfrac{m^2-4}{16-8m} \div \dfrac{m+2}{8}$

38. $\dfrac{r^2-36}{54-9r} \div \dfrac{r+6}{9}$

39. $\dfrac{p^2+4p-5}{p^2+7p+10} \div \dfrac{p-1}{p+4}$

40. $\dfrac{z^2-3z+2}{z^2+4z+3} \div \dfrac{z-1}{z+1}$

41. $\dfrac{2k^2-k-1}{2k^2+5k+3} \div \dfrac{4k^2-1}{2k^2+k-3}$

42. $\dfrac{3t^2-4t-4}{3t^2+10t+8} \div \dfrac{9t^2+21t+10}{3t^2-t-10}$

43. $\dfrac{2k^2+3k-2}{6k^2-7k+2} \cdot \dfrac{4k^2-5k+1}{k^2+k-2}$

44. $\dfrac{2m^2-5m-12}{m^2-10m+24} \cdot \dfrac{m^2-9m+18}{4m^2-9}$

45. $\dfrac{m^2+2mp-3p^2}{m^2-3mp+2p^2} \div \dfrac{m^2+4mp+3p^2}{m^2+2mp-8p^2}$

46. $\dfrac{r^2+rs-12s^2}{r^2-rs-20s^2} \div \dfrac{r^2-2rs-3s^2}{r^2+rs-30s^2}$

47. $\left(\dfrac{x^2+10x+25}{x^2+10x} \cdot \dfrac{10x}{x^2+15x+50}\right) \div \dfrac{x+5}{x+10}$

48. $\left(\dfrac{m^2-12m+32}{8m} \cdot \dfrac{m^2-8m}{m^2-8m+16}\right) \div \dfrac{m-8}{m-4}$

Solve each problem. (Assume all measures are given in appropriate units.)

49. If the rational expression $\frac{5x^2y^3}{2pq}$ represents the area of a rectangle and $\frac{2xy}{p}$ represents the length, what rational expression represents the width?

Width = ?

Length = $\dfrac{2xy}{p}$

The area is $\dfrac{5x^2y^3}{2pq}$.

50. If the rational expression $\frac{12a^3b^4}{5cd}$ represents the area of a rectangle and $\frac{9ab^2}{c}$ represents the width, what rational expression represents the length?

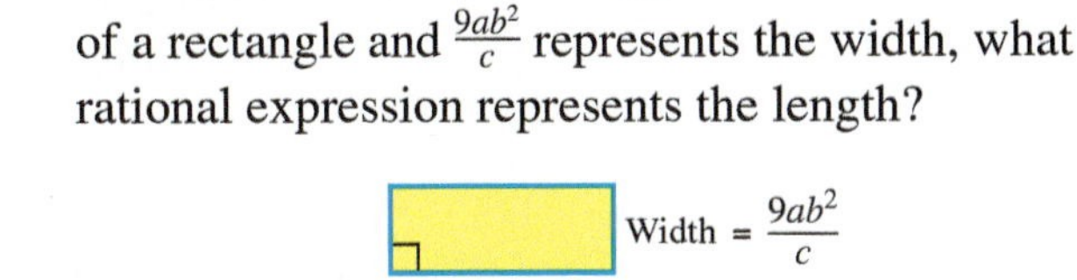

The area is $\dfrac{12a^3b^4}{5cd}$.

7.3 Least Common Denominators

OBJECTIVES

1. Find the least common denominator for a list of fractions.
2. Write equivalent rational expressions.

OBJECTIVE 1 Find the least common denominator for a list of fractions.

Adding or subtracting rational expressions often requires finding a **least common denominator (LCD)**. The LCD is the simplest expression that is divisible by all of the denominators in all of the expressions. For example,

$$\frac{2}{9} \text{ and } \frac{5}{12} \text{ have LCD } 36,$$

because 36 is the least positive number divisible by both 9 and 12.

We can often find least common denominators by inspection. In other cases, we find the LCD using the following procedure.

Finding the Least Common Denominator (LCD)

Step 1 **Factor** each denominator into prime factors.

Step 2 **List each different denominator factor** the *greatest* number of times it appears in any of the denominators.

Step 3 **Multiply** the denominator factors from Step 2 to find the LCD.

When each denominator is factored into prime factors, every prime factor must be a factor of the least common denominator.

EXAMPLE 1 Finding Least Common Denominators

Find the LCD for each pair of fractions.

(a) $\frac{1}{24}, \frac{7}{15}$ Rational numbers

(b) $\frac{1}{8x}, \frac{3}{10x}$ Rational expressions

Step 1 Write each denominator in factored form with numerical coefficients in prime factored form.

(a) $24 = 2 \cdot 2 \cdot 2 \cdot 3 = 2^3 \cdot 3$

$15 = 3 \cdot 5$

(b) $8x = 2 \cdot 2 \cdot 2 \cdot x = 2^3 \cdot x$

$10x = 2 \cdot 5 \cdot x$

Step 2 We find the LCD by taking each different factor the *greatest* number of times it appears as a factor in any of the denominators.

(a) The factor 2 appears three times in one product and not at all in the other, so the greatest number of times 2 appears is three. The greatest number of times both 3 and 5 appear in either product is one.

(b) Here, 2 appears three times in one product and once in the other, so the greatest number of times 2 appears is three. The greatest number of times 5 appears is one. The greatest number of times x appears in either product is one.

Step 3

(a) $\text{LCD} = 2 \cdot 2 \cdot 2 \cdot 3 \cdot 5$

$= 2^3 \cdot 3 \cdot 5$

$= 120$

(b) $\text{LCD} = 2 \cdot 2 \cdot 2 \cdot 5 \cdot x$

$= 2^3 \cdot 5 \cdot x$

$= 40x$

◄ Work Problem 1 at the Side.

---

1. Find the LCD for each pair of fractions.

(a) $\frac{7}{10}, \frac{1}{25}$

(b) $\frac{7}{20p}, \frac{11}{30p}$

(c) $\frac{4}{5x}, \frac{12}{10x}$

Answers

1. (a) 50 (b) $60p$ (c) $10x$

EXAMPLE 2 **Finding the LCD**

Find the LCD for $\frac{5}{6r^2}$ and $\frac{3}{4r^3}$.

Step 1

$$6r^2 = 2 \cdot 3 \cdot r^2$$
$$4r^3 = 2^2 \cdot r^3$$
Factor each denominator.

Step 2 The greatest number of times 2 appears is two, the greatest number of times 3 appears is one, and the greatest number of times r appears is three.

Step 3

$$\text{LCD} = 2^2 \cdot 3 \cdot r^3 = 12r^3$$

Work Problem 2 at the Side. ▶

CAUTION

When finding the LCD, use each factor the ***greatest*** number of times it appears in any *single* denominator, not the ***total*** number of times it appears. For instance, the greatest number of times r appears as a factor in one denominator in **Example 2** is 3, *not* 5.

EXAMPLE 3 **Finding LCDs**

Find the LCD for the fractions in each list.

(a) $\frac{6}{5m}, \frac{4}{m^2 - 3m}$

$$5m = 5 \cdot m$$
$$m^2 - 3m = m(m - 3)$$
Factor each denominator.

Use each different factor the greatest number of times it appears.

$$\text{LCD} = 5 \cdot m \cdot (m - 3) = 5m(m - 3)$$

Be sure to include m as a factor in the LCD.

Because m is not a *factor* of $m - 3$, ***both*** factors, m and $m - 3$ must appear in the LCD.

(b) $\frac{1}{r^2 - 4r - 5}, \frac{3}{r^2 - r - 20}, \frac{1}{r^2 - 10r + 25}$

$$r^2 - 4r - 5 = (r - 5)(r + 1)$$
$$r^2 - r - 20 = (r - 5)(r + 4)$$
$$r^2 - 10r + 25 = (r - 5)^2$$
Factor each denominator.

Use each different factor the greatest number of times it appears as a factor.

$$\text{LCD} = (r + 1)(r + 4)(r - 5)^2$$

Be sure to include the exponent 2 on the factor $r - 5$.

(c) $\frac{1}{q - 5}, \frac{3}{5 - q}$

The expressions $q - 5$ and $5 - q$ are opposites of each other because

$$-(q - 5) = -q + 5 = 5 - q$$

Therefore, either $q - 5$ or $5 - q$ can be used as the LCD.

Work Problem 3 at the Side. ▶

2 Find the LCD for each pair of fractions.

(a) $\frac{4}{16m^3}, \frac{5}{9m^5}$

(b) $\frac{3}{25a^2}, \frac{2}{10a^3b}$

3 Find the LCD for the fractions in each list.

(a) $\frac{7}{3a}, \frac{11}{a^2 - 4a}$

(b) $\frac{1}{x^2 + 7x + 12}, \frac{2}{x^2 + 6x + 9}, \frac{5}{x^2 + 2x - 8}$

(c) $\frac{6}{x - 4}, \frac{3x - 1}{4 - x}$

Answers

2. **(a)** $144m^5$ **(b)** $50a^3b$

3. **(a)** $3a(a - 4)$
(b) $(x + 3)^2(x + 4)(x - 2)$
(c) either $x - 4$ or $4 - x$

OBJECTIVE 2 **Write equivalent rational expressions.** Once the LCD has been found, the next step in preparing to add or subtract two rational expressions is to use the fundamental property to write equivalent rational expressions.

Writing a Rational Expression with a Specified Denominator

Step 1 **Factor** both denominators.

Step 2 **Decide what factor(s) the denominator must be multiplied by** in order to equal the specified denominator.

Step 3 **Multiply** the rational expression by that factor divided by itself. (That is, multiply by 1.)

EXAMPLE 4 Writing Equivalent Rational Expressions

Write each fraction as an equivalent expression with the indicated denominator.

(a) $\frac{3}{8} = \frac{?}{40}$ Rational numbers

(b) $\frac{9k}{25} = \frac{?}{50k}$ Rational expressions

Step 1 For each example, first factor the denominator on the right. Then compare the denominator on the left with the one on the right to decide what factors are missing.

(a) $\frac{3}{8} = \frac{?}{5 \cdot 8}$

(b) $\frac{9k}{25} = \frac{?}{25 \cdot 2k}$

Step 2 (a) A factor of 5 is missing.

(b) Factors of 2 and k are missing.

Step 3 (a) Multiply $\frac{3}{8}$ by $\frac{5}{5}$.

$$\frac{3}{8} = \frac{3}{8} \cdot \underset{\frac{5}{5} = 1}{\frac{5}{5}} = \frac{15}{40}$$

(b) Multiply $\frac{9k}{25}$ by $\frac{2k}{2k}$.

$$\frac{9k}{25} = \frac{9k}{25} \cdot \underset{\frac{2k}{2k} = 1}{\frac{2k}{2k}} = \frac{18k^2}{50k}$$

◄ **Work Problem 4 at the Side.**

4 Write each rational expression as an equivalent expression with the indicated denominator.

(a) $\frac{3}{4} = \frac{?}{36}$

GS **(b)** $\frac{7k}{5} = \frac{?}{30k}$

Step 1 Factor the denominator on the right.

$$\frac{7k}{5} = \frac{?}{5 \cdot \underline{\qquad}}$$

Step 2 Factors of ____ and ____ are missing.

Step 3

$$\frac{7k}{5} = \frac{7k}{5} \cdot \frac{\underline{\qquad}}{\underline{\qquad}} = \frac{\underline{\qquad}}{30k}$$

(c) $\frac{4t}{11} = \frac{?}{33t}$

EXAMPLE 5 Writing Equivalent Rational Expressions

Write each rational expression as an equivalent expression with the indicated denominator.

(a) $\frac{8}{3x+1} = \frac{?}{12x+4}$

$$\frac{8}{3x+1} = \frac{?}{4(3x+1)}$$ Factor the denominator on the right.

The missing factor is 4, so multiply the fraction on the left by $\frac{4}{4}$.

$$\frac{8}{3x+1} \cdot \frac{4}{4} = \frac{32}{12x+4}$$ $\frac{4}{4} = 1$; Fundamental property

Continued on Next Page

Answers

4. (a) $\frac{27}{36}$ (b) $6k$; 6; k; $6k$; $6k$; $42k^2$

(c) $\frac{12t^2}{33t}$

(b) $\dfrac{12p}{p^2+8p} = \dfrac{?}{p^3+4p^2-32p}$

Factor the denominator in each rational expression.

$$\frac{12p}{p(p+8)} = \frac{?}{p(p+8)(p-4)}$$

$p^3 + 4p^2 - 32p$
$= p(p^2 + 4p - 32)$
$= p(p+8)(p-4)$

The factor $p - 4$ is missing, so multiply $\frac{12p}{p(p+8)}$ by $\frac{p-4}{p-4}$.

$$= \frac{12p}{p(p+8)} \cdot \frac{p-4}{p-4} \qquad \frac{p-4}{p-4} = 1\text{; Fundamental property}$$

$$= \frac{12p(p-4)}{p(p+8)(p-4)} \qquad \begin{array}{l}\text{Multiply numerators.}\\ \text{Multiply denominators.}\end{array}$$

$$= \frac{12p^2 - 48p}{p^3+4p^2-32p} \qquad \text{Multiply the factors.}$$

Work Problem 5 at the Side. ▶

Note

In the last step in **Example 5(b),** we multiplied the factors of the numerator and denominator in

$$\frac{12p(p-4)}{p(p+8)(p-4)} \quad \text{to obtain} \quad \frac{12p^2-48p}{p^3+4p^2-32p}.$$

We did this to match the form of the fractions in the original statement of the problem. In actuality, these are equivalent expressions, and either form is acceptable as the answer.

5 Write each rational expression as an equivalent expression with the indicated denominator.

(a) $\dfrac{9}{2a+5} = \dfrac{?}{6a+15}$

(b) $\dfrac{5k+1}{k^2+2k} = \dfrac{?}{k^3+k^2-2k}$

Answers

5. (a) $\dfrac{27}{6a+15}$

(b) $\dfrac{(5k+1)(k-1)}{k^3+k^2-2k}$, or $\dfrac{5k^2-4k-1}{k^3+k^2-2k}$

7.3 Exercises

FOR EXTRA HELP *Go to MyMathLab for worked-out, step-by-step solutions to exercises enclosed in a square and video solutions to ▶ exercises.*

CONCEPT CHECK *Choose the correct response.*

1. The least common denominator for $\frac{11}{20}$ and $\frac{1}{2}$ is

   A. 40 **B.** 2 **C.** 20 **D.** none of these.

2. The least common denominator for $\frac{1}{a}$ and $\frac{1}{5a}$ is

   A. a **B.** $5a$ **C.** $5a^2$ **D.** 5.

3. **CONCEPT CHECK** To find the LCD for $\frac{4}{25x^2}$ and $\frac{7}{10x^4}$, a student factored each denominator as follows.

$$25x^2 = 5^2 \cdot x^2$$
$$10x^4 = 2 \cdot 5 \cdot x^4$$

He multiplied the factors $2 \cdot 5^2 \cdot x^6$ to obtain $50x^6$ as the LCD. ***What Went Wrong?*** Give the correct LCD.

4. **CONCEPT CHECK** A student was asked to find the LCD for

$$\frac{2}{x-1} \text{ and } \frac{5}{x+1}.$$

She answered that because the denominator expressions are opposites, either $x - 1$ or $x + 1$ could be used as the LCD. ***What Went Wrong?*** Give the correct LCD.

Find the LCD for the fractions in each list. ***See Examples 1 and 2.***

5. $\frac{7}{15}, \frac{21}{20}$ ▶

6. $\frac{9}{10}, \frac{13}{25}$

7. $\frac{2}{15}, \frac{3}{10}, \frac{7}{30}$

8. $\frac{5}{24}, \frac{7}{12}, \frac{9}{28}$

9. $\frac{3}{x^4}, \frac{5}{x^7}$

10. $\frac{2}{y^5}, \frac{3}{y^6}$

11. $\frac{2}{5p}, \frac{13}{6p}$

12. $\frac{14}{15k}, \frac{11}{4k}$

13. $\frac{5}{36q}, \frac{17}{24q}$

14. $\frac{4}{30p}, \frac{9}{50p}$

15. $\frac{6}{21r^3}, \frac{8}{12r^5}$ ▶

16. $\frac{9}{35t^2}, \frac{5}{49t^6}$

17. $\frac{13}{5a^2b^3}, \frac{29}{15a^5b}$

18. $\frac{7}{3r^4s^5}, \frac{23}{9r^6s^8}$

19. $\frac{5}{12x^2y}, \frac{7}{24x^3y^2}, \frac{-11}{6xy^4}$

20. $\frac{3}{10a^4b}, \frac{7}{15ab}, \frac{-7}{30ab^7}$

Find the LCD for the fractions in each list. ***See Examples 1–3.***

21. $\frac{7}{6p}, \frac{15}{4p-8}$ ▶

22. $\frac{7}{8k}, \frac{28}{12k-24}$

23. $\frac{9}{28m^2}, \frac{3}{12m-20}$

24. $\frac{15}{27a^3}, \frac{8}{9a-45}$

25. $\frac{7}{5b-10}, \frac{11}{6b-12}$ ▶

26. $\frac{3}{7x+21}, \frac{1}{5x+15}$

27. $\frac{5}{c-d}, \frac{8}{d-c}$

28. $\frac{4}{y-x}, \frac{7}{x-y}$

29. $\frac{13}{x^2-1}, \frac{-5}{2x+2}$

30. $\frac{9}{y^2-9}, \frac{-2}{2y+6}$

31. $\frac{3}{k^2+5k}, \frac{2}{k^2+3k-10}$

32. $\frac{1}{z^2-4z}, \frac{4}{z^2-3z-4}$

33. $\frac{6}{a^2+6a}, \frac{-5}{a^2+3a-18}$

34. $\frac{8}{y^2-5y}, \frac{-5}{y^2-2y-15}$

35. $\frac{5}{p^2+8p+15}, \frac{3}{p^2-3p-18}, \frac{2}{p^2-p-30}$

36. $\frac{10}{y^2-10y+21}, \frac{2}{y^2-2y-3}, \frac{5}{y^2-6y-7}$

Write each rational expression as an equivalent expression with the indicated denominator. ***See Examples 4 and 5.***

37. $\frac{4}{11} = \frac{?}{55}$

38. $\frac{6}{7} = \frac{?}{42}$

39. $\frac{-5}{k} = \frac{?}{9k}$

40. $\frac{-3}{q} = \frac{?}{6q}$

41. $\frac{13}{40y} = \frac{?}{80y^3}$

42. $\frac{5}{27p} = \frac{?}{108p^4}$

43. $\frac{5t^2}{6r} = \frac{?}{42r^4}$

44. $\frac{8y^2}{3x} = \frac{?}{30x^3}$

45. $\frac{5}{2(m+3)} = \frac{?}{8(m+3)}$

46. $\frac{7}{4(y-1)} = \frac{?}{16(y-1)}$

47. $\frac{19z}{2z-6} = \frac{?}{6z-18}$

48. $\frac{3r}{5r-5} = \frac{?}{15r-15}$

49. $\frac{-4t}{3t-6} = \frac{?}{12-6t}$

50. $\frac{-7k}{5k-20} = \frac{?}{40-10k}$

51. $\frac{14}{z^2-3z} = \frac{?}{z(z-3)(z-2)}$

52. $\frac{12}{x(x+4)} = \frac{?}{x(x+4)(x-9)}$

53. $\frac{2(b-1)}{b^2+b} = \frac{?}{b^3+3b^2+2b}$

54. $\frac{3(c+2)}{c(c-1)} = \frac{?}{c^3-5c^2+4c}$

7.4 Adding and Subtracting Rational Expressions

OBJECTIVES

1. Add rational expressions having the same denominator.
2. Add rational expressions having different denominators.
3. Subtract rational expressions.

OBJECTIVE 1 Add rational expressions having the same denominator. We find the sum of two such rational expressions using the procedure for adding two common fractions having the same denominator.

Adding Rational Expressions (Same Denominator)

The rational expressions $\frac{P}{Q}$ and $\frac{R}{Q}$ (where $Q \neq 0$) are added as follows.

$$\frac{P}{Q} + \frac{R}{Q} = \frac{P + R}{Q}$$

In words: To add rational expressions with the same denominator, add the numerators and keep the same denominator.

1 Add. Write each answer in lowest terms.

(a) $\frac{7}{15} + \frac{3}{15}$

(b) $\frac{3}{y+4} + \frac{2}{y+4}$

(c) $\frac{a}{a+b} + \frac{b}{a+b}$

(d) $\frac{x^2}{x+1} + \frac{x}{x+1}$

EXAMPLE 1 Adding Rational Expressions (Same Denominator)

Add. Write each answer in lowest terms.

(a) $\frac{4}{9} + \frac{2}{9}$ Rational numbers

The denominators are the same, so the sum is found by adding the two numerators and keeping the same (common) denominator.

$= \frac{4+2}{9}$ Add.

$= \frac{6}{9}$

$= \frac{2 \cdot 3}{3 \cdot 3}$ Factor.

$= \frac{2}{3}$ Lowest terms

(b) $\frac{3x}{x+1} + \frac{3}{x+1}$ Rational expressions

$= \frac{3x+3}{x+1}$ Add.

$= \frac{3(x+1)}{x+1}$ Factor.

$= 3$ Lowest terms

◀ Work Problem 1 at the Side.

OBJECTIVE 2 Add rational expressions having different denominators. We use the following steps to add two rational expressions having different denominators.

Adding Rational Expressions (Different Denominators)

Step 1 **Find the least common denominator (LCD).**

Step 2 **Write each rational expression** as an equivalent rational expression with the LCD as the denominator.

Step 3 **Add** the numerators to obtain the numerator of the sum. The LCD is the denominator of the sum.

Step 4 **Write in lowest terms** using the fundamental property.

Answers

1. (a) $\frac{2}{3}$ (b) $\frac{5}{y+4}$ (c) 1 (d) x

EXAMPLE 2 **Adding Rational Expressions (Different Denominators)**

Add. Write each answer in lowest terms.

(a) $\frac{1}{12}+\frac{7}{15}$ **(b)** $\frac{2}{3y}+\frac{1}{4y}$

Step 1 First find the LCD, using the methods of the previous section.

$$12 = 2 \cdot 2 \cdot 3 = 2^2 \cdot 3$$
$$15 = 3 \cdot 5$$
$$\text{LCD} = 2^2 \cdot 3 \cdot 5 = 60$$

$$3y = 3 \cdot y$$
$$4y = 2 \cdot 2 \cdot y = 2^2 \cdot y$$
$$\text{LCD} = 2^2 \cdot 3 \cdot y = 12y$$

Step 2 Now write each rational expression as an equivalent expression with the LCD (60 or $12y$, respectively) as the denominator.

$\frac{1}{12}+\frac{7}{15}$ The LCD is 60.

$$=\frac{1}{12}\cdot\frac{5}{5}+\frac{7}{15}\cdot\frac{4}{4}$$
$$=\frac{5}{60}+\frac{28}{60}$$

$\frac{2}{3y}+\frac{1}{4y}$ The LCD is $12y$.

$$=\frac{2}{3y}\cdot\frac{4}{4}+\frac{1}{4y}\cdot\frac{3}{3}$$
$$=\frac{8}{12y}+\frac{3}{12y}$$

Step 3 Add the numerators. The LCD is the denominator.

Step 4 Write in lowest terms if necessary.

$$=\frac{5+28}{60}$$
$$=\frac{33}{60}$$
$$=\frac{11}{20}$$

$$=\frac{8+3}{12y}$$
$$=\frac{11}{12y}$$

Work Problem 2 at the Side. ▶

2 Add. Write each answer in lowest terms.

(a) $\frac{1}{10}+\frac{1}{15}$

GS **(b)** $\frac{6}{5x}+\frac{9}{2x}$

$$=\frac{6}{5x}\cdot\frac{\_\_\_}{\_\_\_}+\frac{9}{2x}\cdot\frac{\_\_\_}{\_\_\_}$$
$$=\frac{12+45}{\_\_\_}$$
$$=\_\_\_$$

(c) $\frac{m}{3n}+\frac{2}{7n}$

EXAMPLE 3 **Adding Rational Expressions**

Add. Write the answer in lowest terms.

$$\frac{2x}{x^2-1}+\frac{-1}{x+1}$$

Step 1 The denominators are different, so find the LCD.

$x^2-1=(x+1)(x-1)$
$x+1$ is prime.
} The LCD is $(x+1)(x-1)$.

Step 2 Write each rational expression with the LCD as the denominator.

$\frac{2x}{x^2-1}+\frac{-1}{x+1}$ The LCD is $(x+1)(x-1)$.

$=\frac{2x}{(x+1)(x-1)}+\frac{-1(x-1)}{(x+1)(x-1)}$ Multiply the second fraction by $\frac{x-1}{x-1}$.

$=\frac{2x}{(x+1)(x-1)}+\frac{-x+1}{(x+1)(x-1)}$ Distributive property

Continued on Next Page

Answers

2. (a) $\frac{1}{6}$ **(b)** 2; 2; 5; 5; $10x$; $\frac{57}{10x}$
(c) $\frac{7m+6}{21n}$

3 Add. Write each answer in lowest terms.

(a) $\frac{2p}{3p+3} + \frac{5p}{2p+2}$

(b) $\frac{4}{y^2-1} + \frac{6}{y+1}$

(c) $\frac{-2}{p+1} + \frac{4p}{p^2-1}$

4 Add. Write each answer in lowest terms.

(a) $\frac{2k}{k^2-5k+4} + \frac{3}{k^2-1}$

(b) $\frac{4m}{m^2+3m+2} + \frac{2m-1}{m^2+6m+5}$

5 Add. Write the answer in lowest terms.

$$\frac{2k}{k-7} + \frac{5}{7-k}$$

Answers

3. (a) $\frac{19p}{6(p+1)}$ (b) $\frac{2(3y-1)}{(y+1)(y-1)}$
(c) $\frac{2}{p-1}$

4. (a) $\frac{(2k-3)(k+4)}{(k-4)(k-1)(k+1)}$
(b) $\frac{6m^2+23m-2}{(m+2)(m+1)(m+5)}$

5. $\frac{2k-5}{k-7}$, or $\frac{5-2k}{7-k}$

Step 3 $= \frac{2x-x+1}{(x+1)(x-1)}$ Add numerators. Keep the same denominator.

$= \frac{x+1}{(x+1)(x-1)}$ Combine like terms in the numerator.

Step 4 $= \frac{1(x+1)}{(x+1)(x-1)}$ Identity property of multiplication

$= \frac{1}{x-1}$ Divide out the common factors. The result is in lowest terms.

Remember to write 1 in the numerator.

◀ **Work Problem 3 at the Side.**

EXAMPLE 4 Adding Rational Expressions

Add. Write the answer in lowest terms.

$$\frac{2x}{x^2+5x+6} + \frac{x+1}{x^2+2x-3}$$

$= \frac{2x}{(x+2)(x+3)} + \frac{x+1}{(x+3)(x-1)}$ Factor the denominators.

$= \frac{2x(x-1)}{(x+2)(x+3)(x-1)} + \frac{(x+1)(x+2)}{(x+2)(x+3)(x-1)}$ The LCD is $(x+2)(x+3)(x-1)$.

$= \frac{2x(x-1)+(x+1)(x+2)}{(x+2)(x+3)(x-1)}$ Add numerators. Keep the same denominator.

$= \frac{2x^2-2x+x^2+3x+2}{(x+2)(x+3)(x-1)}$ Multiply.

$= \frac{3x^2+x+2}{(x+2)(x+3)(x-1)}$ Combine like terms.

The numerator cannot be factored here, so the expression is in lowest terms.

◀ **Work Problem 4 at the Side.**

EXAMPLE 5 Adding Rational Expressions (Denominators Are Opposites)

Add. Write the answer in lowest terms.

$\frac{y}{y-2} + \frac{8}{2-y}$ The denominators are opposites.

$= \frac{y}{y-2} + \frac{8(-1)}{(2-y)(-1)}$ Multiply $\frac{8}{2-y}$ by 1 in the form $\frac{-1}{-1}$ to find a common denominator.

$= \frac{y}{y-2} + \frac{-8}{-2+y}$ Multiply. Apply the distributive property in the denominator.

$= \frac{y}{y-2} + \frac{-8}{y-2}$ Rewrite $-2+y$ as $y-2$.

$= \frac{y-8}{y-2}$ Add numerators. Keep the same denominator.

◀ **Work Problem 5 at the Side.**

Note

If we had chosen to use $2 - y$ as the common denominator in **Example 5,** we would have obtained a different, yet equivalent, form of the answer $\frac{y-8}{y-2}$.

$$\frac{y}{y-2} + \frac{8}{2-y} \quad \text{See Example 5.}$$

$$= \frac{y(-1)}{(y-2)(-1)} + \frac{8}{2-y} \quad \text{Multiply } \tfrac{y}{y-2} \text{ by 1 in the form } \tfrac{-1}{-1}.$$

$$= \frac{-y}{2-y} + \frac{8}{2-y} \quad \text{In the first denominator, } (y-2)(-1) = -y + 2 = 2 - y.$$

$$= \frac{-y+8}{2-y} \quad \text{Add numerators. Keep the same denominator.}$$

$$= \frac{8-y}{2-y} \quad \text{Multiply } \tfrac{8-y}{2-y} \text{ by } \tfrac{-1}{-1} \text{ to confirm that it is equivalent to } \tfrac{y-8}{y-2}.$$

OBJECTIVE 3 Subtract rational expressions.

Subtracting Rational Expressions (Same Denominator)

The rational expressions $\frac{P}{Q}$ and $\frac{R}{Q}$ (where $Q \neq 0$) are subtracted as follows.

$$\frac{P}{Q} - \frac{R}{Q} = \frac{P-R}{Q}$$

In words:* *To subtract rational expressions with the same denominator, subtract the numerators and keep the same denominator.

EXAMPLE 6 Subtracting Rational Expressions (Same Denominator)

Subtract. Write the answer in lowest terms.

$$\frac{2m}{m-1} - \frac{m+3}{m-1}$$

Use parentheses around the numerator of the subtrahend.

$$= \frac{2m - (m+3)}{m-1} \quad \text{Subtract numerators. Keep the same denominator.}$$

Be careful with signs.

$$= \frac{2m - m - 3}{m-1} \quad \text{Distributive property}$$

$$= \frac{m-3}{m-1} \quad \text{Combine like terms.}$$

Work Problem 6 at the Side. ▶

CAUTION

In subtraction problems like the one in **Example 6,** the numerator of the fraction being subtracted must be treated as a single quantity. ***Be sure to use parentheses after the subtraction symbol. Subtract each term within the parentheses, or a sign error may occur.***

6 Subtract. Write each answer in lowest terms.

(a) $\frac{3}{m^2} - \frac{2}{m^2}$

GS **(b)** $\frac{x}{2x+3} - \frac{3x+4}{2x+3}$

$= \frac{x - (\_\_\_\_)}{2x+3}$

$= \frac{x - \_\_\_\_ - \_\_\_\_}{2x+3}$

$= \frac{\_\_\_\_}{2x+3}$

If the numerator is factored, the answer can be written as ____.

(c) $\frac{5t}{t-1} - \frac{5+t}{t-1}$

Answers

6. (a) $\frac{1}{m^2}$

(b) $3x + 4$; $3x$; 4; $-2x - 4$; $\frac{-2(x+2)}{2x+3}$

(c) $\frac{4t-5}{t-1}$

7 Subtract. Write each answer in lowest terms.

(a) $\frac{1}{k+4} - \frac{2}{k}$

(b) $\frac{6}{a+2} - \frac{1}{a-3}$

EXAMPLE 7 **Subtracting Rational Expressions (Different Denominators)**

Subtract. Write the answer in lowest terms.

$\frac{9}{x-2} - \frac{3}{x}$ — The LCD is $x(x-2)$.

$= \frac{9x}{x(x-2)} - \frac{3(x-2)}{x(x-2)}$ — Write each expression with the LCD.

$= \frac{9x - 3(x-2)}{x(x-2)}$ — Subtract numerators. Keep the same denominator.

Be careful with signs.

$= \frac{9x - 3x + 6}{x(x-2)}$ — Distributive property

$= \frac{6x+6}{x(x-2)}$, or $\frac{6(x+1)}{x(x-2)}$ — Combine like terms. Factor the numerator.

We factored in the last step to determine whether there were any common factors to divide out. There were not, so the expression is in lowest terms. The two final forms are equivalent. Either form can be given as the answer.

◀ Work Problem **7** at the Side.

8 Subtract. Write each answer in lowest terms.

(a) $\frac{5}{x-1} - \frac{3x}{1-x}$

(b) $\frac{2y}{y-2} - \frac{1+y}{2-y}$

EXAMPLE 8 **Subtracting Rational Expressions (Denominators Are Opposites)**

Subtract. Write the answer in lowest terms.

$\frac{3x}{x-5} - \frac{2x-25}{5-x}$ — The denominators are opposites. We choose $x-5$ as the common denominator.

$= \frac{3x}{x-5} - \frac{(2x-25)(-1)}{(5-x)(-1)}$ — Multiply $\frac{2x-25}{5-x}$ by $\frac{-1}{-1}$ to find a common denominator.

$= \frac{3x}{x-5} - \frac{-2x+25}{x-5}$ — $(2x-25)(-1) = -2x+25$; $(5-x)(-1) = -5+x = x-5$

Use parentheses.

$= \frac{3x - (-2x+25)}{x-5}$ — Subtract numerators.

Be careful with signs.

$= \frac{3x + 2x - 25}{x-5}$ — Distributive property

$= \frac{5x-25}{x-5}$ — Combine like terms.

$= \frac{5(x-5)}{x-5}$ — Factor.

$= 5$ — Lowest terms

◀ Work Problem **8** at the Side.

Answers

7. (a) $\frac{-k-8}{k(k+4)}$ (b) $\frac{5(a-4)}{(a+2)(a-3)}$

8. (a) $\frac{5+3x}{x-1}$, or $\frac{-5-3x}{1-x}$

(b) $\frac{3y+1}{y-2}$, or $\frac{-3y-1}{2-y}$

EXAMPLE 9 **Subtracting Rational Expressions**

Subtract. Write the answer in lowest terms.

$$\frac{6x}{x^2 - 2x + 1} - \frac{1}{x^2 - 1}$$

$$= \frac{6x}{(x-1)^2} - \frac{1}{(x-1)(x+1)} \quad \text{Factor the denominators.}$$

From the factored denominators, we identify the LCD, $(x-1)^2(x+1)$. ***We use the factor $x - 1$ twice*** because it appears twice in the first denominator.

$$= \frac{6x(x+1)}{(x-1)^2(x+1)} - \frac{1(x-1)}{(x-1)(x-1)(x+1)} \quad \text{Fundamental property}$$

$$= \frac{6x(x+1) - 1(x-1)}{(x-1)^2(x+1)} \quad \text{Subtract numerators.}$$

$$= \frac{6x^2 + 6x - x + 1}{(x-1)^2(x+1)} \quad \text{Distributive property}$$

$$= \frac{6x^2 + 5x + 1}{(x-1)^2(x+1)}, \quad \text{or} \quad \frac{(3x+1)(2x+1)}{(x-1)^2(x+1)} \quad \text{Combine like terms. Factor the numerator.}$$

The two final forms are equivalent. The factored form indicates that the expression is in lowest terms—there are no common factors to divide out. Either form can be given as the answer.

Work Problem 9 at the Side. ▶

EXAMPLE 10 **Subtracting Rational Expressions**

Subtract. Write the answer in lowest terms.

$$\frac{q}{q^2 - 4q - 5} - \frac{3}{2q^2 - 13q + 15}$$

$$= \frac{q}{(q+1)(q-5)} - \frac{3}{(q-5)(2q-3)} \quad \text{Factor the denominators. The LCD is } (q+1)(q-5)(2q-3).$$

$$= \frac{q(2q-3)}{(q+1)(q-5)(2q-3)} - \frac{3(q+1)}{(q+1)(q-5)(2q-3)} \quad \text{Fundamental property}$$

$$= \frac{q(2q-3) - 3(q+1)}{(q+1)(q-5)(2q-3)} \quad \text{Subtract numerators.}$$

$$= \frac{2q^2 - 3q - 3q - 3}{(q+1)(q-5)(2q-3)} \quad \text{Distributive property}$$

$$= \frac{2q^2 - 6q - 3}{(q+1)(q-5)(2q-3)} \quad \text{Combine like terms.}$$

The numerator cannot be factored, so the final answer is in lowest terms.

Work Problem 10 at the Side. ▶

9 Subtract. Write each answer in lowest terms.

(a) $\frac{4y}{y^2 - 1} - \frac{5}{y^2 + 2y + 1}$

GS **(b)** $\frac{3r}{r - 5} - \frac{4}{r^2 - 10r + 25}$

The LCD is ________.

Complete the subtraction.

10 Subtract. Write each answer in lowest terms.

(a) $\frac{2}{p^2 - 5p + 4} - \frac{3}{p^2 - 1}$

(b)

$$\frac{q}{2q^2 + 5q - 3} - \frac{3q + 4}{3q^2 + 10q + 3}$$

Answers

9. (a) $\frac{4y^2 - y + 5}{(y+1)^2(y-1)}$

(b) $(r-5)^2$; $\frac{3r^2 - 15r - 4}{(r-5)^2}$

10. (a) $\frac{14 - p}{(p-4)(p-1)(p+1)}$

(b) $\frac{-3q^2 - 4q + 4}{(2q-1)(q+3)(3q+1)}$

7.4 Exercises

FOR EXTRA HELP — *Go to MyMathLab for worked-out, step-by-step solutions to exercises enclosed in a square and video solutions to ▶ exercises.*

CONCEPT CHECK *Match each expression in Column I with the correct sum or difference in Column II.*

I

1. $\frac{x}{x+6}+\frac{6}{x+6}$
2. $\frac{2x}{x-6}-\frac{12}{x-6}$
3. $\frac{6}{x-6}-\frac{x}{x-6}$
4. $\frac{6}{x+6}-\frac{x}{x+6}$
5. $\frac{x}{x+6}-\frac{6}{x+6}$
6. $\frac{1}{x}+\frac{1}{6}$
7. $\frac{1}{6}-\frac{1}{x}$
8. $\frac{1}{6x}-\frac{1}{6x}$

II

A. 2 **B.** $\frac{x-6}{x+6}$

C. -1 **D.** $\frac{6+x}{6x}$

E. 1 **F.** 0

G. $\frac{x-6}{6x}$ **H.** $\frac{6-x}{x+6}$

9. **CONCEPT CHECK** A student subtracted the following rational expressions incorrectly as shown.

$$\frac{2x}{x+5}-\frac{x+1}{x+5}$$
$$=\frac{2x-x+1}{x+5}$$
$$=\frac{x+1}{x+5}$$

What Went Wrong? Give the correct answer.

10. **CONCEPT CHECK** A student subtracted the following rational expressions incorrectly as shown.

$$\frac{7x}{2x-3}-\frac{3x-5}{2x-3}$$
$$=\frac{7x-(3x-5)}{2x-3}$$
$$=\frac{7x-3x-5}{2x-3}$$
$$=\frac{4x-5}{2x-3}$$

What Went Wrong? Give the correct answer.

Add or subtract. Write each answer in lowest terms. ***See Examples 1 and 6.***

11. $\frac{5}{18}+\frac{7}{18}$
12. $\frac{11}{24}+\frac{7}{24}$
13. ▶ $\frac{4}{m}+\frac{7}{m}$
14. $\frac{5}{p}+\frac{11}{p}$
15. $\frac{5}{y+4}-\frac{1}{y+4}$
16. $\frac{4}{y+3}-\frac{1}{y+3}$
17. $\frac{4y}{y+3}+\frac{12}{y+3}$
18. $\frac{2x}{x+4}+\frac{8}{x+4}$
19. $\frac{a+b}{2}-\frac{a-b}{2}$
20. $\frac{x-y}{2}-\frac{x+y}{2}$
21. ▶ $\frac{5m}{m+1}-\frac{1+4m}{m+1}$
22. $\frac{4x}{x+2}-\frac{2+3x}{x+2}$

23. $\dfrac{6x}{x-4} - \dfrac{4x-3}{x-4}$

24. $\dfrac{8x}{x-2} - \dfrac{5x-4}{x-2}$

25. $\dfrac{x^2}{x+5} + \dfrac{5x}{x+5}$

26. $\dfrac{t^2}{t-3} + \dfrac{-3t}{t-3}$

27. $\dfrac{y^2-3y}{y+3} + \dfrac{-18}{y+3}$

28. $\dfrac{r^2-8r}{r-5} + \dfrac{15}{r-5}$

Add or subtract. Write each answer in lowest terms. ***See Examples 2, 3, 4, and 7.***

29. $\dfrac{5}{12} + \dfrac{3}{20}$

30. $\dfrac{7}{30} + \dfrac{2}{45}$

31. $\dfrac{z}{5} + \dfrac{1}{3}$

32. $\dfrac{p}{8} + \dfrac{3}{5}$

33. $\dfrac{5}{7} - \dfrac{r}{2}$

34. $\dfrac{10}{9} - \dfrac{z}{3}$

35. $-\dfrac{3}{4} - \dfrac{1}{2x}$

36. $-\dfrac{5}{8} - \dfrac{3}{2a}$

37. $\dfrac{3}{5x} + \dfrac{9}{4x}$

38. $\dfrac{3}{2x} + \dfrac{4}{7x}$

39. $\dfrac{x+1}{6} + \dfrac{3x+3}{9}$

40. $\dfrac{2x-6}{4} + \dfrac{x+5}{6}$

41. $\dfrac{x+3}{3x} + \dfrac{2x+2}{4x}$

42. $\dfrac{x+2}{5x} + \dfrac{6x+3}{3x}$

43. $\dfrac{2}{x+3} + \dfrac{1}{x}$

44. $\dfrac{3}{x-4} + \dfrac{2}{x}$

45. $\dfrac{1}{k+4} - \dfrac{2}{k}$

46. $\dfrac{3}{m+1} - \dfrac{4}{m}$

47. $\dfrac{x}{x-2} + \dfrac{-8}{x^2-4}$

48. $\dfrac{2x}{x-1} + \dfrac{-4}{x^2-1}$

49. $\dfrac{x}{x-2} + \dfrac{4}{x+2}$

50. $\dfrac{2x}{x-1} + \dfrac{3}{x+1}$

51. $\dfrac{4m}{m^2+3m+2} + \dfrac{2m-1}{m^2+6m+5}$

52. $\dfrac{a}{a^2+3a-4} + \dfrac{4a}{a^2+7a+12}$

53. $\dfrac{t}{t+2} + \dfrac{5-t}{t} - \dfrac{4}{t^2+2t}$

54. $\dfrac{2p}{p-3} + \dfrac{2+p}{p} - \dfrac{-6}{p^2-3p}$

55. CONCEPT CHECK What are the two possible LCDs that could be used for the following sum?

$$\frac{10}{m-2}+\frac{5}{2-m}$$

56. CONCEPT CHECK If one form of the correct answer to a sum or difference of rational expressions is $\frac{4}{k-3}$, what would be an alternative form of the answer if the denominator is $3-k$?

Add or subtract. Write each answer in lowest terms. ***See Examples 5 and 8.***

57. $\frac{4}{x-5}+\frac{6}{5-x}$

58. $\frac{10}{m-2}+\frac{5}{2-m}$

59. $\frac{-1}{1-y}-\frac{4y-3}{y-1}$

60. $\frac{-4}{p-3}-\frac{p+1}{3-p}$

61. $\frac{2}{x-y^2}+\frac{7}{y^2-x}$

62. $\frac{-8}{p-q^2}+\frac{3}{q^2-p}$

63. $\frac{x}{5x-3y}-\frac{y}{3y-5x}$

64. $\frac{t}{8t-9s}-\frac{s}{9s-8t}$

65. $\frac{3}{4p-5}+\frac{9}{5-4p}$

66. $\frac{8}{3-7y}-\frac{2}{7y-3}$

67. $\frac{15x}{5x-7}-\frac{-21}{7-5x}$

68. $\frac{24y}{6y-5}-\frac{-20}{5-6y}$

In each subtraction problem, the rational expression that follows the subtraction sign has a numerator with more than one term. ***Be careful with signs*** *and find each difference.* ***See Examples 6–10.***

69. $\frac{2m}{m-n}-\frac{5m+n}{2m-2n}$

70. $\frac{5p}{p-q}-\frac{3p+1}{4p-4q}$

71. $\frac{y^2}{y-2}-\frac{9y-14}{y-2}$

72. $\frac{y^2}{y-4}-\frac{y+12}{y-4}$

73. $\frac{5}{x^2-9}-\frac{x+2}{x^2+4x+3}$

74. $\frac{1}{a^2-1}-\frac{a-1}{a^2+3a-4}$

75. $\dfrac{2q+1}{3q^2+10q-8} - \dfrac{3q+5}{2q^2+5q-12}$

76. $\dfrac{4y-1}{2y^2+5y-3} - \dfrac{y+3}{6y^2+y-2}$

Perform the indicated operations. **See Examples 1–10.**

77. $\dfrac{4}{r^2-r} + \dfrac{6}{r^2+2r} - \dfrac{1}{r^2+r-2}$

78. $\dfrac{6}{k^2+3k} - \dfrac{1}{k^2-k} + \dfrac{2}{k^2+2k-3}$

79. $\dfrac{x+3y}{x^2+2xy+y^2} + \dfrac{x-y}{x^2+4xy+3y^2}$

80. $\dfrac{m}{m^2-1} + \dfrac{m-1}{m^2+2m+1}$

81. $\dfrac{r+y}{18r^2+9ry-2y^2} + \dfrac{3r-y}{36r^2-y^2}$

82. $\dfrac{2x-z}{2x^2+xz-10z^2} - \dfrac{x+z}{x^2-4z^2}$

Find an expression that represents **(a)** *the perimeter and* **(b)** *the area of each figure. Give answers in simplified form. (Assume all measures are given in appropriate units. If necessary, refer to the formulas at the back of this text.)*

83.

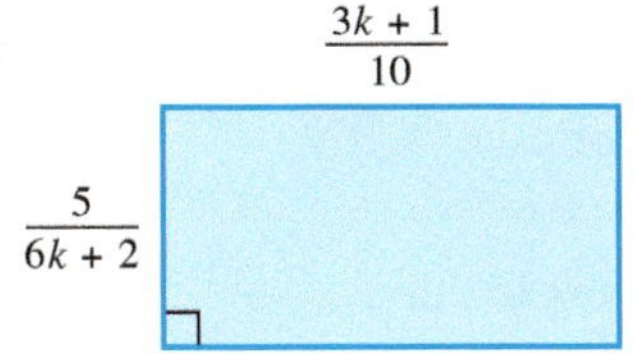

84.

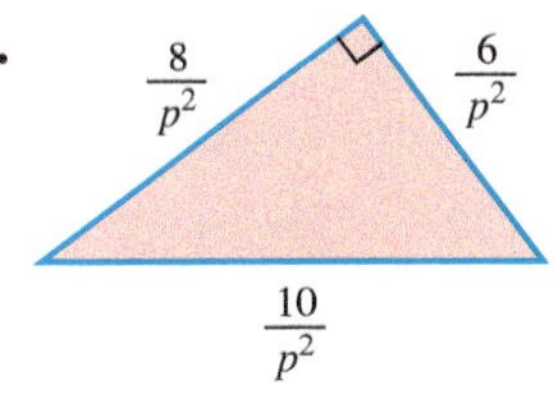

A Concours d'Elégance is a competition in which a maximum of 100 *points is awarded to a car based on its general attractiveness. The rational expression*

$$\frac{9010}{49(101-x)} - \frac{10}{49}$$

approximates the cost, in thousands of dollars, of restoring a car so that it will win x points.

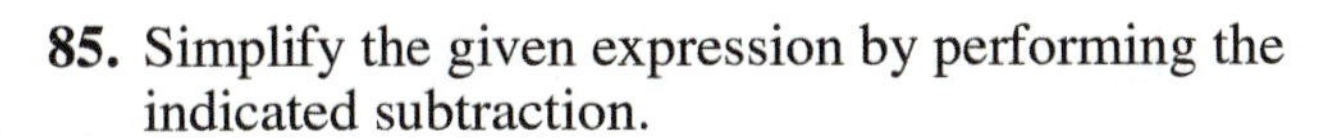

85. Simplify the given expression by performing the indicated subtraction.

86. Use the simplified expression from **Exercise 85** to determine, to two decimal places, how much it would cost to win 95 points.

7.5 Complex Fractions

OBJECTIVES

1. Define and recognize a complex fraction.
2. Simplify a complex fraction by writing it as a division problem (Method 1).
3. Simplify a complex fraction by multiplying numerator and denominator by the least common denominator (Method 2).

OBJECTIVE 1 Define and recognize a complex fraction. The quotient of two mixed numbers in arithmetic, such as $2\frac{1}{2} \div 3\frac{1}{4}$, can be written as a fraction.

$$2\frac{1}{2} \div 3\frac{1}{4}$$

$$= \frac{2\frac{1}{2}}{3\frac{1}{4}} \qquad a \div b = \frac{a}{b}$$

$$= \frac{2 + \frac{1}{2}}{3 + \frac{1}{4}} \qquad \text{Definition of mixed number}$$

We do this to illustrate a *complex fraction*.

In algebra, some rational expressions also have fractions in the numerator, or denominator, or both.

Complex Fraction

A quotient with one or more fractions in the numerator, or denominator, or both, is a **complex fraction.**

Examples: $\frac{2 + \frac{1}{2}}{3 + \frac{1}{4}}$, $\frac{\frac{3x^2 - 5x}{6x^2}}{2x - \frac{1}{x}}$, and $\frac{3 + x}{5 - \frac{2}{x}}$ Complex fractions

The parts of a complex fraction are named as follows.

$$\frac{\frac{2}{p} - \frac{1}{q}}{\frac{3}{p} + \frac{5}{q}}$$

← Numerator of complex fraction
← Main fraction bar
← Denominator of complex fraction

OBJECTIVE 2 Simplify a complex fraction by writing it as a division problem (Method 1). The main fraction bar represents division in a complex fraction, so one method of simplifying a complex fraction involves division.

Method 1 for Simplifying a Complex Fraction

Step 1 Write both the numerator and denominator as single fractions.

Step 2 Change the complex fraction to a division problem.

Step 3 Perform the indicated division.

EXAMPLE 1 Simplifying Complex Fractions (Method 1)

Simplify each complex fraction.

(a) $\dfrac{\frac{2}{3}+\frac{5}{9}}{\frac{1}{4}+\frac{1}{12}}$ **(b)** $\dfrac{6+\frac{3}{x}}{\frac{x}{4}+\frac{1}{8}}$

Step 1 First, write each numerator as a single fraction.

$$\frac{2}{3}+\frac{5}{9} = \frac{2}{3}\cdot\frac{3}{3}+\frac{5}{9} = \frac{6}{9}+\frac{5}{9} = \frac{11}{9}$$

$$6+\frac{3}{x} = \frac{6}{1}\cdot\frac{x}{x}+\frac{3}{x} = \frac{6x}{x}+\frac{3}{x} = \frac{6x+3}{x}$$

Repeat the process for each denominator.

$$\frac{1}{4}+\frac{1}{12} = \frac{1}{4}\cdot\frac{3}{3}+\frac{1}{12} = \frac{3}{12}+\frac{1}{12} = \frac{4}{12}$$

$$\frac{x}{4}+\frac{1}{8} = \frac{x}{4}\cdot\frac{2}{2}+\frac{1}{8} = \frac{2x}{8}+\frac{1}{8} = \frac{2x+1}{8}$$

Step 2 Write the equivalent complex fraction as a division problem.

$$\frac{\frac{11}{9}}{\frac{4}{12}} = \frac{11}{9}\div\frac{4}{12}$$

$$\frac{\frac{6x+3}{x}}{\frac{2x+1}{8}} = \frac{6x+3}{x}\div\frac{2x+1}{8}$$

Step 3 Now use the definition of division and multiply by the reciprocal. Then write in lowest terms using the fundamental property.

$$= \frac{11}{9}\cdot\frac{12}{4} = \frac{11\cdot 3\cdot 4}{3\cdot 3\cdot 4} = \frac{11}{3}$$

$$= \frac{6x+3}{x}\cdot\frac{8}{2x+1} = \frac{3(2x+1)}{x}\cdot\frac{8}{2x+1} = \frac{24}{x}$$

Work Problem 1 at the Side. ▶

---

1 Simplify each complex fraction using Method 1.

GS **(a)** $\dfrac{\frac{2}{5}+\frac{1}{4}}{\frac{1}{2}+\frac{1}{3}}$

Step 1
Write the numerator as a single fraction. ____

Write the denominator as a single fraction. ____

Step 2
Write the equivalent fraction as a division problem.

____

Step 3
Write the division problem as a multiplication problem.

____

Multiply and write the answer in lowest terms. ____

(b) $\dfrac{6+\frac{1}{x}}{5-\frac{2}{x}}$

(c) $\dfrac{9-\frac{4}{p}}{\frac{2}{p}+1}$

Answers

1. (a) $\frac{13}{20}$; $\frac{5}{6}$; $\frac{13}{20}\div\frac{5}{6}$; $\frac{13}{20}\cdot\frac{6}{5}$; $\frac{39}{50}$
(b) $\frac{6x+1}{5x-2}$ (c) $\frac{9p-4}{2+p}$

2 Simplify each complex fraction using Method 1.

(a) $\dfrac{\frac{rs^2}{t}}{\frac{r^2s}{t^2}}$

(b) $\dfrac{\frac{m^2n^3}{p}}{\frac{m^4n}{p^2}}$

3 Simplify using Method 1.

$$\frac{5 + \frac{2}{a-3}}{\frac{1}{a-3} - 2}$$

Answers

2. **(a)** $\frac{st}{r}$ **(b)** $\frac{n^2p}{m^2}$

3. $\frac{5a-13}{7-2a}$

EXAMPLE 2 Simplifying a Complex Fraction (Method 1)

Simplify the complex fraction.

$$\frac{\frac{xp}{q^3}}{\frac{p^2}{qx^2}}$$

The numerator and denominator are single fractions, so use the definition of division and then the fundamental property.

$$\frac{xp}{q^3} \div \frac{p^2}{qx^2}$$

$$= \frac{xp}{q^3} \cdot \frac{qx^2}{p^2} \quad (*)$$

$$= \frac{x \cdot p \cdot q \cdot x^2}{q \cdot q^2 \cdot p \cdot p}$$

$$= \frac{x^3}{pq^2}$$

◀ Work Problem **2** at the Side.

Note

Alternatively, we can simplify equation (*) in **Example 2** as follows.

$$\frac{xp}{q^3} \cdot \frac{qx^2}{p^2} = x^{1+2}p^{1-2}q^{1-3}$$ Product and quotient rules for exponents; Confirm that the same answer results.

EXAMPLE 3 Simplifying a Complex Fraction (Method 1)

Simplify the complex fraction.

$$\frac{\frac{3}{x+2} - 4}{\frac{2}{x+2} + 1}$$ Find a common denominator before subtracting in the numerator or adding in the denominator.

$$= \frac{\frac{3}{x+2} - \frac{4(x+2)}{x+2}}{\frac{2}{x+2} + \frac{1(x+2)}{x+2}}$$ Write both second terms with a denominator of $x + 2$.

$$= \frac{\frac{3 - 4(x+2)}{x+2}}{\frac{2 + 1(x+2)}{x+2}}$$ Subtract in the numerator. Add in the denominator.

$$= \frac{\frac{3 - 4x - 8}{x+2}}{\frac{2 + x + 2}{x+2}}$$ Be careful with signs. Distributive property

$$= \frac{\frac{-5 - 4x}{x+2}}{\frac{4 + x}{x+2}}$$ Combine like terms.

$$= \frac{-5 - 4x}{x+2} \cdot \frac{x+2}{4+x}$$ Multiply by the reciprocal of the denominator (divisor).

$$= \frac{-5 - 4x}{4 + x}$$ Divide out the common factor.

◀ Work Problem **3** at the Side.

OBJECTIVE 3 Simplify a complex fraction by multiplying numerator and denominator by the least common denominator (Method 2). If we multiply both the numerator and the denominator of a complex fraction by the LCD of all the fractions within the complex fraction, the result will no longer be complex. This is Method 2.

Method 2 for Simplifying a Complex Fraction

Step 1 Find the LCD of all fractions within the complex fraction.

Step 2 Multiply both the numerator and the denominator of the complex fraction by this LCD using the distributive property as necessary. Write in lowest terms.

EXAMPLE 4 Simplifying Complex Fractions (Method 2)

Simplify each complex fraction.

(a) $\dfrac{\frac{2}{3}+\frac{5}{9}}{\frac{1}{4}+\frac{1}{12}}$ (In **Example 1,** we simplified these same complex fractions using Method 1.) **(b)** $\dfrac{6+\frac{3}{x}}{\frac{x}{4}+\frac{1}{8}}$

Step 1 Find the LCD for all denominators in the complex fraction.

The LCD for 3, 9, 4, and 12 is 36. | The LCD for x, 4, and 8 is $8x$.

Step 2 Multiply both numerator and denominator by the LCD.

$$\dfrac{\frac{2}{3}+\frac{5}{9}}{\frac{1}{4}+\frac{1}{12}}$$

$$=\dfrac{36\left(\frac{2}{3}+\frac{5}{9}\right)}{36\left(\frac{1}{4}+\frac{1}{12}\right)}$$

$$=\dfrac{36\left(\frac{2}{3}\right)+36\left(\frac{5}{9}\right)}{36\left(\frac{1}{4}\right)+36\left(\frac{1}{12}\right)} \quad \text{Multiply each term by 36.}$$

$$=\dfrac{24+20}{9+3} \quad \text{Multiply.}$$

$$=\dfrac{44}{12} \quad \text{Add.}$$

$$=\dfrac{4\cdot 11}{4\cdot 3} \quad \text{Factor.}$$

$$=\dfrac{11}{3} \quad \text{Same answer as in Example 1(a)}$$

$$\dfrac{6+\frac{3}{x}}{\frac{x}{4}+\frac{1}{8}}$$

$$=\dfrac{8x\left(6+\frac{3}{x}\right)}{8x\left(\frac{x}{4}+\frac{1}{8}\right)}$$

$$=\dfrac{8x(6)+8x\left(\frac{3}{x}\right)}{8x\left(\frac{x}{4}\right)+8x\left(\frac{1}{8}\right)} \quad \text{Multiply each term by } 8x.$$

$$=\dfrac{48x+24}{2x^2+x} \quad \text{Multiply.}$$

$$=\dfrac{24(2x+1)}{x(2x+1)} \quad \text{Factor.}$$

$$=\dfrac{24}{x} \quad \text{Same answer as in Example 1(b)}$$

Work Problem 4 at the Side. ▶

---

4 Simplify each complex fraction using Method 2.

GS **(a)** $\dfrac{\frac{2}{3}-\frac{1}{4}}{\frac{4}{9}+\frac{1}{2}}$

Step 1
The LCD for ____, ____, ____, and ____ is ____.

Refer to Step 2 in **Example 4,** and simplify the complex fraction.

(b) $\dfrac{2-\frac{6}{a}}{3+\frac{4}{a}}$

(c) $\dfrac{9-\frac{4}{p}}{\frac{2}{p}+1}$

(This is the same complex fraction simplified in **Margin Problem 1(c)** using Method 1. Compare the answers.)

Answers

4. (a) 3; 4; 9; 2; 36; $\frac{15}{34}$
(b) $\frac{2a-6}{3a+4}$ (c) $\frac{9p-4}{2+p}$

5 Simplify using Method 2.

$$\frac{\frac{2}{5x} - \frac{3}{x^2}}{\frac{7}{4x} + \frac{1}{2x^2}}$$

The LCD for ____, ____, ____, and ____ is ____.

Simplify the complex fraction.

Answer

5. $5x$; x^2; $4x$; $2x^2$; $20x^2$;
$\frac{8x - 60}{35x + 10}$, or $\frac{4(2x - 15)}{5(7x + 2)}$

EXAMPLE 5 Simplifying a Complex Fraction (Method 2)

Simplify the complex fraction.

$$\frac{\frac{3}{5m} - \frac{2}{m^2}}{\frac{9}{2m} + \frac{3}{4m^2}}$$ The LCD for $5m$, m^2, $2m$, and $4m^2$ is $20m^2$.

$$= \frac{20m^2\left(\frac{3}{5m} - \frac{2}{m^2}\right)}{20m^2\left(\frac{9}{2m} + \frac{3}{4m^2}\right)}$$ Multiply numerator and denominator by $20m^2$.

$$= \frac{20m^2\left(\frac{3}{5m}\right) - 20m^2\left(\frac{2}{m^2}\right)}{20m^2\left(\frac{9}{2m}\right) + 20m^2\left(\frac{3}{4m^2}\right)}$$ Distributive property

$$= \frac{12m - 40}{90m + 15}, \quad \text{or} \quad \frac{4(3m - 10)}{5(18m + 3)}$$ Multiply and factor.

The factored form indicates that there are no common factors to divide out. The two forms are equivalent, so either can be given as the answer.

◀ Work Problem 5 at the Side.

Note

Either method can be used to simplify a complex fraction. A little more or less work may be involved based on the method selected, but the same answer will result if the method is applied correctly **(Examples 1 and 3).**

- We prefer Method 1 for problems that involve the quotient of two fractions, like **Example 2.**
- We prefer Method 2 for complex fractions that have sums or differences in the numerators or denominators, like **Examples 1, 3, 4, and 5.**

EXAMPLE 6 Simplifying Complex Fractions

Simplify each complex fraction. Use either method.

(a) $$\frac{\frac{x+2}{x-3}}{\frac{x^2-4}{x^2-9}}$$ This is a quotient of two rational expressions. We use Method 1.

$$= \frac{x+2}{x-3} \div \frac{x^2-4}{x^2-9}$$ Write as a division problem.

$$= \frac{x+2}{x-3} \cdot \frac{x^2-9}{x^2-4}$$ Multiply by the reciprocal.

$$= \frac{(x+2)(x+3)(x-3)}{(x-3)(x+2)(x-2)}$$ Multiply, and then factor.

$$= \frac{x+3}{x-2}$$ Divide out the common factors.

Continued on Next Page

(b) $\dfrac{\frac{1}{y}+\frac{2}{y+2}}{\frac{4}{y}-\frac{3}{y+2}}$ There are sums and differences in the numerator and denominator. We use Method 2.

$$=\frac{\left(\frac{1}{y}+\frac{2}{y+2}\right)y(y+2)}{\left(\frac{4}{y}-\frac{3}{y+2}\right)y(y+2)}$$ Multiply numerator and denominator by the LCD, $y(y+2)$. Because y appears in two denominators, it must be a factor in the LCD.

$$=\frac{\left(\frac{1}{y}\right)y(y+2)+\left(\frac{2}{y+2}\right)y(y+2)}{\left(\frac{4}{y}\right)y(y+2)-\left(\frac{3}{y+2}\right)y(y+2)}$$ Distributive property

$$=\frac{1(y+2)+2y}{4(y+2)-3y}$$ Multiply and simplify.

$$=\frac{y+2+2y}{4y+8-3y}$$ Distributive property

$$=\frac{3y+2}{y+8}$$ Combine like terms.

(c) $\dfrac{1-\frac{2}{x}-\frac{3}{x^2}}{1-\frac{5}{x}+\frac{6}{x^2}}$ As in **Example 6(b),** there are sums and differences in the numerator and denominator. We use Method 2.

$$=\frac{\left(1-\frac{2}{x}-\frac{3}{x^2}\right)x^2}{\left(1-\frac{5}{x}+\frac{6}{x^2}\right)x^2}$$ Multiply numerator and denominator by the LCD, x^2.

$$=\frac{x^2-2x-3}{x^2-5x+6}$$ Distributive property

$$=\frac{(x-3)(x+1)}{(x-3)(x-2)}$$ Factor.

$$=\frac{x+1}{x-2}$$ Divide out the common factor.

Work Problem 6 at the Side. ▶

6 Simplify each complex fraction. Use either method.

(a) $\dfrac{\frac{2x+3}{x-4}}{\frac{4x^2-9}{x^2-16}}$

(b) $\dfrac{\frac{1}{x}+\frac{2}{x-1}}{\frac{2}{x}-\frac{4}{x-1}}$

(c) $\dfrac{1-\frac{2}{x}-\frac{15}{x^2}}{1+\frac{5}{x}+\frac{6}{x^2}}$

Answers

6. **(a)** $\dfrac{x+4}{2x-3}$ **(b)** $\dfrac{3x-1}{-2x-2}$ **(c)** $\dfrac{x-5}{x+2}$

7.5 Exercises

FOR EXTRA HELP — *Go to MyMathLab for worked-out, step-by-step solutions to exercises enclosed in a square and video solutions to ▶ exercises.*

CONCEPT CHECK *Answer each question.*

1. In a fraction, what operation does the fraction bar represent?

2. What property of real numbers justifies Method 2 of simplifying complex fractions?

GS *Consider the following complex fraction.*

$$\frac{\frac{1}{2}-\frac{1}{3}}{\frac{5}{6}-\frac{1}{12}}$$

3. Answer each part, outlining Method 1 for simplifying this complex fraction.

(a) To combine the terms in the numerator, we must find the LCD of $\frac{1}{2}$ and $\frac{1}{3}$.
What is this LCD? ______
Determine the simplified form of the numerator of the complex fraction. ______

(b) To combine the terms in the denominator, we must find the LCD of $\frac{5}{6}$ and $\frac{1}{12}$.
What is this LCD? ______
Determine the simplified form of the denominator of the complex fraction. ______

(c) Now use the results from parts (a) and (b) to write the complex fraction as a division problem using the symbol ÷. ______

(d) Perform the operation from part (c) to obtain the final simplification. ______

4. Answer each part, outlining Method 2 for simplifying this complex fraction.

(a) We must determine the LCD of all the fractions within the complex fraction.
What is this LCD? ______

(b) Multiply every term in the complex fraction by the LCD found in part (a), but at this time do not combine the terms in the numerator and the denominator. ______

(c) Now combine the terms from part (b) to obtain the simplified form of the complex fraction. ______

Simplify each complex fraction. Use either method. ***See Examples 1–6.***

5. $\dfrac{-\frac{4}{3}}{\frac{2}{9}}$

6. $\dfrac{-\frac{5}{6}}{\frac{5}{4}}$

7. $\dfrac{\frac{5}{8}+\frac{2}{3}}{\frac{7}{3}-\frac{1}{4}}$

8. $\dfrac{\frac{6}{5}-\frac{1}{9}}{\frac{2}{5}+\frac{5}{3}}$

9. ▶ $\dfrac{\frac{x}{y^2}}{\frac{x^2}{y}}$

10. $\dfrac{\frac{p^4}{r}}{\frac{p^2}{r^2}}$

11. $\dfrac{\frac{p}{6q^2}}{\frac{p^2}{q}}$

12. $\dfrac{\frac{a}{x}}{\frac{a^2}{2x}}$

13. $\dfrac{\frac{4a^4b^3}{3a}}{\frac{2ab^4}{b^2}}$

14. $\dfrac{\frac{2r^4t^2}{3t}}{\frac{5r^2t^5}{3r}}$

15. $\dfrac{\frac{m+2}{3}}{\frac{m-4}{m}}$

16. $\dfrac{\frac{q-5}{q}}{\frac{q+5}{3}}$

17. $\dfrac{\frac{2}{x}-3}{\frac{2-3x}{2}}$

18. $\dfrac{6+\frac{2}{r}}{\frac{3r+1}{4}}$

19. $\dfrac{\frac{1}{x}+x}{\frac{x^2+1}{8}}$

20. $\dfrac{\frac{3}{m}-m}{\frac{3-m^2}{4}}$

21. $\dfrac{a-\frac{5}{a}}{a+\frac{1}{a}}$

22. $\dfrac{q+\frac{1}{q}}{q+\frac{4}{q}}$

23. $\dfrac{\frac{1}{2}+\frac{1}{p}}{\frac{2}{3}+\frac{1}{p}}$

24. $\dfrac{\frac{3}{4}-\frac{1}{r}}{\frac{1}{5}+\frac{1}{r}}$

25. $\dfrac{\frac{1}{4}-\frac{1}{a^2}}{\frac{1}{2}+\frac{1}{a}}$

26. $\dfrac{\frac{1}{9}-\frac{1}{m^2}}{\frac{1}{3}+\frac{1}{m}}$

27. $\dfrac{\frac{2}{p^2}-\frac{3}{5p}}{\frac{4}{p}+\frac{1}{4p}}$

28. $\dfrac{\frac{2}{m^2}-\frac{3}{m}}{\frac{2}{5m^2}+\frac{1}{3m}}$

29. $\dfrac{\frac{t}{t+2}}{\frac{4}{t^2-4}}$

30. $\dfrac{\frac{m}{m+1}}{\frac{3}{m^2-1}}$

31. $\dfrac{\frac{1}{m+1}-1}{\frac{1}{m+1}+1}$

32. $\dfrac{\frac{2}{p-1}+2}{\frac{3}{p-1}-2}$

33. $\dfrac{2+\frac{1}{x}-\frac{28}{x^2}}{3+\frac{13}{x}+\frac{4}{x^2}}$

34. $\dfrac{4-\frac{11}{x}-\frac{3}{x^2}}{2-\frac{1}{x}-\frac{15}{x^2}}$

35. $\dfrac{\frac{1}{m-1}+\frac{2}{m+2}}{\frac{2}{m+2}-\frac{1}{m-3}}$

36. $\dfrac{\frac{5}{r+3}-\frac{1}{r-1}}{\frac{2}{r+2}+\frac{3}{r+3}}$

Relating Concepts (Exercises 37–40) For Individual or Group Work

Recall that slope measures the steepness of a line and is calculated using the formula

$$\textbf{slope } m = \frac{\textbf{change in } y}{\textbf{change in } x} = \frac{y_2 - y_1}{x_2 - x_1} \quad \textbf{(where } x_1 \neq x_2\textbf{).}$$

Find the slope of the line that passes through each pair of points. This will involve simplifying complex fractions.

37. $\left(\frac{3}{4}, \frac{1}{3}\right)$ and $\left(\frac{5}{4}, \frac{10}{3}\right)$

38. $\left(\frac{1}{2}, \frac{5}{12}\right)$ and $\left(\frac{1}{4}, \frac{1}{3}\right)$

39. $\left(-\frac{2}{9}, \frac{5}{18}\right)$ and $\left(\frac{1}{18}, -\frac{5}{9}\right)$

40. $\left(-\frac{4}{5}, \frac{1}{2}\right)$ and $\left(-\frac{3}{10}, -\frac{1}{5}\right)$

7.6 Solving Equations with Rational Expressions

OBJECTIVES

1. Distinguish between operations with rational expressions and equations with terms that are rational expressions.
2. Solve equations with rational expressions.
3. Solve a formula for a specified variable.

OBJECTIVE 1 Distinguish between operations with rational expressions and equations with terms that are rational expressions. We emphasize the distinction between sums and differences of terms with rational coefficients—that is, rational *expressions*—and *equations* with terms that are rational expressions.

Sums and differences are expressions to simplify. Equations are solved.

EXAMPLE 1 Distinguishing between an Expression and an Equation

Identify each of the following as an *expression* or an *equation*. Then *simplify the expression* or *solve the equation*.

(a) $\frac{3}{4}x - \frac{2}{3}x$ This is a difference of two terms. It represents an *expression* to simplify—there is no equality symbol.

$= \frac{3 \cdot 3}{3 \cdot 4}x - \frac{4 \cdot 2}{4 \cdot 3}x$ The LCD is 12. Write each coefficient with this LCD.

$= \frac{9}{12}x - \frac{8}{12}x$ Multiply.

$= \frac{1}{12}x$ Combine like terms, using the distributive property: $\frac{9}{12}x - \frac{8}{12}x = \left(\frac{9}{12} - \frac{8}{12}\right)x$.

(b) $\frac{3}{4}x - \frac{2}{3}x = \frac{1}{2}$ Because there is an equality symbol, this is an *equation* to be solved.

$12\left(\frac{3}{4}x - \frac{2}{3}x\right) = 12\left(\frac{1}{2}\right)$ Use the multiplication property of equality to clear the fractions. Multiply by 12, the LCD.

$12\left(\frac{3}{4}x\right) - 12\left(\frac{2}{3}x\right) = 12\left(\frac{1}{2}\right)$ Distributive property

Multiply *each* term by 12.

$9x - 8x = 6$ Multiply.

$x = 6$ Combine like terms.

CHECK $\frac{3}{4}x - \frac{2}{3}x = \frac{1}{2}$ Original equation

$\frac{3}{4}(6) - \frac{2}{3}(6) \stackrel{?}{=} \frac{1}{2}$ Let $x = 6$.

$\frac{9}{2} - 4 \stackrel{?}{=} \frac{1}{2}$ Multiply.

$\frac{1}{2} = \frac{1}{2}$ ✓ True

A true statement results, so $\{6\}$ is the solution set of the equation.

◀ Work Problem 1 at the Side.

1 Identify each of the following as an *expression* or an *equation*. Then *simplify the expression* or *solve the equation*.

(a) $\frac{2}{3}x - \frac{4}{9}x$

(b) $\frac{2}{3}x - \frac{4}{9}x = 2$

Answers

1. (a) expression; $\frac{2}{9}x$
(b) equation; $\{9\}$

The ideas of **Example 1** can be summarized as follows.

Uses of the LCD

When adding or subtracting rational expressions, keep the LCD throughout the simplification. **(See Example 1(a).)**

When solving an equation with terms that are rational expressions, multiply each side by the LCD so that denominators are eliminated. **(See Example 1(b).)**

OBJECTIVE 2 Solve equations with rational expressions. When an equation involves fractions as in **Example 1(b),** we use the multiplication property of equality to clear the fractions. We choose the LCD of all denominators as the multiplier so the resulting equation contains no fractions.

EXAMPLE 2 Solving an Equation with Rational Expressions

Solve, and check the solution.

$$\frac{x}{3}+\frac{x}{4}=10+x$$

$$12\left(\frac{x}{3}+\frac{x}{4}\right)=12(10+x) \quad \text{Multiply by the LCD, 12, to clear the fractions.}$$

$$12\left(\frac{x}{3}\right)+12\left(\frac{x}{4}\right)=12(10)+12x \quad \text{Distributive property}$$

$$4x+3x=120+12x \quad \text{Multiply.}$$

$$7x=120+12x \quad \text{Combine like terms.}$$

$$-5x=120 \quad \text{Subtract } 12x.$$

$$x=-24 \quad \text{Divide by } -5.$$

CHECK

$$\frac{x}{3}+\frac{x}{4}=10+x \quad \text{Original equation}$$

$$\frac{-24}{3}+\frac{-24}{4}\stackrel{?}{=}10+(-24) \quad \text{Let } x=-24.$$

$$-8+(-6)\stackrel{?}{=}-14 \quad \text{Divide. Add.}$$

$$-14=-14 \ \checkmark \quad \text{True}$$

A true statement results, so the solution set is $\{-24\}$.

Work Problem 2 at the Side. ▶

CAUTION

Be careful not to confuse the following procedures.

- In **Examples 2 and 3,** we use the multiplication property of equality to multiply each side of an ***equation*** by the LCD.
- In our work with complex fractions, we used the fundamental property to multiply a *fraction* (an ***expression***) by another fraction that had the LCD as both its numerator and denominator.

2 Solve each equation, and check the solutions.

(a) $\frac{x}{2}-\frac{x}{3}=\frac{5}{6}$

(b) $\frac{x}{6}+\frac{x}{3}=6+x$

Answers

2. **(a)** $\{5\}$ **(b)** $\{-12\}$

3 Solve each equation, and check the solutions.

GS **(a)** $\frac{k}{6} - \frac{k+1}{4} = -\frac{1}{2}$

Multiply by the LCD, ____.

$$\_\_\left(\frac{k}{6} - \frac{k+1}{4}\right) = \_\_\left(-\frac{1}{2}\right)$$

Complete the solution.

(b) $\frac{2m-3}{5} - \frac{m}{3} = -\frac{6}{5}$

4 Solve the equation, and check the proposed solution.

$$1 - \frac{2}{x+1} = \frac{2x}{x+1}$$

Answers

3. **(a)** 12; 12; 12; $\{3\}$ **(b)** $\{-9\}$

4. $\varnothing$ (When the equation is solved, -1 is a proposed solution. However, because $x = -1$ leads to a 0 denominator in the original equation, there is no solution.)

EXAMPLE 3 Solving an Equation with Rational Expressions

Solve, and check the solution.

$$\frac{p}{2} - \frac{p-1}{3} = 1$$

$$6\left(\frac{p}{2} - \frac{p-1}{3}\right) = 6(1) \quad \text{Multiply by the LCD, 6, to clear the fractions.}$$

$$6\left(\frac{p}{2}\right) - 6\left(\frac{p-1}{3}\right) = 6(1) \quad \text{Distributive property}$$

$$3p - 2(p-1) = 6 \quad \text{(Use parentheses around } p-1 \text{ to avoid errors.)}$$

$$3p - 2(p) - 2(-1) = 6 \quad \text{Distributive property}$$

Be careful with signs.

$$3p - 2p + 2 = 6 \quad \text{Multiply.}$$

$$p = 4 \quad \text{Combine like terms. Subtract 2.}$$

Check that $\{4\}$ is the solution set by replacing p with 4 in the original equation.

◀ Work Problem 3 at the Side.

Recall that division by 0 is undefined. ***Therefore, when solving an equation with rational expressions that have variables in the denominator, the solution cannot be a number that makes the denominator equal 0.***

A value of the variable that appears to be a solution after both sides of an equation with rational expressions are multiplied by a variable expression is a **proposed solution.** ***All proposed solutions must be checked.***

EXAMPLE 4 Solving an Equation with Rational Expressions

Solve, and check the proposed solution.

$$\frac{x}{x-2} = \frac{2}{x-2} + 2 \quad x \text{ cannot equal 2 because 2 causes both denominators to equal 0.}$$

$$(x-2)\left(\frac{x}{x-2}\right) = (x-2)\left(\frac{2}{x-2} + 2\right) \quad \text{Multiply each side by the LCD, } x-2.$$

$$(x-2)\left(\frac{x}{x-2}\right) = (x-2)\left(\frac{2}{x-2}\right) + (x-2)(2) \quad \text{Distributive property}$$

$$x = 2 + 2x - 4 \quad \text{Simplify.}$$

$$x = -2 + 2x \quad \text{Combine like terms.}$$

$$-x = -2 \quad \text{Subtract } 2x.$$

Proposed solution → $x = 2$ Multiply by -1.

CHECK

$$\frac{x}{x-2} = \frac{2}{x-2} + 2 \quad \text{Original equation}$$

$$\frac{2}{2-2} \stackrel{?}{=} \frac{2}{2-2} + 2 \quad \text{Let } x = 2.$$

$$\frac{2}{0} \stackrel{?}{=} \frac{2}{0} + 2 \quad \text{Subtract in the denominators.}$$

Division by 0 is undefined.

Thus, the proposed solution 2 must be rejected. The solution set is $\varnothing$.

◀ Work Problem 4 at the Side.

A proposed solution that is not an actual solution of the original equation, such as 2 in **Example 4,** is an **extraneous solution,** or **extraneous value.** Some students like to determine which numbers cannot be solutions *before* solving the equation, as we did at the beginning of **Example 4.**

Solving an Equation with Rational Expressions

Step 1 **Multiply each side of the equation by the LCD** (This clears the equation of fractions.) Be sure to distribute to *every* term on *both* sides of the equation.

Step 2 **Solve** the resulting equation for proposed solutions.

Step 3 **Check** each proposed solution by substituting it in the original equation. Reject any value that causes a denominator to equal 0.

EXAMPLE 5 Solving an Equation with Rational Expressions

Solve, and check the proposed solution.

$$\frac{2}{x^2 - x} = \frac{1}{x^2 - 1}$$

Step 1

$$\frac{2}{x(x-1)} = \frac{1}{(x+1)(x-1)}$$ Factor the denominators to find the LCD.

The LCD is $x(x+1)(x-1)$. ***Notice that* 0, −1, *and* 1 *cannot be solutions of this equation.*** Otherwise, a denominator will equal 0.

$$x(x+1)(x-1)\frac{2}{x(x-1)} = x(x+1)(x-1)\frac{1}{(x+1)(x-1)}$$ Multiply by the LCD to clear the fractions.

Step 2

$2(x+1) = x$ Divide out the common factors.

$2x + 2 = x$ Distributive property

$x + 2 = 0$ Subtract x.

Proposed solution → $x = -2$ Subtract 2.

Step 3 The proposed solution is −2, which does not make any denominator equal 0.

CHECK $\frac{2}{x^2 - x} = \frac{1}{x^2 - 1}$ Original equation

$\frac{2}{(-2)^2 - (-2)} \stackrel{?}{=} \frac{1}{(-2)^2 - 1}$ Let $x = -2$.

$\frac{2}{4+2} \stackrel{?}{=} \frac{1}{4-1}$ Apply the exponents; Definition of subtraction

$\frac{1}{3} = \frac{1}{3}$ ✓ True

A true statement results, so the solution set is $\{-2\}$.

Work Problem 5 at the Side. ▶

5 Solve each equation, and check the proposed solutions.

(a) $\frac{4}{x^2 - 3x} = \frac{1}{x^2 - 9}$

GS **(b)** $\frac{2}{p^2 - 2p} = \frac{3}{p^2 - p}$

Step 1
Factor the denominators.

$p^2 - 2p = p(\_\_\_\_)$

$p^2 - \ \ p = p(\_\_\_\_)$

The LCD is _____.

The numbers ___, ___, and ___ cannot be solutions.

Complete the steps to solve the equation.

Answers

5. **(a)** $\{-4\}$
(b) $p-2$; $p-1$; $p(p-2)(p-1)$; 0; 1; 2 (Order of the numbers 0, 1, and 2 does not matter.); $\{4\}$

6 Solve each equation, and check the proposed solutions.

(a) $\frac{2y}{y^2-25}=\frac{8}{y+5}-\frac{1}{y-5}$

(b) $\frac{8r}{4r^2-1}=\frac{3}{2r+1}+\frac{3}{2r-1}$

Answers

6. **(a)** $\{9\}$ **(b)** $\{0\}$

EXAMPLE 6 **Solving an Equation with Rational Expressions**

Solve, and check the proposed solution.

$$\frac{2m}{m^2-4}+\frac{1}{m-2}=\frac{2}{m+2}$$

$$\frac{2m}{(m+2)(m-2)}+\frac{1}{m-2}=\frac{2}{m+2}$$ Factor the first denominator on the left to find the LCD, $(m+2)(m-2)$.

Notice that -2 and 2 cannot be solutions of the equation.

$$(m+2)(m-2)\left(\frac{2m}{(m+2)(m-2)}+\frac{1}{m-2}\right)$$
$$=(m+2)(m-2)\frac{2}{m+2}$$ Multiply by the LCD.

$$(m+2)(m-2)\frac{2m}{(m+2)(m-2)}+(m+2)(m-2)\frac{1}{m-2}$$
$$=(m+2)(m-2)\frac{2}{m+2}$$ Distributive property

$$2m+m+2=2(m-2)$$ Divide out the common factors.

$$3m+2=2m-4$$ Combine like terms; distributive property

$$m+2=-4$$ Subtract $2m$.

$$m=-6$$ Subtract 2.

CHECK

$$\frac{2m}{m^2-4}+\frac{1}{m-2}=\frac{2}{m+2}$$ Original equation

$$\frac{2(-6)}{(-6)^2-4}+\frac{1}{-6-2}\stackrel{?}{=}\frac{2}{-6+2}$$ Let $m=-6$.

$$\frac{-12}{32}+\frac{1}{-8}\stackrel{?}{=}\frac{2}{-4}$$ Apply the exponent. Subtract and add.

$$-\frac{1}{2}=-\frac{1}{2}\ ✓$$ True

The solution set is $\{-6\}$.

◀ Work Problem **6** at the Side.

EXAMPLE 7 **Solving an Equation with Rational Expressions**

Solve, and check the proposed solution(s).

$$\frac{1}{x-1}+\frac{1}{2}=\frac{2}{x^2-1}$$

$$\frac{1}{x-1}+\frac{1}{2}=\frac{2}{(x+1)(x-1)}$$

$x\neq 1, -1$ or a denominator is 0.

Factor the denominator on the right. The LCD is $2(x+1)(x-1)$.

Continued on Next Page

$$2(x+1)(x-1)\left(\frac{1}{x-1}+\frac{1}{2}\right)=2(x+1)(x-1)\frac{2}{(x+1)(x-1)}$$
Multiply by the LCD.

$$2(x+1)(x-1)\frac{1}{x-1}+2(x+1)(x-1)\frac{1}{2}$$

$$=2(x+1)(x-1)\frac{2}{(x+1)(x-1)}$$
Distributive property

$$2(x+1)+(x+1)(x-1)=2(2)$$ Divide out the common factors.

$$2x+2+x^2-1=4$$ Distributive property; Multiply.

Write in standard form. $$x^2+2x-3=0$$ Subtract 4. Combine like terms.

$$(x+3)(x-1)=0$$ Factor.

$$x+3=0 \quad \text{or} \quad x-1=0$$ Zero-factor property

$$x=-3 \quad \text{or} \quad x=1$$ ⟵ Proposed solutions

Because 1 makes an original denominator equal 0, the proposed solution 1 is an extraneous value. Check that -3 is a solution.

CHECK

$$\frac{1}{x-1}+\frac{1}{2}=\frac{2}{x^2-1}$$ Original equation

$$\frac{1}{-3-1}+\frac{1}{2}\stackrel{?}{=}\frac{2}{(-3)^2-1}$$ Let $x=-3$.

$$\frac{1}{-4}+\frac{1}{2}\stackrel{?}{=}\frac{2}{9-1}$$ Subtract. Apply the exponent.

$$\frac{1}{4}=\frac{1}{4}\ \checkmark$$ True

The check shows that the solution set is $\{-3\}$.

Work Problem 7 at the Side. ▶

EXAMPLE 8 Solving an Equation with Rational Expressions

Solve, and check the proposed solution(s).

$$\frac{1}{k^2+4k+3}+\frac{1}{2k+2}=\frac{3}{4k+12}$$

$$\frac{1}{(k+1)(k+3)}+\frac{1}{2(k+1)}=\frac{3}{4(k+3)}$$
Factor each denominator. The LCD is $4(k+1)(k+3)$.

$k \neq -1, -3$

$$4(k+1)(k+3)\left(\frac{1}{(k+1)(k+3)}+\frac{1}{2(k+1)}\right)$$

$$=4(k+1)(k+3)\frac{3}{4(k+3)}$$
Multiply by the LCD.

Continued on Next Page

7 Solve each equation, and check the proposed solution(s).

(a) $$\frac{3}{m^2-9}=\frac{1}{2(m-3)}-\frac{1}{4}$$

(b) $$\frac{1}{x-1}-\frac{2}{x+3}=\frac{x^2+3x}{x^2+2x-3}$$

Answers

7. **(a)** $\{-1\}$ **(b)** $\{-5\}$

$$4(k+1)(k+3)\frac{1}{(k+1)(k+3)} + 2 \cdot 2(k+1)(k+3)\frac{1}{2(k+1)}$$

Do *not* add 4 + 2 here.

$$= 4(k+1)(k+3)\frac{3}{4(k+3)} \quad \text{Distributive property}$$

$$4 + 2(k+3) = 3(k+1) \quad \text{Divide out the common factors.}$$

$$4 + 2k + 6 = 3k + 3 \quad \text{Distributive property}$$

$$2k + 10 = 3k + 3 \quad \text{Combine like terms.}$$

$$10 = k + 3 \quad \text{Subtract } 2k.$$

$$7 = k \quad \text{Subtract 3.}$$

The proposed solution, 7, does not make an original denominator equal 0. A check shows that the algebra is correct, so the solution set is $\{7\}$.

◀ Work Problem 8 at the Side.

8 Solve each equation, and check the proposed solution(s).

(a) $\dfrac{5}{k^2+k-2} = \dfrac{1}{3k-3} - \dfrac{1}{k+2}$

(b) $\dfrac{1}{x-2} + \dfrac{1}{5} = \dfrac{2}{5(x^2-4)}$

OBJECTIVE 3 Solve a formula for a specified variable.

When solving a formula for a specified variable, remember to treat the variable for which you are solving as if it were the only variable, and all others as if they were constants.

EXAMPLE 9 Solving for a Specified Variable

Solve the following formula for v.

$$a = \frac{v-w}{t}$$

Our goal is to isolate v.

$$at = \left(\frac{v-w}{t}\right)t \quad \text{Multiply by } t \text{ to clear the fraction.}$$

$$at = v - w \quad \text{Divide out the common factor.}$$

$$at + w = v \quad \text{Add } w.$$

$$v = at + w \quad \text{Interchange sides.}$$

CHECK Substitute $at + w$ for v in the original equation.

$$a = \frac{v-w}{t} \quad \text{Original equation}$$

$$a = \frac{at+w-w}{t} \quad \text{Let } v = at + w.$$

$$a = \frac{at}{t} \quad \text{Combine like terms.}$$

$$a = a \ \checkmark \quad \text{True}$$

A true statement results, so $v = at + w$.

◀ Work Problem 9 at the Side.

9 Solve each formula for the specified variable.

GS **(a)** $r = \dfrac{A-p}{pt}$ for A

The goal is to isolate ____.
Multiply by pt to obtain the equation ________.

Add ____ to obtain the equation

________ $= A$, or ________.

(b) $p = \dfrac{x-y}{z}$ for y

Answers

8. **(a)** $\{-5\}$ **(b)** $\{-4, -1\}$

9. **(a)** A; $rpt = A - p$; p; $p + prt$; $A = p + prt$
(b) $y = x - pz$

EXAMPLE 10 **Solving for a Specified Variable**

Solve the following formula for d.

$$F = \frac{k}{d - D}$$ We must isolate d.

$$F(d - D) = \frac{k}{d - D}(d - D)$$ Multiply by $d - D$ to clear the fraction.

$$F(d - D) = k$$ Divide out the common factor.

$$Fd - FD = k$$ Distributive property

$$Fd = k + FD$$ Add FD.

$$d = \frac{k + FD}{F}$$ Divide by F.

We can write an equivalent form of this answer as follows.

$$d = \frac{k + FD}{F}$$ Answer from above

$$d = \frac{k}{F} + \frac{FD}{F}$$ Definition of addition of fractions: $\frac{a+b}{c} = \frac{a}{c} + \frac{b}{c}$

This form of the answer is also correct.

$$d = \frac{k}{F} + D$$ Divide out the common factor from $\frac{FD}{F}$.

Work Problem 10 at the Side. ▶

EXAMPLE 11 **Solving for a Specified Variable**

Solve the following formula for c.

$$\frac{1}{a} = \frac{1}{b} + \frac{1}{c}$$ Goal: Isolate c, the specified variable.

$$abc\left(\frac{1}{a}\right) = abc\left(\frac{1}{b} + \frac{1}{c}\right)$$ Multiply by the LCD, abc, to clear the fractions.

$$abc\left(\frac{1}{a}\right) = abc\left(\frac{1}{b}\right) + abc\left(\frac{1}{c}\right)$$ Distributive property

$$bc = ac + ab$$ Divide out the common factors.

Pay careful attention here.

$$bc - ac = ab \quad (*)$$ Subtract ac so that both terms with c are on the same side.

$$c(b - a) = ab$$ Factor out c.

$$c = \frac{ab}{b - a}$$ Divide by $b - a$.

Work Problem 11 at the Side. ▶

! CAUTION

In **Example 11,** we transformed to obtain equation (*) that has *both* terms with c on the same side of the equality symbol. This key step enabled us to factor out c on the left and ultimately isolate it.

When solving an equation for a specified variable, be sure that the specified variable appears alone on only one side of the equality symbol in the final equation.

10 Solve the following formula for y.

$$z = \frac{x}{x + y}$$

11 Solve the formula

$$\frac{2}{x} = \frac{1}{y} + \frac{1}{z}$$

for the specified variable.

(a) for z

(b) for y

Answers

10. $y = \frac{x - zx}{z}$, or $y = \frac{x}{z} - x$

11. (a) $z = \frac{xy}{2y - x}$ **(b)** $y = \frac{xz}{2z - x}$

7.6 Exercises

FOR EXTRA HELP — *Go to MyMathLab for worked-out, step-by-step solutions to exercises enclosed in a square and video solutions to ▶ exercises.*

CONCEPT CHECK *Fill in each blank with the correct response.*

1. A value of the variable that appears to be a solution after both sides of an equation with rational expressions are multiplied by a variable expression is a(n) ________ solution. It must be checked in the ________ equation to determine whether it is an actual solution.

2. A proposed solution that is not an actual solution of an original equation is a(n) ________ solution, or ________ value.

Identify each as an expression *or an* equation. *Then* simplify the expression *or* solve the equation. ***See Example 1.***

3. $\frac{7}{8}x + \frac{1}{5}x$ **4.** $\frac{4}{7}x + \frac{3}{5}x$ **5.** $\frac{7}{8}x + \frac{1}{5}x = 1$ **6.** $\frac{4}{7}x + \frac{3}{5}x = 1$

7. $\frac{3}{5}y - \frac{7}{10}y$ **8.** $\frac{3}{5}y - \frac{7}{10}y = 1$ **9.** $\frac{2}{3}x - \frac{9}{4}x = -19$ **10.** $\frac{2}{3}x - \frac{9}{4}x$

Solve each equation, and check the solutions. ***See Examples 2 and 3.***

11. $\frac{2}{3}x + \frac{1}{2}x = -7$ **12.** $\frac{1}{4}x - \frac{1}{3}x = 1$ **13.** $\frac{3x}{5} - 6 = x$ **14.** $\frac{5t}{4} + t = 9$

15. $\frac{4m}{7} + m = 11$ **16.** $a - \frac{3a}{2} = 1$ **17.** $\frac{z-1}{4} = \frac{z+3}{3}$ **18.** $\frac{r-5}{2} = \frac{r+2}{3}$

19. $\frac{3p+6}{8} = \frac{3p-3}{16}$ **20.** $\frac{2z+1}{5} = \frac{7z+5}{15}$ **21.** $\frac{2x+3}{-6} = \frac{3}{2}$ **22.** $\frac{4x+3}{6} = \frac{5}{2}$

23. $\frac{r}{6} - \frac{r-2}{3} = -\frac{4}{3}$ **24.** $\frac{p}{2} - \frac{p-1}{4} = \frac{5}{4}$ **25.** $\frac{q+2}{3} + \frac{q-5}{5} = \frac{7}{3}$ **26.** $\frac{x-6}{6} + \frac{x+2}{8} = \frac{11}{4}$

27. $\frac{a+7}{8} - \frac{a-2}{3} = \frac{4}{3}$

28. $\frac{x+3}{7} - \frac{x+2}{6} = \frac{1}{6}$

29. $\frac{3m}{5} - \frac{3m-2}{4} = \frac{1}{5}$

30. $\frac{8p}{5} - \frac{3p-4}{2} = \frac{5}{2}$

When solving an equation with variables in denominators, we must determine the values that cause these denominators to equal 0, so that we can reject these values if they appear as proposed solutions. Find all values for which at least one denominator is equal to 0. Write answers using the symbol ≠. Do not solve. ***See Examples 4–8.***

31. $\frac{3}{x+2} - \frac{5}{x} = 1$

32. $\frac{7}{x} + \frac{9}{x-4} = 5$

33. $\frac{-1}{(x+3)(x-4)} = \frac{1}{2x+1}$

34. $\frac{8}{(x-7)(x+3)} = \frac{7}{3x-10}$

35. $\frac{4}{x^2+8x-9} + \frac{1}{x^2-4} = 0$

36. $\frac{-3}{x^2+9x-10} - \frac{12}{x^2-49} = 0$

Solve each equation, and check the solutions. ***See Examples 4–8.***

37. $\frac{5}{m} - \frac{3}{m} = 8$

38. $\frac{4}{y} + \frac{1}{y} = 2$

39. $\frac{5}{y} + 4 = \frac{2}{y}$

40. $\frac{11}{q} - 3 = \frac{1}{q}$

41. $\frac{5-2x}{x} = \frac{1}{4}$

42. $\frac{2x+3}{x} = \frac{3}{2}$

43. $\frac{k}{k-4} - 5 = \frac{4}{k-4}$

44. $\frac{-5}{a+5} = \frac{a}{a+5} + 2$

45. $\frac{3}{x-1} + \frac{2}{4x-4} = \frac{7}{4}$

46. $\frac{2}{p+3} + \frac{3}{8} = \frac{5}{4p+12}$

47. $\frac{x}{3x+3} = \frac{2x-3}{x+1} - \frac{2x}{3x+3}$

48. $\frac{2k+3}{k+1} - \frac{3k}{2k+2} = \frac{-2k}{2k+2}$

49. $\frac{2}{m} = \frac{m}{5m+12}$

50. $\frac{x}{4-x} = \frac{2}{x}$

51. $\frac{5x}{14x+3} = \frac{1}{x}$

52. $\frac{m}{8m+3} = \frac{1}{3m}$

53. $\frac{2}{z-1} - \frac{5}{4} = \frac{-1}{z+1}$

54. $\frac{5}{p-2} = 7 - \frac{10}{p+2}$

55. $\frac{4}{x^2-3x}=\frac{1}{x^2-9}$

56. $\frac{2}{t^2-4}=\frac{3}{t^2-2t}$

57. $\frac{-2}{z+5}+\frac{3}{z-5}=\frac{20}{z^2-25}$

58. $\frac{3}{r+3}-\frac{2}{r-3}=\frac{-12}{r^2-9}$

59. $\frac{1}{x+4}+\frac{x}{x-4}=\frac{-8}{x^2-16}$

60. $\frac{x}{x-3}+\frac{4}{x+3}=\frac{18}{x^2-9}$

61. $\frac{2p}{p^2-1}=\frac{2}{p+1}-\frac{1}{p-1}$

62. $\frac{2x}{x^2-16}-\frac{2}{x-4}=\frac{4}{x+4}$

63. $\frac{4}{3x+6}-\frac{3}{x+3}=\frac{8}{x^2+5x+6}$

64. $\frac{-13}{t^2+6t+8}+\frac{4}{t+2}=\frac{3}{2t+8}$

65. $\frac{3x}{x^2+5x+6}=\frac{5x}{x^2+2x-3}-\frac{2}{x^2+x-2}$

66. $\frac{m}{m^2+m-2}=\frac{m}{m^2+3m+2}-\frac{m}{m^2-1}$

67. **CONCEPT CHECK** A student simplified the following expression as shown.

$$\frac{3}{2}t+\frac{5}{7}t$$
$$=14\left(\frac{3}{2}t+\frac{5}{7}t\right)$$
$$=21t+10t$$
$$=31t \quad \text{Incorrect}$$

What Went Wrong? Give the correct answer.

68. **CONCEPT CHECK** A student solved the following formula for r as shown.

$$\frac{1}{r}-\frac{1}{m}=\frac{1}{k}$$
$$rmk\left(\frac{1}{r}-\frac{1}{m}\right)=rmk\left(\frac{1}{k}\right)$$
$$mk-rk=rm$$
$$\frac{mk-rk}{m}=r \quad \text{Incorrect}$$

What Went Wrong? Give the correct answer.

Solve each formula for the specified variable. **See Examples 9–11.**

69. $m=\frac{kF}{a}$ for F

70. $I=\frac{kE}{R}$ for E

71. $m=\frac{kF}{a}$ for a

72. $I=\frac{kE}{R}$ for R

73. $m=\frac{y-b}{x}$ for y

74. $y=\frac{C-Ax}{B}$ for C

75. $I = \dfrac{E}{R + r}$ for R

76. $I = \dfrac{E}{R + r}$ for r

77. $h = \dfrac{2\mathcal{A}}{B + b}$ for b

78. $h = \dfrac{2\mathcal{A}}{B + b}$ for B

79. $d = \dfrac{2S}{n(a + L)}$ for a

80. $d = \dfrac{2S}{n(a + L)}$ for L

81. $\dfrac{1}{x} = \dfrac{1}{y} - \dfrac{1}{z}$ for y

82. $\dfrac{3}{k} = \dfrac{1}{p} + \dfrac{1}{q}$ for q

83. $\dfrac{2}{r} + \dfrac{3}{s} + \dfrac{1}{t} = 1$ for t

84. $\dfrac{5}{p} + \dfrac{2}{q} + \dfrac{3}{r} = 1$ for r

85. $\dfrac{1}{a} - \dfrac{1}{b} - \dfrac{1}{c} = 2$ for c

86. $\dfrac{-1}{x} + \dfrac{1}{y} + \dfrac{1}{z} = 4$ for y

87. $9x + \dfrac{3}{z} = \dfrac{5}{y}$ for z

88. $-3t - \dfrac{4}{p} = \dfrac{6}{s}$ for p

Relating Concepts (Exercises 89–94) For Individual or Group Work

In these exercises, we summarize various concepts involving rational expressions.
Work Exercises 89–94 in order.

Let P, Q, and R be rational expressions defined as follows.

$$P = \frac{6}{x + 3}, \qquad Q = \frac{5}{x + 1}, \qquad R = \frac{4x}{x^2 + 4x + 3}$$

89. Find the values for which each expression is undefined. Write answers using the symbol $\neq$.

(a) P **(b)** Q **(c)** R

90. Find and express $(P \cdot Q) \div R$ in lowest terms.

91. Find the LCD for P, Q, and R.

92. Perform the operations and express $P + Q - R$ in lowest terms.

93. Simplify the complex fraction $\dfrac{P + Q}{R}$.

94. Solve the equation $P + Q = R$.

Summary Exercises *Simplifying Rational Expressions vs. Solving Rational Equations*

Students often confuse *simplifying expressions* with *solving equations*. We review the four operations to ***simplify*** the rational expressions $\frac{1}{x}$ and $\frac{1}{x-2}$ as follows.

Add:

$$\frac{1}{x}+\frac{1}{x-2}$$

$$=\frac{1(x-2)}{x(x-2)}+\frac{x(1)}{x(x-2)}$$ Write with a common denominator.

$$=\frac{x-2+x}{x(x-2)}$$ Add numerators. Keep the same denominator.

$$=\frac{2x-2}{x(x-2)}$$ Combine like terms.

Subtract:

$$\frac{1}{x}-\frac{1}{x-2}$$

$$=\frac{1(x-2)}{x(x-2)}-\frac{x(1)}{x(x-2)}$$ Write with a common denominator.

$$=\frac{x-2-x}{x(x-2)}$$ Subtract numerators. Keep the same denominator.

$$=\frac{-2}{x(x-2)}$$ Combine like terms.

Multiply:

$$\frac{1}{x}\cdot\frac{1}{x-2}$$

$$=\frac{1}{x(x-2)}$$ Multiply numerators. Multiply denominators.

Divide:

$$\frac{1}{x}\div\frac{1}{x-2}$$

$$=\frac{1}{x}\cdot\frac{x-2}{1}$$ Multiply by the reciprocal of the divisor.

$$=\frac{x-2}{x}$$ Multiply numerators. Multiply denominators.

By contrast, consider the following ***equation***.

$$\frac{1}{x}+\frac{1}{x-2}=\frac{3}{4}$$ $x \neq 0, 2$ because a denominator is 0 for these values.

$$4x(x-2)\left(\frac{1}{x}+\frac{1}{x-2}\right)=4x(x-2)\frac{3}{4}$$ Multiply each side by the LCD, $4x(x-2)$, to clear the fractions.

$$4x(x-2)\frac{1}{x}+4x(x-2)\frac{1}{x-2}=4x(x-2)\frac{3}{4}$$ Distributive property

$$4(x-2)+4x=3x(x-2)$$ Divide out the common factors.

$$4x-8+4x=3x^2-6x$$ Distributive property

$$3x^2-14x+8=0$$ Standard form

$$(3x-2)(x-4)=0$$ Factor.

$$3x-2=0 \quad \text{or} \quad x-4=0$$ Zero-factor property

Proposed solutions → $x=\frac{2}{3}$ or $x=4$ Solve for x.

The proposed solutions are $\frac{2}{3}$ and 4. Neither makes a denominator equal 0. Check by substituting each proposed solution into the original equation to confirm that the solution set is $\left\{\frac{2}{3}, 4\right\}$.

Points to Remember when Working with Rational Expressions and Equations

1. When simplifying rational expressions, the fundamental property is applied only after numerators and denominators have been *factored*.
2. When adding and subtracting rational expressions, the common denominator must be kept throughout the problem and in the final result.
3. When simplifying rational expressions, check to see if the answer is in lowest terms. If it is not, use the fundamental property.
4. When solving equations with rational expressions, the LCD is used to clear fractions. Multiply each side by the LCD. (Notice how this use differs from that of the common denominator in Point 2.)
5. When solving equations with rational expressions, reject any proposed solution that causes an original denominator to equal 0.

For each exercise, indicate "expression" *if an expression is to be* simplified *or* "equation" *if an equation is to be* solved. *Then simplify the expression or solve the equation.*

1. $\frac{4}{p}+\frac{6}{p}$

2. $\frac{3x}{x+7}-\frac{x+1}{x+7}$

3. $\frac{1}{x^2+x-2}\div\frac{4x^2}{2x-2}$

4. $\frac{8}{m-5}=2$

5. $\frac{2x^2+x-6}{2x^2-9x+9}\cdot\frac{x^2-2x-3}{x^2-1}$

6. $\frac{2}{k^2-4k}+\frac{3}{k^2-16}$

7. $\frac{x-4}{5}=\frac{x+3}{6}$

8. $\frac{3t^2-t}{6t^2+15t}\div\frac{6t^2+t-1}{2t^2-5t-25}$

9. $\frac{4}{p+2}+\frac{1}{3p+6}$

10. $\frac{1}{x}+\frac{1}{x-3}=-\frac{5}{4}$

11. $\frac{3}{t-1}+\frac{1}{t}=\frac{7}{2}$

12. $\frac{5}{4z}-\frac{2}{3z}$

13. $\frac{2m}{m-4}-\frac{m-12}{4-m}$

14. $\frac{k+2}{3}=\frac{2k-1}{5}$

15. $\frac{1}{m^2+5m+6}+\frac{2}{m^2+4m+3}$

16. $\frac{2k^2-3k}{20k^2-5k}\cdot\frac{4k^2+11k-3}{2k^2-5k+3}$

17. $\frac{2}{x+1}+\frac{5}{x-1}=\frac{10}{x^2-1}$

18. $\frac{x}{x-2}+\frac{3}{x+2}=\frac{8}{x^2-4}$

7.7 Applications of Rational Expressions

OBJECTIVES

1. Solve problems about numbers.
2. Solve problems about distance, rate, and time.
3. Solve problems about work.

We continue to use the six-step problem-solving method introduced earlier.

Solving an Applied Problem

Step 1 **Read** the problem, several times if necessary. *What information is given? What is to be found?*

Step 2 **Assign a variable** to represent the unknown value. Use a sketch, diagram, or table, as needed. Express any other unknown values in terms of the variable.

Step 3 **Write an equation** using the variable expression(s).

Step 4 **Solve** the equation.

Step 5 **State the answer.** Label it appropriately. *Does the answer seem reasonable?*

Step 6 **Check** the answer in the words of the *original* problem.

OBJECTIVE 1 Solve problems about numbers.

EXAMPLE 1 Solving a Problem about an Unknown Number

If the same number is added to both the numerator and the denominator of the fraction $\frac{2}{5}$, the result is equivalent to $\frac{2}{3}$. Find the number.

Step 1 **Read** the problem carefully. We are trying to find a number.

Step 2 **Assign a variable.**

Let $x =$ the number added to the numerator and the denominator.

Step 3 **Write an equation.** The fraction $\frac{2}{5}$ is given. The expression

$$\frac{2+x}{5+x}$$

represents the result of adding the same number x to both the numerator and the denominator. This result is equivalent to $\frac{2}{3}$.

$$\frac{2+x}{5+x} = \frac{2}{3}$$

Step 4 **Solve.**

$$3(5+x)\frac{2+x}{5+x} = 3(5+x)\frac{2}{3} \quad \text{Multiply by the LCD, } 3(5+x).$$

$$3(2+x) = (5+x)2 \quad \text{Divide out the common factors.}$$

$$6 + 3x = 10 + 2x \quad \text{Distributive property}$$

$$x = 4 \quad \text{Subtract } 2x. \text{ Subtract } 6.$$

Step 5 **State the answer.** The number is 4.

Step 6 **Check.** If 4 is added to both the numerator and the denominator of $\frac{2}{5}$, the result is $\frac{2+4}{5+4} = \frac{6}{9}$, which in lowest terms is $\frac{2}{3}$, as required.

◀ Work Problem 1 at the Side.

1 (GS) A certain number is added to the numerator and subtracted from the denominator of $\frac{5}{8}$. The new fraction equals the reciprocal of $\frac{5}{8}$. Find the number.

Step 1
We are trying to find a(n) ________.

Step 2
Let $x =$ the number added to the numerator and ________ from the denominator.

Step 3
The expression ________ represents the new fraction. The reciprocal of $\frac{5}{8}$ is ____. Write an equation.

____

Complete Steps 4–6 to solve the problem. Give the answer.

Answer

1. number; subtracted; $\frac{5+x}{8-x}$; $\frac{8}{5}$; $\frac{5+x}{8-x} = \frac{8}{5}$; The number is 3.

OBJECTIVE 2 Solve problems about distance, rate, and time. If an automobile travels at an average rate of 65 mph for 2 hr, then it travels

$65 \times 2 = 130$ mi. $rt = d$, or $d = rt$ (Relationship between distance, rate, and time)

By solving, in turn, for r and t in the distance formula $d = rt$, we obtain two other equivalent forms of the formula.

Forms of the Distance Formula

$$d = rt \qquad r = \frac{d}{t} \qquad t = \frac{d}{r}$$

EXAMPLE 2 Finding Distance, Rate, or Time

Solve each problem using a form of the distance formula.

(a) The speed (rate) of sound is 1088 ft per sec at sea level at 32°F. Find the distance sound travels in 5 sec under these conditions.

We must find distance, given rate and time, using $d = rt$ (or $rt = d$).

$$\underset{\text{Rate}}{1088} \cdot \underset{\text{Time}}{5} = \underset{\text{Distance}}{5440 \text{ ft}}$$

(b) Ray Harroun won the first Indianapolis 500 race (in 1911), driving a Marmon Wasp at an average rate of 74.60 mph. How long did it take him to complete the 500 mi? (Data from *The World Almanac and Book of Facts*.)

We must find time, given rate and distance, using $t = \frac{d}{r}$ (or $\frac{d}{r} = t$).

$$\frac{500 \leftarrow \text{Distance}}{74.60 \leftarrow \text{Rate}} = 6.70 \text{ hr (rounded)} \leftarrow \text{Time}$$

To convert 0.70 hr to minutes, we multiply by 60 to obtain $0.70(60) = 42$. It took Harroun about 6 hr, 42 min, to complete the race.

(c) At the 2016 Olympic Games, U.S. swimmer Michael Phelps won the men's 100-m butterfly swimming event in 51.14 sec. Find his rate. (Data from www.olympic.org)

We must find rate, given distance and time, using $r = \frac{d}{t}$ (or $\frac{d}{t} = r$).

$$\frac{100 \leftarrow \text{Distance}}{51.14 \leftarrow \text{Time}} = 1.96 \text{ m per sec (rounded)} \leftarrow \text{Rate}$$

Work Problem 2 at the Side. ▶

Problem-Solving Hint

Many applied problems use forms of the distance formula. The following two strategies are especially helpful in setting up equations to solve such problems.

- ***Make a sketch*** to visualize what is happening in the problem.
- ***Make a table*** to organize the information given in the problem and the unknown quantities.

---

2 Solve each problem using a form of the distance formula.

(a) A small plane flew from Chicago to St. Louis averaging 145 mph. The trip took 2 hr. What is the distance between Chicago and St. Louis?

(b) Usain Bolt of Jamaica ran the men's 100-m dash in 9.63 sec. What was his rate in meters per second, to the nearest hundredth? (Data from *The World Almanac and Book of Facts*.)

(c) The world record for the women's 3000-m steeplechase is held by Gulnara Samitova of Russia. Her rate was 5.568 m per sec. To the nearest second, what was her time? (Data from *The World Almanac and Book of Facts*.)

Answers

2. (a) 290 mi **(b)** 10.38 m per sec **(c)** 539 sec, or 8 min, 59 sec

EXAMPLE 3 Solving a Distance-Rate-Time Problem

Two cars leave Iowa City, Iowa, at the same time and travel east on Interstate 80. One travels at a constant rate of 55 mph. The other travels at a constant rate of 63 mph. In how many hours will the distance between them be 24 mi?

Step 1 **Read** the problem again.

Step 2 **Assign a variable.** We are looking for time.

Let $t =$ the number of hours until the distance between the cars is 24 mi.

The sketch in **Figure 1** shows what is happening in the problem.

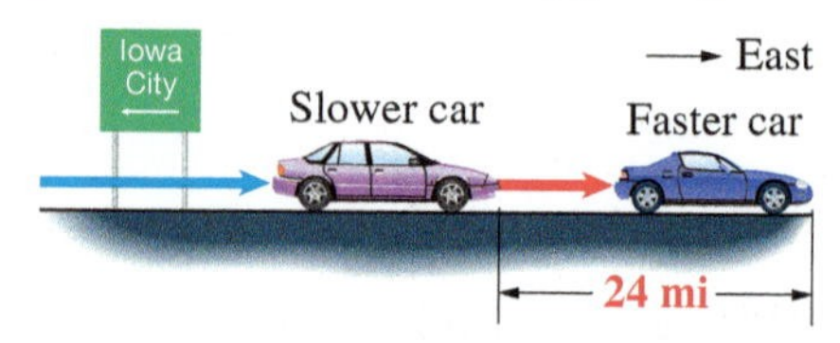

Figure 1

To construct a table, we fill in the rates given in the problem, using t for the time traveled by each car. Because $rt = d$, we multiply rate by time to find expressions for the distances traveled.

	Rate	Time	Distance
Faster Car	63	t	$63t$
Slower Car	55	t	$55t$

The quantities $63t$ and $55t$ represent the two distances.

Step 3 **Write an equation.**

$$63t - 55t = 24$$ The *difference* between the larger distance and the smaller distance is 24 mi.

Step 4 **Solve.**

$$8t = 24$$ Combine like terms.

$$t = 3$$ Divide by 8.

Step 5 **State the answer.** It will take the cars 3 hr to be 24 mi apart.

Step 6 **Check.** After 3 hr, the faster car will have traveled

$$63 \cdot 3 = 189 \text{ mi}$$

and the slower car will have traveled

$$55 \cdot 3 = 165 \text{ mi.}$$

The difference is

$$189 - 165 = 24, \quad \text{as required.}$$

◀ Work Problem 3 at the Side.

Problem-Solving Hint

In distance-rate-time problems like the one in **Example 3,** once we have filled in two pieces of information in each row of a table, we can automatically fill in the third piece of information, using the appropriate form of the distance formula. Then we set up an equation based on our sketch and the information in the table.

---

3 Solve each problem.

GS **(a)** From a point on a straight road, Lupe and Maria ride bicycles in *opposite* directions. Lupe rides 10 mph and Maria rides 12 mph. In how many hours will they be 55 mi apart?

Steps 1 and 2
Let $x =$ the number of ________ until the distance between Lupe and Maria is ________.

Step 3
Complete the table.

	Rate	Time	Distance
Maria	10	t	____
Lupe	____	t	____

Because Lupe and Maria are traveling in *opposite* directions, we must (*add*/*subtract*) the distances they travel to find the distance between them. Write an equation.

____

Complete Steps 4–6 to solve the problem. Give the answer.

(b) At a given hour, two steamboats leave a city in the same direction on a straight canal. One travels at 18 mph, and the other travels at 25 mph. In how many hours will the boats be 35 mi apart?

Answers

3. (a) hours; 55 mi

	Rate	Time	Distance
Maria	10	t	$10t$
Lupe	12	t	$12t$

add; $10t + 12t = 55$;

$2\frac{1}{2}$ hr

(b) 5 hr

EXAMPLE 4 Solving a Distance-Rate-Time Problem

The Tickfaw River has a current of 3 mph. A motorboat takes as long to travel 12 mi downstream as to travel 8 mi upstream. What is the rate of the boat in still water?

Step 1 **Read** the problem. We want the rate (speed) of the boat in still water.

Step 2 **Assign a variable.**

Let $x =$ the rate of the boat in still water.

Because the current pushes the boat when the boat is going downstream, the rate of the boat downstream will be the *sum* of the rate of the boat and the rate of the current, $(x + 3)$ mph.

Because the current slows down the boat when the boat is going upstream, the boat's rate upstream will be the *difference* between the rate of the boat in still water and the rate of the current, $(x - 3)$ mph. See **Figure 2.**

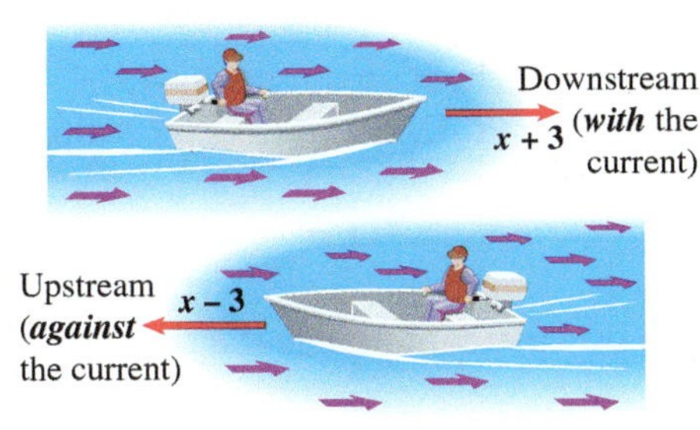

Figure 2

This information is summarized in the following table.

	d	r	t
Downstream	12	$x + 3$	
Upstream	8	$x - 3$	

Fill in the times using the formula $t = \frac{d}{r}$.

The time downstream is the distance divided by the rate.

$$t = \frac{d}{r} = \frac{12}{x + 3} \quad \text{Time downstream}$$

The time upstream is also the distance divided by the rate.

$$t = \frac{d}{r} = \frac{8}{x - 3} \quad \text{Time upstream}$$

	d	r	t
Downstream	12	$x + 3$	$\frac{12}{x + 3}$
Upstream	8	$x - 3$	$\frac{8}{x - 3}$

Enter each time using $t = \frac{d}{r}$.

Step 3 **Write an equation.**

$$\frac{12}{x + 3} = \frac{8}{x - 3}$$

The time downstream equals the time upstream, so the two times from the table must be equal.

Continued on Next Page

4 Solve each problem.

GS **(a)** A boat travels 10 mi against the current in the same time it travels 30 mi with the current. The current is flowing at 4 mph. Find the rate of the boat in still water.

Steps 1 and 2
Let $x =$ the ________ of the boat with no current.

Step 3
Complete the table.

	d	r	t
Against the Current	10	____	____
With the Current	30	____	____

How are the times traveling with the current and against the current related?

____

Write an equation.

____

Complete Steps 4–6 to solve the problem. Give the answer.

(b) An airplane, maintaining a constant airspeed, takes as long to travel 450 mi with the wind as it does to travel 375 mi against the wind. If the wind is blowing at 15 mph, what is the rate of the plane?

Answers

4. (a) rate

	d	r	t
Against the Current	10	$x-4$	$\frac{10}{x-4}$
With the Current	30	$x+4$	$\frac{30}{x+4}$

They are the same; $\frac{10}{x-4} = \frac{30}{x+4}$; 8 mph

(b) 165 mph

Step 4 **Solve.**

$$\frac{12}{x+3} = \frac{8}{x-3} \quad \text{Equation from Step 3}$$

$$(x+3)(x-3)\frac{12}{x+3} = (x+3)(x-3)\frac{8}{x-3} \quad \text{Multiply by the LCD, } (x+3)(x-3).$$

$$12(x-3) = 8(x+3) \quad \text{Divide out the common factors.}$$

$$12x - 36 = 8x + 24 \quad \text{Distributive property}$$

$$4x = 60 \quad \text{Subtract } 8x. \text{ Add } 36.$$

$$x = 15 \quad \text{Divide by 4.}$$

Step 5 **State the answer.** The rate of the boat in still water is 15 mph.

Step 6 **Check.** The rate of the boat downstream is $(x+3)$ mph, which would be $15 + 3 = 18$ mph. Divide 12 mi by 18 mph to find the time.

$$t = \frac{d}{r} = \frac{12}{18} = \frac{2}{3}\text{ hr}$$

The rate of the boat upstream is $(x-3)$ mph, which would be $15 - 3 = 12$ mph. Divide 8 mi by 12 mph to find the time.

$$t = \frac{d}{r} = \frac{8}{12} = \frac{2}{3}\text{ hr}$$

The time downstream equals the time upstream, as required.

◀ Work Problem **4** at the Side.

OBJECTIVE 3 **Solve problems about work.** Suppose that we can mow a lawn in 4 hr. Then after 1 hr, we will have mowed $\frac{1}{4}$ of the lawn. After 2 hr, we will have mowed $\frac{2}{4}$, or $\frac{1}{2}$, of the lawn, and so on. This idea is generalized as follows.

Rate of Work

If a job can be completed in t units of time, then the rate of work is

$$\frac{1}{t} \textbf{ job per unit of time.}$$

Problem-Solving Hint

The amount of work accomplished W is equal to rate of work r multiplied by time worked t—that is, $\boldsymbol{W = rt}$. Note the similarity to the distance formula $d = rt$.

In the lawn mowing example, the amount of work done after 3 hr, is found as follows.

$$\underbrace{\frac{1}{4}}_{\text{Rate of work}} \cdot \underbrace{3}_{\text{Time worked}} = \underbrace{\frac{3}{4}}_{\text{Fractional part of job done}}$$

After 4 hr, $\frac{1}{4}(4) = 1$ whole job has been done.

EXAMPLE 5 Solving a Problem about Work Rates

With spraying equipment, Mateo can paint the trim on a small house in 10 hr. Chet needs 15 hr to complete the same job by hand. If both Mateo and Chet work together, how long will it take them to paint the trim?

Step 1 **Read** the problem again. We are looking for time working together.

Step 2 **Assign a variable.**

Let $x =$ the number of hours it will take for Mateo and Chet to paint the trim, working together.

Making a table is helpful. Based on the previous discussion on work rates, Mateo's rate alone is $\frac{1}{10}$ job per hour, and Chet's rate alone is $\frac{1}{15}$ job per hour.

	Rate	Time Working Together	Fractional Part of the Job Done When Working Together
Mateo	$\frac{1}{10}$	x	$\frac{1}{10}x$
Chet	$\frac{1}{15}$	x	$\frac{1}{15}x$

Because $rt = W$, the quantities $\frac{1}{10}x$ and $\frac{1}{15}x$ represent the two amounts of work.

Step 3 **Write an equation.**

$$\underbrace{\text{Fractional part done by Mateo}} + \underbrace{\text{Fractional part done by Chet}} = \underbrace{\textbf{1 whole job}}$$

$$\frac{1}{10}x + \frac{1}{15}x = 1$$

Together, Mateo and Chet complete **1 whole job.** Add the fractional parts and set the sum equal to **1.**

Step 4 **Solve.**

$$30\left(\frac{1}{10}x + \frac{1}{15}x\right) = 30(1) \qquad \text{Multiply by the LCD, 30.}$$

$$30\left(\frac{1}{10}x\right) + 30\left(\frac{1}{15}x\right) = 30(1) \qquad \text{Distributive property}$$

$$3x + 2x = 30 \quad (*) \qquad \text{Multiply.}$$

$$5x = 30 \qquad \text{Combine like terms.}$$

$$x = 6 \qquad \text{Divide by 6.}$$

Step 5 **State the answer.** Working together, Mateo and Chet can paint the trim in 6 hr.

Step 6 **Check.** The value of x must be *less than* 10 due to the fact that Mateo can complete the job *alone* in 10 hr. So 6 hr seems reasonable.

In 6 hr, Mateo completes

$$\frac{1}{10}x = \frac{1}{10}(6) = \frac{6}{10} = \frac{3}{5} \text{ of the job.}$$

In 6 hr, Chet completes

$$\frac{1}{15}x = \frac{1}{15}(6) = \frac{6}{15} = \frac{2}{5} \text{ of the job.}$$

Working together, they complete $\frac{3}{5} + \frac{2}{5} =$ **1 whole job,** as required. The answer, 6 hr, is correct.

Work Problem 5 at the Side. ▶

5 Solve the problem.

Michael can paint a room, working alone, in 6 hr. Lindsay can paint the same room, working alone, in 12 hr. How long will it take them if they work together?

Steps 1 and 2
Let $x =$ the number of ________ it will take for Michael and Lindsay to paint the room, working ________.

Step 3
Complete the table.

	Rate	Time Working Together	Fractional Part of the Job Done When Working Together
Michael	____	x	____
Lindsay	____	x	____

Together, Michael and Lindsay complete ____ whole job(s). Write an equation.

____

Complete Steps 4–6 to solve the problem. Give the answer.

Answer

5. hours; together

	Rate	Time Working Together	Fractional Part of the Job Done When Working Together
Michael	$\frac{1}{6}$	x	$\frac{1}{6}x$
Lindsay	$\frac{1}{12}$	x	$\frac{1}{12}x$

$1; \frac{1}{6}x + \frac{1}{12}x = 1;$

4 hr

❗ CAUTION

A common error students make when solving a work problem like that in **Example 5** is to add the two times.

$$10 \text{ hr} + 15 \text{ hr} = 25 \text{ hr} \leftarrow \text{Incorrect answer}$$

The answer 25 hr is unreasonable, because the slower worker (Chet) can do the job *alone* in 15 hr.

Another common error students make is to add the two times and divide by 2—that is, average the times.

$$\frac{10 + 15}{2} = \frac{25}{2} = 12\frac{1}{2} \text{ hr} \leftarrow \text{Incorrect answer}$$

The answer $12\frac{1}{2}$ hr is also unreasonable, because the faster worker (Mateo) can do the job *alone* in 10 hr.

The correct time for the two workers together must be *less than* the time for the faster worker alone (in this case Mateo, at 10 hr).

6 Solve the problem using either the method of **Example 5** or the alternative approach discussed at the right.

Roberto can detail his Camaro in 2 hr working alone. His brother Marco can do the job in 3 hr working alone. How long would it take them if they worked together?

An alternative approach when solving work problems is to consider the part of the job that can be done in 1 hr. For instance, in **Example 5** Mateo can do the entire job in 10 hr, and Chet can do it in 15 hr. Thus, their work rates, as we saw in **Example 5,** are $\frac{1}{10}$ and $\frac{1}{15}$, respectively. Since it takes them x hours to complete the job working together, in 1 hr they can paint $\frac{1}{x}$ of the trim.

The amount painted by Mateo in 1 hr plus the amount painted by Chet in 1 hr must equal the amount they can paint *together* in 1 hr. This leads to the following alternative equation.

Amount by Chet ↓

$$\text{Amount by Mateo} \rightarrow \frac{1}{10} + \frac{1}{15} = \frac{1}{x} \leftarrow \text{Amount together}$$

Compare this alternative equation with the equation

$$\frac{1}{10}x + \frac{1}{15}x = 1$$

in Step 3 of **Example 5.** If we multiply each side of the alternative equation by the LCD $30x$, we obtain the following.

$$\frac{1}{10} + \frac{1}{15} = \frac{1}{x} \qquad \text{Alternative equation}$$

$$30x\left(\frac{1}{10} + \frac{1}{15}\right) = 30x\left(\frac{1}{x}\right) \qquad \text{Multiply by the LCD, } 30x.$$

$$30x\left(\frac{1}{10}\right) + 30x\left(\frac{1}{15}\right) = 30x\left(\frac{1}{x}\right) \qquad \text{Distributive property}$$

$$3x + 2x = 30 \qquad \text{Multiply.}$$

This is equation (*) in **Example 5.** The same solution, $x = 6$, results.

◀ **Work Problem 6 at the Side.**

Answer

6. $\frac{6}{5}$ hr, or $1\frac{1}{5}$ hr (that is, 1 hr, 12 min)

7.7 Exercises

FOR EXTRA HELP — *Go to MyMathLab for worked-out, step-by-step solutions to exercises enclosed in a square and video solutions to ▶ exercises.*

1. **CONCEPT CHECK** If a migrating hawk travels m mph in still air, what is its rate when it flies into a steady headwind of 5 mph? What is its rate with a tailwind of 5 mph?

2. **CONCEPT CHECK** Suppose Stephanie walks D miles at R mph in the same time that Wally walks d miles at r mph. Give an equation relating D, R, d, and r.

3. **CONCEPT CHECK** If it takes Elayn 10 hr to do a job, what is her rate?

4. **CONCEPT CHECK** If it takes Clay 12 hr to do a job, how much of the job does he do in 8 hr?

GS *Use Steps 2 and 3 of the six-step problem solving method to set up an equation to use to solve each problem. (Remember that Step 1 is to read the problem carefully.) Do not actually solve the equation.* ***See Example 1.***

5. The numerator of the fraction $\frac{5}{6}$ is increased by an amount so that the value of the resulting fraction is equivalent to $\frac{13}{3}$. By what amount was the numerator increased?

(a) Let $x =$ ___________. *(Step 2)*

(b) Write an expression for "the numerator of the fraction $\frac{5}{6}$ is increased by an amount." ___________

(c) Set up an equation to solve the problem.

(Step 3) ___________

6. If the same number is added to the numerator and subtracted from the denominator of $\frac{23}{12}$, the resulting fraction is equivalent to $\frac{3}{2}$. What is the number?

(a) Let $x =$ ___________. *(Step 2)*

(b) Write an expression for "a number is added to the numerator of $\frac{23}{12}$." ___________ Then write an expression for "the same number is subtracted from the denominator of $\frac{23}{12}$." ___________

(c) Set up an equation to solve the problem.

(Step 3) ___________

In each problem, state what x represents, write an equation, and answer the question. ***See Example 1.***

7. ▶ In a certain fraction, the denominator is 6 more than the numerator. If 3 is added to both the numerator and the denominator, the resulting fraction is equivalent to $\frac{5}{7}$. What was the original fraction (*not* written in lowest terms)?

8. In a certain fraction, the denominator is 4 less than the numerator. If 3 is added to both the numerator and the denominator, the resulting fraction is equivalent to $\frac{3}{2}$. What was the original fraction?

9. The denominator of a certain fraction is three times the numerator. If 2 is added to the numerator and subtracted from the denominator, the resulting fraction is equivalent to 1. What was the original fraction (*not* written in lowest terms)?

10. The numerator of a certain fraction is four times the denominator. If 6 is added to both the numerator and the denominator, the resulting fraction is equivalent to 2. What was the original fraction (*not* written in lowest terms)?

11. One-sixth of a number is 5 more than the same number. What is the number?

12. One-third of a number is 2 more than one-sixth of the same number. What is the number?

13. "A quantity, its $\frac{3}{4}$, its $\frac{1}{2}$, and its $\frac{1}{3}$, added together, become 93." What is the quantity? (Data from *Rhind Mathematical Papyrus.*)

14. "A quantity, its $\frac{2}{3}$, its $\frac{1}{2}$, and its $\frac{1}{7}$, added together, become 33." What is the quantity? (Data from *Rhind Mathematical Papyrus.*)

Solve each problem. ***See Example 2.***

15. British explorer and endurance swimmer Lewis Gordon Pugh was the first person to swim at the North Pole. He swam 0.6 mi at 0.0319 mi per min in waters created by melted sea ice. What was his time (to three decimal places)? (Data from *The Gazette.*)

16. In the 2012 Olympics, Missy Franklin of the United States won the women's 100-m backstroke swimming event. Her rate was 1.7143 m per sec. What was her time (to two decimal places)? (Data from *The World Almanac and Book of Facts.*)

17. Caroline Rotich of Kenya won the women's 26.2 mi Boston Marathon in 2015 with a time of 2.4153 hr. What was her rate (to three decimal places)? (Data from *The World Almanac and Book of Facts.*)

18. Ireen Wüst of the Netherlands won the women's 3000-m speed skating event in the 2014 Olympics with a time of 4.009 min. What was her rate (to three decimal places)? (Data from *The World Almanac and Book of Facts.*)

19. The winner of the 2015 Daytona 500 (mile) race was Joey Logano, who drove his Ford to victory with a rate of 161.939 mph. What was his time (to the nearest thousandth of an hour)? (Data from *The World Almanac and Book of Facts.*)

20. In 2015, Kyle Busch drove his Toyota to victory in the Brickyard 400 (mile) race. His rate was 131.656 mph. What was his time (to the nearest thousandth of an hour)? (Data from *The World Almanac and Book of Facts.*)

Complete the table and write an equation to use to solve each problem. Do not actually solve the equation. ***See Examples 3 and 4.***

21. Luvenia can row 4 mph in still water. She takes as long to row 8 mi upstream as 24 mi downstream. How fast is the current?

Let $x =$ the rate of the current.

	d	r	t
Upstream	8	$4 - x$	____
Downstream	24	$4 + x$	____

22. Julio flew his airplane 500 mi against the wind in the same time it took him to fly it 600 mi with the wind. If the rate of the wind was 10 mph, what was the average rate of his plane in still air?

Let $x =$ the rate of the plane in still air.

	d	r	t
Against the Wind	500	$x - 10$	____
With the Wind	600	$x + 10$	____

Solve each problem. **See Examples 3 and 4.**

23. From a point on a straight road, Marco and Celeste ride bicycles in the same direction. Marco rides at 10 mph and Celeste rides at 12 mph. In how many hours will they be 15 mi apart?

24. Two steamboats leaves a city on a river at the same time, traveling in the same direction. One travels at 18 mph and the other travels at 24 mph. In how many hours will the boats be 9 mi apart?

25. A train leaves Kansas City and travels north at 85 km per hr. Another train leaves at the same time and travels south at 95 km per hour. How long will it take before they are 315 km apart? (*Hint*: **See Margin Problem 3(a).**)

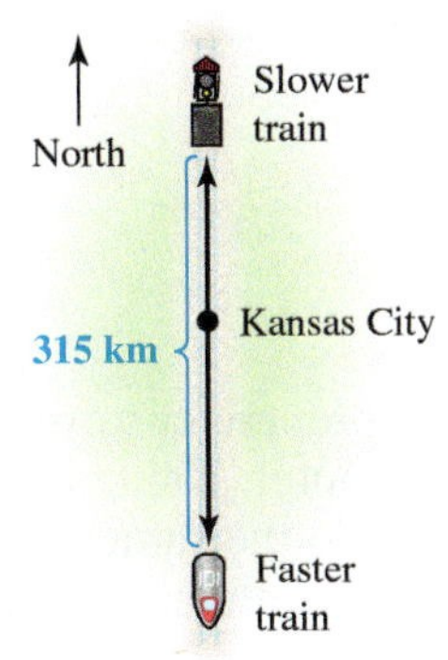

26. Two planes leave Boston at 12:00 noon and fly in opposite directions. If one flies at 410 mph and the other flies at 530 mph, how long will it take them to be 3290 mi apart? (*Hint*: **See Margin Problem 3(a).**)

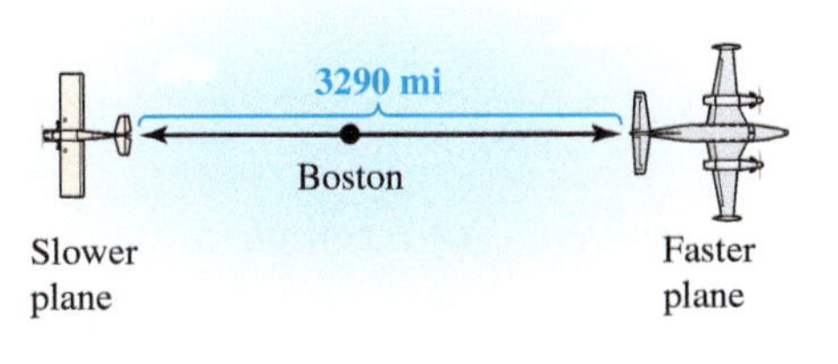

27. A boat can travel 20 mi against a current in the same time that it can travel 60 mi with the current. The rate of the current is 4 mph. Find the rate of the boat in still water.

28. Vince can fly his plane 200 mi against the wind in the same time it takes him to fly 300 mi with the wind. The wind blows at 30 mph. Find the rate of his plane in still air.

29. The sanderling is a small shorebird about 6.5 in. long, with a thin, dark bill and a wide, white wing stripe. If a sanderling can fly 30 mi with the wind in the same time it can fly 18 mi against the wind when the wind speed is 8 mph, what is the rate of the bird in still air? (Data from U.S. Geological Survey.)

30. Airplanes usually fly faster from west to east than from east to west because the prevailing winds go from west to east. The air distance between Chicago and London is about 4000 mi, while the air distance between New York and London is about 3500 mi. If a jet can fly eastbound from Chicago to London in the same time it can fly westbound from London to New York in a 35-mph wind, what is the rate of the plane in still air? (Data from www.geobytes.com)

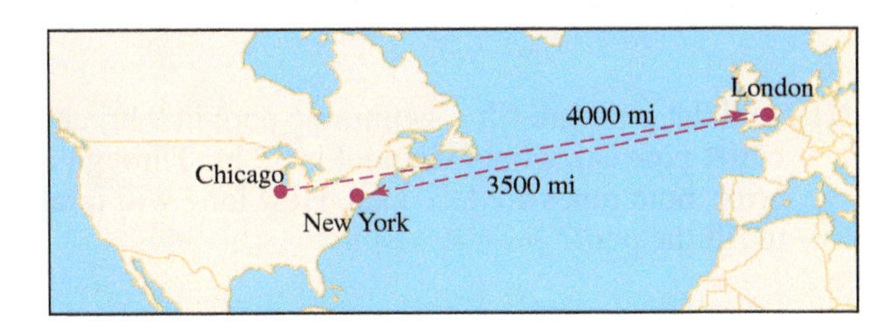

31. Perian's boat travels at 12 mph. Find the rate of the current of the river if she can travel 6 mi upstream in the same amount of time she can travel 10 mi downstream.

32. Bridget can travel 8 mi upstream in the same time it takes her to travel 12 mi downstream. Her boat travels 15 mph in still water. What is the rate of the current?

Complete the table and write an equation to use to solve each problem. Do not actually solve the equation. ***See Example 5.***

33. Eric can tune up his Chevy in 2 hr working alone. His son Oscar can do the job in 3 hr working alone. How long would it take them if they worked together?

Let t = the time working together.

	Rate	Time Working Together	Fractional Part of the Job Done When Working Together
Eric	____	t	____
Oscar	____	t	____

34. Working alone, Kyle can paint a room in 8 hr. Julianne can paint the same room working alone in 6 hr. How long will it take them if they work together?

Let t = the time working together.

	Rate	Time Working Together	Fractional Part of the Job Done When Working Together
Kyle	____	t	____
Julianne	____	t	____

Solve each problem. ***See Example 5.***

35. A copier can do a large printing job in 20 hr. An older model can do the same job in 30 hr. How long would it take to do the job using both copiers?

36. A company can prepare customer statements in 8 hr using a new computer. Using an older computer requires 24 hr to do the same job. How long would it take to prepare the statements using both computers?

37. A high school mathematics teacher gave a geometry test. Working alone, it would take her 4 hr to grade the tests. Her student teacher would take 6 hr to grade the same tests. How long would it take them to grade these tests if they work together?

38. A pump can pump the water out of a flooded basement in 10 hr. A smaller pump takes 12 hr. How long would it take to pump the water from the basement using both pumps?

39. Hilda can paint a room in 6 hr. Working together with Brenda, they can paint the room in $3\frac{3}{4}$ hr. How long would it take Brenda to paint the room by herself?

40. Grant can completely mess up his room in 15 min. If his cousin Wade helps him, they can completely mess up the room in $8\frac{4}{7}$ min. How long would it take Wade to mess up the room by himself?

41. An inlet pipe can fill a swimming pool in 9 hr, and an outlet pipe can empty the pool in 12 hr. Through an error, both pipes are left open. How long will it take to fill the pool?

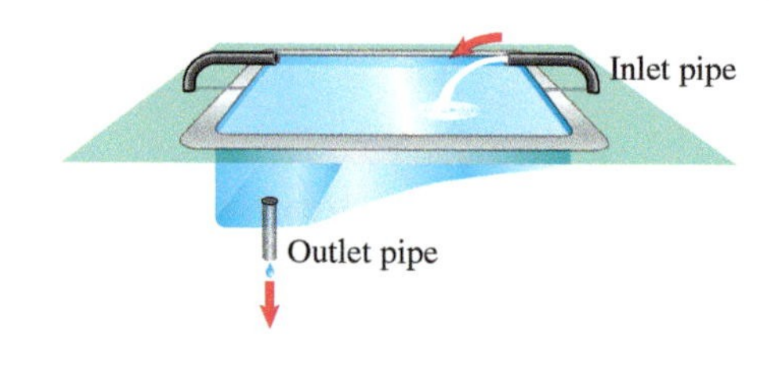

42. One pipe can fill a swimming pool in 6 hr, and another pipe can do it in 9 hr. How long will it take the two pipes working together to fill the pool $\frac{3}{4}$ full?

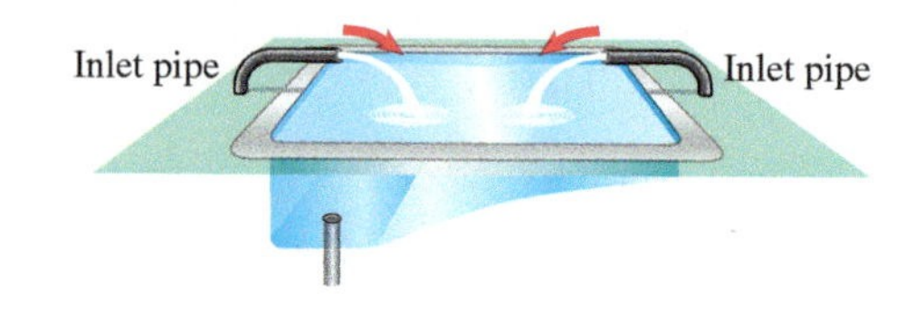

Solve each problem. (Data from Mary Jo Boyer, Math for Nurses, *Wolter Kluwer.)*

43. Nurses use Young's Rule to calculate the pediatric (child's) dose P of a medication, given a child's age c in years and a normal adult dose a.

$$P = \frac{c}{c + 12} \cdot a$$

The normal adult dose for milk of magnesia is 30 mL. Use Young's Rule to calculate the correct dose to give a 6-yr-old boy.

44. Nurses use Clark's Rule to calculate the pediatric (child's) dose P of a medication, given a child's weight w in pounds and a normal adult dose a.

$$P = \frac{w}{150} \cdot a$$

The normal adult dose for ibuprofen is 200 mg. Use Clark's Rule to calculate the correct dose to give a 4-yr-old girl who weighs 30 lb.

Relating Concepts (Exercises 45–48) For Individual or Group Work

In the movie Little Big League, *young Billy Heywood inherits the Minnesota Twins baseball team and becomes its manager. Before the biggest game of the year, he can't keep his mind on his job because a homework problem is giving him trouble.*

If Joe can paint a house in 3 hr, *and Sam can paint the same house in* 5 hr, *how long does it take for them to do it together?*

With the help of one of his players, Billy solves the problem.

45. Use the method of **Example 5** of this section to solve this problem.

46. Billy got "help" from some of the other players. The incorrect answers they gave him follow. Explain the faulty reasoning behind each of these answers.

(a) 15 hr **(b)** 8 hr **(c)** 4 hr

47. The player who gave Billy the correct answer solved the problem as follows:

Using the simple formula a times b over a plus b,
we get our answer of one and seven-eighths.

Show that if it takes one person a hours to complete a job and another b hours to complete the same job, then the expression stated by the player,

$$\frac{a \cdot b}{a + b},$$

actually does give the number of hours it would take them to do the job together. (*Hint:* Refer to **Example 5** and use a and b rather than 10 and 15 to write a formula. Then solve the formula for x.)

48. Solve the following problem using the method of **Example 5.** Then solve it using the formula obtained in **Exercise 47.** How do the answers compare?

A screen printer can complete a t-shirt order for a Little League baseball organization in 15 hr *using a large machine. The same order would take* 30 hr *using a smaller machine. How long would it take to complete the order using both machines together?*

7.8 Variation

OBJECTIVES

1. Solve direct variation problems.
2. Solve inverse variation problems.

OBJECTIVE 1 Solve direct variation problems. Suppose that gasoline costs \$3.00 per gal. Then 1 gal costs \$3.00, 2 gal cost $2(\$3.00) = \$6.00$, 3 gal cost $3(\$3.00) = \$9.00$, and so on. Each time, the total cost is obtained by multiplying the number of gallons by the price per gallon. In general, if k equals the price per gallon and x equals the number of gallons, then the total cost y is given by the equation $y = kx$.

As the *number of gallons* **increases,** the *total cost* **increases.**

As the *number of gallons* **decreases,** the *total cost* **decreases.**

The preceding discussion is an example of *variation.* In both cases, the rate of change is constant. ***Two variables vary directly if one is a constant multiple of the other.***

Direct Variation

y* varies directly as *x if there is a constant k such that the following holds true.

$$y = kx$$

← Constant of variation

Stated another way, ***y* is proportional to *x*.** The constant k is a numerical value (such as 3.00 in the gasoline price discussion above) called the **constant of variation.**

Another example of direct variation is the formula for the diameter of a circle,

$$d = 2r.$$

See **Figure 3.** Here, 2 is the constant of variation, and the diameter d varies directly as the length of the radius r.

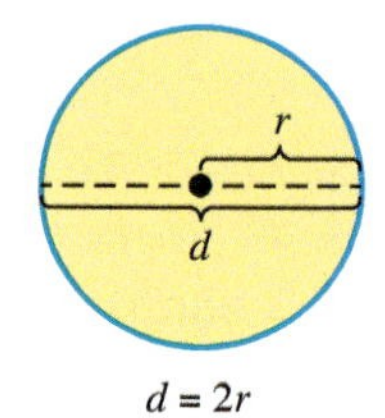

$d = 2r$

Figure 3

Solving a Variation Problem

Step 1 Write a variation equation. For direct variation, this equation will be of the form

"first variable" $= k \cdot$ "second variable,"

where k is a constant.

Step 2 Substitute the appropriate given values and solve for k.

Step 3 Write the variation equation with the value of k from Step 2.

Step 4 Substitute the remaining values, solve for the unknown, and find the required answer.

◀ Work Problem 1 at the Side.

1 Write a variation equation for each situation. Use k as the constant of variation.

GS **(a)** A varies directly as b.

$\_\_\_\_ = kb$

(b) m varies directly as d.

(c) F is proportional to a.

Answers

1. (a) A (b) $m = kd$ (c) $F = ka$

EXAMPLE 1 **Using Direct Variation**

Suppose y varies directly as x, and $y = 20$ when $x = 4$. Find y when $x = 9$.

Step 1 Here y varies directly as x, so there is a constant k such that $y = kx$.

Step 2 We let $y = 20$ and $x = 4$ in this equation and solve for k.

$y = kx$ Equation for direct variation

$20 = k \cdot 4$ Substitute the given values.

$k = 5 \longleftarrow$ Constant of variation

Step 3 Because $y = kx$ and $k = 5$, we have the following.

$y = 5x$ Let $k = 5$.

Step 4 Now we can find the value of y when $x = 9$.

$y = 5x$ Variation equation

$y = 5 \cdot 9$ Let $x = 9$.

$y = 45$ Multiply.

Thus, $y = 45$ when $x = 9$.

Work Problem 2 at the Side. ▶

EXAMPLE 2 **Solving a Direct Variation Problem**

The perimeter of a square varies directly as the length of a side. A square with a side of length 9 in. has a perimeter of 36 in. Find the perimeter of a square with a side of length 19 in.

Step 1 If we select P to represent the perimeter and s the length of a side, then for some constant k, the variation equation is

$P = ks$ ← Choose variables that "connect" to the problem. Here, we choose P for perimeter and s for side.

Step 2 To find the value of k, we substitute the given values for P and s.

$P = ks$ Equation for direct variation

$36 = k \cdot 9$ Let $P = 36$ and $s = 9$.

$k = 4$ Divide by 9.

Step 3 Substitute 4 for k in the variation equation.

$P = 4s$ Intuitively this makes sense. This is the formula for finding the perimeter of a square.

Step 4 Now we can find the value of P when $s = 19$.

$P = 4s$ Variation equation

$P = 4 \cdot 19$ Let $s = 19$.

$P = 76$ Multiply.

Thus, $P = 76$ when $s = 19$. The perimeter is 76 in.

Work Problem 3 at the Side. ▶

OBJECTIVE ▶ 2 Solve inverse variation problems. In direct variation, where $k > 0$, as x increases, y increases, and as x decreases, y decreases. With *inverse variation,* where $k > 0$, we have the following.

As one variable ***increases,*** the other variable ***decreases.***

For example, in a closed space, volume decreases as pressure increases, as illustrated by a trash compactor. See **Figure 4** on the next page.

2 Solve each problem.

GS **(a)** If W varies directly as r, and $W = 40$ when $r = 5$, find W when $r = 10$.

Step 1
The problem states that "W varies directly as r." Write a variation equation using k as the constant of variation.

____

Step 2
Let $W = 40$ and $r = 5$ in this equation, and solve for k.

$40 = k \cdot 5$

$k =$ ____

Step 3
Substitute 8 for k in the variation equation to obtain

$W =$ ____ r.

Step 4
Let $r = 10$ in this equation, and find the value of W.

$W = 8 \cdot$ ____

$W =$ ____

Thus, $W = 80$ when $r = 10$.

(b) If z varies directly as t, and $z = 11$ when $t = 4$, find z when $t = 32$.

3 Solve the problem.
The amount a student earns at a part-time job varies directly as the number of hours worked. If the student earns \$225 working 18 hr, find the paycheck amount in dollars if the student works 22 hr.

Answers

2. **(a)** $W = kr$; 8; 8; 10; 80 **(b)** 88

3. \$275

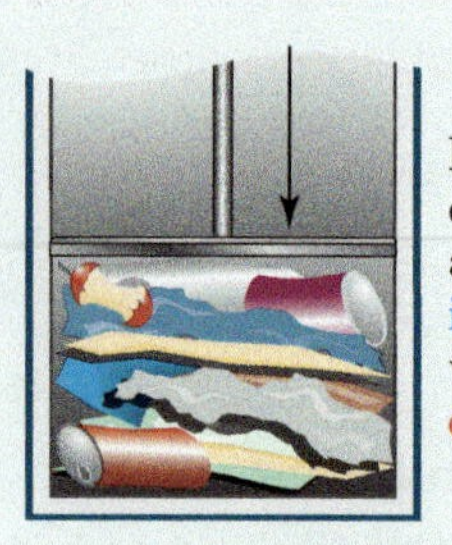

In a trash compactor, as pressure increases, volume decreases.

Figure 4

4 Write a variation equation for each situation. Use k as the constant of variation.

GS **(a)** h varies inversely as w.

$h =$ ____

(b) P varies inversely as d.

(c) d is inversely proportional to t.

5 Solve the problem.

If t varies inversely as r, and $t = 12$ when $r = 3$, find t when $r = 6$.

6 Solve the problem.

The unit cost of producing electric cables varies inversely as the number of units produced. If 5000 units are produced, the cost is \$0.50 per unit. Find the cost per unit to produce 10,000 units.

Answers

4. (a) $\frac{k}{w}$ (b) $P = \frac{k}{d}$ (c) $d = \frac{k}{t}$

5. 6

6. \$0.25

Inverse Variation

y* varies inversely as *x (or ***y* is inversely proportional to *x***) if there exists a constant k such that the following holds true.

$$y = \frac{k}{x}$$

EXAMPLE 3 Using Inverse Variation

Suppose y varies inversely as x, and $y = 3$ when $x = 8$. Find y when $x = 6$.

Step 1 Because y varies inversely as x, there is a constant k such that $y = \frac{k}{x}$.

Step 2 We know that $y = 3$ when $x = 8$, so we can find k.

$$y = \frac{k}{x} \quad \text{Equation for inverse variation}$$

$$3 = \frac{k}{8} \quad \text{Substitute the given values.}$$

$$k = 24 \quad \text{Multiply by 8. Rewrite } 24 = k \text{ as } k = 24.$$

Step 3 Because $y = \frac{k}{x}$ and $k = 24$, we have the following.

$$y = \frac{24}{x} \quad \text{Let } k = 24 \text{ in the variation equation.}$$

Step 4

$$y = \frac{24}{6} \quad \text{Now let } x = 6.$$

$$y = 4 \quad \text{Divide.}$$

Therefore, $y = 4$ when $x = 6$.

◀ **Work Problems 4 and 5 at the Side.**

EXAMPLE 4 Solving an Inverse Variation Problem

The unit cost of producing a phone-charging device varies inversely as the number of units produced. If 10,000 units are produced, the cost is \$2 per unit. Find the cost per unit to produce 25,000 units.

Let $x =$ the number of units produced and $c =$ the cost per unit.

Because c varies inversely as x, there is a constant k such that $c = \frac{k}{x}$.

$$c = \frac{k}{x} \quad \text{Equation for inverse variation}$$

$$2 = \frac{k}{10{,}000} \quad \text{Substitute the given values.}$$

$$k = 20{,}000 \quad \text{Multiply by 10,000. Rewrite.}$$

Now use $c = \frac{20{,}000}{x}$ to find the value of c when $x = 25{,}000$.

$$c = \frac{20{,}000}{25{,}000} = 0.80 \quad \text{Let } x = 25{,}000.$$

The cost per unit to make 25,000 devices is \$0.80.

◀ **Work Problem 6 at the Side.**

7.8 Exercises

FOR EXTRA HELP Go to MyMathLab for worked-out, step-by-step solutions to exercises enclosed in a square and video solutions to ▶ exercises.

CONCEPT CHECK *Use personal experience or intuition to determine whether the situation suggests* direct *or* inverse *variation.**

1. The rate and the distance traveled by a pickup truck in 3 hr

2. The amount of pressure placed on the accelerator of a truck and the rate of the truck

3. The amount of gasoline pumped and the amount of empty space in the tank

4. A person's age and the probability that the person believes in the tooth fairy

5. The surface area of a balloon and its diameter

6. The intensity of a light source (such as a light bulb) and the distance from which a person views the light

7. The loudness of a sound source (such as a car horn) and the distance from which a person hears the sound

8. The number of different lottery tickets purchased and the probability of winning that lottery

9. CONCEPT CHECK Refer to **Exercises 1–8.** Give an example of direct variation from everyday life.

10. CONCEPT CHECK Refer to **Exercises 1–8.** Give an example of inverse variation from everyday life.

11. CONCEPT CHECK Make the correct choice.

(a) If the constant of variation is positive and y varies directly as x, then as x increases, y (*increases* / *decreases*).

(b) If the constant of variation is positive and y varies inversely as x, then as x increases, y (*increases* / *decreases*).

12. Bill Veeck was the owner of several major league baseball teams in the 1950s and 1960s. He was known to often sit in the stands and enjoy games with his paying customers. Here is a quote attributed to him:

"I have discovered in 20 years of moving around a ballpark, that the knowledge of the game is usually in inverse proportion to the price of the seats."

Explain the meaning of this statement.

CONCEPT CHECK *Determine whether each equation represents* direct *or* inverse *variation. Give the constant of variation.*

13. $y = \frac{3}{x}$

14. $y = \frac{5}{x}$

15. $y = 50x$

16. $y = 200x$

*The authors thank Linda Kodama for suggesting these exercises.

Solve each problem involving variation. ***See Examples 1 and 3.***

17. If z varies directly as x, and $z = 30$ when $x = 8$, find z when $x = 4$.

18. If y varies directly as x, and $x = 27$ when $y = 6$, find x when $y = 2$.

19. If d varies directly as t, and $d = 150$ when $t = 3$, find d when $t = 5$.

20. If d varies directly as r, and $d = 200$ when $r = 40$, find d when $r = 60$.

21. If z varies inversely as x, and $z = 50$ when $x = 2$, find z when $x = 25$.

22. If x varies inversely as y, and $x = 3$ when $y = 8$, find y when $x = 4$.

23. If p varies inversely as q, and $p = 7$ when $q = 6$, find p when $q = 2$.

24. If m varies inversely as r, and $m = 12$ when $r = 8$, find m when $r = 16$.

25. Suppose a is proportional to b, and $a = 20$ when $b = 14$. Find b when $a = 50$.

26. Suppose p is proportional to q, and $p = 24$ when $q = 18$. Find q when $p = 16$.

27. Suppose z is inversely proportional to t, and $z = 8$ when $t = 2$. Find z when $t = 32$.

28. Suppose y is inversely proportional to x, and $y = 10$ when $x = 3$. Find y when $x = 20$.

Solve each problem. ***See Examples 1–4.***

29. For a given base, the area of a triangle varies directly as its height. Find the area of a triangle with a height of 6 in., if the area is 10 in.^{2} when the height is 4 in.

30. The interest on an investment varies directly as the rate of interest. If the interest is \$60 when the interest rate is 4%, find the interest when the rate is 1.5%.

31. Hooke's law for an elastic spring states that the distance a spring stretches varies directly with the force applied. If a force of 75 lb stretches a certain spring 16 in., how much will a force of 200 lb stretch the spring?

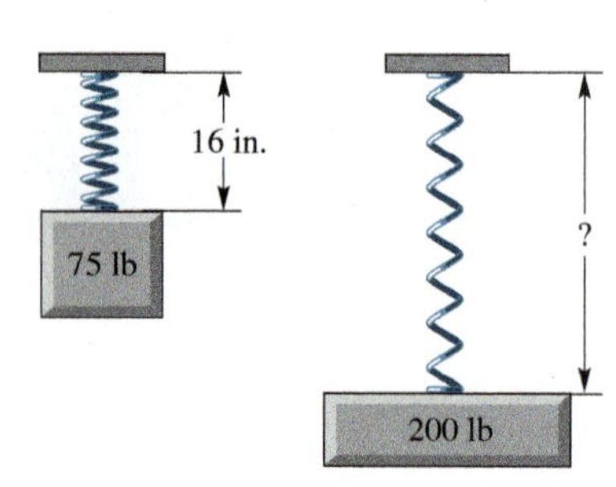

32. The pressure exerted by water at a given point varies directly with the depth of the point beneath the surface of the water. Water exerts 4.34 lb per in.^{2} for every 10 ft traveled below the water's surface. What is the pressure exerted on a scuba diver at 20 ft?

33. For a constant area, the length of a rectangle varies inversely as the width. The length of a rectangle is 27 ft when the width is 10 ft. Find the width of a rectangle with the same area if the length is 18 ft.

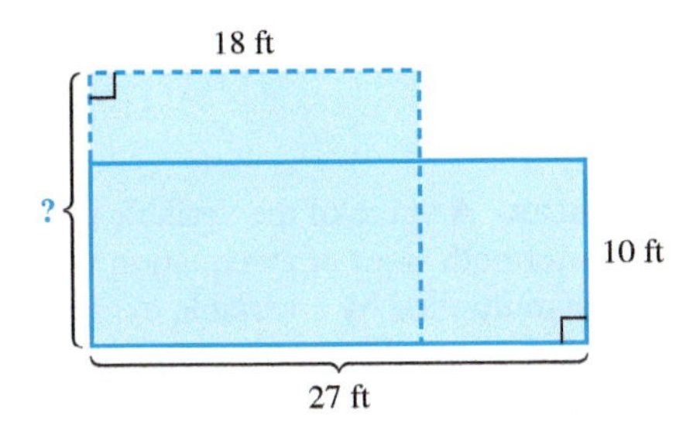

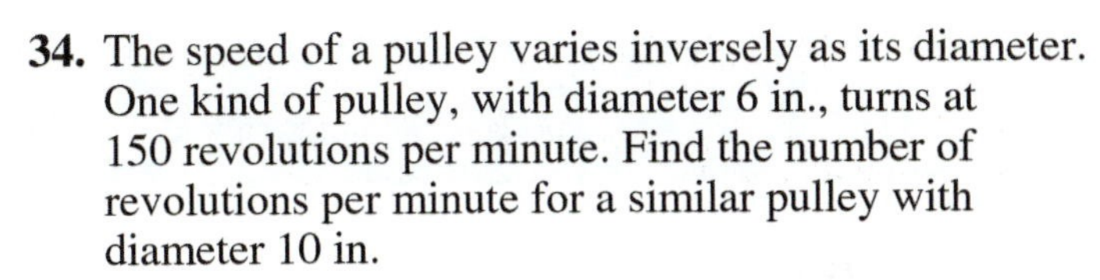

34. The speed of a pulley varies inversely as its diameter. One kind of pulley, with diameter 6 in., turns at 150 revolutions per minute. Find the number of revolutions per minute for a similar pulley with diameter 10 in.

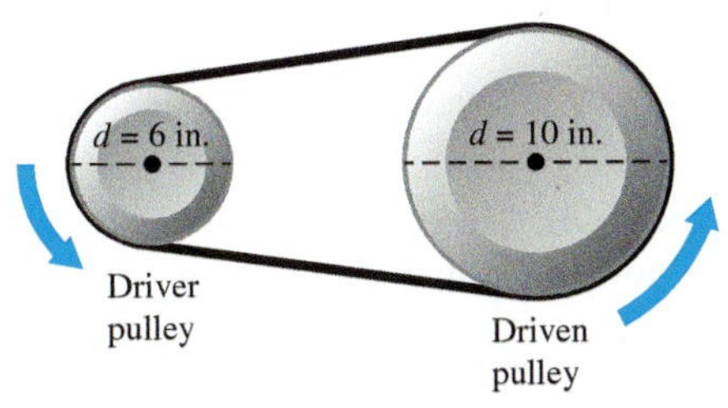

35. If the temperature is constant, the pressure of a gas in a container varies inversely as the volume of the container. If the pressure is 10 lb per ft^2 in a container with volume 3 ft^3, what is the pressure in a container with volume 1.5 ft^3?

36. In rectangles of constant area, length and width vary inversely. When the length is 24 cm, the width is 2 cm. What is the width when the length is 12 cm?

37. In the inversion of raw sugar, the rate of change of the amount of raw sugar varies directly as the amount of raw sugar remaining. The rate is 200 kg per hr when there are 800 kg left. What is the rate of change per hour when only 100 kg are left?

38. The force required to compress a spring varies directly as the change in the length of the spring. If a force of 12 lb is required to compress a certain spring 3 in., how much force is required to compress the spring 5 in.?

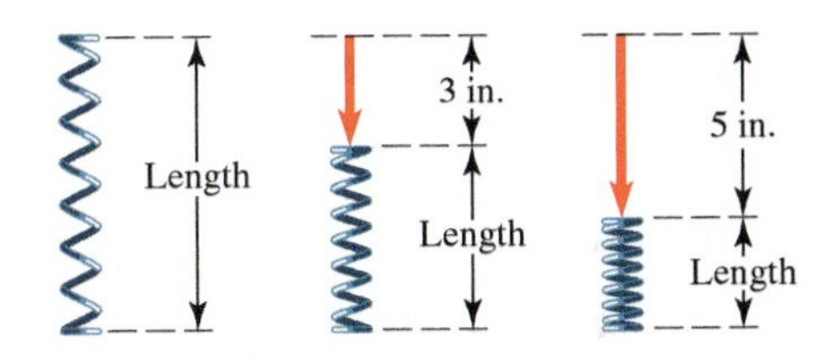

39. The current in a simple electrical circuit varies inversely as the resistance. If the current is 20 amps when the resistance is 5 ohms, find the current when the resistance is 8 ohms.

40. Over a specified distance, rate varies inversely with time. If a Dodge Viper on a test track goes a certain distance in one-half minute at 160 mph, what rate is needed to go the same distance in three-fourths minute?

41. The weight of an object on Earth is directly proportional to the weight of that same object on the moon. A 200-lb astronaut would weigh 32 lb on the moon. How much would a 50-lb dog weigh on the moon?

42. The pressure exerted by a certain liquid at a given point is directly proportional to the depth of the point beneath the surface of the liquid. The pressure at 30 m is 80 newtons. What pressure is exerted at 50 m?

Chapter 7 Summary

Key Terms

7.1

rational expression The quotient of two polynomials with denominator not 0 is a rational expression.

lowest terms A rational expression is written in lowest terms if the greatest common factor of its numerator and denominator is 1.

7.3

least common denominator (LCD) The simplest expression that is divisible by all denominators is the least common denominator.

7.5

complex fraction A quotient with one or more fractions in the numerator, or denominator, or both, is a complex fraction.

7.6

proposed solution A value of the variable that appears to be a solution after both sides of an equation with rational expressions are multiplied by a variable expression is a proposed solution.

extraneous solution (extraneous value) A proposed solution that is not an actual solution of a given equation is an extraneous solution, or extraneous value.

7.8

direct variation y varies directly as x (or y is proportional to x) if there is a constant k such that $y = kx$.

constant of variation In the equation $y = kx$ or $y = \frac{k}{x}$, the number k is the constant of variation.

inverse variation y varies inversely as x (or y is inversely proportional to x) if there is a constant k such that $y = \frac{k}{x}$.

Test Your Word Power

See how well you have learned the vocabulary in this chapter.

1. A **rational expression** is
   A. an algebraic expression made up of a term or the sum of a finite number of terms with real coefficients and whole number exponents
   B. a polynomial equation of degree 2
   C. an expression with one or more fractions in the numerator, or denominator, or both
   D. the quotient of two polynomials with denominator not 0.

2. A **complex fraction** is
   A. an algebraic expression made up of a term or the sum of a finite number of terms with real coefficients and whole number exponents
   B. a polynomial equation of degree 2
   C. a rational expression with one or more fractions in the numerator, or denominator, or both
   D. the quotient of two polynomials with denominator not 0.

3. If y varies directly as x, and the constant of variation is positive, then
   A. as x increases, y decreases
   B. as x increases, y increases
   C. as x increases, y remains constant
   D. as x decreases, y remains constant.

4. If y varies inversely as x, and the constant of variation is positive, then
   A. as x increases, y decreases
   B. as x increases, y increases
   C. as x increases, y remains constant
   D. as x decreases, y remains constant.

Answers to Test Your Word Power

1. D; *Examples:* $-\frac{3}{4y}, \frac{5x^3}{x+2}, \frac{a+3}{a^2-4a-5}$

2. C; *Examples:* $\frac{\frac{2}{3}}{\frac{4}{7}}, \frac{x-\frac{1}{y}}{x+\frac{1}{y}}, \frac{\frac{2}{a+1}}{a^2-1}$

3. B; *Example:* The equation $y = 3x$ represents direct variation. When $x = 2$, $y = 6$. If x increases to 3, then y increases to $3(3) = 9$.

4. A; *Example:* The equation $y = \frac{3}{x}$ represents inverse variation. When $x = 1$, $y = 3$. If x increases to 2, then y decreases to $\frac{3}{2}$, or $1\frac{1}{2}$.

Quick Review

Concepts	Examples

7.1 The Fundamental Property of Rational Expressions

Determining When a Rational Expression Is Undefined

Step 1 Set the denominator equal to 0.

Step 2 Solve this equation.

Step 3 The solutions of the equation are the values that make the rational expression undefined. The variable *cannot* equal these values.

Find the values for which $\frac{x-4}{x^2-16}$ is undefined.

$$x^2 - 16 = 0$$

$$(x-4)(x+4) = 0 \quad \text{Factor.}$$

$$x - 4 = 0 \quad \text{or} \quad x + 4 = 0 \quad \text{Zero-factor property}$$

$$x = 4 \quad \text{or} \quad x = -4 \quad \text{Solve for } x.$$

The rational expression is undefined for 4 and -4, so $x \neq 4$ and $x \neq -4$.

Writing a Rational Expression in Lowest Terms

Step 1 Factor the numerator and denominator completely.

Step 2 Use the fundamental property to divide out any common factors.

Write the rational expression in lowest terms.

$$\frac{x^2-1}{(x-1)^2}$$

$$= \frac{(x-1)(x+1)}{(x-1)(x-1)} \quad \text{Factor.}$$

$$= \frac{x+1}{x-1} \quad \text{Lowest terms}$$

Writing Equivalent Forms of a Rational Expression

There are often several different equivalent forms of a rational expression.

Write four equivalent forms of $-\frac{x-1}{x+2}$.

$$① \rightarrow \frac{-(x-1)}{x+2}, \quad \text{or} \quad \frac{-x+1}{x+2} \leftarrow ②$$

Distribute the negative sign in the numerator.

$$③ \rightarrow \frac{x-1}{-(x+2)}, \quad \text{or} \quad \frac{x-1}{-x-2} \leftarrow ④$$

Distribute the negative sign in the denominator.

7.2 Multiplying and Dividing Rational Expressions

Multiplying or Dividing Rational Expressions

Step 1 Note the operation. If the operation is division, use the definition of division to rewrite it as multiplication.

Step 2 Multiply numerators and multiply denominators.

Step 3 Factor numerators and denominators completely.

Step 4 Write in lowest terms using the fundamental property.

Steps 2 and 3 may be interchanged based on personal preference.

Multiply. $\frac{3x+9}{x-5} \cdot \frac{x^2-3x-10}{x^2-9}$

$$= \frac{(3x+9)(x^2-3x-10)}{(x-5)(x^2-9)} \quad \text{Multiply numerators and denominators.}$$

$$= \frac{3(x+3)(x-5)(x+2)}{(x-5)(x+3)(x-3)} \quad \text{Factor.}$$

$$= \frac{3(x+2)}{x-3} \quad \text{Lowest terms}$$

Divide. $\frac{2x+1}{x+5} \div \frac{6x^2-x-2}{x^2-25}$

$$= \frac{2x+1}{x+5} \cdot \frac{x^2-25}{6x^2-x-2} \quad \text{Multiply by the reciprocal of the divisor.}$$

$$= \frac{(2x+1)(x+5)(x-5)}{(x+5)(2x+1)(3x-2)} \quad \text{Multiply and factor.}$$

$$= \frac{x-5}{3x-2} \quad \text{Lowest terms}$$

Concepts	Examples

7.3 Least Common Denominators

Finding the LCD

Step 1 Factor each denominator into prime factors.

Step 2 List each different denominator factor the *greatest* number of times it appears in any of the denominators.

Step 3 Multiply the factors from Step 2 to find the LCD.

Find the LCD for $\frac{3}{k^2 - 8k + 16}$ and $\frac{1}{4k^2 - 16k}$.

$$k^2 - 8k + 16 = (k-4)^2$$
$$4k^2 - 16k = 4k(k-4)$$
Factor each denominator.

$$\text{LCD} = (k-4)^2 \cdot 4 \cdot k$$
$$= 4k(k-4)^2$$

Writing a Rational Expression with a Specified Denominator

Step 1 Factor both denominators.

Step 2 Decide what factors the denominator must be multiplied by in order to equal the specified denominator.

Step 3 Multiply the rational expression by that factor divided by itself. (That is, multiply by 1.)

Write the rational expression as an equivalent expression with the indicated denominator.

$$\frac{5}{2z^2 - 6z} = \frac{?}{4z^3 - 12z^2}$$

$$\frac{5}{2z(z-3)} = \frac{?}{4z^2(z-3)}$$

$2z(z-3)$ must be multiplied by $2z$ to obtain $4z^2(z-3)$.

$$\frac{5}{2z(z-3)} \cdot \frac{2z}{2z} \qquad \frac{2z}{2z} = 1$$

$$= \frac{10z}{4z^2(z-3)}, \quad \text{or} \quad \frac{10z}{4z^3 - 12z^2}$$

7.4 Adding and Subtracting Rational Expressions

Adding Rational Expressions

Step 1 Find the LCD.

Step 2 Write each rational expression as an equivalent rational expression with the LCD as denominator.

Step 3 Add the numerators to obtain the numerator of the sum. The LCD is the denominator of the sum.

Step 4 Write in lowest terms using the fundamental property.

Add. $\frac{2}{3m+6} + \frac{m}{m^2-4}$

$$3m + 6 = 3(m+2)$$
$$m^2 - 4 = (m+2)(m-2)$$
The LCD is $3(m+2)(m-2)$.

$$= \frac{2(m-2)}{3(m+2)(m-2)} + \frac{3m}{3(m+2)(m-2)}$$
Write with the LCD.

$$= \frac{2m - 4 + 3m}{3(m+2)(m-2)}$$
Add numerators. Keep the same denominator.

$$= \frac{5m-4}{3(m+2)(m-2)}$$
Combine like terms.

Subtracting Rational Expressions

Follow the steps above for addition, but subtract in Step 3.

Subtract. $\frac{6}{k+4} - \frac{2}{k}$ The LCD is $k(k+4)$.

$$= \frac{6k}{(k+4)k} - \frac{2(k+4)}{k(k+4)}$$
Write with the LCD.

$$= \frac{6k - 2(k+4)}{k(k+4)}$$
Subtract numerators. Keep the same denominator.

$$= \frac{6k - 2k - 8}{k(k+4)}$$
Distributive property

$$= \frac{4k-8}{k(k+4)}, \quad \text{or} \quad \frac{4(k-2)}{k(k+4)}$$
Either form is correct.

Concepts

7.5 Complex Fractions

Simplifying a Complex Fraction

Method 1

Step 1 Write both the numerator and the denominator as single fractions.

Step 2 Change the complex fraction to a division problem.

Step 3 Perform the indicated division.

Method 2

Step 1 Find the LCD of all fractions within the complex fraction.

Step 2 Multiply both the numerator and the denominator of the complex fraction by this LCD using the distributive property as necessary. Write in lowest terms.

Which method to use is a matter of individual preference.

Examples

Simplify the complex fraction.

Method 1

$$\frac{\frac{1}{a} - a}{1 - a}$$

$$= \frac{\frac{1}{a} - \frac{a^2}{a}}{1 - a}$$

$$= \frac{\frac{1 - a^2}{a}}{1 - a}$$

$$= \frac{1 - a^2}{a} \div (1 - a)$$

$$= \frac{1 - a^2}{a} \cdot \frac{1}{1 - a}$$

$$= \frac{(1 - a)(1 + a)}{a(1 - a)}$$

$$= \frac{1 + a}{a}$$

Method 2

$$\frac{\frac{1}{a} - a}{1 - a}$$

$$= \frac{a\left(\frac{1}{a} - a\right)}{a(1 - a)}$$

$$= \frac{\frac{a}{a} - a^2}{(1 - a)a}$$

$$= \frac{1 - a^2}{(1 - a)a}$$

$$= \frac{(1 + a)(1 - a)}{(1 - a)a}$$

$$= \frac{1 + a}{a}$$

The same answer results using either method.

7.6 Solving Equations with Rational Expressions

Solving an Equation with Rational Expressions

Step 1 Multiply each side of the equation by the LCD. (This clears the equation of fractions.) Be sure to distribute to *every* term on *both* sides of the equation.

Step 2 Solve the resulting equation for proposed solutions.

Step 3 Check each proposed solution by substituting it in the original equation. Reject any value that causes a denominator to equal 0.

Solve.

$$\frac{x}{x - 3} + \frac{4}{x + 3} = \frac{18}{x^2 - 9}$$

$$\frac{x}{x - 3} + \frac{4}{x + 3} = \frac{18}{(x - 3)(x + 3)} \quad \text{Factor.}$$

The LCD is $(x - 3)(x + 3)$. ***Note that 3 and −3 cannot be solutions, as they cause a denominator to equal 0.***

$$(x - 3)(x + 3)\left(\frac{x}{x - 3} + \frac{4}{x + 3}\right)$$

$$= (x - 3)(x + 3)\frac{18}{(x - 3)(x + 3)} \quad \text{Multiply by the LCD.}$$

$$x(x + 3) + 4(x - 3) = 18 \quad \text{Distributive property}$$

$$x^2 + 3x + 4x - 12 = 18 \quad \text{Distributive property}$$

$$x^2 + 7x - 30 = 0 \quad \text{Standard form}$$

$$(x - 3)(x + 10) = 0 \quad \text{Factor.}$$

$$x - 3 = 0 \quad \text{or} \quad x + 10 = 0 \quad \text{Zero-factor property}$$

Reject → $x = 3$ or $x = -10$ Solve for x.

Because 3 causes denominators to equal 0, it is an extraneous value. Check that the only solution is -10. Thus, $\{-10\}$ is the solution set.

Concepts

7.7 Applications of Rational Expressions

Solving a Distance-Rate-Time Problem

Use the formulas relating d, r, and t.

$$d = rt, \quad r = \frac{d}{t}, \quad t = \frac{d}{r}$$

Examples

A small plane flew from Chicago to Kansas City averaging 145 mph. The trip took 3.5 hr. What is the distance between Chicago and Kansas City?

$$\underset{\text{Rate}}{145} \cdot \underset{\text{Time}}{3.5} = \underset{\text{Distance}}{507.5 \text{ mi}}$$

Solving a Work Problem

Step 1 Read the problem, several times if necessary.

Step 2 Assign a variable. State what the variable represents. Organize the information from the problem in a table.

If a job can be completed in t units of time, then the rate of work is

$$\frac{1}{t} \text{ job per unit of time.}$$

Step 3 Write an equation. The sum of the fractional parts should equal 1 (whole job).

Step 4 Solve the equation.

Step 5 State the answer.

Step 6 Check.

It takes the regular mail carrier 6 hr to cover her route. A substitute takes 8 hr to cover the same route. How long would it take them to cover the route together?

Let x = the number of hours to cover the route together.

The rate of the regular carrier is $\frac{1}{6}$ job per hour, and the rate of the substitute is $\frac{1}{8}$ job per hour.

	Rate	Time	Part of the Job Done
Regular	$\frac{1}{6}$	x	$\frac{1}{6}x$
Substitute	$\frac{1}{8}$	x	$\frac{1}{8}x$

Multiply rate by time to find the fractional part of the job done.

$$\frac{1}{6}x + \frac{1}{8}x = 1 \quad \text{The parts add to 1 whole job.}$$

$$24\left(\frac{1}{6}x + \frac{1}{8}x\right) = 24(1) \quad \text{Multiply by the LCD, 24.}$$

$$4x + 3x = 24 \quad \text{Distributive property}$$

$$7x = 24 \quad \text{Combine like terms.}$$

$$x = \frac{24}{7} \quad \text{Divide by 7.}$$

It would take them $\frac{24}{7}$ hr, or $3\frac{3}{7}$ hr, to cover the route together.

This makes sense. The time together is *less than* the time of the regular carrier working alone. Also, $\frac{1}{6}\left(\frac{24}{7}\right) + \frac{1}{8}\left(\frac{24}{7}\right) = 1.$

7.8 Variation

Solving a Variation Problem

Step 1 Write a variation equation using

Direct variation $\longrightarrow y = kx$ or $y = \frac{k}{x}$. $\longleftarrow$ Inverse variation

Step 2 Substitute the given values and solve for k.

Step 3 Write the equation with the value of k from Step 2.

Step 4 Substitute the remaining values, solve for the unknown, and find the required answer.

If y varies inversely as x, and $y = 4$ when $x = 9$, find y when $x = 3$.

$$y = \frac{k}{x} \quad \text{Equation for inverse variation}$$

$$4 = \frac{k}{9} \quad \text{Substitute the given values.}$$

$$k = 36 \quad \text{Solve for } k.$$

Because $y = \frac{k}{x}$ and $k = 36$, we have the following when $x = 3$.

$$y = \frac{36}{x} = \frac{36}{3} = 12.$$

Thus, $y = 12$ when $x = 3$.

Chapter 7 Review Exercises

7.1 *Find the numerical value of each rational expression for* **(a)** $x = -2$ *and* **(b)** $x = 4$.

1. $\dfrac{x^2}{x-5}$ **2.** $\dfrac{4x-3}{5x+2}$ **3.** $\dfrac{3x}{x^2-4}$ **4.** $\dfrac{x-1}{x+2}$

Find any value(s) of the variable for which each rational expression is undefined. Write answers with the symbol $\neq$.

5. $\dfrac{4}{x-3}$ **6.** $\dfrac{x+3}{2x}$ **7.** $\dfrac{m-2}{m^2-2m-3}$ **8.** $\dfrac{2k+1}{3k^2+17k+10}$

Write each rational expression in lowest terms.

9. $\dfrac{5a^3b^3}{15a^4b^2}$ **10.** $\dfrac{m-4}{4-m}$ **11.** $\dfrac{4x^2-9}{6-4x}$ **12.** $\dfrac{4p^2+8pq-5q^2}{10p^2-3pq-q^2}$

Write four equivalent forms of each rational expression.

13. $-\dfrac{4x-9}{2x+3}$ **14.** $-\dfrac{8-3x}{3-6x}$

7.2 *Multiply or divide. Write each answer in lowest terms.*

15. $\dfrac{8x^2}{12x^5} \cdot \dfrac{6x^4}{2x}$ **16.** $\dfrac{9m^2}{(3m)^4} \div \dfrac{6m^5}{36m}$ **17.** $\dfrac{x-3}{4} \cdot \dfrac{5}{2x-6}$

18. $\dfrac{2r+3}{r-4} \cdot \dfrac{r^2-16}{6r+9}$ **19.** $\dfrac{3q+3}{5-6q} \div \dfrac{4q+4}{2(5-6q)}$ **20.** $\dfrac{y^2-6y+8}{y^2+3y-18} \div \dfrac{y-4}{y+6}$

21. $\dfrac{2p^2+13p+20}{p^2+p-12} \cdot \dfrac{p^2+2p-15}{2p^2+7p+5}$ **22.** $\dfrac{3z^2+5z-2}{9z^2-1} \cdot \dfrac{9z^2+6z+1}{z^2+5z+6}$

7.3 *Find the LCD for the fractions in each list.*

23. $\frac{1}{8}, \frac{5}{12}, \frac{7}{32}$

24. $\frac{4}{9y}, \frac{7}{12y^2}, \frac{5}{27y^4}$

25. $\frac{1}{m^2 + 2m}, \frac{4}{m^2 + 7m + 10}$

26. $\frac{3}{x^2 + 4x + 3}, \frac{5}{x^2 + 5x + 4}, \frac{2}{x^2 + 7x + 12}$

Write each rational expression as an equivalent expression with the indicated denominator.

27. $\frac{5}{8} = \frac{?}{56}$

28. $\frac{10}{k} = \frac{?}{4k}$

29. $\frac{3}{2a^3} = \frac{?}{10a^4}$

30. $\frac{9}{x - 3} = \frac{?}{18 - 6x}$

31. $\frac{-3y}{2y - 10} = \frac{?}{50 - 10y}$

32. $\frac{4b}{b^2 + 2b - 3} = \frac{?}{(b + 3)(b - 1)(b + 2)}$

7.4 *Add or subtract. Write each answer in lowest terms.*

33. $\frac{10}{x} + \frac{5}{x}$

34. $\frac{6}{3p} - \frac{12}{3p}$

35. $\frac{9}{k} - \frac{5}{k - 5}$

36. $\frac{4}{y} + \frac{7}{7 + y}$

37. $\frac{m}{3} - \frac{2 + 5m}{6}$

38. $\frac{12}{x^2} - \frac{3}{4x}$

39. $\frac{5}{a - 2b} + \frac{2}{a + 2b}$

40. $\frac{4}{k^2 - 9} - \frac{k + 3}{3k - 9}$

41. $\frac{8}{z^2 + 6z} - \frac{3}{z^2 + 4z - 12}$

42. $\frac{11}{2p - p^2} - \frac{2}{p^2 - 5p + 6}$

7.5 *Simplify each complex fraction.*

43. $\dfrac{\frac{a^4}{b^2}}{\frac{a^3}{b}}$

44. $\dfrac{\frac{y - 3}{y}}{\frac{y + 3}{4y}}$

45. $\dfrac{\frac{3m + 2}{m}}{\frac{2m - 5}{6m}}$

46. $\dfrac{\frac{1}{p} - \frac{1}{q}}{\frac{1}{q - p}}$

47. $\dfrac{x + \frac{1}{w}}{x - \frac{1}{w}}$

48. $\dfrac{\frac{1}{r + t} - 1}{\frac{1}{r + t} + 1}$

7.6 *Solve each equation, and check the solutions.*

49. $\frac{k}{5} - \frac{2}{3} = \frac{1}{2}$

50. $\frac{4 - z}{z} + \frac{3}{2} = \frac{-4}{z}$

51. $\frac{x}{2} - \frac{x - 3}{7} = -1$

52. $\frac{3y - 1}{y - 2} = \frac{5}{y - 2} + 1$

53. $\frac{3}{x + 4} - \frac{2x}{5} = \frac{3}{x + 4}$

54. $\frac{3}{m - 2} + \frac{1}{m - 1} = \frac{7}{m^2 - 3m + 2}$

Solve each formula for the specified variable.

55. $m = \frac{Ry}{t}$ for t

56. $b = \frac{s + t}{r}$ for s

57. $a = \frac{b}{c + d}$ for d

58. $\frac{1}{r} - \frac{1}{s} = \frac{1}{t}$ for t

7.7 *Solve each problem.*

59. In a certain fraction, the denominator is 5 less than the numerator. If 5 is added to both the numerator and the denominator, the resulting fraction is equivalent to $\frac{5}{4}$. Find the original fraction (*not* written in lowest terms).

60. The denominator of a certain fraction is six times the numerator. If 3 is added to the numerator and subtracted from the denominator, the resulting fraction is equivalent to $\frac{2}{5}$. Find the original fraction (*not* written in lowest terms).

61. Ryan Hunter-Reay won the Iowa Corn Indy 300. He drove a Dallara-Honda the 262.5 mi distance with an average rate of 129.943 mph. What was his time (to the nearest thousandth of an hour)? (Data from www.indycar.com)

62. In the 2014 Winter Olympics in Sochi, Russia, Sven Kramer of the Netherlands won the men's 5000-m speed skating event in 6.179 min. What was his rate (to three decimal places)? (Data from *The World Almanac and Book of Facts.*)

63. Zachary and Samuel are brothers who share a bedroom. By himself, Zachary can completely mess up their room in 20 min, while it would take Samuel only 12 min to do the same thing. How long would it take them to mess up the room together?

64. A man can plant his garden in 5 hr, working alone. His daughter can do the same job in 8 hr. How long would it take them if they worked together?

7.8 *Solve each problem.*

65. If y varies directly as x, and $x = 12$ when $y = 5$, find x when $y = 3$.

66. If y varies inversely as x, and $y = 5$ when $x = 2$, find y when $x = 20$.

67. If a parallelogram has a fixed area, the height varies inversely as the base. A parallelogram has a height of 8 cm and a base of 12 cm. Find the height if the base is changed to 24 cm.

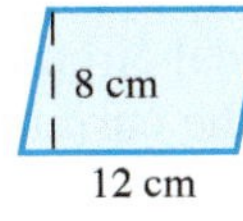

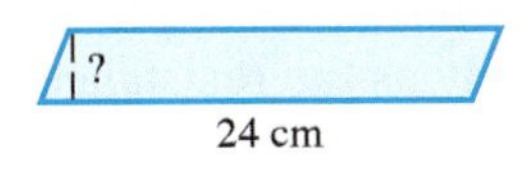

68. The circumference of a circle varies directly as its radius. A circle with circumference 9.42 in. has radius approximately 1.5 in. Find the circumference of a circle with radius 5.25 in. (to the nearest hundredth).

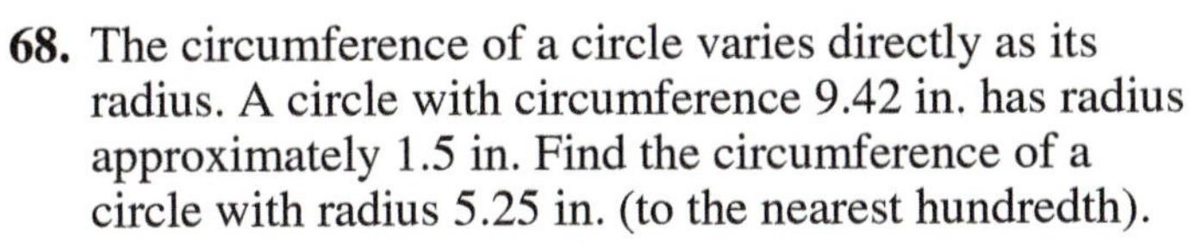

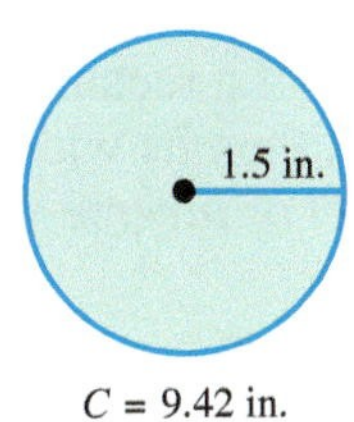

Chapter 7 Mixed Review Exercises

Perform the indicated operations. Write each answer in lowest terms.

1. $\frac{4}{m-1} - \frac{3}{m+1}$

2. $\frac{8p^5}{5} \div \frac{2p^3}{10}$

3. $\frac{r-3}{8} \div \frac{3r-9}{4}$

4. $\frac{\frac{5}{x}-1}{\frac{5-x}{3x}}$

5. $\frac{4}{z^2-2z+1} - \frac{3}{z^2-1}$

6. $\frac{2x^2+5x-12}{4x^2-9} \cdot \frac{x^2-3x-28}{x^2+8x+16}$

Solve each equation, and check the solutions.

7. $\frac{5t}{6} = \frac{2t-1}{3} + 1$

8. $\frac{2}{z} - \frac{z}{z+3} = \frac{1}{z+3}$

9. $\frac{2x}{x^2-16} - \frac{2}{x-4} = \frac{4}{x+4}$

10. Solve $a = \frac{v-w}{t}$ for w.

Solve each problem.

11. If the same number is added to both the numerator and denominator of the fraction $\frac{4}{11}$, the result is equivalent to $\frac{1}{2}$. Find the number.

12. Seema can clean the house in 3 hr. Satish can clean the house in 6 hr. Working together, how long will it take them to clean the house?

13. Anne flew her plane 400 km with the wind in the same time it took her to go 200 km against the wind. The wind speed is 50 km per hr. Find the rate of the plane in still air. Use x as the variable.

	d	r	t
With the Wind	400	______	______
Against the Wind	200	______	______

14. At a given hour, two steamboats leave a city in the same direction on a straight canal. One travels at 18 mph, and the other travels at 25 mph. In how many hours will the boats be 70 mi apart?

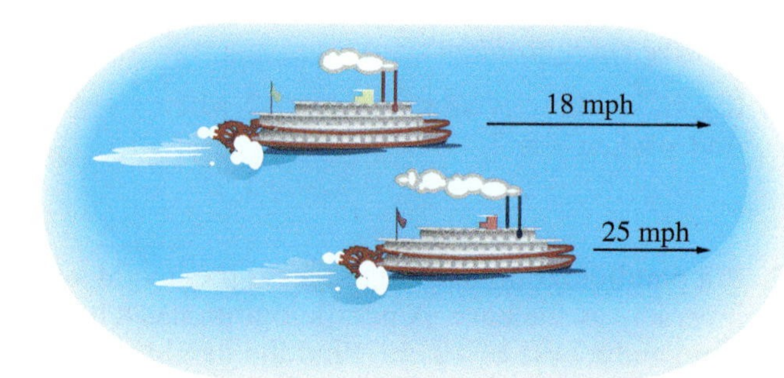

15. **CONCEPT CHECK** If a rectangle of fixed area has its length increased, then its width decreases. Is this an example of *direct* or *inverse* variation?

16. The current in a simple electrical circuit varies inversely as the resistance. If the current is 80 amps when the resistance is 10 ohms, find the current if the resistance is 16 ohms.

Chapter 7 Test

The Chapter Test Prep Videos with step-by-step solutions are available in MyMathLab or on YouTube at https://goo.gl/8aAsWP

1. Find the numerical value of $\frac{6r+1}{2r^2-3r-20}$ for each value of r.

(a) $r=-2$ (b) $r=4$

2. Find any values for which $\frac{3x-1}{x^2-2x-8}$ is undefined. Write the answer with the symbol $\neq$.

3. Write four equivalent forms of the rational expression $-\frac{6x-5}{2x+3}$.

Write each rational expression in lowest terms.

4. $\frac{-15x^6y^4}{5x^4y}$

5. $\frac{6a^2+a-2}{2a^2-3a+1}$

Multiply or divide. Write each answer in lowest terms.

6. $\frac{5(d-2)}{9} \div \frac{3(d-2)}{5}$

7. $\frac{6k^2-k-2}{8k^2+10k+3} \cdot \frac{4k^2+7k+3}{3k^2+5k+2}$

8. $\frac{4a^2+9a+2}{3a^2+11a+10} \div \frac{4a^2+17a+4}{3a^2+2a-5}$

9. $\frac{x^2-10x+25}{9-6x+x^2} \cdot \frac{x-3}{5-x}$

Find the LCD for the fractions in each list.

10. $\frac{-3}{10p^2}, \frac{21}{25p^3}, \frac{-7}{30p^5}$

11. $\frac{r+1}{2r^2+7r+6}, \frac{-2r+1}{2r^2-7r-15}$

Write each rational expression as an equivalent expression with the indicated denominator.

12. $\frac{15}{4p} = \frac{?}{64p^3}$

13. $\frac{3}{6m-12} = \frac{?}{42m-84}$

Add or subtract. Write each answer in lowest terms.

14. $\frac{4x+2}{x+5} + \frac{-2x+8}{x+5}$

15. $\frac{-4}{y+2} + \frac{6}{5y+10}$

16. $\frac{x+1}{3-x} - \frac{x^2}{x-3}$

17. $\frac{3}{2m^2-9m-5} - \frac{m+1}{2m^2-m-1}$

Simplify each complex fraction.

18. $\dfrac{\frac{2p}{k^2}}{\frac{3p^2}{k^3}}$

19. $\dfrac{\frac{x^2-25}{x+3}}{\frac{x+5}{x^2-9}}$

20. $\dfrac{\frac{1}{x+3}-1}{1+\frac{1}{x+3}}$

Solve each equation.

21. $\dfrac{3x}{x+1}=\dfrac{3}{2x}$

22. $\dfrac{2}{x-1}-\dfrac{2}{3}=\dfrac{-1}{x+1}$

23. $4+\dfrac{6}{x-3}=\dfrac{2x}{x-3}$

24. $\dfrac{2x}{x-3}+\dfrac{1}{x+3}=\dfrac{-6}{x^2-9}$

25. Solve the formula $F=\dfrac{k}{d-D}$ for D.

Solve each problem.

26. A man can paint a room in his house, working alone, in 5 hr. His wife can do the job in 4 hr. How long will it take them to paint the room if they work together?

	Rate	Time Working Together	Fractional Part of the Job Done When Working Together
Man			
Wife			

27. A boat travels 7 mph in still water. It takes as long to go 20 mi upstream as 50 mi downstream. Find the rate of the current.

	d	r	t
Upstream			
Downstream			

28. If the same number is added to the numerator and subtracted from the denominator of $\frac{5}{6}$, the resulting fraction is equivalent to $\frac{1}{10}$. What is the number?

29. If x varies directly as y, and $x = 12$ when $y = 4$, find x when $y = 9$.

30. Under certain conditions, the length of time that it takes for fruit to ripen during the growing season varies inversely as the average maximum temperature during the season. If it takes 25 days for fruit to ripen with an average maximum temperature of 80°F, find the number of days it would take at 75°F. Round the answer to the nearest whole number.

Chapters R–7 Cumulative Review Exercises

1. Evaluate $3 + 4\left(\frac{1}{2} - \frac{3}{4}\right)$.

Solve each equation.

2. $3(2t - 5) = 2 + 5t$

3. $\mathcal{A} = \frac{1}{2}bh$ for b

4. $\frac{2 + m}{2 - m} = \frac{3}{4}$

Solve each inequality. Write the solution set in interval notation, and graph it.

5. $5x \le 6x + 8$

6. $5m - 9 > 2m + 3$

Graph each equation or inequality.

7. $y = -3x + 2$

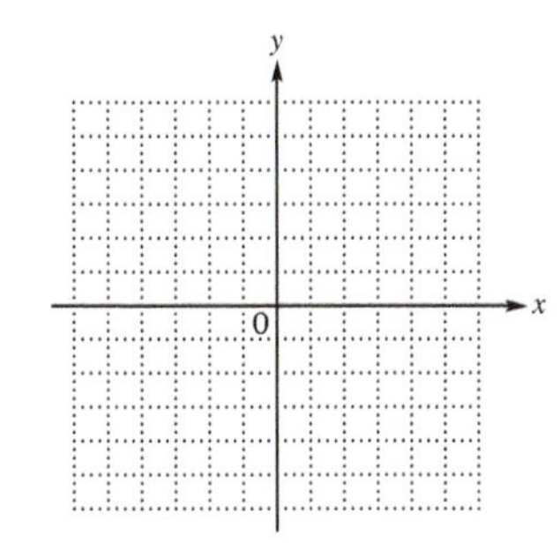

8. $y \ge 2x + 3$

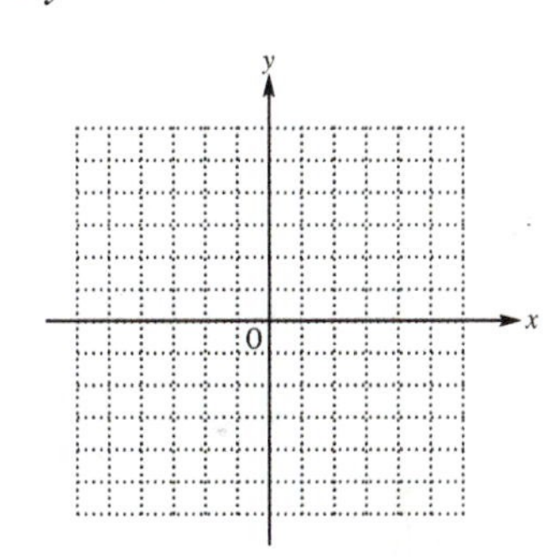

Solve each system using the method indicated.

9. $3x + 4y = 5$
$6x + 7y = 8$ (elimination)

10. $y = -3x + 1$
$x + 2y = -3$ (substitution)

Simplify each expression.

11. $\frac{(2x^3)^{-1} \cdot x}{2^3x^5}$

12. $\frac{(m^{-2})^3m}{m^5m^{-4}}$

13. $\frac{2p^3q^4}{8p^5q^3}$

Perform the indicated operations.

14. $(2k^2 + 3k) - (k^2 + k - 1)$

15. $8x^2y^2(9x^4y^5)$

16. $(2a - b)^2$

17. $(y^2 + 3y + 5)(3y - 1)$

18. $\frac{12p^3 + 2p^2 - 12p + 4}{2p - 2}$

Factor completely.

19. $8t^2 + 10tv + 3v^2$

20. $8r^2 - 9rs + 12s^2$

21. $16x^4 - 1$

Solve each equation.

22. $r^2 = 2r + 15$

23. $y^2 - 121 = 0$

24. $(r - 5)(2r + 1)(3r - 2) = 0$

Solve each problem.

25. One number is 4 greater than another. The product of the numbers is 2 less than the lesser number. Find the lesser number.

26. The length of a rectangle is 2 m less than twice the width. The area is 60 m^2. Find the width of the rectangle.

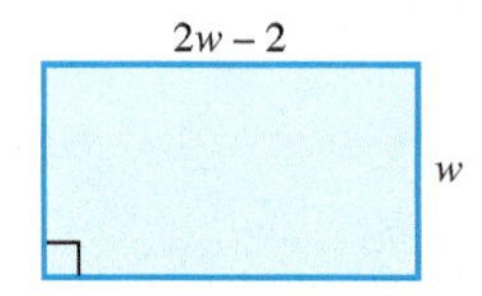

27. For what value(s) of t is $\frac{2+t}{t^2-4}$ undefined? Write the answer with the symbol $\neq$.

28. Which one of the following rational expressions is *not* equivalent to $\frac{4-3x}{7}$?

A. $-\dfrac{-4+3x}{7}$ **B.** $-\dfrac{4-3x}{-7}$ **C.** $\dfrac{-4+3x}{-7}$ **D.** $\dfrac{-(3x+4)}{7}$

Perform the indicated operations. Write each answer in lowest terms.

29. $\dfrac{5}{q} - \dfrac{1}{q}$

30. $\dfrac{3}{7} + \dfrac{4}{r}$

31. $\dfrac{4}{5q-20} - \dfrac{1}{3q-12}$

32. $\dfrac{2}{k^2+k} - \dfrac{3}{k^2-k}$

33. $\dfrac{7z^2+49z+70}{16z^2+72z-40} \div \dfrac{3z+6}{4z^2-1}$

34. $\dfrac{\frac{4}{a}+\frac{5}{2a}}{\frac{7}{6a}-\frac{1}{5a}}$

Solve each equation, and check the solutions.

35. $\dfrac{r+2}{5} = \dfrac{r-3}{3}$

36. $\dfrac{1}{x} = \dfrac{1}{x+1} + \dfrac{1}{2}$

Solve each problem.

37. Len can weed the yard in 3 hr. Bruno can weed the yard in 2 hr. How long would it take them if they worked together?

38. If the area is constant, the height of a triangle varies inversely as the base. If the height is 6 ft when the base is 4 ft, find the height when the base is 12 ft.

8 Roots and Radicals

Many formulas in mathematics and science, such as that for finding the time it takes for a pendulum to swing from one extreme to the other and back again, include *square roots,* the topic of this chapter.

8.1 Evaluating Roots

OBJECTIVES

1. Find square roots.
2. Decide whether a given root is rational, irrational, or not a real number.
3. Find decimal approximations for irrational square roots.
4. Use the Pythagorean theorem.
5. Find cube, fourth, and other roots.

OBJECTIVE 1 Find square roots. Recall that *squaring* a number means multiplying the number by itself.

$$7^2 \quad \text{means} \quad 7 \cdot 7, \quad \text{which equals} \quad 49. \qquad \text{The square of 7 is 49.}$$

The opposite (inverse) of squaring a number is taking its *square root.* This is equivalent to asking

"*What number when multiplied by itself equals* 49?"

For the example above, one answer is 7 because $7 \cdot 7 = 49$.

Square Root

A number b is a **square root** of a if $b^2 = a$ (that is, $b \cdot b = a$).

EXAMPLE 1 Finding All Square Roots of a Number

Find all square roots of 49.

We ask, "*What number when multiplied by itself equals* 49?" As mentioned above, one square root is 7. Another square root of 49 is −7, because

$$(-7)(-7) = 49.$$

Thus, the number 49 has *two* square roots: 7 and −7. One square root is positive, and one is negative.

◀ Work Problem 1 at the Side.

1 Find all square roots of each number.

(a) 100

Ask "*What number when multiplied by itself equals* 100?" There are two answers.

(b) 25 (c) 36 (d) $\frac{25}{36}$

Leonardo of Pisa (c. 1170–1250)

The **positive** or **principal square root** of a number is written with the symbol $\sqrt{}$. For example, the positive square root of 121 is 11.

$$\sqrt{121} = 11 \qquad 11^2 = 121$$

The symbol $-\sqrt{}$ is used for the **negative square root** of a number. For example, the negative square root of 121 is −11.

$$-\sqrt{121} = -11 \qquad (-11)^2 = 121$$

The **radical symbol** $\sqrt{}$ always represents the positive square root (except that $\sqrt{0} = 0$). The number inside the radical symbol is the **radicand,** and the entire expression—radical symbol and radicand—is a **radical.**

Radical symbol → $\sqrt{a}$ ← Radicand

Radical

An algebraic expression containing a radical is a **radical expression.**

The radical symbol $\sqrt{}$ has been used since sixteenth-century Germany and was probably derived from the letter *R*. The radical symbol at the right comes from the Latin word *radix,* for root. It was first used by Leonardo of Pisa (Fibonacci) in 1220.

Early radical symbol

Answers

1. (a) 10, −10 (b) 5, −5 (c) 6, −6 (d) $\frac{5}{6}, -\frac{5}{6}$

We summarize our discussion of square roots as follows.

Square Roots of a

Let a be a positive real number.

$\sqrt{a}$ is the positive or principal square root of a.

$-\sqrt{a}$ is the negative square root of a.

For nonnegative a, the following hold true.

$$\sqrt{a} \cdot \sqrt{a} = \left(\sqrt{a}\right)^2 = a \quad \text{and} \quad -\sqrt{a} \cdot \left(-\sqrt{a}\right) = \left(-\sqrt{a}\right)^2 = a$$

Also, $\sqrt{0} = 0$.

EXAMPLE 2 Finding Square Roots

Find each square root.

(a) $\sqrt{144}$

The radical $\sqrt{144}$ represents the positive or principal square root of 144. Think of a positive number whose square is 144.

$$12^2 = 144, \quad \text{so} \quad \sqrt{144} = 12.$$

(b) $-\sqrt{1024}$

This symbol represents the negative square root of 1024. A calculator with a square root key can be used to find $\sqrt{1024} = 32$. Therefore,

$$-\sqrt{1024} = -32.$$

(c) $\sqrt{\frac{4}{9}} = \frac{2}{3}$ **(d)** $-\sqrt{\frac{16}{49}} = -\frac{4}{7}$ **(e)** $\sqrt{0.81} = 0.9$

$(0.9)^2 = 0.9 \cdot 0.9 = 0.81$

Work Problem 2 at the Side. ▶

! CAUTION

By definition, $\sqrt{4} = 2$ because $2^2 = 4$. ***In general, however, the square root of a number is not half the number.***

As noted above, when the square root of a positive real number is squared, the result is that positive real number. $\left(\text{Also, } \left(\sqrt{0}\right)^2 = 0.\right)$

EXAMPLE 3 Squaring Radical Expressions

Find the *square* of each radical expression.

(a) $\sqrt{13}$ The square of $\sqrt{13}$ is $\left(\sqrt{13}\right)^2 = 13$. Definition of square root

(b) $-\sqrt{29}$

$\left(-\sqrt{29}\right)^2 = 29$

The square of a *negative* number is positive.

(c) $\sqrt{p^2 + 1}$

$\left(\sqrt{p^2 + 1}\right)^2 = p^2 + 1$

Work Problem 3 at the Side. ▶

---

2 Find each square root.

GS **(a)** $\sqrt{16}$

$4^2 =$ ____, so

$\sqrt{16} =$ ____.

GS **(b)** $-\sqrt{169}$

$13^2 =$ ____, so

$-\sqrt{169} =$ ____.

(c) $-\sqrt{225}$ **(d)** $\sqrt{729}$

(e) $-\sqrt{\frac{36}{25}}$ **(f)** $\sqrt{0.49}$

3 Find the *square* of each radical expression.

GS **(a)** $\sqrt{41}$

$\left(\sqrt{41}\right)^2 =$ ____

GS **(b)** $-\sqrt{39}$

$\left(-\sqrt{39}\right)^2 =$ ____

(c) $\sqrt{120}$

(d) $\sqrt{2x^2 + 3}$

Answers

2. **(a)** 16; 4 **(b)** 169; −13 **(c)** −15 **(d)** 27 **(e)** $-\frac{6}{5}$ **(f)** 0.7

3. **(a)** 41 **(b)** 39 **(c)** 120 **(d)** $2x^2 + 3$

4 Determine whether each square root is *rational, irrational,* or *not a real number.*

(a) $\sqrt{9}$

(b) $\sqrt{7}$

(c) $\sqrt{\frac{9}{16}}$

(d) $\sqrt{72}$

(e) $\sqrt{-43}$

Answers

4. (a) rational (b) irrational (c) rational (d) irrational (e) not a real number

OBJECTIVE 2 Decide whether a given root is rational, irrational, or not a real number. Numbers with rational square roots are **perfect squares.**

Perfect squares: 25, 144, $\frac{4}{9}$ are perfect squares because $\sqrt{25} = 5$, $\sqrt{144} = 12$, $\sqrt{\frac{4}{9}} = \frac{2}{3}$ (Rational square roots)

A number that is not a perfect square has a square root that is not a rational number. For example, $\sqrt{5}$ is not a rational number because it cannot be written as the ratio of two integers. Its decimal equivalent (or approximation) neither terminates nor repeats. However, $\sqrt{5}$ is a real number and corresponds to a point on the number line.

A real number that is not rational is an **irrational number.** The number $\sqrt{5}$ is irrational. ***Many square roots of integers are irrational.***

If a is a ***positive*** real number that is ***not*** a perfect square, then $\sqrt{a}$ is irrational.

Not every number has a real number square root. For example, there is no real number that can be squared to obtain -36. (The square of a real number can never be negative.) Because of this, $\sqrt{-36}$ ***is not a real number.***

If a is a ***negative*** real number, then $\sqrt{a}$ is *not* a real number.

CAUTION

Do not confuse $\sqrt{-36}$ and $-\sqrt{36}$. $\sqrt{-36}$ is not a real number because there is no real number that can be squared to obtain -36. However, $-\sqrt{36}$ is the negative square root of 36, which is -6.

EXAMPLE 4 Identifying Types of Square Roots

Determine whether each square root is *rational, irrational,* or *not a real number.*

(a) $\sqrt{17}$ Because 17 is not a perfect square, $\sqrt{17}$ is irrational.

(b) $\sqrt{64}$ 64 is a perfect square, 8^2, so $\sqrt{64} = 8$ is a rational number.

(c) $\sqrt{-25}$ There is no real number whose square is -25. Therefore, $\sqrt{-25}$ is not a real number.

◀ Work Problem 4 at the Side.

Note

Not all irrational numbers are square roots of integers. For example, the number π (approximately 3.14159) is an irrational number that is not a square root of any integer.

OBJECTIVE 3 Find decimal approximations for irrational square roots. Even if a number is irrational, a decimal that approximates the number can be found using a calculator.

EXAMPLE 5 Approximating Irrational Square Roots

Find a decimal approximation for each square root. Round answers to the nearest thousandth.

(a) $\sqrt{11}$

Using the square root key of a calculator gives $3.31662479 \approx 3.317$, where $\approx$ means **"is approximately equal to."**

(b) $\sqrt{39} \approx 6.245$ Use a calculator. **(c)** $-\sqrt{741} \approx -27.221$

Work Problem 5 at the Side. ▶

5 Find a decimal approximation for each square root. Round answers to the nearest thousandth.

(a) $\sqrt{28}$ **(b)** $\sqrt{63}$

(c) $-\sqrt{190}$ **(d)** $\sqrt{1000}$

OBJECTIVE 4 Use the Pythagorean theorem. Many applications of square roots use the Pythagorean theorem. Recall that if c is the length of the hypotenuse of a right triangle, and a and b are the lengths of the two legs, then

$$a^2 + b^2 = c^2.$$

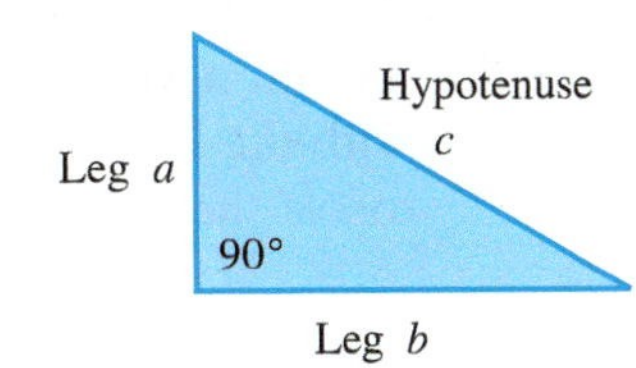

Figure 1

In the next example, we use the fact that if $k > 0$, then the positive solution of the equation $x^2 = k$ is $\sqrt{k}$.

EXAMPLE 6 Using the Pythagorean Theorem

Find the length of the unknown side of each right triangle with sides a, b, and c, where c is the hypotenuse.

(a) $a = 3, b = 4$

$a^2 + b^2 = c^2$ Use the Pythagorean theorem.

$3^2 + 4^2 = c^2$ Let $a = 3$ and $b = 4$.

$9 + 16 = c^2$ Square.

$25 = c^2$ Add.

Because the length of a side of a triangle must be a positive number, find the positive square root of 25 to obtain c.

$$c = \sqrt{25} = 5$$

(b) $b = 5, c = 9$

$a^2 + b^2 = c^2$ Use the Pythagorean theorem.

Solve for a^2. → $a^2 + 5^2 = 9^2$ Let $b = 5$ and $c = 9$.

$a^2 + 25 = 81$ Square.

$a^2 = 56$ Subtract 25.

Use a calculator to find the positive square root of 56 to approximate a.

$$a = \sqrt{56} \approx 7.483 \quad \text{Nearest thousandth}$$

Work Problem 6 at the Side. ▶

6 Find the length of the unknown side of each right triangle with sides a, b, and c, where c is the hypotenuse. Give any decimal approximations to the nearest thousandth.

(a) $a = 7, b = 24$

(b) $b = 13, c = 15$

(c)

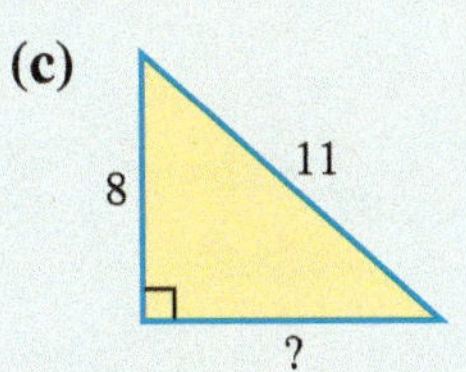

Answers

5. (a) 5.292 **(b)** 7.937 **(c)** −13.784 **(d)** 31.623

6. (a) 25 **(b)** 7.483 **(c)** 7.550

7 A rectangle has dimensions 5 ft by 12 ft. Find the length of its diagonal.

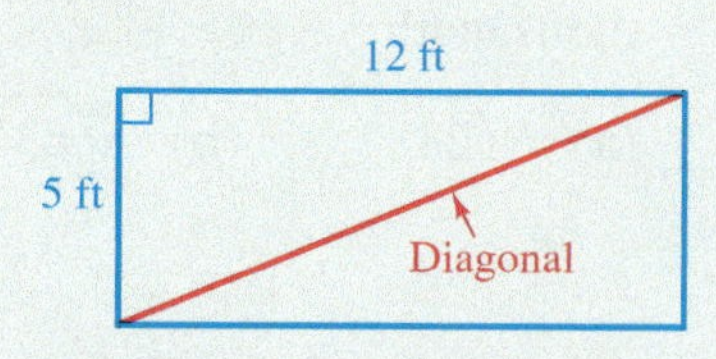

(Note that the diagonal divides the rectangle into two right triangles with itself as the hypotenuse.)

Answer

7. 13 ft

! CAUTION

Be careful not to make the common mistake of thinking that $\sqrt{a^2 + b^2}$ equals $a + b$. Consider the following.

$$\sqrt{9 + 16} = \sqrt{25} = 5 \quad \text{Correct}$$

$$\sqrt{9 + 16} = \sqrt{9} + \sqrt{16} = 3 + 4 = 7 \quad \text{Incorrect}$$

In general, $\sqrt{a^2 + b^2} \neq a + b$.

EXAMPLE 7 Using the Pythagorean Theorem to Solve an Application

A ladder 10 ft long leans against a wall. The foot of the ladder is 6 ft from the base of the wall. How high up the wall does the top of the ladder rest?

Step 1 **Read** the problem again.

Step 2 **Assign a variable.** As shown in **Figure 2,** a right triangle is formed with the ladder as the hypotenuse. Let a represent the height of the top of the ladder when measured straight down to the ground.

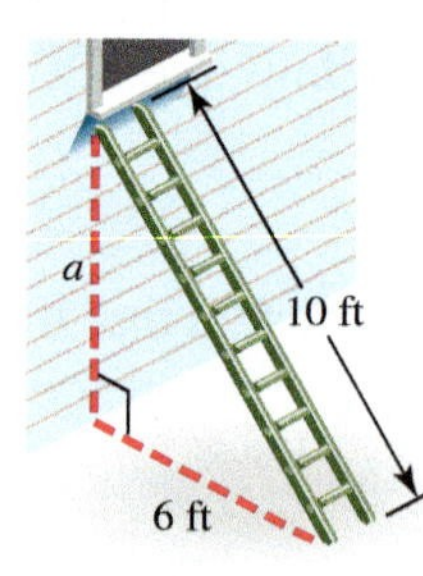

Figure 2

Recall that the symbol ⌉ indicates a 90° or right angle.

Step 3 **Write an equation** using the Pythagorean theorem, and substitute the known quantities. Because the hypotenuse is the longest side of a right triangle, the greatest value is *always* substituted for c.

$$a^2 + b^2 = c^2$$

$$a^2 + 6^2 = 10^2 \quad \text{Let } b = 6 \text{ and } c = 10.$$

Substitute carefully.

Step 4 **Solve.**

$$a^2 + 36 = 100 \quad \text{Square.}$$

$$a^2 = 64 \quad \text{Subtract 36.}$$

$$a = \sqrt{64} \quad \text{Solve for } a.$$

$$a = 8 \quad \sqrt{64} = 8$$

Choose the positive square root of 64 because a represents a length.

Step 5 **State the answer.** The top of the ladder rests 8 ft up the wall.

Step 6 **Check.** From **Figure 2,** we have the following.

$$8^2 + 6^2 \stackrel{?}{=} 10^2 \quad a^2 + b^2 = c^2$$

$$64 + 36 = 100 \checkmark \quad \text{True}$$

The check confirms that the top of the ladder rests 8 ft up the wall.

◀ **Work Problem 7 at the Side.**

OBJECTIVE 5 Find cube, fourth, and other roots. Finding the square root of a number is the inverse (opposite) of squaring a number. There are inverses to finding the cube of a number and to finding the fourth or greater power of a number. These inverses are, respectively, the **cube root,** $\sqrt[3]{a}$**,** and the **fourth root,** $\sqrt[4]{a}$**.** Similar symbols are used for other roots.

> **$\sqrt[n]{a}$**
>
> The nth root of a, written $\sqrt[n]{a}$, is a number whose nth power equals a. That is,
>
> $$\sqrt[n]{a} = b \quad \text{means} \quad b^n = a.$$

In $\sqrt[n]{a}$, the number n is the **index,** or **order,** of the radical.

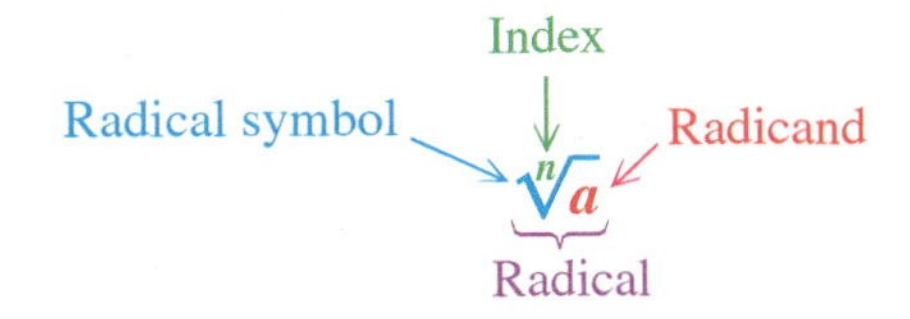

We could write $\sqrt[2]{a}$ instead of $\sqrt{a}$, but the simpler symbol $\sqrt{a}$ is customary since the square root is the most commonly used root.

When working with cube roots or fourth roots, it is helpful to memorize the first few **perfect cubes** ($1^3 = 1$, $2^3 = 8$, $3^3 = 27$, and so on) and the first few **perfect fourth powers** ($1^4 = 1$, $2^4 = 16$, $3^4 = 81$, and so on).

Work Problem 8 at the Side. ▶

EXAMPLE 8 Finding Cube Roots

Find each cube root.

(a) $\sqrt[3]{8}$

Ask "*What number can be cubed to give* 8?" Because $2^3 = 8$, $\sqrt[3]{8} = 2$.

$2^3 = 2 \cdot 2 \cdot 2$

(b) $\sqrt[3]{-8}$ Because $(-2)^3 = -8$, $\sqrt[3]{-8} = -2$.

(c) $\sqrt[3]{216}$ Because $6^3 = 216$, $\sqrt[3]{216} = 6$.

Work Problem 9 at the Side. ▶

Notice in **Example 8(b)** that we can find the cube root of a negative number. (Contrast this with the square root of a negative number, which is not real.) In fact, the cube root of a positive number is positive, and the cube root of a negative number is negative. ***There is only one real number cube root for each real number.***

When a radical has an ***even index*** (square root, fourth root, and so on), ***the radicand must be nonnegative*** to yield a real number root. Also, for $a > 0$,

$$\sqrt{a},\ \sqrt[4]{a},\ \sqrt[6]{a}, \text{ and so on are positive (principal) roots.}$$

$$-\sqrt{a},\ -\sqrt[4]{a},\ -\sqrt[6]{a}, \text{ and so on are negative roots.}$$

8 Complete the following list of perfect cubes and perfect fourth powers.

Perfect Cubes	Perfect Fourth Powers
$1^3 = 1$	$1^4 = 1$
$2^3 = 8$	$2^4 = 16$
$3^3 = 27$	$3^4 = 81$
$4^3 =$ ____	$4^4 =$ ____
$5^3 =$ ____	$5^4 =$ ____
$6^3 =$ ____	$6^4 =$ ____
$7^3 =$ ____	$7^4 =$ ____
$8^3 =$ ____	$8^4 =$ ____
$9^3 =$ ____	$9^4 =$ ____
$10^3 =$ ____	$10^4 =$ ____

9 Find each cube root.

GS **(a)** $\sqrt[3]{27}$

____$^3 = 27$, so

$\sqrt[3]{27} =$ ____.

(b) $\sqrt[3]{1}$

(c) $\sqrt[3]{-125}$

Answers

8. Perfect cubes: 64; 125; 216; 343; 512; 729; 1000
Perfect fourth powers: 256; 625; 1296; 2401; 4096; 6561; 10,000

9. (a) 3; 3 (b) 1 (c) −5

10 Find each root.

(a) $\sqrt[4]{81}$

(b) $\sqrt[4]{-81}$

(c) $-\sqrt[4]{81}$

(d) $\sqrt[5]{243}$

(e) $\sqrt[5]{-243}$

Answers

10. (a) 3 (b) not a real number (c) −3 (d) 3 (e) −3

EXAMPLE 9 Finding Other Roots

Find each root.

$2^4 = 2 \cdot 2 \cdot 2 \cdot 2$

(a) $\sqrt[4]{16}$ Because 2 is positive and $2^4 = 16$, $\sqrt[4]{16} = 2$.

(b) $-\sqrt[4]{16}$

From part (a), $\sqrt[4]{16} = 2$, so the negative root is $-\sqrt[4]{16} = -2$.

(c) $\sqrt[4]{-16}$

For a real number fourth root, the radicand must be nonnegative. There is no real number that equals $\sqrt[4]{-16}$.

(d) $-\sqrt[5]{32}$

First find $\sqrt[5]{32}$. Because 2 is the number whose fifth power is 32, $\sqrt[5]{32} = 2$. Since $\sqrt[5]{32} = 2$, it follows that

$$-\sqrt[5]{32} = -2.$$

(e) $\sqrt[5]{-32}$ Because $(-2)^5 = -32$, $\sqrt[5]{-32} = -2$.

◀ Work Problem **10** at the Side.

8.1 Exercises

FOR EXTRA HELP *Go to* MyMathLab *for worked-out, step-by-step solutions to exercises enclosed in a square*  *and video solutions to ▶ exercises.*

CONCEPT CHECK *Decide whether each statement is* true *or* false. *If false, tell why.*

1. Every positive number has two real square roots.

2. A negative number has negative square roots.

3. Every nonnegative number has two real square roots.

4. The positive square root of a positive number is its principal square root.

5. The cube root of every real number has the same sign as the number itself.

6. Every positive number has three real cube roots.

CONCEPT CHECK *What must be true about the value of the variable a for each statement to be true?*

7. $\sqrt{a}$ represents a positive number.

8. $-\sqrt{a}$ represents a negative number.

9. $\sqrt{a}$ is not a real number.

10. $-\sqrt{a}$ is not a real number.

Find all square roots of each number. **See Example 1.**

11. 9 ▶ **12.** 16 **13.** 64 **14.** 121 **15.** 169

16. 225 **17.** $\frac{25}{196}$ **18.** $\frac{81}{400}$ **19.** 900 **20.** 1600

Find each square root. **See Examples 2 and 4(c).**

21. $\sqrt{1}$ **22.** $\sqrt{4}$ **23.** $\sqrt{64}$ **24.** $\sqrt{9}$

25. $\sqrt{49}$ ▶ **26.** $\sqrt{81}$ **27.** $-\sqrt{256}$ **28.** $-\sqrt{196}$

29. $-\sqrt{\frac{144}{121}}$ ▶ **30.** $-\sqrt{\frac{49}{36}}$ **31.** $\sqrt{0.64}$ **32.** $\sqrt{0.16}$

33. $\sqrt{-121}$ **34.** $\sqrt{-64}$ **35.** $-\sqrt{-49}$ **36.** $-\sqrt{-100}$

Find the square of each radical expression. **See Example 3.**

37. $\sqrt{100}$
GS $(\sqrt{100})^2 =$ ____

38. $\sqrt{36}$
GS $(\sqrt{36})^2 =$ ____

39. $-\sqrt{19}$

40. $-\sqrt{99}$

41. $\sqrt{\frac{2}{3}}$ **42.** $\sqrt{\frac{5}{7}}$ **43.** $\sqrt{3x^2 + 4}$ ▶ **44.** $\sqrt{9y^2 + 3}$

CONCEPT CHECK *Without using a calculator, determine between which two consecutive integers each square root lies. For example,*

$\sqrt{75}$ *is between* 8 *and* 9, *because* $\sqrt{64} = 8$, $\sqrt{81} = 9$, *and* $64 < 75 < 81$.

45. $\sqrt{94}$ **46.** $\sqrt{43}$ **47.** $\sqrt{51}$ **48.** $\sqrt{30}$

49. $-\sqrt{40}$ **50.** $-\sqrt{63}$ **51.** $\sqrt{23.2}$ **52.** $\sqrt{10.3}$

Determine whether each number is rational, irrational, *or* not a real number. *If a number is rational, give its exact value. If a number is irrational, give a decimal approximation to the nearest thousandth. Use a calculator as necessary.* ***See Examples 4 and 5.***

53. $\sqrt{25}$ **54.** $\sqrt{169}$ **55.** $\sqrt{29}$ **56.** $\sqrt{33}$

57. $-\sqrt{64}$ **58.** $-\sqrt{81}$ **59.** $-\sqrt{300}$ **60.** $-\sqrt{500}$

61. $\sqrt{-29}$ **62.** $\sqrt{-47}$ **63.** $\sqrt{1200}$ **64.** $\sqrt{1500}$

Work each exercise without using a calculator.

65. Choose the best estimate for the length and width (in meters) of this rectangle.

A. 11 by 6 **B.** 11 by 7 **C.** 10 by 7 **D.** 10 by 6

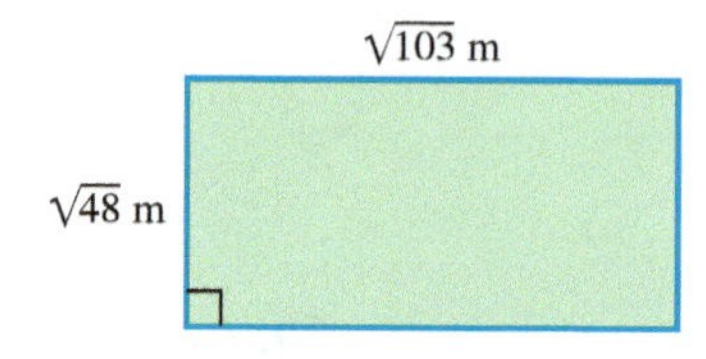

66. Choose the best estimate for the base and height (in feet) of this triangle.

A. $b = 8, h = 5$ **B.** $b = 8, h = 4$

C. $b = 9, h = 5$ **D.** $b = 9, h = 4$

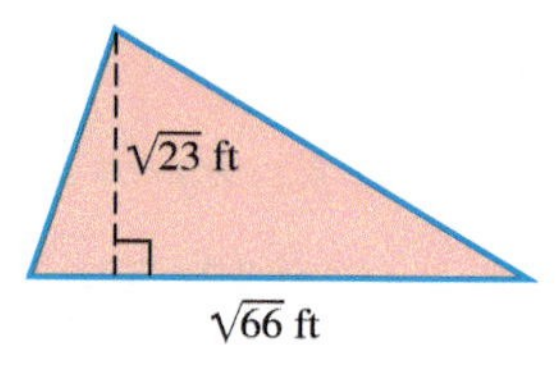

Find the length of the unknown side of each right triangle with sides a, b, and c, where c is the hypotenuse. Give any decimal approximations to the nearest thousandth. ***See Figure 1 and Example 6.***

67. $a = 8, b = 15$ **68.** $a = 24, b = 10$ **69.** $a = 6, c = 10$ **70.** $a = 5, c = 13$

71. $a = 11, b = 4$ **72.** $a = 13, b = 9$ **73.** $b = 10, c = 15$ **74.** $b = 13, c = 20$

Solve each problem. Give any decimal approximations to the nearest tenth. ***See Example 7.***

75. The diagonal of a rectangle measures 25 cm. The width of the rectangle is 7 cm. Find the length of the rectangle.

25 cm
7 cm

76. The length of a rectangle is 40 m, and the width is 9 m. Find the measure of the diagonal of the rectangle.

9 m
40 m

77. Tyler is flying a kite on 100 ft of string. How high is it above his hand (vertically) if the horizontal distance between Tyler and the kite is 60 ft?

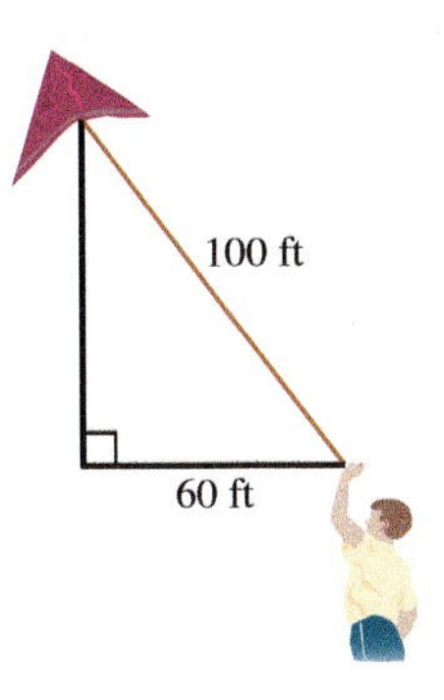

78. A guy wire is attached to the mast of a transmitting antenna at a point 96 ft above ground level. If the wire is staked to the ground 72 ft from the base of the mast, how long is the wire?

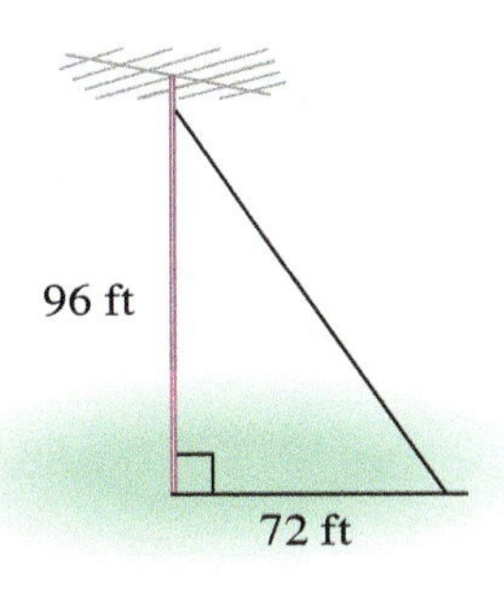

79. A surveyor measured the distances shown in the figure. Find the distance across the lake between points R and S.

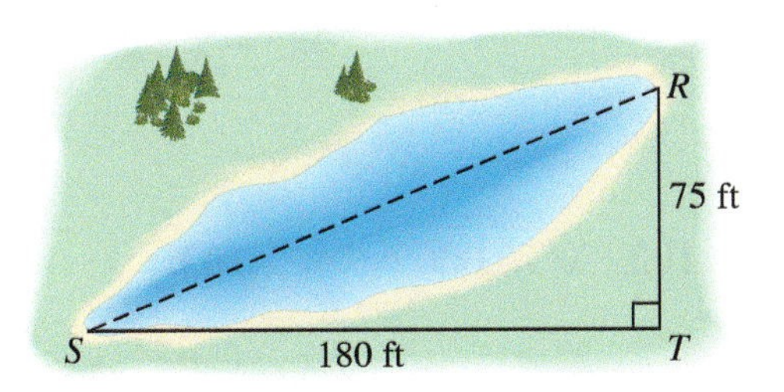

80. A boat is being pulled toward a dock with a rope attached at water level. When the boat is 24 ft from the dock, 30 ft of rope is extended. What is the height of the dock above the water?

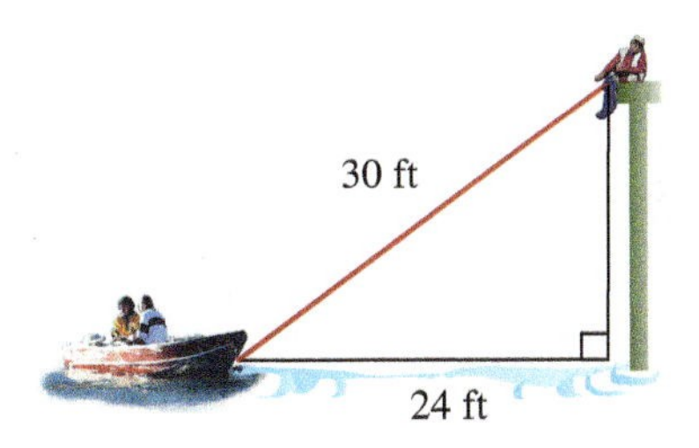

81. A surveyor wants to find the height of a building. At a point 110.0 ft from the base of the building he sights to the top of the building and finds the distance to be 193.0 ft. How high is the building?

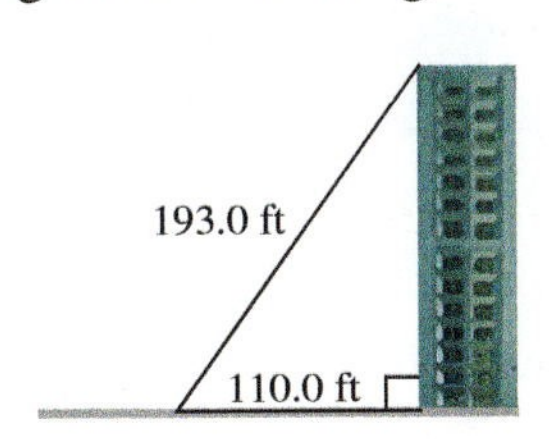

82. Two towns are separated by a dense forest. To go from Town B to Town A, it is necessary to travel due west for 19.0 mi, then turn due north and travel for 14.0 mi. How far apart are the towns?

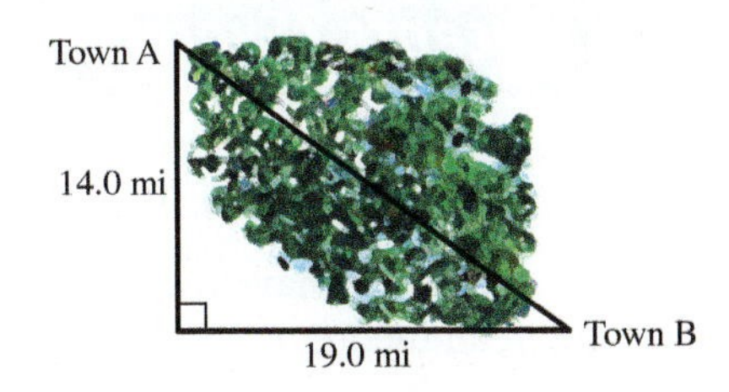

83. During Hurricane Katrina, thousands of pine trees snapped off to form right triangles. Suppose that for one such tree, the vertical distance from the base of the broken tree to the point of the break was 4.5 ft. The length of the broken part was 12.0 ft. How far along the ground was it from the base of the tree to the point where the broken part touched the ground?

84. A television set is "sized" according to the diagonal measurement of the viewing screen. A rectangular 46-in. TV measures 46 in. from one corner of the viewing screen diagonally to the other corner. The viewing screen is 40 in. wide. Find the height of the viewing screen.

85. What is the value of x (to the nearest thousandth) in the figure?

86. What is the value of y (to the nearest thousandth) in the figure?

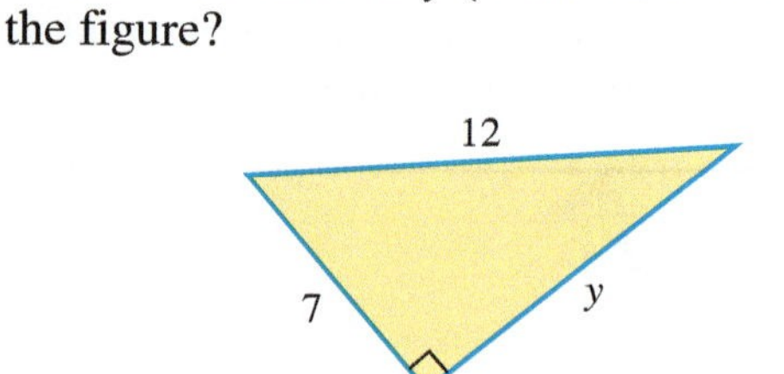

Find each root. ***See Examples 8 and 9.***

87. $\sqrt[3]{64}$

88. $\sqrt[3]{343}$

89. $\sqrt[3]{125}$

90. $\sqrt[3]{729}$

91. $\sqrt[3]{512}$

92. $\sqrt[3]{1000}$

93. $\sqrt[3]{-27}$

94. $\sqrt[3]{-64}$

95. $\sqrt[3]{-216}$

96. $\sqrt[3]{-343}$

97. $-\sqrt[3]{-8}$

98. $-\sqrt[3]{-216}$

99. $\sqrt[4]{625}$

100. $\sqrt[4]{256}$

101. $\sqrt[4]{1296}$

102. $\sqrt[4]{10{,}000}$

103. $\sqrt[4]{-1}$

104. $\sqrt[4]{-625}$

105. $-\sqrt[4]{625}$

106. $-\sqrt[4]{256}$

107. $\sqrt[5]{-1024}$

108. $\sqrt[5]{-100{,}000}$

Relating Concepts (Exercises 109 and 110) For Individual or Group Work

Although Pythagoras may have written the first proof of the Pythagorean relationship, many other proofs have been given. Two of them follow.

109. The Babylonians may have used a tile pattern like that shown here to illustrate the Pythagorean theorem.

(a) The side of the square along the hypotenuse measures 5 units. What are the measures of the sides along the legs?

(b) Using the measures from part (a), show that the Pythagorean theorem is satisfied.

110. The diagram shown here can be used to verify the Pythagorean theorem.

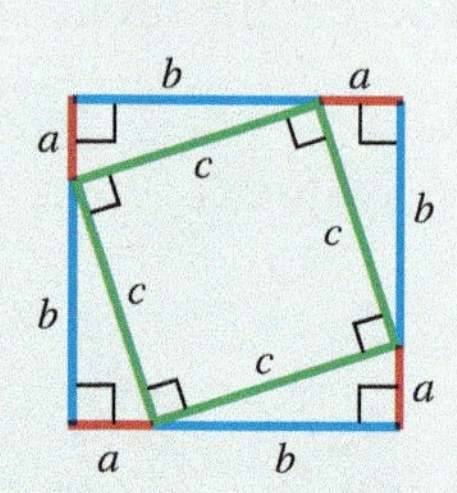

(a) Find the area of the large square.

(b) Find the sum of the areas of the smaller square and the four right triangles.

(c) Set the areas equal and write a simpler, equivalent equation.

8.2 Multiplying, Dividing, and Simplifying Radicals

OBJECTIVES

1. Multiply square root radicals.
2. Simplify radicals using the product rule.
3. Simplify radicals using the quotient rule.
4. Simplify radicals involving variables.
5. Simplify other roots.

OBJECTIVE 1 Multiply square root radicals. Consider the following.

$$\sqrt{4} \cdot \sqrt{9} = 2 \cdot 3 = 6 \quad \text{and} \quad \sqrt{4 \cdot 9} = \sqrt{36} = 6$$

This shows that $\sqrt{4} \cdot \sqrt{9} = \sqrt{4 \cdot 9}$.

The result here is a particular case of the **product rule for radicals.**

Product Rule for Radicals

If a and b are nonnegative real numbers, then the following hold true.

$$\sqrt{a} \cdot \sqrt{b} = \sqrt{a \cdot b} \quad \text{and} \quad \sqrt{a \cdot b} = \sqrt{a} \cdot \sqrt{b}$$

In words, the product of two square roots is the square root of the product. The square root of a product is the product of the two square roots.

EXAMPLE 1 Using the Product Rule to Multiply Radicals

Use the product rule for radicals to find each product.

(a) $\sqrt{2} \cdot \sqrt{3}$
$= \sqrt{2 \cdot 3}$
$= \sqrt{6}$

(b) $\sqrt{7} \cdot \sqrt{5}$
$= \sqrt{35}$

(c) $\sqrt{11} \cdot \sqrt{a}$ $(a \geq 0)$
$= \sqrt{11a}$

Work Problem 1 at the Side. ▶

OBJECTIVE 2 Simplify radicals using the product rule. ***A square root radical is simplified when no perfect square factor other than 1 remains under the radical symbol.*** This is accomplished using the product rule.

EXAMPLE 2 Using the Product Rule to Simplify Radicals

Simplify each radical.

(a) $\sqrt{20}$ (20 has a perfect square factor of 4.)

$= \sqrt{4 \cdot 5}$ Factor; 4 is a perfect square.

$= \sqrt{4} \cdot \sqrt{5}$ Product rule in the form $\sqrt{a \cdot b} = \sqrt{a} \cdot \sqrt{b}$

$= 2\sqrt{5}$ $\sqrt{4} = 2$

Thus, $\sqrt{20} = 2\sqrt{5}$. Because 5 has no perfect square factor (other than 1), $2\sqrt{5}$ is the **simplified form** of $\sqrt{20}$. Note that $2\sqrt{5}$ represents a product whose factors are 2 and $\sqrt{5}$.

Alternatively, we could begin by factoring 20 into prime factors.

$\sqrt{20}$

$= \sqrt{2 \cdot 2 \cdot 5}$ Look for pairs of like factors. Each pair produces one factor outside the radical.

$= 2\sqrt{5}$ The result is the same.

Continued on Next Page

1 Use the product rule for radicals to find each product.

GS **(a)** $\sqrt{6} \cdot \sqrt{11}$
$= \sqrt{\_\_\_ \cdot \_\_\_}$
$= \sqrt{\_\_\_}$

(b) $\sqrt{2} \cdot \sqrt{5}$

(c) $\sqrt{10} \cdot \sqrt{r}$ $(r \geq 0)$

Answers

1. (a) 6; 11; 66 (b) $\sqrt{10}$ (c) $\sqrt{10r}$

2 Simplify each radical.

GS (a) $\sqrt{8}$

$= \sqrt{\_\_\_ \cdot 2}$

$= \sqrt{\_\_\_} \cdot \sqrt{2}$

$= \_\_\_$

(b) $\sqrt{27}$

(c) $\sqrt{50}$

(d) $-\sqrt{60}$

(e) $\sqrt{30}$

Answers

2. (a) 4; 4; $2\sqrt{2}$ (b) $3\sqrt{3}$ (c) $5\sqrt{2}$ (d) $-2\sqrt{15}$ (e) It cannot be simplified.

(b) $\sqrt{72}$ Look for the *greatest* perfect square factor of 72.

$= \sqrt{36 \cdot 2}$ Factor; 36 is a perfect square.

$= \sqrt{36} \cdot \sqrt{2}$ Product rule

$= 6\sqrt{2}$ $\sqrt{36} = 6$

We could also factor 72 into prime factors and look for pairs of like factors.

$\sqrt{72}$

$= \sqrt{2 \cdot 2 \cdot 2 \cdot 3 \cdot 3}$ Factor into primes.

$= 2 \cdot 3 \cdot \sqrt{2}$ $\sqrt{2 \cdot 2} = 2$; $\sqrt{3 \cdot 3} = 3$

$= 6\sqrt{2}$ The result is the same.

(c) $-\sqrt{300}$

$= -\sqrt{100 \cdot 3}$ 100 is a perfect square.

$= -\sqrt{100} \cdot \sqrt{3}$ Product rule

$= -10\sqrt{3}$ $\sqrt{100} = 10$

(d) $\sqrt{15}$ Because 15 has no perfect square factors (except 1), $\sqrt{15}$ cannot be simplified.

◀ Work Problem 2 at the Side.

EXAMPLE 3 Multiplying and Simplifying Radicals

Find each product and simplify.

(a) $\sqrt{9} \cdot \sqrt{75}$

$= 3\sqrt{75}$ $\sqrt{9} = 3$

$= 3\sqrt{25 \cdot 3}$ Factor; 25 is a perfect square.

$= 3\sqrt{25} \cdot \sqrt{3}$ Product rule

$= 3 \cdot 5 \cdot \sqrt{3}$ $\sqrt{25} = 5$

$= 15\sqrt{3}$ Multiply.

We could have used the product rule to obtain $\sqrt{9} \cdot \sqrt{75} = \sqrt{675}$, and then simplified. The method above is preferred because it uses smaller numbers.

(b) $\sqrt{8} \cdot \sqrt{12}$

$= \sqrt{8 \cdot 12}$ Product rule

$= \sqrt{4 \cdot 2 \cdot 4 \cdot 3}$ Factor; 4 is a perfect square.

$= \sqrt{4} \cdot \sqrt{4} \cdot \sqrt{2 \cdot 3}$ Commutative property; product rule

$= 2 \cdot 2 \cdot \sqrt{6}$ $\sqrt{4} = 2$

$= 4\sqrt{6}$ Multiply.

Continued on Next Page

(c) $2\sqrt{3} \cdot 3\sqrt{6}$

$= 2 \cdot 3 \cdot \sqrt{3 \cdot 6}$ Commutative property; product rule

$= 6\sqrt{18}$ Multiply.

$= 6\sqrt{9 \cdot 2}$ Factor; 9 is a perfect square.

$= 6\sqrt{9} \cdot \sqrt{2}$ Product rule

$= 6 \cdot 3 \cdot \sqrt{2}$ $\sqrt{9} = 3$

$= 18\sqrt{2}$ Multiply.

Work Problem 3 at the Side. ▶

Note

There is often more than one way to find a product of radicals.

$\sqrt{8} \cdot \sqrt{12}$ See Example 3(b).

$= \sqrt{4 \cdot 2} \cdot \sqrt{4 \cdot 3}$ Factor.

$= 2\sqrt{2} \cdot 2\sqrt{3}$ $\sqrt{4} = 2$

$= 2 \cdot 2 \cdot \sqrt{2} \cdot \sqrt{3}$ Commutative property

Same result → $= 4\sqrt{6}$ Multiply; product rule

OBJECTIVE 3 Simplify radicals using the quotient rule. The **quotient rule for radicals** is very similar to the product rule.

Quotient Rule for Radicals

If a and b are nonnegative real numbers and $b \neq 0$, then the following hold true.

$$\sqrt{\frac{a}{b}} = \frac{\sqrt{a}}{\sqrt{b}} \quad \text{and} \quad \frac{\sqrt{a}}{\sqrt{b}} = \sqrt{\frac{a}{b}}$$

In words, the square root of a quotient is the quotient of the two square roots. The quotient of two square roots is the square root of the quotient.

EXAMPLE 4 **Using the Quotient Rule to Simplify Radicals**

Use the quotient rule to simplify each radical.

(a) $\sqrt{\frac{25}{9}}$

$= \frac{\sqrt{25}}{\sqrt{9}}$

$= \frac{5}{3}$

(b) $\frac{\sqrt{288}}{\sqrt{2}}$

$= \sqrt{\frac{288}{2}}$

$= \sqrt{144}$

$= 12$

(c) $\sqrt{\frac{3}{4}}$

$= \frac{\sqrt{3}}{\sqrt{4}}$

$= \frac{\sqrt{3}}{2}$

Work Problem 4 at the Side. ▶

3 Find each product and simplify.

GS **(a)** $\sqrt{16} \cdot \sqrt{50}$

$= \_\_\_\_ \cdot \sqrt{50}$

$= 4\sqrt{\_\_\_\_ \cdot 2}$

$= 4\sqrt{\_\_\_\_} \cdot \sqrt{2}$

$= 4 \cdot \_\_\_\_ \cdot \sqrt{2}$

$= \_\_\_\_$

(b) $\sqrt{10} \cdot \sqrt{50}$

(c) $\sqrt{12} \cdot \sqrt{2}$

(d) $3\sqrt{5} \cdot 4\sqrt{10}$

4 Use the quotient rule to simplify each radical.

GS **(a)** $\sqrt{\frac{81}{16}}$

$= \frac{\sqrt{\_\_\_\_}}{\sqrt{\_\_\_\_}}$

$= \_\_\_\_$

(b) $\frac{\sqrt{48}}{\sqrt{3}}$

(c) $\sqrt{\frac{10}{49}}$

Answers

3. **(a)** 4; 25; 25; 5; $20\sqrt{2}$
(b) $10\sqrt{5}$ **(c)** $2\sqrt{6}$ **(d)** $60\sqrt{2}$

4. **(a)** 81; 16; $\frac{9}{4}$ **(b)** 4 **(c)** $\frac{\sqrt{10}}{7}$

5 Simplify.

(a) $\dfrac{8\sqrt{50}}{4\sqrt{5}}$

(b) $\dfrac{24\sqrt{39}}{4\sqrt{13}}$

EXAMPLE 5 **Using the Quotient Rule to Divide Radicals**

Simplify.

$$\frac{27\sqrt{15}}{9\sqrt{3}}$$

$= \dfrac{27}{9} \cdot \dfrac{\sqrt{15}}{\sqrt{3}}$ Multiplication of fractions

$= \dfrac{27}{9} \cdot \sqrt{\dfrac{15}{3}}$ Quotient rule

$= 3\sqrt{5}$ Divide.

◀ Work Problem 5 at the Side.

6 Simplify.

(a) $\sqrt{\dfrac{3}{8}} \cdot \sqrt{\dfrac{7}{2}}$

GS **(b)** $\sqrt{\dfrac{5}{6}} \cdot \sqrt{120}$

$= \sqrt{\dfrac{5}{6} \cdot \_\_\_}$

$= \sqrt{\_\_\_}$

$= \_\_\_$

EXAMPLE 6 **Using Both the Product and Quotient Rules**

Simplify.

$$\sqrt{\frac{3}{5}} \cdot \sqrt{\frac{1}{5}}$$

$= \sqrt{\dfrac{3}{5} \cdot \dfrac{1}{5}}$ Product rule

$= \sqrt{\dfrac{3}{25}}$ Multiply fractions.

$= \dfrac{\sqrt{3}}{\sqrt{25}}$ Quotient rule

$= \dfrac{\sqrt{3}}{5}$ $\sqrt{25} = 5$

◀ Work Problem 6 at the Side.

OBJECTIVE 4 **Simplify radicals involving variables.** Consider a radical with variable radicand, such as $\sqrt{x^2}$.

If x represents a nonnegative number, then $\sqrt{x^2} = x$.

If x represents a negative number, then $\sqrt{x^2} = -x$, the *opposite* of x (which is positive).

Example: $\sqrt{5^2} = 5$, but $\sqrt{(-5)^2} = \sqrt{25} = 5$, the *opposite* of -5.

This means that the square root of a squared number is always nonnegative. We can use absolute value to express this.

$\sqrt{a^2}$

For any real number a, the following holds true.

$$\sqrt{a^2} = |a|$$

Answers

5. **(a)** $2\sqrt{10}$ **(b)** $6\sqrt{3}$

6. **(a)** $\dfrac{\sqrt{21}}{4}$ **(b)** 120; 100; 10

The product and quotient rules apply when variables appear under radical symbols, as long as the variables represent only *nonnegative* real numbers. ***To avoid negative radicands, we assume variables under radical symbols are nonnegative unless otherwise specified.*** In such cases, absolute value bars are not necessary because for all $x \geq 0$, $|x| = x$.

EXAMPLE 7 **Simplifying Radicals Involving Variables**

Simplify each radical. Assume that all variables represent nonnegative real numbers.

(a) $\sqrt{x^4}$ Because $(x^2)^2 = x^4$, $\sqrt{x^4} = x^2$.

(b) $\sqrt{25m^6}$

$= \sqrt{25} \cdot \sqrt{m^6}$ Product rule

$= 5m^3$ $(m^3)^2 = m^6$

(c) $\sqrt{8p^{10}}$

$= \sqrt{4 \cdot 2 \cdot p^{10}}$ Factor; 4 is a perfect square.

$= \sqrt{4} \cdot \sqrt{2} \cdot \sqrt{p^{10}}$ Product rule

$= 2 \cdot \sqrt{2} \cdot p^5$ $\sqrt{4} = 2$; $(p^5)^2 = p^{10}$

$= 2p^5\sqrt{2}$ Commutative property

(d) $\sqrt{r^9}$ (r^8 is a perfect square.)

$= \sqrt{r^8 \cdot r}$

$= \sqrt{r^8} \cdot \sqrt{r}$ Product rule

$= r^4\sqrt{r}$ $(r^4)^2 = r^8$

(e) $\sqrt{\dfrac{5}{x^2}}$ $(x \neq 0)$

$= \dfrac{\sqrt{5}}{\sqrt{x^2}}$ Quotient rule

$= \dfrac{\sqrt{5}}{x}$

Work Problem **7** at the Side. ▶

Note

A quick way to find the square root of a variable raised to an even power is to divide the exponent by the index, 2.

Examples: $\sqrt{x^6} = x^3$ and $\sqrt{x^{10}} = x^5$

$6 \div 2 = 3$ $10 \div 2 = 5$

OBJECTIVE 5 **Simplify other roots.** The product and quotient rules for radicals also apply to other roots.

Properties of Radicals

For all real numbers a and b where the indicated roots exist, the following hold true.

$$\sqrt[n]{a} \cdot \sqrt[n]{b} = \sqrt[n]{ab} \quad \text{and} \quad \frac{\sqrt[n]{a}}{\sqrt[n]{b}} = \sqrt[n]{\frac{a}{b}} \quad (b \neq 0)$$

7 Simplify each radical. Assume that all variables represent nonnegative real numbers.

GS **(a)** $\sqrt{x^8}$

$(\_\_\_\_)^2 = x^8$, so

$\sqrt{x^8} = \_\_\_\_$.

GS **(b)** $\sqrt{36y^6}$

$= \sqrt{36} \cdot \sqrt{\_\_\_\_}$

$= \_\_\_\_$

(c) $\sqrt{100p^{12}}$

(d) $\sqrt{12z^2}$

(e) $\sqrt{a^5}$

(f) $\sqrt{\dfrac{10}{n^4}}$ $(n \neq 0)$

Answers

7. **(a)** x^4; x^4 **(b)** y^6; $6y^3$ **(c)** $10p^6$ **(d)** $2z\sqrt{3}$ **(e)** $a^2\sqrt{a}$ **(f)** $\dfrac{\sqrt{10}}{n^2}$

8 Simplify each radical.

GS **(a)** $\sqrt[3]{108}$

$= \sqrt[3]{\_\_\_ \cdot 4}$

$= \sqrt[3]{\_\_\_} \cdot \sqrt[3]{4}$

$= \_\_\_$

(b) $\sqrt[3]{250}$

(c) $\sqrt[4]{160}$

(d) $\sqrt[4]{\frac{16}{625}}$

9 Simplify each radical.

(a) $\sqrt[3]{z^9}$

(b) $\sqrt[3]{8x^6}$

(c) $\sqrt[3]{54t^5}$

(d) $\sqrt[3]{\frac{a^{15}}{64}}$

Answers

8. (a) 27; 27; $3\sqrt[3]{4}$ (b) $5\sqrt[3]{2}$ (c) $2\sqrt[4]{10}$ (d) $\frac{2}{5}$

9. (a) z^3 (b) $2x^2$ (c) $3t\sqrt[3]{2t^2}$ (d) $\frac{a^5}{4}$

EXAMPLE 8 Simplifying Other Roots

Simplify each radical.

Because the index is 3, look for factors that are perfect cubes.

(a) $\sqrt[3]{32}$

Remember to write the root index 3 in each radical.

$= \sqrt[3]{8 \cdot 4}$ Factor; 8 is a perfect cube.

$= \sqrt[3]{8} \cdot \sqrt[3]{4}$ Product rule

$= 2\sqrt[3]{4}$ Take the cube root.

(b) $\sqrt[4]{32}$

Remember to write the root index 4 in each radical.

$= \sqrt[4]{16 \cdot 2}$ Factor; 16 is a perfect fourth power.

$= \sqrt[4]{16} \cdot \sqrt[4]{2}$ Product rule

$= 2\sqrt[4]{2}$ Take the fourth root.

(c) $\sqrt[3]{\frac{27}{125}}$

$= \frac{\sqrt[3]{27}}{\sqrt[3]{125}}$ Quotient rule

$= \frac{3}{5}$ Take cube roots.

◀ Work Problem **8** at the Side.

Other roots of radicals involving variables can also be simplified. To simplify cube roots with variables, use the fact that for *any* real number a,

$$\sqrt[3]{a^3} = a.$$

This is true whether a is positive, negative, or 0.

EXAMPLE 9 Simplifying Cube Roots Involving Variables

Simplify each radical.

(a) $\sqrt[3]{m^6}$

$= m^2$ $(m^2)^3 = m^6$

(b) $\sqrt[3]{27x^{12}}$

$= \sqrt[3]{27} \cdot \sqrt[3]{x^{12}}$ Product rule

$= 3x^4$ $3^3 = 27$; $(x^4)^3 = x^{12}$

(c) $\sqrt[3]{32a^4}$

$= \sqrt[3]{8a^3 \cdot 4a}$ Factor; $8a^3$ is a perfect cube.

$= \sqrt[3]{8a^3} \cdot \sqrt[3]{4a}$ Product rule

$= 2a\sqrt[3]{4a}$ $(2a)^3 = 8a^3$

(d) $\sqrt[3]{\frac{y^3}{125}}$

$= \frac{\sqrt[3]{y^3}}{\sqrt[3]{125}}$ Quotient rule

$= \frac{y}{5}$ Take cube roots.

◀ Work Problem **9** at the Side.

8.2 Exercises

FOR EXTRA HELP — *Go to MyMathLab for worked-out, step-by-step solutions to exercises enclosed in a square and video solutions to ▶ exercises.*

CONCEPT CHECK *Decide whether each statement is* true *or* false. *If false, explain why.*

1. $\sqrt{(-6)^2} = -6$

2. $\sqrt[3]{(-6)^3} = -6$

3. The radical $2\sqrt{7}$ represents a sum.

4. In the radical $3\sqrt{11}$, the numbers 3 and $\sqrt{11}$ are factors.

5. CONCEPT CHECK Which one of the following radicals is simplified?

A. $\sqrt{47}$ **B.** $\sqrt{45}$ **C.** $\sqrt{48}$ **D.** $\sqrt{44}$

6. If p is a prime number, is $\sqrt{p}$ in simplified form? Explain.

Use the product rule for radicals to find each product. **See Example 1.**

7. $\sqrt{3} \cdot \sqrt{5}$ ▶

8. $\sqrt{3} \cdot \sqrt{7}$

9. $\sqrt{2} \cdot \sqrt{11}$

10. $\sqrt{2} \cdot \sqrt{15}$

11. $\sqrt{6} \cdot \sqrt{7}$ ▶

12. $\sqrt{5} \cdot \sqrt{6}$

13. $\sqrt{13} \cdot \sqrt{r}\ (r \geq 0)$ ▶

14. $\sqrt{19} \cdot \sqrt{k}\ (k \geq 0)$

Simplify each radical. **See Example 2.**

15. $\sqrt{24}$ ▶

16. $\sqrt{63}$

17. $\sqrt{45}$

18. $\sqrt{48}$

19. $\sqrt{90}$

20. $\sqrt{80}$

21. $\sqrt{75}$

22. $\sqrt{18}$

23. $\sqrt{125}$

24. $\sqrt{56}$

25. $\sqrt{145}$ ▶

26. $\sqrt{110}$

27. $-\sqrt{160}$

28. $-\sqrt{128}$

29. $-\sqrt{700}$ ▶

30. $-\sqrt{600}$

31. $3\sqrt{52}$

32. $9\sqrt{8}$

33. $5\sqrt{50}$

34. $6\sqrt{90}$

Use the Pythagorean theorem to find the length of the unknown side of each right triangle. Express answers as simplified radicals.

35. 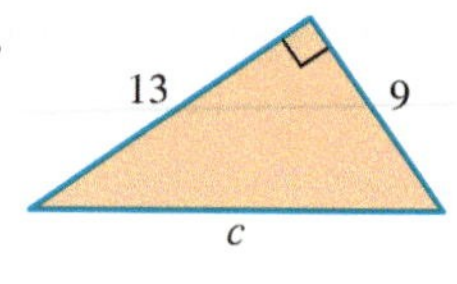

36. 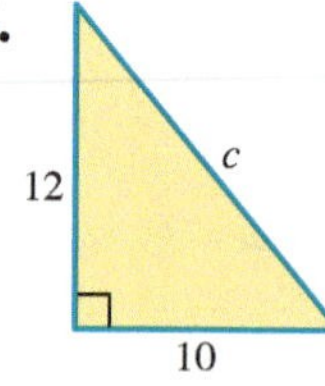

37. 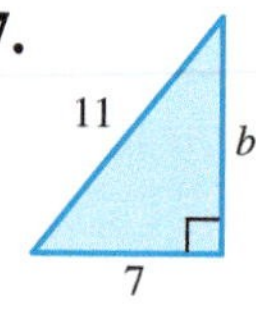

38. 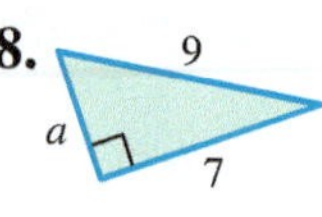

Find each product and simplify. **See Example 3.**

39. $\sqrt{9} \cdot \sqrt{32}$

40. $\sqrt{16} \cdot \sqrt{50}$

41. $\sqrt{3} \cdot \sqrt{18}$

42. $\sqrt{3} \cdot \sqrt{21}$

43. $\sqrt{12} \cdot \sqrt{48}$

44. $\sqrt{50} \cdot \sqrt{72}$

45. $\sqrt{12} \cdot \sqrt{30}$

46. $\sqrt{42} \cdot \sqrt{24}$

47. $2\sqrt{10} \cdot 3\sqrt{2}$

48. $5\sqrt{6} \cdot 2\sqrt{10}$

49. $5\sqrt{3} \cdot 2\sqrt{15}$

50. $4\sqrt{6} \cdot 3\sqrt{2}$

51. Simplify the product $\sqrt{8} \cdot \sqrt{32}$ in two ways.

Method 1: Multiply 8 by 32 and simplify the square root of this product.

Method 2: Simplify $\sqrt{8}$, simplify $\sqrt{32}$, and then multiply.

How do the answers compare? Make a conjecture (an educated guess) about using these methods when simplifying a product such as this.

52. Simplify the radical $\sqrt{288}$ in two ways.

Method 1: Factor 288 as $144 \cdot 2$ and then simplify.

Method 2: Factor 288 as $16 \cdot 18$ and then simplify.

How do the answers compare? Make a conjecture concerning the quickest way to simplify such a radical.

Simplify each radical expression. **See Examples 4–6.**

53. $\sqrt{\frac{16}{225}}$

54. $\sqrt{\frac{9}{100}}$

55. $\sqrt{\frac{7}{16}}$

56. $\sqrt{\frac{13}{25}}$

57. $\sqrt{\frac{4}{50}}$

58. $\sqrt{\frac{14}{72}}$

59. $\frac{\sqrt{75}}{\sqrt{3}}$

60. $\frac{\sqrt{200}}{\sqrt{2}}$

61. $\frac{30\sqrt{10}}{5\sqrt{2}}$

62. $\frac{50\sqrt{20}}{2\sqrt{10}}$

63. $\sqrt{\frac{5}{2}} \cdot \sqrt{\frac{125}{8}}$

64. $\sqrt{\frac{8}{3}} \cdot \sqrt{\frac{512}{27}}$

Simplify each radical. Assume that all variables represent nonnegative real numbers. ***See Example 7.***

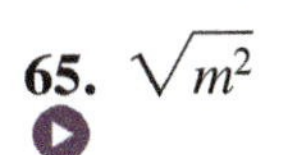

65. $\sqrt{m^2}$ **66.** $\sqrt{k^2}$ **67.** $\sqrt{y^4}$ **68.** $\sqrt{s^4}$

69. $\sqrt{36z^2}$ **70.** $\sqrt{49n^2}$ **71.** $\sqrt{400x^6}$ **72.** $\sqrt{900y^8}$

73. $\sqrt{18x^8}$ **74.** $\sqrt{20r^{10}}$ **75.** $\sqrt{45c^{14}}$ **76.** $\sqrt{50d^{20}}$

77. $\sqrt{z^5}$ **78.** $\sqrt{y^3}$ **79.** $\sqrt{a^{13}}$ **80.** $\sqrt{p^{17}}$

81. $\sqrt{64x^7}$ **82.** $\sqrt{25t^{11}}$ **83.** $\sqrt{x^6y^{12}}$ **84.** $\sqrt{a^8b^{10}}$

85. $\sqrt{81m^4n^2}$ **86.** $\sqrt{100c^4d^6}$ **87.** $\sqrt{\frac{7}{x^{10}}}$ $(x \neq 0)$ **88.** $\sqrt{\frac{14}{z^{12}}}$ $(z \neq 0)$

89. $\sqrt{\frac{y^4}{100}}$ **90.** $\sqrt{\frac{w^8}{144}}$ **91.** $\sqrt{\frac{x^6}{y^8}}$ $(y \neq 0)$ **92.** $\sqrt{\frac{a^4}{b^6}}$ $(b \neq 0)$

Simplify each radical. ***See Example 8.***

93. $\sqrt[3]{40}$ **94.** $\sqrt[3]{48}$ **95.** $\sqrt[3]{54}$ **96.** $\sqrt[3]{135}$

97. $\sqrt[3]{128}$ **98.** $\sqrt[3]{192}$ **99.** $\sqrt[4]{80}$ **100.** $\sqrt[4]{243}$

101. $\sqrt[3]{\frac{8}{27}}$ **102.** $\sqrt[3]{\frac{64}{125}}$ **103.** $\sqrt[3]{-\frac{216}{125}}$ **104.** $\sqrt[3]{-\frac{1}{64}}$

Simplify each radical. **See Example 9.**

105. $\sqrt[3]{p^3}$ **106.** $\sqrt[3]{w^3}$ **107.** $\sqrt[3]{x^9}$ **108.** $\sqrt[3]{y^{18}}$

109. $\sqrt[3]{64z^6}$ **110.** $\sqrt[3]{125a^{15}}$ **111.** $\sqrt[3]{343a^9b^3}$ **112.** $\sqrt[3]{216m^3n^6}$

113. $\sqrt[3]{16t^5}$ **114.** $\sqrt[3]{24x^4}$ **115.** $\sqrt[3]{\frac{m^{12}}{8}}$ **116.** $\sqrt[3]{\frac{n^9}{27}}$

The volume V of a cube is found with the formula $V = s^3$, where s is the length of an edge of the cube. Use this information in each problem.

117. A container in the shape of a cube has a volume of 216 cm^3. What is the length of each side of the container?

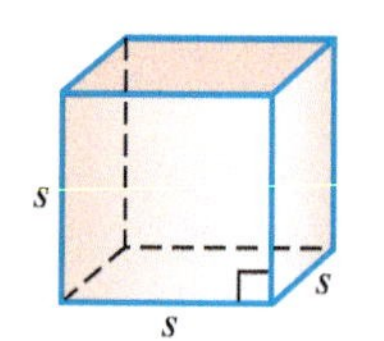

118. A cube-shaped box must be constructed to contain 128 ft^3. What should the dimensions (height, width, and length) of the box be?

The volume V of a sphere is found with the formula $V = \frac{4}{3}\pi r^3$, where r is the length of the radius of the sphere. Use this information in each problem.

119. A ball in the shape of a sphere has a volume of 288π $in.^3$. What is the radius of the ball?

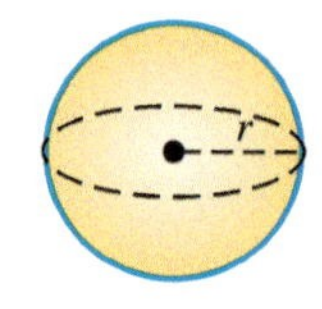

120. Suppose that the volume of the ball described in **Exercise 119** is multiplied by 8. How is the radius affected?

Work each problem without using a calculator.

121. Choose the best estimate for the area (in square inches) of this rectangle.

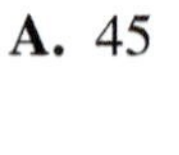

A. 45 **B.** 72 **C.** 80 **D.** 90

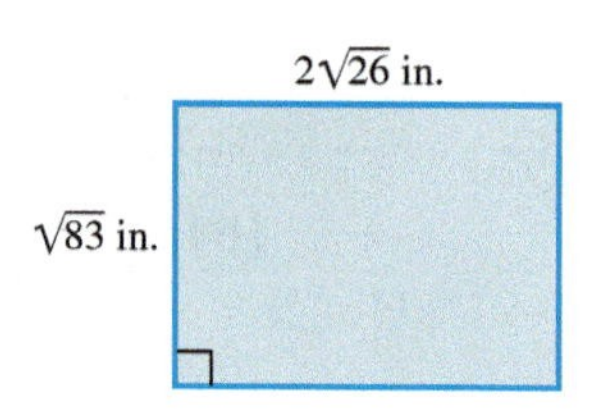

122. Choose the best estimate for the area (in square feet) of the triangle.

A. 20 **B.** 40 **C.** 60 **D.** 80

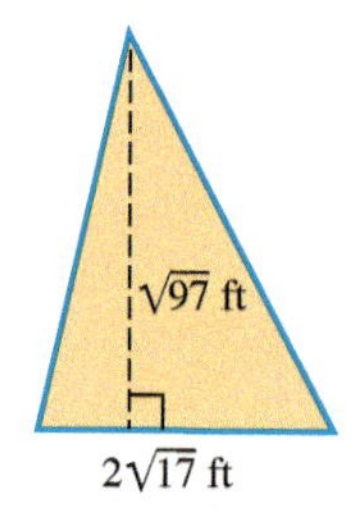

8.3 Adding and Subtracting Radicals

OBJECTIVES

1. Add and subtract radicals.
2. Simplify radical sums and differences.
3. Simplify more complicated radical expressions.

OBJECTIVE 1 Add and subtract radicals. We add or subtract radicals using the distributive property, $ac + bc = (a + b)c$.

$8\sqrt{3} + 6\sqrt{3}$	Distributive property	$2\sqrt{11} - 7\sqrt{11}$	Distributive property
$= (8 + 6)\sqrt{3}$		$= (2 - 7)\sqrt{11}$	
$= 14\sqrt{3}$	Add.	$= -5\sqrt{11}$	Subtract.

Only **like radicals**—those that are ***multiples of the same root of the same number***—can be combined in this way. Examples of **unlike radicals** are

$2\sqrt{5}$ and $2\sqrt{3}$, Radicands are different.

as well as $2\sqrt{3}$ and $2\sqrt[3]{3}$. Indexes are different.

Work Problem 1 at the Side. ▶

1 Are the radicals in each pair *like* or *unlike*? Explain why.

(a) $5\sqrt{6}$ and $4\sqrt{6}$

(b) $2\sqrt{3}$ and $3\sqrt{2}$

(c) $\sqrt{10}$ and $\sqrt[3]{10}$

(d) $7\sqrt{2x}$ and $8\sqrt{2x}$

(e) $\sqrt{3y}$ and $\sqrt{6y}$

EXAMPLE 1 Adding and Subtracting Like Radicals

Add or subtract, as indicated.

(a) $3\sqrt{6} + 5\sqrt{6}$ These are like radicals.

$= (3 + 5)\sqrt{6}$ We are factoring out $\sqrt{6}$ here.

$= 8\sqrt{6}$

(b) $5\sqrt{10} - 7\sqrt{10}$ These are like radicals.

$= (5 - 7)\sqrt{10}$

$= -2\sqrt{10}$

(c) $\sqrt{7} + 2\sqrt{7}$

$= 1\sqrt{7} + 2\sqrt{7}$

$= (1 + 2)\sqrt{7}$

$= 3\sqrt{7}$

(d) $\sqrt{5} + \sqrt{5}$

$= 1\sqrt{5} + 1\sqrt{5}$

$= (1 + 1)\sqrt{5}$

$= 2\sqrt{5}$

(e) $\sqrt{3} + \sqrt{7}$ cannot be combined. They are unlike radicals.

Work Problem 2 at the Side. ▶

2 Add or subtract, as indicated.

(a) $8\sqrt{5} + 2\sqrt{5}$

(b) $4\sqrt{3} - 9\sqrt{3}$

(c) $4\sqrt{11} - 3\sqrt{11}$

(d) $\sqrt{15} + \sqrt{15}$

(e) $2\sqrt{7} + 2\sqrt{10}$

OBJECTIVE 2 Simplify radical sums and differences.

EXAMPLE 2 Adding and Subtracting Radicals

Add or subtract, as indicated.

(a) $3\sqrt{2} + \sqrt{8}$

$= 3\sqrt{2} + \sqrt{4 \cdot 2}$ Factor; 4 is a perfect square.

$= 3\sqrt{2} + \sqrt{4} \cdot \sqrt{2}$ Product rule

$= 3\sqrt{2} + 2\sqrt{2}$ $\sqrt{4} = 2$

$= (3 + 2)\sqrt{2}$ Distributive property

$= 5\sqrt{2}$ Add.

Continued on Next Page

Answers

1. (a) like; Both are square roots and have the same radicand, 6.
   (b) unlike; The radicands are different—one is 3 and one is 2.
   (c) unlike; The indexes are different—one is a square root and one is a cube root.
   (d) like; Both are square roots and have the same radicand, $2x$.
   (e) unlike; The radicands are different—one is $3y$ and one is $6y$.
2. (a) $10\sqrt{5}$ (b) $-5\sqrt{3}$ (c) $\sqrt{11}$
   (d) $2\sqrt{15}$
   (e) These are unlike radicals and cannot be combined.

3 Add or subtract, as indicated.

(a) $\sqrt{8} + 4\sqrt{2}$

(b) $\sqrt{18} - \sqrt{2}$

(c) $3\sqrt{48} - 2\sqrt{75}$

(d) $8\sqrt[3]{5} + 10\sqrt[3]{40}$

Answers

3. (a) $6\sqrt{2}$ (b) $2\sqrt{2}$ (c) $2\sqrt{3}$ (d) $28\sqrt[3]{5}$

(b) $\sqrt{18} - \sqrt{27}$

$= \sqrt{9 \cdot 2} - \sqrt{9 \cdot 3}$ Factor; 9 is a perfect square.

$= \sqrt{9} \cdot \sqrt{2} - \sqrt{9} \cdot \sqrt{3}$ Product rule

$= 3\sqrt{2} - 3\sqrt{3}$ $\sqrt{9} = 3$

Stop here. **These are unlike radicals.** They cannot be combined.

(c) $2\sqrt{12} + 3\sqrt{75}$

$= 2\sqrt{4 \cdot 3} + 3\sqrt{25 \cdot 3}$ Factor; 4 and 25 are perfect squares.

$= 2(\sqrt{4} \cdot \sqrt{3}) + 3(\sqrt{25} \cdot \sqrt{3})$ Product rule

$= 2(2\sqrt{3}) + 3(5\sqrt{3})$ $\sqrt{4} = 2$; $\sqrt{25} = 5$

$= 4\sqrt{3} + 15\sqrt{3}$ Multiply.

Think: $(4 + 15)\sqrt{3}$

$= 19\sqrt{3}$ Add like radicals.

(d) $3\sqrt[3]{16} + 5\sqrt[3]{2}$

$= 3\sqrt[3]{8 \cdot 2} + 5\sqrt[3]{2}$ Factor; 8 is a perfect cube.

$= 3(\sqrt[3]{8} \cdot \sqrt[3]{2}) + 5\sqrt[3]{2}$ Product rule

$= 3(2\sqrt[3]{2}) + 5\sqrt[3]{2}$ $\sqrt[3]{8} = 2$

$= 6\sqrt[3]{2} + 5\sqrt[3]{2}$ Multiply.

$= 11\sqrt[3]{2}$ Add like radicals.

◀ **Work Problem 3 at the Side.**

CAUTION

Only like radicals can be combined.

$\sqrt{5} + 3\sqrt{5} = 4\sqrt{5}$ — Add like radicals.	$\sqrt{5} + 5\sqrt{3}$, $2\sqrt{3} + 5\sqrt[3]{3}$ — Unlike radicals cannot be combined.

OBJECTIVE 3 Simplify more complicated radical expressions.

EXAMPLE 3 Simplifying Radical Expressions

Simplify. Assume that all variables represent nonnegative real numbers.

(a) $\sqrt{5} \cdot \sqrt{15} + 4\sqrt{3}$ We simplify this expression by performing the operations.

$= \sqrt{5 \cdot 15} + 4\sqrt{3}$ Product rule

$= \sqrt{75} + 4\sqrt{3}$ Multiply.

$= \sqrt{25 \cdot 3} + 4\sqrt{3}$ Factor; 25 is a perfect square.

$= \sqrt{25} \cdot \sqrt{3} + 4\sqrt{3}$ Product rule

$= 5\sqrt{3} + 4\sqrt{3}$ $\sqrt{25} = 5$

$= 9\sqrt{3}$ Add like radicals.

Continued on Next Page

(b) $\sqrt{12k} + \sqrt{27k}$

$= \sqrt{4 \cdot 3k} + \sqrt{9 \cdot 3k}$	Factor.
$= \sqrt{4} \cdot \sqrt{3k} + \sqrt{9} \cdot \sqrt{3k}$	Product rule
$= 2\sqrt{3k} + 3\sqrt{3k}$	$\sqrt{4} = 2;\ \sqrt{9} = 3$
$= 5\sqrt{3k}$	Add like radicals.

(c) $3x\sqrt{50} + \sqrt{2x^2}$

$= 3x\sqrt{25 \cdot 2} + \sqrt{x^2 \cdot 2}$	Factor.
$= 3x\sqrt{25} \cdot \sqrt{2} + \sqrt{x^2} \cdot \sqrt{2}$	Product rule
$= 3x \cdot 5\sqrt{2} + x\sqrt{2}$	$\sqrt{25} = 5;\ \sqrt{x^2} = x$
$= 15x\sqrt{2} + x\sqrt{2}$	Multiply.
$= 16x\sqrt{2}$	Add like radicals.

Think: $(15x + 1x)\sqrt{2}$

(d) $2\sqrt[3]{32m^3} - \sqrt[3]{108m^3}$

$= 2\sqrt[3]{8m^3 \cdot 4} - \sqrt[3]{27m^3 \cdot 4}$	Factor.
$= 2\sqrt[3]{8m^3} \cdot \sqrt[3]{4} - \sqrt[3]{27m^3} \cdot \sqrt[3]{4}$	Product rule
$= 2 \cdot 2m\sqrt[3]{4} - 3m\sqrt[3]{4}$	$\sqrt[3]{8m^3} = 2m;\ \sqrt[3]{27m^3} = 3m$
$= 4m\sqrt[3]{4} - 3m\sqrt[3]{4}$	Multiply.
$= m\sqrt[3]{4}$	Subtract like radicals.

Work Problem 4 at the Side. ▶

4 Simplify. Assume that all variables represent nonnegative real numbers.

(a) $\sqrt{7} \cdot \sqrt{21} + 2\sqrt{27}$

(b) $\sqrt{3r} \cdot \sqrt{6} + \sqrt{8r}$

(c) $y\sqrt{72} - \sqrt{18y^2}$

(d) $\sqrt[3]{81x^4} + 5\sqrt[3]{24x^4}$

Answers

4. **(a)** $13\sqrt{3}$ **(b)** $5\sqrt{2r}$ **(c)** $3y\sqrt{2}$ **(d)** $13x\sqrt[3]{3x}$

8.3 Exercises

FOR EXTRA HELP

Go to MyMathLab for worked-out, step-by-step solutions to exercises enclosed in a square and video solutions to ▶ exercises.

CONCEPT CHECK *Complete each statement.*

1. Like radicals have the same ______________ and the same ______________, or order. For example, $5\sqrt{2}$ and $-3\sqrt{2}$ are (*like* / *unlike*) radicals.

2. The radicals $\sqrt[4]{3xy^3}$ and $-6\sqrt[4]{3xy^3}$ are (*like* / *unlike*) radicals because both have the same root index, _____, and the same radicand, _____.

3. $\sqrt{5} + 5\sqrt{3}$ cannot be simplified because the ______________ are different. They are (*like* / *unlike*) radicals.

4. $4\sqrt[3]{2} + 3\sqrt{2}$ cannot be simplified because the ______________ are different. They are (*like* / *unlike*) radicals.

Perform the indicated operations. ***See Examples 1 and 2.***

5. $2\sqrt{3} + 5\sqrt{3}$ ▶

6. $6\sqrt{5} + 8\sqrt{5}$

7. $14\sqrt{7} - 19\sqrt{7}$ ▶

8. $16\sqrt{2} - 18\sqrt{2}$

9. $\sqrt{17} + 4\sqrt{17}$ ▶

10. $5\sqrt{19} + \sqrt{19}$

11. $6\sqrt{7} - \sqrt{7}$

12. $11\sqrt{14} - \sqrt{14}$

13. $\sqrt{6} + \sqrt{6}$

14. $\sqrt{11} + \sqrt{11}$

15. $\sqrt{6} + \sqrt{7}$

16. $\sqrt{14} + \sqrt{17}$

17. $5\sqrt{3} + \sqrt{12}$ ▶

18. $3\sqrt{2} + \sqrt{50}$

19. $\sqrt{45} + 4\sqrt{20}$ ▶

20. $\sqrt{24} + 6\sqrt{54}$

21. $5\sqrt{72} - 3\sqrt{50}$

22. $6\sqrt{18} - 5\sqrt{32}$

23. $-5\sqrt{32} + 2\sqrt{98}$ ▶

24. $-4\sqrt{75} + 3\sqrt{12}$

25. $\frac{1}{4}\sqrt{288} + \frac{1}{6}\sqrt{72}$

26. $\frac{2}{3}\sqrt{27} + \frac{3}{4}\sqrt{48}$

27. $4\sqrt[3]{16} - 3\sqrt[3]{54}$

28. $3\sqrt[3]{250} - 4\sqrt[3]{128}$

29. $3\sqrt[3]{24} + 6\sqrt[3]{81}$

30. $2\sqrt[4]{48} - \sqrt[4]{243}$

31. $5\sqrt{7} - 3\sqrt{28} + 6\sqrt{63}$

32. $3\sqrt{11} + 5\sqrt{44} - 8\sqrt{99}$

33. $2\sqrt{8} - 5\sqrt{32} - 2\sqrt{48}$

34. $5\sqrt{72} - 3\sqrt{48} + 4\sqrt{128}$

Find the perimeter of each figure.

35.

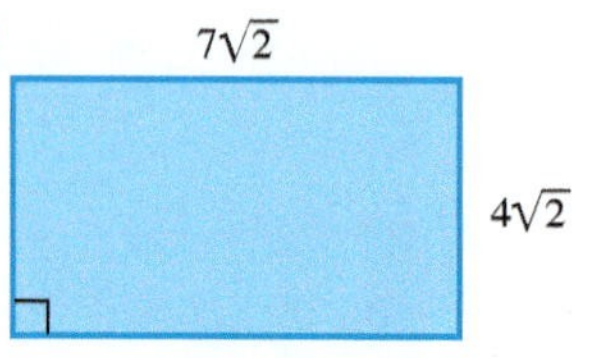

36.

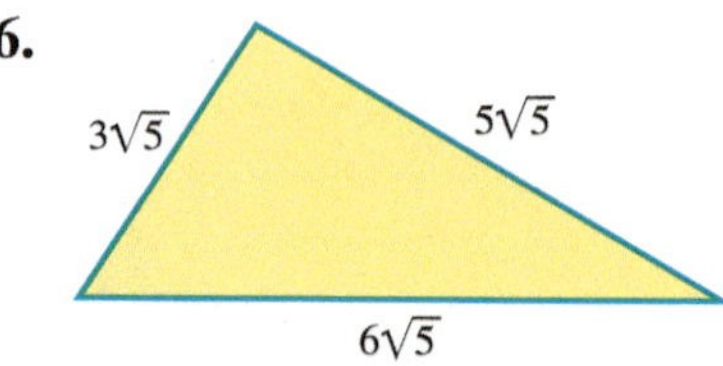

Simplify. Assume that all variables represent nonnegative real numbers.
See Example 3.

37. $\sqrt{6} \cdot \sqrt{2} + 9\sqrt{3}$

38. $4\sqrt{15} \cdot \sqrt{3} + 4\sqrt{5}$

39. $\sqrt{3} \cdot \sqrt{7} + 2\sqrt{21}$

40. $\sqrt{13} \cdot \sqrt{2} + 3\sqrt{26}$

41. $\sqrt{32x} - \sqrt{18x}$

42. $\sqrt{125t} - \sqrt{80t}$

43. $\sqrt{27r} + \sqrt{48r}$

44. $\sqrt{24x} + \sqrt{54x}$

45. $\sqrt{9x} + \sqrt{49x} - \sqrt{25x}$

46. $\sqrt{4a} - \sqrt{16a} + \sqrt{100a}$

47. $\sqrt{6x^2} + x\sqrt{24}$

48. $\sqrt{75x^2} + x\sqrt{108}$

49. $3\sqrt{8x^2} - 4x\sqrt{2} - x\sqrt{8}$

50. $\sqrt{2b^2} + 3b\sqrt{18} - b\sqrt{200}$

51. $-8\sqrt{32k} + 6\sqrt{8k}$

52. $4\sqrt{12x} + 2\sqrt{27x}$

53. $2\sqrt{125x^2z} + 8x\sqrt{80z}$

54. $\sqrt{48x^2y} + 5x\sqrt{27y}$

55. $4\sqrt[3]{16} - 3\sqrt[3]{54}$

56. $5\sqrt[3]{128} + 3\sqrt[3]{250}$

57. $6\sqrt[3]{8p^2} - 2\sqrt[3]{27p^2}$

58. $8k\sqrt[3]{54k} + 6\sqrt[3]{16k^4}$

59. $5\sqrt[4]{m^3} + 8\sqrt[4]{16m^3}$

60. $5\sqrt[4]{m^5} + 3\sqrt[4]{81m^5}$

Relating Concepts (Exercises 61–64) For Individual or Group Work

Work Exercises 61–64 in order.

61. Combine like terms: $5x^2y + 3x^2y - 14x^2y$.

62. Combine like terms: $5(p - 2q)^2(a + b) + 3(p - 2q)^2(a + b) - 14(p - 2q)^2(a + b)$.

63. Combine like radicals: $5a^2\sqrt{x} + 3a^2\sqrt{x} - 14a^2\sqrt{x}$.

64. Compare the work done in **Exercises 61–63.** What can we conclude?

8.4 Rationalizing the Denominator

OBJECTIVES

1. Rationalize denominators with square roots.
2. Write radicals in simplified form.
3. Rationalize denominators with cube roots.

OBJECTIVE 1 Rationalize denominators with square roots. Although calculators now make it fairly easy to divide by a radical in an expression such as $\frac{1}{\sqrt{2}}$, it is sometimes easier to work with radical expressions if the denominators do not contain any radicals.

The radical in the denominator of $\frac{1}{\sqrt{2}}$ can be eliminated by multiplying the numerator and denominator by $\sqrt{2}$ because $\sqrt{2} \cdot \sqrt{2} = \sqrt{4} = 2$.

$$\frac{1}{\sqrt{2}} = \frac{1 \cdot \sqrt{2}}{\sqrt{2} \cdot \sqrt{2}} = \frac{\sqrt{2}}{2} \quad \text{Multiply by } \frac{\sqrt{2}}{\sqrt{2}} = 1.$$

Rationalizing a Denominator

The process of changing a denominator from one with a radical to one without a radical is called **rationalizing the denominator.**

The value of the radical expression is not changed. Only the form is changed, because the expression has been multiplied by a form of 1.

EXAMPLE 1 Rationalizing Denominators

Rationalize each denominator.

(a) $\frac{9}{\sqrt{6}}$

$= \frac{9 \cdot \sqrt{6}}{\sqrt{6} \cdot \sqrt{6}}$ Multiply by $\frac{\sqrt{6}}{\sqrt{6}} = 1$.

$= \frac{9\sqrt{6}}{6}$ In the denominator, $\sqrt{6} \cdot \sqrt{6} = \sqrt{36} = 6$.

$= \frac{3\sqrt{6}}{2}$ Write in lowest terms; $\frac{9}{6} = \frac{3}{2}$.

The denominator is now a rational number.

(b) $\frac{12}{\sqrt{8}}$ While the denominator could be rationalized by multiplying by $\sqrt{8}$, simplifying the denominator first is more direct.

$= \frac{12}{2\sqrt{2}}$ $\sqrt{8} = \sqrt{4} \cdot \sqrt{2} = 2\sqrt{2}$

$= \frac{12 \cdot \sqrt{2}}{2\sqrt{2} \cdot \sqrt{2}}$ Multiply by $\frac{\sqrt{2}}{\sqrt{2}} = 1$.

$= \frac{12 \cdot \sqrt{2}}{2 \cdot 2}$ $\sqrt{2} \cdot \sqrt{2} = \sqrt{4} = 2$

$= \frac{12\sqrt{2}}{4}$ Multiply.

$= 3\sqrt{2}$ Write in lowest terms; $\frac{12}{4} = 3$.

◀ Work Problem 1 at the Side.

---

1 Rationalize each denominator.

GS (a) $\frac{3}{\sqrt{5}}$

$= \frac{3 \cdot \_\_\_}{\sqrt{5} \cdot \_\_\_}$

$= \_\_\_$

(b) $\frac{-6}{\sqrt{11}}$

(c) $-\frac{20}{\sqrt{18}}$

Answers

1. (a) $\sqrt{5}$; $\sqrt{5}$; $\frac{3\sqrt{5}}{5}$ (b) $\frac{-6\sqrt{11}}{11}$ (c) $-\frac{10\sqrt{2}}{3}$

Note

In **Example 1(b)**, we could also have rationalized the original denominator, $\sqrt{8}$, by multiplying by $\sqrt{2}$ because $\sqrt{8} \cdot \sqrt{2} = \sqrt{16} = 4$.

$$\frac{12}{\sqrt{8}} = \frac{12 \cdot \sqrt{2}}{\sqrt{8} \cdot \sqrt{2}} = \frac{12\sqrt{2}}{\sqrt{16}} = \frac{12\sqrt{2}}{4} = 3\sqrt{2}$$

Either approach yields the same correct answer.

OBJECTIVE 2 Write radicals in simplified form. A radical is considered to be in simplified form if the following three conditions are met.

Conditions for Simplified Form of a Radical

1. The radicand contains no factor (except 1) that is a perfect square (when dealing with square roots), a perfect cube (when dealing with cube roots), and so on.
2. The radicand has no fractions.
3. No denominator contains a radical.

EXAMPLE 2 Simplifying a Radical

Simplify.

$$\sqrt{\frac{27}{5}}$$

$$= \frac{\sqrt{27}}{\sqrt{5}} \quad \text{Quotient rule}$$

$$= \frac{\sqrt{27} \cdot \sqrt{5}}{\sqrt{5} \cdot \sqrt{5}} \quad \text{Rationalize the denominator.}$$

$$= \frac{\sqrt{27} \cdot \sqrt{5}}{5} \quad \sqrt{5} \cdot \sqrt{5} = \sqrt{25} = 5$$

$$= \frac{\sqrt{9 \cdot 3} \cdot \sqrt{5}}{5} \quad \text{Factor.}$$

$$= \frac{\sqrt{9} \cdot \sqrt{3} \cdot \sqrt{5}}{5} \quad \text{Product rule}$$

$$= \frac{3 \cdot \sqrt{3} \cdot \sqrt{5}}{5} \quad \sqrt{9} = 3$$

$$= \frac{3\sqrt{15}}{5} \quad \text{Product rule}$$

The three conditions for a simplified radical are met.

Work Problem 2 at the Side. ▶

2 Simplify.

GS (a) $\sqrt{\frac{16}{11}}$

$= \frac{\sqrt{16}}{\_\_\_\_}$

$= \frac{\sqrt{16} \cdot \_\_\_\_}{\sqrt{11} \cdot \_\_\_\_}$

$= \frac{\_\_\_\_\sqrt{11}}{\_\_\_\_}$

(b) $\sqrt{\frac{5}{18}}$

(c) $\sqrt{\frac{8}{32}}$

Answers

2. (a) $\sqrt{11}$; $\sqrt{11}$; $\sqrt{11}$; 4; 11

(b) $\frac{\sqrt{10}}{6}$ (c) $\frac{1}{2}$

3 Simplify.

(a) $\sqrt{\frac{1}{2}} \cdot \sqrt{\frac{5}{6}}$

GS (b) $\sqrt{\frac{1}{10}} \cdot \sqrt{20}$

$= \sqrt{\frac{1}{10}} \cdot \sqrt{\frac{20}{1}}$

$= \sqrt{\frac{\_\_\_}{10}}$

$= \_\_\_$

(c) $\sqrt{\frac{5}{8}} \cdot \sqrt{\frac{24}{10}}$

4 Simplify. Assume that all variables represent positive real numbers.

(a) $\frac{\sqrt{5p}}{\sqrt{q}}$

(b) $\sqrt{\frac{5r^2t^2}{7}}$

Answers

3. (a) $\frac{\sqrt{15}}{6}$ (b) 20; $\sqrt{2}$ (c) $\frac{\sqrt{6}}{2}$

4. (a) $\frac{\sqrt{5pq}}{q}$ (b) $\frac{rt\sqrt{35}}{7}$

EXAMPLE 3 Simplifying a Product of Radicals

Simplify.

$\sqrt{\frac{5}{8}} \cdot \sqrt{\frac{1}{6}}$

$= \sqrt{\frac{5}{8} \cdot \frac{1}{6}}$ Product rule

$= \sqrt{\frac{5}{48}}$ Multiply fractions.

$= \frac{\sqrt{5}}{\sqrt{48}}$ Quotient rule

$= \frac{\sqrt{5}}{\sqrt{16} \cdot \sqrt{3}}$ Product rule

$= \frac{\sqrt{5}}{4\sqrt{3}}$ $\sqrt{16} = 4$

$= \frac{\sqrt{5} \cdot \sqrt{3}}{4\sqrt{3} \cdot \sqrt{3}}$ Rationalize the denominator.

$= \frac{\sqrt{15}}{4 \cdot 3}$ Product rule; $\sqrt{3} \cdot \sqrt{3} = 3$

$= \frac{\sqrt{15}}{12}$ Multiply.

◄ Work Problem **3** at the Side.

EXAMPLE 4 Simplifying Quotients Involving Radicals

Simplify. Assume that x and y represent positive real numbers.

(a) $\frac{\sqrt{4x}}{\sqrt{y}}$

$= \frac{\sqrt{4x} \cdot \sqrt{y}}{\sqrt{y} \cdot \sqrt{y}}$ Rationalize the denominator.

$= \frac{\sqrt{4xy}}{y}$ Product rule; $\sqrt{y} \cdot \sqrt{y} = y$

$= \frac{\sqrt{4} \cdot \sqrt{xy}}{y}$ Product rule

$= \frac{2\sqrt{xy}}{y}$ $\sqrt{4} = 2$

(b) $\sqrt{\frac{2x^2y}{3}}$

$= \frac{\sqrt{2x^2y}}{\sqrt{3}}$ Quotient rule

$= \frac{\sqrt{2x^2y} \cdot \sqrt{3}}{\sqrt{3} \cdot \sqrt{3}}$ Rationalize the denominator.

$= \frac{\sqrt{6x^2y}}{3}$ Product rule; $\sqrt{3} \cdot \sqrt{3} = 3$

$= \frac{\sqrt{x^2} \cdot \sqrt{6y}}{3}$ Product rule

$= \frac{x\sqrt{6y}}{3}$ $\sqrt{x^2} = x$, because $x > 0$.

◄ Work Problem **4** at the Side.

OBJECTIVE 3 **Rationalize denominators with cube roots.**

EXAMPLE 5 **Rationalizing Denominators with Cube Roots**

Rationalize each denominator.

(a) $\sqrt[3]{\frac{3}{2}}$

First write the expression as a quotient of radicals. Then multiply the numerator and denominator by the appropriate number of factors of 2 to make the radicand in the denominator a *perfect cube*. This will eliminate the radical in the denominator. Here, multiply by $\sqrt[3]{2 \cdot 2}$, or $\sqrt[3]{2^2}$.

$$\sqrt[3]{\frac{3}{2}} = \frac{\sqrt[3]{3}}{\sqrt[3]{2}} = \frac{\sqrt[3]{3} \cdot \sqrt[3]{2 \cdot 2}}{\sqrt[3]{2} \cdot \sqrt[3]{2 \cdot 2}} = \frac{\sqrt[3]{3 \cdot 2 \cdot 2}}{\sqrt[3]{2 \cdot 2 \cdot 2}} = \frac{\sqrt[3]{12}}{2}$$

We need 3 factors of 2 in the radicand in the denominator.

$\sqrt[3]{2 \cdot 2 \cdot 2} = \sqrt[3]{2^3} = 2$
Denominator radicand is a perfect cube.

(b) $\frac{\sqrt[3]{3}}{\sqrt[3]{4}}$

Because $\sqrt[3]{4} = \sqrt[3]{2 \cdot 2}$, multiply the numerator and denominator by a radical with a sufficient number of factors of 2 to obtain a perfect cube in the radicand in the denominator. Here, multiply by $\sqrt[3]{2}$.

$$\frac{\sqrt[3]{3}}{\sqrt[3]{4}} = \frac{\sqrt[3]{3} \cdot \sqrt[3]{2}}{\sqrt[3]{2 \cdot 2} \cdot \sqrt[3]{2}} = \frac{\sqrt[3]{6}}{\sqrt[3]{2 \cdot 2 \cdot 2}} = \frac{\sqrt[3]{6}}{2}$$

(c) $\frac{\sqrt[3]{2}}{\sqrt[3]{3x^2}} \quad (x \neq 0)$

Multiply the numerator and denominator by a radical with a sufficient number of factors of 3 and of x to obtain a perfect cube in the radicand in the denominator.

$$\frac{\sqrt[3]{2}}{\sqrt[3]{3x^2}} = \frac{\sqrt[3]{2} \cdot \sqrt[3]{3 \cdot 3 \cdot x}}{\sqrt[3]{3 \cdot x \cdot x} \cdot \sqrt[3]{3 \cdot 3 \cdot x}} = \frac{\sqrt[3]{18x}}{\sqrt[3]{(3x)^3}} = \frac{\sqrt[3]{18x}}{3x}$$

We need 3 factors of 3 and 3 factors of x in the radicand in the denominator.

Denominator radicand is a perfect cube.

Work Problem 5 at the Side. ▶

CAUTION

A common error is to multiply the numerator and denominator of

$$\frac{\sqrt[3]{3}}{\sqrt[3]{2}}$$ See Example 5(a).

by $\sqrt[3]{2}$ instead of $\sqrt[3]{2^2}$. Doing this would give a denominator of

$$\sqrt[3]{2} \cdot \sqrt[3]{2} = \sqrt[3]{4}.$$

Because 4 is not a perfect cube, the denominator is still not rationalized.

5 Rationalize each denominator.

GS **(a)** $\sqrt[3]{\frac{5}{7}}$

$$= \frac{\sqrt[3]{5}}{\sqrt[3]{7}}$$

To obtain a perfect cube in the denominator radicand, multiply the ______ and denominator by

$\sqrt[3]{\_\_\_ \cdot \_\_\_}$.

Complete the solution.

(b) $\frac{\sqrt[3]{5}}{\sqrt[3]{16}}$

(c) $\frac{\sqrt[3]{4}}{\sqrt[3]{25y}} \quad (y \neq 0)$

Answers

5. **(a)** numerator; 7; 7; $\frac{\sqrt[3]{245}}{7}$ **(b)** $\frac{\sqrt[3]{20}}{4}$ **(c)** $\frac{\sqrt[3]{20y^2}}{5y}$

8.4 Exercises

FOR EXTRA HELP *Go to MyMathLab for worked-out, step-by-step solutions to exercises enclosed in a square and video solutions to ▶ exercises.*

1. CONCEPT CHECK A student incorrectly rationalized the denominator of $\frac{4}{\sqrt{3}}$ as follows.

$$\frac{4}{\sqrt{3}} = \frac{4}{\sqrt{3}} \cdot \frac{1}{\sqrt{3}} = \frac{4}{3} \quad \text{Incorrect}$$

What Went Wrong? Rationalize the denominator correctly.

2. CONCEPT CHECK A student incorrectly rationalized the denominator of $\frac{2}{\sqrt[3]{5}}$ as follows.

$$\frac{2}{\sqrt[3]{5}} = \frac{2}{\sqrt[3]{5}} \cdot \frac{\sqrt[3]{5}}{\sqrt[3]{5}} = \frac{2\sqrt[3]{5}}{5} \quad \text{Incorrect}$$

What Went Wrong? Rationalize the denominator correctly.

Rationalize each denominator. ***See Examples 1 and 2.***

3. $\frac{6}{\sqrt{5}}$ **4.** $\frac{3}{\sqrt{2}}$ **5.** $\frac{5}{\sqrt{5}}$ **6.** $\frac{15}{\sqrt{15}}$

7. $\frac{4}{\sqrt{6}}$ **8.** $\frac{15}{\sqrt{10}}$ **9.** $\frac{18}{\sqrt{27}}$ **10.** $\frac{24}{\sqrt{18}}$ **11.** $\frac{-3}{\sqrt{50}}$

12. $\frac{-5}{\sqrt{75}}$ **13.** $\frac{63}{\sqrt{45}}$ **14.** $\frac{27}{\sqrt{32}}$ **15.** $\sqrt{\frac{13}{5}}$ **16.** $\sqrt{\frac{17}{11}}$

17. $\sqrt{\frac{1}{2}}$ **18.** $\sqrt{\frac{1}{3}}$ **19.** $\frac{\sqrt{24}}{\sqrt{8}}$ **20.** $\frac{\sqrt{36}}{\sqrt{18}}$ **21.** $\frac{-\sqrt{11}}{\sqrt{3}}$

22. $\frac{-\sqrt{13}}{\sqrt{5}}$ **23.** $\frac{7\sqrt{3}}{\sqrt{5}}$ **24.** $\frac{4\sqrt{6}}{\sqrt{5}}$ **25.** $\frac{24\sqrt{10}}{16\sqrt{3}}$ **26.** $\frac{18\sqrt{15}}{12\sqrt{2}}$

Simplify. ***See Example 3.***

27. $\sqrt{\frac{3}{4}} \cdot \sqrt{\frac{1}{5}}$ **28.** $\sqrt{\frac{7}{8}} \cdot \sqrt{\frac{1}{3}}$ **29.** $\sqrt{\frac{1}{6}} \cdot \sqrt{\frac{7}{2}}$ **30.** $\sqrt{\frac{3}{7}} \cdot \sqrt{\frac{5}{14}}$

31. $\sqrt{\frac{7}{13}} \cdot \sqrt{\frac{13}{3}}$ **32.** $\sqrt{\frac{19}{20}} \cdot \sqrt{\frac{20}{3}}$ **33.** $\sqrt{\frac{21}{7}} \cdot \sqrt{\frac{21}{8}}$ **34.** $\sqrt{\frac{5}{8}} \cdot \sqrt{\frac{5}{6}}$

35. $\sqrt{\frac{1}{12}} \cdot \sqrt{\frac{1}{3}}$ **36.** $\sqrt{\frac{1}{8}} \cdot \sqrt{\frac{1}{2}}$ **37.** $\sqrt{\frac{2}{9}} \cdot \sqrt{\frac{9}{2}}$ **38.** $\sqrt{\frac{4}{3}} \cdot \sqrt{\frac{3}{4}}$

Simplify. Assume that all variables represent positive real numbers. **See Example 4.**

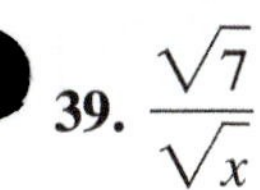

39. $\frac{\sqrt{7}}{\sqrt{x}}$ **40.** $\frac{\sqrt{19}}{\sqrt{y}}$ **41.** $\frac{\sqrt{3p^2}}{\sqrt{q}}$ **42.** $\frac{\sqrt{5a^2}}{\sqrt{b}}$

43. $\sqrt{\frac{9a^2r}{5}}$ **44.** $\sqrt{\frac{4x^2z}{3}}$ **45.** $\frac{\sqrt{4x^3}}{\sqrt{y}}$ **46.** $\frac{\sqrt{9t^3}}{\sqrt{s}}$

47. $\sqrt{\frac{5x^3z}{6}}$ **48.** $\sqrt{\frac{3st^3}{5}}$ **49.** $\sqrt{\frac{9a^2r^5}{7t}}$ **50.** $\sqrt{\frac{16x^3y^2}{13z}}$

Rationalize each denominator. Assume that variables in the denominator represent nonzero real numbers. **See Example 5.**

51. $\sqrt[3]{\frac{1}{2}}$ **52.** $\sqrt[3]{\frac{1}{3}}$ **53.** $\sqrt[3]{\frac{2}{5}}$ **54.** $\sqrt[3]{\frac{3}{2}}$

55. $\frac{\sqrt[3]{4}}{\sqrt[3]{7}}$ **56.** $\frac{\sqrt[3]{5}}{\sqrt[3]{10}}$ **57.** $\sqrt[3]{\frac{5}{9}}$ **58.** $\sqrt[3]{\frac{7}{100}}$

59. $\sqrt[3]{\frac{3}{4y^2}}$ **60.** $\sqrt[3]{\frac{3}{25x^2}}$ **61.** $\frac{\sqrt[3]{7m}}{\sqrt[3]{36n}}$ **62.** $\frac{\sqrt[3]{11p}}{\sqrt[3]{49q}}$

In each problem, **(a)** *give the answer as a simplified radical and* **(b)** *use a calculator to give the answer correct to the nearest thousandth.*

63. The period p of a pendulum is the time it takes for it to swing from one extreme to the other and back again. The value of p in seconds is given by

$$p = k \cdot \sqrt{\frac{L}{g}},$$

where L is length of the pendulum, g is acceleration due to gravity, and k is a constant. Find the period when $k = 6$, $L = 9$ ft, and $g = 32$ ft per sec².

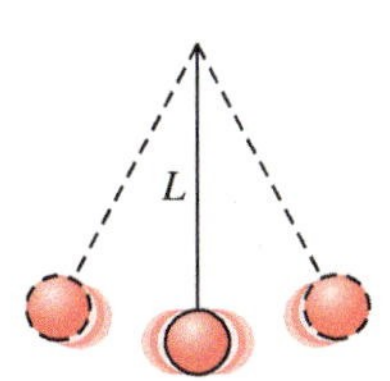

64. The velocity v in kilometers per second of a meteor approaching Earth is given by

$$v = \frac{k}{\sqrt{d}},$$

where d is its distance from the center of Earth and k is a constant. Find the velocity of a meteor that is 6000 km away from the center of Earth if $k = 450$.

8.5 More Simplifying and Operations with Radicals

OBJECTIVES

1. Simplify products of radical expressions.
2. Use conjugates to rationalize denominators of radical expressions.
3. Write radical expressions with quotients in lowest terms.

Apply these guidelines when simplifying radical expressions.

Guidelines for Simplifying Radical Expressions

1. If a radical represents a rational number, use that rational number in place of the radical.

   Examples: $\sqrt{49} = 7$, $\sqrt{\frac{169}{9}} = \frac{13}{3}$, $\sqrt[3]{27} = 3$

2. If a radical expression contains products of radicals, use the product rule for radicals, $\sqrt[n]{a} \cdot \sqrt[n]{b} = \sqrt[n]{ab}$, to obtain a single radical.

   Examples: $\sqrt{3} \cdot \sqrt{2} = \sqrt{6}$, $\sqrt[3]{5} \cdot \sqrt[3]{x} = \sqrt[3]{5x}$

3. If a radicand of a square root radical has a factor that is a perfect square, express the radical as the product of the positive square root of the perfect square and the remaining radical factor. Higher roots are treated similarly.

   Examples: $\sqrt{20} = \sqrt{4 \cdot 5} = \sqrt{4} \cdot \sqrt{5} = 2\sqrt{5}$

   $\sqrt[3]{16} = \sqrt[3]{8 \cdot 2} = \sqrt[3]{8} \cdot \sqrt[3]{2} = 2\sqrt[3]{2}$

4. If a radical expression contains sums or differences of radicals, use the distributive property to combine like radicals.

   Examples: $3\sqrt{2} + 4\sqrt{2}$ can be combined to obtain $7\sqrt{2}$.

   $3\sqrt{2} + 4\sqrt{3}$ cannot be combined in this way.

5. Rationalize any denominator containing a radical.

   Examples: $\frac{5}{\sqrt{3}} = \frac{5 \cdot \sqrt{3}}{\sqrt{3} \cdot \sqrt{3}} = \frac{5\sqrt{3}}{3}$

   $\sqrt[3]{\frac{1}{4}} = \frac{\sqrt[3]{1}}{\sqrt[3]{4}} = \frac{\sqrt[3]{1} \cdot \sqrt[3]{2}}{\sqrt[3]{4} \cdot \sqrt[3]{2}} = \frac{\sqrt[3]{2}}{\sqrt[3]{8}} = \frac{\sqrt[3]{2}}{2}$

OBJECTIVE 1 Simplify products of radical expressions.

EXAMPLE 1 Multiplying Radical Expressions

Find each product and simplify.

(a) $\sqrt{5}(\sqrt{8} - \sqrt{32})$ — Simplify inside the parentheses.

$= \sqrt{5}(2\sqrt{2} - 4\sqrt{2})$ — $\sqrt{8} = 2\sqrt{2}$; $\sqrt{32} = 4\sqrt{2}$

$= \sqrt{5}(-2\sqrt{2})$ — Subtract like radicals.

$= -2\sqrt{5 \cdot 2}$ — Product rule

$= -2\sqrt{10}$ — Multiply.

Continued on Next Page

(b) $(\sqrt{3}+2\sqrt{5})(\sqrt{3}-4\sqrt{5})$ — Use the FOIL method to multiply.

$$= \underbrace{\sqrt{3}(\sqrt{3})}_{\text{First}} + \underbrace{\sqrt{3}(-4\sqrt{5})}_{\text{Outer}} + \underbrace{2\sqrt{5}(\sqrt{3})}_{\text{Inner}} + \underbrace{2\sqrt{5}(-4\sqrt{5})}_{\text{Last}}$$

$= 3 - 4\sqrt{15} + 2\sqrt{15} - 8 \cdot 5$ Product rule

$= 3 - 2\sqrt{15} - 40$ Add like radicals. Multiply.

$= -37 - 2\sqrt{15}$ Add 3 and −40.

This does *not* equal $-39\sqrt{15}$.

(c) $(\sqrt{3}+\sqrt{21})(\sqrt{3}-\sqrt{7})$

$= \sqrt{3}(\sqrt{3}) + \sqrt{3}(-\sqrt{7}) + \sqrt{21}(\sqrt{3}) + \sqrt{21}(-\sqrt{7})$ FOIL method

$= 3 - \sqrt{21} + \sqrt{63} - \sqrt{147}$ Product rule

$= 3 - \sqrt{21} + \sqrt{9}\cdot\sqrt{7} - \sqrt{49}\cdot\sqrt{3}$ Factor; 9 and 49 are perfect squares.

$= 3 - \sqrt{21} + 3\sqrt{7} - 7\sqrt{3}$ $\sqrt{9}=3$; $\sqrt{49}=7$

Because there are no like radicals, no terms can be combined.

Work Problem 1 at the Side. ►

Example 2 uses the following rules for squaring binomials.

$$(x+y)^2 = x^2 + 2xy + y^2 \quad \text{and} \quad (x-y)^2 = x^2 - 2xy + y^2$$

EXAMPLE 2 Using Special Products with Radicals

Find each product. In part (c), assume that $x \geq 0$.

(a) $(\sqrt{10}-7)^2$

$= (\sqrt{10})^2 - 2(\sqrt{10})(7) + 7^2$ $(x-y)^2 = x^2 - 2xy + y^2$; Let $x = \sqrt{10}$ and $y = 7$.

$= 10 - 14\sqrt{10} + 49$ $(\sqrt{10})^2 = 10$; $7^2 = 49$

$= 59 - 14\sqrt{10}$ Add 10 and 49.

Do *not* try to combine further here. $59 - 14\sqrt{10} \neq 45\sqrt{10}$

(b) $(2\sqrt{3}+4)^2$

$= (2\sqrt{3})^2 + 2(2\sqrt{3})(4) + 4^2$ $(x+y)^2 = x^2 + 2xy + y^2$; Let $x = 2\sqrt{3}$ and $y = 4$.

$= 12 + 16\sqrt{3} + 16$ $(2\sqrt{3})^2 = 4 \cdot 3 = 12$

$= 28 + 16\sqrt{3}$ Add 12 and 16.

Do *not* try to combine further here. $28 + 16\sqrt{3} \neq 44\sqrt{3}$

(c) $(5-\sqrt{x})^2$

$= 5^2 - 2(5)(\sqrt{x}) + (\sqrt{x})^2$ Square the binomial.

$= 25 - 10\sqrt{x} + x$ Apply the exponents. Multiply.

Work Problem 2 at the Side. ►

---

1 Find each product and simplify.

GS **(a)** $\sqrt{7}(\sqrt{2}-\sqrt{5})$

$= \_\_\_ \cdot \sqrt{2} - \sqrt{7} \cdot \_\_\_$

$= \_\_\_\_$

(b) $\sqrt{2}(\sqrt{8}+\sqrt{20})$

(c) $(\sqrt{2}+5\sqrt{3})(\sqrt{3}-2\sqrt{2})$

(d) $(\sqrt{2}-\sqrt{5})(\sqrt{10}+\sqrt{2})$

2 Find each product. Simplify the answers. In part (c), assume that $x \geq 0$.

GS **(a)** $(\sqrt{5}-3)^2$

$= (\_\_\_)^2 - 2(\_\_\_)(3) + \_\_\_^2$

$= \_\_\_ - \_\_\_ + 9$

$= \_\_\_\_$

(b) $(4\sqrt{2}+5)^2$

(c) $(6+\sqrt{x})^2$

Answers

1. **(a)** $\sqrt{7}$; $\sqrt{5}$; $\sqrt{14}-\sqrt{35}$
(b) $4+2\sqrt{10}$ **(c)** $11-9\sqrt{6}$
(d) $2\sqrt{5}+2-5\sqrt{2}-\sqrt{10}$

2. **(a)** $\sqrt{5}$; $\sqrt{5}$; 3; 5; $6\sqrt{5}$; $14-6\sqrt{5}$
(b) $57+40\sqrt{2}$ **(c)** $36+12\sqrt{x}+x$

3 Find each product. In part (d), assume that $x \geq 0$.

GS (a) $(3+\sqrt{5})(3-\sqrt{5})$

$= \_\_\_^2 - (\_\_\_)^2$

$= \_\_\_ - \_\_\_$

$= \_\_\_$

(b) $(\sqrt{3}-2)(\sqrt{3}+2)$

(c) $(\sqrt{5}+\sqrt{3})(\sqrt{5}-\sqrt{3})$

(d) $(\sqrt{10}-\sqrt{x})(\sqrt{10}+\sqrt{x})$

Answers

3. (a) 3; $\sqrt{5}$; 9; 5; 4 (b) -1
(c) 2 (d) $10-x$

Example 3 involves the product of the sum and difference of two terms.

$$(x+y)(x-y) = x^2 - y^2$$

EXAMPLE 3 Using a Special Product with Radicals

Find each product. In part (b), assume that $x \geq 0$.

(a) $(4+\sqrt{3})(4-\sqrt{3})$ — $(x+y)(x-y)=x^2-y^2$; Let $x=4$ and $y=\sqrt{3}$.

$= 4^2 - (\sqrt{3})^2$

$= 16 - 3$ — $4^2 = 16$; $(\sqrt{3})^2 = 3$

$= 13$ — Subtract.

(b) $(\sqrt{x}-\sqrt{6})(\sqrt{x}+\sqrt{6})$

$= (\sqrt{x})^2 - (\sqrt{6})^2$ — Product of the sum and difference of two terms

$= x - 6$ — $(\sqrt{x})^2 = x$; $(\sqrt{6})^2 = 6$

◀ Work Problem 3 at the Side.

OBJECTIVE 2 Use conjugates to rationalize denominators of radical expressions. Recall that the expressions $x+y$ and $x-y$ are **conjugates.** We rationalize the denominator in a quotient such as $\frac{2}{4+\sqrt{3}}$ as follows.

$$\frac{2}{4+\sqrt{3}} = \frac{2(4-\sqrt{3})}{(4+\sqrt{3})(4-\sqrt{3})} = \frac{2(4-\sqrt{3})}{4^2-(\sqrt{3})^2} = \frac{2(4-\sqrt{3})}{13}$$

Rationalizing a Binomial Denominator

To rationalize a binomial denominator, where at least one of those terms is a square root radical, multiply the numerator and denominator by the conjugate of the denominator.

EXAMPLE 4 Using Conjugates to Rationalize Denominators

Rationalize each denominator. In part (c), assume that $x \geq 0$.

(a) $\frac{5}{3+\sqrt{5}}$

$= \frac{5(3-\sqrt{5})}{(3+\sqrt{5})(3-\sqrt{5})}$ — Multiply the numerator and denominator by the conjugate of the denominator.

$= \frac{5(3-\sqrt{5})}{3^2-(\sqrt{5})^2}$ — $(x+y)(x-y)=x^2-y^2$

$= \frac{5(3-\sqrt{5})}{9-5}$ — $3^2 = 9$ and $(\sqrt{5})^2 = 5$

$= \frac{5(3-\sqrt{5})}{4}$ — Subtract in the denominator.

Continued on Next Page

(b) $\dfrac{6+\sqrt{2}}{\sqrt{2}-5}$

$=\dfrac{(6+\sqrt{2})(\sqrt{2}+5)}{(\sqrt{2}-5)(\sqrt{2}+5)}$ Multiply the numerator and denominator by the conjugate of the denominator.

$=\dfrac{6\sqrt{2}+30+2+5\sqrt{2}}{2-25}$ FOIL method; $(x+y)(x-y)=x^2-y^2$

$=\dfrac{11\sqrt{2}+32}{-23}$ Combine like terms.

Be careful. Distribute the − sign to *both* terms in the numerator.

$=\dfrac{-11\sqrt{2}-32}{23}$ $\frac{a}{-b}=\frac{-a}{b}$; Distributive property

(c) $\dfrac{4}{3+\sqrt{x}}$

$=\dfrac{4(3-\sqrt{x})}{(3+\sqrt{x})(3-\sqrt{x})}$ Multiply by $\frac{3-\sqrt{x}}{3-\sqrt{x}}=1$.

We assume here that $x \neq 9$.

$=\dfrac{4(3-\sqrt{x})}{9-x}$ $3^2=9$; $(\sqrt{x})^2=x$

Work Problem 4 at the Side. ▶

4 Rationalize each denominator. In part (c), assume that $x \geq 0$.

(a) $\dfrac{5}{4+\sqrt{2}}$

(b) $\dfrac{\sqrt{5}+3}{2-\sqrt{5}}$

(c) $\dfrac{7}{5+\sqrt{x}}$

OBJECTIVE 3 Write radical expressions with quotients in lowest terms.

EXAMPLE 5 Writing a Radical Quotient in Lowest Terms

Write the quotient in lowest terms.

$\dfrac{3\sqrt{3}+9}{12}$ Don't simplify yet.

$=\dfrac{3(\sqrt{3}+3)}{3(4)}$ Factor first.

$=1\cdot\dfrac{\sqrt{3}+3}{4}$ Now divide out the common factor; $\frac{3}{3}=1$.

$=\dfrac{\sqrt{3}+3}{4}$ Identity property; lowest terms

Work Problem 5 at the Side. ▶

5 Write each quotient in lowest terms.

GS **(a)** $\dfrac{5\sqrt{3}-15}{10}$

$=\dfrac{\_\_\_(\sqrt{3}-\_\_\_)}{\_\_\_\cdot 2}$

$=\_\_\_\_$

(b) $\dfrac{12+8\sqrt{5}}{16}$

CAUTION

An expression like the one in **Example 5** can be simplified only by factoring a common factor from the denominator and *each* term of the numerator. ***First factor, and then divide out the common factor.***

Factor $\dfrac{4+8\sqrt{5}}{4}$ as $\dfrac{4(1+2\sqrt{5})}{4}$ to obtain $1+2\sqrt{5}$.

Answers

4. **(a)** $\dfrac{5(4-\sqrt{2})}{14}$ **(b)** $-11-5\sqrt{5}$

(c) $\dfrac{7(5-\sqrt{x})}{25-x}$ $(x \neq 25)$

5. **(a)** 5; 3; 5; $\dfrac{\sqrt{3}-3}{2}$ **(b)** $\dfrac{3+2\sqrt{5}}{4}$

8.5 Exercises

FOR EXTRA HELP *Go to MyMathLab for worked-out, step-by-step solutions to exercises enclosed in a square and video solutions to ▶ exercises.*

In this exercise set, we assume that variables are such that no negative numbers appear as radicals in square roots and no denominators are 0.

CONCEPT CHECK *Perform the operations mentally, and write the answers without doing intermediate steps.*

1. $\sqrt{49}+\sqrt{36}$

2. $\sqrt{2}\cdot\sqrt{8}$

3. $\sqrt{8}\cdot\sqrt{8}$

4. $(\sqrt{28}-\sqrt{14})(\sqrt{28}+\sqrt{14})$

5. **CONCEPT CHECK** A student simplified

$$-37-2\sqrt{15}$$

by combining the -37 and the -2 to obtain $-39\sqrt{15}$, which is incorrect. ***What Went Wrong?***

6. **CONCEPT CHECK** A student incorrectly squared a binomial as follows.

$$(\sqrt{7}+5)^2=(\sqrt{7})^2+5^2=7+25=32$$

What Went Wrong? Square the binomial correctly.

Find each product. Refer to the five guidelines given in this section to be sure answers are simplified. ***See Examples 1–3.***

7. $\sqrt{5}(\sqrt{3}-\sqrt{7})$ ▶

8. $\sqrt{7}(\sqrt{10}+\sqrt{3})$

9. $\sqrt{3}(\sqrt{2}+\sqrt{32})$

10. $\sqrt{7}(\sqrt{45}-\sqrt{20})$

11. $2\sqrt{5}(\sqrt{2}+3\sqrt{5})$

12. $3\sqrt{7}(2\sqrt{7}+4\sqrt{5})$

13. $3\sqrt{14}\cdot\sqrt{2}-\sqrt{28}$ ▶

14. $7\sqrt{6}\cdot\sqrt{3}-2\sqrt{18}$

15. $(2\sqrt{6}+3)(3\sqrt{6}+7)$ ▶

16. $(4\sqrt{5}-2)(2\sqrt{5}-4)$

17. $(5\sqrt{7}-2\sqrt{3})(3\sqrt{7}+4\sqrt{3})$

18. $(2\sqrt{10}+5\sqrt{2})(3\sqrt{10}-3\sqrt{2})$

19. $(\sqrt{15}-4)^2$

20. $(\sqrt{13}-1)^2$

21. $(8-\sqrt{7})^2$ ▶

22. $(6-\sqrt{11})^2$

23. $(2\sqrt{7}+3)^2$ ▶

24. $(4\sqrt{5}+5)^2$

25. $(5\sqrt{7}-2\sqrt{3})^2$

26. $(8\sqrt{2}-3\sqrt{3})^2$

27. $(\sqrt{a}+1)^2$

28. $(\sqrt{y}+4)^2$

29. $(5-\sqrt{2})(5+\sqrt{2})$

30. $(3-\sqrt{5})(3+\sqrt{5})$

31. $(\sqrt{8}-\sqrt{7})(\sqrt{8}+\sqrt{7})$

32. $(\sqrt{12}-\sqrt{11})(\sqrt{12}+\sqrt{11})$

33. $(\sqrt{y}-\sqrt{10})(\sqrt{y}+\sqrt{10})$

34. $(\sqrt{t}-\sqrt{13})(\sqrt{t}+\sqrt{13})$

35. $(\sqrt{2}+\sqrt{3})(\sqrt{6}-\sqrt{2})$

36. $(\sqrt{3}+\sqrt{5})(\sqrt{15}-\sqrt{5})$

37. $(\sqrt{10}-\sqrt{5})(\sqrt{5}+\sqrt{20})$

38. $(\sqrt{6}-\sqrt{3})(\sqrt{3}+\sqrt{18})$

39. $(\sqrt{5}+\sqrt{30})(\sqrt{6}+\sqrt{3})$

40. $(\sqrt{10}-\sqrt{20})(\sqrt{2}-\sqrt{5})$

41. $(\sqrt{5}-\sqrt{10})(\sqrt{x}-\sqrt{2})$

42. $(\sqrt{x}+\sqrt{6})(\sqrt{10}+\sqrt{3})$

43. CONCEPT CHECK A student tried to rationalize the denominator of

$$\frac{2}{4+\sqrt{3}} \text{ by multiplying by } \frac{4+\sqrt{3}}{4+\sqrt{3}}.$$

What Went Wrong? By what should he multiply?

44. CONCEPT CHECK A student simplified the expression

$$\frac{6+30\sqrt{11}}{6}$$

by dividing out 6 from the first term of the numerator and the denominator to obtain $1+30\sqrt{11}$.

What Went Wrong? Give the correct simplified form.

Rationalize each denominator. Write quotients in lowest terms. ***See Examples 4 and 5.***

45. $\frac{1}{3+\sqrt{2}}$

46. $\frac{1}{4+\sqrt{3}}$

47. $\frac{14}{2-\sqrt{11}}$

48. $\frac{19}{5-\sqrt{6}}$

49. $\frac{\sqrt{2}}{2-\sqrt{2}}$

50. $\frac{\sqrt{7}}{7-\sqrt{7}}$

51. $\frac{\sqrt{5}}{\sqrt{2}+\sqrt{3}}$

52. $\frac{\sqrt{3}}{\sqrt{2}+\sqrt{3}}$

53. $\frac{\sqrt{5}+2}{2-\sqrt{3}}$

54. $\frac{\sqrt{7}+3}{4-\sqrt{5}}$

55. $\frac{6-\sqrt{5}}{\sqrt{2}+2}$

56. $\frac{3+\sqrt{2}}{\sqrt{2}+1}$

57. $\frac{12}{\sqrt{x}+1}$

58. $\frac{10}{\sqrt{x}-4}\quad (x \neq 16)$

59. $\frac{3}{7-\sqrt{x}}\quad (x \neq 49)$

60. $\frac{1}{6-\sqrt{z}}\quad (x \neq 36)$

Write each quotient in lowest terms. ***See Example 5.***

61. $\frac{6\sqrt{11}-12}{6}$

62. $\frac{12\sqrt{5}-24}{12}$

63. $\frac{2\sqrt{3}+10}{16}$

64. $\frac{4\sqrt{6}+24}{20}$

65. $\frac{12-\sqrt{40}}{4}$

66. $\frac{9-\sqrt{72}}{12}$

67. $\frac{16+\sqrt{128}}{24}$

68. $\frac{25+\sqrt{75}}{10}$

Solve each problem.

69. The radius r of the circular top or bottom of a tin can with surface area S and height h is given by

$$r = \frac{-h+\sqrt{h^2+0.64S}}{2}.$$

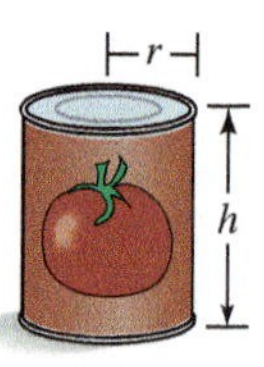

(a) What radius should be used to make a can with height 12 in. and surface area 400 in.^{2}?

(b) What radius should be used to make a can with height 6 in. and surface area 200 in.^{2}? Round the answer to the nearest tenth.

70. If an investment of P dollars grows to A dollars in 2 yr, the annual rate of return on the investment is given by

$$r = \frac{\sqrt{A}-\sqrt{P}}{\sqrt{P}}.$$

(a) Rationalize the denominator to simplify the expression.

(b) Find the annual rate of return r (as a percent) if \$50,000 increases to \$54,080.

Summary Exercises *Applying Operations with Radicals*

Perform all indicated operations, and express each answer in simplest form. Assume that all variables represent positive real numbers.

1. $5\sqrt{10} - 8\sqrt{10}$

2. $\sqrt{5}(\sqrt{5} - \sqrt{3})$

3. $(1 + \sqrt{3})(2 - \sqrt{6})$

4. $\sqrt{98} - \sqrt{72} + \sqrt{50}$

5. $(3\sqrt{5} - 2\sqrt{7})^2$

6. $\dfrac{3}{\sqrt{6}}$

7. $\sqrt[3]{16t^2} - \sqrt[3]{54t^2} + \sqrt[3]{128t^2}$

8. $\dfrac{8}{\sqrt{7} - \sqrt{5}}$

9. $\dfrac{1 + \sqrt{2}}{1 - \sqrt{2}}$

10. $(1 + \sqrt[3]{3})(1 - \sqrt[3]{3} + \sqrt[3]{9})$

11. $(\sqrt{3} + 6)(\sqrt{3} - 6)$

12. $\dfrac{1}{\sqrt{t} + \sqrt{3}}$

13. $\sqrt[3]{8x^3y^5z^6}$

14. $\dfrac{12}{\sqrt[3]{9}}$

15. $\dfrac{5}{\sqrt{6} - 1}$

16. $\sqrt{\dfrac{2}{3x}}$

17. $\dfrac{6\sqrt{3}}{5\sqrt{12}}$

18. $\dfrac{8\sqrt{50}}{2\sqrt{25}}$

19. $\dfrac{-4}{\sqrt[3]{4}}$

20. $\dfrac{\sqrt{6} - \sqrt{5}}{\sqrt{6} + \sqrt{5}}$

21. $\sqrt{75x} - \sqrt{12x}$

22. $(5 + 3\sqrt{3})^2$

23. $(\sqrt{7} - \sqrt{6})(\sqrt{7} + \sqrt{6})$

24. $\sqrt[3]{\frac{16}{81}}$

25. $x\sqrt[4]{x^5} - 3\sqrt[4]{x^9} + x^2\sqrt[4]{x}$

26. $\sqrt{7} + \sqrt{7}$

27. $\sqrt{14} + \sqrt{5}$

28. $9\sqrt{24} - 2\sqrt{54} + 3\sqrt{20}$

29. $\sqrt{\frac{3}{4}} \cdot \sqrt{\frac{1}{5}}$

30. $\frac{5}{\sqrt{5}}$

31. $\sqrt[3]{24} + 6\sqrt[3]{81}$

32. $\frac{8}{4 - \sqrt{x}} \quad (x \neq 16)$

33. $\sqrt[3]{4}(\sqrt[3]{2} - 3)$

34. $\sqrt{32x} - \sqrt{18x}$

35. $\sqrt{\frac{5}{8}}$

36. $(7 + \sqrt{x})^2$

A biologist found that the number of different plant species S on a Galápagos Island is related to the area of the island, $\mathcal{A}$ (in square miles), by the following formula.

$$S = 28.6\sqrt[3]{\mathcal{A}}$$

How many plant species (to the nearest whole number) would exist on such an island with the following areas?

37. 8 mi^2

38. 27,000 mi^2

8.6 Solving Equations with Radicals

OBJECTIVE 1 Solve radical equations having square root radicals. A **radical equation** is an equation that has a variable in a radicand.

$$\sqrt{x} = 6, \quad \sqrt{x+1} = 3, \quad \text{and} \quad 3\sqrt{x} = \sqrt{8x+9} \qquad \text{Radical equations}$$

To solve radical equations, we use the ***squaring property of equality.***

Squaring Property of Equality

If each side of a given equation is squared, then all solutions of the original equation are *among* the solutions of the squared equation.

CAUTION

Using the squaring property can give a new equation with *more* solutions than the original equation. For example, starting with the equation $x = 4$ and squaring each side gives

$$x^2 = 4^2, \quad \text{or} \quad x^2 = 16.$$

This last equation, $x^2 = 16$, has *two* solutions, 4 or -4, while the original equation, $x = 4$, has only *one* solution, 4.

Because of this possibility, checking is more than just a guard against algebraic errors when solving an equation with radicals. It is an essential part of the solution process. ***All proposed solutions from the squared equation must be checked in the original equation.***

EXAMPLE 1 Using the Squaring Property of Equality

Solve $\sqrt{p+1} = 3$.

$\sqrt{p+1} = 3$ — To eliminate the radical, use the squaring property and square each side.

$\left(\sqrt{p+1}\right)^2 = 3^2$

This equation is linear. $p + 1 = 9$ — $\left(\sqrt{p+1}\right)^2 = p + 1$

Proposed solution ⟶ $p = 8$ — Subtract 1.

CHECK (A check is essential.)

$\sqrt{p+1} = 3$ — Original equation

$\sqrt{8+1} \stackrel{?}{=} 3$ — Let $p = 8$.

$\sqrt{9} \stackrel{?}{=} 3$ — Add.

$3 = 3$ ✓ True

Because this statement is true, $\{8\}$ is the solution set of $\sqrt{p+1} = 3$. In this case the equation obtained by squaring had just one solution, which also satisfied the original equation.

Work Problem 1 at the Side. ▶

OBJECTIVES

1. Solve radical equations having square root radicals.
2. Identify equations with no solutions.
3. Solve equations by squaring a binomial.
4. Solve problems using formulas that involve radicals.

1 Solve each equation, and check the solution.

GS (a)
$\sqrt{k} = 3$

$\left(\sqrt{k}\right)^{\_\_} = 3^2$

$\_\_\_\_ = 9$

CHECK $\sqrt{k} = 3$

$\sqrt{\_\_\_\_} \stackrel{?}{=} 3$

$\_\_\_\_ = 3$

(True / False)

The solution set is ____.

(b) $\sqrt{x-2} = 4$

(c) $\sqrt{9-t} = 4$

Answers

1. (a) 2; k; 9; 3; True; $\{9\}$
(b) $\{18\}$ (c) $\{-7\}$

2 Solve each equation.

(a) $\sqrt{3x+9} = 2\sqrt{x}$

(b) $5\sqrt{x} = \sqrt{20x+5}$

EXAMPLE 2 Using the Squaring Property with a Radical on Each Side

Solve $3\sqrt{x} = \sqrt{x+8}$.

$3\sqrt{x} = \sqrt{x+8}$ — We need to eliminate *both* radicals.

$(3\sqrt{x})^2 = (\sqrt{x+8})^2$ — Squaring property

$3^2(\sqrt{x})^2 = (\sqrt{x+8})^2$ — On the left, $(ab)^2 = a^2b^2$.

Be careful here.

$9x = x + 8$ — $(\sqrt{x})^2 = x$; $(\sqrt{x+8})^2 = x+8$

$8x = 8$ — Subtract x.

Proposed solution → $x = 1$ — Divide by 8.

CHECK

$3\sqrt{x} = \sqrt{x+8}$ — Original equation

$3\sqrt{1} \stackrel{?}{=} \sqrt{1+8}$ — Let $x = 1$.

$3(1) \stackrel{?}{=} \sqrt{9}$ — Simplify.

$3 = 3$ ✓ — True (This is *not* the solution.)

Because a true statement results, the solution set is $\{1\}$.

◀ Work Problem **2** at the Side.

3 Solve each equation.

(a) $\sqrt{x} = -4$

GS (b) $\sqrt{x} + 6 = 0$

$\sqrt{x} =$ ____

$(\sqrt{x})^2 = (\_\_\_\_)^2$

$x =$ ____

CHECK $\sqrt{x} + 6 = 0$

$\sqrt{\_\_\_\_} + 6 \stackrel{?}{=} 0$

____ $= 0$

(*True / False*)

The solution set is ____.

OBJECTIVE 2 Identify equations with no solutions. Not all radical equations have real number solutions.

EXAMPLE 3 Using the Squaring Property When One Side Is Negative

Solve $\sqrt{x} = -3$.

$\sqrt{x} = -3$

$(\sqrt{x})^2 = (-3)^2$ — Squaring property

Proposed solution → $x = 9$ — Apply the exponents.

CHECK

$\sqrt{x} = -3$ — Original equation

$\sqrt{9} \stackrel{?}{=} -3$ — Let $x = 9$.

$3 = -3$ — False

Because the statement $3 = -3$ is false, the number 9 is *not* a solution of the given equation. It is an **extraneous solution** and must be rejected. In fact, $\sqrt{x} = -3$ has no real number solution. The solution set is $\emptyset$.

◀ Work Problem **3** at the Side.

Note

Because $\sqrt{x}$ represents the *principal* or *nonnegative* square root of x, we might have seen immediately in **Example 3** that there is no real number solution.

Answers

2. (a) $\{9\}$ (b) $\{1\}$

3. (a) $\emptyset$ (b) -6; -6; 36; 36; 12; False; $\emptyset$

We use the following steps when solving an equation with radicals.

Solving a Radical Equation

Step 1 **Isolate a radical.** Arrange the terms so that a radical is isolated on one side of the equation.

Step 2 **Square each side.**

Step 3 **Combine like terms.**

Step 4 **Repeat Steps 1–3,** if there is still a term with a radical.

Step 5 **Solve the equation.** Find all proposed solutions.

Step 6 **Check all proposed solutions** in the original equation. Write the solution set.

EXAMPLE 4 Using the Squaring Property with a Quadratic Expression

Solve $x = \sqrt{x^2 + 5x + 10}$.

Step 1 The radical is already isolated on the right side of the equation.

Step 2 Square each side.

$$x^2 = \left(\sqrt{x^2 + 5x + 10}\right)^2 \quad \text{Squaring property}$$

$$x^2 = x^2 + 5x + 10 \quad \left(\sqrt{a}\right)^2 = a$$

Step 3 $\quad 0 = 5x + 10 \quad$ Subtract x^2.

Step 4 This step is not needed.

Step 5 $\quad -10 = 5x \quad$ Subtract 10.

Proposed solution $\longrightarrow$ $\mathbf{-2 = x}$ $\quad$ Divide by 5.

Step 6 **CHECK** $\quad x = \sqrt{x^2 + 5x + 10} \quad$ Original equation

$$-2 \stackrel{?}{=} \sqrt{(-2)^2 + 5(-2) + 10} \quad \text{Let } x = -2.$$

$$-2 \stackrel{?}{=} \sqrt{4 - 10 + 10} \quad \text{Multiply.}$$

$$-2 = 2 \quad \text{False}$$

The principal square root of a quantity *cannot* be negative.

Because substituting -2 for x leads to a false result, the equation has no solution. The solution set is $\emptyset$.

Work Problem 4 at the Side. ▶

OBJECTIVE 3 **Solve equations by squaring a binomial.** Recall the rules for squaring binomials.

$$(x + y)^2 = x^2 + 2xy + y^2 \quad \text{and} \quad (x - y)^2 = x^2 - 2xy + y^2$$

We apply the second rule in **Example 5** on the next page with $(x - 3)^2$.

$$(x - 3)^2$$
$$= x^2 - 2x(3) + 3^2$$
$$= x^2 - 6x + 9$$

Remember the middle term when squaring.

Work Problem 5 at the Side. ▶

4 Solve $x = \sqrt{x^2 - 4x - 16}$.

5 Square each binomial.

GS **(a)** $w - 5$

$(w - 5)^2$

$= \_\_\_^2 - \_\_\_ w(\_\_\_) + \_\_\_^2$

$= \_\_\_\_\_\_$

(b) $2k - 5$

(c) $3m - 2p$

Answers

4. $\emptyset$

5. **(a)** w; 2; 5; 5; $w^2 - 10w + 25$
(b) $4k^2 - 20k + 25$
(c) $9m^2 - 12mp + 4p^2$

6 Solve each equation.

GS **(a)** $\sqrt{6w+6} = w + 1$

$$(\sqrt{6w+6})^{\_\_} = (w+1)^2$$

$$\_\_\_\_ = w^2 + 2w + 1$$

$$w^2 - \_\_\_\_ = 0$$

Complete the solution.

(b) $2u - 1 = \sqrt{10u + 9}$

EXAMPLE 5 Using the Squaring Property When One Side Has Two Terms

Solve $\sqrt{2x-3} = x - 3$.

$$\sqrt{2x-3} = x - 3$$

$$(\sqrt{2x-3})^2 = (x-3)^2 \quad \text{Square each side.}$$

Be careful squaring. $2x - 3 = x^2 - 6x + 9$ $\quad (\sqrt{a})^2 = a;\ (x-y)^2 = x^2 - 2xy + y^2$

This equation is quadratic because of the presence of the x^2-term.

Standard form → $x^2 - 8x + 12 = 0$ $\quad$ Subtract $2x$, add 3, and interchange sides.

$(x-6)(x-2) = 0$ $\quad$ Factor.

$x - 6 = 0$ or $x - 2 = 0$ $\quad$ Zero-factor property

Proposed solutions → $x = 6$ or $x = 2$ $\quad$ Solve each equation.

CHECK

$\sqrt{2x-3} = x - 3$		$\sqrt{2x-3} = x - 3$	
$\sqrt{2(6)-3} \stackrel{?}{=} 6 - 3$	Let $x = 6$.	$\sqrt{2(2)-3} \stackrel{?}{=} 2 - 3$	Let $x = 2$.
$\sqrt{12-3} \stackrel{?}{=} 3$		$\sqrt{4-3} \stackrel{?}{=} -1$	
$\sqrt{9} \stackrel{?}{=} 3$		$\sqrt{1} \stackrel{?}{=} -1$	
$3 = 3$ ✓	True	$1 = -1$	False

Only 6 is a valid solution. (2 is extraneous.) The solution set is $\{6\}$.

◀ Work Problem **6** at the Side.

EXAMPLE 6 Rewriting an Equation before Using the Squaring Property

Solve $\sqrt{9x} - 1 = 2x$.

We must apply Step 1 to isolate the radical *before* squaring each side. If we skip Step 1 and begin by squaring, we obtain the following.

$$(\sqrt{9x} - 1)^2 = (2x)^2 \quad \text{Don't do this.}$$

$$9x - 2\sqrt{9x} + 1 = 4x^2$$

This equation still contains a radical. Follow the steps below instead.

$$\sqrt{9x} - 1 = 2x$$

This is a key step. $\sqrt{9x} = 2x + 1$ $\quad$ Add 1 to isolate the radical. (Step 1)

$(\sqrt{9x})^2 = (2x+1)^2$ $\quad$ Square each side. (Step 2)

No terms contain radicals. $9x = 4x^2 + 4x + 1$ $\quad (\sqrt{a})^2 = a;\ (x+y)^2 = x^2 + 2xy + y^2$

$4x^2 - 5x + 1 = 0$ $\quad$ Standard form (Step 3)

$(4x-1)(x-1) = 0$ $\quad$ Factor. (Step 5; Step 4 is not needed.)

$4x - 1 = 0$ or $x - 1 = 0$ $\quad$ Zero-factor property

$x = \frac{1}{4}$ or $x = 1$ ← Proposed solutions

Continued on Next Page

Answers

6. (a) 2; $6w + 6$; $4w - 5$; $\{-1, 5\}$
(b) $\{4\}$

CHECK Let $x = \frac{1}{4}$. (Step 6)

$$\sqrt{9x} - 1 = 2x$$

$$\sqrt{9\left(\frac{1}{4}\right)} - 1 \stackrel{?}{=} 2\left(\frac{1}{4}\right)$$

$$\frac{3}{2} - 1 \stackrel{?}{=} \frac{1}{2}$$

$$\frac{1}{2} = \frac{1}{2} \checkmark \quad \text{True}$$

Let $x = 1$. (Step 6)

$$\sqrt{9x} - 1 = 2x$$

$$\sqrt{9(1)} - 1 \stackrel{?}{=} 2(1)$$

$$3 - 1 \stackrel{?}{=} 2$$

$$2 = 2 \checkmark \quad \text{True}$$

Both proposed solutions check, so the solution set is $\left\{\frac{1}{4}, 1\right\}$.

Work Problem 7 at the Side. ▶

CAUTION

When squaring each side of

$$\sqrt{9x} = 2x + 1$$

in **Example 6,** the *entire* binomial $2x + 1$ must be squared to obtain $4x^2 + 4x + 1$. It is incorrect to square the $2x$ and the 1 separately and write $4x^2 + 1$.

Some radical equations require squaring twice, as in the next example.

EXAMPLE 7 Using the Squaring Property Twice

Solve $\sqrt{21 + x} = 3 + \sqrt{x}$.

$$\sqrt{21 + x} = 3 + \sqrt{x} \qquad \text{A radical is isolated on the left. (Step 1)}$$

$$\left(\sqrt{21 + x}\right)^2 = \left(3 + \sqrt{x}\right)^2 \qquad \text{Square each side. (Step 2)}$$

$$21 + x = 9 + 6\sqrt{x} + x \qquad \text{Be careful here.}$$

$$12 = 6\sqrt{x} \qquad \text{Subtract 9. Subtract } x. \text{ (Step 3)}$$

$$2 = \sqrt{x} \qquad \text{Divide by 6.}$$

$$2^2 = \left(\sqrt{x}\right)^2 \qquad \text{Square each side again. (Step 4)}$$

Proposed solution ⟶ $4 = x$ Apply the exponents. (Step 5)

CHECK

$$\sqrt{21 + x} = 3 + \sqrt{x} \qquad \text{Original equation (Step 6)}$$

$$\sqrt{21 + 4} \stackrel{?}{=} 3 + \sqrt{4} \qquad \text{Let } x = 4.$$

$$\sqrt{25} \stackrel{?}{=} 3 + 2 \qquad \text{Simplify.}$$

$$5 = 5 \checkmark \qquad \text{True}$$

The solution set is $\{4\}$.

Work Problem 8 at the Side. ▶

7 Solve each equation.

GS **(a)** $\sqrt{x} - 3 = x - 15$

$$\sqrt{x} = x - \_\_\_\_$$

$$\left(\sqrt{x}\right)^{\_\_} = (x - \_\_\_\_)^2$$

$$\_\_\_\_ = x^2 - \_\_\_\_\_\_$$

$$\_\_\_\_\_\_\_\_ = 0$$

Complete the solution.

(b) $\sqrt{z + 5} + 2 = z + 5$

8 Solve each equation.

(a) $\sqrt{p + 1} - \sqrt{p - 4} = 1$

(b) $\sqrt{2x + 1} + \sqrt{x + 4} = 3$

Answers

7. **(a)** 12; 2; 12; x; $24x + 144$; $x^2 - 25x + 144$; $\{16\}$
(b) $\{-1\}$

8. **(a)** $\{8\}$ **(b)** $\{0\}$

Heron of Alexandria
First-century Greek mathematician

9 Find the area of a triangle with sides of lengths 7 cm, 15 cm, and 20 cm.

OBJECTIVE 4 Solve problems using formulas that involve radicals. The most common formula for the area $\mathcal{A}$ of a triangle is $\mathcal{A} = \frac{1}{2}bh$, where b is the length of the base and h is the height. *What if the height h is not known?* **Heron's formula** allows us to calculate the area of a triangle if we know only the lengths of the sides a, b, and c. To use Heron's formula, we first find the **semiperimeter** s, which is one-half the perimeter.

$$s = \frac{1}{2}(a + b + c) \quad \text{Semiperimeter}$$

The area $\mathcal{A}$ is then given by the following formula.

$$\mathcal{A} = \sqrt{s(s-a)(s-b)(s-c)} \quad \text{Heron's formula}$$

Using the familiar formula, we find the area of a 3-4-5 right triangle as follows.

$$\mathcal{A} = \frac{1}{2}(3)(4) = 6 \text{ square units} \quad \mathcal{A} = \tfrac{1}{2}bh$$

We can also use Heron's formula, with $s = \frac{1}{2}(3 + 4 + 5) = 6$.

$$\mathcal{A} = \sqrt{6(6-3)(6-4)(6-5)} \quad \text{Let } s = 6, a = 3, b = 4, \text{ and } c = 5.$$

$$\mathcal{A} = \sqrt{6 \cdot 3 \cdot 2 \cdot 1} \quad \text{Subtract.}$$

$$\mathcal{A} = \sqrt{36} \quad \text{Multiply.}$$

$$\mathcal{A} = 6 \text{ square units} \quad \text{Take the positive root.}$$

EXAMPLE 8 Using Heron's Formula to Find the Area of a Triangle

The sides of a triangular garden plot have lengths 20 ft, 34 ft, and 42 ft. See **Figure 3.** Find its area.

First find the semiperimeter of the triangle.

$$s = \frac{1}{2}(a + b + c)$$

$$s = \frac{1}{2}(20 + 34 + 42) \quad \text{Let } a = 20, b = 34, \text{ and } c = 42.$$

$$s = \frac{1}{2}(96) \quad \text{Add.}$$

$$s = 48 \quad \text{Multiply.}$$

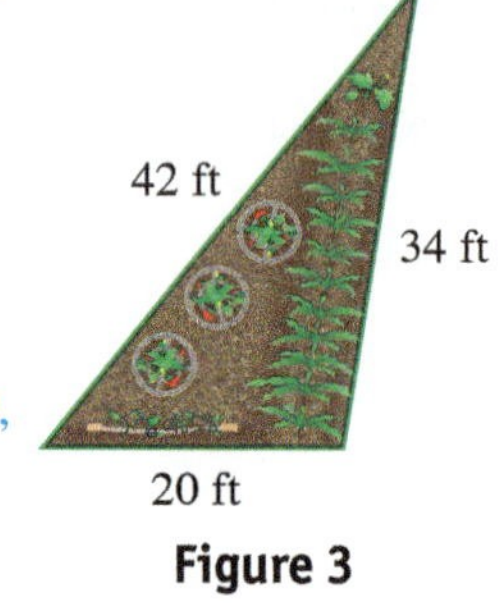

Figure 3

Now use Heron's formula to find the area.

$$\mathcal{A} = \sqrt{s(s-a)(s-b)(s-c)} \quad \text{Heron's formula}$$

$$\mathcal{A} = \sqrt{48(48-20)(48-34)(48-42)} \quad \text{Substitute.}$$

$$\mathcal{A} = \sqrt{48(28)(14)(6)} \quad \text{Subtract.}$$

$$\mathcal{A} = \sqrt{112{,}896} \quad \text{Multiply.}$$

$$\mathcal{A} = 336 \quad \text{Use a calculator.}$$

The area of the garden plot is 336 ft^2.

◀ Work Problem **9** at the Side.

Answer

9. 42 cm^2

8.6 Exercises

FOR EXTRA HELP — Go to MyMathLab for worked-out, step-by-step solutions to exercises enclosed in a square and video solutions to ▶ exercises.

CONCEPT CHECK *Fill in the blanks to complete the following.*

1. To solve an equation involving a radical, such as $\sqrt{2x-1}=5$, use the __________ property of equality. This property says that if each side of an equation is __________, all solutions of the __________ equation are among the solutions of the squared equation.

2. Solving some radical equations involves squaring a binomial using the following rules.

$$(x+y)^2 = \_\_\_\_\_\_\_\_\_\_$$

$$(x-y)^2 = \_\_\_\_\_\_\_\_\_\_$$

Solve each equation. **See Examples 1–4.**

3. $\sqrt{x}=7$

4. $\sqrt{k}=10$

5. $\sqrt{x+2}=3$ ▶

6. $\sqrt{x+7}=5$

7. $\sqrt{r-4}=9$

8. $\sqrt{k-12}=3$

9. $\sqrt{t}=-5$ ▶

10. $\sqrt{x}=-8$

11. $\sqrt{4-t}=7$ ▶

12. $\sqrt{9-s}=5$

13. $\sqrt{2t+3}=0$

14. $\sqrt{5x-4}=0$

15. $\sqrt{3x-8}=-2$

16. $\sqrt{6x+4}=-3$

17. $\sqrt{w}-4=7$

18. $\sqrt{t}+3=10$

19. $\sqrt{10x-8}=3\sqrt{x}$ ▶

20. $\sqrt{17t-4}=4\sqrt{t}$

21. $5\sqrt{x}=\sqrt{10x+15}$

22. $4\sqrt{z}=\sqrt{20z-16}$

23. $\sqrt{3x-5}=\sqrt{2x+1}$ ▶

24. $\sqrt{5x+2}=\sqrt{3x+8}$

25. $k=\sqrt{k^2-5k-15}$ ▶

26. $s=\sqrt{s^2-2s-6}$

27. $7x=\sqrt{49x^2+2x-10}$

28. $6x=\sqrt{36x^2+5x-5}$

29. CONCEPT CHECK Consider the following incorrect "solution." ***What Went Wrong?***

$$-\sqrt{x-1} = -4$$
$$-(x-1) = 16 \quad \text{Square each side.}$$
$$-x+1 = 16 \quad \text{Distributive property}$$
$$-x = 15 \quad \text{Subtract 1.}$$
$$x = -15 \quad \text{Multiply by } -1.$$

Solution set: $\{-15\}$

30. CONCEPT CHECK The first step in solving the equation

$$\sqrt{2x+1} = x-7$$

is to square each side of the equation. When a student did this, he incorrectly obtained

$$2x + 1 = x^2 + 49.$$

What Went Wrong? Square each side of the original equation correctly.

Solve each equation. ***See Examples 5 and 6.***

31. $\sqrt{2x+1} = x-7$

32. $\sqrt{3x+3} = x-5$

33. $\sqrt{3k+10} + 5 = 2k$

34. $\sqrt{4t+13} + 1 = 2t$

35. $\sqrt{5x+1} - 1 = x$

36. $\sqrt{x+1} - x = 1$

37. $\sqrt{6t+7} + 3 = t+5$

38. $\sqrt{10x+24} = x+4$

39. $x - 4 - \sqrt{2x} = 0$

40. $x - 3 - \sqrt{4x} = 0$

41. $\sqrt{x} + 6 = 2x$

42. $\sqrt{k} + 12 = k$

Solve each equation. ***See Example 7.***

43. $\sqrt{x+1} - \sqrt{x-4} = 1$

44. $\sqrt{2x+3} + \sqrt{x+1} = 1$

45. $\sqrt{x} = \sqrt{x-5} + 1$

46. $\sqrt{2x} = \sqrt{x+7} - 1$

47. $\sqrt{3x+4} - \sqrt{2x-4} = 2$

48. $\sqrt{1-x} + \sqrt{x+9} = 4$

49. $\sqrt{2x+11} + \sqrt{x+6} = 2$

50. $\sqrt{x+9} + \sqrt{x+16} = 7$

Solve each problem.

51. The square root of the sum of a number and 4 is 5. Find the number.

52. A certain number is the same as the square root of the product of 8 and the number. Find the number.

53. Three times the square root of 2 equals the square root of the sum of some number and 10. Find the number.

54. The negative square root of a number equals that number decreased by 2. Find the number.

Use Heron's formula to solve each problem. ***See Example 8.***

55. A surveyor has measured the lengths of the sides of a triangular plot of land as 180 ft, 200 ft, and 240 ft. Find the area of the plot. (Round to the nearest whole number.)

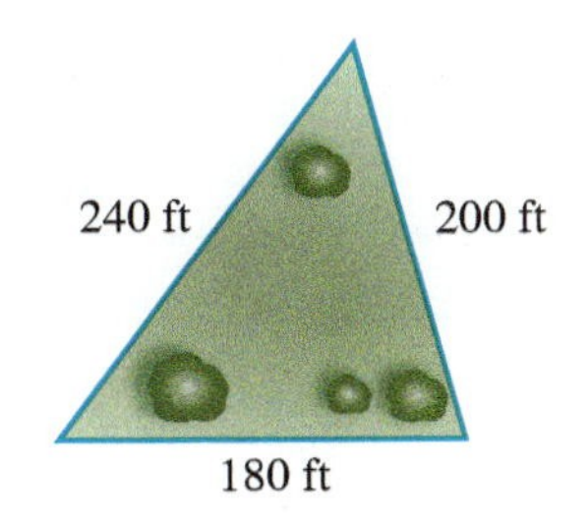

56. A triangular indoor play space in a day care center has sides of length 30 ft, 45 ft, and 65 ft. How much carpeting would be needed to cover this space? (Round to the nearest whole number.)

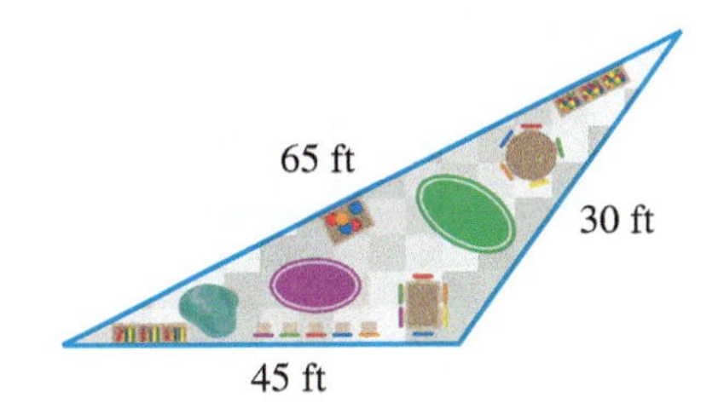

Solve each problem. Give answers to the nearest whole number.

57. To estimate the speed at which a car was traveling at the time of an accident, a police officer drives the car under conditions similar to those during which the accident took place and then skids to a stop. If the car is driven at 30 mph, then the speed at the time of the accident is given by

$$s = 30\sqrt{\frac{a}{p}},$$

where a is the length of the skid marks left at the time of the accident and p is the length of the skid marks in the police test.

Find s for the following values of a and p.

(a) $a = 862$ ft; $p = 156$ ft

(b) $a = 382$ ft; $p = 96$ ft

(c) $a = 84$ ft; $p = 26$ ft

58. A formula for calculating the distance d in miles one can see from an airplane to the horizon on a clear day is

$$d = 1.22\sqrt{x},$$

where x is the altitude of the plane in feet.

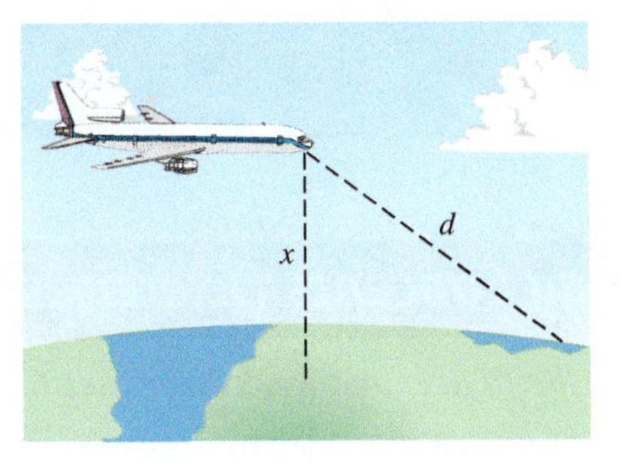

How far can one see to the horizon in a plane flying at the following altitudes?

(a) 15,000 ft

(b) 18,000 ft

(c) 24,000 ft

On a clear day, the maximum distance in kilometers that can be seen from a tall building is given by the formula

$$\text{sight distance} = 111.7\sqrt{\text{height of structure in kilometers}}.$$

Use this formula and the conversion equations $1 \text{ ft} \approx 0.3048 \text{ m}$ *and* $1 \text{ km} \approx 0.621371 \text{ mi}$ *as necessary to solve each problem. Round answers to the nearest mile. (Data from* A Sourcebook of Applications of School Mathematics, *NCTM.)*

59. The London Eye is a unique structure that features 32 observation capsules, and has a diameter of 135 m. Does the formula justify the claim that on a clear day passengers on the London Eye can see Windsor Castle, 25 mi away? (Data from www.londoneye.com)

60. The Empire State Building in New York City is 1250 ft high (excluding the antenna). The observation deck, located on the 102nd floor, is at a height of 1050 ft. How far could we see on a clear day from the observation deck? (Data from www.esbnyc.com)

61. The twin Petronas Towers in Kuala Lumpur, Malaysia, are 1483 ft high (including the spires). How far would one of the builders have been able to see on a clear day from the top of a spire? (Data from *The World Almanac and Book of Facts.*)

62. The Khufu Pyramid in Giza was built in about 2566 B.C. to a height, at that time, of 481 ft. It is now only about 449 ft high. How far would one of the original builders of the pyramid have been able to see from the top of the pyramid? (Data from www.archaeology.com)

Relating Concepts (Exercises 63–68) For Individual or Group Work

Consider the figure, and **work Exercises 63–68 in order.**

63. The lengths of the sides of the entire triangle are 7, 7, and 12. Find the semiperimeter s.

64. Now use Heron's formula to find the area of the entire triangle. Write it as a simplified radical.

65. Find the value of h using the Pythagorean theorem.

66. Find the area of each of the congruent right triangles forming the entire triangle by using the formula $\mathcal{A} = \frac{1}{2}bh$.

67. Double the result from **Exercise 66** to determine the area of the entire triangle.

68. How do the answers in **Exercises 64 and 67** compare?

Chapter 8 Summary

Key Terms

8.1

square root A number b is a square root of a if $b^2 = a$.

principal square root The positive square root of a number is its principal square root.

Radical symbol → $\sqrt[n]{a}$ ← Radicand; Index; Radical

radicand The number or expression inside a radical symbol is the radicand.

radical A radical symbol with a radicand is a radical.

radical expression An algebraic expression containing a radical is a radical expression.

perfect square A number with a rational square root is a perfect square.

irrational number A real number that is not rational is an irrational number.

cube root A number b is a cube root of a if $b^3 = a$.

index (order) In a radical of the form $\sqrt[n]{a}$, the number n is the index, or order.

perfect cube A number with a rational cube root is a perfect cube.

8.3

like radicals Like radicals are multiples of the same root of the same number.

8.4

rationalizing the denominator The process of changing a denominator from one with a radical to one without a radical is called rationalizing the denominator.

8.5

conjugate The conjugate of $x + y$ is $x - y$.

8.6

radical equation An equation with a variable in a radicand is a radical equation.

extraneous solution A proposed solution that does not satisfy the given equation is an extraneous solution.

New Symbols

 radical symbol

$\approx$ is approximately equal to

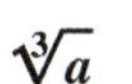 cube root of a

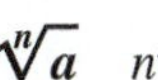 nth root of a

Test Your Word Power

See how well you have learned the vocabulary in this chapter.

1. A **square root** of a number is
   A. the number raised to the second power
   B. the number under a radical symbol
   C. a number that when multiplied by itself gives the original number
   D. the inverse of the number.

2. A **radicand** is
   A. the index of a radical
   B. the number or expression inside the radical symbol
   C. the positive root of a number
   D. the radical symbol.

3. A **radical** is
   A. a symbol that indicates the nth root
   B. an algebraic expression containing a square root
   C. the positive nth root of a number
   D. a radical symbol and the number or expression inside it.

4. The **principal root** of a positive number with even index n is
   A. the positive nth root of the number
   B. the negative nth root of the number
   C. the square root of the number
   D. the cube root of the number.

5. An **irrational number** is
   A. the quotient of two integers, with denominator not 0
   B. a decimal number that neither terminates nor repeats
   C. the principal square root of a number
   D. a nonreal number.

Test Your Word Power (*continued*)

6. The **Pythagorean theorem** states that, in a right triangle,
   A. the sum of the measures of the angles is 180°
   B. the sum of the lengths of the two shorter sides equals the length of the longest side
   C. the longest side is opposite the right angle
   D. the square of the length of the longest side equals the sum of the squares of the lengths of the two shorter sides.

7. **Like radicals** are
   A. radicals in simplest form
   B. expressions containing radicals
   C. multiples of the same root of the same number
   D. radicals with the same index.

8. The **conjugate** of $a + b$ is
   A. $a - b$
   B. $a \cdot b$
   C. $a \div b$
   D. $(a + b)^2$.

9. **Rationalizing the denominator** is the process of
   A. eliminating fractions from a radical expression
   B. changing the denominator of a fraction from one with a radical to one without a radical
   C. clearing a radical expression of radicals
   D. multiplying radical expressions.

10. An **extraneous solution** is a value
    A. that makes an equation false and must be discarded
    B. that makes an equation true
    C. that makes an expression equal 0
    D. that checks in the original equation.

Answers to Test Your Word Power

1. C; *Examples:* 6 is a square root of 36 because $6^2 = 6 \cdot 6 = 36$. -6 is also a square root of 36.
2. B; *Example:* In $\sqrt{3xy}$, $3xy$ is the radicand.
3. D; *Examples:* $\sqrt{144}$, $\sqrt{4xy^2}$, and $\sqrt{4 + t^2}$
4. A; *Examples:* $\sqrt{36} = 6$, $\sqrt[4]{81} = 3$, and $\sqrt[6]{64} = 2$
5. B; *Examples:* π, $\sqrt{2}, -\sqrt{5}$
6. D; *Example:* In a right triangle where $a = 6$, $b = 8$, and $c = 10$, $6^2 + 8^2 = 10^2$.
7. C; *Examples:* $\sqrt{7}$ and $3\sqrt{7}$ are like radicals, as are $2\sqrt[3]{6k}$ and $5\sqrt[3]{6k}$.
8. A; *Example:* The conjugate of $\sqrt{3} + 1$ is $\sqrt{3} - 1$.
9. B; *Example:* To rationalize the denominator of $\frac{5}{\sqrt{3}+1}$, multiply the numerator and denominator by $\sqrt{3} - 1$ (the conjugate of the denominator) to obtain $\frac{5(\sqrt{3}-1)}{2}$.
10. A; *Example:* The proposed solution 2 is extraneous when $\sqrt{5q - 1} + 3 = 0$ is solved using the squaring property of equality.

Quick Review

Concepts	Examples

8.1 Evaluating Roots

Concepts

Let a be a positive real number.

$\sqrt{a}$ is the positive or principal square root of a.

$-\sqrt{a}$ is the negative square root of a. Also, $\sqrt{0} = 0$.

If a is a negative real number, then $\sqrt{a}$ is not a real number.

If a is a positive rational number, then $\sqrt{a}$ is rational if a is a perfect square. $\sqrt{a}$ is irrational if a is not a perfect square.

Every real number has exactly one real cube root.

Examples

$$\sqrt{81} = 9$$

$$-\sqrt{81} = -9$$

$\sqrt{-25}$ is not a real number.

$\sqrt{\frac{4}{9}}$ and $\sqrt{16}$ are rational. $\sqrt{\frac{2}{3}}$ and $\sqrt{21}$ are irrational.

$$\sqrt[3]{27} = 3 \qquad \sqrt[3]{-8} = -2$$

Pythagorean Theorem

If c is the length of the longest side (hypotenuse) of a right triangle and a and b are the lengths of the shorter sides (legs), then

$$a^2 + b^2 = c^2.$$

Find a for the triangle in the figure.

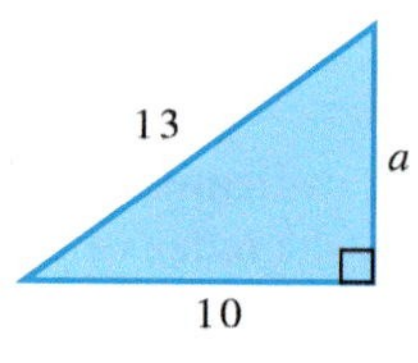

$$a^2 + 10^2 = 13^2$$
$$a^2 + 100 = 169$$
$$a^2 = 69$$
$$a = \sqrt{69}$$
$$a \approx 8.3$$

Concepts	Examples
8.2 Multiplying, Dividing, and Simplifying Radicals	
Product Rule for Radicals If a and b are nonnegative real numbers, then the following hold true. $\sqrt{a}\cdot\sqrt{b}=\sqrt{ab}$ and $\sqrt{ab}=\sqrt{a}\cdot\sqrt{b}$	$\sqrt{5}\cdot\sqrt{7}=\sqrt{5\cdot 7}=\sqrt{35}$ $\sqrt{48}=\sqrt{16\cdot 3}=\sqrt{16}\cdot\sqrt{3}=4\sqrt{3}$
Quotient Rule for Radicals If a and b are nonnegative real numbers and $b\neq 0$, then the following hold true. $\sqrt{\frac{a}{b}}=\frac{\sqrt{a}}{\sqrt{b}}$ and $\frac{\sqrt{a}}{\sqrt{b}}=\sqrt{\frac{a}{b}}$	$\sqrt{\frac{25}{64}}=\frac{\sqrt{25}}{\sqrt{64}}=\frac{5}{8}$ $\quad\frac{\sqrt{8}}{\sqrt{2}}=\sqrt{\frac{8}{2}}=\sqrt{4}=2$
For all real numbers a and b where the indicated roots exist, the following hold true. $\sqrt[n]{a}\cdot\sqrt[n]{b}=\sqrt[n]{ab}$ and $\frac{\sqrt[n]{a}}{\sqrt[n]{b}}=\sqrt[n]{\frac{a}{b}}$ $(b\neq 0)$	$\sqrt[3]{5}\cdot\sqrt[3]{3}=\sqrt[3]{15}$ $\quad\frac{\sqrt[4]{12}}{\sqrt[4]{4}}=\sqrt[4]{\frac{12}{4}}=\sqrt[4]{3}$
8.3 Adding and Subtracting Radicals Add and subtract like radicals using the distributive property. ***Only like radicals can be combined in this way.***	$2\sqrt{5}+4\sqrt{5}$ $=(2+4)\sqrt{5}$ $=6\sqrt{5}$ $\sqrt{8}-\sqrt{32}$ $=2\sqrt{2}-4\sqrt{2}$ $=-2\sqrt{2}$
8.4 Rationalizing the Denominator To rationalize the denominator of a radical expression, multiply both the numerator and denominator by a number that will eliminate the radical from the denominator.	$\frac{2}{\sqrt{3}}=\frac{2\cdot\sqrt{3}}{\sqrt{3}\cdot\sqrt{3}}=\frac{2\sqrt{3}}{3}$ $\sqrt[3]{\frac{5}{121}}=\frac{\sqrt[3]{5}\cdot\sqrt[3]{11}}{\sqrt[3]{11^2}\cdot\sqrt[3]{11}}=\frac{\sqrt[3]{55}}{11}$
8.5 More Simplifying and Operations with Radicals When appropriate, use the rules for adding and multiplying polynomials to simplify radical expressions.	$\sqrt{6}(\sqrt{5}-\sqrt{7})=\sqrt{30}-\sqrt{42}$ $(\sqrt{3}+1)(\sqrt{3}-2)$ $=\sqrt{3}\cdot\sqrt{3}-2\sqrt{3}+\sqrt{3}-2$ FOIL method $=3-2\sqrt{3}+\sqrt{3}-2$ $(\sqrt{a})^2=a$ $=1-\sqrt{3}$ Combine like terms.
The following rules are useful in simplifying radical expressions. $(x+y)^2=x^2+2xy+y^2$ $(x-y)^2=x^2-2xy+y^2$ $(x+y)(x-y)=x^2-y^2$	$(\sqrt{13}-\sqrt{2})^2$ $=(\sqrt{13})^2-2(\sqrt{13})(\sqrt{2})+(\sqrt{2})^2$ $=13-2\sqrt{26}+2$ $=15-2\sqrt{26}$ $(\sqrt{5}+\sqrt{3})(\sqrt{5}-\sqrt{3})$ $=5-3$ $=2$

Concepts	Examples

Any denominators with radicals should be rationalized.

$$\frac{3}{\sqrt{6}} = \frac{3 \cdot \sqrt{6}}{\sqrt{6} \cdot \sqrt{6}} = \frac{3\sqrt{6}}{6} = \frac{\sqrt{6}}{2}$$

If a radical expression contains two terms in the denominator and at least one of those terms is a square root radical, multiply both the numerator and denominator by the conjugate of the denominator. In general, the expressions $x + y$ and $x - y$ are conjugates.

$$\frac{6}{\sqrt{7} - \sqrt{2}}$$

$$= \frac{6(\sqrt{7} + \sqrt{2})}{(\sqrt{7} - \sqrt{2})(\sqrt{7} + \sqrt{2})}$$ Multiply the numerator and denominator by the conjugate of the denominator.

$$= \frac{6(\sqrt{7} + \sqrt{2})}{7 - 2}$$ Multiply.

$$= \frac{6(\sqrt{7} + \sqrt{2})}{5}$$ Subtract.

To write a radical expression with a quotient in lowest terms, first factor, and then divide out the common factor.

Write in lowest terms.

$$\frac{4\sqrt{5} + 20}{8}$$

$$= \frac{4(\sqrt{5} + 5)}{4(2)}$$ Factor.

$$= \frac{\sqrt{5} + 5}{2}$$ Divide out 4.

8.6 Solving Equations with Radicals

Solving a Radical Equation

Step 1 Isolate a radical.

Step 2 Square each side. (By the squaring property of equality, all solutions of the original equation are *among* the solutions of the squared equation.)

Step 3 Combine like terms.

Step 4 If there is still a term with a radical, repeat Steps 1–3.

Step 5 Solve the equation for all proposed solutions.

Step 6 Check all proposed solutions in the original equation. Write the solution set.

Solve.

$$\sqrt{2x - 3} + x = 3$$

$$\sqrt{2x - 3} = 3 - x$$ Isolate the radical.

$$(\sqrt{2x - 3})^2 = (3 - x)^2$$ Square each side.

$$2x - 3 = 9 - 6x + x^2$$ Remember the middle term when squaring $3 - x$.

$$x^2 - 8x + 12 = 0$$ Standard form

$$(x - 2)(x - 6) = 0$$ Factor.

$x - 2 = 0$ or $x - 6 = 0$ Zero-factor property

$x = 2$ or $x = 6$ ← Proposed solutions

CHECK $\sqrt{2x - 3} + x = 3$ Original equation

Let $x = 2$.	Let $x = 6$.
$\sqrt{2(2) - 3} + 2 \stackrel{?}{=} 3$	$\sqrt{2(6) - 3} + 6 \stackrel{?}{=} 3$
$\sqrt{1} + 2 \stackrel{?}{=} 3$	$\sqrt{9} + 6 \stackrel{?}{=} 3$
$3 = 3$ ✓ True	$9 = 3$ False

Only 2 is a valid solution. (6 is extraneous.)

Solution set: $\{2\}$

Chapter 8 Review Exercises

8.1 *Find all square roots of each number.*

1. 49 **2.** 81 **3.** 196 **4.** 121 **5.** 256 **6.** 729

Find each root.

7. $\sqrt{16}$ **8.** $-\sqrt{0.36}$ **9.** $\sqrt{-8100}$ **10.** $-\sqrt{4225}$ **11.** $\sqrt{\frac{144}{169}}$

12. $-\sqrt{\frac{100}{81}}$ **13.** $\sqrt[3]{216}$ **14.** $\sqrt[3]{-512}$ **15.** $\sqrt[4]{81}$ **16.** $-\sqrt[4]{625}$

17. Find the value of x (to the nearest thousandth as necessary) in each triangle.

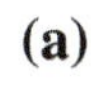

(a)

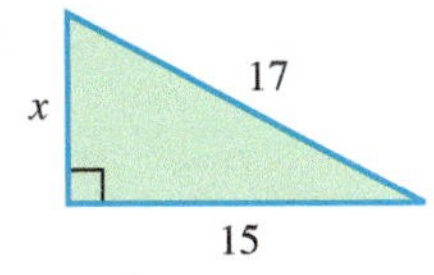

(b)

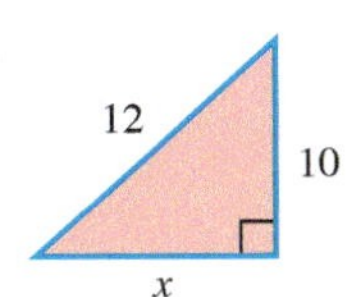

18. An HP f1905 computer monitor has viewing screen dimensions as shown in the figure. Find the diagonal measure of the viewing screen to the nearest tenth.

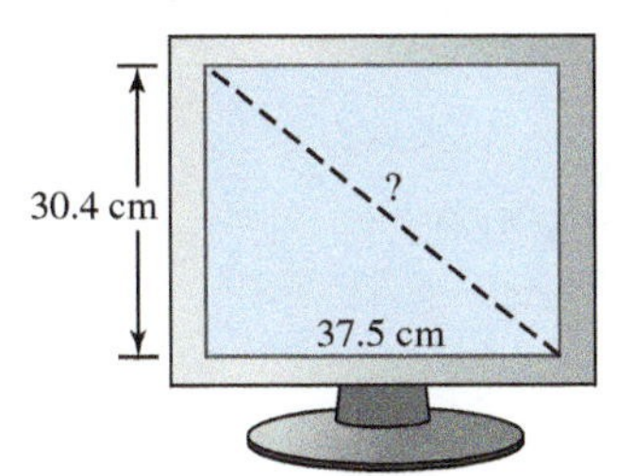

Determine whether each number is rational, irrational, *or* not a real number. *If a number is rational, give its exact value. If a number is irrational, give a decimal approximation to the nearest thousandth.*

19. $\sqrt{73}$ **20.** $\sqrt{169}$ **21.** $-\sqrt{625}$ **22.** $\sqrt{-19}$

8.2 *Simplify each radical expression.*

23. $\sqrt{48}$ **24.** $-\sqrt{288}$ **25.** $\sqrt[3]{16}$ **26.** $\sqrt[3]{375}$

27. $\sqrt{12} \cdot \sqrt{27}$ **28.** $\sqrt{32} \cdot \sqrt{48}$ **29.** $-\sqrt{\frac{121}{400}}$ **30.** $\sqrt{\frac{7}{169}}$

31. $\sqrt{\frac{1}{6}} \cdot \sqrt{\frac{5}{6}}$

32. $\sqrt{\frac{2}{5}} \cdot \sqrt{\frac{2}{45}}$

33. $\frac{3\sqrt{10}}{\sqrt{5}}$

34. $\frac{8\sqrt{150}}{4\sqrt{75}}$

Simplify. Assume that all variables represent nonnegative real numbers.

35. $\sqrt{r^{18}}$

36. $\sqrt{x^{10}y^{16}}$

37. $\sqrt{162x^9}$

38. $\sqrt{\frac{36}{p^2}} \quad (p \neq 0)$

39. $\sqrt{a^{15}b^{21}}$

40. $\sqrt{121x^6y^{10}}$

41. $\sqrt[3]{y^6}$

42. $\sqrt[3]{216x^{15}}$

8.3 *Perform the indicated operations.*

43. $7\sqrt{11} + \sqrt{11}$

44. $3\sqrt{2} + 6\sqrt{2}$

45. $3\sqrt{75} + 2\sqrt{27}$

46. $4\sqrt{12} + \sqrt{48}$

47. $4\sqrt{24} - 3\sqrt{54} + \sqrt{6}$

48. $2\sqrt{7} - 4\sqrt{28} + 3\sqrt{45}$

49. $\frac{2}{5}\sqrt{75} + \frac{3}{4}\sqrt{160}$

50. $\frac{1}{3}\sqrt{18} + \frac{1}{4}\sqrt{32}$

51. $\sqrt{15} \cdot \sqrt{2} + 5\sqrt{30}$

Simplify. Assume that all variables represent nonnegative real numbers.

52. $\sqrt{4x} + \sqrt{36x} - \sqrt{9x}$

53. $\sqrt{16p} + 3\sqrt{p} - \sqrt{49p}$

54. $3k\sqrt{8k^2n} + 5k^2\sqrt{2n}$

8.4 *Simplify. Assume that all variables represent positive real numbers.*

55. $\frac{10}{\sqrt{3}}$

56. $\frac{8\sqrt{2}}{\sqrt{5}}$

57. $\frac{12}{\sqrt{24}}$

58. $\sqrt{\frac{2}{5}}$

59. $\sqrt{\frac{5}{14}} \cdot \sqrt{28}$

60. $\sqrt{\frac{2}{7}} \cdot \sqrt{\frac{1}{3}}$

61. $\sqrt{\frac{r^2}{16x}}$

62. $\sqrt[3]{\frac{1}{3}}$

8.5 *Find each product.*

63. $-\sqrt{3}\left(\sqrt{5}+\sqrt{27}\right)$

64. $3\sqrt{2}\left(\sqrt{3}+2\sqrt{2}\right)$

65. $\left(2\sqrt{3}-4\right)\left(5\sqrt{3}+2\right)$

66. $\left(\sqrt{7}+2\sqrt{6}\right)\left(\sqrt{12}-\sqrt{2}\right)$

67. $\left(2\sqrt{3}+5\right)\left(2\sqrt{3}-5\right)$

68. $\left(\sqrt{x}+2\right)^2 \quad (x \geq 0)$

Rationalize each denominator.

69. $\dfrac{1}{2+\sqrt{5}}$

70. $\dfrac{3}{4-\sqrt{3}}$

71. $\dfrac{\sqrt{5}-1}{\sqrt{2}+3}$

Write each quotient in lowest terms.

72. $\dfrac{15+10\sqrt{6}}{15}$

73. $\dfrac{3+9\sqrt{7}}{12}$

74. $\dfrac{6-\sqrt{192}}{2}$

8.6 *Solve each equation.*

75. $\sqrt{x}+5=0$

76. $\sqrt{k+1}=7$

77. $\sqrt{5t+4}=3\sqrt{t}$

78. $\sqrt{2p+3}=\sqrt{5p-3}$

79. $\sqrt{4x+1}=x-1$

80. $\sqrt{13+4t}=t+4$

81. $\sqrt{2-x}+3=x+7$

82. $\sqrt{x}-x+2=0$

83. $\sqrt{x+2}-\sqrt{x-3}=1$

84. A surveyor measured the lengths of the sides of a triangular plot of land as 190 ft, 210 ft, and 250 ft. Use Heron's formula,

$$\mathcal{A}=\sqrt{s(s-a)(s-b)(s-c)}, \quad \text{where} \quad s=\frac{1}{2}(a+b+c),$$

to find the area of the plot. Round to the nearest whole number.

Chapter 8 Mixed Review Exercises

Simplify. Assume that all variables represent positive real numbers.

1. $2\sqrt{27} + 3\sqrt{75} - \sqrt{300}$

2. $\dfrac{1}{5 + \sqrt{2}}$

3. $\sqrt{\dfrac{1}{3}} \cdot \sqrt{\dfrac{24}{5}}$

4. $\sqrt[3]{-125}$

5. $\sqrt{50y^2}$

6. $-\sqrt{121}$

7. $\sqrt[3]{54a^7b^{10}}$

8. $-\sqrt{5}\left(\sqrt{2} + \sqrt{75}\right)$

9. $\sqrt{\dfrac{16r^3}{3s}}$

10. $\dfrac{12 + 6\sqrt{13}}{12}$

11. $\left(\sqrt{5} - \sqrt{2}\right)^2$

12. $\left(6\sqrt{7} + 2\right)\left(4\sqrt{7} - 1\right)$

Solve each equation.

13. $\sqrt{x + 2} = x - 4$

14. $\sqrt{k} + 3 = 0$

15. $\sqrt{1 + 3t} - t = -3$

16. The period p, in seconds, of the swing of a pendulum is given by the formula

$$p = 2\pi\sqrt{\frac{L}{32}},$$

where L is the length of the pendulum in feet. Find the period of a pendulum that is 4 ft long. Give the answer in simplified radical form.

Chapter 8 Test

The Chapter Test Prep Videos with step-by-step solutions are available in MyMathLab or on YouTube at https://goo.gl/8aAsWP

1. Find all square roots of 400.

2. Consider $\sqrt{142}$.

   (a) Determine whether this number is *rational* or *irrational*.

   (b) Find a decimal approximation to the nearest thousandth.

3. **CONCEPT CHECK** Match each radical in Column I with the equivalent choice in Column II. Choices may be used once, more than once, or not at all.

I		II	
(a) $\sqrt{64}$	(b) $-\sqrt{64}$	A. 4	B. 8
(c) $\sqrt{-64}$	(d) $\sqrt[3]{64}$	C. -4	D. Not a real number
(e) $\sqrt[3]{-64}$	(f) $-\sqrt[3]{-64}$	E. 16	F. -8

Simplify. Assume that all variables represent positive real numbers.

4. $\sqrt[3]{216}$

5. $-\sqrt{54}$

6. $\sqrt{\frac{128}{25}}$

7. $\sqrt[3]{32}$

8. $\frac{20\sqrt{18}}{5\sqrt{3}}$

9. $3\sqrt{28} + \sqrt{63}$

10. $(\sqrt{5} + \sqrt{6})^2$

11. $\sqrt[3]{32x^2y^3}$

12. $(6 - \sqrt{5})(6 + \sqrt{5})$

13. $(2 - \sqrt{7})(3\sqrt{2} + 1)$

14. $3\sqrt{27x} - 4\sqrt{48x} + 2\sqrt{3x}$

Solve each problem.

15. Find the measure of the unknown leg of this right triangle.

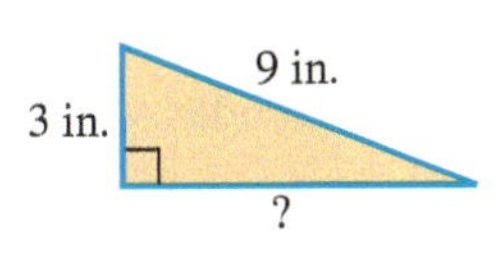

   (a) Give its length in simplified radical form.

   (b) Give its length to the nearest thousandth.

16. In electronics, the impedance Z of an alternating current circuit is given by the following formula.

$$Z = \sqrt{R^2 + X^2}$$

Here, R is the resistance and X is the reactance, both in ohms. Find the value of the impedance Z if $R = 40$ ohms and $X = 30$ ohms. (Data from Cooke, N., and J. Orleans, *Mathematics Essential to Electricity and Radio*, McGraw-Hill.)

Rationalize each denominator.

17. $\dfrac{5\sqrt{2}}{\sqrt{7}}$

18. $\sqrt{\dfrac{2}{3x}}\quad (x > 0)$

19. $\dfrac{-2}{\sqrt[3]{4}}$

20. $\dfrac{-3}{4 - \sqrt{3}}$

21. Write $\dfrac{\sqrt{12} + 3\sqrt{128}}{6}$ in lowest terms.

Solve each equation.

22. $\sqrt{p} + 4 = 0$

23. $\sqrt{x + 1} = 5 - x$

24. $3\sqrt{x} - 2 = x$

25. $\sqrt{x + 7} - \sqrt{x} = 1$

26. Consider the following incorrect "solution." ***What Went Wrong?*** Give the correct solution set.

$$\sqrt{2x + 1} + 5 = 0$$

$$\sqrt{2x + 1} = -5 \quad \text{Subtract 5.}$$

$$2x + 1 = 25 \quad \text{Square each side.}$$

$$2x = 24 \quad \text{Subtract 1.}$$

$$x = 12 \quad \text{Divide by 2.}$$

The solution set is $\{12\}$.

Chapters R–8 Cumulative Review Exercises

Simplify each expression.

1. $3(6+7)+6\cdot 4-3^2$

2. $\dfrac{3(6+7)+3}{2(4)-1}$

3. $|-6|-|-3|$

Solve each equation or inequality.

4. $5(k-4)-k=k-11$

5. $-\frac{3}{4}x \le 12$

6. $5z+3-4>2z+9+z$

7. Trevor Brazile won the ProRodeo All-Around Championship in both 2012 and 2013. He won a total of \$724,636 in these two years, winning \$127,384 more in 2013 than in 2012. How much did he win each year? (Data from *The World Almanac and Book of Facts*.)

Graph.

8. $-4x+5y=-20$

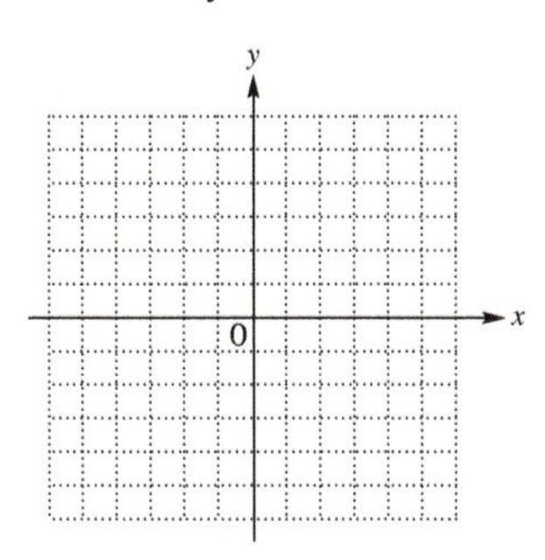

9. $x=2$

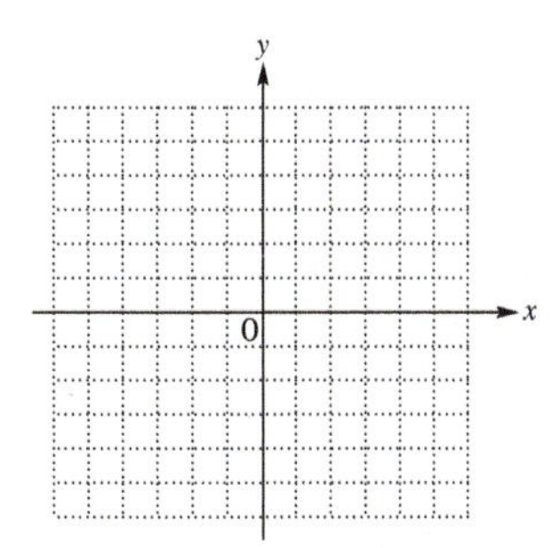

10. $2x-5y>10$

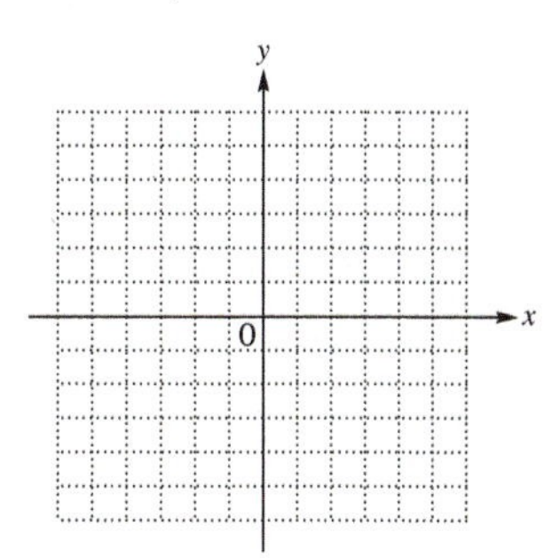

The graph shows a line that models the number of cell phone subscribers in millions in the United States from 2000 to 2014.

11. Use the ordered pairs shown on the graph to find the slope of the line to the nearest tenth. Interpret the slope.

12. Use the slope from **Exercise 11** and the ordered pair $(0, 109.5)$ to find the equation of the line that models the data, where $x=0$ represents 2000.

13. Use the equation from **Exercise 12** to estimate the number of cell phone subscribers in 2016. Round to the nearest tenth of a million.

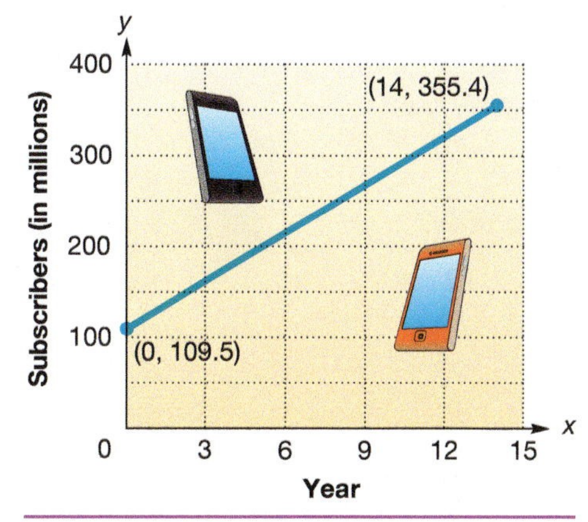

Data from CTIA–The Wireless Association.

Solve each system of equations.

14. $4x - y = 19$
$3x + 2y = -5$

15. $2x - y = 6$
$3y = 6x - 18$

16. Des Moines and Chicago are 345 mi apart. Two cars start from these cities traveling toward each other. They meet after 3 hr. The car from Chicago has an average rate 7 mph faster than the other car. Find the average rate of each car. (Data from *State Farm Road Atlas.*)

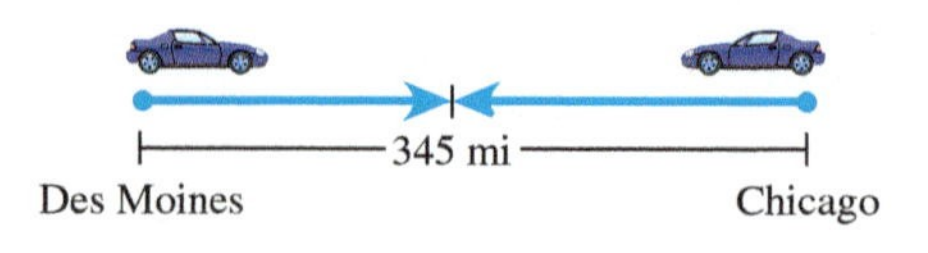

Simplify and write each expression without negative exponents. Assume that variables represent positive numbers.

17. $(3x^6)(2x^2y)^2$

18. $\left(\dfrac{3^2y^{-2}}{2^{-1}y^3}\right)^{-3}$

19. Subtract $7x^3 - 8x^2 + 4$ from $10x^3 + 3x^2 - 9$.

20. Divide $\dfrac{8t^3 - 4t^2 - 14t + 15}{2t + 3}$.

Factor each polynomial completely.

21. $m^2 + 12m + 32$

22. $25t^4 - 36$

23. $12a^2 + 4ab - 5b^2$

24. $81z^2 + 72z + 16$

Perform the indicated operations. Express answers in lowest terms.

25. $\dfrac{x^2 - 3x - 4}{x^2 + 3x} \cdot \dfrac{x^2 + 2x - 3}{x^2 - 5x + 4}$

26. $\dfrac{t^2 + 4t - 5}{t + 5} \div \dfrac{t - 1}{t^2 + 8t + 15}$

27. $\dfrac{2}{x + 3} - \dfrac{4}{x - 1}$

Simplify each expression.

28. $\sqrt{27} - 2\sqrt{12} + 6\sqrt{75}$

29. $\dfrac{2}{\sqrt{3} + \sqrt{5}}$

30. $(3\sqrt{2} + 1)(4\sqrt{2} - 3)$

Solve each equation.

31. $x^2 - 7x = -12$

32. $(x + 4)(x - 1) = -6$

33. $z^2 + 144 = 24z$

34. $\dfrac{x}{x + 8} - \dfrac{3}{x - 8} = \dfrac{128}{x^2 - 64}$

35. $\sqrt{x + 2} = -8$

36. $\sqrt{x} + 2 = x - 10$

9

Quadratic Equations

The trajectory of a water spout follows a *parabolic* path, which can be described by a *quadratic* (second-degree) *equation*, the topic of this chapter.

9.1 Solving Quadratic Equations by the Square Root Property

OBJECTIVES

1. Review the zero-factor property.
2. Solve equations of the form $x^2 = k$, where $k > 0$.
3. Solve equations of the form $(ax + b)^2 = k$, where $k > 0$.
4. Use formulas involving second-degree variables.

OBJECTIVE 1 Review the zero-factor property. Recall that a *quadratic equation* is defined as follows.

Quadratic Equation

A **quadratic equation** (in x here) can be written in the form

$$ax^2 + bx + c = 0,$$

where a, b, and c are real numbers and $a \neq 0$. The given form is called **standard form.**

Examples: $4x^2 + 4x - 5 = 0$ and $3x^2 = 4x - 8$ Quadratic equations (The first equation is in standard form.)

A quadratic equation is a *second-degree equation*—that is, an equation with a squared variable term and no terms of greater degree.

The **zero-factor property** can be used to solve some quadratic equations.

Zero-Factor Property

If a and b are real numbers and if $ab = 0$, then $a = 0$ or $b = 0$.

In words, if the product of two numbers is 0, then at least one of the numbers must be 0. One number must be 0, but both *may* be 0.

EXAMPLE 1 Solving Quadratic Equations by the Zero-Factor Property

Solve each equation using the zero-factor property.

(a)

$$x^2 + 4x + 3 = 0$$

$$(x + 3)(x + 1) = 0 \quad \text{Factor.}$$

$$x + 3 = 0 \quad \text{or} \quad x + 1 = 0 \quad \text{Zero-factor property}$$

$$x = -3 \quad \text{or} \quad x = -1 \quad \text{Solve each equation.}$$

The solution set is $\{-3, -1\}$.

(b)

$$x^2 = 9$$

$$x^2 - 9 = 0 \quad \text{Subtract 9.}$$

$$(x + 3)(x - 3) = 0 \quad \text{Factor.}$$

$$x + 3 = 0 \quad \text{or} \quad x - 3 = 0 \quad \text{Zero-factor property}$$

$$x = -3 \quad \text{or} \quad x = 3 \quad \text{Solve each equation.}$$

The solution set is $\{-3, 3\}$.

◀ Work Problem 1 at the Side.

Not all quadratic expressions can easily be factored, so we develop other methods for solving quadratic equations.

1 Solve each equation using the zero-factor property.

GS **(a)** $x^2 - 2x - 15 = 0$

$(x + \_\_\_)(x - \_\_\_) = 0$

$x + \_\_\_ = 0$ or $x - 5 = 0$

$x = -3$ or $x = \_\_\_$

The solution set is $\{\_\_\_, \_\_\_\}$.

(b) $2x^2 - 3x + 1 = 0$

(c) $x^2 = 4$

Answers

1. (a) 3; 5; 3; 5; -3; 5

(b) $\left\{\frac{1}{2}, 1\right\}$ (c) $\{-2, 2\}$

OBJECTIVE 2 Solve equations of the form $x^2 = k$, where $k > 0$. In **Example 1(b),** we might also have solved $x^2 = 9$ by noticing that x must be a number whose square is 9. Thus

$$x = \sqrt{9} = 3 \quad \text{or} \quad x = -\sqrt{9} = -3.$$

This is generalized as the **square root property.**

Square Root Property

If k is a positive number and if $x^2 = k$, then

$$x = \sqrt{k} \quad \text{or} \quad x = -\sqrt{k}.$$

The solution set is $\{-\sqrt{k}, \sqrt{k}\}$, which can be written $\{\pm\sqrt{k}\}$. (The symbol $\pm$ is read "*positive or negative*" or "*plus or minus.*")

EXAMPLE 2 Solving Quadratic Equations of the Form $x^2 = k$

Solve each equation. Express radicals in simplified form.

(a) $x^2 = 16$

By the square root property, if $x^2 = 16$, then

$$x = \sqrt{16} = 4 \quad \text{or} \quad x = -\sqrt{16} = -4.$$

Check each value by substituting it for x in the original equation. The solution set is $\{-4, 4\}$, or $\{\pm 4\}$.

This $\pm$ notation indicates ***two*** solutions, one ***positive*** and one ***negative***.

(b) $x^2 = 5$

By the square root property, if $x^2 = 5$, then

$$x = \sqrt{5} \quad \text{or} \quad x = -\sqrt{5}.$$

Don't forget the negative solution.

The solution set is $\{-\sqrt{5}, \sqrt{5}\}$, or $\{\pm\sqrt{5}\}$.

(c)

$5m^2 - 32 = 8$

$5m^2 = 40$ Add 32.

$m^2 = 8$ Divide by 5.

$m = \sqrt{8}$ or $m = -\sqrt{8}$ Square root property

$m = 2\sqrt{2}$ or $m = -2\sqrt{2}$ $\sqrt{8} = \sqrt{4} \cdot \sqrt{2} = 2\sqrt{2}$

Don't stop here. Simplify the radicals.

CHECK

$5m^2 - 32 = 8$ Original equation

$5(2\sqrt{2})^2 - 32 \stackrel{?}{=} 8$ Let $m = 2\sqrt{2}$.

$5(8) - 32 \stackrel{?}{=} 8$ $(2\sqrt{2})^2 = 2^2 \cdot (\sqrt{2})^2 = 4 \cdot 2 = 8$

$40 - 32 \stackrel{?}{=} 8$ Multiply.

$8 = 8$ ✓   True

The check of the other value is similar. The solution set is

$$\{-2\sqrt{2}, 2\sqrt{2}\}, \quad \text{or} \quad \{\pm 2\sqrt{2}\}.$$

Work Problem 2 at the Side. ▶

2 Solve each equation. Express radicals in simplified form.

(a) $x^2 = 49$

(b) $x^2 = 11$

(c) $3x^2 - 8 = 88$

Answers

2. (a) $\{-7, 7\}$ **(b)** $\{-\sqrt{11}, \sqrt{11}\}$ **(c)** $\{-4\sqrt{2}, 4\sqrt{2}\}$

3 Solve each equation.

(a) $x^2 = -9$

(b) $x^2 + 25 = 0$

4 Solve each equation.

(a) $(x + 2)^2 = 36$

GS (b) $(x - 4)^2 = 3$

$x - 4 =$ ____ or $x - 4 =$ ____

$x =$ ____ or $x =$ ____

The solution set is ________.

Answers

3. (a) no real solution
   (b) no real solution
4. (a) $\{-8, 4\}$
   (b) $\sqrt{3}; -\sqrt{3}; 4 + \sqrt{3}; 4 - \sqrt{3};$ $\{4 + \sqrt{3}, 4 - \sqrt{3}\}$

EXAMPLE 3 Recognizing When There Is No Real Solution

Solve $x^2 = -4$.

Because -4 is a negative number and because the square of a real number cannot be negative, ***there is no real solution*** of this equation.*

◀ Work Problem 3 at the Side.

OBJECTIVE 3 Solve equations of the form $(ax + b)^2 = k$, where $k > 0$.

EXAMPLE 4 Solving Quadratic Equations of the Form $(x + b)^2 = k$

Solve each equation.

(a) (Use $x - 3$ as the base.) $(x - 3)^2 = 16$

$x - 3 = \sqrt{16}$ or $x - 3 = -\sqrt{16}$ — Square root property

$x - 3 = 4$ or $x - 3 = -4$ — $\sqrt{16} = 4$

$x = 7$ or $x = -1$ — Add 3.

CHECK

$(x - 3)^2 = 16$

$(7 - 3)^2 \stackrel{?}{=} 16$ — Let $x = 7$.

$4^2 \stackrel{?}{=} 16$ — Subtract.

$16 = 16$ ✓ True

$(x - 3)^2 = 16$

$(-1 - 3)^2 \stackrel{?}{=} 16$ — Let $x = -1$.

$(-4)^2 \stackrel{?}{=} 16$ — Subtract.

$16 = 16$ ✓ True

True statements result, so the solution set is $\{-1, 7\}$.

(b) $(x + 1)^2 = 6$

$x + 1 = \sqrt{6}$ or $x + 1 = -\sqrt{6}$ — Square root property

$x = -1 + \sqrt{6}$ or $x = -1 - \sqrt{6}$ — Add -1.

CHECK

$\left(-1 + \sqrt{6} + 1\right)^2 = \left(\sqrt{6}\right)^2 = 6$ ✓ Let $x = -1 + \sqrt{6}$.

$\left(-1 - \sqrt{6} + 1\right)^2 = \left(-\sqrt{6}\right)^2 = 6$ ✓ Let $x = -1 - \sqrt{6}$.

The solution set is $\{-1 + \sqrt{6}, -1 - \sqrt{6}\}$, or $\{-1 \pm \sqrt{6}\}$.

◀ Work Problem 4 at the Side.

EXAMPLE 5 Solving a Quadratic Equation of the Form $(ax + b)^2 = k$

Solve $(3r - 2)^2 = 27$.

$$(3r - 2)^2 = 27$$

$3r - 2 = \sqrt{27}$ or $3r - 2 = -\sqrt{27}$ — Square root property

$3r - 2 = 3\sqrt{3}$ or $3r - 2 = -3\sqrt{3}$ — $\sqrt{27} = \sqrt{9} \cdot \sqrt{3} = 3\sqrt{3}$

$3r = 2 + 3\sqrt{3}$ or $3r = 2 - 3\sqrt{3}$ — Add 2.

$r = \frac{2 + 3\sqrt{3}}{3}$ or $r = \frac{2 - 3\sqrt{3}}{3}$ — Divide by 3.

Continued on Next Page

*The equation in **Example 3** has no solution over the *real number system*. In the **complex number system,** which includes numbers whose squares are negative, this equation does have solutions. Such numbers are discussed in intermediate and college algebra courses.

CHECK

$$(3r-2)^2 = 27 \quad \text{Original equation}$$

$$\left(3 \cdot \frac{2+3\sqrt{3}}{3} - 2\right)^2 \stackrel{?}{=} 27 \quad \text{Let } r = \frac{2+3\sqrt{3}}{3}.$$

$$\left(2+3\sqrt{3}-2\right)^2 \stackrel{?}{=} 27 \quad \text{Multiply.}$$

$(ab)^2 = a^2b^2$

$$\left(3\sqrt{3}\right)^2 \stackrel{?}{=} 27 \quad \text{Subtract.}$$

$$27 = 27 \checkmark \quad \text{True}$$

The check of the other value is similar. The solution set is

$$\left\{\frac{2+3\sqrt{3}}{3}, \frac{2-3\sqrt{3}}{3}\right\}.$$

These fractions *cannot* be simplified. 3 is *not* a common factor in the numerator.

Work Problem 5 at the Side. ▶

5 Solve $(2x-5)^2 = 18$.

6 Solve each equation.

(a) $(x-1)^2 = -1$

(b) $(2x+1)^2 = -5$

(c) $(5x+1)^2 = 7$

EXAMPLE 6 Recognizing When There Is No Real Solution

Solve $(x+3)^2 = -9$.

The square root of -9 is not a real number, so there is no real solution.

Work Problem 6 at the Side. ▶

OBJECTIVE 4 Use formulas involving second-degree variables.

EXAMPLE 7 Finding the Length of a Bass

The weight w in pounds, length L in inches, and girth g in inches of a bass are related by the formula

$$w = \frac{L^2 g}{1200}.$$

Approximate, to the nearest inch, the length of a bass weighing 2.20 lb and having girth 10 in. (Data from *The Sacramento Bee*.)

$$w = \frac{L^2 g}{1200} \quad \text{Given formula}$$

$$2.20 = \frac{L^2 \cdot 10}{1200} \quad \text{Substitute } w = 2.20 \text{ and } g = 10.$$

$$2640 = 10L^2 \quad \text{Multiply by 1200.}$$

$$L^2 = 264 \quad \text{Divide by 10. Interchange sides.}$$

$$L = \sqrt{264} \quad \text{or} \quad L = -\sqrt{264} \quad \text{Square root property}$$

A calculator gives $\sqrt{264} \approx 16.25$, so the length of the bass, to the nearest inch, is 16 in. (We must reject the negative value $-\sqrt{264} \approx -16.25$, because L represents length.)

Work Problem 7 at the Side. ▶

7 Use the formula in **Example 7** to approximate, to the nearest inch, the length of a bass weighing 2.80 lb and having girth 11 in.

Answers

5. $\left\{\frac{5+3\sqrt{2}}{2}, \frac{5-3\sqrt{2}}{2}\right\}$

6. (a) no real solution
(b) no real solution
(c) $\left\{\frac{-1+\sqrt{7}}{5}, \frac{-1-\sqrt{7}}{5}\right\}$

7. 17 in.

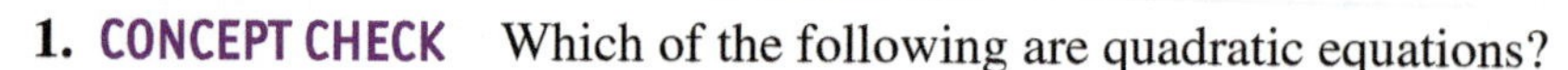

9.1 Exercises

FOR EXTRA HELP

Go to MyMathLab for worked-out, step-by-step solutions to exercises enclosed in a square and video solutions to ▶ exercises.

1. **CONCEPT CHECK** Which of the following are quadratic equations?

   A. $x + 2y = 0$ **B.** $x^2 - 8x + 16 = 0$ **C.** $2x^2 - 5x = 3$ **D.** $x^3 + x^2 + 4 = 0$

2. **CONCEPT CHECK** Which quadratic equation identified in **Exercise 1** is in standard form?

CONCEPT CHECK *Match each equation in Column I with the correct description of its solution(s) in Column II.*

I		II	
3. $x^2 = 12$	**4.** $x^2 = -36$	**A.** No real solution	**B.** Two integer solutions
5. $x^2 = \frac{25}{36}$	**6.** $x^2 = 25$	**C.** Two irrational solutions	**D.** Two rational solutions that are not integers

7. **CONCEPT CHECK** When a student was asked to solve

$$x^2 = 81,$$

she wrote $\{9\}$ as the solution set. Her teacher did not give her full credit. The student argued that because $9^2 = 81$, her answer had to be correct. ***What Went Wrong?*** Give the correct solution set.

8. **CONCEPT CHECK** When solving a quadratic equation, a student obtained the solutions

$$x = \frac{3 + 2\sqrt{5}}{2} \quad \text{or} \quad x = \frac{3 - 2\sqrt{5}}{2},$$

and wrote the solution set incorrectly as

$$\{3 + \sqrt{5}, 3 - \sqrt{5}\}.$$

What Went Wrong? Give the correct solution set.

Solve each equation using the zero-factor property. ***See Example 1.***

9. $x^2 - x - 56 = 0$ **10.** $x^2 - 2x - 99 = 0$ **11.** $x^2 - 8x + 15 = 0$ **12.** $x^2 - 6x + 5 = 0$

13. $x^2 = 121$ **14.** $x^2 = 144$ **15.** $3x^2 - 13x = 30$ **16.** $5x^2 - 14x = 3$

Solve each equation using the square root property. Express radicals in simplified form. ***See Examples 2 and 3.***

17. $x^2 = 81$ ▶ **18.** $x^2 = 36$ **19.** $k^2 = 14$ ▶ **20.** $m^2 = 22$

21. $t^2 = 48$ **22.** $x^2 = 54$ **23.** $x^2 = -100$ **24.** $x^2 = -64$

25. $x^2 = \frac{25}{4}$ **26.** $m^2 = \frac{36}{121}$ **27.** $z^2 = 0.25$ **28.** $x^2 = 0.49$

29. $x^2 - 64 = 0$
30. $x^2 - 100 = 0$
31. $r^2 - 3 = 0$
32. $x^2 - 13 = 0$

33. $x^2 + 16 = 0$
34. $x^2 + 4 = 0$
35. $4x^2 - 72 = 0$
36. $5z^2 - 200 = 0$

37. $2x^2 + 7 = 61$
38. $2x^2 + 8 = 32$
39. $3x^2 - 8 = 64$
40. $2x^2 - 5 = 35$

41. $7x^2 = 4$
42. $3x^2 = 10$
43. $5x^2 + 4 = 8$
44. $4x^2 - 3 = 7$

Solve each equation using the square root property. Express radicals in simplified form. ***See Examples 4–6.***

45. $(x - 3)^2 = 25$
46. $(x - 7)^2 = 16$
47. $(x - 8)^2 = 27$

48. $(x - 5)^2 = 40$
49. $(x + 5)^2 = -13$
50. $(x + 2)^2 = -17$

51. $(3x + 2)^2 = 49$
52. $(5x + 3)^2 = 36$
53. $(4x - 3)^2 = 9$

54. $(7x - 5)^2 = 25$
55. $(5 - 2x)^2 = 30$
56. $(3 - 2x)^2 = 70$

57. $(3k + 1)^2 = 18$
58. $(5x + 6)^2 = 75$
59. $(4x - 1)^2 - 48 = 0$

60. $(2x - 5)^2 - 180 = 0$
61. $\left(\frac{1}{2}x + 5\right)^2 = 12$
62. $\left(\frac{1}{3}x + 4\right)^2 = 27$

Solve each problem. **See Example 7.**

63. Find the lengths of the three sides of the right triangle. (*Hint:* Set up an equation using the Pythagorean theorem.)

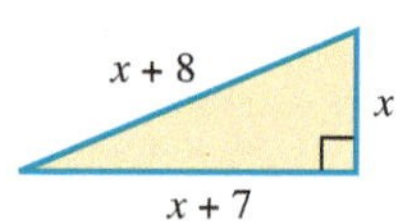

64. Find the lengths of the three sides of the right triangle. (*Hint:* Set up an equation using the Pythagorean theorem.)

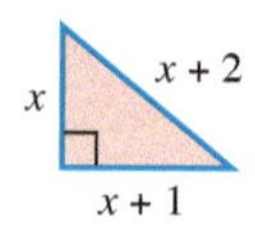

65. An expert marksman can hold a silver dollar at forehead level, drop it, draw his gun, and shoot the coin as it passes waist level. The distance traveled by a falling object is given by

$$d = 16t^2,$$

where d is the distance (in feet) the object falls in t seconds. If the coin falls 4 ft, use the formula to find the time that elapses between the dropping of the coin and the shot.

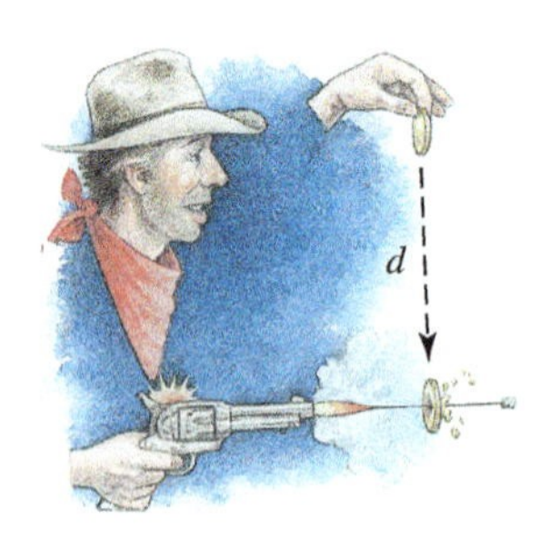

66. The illumination produced by a light source depends on the distance from the source. For a particular light source, this relationship can be expressed as

$$I = \frac{4050}{d^2},$$

where I is the amount of illumination in footcandles and d is the distance from the light source (in feet). How far from the source is the illumination equal to 50 footcandles?

67. The area $\mathcal{A}$ of a circle with radius r is given by the formula

$$\mathcal{A} = \pi r^2.$$

If a circle has area 81π in.^{2}, what is its radius?

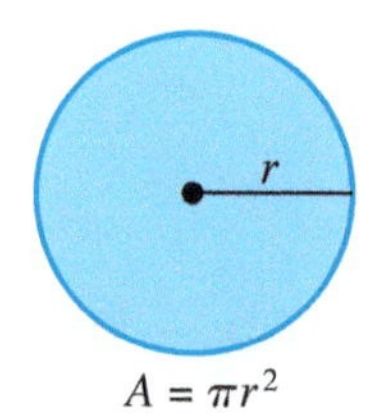

$A = \pi r^2$

68. The surface area S of a sphere with radius r is given by the formula

$$S = 4\pi r^2.$$

If a sphere has surface area 36π ft^2, what is its radius?

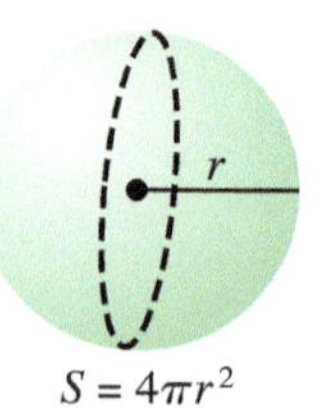

$S = 4\pi r^2$

The amount A that P dollars invested at an annual rate of interest r will grow to in 2 yr is

$$A = P(1 + r)^2.$$

69. At what interest rate will \$100 grow to \$104.04 in 2 yr?

70. At what interest rate will \$500 grow to \$530.45 in 2 yr?

9.2 Solving Quadratic Equations by Completing the Square

OBJECTIVES

1. Solve quadratic equations by completing the square when the coefficient of the second-degree term is 1.
2. Solve quadratic equations by completing the square when the coefficient of the second-degree term is not 1.
3. Simplify the terms of an equation before solving.
4. Solve applied problems that require quadratic equations.

OBJECTIVE 1 Solve quadratic equations by completing the square when the coefficient of the second-degree term is 1. The methods we have studied so far are not enough to solve an equation such as

$$x^2 + 6x + 7 = 0.$$

If we could write the equation in the form $(x + 3)^2$ equals a constant, we could solve it with the square root property. To do that, we need a perfect square trinomial on one side of the equation.

Recall that a perfect square trinomial has the form

$$x^2 + 2kx + k^2 \quad \text{or} \quad x^2 - 2kx + k^2,$$

where k represents a real number.

EXAMPLE 1 Creating Perfect Square Trinomials

Complete each trinomial so that it is a perfect square. Then factor the trinomial.

(a) $x^2 + 8x +$ ______

The perfect square trinomial will have the form $x^2 + 2kx + k^2$. Thus, the middle term, $8x$, must equal $2kx$.

$$8x = 2kx \quad \leftarrow \text{Solve this equation for } k.$$

$$4 = k \quad \text{Divide each side by } 2x.$$

Therefore, $k = 4$ and $k^2 = 4^2 = 16$. The required perfect square trinomial is

$$x^2 + 8x + 16, \quad \text{which factors as} \quad (x + 4)^2.$$

(b) $x^2 - 18x +$ ______

Here the perfect square trinomial will have the form $x^2 - 2kx + k^2$. The middle term, $-18x$, must equal $-2kx$.

$$-18x = -2kx \quad \leftarrow \text{Solve this equation for } k.$$

$$9 = k \quad \text{Divide each side by } -2x.$$

Thus, $k = 9$ and $k^2 = 9^2 = 81$. The required perfect square trinomial is

$$x^2 - 18x + 81, \quad \text{which factors as} \quad (x - 9)^2.$$

Work Problem 1 at the Side. ▶

1 Complete each trinomial so that it is a perfect square. Then factor the trinomial.

(a) $x^2 + 12x +$ ____ factors as ________.

(b) $x^2 - 14x +$ ____ factors as ________.

(c) $x^2 - 2x +$ ____ factors as ________.

EXAMPLE 2 Rewriting an Equation to Use the Square Root Property

Solve $x^2 + 6x + 7 = 0$.

$$x^2 + 6x = -7 \quad \text{Subtract 7 from each side.}$$

To use the square root property, the expression on the left, $x^2 + 6x$, must be written as a perfect square trinomial in the form $x^2 + 2kx + k^2$.

$$x^2 + 6x + \text{\_\_\_\_\_\_}$$

A square must go here.

Here, $2kx = 6x$, so $k = 3$ and $k^2 = 9$. The required perfect square trinomial is

$$x^2 + 6x + 9, \quad \text{which factors as} \quad (x + 3)^2.$$

Continued on Next Page

Answers

1. (a) 36; $(x + 6)^2$
   (b) 49; $(x - 7)^2$
   (c) 1; $(x - 1)^2$

2 Solve $x^2 - 4x - 1 = 0$.

3 Solve each equation by completing the square.

GS **(a)** $x^2 + 4x = 1$

Take half the coefficient of x and square it.

$\frac{1}{2} \cdot$ ____ $=$ ____,

and ____^{2} $=$ ____.

Add ____ to each side of the equation.

$x^2 + 4x +$ ____ $= 1 + 4$

Factor and add.

____

Complete the solution.

(b) $z^2 + 6z = 3$

Answers

2. $\{2 + \sqrt{5}, 2 - \sqrt{5}\}$

3. **(a)** 4; 2; 2; 4; 4; 4; $(x + 2)^2 = 5$; $\{-2 + \sqrt{5}, -2 - \sqrt{5}\}$

(b) $\{-3 + 2\sqrt{3}, -3 - 2\sqrt{3}\}$

Therefore, if we add 9 to *each* side of $x^2 + 6x = -7$ the equation will have a perfect square trinomial on the left side, as needed.

This is a key step.

$$x^2 + 6x = -7 \quad \text{Transformed equation}$$

$$x^2 + 6x + 9 = -7 + 9 \quad \text{Add 9 to each side.}$$

$$(x + 3)^2 = 2 \quad \text{Factor on the left. Add on the right.}$$

Now use the square root property to complete the solution.

$$x + 3 = \sqrt{2} \quad \text{or} \quad x + 3 = -\sqrt{2}$$

$$x = -3 + \sqrt{2} \quad \text{or} \quad x = -3 - \sqrt{2} \quad \text{Add } -3.$$

Check by first substituting $-3 + \sqrt{2}$ and then substituting $-3 - \sqrt{2}$ for x in the original equation. The solution set is $\{-3 + \sqrt{2}, -3 - \sqrt{2}\}$.

◀ **Work Problem 2 at the Side.**

The process of changing the form of the equation in **Example 2** from

$$x^2 + 6x + 7 = 0 \quad \text{to} \quad (x + 3)^2 = 2$$

is called **completing the square.** Completing the square changes only the form of the equation. To see this, multiply out the left side of $(x + 3)^2 = 2$, and write the equation in standard form to obtain $x^2 + 6x + 7 = 0$.

Look again at the original equation in **Example 2.**

$$x^2 + 6x + 7 = 0$$

If we take half the coefficient of x, which is 6 here, and square it, we get 9.

$$\frac{1}{2} \cdot 6 = 3 \quad \text{and} \quad 3^2 = 9$$

Coefficient of x — Quantity added to each side

To complete the square in **Example 2,** we added 9 to each side.

EXAMPLE 3 Completing the Square to Solve a Quadratic Equation

Solve $x^2 - 8x = 5$ by completing the square.

To complete the square on $x^2 - 8x$, take half the coefficient of x and square it.

$$\frac{1}{2}(-8) = -4 \quad \text{and} \quad (-4)^2 = 16$$

Coefficient of x

Add the result, 16, to each side of the equation.

$$x^2 - 8x = 5 \quad \text{Given equation}$$

$$x^2 - 8x + 16 = 5 + 16 \quad \text{Add 16.}$$

$$(x - 4)^2 = 21 \quad \text{Factor. Add.}$$

$$x - 4 = \sqrt{21} \quad \text{or} \quad x - 4 = -\sqrt{21} \quad \text{Square root property}$$

$$x = 4 + \sqrt{21} \quad \text{or} \quad x = 4 - \sqrt{21} \quad \text{Add 4.}$$

A check confirms that the solution set is $\{4 + \sqrt{21}, 4 - \sqrt{21}\}$.

◀ **Work Problem 3 at the Side.**

OBJECTIVE 2 Solve quadratic equations by completing the square when the coefficient of the second-degree term is not 1. If a quadratic equation has the form

$$ax^2 + bx + c = 0, \quad \text{where} \quad a \neq 1,$$

we obtain 1 as the coefficient of x^2 by dividing each side by a.

Completing the Square

To solve $ax^2 + bx + c = 0$ (where $a \neq 0$) by completing the square, follow these steps.

Step 1 **Be sure the second-degree term has coefficient 1.**
- If the coefficient a of the second-degree term is 1, go to Step 2.
- If the coefficient is not 1, but some other nonzero number a, divide each side of the equation by a.

Step 2 **Write the equation in correct form.** Make sure that all variable terms are on one side of the equality symbol and the constant term is on the other side.

Step 3 **Complete the square.**
- Take half the coefficient of the first-degree term, and square it—that is, find $\left(\frac{1}{2}b\right)^2$.
- Add the square to each side of the equation.
- Factor the variable side, which should be a perfect square trinomial, as the square of a binomial. Add on the other side.

Step 4 **Solve** the equation using the square root property.

EXAMPLE 4 **Solving a Quadratic Equation by Completing the Square**

Solve $4x^2 + 16x - 9 = 0$.

Step 1 ***Before completing the square, the coefficient of x^2 must be 1,*** not 4.

$$4x^2 + 16x - 9 = 0 \qquad \text{Given equation}$$

$$x^2 + 4x - \frac{9}{4} = 0 \qquad \text{Divide by 4 to obtain 1 as the coefficient of } x^2.$$

Step 2 Write the equation so that all variable terms are on one side of the equation and all constant terms are on the other side.

$$x^2 + 4x = \frac{9}{4} \qquad \text{Add } \tfrac{9}{4}.$$

Step 3 Complete the square. Take half the coefficient of x and square it.

$$\left[\frac{1}{2}(4)\right]^2 = 2^2 = 4$$

Add the result, 4, to each side of the equation.

$$x^2 + 4x + 4 = \frac{9}{4} + 4 \qquad \text{Add 4.}$$

$$(x + 2)^2 = \frac{25}{4} \qquad \text{Factor; } \tfrac{9}{4} + 4 = \tfrac{9}{4} + \tfrac{16}{4} = \tfrac{25}{4}.$$

Continued on Next Page

Step 4 Solve the equation $(x+2)^2 = \frac{25}{4}$ using the square root property.

$$x+2=\sqrt{\frac{25}{4}} \quad \text{or} \quad x+2=-\sqrt{\frac{25}{4}} \qquad \text{Square root property}$$

$$x+2=\frac{5}{2} \quad \text{or} \quad x+2=-\frac{5}{2} \qquad \text{Take square roots.}$$

$$x=\frac{5}{2}-2 \quad \text{or} \quad x=-\frac{5}{2}-2 \qquad \text{Subtract 2.}$$

$$x=\frac{5}{2}-\frac{4}{2} \quad \text{or} \quad x=-\frac{5}{2}-\frac{4}{2} \qquad 2=\frac{4}{2}$$

$$x=\frac{1}{2} \quad \text{or} \quad x=-\frac{9}{2} \qquad \text{Subtract fractions.}$$

CHECK $4x^2+16x-9=0$

$$4\left(\frac{1}{2}\right)^2+16\left(\frac{1}{2}\right)-9 \stackrel{?}{=} 0 \qquad \text{Let } x=\tfrac{1}{2}.$$

$$4\left(\frac{1}{4}\right)+8-9 \stackrel{?}{=} 0$$

$$1+8-9 \stackrel{?}{=} 9$$

$$0=0 \ ✓ \ \text{True}$$

$$4\left(-\frac{9}{2}\right)^2+16\left(-\frac{9}{2}\right)-9 \stackrel{?}{=} 0 \qquad \text{Let } x=-\tfrac{9}{2}.$$

$$4\left(\frac{81}{4}\right)-72-9 \stackrel{?}{=} 0$$

$$81-72-9 \stackrel{?}{=} 0$$

$$0=0 \ ✓ \ \text{True}$$

The two values $\frac{1}{2}$ and $-\frac{9}{2}$ check, so the solution set is $\left\{-\frac{9}{2}, \frac{1}{2}\right\}$.

◀ **Work Problem 4 at the Side.**

4 Solve each equation by completing the square.

(a) $9x^2+18x+5=0$

(b) $4t^2-24t+11=0$

EXAMPLE 5 Solving a Quadratic Equation by Completing the Square

Solve $2x^2-7x-9=0$.

Step 1 Transform the equation so that 1 is the coefficient of the x^2-term.

$$2x^2-7x-9=0 \qquad \text{Given equation}$$

$$x^2-\frac{7}{2}x-\frac{9}{2}=0 \qquad \text{Divide by 2.}$$

Step 2 Add $\frac{9}{2}$ to each side so that the variable terms are on the left and the constant is on the right.

$$x^2-\frac{7}{2}x=\frac{9}{2} \qquad \text{Add } \tfrac{9}{2}.$$

Step 3 To complete the square, take half the coefficient of x and square it.

$$\left[\frac{1}{2}\left(-\frac{7}{2}\right)\right]^2=\left(-\frac{7}{4}\right)^2=\frac{49}{16}$$

Add the result, $\frac{49}{16}$, to each side of the equation.

$$x^2-\frac{7}{2}x+\frac{49}{16}=\frac{9}{2}+\frac{49}{16}$$

Be sure to add $\frac{49}{16}$ to *each* side.

$$\left(x-\frac{7}{4}\right)^2=\frac{121}{16} \qquad \text{Factor; } \tfrac{9}{2}+\tfrac{49}{16}=\tfrac{72}{16}+\tfrac{49}{16}=\tfrac{121}{16}.$$

Continued on Next Page

Answers

4. (a) $\left\{-\frac{1}{3}, -\frac{5}{3}\right\}$ (b) $\left\{\frac{11}{2}, \frac{1}{2}\right\}$

Step 4 Solve the equation $\left(x-\frac{7}{4}\right)^2=\frac{121}{16}$ using the square root property.

$$x-\frac{7}{4}=\sqrt{\frac{121}{16}} \quad \text{or} \quad x-\frac{7}{4}=-\sqrt{\frac{121}{16}} \qquad \text{Square root property}$$

$$x=\frac{7}{4}+\frac{11}{4} \quad \text{or} \quad x=\frac{7}{4}-\frac{11}{4} \qquad \text{Add } \tfrac{7}{4};\ \sqrt{\tfrac{121}{16}}=\tfrac{11}{4}.$$

$$x=\frac{18}{4} \quad \text{or} \quad x=-\frac{4}{4} \qquad \text{Add and subtract the fractions.}$$

$$x=\frac{9}{2} \quad \text{or} \quad x=-1 \qquad \text{Lowest terms}$$

A check confirms that the solution set is $\left\{-1,\frac{9}{2}\right\}$.

Work Problem 5 at the Side. ▶

EXAMPLE 6 Recognizing When There Is No Real Solution

Solve $4p^2+8p+5=0$.

$$4p^2+8p+5=0$$

The coefficient of the second-degree term must be 1.

$$p^2+2p+\frac{5}{4}=0 \qquad \text{Divide by 4.}$$

$$p^2+2p=-\frac{5}{4} \qquad \text{Subtract } \tfrac{5}{4}.$$

$$p^2+2p+1=-\frac{5}{4}+1 \qquad \text{Complete the square. Add } \left[\tfrac{1}{2}(2)\right]^2=1^2=1 \text{ to each side.}$$

$$(p+1)^2=-\frac{1}{4} \qquad \text{Factor; } -\tfrac{5}{4}+1=-\tfrac{5}{4}+\tfrac{4}{4}=-\tfrac{1}{4}.$$

If we apply the square root property to solve this equation, we obtain the square root of $-\frac{1}{4}$, which is not a real number. There is no real solution.

Work Problem 6 at the Side. ▶

OBJECTIVE 3 Simplify the terms of an equation before solving.

EXAMPLE 7 Simplifying before Completing the Square

Solve $(x+3)(x-1)=2$.

$$(x+3)(x-1)=2$$

$$x^2+2x-3=2 \qquad \text{Multiply using the FOIL method.}$$

$$x^2+2x=5 \qquad \text{Add 3.}$$

$$x^2+2x+1=5+1 \qquad \text{Add } \left[\tfrac{1}{2}(2)\right]^2=1^2=1.$$

$$(x+1)^2=6 \qquad \text{Factor on the left. Add on the right.}$$

$$x+1=\sqrt{6} \quad \text{or} \quad x+1=-\sqrt{6} \qquad \text{Square root property}$$

$$x=-1+\sqrt{6} \quad \text{or} \quad x=-1-\sqrt{6} \qquad \text{Subtract 1.}$$

The solution set is $\left\{-1+\sqrt{6},-1-\sqrt{6}\right\}$.

Work Problem 7 at the Side. ▶

---

5 Solve $3x^2+5x-2=0$ by completing the square.

6 Solve $5x^2+3x+1=0$.

7 Solve each equation.

(a) $(x-5)(x+1)=2$

(b) $(x+6)(x+2)=1$

Answers

5. $\left\{-2,\frac{1}{3}\right\}$

6. no real solution

7. (a) $\left\{2+\sqrt{11},2-\sqrt{11}\right\}$
(b) $\left\{-4+\sqrt{5},-4-\sqrt{5}\right\}$

8 Solve each equation.

(a) $r(r-3)=-1$

(b) $x(x+5)=3$

9 Solve each problem.

(a) If a ball is projected upward from ground level with an initial velocity of 128 ft per sec, its altitude (height) s, in feet, in t seconds is given by the formula

$$s=-16t^2+128t.$$

At what times will the ball be 48 ft above the ground? Give answers to the nearest tenth.

(b) At what times will the ball in **Example 9** be 28 ft above the ground?

Answers

8. **(a)** $\left\{\frac{3+\sqrt{5}}{2}, \frac{3-\sqrt{5}}{2}\right\}$

(b) $\left\{\frac{-5+\sqrt{37}}{2}, \frac{-5-\sqrt{37}}{2}\right\}$

9. **(a)** 0.4 sec and 7.6 sec

(b) 0.5 sec and 3.5 sec

EXAMPLE 8 Simplifying before Completing the Square

Solve $x(x+7)=2$.

$$x(x+7)=2$$

$$x^2+7x=2 \quad \text{Multiply.}$$

$$x^2+7x+\frac{49}{4}=2+\frac{49}{4} \quad \text{Add } \left[\tfrac{1}{2}(7)\right]^2=\tfrac{49}{4}.$$

$$\left(x+\frac{7}{2}\right)^2=\frac{57}{4} \quad \text{Factor; } 2+\tfrac{49}{4}=\tfrac{8}{4}+\tfrac{49}{4}=\tfrac{57}{4}.$$

$$x+\frac{7}{2}=\sqrt{\frac{57}{4}} \quad \text{or} \quad x+\frac{7}{2}=-\sqrt{\frac{57}{4}} \quad \text{Square root property}$$

$$x+\frac{7}{2}=\frac{\sqrt{57}}{2} \quad \text{or} \quad x+\frac{7}{2}=-\frac{\sqrt{57}}{2} \quad \text{Quotient rule for radicals, } \sqrt{\tfrac{a}{b}}=\tfrac{\sqrt{a}}{\sqrt{b}}$$

$$x=-\frac{7}{2}+\frac{\sqrt{57}}{2} \quad \text{or} \quad x=-\frac{7}{2}-\frac{\sqrt{57}}{2} \quad \text{Add } -\tfrac{7}{2}.$$

$$x=\frac{-7+\sqrt{57}}{2} \quad \text{or} \quad x=\frac{-7-\sqrt{57}}{2} \quad \text{Add and subtract the fractions.}$$

The solution set is $\left\{\frac{-7+\sqrt{57}}{2}, \frac{-7-\sqrt{57}}{2}\right\}$.

◀ **Work Problem 8 at the Side.**

OBJECTIVE 4 Solve applied problems that require quadratic equations.

EXAMPLE 9 Solving a Velocity Problem

If a ball is projected upward from ground level with an initial velocity of 64 ft per sec, its altitude (height) s, in feet, in t seconds is given by the formula

$$s=-16t^2+64t.$$

At what times will the ball be 48 ft above the ground?

Because s represents the height, we substitute 48 for s in the formula and then solve this equation for time t by completing the square.

$$s=-16t^2+64t$$

$$48=-16t^2+64t \quad \text{Let } s=48.$$

$$-3=t^2-4t \quad \text{Divide by } -16.$$

$$t^2-4t=-3 \quad \text{Interchange sides.}$$

$$t^2-4t+4=-3+4 \quad \text{Add } \left[\tfrac{1}{2}(-4)\right]^2=(-2)^2=4.$$

$$(t-2)^2=1 \quad \text{Factor. Add.}$$

$$t-2=1 \quad \text{or} \quad t-2=-1 \quad \text{Square root property}$$

$$t=3 \quad \text{or} \quad t=1 \quad \text{Add 2.}$$

The ball reaches a height of 48 ft twice, once on the way up and again on the way down. It takes 1 sec to reach 48 ft on the way up, and then after 3 sec, the ball reaches 48 ft again on the way down.

◀ **Work Problem 9 at the Side.**

9.2 Exercises

FOR EXTRA HELP — *Go to MyMathLab for worked-out, step-by-step solutions to exercises enclosed in a square and video solutions to ▶ exercises.*

1. CONCEPT CHECK Which step is an appropriate way to begin solving the quadratic equation

$$2x^2 - 4x = 9$$

by completing the square?

A. Add 4 to each side of the equation.

B. Factor the left side as $2x(x - 2)$.

C. Factor the left side as $x(2x - 4)$.

D. Divide each side by 2.

2. CONCEPT CHECK Solve the following quadratic equation using the zero-factor property.

$$2p^2 - 13p + 20 = 0$$

If we were to solve this equation by completing the square, what would the solution set be? Explain.

Complete each trinomial so that it is a perfect square. Then factor the trinomial. **See Example 1.**

3. $x^2 + 10x +$ ____ ▶

4. $x^2 + 16x +$ ____

5. $z^2 - 20z +$ ____ ▶

6. $x^2 - 32x +$ ____

7. $x^2 + 2x +$ ____

8. $m^2 - 2m +$ ____

9. $p^2 - 5p +$ ____

10. $x^2 + 3x +$ ____

Solve each equation by completing the square. **See Examples 2 and 3.**

11. $x^2 - 4x = -3$ ▶

12. $x^2 - 2x = 8$

13. $x^2 + 5x + 6 = 0$

14. $x^2 + 6x + 5 = 0$

15. $x^2 + 2x - 5 = 0$ ▶

16. $x^2 + 4x + 1 = 0$

17. $x^2 - 8x = -4$

18. $m^2 - 4m = 14$

19. $t^2 + 6t + 9 = 0$

20. $k^2 - 8k + 16 = 0$

21. $x^2 + x - 1 = 0$

22. $x^2 + x - 3 = 0$

Solve each equation by completing the square. **See Examples 4–8.**

23. $4x^2 + 4x = 3$ ▶

24. $9x^2 + 3x = 2$

25. $2p^2 - 2p + 3 = 0$ ▶

26. $3q^2 - 3q + 4 = 0$

27. $3k^2 + 7k = 4$ ▶

28. $2k^2 + 5k = 1$

29. $(x + 3)(x - 1) = 5$

30. $(x - 8)(x + 2) = 24$

31. $(r - 3)(r - 5) = 2$

32. $(k - 1)(k - 7) = 1$

33. $x(x - 3) = 1$

34. $x(x - 5) = 2$

35. $x(x + 3) = -1$

36. $x(x + 7) = -2$

37. $-x^2 + 2x = -5$

38. $-x^2 + 4x = 1$

39. $2x^2 - 4x = 5$

40. $2x^2 - 4x = 1$

Solve each problem. ***See Example 9.***

41. If an object is projected upward on the surface of Mars from ground level with an initial velocity of 104 ft per sec, its altitude (height) s, in feet, in t seconds is given by the formula

$$s = -13t^2 + 104t.$$

At what times will the object be 195 ft above the surface?

42. If an object is projected upward from ground level on Earth with an initial velocity of 96 ft per sec, its altitude (height) s, in feet, in t seconds is given by the formula

$$s = -16t^2 + 96t.$$

At what times will the object be 80 ft above the ground?

43. A farmer has a rectangular cattle pen with perimeter 350 ft and area 7500 ft^2. What are the dimensions of the pen? (*Hint:* Use the figure to set up the equation.)

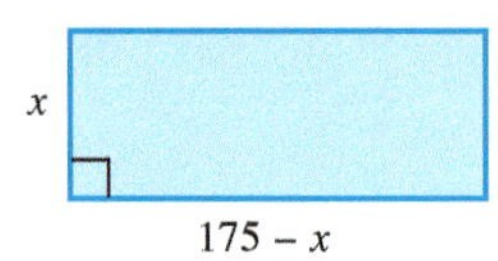

44. The base of a triangle measures 1 m more than three times the height of the triangle. Its area is 15 m^2. Find the lengths of the base and the height.

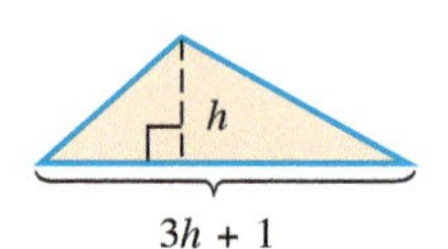

Relating Concepts (Exercises 45–48) For Individual or Group Work

We have discussed "completing the square" in an algebraic sense. This procedure can literally be applied to a geometric figure so that it becomes a square.

For example, to complete the square for $x^2 + 8x$, begin with a square having a side of length x. Add four rectangles of width 1 to the right side and to the bottom, as shown in the top figure. To "complete the square," fill in the bottom right corner with 16 squares of area 1, as shown in the bottom figure.

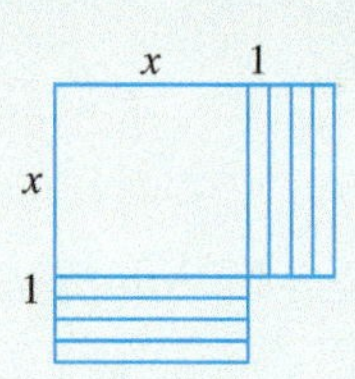

Work Exercises 45–48 in order.

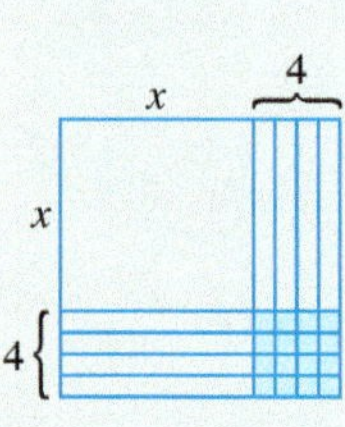

45. What is the area of the original square?

46. What is the area of the figure after the 8 rectangles are added?

47. What is the area of the figure after the 16 small squares are added?

48. At what point did we "complete the square"?

9.3 Solving Quadratic Equations by the Quadratic Formula

OBJECTIVES

1. Identify the values of a, b, and c in a quadratic equation.
2. Use the quadratic formula to solve quadratic equations.
3. Solve quadratic equations with one distinct solution.
4. Solve quadratic equations with fractional coefficients.

Although we can solve any quadratic equation by completing the square, the method can be tedious. By completing the square on the general quadratic equation

$$ax^2 + bx + c = 0, \quad \text{where} \quad a \neq 0, \qquad \text{Standard form}$$

we obtain the *quadratic formula,* which gives the solution(s) of any quadratic equation.

Note

In $ax^2 + bx + c = 0$, there is a restriction that a not equal zero. If it did, the equation would be linear, not quadratic.

OBJECTIVE 1 Identify the values of a, b, and c in a quadratic equation. To solve a quadratic equation using the quadratic formula, we must first identify the values of a, b, and c in the standard form.

EXAMPLE 1 Identifying Values of a, b, and c in Quadratic Equations

Identify the values of a, b, and c in each quadratic equation

$$ax^2 + bx + c = 0.$$

(a) $2x^2 + 3x - 5 = 0$ (This equation is in standard form.)

Here, $a = 2$, $b = 3$, and $c = -5$.

(b) $-x^2 + 2 = 6x$

Subtract $6x$ to write the equation in standard form $ax^2 + bx + c = 0$.

$$-x^2 - 6x + 2 = 0 \qquad \text{Subtract } 6x.$$

($-x^2$ means $-1x^2$.)

Here, $a = -1$, $b = -6$, and $c = 2$.

(c) $5x^2 - 12 = 0$

There is no x-term, so we write the equation as follows.

$$5x^2 + 0x - 12 = 0 \qquad \text{Standard form}$$

Then, $a = 5$, $b = 0$, and $c = -12$.

(d) $-4x^2 = -x$

In standard form, this equation is written $-4x^2 + x = 0$. There is no constant term c, so $a = -4$, $b = 1$, and $c = 0$.

(e)

$$(2x - 7)(x + 4) = -23$$

$$2x^2 + x - 28 = -23 \qquad \text{Multiply using the FOIL method.}$$

$$2x^2 + x - 5 = 0 \qquad \text{Add 23. The equation is now in standard form.}$$

(The equation is not in standard form.)

Identify the required values. We see that $a = 2$, $b = 1$, and $c = -5$.

Work Problem 1 at the Side. ▶

1 Identify the values of a, b, and c in each quadratic equation

$$ax^2 + bx + c = 0.$$

(a) $5x^2 + 2x - 1 = 0$

(b) $3x^2 = x - 2$

(c) $9x^2 - 13 = 0$

(d) $-x^2 + x = 0$

(e) $(3x + 2)(x - 1) = 8$

Answers

1. (a) $a = 5, b = 2, c = -1$
(b) $a = 3, b = -1, c = 2$
(c) $a = 9, b = 0, c = -13$
(d) $a = -1, b = 1, c = 0$
(e) $a = 3, b = -1, c = -10$

OBJECTIVE 2 Use the quadratic formula to solve quadratic equations. To develop the quadratic formula, we follow the steps to complete the square on $ax^2 + bx + c = 0$ (where $a > 0$). For comparison, we also show the corresponding steps for solving $2x^2 + x - 5 = 0$ (from **Example 1(e)**).

Step 1 Transform so that the coefficient of x^2 is equal to 1.

$2x^2 + x - 5 = 0$	Standard form	$ax^2 + bx + c = 0$	Standard form
$x^2 + \frac{1}{2}x - \frac{5}{2} = 0$	Divide by 2.	$x^2 + \frac{b}{a}x + \frac{c}{a} = 0$	Divide by a.

Step 2 Write so that the variable terms with x are alone on the left side.

$x^2 + \frac{1}{2}x = \frac{5}{2}$	Add $\frac{5}{2}$.	$x^2 + \frac{b}{a}x = -\frac{c}{a}$	Subtract $\frac{c}{a}$.

Step 3 Add the square of half the coefficient of x to each side, factor the left side, and combine terms on the right.

$x^2 + \frac{1}{2}x + \frac{1}{16} = \frac{5}{2} + \frac{1}{16}$	Add $\frac{1}{16}$.	$x^2 + \frac{b}{a}x + \frac{b^2}{4a^2} = -\frac{c}{a} + \frac{b^2}{4a^2}$	Add $\frac{b^2}{4a^2}$.
$\left(x + \frac{1}{4}\right)^2 = \frac{41}{16}$	Factor. Add.	$\left(x + \frac{b}{2a}\right)^2 = \frac{b^2 - 4ac}{4a^2}$	Factor. Add.

Step 4 Use the square root property to complete the solution.

$x + \frac{1}{4} = \pm\sqrt{\frac{41}{16}}$	$x + \frac{b}{2a} = \pm\sqrt{\frac{b^2 - 4ac}{4a^2}}$
$x + \frac{1}{4} = \pm\frac{\sqrt{41}}{4}$	$x + \frac{b}{2a} = \pm\frac{\sqrt{b^2 - 4ac}}{2a}$
$x = -\frac{1}{4} \pm \frac{\sqrt{41}}{4}$	$x = -\frac{b}{2a} \pm \frac{\sqrt{b^2 - 4ac}}{2a}$
$x = \frac{-1 \pm \sqrt{41}}{4}$	$\boldsymbol{x = \frac{-b \pm \sqrt{b^2 - 4ac}}{2a}}$

The final result on the right is the **quadratic formula** (which is also valid for $a < 0$). ***It gives two values: one for the + sign and one for the − sign.***

Quadratic Formula

The quadratic equation $ax^2 + bx + c = 0$ (where $a \neq 0$) has solutions

$$x = \frac{-b + \sqrt{b^2 - 4ac}}{2a} \quad \text{and} \quad x = \frac{-b - \sqrt{b^2 - 4ac}}{2a},$$

or in compact form, $$x = \frac{-b \pm \sqrt{b^2 - 4ac}}{2a}.$$

CAUTION

In the quadratic formula, the fraction bar extends under $-b$ as well as the radical. ***Be sure to find the values of $-b \pm \sqrt{b^2 - 4ac}$ first, and then divide those results by the value of $2a$.***

EXAMPLE 2 Solving a Quadratic Equation Using the Quadratic Formula

Solve $2x^2 - 7x - 9 = 0$.

$$x = \frac{-b \pm \sqrt{b^2 - 4ac}}{2a}$$ Quadratic formula

Be sure to write $-b$ in the numerator.

$$x = \frac{-(-7) \pm \sqrt{(-7)^2 - 4(2)(-9)}}{2(2)}$$ Substitute $a = 2$, $b = -7$, and $c = -9$.

$$x = \frac{7 \pm \sqrt{49 + 72}}{4}$$ Simplify.

$$x = \frac{7 \pm \sqrt{121}}{4}$$ Add.

The $\pm$ represents two solutions.

$$x = \frac{7 \pm 11}{4}$$ $\sqrt{121} = 11$

Find two values using the plus symbol and then the minus symbol.

$$x = \frac{7 + 11}{4} = \frac{18}{4} = \frac{9}{2} \quad \text{or} \quad x = \frac{7 - 11}{4} = \frac{-4}{4} = -1$$

Check each value in the original equation. The solution set is $\left\{-1, \frac{9}{2}\right\}$.

Work Problem 2 at the Side. ▶

2 Solve each equation using the quadratic formula.

(a) $2x^2 + 3x - 5 = 0$

(b) $6x^2 + x - 1 = 0$

EXAMPLE 3 Rewriting a Quadratic Equation before Solving

Solve $x^2 = 2x + 1$.

Write the equation in standard form as $x^2 - 2x - 1 = 0$.

$$x = \frac{-b \pm \sqrt{b^2 - 4ac}}{2a}$$ Quadratic formula

Be careful substituting negative values.

$$x = \frac{-(-2) \pm \sqrt{(-2)^2 - 4(1)(-1)}}{2(1)}$$ Substitute $a = 1$, $b = -2$, and $c = -1$.

$$x = \frac{2 \pm \sqrt{4 + 4}}{2}$$ Simplify.

$$x = \frac{2 \pm \sqrt{8}}{2}$$ Add.

$$x = \frac{2 \pm 2\sqrt{2}}{2}$$ $\sqrt{8} = \sqrt{4} \cdot \sqrt{2} = 2\sqrt{2}$

Factor first. Then divide out the common factor.

$$x = \frac{2(1 \pm \sqrt{2})}{2}$$ Factor.

$$x = 1 \pm \sqrt{2}$$ Divide out the common factor to write in lowest terms.

The solution set is $\left\{1 + \sqrt{2}, 1 - \sqrt{2}\right\}$.

Work Problem 3 at the Side. ▶

3 Solve $x^2 + 1 = -8x$.

GS Write the equation in standard form.

____

Substitute $a =$ ____, $b =$ ____, and $c =$ ____ in the quadratic formula and complete the solution.

Answers

2. **(a)** $\left\{1, -\frac{5}{2}\right\}$ **(b)** $\left\{-\frac{1}{2}, \frac{1}{3}\right\}$

3. $x^2 + 8x + 1 = 0$; 1; 8; 1; $\left\{-4 + \sqrt{15}, -4 - \sqrt{15}\right\}$

4 Solve $2t^2 - 4t + 5 = 0$.

5 Solve $9x^2 + 4 = 12x$.

Answers

4. no real solution

5. $\left\{\frac{2}{3}\right\}$

EXAMPLE 4 Recognizing When There Is No Real Solution

Solve $3x^2 - 4x + 2 = 0$.

$$x = \frac{-b \pm \sqrt{b^2 - 4ac}}{2a} \quad \text{Quadratic formula}$$

$$x = \frac{-(-4) \pm \sqrt{(-4)^2 - 4(3)(2)}}{2(3)} \quad \text{Substitute } a = 3, b = -4, \text{ and } c = 2.$$

$$x = \frac{4 \pm \sqrt{16 - 24}}{6} \quad \text{Simplify.}$$

$$x = \frac{4 \pm \sqrt{-8}}{6} \quad \text{Subtract.}$$

Because $\sqrt{-8}$ does not represent a real number, there is no real solution.

◀ Work Problem **4** at the Side.

OBJECTIVE 3 Solve quadratic equations with one distinct solution.

EXAMPLE 5 Solving a Quadratic Equation with One Solution

Solve $4x^2 + 25 = 20x$.

Write the equation in standard form as $4x^2 - 20x + 25 = 0$.

$$x = \frac{-b \pm \sqrt{b^2 - 4ac}}{2a} \quad \text{Quadratic formula}$$

$$x = \frac{-(-20) \pm \sqrt{(-20)^2 - 4(4)(25)}}{2(4)} \quad \text{Substitute } a = 4, b = -20, \text{ and } c = 25.$$

$$x = \frac{20 \pm \sqrt{400 - 400}}{8} \quad \text{Simplify.}$$

$$x = \frac{20 \pm 0}{8} \quad \sqrt{400 - 400} = \sqrt{0} = 0$$

$$x = \frac{20}{8} \quad \text{Simplify.}$$

$$x = \frac{5}{2} \quad \text{Write in lowest terms.}$$

Here $b^2 - 4ac = 0$, and the trinomial

$$4x^2 - 20x + 25 \quad \text{is a perfect square,} \quad (2x - 5)^2.$$

This would lead to two solutions, each of which is $\frac{5}{2}$. Thus, there is just *one* distinct solution, $\frac{5}{2}$. A check confirms that the solution set is $\left\{\frac{5}{2}\right\}$.

◀ Work Problem **5** at the Side.

Note

The single solution of the equation in **Example 5** is a rational number. If all solutions of a quadratic equation are rational, the equation can be solved using the zero-factor property.

$$4x^2 - 20x + 25 = 0 \quad \text{See Example 5.}$$

$$(2x - 5)^2 = 0 \quad \text{Factor.}$$

$$2x - 5 = 0 \quad \text{Zero-factor property}$$

The same solution results. ⟶ $x = \frac{5}{2}$ Solve for x.

OBJECTIVE 4 **Solve quadratic equations with fractional coefficients.**

EXAMPLE 6 **Solving a Quadratic Equation with Fractions**

Solve $\frac{1}{10}t^2 = \frac{2}{5}t + \frac{1}{5}$.

$$\frac{1}{10}t^2 = \frac{2}{5}t + \frac{1}{5}$$

$$10\left(\frac{1}{10}t^2\right) = 10\left(\frac{2}{5}t + \frac{1}{5}\right) \quad \text{To clear fractions, multiply by the LCD, 10.}$$

$$10\left(\frac{1}{10}t^2\right) = 10\left(\frac{2}{5}t\right) + 10\left(\frac{1}{5}\right) \quad \text{Distributive property}$$

$$t^2 = 4t + 2 \quad \text{Multiply.}$$

$$t^2 - 4t - 2 = 0 \quad \text{Subtract } 4t \text{ and } 2.$$

Identify $a = 1$, $b = -4$, and $c = -2$.

$$t = \frac{-b \pm \sqrt{b^2 - 4ac}}{2a} \quad \text{Quadratic formula}$$

$$t = \frac{-(-4) \pm \sqrt{(-4)^2 - 4(1)(-2)}}{2(1)} \quad \text{Substitute into the quadratic formula.}$$

$$t = \frac{4 \pm \sqrt{16 + 8}}{2} \quad \text{Simplify.}$$

$$t = \frac{4 \pm \sqrt{24}}{2} \quad \text{Add.}$$

$$t = \frac{4 \pm 2\sqrt{6}}{2} \quad \sqrt{24} = \sqrt{4 \cdot 6} = 2\sqrt{6}$$

$$t = \frac{2(2 \pm \sqrt{6})}{2} \quad \text{Factor.}$$

Be careful here.

$$t = 2 \pm \sqrt{6} \quad \text{Divide out the common factor to write in lowest terms.}$$

The solution set is $\{2 + \sqrt{6}, 2 - \sqrt{6}\}$.

Work Problem 6 at the Side. ▶

6 Solve each equation.

(a) $x^2 - \frac{9}{5}x = \frac{2}{5}$

(b) $\frac{1}{12}x^2 = \frac{1}{2}x - \frac{1}{3}$

Answers

6. **(a)** $\left\{-\frac{1}{5}, 2\right\}$ **(b)** $\{3 + \sqrt{5}, 3 - \sqrt{5}\}$

9.3 Exercises

FOR EXTRA HELP

Go to MyMathLab for worked-out, step-by-step solutions to exercises enclosed in a square and video solutions to ▶ exercises.

1. **CONCEPT CHECK** A student writes the quadratic formula as

$$x = -b \pm \frac{\sqrt{b^2 - 4ac}}{2a}. \quad \text{Incorrect}$$

What Went Wrong? Explain the error, and give the correct formula.

2. **CONCEPT CHECK** A student writes the quadratic formula as

$$x = -b \pm \sqrt{\frac{b^2 - 4ac}{2a}}. \quad \text{Incorrect}$$

What Went Wrong? Explain the error, and give the correct formula.

Write each equation in standard form $ax^2 + bx + c = 0$, if necessary. Then identify the values of a, b, and c. Do not actually solve the equation. ***See Example 1.***

3. $4x^2 + 5x - 9 = 0$

$a =$ ____, $b =$ ____, $c =$ ____

4. $8x^2 + 3x - 4 = 0$

$a =$ ____, $b =$ ____, $c =$ ____

5. $3x^2 = 4x + 2$

$a =$ ____, $b =$ ____, $c =$ ____

6. $5x^2 = 3x - 6$

$a =$ ____, $b =$ ____, $c =$ ____

7. $3x^2 = -7x$

$a =$ ____, $b =$ ____, $c =$ ____

8. $9x^2 = 8x$

$a =$ ____, $b =$ ____, $c =$ ____

Solve each equation using the quadratic formula. Express radicals in simplified form and answers in lowest terms. ***See Examples 2–5.***

9. $k^2 + 12k - 13 = 0$ ▶

10. $r^2 - 8r - 9 = 0$

11. $2x^2 = 5 + 3x$

12. $2z^2 = 30 + 7z$

13. $p^2 - 4p + 4 = 0$ ▶

14. $x^2 - 10x + 25 = 0$

15. $2x^2 + 12x = -5$ ▶

16. $5m^2 + m = 1$

17. $3x^2 + 5x + 1 = 0$ ▶

18. $6x^2 - 6x + 1 = 0$

19. $2x^2 + x + 5 = 0$

20. $3x^2 + 2x + 8 = 0$

21. $6x^2 + 6x = 0$

22. $4n^2 - 12n = 0$

23. $7x^2 = 12x$

24. $9r^2 = 11r$

25. $x^2 - 24 = 0$

26. $z^2 - 96 = 0$

27. $25x^2 - 4 = 0$

28. $16x^2 - 9 = 0$

29. $-2x^2 = -3x + 2$

30. $-x^2 = -5x + 20$

31. $3x^2 - 2x + 5 = 10x + 1$

32. $4x^2 - x + 4 = x + 7$

Solve each equation using the quadratic formula. ***See Example 6.***

33. $\frac{3}{2}k^2 - k - \frac{4}{3} = 0$

34. $\frac{2}{5}x^2 - \frac{3}{5}x - 1 = 0$

35. $\frac{1}{2}x^2 + \frac{1}{6}x = 1$

36. $\frac{2}{3}t^2 - \frac{4}{9}t = \frac{1}{3}$

37. $\frac{3}{8}x^2 - x + \frac{17}{24} = 0$

38. $\frac{1}{3}x^2 + \frac{8}{9}x + \frac{7}{9} = 0$

39. $0.6x - 0.4x^2 = -1$

40. $0.25x + 0.5x^2 = 1.5$

41. $0.25x^2 = -1.5x - 1$

42. $0.5x^2 = x + 0.5$

Solve each problem.

43. A frog is sitting on a stump 3 ft above the ground. He hops off the stump and lands on the ground 4 ft away. During his leap, his height h in feet is given by the equation

$$h = -0.5x^2 + 1.25x + 3,$$

where x is the distance in feet from the base of the stump. How far from the base of the stump was the frog when he was 1.25 ft above the ground?

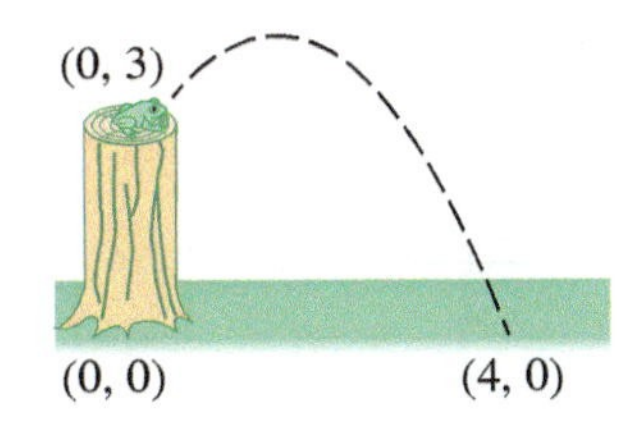

44. An astronaut on the moon throws a baseball upward. The altitude (height) h of the ball, in feet, x seconds after he throws it is given by the equation

$$h = -2.7x^2 + 30x + 6.5.$$

At what times is the ball 12 ft above the moon's surface? Give answer(s) to the nearest tenth.

45. A rule for estimating the number of board feet of lumber that can be cut from a log depends on the diameter of the log. To find the diameter d required to get 9 board feet of lumber, we use the equation

$$\left(\frac{d - 4}{4}\right)^2 = 9.$$

Solve this equation for d. Are both answers reasonable?

46. An old Babylonian problem asks for the length of the side of a square, given that the area of the square minus the length of a side is 870. Find the length of the side. (Data from Eves, H., *An Introduction to the History of Mathematics*, Saunders College Publishing.)

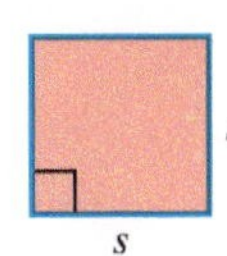

Summary Exercises *Applying Methods for Solving Quadratic Equations*

The table summarizes four methods for solving a quadratic equation $ax^2 + bx + c = 0$.

Method	Advantages	Disadvantages
1. Zero-factor property	It is usually the fastest method.	Not all equations can be solved using this property. Some factorable polynomials are difficult to factor.
2. Square root property	It is the simplest method for solving equations of the form $(ax + b)^2 = k$.	Few equations are given in this form.
3. Completing the square	It can always be used. (Also, the procedure is useful in other areas of mathematics.)	It requires more steps than other methods and can be tedious.
4. Quadratic formula	It can always be used.	Sign errors are common because of the presence of the expression $\sqrt{b^2 - 4ac}$.

A quadratic equation can be solved using more than one method, although one may be more direct than another. The following example compares several of the methods.

EXAMPLE Solving a Quadratic Equation

Solve $x^2 - 64 = 0$ using the method specified.

Zero-factor property

$$x^2 - 64 = 0$$

$$(x + 8)(x - 8) = 0$$

$$x + 8 = 0 \quad \text{or} \quad x - 8 = 0$$

$$x = -8 \quad \text{or} \quad x = 8$$

Solution set: $\{-8, 8\}$

Square root property

$$x^2 - 64 = 0$$

$$x^2 = 64$$

$$x = \sqrt{64} \quad \text{or} \quad x = -\sqrt{64}$$

$$x = 8 \quad \text{or} \quad x = -8$$

Solution set: $\{-8, 8\}$

Quadratic formula

$$x = \frac{-b \pm \sqrt{b^2 - 4ac}}{2a}$$

$$x = \frac{-0 \pm \sqrt{0^2 - 4(1)(-64)}}{2(1)}$$

In $x^2 - 64 = 0$, $a = 1$, $b = 0$, and $c = -64$.

$$x = \frac{\pm\sqrt{256}}{2}$$

$$x = \frac{\pm 16}{2}$$

$$x = \pm 8$$

Solution set: $\{-8, 8\}$

Solve each quadratic equation using the method of your choice.

1. $x^2 = 36$

2. $x^2 + 3x = -1$

3. $x^2 - \frac{100}{81} = 0$

4. $81t^2 = 49$

5. $z^2 - 4z + 3 = 0$

6. $w^2 + 3w + 2 = 0$

7. $z(z-9)=-20$

8. $x^2+3x-2=0$

9. $(3k-2)^2=9$

10. $(2s-1)^2=10$

11. $(x+6)^2=121$

12. $(5k+1)^2=36$

13. $(3r-7)^2=24$

14. $(7p-1)^2=32$

15. $(5x-8)^2=-6$

16. $2t^2+1=t$

17. $-2x^2=-3x-2$

18. $-2x^2+x=-1$

19. $8z^2=15+2z$

20. $3k^2=3-8k$

21. $0=-x^2+2x+1$

22. $3x^2+5x=-1$

23. $5x^2-22x=-8$

24. $x(x+6)+4=0$

25. $(x+2)(x+1)=10$

26. $16x^2+40x+25=0$

27. $4x^2=-1+5x$

28. $2p^2=2p+1$

29. $3x(3x+4)=7$

30. $5x-1+4x^2=0$

31. $\frac{x^2}{2}+\frac{7x}{4}+\frac{11}{8}=0$

32. $t(15t+58)=-48$

33. $9k^2=16(3k+4)$

34. $\frac{1}{5}x^2+x+1=0$

35. $x^2-x+3=0$

36. $4x^2-11x+8=-2$

37. $-3x^2+4x=-4$

38. $z^2-\frac{5}{12}z=\frac{1}{6}$

39. $5k^2+19k=2k+12$

40. $\frac{1}{2}x^2-x=\frac{15}{2}$

41. $x^2-\frac{4}{15}=-\frac{4}{15}x$

42. $(x+2)(x-4)=16$

9.4 Graphing Quadratic Equations

OBJECTIVES

1. Graph quadratic equations.
2. Find the vertex of a parabola.

OBJECTIVE 1 Graph quadratic equations. In this section, we graph quadratic equations in two variables, of the form

$$y = ax^2 + bx + c.$$

The simplest quadratic equation is $\mathbf{y = x^2}$ (or $y = 1x^2 + 0x + 0$).

1 Complete the table of values for $y = x^2$.

x	y
3	
2	4
1	
0	0
−1	1
−2	
−3	

2 Graph $y = \frac{1}{2}x^2$ by first completing the table of values.

x	y
−2	
−1	
0	
1	
2	

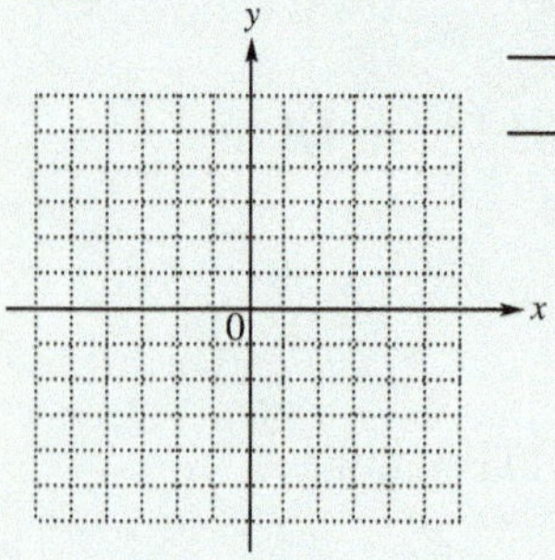

EXAMPLE 1 Graphing a Quadratic Equation

Graph $y = x^2$.

To find ordered pairs that satisfy the equation, we select several values for x and find the corresponding y-values. For example, we select $x = 2$, $x = 0$, and $x = -1$.

$y = x^2$	$y = x^2$	$y = x^2$
$y = 2^2$ Let $x = 2$.	$y = 0^2$ Let $x = 0$.	$y = (-1)^2$ Let $x = -1$.
$y = 4$	$y = 0$	$y = 1$ (Be careful substituting.)

The points $(2, 4)$, $(0, 0)$, and $(-1, 1)$ are on the graph of $y = x^2$. ***(Recall that in an ordered pair, the x-value comes first and the y-value second.)***

◀ **Work Problem 1 at the Side.**

If we plot the points from **Margin Problem 1** on a coordinate system and draw a smooth curve through them, we obtain the graph in **Figure 1.**

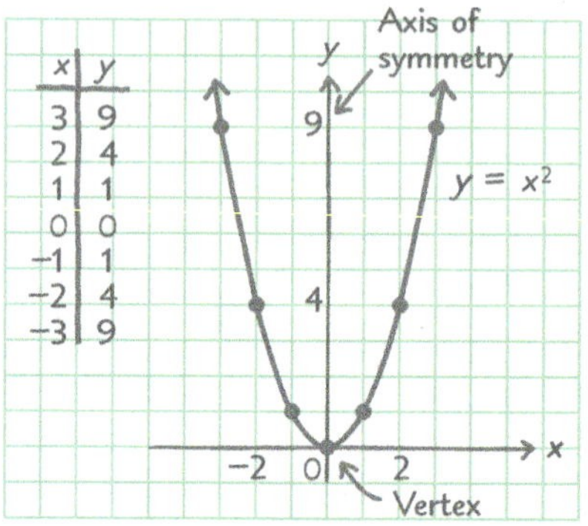

Figure 1

◀ **Work Problem 2 at the Side.**

The curve in **Figure 1** is a **parabola.** Every equation of the form

$$y = ax^2 + bx + c, \quad \text{where} \quad a \neq 0,$$

has a graph that is a parabola. The point $(0, 0)$, the *lowest* point on the graph of $y = x^2$, is the **vertex** of the parabola. (If a parabola opens downward, then the vertex will be the *highest* point. See **Figure 2** on the next page.)

The vertical line through the vertex of a parabola that opens upward or downward is the **axis of symmetry.** The two halves of the parabola are mirror images of each other across this line. For the parabola shown in **Figure 1**, the axis of symmetry is the y-axis.

Answers

1. See the table beside the graph in **Figure 1.**
2.

x	y
−2	2
−1	$\frac{1}{2}$
0	0
1	$\frac{1}{2}$
2	2

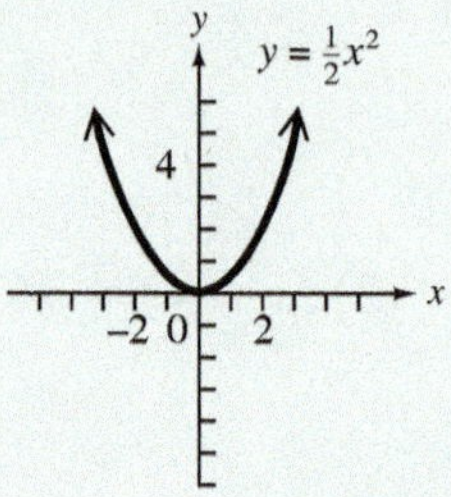

EXAMPLE 2 **Graphing a Parabola by Plotting Points**

Graph $y = -x^2 + 3$. Identify the vertex.

We find several ordered pairs by substituting values for x.

Let $x = 2$.	Let $x = 0$.	Let $x = -2$.
$y = -x^2 + 3$	$y = -x^2 + 3$	$y = -x^2 + 3$
$y = -2^2 + 3$	$y = -0^2 + 3$	$y = -(-2)^2 + 3$
$y = -4 + 3$	$y = 0 + 3$	$y = -4 + 3$
$y = -1$	$y = 3$	$y = -1$
Ordered pair: $(2, -1)$	Ordered pair: $(0, 3)$	Ordered pair: $(-2, -1)$

We plot these three ordered pairs and several others and join them with a smooth curve as shown in **Figure 2.** The vertex of this parabola is $(0, 3)$. The graph opens downward because x^2 has a negative coefficient, so the vertex is the *highest* point on the graph.

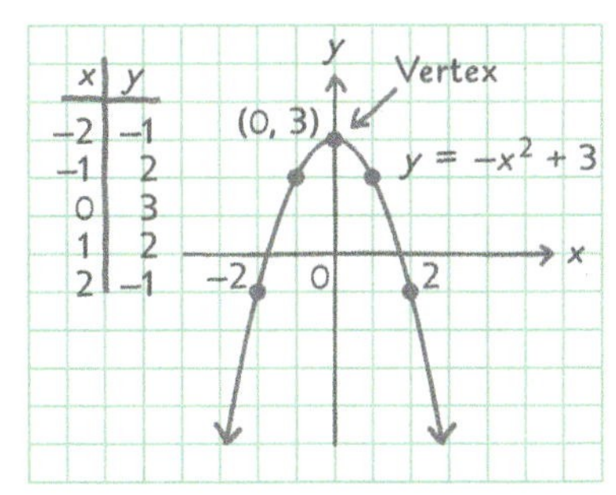

Figure 2

Work Problem 3 at the Side. ▶

OBJECTIVE 2 **Find the vertex of a parabola.** The vertex is an important point to locate when graphing a quadratic equation.

EXAMPLE 3 **Graphing a Parabola**

Graph $y = x^2 - 2x - 3$.

Because of its symmetry, if a parabola has two x-intercepts, the x-value of the vertex is exactly halfway between them. Therefore, we begin by finding the x-intercepts. We let $y = 0$ in the equation, and solve for x.

$$0 = x^2 - 2x - 3 \quad \text{Let } y = 0.$$

$$0 = (x + 1)(x - 3) \quad \text{Factor.}$$

$$x + 1 = 0 \quad \text{or} \quad x - 3 = 0 \quad \text{Zero-factor property}$$

$$x = -1 \quad \text{or} \quad x = 3 \quad \text{Solve each equation.}$$

There are two x-intercepts, $(-1, 0)$ and $(3, 0)$. Since the x-value of the vertex is halfway between the x-values of the two x-intercepts, it is half their sum.

$$x = \frac{1}{2}(-1 + 3)$$

$$x = 1 \leftarrow x\text{-value of the vertex}$$

We find the y-value of the vertex by substituting 1 for x in the equation.

$$y = x^2 - 2x - 3$$

$$y = 1^2 - 2(1) - 3 \quad \text{Let } x = 1.$$

$$y = -4 \leftarrow y\text{-value of the vertex}$$

The vertex is $(1, -4)$. The axis of symmetry is the vertical line $x = 1$.

Continued on Next Page

3 Graph each equation. Identify the vertex.

(a) $y = -x^2 - 3$

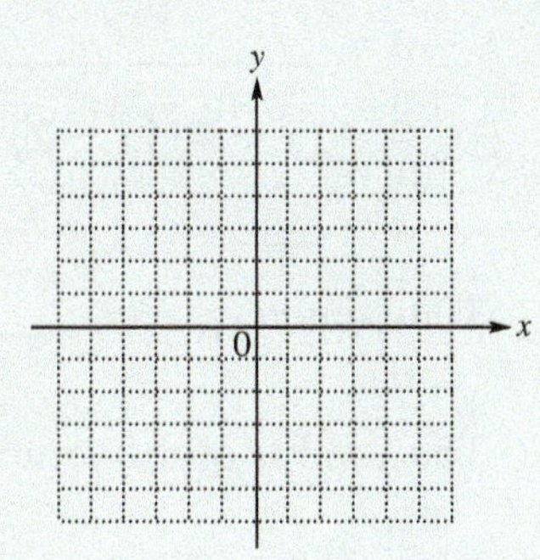

(b) $y = x^2 + 3$

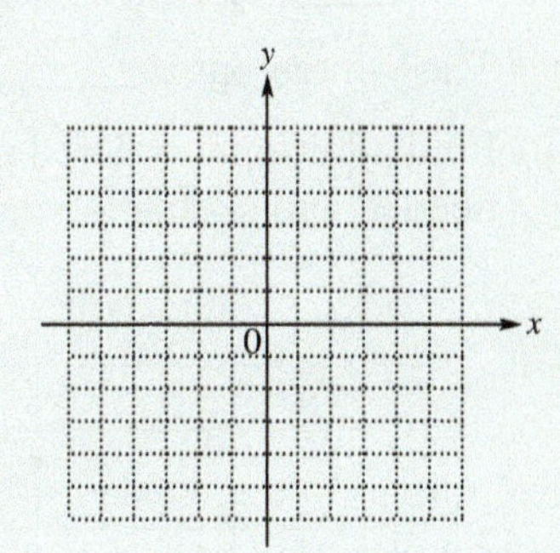

Answers

3. (a)

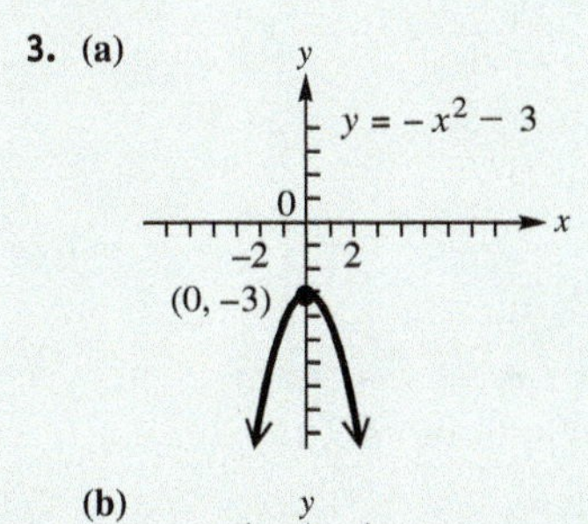

(b)

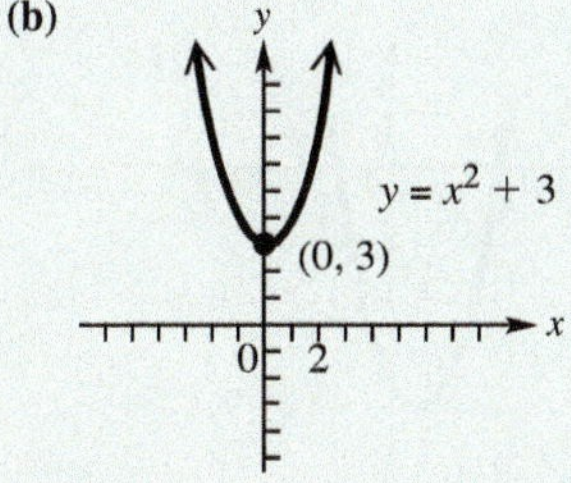

4 Graph $y = x^2 + 2x - 8$.

GS To find the x-intercepts, solve the equation.

$0 = \_\_\_\_\_\_\_\_$

$0 = (x + 4)(x - 2)$

$x = \_\_\_\_$ or $x = 2$

The x-intercepts are ____ and $(2, 0)$.

The x-value of the vertex is

$\frac{1}{2}(-4 + \_\_\_\_) = \_\_\_\_.$

The y-value of the vertex is

____.

The vertex is ____.

The axis of symmetry is the line $x = \_\_\_\_$.

The y-intercept is ____.

Find additional ordered pairs as needed and sketch the graph.

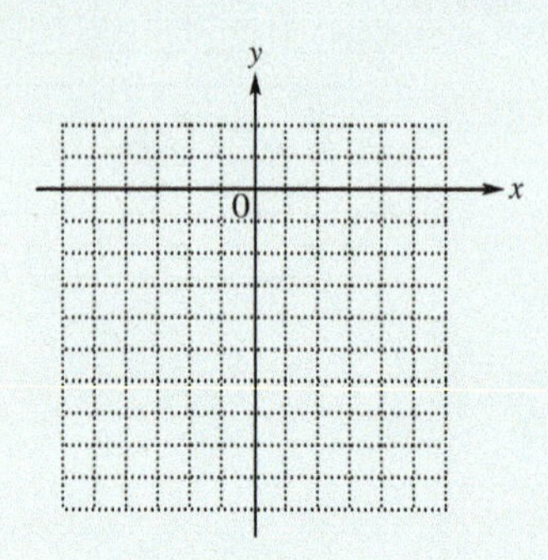

Answer

4. $x^2 + 2x - 8$; -4; $(-4, 0)$; 2; -1; -9; $(-1, -9)$; -1; $(0, -8)$

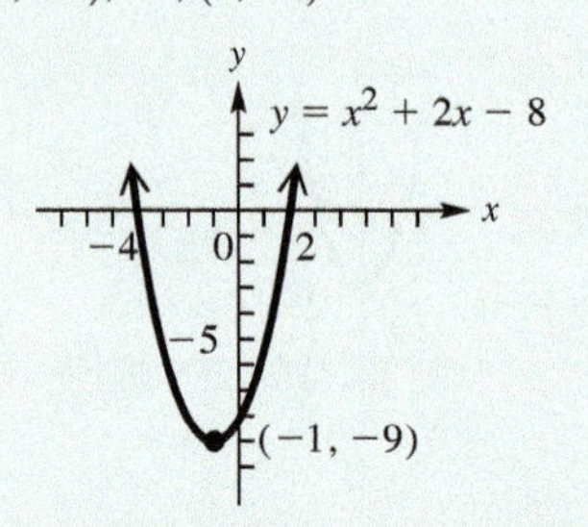

To find the y-intercept, we substitute 0 for x in the equation.

$$y = x^2 - 2x - 3$$

$$y = 0^2 - 2(0) - 3 \quad \text{Let } x = 0.$$

$$y = -3 \quad \text{Simplify.}$$

The y-intercept is $(0, -3)$.

We plot the three intercepts and the vertex, and find additional ordered pairs as needed. For example, we let $x = -2$.

$$y = (-2)^2 - 2(-2) - 3 \quad \text{Let } x = -2.$$

$$y = 5 \quad \text{Simplify.}$$

This leads to the ordered pair $(-2, 5)$. A table that includes the ordered pairs we found is shown with the graph in **Figure 3.**

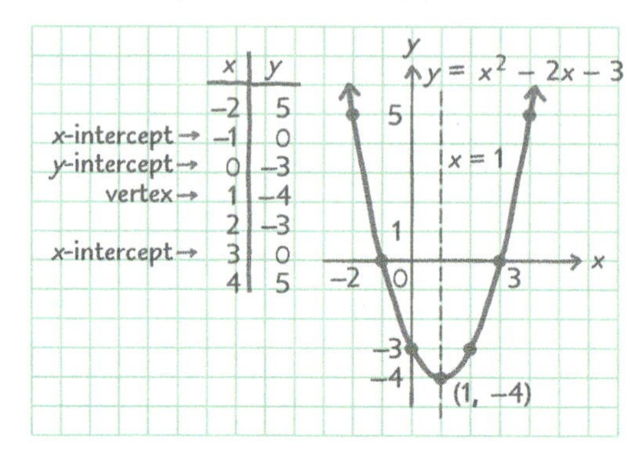

Figure 3

◀ **Work Problem 4 at the Side.**

We can generalize from **Example 3.** By the quadratic formula, the x-coordinates of the x-intercepts for the equation $y = ax^2 + bx + c$ are

$$x = \frac{-b + \sqrt{b^2 - 4ac}}{2a} \quad \text{and} \quad x = \frac{-b - \sqrt{b^2 - 4ac}}{2a}.$$

Thus, the x-value of the vertex is half their sum.

$$x = \frac{1}{2}\left(\frac{-b + \sqrt{b^2 - 4ac}}{2a} + \frac{-b - \sqrt{b^2 - 4ac}}{2a}\right)$$

$$x = \frac{1}{2}\left(\frac{-b + \sqrt{b^2 - 4ac} - b - \sqrt{b^2 - 4ac}}{2a}\right)$$

$$x = \frac{1}{2}\left(\frac{-2b}{2a}\right) \quad \text{Combine like terms.}$$

$$\mathbf{x = -\frac{b}{2a}} \quad \text{Multiply. Write in lowest terms.}$$

For the equation in **Example 3,** $y = x^2 - 2x - 3$, we have $a = 1$ and $b = -2$. Thus, the x-value of the vertex is found as follows.

$$x = -\frac{b}{2a} = -\frac{-2}{2(1)} = 1 \leftarrow \text{This is the same } x\text{-value for the vertex found in Example 3.}$$

(It can be shown that the x-value of the vertex is $x = -\frac{b}{2a}$, even if the graph has no x-intercepts.)

EXAMPLE 4 Graphing a Parabola

Graph $y = x^2 - 4x + 1$.

Here, $a = 1$ and $b = -4$, so we substitute to find the x-value of the vertex.

$$x = -\frac{b}{2a} = -\frac{-4}{2(1)} = 2$$

The y-value of the vertex is found by substituting 2 for x in $y = x^2 - 4x + 1$.

$$y = x^2 - 4x + 1$$

$$y = 2^2 - 4(2) + 1 \quad \text{Let } x = 2.$$

$$y = -3 \quad \text{Simplify.}$$

The vertex is $(2, -3)$. The axis of symmetry is the line $x = 2$.

To find the y-intercept, we let $x = 0$ in $y = x^2 - 4x + 1$.

$$y = 0^2 - 4(0) + 1 \quad \text{Let } x = 0.$$

$$y = 1$$

The y-intercept is $(0, 1)$. We let $y = 0$ to find the x-intercepts. If $y = 0$, the equation becomes $0 = x^2 - 4x + 1$, which cannot be solved using the zero-factor property, so we use the quadratic formula to solve for x.

$$x = \frac{-(-4) \pm \sqrt{(-4)^2 - 4(1)(1)}}{2(1)} \quad \text{Let } a = 1, b = -4, c = 1 \text{ in the quadratic formula.}$$

$$x = \frac{4 \pm \sqrt{12}}{2} \quad \text{Simplify.}$$

$$x = \frac{4 \pm 2\sqrt{3}}{2} \quad \sqrt{12} = \sqrt{4} \cdot \sqrt{3} = 2\sqrt{3}$$

Factor first. Then divide out the common factor.

$$x = \frac{2(2 \pm \sqrt{3})}{2} \quad \text{Factor.}$$

$$x = 2 \pm \sqrt{3} \quad \text{Divide out 2.}$$

$$x \approx 3.7 \quad \text{and} \quad x \approx 0.3 \quad \text{Use a calculator.}$$

The x-intercepts are $(3.7, 0)$ and $(0.3, 0)$, to the nearest tenth.

Work Problem 5 at the Side. ▶

We plot the intercepts, vertex, and the points found in **Margin Problem 5** and join these points with a smooth curve, as shown in **Figure 4**.

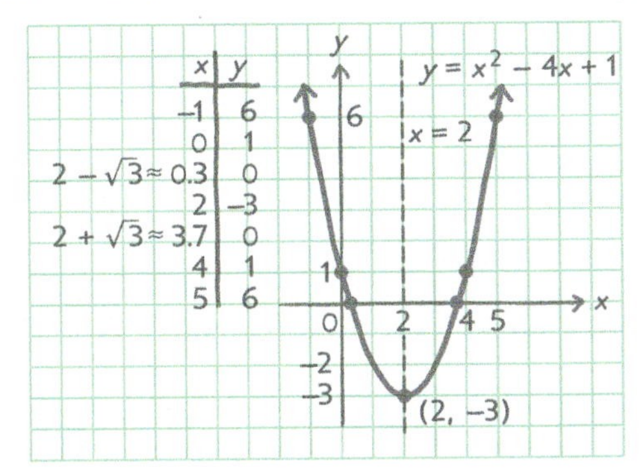

Figure 4

Work Problem 6 at the Side. ▶

5 Complete each ordered pair for

$$y = x^2 - 4x + 1.$$

$(5, \_\_)$, $(4, \_\_)$, $(-1, \_\_)$

6 Graph each parabola.

(a) $y = x^2 - 3x - 3$

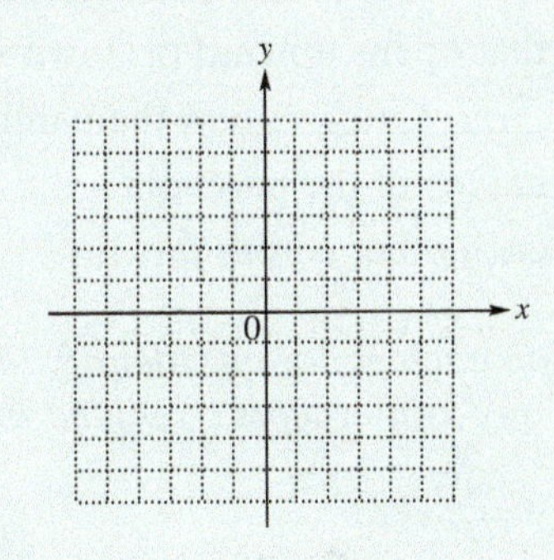

(b) $y = -x^2 + 2x + 4$

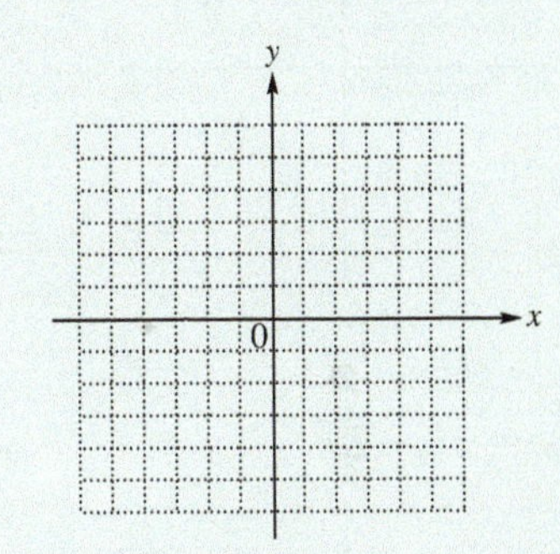

Answers

5. $(5, 6)$, $(4, 1)$, $(-1, 6)$

6. (a)

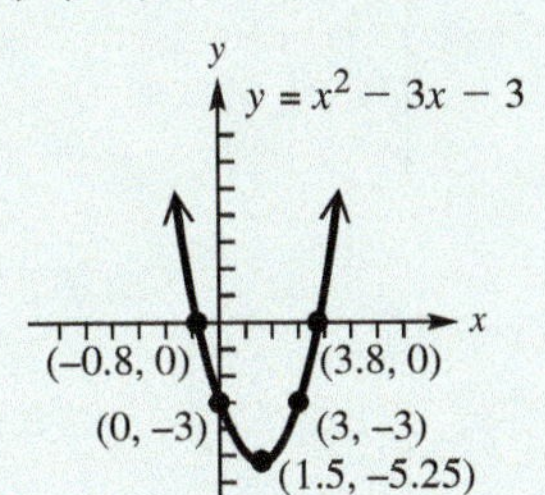

(b)

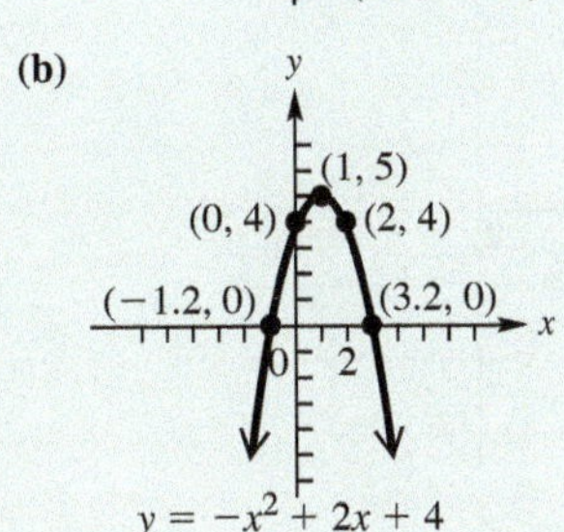

9.4 Exercises

FOR EXTRA HELP — *Go to MyMathLab for worked-out, step-by-step solutions to exercises enclosed in a square and video solutions to ▶ exercises.*

CONCEPT CHECK *Fill in each blank with the correct response.*

1. Every equation of the form $y = ax^2 + bx + c$, with $a \neq 0$, has a graph that is a(n) ______.

2. The ______ of a parabola is the lowest point of the graph if the parabola opens upward and the ______ point of the graph if the parabola opens downward.

3. The vertical line through the vertex of a parabola that opens upward or downward is the ____________ of the parabola. The two halves of the parabola are ______ images of each other across this line.

4. The graph of an equation of the form $y = ax^2 + bx + c$, where $a \neq 0$, has exactly ______ y-intercept(s), but may have either two, ______, or ______ x-intercepts.

Graph each equation. Identify the vertex. In Exercises 13–19, also give the y-intercept and any x-intercepts (rounded to the nearest tenth, if necessary). **See Examples 1–4.**

5. $y = 2x^2$

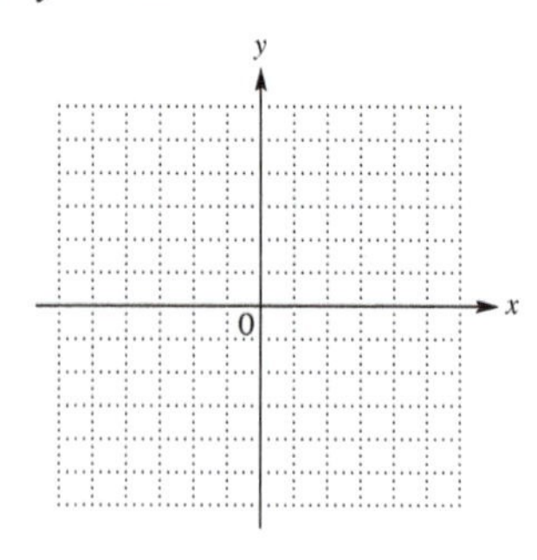

6. $y = \frac{1}{4}x^2$

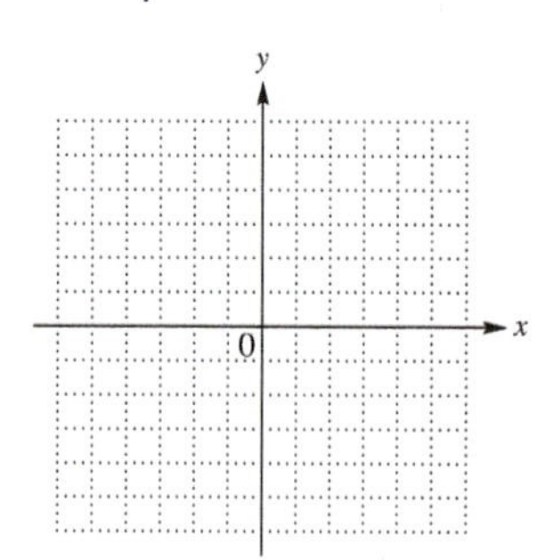

7. $y = x^2 - 4$

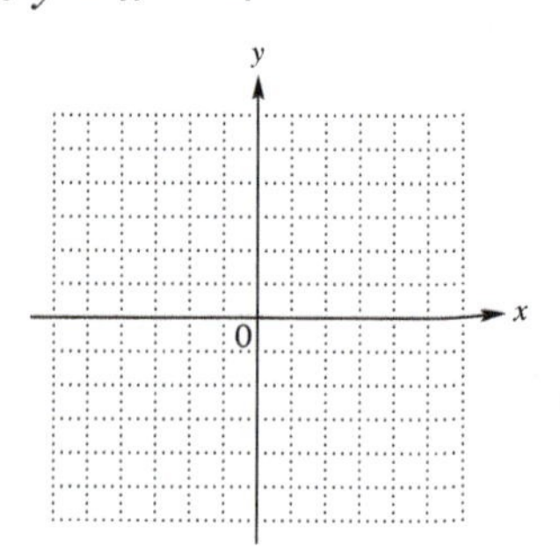

8. $y = x^2 - 2$

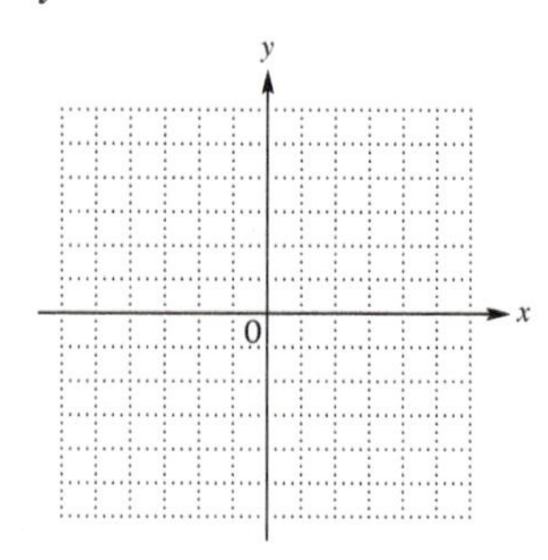

9. $y = -x^2 + 2$

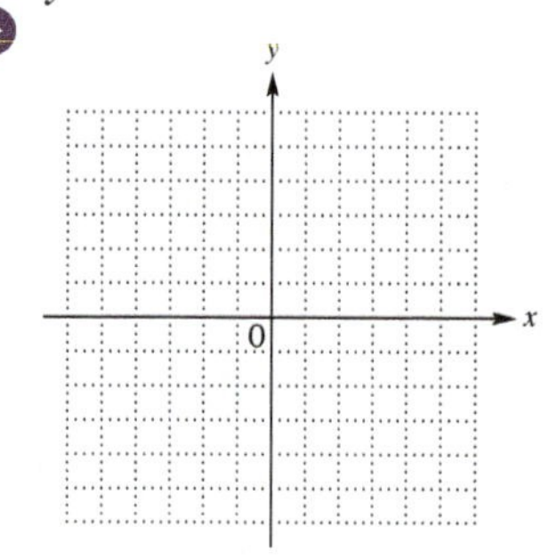

10. $y = -x^2 + 4$

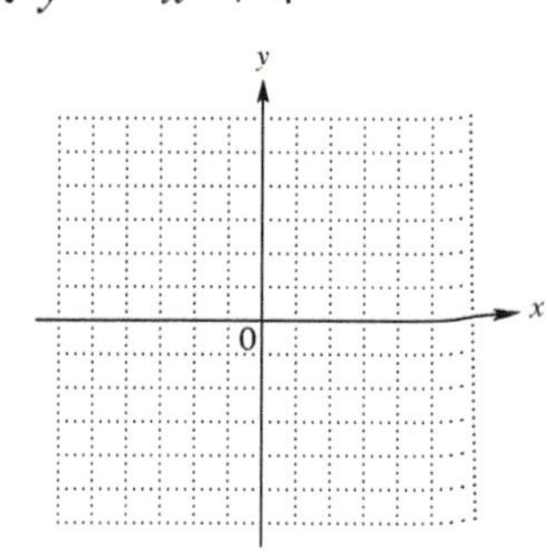

11. $y = (x + 1)^2$

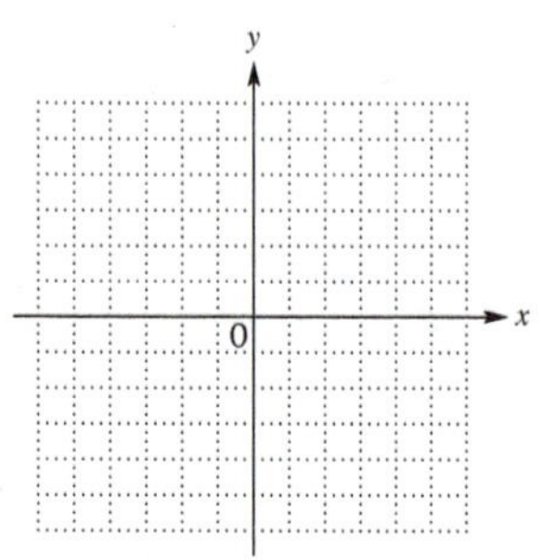

12. $y = (x - 2)^2$

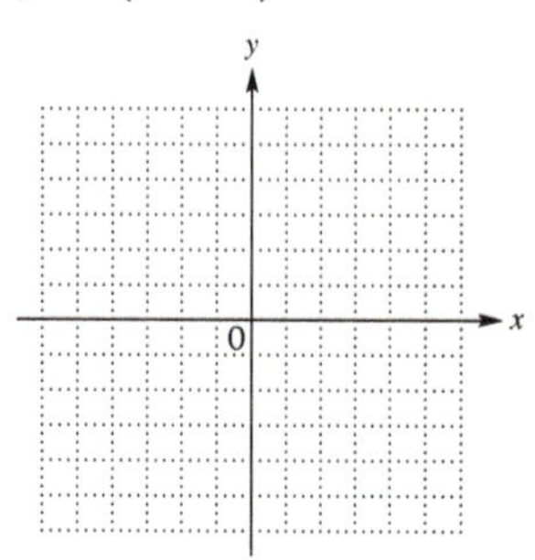

13. $y = x^2 + 2x + 3$

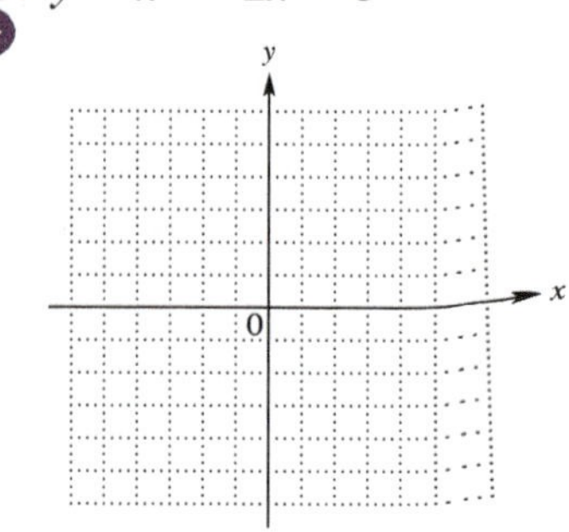

14. $y = x^2 - 4x + 3$

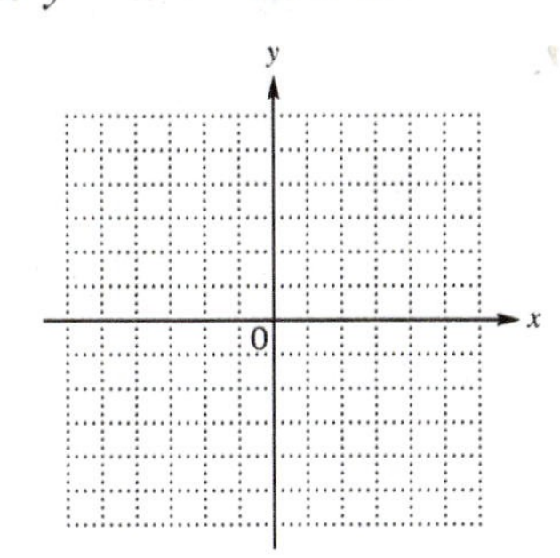

15. $y = -x^2 + 6x - 5$

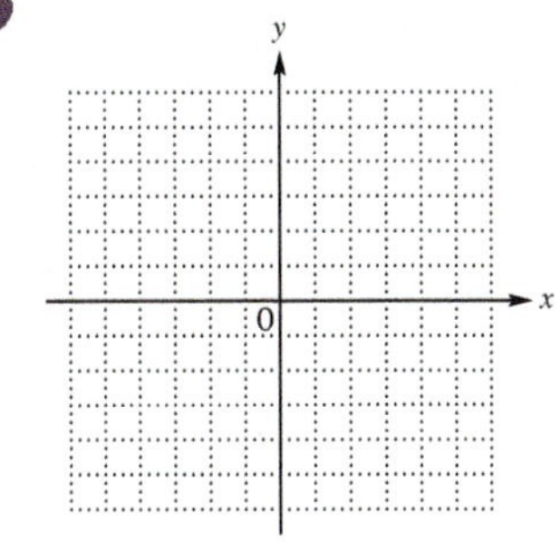

16. $y = -x^2 - 4x - 3$

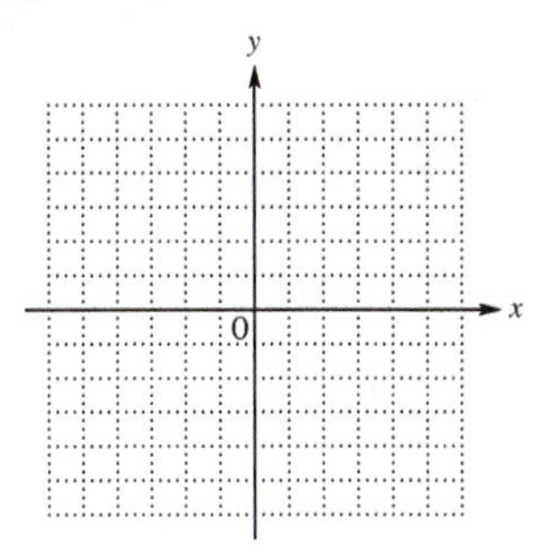

17. $y = x^2 - 4x - 2$

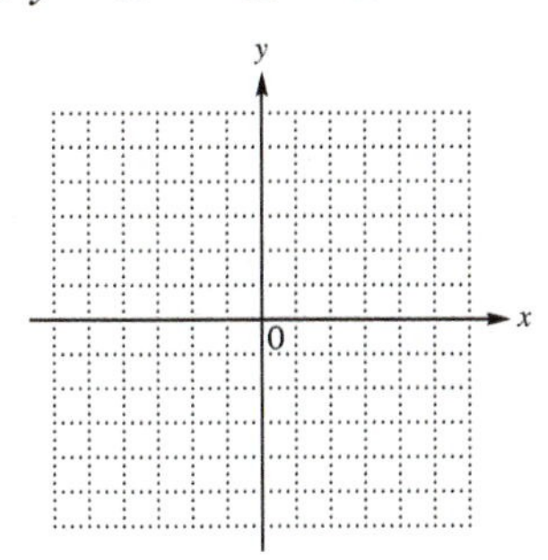

18. $y = x^2 - 2x - 4$

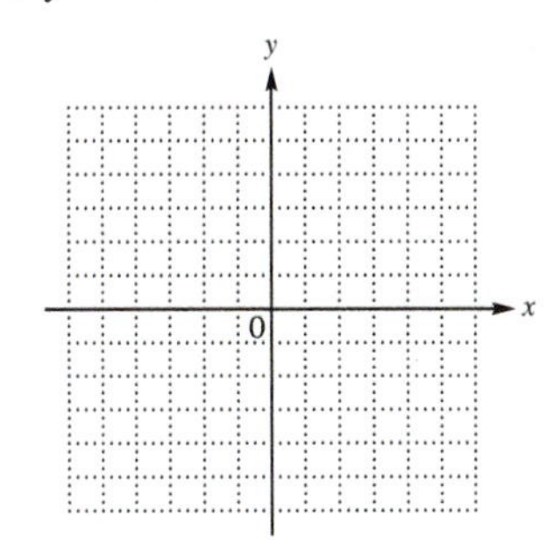

19. $y = -x^2 + 2x + 5$

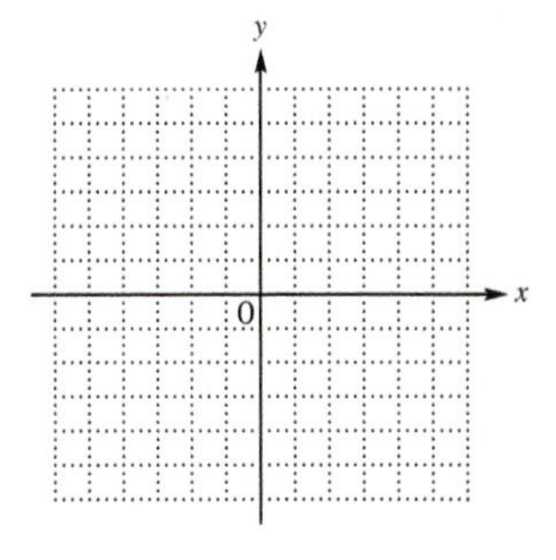

20. CONCEPT CHECK Based on your work in **Exercises 5–19,** what seems to be the direction in which the parabola

$$y = ax^2 + bx + c$$

opens if $a > 0$? If $a < 0$?

Solve each problem.

21. The U.S. Naval Research Laboratory designed a giant radio telescope that had a diameter of 300 ft and maximum depth of 44 ft. The graph on the right describes a cross section of this telescope. Find the equation of this parabola. (Data from Mar, J. and Liebowitz, H., *Structure Technology for Large Radio and Radar Telescope Systems,* The MIT Press.)

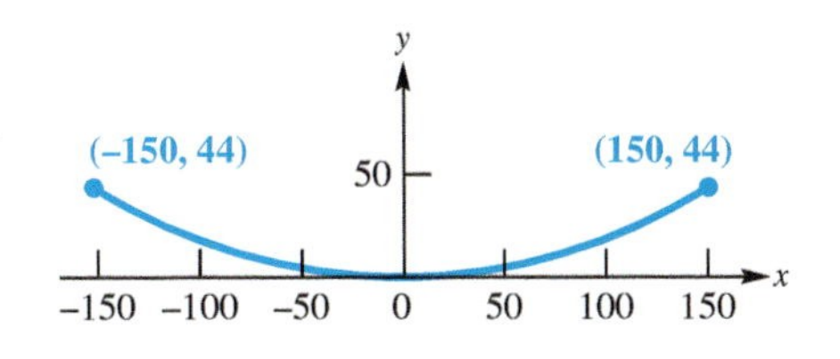

22. Suppose the telescope in **Exercise 21** had a diameter of 400 ft and maximum depth of 50 ft. Find the equation of this parabola.

9.5 Introduction to Functions

OBJECTIVES

1. Understand the definition of a relation.
2. Understand the definition of a function.
3. Determine whether a graph or equation represents a function.
4. Use function notation.
5. Apply the function concept in an application.

OBJECTIVE 1 Understand the definition of a relation. If gasoline costs \$3.00 per gal and we purchase 1 gal, then we must pay $\$3.00(1) = \$3.00$. For 2 gal, the cost is $\$3.00(2) = \$6.00$. For 3 gal, the cost is $\$3.00(3) = \$9.00$, and so on. Generalizing, if x represents the number of gallons, then the cost y is $\$3.00x$. The equation

$$y = 3.00x$$

relates the number of gallons, x, to the cost in dollars, y. The set of ordered pairs (x, y) that satisfy this equation forms a *relation*. In an ordered pair (x, y), x and y are the **components** of the ordered pair.

Relation

Any set of ordered pairs is a **relation**.

- The set of all first components in the ordered pairs of a relation is the **domain** of the relation.
- The set of all second components in the ordered pairs is the **range** of the relation.

1 Give the domain and range of each relation.

(a) $\{(5, 10), (15, 20), (25, 30), (35, 40)\}$

(b) $\{(1, 4), (2, 4), (3, 4)\}$

EXAMPLE 1 Identifying Domains and Ranges of Relations

Give the domain and range of each relation.

(a) $\{(0, 1), (2, 5), (3, 8), (4, 2)\}$

Domain: $\{0, 2, 3, 4\}$ ← Set of first components

Range: $\{1, 5, 8, 2\}$ ← Set of second components

The correspondence between the elements of the domain and the elements of the range is shown in **Figure 5.**

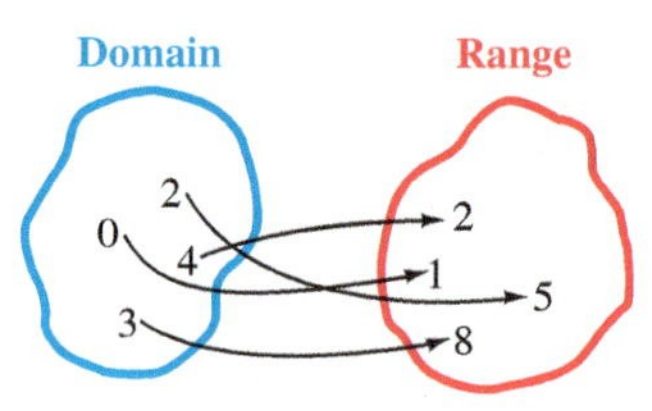

Figure 5

(b) $\{(3, 5), (3, 6), (3, 7), (3, 8)\}$

The relation has domain $\{3\}$ and range $\{5, 6, 7, 8\}$.

◀ Work Problem 1 at the Side.

OBJECTIVE 2 Understand the definition of a function. A *function* is a special type of relation.

Function

A **function** is a set of ordered pairs (a relation) in which each distinct first component corresponds to exactly one second component.

Answers

1. (a) domain: $\{5, 15, 25, 35\}$; range: $\{10, 20, 30, 40\}$
   (b) domain: $\{1, 2, 3\}$; range: $\{4\}$

Based on this definition, the relation in **Example 1(a)**

$$\{(0, 1), (2, 5), (3, 8), (4, 2)\} \text{ is a function.}$$

However, the relation in **Example 1(b)**

$$\{(3, 5), (3, 6), (3, 7), (3, 8)\} \text{ is } \textbf{\textit{not}} \text{ a function.}$$

The ***same*** first component, 3, corresponds to more than one second component. See **Figure 6.** If the ordered pairs in **Example 1(b)** were interchanged, giving the relation

$$\{(5, 3), (6, 3), (7, 3), (8, 3)\},$$

then this relation *would* be a function. ***In that case, each domain element (first component) is distinct and corresponds to exactly one range element (second component).***

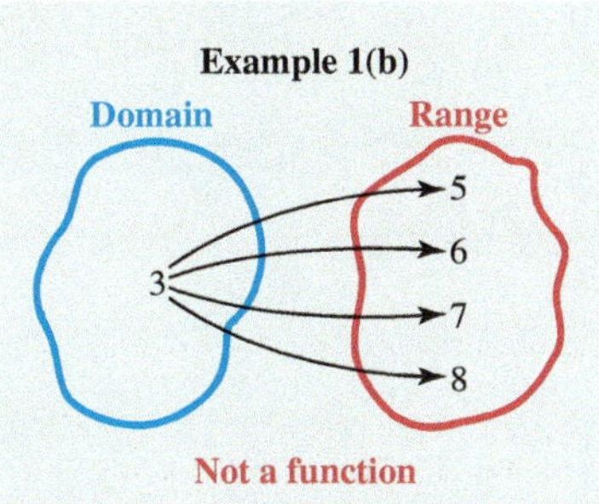

Figure 6

2 Determine whether each relation is a function.

(a) $\{(-2, 8), (-1, 1), (0, 0), (1, 1), (2, 8)\}$

(b) $\{(5, 2), (5, 1), (5, 0)\}$

(c) $\{(-1, -3), (0, 2), (3, 1), (8, 1)\}$

EXAMPLE 2 Determining Whether Relations Are Functions

Determine whether each relation is a function.

(a) $\{(-2, 4), (-1, 1), (0, 0), (1, 1), (2, 4)\}$

Each first component appears once and only once. The relation is a function.

(b) $\{(9, 3), (9, -3), (4, 2)\}$

The first component 9 appears in two ordered pairs and corresponds to two different second components. Therefore, this relation is not a function.

Work Problem **2** at the Side. ▶

Functions often have an infinite number of ordered pairs and are represented by equations that tell how to find the second components (**outputs**), given the first components (**inputs**).

Here are some everyday examples of functions.

1. The **cost** y in dollars charged by an express mail company is a function of the **weight** x in pounds determined by the equation $y = 1.5(x - 1) + 9$.
2. In Cedar Rapids, Iowa, sales tax is 7% of the price of an item. The **tax** y on a particular item is a function of the **price** x, because $y = 0.07x$.
3. The **distance** d traveled by a car moving at a constant rate of 45 mph is a function of the **time** t. Thus, $d = 45t$.

The function concept can be illustrated by an input-output "machine," as seen in **Figure 7.** The express mail company equation $y = 1.5(x - 1) + 9$ provides an output (cost y in dollars) for a given input (weight x in pounds).

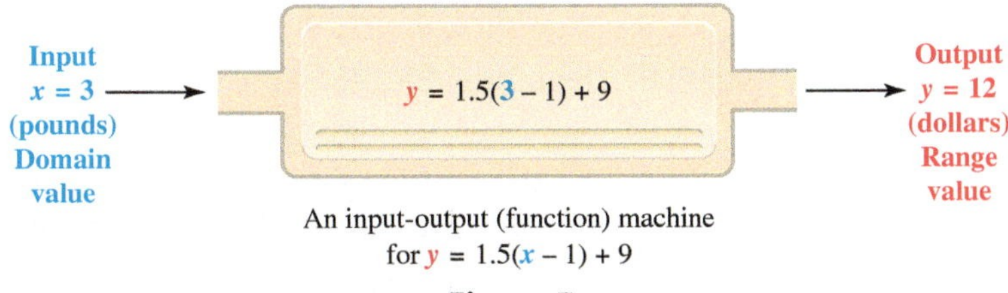

An input-output (function) machine for $y = 1.5(x - 1) + 9$

Figure 7

Answers

2. **(a)** function **(b)** not a function **(c)** function

OBJECTIVE 3 Determine whether a graph or equation represents a function. Given the graph of an equation, the definition of a function can be used to decide whether or not the graph represents a function. By definition, each x-value of a function must lead to exactly one y-value.

In **Figure 8(a)**, the indicated x-value leads to two y-values, so this graph is not the graph of a function. A vertical line can be drawn that intersects this graph in more than one point.

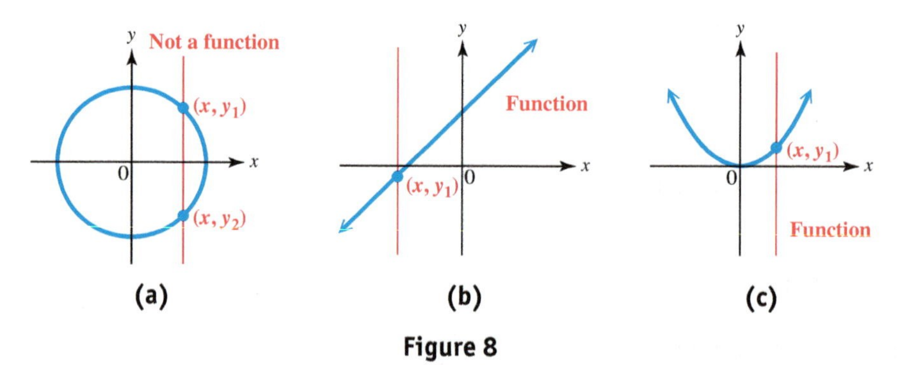

Figure 8

By contrast, in **Figure 8(b)** and **Figure 8(c)** any vertical line will intersect each graph in no more than one point, so these graphs are graphs of functions. This idea leads to the **vertical line test** for a function.

Vertical Line Test

If a vertical line intersects a graph in more than one point, then the graph is not the graph of a function.

As **Figure 8(b)** suggests, any nonvertical line is the graph of a function.

Thus, any linear equation of the form $y = mx + b$ defines a function.

(Recall that a vertical line has undefined slope.) Also, any vertical parabola, as in **Figure 8(c)**, is the graph of a function.

Thus, any quadratic equation of the form $y = ax^2 + bx + c$ (where $a \neq 0$) defines a function.

EXAMPLE 3 Determining Whether Relations Are Functions

Determine whether each relation represented by a graph or an equation is a function.

(a)

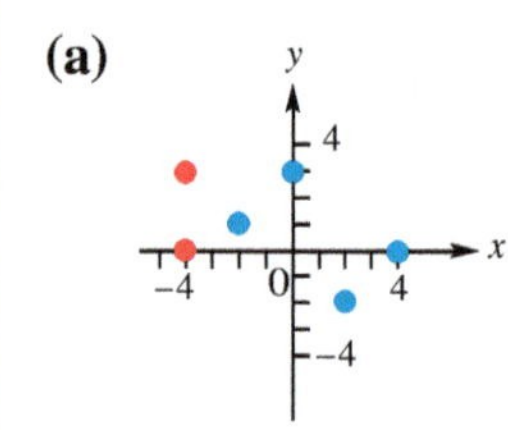

We can apply the vertical line test to decide whether this relation is a function. The two ordered pairs shown in red, $(-4, 3)$ and $(-4, 0)$, have first component -4. A vertical line drawn through these points would intersect the graph *twice*, so the vertical line test is *not* satisfied. Therefore, this relation is *not* a function.

(b)

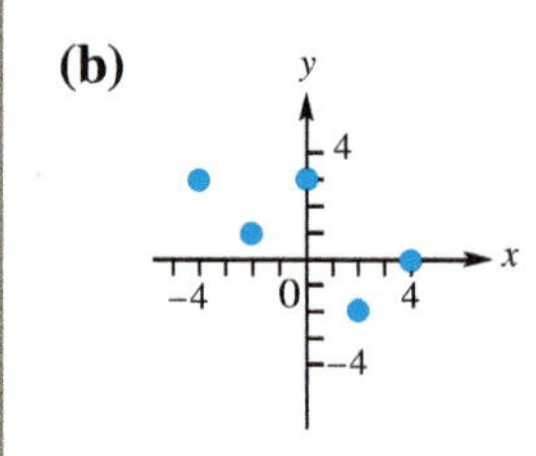

Applying the vertical line test reveals that no vertical line intersects the graph in more than one point. Every first component of the ordered pairs represented by the graph is paired with one and only one second component. Therefore, this relation is a function.

Continued on Next Page

(c)

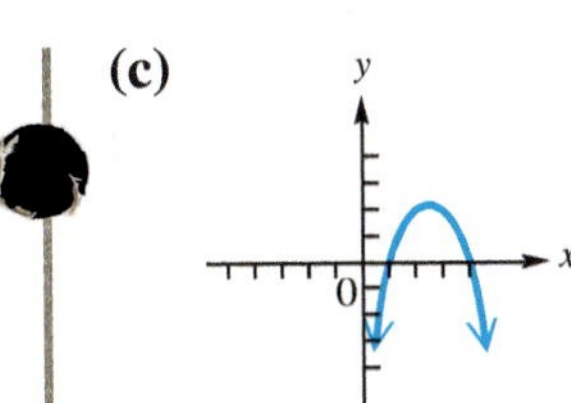

Apply the vertical line test. Any vertical line intersects the graph of a vertical parabola just once, so this relation is a function.

(d)

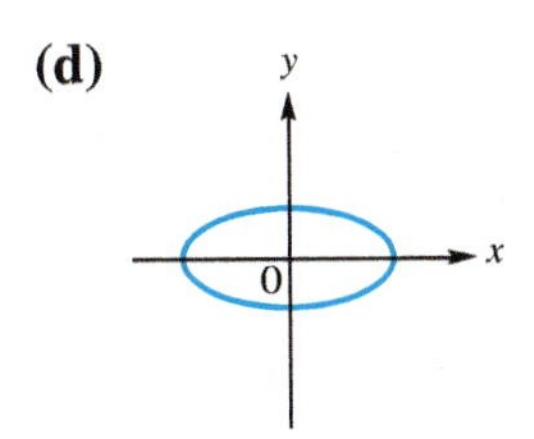

Applying the vertical line test shows that this relation is *not* a function—a vertical line could intersect the graph twice.

(e) $y = 2x - 9$

This linear equation is in the form $y = mx + b$. Because the graph of this equation is a line that is not vertical, the equation defines a function.

(f) $x = 4$

The graph of $x = 4$ is a vertical line, so the equation does *not* define a function. (Every ordered pair has x-value 4.)

Work Problem 3 at the Side. ▶

OBJECTIVE 4 Use function notation. The letters f, g, and h are commonly used to name functions. For example, the function

$$y = 3x + 5$$

may be named f and written

$$f(x) = 3x + 5,$$

where $f(x)$, which represents the value of f at x, is read **"f of x."** The notation $f(x)$, called **function notation,** is another way of expressing the range element y for a function f. For the function $f(x) = 3x + 5$, if $x = 7$ then we find $f(7)$ as follows.

$$f(x) = 3x + 5$$

$$f(7) = 3(7) + 5 \quad \text{Let } x = 7.$$

$$f(7) = 26 \quad \text{Multiply, and then add.}$$

We read $f(7) = 26$ as "f *of* 7 *equals* 26." The notation $f(7)$ represents the value of y when x is 7. The statement $f(7) = 26$ says that the value of y is 26 when x is 7. It also indicates that the point $(7, 26)$ lies on the graph of f.

CAUTION

The notation $f(x)$ does *not* mean f times x. ***It represents the y-value that corresponds to x in function f.***

Function Notation

In the notation $f(x)$, remember the following.

f	is the name of the function.
x	is the domain value.
$f(x)$	is the range value y for the domain value x.

3 Determine whether each relation represented by a graph or an equation is a function.

(a)

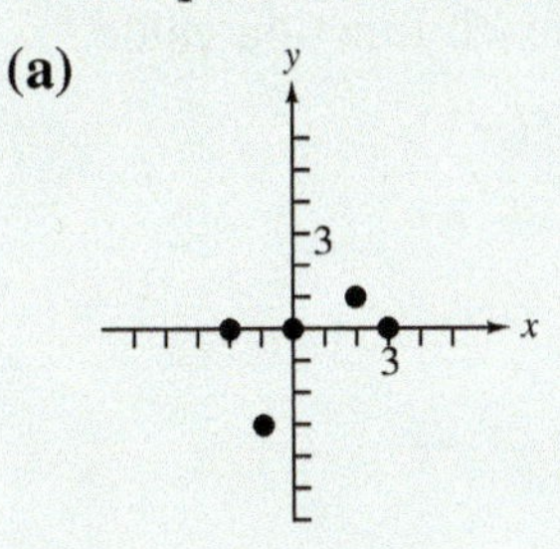

(b)

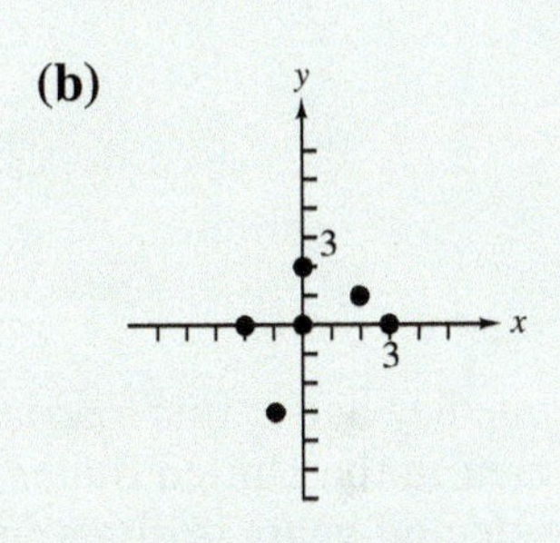

(c)

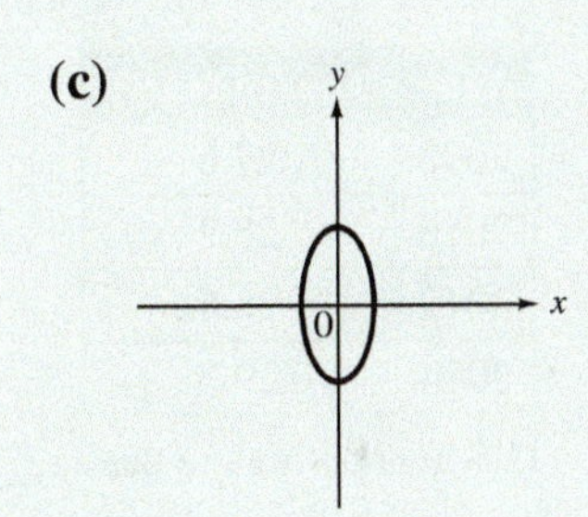

(d)

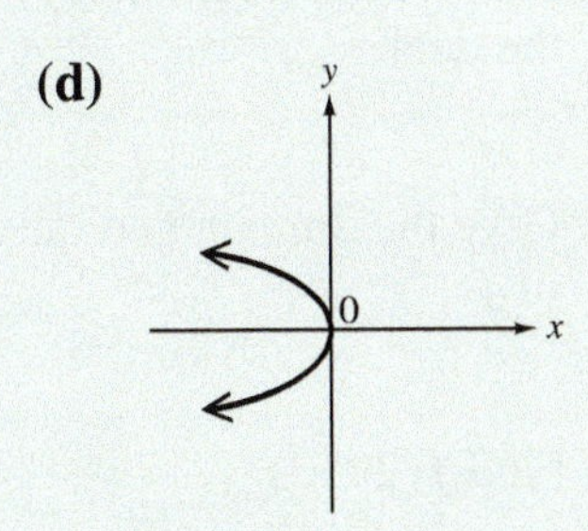

(e) $y = 3$

(f) $x = -3$

Answers

3. **(a)** function **(b)** not a function **(c)** not a function **(d)** not a function **(e)** function **(f)** not a function

4 For the function

$$f(x) = 6x - 2,$$

find each function value.

GS **(a)** $f(-1)$

$= 6(\_\_\_\_) - 2$

$= \_\_\_\_ - 2$

$= \_\_\_\_$

(b) $f(0)$ **(c)** $f(1)$

5 The median age at first marriage for men in the United States for selected years is given in the table.

Year	Age (in Years)
2008	27.6
2010	28.2
2012	28.6
2014	29.3

Data from U.S. Census Bureau.

(a) Write a set of ordered pairs that defines a function f for the data.

(b) Give the domain and range of f.

(c) Find $f(2010)$.

(d) For what x-value does $f(x)$ equal 28.6?

Answers

4. (a) -1; -6; -8
(b) -2 (c) 4

5. (a) $f = \{(2008, 27.6), (2010, 28.2), (2012, 28.6), (2014, 29.3)\}$
(b) domain: $\{2008, 2010, 2012, 2014\}$; range: $\{27.6, 28.2, 28.6, 29.3\}$
(c) 28.2 (d) 2012

EXAMPLE 4 Using Function Notation

For the function $f(x) = x^2 - 3$, find each function value.

(a) $f(4)$

$f(x) = x^2 - 3$ Given function

$f(4) = 4^2 - 3$ Substitute 4 for x.

$= 16 - 3$ Apply the exponent. (4^2 means $4 \cdot 4$.)

$f(4) = 13$ Subtract.

(b) $f(0)$

$f(x) = x^2 - 3$

$f(0) = 0^2 - 3$ Let $x = 0$.

$f(0) = 0 - 3$

$f(0) = -3$

(c) $f(-3)$

$f(x) = x^2 - 3$

$f(-3) = (-3)^2 - 3$ Let $x = -3$.

$f(-3) = 9 - 3$ ($(-3)^2$ means $-3 \cdot (-3)$.)

$f(-3) = 6$

◀ Work Problem 4 at the Side.

OBJECTIVE 5 Apply the function concept in an application.

EXAMPLE 5 Applying the Function Concept to Population

Asian-American populations (in millions) are shown in the table.

Year	Population (in millions)
2004	13.1
2006	14.9
2012	15.6
2014	16.7

Data from U.S. Census Bureau.

(a) Write a set of ordered pairs that defines a function f for the data.

If we choose the years as the domain elements and the populations in millions as the range elements, the information in the table can be written as a set of four ordered pairs. In set notation, the function f is

$$f = \{(2004, 13.1), (2006, 14.9), (2012, 15.6), (2014, 16.7)\}.$$

y-values are in millions.

(b) What is the domain of f? What is the range?

$\{2004, 2006, 2012, 2014\}$ The domain is the set of years, or x-values.

$\{13.1, 14.9, 15.6, 16.7\}$ The range is the set of populations, or y-values.

(c) Find $f(2004)$ and $f(2012)$.

We refer to the table or the ordered pairs in part (a).

$$f(2004) = 13.1 \text{ million} \quad \text{and} \quad f(2012) = 15.6 \text{ million}$$

(d) For what x-value does $f(x)$ equal 16.7 million? 14.9 million?

Again, we refer to the table or the ordered pairs found in part (a).

$$f(2014) = 16.7 \text{ million} \quad \text{and} \quad f(2006) = 14.9 \text{ million}$$

◀ Work Problem 5 at the Side.

9.5 Exercises

FOR EXTRA HELP *Go to MyMathLab for worked-out, step-by-step solutions to exercises enclosed in a square and video solutions to ▶ exercises.*

CONCEPT CHECK *Complete each statement.*

1. Any set of ordered pairs is a(n) ________. The set of first components of the ordered pairs of a relation is the ________. The set of second components of the ordered pairs of a relation is the ________.

2. A(n) ________ is a special type of relation in which each first component corresponds to exactly (*one / two / unlimited*) second ________.

Determine whether each relation is a function. In Exercises 3–12, give the domain and range of each relation. **See Examples 1–3(d).**

3. $\{(3, 7), (1, 4), (0, -2), (-1, -1)\}$

4. $\{(-2, 6), (0, 5), (2, 4), (3, 3)\}$

5. $\{(1, -1), (2, -2), (3, -1)\}$

6. $\{(4, 2), (0, 0), (-4, 2)\}$

7. ▶ $\{(-4, 3), (-2, 1), (0, 5), (-2, -8)\}$

8. $\{(6, 0), (3, -2), (3, -6), (0, -4)\}$

9.

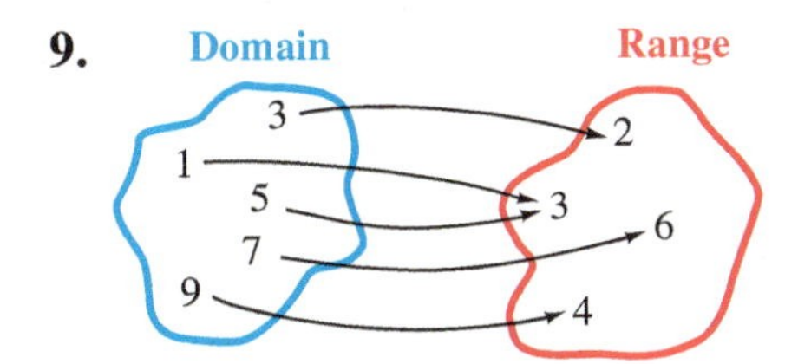

10.

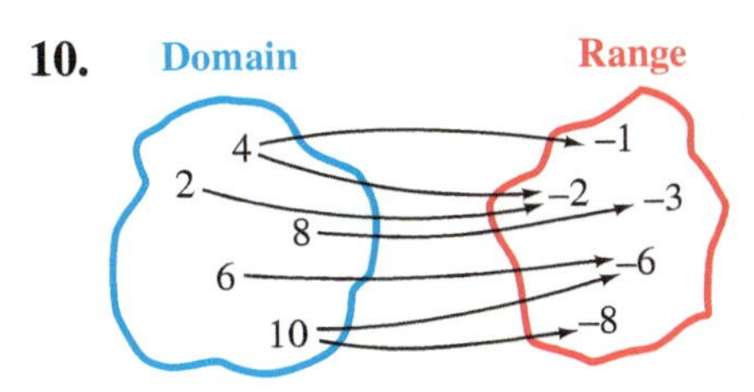

11. ▶

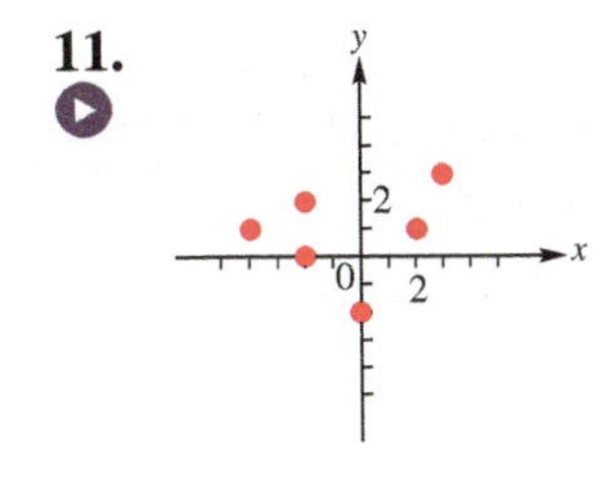

12.

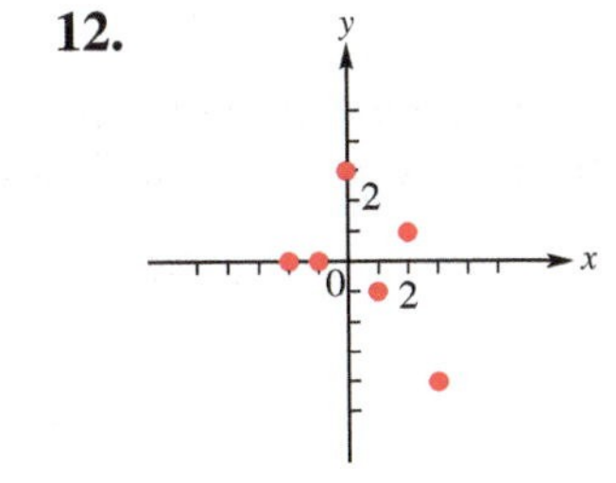

13. ▶

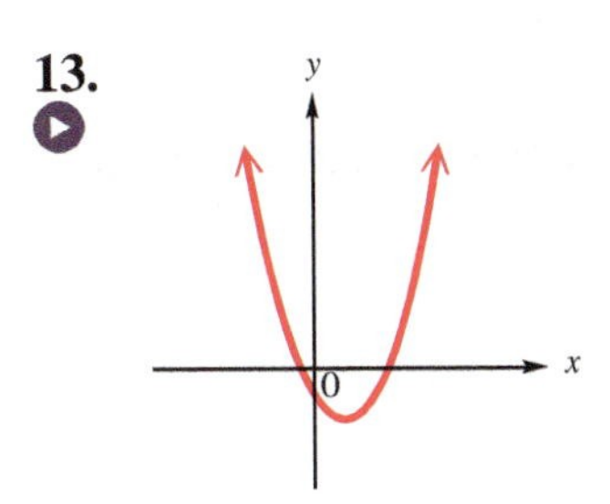

14.

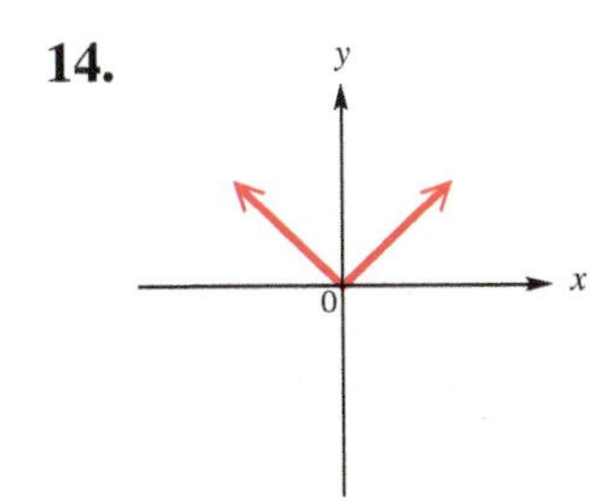

15. ▶

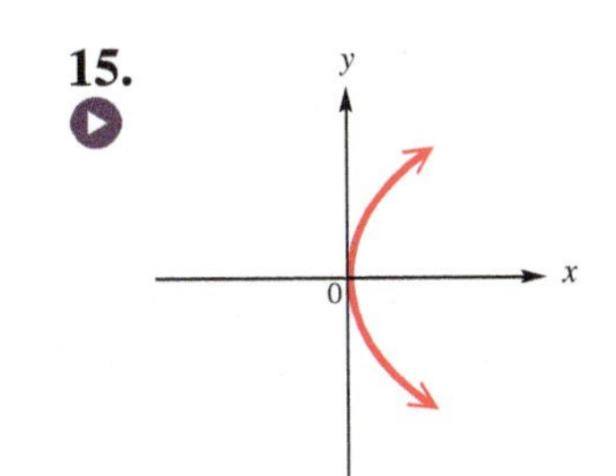

16.

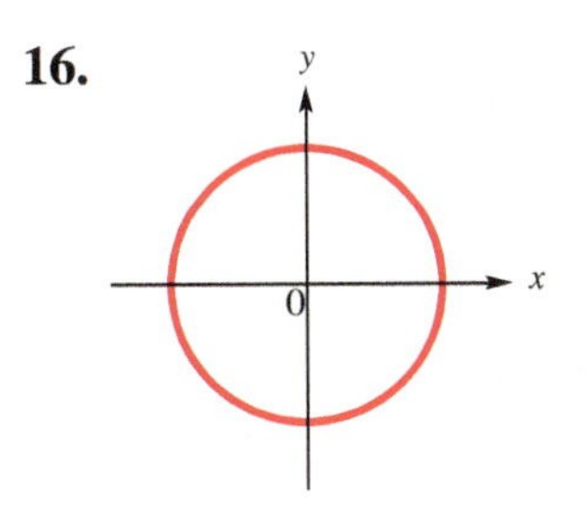

Determine whether each relation represented by an equation is a function. ***See Examples 3(e) and (f).***

17. $y = 5x + 3$

18. $y = -7x + 12$

19. $y = x$

20. $y = -x$

21. $x = -7$

22. $x = -4$

23. $y = 5$

24. $y = -8$

For each function f, find **(a)** $f(2)$, **(b)** $f(0)$, *and* **(c)** $f(-3)$. ***See Example 4.***

25. $f(x) = 4x + 3$

26. $f(x) = -3x + 5$

27. $f(x) = -x - 2$

28. $f(x) = -2x - 1$

29. $f(x) = x^2 - x + 2$

30. $f(x) = x^3 + x$

The table shows the number of foreign-born U.S. residents (in millions) for selected years. Use the information to work each problem. ***See Example 5.***

Year	Population (in millions)
1980	14.1
1990	19.8
2000	28.4
2010	40.0
2014	42.4

Data from U.S. Census Bureau.

31. Write the information in the table as a set of ordered pairs. Does this set represent a function?

32. Suppose that g is the name given to this relation. Give the domain and range of g.

33. Find $g(1980)$ and $g(2000)$.

34. For what value of x does $g(x) = 40.0$ (million)?

Relating Concepts (Exercises 35–38) For Individual or Group Work

A ***linear function*** *is a function defined by the equation of a line. Let us assume that a function is written in the form*

$$f(x) = mx + b, \quad \text{Linear function}$$

for particular values of m and b. ***Work Exercises 35–38 in order.***

35. Name the coordinates of the point on a line that corresponds to the statement $f(0) = -4$.

36. Name the coordinates of the point on a line that corresponds to the statement $f(3) = -2$.

37. Use the results of **Exercises 35 and 36** to find the slope of the line.

38. Use the slope-intercept form of the equation of a line and the results of **Exercises 35 and 37** to write a linear function $f(x) = mx + b$.

Chapter 9 Summary

Key Terms

9.1

quadratic equation A quadratic equation can be written in the form $ax^2 + bx + c = 0$ (called **standard form**), where a, b, and c are real numbers and $a \neq 0$.

9.4

parabola The graph of a quadratic equation $y = ax^2 + bx + c$ is a parabola.

vertex The vertex of a parabola that opens upward or downward is the lowest or highest point on the graph.

axis of symmetry The axis of symmetry of a parabola that opens upward or downward is the vertical line through the vertex.

9.5

components In an ordered pair (x, y), x and y are the components.

relation Any set of ordered pairs is a relation.

domain The set of all first components in the ordered pairs of a relation is the domain of the relation.

range The set of all second components in the ordered pairs of a relation is the range of the relation.

function A function is a set of ordered pairs in which each distinct first component corresponds to exactly one second component.

New Symbols

$\pm$ positive or negative (plus or minus)

$f(x)$ function notation; function of x (read "f *of* x")

Test Your Word Power

See how well you have learned the vocabulary in this chapter.

1 A **quadratic equation** is an equation that can be written in the form
- **A.** $Ax + By = C$
- **B.** $ax^2 + bx + c = 0$
- **C.** $Ax + B = 0$
- **D.** $y = mx + b$.

2 A **parabola** is the graph of
- **A.** any equation in two variables
- **B.** a linear equation
- **C.** an equation of degree three
- **D.** a quadratic equation in two variables.

3 The **vertex** of a parabola is
- **A.** the point where the graph intersects the y-axis
- **B.** the point where the graph intersects the x-axis
- **C.** the lowest point on a parabola that opens up or the highest point on a parabola that opens down
- **D.** the origin.

4 A **relation** is
- **A.** any set of ordered pairs
- **B.** a set of ordered pairs in which each distinct first component corresponds to exactly one second component
- **C.** two sets of ordered pairs that are related
- **D.** a graph of any ordered pairs.

5 A **function** is
- **A.** any set of ordered pairs
- **B.** a set of ordered pairs in which each distinct first component corresponds to exactly one second component
- **C.** two sets of ordered pairs that are related
- **D.** a graph of any ordered pairs.

Answers to Test Your Word Power

1. B; *Examples:* $x^2 + 6x + 9 = 0$, $x^2 - 2x = 8$
2. D; *Example:* See the figure shown above in the Key Terms.
3. C; *Example:* The graph of $y = (x + 3)^2$ has vertex $(-3, 0)$, which is the lowest point on the graph.
4. A; *Example:* $\{(0, 2), (2, 4), (3, 6), (-1, 3)\}$
5. B; *Example:* The relation $\{(0, 2), (2, 4), (3, 6), (-1, 3)\}$ is a function because each x-value corresponds to exactly one y-value.

Quick Review

Concepts

9.1 Solving Quadratic Equations by the Square Root Property

Zero-Factor Property
If a and b are real numbers and if $ab = 0$, then $a = 0$ or $b = 0$.

Square Root Property
If k is a positive number and if $x^2 = k$, then

$$x = \sqrt{k} \quad \text{or} \quad x = -\sqrt{k}.$$

The solution set $\{-\sqrt{k}, \sqrt{k}\}$ can be written $\{\pm\sqrt{k}\}$.

Examples

Solve $x^2 - x - 72 = 0$.

$$x^2 - x - 72 = 0$$
$$(x + 8)(x - 9) = 0 \quad \text{Factor.}$$
$$x + 8 = 0 \quad \text{or} \quad x - 9 = 0 \quad \text{Zero-factor property}$$
$$x = -8 \quad \text{or} \quad x = 9 \quad \text{Solve each equation.}$$

Solution set: $\{-8, 9\}$

Solve $(2x + 1)^2 = 5$.

$$2x + 1 = \sqrt{5} \quad \text{or} \quad 2x + 1 = -\sqrt{5}$$
$$2x = -1 + \sqrt{5} \quad \text{or} \quad 2x = -1 - \sqrt{5}$$
$$x = \frac{-1 + \sqrt{5}}{2} \quad \text{or} \quad x = \frac{-1 - \sqrt{5}}{2}$$

Solution set: $\left\{\frac{-1 + \sqrt{5}}{2}, \frac{-1 - \sqrt{5}}{2}\right\}$

Concepts

9.2 Solving Quadratic Equations by Completing the Square

Solving a Quadratic Equation by Completing the Square

Step 1 Be sure the second-degree term has coefficient 1.

- If the coefficient of the second-degree term is 1, go to Step 2.
- If the coefficient is not 1, divide each side of the equation by this coefficient.

Step 2 Make sure that all variable terms are on one side of the equality symbol and the constant term is on the other side.

Step 3 Complete the square.

- Take half the coefficient of the first-degree term, and square it.
- Add the square to each side of the equation.
- Factor the variable side, which should be a perfect square trinomial, as the square of a binomial. Add on the other side.

Step 4 Solve the equation using the square root property.

Examples

Solve $4x^2 + 8x - 1 = 0$.

$$4x^2 + 8x - 1 = 0$$
$$x^2 + 2x - \frac{1}{4} = 0 \quad \text{Divide by 4.}$$
$$x^2 + 2x = \frac{1}{4} \quad \text{Add } \tfrac{1}{4}.$$
$$x^2 + 2x + 1 = \frac{1}{4} + 1 \quad \left[\tfrac{1}{2}(2)\right]^4 = 1^2 = 1$$
$$(x + 1)^2 = \frac{5}{4} \quad \text{Factor. Add.}$$
$$x + 1 = \sqrt{\frac{5}{4}} \quad \text{or} \quad x + 1 = -\sqrt{\frac{5}{4}}$$
$$x + 1 = \frac{\sqrt{5}}{2} \quad \text{or} \quad x + 1 = -\frac{\sqrt{5}}{2}$$
$$x = -1 + \frac{\sqrt{5}}{2} \quad \text{or} \quad x = -1 - \frac{\sqrt{5}}{2}$$
$$x = \frac{-2}{2} + \frac{\sqrt{5}}{2} \quad \text{or} \quad x = \frac{-2}{2} - \frac{\sqrt{5}}{2}$$
$$x = \frac{-2 + \sqrt{5}}{2} \quad \text{or} \quad x = \frac{-2 - \sqrt{5}}{2}$$

Solution set: $\left\{\frac{-2 + \sqrt{5}}{2}, \frac{-2 - \sqrt{5}}{2}\right\}$

Concepts	*Examples*

9.3 Solving Quadratic Equations by the Quadratic Formula

Quadratic Formula

The quadratic equation $ax^2 + bx + c = 0$ (where $a \neq 0$) has solutions

$$x = \frac{-b \pm \sqrt{b^2 - 4ac}}{2a}.$$ $\pm$ indicates two solutions.

Solve $3x^2 - 4x - 2 = 0$.

$$x = \frac{-(-4) \pm \sqrt{(-4)^2 - 4(3)(-2)}}{2(3)}$$ Let $a = 3$, $b = -4$, $c = -2$.

$$x = \frac{4 \pm \sqrt{40}}{6}$$ Simplify.

$$x = \frac{4 \pm 2\sqrt{10}}{6}$$ $\sqrt{40} = \sqrt{4} \cdot \sqrt{10} = 2\sqrt{10}$

$$x = \frac{2(2 \pm \sqrt{10})}{2(3)}$$ Factor.

$$x = \frac{2 \pm \sqrt{10}}{3}$$ Divide out 2.

Solution set: $\left\{\frac{2 + \sqrt{10}}{3}, \frac{2 - \sqrt{10}}{3}\right\}$

9.4 Graphing Quadratic Equations

Graph $y = ax^2 + bx + c$ as follows.

- Find the vertex. Use $x = -\frac{b}{2a}$ and find y by substituting this value for x in the equation.
- Let $x = 0$ to find the y-intercept.
- Let $y = 0$ and solve to find any x-intercepts.
- Plot the intercepts, the vertex, and any additional ordered pairs as needed, and join these points with a smooth curve.

Graph $y = x^2 + 4x + 3$.

$$x = -\frac{b}{2a} = -\frac{4}{2(1)} = -2$$

$$y = (-2)^2 + 4(-2) + 3$$

$$y = -1$$

Vertex: $(-2, -1)$

$y = x^2 + 4x + 3$

$y = 0^2 + 4(0) + 3$

$y = 3$

y-intercept: $(0, 3)$

$0 = x^2 + 4x + 3$

$0 = (x + 3)(x + 1)$

$x + 3 = 0$ or $x + 1 = 0$

$x = -3$ or $x = -1$

x-intercepts: $(-3, 0)$ and $(-1, 0)$

9.5 Introduction to Functions

The set of all first components in the ordered pairs of a relation is the **domain**, and the set of all second components in the ordered pairs is the **range.**

The relation $\{(10, 5), (20, 15), (30, 25)\}$ has

domain $\{10, 20, 30\}$, and range $\{5, 15, 25\}$.

Vertical Line Test

If a vertical line intersects a graph in more than one point, then the graph is not the graph of a function.

By the vertical line test, the graph shown is not the graph of a function.

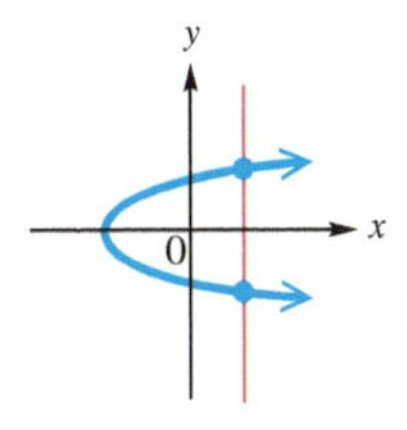

Function Notation

To find $f(x)$ for a specific value of x, replace x by that value in the expression for the function f.

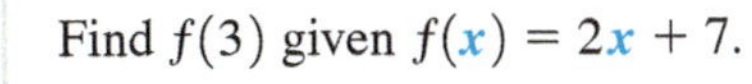

Find $f(3)$ given $f(x) = 2x + 7$.

$f(3) = 2(3) + 7$ Let $x = 3$.

$f(3) = 13$ Multiply. Add.

Chapter 9 Review Exercises

9.1 *Solve each equation using the zero-factor property.*

1. $x^2 + 3x - 28 = 0$
2. $x^2 + 14x + 45 = 0$
3. $2z^2 + 7z = 15$
4. $3x^2 - 5x + 2 = 0$
5. $x^2 = 100$
6. $r^2 - 169 = 0$

Solve each equation using the square root property. Express radicals in simplified form.

7. $x^2 = 144$
8. $x^2 = 37$
9. $m^2 = 128$
10. $(k + 2)^2 = 25$
11. $(r - 3)^2 = 10$
12. $(2p + 1)^2 = 14$
13. $(3k + 2)^2 = -3$
14. $(3x + 5)^2 = 0$

9.2 *Solve each equation by completing the square.*

15. $m^2 + 6m + 5 = 0$
16. $p^2 + 4p = 7$
17. $-x^2 + 5 = 2x$
18. $4x^2 - 3 = -8x$
19. $4(x^2 + 7x) + 29 = -20$
20. $(4x + 1)(x - 1) = -7$

Solve each problem. Give answers to the nearest tenth as needed.

21. If an object is projected upward on Earth from a height of 50 ft, with an initial velocity of 32 ft per sec, then its altitude (height) s, in feet, in t seconds is given by
$$s = -16t^2 + 32t + 50.$$
At what time(s) will the object be at a height of 30 ft?

22. If an object is projected upward from ground level on Earth with an initial velocity of 96 ft per sec, its altitude (height) s, in feet, in t seconds is given by
$$s = -16t^2 + 96t.$$
At what time(s) will the object be 100 ft above the ground?

9.3 *Solve each equation using the quadratic formula.*

23. $m^2 - 5m - 36 = 0$

24. $5r^2 = 14r$

25. $2x^2 - x + 3 = 0$

26. $4w^2 - 12w + 9 = 0$

27. $-4x^2 - 2x + 7 = 0$

28. $2x^2 + 8 = 4x + 11$

29. $x(5x - 1) = 1$

30. $\frac{1}{4}x^2 = 2 - \frac{3}{4}x$

31. $\frac{1}{2}x^2 + 3x = 5$

32. $0.2x^2 = 0.4x - 0.1$

9.4 *Graph each equation. Identify the vertex. In Exercises 35 and 36, also give the y-intercept and any x-intercepts (rounded to the nearest tenth, if necessary).*

33. $y = -x^2 + 5$

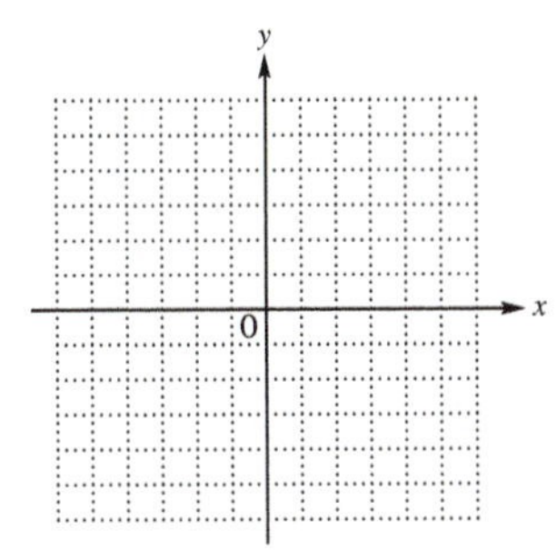

34. $y = (x - 1)^2$

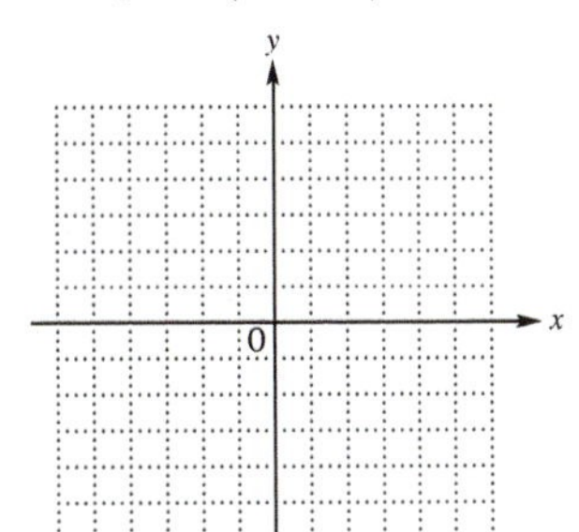

35. $y = -x^2 + 2x + 3$

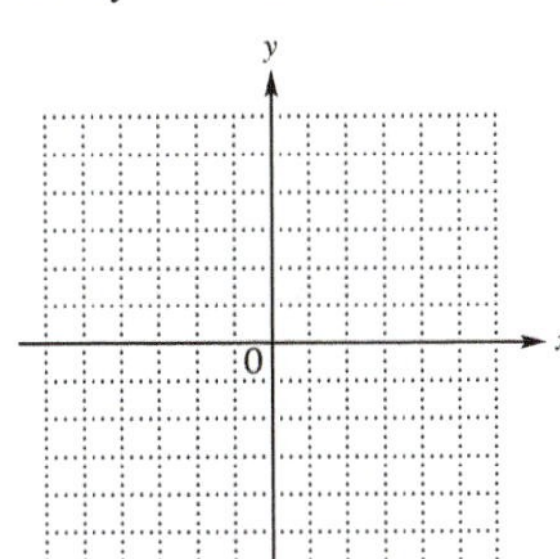

36. $y = x^2 + 4x + 2$

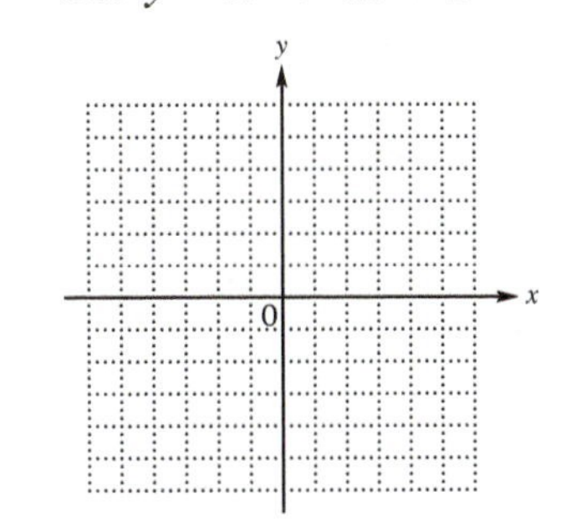

9.5 *Determine whether each relation is a function. In Exercises 37 and 38, give the domain and range of each relation.*

37. $\{(-2, 4), (0, 8), (2, 5), (2, 3)\}$

38. $\{(8, 3), (7, 4), (6, 5), (5, 6), (4, 7)\}$

39.

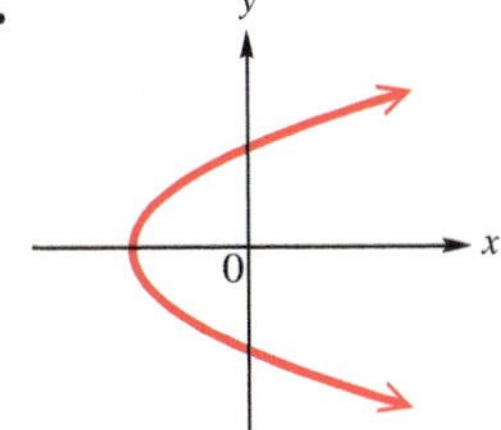

40.

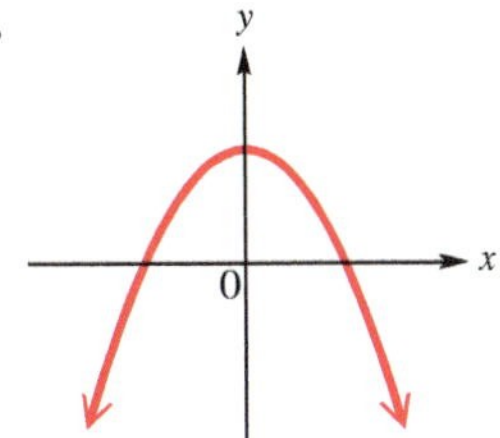

41. $2x + 3y = 12$

42. $x = -5$

For each function f, find **(a)** $f(2)$ *and* **(b)** $f(-1)$.

43. $f(x) = 3x + 2$

44. $f(x) = x - 3$

45. $f(x) = 2x^2 - 1$

46. $f(x) = -x^2 + 2x - 3$

Chapter 9 Mixed Review Exercises

Solve each quadratic equation using the method of your choice.

1. $(2t - 1)(t + 1) = 54$

2. $(2p + 1)^2 = 100$

3. $(k + 2)(k - 1) = 3$

4. $6t^2 + 7t - 3 = 0$

5. $2x^2 + 3x + 2 = x^2 - 2x$

6. $x^2 + 2x + 5 = 7$

7. $m^2 - 4m + 10 = 0$

8. $k^2 - 9k + 10 = 0$

9. $(5x + 6)^2 = 0$

10. $\frac{1}{2}r^2 = \frac{7}{2} - r$

11. $x^2 + 4x = 1$

12. $7x^2 - 8 = 5x^2 + 8$

Solve each problem.

13. Consider the quadratic equation $x^2 - 9 = 0$.

(a) Solve the equation using the zero-factor property.

(b) Solve the equation using the square root property.

(c) Solve the equation using the quadratic formula.

(d) Compare the answers in parts (a)–(c). If a quadratic equation can be solved using the zero-factor property and the quadratic formula, should we always obtain the same results? Explain.

14. Find the lengths of the three sides of the right triangle.

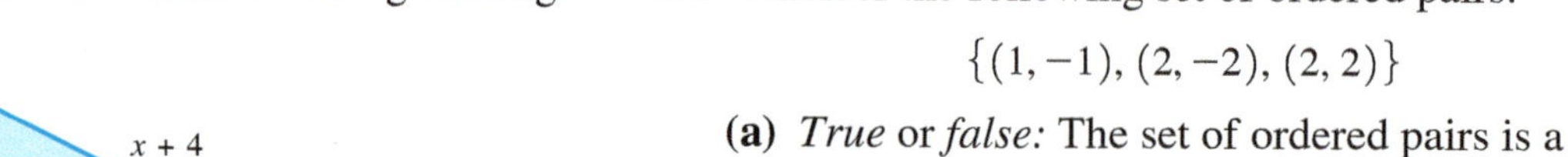

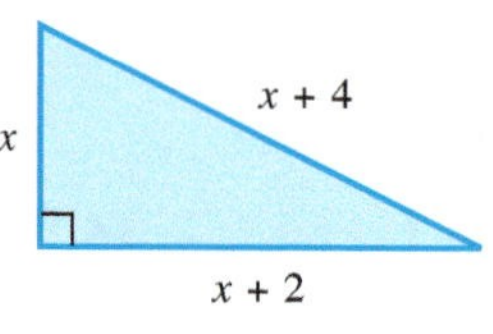

15. Consider the following set of ordered pairs.

$$\{(1, -1), (2, -2), (2, 2)\}$$

(a) *True* or *false:* The set of ordered pairs is a relation.

(b) *True* or *false:* The set of ordered pairs is a function.

(c) Give the domain and range.

Chapter 9 Test

The Chapter Test Prep Videos with step-by-step solutions are available in MyMathLab or on YouTube at https://goo.gl/8aAsWP

Solve using the zero-factor property.

1. $x^2 - 2x - 48 = 0$

2. $x^2 = 49$

3. $3x^2 - 7x + 4 = 0$

Solve using the square root property.

4. $x^2 = 39$

5. $(x + 3)^2 = 64$

6. $(4x + 3)^2 = 24$

Solve by completing the square.

7. $x^2 - 4x = 6$

8. $4x^2 + 12x + 3 = 0$

Solve using the quadratic formula.

9. $2x^2 + 5x - 3 = 0$

10. $3w^2 + 2 = 6w$

11. $4x^2 + 8x + 11 = 0$

12. $t^2 - \frac{5}{3}t + \frac{1}{3} = 0$

Solve using the method of your choice.

13. $p^2 - 2p - 1 = 0$

14. $(2x + 1)^2 = 18$

15. $(x - 5)(2x - 1) = 1$

16. $t^2 + 25 = 10t$

Solve each problem.

17. If an object is projected into the air from ground level with an initial velocity of 64 ft per sec, its altitude (height) s, in feet, in t seconds is given by the formula

$$s = -16t^2 + 64t.$$

At what time(s) will the object be at a height of 64 ft?

18. Find the lengths of the three sides of the right triangle.

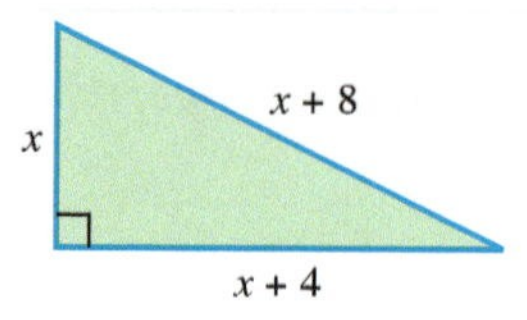

Graph each equation. Identify the vertex. In Exercise 21, also give the y-intercept and any x-intercepts.

19. $y = \frac{1}{3}x^2$

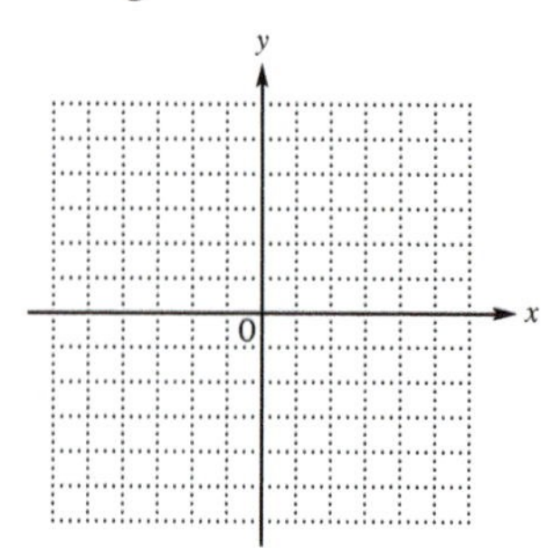

20. $y = (x - 3)^2$

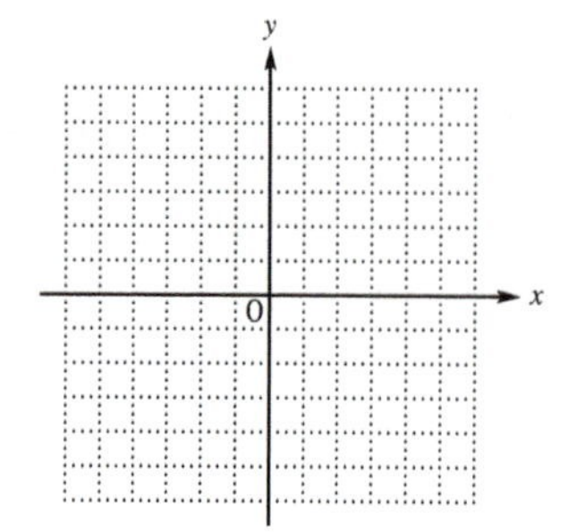

21. $y = -x^2 - 2x - 4$

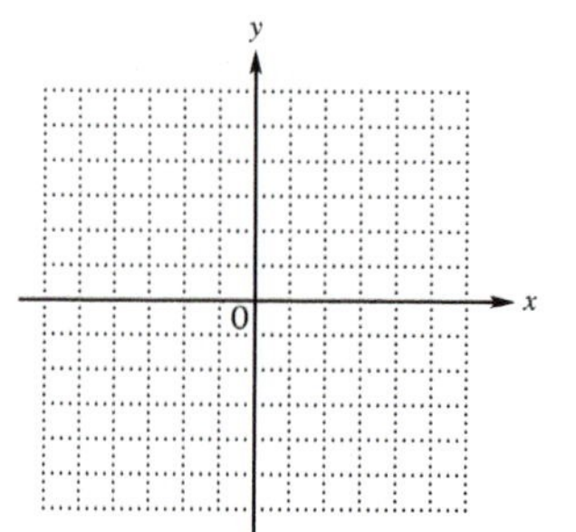

Determine whether each relation is a function. Give the domain and range.

22. $\{(2, 3), (2, 4), (2, 5)\}$

23. $\{(0, 2), (1, 2), (2, 2)\}$

24. Use the vertical line test to determine whether the graph is that of a function.

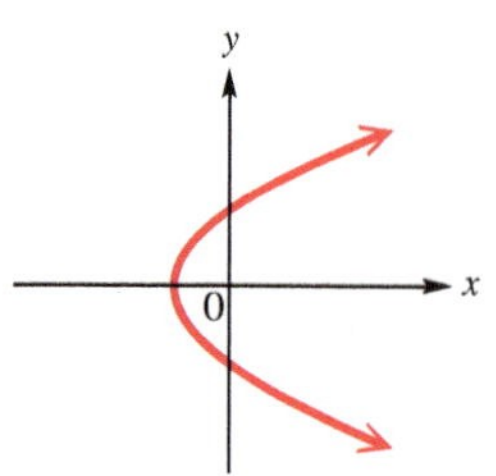

25. If $f(x) = 3x + 7$, find $f(-2)$.

Chapters R–9 Cumulative Review Exercises

Perform the indicated operations.

1. $\dfrac{-4 \cdot 3^2 + 2 \cdot 3}{2 - 4 \cdot 1}$

2. $|-3| - |1 - 6|$

3. $5(4m - 2) - (m + 7)$

Solve each equation.

4. $6x - 5 = 13$

5. $3k - 9k - 8k + 6 = -64$

6. $2(m - 1) - 6(3 - m) = -4$

7. The perimeter of a basketball court is 288 ft. The width of the court is 44 ft less than the length. What are the dimensions of the court?

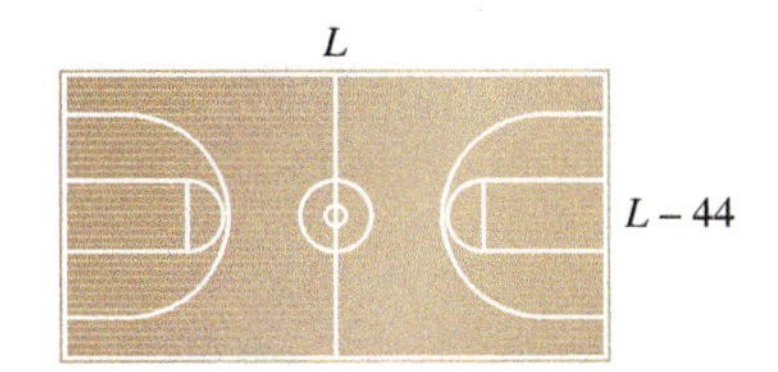

8. Find the measures of the marked angles.

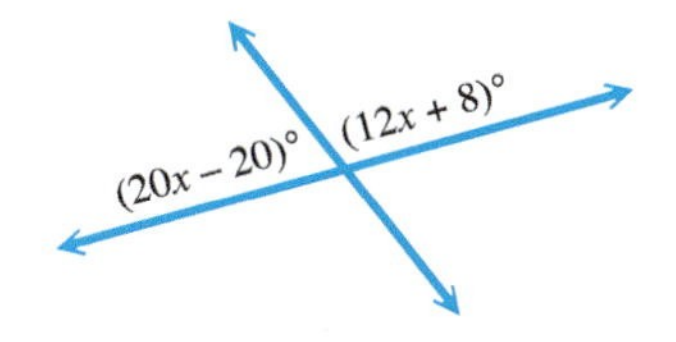

9. Solve the formula $P = 2L + 2W$ for L.

10. Solve. Give the solution set in interval and graph forms.

$$-9p + 2(8 - p) - 6 \geq 4p - 50$$

Graph each equation or inequality.

11. $2x + 3y = 6$

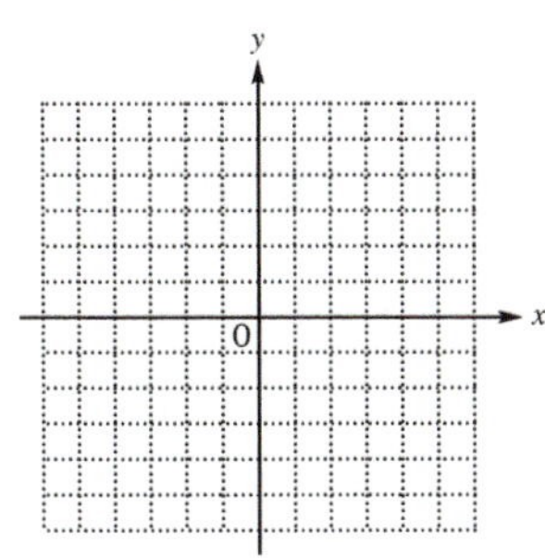

12. $y = 3$

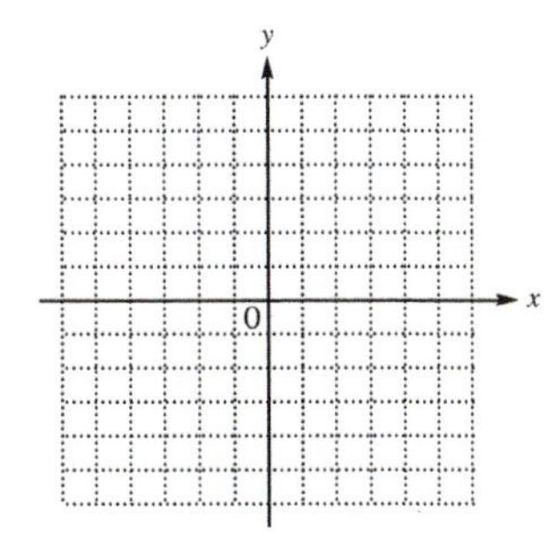

13. $2x - 5y < 10$

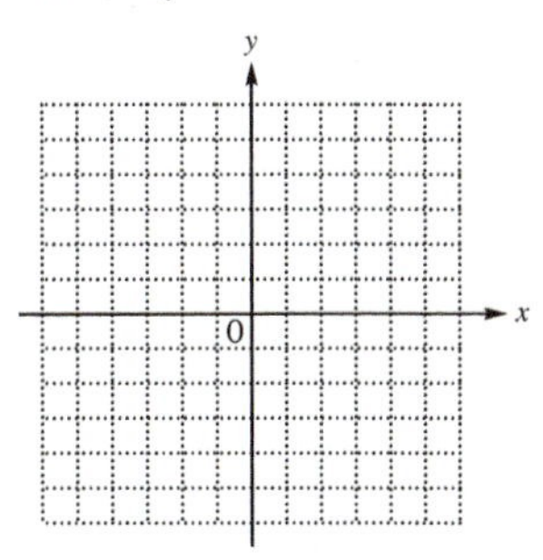

14. Find the slope of the line passing through $(-1, 4)$ and $(5, 2)$.

15. Write an equation of a line with slope 2 and y-intercept $(0, 3)$. Give it in standard form $Ax + By = C$.

Solve each system of equations.

16. $2x + y = -4$
$-3x + 2y = 13$

17. $3x - 5y = 8$
$-6x + 10y = 16$

18. Three Motorola Theory cell phones and two Motorola i412 cell phones cost \$379.95. Two Theory cell phones and three i412 phones cost \$369.95. Find the price for a single phone of each model. (Data from www.boostmobile.com)

Simplify each expression.

19. $2^0 + 2^{-1}$

20. $\left(\dfrac{b^{-3}c^4}{b^5c^3}\right)^2$

21. $\left(\dfrac{5}{3}\right)^{-3}$

Perform the indicated operations.

22. $(5x^5 - 9x^4 + 8x^2) - (9x^2 + 8x^4 - 3x^5)$

23. $(2x - 5)(x^3 + 3x^2 - 2x - 4)$

Factor completely.

24. $16x^3 - 48x^2y$

25. $2a^2 - 5a - 3$

26. $25m^2 - 20m + 4$

27. Solve $3x^2 = x + 4$.

28. The length of a rectangle is 2.5 times its width. The area is 1000 m^2. Find the length.

Perform the indicated operations. Write all answers in lowest terms.

29. $\dfrac{2}{a-3} \div \dfrac{5}{2a-6}$

30. $\dfrac{1}{k} - \dfrac{2}{k-1}$

31. $\dfrac{6 + \dfrac{1}{x}}{3 - \dfrac{1}{x}}$

Solve each equation.

32. $\dfrac{1}{x+3} + \dfrac{1}{x} = \dfrac{7}{10}$

33. $\sqrt{x+2} = x - 4$

34. $2x^2 - 2x = 1$

Simplify each expression.

35. $\dfrac{6\sqrt{6}}{\sqrt{5}}$

36. $3\sqrt{5} - 2\sqrt{20} + \sqrt{125}$

37. $\sqrt[3]{16a^3b^4} - \sqrt[3]{54a^3b^4}$

38. Graph the equation $y = x^2 - 4x$. Identify the vertex. What are the x-intercepts?

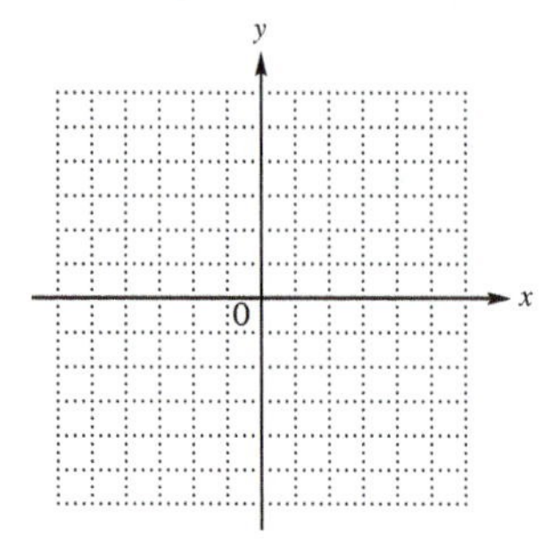

39. Consider the relation

$$\{(0, 4), (1, 2), (3, 5)\}.$$

(a) Is this relation a function?

(b) Give the domain and range.

40. If $f(x) = -2x + 7$, find $f(-2)$.

Appendix: Factoring Sums and Differences of Cubes

OBJECTIVES

1. Factor a difference of cubes.
2. Factor a sum of cubes.

OBJECTIVE 1 Factor a difference of cubes. In a *difference of cubes* $x^3 - y^3$, both terms of the binomial must be **perfect cubes**, such as

$$x^3, \quad 8p^3 = (2p)^3, \quad s^6 = (s^2)^3, \quad 1 = 1^3, \quad 27 = 3^3, \quad 216 = 6^3.$$

We can factor a **difference of cubes** using the following rule.

Factoring a Difference of Cubes

$$x^3 - y^3 = (x - y)(x^2 + xy + y^2)$$

To see that the rule for factoring a difference of cubes is correct, multiply

$$(x - y)(x^2 + xy + y^2).$$

$$\begin{array}{rl} x^2 + xy + y^2 & \text{Multiply} \\ x - y & \text{vertically.} \\ \hline -x^2y - xy^2 - y^3 & \\ x^3 + x^2y + xy^2 \qquad & \\ \hline x^3 \qquad\qquad\quad - y^3 & \text{Add.} \end{array}$$

Notice the pattern of the terms in the factored form of $x^3 - y^3$.

- $x^3 - y^3$ factors as (a binomial factor) • (a trinomial factor).
- The binomial factor has the difference of the cube roots of the given terms. (*Note*: A **cube root** of 1 is 1 because $1^3 = 1$, a **cube root** of 8 is 2 because $2^3 = 8$, and so on.)
- The terms in the trinomial factor are all positive.
- The terms in the binomial factor help determine the trinomial factor.

$$x^3 - y^3 = (x - y)(\underbrace{x^2}_{\text{First term squared}} + \underbrace{xy}_{\text{Positive product of the terms}} + \underbrace{y^2}_{\text{Second term squared}})$$

EXAMPLE 1 Factoring Differences of Cubes

Factor each binomial.

$$x^3 - y^3 = (x - y)(x^2 + xy + y^2)$$

(a) $m^3 - 125 = m^3 - 5^3 = (m - 5)(m^2 + 5m + 5^2)$ Let $x = m$ and $y = 5$.

$= (m - 5)(m^2 + 5m + 25)$ $5^2 = 25$

(b) $8p^3 - 27$

$= (2p)^3 - 3^3$ $8p^3 = (2p)^3$ and $27 = 3^3$.

$= (2p - 3)[(2p)^2 + (2p)3 + 3^2]$ Let $x = 2p$ and $y = 3$.

$= (2p - 3)(4p^2 + 6p + 9)$ Apply the exponents. Multiply.

$(2p)^2 = 2^2p^2 = 4p^2$, **not** $2p^2$.

Continued on Next Page

(c) $4m^3 - 32n^3$

$= 4(m^3 - 8n^3)$ Factor out the common factor.

$= 4[m^3 - (2n)^3]$ $8n^3 = (2n)^3$

$= 4(m - 2n)[m^2 + m(2n) + (2n)^2]$ Let $x = m$ and $y = 2n$.

$= 4(m - 2n)(m^2 + 2mn + 4n^2)$ Apply the exponents. Multiply.

◀ **Work Problem 1 at the Side.**

CAUTION

A common error in factoring $x^3 - y^3 = (x - y)(x^2 + xy + y^2)$ is to try to factor $x^2 + xy + y^2$. This is usually not possible.

OBJECTIVE 2 Factor a sum of cubes. A *sum of squares,* such as $m^2 + 25$, *cannot* be factored using real numbers, but **a sum of cubes** can.

Factoring a Sum of Cubes

$$x^3 + y^3 = (x + y)(x^2 - xy + y^2)$$

Compare the rule for the *sum* of cubes with that for the *difference* of cubes.

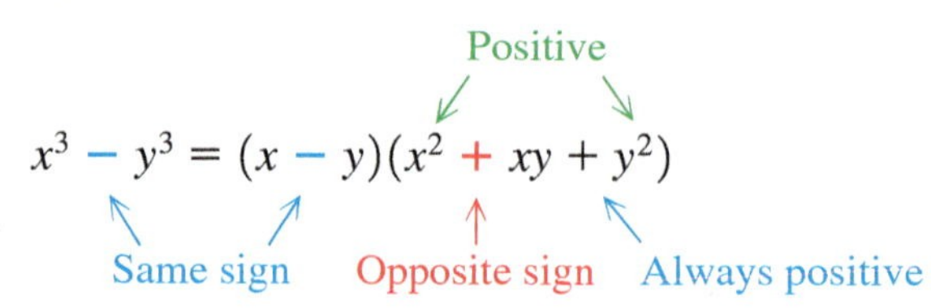

Difference of cubes

The only difference between the rules is the positive and negative signs.

Positive

$x^3 + y^3 = (x + y)(x^2 - xy + y^2)$

Same sign Opposite sign Always positive

Sum of cubes

EXAMPLE 2 Factoring Sums of Cubes

Factor each binomial.

(a) $k^3 + 27$

$= k^3 + 3^3$ $27 = 3^3$

$= (k + 3)(k^2 - 3k + 3^2)$ Let $x = k$ and $y = 3$.

$= (k + 3)(k^2 - 3k + 9)$ Apply the exponent.

(b) $8m^3 + 125p^3$

$= (2m)^3 + (5p)^3$ $8m^3 = (2m)^3$; $125p^3 = (5p)^3$

$= (2m + 5p)[(2m)^2 - (2m)(5p) + (5p)^2]$ Let $x = 2m$ and $y = 5p$.

$= (2m + 5p)(4m^2 - 10mp + 25p^2)$ Apply the exponents. Multiply.

◀ **Work Problem 2 at the Side.**

1 Factor each binomial.

GS **(a)** $t^3 - 64$

$= \_\_\_^3 - \_\_\_^3$

$= (\_\_\_ - \_\_\_) \cdot (\_\_\_^2 + \_\_\_ + \_\_\_^2)$

$= \_\_\_\_\_\_$

(b) $27x^3 - 8$

(c) $2x^3 - 432y^3$

2 Factor each binomial.

GS **(a)** $x^3 + 8$

$= \_\_\_^3 + \_\_\_^3$

$= (\_\_\_ + \_\_\_) \cdot (\_\_\_^2 - \_\_\_ + \_\_\_^2)$

$= \_\_\_\_\_\_$

(b) $64y^3 + 1$

(c) $27m^3 + 343n^3$

Answers

1. (a) t; 4; t; 4; t; $4t$; 4; $(t - 4)(t^2 + 4t + 16)$
   (b) $(3x - 2)(9x^2 + 6x + 4)$
   (c) $2(x - 6y)(x^2 + 6xy + 36y^2)$
2. (a) x; 2; x; 2; x; $2x$; 2; $(x + 2)(x^2 - 2x + 4)$
   (b) $(4y + 1)(16y^2 - 4y + 1)$
   (c) $(3m + 7n)(9m^2 - 21mn + 49n^2)$

Appendix Exercises

FOR EXTRA HELP — *Go to* MyMathLab *for worked-out, step-by-step solutions to exercises enclosed in a square and video solutions to ▶ exercises.*

CONCEPT CHECK *Work each problem.*

1. To help factor the sum or difference of cubes, complete the list of perfect cubes.

$1^3 =$ ____ $2^3 =$ ____ $3^3 =$ ____ $4^3 =$ ____ $5^3 =$ ____

$6^3 =$ ____ $7^3 =$ ____ $8^3 =$ ____ $9^3 =$ ____ $10^3 =$ ____

2. The following powers of x are all perfect cubes: $x^3, x^6, x^9, x^{12}, x^{15}$. Based on this observation, we may make a conjecture that if the power of a variable is divisible by ____ (with 0 remainder), then we have a perfect cube.

3. Which of the following are differences of cubes?

A. $9x^3 - 125$ **B.** $x^3 - 16$ **C.** $x^3 - 1$ **D.** $8x^3 - 27y^3$

4. Which of the following are sums of cubes?

A. $x^3 - 1$ **B.** $x^3 + 36$ **C.** $12x^3 + 27$ **D.** $64x^3 + 216y^3$

Factor each binomial completely. ***See Examples 1 and 2.***

5. $a^3 - 1$ ▶

6. $m^3 - 8$

7. $m^3 + 8$ ▶

8. $x^3 + 1$

9. $y^3 - 216$

10. $x^3 - 343$

11. $k^3 + 1000$

12. $p^3 + 512$

13. $27x^3 - 1$

14. $64y^3 - 27$

15. $125x^3 + 8$

16. $216x^3 + 125$

17. $y^3 - 8x^3$

18. $w^3 - 216z^3$

19. $27x^3 - 64y^3$

20. $125m^3 - 8n^3$

21. $8p^3 + 729q^3$

22. $27x^3 + 1000y^3$

23. $16t^3 - 2$

24. $3p^3 - 81$

25. $40w^3 + 135$

26. $32z^3 + 500$

27. $x^3 + y^6$

28. $p^9 + q^3$

29. $125k^3 - 8m^9$

30. $125c^6 - 216d^3$

Answers to Selected Exercises

In this section we provide the answers that we think most students will obtain when they work the exercises using the methods explained in the text. If your answer does not look exactly like the one given here, it is not necessarily wrong. In many cases there are equivalent forms of the answer that are correct. For example, if the answer section shows $\frac{3}{4}$ and your answer is 0.75, you have obtained the correct answer but written it in a different (yet equivalent) form. Unless the directions specify otherwise, 0.75 is just as valid an answer as $\frac{3}{4}$.

In general, if your answer does not agree with the one given in the text, see whether it can be transformed into the other form. If it can, then it is the correct answer. If you still have doubts, talk with your instructor.

CHAPTER R Prealgebra Review

SECTION R.1

1. true **2.** true **3.** false; This is an improper fraction. Its value is 1. **4.** false; The number 1 is neither prime nor composite. **5.** false; The fraction $\frac{17}{51}$ can be simplified to $\frac{1}{3}$. **6.** false; The reciprocal of $\frac{8}{2} = 4$ is $\frac{2}{8} = \frac{1}{4}$. **7.** false; *Product* indicates multiplication, so the product of 8 and 2 is 16. **8.** false; *Difference* indicates subtraction, so the difference of 12 and 2 is 10. **9.** prime **11.** composite **13.** composite **15.** prime **17.** $2 \cdot 3 \cdot 5$ **19.** $3 \cdot 19$ **21.** $2 \cdot 2 \cdot 31$ **23.** $2 \cdot 2 \cdot 3 \cdot 3 \cdot 7$ **25.** $2 \cdot 2 \cdot 5 \cdot 5 \cdot 5$ **27.** $\frac{5}{6}$ **29.** $\frac{1}{2}$ **31.** $\frac{3}{10}$ **33.** $\frac{1}{5}$ **35.** $\frac{3}{5}$ **37.** $6\frac{5}{12}$ **39.** $7\frac{6}{11}$ **41.** $1\frac{5}{7}$ **43.** $\frac{13}{5}$ **45.** $\frac{83}{8}$ **47.** $\frac{51}{5}$ **49.** A **50.** C **51.** $\frac{24}{35}$ **53.** $\frac{1}{8}$ **55.** $\frac{6}{25}$ **57.** $\frac{6}{5}$, or $1\frac{1}{5}$ **59.** $\frac{65}{12}$, or $5\frac{5}{12}$ **61.** $\frac{38}{5}$, or $7\frac{3}{5}$ **63.** $\frac{18}{35}$ **65.** $\frac{10}{3}$, or $3\frac{1}{3}$ **67.** 12 **69.** $\frac{1}{16}$ **71.** $\frac{24}{35}$ **73.** $\frac{84}{47}$, or $1\frac{37}{47}$ **75.** $\frac{11}{15}$ **77.** $\frac{2}{3}$ **79.** $\frac{8}{9}$ **81.** $\frac{29}{24}$, or $1\frac{5}{24}$ **83.** $\frac{43}{8}$, or $5\frac{3}{8}$ **85.** $\frac{5}{9}$ **87.** $\frac{2}{3}$ **89.** $\frac{17}{36}$ **91.** $\frac{11}{12}$ **93.** $\frac{4}{3}$, or $1\frac{1}{3}$ **95.** 2 cups **97.** $618\frac{3}{4}$ ft **99.** $\frac{9}{16}$ in. **101.** $\frac{5}{16}$ in. **103.** 8 cakes (There will be some sugar left over.) **105.** $16\frac{5}{8}$ yd **107.** 10 million, or 10,000,000 **109.** $4\frac{4}{5}$ million, or 4,800,000 **111.** $\frac{3}{50}$

SECTION R.2

1. **(a)** 6 **(b)** 9 **(c)** 1 **(d)** 7 **(e)** 4 **2.** **(a)** 46.25 **(b)** 46.2 **(c)** 46 **(d)** 50 **3.** **(a)** 0.889 **(b)** 0.556 **(c)** 0.976 **(d)** 0.864 **4.** **(a)** 254 **(b)** 2540 **(c)** 0.254 **(d)** 0.0254 **5.** $\frac{4}{10}$ **7.** $\frac{64}{100}$ **9.** $\frac{138}{1000}$ **11.** $\frac{43}{1000}$ **13.** $\frac{3805}{1000}$ **15.** 143.094 **17.** 25.61 **19.** 15.33 **21.** 21.77 **23.** 81.716 **25.** 15.211 **27.** 116.48 **29.** 0.006 **31.** 7.15 **33.** 2.05 **35.** 5711.6 **37.** 94 **39.** 0.162 **41.** 0.2403 **43.** 1% **44.** 0.02 **45.** $\frac{1}{20}$ **46.** 0.1; 10% **47.** $12\frac{1}{2}$%, or 12.5% **48.** $\frac{1}{5}$; 0.2 **49.** 0.25; 25% **50.** $0.\overline{3}$; $33\frac{1}{3}$%, or $33.\overline{3}$% **51.** $\frac{1}{2}$; 0.5 **52.** $0.\overline{6}$ **53.** $\frac{3}{4}$; 75% **54.** 100% **55.** 0.125 **57.** 1.25 **59.** $0.\overline{5}$; 0.556 **61.** $0.1\overline{6}$; 0.167 **63.** 0.54 **65.** 0.07 **67.** 0.9 **69.** 1.17 **71.** 0.024 **73.** 0.0625 **75.** 0.008 **77.** 73% **79.** 2% **81.** 0.4% **83.** 128% **85.** 30% **87.** 600% **89.** $\frac{51}{100}$ **91.** $\frac{3}{20}$ **93.** $\frac{1}{50}$ **95.** $\frac{7}{5}$, or $1\frac{2}{5}$ **97.** $\frac{3}{40}$ **99.** 80% **101.** 14% **103.** $18\frac{2}{11}$%, or $18.\overline{18}$% **105.** 225% **107.** $216\frac{2}{3}$%, or $216.\overline{6}$% **109.** 160 **111.** 4.8 **113.** 109.2 **115.** \$17.80; \$106.80 **117.** \$119.25; \$675.75 **119.** 19.8 million, or 19,800,000 **121.** 13%

CHAPTER 1 The Real Number System

SECTION 1.1

1. false; $3^2 = 3 \cdot 3 = 9$ **2.** false; 1 raised to *any* power is 1. Here, $1^3 = 1 \cdot 1 \cdot 1 = 1$. **3.** false; A number raised to the first power is that number, so $3^1 = 3$. **4.** false; 6^2 means that 6 is used as a factor 2 times, so $6^2 = 6 \cdot 6 = 36$. **5.** 49 **7.** 9 **9.** 144 **11.** 64 **13.** 1000 **15.** 81 **17.** 1024 **19.** $\frac{16}{81}$ **21.** $\frac{1}{36}$ **23.** 0.000064 **25.** ②①; 45; 58 **27.** ②①③; 12; 8; 13 **29.** ①②; 32 **31.** ①②; 19 **33.** 32 **35.** 12 **37.** $\frac{49}{30}$ **39.** 22.2 **41.** 26 **43.** 4 **45.** 42 **47.** 5 **49.** 95 **51.** 90 **53.** 41 **55.** 14 **57.** $\frac{19}{2}$ **59.** $3 \cdot (6 + 4) \cdot 2$ **60.** $2 \cdot (8 - 1) \cdot 3$ **61.** $10 - (7 - 3)$ **62.** $15 - (10 - 2)$ **63.** $(8 + 2)^2$ **64.** $(4 + 2)^2$ **65.** false **67.** true **69.** true **71.** true **73.** true **75.** false **77.** false **79.** true **81.** $15 = 5 + 10$ **83.** $9 > 5 - 4$ **85.** $16 \neq 19$ **87.** $2 \leq 3$ **89.** Seven is less than nineteen; true **91.** Eight is greater than or equal to eleven; false **93.** One-third is not equal to three-tenths; true **95.** $30 > 5$ **97.** $3 \leq 12$ **99.** $1.3 \leq 2.5$ **101.** $\frac{3}{4} < \frac{4}{5}$ **103.** **(a)** $14.7 - 40 \cdot 0.13$ **(b)** 9.5 **(c)** 8.075; walking (5 mph) **(d)** $14.7 - 55 \cdot 0.11$; 8.65; 7.3525, swimming **105.** Alaska, Texas, California, Idaho **107.** Alaska, Texas, California, Idaho, Missouri

SECTION 1.2

1. B **2.** C **3.** A **4.** B, C **5.** 11 **6.** 10 **7.** $13 + x$; 16 **8.** expression; equation **9.** **(a)** 11 **(b)** 13 **11.** **(a)** 16 **(b)** 24 **13.** **(a)** 22 **(b)** 32 **15.** **(a)** 64 **(b)** 144 **17.** **(a)** $\frac{5}{3}$ **(b)** $\frac{7}{3}$ **19.** **(a)** $\frac{7}{8}$ **(b)** $\frac{13}{12}$ **21.** **(a)** 25.836 **(b)** 38.754 **23.** **(a)** 52 **(b)** 114 **25.** **(a)** 24 **(b)** 3 **27.** **(a)** 24 **(b)** 28 **29.** **(a)** 12 **(b)** 33 **31.** **(a)** $\frac{4}{3}$ **(b)** $\frac{13}{6}$ **33.** **(a)** $\frac{2}{7}$ **(b)** $\frac{16}{27}$ **35.** **(a)** 12 **(b)** 55 **37.** **(a)** 1 **(b)** $\frac{28}{17}$ **39.** **(a)** 3.684 **(b)** 8.841 **41.** $12x$ **43.** $x + 13$ **45.** $x - 2$ **47.** $2x - 6$ **49.** $7 - \frac{1}{3}x$ **51.** $\frac{12}{x + 3}$ **53.** $6(x - 4)$

55. no **57.** yes **59.** yes **61.** no **63.** yes **65.** yes **67.** $x + 8 = 18$ **69.** $2x + 5 = 5$ **71.** $16 - \frac{3}{4}x = 13$ **73.** $3x = 2x + 8$ **75.** $\frac{x}{3} = x - 4$ **77.** expression **79.** equation **81.** equation **83.** expression **85.** 70 yr **86.** 73 yr **87.** 76 yr **88.** 79 yr

SECTION 1.3

1. 0 **2.** integers **3.** positive **4.** right **5.** quotient; denominator **6.** irrational **7.** 1,212,795 **9.** −3413 **11.** −6 −5 0 3

13. **15.**

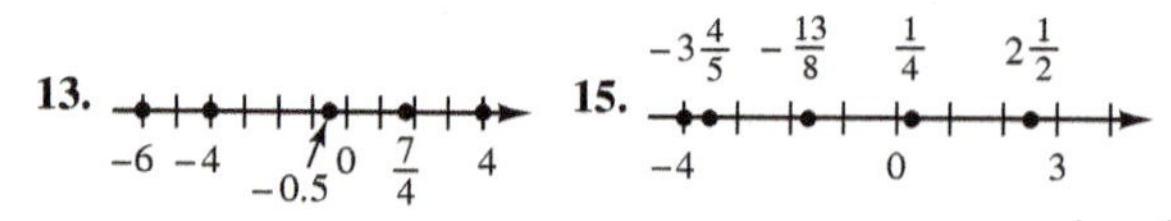

17. **(a)** 3, 7 **(b)** 0, 3, 7 **(c)** −9, 0, 3, 7 **(d)** $-9, -1\frac{1}{4}, -\frac{3}{5}, 0, 3, 5.9, 7$ **(e)** $-\sqrt{7}, \sqrt{5}$ **(f)** All are real numbers. **19.** **(a)** 11 **(b)** 0, 11 **(c)** 0, 11, −6 **(d)** $\frac{7}{9}, -2.\overline{3}, 0, -8\frac{3}{4}, 11, -6$ **(e)** $\sqrt{3}, \pi$ **(f)** All are real numbers. **21.** 4 **22.** One example is 2.85. There are others. **23.** 0 **24.** One example is 4. There are others. **25.** One example is $\sqrt{13}$. There are others. **26.** 0 **27.** true **28.** false **29.** true **30.** true **31.** false **32.** true *In Exercises 33–38, answers will vary.* **33.** $\frac{1}{2}, \frac{5}{8}, 1\frac{3}{4}$ **34.** $-1, -\frac{3}{4}, -5$ **35.** $-3\frac{1}{2}, -\frac{2}{3}, \frac{3}{7}$ **36.** $\frac{1}{2}, -\frac{2}{3}, \frac{2}{7}$ **37.** $\sqrt{5}, \pi, -\sqrt{3}$ **38.** $\frac{2}{3}, \frac{5}{6}, \frac{5}{2}$ **39.** −11 **41.** −21 **43.** −100 **45.** $-\frac{2}{3}$ **47.** false **49.** true **51.** false **53.** false **55.** **(a)** 2 **(b)** 2 **57.** **(a)** −6 **(b)** 6 **59.** **(a)** $\frac{3}{4}$ **(b)** $\frac{3}{4}$ **61.** **(a)** −4.95 **(b)** 4.95 **63.** **(a)** A **(b)** A **(c)** B **(d)** B **64.** 5; 5; 5; −5 **65.** 6 **67.** −12 **69.** $-\frac{2}{3}$ **71.** 9 **73.** −3 **75.** false **77.** true **79.** true **81.** false **83.** Energy, 2014 to 2015 **85.** Apparel, 2013 to 2014

SECTION 1.4

1. negative; −5 **2.** negative; −2 **3.** zero (0)

4. −3; 5 **5.** −10 **7.** −13 **9.** 2 **11.** −3 **13.** −9 **15.** 0 **17.** $-\frac{3}{5}$ **19.** $\frac{1}{2}$ **21.** $-\frac{19}{40}$ **23.** $-\frac{3}{4}$ **25.** −15.9 **27.** −1.6 **29.** 5 **31.** 13 **33.** 0 **35.** −8 **37.** −8.7 **39.** −11; −14; −25 **41.** 6 **43.** 22 **45.** $-\frac{1}{4}$, or −0.25 **47.** false **49.** true **51.** false **53.** true **55.** false **57.** $-5 + 12 + 6$; 13 **59.** $[-19 + (-4)] + 14$; −9 **61.** $[-4 + (-10)] + 12$; −2 **63.** $\left[\frac{5}{7} + \left(-\frac{9}{7}\right)\right] + \frac{2}{7}; -\frac{2}{7}$ **65.** −12 **67.** +4 **69.** −184 m **71.** −12 **73.** 17 **75.** 37 yd **77.** 120°F

SECTION 1.5

1. −8; −6; 2 **2.** additive; inverse; opposite **3.** 5; −4 **4.** $7 - 12$; $12 - 7$ **5.** −4 **7.** −3 **9.** −3; −10 **11.** −16 **13.** 4; 11 **15.** 19 **17.** −4 **19.** 9 **21.** $\frac{3}{4}$ **23.** $-\frac{11}{8}$ **25.** 13.6 **27.** −11.9 **29.** 10 **31.** −5 **33.** 5 **35.** 0 **37.** 8 **39.** −4 **41.** −20 **43.** 22 **45.** −2 **47.** −21 **49.** $-\frac{5}{8}$ **51.** $\frac{37}{12}$ **53.** −2.8 **55.** −6.3 **57.** −28 **59.** −42.04 **61.** negative **62.** positive **63.** positive **64.** negative **65.** positive **66.** negative **67.** $4 - (-8)$; 12 **69.** $-2 - 8$; −10 **71.** $[9 + (-4)] - 7$; −2 **73.** $[8 - (-5)] - 12$; 1 **75.** −69°F **77.** 14,776 ft **79.** −176.9°F **81.** $1045.55 **83.** $323.83 **85.** 14 ft **87.** $710 billion **89.** −$22,600 **91.** −$6900 **93.** 50,395 ft **95.** 1345 ft **97.** 136 ft

SECTION 1.6

1. greater than 0 **2.** less than 0 **3.** less than 0 **4.** less than 0 **5.** greater than 0 **6.** less than 0 **7.** undefined; 0; Examples will vary. For instance, $\frac{1}{0}$ is undefined, and $\frac{0}{1}$ equals 0. **8.** A **9.** −28 **11.** 30 **13.** 0 **15.** $\frac{5}{6}$ **17.** −2.38 **19.** $\frac{3}{2}$ **21.** −3 **23.** −6 **25.** −4 **27.** 16 **29.** 0 **31.** undefined **33.** $-\frac{15}{16}$ **35.** $\frac{3}{2}$ **37.** −11 **39.** −2 **41.** 35 **43.** 13 **45.** −22 **47.** −8 **49.** 18 **51.** 3 **53.** 7 **55.** 4 **57.** −2 **59.** −3 **61.** −1 **63.** $\frac{7}{4}$ **65.** 68 **67.** 72 **69.** 1 **71.** 0 **73.** −6 **75.** undefined **77.** $-12 + 4(-7)$; −40 **79.** $-4 - 2(-1)(6)$; 8 **81.** $-3[3 - (-7)]$; −30 **83.** $\frac{3}{10}[-2 + (-28)]$; −9 **85.** $0.20(-5 \cdot 6)$; −6 **87.** $\frac{-12}{-5 + (-1)}$; 2 **89.** $\frac{-18 + (-6)}{2(-4)}$; 3 **91.** $\frac{-\frac{2}{3}\left(-\frac{1}{5}\right)}{\frac{1}{7}}; \frac{14}{15}$ **93.** $9x = -36$ **95.** $\frac{x}{4} = -1$ **97.** $x - \frac{9}{11} = 5$ **99.** $\frac{6}{x} = -3$ **101.** 29 **102.** The incorrect answer, 92, was obtained by performing all of the operations in order from left to right rather than following the rules for order of operations. The multiplications and divisions need to be done in order, before the additions and subtractions. **103.** 42 **104.** 5 **105.** $8\frac{2}{5}$ **106.** $8\frac{2}{5}$ **107.** 2 **108.** $-12\frac{1}{2}$

SUMMARY EXERCISES Performing Operations With Real Numbers

1. −16 **2.** 4 **3.** 0 **4.** −24 **5.** −17 **6.** 76 **7.** −18 **8.** 90 **9.** 38 **10.** 4 **11.** −5 **12.** 5 **13.** $-\frac{7}{2}$, or $-3\frac{1}{2}$ **14.** 4 **15.** 13 **16.** $\frac{5}{4}$, or $1\frac{1}{4}$ **17.** 9 **18.** $\frac{37}{10}$, or $3\frac{7}{10}$ **19.** 0 **20.** 25 **21.** 14 **22.** 0 **23.** −4 **24.** $\frac{6}{5}$, or $1\frac{1}{5}$ **25.** −1 **26.** $\frac{52}{37}$, or $1\frac{15}{37}$ **27.** $\frac{17}{16}$, or $1\frac{1}{16}$ **28.** $-\frac{2}{3}$ **29.** 3.33 **30.** 1.02 **31.** 0 **32.** 24 **33.** −7 **34.** −3 **35.** −1 **36.** $\frac{1}{2}$ **37.** $-\frac{5}{13}$ **38.** 5 **39.** undefined **40.** 0

SECTION 1.7

1. **(a)** B **(b)** F **(c)** C **(d)** I **(e)** B **(f)** D, F **(g)** B **(h)** A **(i)** G **(j)** H **2.** order; grouping **3.** yes **4.** yes **5.** no **6.** no **7.** no **8.** no **9.** (foreign sales) clerk; foreign (sales clerk) **10.** (defective merchandise) counter; defective (merchandise counter) **11.** −15; commutative property **13.** 3; commutative property **15.** 6; associative property **17.** 7; associative property **19.** Subtraction is not associative.

ANSWERS

20. Division is not associative. **21.** row 1: $-5, \frac{1}{5}$; row 2: $10, -\frac{1}{10}$; row 3: $\frac{1}{2}, -2$; row 4: $-\frac{3}{8}, \frac{8}{3}$; row 5: $-x, \frac{1}{x}\ (x \neq 0)$; row 6: $y, -\frac{1}{y}\ (y \neq 0)$; opposite; the same **22.** identity property **23.** commutative property **25.** inverse property **27.** inverse property **29.** identity property **31.** distributive property **33.** identity property **35.** associative property **37.** commutative property **39.** distributive property **41.** $7 + r$ **43.** s **45.** $-6x + (-6)7; -6x - 42$ **47.** $w + [5 + (-3)]; w + 2$ **49.** 2010 **51.** 6700 **53.** 50 **55.** 2 **57.** 0.77 **59.** 11 **61.** The expression following the first equality symbol should be $-3(4) - 3(-6)$. This simplifies to $-12 + 18$, which equals 6. **62.** The expression following the second equality symbol should be $-1(3x) + (-1)4$. This simplifies to $-3x - 4$. **63.** We must multiply $\frac{3}{4}$ by 1 in the form of a fraction, $\frac{3}{3}$: $\frac{3}{4} \cdot \frac{3}{3} = \frac{9}{12}$. **64.** This is the reverse of the procedure in **Exercise 63.** We factor the numerator and denominator, writing the identity element 1 as $\frac{3}{3}$: $\frac{9}{12} = \frac{3}{4} \cdot \frac{3}{3} = \frac{3}{4} \cdot 1 = \frac{3}{4}$.
65. 85 **67.** $4t + 12$ **69.** $7z - 56$ **71.** $-8r - 24$ **73.** $-2x - \frac{3}{4}$ **75.** $y; -4; -5y + 20$ **77.** $12x + 10$ **79.** $-6x + 15$ **81.** $-48x - 6$ **83.** $-16y - 20z$ **85.** $24r + 32s - 40y$ **87.** $-24x - 9y - 12z$ **89.** $-6x - 5$ **91.** $-4t - 3m$ **93.** $5c + 4d$ **95.** $3q - 5r + 8s$

SECTION 1.8

1. B **2.** C **3.** A **4.** B **5.** $4r + 11$ **7.** $21x - 28y$ **9.** $5 + 2x - 6y$ **11.** $-7 + 3p$ **13.** $2 - 3x$ **15.** -12 **17.** 5 **19.** 1 **21.** -1 **23.** 10 **25.** 28 **27.** $-\frac{3}{8}$ **29.** $\frac{1}{2}$ **31.** $\frac{2}{5}$ **33.** -1.28 **35.** like **37.** unlike **39.** unlike **41.** like **43.** unlike **45.** $15x$ **47.** $-9x$ **49.** $13b$ **51.** $7k + 15$ **53.** $2x + 6$ **55.** $-\frac{1}{3}t - \frac{28}{3}$ **57.** $9y^2$ **59.** $5p^2 - 14p^3$ **61.** $-2y^2 + 3y^3$ **63.** $8x + 15$ **65.** $-\frac{4}{3}y - 10$ **67.** $-\frac{3}{2}y + 16$ **69.** $x; -3; 2x; 6; 1 - 2x$ **71.** $-19p + 16$ **73.** $-16y + 63$ **75.** $4r + 15$ **77.** $12k - 5$ **79.** $-2x + 4$ **81.** $-\frac{14}{3}x - \frac{22}{3}$ **83.** $-2k - 3$ **85.** $4x - 7$ **87.** $-4.1r + 4.2$ **89.** $-23.7y - 12.6$ **91.** The student made a sign error when applying the distributive property. $7x - 2(3 - 2x)$ means $7x - 2(3) - 2(-2x)$, which simplifies to $7x - 6 + 4x$, or $11x - 6$. **92.** The student incorrectly started by adding $3 + 2$. As the first step, 2 must be multiplied by $4x - 5$. Thus, $3 + 2(4x - 5)$ equals $3 + 8x - 10$, which simplifies to $8x - 7$. **93.** $(x + 3) + 5x; 6x + 3$ **95.** $(13 + 6x) - (-7x); 13 + 13x$ **97.** $2(3x + 4) - (-4 + 6x); 12$ **99.** $1000 + 5x$ (dollars) **100.** $750 + 3y$ (dollars) **101.** $1000 + 5x + 750 + 3y$ (dollars) **102.** $1750 + 5x + 3y$ (dollars)

Chapter 1 REVIEW EXERCISES

1. 625 **2.** 0.00000081 **3.** 0.009261 **4.** $\frac{125}{8}$ **5.** 27 **6.** 200 **7.** 17 **8.** 4 **9.** 7 **10.** 4 **11.** $13 < 17$ **12.** $5 + 2 \neq 10$ **13.** Six is less than fifteen; true **14.** Two-fourths is not equal to three-sixths; false **15.** 30 **16.** 60 **17.** 14 **18.** 13 **19.** $x + 6$ **20.** $8 - x$ **21.** $6x - 9$ **22.** $12 + \frac{3}{5}x$ **23.** yes **24.** no **25.** $2x - 6 = 10$ **26.** $4x = 8$ **27.** equation **28.** expression

29. 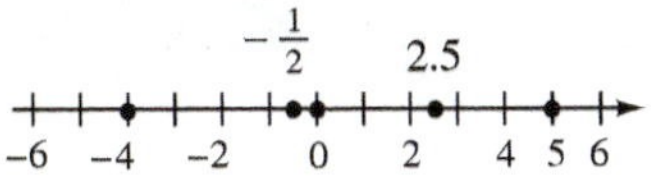

30. 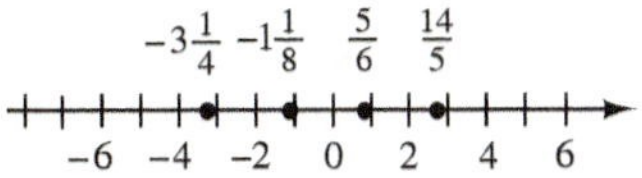

31. rational numbers, real numbers **32.** natural numbers, whole numbers, integers, rational numbers, real numbers **33.** -10 **34.** -9 **35.** $-\frac{3}{4}$ **36.** $-|23|$ **37.** true **38.** true **39.** true **40.** false **41.** -3 **42.** -19 **43.** -7 **44.** 9 **45.** -6 **46.** -4 **47.** -17 **48.** $-\frac{29}{36}$ **49.** 0 **50.** 15 **51.** $(-31 + 12) + 19; 0$ **52.** $[-4 + (-8)] + 13; 1$ **53.** \$26.25 **54.** -10°F **55.** -11 **56.** -1 **57.** 7 **58.** $-\frac{43}{35}$ **59.** 10.31 **60.** -12 **61.** 2 **62.** 1 **63.** $-4 - (-6); 2$ **64.** $[4 + (-8)] - 5; -9$ **65.** $[18 - (-23)] - 15; 26$ **66.** $19 - (-7 - 12); 38$ **67.** 38 yd **68.** 17,477.63 **69.** -308 thousand **70.** 72 thousand **71.** 14 thousand **72.** -122 thousand **73.** 36 **74.** -105 **75.** $\frac{1}{2}$ **76.** 10.08 **77.** -20 **78.** -10 **79.** -24 **80.** -35 **81.** 4 **82.** -20 **83.** $-\frac{3}{4}$ **84.** 11.3 **85.** -1 **86.** undefined **87.** 1 **88.** 0 **89.** -18 **90.** -18 **91.** 125 **92.** -423 **93.** $-4(5) - 9; -29$ **94.** $\frac{5}{6}[12 + (-6)]; 5$ **95.** $\frac{12}{8 + (-4)}; 3$ **96.** $\frac{-20(12)}{15 - (-15)}; -8$ **97.** $\frac{x}{x + 5} = -2$ **98.** $8x - 3 = -7$ **99.** identity property **100.** identity property **101.** inverse property **102.** inverse property **103.** distributive property **104.** associative property **105.** associative property **106.** commutative property **107.** $7y + 14$ **108.** $-48 + 12t$ **109.** $6s + 15y$ **110.** $4r - 5s$ **111.** $8y$ **112.** $17p^2$ **113.** $16r^2 + 7r$ **114.** $-19k + 54$ **115.** $5s - 6$ **116.** $-45t - 23$ **117.** $-2(3x) - 7x; -13x$ **118.** $(5 + 4x) + 8x; 5 + 12x$

Chapter 1 MIXED REVIEW EXERCISES

1. $3; 3; -\frac{1}{3}$ **2.** $12; -12; \frac{1}{12}$ **3.** $-\frac{2}{3}; \frac{2}{3}; \frac{2}{3}$ **4.** 0.2; 0.2; 5 **5.** rational numbers, real numbers **6.** 37 **7.** -6 **8.** $\frac{25}{36}$ **9.** -26 **10.** $\frac{8}{3}$ **11.** $-\frac{1}{24}$ **12.** $\frac{7}{2}$ **13.** 2 **14.** 77.6 **15.** $-1\frac{1}{2}$ **16.** 11 **17.** $-\frac{28}{15}$ **18.** 24 **19.** -47°F **20.** 27 ft

Chapter 1 TEST

1. true **2.** false **3.** 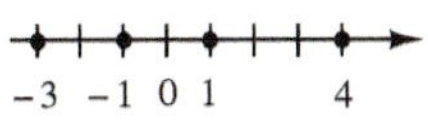

4. rational numbers, real numbers **5.** $-|-8|$ (or -8) **6.** -1.277 **7.** $\frac{-6}{2 + (-8)}; 1$ **8.** negative **9.** 4 **10.** $-2\frac{5}{6}$ **11.** 6 **12.** 2 **13.** 108 **14.** 11 **15.** $\frac{30}{7}$ **16.** -70 **17.** 3 **18.** 178°F **19.** $-$\$0.49 trillion **20.** 15 **21.** D **22.** A **23.** E **24.** B **25.** C **26.** $21x$ **27.** $-3x + 1$ **28.** $-9x^2 - 6x - 8$ **29.** $15x - 3$ **30.** **(a)** -18 **(b)** -18 **(c)** The distributive property tells us that the answers must be the same because $a(b + c) = ab + ac$ for all a, b, c.

ANSWERS

CHAPTER 2 Equations, Inequalities, and Applications

SECTION 2.1

1. equation; expression **2.** linear; variable; = **3.** equivalent equations **4.** addition; solution set **5.** A, B **6. (a)** expression; $x + 15$ **(b)** expression; $m + 7$ **(c)** equation; $\{-1\}$ **(d)** equation; $\{-17\}$ **7.** $\{12\}$ **9.** $\{-3\}$ **11.** $\{4\}$ **13.** $\{4\}$ **15.** $\{-9\}$ **17.** $\{6.3\}$ **19.** $\{-6.5\}$ **21.** $\{-16.9\}$ **23.** $\left\{-\frac{3}{4}\right\}$ **25.** $\{-10\}$ **27.** $\{-13\}$ **29.** $\{10\}$ **31.** $\{10.1\}$ **33.** $\left\{\frac{4}{15}\right\}$ **35.** $\{-3\}$ **37.** $\{7\}$ **39.** $\{-4\}$ **41.** $\{12\}$ **43.** $\{-5\}$ **45.** $\{-6\}$ **47.** $\{-2\}$ **49.** $\{4\}$ **51.** $\{-16\}$ **53.** $\{2\}$ **55.** $\{2\}$ **57.** $\{-4\}$ **59.** $\{0\}$ **61.** $\{0\}$ **63.** $\left\{\frac{7}{15}\right\}$ **65.** $\{13\}$ **67.** $\{-2\}$ **69.** $\{7\}$ **71.** $\{-4\}$ **73.** $\{0\}$ **75.** $\{29\}$ **77.** $\{18\}$

ANSWERS

SECTION 2.2

1. (a) and **(c)**: multiplication property of equality; **(b)** and **(d)**: addition property of equality **2.** C **3.** $\frac{3}{2}$ **4.** $\frac{5}{4}$ **5.** 10 **6.** 100 **7.** $-\frac{2}{9}$ **8.** $-\frac{3}{8}$ **9.** -1 **10.** -1 **11.** 6 **12.** 7 **13.** -4 **14.** -13 **15.** 0.12 **16.** 0.21 **17.** -1 **18.** -1 **19.** B **20.** A **21.** $\{6\}$ **23.** $\left\{\frac{15}{2}\right\}$ **25.** $\{-5\}$ **27.** $\left\{-\frac{18}{5}\right\}$ **29.** $\{12\}$ **31.** 2; 2; 0; $\{0\}$ **33.** $\{-12\}$ **35.** $\{40\}$ **37.** $\{-12.2\}$ **39.** $\{-48\}$ **41.** $\{72\}$ **43.** $\{-35\}$ **45.** $\{14\}$ **47.** $\left\{-\frac{27}{35}\right\}$ **49.** $\{3\}$ **51.** $\{-5\}$ **53.** $\{20\}$ **55.** $\{7\}$ **57.** $\{-12\}$ **59.** $\{0\}$ **61.** $\{-3\}$ **63.** $\{-4\}$ **65.** $\{-5\}$ **67.** $\{1\}$ **69.** $-4x = 10; -\frac{5}{2}$ **71.** $\frac{x}{-5} = 2; -10$

SECTION 2.3

1. addition; subtract **2.** left; like **3.** distributive; parentheses **4.** multiplication; $\frac{4}{3}$ **5.** fractions; 6 **6.** decimals; 10 **7. (a)** identity; B **(b)** conditional; A **(c)** contradiction; C **8.** D **9.** A **10.** D **11.** $\{4\}$ **13.** $\{-5\}$ **15.** $\left\{\frac{5}{2}\right\}$ **17.** $\left\{-\frac{1}{2}\right\}$ **19.** $\{5\}$ **21.** $\{1\}$ **23.** $\{5\}$ **25.** $\left\{-\frac{5}{3}\right\}$ **27.** $\left\{\frac{4}{3}\right\}$ **29.** $\left\{-\frac{5}{3}\right\}$ **31.** $\{5\}$ **33.** $\{0\}$ **35.** $\{0\}$ **37.** {all real numbers} **39.** {all real numbers} **41.** $\emptyset$ **43.** $\emptyset$ **45.** $\{5\}$ **47.** $\{12\}$ **49.** $\{11\}$ **51.** $\{0\}$ **53.** $\left\{\frac{3}{25}\right\}$ **55.** 100; $\{60\}$ **57.** $\{4\}$ **59.** $\{5000\}$ **61.** $\left\{-\frac{72}{11}\right\}$ **63.** $\{0\}$ **65.** $\emptyset$ **67.** {all real numbers} **69.** $\{-6\}$ **71.** $\{15\}$ **73.** $12 - q$ **75.** $\frac{9}{z}$ **77.** $x + 29$ **79.** $m + 12; m - 2$ **81.** $25r$ **83.** $\frac{t}{5}$ **85.** $3x + 2y$

SUMMARY EXERCISES Applying Methods for Solving Linear Equations

1. equation; $\{-5\}$ **2.** expression; $7p - 14$ **3.** expression; $-3m - 2$ **4.** equation; $\{7\}$ **5.** equation; $\{0\}$ **6.** expression; $-\frac{1}{6}x + 5$ **7.** $\left\{\frac{7}{3}\right\}$ **8.** $\{4\}$ **9.** $\{-5.1\}$ **10.** $\{12\}$ **11.** $\{-25\}$ **12.** $\{-6\}$ **13.** $\{-6\}$ **14.** $\{-16\}$ **15.** {all real numbers} **16.** $\{23.7\}$ **17.** $\{6\}$ **18.** $\{0\}$ **19.** $\{7\}$ **20.** $\{1\}$ **21.** $\{5\}$ **22.** $\emptyset$ **23.** $\emptyset$ **24.** $\{-10.8\}$ **25.** $\{25\}$ **26.** {all real numbers} **27.** $\{3\}$ **28.** $\{70\}$ **29.** $\{-2\}$ **30.** $\left\{\frac{14}{17}\right\}$

SECTION 2.4

1. *Step 1:* Read the problem carefully; *Step 2:* Assign a variable to represent the unknown; *Step 3:* Write an equation; *Step 4:* Solve the equation; *Step 5:* State the answer; *Step 6:* Check the answer. **2.** Some examples are *is, are, was,* and *were.* **3.** D; There cannot be a fractional number of cars. **4.** D; A day cannot have more than 24 hr. **5.** A; Distance cannot be negative. **6.** C; Time cannot be negative. **7.** 1; 16 (or 14); -7 (or -9); $x + 1$ **8.** odd; 2; 11 (or 15); even; 2; 14 (or 10) **9.** complementary; supplementary **10.** 90°; 180° **11.** yes, 90°; yes, 45° **12.** $x - 1; x - 2$ **13.** $8 \cdot (x + 6) = 104; 7$ **15.** $5x + 2 = 4x + 5; 3$ **17.** $3x - 2 = 5x + 14; -8$ **19.** $3(x - 2) = x + 6; 6$ **21.** $3x + (x + 7) = -11 - 2x; -3$ **23.** *Step 1:* We are asked to find the number of drive-in movie screens in the two states; *Step 2:* the number of screens in Pennsylvania; *Step 3:* $x; x - 1$; *Step 4:* 28; *Step 5:* 28; 28; 27; *Step 6:* 1; screens in New York; 27; 55 **25.** Democrats: 188; Republicans: 247 **27.** Kenny Chesney: \$116.4 million; Taylor Swift: \$199.4 million **29.** wins: 73; losses: 9 **31.** orange: 97 mg; pineapple: 25 mg **33.** whole wheat: 25.6 oz; rye: 6.4 oz **35.** active: 225 mg; inert: 25 mg **37.** 1950 Denver nickel: \$16.00; 1944 Philadelphia nickel: \$12.00 **39.** onions: 81.3 kg; grilled steak: 536.3 kg **41.** $x + 5; x + 9$; shortest piece: 15 in.; middle piece: 20 in.; longest piece: 24 in. **43.** American: 18; United: 11; Southwest: 26 **45.** gold: 9; silver: 7; bronze: 12 **47.** 36 million mi **49.** *A* and *B*: 40°; *C*: 100° **51.** 68, 69 **53.** 101, 102 **55.** 10, 12 **57.** 17, 19 **59.** 10, 11 **61.** 18 **63.** 18° **65.** 20° **67.** 39° **69.** 50°

SECTION 2.5

1. The perimeter of a plane geometric figure is the measure of the outer boundary of the figure. **2.** The area of a plane geometric figure is the measure of the surface covered or enclosed by the figure. **3.** area **4.** area **5.** perimeter **6.** perimeter **7.** area **8.** area **9.** area **10.** area **11.** $P = 26$ **13.** $\mathcal{A} = 64$ **15.** $b = 4$ **17.** $t = 5.6$ **19.** $h = 7$ **21.** $r = 2.6$ **23.** $\mathcal{A} = 50.24$ **25.** $V = 150$ **27.** $V = 52$ **29.** $V = 7234.56$ **31.** $I = \$600$ **33.** $p = \$550$ **35.** 0.025; $t = 1.5$ yr **37.** length: 18 in.; width: 9 in. **39.** length: 14 m; width: 4 m **41.** shortest: 5 in.; medium: 7 in.; longest: 8 in. **43.** two equal sides: 7 m; third side: 10 m **45.** perimeter: 5.4 m; area: 1.8 m^2 **47.** 10 ft **49.** 154,000 ft^2 **51.** 194.48 ft^2; 49.42 ft **53.** 23,800.10 ft^2 **55.** length: 36 in.; maximum volume: 11,664 $in.^3$ **57.** 48°, 132° **59.** 70°, 110° **61.** 55°, 35° **63.** 30°, 60° **65.** 51°, 51° **67.** 105°, 105° **69.** $t = \frac{d}{r}$ **71.** $H = \frac{V}{LW}$ **73.** $b = P - a - c$

75. $r=\frac{C}{2\pi}$ **77.** $r=\frac{I}{pt}$ **79.** $h=\frac{2\mathcal{A}}{b}$ **81.** $h=\frac{3V}{\pi r^2}$ **83.** $W=\frac{P-2L}{2}$ **85.** $m=\frac{y-b}{x}$ **87.** $y=\frac{C-Ax}{B}$ **89.** $r=\frac{M-C}{C}$ **91.** $a=\frac{P-2b}{2}$ **93.** $x=\frac{f+ah}{a}$ **95.** $b=2S-a-c$ **97.** $F=\frac{9}{5}C+32$

We give one possible answer for Exercises 99–105. There are other correct forms.

99. $y=-6x+4$ **101.** $y=5x-2$ **103.** $y=\frac{3}{5}x-3$ **105.** $y=\frac{1}{3}x-4$ **107.** $W_1+M=W_2+N_2+1$ **108.** $M=W_2+N_2+1-W_1$ **109.** **(a)** 12 **(b)** 3 **(c)** 12 **110.** $M=-3$; A negative magic number indicates that Oakland has been eliminated from winning the division.

SECTION 2.6

1. compare; A, D **2.** ratios; proportion; cross products **3.** **(a)** C **(b)** D **(c)** B **(d)** A **4.** C, E **5.** $\frac{6}{7}$ **7.** $\frac{18}{55}$ **9.** $\frac{5}{16}$ **11.** $\frac{4}{15}$ **13.** $\frac{6}{5}$ **15.** 10 lb; \$0.749 **17.** 64 oz; \$0.047 **19.** 32 oz; \$0.531 **21.** 32 oz; \$0.056 **23.** $\{35\}$ **25.** $\{7\}$ **27.** $\left\{\frac{8}{5}\right\}$ **29.** $\{1\}$ **31.** $\{2\}$ **33.** $\{-1\}$ **35.** $\{5\}$ **37.** $\left\{-\frac{31}{5}\right\}$ **39.** \$30.00 **41.** \$56.85 **43.** 50,000 fish **45.** 4 ft **47.** 17.0 in. **49.** $2\frac{5}{8}$ cups **51.** \$326.67 **53.** 9; $x=4$ **55.** $x=8$ **57.** $x=3$; $y=5.5$ **59.** side of triangle labeled Chair: 18 ft; side of triangle labeled Pole: 12 ft; One proportion is $\frac{x}{12}=\frac{18}{4}$; 54 ft **61.** **(a)** 2625 mg **(b)** $\frac{125\text{ mg}}{5\text{ mL}}=\frac{2625\text{ mg}}{x\text{ mL}}$ **(c)** 105 mL **63.** \$236 **65.** \$273 **67.** 109.2 **69.** 54 **71.** 700 **73.** 425 **75.** 8% **77.** 120% **79.** 80% **81.** 28% **83.** 32% **85.** \$3000 **87.** \$304 **89.** 9,617,000; · ; 155,922,000; 6.2% **91.** 75.1% **93.** 860% **95.** 30 **96.** **(a)** $5x=12$ **(b)** $\left\{\frac{12}{5}\right\}$ **97.** $\left\{\frac{12}{5}\right\}$ **98.** Both methods give the same solution set.

SUMMARY EXERCISES Applying Problem-Solving Techniques

1. 48 **2.** 35° **3.** 3 **4.** 18, 20 **5.** 140°, 40° **6.** 104°, 104° **7.** 4 **8.** 36 quart cartons **9.** $16\frac{2}{3}\%$ **10.** 510 calories **11.** 4000 calories **12.** U.S.: 104; China: 88; Russia: 82 **13.** 24 oz; \$0.074 **14.** 12.42 cm

SECTION 2.7

1. $<$ (or $>$); $>$ (or $<$); $\le$ (or $\ge$); $\ge$ (or $\le$) **2.** false **3.** $(0, \infty)$ **4.** $(-\infty, \infty)$ **5.** $x>-4$ **6.** $x\ge -4$ **7.** $x\le 4$ **8.** $x<4$ **9.** $-1<x\le 2$ **10.** $-1\le x<2$

11. $(-\infty, 4]$

13. $(-3, \infty)$

15. $[8, 10]$

17. $(0, 10]$

19. $(1, \infty)$

21. $[5, \infty)$

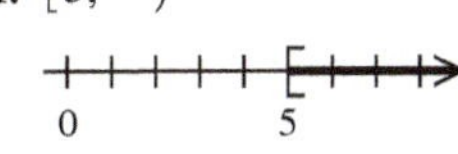

23. $(-\infty, -6)$

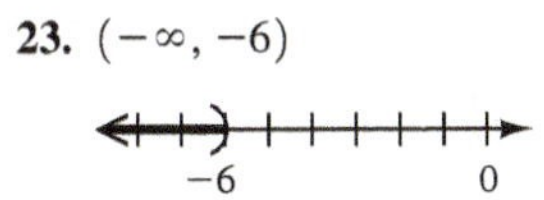

25. The inequality symbol must be reversed when multiplying or dividing by a negative number.

26. Divide by -5 and reverse the direction of the inequality symbol to get $x<-4$.

27. $(-\infty, 6)$

29. $[-10, \infty)$

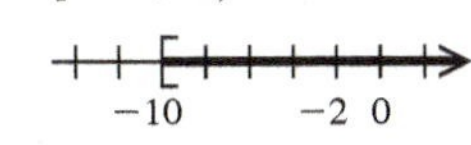

31. $(-\infty, -3)$

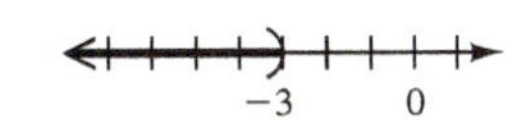

33. $(-\infty, 0]$

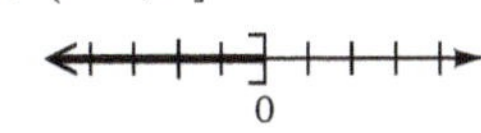

35. $(20, \infty)$

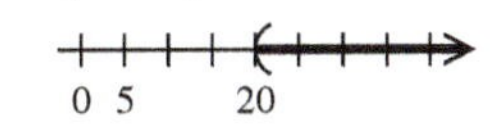

37. $[-3, \infty)$

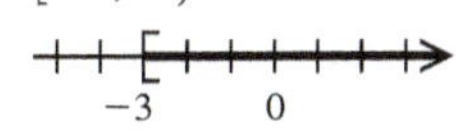

39. $(-\infty, -3]$

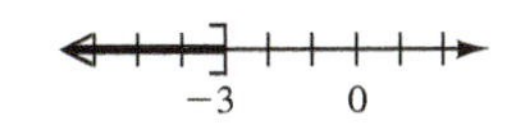

41. $(-1, \infty)$

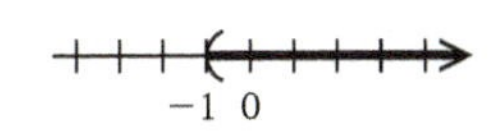

43. $[-5, \infty)$

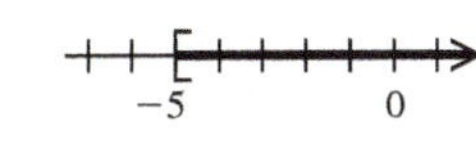

45. $(-\infty, 1)$

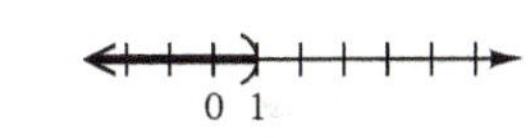

47. $(-\infty, 0]$

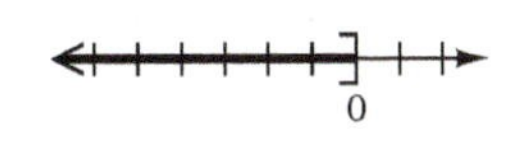

49. $\left(-\frac{1}{2}, \infty\right)$

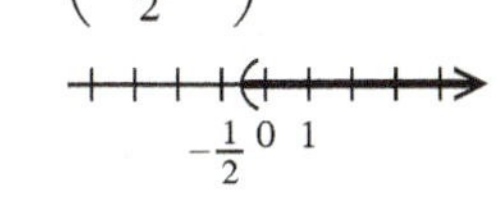

51. $[4, \infty)$

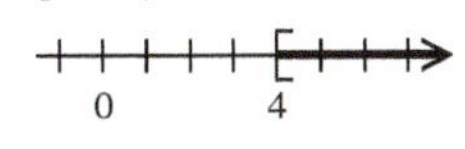

53. $(-\infty, 32)$

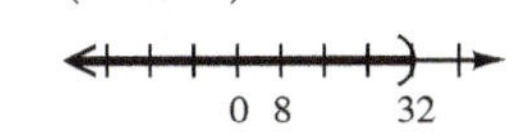

55. $[2, \infty)$

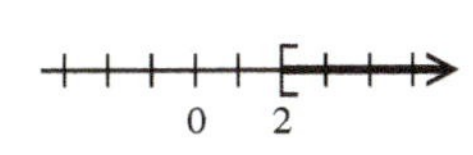

57. $\left[\frac{5}{12}, \infty\right)$

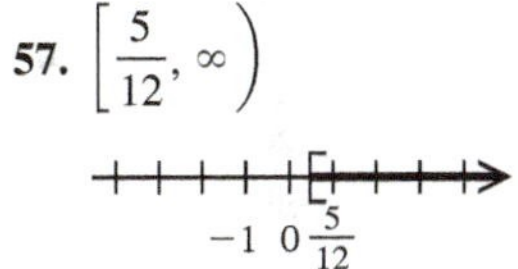

59. $(-21, \infty)$

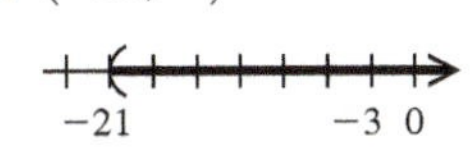

61. $x\ge 16$ **62.** $x<1$ **63.** $x>8$ **64.** $x\ge 12$ **65.** $x\le 20$ **66.** $x>40$ **67.** 88 or more **69.** 80 or more **71.** all numbers greater than 16 **73.** It has never exceeded 40°C. **75.** 32 or greater **77.** 12 min **79.** more than 3.8 in.

81. $[-1, 6]$

83. $(1, 3)$

85. $\left(-\frac{11}{6}, -\frac{2}{3}\right]$

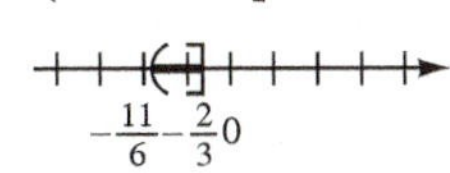

87. $[3, 7)$

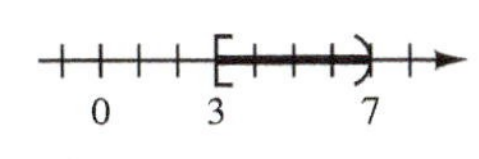

89. $[-26, 6]$

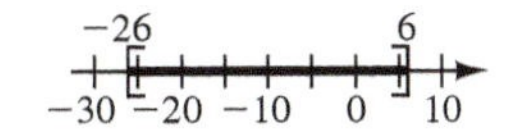

91. $[-3, 6]$

93. $\{4\}$

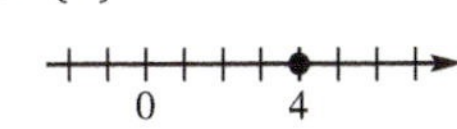

94. $(-\infty, 4)$

ANSWERS

95. $(4, \infty)$

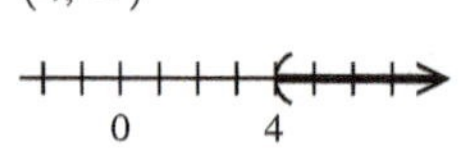

96. The graph would be the set of all real numbers; $(-\infty, \infty)$

Chapter 2 REVIEW EXERCISES

1. $\{9\}$ **2.** $\{4\}$ **3.** $\{-6\}$ **4.** $\left\{\frac{3}{2}\right\}$ **5.** $\{0\}$ **6.** $\left\{-\frac{61}{2}\right\}$ **7.** $\{15\}$ **8.** $\{-21\}$ **9.** $\varnothing$ **10.** {all real numbers} **11.** $-\frac{7}{2}$ **12.** 20 **13.** Hawaii: 6425 mi^2; Rhode Island: 1212 mi^2 **14.** Seven Falls: 300 ft; Twin Falls: 120 ft **15.** 80° **16.** 11, 13 **17.** oil: 4 oz; gasoline: 128 oz **18.** shortest piece: 13 in.; middle-sized piece: 20 in.; longest piece: 39 in. **19.** $h = 11$ **20.** $\mathcal{A} = 28$ **21.** $r = 4.75$ **22.** $V = 3052.08$ **23.** $h = \frac{\mathcal{A}}{b}$ **24.** $h = \frac{2\mathcal{A}}{b + B}$

We give one possible answer for Exercises 25 and 26. There are other correct forms.

25. $y = -x + 11$ **26.** $y = \frac{3}{2}x - 6$ **27.** 135°, 45° **28.** 100°, 100° **29.** perimeter: 428 ft; area: 11,349 ft^2 **30.** diameter: 27.07 ft; radius: 13.54 ft **31.** $\frac{3}{2}$ **32.** $\frac{5}{14}$ **33.** $\frac{3}{4}$ **34.** $\left\{\frac{7}{2}\right\}$ **35.** $\left\{-\frac{8}{3}\right\}$ **36.** $\left\{\frac{25}{19}\right\}$ **37.** $6\frac{2}{3}$ lb **38.** 36 oz **39.** 375 km **40.** 18 oz; \$0.249 **41.** 6 **42.** 175% **43.** $33\frac{1}{3}\%$ **44.** 2500 **45.** \$21,575 **46.** \$350.46 **47.** 25% **48.** 17.5%

49. $[-4, \infty)$

50. $(-\infty, 7)$

51. $[-5, 6)$

52. $\left[\frac{1}{2}, \infty\right)$

53. $[-3, \infty)$

54. $(-\infty, 2)$

55. $[3, \infty)$

56. $[46, \infty)$

57. $(-\infty, -5)$

58. $(-\infty, -37)$

59. $\left[-2, \frac{3}{2}\right)$

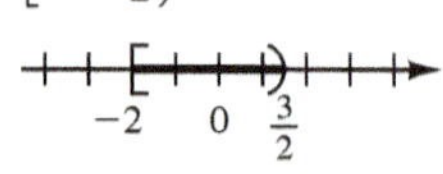

60. $(1, 5]$

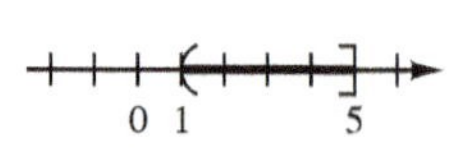

61. 88 or more **62.** all numbers less than or equal to $-\frac{1}{3}$

Chapter 2 MIXED REVIEW EXERCISES

1. $\{7\}$ **2.** $r = \frac{d}{2}$ **3.** $(-\infty, 2)$ **4.** $\{-9\}$ **5.** $\{70\}$ **6.** $\left\{\frac{13}{4}\right\}$ **7.** $\varnothing$ **8.** {all real numbers} **9.** 2.0 in. **10.** D: 22°; E: 44°; F: 114° **11.** 44 m **12.** 70 ft **13.** 160 oz; \$0.062 **14.** 24°, 66° **15.** 92 or more **16.** \$197.50

Chapter 2 TEST

1. $\{6\}$ **2.** $\{-6\}$ **3.** $\left\{\frac{13}{4}\right\}$ **4.** $\{-10.8\}$ **5.** $\{21\}$ **6.** {all real numbers} **7.** $\{30\}$ **8.** $\varnothing$ **9.** Chamberlain: 4029; Jordan: 3041 **10.** Hawaii: 4021 mi^2; Maui: 728 mi^2; Kauai: 551 mi^2 **11.** 24, 26 **12.** 50° **13.** **(a)** $W = \frac{P - 2L}{2}$ **(b)** 18 **14.** $y = \frac{5}{4}x - 2$ (Other correct forms of the answer are possible.) **15.** 100°, 80° **16.** 75°, 75° **17.** $\{6\}$ **18.** $\{-29\}$ **19.** 16 oz; \$0.249 **20.** 2300 mi **21.** \$264 **22.** 40% **23.** **(a)** $x < 0$ **(b)** $-2 < x \le 3$

24. $(-\infty, 11)$

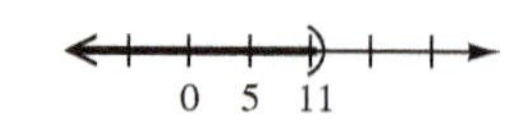

25. $[-3, \infty)$

26. $(-\infty, 4]$

27. $(-2, 6]$

28. 83 or more

Chapters R–2 CUMULATIVE REVIEW EXERCISES

1. $\frac{3}{8}$ **2.** $\frac{3}{4}$ **3.** $\frac{31}{20}$ **4.** $\frac{551}{40}$, or $13\frac{31}{40}$ **5.** 6 **6.** $\frac{6}{5}$ **7.** 34.03 **8.** 27.31 **9.** 30.51 **10.** 56.3 **11.** 35 yd **12.** $7\frac{1}{2}$ cups **13.** $3\frac{3}{8}$ in. **14.** \$2599.94 **15.** true **16.** false **17.** 7 **18.** 1 **19.** 13 **20.** −40 **21.** −12 **22.** undefined **23.** −6 **24.** 28 **25.** 1 **26.** 0 **27.** $\frac{73}{18}$ **28.** −64 **29.** −134 **30.** $-\frac{29}{6}$ **31.** distributive property **32.** commutative property **33.** inverse property **34.** identity property **35.** $7p - 14$ **36.** $2k - 11$ **37.** $\{7\}$ **38.** $\{-4\}$ **39.** $\{-1\}$ **40.** $\left\{-\frac{3}{5}\right\}$ **41.** $\{2\}$ **42.** $\{-13\}$ **43.** $\{26\}$ **44.** $\{-12\}$ **45.** $c = P - a - b - B$ **46.** $s = \frac{P}{4}$

47. $(-\infty, 2]$

48. $(-\infty, 1)$

49. \$260.50 **50.** \$8625 **51.** 30 cm **52.** 16 in.

CHAPTER 3 Graphs of Linear Equations and Inequalities in Two Variables

SECTION 3.1

1. does; do not **2.** (0, 0) **3.** 4; −1; is not **4.** II **5.** y **6.** 3 **7.** 6 **8.** −4 **9.** negative; negative **10.** negative; positive **11.** positive; negative **12.** positive; positive **13.** 2009 and 2010 **14.** The unemployment rate was decreasing. **15.** 2010: 9.5%; 2015: 5%; 4.5% **17.** 2009 and 2010; 1.5 million **19.** Sales decreased slightly between 2012 and 2013 and then increased slightly between 2013 and 2014. **21.** yes **23.** yes **25.** no **27.** yes **29.** no **31.** no **33.** 17 **35.** $-\frac{7}{2}$ **37.** −4 **39.** −5 **41.** 8, 6, 3; (0, 8), (6, 0), (3, 4) **43.** 3, −5, −15; (0, 3), (−5, 0), (−15, −6) **45.** −9, −9, −9; (−9, 6), (−9, 2), (−9, −3) **47.** 6, 6, 6; (8, 6), (4, 6), (−2, 6) **49.** 8, 8, 8; (8, 8), (8, 3), (8, 0) **51.** −2, −2, −2; (9, −2), (2, −2), (0, −2) **53.** (2, 4); I

55. $(-5, 4)$; II **57.** $(3, 0)$; no quadrant **59.** $(4, -4)$; IV **61.** If $xy < 0$, then either $x < 0$ and $y > 0$ or $x > 0$ and $y < 0$. If $x < 0$ and $y > 0$, then the point lies in quadrant II. If $x > 0$ and $y < 0$, then the point lies in quadrant IV. **62.** If $xy > 0$, then either $x > 0$ and $y > 0$ or $x < 0$ and $y < 0$. If $x > 0$ and $y > 0$, then the point lies in quadrant I. If $x < 0$ and $y < 0$, then the point lies in quadrant III.

63.–74.

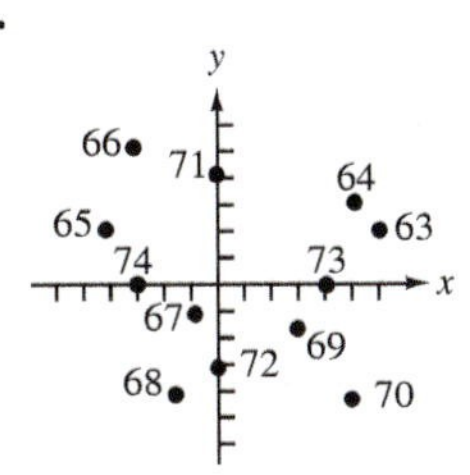

75. $-3, 6, -2, 4$

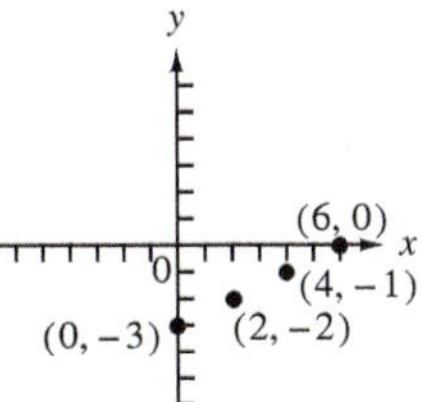

77. $-3, 4, -6, -\frac{4}{3}$

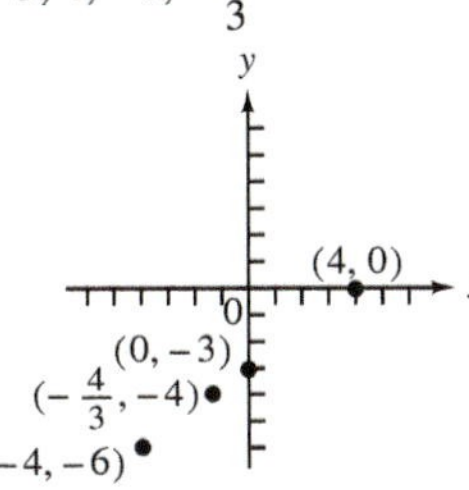

79. $-4, -4, -4, -4$

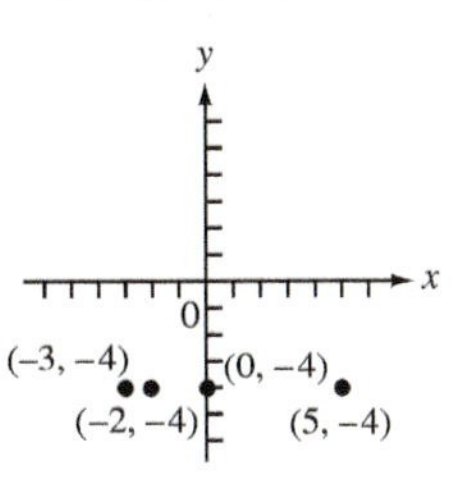

81. The points in each graph appear to lie on a straight line.
82. (a) horizontal; They are the same (all -4). **(b)** vertical; They are the same (all 5). **83. (a)** $(5, 45)$ **(b)** $(6, 50)$ **85. (a)** $(2009, 28.3)$, $(2010, 28.0)$, $(2011, 26.9)$, $(2012, 25.4)$, $(2013, 22.5)$, $(2014, 21.9)$
(b) $(2000, 32.4)$ means that 32.4 percent of 2-yr college students in 2000 received a degree within 3 yr.

(c) 2-YEAR COLLEGE STUDENTS COMPLETING A DEGREE WITHIN 3 YEARS

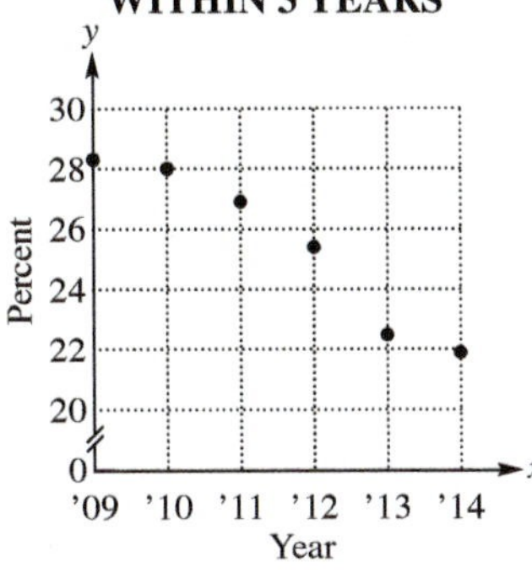

(d) The points lie approximately in a linear pattern. Rates at which 2-yr college students complete a degree within 3 yr were decreasing.

87. (a) 98, 88, 78, 68 **(b)** $(20, 98)$, $(40, 88)$, $(60, 78)$, $(80, 68)$

(c) TARGET HEART RATE ZONE (Lower Limit)

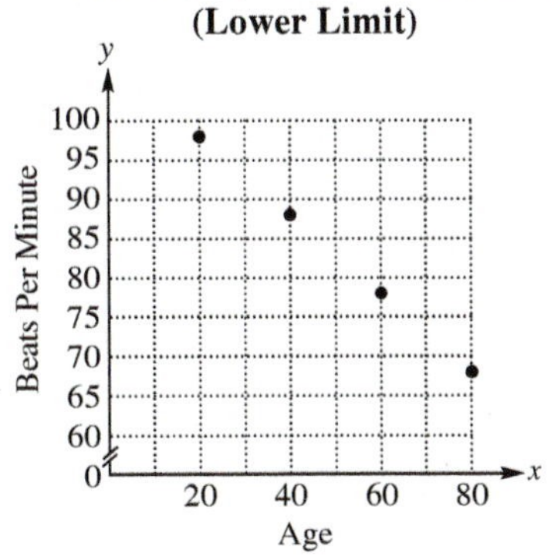

Yes, the points lie in a linear pattern.

89. between 98 and 157 beats per minute; between 88 and 141 beats per minute

SECTION 3.2

1. By; C; 0 **2.** line; solution **3. (a)** A **(b)** C **(c)** D **(d)** B **4.** A, C, D **5.** x-intercept: $(4, 0)$; y-intercept: $(0, -4)$ **6.** x-intercept: $(-5, 0)$; y-intercept: $(0, 5)$ **7.** x-intercept: $(-2, 0)$; y-intercept: $(0, -3)$ **8.** x-intercept: $(4, 0)$; y-intercept: $(0, 3)$

9. 5, 5, 3

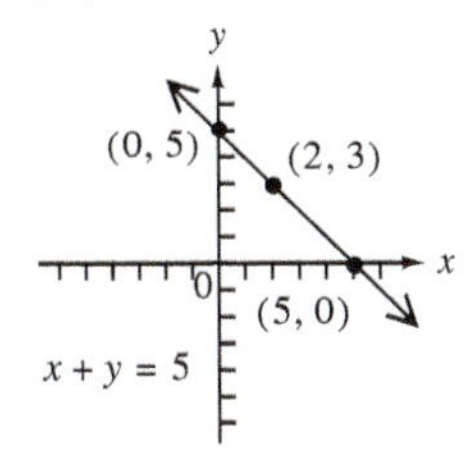

11. 1, 3, −1

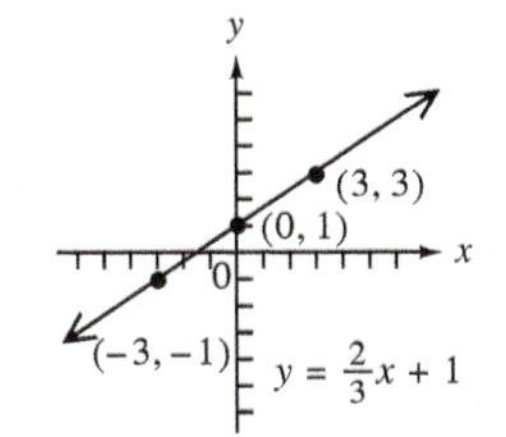

13. $-6, -2, -5$

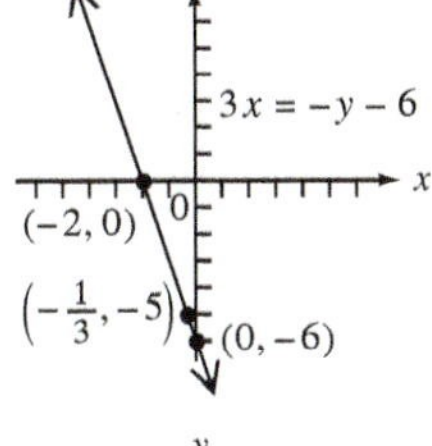
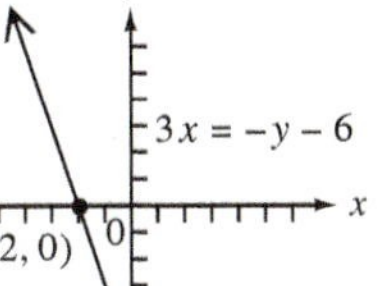

15. 0; $(0, -8)$; 0; $(8, 0)$
17. $(0, -8)$; $(12, 0)$
19. $(0, 0)$; $(0, 0)$

21.

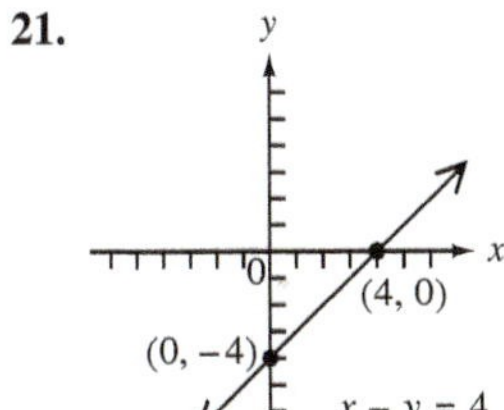

23.

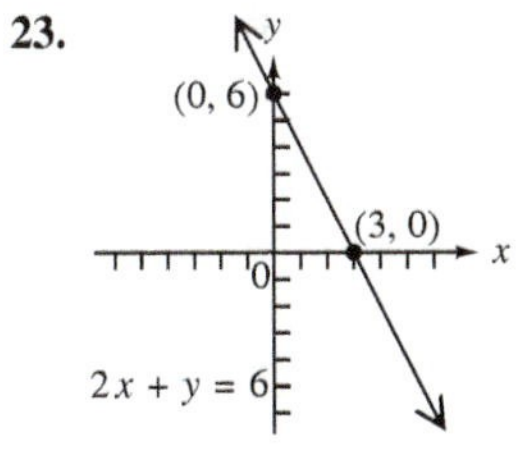

25.

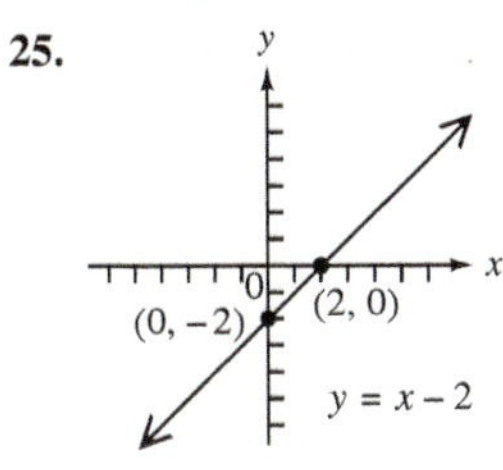

27.

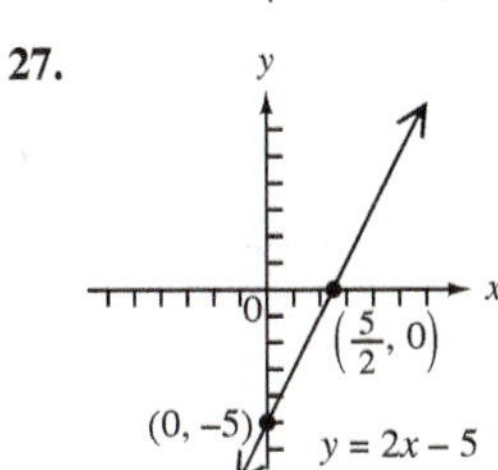
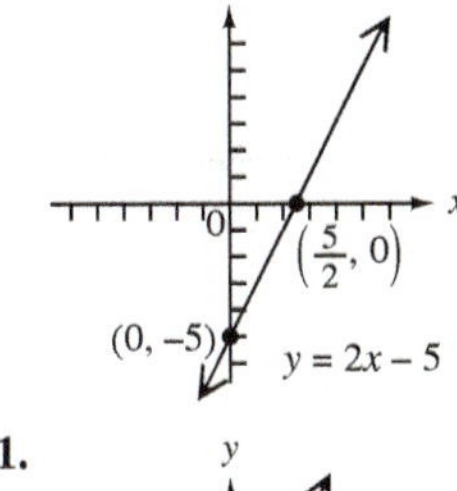

29.

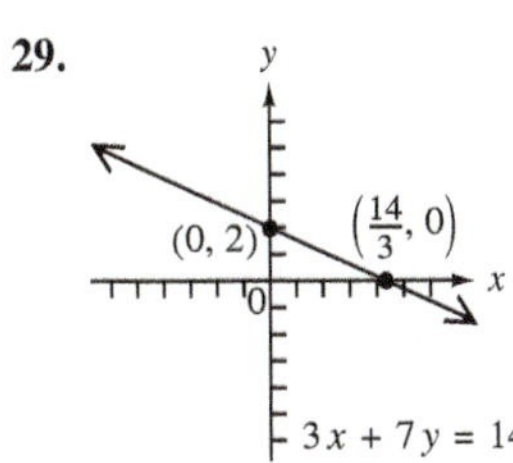
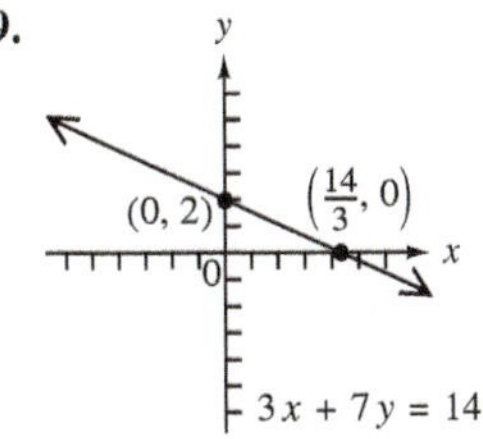

31.

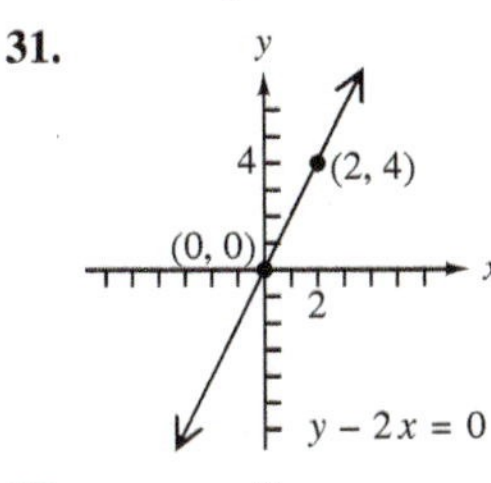

33.

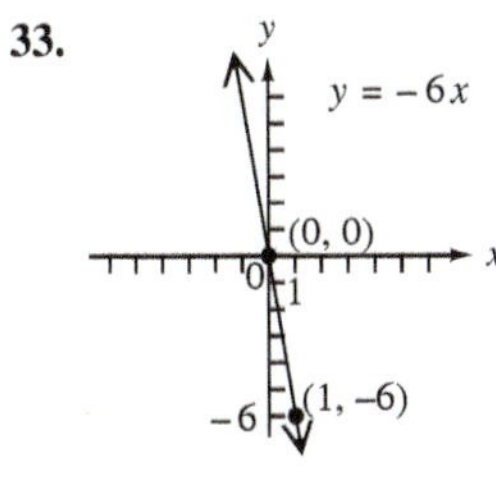

35.

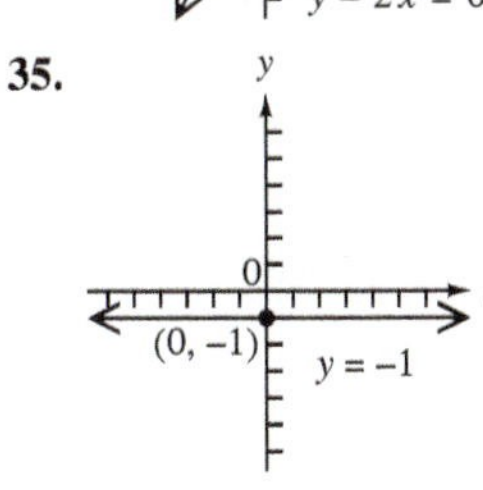

37.

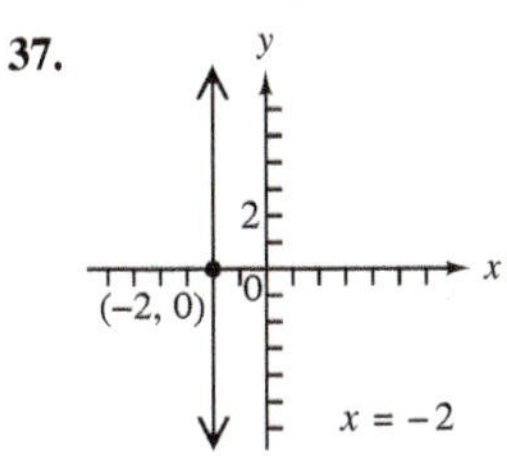

39.

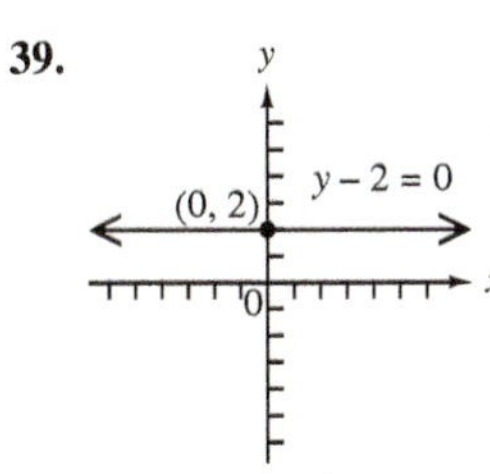

41.

43.

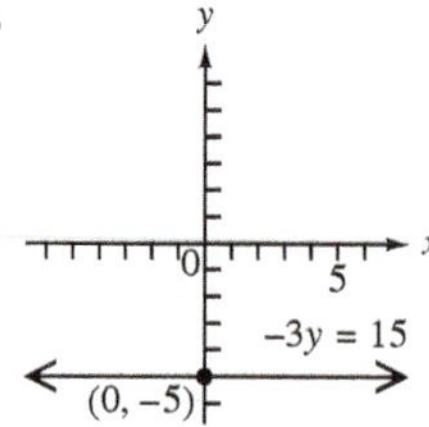

45. **(a)** D **(b)** C **(c)** B **(d)** A **46.** $y = 0; x = 0$

In Exercises 47–50, descriptions may vary.

47. The graph is a line with x-intercept $(-3, 0)$ and y-intercept $(0, 9)$. **48.** The graph is a vertical line with x-intercept $(11, 0)$. **49.** The graph is a horizontal line with y-intercept $(0, -2)$. **50.** The graph passes through the origin $(0, 0)$ and the points $(2, 1)$ and $(4, 2)$.

51. **(a)** 151.5 cm, 159.3 cm, 174.9 cm
(b) $(20, 151.5)$, $(22, 159.3)$, $(26, 174.9)$
(c)  HEIGHTS OF WOMEN 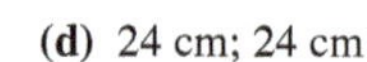 **(d)** 24 cm; 24 cm

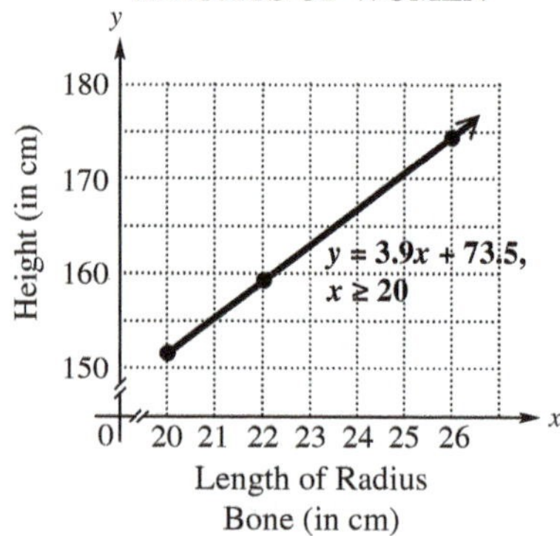

53. **(a)** \$62.50; \$100 **(b)** 200 **(c)** $(50, 62.50)$, $(100, 100)$, $(200, 175)$
(d) POSTER COSTS **(e)** \$120; \$118.75

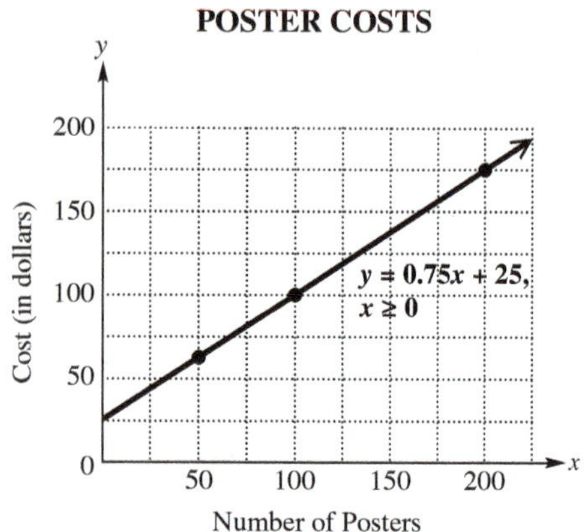

55. **(a)** \$30,000 **(b)** \$20,000 **(c)** \$5000 **(d)** After 5 yr, the SUV has a value of \$5000. **57.** **(a)** 2000: 30.1 lb; 2010: 33.2 lb; 2014: 34.4 lb **(b)** 2000: 30 lb; 2010: 33 lb; 2014: 34 lb **(c)** The values are quite close. **(d)** 36.9 lb; It is very close to the USDA projection.

SECTION 3.3

1. steepness; vertical; horizontal **2.** ratio; y; rise; x; run **3.** **(a)** 6 **(b)** 4 **(c)** $\frac{6}{4}$, or $\frac{3}{2}$; slope of the line **(d)** Yes, it doesn't matter which point we start with. The slope would be expressed as the quotient of -6 and -4, which simplifies to $\frac{3}{2}$. **4.** **(a)** C **(b)** A **(c)** D **(d)** B

5. Answers will vary.

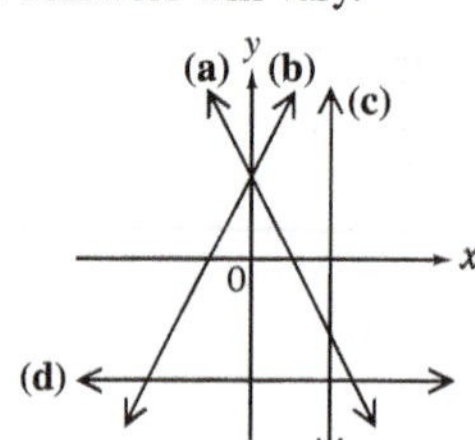

6. **(a)** falls from left to right **(b)** horizontal **(c)** vertical **(d)** rises from left to right **7.** **(a)** negative **(b)** 0 **8.** **(a)** negative **(b)** negative **9.** **(a)** positive **(b)** negative **10.** **(a)** positive **(b)** 0 **11.** **(a)** 0 **(b)** negative **12.** **(a)** 0 **(b)** positive

13. $\frac{8}{27}$ **14.** $\frac{3}{10}$ **15.** $-\frac{2}{3}$ **16.** $-\frac{1}{4}$ **17.** Because the student found the difference $3 - 5 = -2$ in the numerator, he should have subtracted in the *same* order in the denominator, $-1 - 2 = -3$. The correct slope is $\frac{-2}{-3} = \frac{2}{3}$.

18. Slope is defined as $\frac{\text{change in } y}{\text{change in } x}$, but the student found $\frac{\text{change in } x}{\text{change in } y}$. The correct slope is $\frac{5}{-8} = -\frac{5}{8}$. **19.** 4 **21.** $-\frac{1}{2}$ **23.** 0 **25.** $\frac{5}{4}$ **27.** $\frac{3}{2}$ **29.** -3 **31.** 0 **33.** undefined **35.** $\frac{1}{4}$ **37.** $-\frac{1}{2}$ **39.** 5 **41.** $\frac{1}{4}$ **43.** $\frac{3}{2}$ **45.** $-\frac{3}{2}$ **47.** 0 **49.** undefined **51.** 1

In part (a) of Exercises 53 and 55, we used the intercepts. Other points can be used.

53. **(a)** $(5, 0)$ and $(0, 10)$; -2 **(b)** $y = -2x + 10$; -2 **55.** **(a)** $(3, 0)$  and $(0, -5)$; $\frac{5}{3}$ 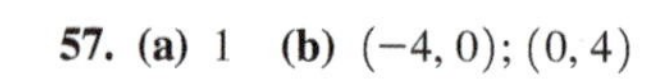 **(b)** $y = \frac{5}{3}x - 5$; $\frac{5}{3}$

57. **(a)** 1 **(b)** $(-4, 0)$; $(0, 4)$ **(c)**

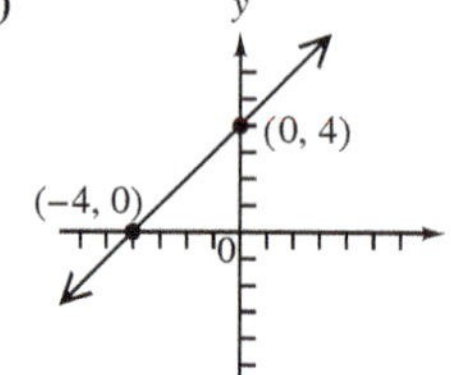

59. **(a)** $-\frac{1}{3}$ **(b)** $(-6, 0)$; $(0, -2)$ **(c)**

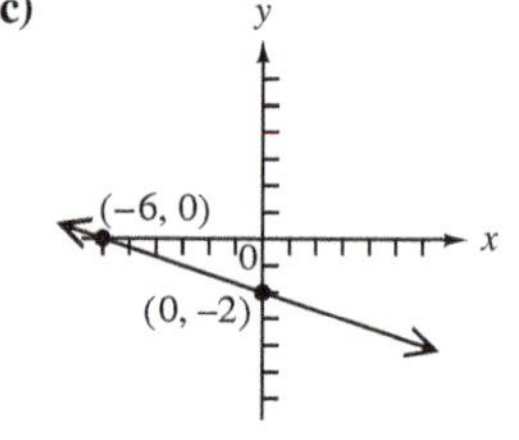

61. **(a)** -3 **(b)** $\frac{1}{3}$ **62.** **(a)** 5 **(b)** $-\frac{1}{5}$ **63.** A **64.** It is horizontal. **65.** $\frac{4}{3}$; $\frac{4}{3}$; parallel **67.** $\frac{5}{3}$; $\frac{3}{5}$; neither **69.** $\frac{3}{5}$; $-\frac{5}{3}$; perpendicular **71.** -6; $-\frac{1}{6}$; neither **73.** 0.5 **74.** positive; increased **75.** 0.5% **76.** -0.2 **77.** negative; decreased **78.** 0.2%

SECTION 3.4

1. m; $(0, b)$ **2.** $-\frac{1}{2}$; $(0, -3)$ **3.** **(a)** C **(b)** B **(c)** A **(d)** D **4.** **(a)** E **(b)** D **(c)** B **(d)** A **5.** y-axis **6.** x-axis **7.** $\frac{5}{2}$; $(0, -4)$ **9.** -1; $(0, 9)$ **11.** -8; $(0, 0)$ **13.** $\frac{1}{5}$; $\left(0, -\frac{3}{10}\right)$

15.

17.

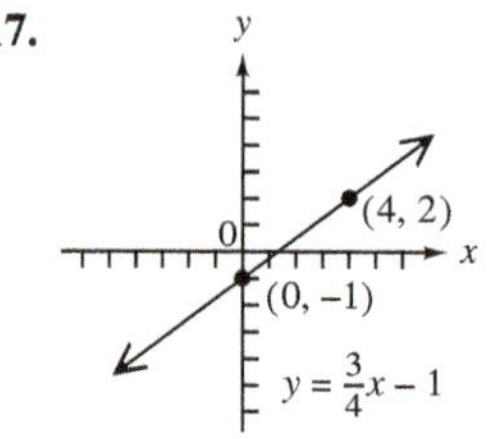

19.

$2x + y = -5$, (0, −5), (1, −7)

21.

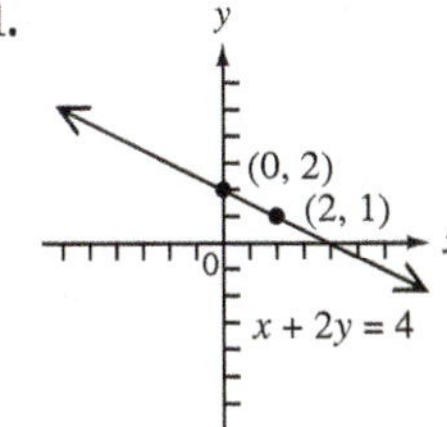

23.

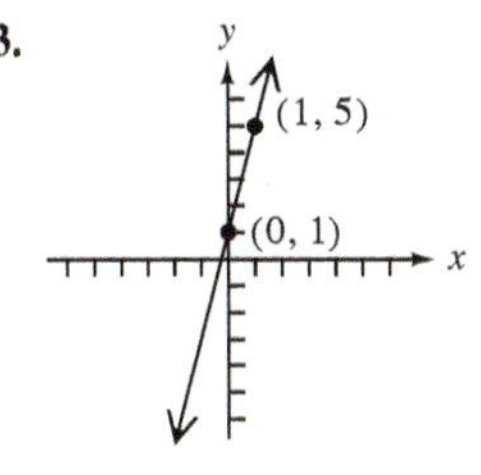

25.

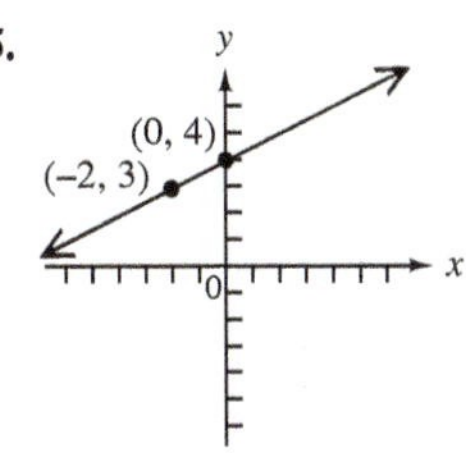

27.

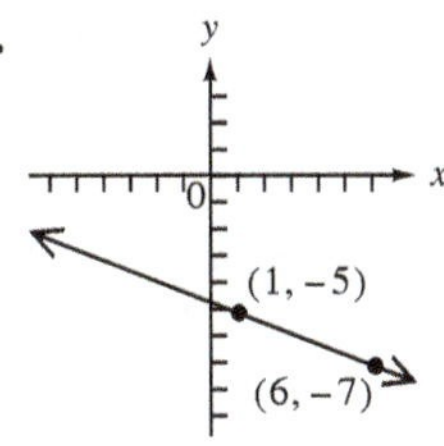

29.

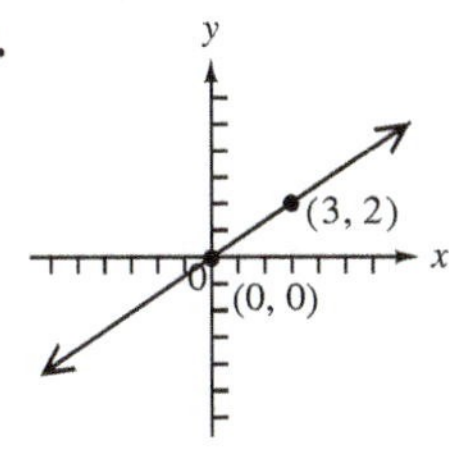

31.

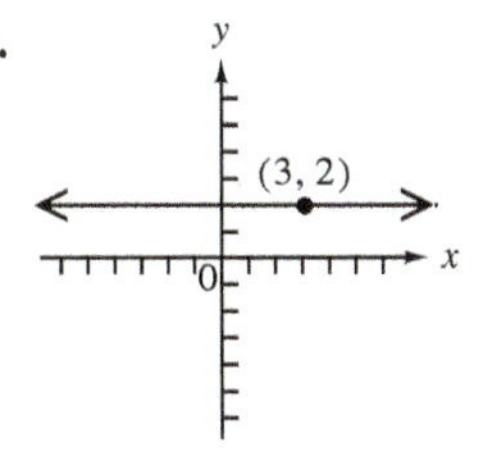

33.

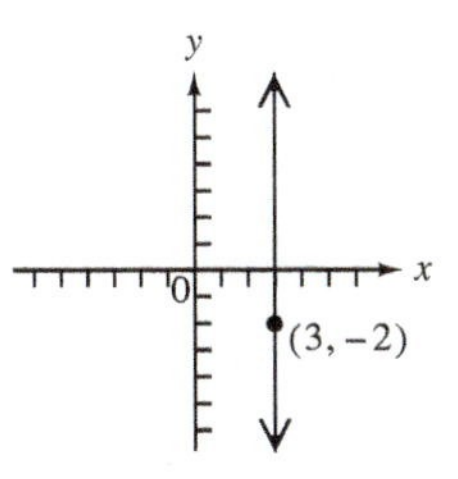

35. $y = 3x - 3$ **36.** $y = 2x - 4$ **37.** $y = -x + 3$ **38.** $y = -x - 2$ **39.** $y = -\frac{1}{2}x + 2$ **40.** $y = \frac{3}{2}x - 3$ **41.** $y = 4x - 3$ **43.** $y = -x - 7$ **45.** $y = 2x - 7$ **47.** $y = -4x - 1$ **49.** $y = -2x - 4$ **51.** $y = \frac{3}{4}x + 4$ **53.** $y = 3$ **55.** $x = 2$ **57.** $y = -6$ **59.** **(a)** 2 **(b)** $(0, -1)$ **(c)** $y = 2x - 1$ **(d)**

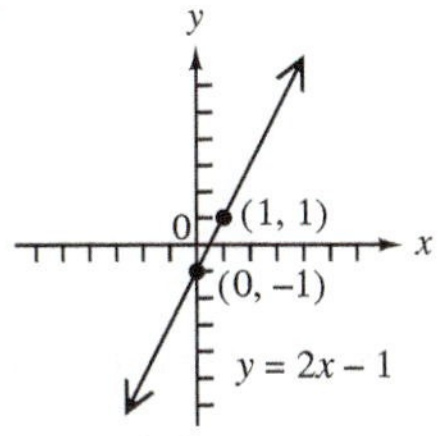

61. **(a)** \$400 **(b)** \$0.25 **(c)** $y = 0.25x + 400$ **(d)** \$425 **(e)** 1500 **63.** **(a)** 0.05; commission rate **(b)** (0, 2000); base salary per month **(c)** \$2500 **(d)** \$30,000 **65.** $y = -\frac{A}{B}x + \frac{C}{B}$ **66.** **(a)** $-\frac{2}{3}$ **(b)** 2 **(c)** $\frac{3}{7}$ **67.** $\left(0, \frac{C}{B}\right)$ **68.** **(a)** $(0, 6)$ **(b)** $\left(0, \frac{1}{2}\right)$ **(c)** $(0, -3)$

SECTION 3.5

1. **(a)** D **(b)** C **(c)** B **(d)** E **(e)** A **2.** $y = -2x + 9$; $2x + y = 9$ **3.** A, B, D **4.** In standard form, both equations are written $3x - 2y = 12$. They are different, but equivalent, forms of the same equation. **5.** $y = 5x + 2$ **7.** $y = x - 9$ **9.** $y = -3x - 4$ **11.** $y = -x + 1$ **13.** $y = \frac{2}{3}x + \frac{19}{3}$ **15.** $y = -\frac{4}{5}x + \frac{9}{5}$ **17.** **(a)** $y = x + 6$ **(b)** $x - y = -6$ **19.** **(a)** $y = -\frac{5}{7}x - \frac{54}{7}$ **(b)** $5x + 7y = -54$ **21.** **(a)** $y = -\frac{2}{3}x - 2$ **(b)** $2x + 3y = -6$ **23.** **(a)** $y = \frac{1}{3}x + \frac{4}{3}$ **(b)** $x - 3y = -4$ **25.** **(a)** $y = 3x - 9$ **(b)** $3x - y = 9$ **27.** **(a)** $y = -\frac{2}{3}x + \frac{4}{3}$ **(b)** $2x + 3y = 4$ **29.** $y = \frac{3}{4}x - 7$ **31.** $y = -2x - 3$ **33.** $y = -3x + 14$ **35.** $y = \frac{3}{4}x - \frac{9}{2}$ **37.** **(a)** (1, 2283), (2, 2441), (3, 2651), (4, 2792), (5, 2882) **(b)** yes

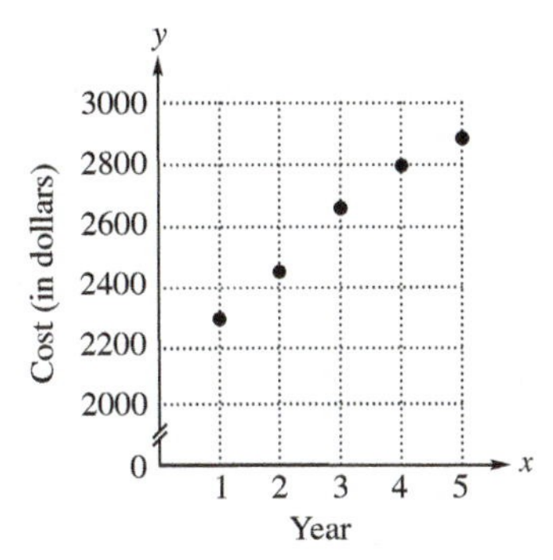

(c) $y = 147x + 2147$ **(d)** \$3176 $(x = 7)$ **39.** $y = 0.2375x + 59.7$ **41.** (0, 32); (100, 212) **42.** $\frac{9}{5}$ **43.** $F - 32 = \frac{9}{5}(C - 0)$ **44.** $F = \frac{9}{5}C + 32$ **45.** $C = \frac{5}{9}(F - 32)$ **46.** 86° **47.** 10° **48.** −40°

SUMMARY EXERCISES Applying Graphing and Equation-Writing Techniques for Lines

1. **(a)** B **(b)** D **(c)** A **(d)** C **2.** A, B

3.

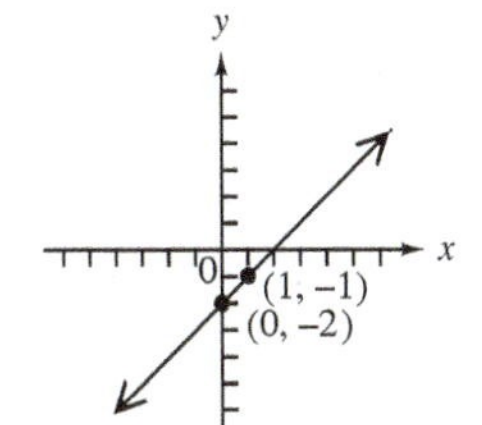

4.

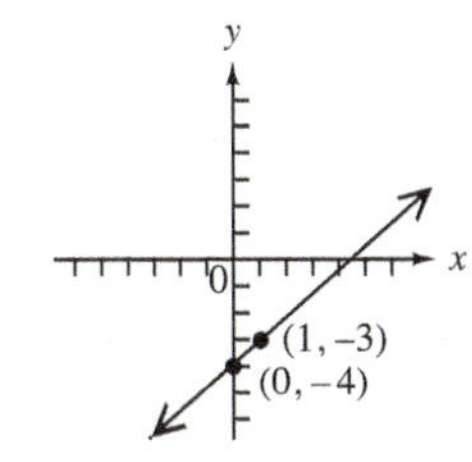

5.

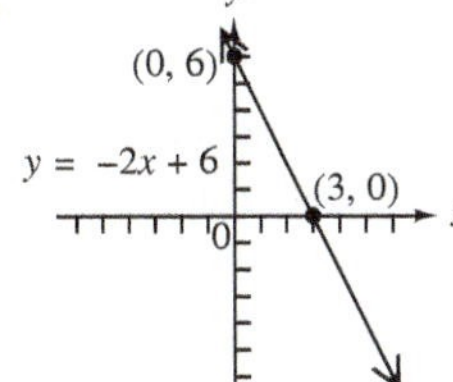

6.

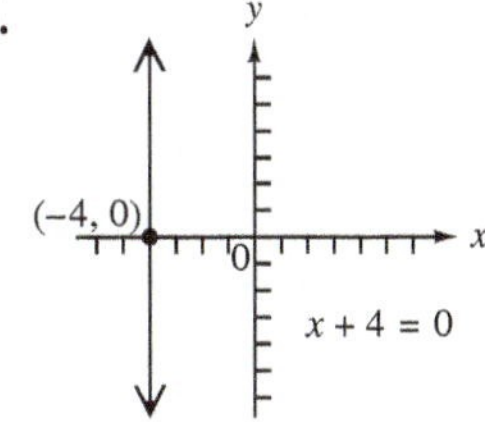

7.

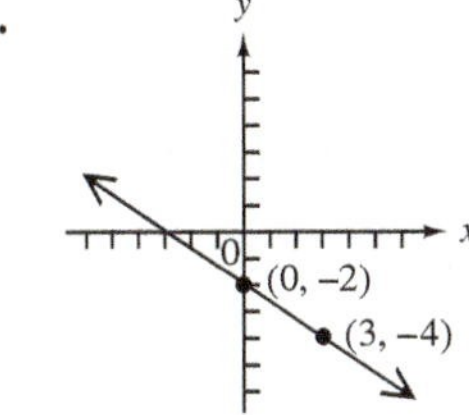

8.

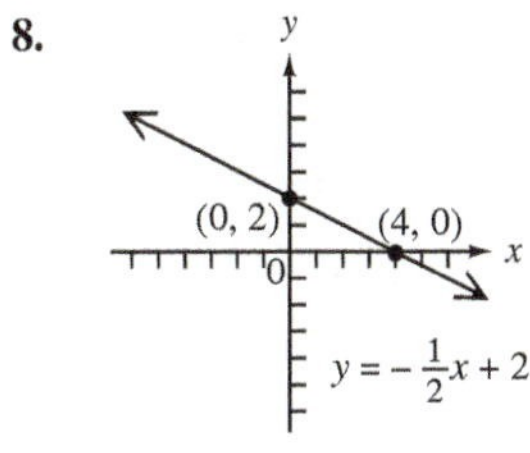

ANSWERS

9.

$y - 4 = -9$ (0, −5)

10.
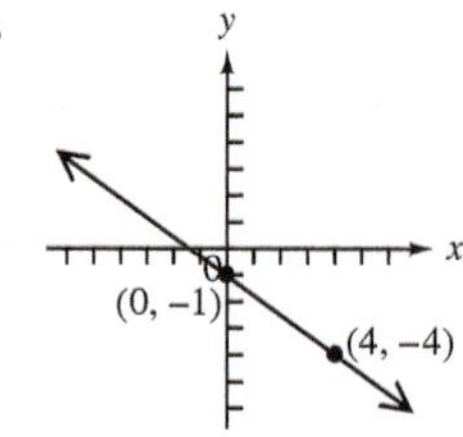

11.
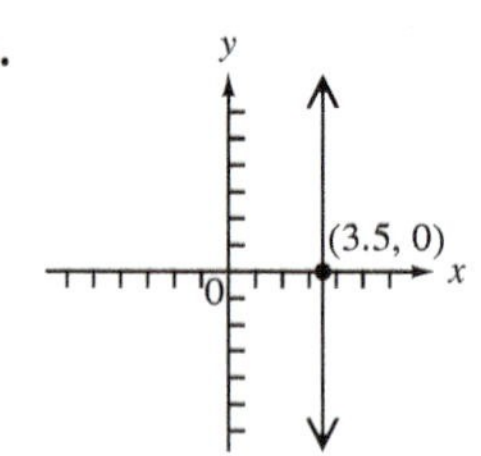

12.
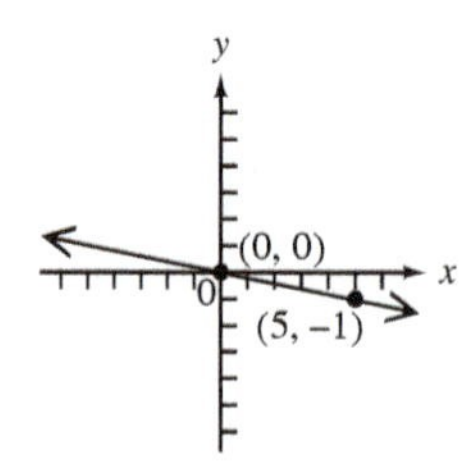

13.
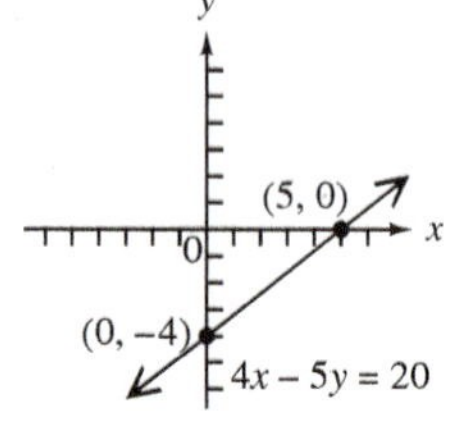

14.
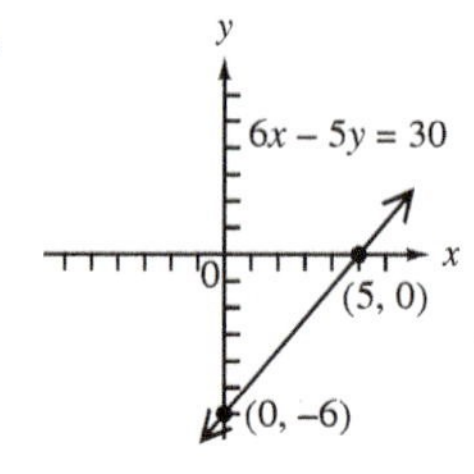

15.
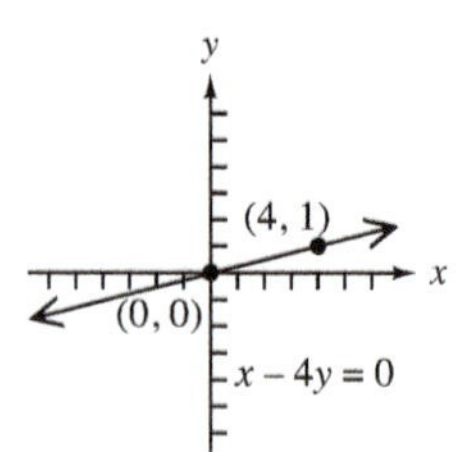

16.
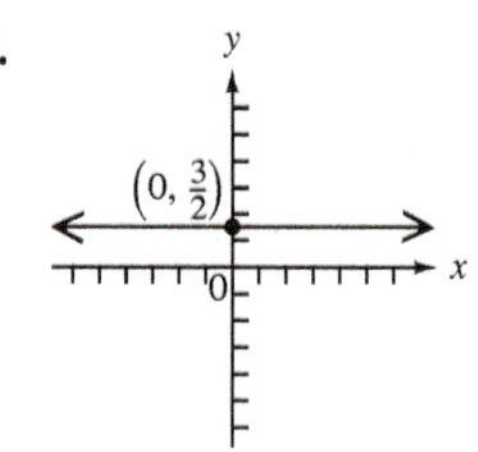

17.
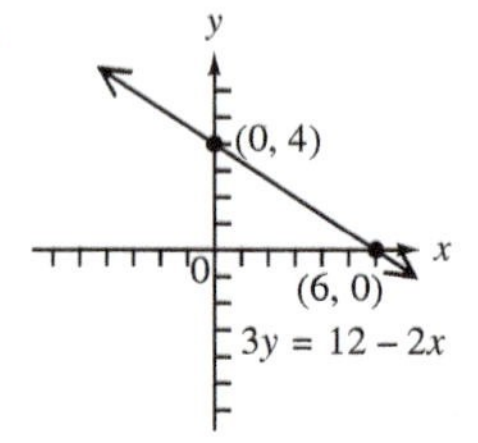

18.
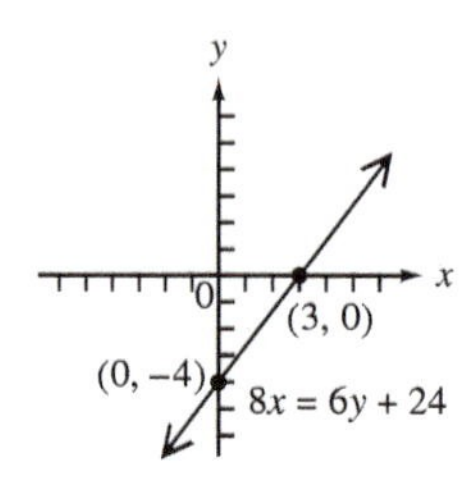

19. $y = -2x + 6$ **20.** $y = \frac{4}{3}x + 8$ **21.** $y = \frac{1}{2}x - 2$ **22.** $y = -\frac{2}{3}x + 5$

23. $y = -3x - 6$ **24.** $y = -4x - 3$ **25.** $y = \frac{3}{5}x$ **26.** $x = 0$

27. $y = 0$ **28.** $y = \frac{5}{3}x + 5$

SECTION 3.6

1. $>, >$ **2.** $<$ **3.** $\leq$ **4.** $\geq$ **5.** false; The point $(4, 0)$ lies on the boundary line $3x - 4y = 12$, which is *not* part of the graph because the symbol $<$ does not involve equality. **6.** true **7.** false; Because $(0, 0)$ is on the boundary line $x + 4y = 0$, it cannot be used as a test point. Use a test point *off* the line. **8.** false; Use a solid line for the boundary line because the symbol $\geq$ involves equality. **9.** true **10.** true **11.** **(a)** no **(b)** no **(c)** yes **(d)** no **12.** **(a)** yes **(b)** no **(c)** yes **(d)** yes **13.** Use a dashed line if the symbol is $<$ or $>$. Use a solid line if the symbol is $\leq$ or $\geq$. **14.** A test point cannot lie on the boundary line. It must lie on one side of the boundary.

15.
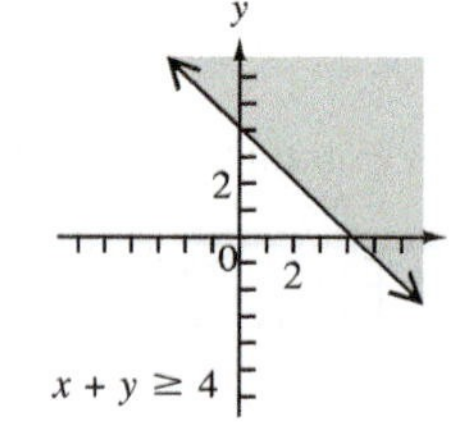

17.
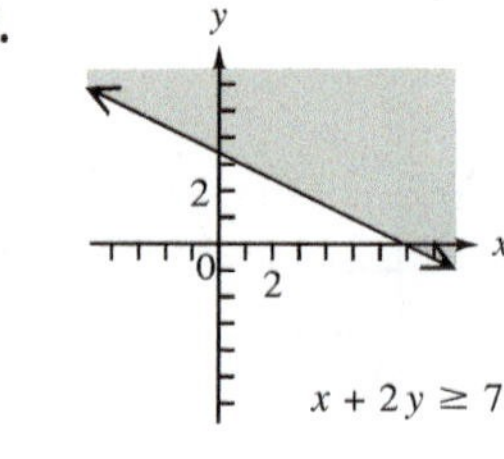

19.
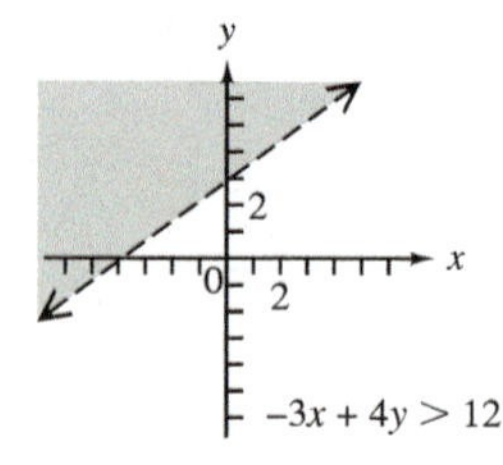

21.
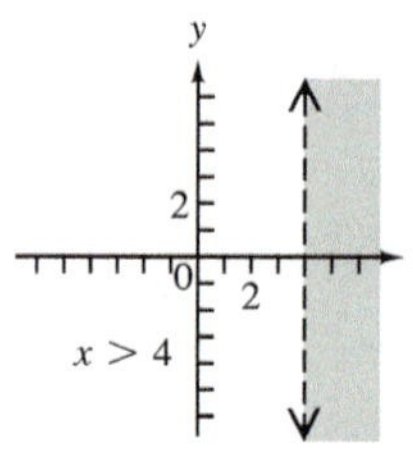

23.
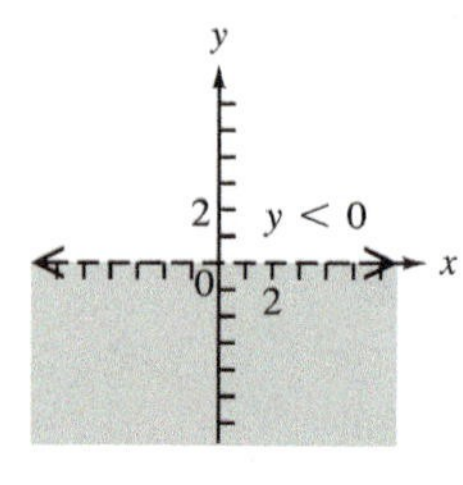

25.
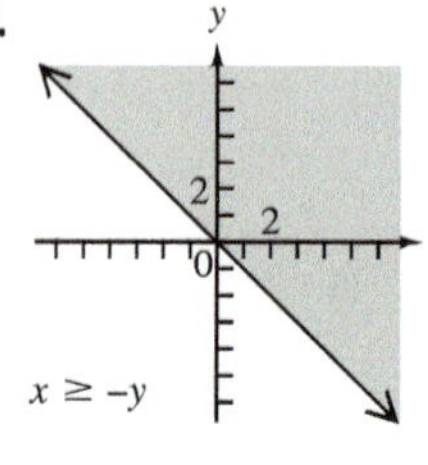

27.
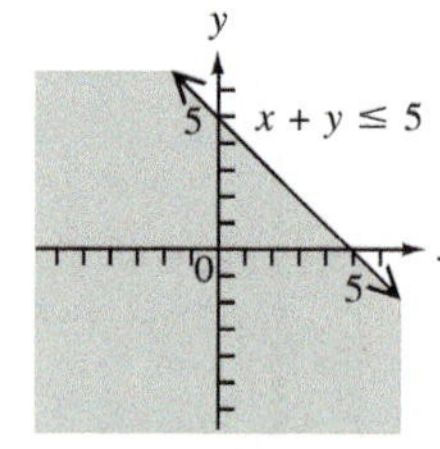

29.
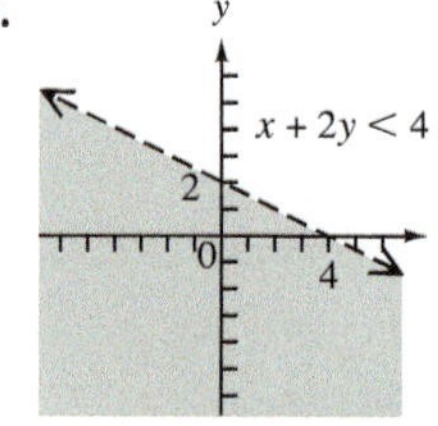

31.
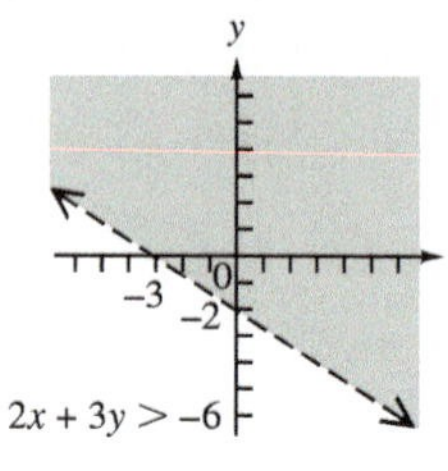

33.
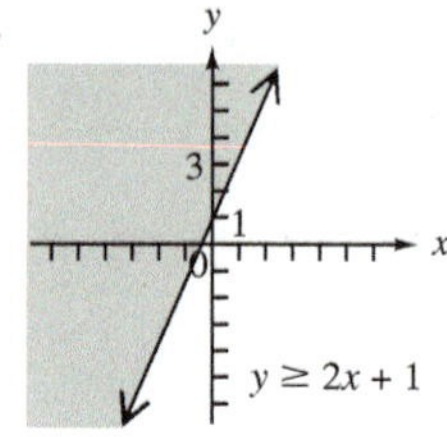

35.
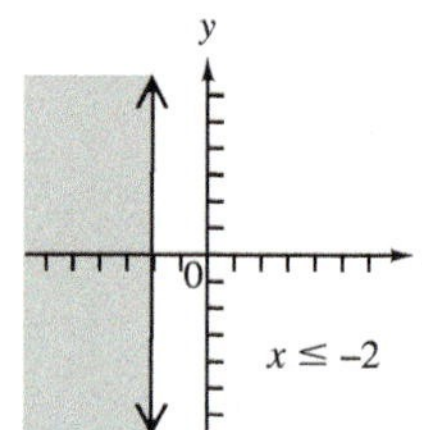

37.
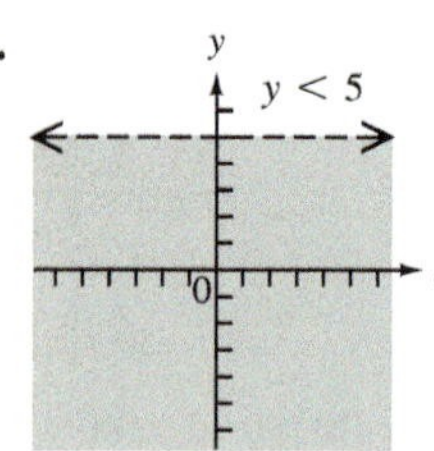

39.
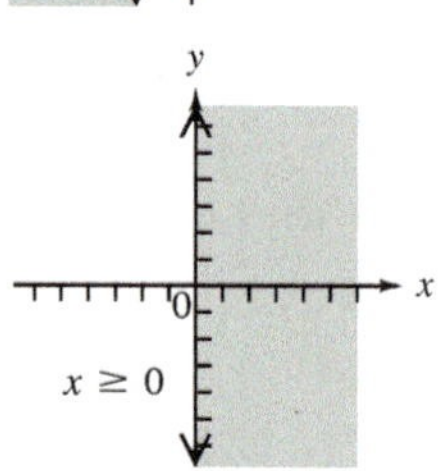

41.
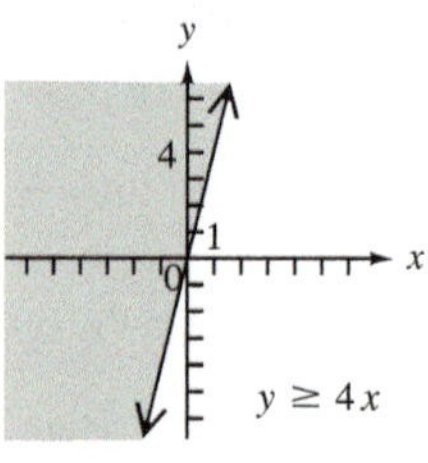

43.
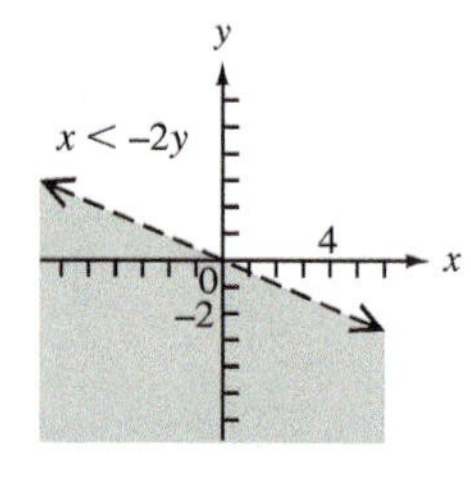

45.
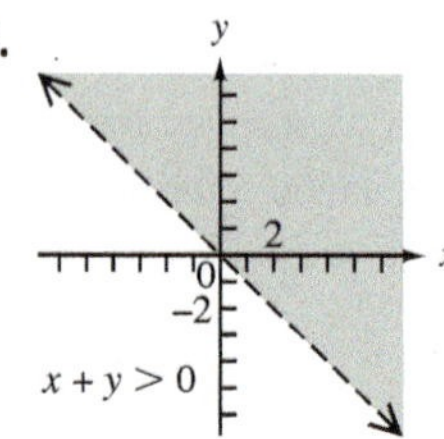

Chapter 3 REVIEW EXERCISES

1. The percent was decreasing. **2.** 2010: 39.5%; 2013: 36% **3.** 3.5% **4.** In the year 2015, 36.4% of students at 4-yr public institutions earned a degree within 5 yr. **5.** -1; 2; 1 **6.** 2; $\frac{3}{2}$; $\frac{14}{3}$ **7.** 0; $\frac{8}{3}$; -9 **8.** 7; 7; 7 **9.** yes **10.** no **11.** yes **12.** no **13.** quadrant I **14.** quadrant II **15.** no quadrant **16.** no quadrant **17.** x is positive in quadrants I and IV; y is negative in quadrants III and IV. Thus, if x is positive and y is negative, (x, y) must lie in quadrant IV. **18.** In the ordered pair $(k, 0)$, the y-value is 0, so the point lies on the x-axis. In the ordered pair $(0, k)$, the x-value is 0, so the point lies on the y-axis.

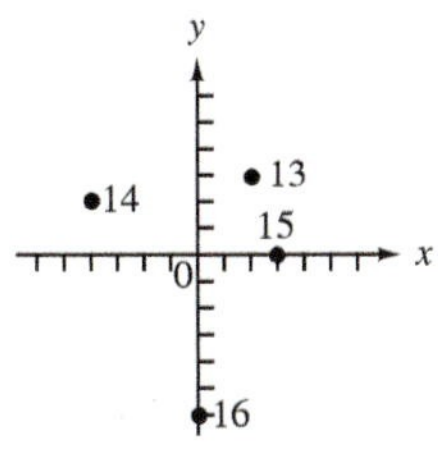

Graph for Exercises 13–16

19.

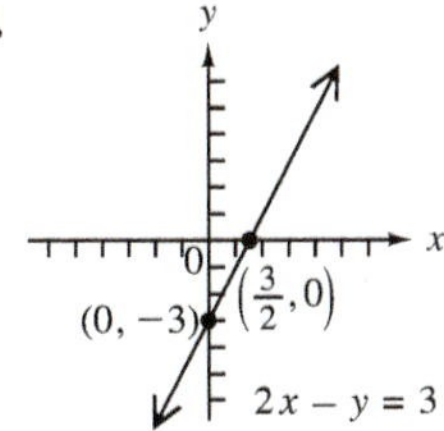

20.

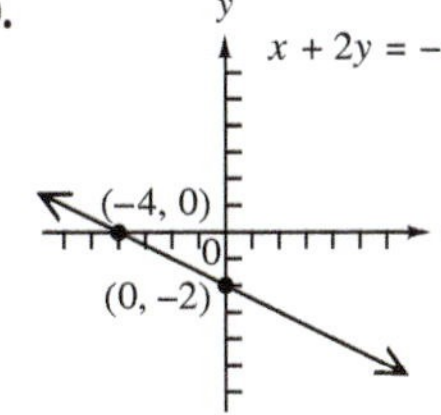

21.

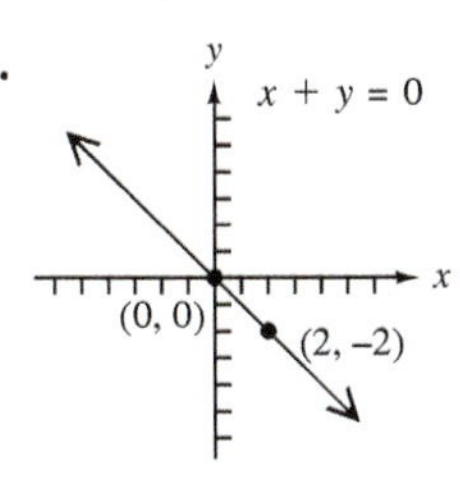

22.

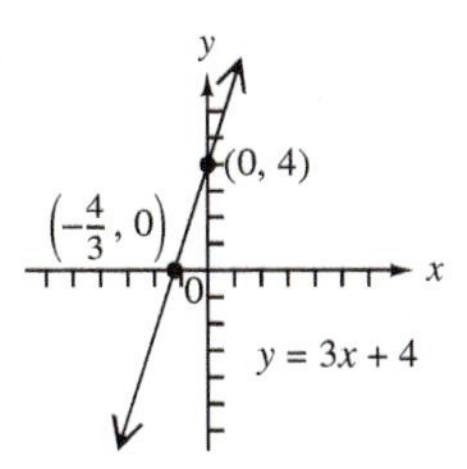

23. $-\frac{1}{2}$ **24.** $-\frac{2}{3}$ **25.** 0 **26.** undefined **27.** 3 **28.** $\frac{2}{3}$ **29.** $\frac{3}{2}$ **30.** $-\frac{1}{3}$ **31.** $\frac{3}{2}$ **32.** 0 **33.** 2 **34.** $\frac{1}{3}$ **35.** parallel **36.** perpendicular **37.** neither **38.** 0 **39.** $y = -x + \frac{2}{3}$ **40.** $y = -\frac{1}{3}x + 1$ **41.** $y = x - 7$ **42.** $y = \frac{2}{3}x + \frac{14}{3}$ **43.** $y = -\frac{3}{4}x - \frac{1}{4}$ **44.** $y = -\frac{1}{4}x + \frac{3}{2}$ **45.** $y = 1$ **46.** $x = \frac{1}{3}$ **47. (a)** $y = -\frac{1}{3}x + 5$ **(b)** slope: $-\frac{1}{3}$; y-intercept: $(0, 5)$

(c)

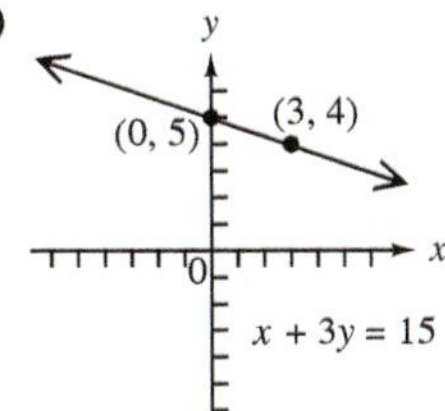

48. positive; The graph of the line rises from left to right, so the slope is positive. **49.** $(1, 240)$, $(4, 1508)$; $y = 423x - 183$ **50.** $663 million $(x = 2)$

51.

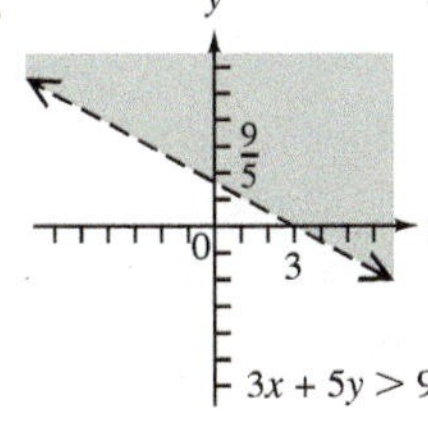

52.

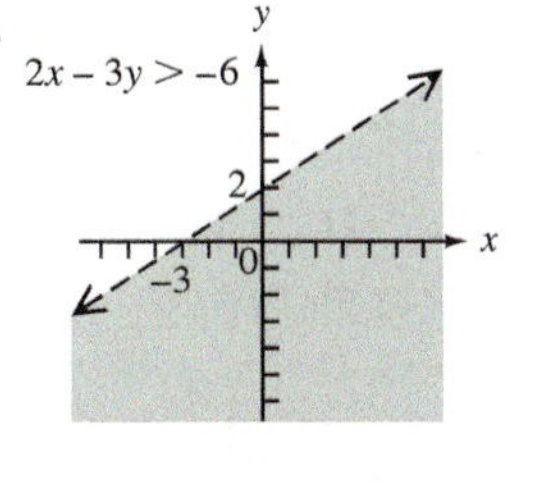

53.

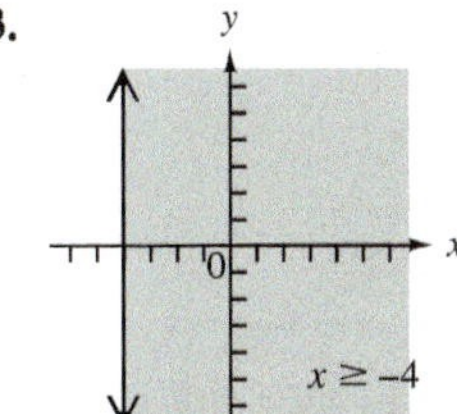

54.

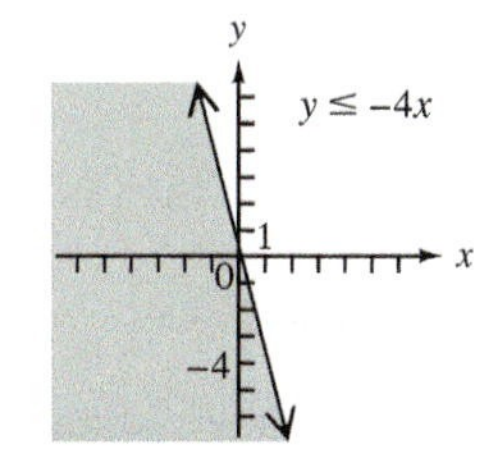

Chapter 3 MIXED REVIEW EXERCISES

1. A **2.** C, D **3.** A, B, D **4.** D **5.** C **6.** B

7. $(0, -5)$; $\left(-\frac{5}{2}, 0\right)$; -2

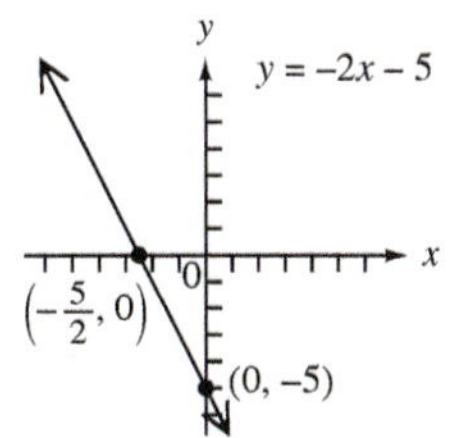

8. $(0, 0)$; $(0, 0)$; $-\frac{1}{3}$

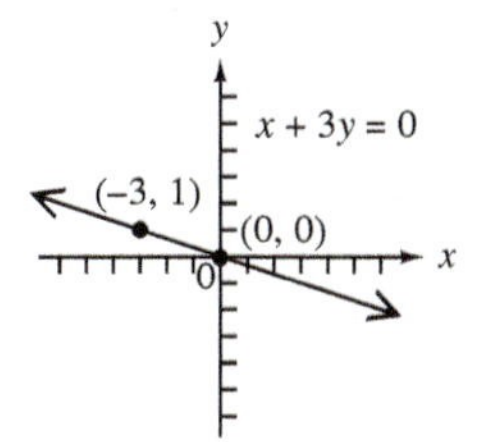

9. $(0, 5)$; no x-intercept; 0

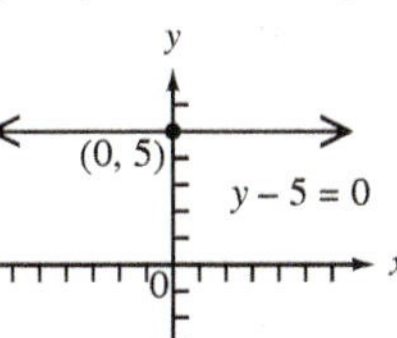
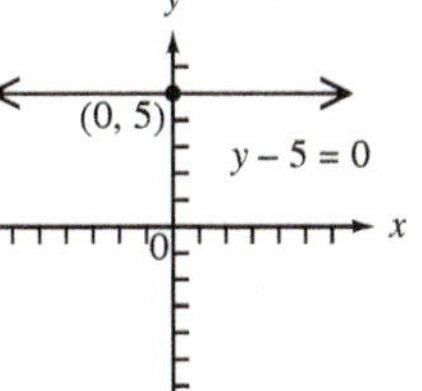

10. no y-intercept; $(-1, 0)$; undefined

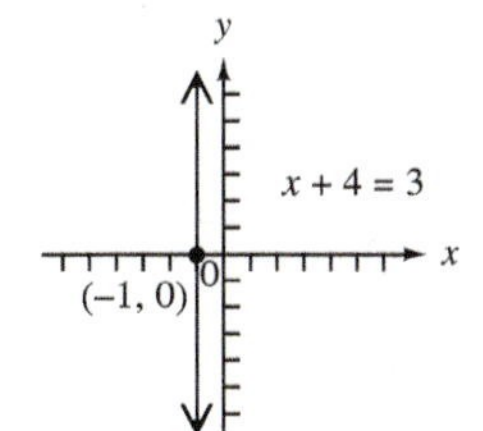

11. (a) $y = -\frac{1}{4}x - \frac{5}{4}$ **(b)** $x + 4y = -5$ **12. (a)** $y = -3x + 30$ **(b)** $3x + y = 30$ **13. (a)** $y = -\frac{4}{7}x - \frac{23}{7}$ **(b)** $4x + 7y = -23$ **14. (a)** $y = -5$ **(b)** $y = -5$

Chapter 3 TEST

1. -6; -10; -5 **2.** no **3.** To find the x-intercept, let $y = 0$, and to find the y-intercept, let $x = 0$. **4.** true

5.

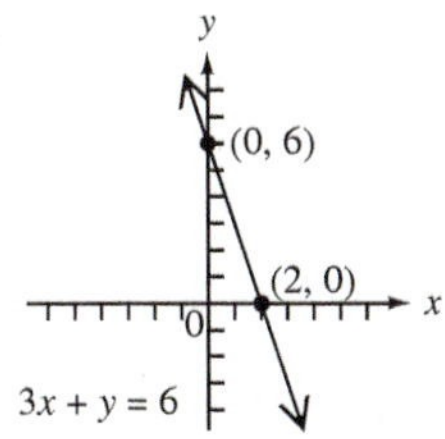

6.

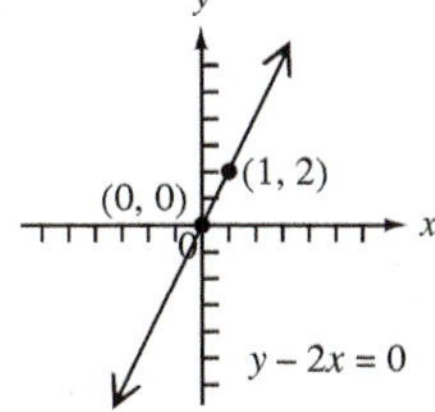

7.

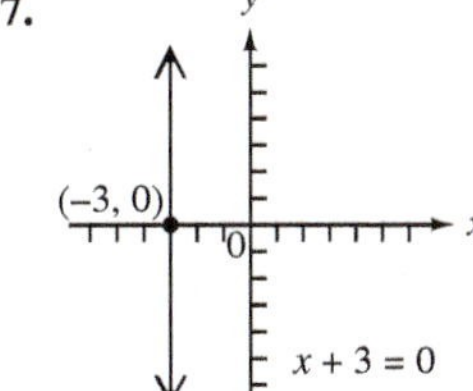

8.

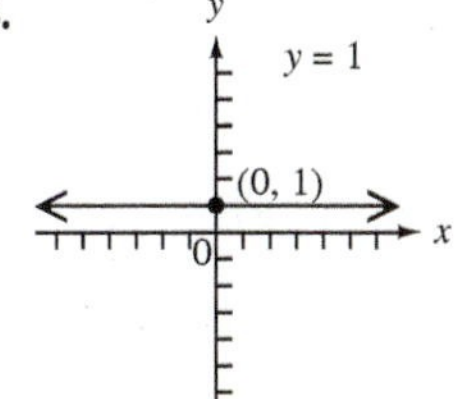

9. 1; $(0, -4)$

10.

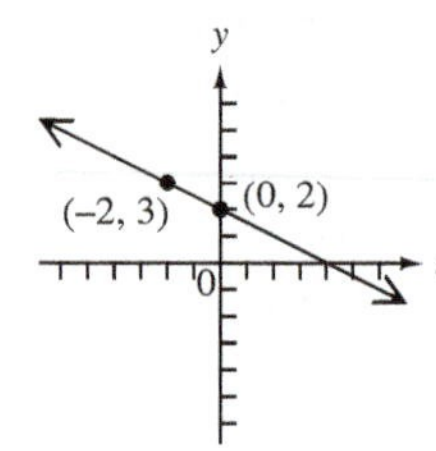

11. $-\frac{8}{3}$ **12.** -2 **13.** undefined **14.** $\frac{5}{2}$ **15.** 0 **16.** $-\frac{1}{4}$

17. $y = 2x + 6$ **18.** $y = \frac{5}{2}x - 4$ **19.** $y = -9x + 12$ **20.** $y = -\frac{3}{2}x + \frac{9}{2}$

21.

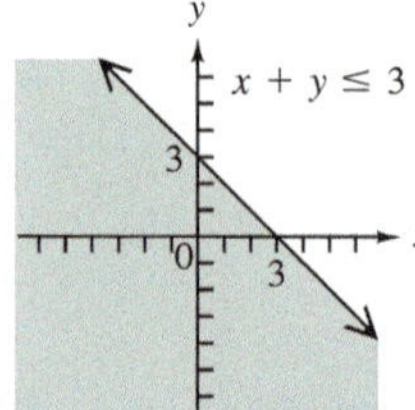

22.

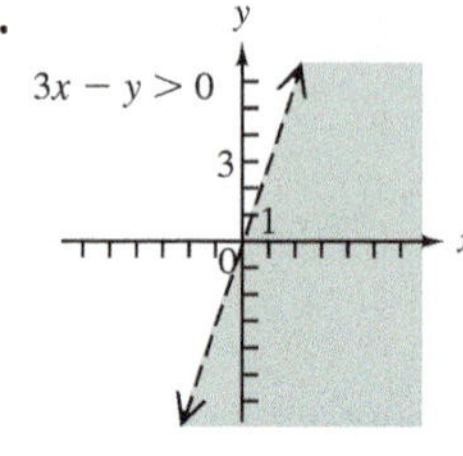

23.

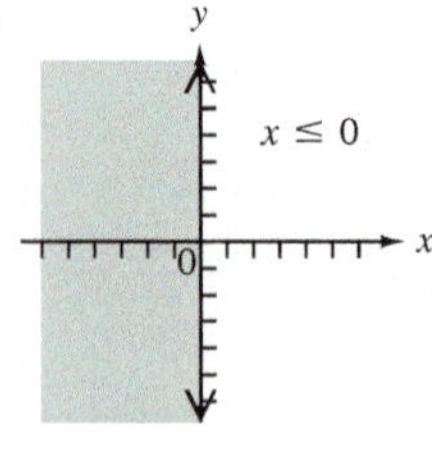

24. The graph falls from left to right, so the slope is negative.
25. $(0, 209), (15, 151); -3.9$ **26.** $y = -3.9x + 209$
27. 170 thousand **28.** In 2015, worldwide snowmobile sales were 151 thousand.

Chapters R–3 CUMULATIVE REVIEW EXERCISES

1. $\frac{301}{40}$, or $7\frac{21}{40}$ **2.** 6 **3.** 7 **4.** $\frac{73}{18}$ **5.** true **6.** -43

7. distributive property **8.** $-p + 2$ **9.** $h = \frac{3V}{\pi r^2}$ **10.** $\{-1\}$ **11.** $\{2\}$

12. $\{-13\}$ **13.** $(-2.6, \infty)$

14. $(0, \infty)$ **15.** $(-\infty, -4]$

16. high school credential: \$30,000; bachelor's degree: \$48,500 **17.** 13 mi **18.** **(a)** 85.63; 77.82; 76.09; 50 **(b)** $(20, 85.63), (38, 77.82), (42, 76.09), (50, 72.62)$ **(c)** In 2014, the winning time was approximately 70.89 sec. **19.** **(a)** 19.8 million **(b)** 13.2 million **(c)** 7.8 million

20. $(-4, 0); (0, 3)$ **21.** $\frac{3}{4}$ **22.** 6

23.

24. perpendicular **25.** $y = 3x - 11$
26. $y = 4$

CHAPTER 4 Systems of Linear Equations and Inequalities

SECTION 4.1

1. system of linear equations; same **2.** ordered pair; true **3.** inconsistent; no; independent **4.** solution; consistent **5.** dependent; consistent; infinitely many **6.** parallel; solution **7.** It is not a solution of the system because it is not a solution of the second equation, $2x + y = 4$.
8. $\{(x, y) \mid 3x - 2y = 4\}$ **9.** B; The ordered pair must be in quadrant II.
10. D; The ordered pair must be on the y-axis, with $y < 0$. **11.** no
13. yes **15.** yes **17.** no **19.** yes **21.** no

We show the graphs here only for Exercises 23–27.

23. $\{(4, 2)\}$

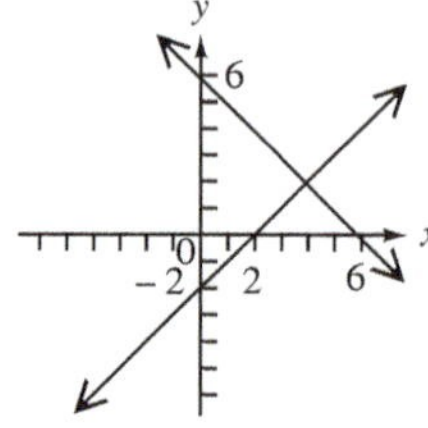

25. $\{(0, 4)\}$

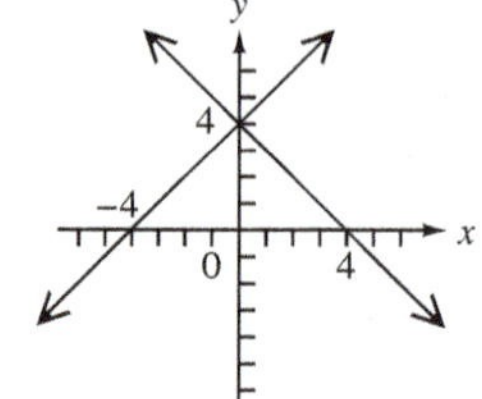

27. $\{(4, -1)\}$

29. $\{(1, 3)\}$ **31.** $\{(0, 2)\}$
33. $\varnothing$ (inconsistent system)
35. $\{(x, y) \mid 5x - 3y = 2\}$ (dependent equations)
37. $\{(4, -3)\}$
39. $\{(x, y) \mid 2x - y = 4\}$ (dependent equations)

41. $\varnothing$ (inconsistent system) **43.** **(a)** neither **(b)** intersecting lines **(c)** one solution **45.** **(a)** dependent **(b)** one line **(c)** infinite number of solutions **47.** **(a)** inconsistent **(b)** parallel lines **(c)** no solution
49. **(a)** neither **(b)** intersecting lines **(c)** one solution
51. **(a)** dependent **(b)** one line **(c)** infinite number of solutions
53. **(a)** inconsistent **(b)** parallel lines **(c)** no solution
55. 40; 30 **57.** Supply exceeds demand. **59.** 1980–2000 **61.** 2006; 600 (million) units **62.** (2006, 600) **63.** The slope would be negative. Sales of CDs were decreasing during this period. **64.** The slope would be positive. Sales of digital downloads were increasing during this period.

SECTION 4.2

1. The student must find the value of y and write the solution as an ordered pair. The solution set is $\{(3, 0)\}$. **2.** The true result $0 = 0$ means that the system has an infinite number of solutions. The solution set is $\{(x, y) \mid x + y = 4\}$. **3.** A false statement, such as $0 = 3$, occurs.
4. A true statement, such as $0 = 0$, occurs. **5.** $\{(3, 9)\}$ **7.** $\{(7, 3)\}$
9. $\{(-4, 8)\}$ **11.** $\{(3, -2)\}$ **13.** $\{(0, 5)\}$ **15.** $\{(1, 5)\}$
17. $\{(x, y) \mid 3x - y = 5\}$ **19.** $\varnothing$ **21.** $\{(x, y) \mid 2x - y = -12\}$

23. $\{(0, 0)\}$ **25.** $\varnothing$ **27.** $\left\{\left(\frac{1}{3}, -\frac{1}{2}\right)\right\}$ **29.** $\{(2, -3)\}$

31. $\{(2, -4)\}$ **33.** $\{(-4, 2)\}$ **35.** $\{(5, 0)\}$ **37.** $\{(7, -3)\}$
39. $\{(2, 3)\}$ **41.** To find the total cost, multiply the number of bicycles (x) by the cost per bicycle (400 dollars) and add the fixed cost (5000 dollars). Thus $y_1 = 400x + 5000$ gives this total cost (in dollars).
42. $y_2 = 600x$ **43.** $y_1 = 400x + 5000$, $y_2 = 600x$; solution set: $\{(25, 15{,}000)\}$ **44.** 25: 15,000; 15,000

SECTION 4.3

1. true **2.** false; Multiply by -3. **3.** The student incorrectly stated the solution set. A false statement indicates that the solution set is $\varnothing$.
4. The student did not multiply *both* sides of equation (1) by 3. The correct solution set is $\{(x, y) \mid 2x - y = 5\}$. **5.** $\{(-1, 3)\}$ **7.** $\{(-1, -3)\}$
9. $\{(5, 3)\}$ **11.** $\{(-2, 3)\}$ **13.** $\left\{\left(\frac{1}{2}, 4\right)\right\}$ **15.** $\{(-3, 4)\}$
17. $\{(3, -6)\}$ **19.** $\{(7, 4)\}$ **21.** $\{(0, 4)\}$ **23.** $\{(-4, 0)\}$
25. $\{(0, 0)\}$ **27.** $\varnothing$ **29.** $\{(x, y) \mid x - 3y = -4\}$ **31.** $\{(0, 7)\}$
33. $\{(-6, 5)\}$ **35.** $\left\{\left(-\frac{5}{7}, -\frac{2}{7}\right)\right\}$ **37.** $\left\{\left(\frac{1}{8}, -\frac{5}{6}\right)\right\}$ **39.** $\varnothing$
41. $\{(x, y) \mid 2x + y = 0\}$ **43.** $\{(11, 15)\}$ **45.** $\left\{\left(13, -\frac{7}{5}\right)\right\}$
47. $\{(6, -4)\}$ **49.** $6.21 = 2004a + b$ **50.** $8.17 = 2014a + b$
51. $2004a + b = 6.21$, $2014a + b = 8.17$; solution set: $\{(0.196, -386.574)\}$ **52. (a)** $y = 0.196x - 386.574$
(b) \$7.97; This is a bit less than the actual figure.

SUMMARY EXERCISES Applying Techniques for Solving Systems of Linear Equations

1. (a) Use substitution because the second equation is solved for y.
(b) Use elimination because the coefficients of the y-terms are opposites.
(c) Use elimination because the equations are in $Ax + By = C$ form with no coefficients of 1 or -1. Solving by substitution would involve fractions.
2. System B is easier to solve by substitution because the second equation is already solved for y. **3. (a)** $\{(1, 4)\}$ **(b)** $\{(1, 4)\}$
(c) Answers will vary. **4. (a)** $\{(-5, 2)\}$ **(b)** $\{(-5, 2)\}$
(c) Answers will vary. **5.** $\{(2, 6)\}$ **6.** $\{(-3, 2)\}$ **7.** $\left\{\left(\frac{1}{3}, \frac{1}{2}\right)\right\}$
8. $\varnothing$ **9.** $\{(3, 0)\}$ **10.** $\left\{\left(\frac{3}{2}, -\frac{3}{2}\right)\right\}$ **11.** $\{(x, y) \mid 3x + y = 7\}$
12. $\{(9, 4)\}$ **13.** $\left\{\left(\frac{45}{31}, \frac{4}{31}\right)\right\}$ **14.** $\{(4, -5)\}$ **15.** $\varnothing$ **16.** $\{(0, 0)\}$
17. $\left\{\left(\frac{19}{3}, -5\right)\right\}$ **18.** $\left\{\left(\frac{22}{13}, -\frac{23}{13}\right)\right\}$ **19.** $\{(-12, -60)\}$
20. $\{(2, -3)\}$ **21.** $\{(24, -12)\}$ **22.** $\{(-2, 1)\}$ **23.** $\{(-35, 13)\}$
24. $\{(10, -9)\}$ **25.** $\{(-4, 6)\}$ **26.** $\{(5, 3)\}$

SECTION 4.4

1. D **2.** A **3.** B **4.** C **5.** D **6.** D **7.** C **8.** C **9.** B **10.** A
11. the second number; $x - y = 48$ (or $y - x = 48$); The two numbers are 73 and 25. **13.** *The Phantom of the Opera*: 11,669; *The Lion King*: 7603
15. *Furious 7*: \$353 million; *Minions*: \$336 million **17.** Terminal Tower: 708 ft; Key Tower: 947 ft **19.** variables; width; Equation (1): $x = 38 + y$; length; twice; Equation (2): $2x + 2y = 188$; length: 66 yd; width: 28 yd
21. (a) 45 units **(b)** Do not produce—the product will lead to a loss.
23. table entries: (third column) $1x$ (or x), $10y$; 46 ones; 28 tens
25. 5 DVDs of *Ant-Man*; 2 Blu-ray discs of *The Martian*
27. table entries: (second column) 4%, or 0.04; (third column) $0.04y$; Equation (1): $x = 2y$; Equation (2): $0.05x + 0.04y = 350$; \$2500 at 4%; \$5000 at 5% **29.** Taylor Swift: \$112; Kenny Chesney: \$85 **31.** table entries: (third column) $0.40x$, $0.70y$, $0.50(120)$, or 60; 80 L of 40% solution; 40 L of 70% solution **33.** table entries: (second column) 3, 4; (third column) $6x$, $3y$, $4(90)$, or 360; 30 lb at \$6 per lb; 60 lb at \$3 per lb
35. nuts: 40 lb; raisins: 20 lb **37.** table entries: (third column) 4.5, 4.5; (fourth column) $4.5x$, $4.5y$; Equation (1): $4.5x + 4.5y = 495$; Equation (2): $x = 10 + y$; 60 mph; 50 mph **39.** bicycle: 13.5 mph; car: 46.5 mph **41.** car leaving Cincinnati: 55 mph; car leaving Toledo: 70 mph **43.** table entries: (third column) 3, 3; (fourth column) 24; boat: 10 mph; current: 2 mph **45.** plane: 470 mph; wind: 30 mph
47. Roberto: 17.5 mph; Juana: 12.5 mph

SECTION 4.5

1. C **2.** A **3.** B **4.** D
5. (a) no **(b)** yes

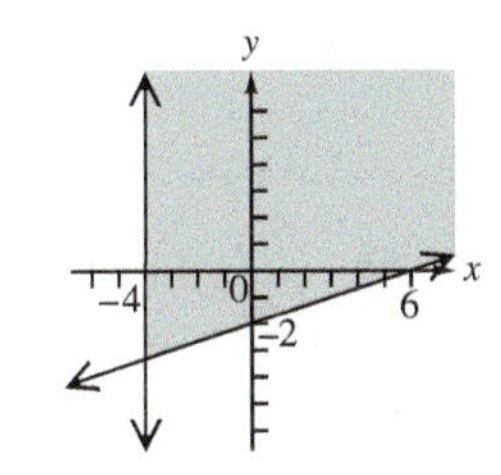

6. (a) no **(b)** yes

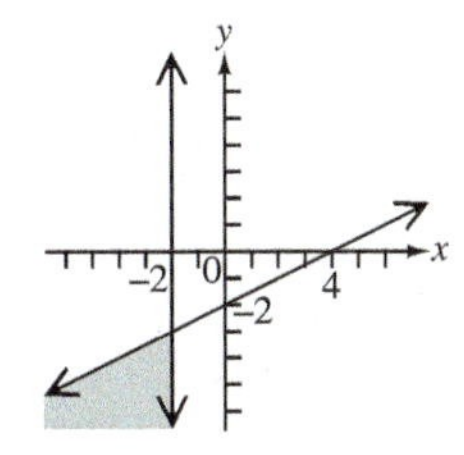

7. (a) yes **(b)** no

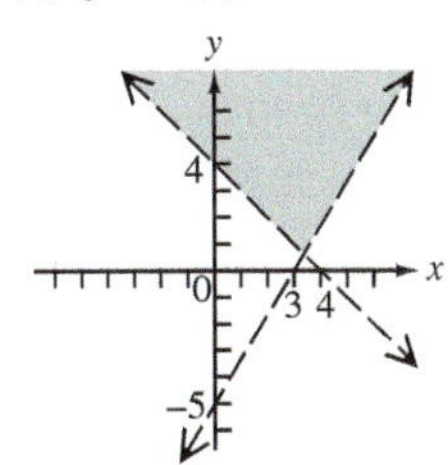

8. (a) yes **(b)** no

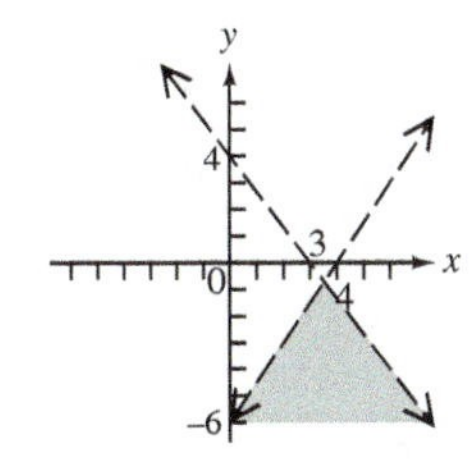

9.

$x + y \le 6$
$x - y \ge 1$

11.

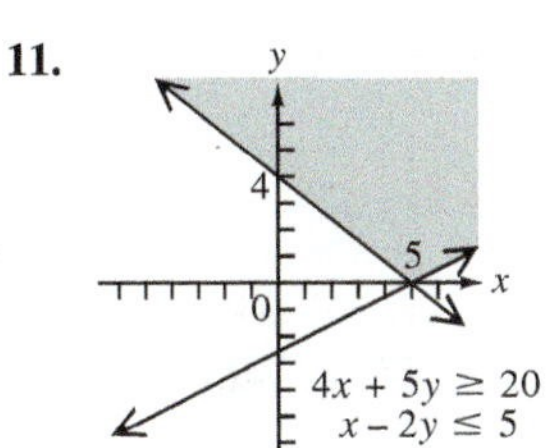

13.

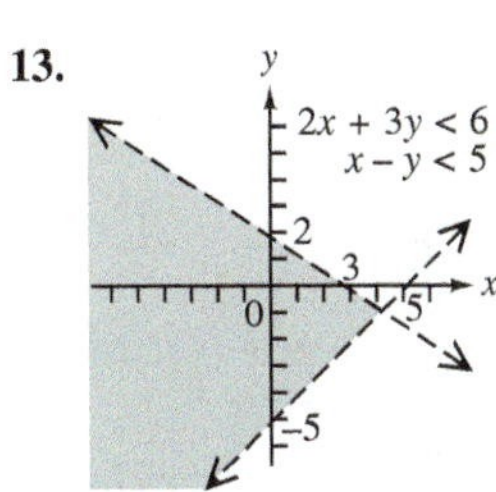

15.

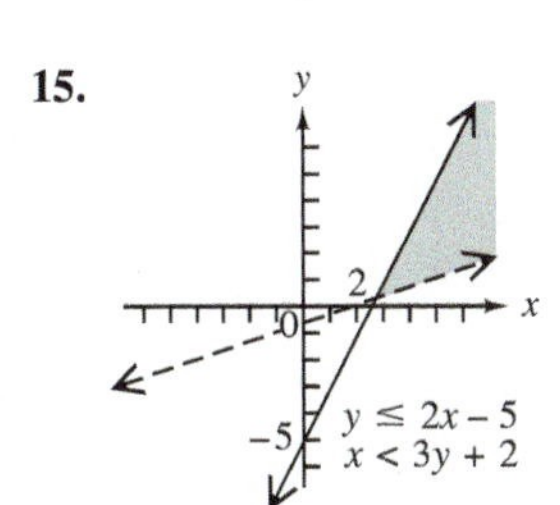

17.

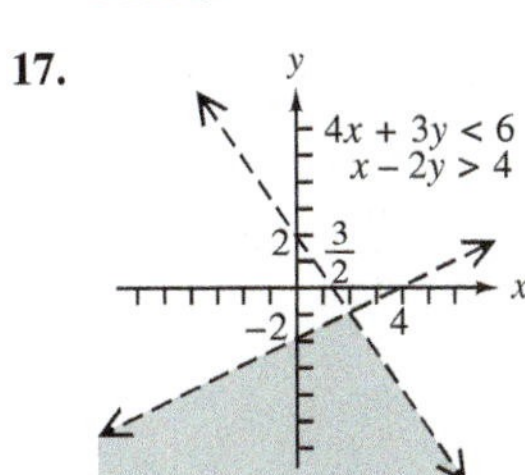

19.

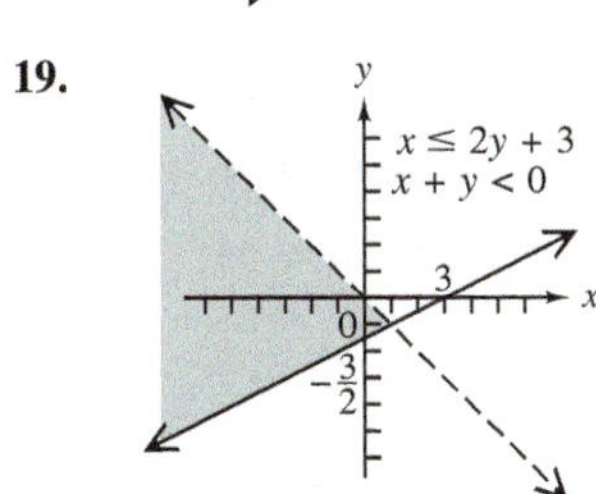

21.

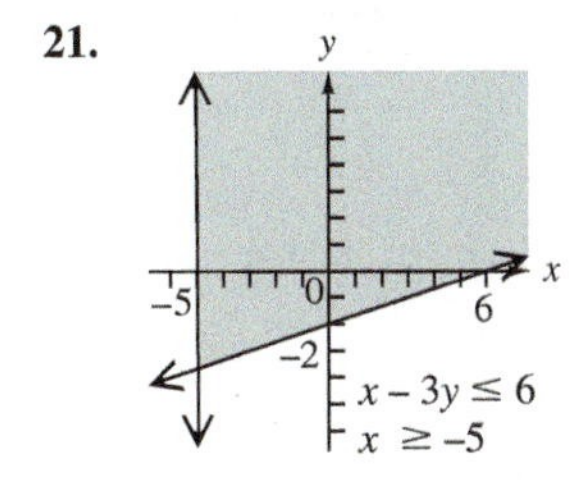

23.

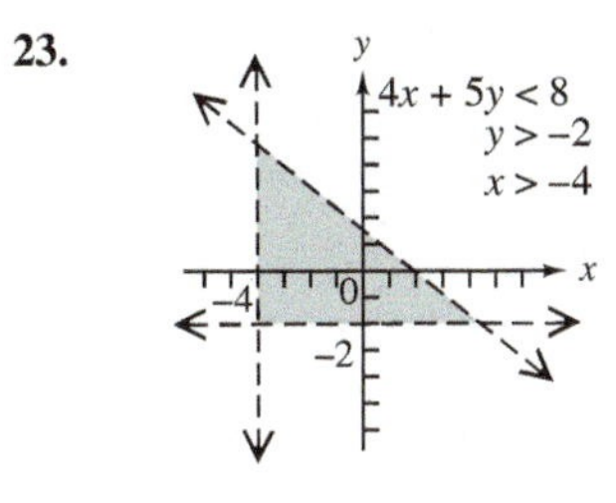

ANSWERS

25.

$x + y \geq -3$
$x - y \leq 3$
$y \leq 3$

27. 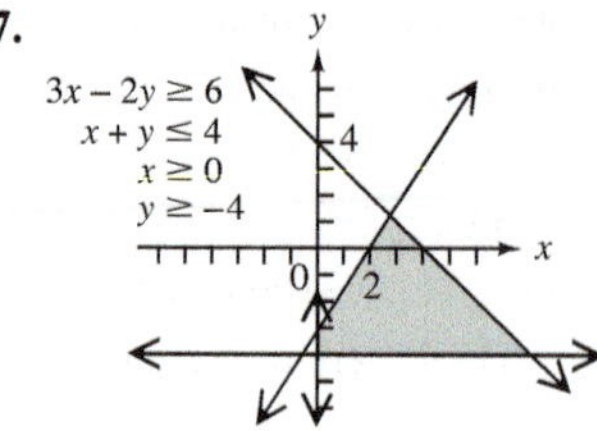

Chapter 4 REVIEW EXERCISES

1. yes **2.** no

We do not show the graphs for Exercises 3–6.

3. $\{(3, 1)\}$ **4.** $\{(0, -2)\}$ **5.** $\{(x, y) \,|\, x - 2y = 2\}$ **6.** $\varnothing$ **7.** $\{(2, 1)\}$ **8.** $\{(3, 5)\}$ **9.** $\{(6, 4)\}$ **10.** $\varnothing$ **11.** $\{(7, 1)\}$ **12.** $\{(-5, -2)\}$ **13.** $\{(-4, 3)\}$ **14.** $\{(x, y) \,|\, 3x - 4y = 9\}$ **15.** $\{(9, 2)\}$ **16.** $\left\{\left(\frac{10}{7}, -\frac{9}{7}\right)\right\}$ **17.** $\{(0, 0)\}$ **18.** $\left\{\left(\frac{3}{2}, 0\right)\right\}$ **19.** $\{(8, 9)\}$ **20.** $\{(2, 1)\}$ **21.** $\{(7, -2)\}$ **22.** $\{(-4, 2)\}$

23. Pizza Hut: 15,605 locations: Domino's: 11,629 locations

24. *Reader's Digest*: 2.7 million; *People*: 3.5 million

25. table entries: (first column) x; (second column) 0.90; (third column) $0.90y$, $100(1)$, or 100; 25 lb of \$1.30 candy; 75 lb of \$0.90 candy

26. table entries: (first column) y, 20; (second column) 20; (third column) $10x$; 13 twenties; 7 tens **27.** length: 27 m; width: 18 m **28.** plane: 250 mph; wind: 20 mph **29.** table entries: (second column) 0.04; (third column) $0.03x$, $0.04y$, 650; \$7000 at 3%; \$11,000 at 4%

30. table entries: (second column) 0.70; (third column) $0.40x$, $0.70y$, $0.50(90)$, or 45; 60 L of 40% solution; 30 L of 70% solution

31. 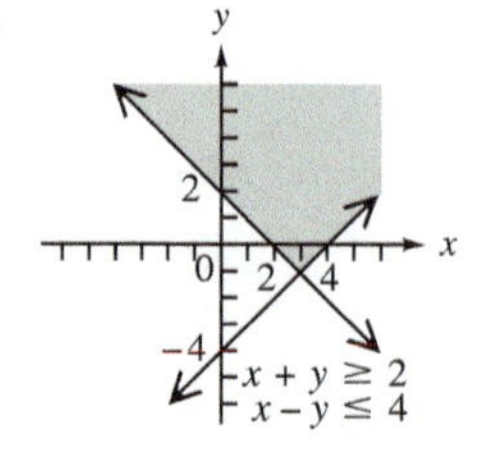

32. 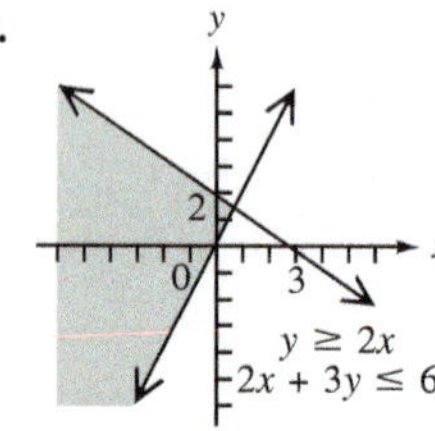

33. 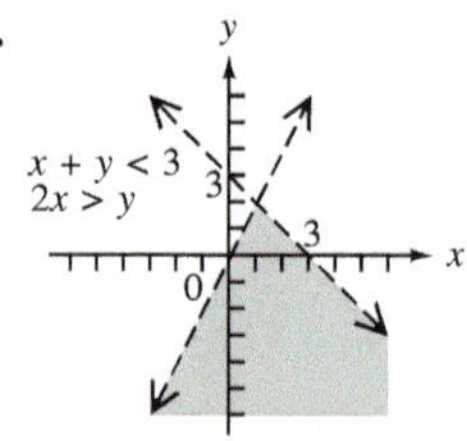

34. 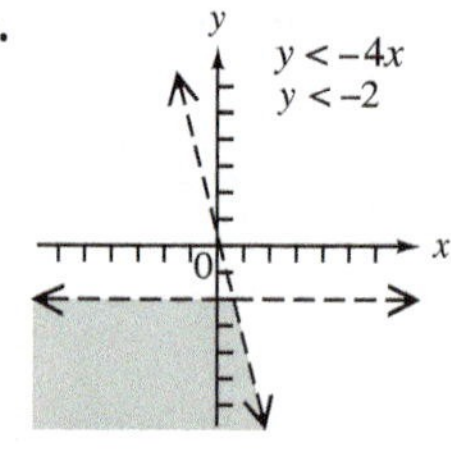

Chapter 4 MIXED REVIEW EXERCISES

1. $\{(2, 0)\}$ **2.** $\{(-4, 15)\}$ **3.** $\varnothing$

4. 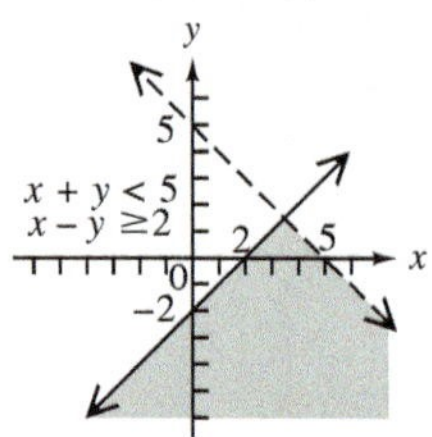

5. 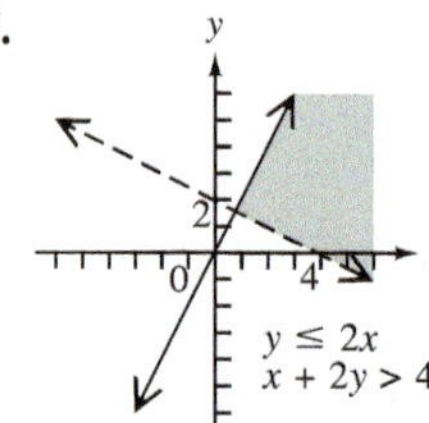

6. 8 in., 8 in., and 13 in. **7.** Broncos: 24; Panthers: 10 **8.** **(a)** years 0–6 **(b)** year 6; \$650 **9.** B **10.** B

Chapter 4 TEST

1. **(a)** no **(b)** no **(c)** yes **2.** $\{(2, -3)\}$ **3.** It has no solution. **4.** $\{(1, -6)\}$ **5.** $\{(-35, 35)\}$ **6.** $\left\{\left(-\frac{1}{3}, -10\right)\right\}$ **7.** $\{(5, 6)\}$ **8.** $\{(-1, 3)\}$ **9.** $\{(-1, 3)\}$ **10.** $\{(0, 0)\}$ **11.** $\varnothing$ **12.** $\{(-2, 2)\}$ **13.** $\{(x, y) \,|\, 3x - y = 6\}$ **14.** $\{(-15, 6)\}$

15. Memphis and Atlanta: 394 mi; Minneapolis and Houston: 1176 mi

16. Statue of Liberty: 4.3 million: Mount Rushmore National Memorial: 2.4 million **17.** 20 L of 15% solution; 30 L of 40% solution

18. slower car: 45 mph; faster car: 60 mph

19. 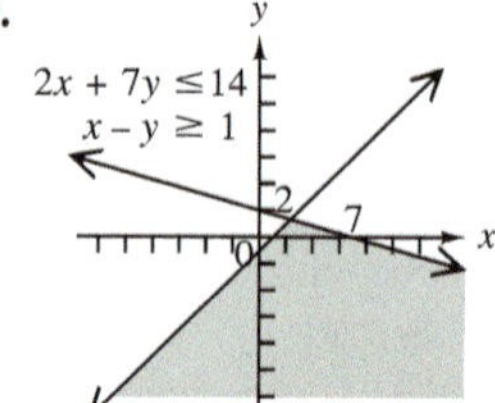

20. 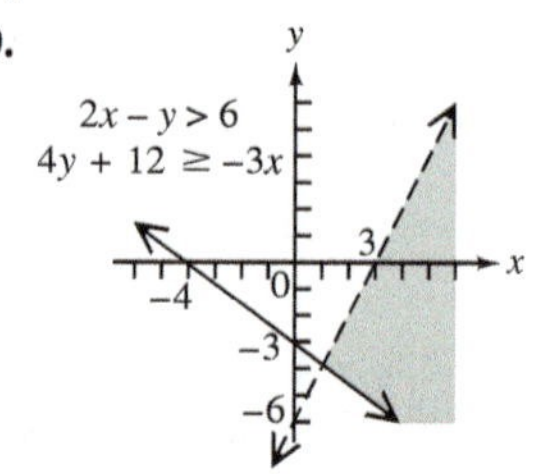

Chapters R-4 CUMULATIVE REVIEW EXERCISES

1. $-1, 1, -2, 2, -4, 4, -5, 5, -8, 8, -10, 10, -20, 20, -40, 40$ **2.** 1 **3.** commutative property **4.** distributive property **5.** inverse property **6.** 46 **7.** $T = \frac{PV}{k}$ **8.** $\left\{-\frac{13}{11}\right\}$ **9.** $\left\{\frac{9}{11}\right\}$ **10.** $(-18, \infty)$ **11.** $\left(-\frac{11}{2}, \infty\right)$ **12.** length: 12 in.; width: 8.7 in.

13. 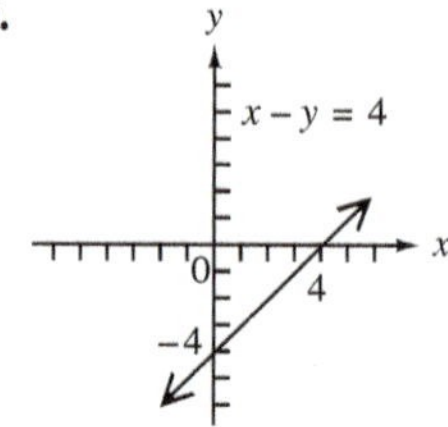

14. 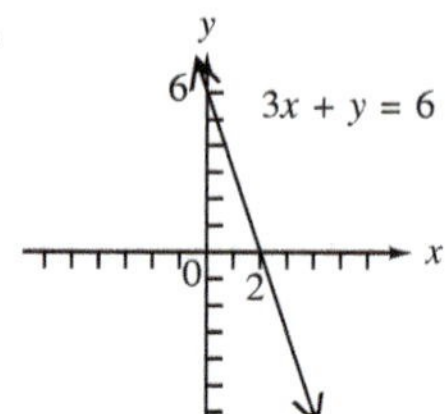

15. $-\frac{4}{3}$ **16.** $-\frac{1}{4}$ **17.** $y = \frac{1}{2}x + 3$ **18.** $y = 2x + 1$ **19.** **(a)** $x = 9$ **(b)** $y = -1$ **20.** $\{(-1, 6)\}$ **21.** $\{(3, -4)\}$ **22.** $\varnothing$

23. table entries: (third column) 2; (fourth column) $2y$, 2528; 405 adults, 49 children **24.** 19 in., 19 in., 15 in.

25. 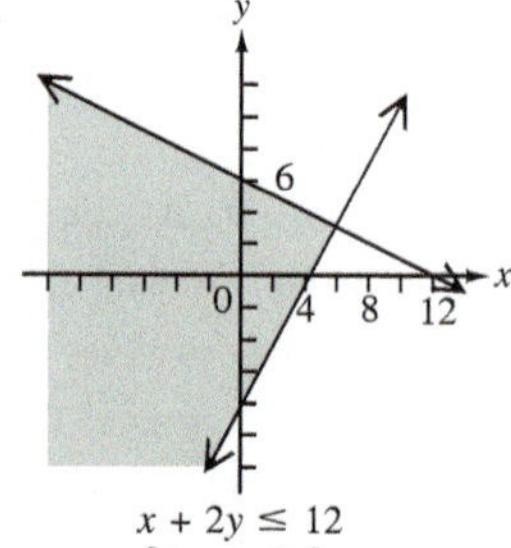

$x + 2y \leq 12$
$2x - y \leq 8$

CHAPTER 5 Exponents and Polynomials

SECTION 5.1

1. false; $3^3 = 3 \cdot 3 \cdot 3 = 27$ **2.** true **3.** false; $(a^2)^3 = a^{2 \cdot 3} = a^6$ **4.** true **5.** false; $-2^2 = -(2 \cdot 2) = -4$ **6.** false; $2^3 \cdot 2^4 = 2^7$ **7.** false; $(3x)^2 = 3^2x^2 = 9x^2$ **8.** true **9.** t^7 **11.** $\left(\frac{1}{2}\right)^5$ **13.** $(-8p)^2$

15. base: 3; exponent: 5; 243 **17.** base: -3; exponent: 5; -243

ANSWERS

19. base: -6; exponent: 2; 36 **21.** base: 6; exponent: 2; -36 **23.** base: $-2x$; exponent: 4; $16x^4$ **25.** base: x; exponent: 4 **27.** 5^8 **29.** 4^{12} **31.** $(-7)^9$ **33.** t^{24} **35.** $-56r^7$ **37.** $42p^{10}$ **39.** The product rule does not apply. **41.** The product rule does not apply. **43.** 4^6 **45.** t^{20} **47.** $343r^3$ **49.** 5^{12} **51.** -8^{15} **53.** $5^5x^5y^5$ **55.** $8q^3r^3$ **57.** $\frac{9^8}{5^8}$ **59.** $\frac{1}{8}$ **61.** $\frac{a^3}{b^3}$ **63.** $-8x^6y^3$ **65.** $9a^6b^4$ **67.** $\frac{5^5}{2^5}$ **69.** $\frac{9^5}{8^3}$ **71.** $2^{12}x^{12}$ **73.** -6^5p^5 **75.** $6^5x^{10}y^{15}$ **77.** x^{21} **79.** $4w^4x^{26}y^7$ **81.** $-r^{18}s^{17}$ **83.** $\frac{64x^6}{125}$ **85.** $\frac{125a^6b^{15}}{c^{18}}$ **87.** $25m^6p^{14}q^5$ **89.** $16x^{10}y^{16}z^{10}$ **91.** Using power rule (a), raise a power to a power by multiplying *exponents*. The base remains the same. Simplify as follows: $(10^2)^3 = 10^{2\cdot3} = 10^6 = 1{,}000{,}000$. **92.** The 4 is used as an *exponent* on 3. It is *not* multiplied by 3. The correct simplification is $\mathbf{3^4} \cdot x^8 \cdot y^{12} = \mathbf{81}x^8y^{12}$. **93.** $30x^7$ **95.** $6p^7$ **97.** $125x^6$

SECTION 5.2

1. negative **2.** positive **3.** negative **4.** negative **5.** positive **6.** positive **7.** 0 **8.** 0 **9.** (a) B (b) C (c) D (d) B (e) E (f) B **10.** (a) C (b) F (c) F (d) B (e) E (f) B **11.** 1 **13.** 1 **15.** -1 **17.** 0 **19.** 0 **21.** $\frac{1}{64}$ **23.** 16 **25.** $\frac{49}{36}$ **27.** $\frac{1}{81}$ **29.** 3 **31.** 1 **33.** 2 **35.** $\frac{8}{15}$ **37.** $-\frac{7}{18}$ **39.** 125 **41.** $\frac{1}{9}$ **43.** $\frac{1}{6^5}$ **45.** 27 **47.** 216 **49.** $2r^4$ **51.** $\frac{125}{9}$ **53.** $-\frac{4}{x^3}$ **55.** $\frac{p^5}{q^8}$ **57.** r^9 **59.** $\frac{x^5}{6}$ **61.** $3y^2$ **63.** x^3 **65.** $\frac{yz^2}{4x^3}$ **67.** $a+b$ **69.** 343 **71.** $\frac{1}{x^2}$ **73.** $\frac{64x}{9}$ **75.** $\frac{x^2z^4}{y^2}$ **77.** $6x$ **79.** $\frac{1}{m^{10}n^5}$ **81.** $\frac{5}{16x^5}$ **83.** $\frac{36q^2}{m^4p^2}$ **85.** The student attempted to use the quotient rule with unequal bases. The correct way to simplify is $\frac{16^3}{2^2} = \frac{(2^4)^3}{2^2} = \frac{2^{12}}{2^2} = 2^{12-2} = 2^{10} = 1024$. **86.** The student incorrectly assumed that the negative exponent indicated a negative number. The correct way to simplify is $5^{-4} = \frac{1}{5^4} = \frac{1}{625}$.

SUMMARY EXERCISES Applying the Rules for Exponents

1. positive **2.** negative **3.** negative **4.** negative **5.** positive **6.** negative **7.** positive **8.** negative **9.** $\frac{6^{12}x^{24}}{5^{12}}$ **10.** $\frac{r^6s^{12}}{729t^6}$ **11.** $100{,}000x^7y^{14}$ **12.** $64a^8b^{14}c^4$ **13.** $\frac{729w^3x^9}{y^{12}}$ **14.** $\frac{x^4y^6}{16}$ **15.** c^{22} **16.** $\frac{1}{k^4t^{12}}$ **17.** $\frac{11}{30}$ **18.** $y^{12}z^3$ **19.** $\frac{x^6}{y^5}$ **20.** 0 **21.** $\frac{1}{z^2}$ **22.** $\frac{9}{r^2s^2t^{10}}$ **23.** $\frac{300x^3}{y^3}$ **24.** $\frac{3}{5x^6}$ **25.** x^8 **26.** $\frac{y^{11}}{x^{11}}$ **27.** $\frac{a^6}{b^4}$ **28.** $6ab$ **29.** $\frac{61}{900}$ **30.** 1 **31.** $\frac{343a^6b^9}{8}$ **32.** 1 **33.** -1 **34.** 0 **35.** $\frac{27y^{18}}{4x^8}$ **36.** $\frac{1}{a^8b^{12}c^{16}}$ **37.** $\frac{x^{15}}{216z^9}$ **38.** $\frac{q}{8p^6r^3}$ **39.** x^6y^6 **40.** 0 **41.** $\frac{343}{x^{15}}$ **42.** $\frac{9}{x^6}$ **43.** $5p^{10}q^9$ **44.** $\frac{7}{24}$ **45.** $\frac{r^{14}t}{2s^2}$ **46.** 1 **47.** $8p^{10}q$ **48.** $\frac{1}{mn^3p^3}$ **49.** -1 **50.** $\frac{3}{40}$

SECTION 5.3

1. (a) C (b) A (c) B (d) D **2.** (a) A (b) C (c) B (d) D **3.** in scientific notation **4.** in scientific notation **5.** not in scientific notation; 5.6×10^6 **6.** not in scientific notation; 3.4×10^4 **7.** not in scientific notation; 4×10^{-3} **8.** not in scientific notation; 7×10^{-4} **9.** not in scientific notation; 8×10^1 **10.** not in scientific notation; 9×10^2 **11.** (a) 6; 4; 6.3; 4 (b) 5; 2; 5.71; -2 **13.** 5.876×10^9 **15.** 8.235×10^4 **17.** 7×10^{-6} **19.** -2.03×10^{-3} **21.** 750,000 **23.** 5,677,000,000,000 **25.** 1,000,000,000,000 **27.** -6.21 **29.** 0.00078 **31.** 0.000000005134 **33.** 0.000002 **35.** 4.2×10^{42} **37.** (a) 6×10^{11} (b) 600,000,000,000 **39.** (a) 1.5×10^7 (b) 15,000,000 **41.** (a) 8×10^{-3} (b) 0.008 **43.** (a) 2.4×10^2 (b) 240 **45.** (a) -6×10^4 (b) $-60{,}000$ **47.** (a) 6.3×10^{-2} (b) 0.063 **49.** (a) 3×10^{-4} (b) 0.0003 **51.** (a) 4×10^1 (b) 40 **53.** (a) 1.3×10^{-5} (b) 0.000013 **55.** (a) 5×10^2 (b) 500 **57.** (a) 9×10^6 (b) 9,000,000 **59.** (a) 2.6×10^{-3} (b) 0.0026 **61.** 1.5×10^{17} mi **63.** \$3097 **65.** \$1,394,668,000,000 **67.** 3.59×10^2 sec, or 359 sec **69.** 6×10^{17}, or 600,000,000,000,000,000; 3.6×10^{19}, or 36,000,000,000,000,000,000 **71.** 4.7E–7 **73.** 2E7 **75.** The Chile earthquake was 10 times as intense as the Southern Sumatra earthquake. **76.** The Solomon Islands earthquake was 10 times as intense as the Falkland Islands earthquake. **77.** The NE Japan earthquake was 100 times as intense as the Falkland Islands earthquake. **78.** The Chile earthquake would be 10,000 times as intense.

SECTION 5.4

1. 7; 5 **2.** two **3.** 8 **4.** is not **5.** 26 **6.** $5x^9$ **7.** 6; 1 **9.** 1; 1 **11.** $\frac{1}{5}$; 1 **13.** $-19, -1$; 2 **15.** $1, -8, \frac{2}{3}$; 3 **17.** $2m^5$ **19.** $-r^5$ **21.** $\frac{2}{3}x^4$ **23.** cannot be simplified; $0.2m^5 - 0.5m^2$ **25.** $-5x^5$ **27.** $5p^9 + 4p^7$ **29.** $-2y^2$ **31.** $-4xy$ **33.** already simplified; 4; binomial **35.** $x^2 + 9x - 3$; 2; trinomial **37.** $6xy$; 2; monomial **39.** already simplified; $6m^5 + 5m^4 - 7m^3 - 3m^2$; 5; none of these **41.** x^4; 4; monomial **43.** 7; 0; monomial **45.** (a) -1 (b) 5 **47.** (a) 19 (b) -2 **49.** (a) 36 (b) -12 **51.** (a) -124 (b) 5 **53.** $5x^2 - 2x$ **55.** $5m^2 + 3m + 2$ **57.** $\frac{7}{6}x^2 - \frac{2}{15}x + \frac{5}{6}$ **59.** $3y^3 - 11y^2$ **61.** $4x^4 - 4x^2 + 4x$ **63.** $15m^3 - 13m^2 + 8m + 11$ **65.** $8r^2 + 5r - 12$ **67.** $5m^2 - 14m + 6$ **69.** $4x^3 + 2x^2 + 5x$ **71.** $-18y^5 + 7y^4 + 5y^3 + 3y^2 + y$ **73.** $-10x^2 + 4x - 6$ **75.** $-2m^3 + 7m^2 + 8m - 9$ **77.** $6m^3 + m^2 + 4m - 14$ **79.** $-11x^2 - 3x - 3$ **81.** $2x^2 + 8x$ **83.** $8x^2 + 8x + 6$ **85.** $8t^2 + 8t + 13$ **87.** $6x - xy - 7$ **89.** $13a^2b - 7a^2 - b$ **91.** $c^4d - 5c^2d^2 + d^2$ **93.** (a) $23y + 5t$ (b) 25°, 67°, 88° **95.** 5; 175 **96.** 9; 63 **97.** 6; 27 **98.** 2.5; 130

SECTION 5.5

1. (a) B (b) D (c) A (d) C **2.** (a) C (b) A (c) B (d) D **3.** distributive **4.** binomials **5.** $12x^3$ **7.** $-15y^{11}$ **9.** $30a^9$ **11.** $-18m^3n^2$ **13.** $20a^2b$ **15.** $-12m^3n^4$ **17.** $6m^2 + 4m$ **19.** $10y^9 + 4y^6 + 6y^5$ **21.** $6y^6 + 4y^4 + 2y^3$ **23.** $6p - \frac{9}{2}p^2 + 9p^4$ **25.** $28r^5 - 32r^4 + 36r^3$ **27.** $6a^4 - 12a^3b + 15a^2b^2$ **29.** $3m^2$; $2mn$; $-n^3$; $21m^5n^2 + 14m^4n^3 - 7m^3n^5$ **31.** $12x^3 + 26x^2 + 10x + 1$ **33.** $6r^3 + 5r^2 - 12r + 4$ **35.** $20m^4 - m^3 - 8m^2 - 17m - 15$ **37.** $5x^4 - 13x^3 + 20x^2 + 7x + 5$ **39.** $3x^5 + 18x^4 - 2x^3 - 8x^2 + 24x$ **41.** first row: x^2; $4x$; second row: $3x$; 12; Product: $x^2 + 7x + 12$

43. first row: $2x^3$; $6x^2$; $4x$; second row: x^2; $3x$; 2; Product: $2x^3+7x^2+7x+2$ **45.** $m^2+12m+35$ **47.** n^2+n-6 **49.** $8r^2-10r-3$ **51.** $9x^2-4$ **53.** $9q^2+6q+1$ **55.** $8xy-4x+6y-3$ **57.** $15x^2+xy-6y^2$ **59.** $6t^2+23st+20s^2$ **61.** $-0.3t^2+0.22t+0.24$ **63.** $x^2-\frac{5}{12}x-\frac{1}{6}$ **65.** $\frac{15}{16}-\frac{1}{4}r-2r^2$ **67.** $2x^3+x^2-15x$ **69.** $6y^5-21y^4-45y^3$ **71.** $-200r^7+32r^3$ **73. (a)** $3y^2+10y+7$ **(b)** $8y+16$ **75.** $14x+49$ **77.** $30x+60$ **78.** $30x+60=600$; $\{18\}$ **79.** 10 ft by 60 ft **80.** 140 ft **81.** \$900 **82.** \$2870

SECTION 5.6

1. square; twice; product; square **2.** difference; squares **3. (a)** $2x$; $4x^2$ **(b)** $2x$; 3; $12x$ **(c)** 3; 9 **(d)** $4x^2+12x+9$ **5.** p^2+4p+4 **7.** $z^2-10z+25$ **9.** $x^2-\frac{3}{2}x+\frac{9}{16}$ **11.** $v^2+0.8v+0.16$ **13.** $16x^2-24x+9$ **15.** $100z^2+120z+36$ **17.** $x^2+4xy+4y^2$ **19.** $4p^2+20pq+25q^2$ **21.** $16a^2-40ab+25b^2$ **23.** $0.64t^2+1.12ts+0.49s^2$ **25.** $36m^2-\frac{48}{5}mn+\frac{16}{25}n^2$ **27.** $9t^3-6t^2+t$ **29.** $48t^3+24t^2+3t$ **31.** $-16r^2+16r-4$ **33. (a)** $7x$; $49x^2$ **(b)** $-21xy$; $21xy$; 0 **(c)** $3y$ and $-3y$; $-9y^2$ **(d)** $49x^2-9y^2$ **35.** k^2-25 **37.** $r^2-\frac{9}{16}$ **39.** $s^2-6.25$ **41.** $4w^2-25$ **43.** $9x^2-16y^2$ **45.** $100x^2-9y^2$ **47.** $49x^2-\frac{9}{49}$ **49.** $4x^4-25$ **51.** $25q^3-q$ **53.** $-5a^2+5b^6$ **55.** x^3+3x^2+3x+1 **57.** $m^3-15m^2+75m-125$ **59.** $8a^3+12a^2+6a+1$ **61.** $256x^4-256x^3+96x^2-16x+1$ **63.** $81r^4-216r^3t+216r^2t^2-96rt^3+16t^4$ **65.** $3x^5-27x^4+81x^3-81x^2$ **67.** $-8x^6y-32x^5y^2-48x^4y^3-32x^3y^4-8x^2y^5$ **69.** $\frac{1}{2}m^2-2n^2$ **71.** $9a^2-4$ **73.** $\pi x^2+4\pi x+4\pi$ **75.** $x^3+6x^2+12x+8$ **77.** $(x+y)^2$ **78.** x^2 **79.** $2xy$ **80.** y^2 **81.** $x^2+2xy+y^2$ **82.** They both represent the area of the entire large square. **83.** 1225 **84.** $30^2+2(30)(5)+5^2$ **85.** 1225 **86.** They are equal.

SECTION 5.7

1. $6x^2+8$; 2; $3x^2+4$ **2.** $3x^2+4$; 2 (These may be reversed.); $6x^2+8$ **3.** is not **4.** 0 **5.** $2m^2-m$ **7.** $30x^3-10x+5$ **9.** $-4m^3+2m^2-1$ **11.** $4t^4-2t^2+2t$ **13.** $a^4-a+\frac{2}{a}$ **15.** $-2x^3+\frac{2x^2}{3}-x$ **17.** $-9x^2+5x+1$ **19.** $7r^2-6+\frac{1}{r}$ **21.** $\frac{4x^2}{3}+x-\frac{2}{3x}$ **23.** $15x^5-35x^4+35x^3$ **24.** $-72y^6+60y^5-24y^4+36y^3-84y^2$ **25.** $27r^4$; $36r^3$; $6r^2$; $3r$; 2; $9r^3-12r^2-2r+1-\frac{2}{3r}$ **27.** $-m^2+3m-\frac{4}{m}$ **29.** $\frac{12}{x}-\frac{6}{x^2}+\frac{14}{x^3}-\frac{10}{x^4}$ **31.** $-4b^2+3ab-\frac{5}{a}$ **33.** $6x^4y^2-4xy+2xy^2-x^4y$ **35.** 1423 **36.** $(1\times10^3)+(4\times10^2)+(2\times10^1)+(3\times10^0)$ **37.** x^3+4x^2+2x+3 **38.** The coefficients of the powers of 10 are equal to the coefficients of the powers of x. One is a constant, while the other is a polynomial. They are equal if $x=10$ (which is the base of our decimal system).

SECTION 5.8

1. dividend; divisor; quotient **2.** Stop when the degree of the remainder is less than the degree of the divisor, or when the remainder is 0. **3.** Divide $12m^2$ by $2m$ to obtain $6m$. **4.** Multiply $6m$ by $2m-3$ to obtain $12m^2-18m$. **5.** $x+2$ **7.** $2y-5$ **9.** $p-4+\frac{44}{p+6}$ **11.** $r-5$ **13.** $2a-14+\frac{74}{2a+3}$ **15.** $4x^2-7x+3$ **17.** $2r^2+r-3+\frac{6}{r-3}$ **19.** $3y^2-2y+2$ **21.** $2x^2-6x+19+\frac{-55}{x+3}$ **23.** $3k-4+\frac{2}{k^2-2}$ **25.** x^2+1 **27.** x^2+1 **29.** $2p^2-5p+4+\frac{6}{3p^2+1}$ **31.** $2x^2+\frac{3}{5}x+\frac{1}{5}$ **33.** $x^2+\frac{8}{3}x-\frac{1}{3}+\frac{4}{3x-3}$ **35.** x^3+6x-7 **37.** $\left(6x-2+\frac{1}{x}\right)$ units **39.** (x^2+x-3) units **41.** $(5x^2-11x+14)$ hours

CHAPTER 5 Review Exercises

1. 4^{11} **2.** -5^{11} **3.** $-72x^7$ **4.** $10x^{14}$ **5.** 19^5x^5 **6.** -4^7y^7 **7.** $5p^4t^4$ **8.** $\frac{7^6}{5^6}$ **9.** $27x^6y^9$ **10.** t^{42} **11.** $36x^{16}y^4z^{16}$ **12.** $\frac{8m^9n^3}{p^6}$ **13.** $27x^6$ **14.** The product rule for exponents does not apply here because we want the sum of 7^2 and 7^3, not their product; $7^2+7^3=49+343=392$ **15.** -1 **16.** 1 **17.** 2 **18.** $\frac{1}{32}$ **19.** $\frac{25}{36}$ **20.** $-\frac{3}{16}$ **21.** 36 **22.** x^2 **23.** $\frac{1}{p^{12}}$ **24.** r^4 **25.** 2^8 **26.** $\frac{1}{9^6}$ **27.** 5^8 **28.** $\frac{1}{8^{12}}$ **29.** $\frac{1}{m^2}$ **30.** y^7 **31.** r^{13} **32.** $25m^6$ **33.** $\frac{y^{12}}{8}$ **34.** $\frac{1}{a^3b^5}$ **35.** $72r^5$ **36.** $\frac{8n^{10}}{3m^{13}}$ **37.** 4.8×10^7 **38.** 2.8988×10^{10} **39.** 6.5×10^{-5} **40.** 8.24×10^{-8} **41.** 24,000 **42.** 78,300,000 **43.** 0.000000897 **44.** 0.00000000000995 **45. (a)** 8×10^2 **(b)** 800 **46. (a)** 4×10^6 **(b)** 4,000,000 **47. (a)** 6×10^{-2} **(b)** 0.06 **48. (a)** 1×10^{-2} **(b)** 0.01 **49.** 2.796×10^{10} calculations; 1.6776×10^{12} calculations **50.** about 3.3 **51.** $22m^2$; 2; monomial **52.** p^3-p^2+4p+2; 3; none of these **53.** already simplified; 5; none of these **54.** $-8y^5-7y^4+9y$; 5; trinomial **55.** $-5a^3+4a^2$ **56.** $2r^3-3r^2+9r$ **57.** $11y^2-10y+9$ **58.** $-13k^4-15k^2-4k-6$ **59.** $10m^3-6m^2-3$ **60.** $-y^2-4y+26$ **61.** $10p^2-3p-11$ **62.** $7r^4-4r^3-1$ **63.** $10x^2+70x$ **64.** $-6p^5+15p^4$ **65.** $6r^3+8r^2-17r+6$ **66.** $8y^3+27$ **67.** $5p^5-2p^4-3p^3+25p^2+15p$ **68.** $x^2+3x-18$ **69.** $6k^2-9k-6$ **70.** $12p^2-48pq+21q^2$ **71.** $2m^4+5m^3-16m^2-28m+9$ **72.** $a^2+8a+16$ **73.** $9p^2-12p+4$ **74.** $4r^2+20rs+25s^2$ **75.** $r^3+6r^2+12r+8$ **76.** $8x^3-12x^2+6x-1$ **77.** $4z^2-49$ **78.** $36m^2-25$ **79.** $25a^2-36b^2$ **80.** $12x^4-75$ **81.** three; two **82.** $(a+b)^2=(a+b)(a+b)=a^2+2ab+b^2$. The term $2ab$ is not in a^2+b^2. **83.** $\frac{5y^2}{3}$ **84.** $-2x^2y$ **85.** $-y^3+2y-3$ **86.** $p-3+\frac{5}{2p}$ **87.** $-x^9+2x^8-4x^3+7x$ **88.** $-2m^2n+mn^2+\frac{6n^3}{5}$ **89.** $2r+7$ **90.** $4m+3+\frac{5}{3m-5}$ **91.** $2a+1+\frac{-8a+12}{5a^2-3}$ **92.** $k^2+2k+4+\frac{-2k-12}{2k^2+1}$

ANSWERS

CHAPTER 5 Mixed Review Exercises

1. 0 **2.** $\frac{243}{p^3}$ **3.** $\frac{1}{49}$ **4.** $4k^2 - 28k + 49$ **5.** $y^2 + 5y + 1$ **6.** $\frac{1296r^8s^4}{625}$ **7.** $-8m^7 - 10m^6 - 6m^5$ **8.** 32 **9.** $5xy^3 - \frac{8y^2}{5} + 3x^2y$ **10.** $\frac{r^2}{6}$ **11.** $8x^3 + 12x^2y + 6xy^2 + y^3$ **12.** $\frac{3}{4}$ **13.** $a^3 - 2a^2 - 7a + 2$ **14.** $8y^3 - 9y^2 + 5$ **15.** $10r^2 + 21r - 10$ **16.** $144a^2 - 1$ **17. (a)** $6x - 2$ **(b)** $2x^2 + x - 6$ **18. (a)** $20x^4 + 8x^2$ **(b)** $25x^8 + 20x^6 + 4x^4$ **19.** The second term of the quotient should be $-2x$, not $-12x$. Simplify as follows: $\frac{6x^2 - 12x}{6} = \frac{6x^2}{6} - \frac{12x}{6} = x^2 - 2x$. **20.** $2mn + 3m^4n^2 - 4n$

CHAPTER 5 Test

1. -32 **2.** $\frac{1}{625}$ **3.** 2 **4.** $\frac{7}{12}$ **5.** $\frac{216}{m^6}$ **6.** $9x^3y^5$ **7.** 8^5 **8.** x^2y^6 **9. (a)** positive **(b)** positive **(c)** negative **(d)** positive **(e)** zero **(f)** negative **10. (a)** 3.44×10^{11} **(b)** 5.57×10^{-6} **11. (a)** 29,600,000 **(b)** 0.0000000607 **12. (a)** 1×10^3; 5.89×10^{12} **(b)** 5.89×10^{15} mi **13.** $-7x^2 + 8x$; 2; binomial **14.** $4n^4 + 13n^3 - 10n^2$; 4; trinomial **15.** $4t^4 + t^3 - 6t^2 - t$ **16.** $-2y^2 - 9y + 17$ **17.** $16r^2 - 19$ **18.** $-12t^2 + 5t + 8$ **19.** $-27x^5 + 18x^4 - 6x^3 + 3x^2$ **20.** $2r^3 + r^2 - 16r + 15$ **21.** $t^2 - 5t - 24$ **22.** $8x^2 + 2xy - 3y^2$ **23.** $25x^2 - 20xy + 4y^2$ **24.** $100v^2 - 9w^2$ **25.** $x^3 + 3x^2 + 3x + 1$ **26.** $12x + 36$; $9x^2 + 54x + 81$ **27.** $4y^2 - 3y + 2 + \frac{5}{y}$ **28.** $-3xy^2 + 2x^3y^2 + 4y^2$ **29.** $2x + 9$ **30.** $3x^2 + 6x + 11 + \frac{26}{x-2}$

CHAPTERS R–5 Cumulative Review Exercises

1. $\frac{19}{24}$ **2.** $-\frac{1}{20}$ **3.** 3.72 **4.** 0.000042 **5.** \$1836 **6.** -8 **7.** 24 **8.** $\frac{1}{2}$ **9.** -4 **10.** associative property **11.** inverse property **12.** distributive property **13.** $\{10\}$ **14.** $\left\{\frac{13}{4}\right\}$ **15.** $\emptyset$ **16.** $r = \frac{d}{t}$ **17.** $\{-5\}$ **18.** $\{-12\}$ **19.** $\{20\}$ **20.** {all real numbers} **21.** mouse: 160; elephant: 10 **22.** 4 **23.** $[10, \infty)$ **24.** $\left(-\infty, -\frac{14}{5}\right)$ **25.** $[-4, 2)$ **26.** $(0, 2)$ and $(-3, 0)$ **27.** $\frac{2}{3}$

28.

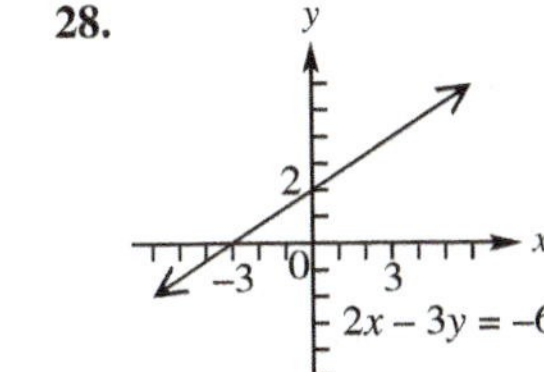

29. 1 **30.** $y = x + 6$ **31.** $\{(-3, -1)\}$ **32.** $\{(4, -5)\}$ **33.** $\frac{5}{4}$, or $1\frac{1}{4}$ **34.** 2 **35.** 1 **36.** $\frac{2b}{a^{10}}$ **37.** 3.45×10^4 **38.** $11x^3 - 14x^2 - x + 14$ **39.** $18x^7 - 54x^6 + 60x^5$ **40.** $63x^2 + 57x + 12$ **41.** $25x^2 + 80x + 64$ **42.** $y^2 - 2y + 6$

CHAPTER 6 Factoring and Applications

SECTION 6.1

1. product; multiplying **2.** common factor; is; divides **3.** 4 **5.** 4 **7.** 6 **9.** 1 **11.** 8 **13.** $10x^3$ **15.** $15m^2$ **17.** xy^2 **19.** 6 **21.** $6m^3n^2$ **23.** factored **24.** factored **25.** not factored **26.** not factored **27.** $18x^3y^2 + 9xy = 9xy(2x^2y + 1)$; If a polynomial has two terms, the product of the factors must have two terms. $9xy(2x^2y) = 18x^3y^2$ is just one term. **28.** $12x^2y - 24xy = 12xy(x - 2)$; The polynomial is *completely* factored when each factor has no common factor greater than 1. **29.** $3m^2$ **31.** $2z^4$ **33.** $2mn^4$ **35.** $y + 2$ **37.** $a - 2$ **39.** $2 + 3xy$ **41.** $x(x - 4)$ **43.** $3t(2t + 5)$ **45.** $m^2(m - 1)$ **47.** $-6x^2(2x + 1)$ **49.** $8z^2(2z^2 + 3)$ **51.** no common factor (except 1) **53.** $8mn^3(1 + 3m)$ **55.** $-2x(2x^2 - 5x + 3)$ **57.** $13y^2(y^6 + 2y^2 - 3)$ **59.** $9qp^3(5q^3p^2 - 4p^3 + 9q)$ **61.** $(x + 2)(c + d)$ **63.** $(2a + b)(a^2 - b)$ **65.** $(p + 4)(q - 1)$ **67.** not in factored form; $(7t + 4)(8 + x)$ **68.** not in factored form; $(5x - 1)(3r + 7)$ **69.** in factored form **70.** in factored form **71.** not in factored form **72.** not in factored form **73.** $(5 + n)(m + 4)$ **75.** $(2y - 7)(3x + 4)$ **77.** $(a - 2)(a + b)$ **79.** $(z + 2)(7z - a)$ **81.** $(3r + 2y)(6r - x)$ **83.** $(w + 1)(w^2 + 9)$ **85.** $(a + 2)(3a^2 - 2)$ **87.** $(4m - p^2)(4m^2 - p)$ **89.** $(y + 3)(y + x)$ **91.** $(z - 2)(2z - 3w)$ **93.** $(5 - 2p)(m + 3)$ **95.** $(3r + 2y)(6r - t)$ **97.** commutative property **98.** $2x(y - 4) - 3(y - 4)$ **99.** No, because it is not a product. It is the difference of $2x(y - 4)$ and $3(y - 4)$. **100.** $(y - 4)(2x - 3)$, or $(2x - 3)(y - 4)$; yes

SECTION 6.2

1. a and b must have different signs. **2.** a and b must have the same sign. **3.** C **4.** Factor out the greatest common factor, $2x$. **5.** $a^2 + 13a + 36$ **6.** $y^2 - 4y - 21$ **7.** The greatest common factor must be included in the factorization; $x^3 + 3x^2 - 28x = x(x + 7)(x - 4)$ **8.** The polynomial cannot be factored. It is prime. **9.** 1 and 12, -1 and -12, 2 and 6, -2 and -6, 3 and 4, -3 and -4; The pair with a sum of 7 is 3 and 4. **11.** 1 and -24, -1 and 24, 2 and -12, -2 and 12, 3 and -8, -3 and 8, 4 and -6, -4 and 6; The pair with a sum of -5 is 3 and -8. **13.** $p + 6$ **15.** $x + 11$ **17.** $x - 8$ **19.** $y - 5$ **21.** $x + 11$ **23.** $y - 9$ **25.** $(y + 8)(y + 1)$ **27.** $(b + 3)(b + 5)$ **29.** $(m + 5)(m - 4)$ **31.** $(x + 8)(x - 5)$ **33.** prime **35.** $(y - 5)(y - 3)$ **37.** $(z - 8)(z - 7)$ **39.** $(r - 6)(r + 5)$ **41.** $(a - 12)(a + 4)$ **43.** $(r + 2a)(r + a)$ **45.** $(x + y)(x + 3y)$ **47.** $(t + 2z)(t - 3z)$ **49.** $(v - 5w)(v - 6w)$ **51.** $(a + 5b)(a - 3b)$ **53.** $(a - 6b)(a - 3b)$ **55.** $4(x + 5)(x - 2)$ **57.** $2t(t + 1)(t + 3)$ **59.** $-2x^4(x - 3)(x + 7)$ **61.** $-a^3(a + 4b)(a - b)$ **63.** $5m^2(m^3 + 5m^2 - 8)$ **65.** $mn(m - 6n)(m - 4n)$

SECTION 6.3

1. B **2.** D **3.** $(m + 6)(m + 2)$ **5.** $(a + 5)(a - 2)$ **7.** $(2t + 1)(5t + 2)$ **9.** $(3z - 2)(5z - 3)$ **11.** $(2s + t)(4s - 3t)$ **13.** $(3a + 2b)(5a + 4b)$ **15. (a)** 2; 12; 24; 11 **(b)** 3; 8 (Order is irrelevant.) **(c)** $3m$; $8m$ **(d)** $2m^2 + 3m + 8m + 12$ **(e)** $(2m + 3)(m + 4)$ **(f)** $(2m + 3)(m + 4) = 2m^2 + 8m + 3m + 12$; Combine like terms to obtain $2m^2 + 11m + 12$. **17.** $(2x + 1)(x + 3)$ **19.** $(4r - 3)(r + 1)$ **21.** $(4m + 1)(2m - 3)$ **23.** $(3m + 1)(7m + 2)$ **25.** $(3a + 7)(a + 1)$ **27.** $(4y - 3)(3y - 1)$ **29.** $(4 + x)(4 + 3x)$, or $(x + 4)(3x + 4)$ **31.** $3(4x - 1)(2x - 3)$ **33.** $2m(m - 4)(m + 5)$ **35.** $-4z^3(z - 1)(8z + 3)$ **37.** $(3p + 4q)(4p - 3q)$ **39.** $(3a - 5b)(2a + b)$ **41.** The student stopped too soon. He needs to factor out the common factor $4x - 1$ to obtain $(4x - 1)(4x - 5)$ as the correct answer. **42.** The student forgot to include the common factor $3k$ in her answer. The correct answer is $3k(k - 5)(k + 1)$.

SECTION 6.4

1. B **2.** A **3.** A **4.** B **5.** A **6.** A **7.** $2a + 5b$
9. $x^2 + 3x - 4$; $x + 4$, $x - 1$ **11.** $2z^2 - 5z - 3$; $2z + 1$, $z - 3$
13. $(4x + 4)$ cannot be a factor because its terms have a common factor of 4, but those of the polynomial do not. The correct factored form is $(4x - 3)(3x + 4)$. **14.** The student forgot to factor out the common factor 2 from the terms of the trinomial. The *completely* factored form is $2(2x - 1)(x + 3)$. **15.** $(3a + 7)(a + 1)$ **17.** $(2y + 3)(y + 2)$
19. $(3m - 1)(5m + 2)$ **21.** $(3s - 1)(4s + 5)$ **23.** $(5m - 4)(2m - 3)$
25. $(4w - 1)(2w - 3)$ **27.** $(4y + 1)(5y - 11)$ **29.** prime
31. $2(5x + 3)(2x + 1)$ **33.** $-q(5m + 2)(8m - 3)$
35. $3n^2(5n - 3)(n - 2)$ **37.** $-y^2(5x - 4)(3x + 1)$
39. $(5a + 3b)(a - 2b)$ **41.** $(4s + 5t)(3s - t)$
43. $m^4n(3m + 2n)(2m + n)$ **45.** $-1(x + 7)(x - 3)$
47. $-1(3x + 4)(x - 1)$ **49.** $-1(a + 2b)(2a + b)$ **51.** $5 \cdot 7$
52. $(-5)(-7)$ **53.** The product of $3x - 4$ and $2x - 1$ is $6x^2 - 11x + 4$.
54. The product of $4 - 3x$ and $1 - 2x$ is $6x^2 - 11x + 4$. **55.** The factors in **Exercise 53** are the opposites of the factors in **Exercise 54.**
56. $(3 - 7t)(5 - 2t)$

SECTION 6.5

1. 1; 4; 9; 16; 25; 36; 49; 64; 81; 100; 121; 144; 169; 196; 225; 256; 289; 324; 361; 400 **2.** 1; 16; 81; 256; 625 **3.** A, D **4.** B, C **5.** The binomial $4x^2 + 16$ can be factored as $4(x^2 + 4)$. After any common factor is removed, a sum of squares (like $x^2 + 4$ here) *cannot* be factored. **6.** $k^2 - 9$ can be factored further as $(k + 3)(k - 3)$. The completely factored form is $(k^2 + 9)(k + 3)(k - 3)$. **7.** $(y + 5)(y - 5)$ **9.** $(x + 12)(x - 12)$
11. prime **13.** prime **15.** $(3r + 2)(3r - 2)$ **17.** $4(3x + 2)(3x - 2)$
19. $(14p + 15)(14p - 15)$ **21.** $(4r + 5a)(4r - 5a)$ **23.** $16(m^2 + 4)$
25. $(p^2 + 7)(p^2 - 7)$ **27.** $(x^2 + 1)(x + 1)(x - 1)$
29. $(p^2 + 16)(p + 4)(p - 4)$ **31.** B, C **32.** This polynomial is prime. It is not a perfect square trinomial because the middle term would have to be $30y$. **33.** $(w + 1)^2$ **35.** $(x - 4)^2$ **37.** prime **39.** $2(x + 6)^2$
41. $(2x + 3)^2$ **43.** $x(4x - 5)^2$ **45.** $(7x - 2y)^2$ **47.** $(8x + 3y)^2$
49. $-2h(5h - 2y)^2$ **51.** $\left(p + \frac{1}{3}\right)\left(p - \frac{1}{3}\right)$ **53.** $\left(2m + \frac{3}{5}\right)\left(2m - \frac{3}{5}\right)$
55. $(x + 0.8)(x - 0.8)$ **57.** $\left(t + \frac{1}{2}\right)^2$ **59.** $\left(a - \frac{2}{7}\right)^2$ **61.** $(x - 0.5)^2$

SUMMARY EXERCISES Recognizing and Applying Factoring Strategies

1. F **2.** G **3.** A **4.** B **5.** D **6.** H **7.** C **8.** E **9.** H
10. D **11.** $8m^3(4m^6 + 2m^2 + 3)$ **12.** $2(m + 3)(m - 8)$
13. $7k(2k + 5)(k - 2)$ **14.** prime **15.** $(6z + 1)(z + 5)$
16. $(m + n)(m - 4n)$ **17.** $(7z + 4y)(7z - 4y)$
18. $10nr(10nr + 3r^2 - 5n)$ **19.** $4x(4x^2 + 25)$ **20.** $(4 + m)(5 + 3n)$
21. $(5y - 6z)(2y + z)$ **22.** $(y^2 + 9)(y + 3)(y - 3)$
23. $(m - 3)(m + 5)$ **24.** $(2y + 1)(3y - 4)$ **25.** $8z(4z - 1)(z + 2)$
26. $(p - 12)^2$ **27.** $(z - 6)^2$ **28.** $(3m + 8)(3m - 8)$
29. $(y - 6k)(y + 2k)$ **30.** $(4z - 1)^2$ **31.** $6(y - 2)(y + 1)$
32. prime **33.** $(p - 6)(p - 11)$ **34.** $(a + 8)(a + 9)$
35. prime **36.** $3(6m - 1)^2$ **37.** $(z + 2a)(z - 5a)$
38. $(2a + 1)(a^2 - 7)$ **39.** $(2k - 3)^2$ **40.** $(a - 7b)(a + 4b)$
41. $(4r + 3m)^2$ **42.** $(3k - 2)(k + 2)$ **43.** prime
44. $(a^2 + 25)(a + 5)(a - 5)$ **45.** $4(2k - 3)^2$
46. $(4k + 1)(2k - 3)$ **47.** $6y^4(3y + 4)(2y - 5)$
48. $5z(z - 2)(z - 7)$ **49.** $(8p - 1)(p + 3)$
50. $(4k - 3h)(2k + h)$ **51.** $6(3m + 2z)(3m - 2z)$
52. $(2k - 5z)^2$ **53.** $2(3a - 1)(a + 2)$ **54.** $(3h - 2g)(5h + 7g)$
55. $7(2a + 3b)(2a - 3b)$ **56.** $(5z - 6)(2z + 1)$
57. $5m^2(5m - 13n)(5m - 3n)$ **58.** $(3y - 1)(3y + 5)$ **59.** $(3u + 11v)^2$
60. prime **61.** $9p^8(3p + 7)(p - 4)$ **62.** $5(2m - 3)(m + 4)$
63. $(2 - q)(2 - 3p)$ **64.** $(k + 11)(k - 11)$
65. $4(4p + 5m)(4p - 5m)$ **66.** $(m + 4)(m^2 - 6)$
67. $(10a + 9y)(10a - 9y)$ **68.** $(8a - b)(a + 3b)$ **69.** $(a + 4)^2$
70. $(2y + 5)(2y - 5)$ **71.** prime **72.** $-3x(x + 2y)(x - 2y)$
73. $(5a - 7b)^2$ **74.** $8(t^2 + 1)(t + 1)(t - 1)$ **75.** $-4(x - 3y)^2$
76. $25(2a + b)(2a - b)$ **77.** $-2(x - 9)(x - 4)$
78. $(m + 3)(2m - 5n)$ **79.** $2(2x + 5)(3x - 2)$ **80.** $(y^2 + 5)(y^4 - 3)$
81. $(y + 8)(y - 8)$ **82.** $6p(2p + 1)(p - 5)$

SECTION 6.6

1. $ax^2 + bx + c$ **2.** standard **3.** factor **4.** 0; zero; factor
5. **(a)** linear **(b)** quadratic **(c)** quadratic **(d)** linear
6. Because $(x - 9)^2 = (x - 9)(x - 9)$, applying the zero-factor property leads to two solutions of 9. Thus, 9 is a double solution. **7.** Set each *variable* factor equal to 0, to obtain $2x = 0$ or $3x - 4 = 0$. The solution set is $\left\{0, \frac{4}{3}\right\}$. **8.** The student should not divide by a variable because this causes the solution 0 to be eliminated. The solution set is $\left\{0, \frac{1}{7}\right\}$.
9. $\{-5, 2\}$ **11.** $\left\{3, \frac{7}{2}\right\}$ **13.** $\left\{-\frac{1}{2}, \frac{1}{6}\right\}$ **15.** $\left\{-\frac{5}{6}, 0\right\}$ **17.** $\left\{0, \frac{4}{3}\right\}$
19. $\{9\}$ **21.** $\{-4, -1\}$ **23.** $\{1, 2\}$ **25.** $\{-8, 3\}$ **27.** $\{-1, 3\}$
29. $\{-2, -1\}$ **31.** $\{-4\}$ **33.** $\left\{\frac{1}{4}\right\}$ **35.** $\left\{-2, \frac{1}{3}\right\}$ **37.** $\left\{-\frac{4}{3}, \frac{1}{2}\right\}$
39. $\left\{-\frac{2}{3}\right\}$ **41.** $\{-3, 3\}$ **43.** $\left\{-\frac{7}{4}, \frac{7}{4}\right\}$ **45.** $\{-13, 13\}$
47. $\{0, 7\}$ **49.** $\left\{0, \frac{1}{2}\right\}$ **51.** $\{2, 5\}$ **53.** $\left\{-4, \frac{1}{2}\right\}$ **55.** $\left\{-12, \frac{11}{2}\right\}$
57. $\left\{-4, \frac{1}{2}\right\}$ **59.** $\{-2, 0, 2\}$ **61.** $\left\{-\frac{7}{3}, 0, \frac{7}{3}\right\}$ **63.** $\left\{-\frac{5}{2}, \frac{1}{3}, 5\right\}$
65. $\left\{-\frac{7}{2}, -3, 1\right\}$ **67.** $\{-5, 0, 4\}$ **69.** $\{-3, 0, 5\}$ **71.** $\{-4, 12\}$
73. $\{-1, 3\}$ **75.** **(a)** 64; 144; 4; 6 **(b)** No time has elapsed, so the object hasn't fallen (been released) yet. **76.** Time cannot be negative.

SECTION 6.7

1. Read; variable; equation; Solve; answer; Check; original **2.** Only 6 is reasonable because a square cannot have a side of negative length.
3. $\mathcal{A} = bh$; *Step 3:* $(2x + 1)(x + 1)$; *Step 4:* 4; $-\frac{11}{2}$; *Step 5:* 9; 5; *Step 6:* $9 \cdot 5$ **5.** $V = LWH$; *Step 3:* 192; $4x$; *Step 4:* 6; -8; *Step 5:* 8; 6; *Step 6:* $8 \cdot 6$; 192 **7.** length: 14 cm; width: 12 cm **9.** base: 12 in.; height: 5 in. **11.** length: 15 in.; width: 12 in. **13.** height: 13 in.; width: 10 in. **15.** mirror: 7 ft; painting: 9 ft **17.** 20, 21 **19.** $-3, -2$ or 4, 5
21. $-3, -1$ or 7, 9 **23.** $-2, 0, 2$ or 6, 8, 10 **25.** 7, 9, 11 **27.** 12 cm
29. 12 mi **31.** length: 20 in.; width: 15 in.; diagonal: 25 in. **33.** 8 ft
35. **(a)** 1 sec **(b)** $\frac{1}{2}$ sec and $1\frac{1}{2}$ sec **(c)** 3 sec **(d)** The negative solution, -1, does not make sense because t represents time, which cannot be negative.

ANSWERS

37. 112 ft **39.** 256 ft **41.** **(a)** 100 million; The result is less than 109 million, the actual number from the table for 2000. **(b)** 10 **(c)** 303 million; The result using the model is the same as the actual number for 2010. **(d)** 393 million **43.** c^2 **44.** b^2 **45.** a^2 **46.** $a^2 + b^2 = c^2$; This is the equation of the Pythagorean theorem.

Chapter 6 REVIEW EXERCISES

1. $15(t+3)$ **2.** $30z(2z^2-1)$ **3.** $11x^2(4x+5)$ **4.** $50m^2n^2(2n-mn^2+3)$ **5.** $(x-4)(2y+3)$ **6.** $(2y+3)(3y+2x)$ **7.** $(x+3)(x+7)$ **8.** $(y-5)(y-8)$ **9.** $(q+9)(q-3)$ **10.** $(r-8)(r+7)$ **11.** prime **12.** $3(x^2+2x+2)$ **13.** $(r+8s)(r-12s)$ **14.** $(p+12q)(p-10q)$ **15.** $-8p(p+2)(p-5)$ **16.** $3x^2(x+2)(x+8)$ **17.** $(m+3n)(m-6n)$ **18.** $(y-3z)(y-5z)$ **19.** $p^5(p-2q)(p+q)$ **20.** $-3r^3(r+3s)(r-5s)$ **21.** r and $6r$, $2r$ and $3r$ **22.** Factor out z. **23.** $(2k-1)(k-2)$ **24.** $(3r-1)(r+4)$ **25.** $(3r+2)(2r-3)$ **26.** $(5z+1)(2z-1)$ **27.** prime **28.** $4x^3(3x-1)(2x-1)$ **29.** $-3(x+2)(2x-5)$ **30.** $rs(5r+6s)(2r+s)$ **31.** $-5y(3y+2)(2y-1)$ **32.** prime **33.** $-mn(3m+5)(m-8)$ **34.** $(2a-5b)(7a+4b)$ **35.** B **36.** D **37.** $(n+8)(n-8)$ **38.** $(5b+11)(5b-11)$ **39.** $(7y+5w)(7y-5w)$ **40.** $9(m^2+9)$ **41.** $36(2p+q)(2p-q)$ **42.** $(v^2+1)(v+1)(v-1)$ **43.** prime **44.** $(z+5)^2$ **45.** $(r-6)^2$ **46.** $(3t-7)^2$ **47.** $(4m+5n)^2$ **48.** $6x(3x-2)^2$ **49.** $\left\{-\frac{3}{4}, 1\right\}$ **50.** $\{-7, -3, 4\}$ **51.** $\left\{0, \frac{5}{2}\right\}$ **52.** $\{-3, -1\}$ **53.** $\{1, 4\}$ **54.** $\{3, 5\}$ **55.** $\left\{-\frac{4}{3}, 5\right\}$ **56.** $\left\{-\frac{8}{9}, \frac{8}{9}\right\}$ **57.** $\{0, 8\}$ **58.** $\{-1, 6\}$ **59.** $\{-7\}$ **60.** $\{6\}$ **61.** $\left\{-\frac{2}{5}, -2, -1\right\}$ **62.** $\{-3, 3\}$ **63.** $\{-1, 6\}$ **64.** $\left\{-\frac{3}{4}, -\frac{1}{2}, \frac{1}{3}\right\}$ **65.** $\left\{\frac{9}{5}\right\}$ **66.** $\{-3, 10\}$ **67.** length: 10 ft; width: 4 ft **68.** 5 ft **69.** 6, 7 or $-5, -4$ **70.** $-5, -4, -3$ or 5, 6, 7 **71.** 26 mi **72.** 6 m **73.** width: 10 m; length: 17 m **74.** **(a)** 256 ft **(b)** 1024 ft **75.** 704 thousand; The result is a little higher than the 696 thousand given in the table. **76.** 1454 thousand

Chapter 6 MIXED REVIEW EXERCISES

1. D **2.** The student forgot to factor out the common factor 2 from the terms of the trinomial. The completely factored form is $2(x+4)(3x-4)$. **3.** $(3m+4p)(5m-4)$ **4.** $8ab(3b^2-7ac^3+9ab)$ **5.** prime **6.** $(z-x)(z-10x)$ **7.** $(3k+5)(k+2)$ **8.** $(y^2+25)(y+5)(y-5)$ **9.** $3m(2m+3)(m-5)$ **10.** prime **11.** $2a^3(a+2)(a-6)$ **12.** $-1(2r+3q)(6r-5q)$ **13.** $(10a+3)(10a-3)$ **14.** $(7t+4)^2$ **15.** $\{0, 7\}$ **16.** $\{-5, 2\}$ **17.** $\left\{-\frac{2}{5}\right\}$ **18.** $\left\{-\frac{3}{8}, 0, \frac{3}{8}\right\}$ **19.** $\left\{-\frac{5}{3}, \frac{3}{2}\right\}$ **20.** $\left\{-6, -1, -\frac{1}{3}\right\}$ **21.** 15 m, 36 m, 39 m **22.** length: 6 m; width: 4 m

Chapter 6 TEST

1. D **2.** $6x(2x-5)$ **3.** $m^2n(2mn+3m-5n)$ **4.** $(2x+y)(a-b)$ **5.** $(x-7)(x-2)$ **6.** $(t+2)(t+5)$ **7.** $(3x+1)(2x-7)$ **8.** $3(x+1)(x-5)$ **9.** $(5z-1)(2z-3)$ **10.** $(2x+3)(x-1)$ **11.** prime **12.** $(y+7)(y-7)$ **13.** $(9a+11b)(9a-11b)$ **14.** $(x+8)^2$ **15.** $(2x-7y)^2$ **16.** $-2(x+1)^2$ **17.** $3t^2(2t+9)(t-4)$ **18.** prime **19.** $4t(t+4)^2$ **20.** $(x^2+9)(x+3)(x-3)$ **21.** $\{-3, 9\}$ **22.** $\left\{\frac{1}{2}, 6\right\}$ **23.** $\left\{-\frac{2}{5}, \frac{2}{5}\right\}$ **24.** $\{10\}$ **25.** $\{0, 3\}$ **26.** $\left\{-8, -\frac{5}{2}, \frac{1}{3}\right\}$ **27.** 6 ft by 9 ft **28.** $-2, -1$ **29.** 17 ft **30.** \$20,744 billion

Chapters R–6 CUMULATIVE REVIEW EXERCISES

1. $\{0\}$ **2.** $\{0.05\}$ **3.** $\{6\}$ **4.** $t = \frac{A-P}{Pr}$ **5.** second column: 38%, 12%; third column: 230, 205 **6.** gold: 9; silver: 7; bronze: 12 **7.** \$39,803 **8.** 110° and 70° **9.** **(a)** negative; positive **(b)** negative; negative **10.** $\left(-\frac{3}{4}, 0\right)$, $(0, 3)$ **11.** 4

12.

y
y = 4x + 3
(0, 3)
x
$(-\frac{3}{4}, 0)$
0

13. **(a)** 14.75; A slope of 14.75 means that revenue increased by about \$14.75 billion per year. **(b)** (2014, 90) **14.** $\{(-1, 2)\}$ **15.** $\varnothing$ **16.** 4 **17.** $\frac{16}{9}$ **18.** 1 **19.** 256 **20.** $\frac{1}{p^2}$ **21.** $\frac{1}{m^6}$ **22.** $-4k^2-4k+8$ **23.** $45x^2+3x-18$ **24.** $9p^2+12p+4$ **25.** $4x^3+6x^2-3x+10$ **26.** $(2a-1)(a+4)$ **27.** $(2m+3)(5m+2)$ **28.** $(4t+3v)(2t+v)$ **29.** $(2p-3)^2$ **30.** $(5r+9t)(5r-9t)$ **31.** $2pq(3p+1)(p+1)$ **32.** $\left\{-\frac{2}{3}, \frac{1}{2}\right\}$ **33.** $\{0, 8\}$ **34.** $\left\{\frac{4}{7}\right\}$ **35.** 5 m, 12 m, 13 m

CHAPTER 7 Rational Expressions and Applications

Note: In work with rational expressions, several different equivalent forms of the answer often exist. If your answer does not look exactly like the one given here, check to see if you have written an equivalent form.

SECTION 7.1

1. $-3; -3; 5$ **2.** $q; q; -1$ **3.** B, C **4.** **(a)** is not **(b)** are **5.** A rational expression is a quotient of polynomials, such as $\frac{x+3}{x^2-4}$, with denominator not equal to 0. **6.** Division by 0 is undefined, so if the denominator of a rational expression equals 0, the expression is undefined. **7.** **(a)** $\frac{7}{10}$ **(b)** $\frac{8}{15}$ **9.** **(a)** 0 **(b)** -1 **11.** **(a)** $\frac{9}{5}$ **(b)** undefined **13.** **(a)** $\frac{2}{7}$ **(b)** $\frac{13}{3}$ **15.** $y \neq 0$ **17.** $x \neq -6$ **19.** $x \neq \frac{5}{3}$ **21.** $m \neq -3, m \neq 2$ **23.** It is never undefined. **25.** It is never undefined. **27.** $\frac{3}{7}$ **29.** $\frac{3}{11}$ **31.** $3r^2$ **33.** $\frac{2}{5}$ **35.** $\frac{x-1}{x+1}$ **37.** $\frac{7}{5}$ **39.** $\frac{8}{7}$ **41.** $m-n$ **43.** $\frac{3(2m+1)}{4}$ **45.** $\frac{3m}{5}$ **47.** $\frac{3r-2s}{3}$ **49.** $\frac{x+1}{x-1}$ **51.** $\frac{z-3}{z+5}$ **53.** $\frac{a+b}{a-b}$ **55.** -1 **57.** $-(m+1)$ **59.** -1 **61.** It is already in lowest terms.

Answers may vary in Exercises 63–67.

63. $\frac{-(x+4)}{x-3}, \frac{-x-4}{x-3}, \frac{x+4}{-(x-3)}, \frac{x+4}{-x+3}$

65. $\frac{-(2x-3)}{x+3}, \frac{-2x+3}{x+3}, \frac{2x-3}{-(x+3)}, \frac{2x-3}{-x-3}$

67. $\frac{-(3x-1)}{5x-6}, \frac{-3x+1}{5x-6}, \frac{3x-1}{-(5x-6)}, \frac{3x-1}{-5x+6}$ **69.** x^2+3

71. **(a)** 80% **(b)** 90% **(c)** 95% **72.** No. If x is 0, then the expression is undefined. **73.** Both yield $2x + 3$. **74.** Both yield $2x + 1$. **75.** Both yield $x^2 + 1$.

SECTION 7.2

1. **(a)** B **(b)** D **(c)** C **(d)** A **2.** **(a)** D **(b)** C **(c)** A **(d)** B **3.** $\frac{5}{12}$ **5.** $\frac{3a}{2}$ **7.** $\frac{40y^2}{3}$ **9.** $\frac{2}{c+d}$ **11.** $4(x-y)$ **13.** $\frac{16q}{3p^3}$ **15.** $\frac{7}{r^2+rp}$ **17.** $\frac{z^2-9}{z^2+7z+12}$ **19.** $\frac{16}{13}$ **21.** 5 **23.** $-\frac{3}{2t^4}$ **25.** $\frac{1}{4}$ **27.** $\frac{x(x-3)}{6}$ **29.** $\frac{10}{9}$ **31.** $-\frac{3}{4}$ **33.** -1 **35.** $\frac{9(m-2)}{-(m+4)}$, or $\frac{-9(m-2)}{m+4}$ **37.** -1 **39.** $\frac{p+4}{p+2}$ **41.** $\frac{(k-1)^2}{(k+1)(2k-1)}$ **43.** $\frac{4k-1}{3k-2}$ **45.** $\frac{m+4p}{m+p}$ **47.** $\frac{10}{x+10}$ **49.** $\frac{5xy^2}{4q}$

SECTION 7.3

1. C **2.** B **3.** The factor x should appear in the LCD the *greatest* number of times it appears in any single denominator, not the *total* number of times. The correct LCD is $50x^4$. **4.** The expressions are not opposites. The opposite of $x - 1$ is $1 - x$. The correct LCD is $(x-1)(x+1)$. **5.** 60 **7.** 30 **9.** x^7 **11.** $30p$ **13.** $72q$ **15.** $84r^5$ **17.** $15a^5b^3$ **19.** $24x^3y^4$ **21.** $12p(p-2)$ **23.** $28m^2(3m-5)$ **25.** $30(b-2)$ **27.** $c - d$ or $d - c$ **29.** $2(x+1)(x-1)$ **31.** $k(k+5)(k-2)$ **33.** $a(a+6)(a-3)$ **35.** $(p+3)(p+5)(p-6)$ **37.** $\frac{20}{55}$ **39.** $\frac{-45}{9k}$ **41.** $\frac{26y^2}{80y^3}$ **43.** $\frac{35t^2r^3}{42r^4}$ **45.** $\frac{20}{8(m+3)}$ **47.** $\frac{57z}{6z-18}$ **49.** $\frac{8t}{12-6t}$ **51.** $\frac{14(z-2)}{z(z-3)(z-2)}$ **53.** $\frac{2(b-1)(b+2)}{b^3+3b^2+2b}$

SECTION 7.4

1. E **2.** A **3.** C **4.** H **5.** B **6.** D **7.** G **8.** F
9. *Each* term in the numerator of the second expression must be subtracted. Using parentheses will help avoid this error.

$$\frac{2x}{x+5} - \frac{x+1}{x+5} = \frac{2x - \mathbf{(x+1)}}{x+5} = \frac{2x - \mathbf{x - 1}}{x+5} = \frac{x-1}{x+5}$$

10. The student did not apply the distributive property correctly. In the third line, the numerator should be $7x - 3x + 5$. The correct answer is $\frac{4x+5}{2x-3}$. **11.** $\frac{2}{3}$ **13.** $\frac{11}{m}$ **15.** $\frac{4}{y+4}$ **17.** 4 **19.** b **21.** $\frac{m-1}{m+1}$ **23.** $\frac{2x+3}{x-4}$ **25.** x **27.** $y - 6$ **29.** $\frac{17}{30}$ **31.** $\frac{3z+5}{15}$ **33.** $\frac{10-7r}{14}$ **35.** $\frac{-3x-2}{4x}$ **37.** $\frac{57}{20x}$ **39.** $\frac{x+1}{2}$ **41.** $\frac{5x+9}{6x}$ **43.** $\frac{3x+3}{x(x+3)}$ **45.** $\frac{-k-8}{k(k+4)}$ **47.** $\frac{x+4}{x+2}$ **49.** $\frac{x^2+6x-8}{(x-2)(x+2)}$ **51.** $\frac{6m^2+23m-2}{(m+2)(m+1)(m+5)}$ **53.** $\frac{3}{t}$ **55.** $m - 2$ or $2 - m$ **56.** $\frac{-4}{3-k}$, or $-\frac{4}{3-k}$ **57.** $\frac{-2}{x-5}$, or $\frac{2}{5-x}$ **59.** -4 **61.** $\frac{-5}{x-y^2}$, or $\frac{5}{y^2-x}$ **63.** $\frac{x+y}{5x-3y}$, or $\frac{-x-y}{3y-5x}$ **65.** $\frac{-6}{4p-5}$, or $\frac{6}{5-4p}$ **67.** 3 **69.** $\frac{-(m+n)}{2(m-n)}$ **71.** $y - 7$ **73.** $\frac{-x^2+6x+11}{(x+3)(x-3)(x+1)}$ **75.** $\frac{-5q^2-13q+7}{(3q-2)(q+4)(2q-3)}$ **77.** $\frac{9r+2}{r(r+2)(r-1)}$ **79.** $\frac{2x^2+6xy+8y^2}{(x+y)(x+y)(x+3y)}$, or $\frac{2x^2+6xy+8y^2}{(x+y)^2(x+3y)}$ **81.** $\frac{15r^2+10ry-y^2}{(3r+2y)(6r-y)(6r+y)}$ **83.** **(a)** $\frac{9k^2+6k+26}{5(3k+1)}$ **(b)** $\frac{1}{4}$ **85.** $\frac{8000+10x}{49(101-x)}$

SECTION 7.5

1. division **2.** identity property of multiplication **3.** **(a)** 6; $\frac{1}{6}$ **(b)** 12; $\frac{3}{4}$ **(c)** $\frac{1}{6} \div \frac{3}{4}$ **(d)** $\frac{2}{9}$ **5.** -6 **7.** $\frac{31}{50}$ **9.** $\frac{1}{xy}$ **11.** $\frac{1}{6pq}$ **13.** $\frac{2a^2b}{3}$ **15.** $\frac{m(m+2)}{3(m-4)}$ **17.** $\frac{2}{x}$ **19.** $\frac{8}{x}$ **21.** $\frac{a^2-5}{a^2+1}$ **23.** $\frac{3(p+2)}{2(2p+3)}$ **25.** $\frac{a-2}{2a}$ **27.** $\frac{40-12p}{85p}$, or $\frac{4(10-3p)}{85p}$ **29.** $\frac{t(t-2)}{4}$ **31.** $\frac{-m}{2+m}$ **33.** $\frac{2x-7}{3x+1}$ **35.** $\frac{3m(m-3)}{(m-1)(m-8)}$ **37.** 6 **38.** $\frac{1}{3}$ **39.** -3 **40.** $-\frac{7}{5}$

SECTION 7.6

1. proposed; original **2.** extraneous; extraneous **3.** expression; $\frac{43}{40}x$ **5.** equation; $\left\{\frac{40}{43}\right\}$ **7.** expression; $-\frac{1}{10}y$ **9.** equation; $\{12\}$ **11.** $\{-6\}$ **13.** $\{-15\}$ **15.** $\{7\}$ **17.** $\{-15\}$ **19.** $\{-5\}$ **21.** $\{-6\}$ **23.** $\{12\}$ **25.** $\{5\}$ **27.** $\{1\}$ **29.** $\{2\}$ **31.** $x \neq -2, x \neq 0$ **33.** $x \neq -3, x \neq 4, x \neq -\frac{1}{2}$ **35.** $x \neq -9, x \neq 1, x \neq -2, x \neq 2$ **37.** $\left\{\frac{1}{4}\right\}$ **39.** $\left\{-\frac{3}{4}\right\}$ **41.** $\left\{\frac{20}{9}\right\}$ **43.** $\emptyset$ **45.** $\{3\}$ **47.** $\{3\}$ **49.** $\{-2, 12\}$ **51.** $\left\{-\frac{1}{5}, 3\right\}$ **53.** $\left\{-\frac{3}{5}, 3\right\}$ **55.** $\{-4\}$ **57.** $\emptyset$ **59.** $\{-1\}$ **61.** $\{-3\}$ **63.** $\{-6\}$ **65.** $\left\{-6, \frac{1}{2}\right\}$

67. This is an expression, *not* an equation. The student multiplied by the LCD, 14, instead of writing each coefficient with the LCD.

$$\frac{7}{7} \cdot \frac{3}{2}t + \frac{2}{2} \cdot \frac{5}{7}t$$
$$= \frac{21}{14}t + \frac{10}{14}t$$
$$= \frac{31}{14}t$$

68. The specified variable r appears on *both* sides of the final equation. Add rk in the third line, factor, and then divide to isolate r.

$$mk - rk = rm$$
$$mk = rm + rk$$
$$mk = r(m+k)$$
$$\frac{mk}{m+k} = r$$

69. $F = \frac{ma}{k}$ **71.** $a = \frac{kF}{m}$ **73.** $y = mx + b$ **75.** $R = \frac{E - Ir}{I}$, or $R = \frac{E}{I} - r$ **77.** $b = \frac{2\mathcal{A} - hB}{h}$, or $b = \frac{2\mathcal{A}}{h} - B$ **79.** $a = \frac{2S - dnL}{dn}$, or $a = \frac{2S}{dn} - L$ **81.** $y = \frac{xz}{x+z}$ **83.** $t = \frac{rs}{rs - 2s - 3r}$, or $t = \frac{-rs}{-rs + 2s + 3r}$ **85.** $c = \frac{ab}{b - a - 2ab}$, or $c = \frac{-ab}{-b + a + 2ab}$

87. $z = \frac{3y}{5 - 9xy}$, or $z = \frac{-3y}{9xy - 5}$ **89. (a)** $x \neq -3$ **(b)** $x \neq -1$ **(c)** $x \neq -3, x \neq -1$ **90.** $\frac{15}{2x}$ **91.** $(x + 3)(x + 1)$ **92.** $\frac{7}{x + 1}$ **93.** $\frac{11x + 21}{4x}$ **94.** $\varnothing$

SUMMARY EXERCISES Simplifying Rational Expressions vs. Solving Rational Equations

1. expression; $\frac{10}{p}$ **2.** expression; $\frac{2x - 1}{x + 7}$ **3.** expression; $\frac{1}{2x^2(x + 2)}$ **4.** equation; $\{9\}$ **5.** expression; $\frac{x + 2}{x - 1}$ **6.** expression; $\frac{5k + 8}{k(k - 4)(k + 4)}$ **7.** equation; $\{39\}$ **8.** expression; $\frac{t - 5}{3(2t + 1)}$ **9.** expression; $\frac{13}{3(p + 2)}$ **10.** equation; $\left\{-1, \frac{12}{5}\right\}$ **11.** equation; $\left\{\frac{1}{7}, 2\right\}$ **12.** expression; $\frac{7}{12z}$ **13.** expression; 3 **14.** equation; $\{13\}$ **15.** expression; $\frac{3m + 5}{(m + 2)(m + 3)(m + 1)}$ **16.** expression; $\frac{k + 3}{5(k - 1)}$ **17.** equation; $\varnothing$ **18.** equation; $\{-7\}$

SECTION 7.7

1. into a headwind: $(m - 5)$ mph; with a tailwind: $(m + 5)$ mph **2.** $\frac{D}{R} = \frac{d}{r}$ **3.** $\frac{1}{10}$ job per hr **4.** $\frac{2}{3}$ of the job **5. (a)** the amount **(b)** $5 + x$ **(c)** $\frac{5 + x}{6} = \frac{13}{3}$ **7.** x represents the original numerator; $\frac{x + 3}{(x + 6) + 3} = \frac{5}{7}$; $\frac{12}{18}$ **9.** x represents the original numerator; $\frac{x + 2}{3x - 2} = 1$; $\frac{2}{6}$ **11.** x represents the number; $\frac{1}{6}x = x + 5$; -6 **13.** x represents the quantity; $x + \frac{3}{4}x + \frac{1}{2}x + \frac{1}{3}x = 93$; 36 **15.** 18.809 min **17.** 10.848 mph **19.** 3.088 hr **21.** table entries: (fourth column) $\frac{8}{4 - x}, \frac{24}{4 + x}$; $\frac{8}{4 - x} = \frac{24}{4 + x}$ **23.** $7\frac{1}{2}$ hr **25.** $1\frac{3}{4}$ hr **27.** 8 mph **29.** 32 mph **31.** 3 mph **33.** table entries: (second column) $\frac{1}{2}, \frac{1}{3}$; (fourth column) $\frac{1}{2}t, \frac{1}{3}t$; $\frac{1}{2}t + \frac{1}{3}t = 1$ **35.** 12 hr **37.** $2\frac{2}{5}$ hr **39.** 10 hr **41.** 36 hr **43.** 10 mL **45.** $\frac{15}{8}$ hr, or $1\frac{7}{8}$ hr **46. (a)** The player multiplied: $5 \cdot 3 = 15$. **(b)** The player added: $5 + 3 = 8$. **(c)** The player added the two times and divided by 2—that is, he averaged the times: $\frac{5 + 3}{2} = 4$.

47.

$$\frac{1}{a}x + \frac{1}{b}x = 1$$

$$ab\left(\frac{1}{a}x + \frac{1}{b}x\right) = ab(1)$$

$$bx + ax = ab$$

$$x(b + a) = ab$$

$$x = \frac{ab}{b + a}$$

$$x = \frac{a \cdot b}{a + b}$$

48. 10 hr; 10 hr; The same answer results.

SECTION 7.8

1. direct **2.** direct **3.** inverse **4.** inverse **5.** direct **6.** inverse **7.** inverse **8.** direct **9.** Answers will vary; for example, number of movie tickets purchased and total price paid for the tickets. **10.** Answers will vary; for example, percentage off an item that is on sale and price paid for the item. **11. (a)** increases **(b)** decreases **12.** The customers in the lower-priced seats know more about the game than those in the higher-priced seats. **13.** inverse; 3 **14.** inverse; 5 **15.** direct; 50 **16.** direct; 200 **17.** 15 **19.** 250 **21.** 4 **23.** 21 **25.** 35 **27.** $\frac{1}{2}$ **29.** 15 in.^{2} **31.** $42\frac{2}{3}$ in. **33.** 15 ft **35.** 20 lb per ft^2 **37.** 25 kg per hr **39.** $12\frac{1}{2}$ amps **41.** 8 lb

Chapter 7 REVIEW EXERCISES

1. (a) $-\frac{4}{7}$ **(b)** -16 **2. (a)** $\frac{11}{8}$ **(b)** $\frac{13}{22}$ **3. (a)** undefined **(b)** 1 **4. (a)** undefined **(b)** $\frac{1}{2}$ **5.** $x \neq 3$ **6.** $x \neq 0$ **7.** $m \neq -1, m \neq 3$ **8.** $k \neq -5, k \neq -\frac{2}{3}$ **9.** $\frac{b}{3a}$ **10.** -1 **11.** $\frac{-(2x + 3)}{2}$ **12.** $\frac{2p + 5q}{5p + q}$

Answers may vary in Exercises 13 and 14.

13. $\frac{-(4x - 9)}{2x + 3}, \frac{-4x + 9}{2x + 3}, \frac{4x - 9}{-(2x + 3)}, \frac{4x - 9}{-2x - 3}$ **14.** $\frac{-(8 - 3x)}{3 - 6x}, \frac{-8 + 3x}{3 - 6x}, \frac{8 - 3x}{-(3 - 6x)}, \frac{8 - 3x}{-3 + 6x}$ **15.** 2 **16.** $\frac{2}{3m^6}$ **17.** $\frac{5}{8}$ **18.** $\frac{r + 4}{3}$ **19.** $\frac{3}{2}$ **20.** $\frac{y - 2}{y - 3}$ **21.** $\frac{p + 5}{p + 1}$ **22.** $\frac{3z + 1}{z + 3}$ **23.** 96 **24.** $108y^4$ **25.** $m(m + 2)(m + 5)$ **26.** $(x + 3)(x + 1)(x + 4)$ **27.** $\frac{35}{56}$ **28.** $\frac{40}{4k}$ **29.** $\frac{15a}{10a^4}$ **30.** $\frac{-54}{18 - 6x}$ **31.** $\frac{15y}{50 - 10y}$ **32.** $\frac{4b(b + 2)}{(b + 3)(b - 1)(b + 2)}$ **33.** $\frac{15}{x}$ **34.** $-\frac{2}{p}$ **35.** $\frac{4k - 45}{k(k - 5)}$ **36.** $\frac{28 + 11y}{y(7 + y)}$ **37.** $\frac{-2 - 3m}{6}$ **38.** $\frac{3(16 - x)}{4x^2}$ **39.** $\frac{7a + 6b}{(a - 2b)(a + 2b)}$ **40.** $\frac{-k^2 - 6k + 3}{3(k + 3)(k - 3)}$ **41.** $\frac{5z - 16}{z(z + 6)(z - 2)}$ **42.** $\frac{-13p + 33}{p(p - 2)(p - 3)}$ **43.** $\frac{a}{b}$ **44.** $\frac{4(y - 3)}{y + 3}$ **45.** $\frac{6(3m + 2)}{2m - 5}$ **46.** $\frac{(q - p)^2}{pq}$ **47.** $\frac{xw + 1}{xw - 1}$ **48.** $\frac{1 - r - t}{1 + r + t}$ **49.** $\left\{\frac{35}{6}\right\}$ **50.** $\{-16\}$ **51.** $\{-4\}$ **52.** $\varnothing$ **53.** $\{0\}$ **54.** $\{3\}$ **55.** $t = \frac{Ry}{m}$ **56.** $s = br - t$ **57.** $d = \frac{b - ac}{a}$, or $d = \frac{b}{a} - c$ **58.** $t = \frac{rs}{s - r}$ **59.** $\frac{20}{15}$ **60.** $\frac{3}{18}$ **61.** 2.020 hr **62.** 809.192 m per min **63.** $7\frac{1}{2}$ min **64.** $3\frac{1}{13}$ hr **65.** $\frac{36}{5}$ **66.** $\frac{1}{2}$ **67.** 4 cm **68.** 32.97 in.

Chapter 7 MIXED REVIEW EXERCISES

1. $\frac{m + 7}{(m - 1)(m + 1)}$ **2.** $8p^2$ **3.** $\frac{1}{6}$ **4.** 3 **5.** $\frac{z + 7}{(z + 1)(z - 1)^2}$ **6.** $\frac{x - 7}{2x + 3}$ **7.** $\{4\}$ **8.** $\{-2, 3\}$ **9.** $\{2\}$ **10.** $w = v - at$ **11.** 3 **12.** 2 hr **13.** table entries: (third column) $x + 50$, $x - 50$; (fourth column) $\frac{400}{x + 50}, \frac{200}{x - 50}$; 150 km per hr **14.** 10 hr **15.** inverse **16.** 50 amps

Chapter 7 TEST

1. (a) $\frac{11}{6}$ **(b)** undefined **2.** $x \neq -2, x \neq 4$ **3.** (Answers may vary.) $\frac{-(6x-5)}{2x+3}, \frac{-6x+5}{2x+3}, \frac{6x-5}{-(2x+3)}, \frac{6x-5}{-2x-3}$ **4.** $-3x^2y^3$ **5.** $\frac{3a+2}{a-1}$ **6.** $\frac{25}{27}$ **7.** $\frac{3k-2}{3k+2}$ **8.** $\frac{a-1}{a+4}$ **9.** $\frac{x-5}{3-x}$ **10.** $150p^5$ **11.** $(2r+3)(r+2)(r-5)$ **12.** $\frac{240p^2}{64p^3}$ **13.** $\frac{21}{42m-84}$ **14.** 2 **15.** $\frac{-14}{5(y+2)}$ **16.** $\frac{x^2+x+1}{3-x}$, or $\frac{-x^2-x-1}{x-3}$ **17.** $\frac{-m^2+7m+2}{(2m+1)(m-5)(m-1)}$ **18.** $\frac{2k}{3p}$ **19.** $(x-5)(x-3)$, or $x^2-8x+15$ **20.** $\frac{-2-x}{4+x}$ **21.** $\left\{-\frac{1}{2}, 1\right\}$ **22.** $\left\{-\frac{1}{2}, 5\right\}$ **23.** $\varnothing$ **24.** $\left\{-\frac{1}{2}\right\}$ **25.** $D = \frac{dF-k}{F}$, or $D = d - \frac{k}{F}$ **26.** $2\frac{2}{9}$ hr **27.** 3 mph **28.** -4 **29.** 27 **30.** 27 days

Chapters R–7 CUMULATIVE REVIEW EXERCISES

1. 2 **2.** $\{17\}$ **3.** $b = \frac{2\mathcal{A}}{h}$ **4.** $\left\{-\frac{2}{7}\right\}$

5. $[-8, \infty)$

6. $(4, \infty)$

7.

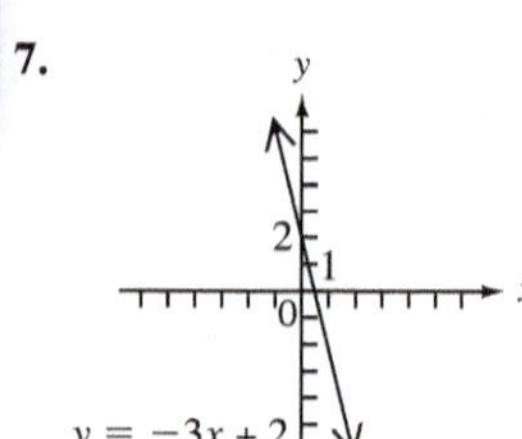

8.

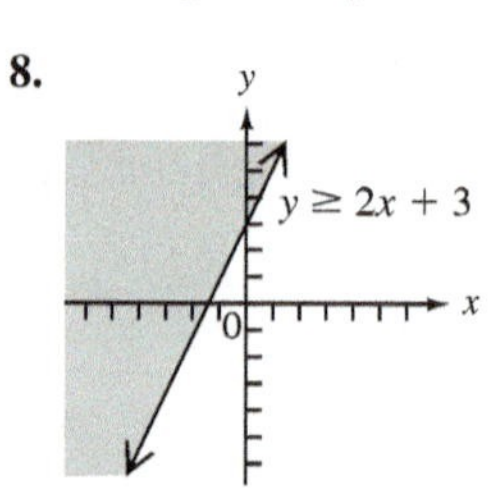

9. $\{(-1, 2)\}$ **10.** $\{(1, -2)\}$ **11.** $\frac{1}{16x^7}$ **12.** $\frac{1}{m^6}$ **13.** $\frac{q}{4p^2}$ **14.** $k^2 + 2k + 1$ **15.** $72x^6y^7$ **16.** $4a^2 - 4ab + b^2$ **17.** $3y^3 + 8y^2 + 12y - 5$ **18.** $6p^2 + 7p + 1 + \frac{3}{p-1}$ **19.** $(4t + 3v)(2t + v)$ **20.** prime **21.** $(4x^2 + 1)(2x + 1)(2x - 1)$ **22.** $\{-3, 5\}$ **23.** $\{-11, 11\}$ **24.** $\left\{-\frac{1}{2}, \frac{2}{3}, 5\right\}$ **25.** -2 or -1 **26.** 6 m **27.** $t \neq -2, t \neq 2$ **28.** D **29.** $\frac{4}{q}$ **30.** $\frac{3r+28}{7r}$ **31.** $\frac{7}{15(q-4)}$ **32.** $\frac{-k-5}{k(k+1)(k-1)}$ **33.** $\frac{7(2z+1)}{24}$ **34.** $\frac{195}{29}$ **35.** $\left\{\frac{21}{2}\right\}$ **36.** $\{-2, 1\}$ **37.** $1\frac{1}{5}$ hr **38.** 2 ft

CHAPTER 8 Roots and Radicals

SECTION 8.1

1. true **2.** false; A negative number has no real square roots. **3.** false; Zero has only one square root. **4.** true **5.** true **6.** false; A positive number has just one real cube root. **7.** a must be positive. **8.** a must be positive. **9.** a must be negative. **10.** a must be negative. **11.** $-3, 3$ **13.** $-8, 8$ **15.** $-13, 13$ **17.** $-\frac{5}{14}, \frac{5}{14}$ **19.** $-30, 30$ **21.** 1 **23.** 8 **25.** 7 **27.** -16 **29.** $-\frac{12}{11}$ **31.** 0.8 **33.** not a real number **35.** not a real number **37.** 100 **39.** 19 **41.** $\frac{2}{3}$ **43.** $3x^2 + 4$ **45.** 9 and 10 **46.** 6 and 7 **47.** 7 and 8 **48.** 5 and 6 **49.** -7 and -6 **50.** -8 and -7 **51.** 4 and 5 **52.** 3 and 4 **53.** rational; 5 **55.** irrational; 5.385 **57.** rational; -8 **59.** irrational; -17.321 **61.** not a real number **63.** irrational; 34.641 **65.** C **67.** $c = 17$ **69.** $b = 8$ **71.** $c \approx 11.705$ **73.** $a \approx 11.180$ **75.** 24 cm **77.** 80 ft **79.** 195 ft **81.** 158.6 ft **83.** 11.1 ft **85.** 9.434 **87.** 4 **89.** 5 **91.** 8 **93.** -3 **95.** -6 **97.** 2 **99.** 5 **101.** 6 **103.** not a real number **105.** -5 **107.** -4 **109. (a)** 3 units, 4 units **(b)** If we let $a = 3$, $b = 4$, and $c = 5$, then the Pythagorean theorem is satisfied. $a^2 + b^2 = c^2$ becomes $3^2 + 4^2 \stackrel{?}{=} 5^2$, which simplifies to $25 = 25$, a true statement. **110. (a)** $(a+b)^2$, or $a^2 + 2ab + b^2$ **(b)** $c^2 + 4\left(\frac{1}{2}ab\right)$, or $c^2 + 2ab$ **(c)** Subtract $2ab$ from each side to obtain $a^2 + b^2 = c^2$.

SECTION 8.2

1. false; $\sqrt{(-6)^2} = \sqrt{36} = 6$ **2.** true **3.** false; $2\sqrt{7}$ represents the product of 2 and $\sqrt{7}$. **4.** true **5.** A **6.** Yes; because a prime number cannot be factored (except as the product of itself and 1), it cannot have any perfect square factors (except 1). **7.** $\sqrt{15}$ **9.** $\sqrt{22}$ **11.** $\sqrt{42}$ **13.** $\sqrt{13r}$ **15.** $2\sqrt{6}$ **17.** $3\sqrt{5}$ **19.** $3\sqrt{10}$ **21.** $5\sqrt{3}$ **23.** $5\sqrt{5}$ **25.** It cannot be simplified. **27.** $-4\sqrt{10}$ **29.** $-10\sqrt{7}$ **31.** $6\sqrt{13}$ **33.** $25\sqrt{2}$ **35.** $5\sqrt{10}$ **37.** $6\sqrt{2}$ **39.** $12\sqrt{2}$ **41.** $3\sqrt{6}$ **43.** 24 **45.** $6\sqrt{10}$ **47.** $12\sqrt{5}$ **49.** $30\sqrt{5}$ **51.** *Method 1:* $\sqrt{8} \cdot \sqrt{32} = \sqrt{8 \cdot 32} = \sqrt{256} = 16$; *Method 2:* $\sqrt{8} = 2\sqrt{2}$ and $\sqrt{32} = 4\sqrt{2}$, so $\sqrt{8} \cdot \sqrt{32} = 2\sqrt{2} \cdot 4\sqrt{2} = 8 \cdot 2 = 16$; The same answer results. Either method can be used to obtain the correct answer. **52.** *Method 1:* $\sqrt{288} = \sqrt{144 \cdot 2} = \sqrt{144} \cdot \sqrt{2} = 12\sqrt{2}$; *Method 2:* $\sqrt{288} = \sqrt{16 \cdot 18} = \sqrt{16} \cdot \sqrt{18} = 4\sqrt{18} = 4\sqrt{9 \cdot 2} = 4 \cdot \sqrt{9} \cdot \sqrt{2} = 4 \cdot 3\sqrt{2} = 12\sqrt{2}$; The same answer results. Method 1 yields the answer more quickly using the greatest perfect square factor first. **53.** $\frac{4}{15}$ **55.** $\frac{\sqrt{7}}{4}$ **57.** $\frac{\sqrt{2}}{5}$ **59.** 5 **61.** $6\sqrt{5}$ **63.** $\frac{25}{4}$ **65.** m **67.** y^2 **69.** $6z$ **71.** $20x^3$ **73.** $3x^4\sqrt{2}$ **75.** $3c^7\sqrt{5}$ **77.** $z^2\sqrt{z}$ **79.** $a^6\sqrt{a}$ **81.** $8x^3\sqrt{x}$ **83.** x^3y^6 **85.** $9m^2n$ **87.** $\frac{\sqrt{7}}{x^5}$ **89.** $\frac{y^2}{10}$ **91.** $\frac{x^3}{y^4}$ **93.** $2\sqrt[3]{5}$ **95.** $3\sqrt[3]{2}$ **97.** $4\sqrt[3]{2}$ **99.** $2\sqrt[4]{5}$ **101.** $\frac{2}{3}$ **103.** $-\frac{6}{5}$ **105.** p **107.** x^3 **109.** $4z^2$ **111.** $7a^3b$ **113.** $2t\sqrt[3]{2t^2}$ **115.** $\frac{m^4}{2}$ **117.** 6 cm **119.** 6 in. **121.** D

SECTION 8.3

1. radicand; index; like **2.** like; 4; $3xy^3$ **3.** radicands; unlike **4.** indexes or roots; unlike **5.** $7\sqrt{3}$ **7.** $-5\sqrt{7}$ **9.** $5\sqrt{17}$ **11.** $5\sqrt{7}$ **13.** $2\sqrt{6}$ **15.** These unlike radicals cannot be combined. **17.** $7\sqrt{3}$ **19.** $11\sqrt{5}$ **21.** $15\sqrt{2}$ **23.** $-6\sqrt{2}$ **25.** $4\sqrt{2}$ **27.** $-\sqrt[3]{2}$

29. $24\sqrt[3]{3}$ **31.** $17\sqrt{7}$ **33.** $-16\sqrt{2}-8\sqrt{3}$ **35.** $22\sqrt{2}$ **37.** $11\sqrt{3}$ **39.** $3\sqrt{21}$ **41.** $\sqrt{2x}$ **43.** $7\sqrt{3r}$ **45.** $5\sqrt{x}$ **47.** $3x\sqrt{6}$ **49.** 0 **51.** $-20\sqrt{2k}$ **53.** $42x\sqrt{5z}$ **55.** $-\sqrt[3]{2}$ **57.** $6\sqrt[3]{p^2}$ **59.** $21\sqrt[4]{m^3}$ **61.** $-6x^2y$ **62.** $-6(p-2q)^2(a+b)$ **63.** $-6a^2\sqrt{x}$ **64.** Combining like terms and combining like radicals is essentially the same process—we use the distributive property to combine the numerical coefficients.

SECTION 8.4

1. The student did not use the identity property of multiplication which requires multiplying by a form of 1. The correct process is $\frac{4\cdot\sqrt{3}}{\sqrt{3}\cdot\sqrt{3}}=\frac{4\sqrt{3}}{3}$.
2. The student did not create a perfect cube in the radicand in the denominator. The correct process is $\frac{2\cdot\sqrt[3]{5\cdot5}}{\sqrt[3]{5}\cdot\sqrt[3]{5\cdot5}}=\frac{2\sqrt[3]{5\cdot5}}{\sqrt[3]{5\cdot5\cdot5}}=\frac{2\sqrt[3]{25}}{5}$.
3. $\frac{6\sqrt{5}}{5}$ **5.** $\sqrt{5}$ **7.** $\frac{2\sqrt{6}}{3}$ **9.** $2\sqrt{3}$ **11.** $\frac{-3\sqrt{2}}{10}$ **13.** $\frac{21\sqrt{5}}{5}$ **15.** $\frac{\sqrt{65}}{5}$ **17.** $\frac{\sqrt{2}}{2}$ **19.** $\sqrt{3}$ **21.** $\frac{-\sqrt{33}}{3}$ **23.** $\frac{7\sqrt{15}}{5}$ **25.** $\frac{\sqrt{30}}{2}$ **27.** $\frac{\sqrt{15}}{10}$ **29.** $\frac{\sqrt{21}}{6}$ **31.** $\frac{\sqrt{21}}{3}$ **33.** $\frac{3\sqrt{14}}{4}$ **35.** $\frac{1}{6}$ **37.** 1 **39.** $\frac{\sqrt{7x}}{x}$ **41.** $\frac{p\sqrt{3q}}{q}$ **43.** $\frac{3a\sqrt{5r}}{5}$ **45.** $\frac{2x\sqrt{xy}}{y}$ **47.** $\frac{x\sqrt{30xz}}{6}$ **49.** $\frac{3ar^2\sqrt{7rt}}{7t}$ **51.** $\frac{\sqrt[3]{4}}{2}$ **53.** $\frac{\sqrt[3]{50}}{5}$ **55.** $\frac{\sqrt[3]{196}}{7}$ **57.** $\frac{\sqrt[3]{15}}{3}$ **59.** $\frac{\sqrt[3]{6y}}{2y}$ **61.** $\frac{\sqrt[3]{42mn^2}}{6n}$ **63. (a)** $\frac{9\sqrt{2}}{4}$ sec **(b)** 3.182 sec

SECTION 8.5

1. 13 **2.** 4 **3.** 8 **4.** 14 **5.** It is incorrect to combine -37 and $-2\sqrt{15}$ because they are not like terms. This expression cannot be simplified further. **6.** Remember the middle term when squaring a binomial: $(x+y)^2=x^2+2xy+y^2$. The correct work is $\left(\sqrt{7}+5\right)^2=\left(\sqrt{7}\right)^2+2\left(\sqrt{7}\right)(5)+5^2=32+10\sqrt{7}$. **7.** $\sqrt{15}-\sqrt{35}$ **9.** $5\sqrt{6}$ **11.** $2\sqrt{10}+30$ **13.** $4\sqrt{7}$ **15.** $57+23\sqrt{6}$ **17.** $81+14\sqrt{21}$ **19.** $31-8\sqrt{15}$ **21.** $71-16\sqrt{7}$ **23.** $37+12\sqrt{7}$ **25.** $187-20\sqrt{21}$ **27.** $a+2\sqrt{a}+1$ **29.** 23 **31.** 1 **33.** $y-10$ **35.** $2\sqrt{3}-2+3\sqrt{2}-\sqrt{6}$ **37.** $15\sqrt{2}-15$ **39.** $\sqrt{30}+\sqrt{15}+6\sqrt{5}+3\sqrt{10}$ **41.** $\sqrt{5x}-\sqrt{10}-\sqrt{10x}+2\sqrt{5}$ **43.** There would still be a radical in the denominator. He should multiply by $\frac{4-\sqrt{3}}{4-\sqrt{3}}$. **44.** This expression must be factored first. Then the common factor 6 can be divided out as follows: $\frac{6+30\sqrt{11}}{6}=\frac{6\left(1+5\sqrt{11}\right)}{6}=1+5\sqrt{11}$. **45.** $\frac{3-\sqrt{2}}{7}$ **47.** $-4-2\sqrt{11}$ **49.** $1+\sqrt{2}$ **51.** $-\sqrt{10}+\sqrt{15}$ **53.** $2\sqrt{5}+\sqrt{15}+4+2\sqrt{3}$ **55.** $\frac{-6\sqrt{2}+12+\sqrt{10}-2\sqrt{5}}{2}$ **57.** $\frac{12\left(\sqrt{x}-1\right)}{x-1}$ **59.** $\frac{3\left(7+\sqrt{x}\right)}{49-x}$ **61.** $\sqrt{11}-2$ **63.** $\frac{\sqrt{3}+5}{8}$ **65.** $\frac{6-\sqrt{10}}{2}$ **67.** $\frac{2+\sqrt{2}}{3}$ **69. (a)** 4 in. **(b)** 3.4 in.

SUMMARY EXERCISES Applying Operations with Radicals

1. $-3\sqrt{10}$ **2.** $5-\sqrt{15}$ **3.** $2-\sqrt{6}+2\sqrt{3}-3\sqrt{2}$ **4.** $6\sqrt{2}$ **5.** $73-12\sqrt{35}$ **6.** $\frac{\sqrt{6}}{2}$ **7.** $3\sqrt[3]{2t^2}$ **8.** $4\sqrt{7}+4\sqrt{5}$ **9.** $-3-2\sqrt{2}$ **10.** 4 **11.** -33 **12.** $\frac{\sqrt{t}-\sqrt{3}}{t-3}\ (t\neq3)$ **13.** $2xyz^2\sqrt[3]{y^2}$ **14.** $4\sqrt[3]{3}$ **15.** $\sqrt{6}+1$ **16.** $\frac{\sqrt{6x}}{3x}$ **17.** $\frac{3}{5}$ **18.** $4\sqrt{2}$ **19.** $-2\sqrt[3]{2}$ **20.** $11-2\sqrt{30}$ **21.** $3\sqrt{3x}$ **22.** $52+30\sqrt{3}$ **23.** 1 **24.** $\frac{2\sqrt[3]{18}}{9}$ **25.** $-x^2\sqrt[4]{x}$ **26.** $2\sqrt{7}$ **27.** It cannot be simplified. **28.** $12\sqrt{6}+6\sqrt{5}$ **29.** $\frac{\sqrt{15}}{10}$ **30.** $\sqrt{5}$ **31.** $20\sqrt[3]{3}$ **32.** $\frac{8\left(4+\sqrt{x}\right)}{16-x}$ **33.** $2-3\sqrt[3]{4}$ **34.** $\sqrt{2x}$ **35.** $\frac{\sqrt{10}}{4}$ **36.** $49+14\sqrt{x}+x$ **37.** 57 species **38.** 858 species

SECTION 8.6

1. squaring; squared; original **2.** $x^2+2xy+y^2$; $x^2-2xy+y^2$ **3.** $\{49\}$ **5.** $\{7\}$ **7.** $\{85\}$ **9.** $\emptyset$ **11.** $\{-45\}$ **13.** $\left\{-\frac{3}{2}\right\}$ **15.** $\emptyset$ **17.** $\{121\}$ **19.** $\{8\}$ **21.** $\{1\}$ **23.** $\{6\}$ **25.** $\emptyset$ **27.** $\{5\}$ **29.** When the left side is squared, the result should be $x-1$, not $-(x-1)$. The correct solution set is $\{17\}$. **30.** Squaring $x-7$ gives $x^2-14x+49$. Remember to include the middle term. **31.** $\{12\}$ **33.** $\{5\}$ **35.** $\{0,3\}$ **37.** $\{-1,3\}$ **39.** $\{8\}$ **41.** $\{4\}$ **43.** $\{8\}$ **45.** $\{9\}$ **47.** $\{4,20\}$ **49.** $\{-5\}$ **51.** 21 **53.** 8 **55.** 17,616 ft^2 **57. (a)** 71 mph **(b)** 60 mph **(c)** 54 mph **59.** yes; 26 mi **61.** 47 mi **63.** $s=13$ units **64.** $6\sqrt{13}$ sq. units **65.** $h=\sqrt{13}$ units **66.** $3\sqrt{13}$ sq. units **67.** $6\sqrt{13}$ sq. units **68.** They are both $6\sqrt{13}$.

Chapter 8 REVIEW EXERCISES

1. $-7,7$ **2.** $-9,9$ **3.** $-14,14$ **4.** $-11,11$ **5.** $-16,16$ **6.** $-27,27$ **7.** 4 **8.** -0.6 **9.** not a real number **10.** -65 **11.** $\frac{12}{13}$ **12.** $-\frac{10}{9}$ **13.** 6 **14.** -8 **15.** 3 **16.** -5 **17. (a)** 8 **(b)** 6.633 **18.** 48.3 cm **19.** irrational; 8.544 **20.** rational; 13 **21.** rational; -25 **22.** not a real number **23.** $4\sqrt{3}$ **24.** $-12\sqrt{2}$ **25.** $2\sqrt[3]{2}$ **26.** $5\sqrt[3]{3}$ **27.** 18 **28.** $16\sqrt{6}$ **29.** $-\frac{11}{20}$ **30.** $\frac{\sqrt{7}}{13}$ **31.** $\frac{\sqrt{5}}{6}$ **32.** $\frac{2}{15}$ **33.** $3\sqrt{2}$ **34.** $2\sqrt{2}$ **35.** r^9 **36.** x^5y^8 **37.** $9x^4\sqrt{2x}$ **38.** $\frac{6}{p}$ **39.** $a^7b^{10}\sqrt{ab}$ **40.** $11x^3y^5$ **41.** y^2 **42.** $6x^5$ **43.** $8\sqrt{11}$ **44.** $9\sqrt{2}$ **45.** $21\sqrt{3}$ **46.** $12\sqrt{3}$ **47.** 0 **48.** $-6\sqrt{7}+9\sqrt{5}$ **49.** $2\sqrt{3}+3\sqrt{10}$ **50.** $2\sqrt{2}$ **51.** $6\sqrt{30}$ **52.** $5\sqrt{x}$ **53.** 0 **54.** $11k^2\sqrt{2n}$ **55.** $\frac{10\sqrt{3}}{3}$ **56.** $\frac{8\sqrt{10}}{5}$ **57.** $\sqrt{6}$ **58.** $\frac{\sqrt{10}}{5}$ **59.** $\sqrt{10}$ **60.** $\frac{\sqrt{42}}{21}$ **61.** $\frac{r\sqrt{x}}{4x}$ **62.** $\frac{\sqrt[3]{9}}{3}$ **63.** $-\sqrt{15}-9$ **64.** $3\sqrt{6}+12$ **65.** $22-16\sqrt{3}$ **66.** $2\sqrt{21}-\sqrt{14}+12\sqrt{2}-4\sqrt{3}$ **67.** -13 **68.** $x+4\sqrt{x}+4$ **69.** $-2+\sqrt{5}$ **70.** $\frac{3\left(4+\sqrt{3}\right)}{13}$

71. $\dfrac{-\sqrt{10}+3\sqrt{5}+\sqrt{2}-3}{7}$ **72.** $\dfrac{3+2\sqrt{6}}{3}$ **73.** $\dfrac{1+3\sqrt{7}}{4}$
74. $3-4\sqrt{3}$ **75.** $\varnothing$ **76.** $\{48\}$ **77.** $\{1\}$ **78.** $\{2\}$ **79.** $\{6\}$
80. $\{-3,-1\}$ **81.** $\{-2\}$ **82.** $\{4\}$ **83.** $\{7\}$ **84.** 19,453 ft^2

Chapter 8 MIXED REVIEW EXERCISES

1. $11\sqrt{3}$ **2.** $\dfrac{5-\sqrt{2}}{23}$ **3.** $\dfrac{2\sqrt{10}}{5}$ **4.** -5 **5.** $5y\sqrt{2}$ **6.** -11
7. $3a^2b^3\sqrt[3]{2ab}$ **8.** $-\sqrt{10}-5\sqrt{15}$ **9.** $\dfrac{4r\sqrt{3rs}}{3s}$ **10.** $\dfrac{2+\sqrt{13}}{2}$
11. $7-2\sqrt{10}$ **12.** $166+2\sqrt{7}$ **13.** $\{7\}$ **14.** $\varnothing$ **15.** $\{8\}$
16. $\dfrac{\pi\sqrt{2}}{2}$ sec

Chapter 8 TEST

1. $-20, 20$ **2.** **(a)** irrational **(b)** 11.916 **3.** **(a)** B **(b)** F **(c)** D **(d)** A **(e)** C **(f)** A **4.** 6 **5.** $-3\sqrt{6}$ **6.** $\dfrac{8\sqrt{2}}{5}$ **7.** $2\sqrt[3]{4}$
8. $4\sqrt{6}$ **9.** $9\sqrt{7}$ **10.** $11+2\sqrt{30}$ **11.** $2y\sqrt[3]{4x^2}$ **12.** 31
13. $6\sqrt{2}+2-3\sqrt{14}-\sqrt{7}$ **14.** $-5\sqrt{3x}$ **15.** **(a)** $6\sqrt{2}$ in. **(b)** 8.485 in. **16.** 50 ohms **17.** $\dfrac{5\sqrt{14}}{7}$ **18.** $\dfrac{\sqrt{6x}}{3x}$ **19.** $-\sqrt[3]{2}$
20. $\dfrac{-3(4+\sqrt{3})}{13}$ **21.** $\dfrac{\sqrt{3}+12\sqrt{2}}{3}$ **22.** $\varnothing$ **23.** $\{3\}$ **24.** $\{1,4\}$
25. $\{9\}$ **26.** 12 is not a solution. A check shows that it does not satisfy the original equation. The solution set is $\varnothing$.

Chapters R–8 CUMULATIVE REVIEW EXERCISES

1. 54 **2.** 6 **3.** 3 **4.** $\{3\}$ **5.** $[-16,\infty)$ **6.** $(5,\infty)$
7. 2012: \$298,626; 2013: \$426,010

8.

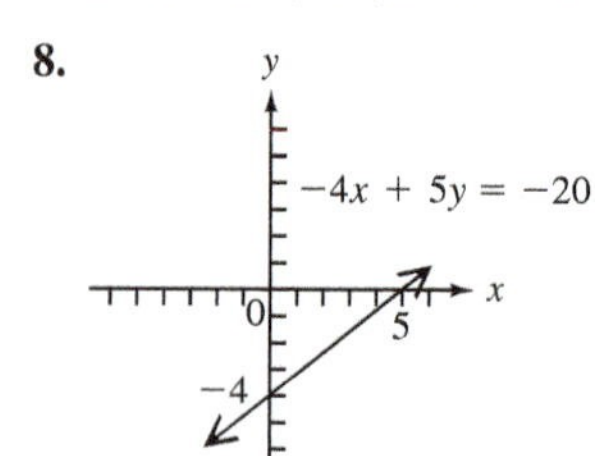

9.

y
x
0
x = 2

10.

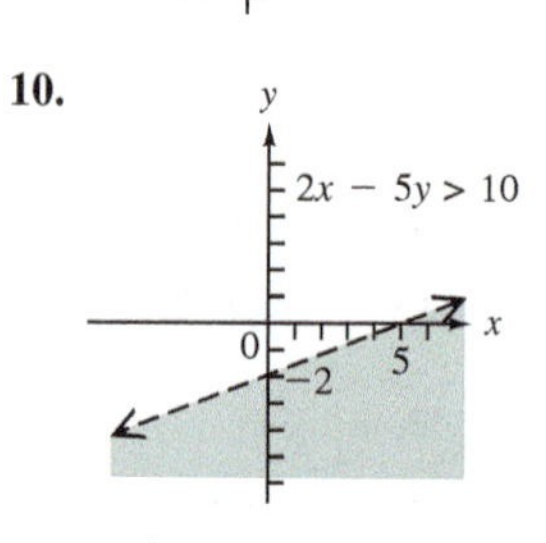

11. 17.6; The number of subscribers increased by an average of 17.6 million per yr. **12.** $y=17.6x+109.5$ **13.** 391.1 million **14.** $\{(3,-7)\}$ **15.** $\{(x,y)\,|\,2x-y=6\}$ **16.** from Chicago: 61 mph; from Des Moines: 54 mph **17.** $12x^{10}y^2$
18. $\dfrac{y^{15}}{5832}$ **19.** $3x^3+11x^2-13$ **20.** $4t^2-8t+5$ **21.** $(m+8)(m+4)$
22. $(5t^2+6)(5t^2-6)$ **23.** $(6a+5b)(2a-b)$ **24.** $(9z+4)^2$
25. $\dfrac{x+1}{x}$ **26.** $(t+5)(t+3)$ **27.** $\dfrac{-2x-14}{(x+3)(x-1)}$ **28.** $29\sqrt{3}$
29. $-\sqrt{3}+\sqrt{5}$ **30.** $21-5\sqrt{2}$ **31.** $\{3,4\}$ **32.** $\{-2,-1\}$
33. $\{12\}$ **34.** $\{19\}$ **35.** $\varnothing$ **36.** $\{16\}$

CHAPTER 9 Quadratic Equations

SECTION 9.1

1. B, C **2.** B **3.** C **4.** A **5.** D **6.** B **7.** According to the square root property, -9 is also a solution, so her answer was not completely correct. The solution set is $\{-9,9\}$, or $\{\pm 9\}$. **8.** The fractions cannot be simplified because 2 is *not* a common factor in the numerator. The correct solution set is $\left\{\dfrac{3+2\sqrt{5}}{2}, \dfrac{3-2\sqrt{5}}{2}\right\}$. **9.** $\{-7,8\}$
11. $\{3,5\}$ **13.** $\{-11,11\}$ **15.** $\left\{-\dfrac{5}{3}, 6\right\}$ **17.** $\{-9,9\}$
19. $\{-\sqrt{14},\sqrt{14}\}$ **21.** $\{-4\sqrt{3},4\sqrt{3}\}$ **23.** no real solution
25. $\left\{-\dfrac{5}{2},\dfrac{5}{2}\right\}$ **27.** $\{-0.5, 0.5\}$ **29.** $\{-8,8\}$ **31.** $\{-\sqrt{3},\sqrt{3}\}$
33. no real solution **35.** $\{-3\sqrt{2},3\sqrt{2}\}$ **37.** $\{-3\sqrt{3},3\sqrt{3}\}$
39. $\{-2\sqrt{6},2\sqrt{6}\}$ **41.** $\left\{-\dfrac{2\sqrt{7}}{7},\dfrac{2\sqrt{7}}{7}\right\}$ **43.** $\left\{-\dfrac{2\sqrt{5}}{5},\dfrac{2\sqrt{5}}{5}\right\}$
45. $\{-2,8\}$ **47.** $\{8+3\sqrt{3},8-3\sqrt{3}\}$ **49.** no real solution
51. $\left\{-3,\dfrac{5}{3}\right\}$ **53.** $\left\{0,\dfrac{3}{2}\right\}$ **55.** $\left\{\dfrac{5+\sqrt{30}}{2},\dfrac{5-\sqrt{30}}{2}\right\}$
57. $\left\{\dfrac{-1+3\sqrt{2}}{3},\dfrac{-1-3\sqrt{2}}{3}\right\}$ **59.** $\left\{\dfrac{1+4\sqrt{3}}{4},\dfrac{1-4\sqrt{3}}{4}\right\}$
61. $\{-10+4\sqrt{3},-10-4\sqrt{3}\}$ **63.** 5, 12, 13 **65.** $\dfrac{1}{2}$ sec
67. 9 in. **69.** 2%

SECTION 9.2

1. D **2.** $\left\{\dfrac{5}{2},4\right\}$; $\left\{\dfrac{5}{2},4\right\}$; This equation can be solved using either method. The same solution set results. **3.** 25; $(x+5)^2$ **5.** 100; $(z-10)^2$
7. 1; $(x+1)^2$ **9.** $\dfrac{25}{4}$; $\left(p-\dfrac{5}{2}\right)^2$ **11.** $\{1,3\}$ **13.** $\{-3,-2\}$
15. $\{-1+\sqrt{6},-1-\sqrt{6}\}$ **17.** $\{4+2\sqrt{3},4-2\sqrt{3}\}$ **19.** $\{-3\}$
21. $\left\{\dfrac{-1+\sqrt{5}}{2},\dfrac{-1-\sqrt{5}}{2}\right\}$ **23.** $\left\{-\dfrac{3}{2},\dfrac{1}{2}\right\}$ **25.** no real solution
27. $\left\{\dfrac{-7+\sqrt{97}}{6},\dfrac{-7-\sqrt{97}}{6}\right\}$ **29.** $\{-4,2\}$ **31.** $\{4+\sqrt{3},4-\sqrt{3}\}$
33. $\left\{\dfrac{3+\sqrt{13}}{2},\dfrac{3-\sqrt{13}}{2}\right\}$ **35.** $\left\{\dfrac{-3+\sqrt{5}}{2},\dfrac{-3-\sqrt{5}}{2}\right\}$
37. $\{1+\sqrt{6},1-\sqrt{6}\}$ **39.** $\left\{\dfrac{2+\sqrt{14}}{2},\dfrac{2-\sqrt{14}}{2}\right\}$
41. 3 sec and 5 sec **43.** 75 ft by 100 ft **45.** x^2 **46.** x^2+8x
47. $x^2+8x+16$ **48.** It occurred when we added the 16 squares.

SECTION 9.3

1. $2a$ should be in the denominator for $-b$ as well. The correct formula is $x=\dfrac{-b\pm\sqrt{b^2-4ac}}{2a}$. **2.** The fraction bar should be under both $-b$ and $\sqrt{b^2-4ac}$. The term $2a$ should not be in the radicand. The correct formula is $x=\dfrac{-b\pm\sqrt{b^2-4ac}}{2a}$. **3.** 4; 5; -9 **5.** 3; -4; -2 **7.** 3; 7; 0
9. $\{-13,1\}$ **11.** $\left\{-1,\dfrac{5}{2}\right\}$ **13.** $\{2\}$ **15.** $\left\{\dfrac{-6+\sqrt{26}}{2},\dfrac{-6-\sqrt{26}}{2}\right\}$

17. $\left\{\frac{-5+\sqrt{13}}{6}, \frac{-5-\sqrt{13}}{6}\right\}$ **19.** no real solution **21.** $\{-1, 0\}$

23. $\left\{0, \frac{12}{7}\right\}$ **25.** $\{-2\sqrt{6}, 2\sqrt{6}\}$ **27.** $\left\{-\frac{2}{5}, \frac{2}{5}\right\}$

29. no real solution **31.** $\left\{\frac{6+2\sqrt{6}}{3}, \frac{6-2\sqrt{6}}{3}\right\}$ **33.** $\left\{-\frac{2}{3}, \frac{4}{3}\right\}$

35. $\left\{\frac{-1+\sqrt{73}}{6}, \frac{-1-\sqrt{73}}{6}\right\}$ **37.** no real solution **39.** $\left\{-1, \frac{5}{2}\right\}$

41. $\{-3+\sqrt{5}, -3-\sqrt{5}\}$ **43.** 3.5 ft **45.** -8, 16; Only 16 board feet is a reasonable answer.

SUMMARY EXERCISES Applying Methods for Solving Quadratic Equations

1. $\{-6, 6\}$ **2.** $\left\{\frac{-3+\sqrt{5}}{2}, \frac{-3-\sqrt{5}}{2}\right\}$ **3.** $\left\{-\frac{10}{9}, \frac{10}{9}\right\}$

4. $\left\{-\frac{7}{9}, \frac{7}{9}\right\}$ **5.** $\{1, 3\}$ **6.** $\{-2, -1\}$ **7.** $\{4, 5\}$

8. $\left\{\frac{-3+\sqrt{17}}{2}, \frac{-3-\sqrt{17}}{2}\right\}$ **9.** $\left\{-\frac{1}{3}, \frac{5}{3}\right\}$

10. $\left\{\frac{1+\sqrt{10}}{2}, \frac{1-\sqrt{10}}{2}\right\}$ **11.** $\{-17, 5\}$ **12.** $\left\{-\frac{7}{5}, 1\right\}$

13. $\left\{\frac{7+2\sqrt{6}}{3}, \frac{7-2\sqrt{6}}{3}\right\}$ **14.** $\left\{\frac{1+4\sqrt{2}}{7}, \frac{1-4\sqrt{2}}{7}\right\}$

15. no real solution **16.** no real solution **17.** $\left\{-\frac{1}{2}, 2\right\}$ **18.** $\left\{-\frac{1}{2}, 1\right\}$

19. $\left\{-\frac{5}{4}, \frac{3}{2}\right\}$ **20.** $\left\{-3, \frac{1}{3}\right\}$ **21.** $\{1+\sqrt{2}, 1-\sqrt{2}\}$

22. $\left\{\frac{-5+\sqrt{13}}{6}, \frac{-5-\sqrt{13}}{6}\right\}$ **23.** $\left\{\frac{2}{5}, 4\right\}$

24. $\{-3+\sqrt{5}, -3-\sqrt{5}\}$ **25.** $\left\{\frac{-3+\sqrt{41}}{2}, \frac{-3-\sqrt{41}}{2}\right\}$

26. $\left\{-\frac{5}{4}\right\}$ **27.** $\left\{\frac{1}{4}, 1\right\}$ **28.** $\left\{\frac{1+\sqrt{3}}{2}, \frac{1-\sqrt{3}}{2}\right\}$

29. $\left\{\frac{-2+\sqrt{11}}{3}, \frac{-2-\sqrt{11}}{3}\right\}$ **30.** $\left\{\frac{-5+\sqrt{41}}{8}, \frac{-5-\sqrt{41}}{8}\right\}$

31. $\left\{\frac{-7+\sqrt{5}}{4}, \frac{-7-\sqrt{5}}{4}\right\}$ **32.** $\left\{-\frac{8}{3}, -\frac{6}{5}\right\}$

33. $\left\{\frac{8+8\sqrt{2}}{3}, \frac{8-8\sqrt{2}}{3}\right\}$ **34.** $\left\{\frac{-5+\sqrt{5}}{2}, \frac{-5-\sqrt{5}}{2}\right\}$

35. no real solution **36.** no real solution **37.** $\left\{-\frac{2}{3}, 2\right\}$ **38.** $\left\{-\frac{1}{4}, \frac{2}{3}\right\}$

39. $\left\{-4, \frac{3}{5}\right\}$ **40.** $\{-3, 5\}$ **41.** $\left\{-\frac{2}{3}, \frac{2}{5}\right\}$ **42.** $\{-4, 6\}$

SECTION 9.4

1. parabola **2.** vertex; highest **3.** axis of symmetry; mirror **4.** one; one; no

5.

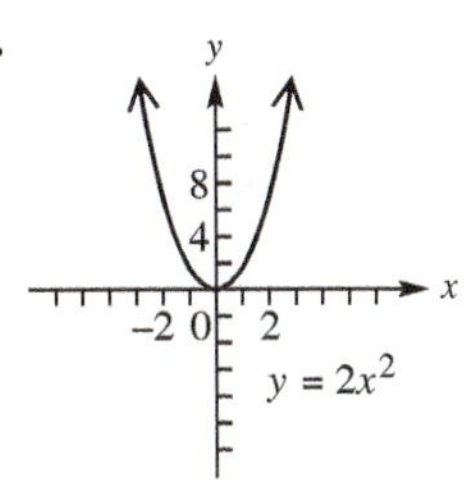

(0, 0)

7.

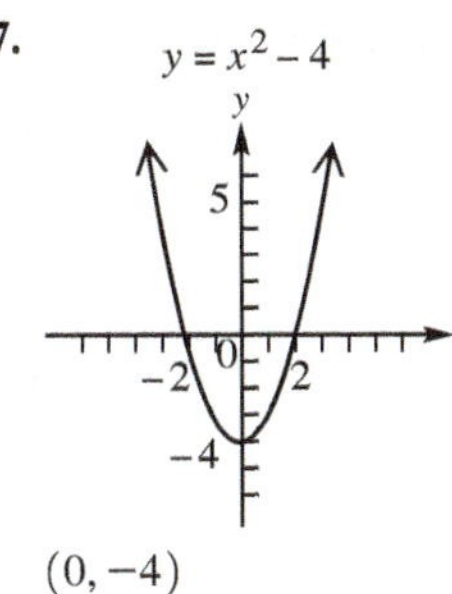

(0, −4)

9.

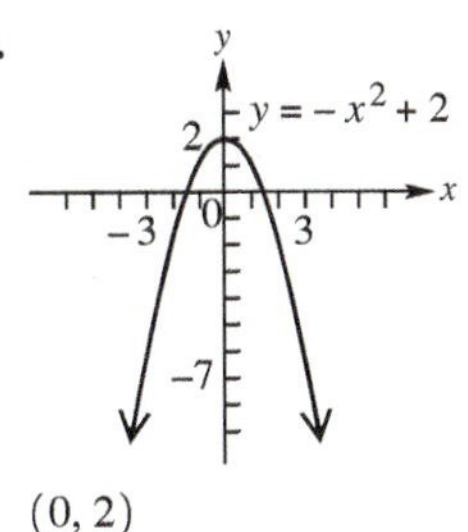

(0, 2)

11.

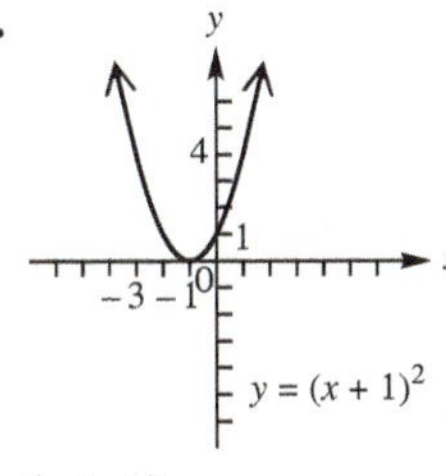

(−1, 0)

13.

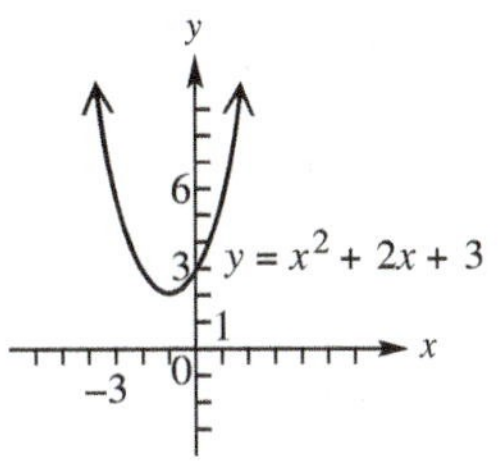

(−1, 2); (0, 3); no x-intercepts

15.

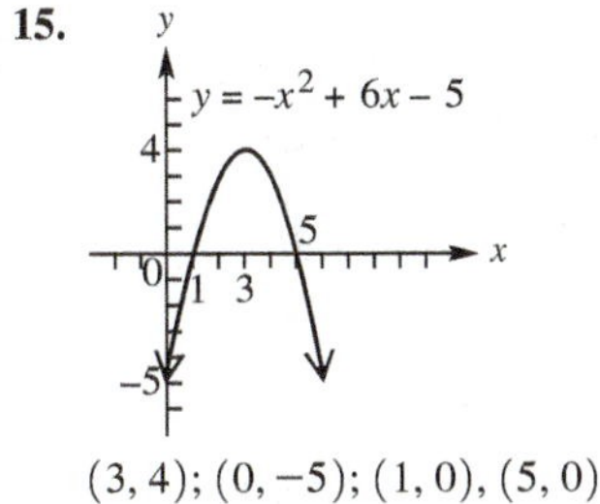

(3, 4); (0, −5); (1, 0), (5, 0)

17. **19.**

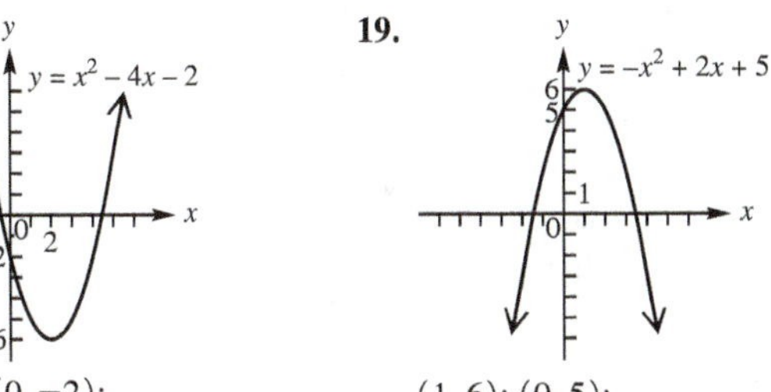

17. (2, −6); (0, −2); (−0.4, 0), (4.4, 0)

19. (1, 6); (0, 5); (−1.4, 0), (3.4, 0)

20. If $a > 0$, then the parabola opens upward. If $a < 0$, then the parabola opens downward. **21.** $y = \frac{11}{5625}x^2$

SECTION 9.5

1. relation; domain; range **2.** function; one; component **3.** function; domain: $\{3, 1, 0, -1\}$; range: $\{7, 4, -2, -1\}$ **5.** function; domain: $\{1, 2, 3\}$; range: $\{-1, -2\}$ **7.** not a function; domain: $\{-4, -2, 0\}$; range: $\{3, 1, 5, -8\}$ **9.** function; domain: $\{3, 1, 5, 7, 9\}$; range: $\{2, 3, 6, 4\}$ **11.** not function; domain: $\{-4, -2, 0, 2, 3\}$; range: $\{-2, 0, 1, 2, 3\}$ **13.** function **15.** not a function **17.** function **19.** function **21.** not a function **23.** function **25.** **(a)** 11 **(b)** 3 **(c)** -9 **27.** **(a)** -4 **(b)** -2 **(c)** 1 **29.** **(a)** 4 **(b)** 2 **(c)** 14 **31.** $\{(1980, 14.1), (1990, 19.8), (2000, 28.4), (2010, 40.0), (2014, 42.4)\}$; yes **33.** $g(1980) = 14.1$ (million); $g(2000) = 28.4$ (million) **35.** $(0, -4)$ **36.** $(3, -2)$ **37.** $\frac{2}{3}$ **38.** $f(x) = \frac{2}{3}x - 4$

ANSWERS

Chapter 9 REVIEW EXERCISES

1. $\{-7, 4\}$ **2.** $\{-9, -5\}$ **3.** $\left\{-5, \frac{3}{2}\right\}$ **4.** $\left\{\frac{2}{3}, 1\right\}$ **5.** $\{-10, 10\}$ **6.** $\{-13, 13\}$ **7.** $\{-12, 12\}$ **8.** $\{-\sqrt{37}, \sqrt{37}\}$ **9.** $\{-8\sqrt{2}, 8\sqrt{2}\}$ **10.** $\{-7, 3\}$ **11.** $\{3 + \sqrt{10}, 3 - \sqrt{10}\}$ **12.** $\left\{\frac{-1 + \sqrt{14}}{2}, \frac{-1 - \sqrt{14}}{2}\right\}$ **13.** no real solution **14.** $\left\{-\frac{5}{3}\right\}$ **15.** $\{-5, -1\}$ **16.** $\{-2 + \sqrt{11}, -2 - \sqrt{11}\}$ **17.** $\{-1 + \sqrt{6}, -1 - \sqrt{6}\}$ **18.** $\left\{\frac{-2 + \sqrt{7}}{2}, \frac{-2 - \sqrt{7}}{2}\right\}$ **19.** $\left\{-\frac{7}{2}\right\}$ **20.** no real solution **21.** 2.5 sec **22.** 1.3 sec and 4.7 sec **23.** $\{-4, 9\}$ **24.** $\left\{0, \frac{14}{5}\right\}$ **25.** no real solution **26.** $\left\{\frac{3}{2}\right\}$ **27.** $\left\{\frac{-1 + \sqrt{29}}{4}, \frac{-1 - \sqrt{29}}{4}\right\}$ **28.** $\left\{\frac{2 + \sqrt{10}}{2}, \frac{2 - \sqrt{10}}{2}\right\}$ **29.** $\left\{\frac{1 + \sqrt{21}}{10}, \frac{1 - \sqrt{21}}{10}\right\}$ **30.** $\left\{\frac{-3 + \sqrt{41}}{2}, \frac{-3 - \sqrt{41}}{2}\right\}$ **31.** $\{-3 + \sqrt{19}, -3 - \sqrt{19}\}$ **32.** $\left\{\frac{2 + \sqrt{2}}{2}, \frac{2 - \sqrt{2}}{2}\right\}$

33.

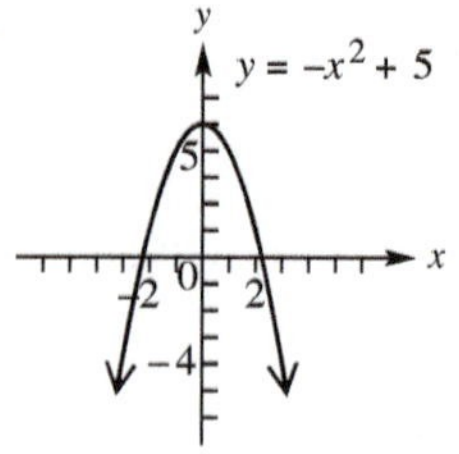

(0, 5)

34.

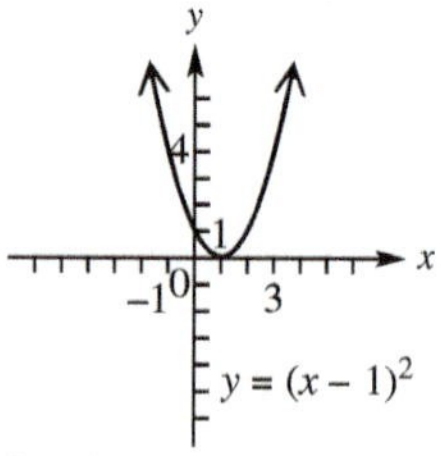

(1, 0)

35.

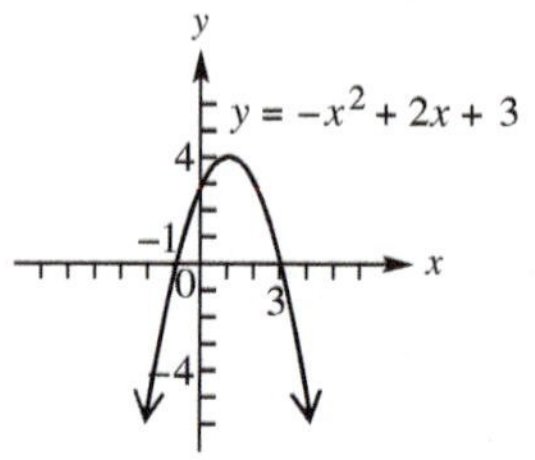

(1, 4); (0, 3); (−1, 0), (3, 0)

36.

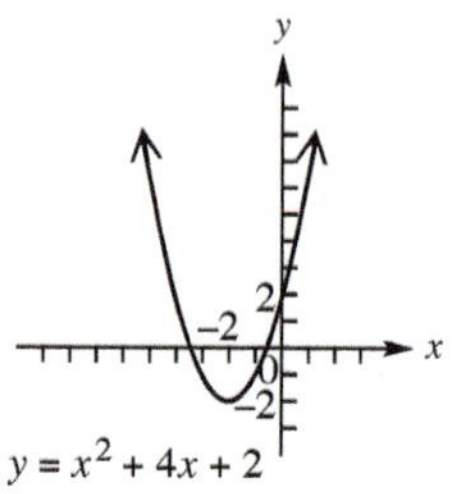

(−2, −2); (0, 2); (−3.4, 0), (−0.6, 0)

37. not a function; domain: $\{-2, 0, 2\}$; range: $\{4, 8, 5, 3\}$ **38.** function; domain: $\{8, 7, 6, 5, 4\}$; range: $\{3, 4, 5, 6, 7\}$ **39.** not a function **40.** function **41.** function **42.** not a function **43.** **(a)** 8 **(b)** −1 **44.** **(a)** −1 **(b)** −4 **45.** **(a)** 7 **(b)** 1 **46.** **(a)** −3 **(b)** −6

Chapter 9 MIXED REVIEW EXERCISES

1. $\left\{-\frac{11}{2}, 5\right\}$ **2.** $\left\{-\frac{11}{2}, \frac{9}{2}\right\}$ **3.** $\left\{\frac{-1 + \sqrt{21}}{2}, \frac{-1 - \sqrt{21}}{2}\right\}$ **4.** $\left\{-\frac{3}{2}, \frac{1}{3}\right\}$ **5.** $\left\{\frac{-5 + \sqrt{17}}{2}, \frac{-5 - \sqrt{17}}{2}\right\}$ **6.** $\{-1 + \sqrt{3}, -1 - \sqrt{3}\}$ **7.** no real solution **8.** $\left\{\frac{9 + \sqrt{41}}{2}, \frac{9 - \sqrt{41}}{2}\right\}$ **9.** $\left\{-\frac{6}{5}\right\}$ **10.** $\{-1 + 2\sqrt{2}, -1 - 2\sqrt{2}\}$ **11.** $\{-2 + \sqrt{5}, -2 - \sqrt{5}\}$ **12.** $\{-2\sqrt{2}, 2\sqrt{2}\}$ **13.** **(a)** $\{-3, 3\}$ **(b)** $\{-3, 3\}$ **(c)** $\{-3, 3\}$ **(d)** Because there is only one solution set, we will always obtain the same results, no matter which method of solution we use. **14.** 6, 8, 10 **15.** **(a)** true **(b)** false **(c)** domain: $\{1, 2\}$; range: $\{-1, -2, 2\}$

Chapter 9 TEST

1. $\{-6, 8\}$ **2.** $\{-7, 7\}$ **3.** $\left\{1, \frac{4}{3}\right\}$ **4.** $\{-\sqrt{39}, \sqrt{39}\}$ **5.** $\{-11, 5\}$ **6.** $\left\{\frac{-3 + 2\sqrt{6}}{4}, \frac{-3 - 2\sqrt{6}}{4}\right\}$ **7.** $\{2 + \sqrt{10}, 2 - \sqrt{10}\}$ **8.** $\left\{\frac{-3 + \sqrt{6}}{2}, \frac{-3 - \sqrt{6}}{2}\right\}$ **9.** $\left\{-3, \frac{1}{2}\right\}$ **10.** $\left\{\frac{3 + \sqrt{3}}{3}, \frac{3 - \sqrt{3}}{3}\right\}$ **11.** no real solution **12.** $\left\{\frac{5 + \sqrt{13}}{6}, \frac{5 - \sqrt{13}}{6}\right\}$ **13.** $\{1 + \sqrt{2}, 1 - \sqrt{2}\}$ **14.** $\left\{\frac{-1 + 3\sqrt{2}}{2}, \frac{-1 - 3\sqrt{2}}{2}\right\}$ **15.** $\left\{\frac{11 + \sqrt{89}}{4}, \frac{11 - \sqrt{89}}{4}\right\}$ **16.** $\{5\}$ **17.** 2 sec **18.** 12, 16, 20

19.

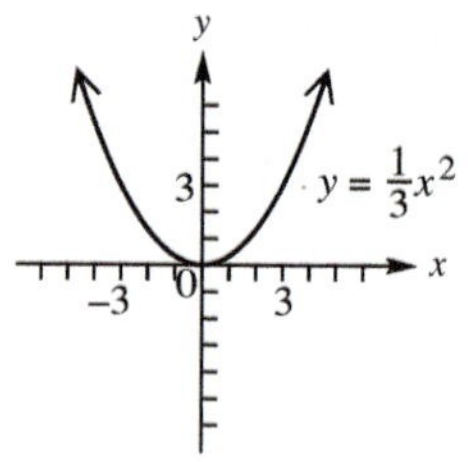

(0, 0)

20.

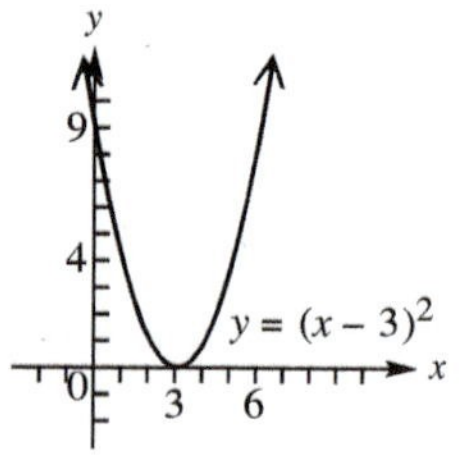

(3, 0)

21.

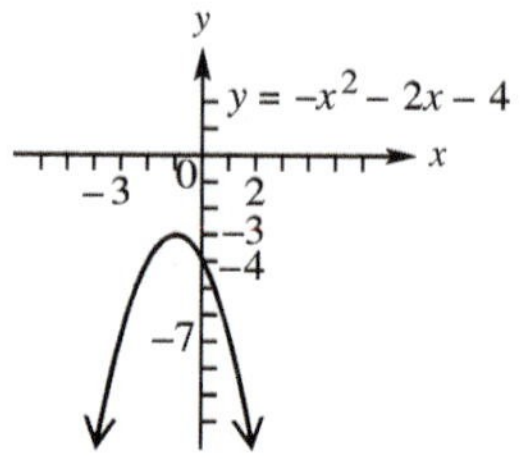

(−1, −3); (0, −4); no x-intercepts

22. not a function; domain: $\{2\}$; range: $\{3, 4, 5\}$ **23.** function; domain: $\{0, 1, 2\}$; range: $\{2\}$ **24.** not a function **25.** 1

Chapters R–9 CUMULATIVE REVIEW EXERCISES

1. 15 **2.** −2 **3.** $19m - 17$ **4.** $\{3\}$ **5.** $\{5\}$ **6.** $\{2\}$ **7.** width: 50 ft; length: 94 ft **8.** 100°; 80° **9.** $L = \frac{P - 2W}{2}$, or $L = \frac{P}{2} - W$ **10.** $(-\infty, 4]$

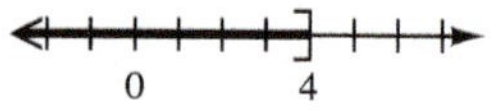

11.

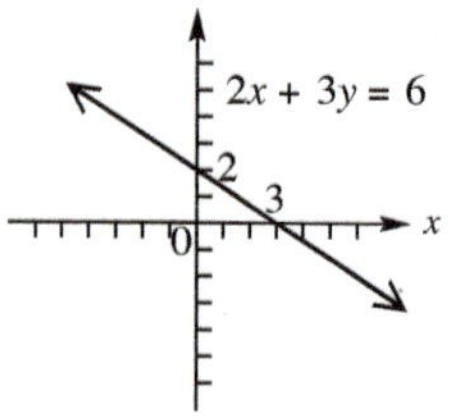

12.

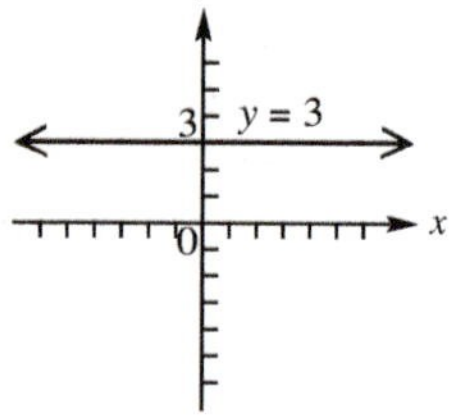

13. $2x - 5y < 10$

14. $-\frac{1}{3}$ **15.** $2x - y = -3$ **16.** $\{(-3, 2)\}$ **17.** $\varnothing$ **18.** Motorola Theory: \$79.99; Motorola i412: \$69.99 **19.** $\frac{3}{2}$ **20.** $\frac{c^2}{b^{16}}$ **21.** $\frac{27}{125}$ **22.** $8x^5 - 17x^4 - x^2$ **23.** $2x^4 + x^3 - 19x^2 + 2x + 20$ **24.** $16x^2(x - 3y)$ **25.** $(2a + 1)(a - 3)$ **26.** $(5m - 2)^2$ **27.** $\left\{-1, \frac{4}{3}\right\}$ **28.** 50 m **29.** $\frac{4}{5}$ **30.** $\frac{-k-1}{k(k-1)}$ **31.** $\frac{6x+1}{3x-1}$ **32.** $\left\{-\frac{15}{7}, 2\right\}$ **33.** $\{7\}$ **34.** $\left\{\frac{1+\sqrt{3}}{2}, \frac{1-\sqrt{3}}{2}\right\}$ **35.** $\frac{6\sqrt{30}}{5}$ **36.** $4\sqrt{5}$ **37.** $-ab\sqrt[3]{2b}$

38.

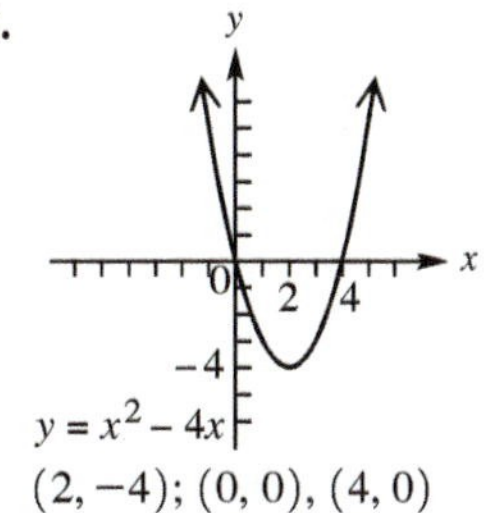

$(2, -4)$; $(0, 0)$, $(4, 0)$

39. (a) yes **(b)** domain: $\{0, 1, 3\}$; range: $\{4, 2, 5\}$ **40.** 11

APPENDIX Factoring Sums and Differences of Cubes

1. 1; 8; 27; 64; 125; 216; 343; 512; 729; 1000 **2.** 3 **3.** C, D **4.** A, D **5.** $(a - 1)(a^2 + a + 1)$ **7.** $(m + 2)(m^2 - 2m + 4)$ **9.** $(y - 6)(y^2 + 6y + 36)$ **11.** $(k + 10)(k^2 - 10k + 100)$ **13.** $(3x - 1)(9x^2 + 3x + 1)$ **15.** $(5x + 2)(25x^2 - 10x + 4)$ **17.** $(y - 2x)(y^2 + 2xy + 4x^2)$ **19.** $(3x - 4y)(9x^2 + 12xy + 16y^2)$ **21.** $(2p + 9q)(4p^2 - 18pq + 81q^2)$ **23.** $2(2t - 1)(4t^2 + 2t + 1)$ **25.** $5(2w + 3)(4w^2 - 6w + 9)$ **27.** $(x + y^2)(x^2 - xy^2 + y^4)$ **29.** $(5k - 2m^3)(25k^2 + 10km^3 + 4m^6)$

Photo Credits

CHAPTER R: **11** Cristovao/Shutterstock; **16** Yevgen Romanenko/Shutterstock

CHAPTER 1: **37** Monkey Business/Fotolia; **45** Monkey Business/Fotolia; **46** WavebreakMediaMicro/Fotolia; **72** (top) Andrey Popov/Shutterstock, (bottom) Brett Bouwer/Fotolia; **73** Gstockstudio/Fotolia; **85** Anjelagr/Fotolia; **89** Maygutyak/Fotolia; **112** Paul Maguire/Fotolia; **116** Sascha Burkhard/Fotolia

CHAPTER 2: **117** Flairimages/Fotolia; **125** Alex_SK/Fotolia; **148** epa european pressphoto agency b.v./Alamy Stock Photos; **157** (top) Emkaplin/Fotolia, (bottom) Denis Nata/Fotolia; **159** Tmass/Fotolia; **173** epa european pressphoto agency b.v./Alamy Stock Photos; **174** Jaimie Duplass/Fotolia; **177** Syda Productions/Shutterstock; **180** Spass/Fotolia; **182** Sumire8/Fotolia; **184** (left) Yanlev/Fotolia, (right) Michael Zhang/123RF; **193** Susan Chiang/iStock/Getty Images; **200** Studio DMM Photography/Design & Art/Shutterstock; **208** WavebreakMediaMicro/Fotolia

CHAPTER 3: **213** Sergunt/Fotolia; **214** ArenaCreative/Fotolia; **217** KingPhoto/Fotolia; **220** Sanjay Goswami/Fotolia; **226** lunamarina/Fotolia; **235** Sumire8/Fotolia; **269** Karen Roach/Fotolia; **273** Rebecca Fisher/Fotolia; **293** Zimmytws/Fotolia

CHAPTER 4: **295** Mmphotos/Fotolia; **313** Vadim Ponomarenko/Fotolia; **321** Sashkin/Fotolia; **324** Julijah/Fotolia; **331** James Steidl/Shutterstock; **333** Dusan Jankovic/Shutterstock; **334** Eric Gevaert/Fotolia; **335** Margo Harrison; **348** Marty/Fotolia

CHAPTER 5: **351** Nikonomad/Fotolia; **376** Koya979/Shutterstock; **377** Africa Studio/Fotolia; **378** Baloncici/Shutterstock; **421** NASA

CHAPTER 6: **425** Alexandre Solodov/Fotolia; **463** Nickolae/Fotolia; **471** Petr Jilek/Shutterstock; **475** Offscreen/Shutterstock; **478** Gina Santa Marie/Fotolia; **491** Giordano Aita/Fotolia; **494** Maxim_Kazmin/Fotolia

CHAPTER 7: **497** Christopher Nolan/Fotolia; **529** Gilles Paire/Fotolia; **558** Johnny Habell/Shutterstock; **560** (left) Pranodhm/Fotolia; (right) epa european pressphoto agency b.v./Alamy Stock Photo; **561** Melinda Fawver/Fotolia; **563** Ilandrea/Fotolia; **564** Ioana Davies/Shutterstock; **567** Richard Coencas/Shutterstock; **568** Dennis Sabol/Shutterstock; **569** Lovegtr35/Fotolia; **580** Verena Matthew/Shutterstock

CHAPTER 8: **583** Danny E. Hooks/Shutterstock; **584** Thepalmer/E+/Getty Images; **593** (left) John Hornsby, (right) AlessandroZocc/Shutterstock; **615** NASA Images; **622** Godsandkings/Fotolia; **624** Kjersti/Fotolia; **630** Betacam SP/Fotolia; **633** Tim Large/Shutterstock; **634** (top left) Lotharingia/Fotolia, (top right) Beboy/Fotolia, (bottom left) Tmax/Fotolia, (bottom right) Sculpies/Fotolia; **642** Amfroey01/Fotolia; **644** Thinkstock/Stockbyte/Getty Images; **645** Tyler Olson/Fotolia

CHAPTER 9: **647** Callie Daniels; **651** Sablin/Fotolia; **669** NASA Images

Index

S

Triangles and Angles

Right Triangle

Triangle has one 90° (right) angle.

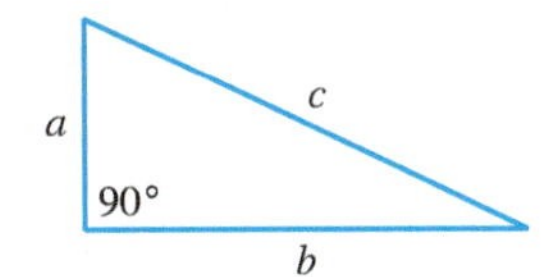

Pythagorean Theorem (for right triangles)

$a^2 + b^2 = c^2$

Isosceles Triangle

Two sides are equal.

$AB = BC$

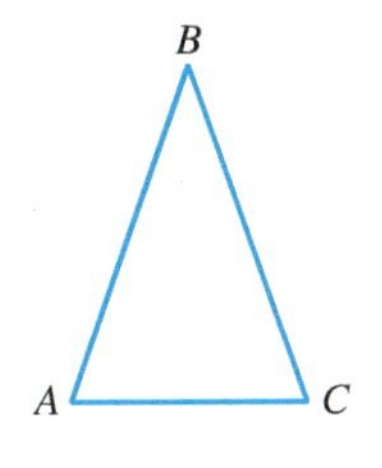

Equilateral Triangle

All sides are equal.

$AB = BC = CA$

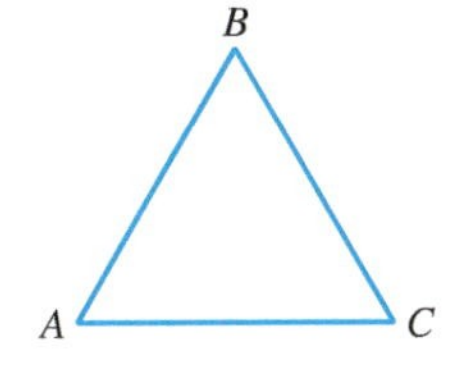

Sum of the Angles of Any Triangle

$A + B + C = 180°$

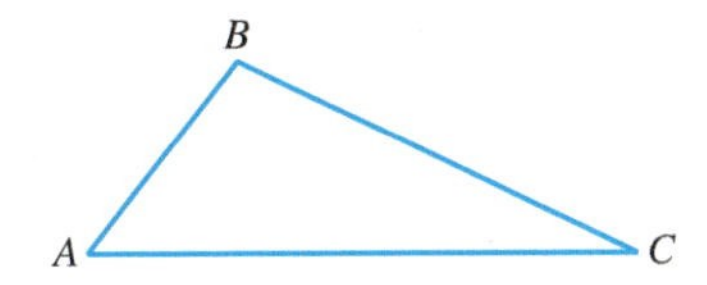

Similar Triangles

Corresponding angles are equal. Corresponding sides are proportional.

$A = D, B = E, C = F$

$$\frac{AB}{DE} = \frac{AC}{DF} = \frac{BC}{EF}$$

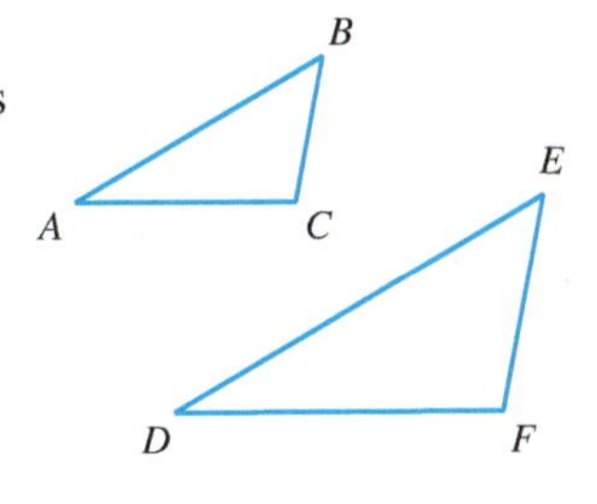

Right Angle

Measure is 90°.

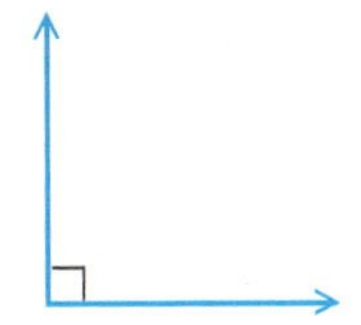

Straight Angle

Measure is 180°.

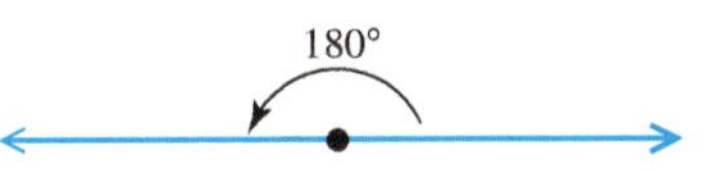

Complementary Angles

The sum of the measures of two complementary angles is 90°.

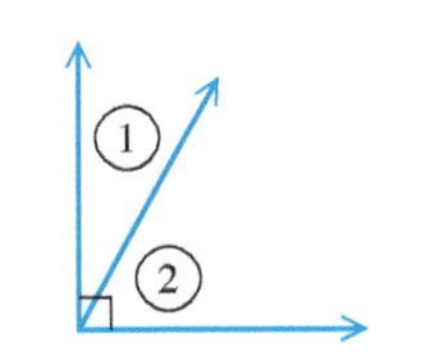

Angles ① and ② are complementary.

Supplementary Angles

The sum of the measures of two supplementary angles is 180°.

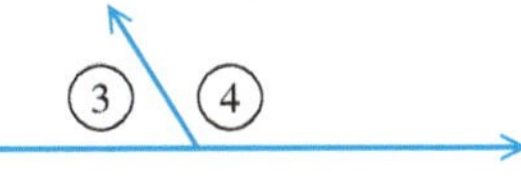

Angles ③ and ④ are supplementary.

Vertical Angles

Vertical angles have equal measures.

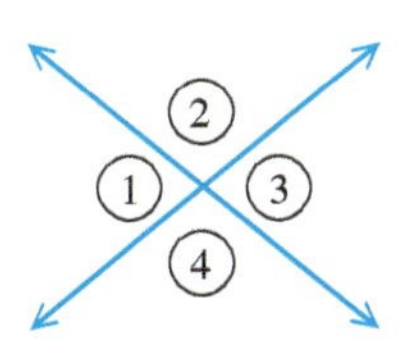

Angle ① = Angle ③

Angle ② = Angle ④

netry Formulas

are
erimeter: $P = 4s$

Area: $\mathcal{A} = s^2$

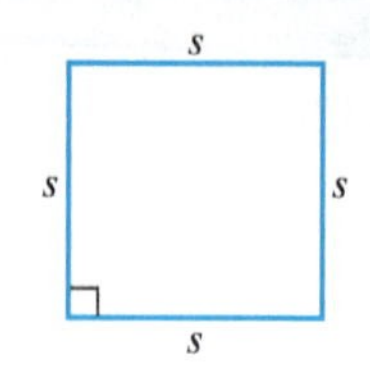

Rectangle
Perimeter: $P = 2L + 2W$

Area: $\mathcal{A} = LW$

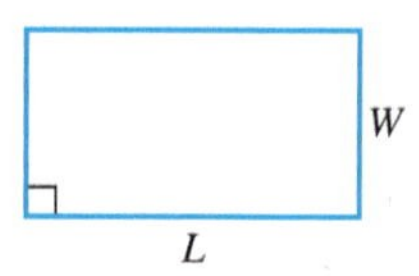

Triangle
Perimeter: $P = a + b + c$

Area: $\mathcal{A} = \frac{1}{2}bh$

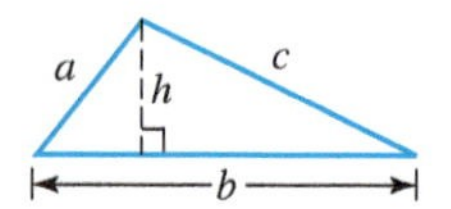

Parallelogram
Perimeter: $P = 2a + 2b$

Area: $\mathcal{A} = bh$

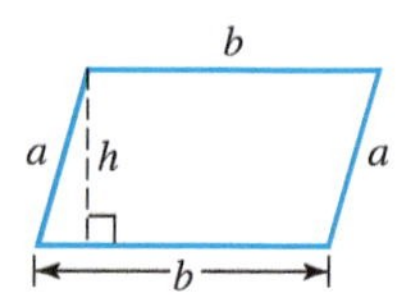

Trapezoid
Perimeter: $P = a + b + c + B$

Area: $\mathcal{A} = \frac{1}{2}h(b + B)$

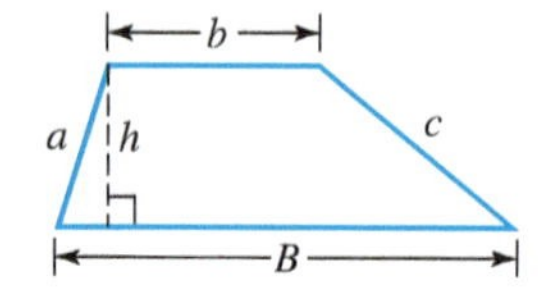

Circle
Diameter: $d = 2r$

Circumference: $C = 2\pi r$

$C = \pi d$

Area: $\mathcal{A} = \pi r^2$

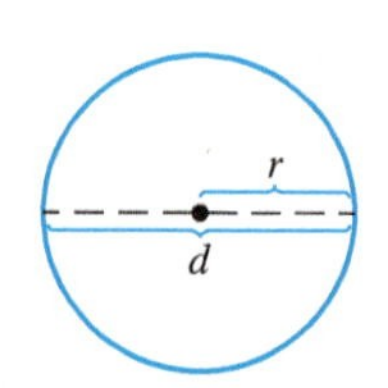

Rectangular Solid
Volume: $V = LWH$

Surface area: $\mathcal{A} = 2HW + 2LW + 2LH$

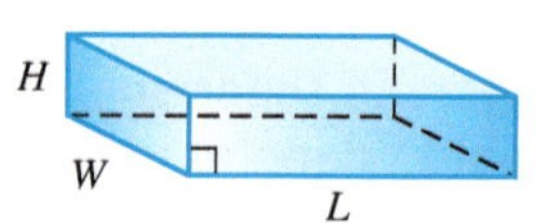

Cube
Volume: $V = e^3$

Surface area: $S = 6e^2$

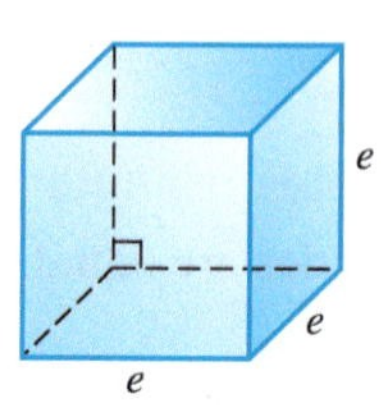

Right Circular Cylinder
Volume: $V = \pi r^2 h$

Surface area: $S = 2\pi rh + 2\pi r^2$
(Includes both circular bases)

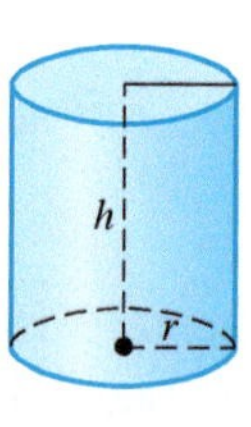

Cone
Volume: $V = \frac{1}{3}\pi r^2 h$

Surface area: $S = \pi r\sqrt{r^2 + h^2} + \pi r^2$
(Includes circular base)

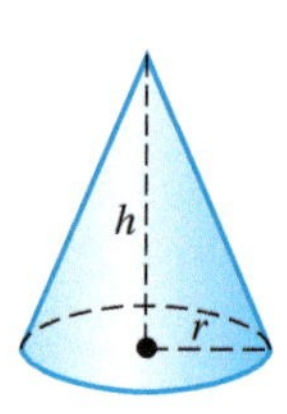

Right Pyramid
Volume: $V = \frac{1}{3}Bh$

B = area of the base

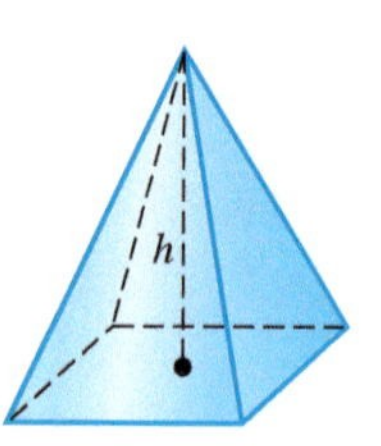

Sphere
Volume: $V = \frac{4}{3}\pi r^3$

Surface area: $S = 4\pi r^2$

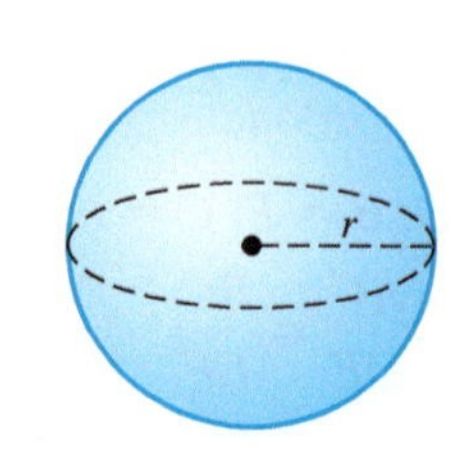